D1288271

KIRK-OTHMER

ENCYCLOPEDIA OF
CHEMICAL
TECHNOLOGY

FOURTH EDITION

VOLUME 22

SILICON COMPOUNDS
TO
SUCCINIC ACID AND SUCCINIC ANHYDRIDE

EXECUTIVE EDITOR
Jacqueline I. Kroschwitz

EDITOR
Mary Howe-Grant

KIRK-OTHMER

ENCYCLOPEDIA OF CHEMICAL TECHNOLOGY

FOURTH EDITION

VOLUME **22**

SILICON COMPOUNDS
TO
SUCCINIC ACID AND SUCCINIC ANHYDRIDE

A Wiley-Interscience Publication
JOHN WILEY & SONS
New York • Chichester • Weinheim • Brisbane • Singapore • Toronto

AEZ9369

ReF
TP
9
.E685
1991
v. 22

This text is printed on acid-free paper.

Copyright © 1997 by John Wiley & Sons, Inc.

All rights reserved. Published simultaneously in Canada.

Reproduction or translation of any part of this work
beyond that permitted by Sections 107 or 108 of the
1976 United States Copyright Act without the permission
of the copyright owner is unlawful. Requests for
permission or further information should be addressed to
the Permissions Department, John Wiley & Sons, Inc.,
605 Third Avenue, New York, NY 10158-0012.

Library of Congress Cataloging-in-Publication Data

Encyclopedia of chemical technology/executive editor, Jacqueline
  I. Kroschwitz; editor, Mary Howe-Grant. — 4th ed.
    p. cm.
  At head of title: Kirk-Othmer.
  "A Wiley-Interscience publication."
  Contents: v. 22, Silicon compounds to succinic acid and succinic anhydride
  ISBN 0471-52691-6 (v. 22)
  1. Chemistry, Technical — Encyclopedias. I. Kirk, Raymond E.
  (Raymond Eller), 1890–1957. II. Othmer, Donald F. (Donald
  Frederick), 1904–1995. III. Kroschwitz, Jacqueline I., 1942– .
  IV. Howe-Grant, Mary, 1943– . V. Title: Kirk-Othmer encyclopedia
  of chemical technology.
  TP9.E685 1992   91-16789
  660′.03 — dc20

Printed in the United States of America

10 9 8 7 6 5 4 3 2 1

OLSON LIBRARY
NORTHERN MICHIGAN UNIVERSITY
MARQUETTE, MICHIGAN 49855

CONTENTS

EDITORIAL STAFF FOR VOLUME 22

Executive Editor: **Jacqueline I. Kroschwitz**
Editor: **Mary Howe-Grant**
Associate Managing Editor: **Lindy Humphreys**
Copy Editors: **Lawrence Altieri**
Jonathan Lee
Assistant Managing Editor: **Brendan A. Vilardo**

CONTRIBUTORS TO VOLUME 22

Steven Arcidaicono, *U.S. Army Natick Research, Development, & [* *Center, Natick, Massachusetts,* Silk

Barry Arkles, *Gelest, Inc., Tullytown, Pennsylvania,* Silanes; Silicor Silicon compounds)

Thomas A. Augurt, *Propper Manufacturing Company, Inc., Lonç York,* Sterilization techniques

Madelyn S. Baggett, *Eastman Chemical Company, Kingsport, Tennessee*, Sorbic acid

Robert G. Bartolo, *The Procter & Gamble Company, Cincinnati, Ohio*, Soap

Alison R. Behling, *Eastman Chemical Company, Kingsport, Tennessee*, Sorbic acid

James Bellows, *Westinghouse Electric Corporation, Orlando, Florida*, Steam

Bruce M. Bertram, *Salt Institute, Alexandria, Virginia*, Sodium halides–sodium chloride (under Sodium compounds)

Walter H. Bortle, *General Chemical Corporation, Claymont, Delaware*, Sodium nitrite (under Sodium compounds)

David R. Bush, *PPG Industries, Inc., New Martinsville, West Virginia*, Sodium sulfides (under Sodium compounds)

David Butts, *Great Salt Lake Minerals Corporation, Ogden, Utah*, Sodium sulfates (under Sodium compounds)

C. Robert Cappel, *Eastman Kodak Company, Rochester, New York*, Silver compounds

C. Edward Capes, *National Research Council of Canada, Ottawa*, Size enlargement

Harold E. Carman, *Eastman Chemical Company, Kingsport, Tennessee*, Sorbic acid

James Cella, *GE Corporate Research and Development, Schenectady, New York*, Silicones (under Silicon compounds)

Shiou-Shan Chen, *Raytheon Engineers & Constructors, East Weymouth, Massachusetts*, Styrene

Ward Collins, *Dow Corning Corporation, Midland, Michigan*, Silicon halides (under Silicon compounds)

Charles R. Craig, *West Virginia University, Morgantown*, Stimulants

K. Darcovich, *National Research Council of Canada, Ottawa*, Size enlargement

James R. Daniel, *Purdue University, Lafayette, Indiana*, Starch

Catherine L. Dorko, *Eastman Chemical Company, Kingsport, Tennessee*, Sorbic acid

Samuel F. Etris, *The Silver Institute, Wayne, Pennsylvania*, Silver and silver alloys

James S. Falcone, Jr., *West Chester University, Pennsylvania*, Synthetic inorganic silicates (under Silicon compounds)

George T. Ford, Jr., *Eastman Chemical Company, Kingsport, Tennessee*, Sorbic acid

Stephen Fossey, *U.S. Army Natick Research, Development, & Engineering Center, Natick, Massachusetts*, Silk

Carlo Fumagalli, *LONZA SpA, Scanzorosciate, Italy*, Succinic acid and succinic anhydride

Larry L. Hench, *University of Florida, Gainesville*, Sol–gel technology

Stephen G. Hibbins, *Timminco Metals, Haley, Ontario, Canada*, Strontium and strontium compounds

David L. Kaplan, *U.S. Army Natick Research, Development, & Engineering Center, Natick, Massachusetts*, Silk

Kaye, *Laurentian University, Ontario, Canada*, Size measurement of particles

Philip C. Kearney, *University of Maryland, College Park*, Soil chemistry of pesticides

James F. Kelly, *Pacific Northwest National Laboratory, Richland, Washington*, Spectroscopy, optical

William C. Koskinen, *USDA-Agricultural Research Service, Beltsville, Maryland*, Soil chemistry of pesticides

Roger N. Kust, *Tetra Technologies, Inc., The Woodlands, Texas*, Sodium halides—sodium bromide (under Sodium compounds)

Richard R. Lattime, *The Goodyear Tire & Rubber Company, Akron, Ohio*, Styrene—butadiene rubber

Charles H. Lemke, *E. I. du Pont de Nemours & Company, Inc., Niagara Falls, New York and the University of Delaware, Newark*, Sodium and sodium alloys

Larry Lewis, *GE Corporate Research and Development, Schenectady, New York*, Silicones (under Silicon compounds)

Matthew L. Lynch, *The Procter & Gamble Company, Cincinnati, Ohio*, Soap

Chien-Pei Mao, *Delavan Inc, West Des Moines, Iowa*, Sprays

Vernon H. Markant, *E. I. du Pont de Nemours & Company, Inc., Niagara Falls, New York*, Sodium and sodium alloys

Ignacio Maturana, *SQM Nitratos SA, Santiago, Chile*, Sodium nitrate (under Sodium compounds)

Robin S. McDowell, *Pacific Northwest National Laboratory, Richland, Washington*, Spectroscopy, optical

Charlene Mello, *U.S. Army Natick Research, Development, & Engineering Center, Natick, Massachusetts*, Silk

Philip H. Merrell, *Mallinckrodt, Inc., St Louis, Missouri*, Sodium halides—sodium iodide (under Sodium compounds)

Bradley P. Morgan, *Pfizer, Inc., Groton, Connecticut*, Steroids

Melinda S. Moynihan, *Pfizer, Inc., Groton, Connecticut*, Steroids

Robert J. Naumann, *University of Alabama in Huntsville*, Space processing

Rodrigo Orefice, *University of Florida, Gainesville*, Sol—gel technology

Peter G. Pape, *Dow Corning Corporation, Midland, Michigan*, Silylating agents (under Silicon compounds)

Harry Paxton, *Carnegie Mellon University, Pittsburgh, Pennsylvania*, Steel

Elizabeth M. Peters, *Mallinckrodt, Inc., St. Louis, Missouri*, Sodium halides—sodium iodide (under Sodium compounds)

Ludwik Pokorny, *SQM Nitratos SA, Santiago, Chile*, Sodium nitrate (under Sodium compounds)

Duane B. Priddy, *The Dow Chemical Company, Midland, Michigan*, Styrene plastics

Michael Prior, *Hosokawa Micron Ltd., Runcorn, Cheshire, England*, Size reduction

Anatol Rabinkin, *AlliedSignal Inc., Parsippany, New Jersey*, Solders and brazing filler metals

Jonathan Rich, *GE Corporate Research and Development, Schenectady, New York*, Silicones (under Silicon compounds)

Slawomir Rubinsztajn, *GE Silicones, Waterford, New York*, Silicones (under Silicon compounds)

Daniel R. Shelton, *USDA-Agricultural Research Service, Beltsville, Maryland,* Soil chemistry of pesticides

Navjot Singh, *GE Corporate Research and Development, Schenectady, New York,* Silicones (under Silicon compounds)

Judith Stein, *GE Corporate Research and Development, Schenectady, New York,* Silicones (under Silicon compounds)

Robert A. Stokes, *Stokes Associates, Golden, Colorado,* Solar energy

Roger Tate, *Delavan, Inc, West Des Moines, Iowa,* Sprays

Remi Trottier, *Aluminum Company of America, Alcoa Center, Pennsylvania,* Size measurement of particles

Ron W. Tucker, *Lilly Industries, Inc., High Point, North Carolina,* Stains, industrial

Roy E. Smith, *Ciba-Geigy Corporation, Greensboro, North Carolina,* Stilbene dyes

Don A. Sullivan, *Shell Chemical Company, Houston, Texas,* Solvents, industrial

Donald Valentine, Jr., *Cytec Industries Inc., Stamford, Connecticut,* Soil stabilization

J.A.A.M. van Asten, *National Institute for Public Health and the Environment, Bilthoven, the Netherlands,* Sterilization techniques

Jeff Wengrovius, *GE Silicones, Waterford, New York,* Silicones (under Silicon compounds)

Leonard A. Wenzel, *Lehigh University, Bethlehem, Pennsylvania,* Simultaneous heat and mass transfer

Roy L. Whistler, *Purdue University, Lafayette, Indiana,* Starch

Walter J. Wolf, *U.S. Department of Agriculture, Peoria, Illinois,* Soybeans and other oilseeds

NOTE ON CHEMICAL ABSTRACTS SERVICE REGISTRY NUMBERS AND NOMENCLATURE

Chemical Abstracts Service (CAS) Registry Numbers are unique numerical identifiers assigned to substances recorded in the CAS Registry System. They appear in brackets in the *Chemical Abstracts* (CA) substance and formula indexes following the names of compounds. A single compound may have synonyms in the chemical literature. A simple compound like phenethylamine can be named β-phenylethylamine or, as in *Chemical Abstracts*, benzeneethanamine. The usefulness of the *Encyclopedia* depends on accessibility through the most common correct name of a substance. Because of this diversity in nomenclature careful attention has been given to the problem in order to assist the reader as much as possible, especially in locating the systematic CA index name by means of the Registry Number. For this purpose, the reader may refer to the CAS Registry Handbook—Number Section which lists in numerical order the Registry Number with the *Chemical Abstracts* index name and the molecular formula; eg, **458-88-8**, Piperidine, 2-propyl-, (*S*)-, $C_8H_{17}N$; in the *Encyclopedia* this compound would be found under its common name, coniine [*458-88-8*]. Alternatively, this information can be retrieved electronically from CAS Online. In many cases molecular formulas have also been provided in the *Encyclopedia* text to facilitate electronic searching. The Registry Number is a valuable link for the reader in retrieving additional published information on substances and also as a point of access for on-line data bases.

In all cases, the CAS Registry Numbers have been given for title compounds in articles and for all compounds in the index. All specific substances indexed in *Chemical Abstracts* since 1965 are included in the CAS Registry System as are a large number of substances derived from a variety of reference works. The CAS Registry System identifies a substance on the basis of an unambiguous computer-language description of its molecular structure including stereochemical detail. The Registry Number is a machine-checkable number (like a Social Security number) assigned in sequential order to each substance as it enters the registry system. The value of the number lies in the fact that it is a concise and unique means of substance identification, which is independent of, and therefore bridges, many systems of chemical nomenclature. For polymers, one Registry Number may be used for the entire family; eg, polyoxyethylene (20) sorbitan monolaurate has the same number as all of its polyoxyethylene homologues.

Cross-references are inserted in the index for many common names and for some systematic names. Trademark names appear in the index. Names that are incorrect, misleading, or ambiguous are avoided. Formulas are given very frequently in the text to help in identifying compounds. The spelling and form used, even for industrial names, follow American chemical usage, but not always the usage of *Chemical Abstracts* (eg, *coniine* is used instead of *(S)-2-propylpiperidine*, *aniline* instead of *benzenamine*, and *acrylic acid* instead of *2-propenoic acid*).

There are variations in representation of rings in different disciplines. The dye industry does not designate aromaticity or double bonds in rings. All double bonds and aromaticity are shown in the *Encyclopedia* as a matter of course. For example, tetralin has an aromatic ring and a saturated ring and its structure

appears in the *Encyclopedia* with its common name, Registry Number enclosed in brackets, and parenthetical CA index name, ie, tetralin [*119-64-2*] (1,2,3,4-tetrahydronaphthalene). With names and structural formulas, and especially with CAS Registry Numbers, the aim is to help the reader have a concise means of substance identification.

CONVERSION FACTORS, ABBREVIATIONS, AND UNIT SYMBOLS

SI Units (Adopted 1960)

The International System of Units (abbreviated SI), is being implemented throughout the world. This measurement system is a modernized version of the MKSA (meter, kilogram, second, ampere) system, and its details are published and controlled by an international treaty organization (The International Bureau of Weights and Measures) (1).

SI units are divided into three classes:

BASE UNITS

length	meter[†] (m)
mass	kilogram (kg)
time	second (s)
electric current	ampere (A)
thermodynamic temperature[‡]	kelvin (K)
amount of substance	mole (mol)
luminous intensity	candela (cd)

SUPPLEMENTARY UNITS

plane angle	radian (rad)
solid angle	steradian (sr)

[†]The spellings "metre" and "litre" are preferred by ASTM; however, "-er" is used in the *Encyclopedia*.

[‡]Wide use is made of Celsius temperature (*t*) defined by

$$t = T - T_0$$

where T is the thermodynamic temperature, expressed in kelvin, and $T_0 = 273.15$ K by definition. A temperature interval may be expressed in degrees Celsius as well as in kelvin.

DERIVED UNITS AND OTHER ACCEPTABLE UNITS

These units are formed by combining base units, supplementary units, and other derived units (2–4). Those derived units having special names and symbols are marked with an asterisk in the list below.

Quantity	Unit	Symbol	Acceptable equivalent
*absorbed dose	gray	Gy	J/kg
acceleration	meter per second squared	m/s^2	
*activity (of a radionuclide)	becquerel	Bq	1/s
area	square kilometer	km^2	
	square hectometer	hm^2	ha (hectare)
	square meter	m^2	
concentration (of amount of substance)	mole per cubic meter	mol/m^3	
current density	ampere per square meter	$A//m^2$	
density, mass density	kilogram per cubic meter	kg/m^3	g/L; mg/cm^3
dipole moment (quantity)	coulomb meter	C·m	
*dose equivalent	sievert	Sv	J/kg
*electric capacitance	farad	F	C/V
*electric charge, quantity of electricity	coulomb	C	A·s
electric charge density	coulomb per cubic meter	C/m^3	
*electric conductance	siemens	S	A/V
electric field strength	volt per meter	V/m	
electric flux density	coulomb per square meter	C/m^2	
*electric potential, potential difference, electromotive force	volt	V	W/A
*electric resistance	ohm	Ω	V/A
*energy, work, quantity of heat	megajoule	MJ	
	kilojoule	kJ	
	joule	J	N·m
	electronvolt[†]	eV[†]	
	kilowatt-hour[†]	kW·h[†]	
energy density	joule per cubic meter	J/m^3	
*force	kilonewton	kN	
	newton	N	$kg·m/s^2$

[†]This non-SI unit is recognized by the CIPM as having to be retained because of practical importance or use in specialized fields (1).

Quantity	Unit	Symbol	Acceptable equivalent
*frequency	megahertz	MHz	
	hertz	Hz	1/s
heat capacity, entropy	joule per kelvin	J/K	
heat capacity (specific), specific entropy	joule per kilogram kelvin	$J/(kg \cdot K)$	
heat-transfer coefficient	watt per square meter kelvin	$W/(m^2 \cdot K)$	
*illuminance	lux	lx	lm/m^2
*inductance	henry	H	Wb/A
linear density	kilogram per meter	kg/m	
luminance	candela per square meter	cd/m^2	
*luminous flux	lumen	lm	$cd \cdot sr$
magnetic field strength	ampere per meter	A/m	
*magnetic flux	weber	Wb	$V \cdot s$
*magnetic flux density	tesla	T	Wb/m^2
molar energy	joule per mole	J/mol	
molar entropy, molar heat capacity	joule per mole kelvin	$J/(mol \cdot K)$	
moment of force, torque	newton meter	$N \cdot m$	
momentum	kilogram meter per second	$kg \cdot m/s$	
permeability	henry per meter	H/m	
permittivity	farad per meter	F/m	
*power, heat flow rate, radiant flux	kilowatt	kW	
	watt	W	J/s
power density, heat flux density, irradiance	watt per square meter	W/m^2	
*pressure, stress	megapascal	MPa	
	kilopascal	kPa	
	pascal	Pa	N/m^2
sound level	decibel	dB	
specific energy	joule per kilogram	J/kg	
specific volume	cubic meter per kilogram	m^3/kg	
surface tension	newton per meter	N/m	
thermal conductivity	watt per meter kelvin	$W/(m \cdot K)$	
velocity	meter per second	m/s	
	kilometer per hour	km/h	
viscosity, dynamic	pascal second	$Pa \cdot s$	
	millipascal second	$mPa \cdot s$	
viscosity, kinematic	square meter per second	m^2/s	
	square millimeter per second	mm^2/s	

Quantity	Unit	Symbol	Acceptable equivalent
volume	cubic meter	m^3	
	cubic diameter	dm^3	L (liter) (5)
	cubic centimeter	cm^3	mL
wave number	1 per meter	m^{-1}	
	1 per centimeter	cm^{-1}	

In addition, there are 16 prefixes used to indicate order of magnitude, as follows:

Multiplication factor	Prefix	Symbol	Note
10^{18}	exa	E	
10^{15}	peta	P	
10^{12}	tera	T	
10^{9}	giga	G	
10^{6}	mega	M	
10^{3}	kilo	k	
10^{2}	hecto	h[a]	[a]Although hecto, deka, deci, and centi
10	deka	da[a]	are SI prefixes, their use should be
10^{-1}	deci	d[a]	avoided except for SI unit-multiples
10^{-2}	centi	c[a]	for area and volume and nontech-
10^{-3}	milli	m	nical use of centimeter, as for body
10^{-6}	micro	μ	and clothing measurement.
10^{-9}	nano	n	
10^{-12}	pico	p	
10^{-15}	femto	f	
10^{-18}	atto	a	

For a complete description of SI and its use the reader is referred to ASTM E380 (4) and the article UNITS AND CONVERSION FACTORS which appears in Vol. 24.

A representative list of conversion factors from non-SI to SI units is presented herewith. Factors are given to four significant figures. Exact relationships are followed by a dagger. A more complete list is given in the latest editions of ASTM E380 (4) and ANSI Z210.1 (6).

Conversion Factors to SI Units

To convert from	To	Multiply by
acre	square meter (m^2)	4.047×10^3
angstrom	meter (m)	1.0×10^{-10}[†]
are	square meter (m^2)	1.0×10^{2}[†]

[†]Exact.

To convert from	To	Multiply by
astronomical unit	meter (m)	1.496×10^{11}
atmosphere, standard	pascal (Pa)	1.013×10^{5}
bar	pascal (Pa)	$1.0 \times 10^{5\dagger}$
barn	square meter (m²)	$1.0 \times 10^{-28\dagger}$
barrel (42 U.S. liquid gallons)	cubic meter (m³)	0.1590
Bohr magneton (μ_B)	J/T	9.274×10^{-24}
Btu (International Table)	joule (J)	1.055×10^{3}
Btu (mean)	joule (J)	1.056×10^{3}
Btu (thermochemical)	joule (J)	1.054×10^{3}
bushel	cubic meter (m³)	3.524×10^{-2}
calorie (International Table)	joule (J)	4.187
calorie (mean)	joule (J)	4.190
calorie (thermochemical)	joule (J)	$4.184^{\dagger}$
centipoise	pascal second (Pa·s)	$1.0 \times 10^{-3\dagger}$
centistokes	square millimeter per second (mm²/s)	$1.0^{\dagger}$
cfm (cubic foot per minute)	cubic meter per second (m³/s)	4.72×10^{-4}
cubic inch	cubic meter (m³)	1.639×10^{-5}
cubic foot	cubic meter (m³)	2.832×10^{-2}
cubic yard	cubic meter (m³)	0.7646
curie	becquerel (Bq)	$3.70 \times 10^{10\dagger}$
debye	coulomb meter (C·m)	3.336×10^{-30}
degree (angle)	radian (rad)	1.745×10^{-2}
denier (international)	kilogram per meter (kg/m)	1.111×10^{-7}
	tex‡	0.1111
dram (apothecaries')	kilogram (kg)	3.888×10^{-3}
dram (avoirdupois)	kilogram (kg)	1.772×10^{-3}
dram (U.S. fluid)	cubic meter (m³)	3.697×10^{-6}
dyne	newton (N)	$1.0 \times 10^{-5\dagger}$
dyne/cm	newton per meter (N/m)	$1.0 \times 10^{-3\dagger}$
electronvolt	joule (J)	1.602×10^{-19}
erg	joule (J)	$1.0 \times 10^{-7\dagger}$
fathom	meter (m)	1.829
fluid ounce (U.S.)	cubic meter (m³)	2.957×10^{-5}
foot	meter (m)	$0.3048^{\dagger}$
footcandle	lux (lx)	10.76
furlong	meter (m)	2.012×10^{-2}
gal	meter per second squared (m/s²)	$1.0 \times 10^{-2\dagger}$
gallon (U.S. dry)	cubic meter (m³)	4.405×10^{-3}
gallon (U.S. liquid)	cubic meter (m³)	3.785×10^{-3}
gallon per minute (gpm)	cubic meter per second (m³/s)	6.309×10^{-5}
	cubic meter per hour (m³/h)	0.2271

†Exact.
‡See footnote on p. xiii.

To convert from	To	Multiply by
gauss	tesla (T)	1.0×10^{-4}
gilbert	ampere (A)	0.7958
gill (U.S.)	cubic meter (m^3)	1.183×10^{-4}
grade	radian	1.571×10^{-2}
grain	kilogram (kg)	6.480×10^{-5}
gram force per denier	newton per tex (N/tex)	8.826×10^{-2}
hectare	square meter (m^2)	$1.0 \times 10^{4\dagger}$
horsepower (550 ft·lbf/s)	watt (W)	7.457×10^2
horsepower (boiler)	watt (W)	9.810×10^3
horsepower (electric)	watt (W)	$7.46 \times 10^{2\dagger}$
hundredweight (long)	kilogram (kg)	50.80
hundredweight (short)	kilogram (kg)	45.36
inch	meter (m)	$2.54 \times 10^{-2\dagger}$
inch of mercury (32°F)	pascal (Pa)	3.386×10^3
inch of water (39.2°F)	pascal (Pa)	2.491×10^2
kilogram-force	newton (N)	9.807
kilowatt hour	megajoule (MJ)	$3.6^\dagger$
kip	newton (N)	4.448×10^3
knot (international)	meter per second (m/S)	0.5144
lambert	candela per square meter (cd/m^3)	3.183×10^3
league (British nautical)	meter (m)	5.559×10^3
league (statute)	meter (m)	4.828×10^3
light year	meter (m)	9.461×10^{15}
liter (for fluids only)	cubic meter (m^3)	$1.0 \times 10^{-3\dagger}$
maxwell	weber (Wb)	$1.0 \times 10^{-8\dagger}$
micron	meter (m)	$1.0 \times 10^{-6\dagger}$
mil	meter (m)	$2.54 \times 10^{-5\dagger}$
mile (statute)	meter (m)	1.609×10^3
mile (U.S. nautical)	meter (m)	$1.852 \times 10^{3\dagger}$
mile per hour	meter per second (m/s)	0.4470
millibar	pascal (Pa)	1.0×10^2
millimeter of mercury (0°C)	pascal (Pa)	$1.333 \times 10^{2\dagger}$
minute (angular)	radian	2.909×10^{-4}
myriagram	kilogram (kg)	10
myriameter	kilometer (km)	10
oersted	ampere per meter (A/m)	79.58
ounce (avoirdupois)	kilogram (kg)	2.835×10^{-2}
ounce (troy)	kilogram (kg)	3.110×10^{-2}
ounce (U.S. fluid)	cubic meter (m^3)	2.957×10^{-5}
ounce-force	newton (N)	0.2780
peck (U.S.)	cubic meter (m^3)	8.810×10^{-3}
pennyweight	kilogram (kg)	1.555×10^{-3}
pint (U.S. dry)	cubic meter (m^3)	5.506×10^{-4}
pint (U.S. liquid)	cubic meter (m^3)	4.732×10^{-4}

†Exact.

To convert from	To	Multiply by
poise (absolute viscosity)	pascal second (Pa·s)	$0.10^{\dagger}$
pound (avoirdupois)	kilogram (kg)	0.4536
pound (troy)	kilogram (kg)	0.3732
poundal	newton (N)	0.1383
pound-force	newton (N)	4.448
pound force per square inch (psi)	pascal (Pa)	6.895×10^{3}
quart (U.S. dry)	cubic meter (m³)	1.101×10^{-3}
quart (U.S. liquid)	cubic meter (m³)	9.464×10^{-4}
quintal	kilogram (kg)	$1.0 \times 10^{2\dagger}$
rad	gray (Gy)	$1.0 \times 10^{-2\dagger}$
rod	meter (m)	5.029
roentgen	coulomb per kilogram (C/kg)	2.58×10^{-4}
second (angle)	radian (rad)	$4.848 \times 10^{-6\dagger}$
section	square meter (m²)	2.590×10^{6}
slug	kilogram (kg)	14.59
spherical candle power	lumen (lm)	12.57
square inch	square meter (m²)	6.452×10^{-4}
square foot	square meter (m²)	9.290×10^{-2}
square mile	square meter (m²)	2.590×10^{6}
square yard	square meter (m²)	0.8361
stere	cubic meter (m³)	$1.0^{\dagger}$
stokes (kinematic viscosity)	square meter per second (m²/s)	$1.0 \times 10^{-4\dagger}$
tex	kilogram per meter (kg/m)	$1.0 \times 10^{-6\dagger}$
ton (long, 2240 pounds)	kilogram (kg)	1.016×10^{3}
ton (metric) (tonne)	kilogram (kg)	$1.0 \times 10^{3\dagger}$
ton (short, 2000 pounds)	kilogram (kg)	9.072×10^{2}
torr	pascal (Pa)	1.333×10^{2}
unit pole	weber (Wb)	1.257×10^{-7}
yard	meter (m)	$0.9144^{\dagger}$

†Exact.

Abbreviations and Unit Symbols

Following is a list of common abbreviations and unit symbols used in the *Encyclopedia*. In general they agree with those listed in *American National Standard Abbreviations for Use on Drawings and in Text* (*ANSI Y1.1*) (6) and *American National Standard Letter Symbols for Units in Science and Technology* (*ANSI Y10*) (6). Also included is a list of acronyms for a number of private and government organizations as well as common industrial solvents, polymers, and other chemicals.

Rules for Writing Unit Symbols (4):

1. Unit symbols are printed in upright letters (roman) regardless of the type style used in the surrounding text.
2. Unit symbols are unaltered in the plural.
3. Unit symbols are not followed by a period except when used at the end of a sentence.
4. Letter unit symbols are generally printed lower-case (for example, cd for candela) unless the unit name has been derived from a proper name, in which case the first letter of the symbol is capitalized (W, Pa). Prefixes and unit symbols retain their prescribed form regardless of the surrounding typography.
5. In the complete expression for a quantity, a space should be left between the numerical value and the unit symbol. For example, write 2.37 lm, *not* 2.37lm, and 35 mm, *not* 35mm. When the quantity is used in an adjectival sense, a hyphen is often used, for example, 35-mm film. *Exception:* No space is left between the numerical value and the symbols of degree, minute, and second of plane angle, degree Celsius, and the percent sign.
6. No space is used between the prefix and unit symbol (for example, kg).
7. Symbols, not abbreviations, should be used for units. For example, use "A," not "amp," for ampere.
8. When multiplying unit symbols, use a raised dot:

$$\text{N·m} \quad \text{for} \quad \text{newton meter}$$

In the case of W·h, the dot may be omitted, thus:

$$\text{Wh}$$

An exception to this practice is made for computer printouts, automatic typewriter work, etc, where the raised dot is not possible, and a dot on the line may be used.
9. When dividing unit symbols, use one of the following forms:

$$\text{m/s} \quad or \quad \text{m·s}^{-1} \quad or \quad \frac{\text{m}}{\text{s}}$$

In no case should more than one slash be used in the same expression unless parentheses are inserted to avoid ambiguity. For example, write:

$$\text{J/(mol·K)} \quad or \quad \text{J·mol}^{-1}\text{·K}^{-1} \quad or \quad \text{(J/mol)/K}$$

but *not*

$$\text{J/mol/K}$$

10. Do not mix symbols and unit names in the same expression. Write:

$$\text{joules per kilogram} \quad or \quad \text{J/kg} \quad or \quad \text{J·kg}^{-1}$$

but *not*

$$\text{joules/kilogram} \quad nor \quad \text{joules/kg} \quad nor \quad \text{joules·kg}^{-1}$$

ABBREVIATIONS AND UNITS

A	ampere	AOAC	Association of Official Analytical Chemists
A	anion (eg, HA)		
A	mass number	AOCS	American Oil Chemists' Society
a	atto (prefix for 10^{-18})		
AATCC	American Association of Textile Chemists and Colorists	APHA	American Public Health Association
		API	American Petroleum Institute
ABS	acrylonitrile–butadiene–styrene	aq	aqueous
abs	absolute	Ar	aryl
ac	alternating current, *n.*	*ar*-	aromatic
a-c	alternating current, *adj.*	*as*-	asymmetric(al)
ac-	alicyclic	ASHRAE	American Society of Heating, Refrigerating, and Air Conditioning Engineers
acac	acetylacetonate		
ACGIH	American Conference of Governmental Industrial Hygienists		
		ASM	American Society for Metals
ACS	American Chemical Society	ASME	American Society of Mechanical Engineers
AGA	American Gas Association	ASTM	American Society for Testing and Materials
Ah	ampere hour		
AIChE	American Institute of Chemical Engineers	at no.	atomic number
AIME	American Institute of Mining, Metallurgical, and Petroleum Engineers	at wt	atomic weight
		av(g)	average
		AWS	American Welding Society
		b	bonding orbital
AIP	American Institute of Physics	bbl	barrel
		bcc	body-centered cubic
AISI	American Iron and Steel Institute	BCT	body-centered tetragonal
		Bé	Baumé
alc	alcohol(ic)	BET	Brunauer-Emmett-Teller (adsorption equation)
Alk	alkyl		
alk	alkaline (not alkali)	bid	twice daily
amt	amount	Boc	*t*-butyloxycarbonyl
amu	atomic mass unit	BOD	biochemical (biological) oxygen demand
ANSI	American National Standards Institute		
		bp	boiling point
AO	atomic orbital	Bq	becquerel

C	coulomb	DIN	Deutsche Industrie
°C	degree Celsius		Normen
C-	denoting attachment to	dl-; DL-	racemic
	carbon	DMA	dimethylacetamide
c	centi (prefix for 10^{-2})	DMF	dimethylformamide
c	critical	DMG	dimethyl glyoxime
ca	circa (approximately)	DMSO	dimethyl sulfoxide
cd	candela; current density;	DOD	Department of Defense
	circular dichroism	DOE	Department of Energy
CFR	Code of Federal	DOT	Department of
	Regulations		Transportation
cgs	centimeter-gram-second	DP	degree of polymerization
CI	Color Index	dp	dew point
cis-	isomer in which	DPH	diamond pyramid
	substituted groups are		hardness
	on same side of double	dstl(d)	distill(ed)
	bond between C atoms	dta	differential thermal
cl	carload		analysis
cm	centimeter	(E)-	entgegen; opposed
cmil	circular mil	ϵ	dielectric constant
cmpd	compound		(unitless number)
CNS	central nervous system	e	electron
CoA	coenzyme A	ECU	electrochemical unit
COD	chemical oxygen demand	ed.	edited, edition, editor
coml	commercial(ly)	ED	effective dose
cp	chemically pure	EDTA	ethylenediaminetetra-
cph	close-packed hexagonal		acetic acid
CPSC	Consumer Product Safety	emf	electromotive force
	Commission	emu	electromagnetic unit
cryst	crystalline	en	ethylene diamine
cub	cubic	eng	engineering
D	debye	EPA	Environmental Protection
D-	denoting configurational		Agency
	relationship	epr	electron paramagnetic
d	differential operator		resonance
d	day; deci (prefix for 10^{-1})	eq.	equation
d	density	esca	electron spectroscopy for
d-	$dextro$-, dextrorotatory		chemical analysis
da	deka (prefix for 10^1)	esp	especially
dB	decibel	esr	electron-spin resonance
dc	direct current, $n.$	est(d)	estimate(d)
d-c	direct current, $adj.$	estn	estimation
dec	decompose	esu	electrostatic unit
detd	determined	exp	experiment, experimental
detn	determination	ext(d)	extract(ed)
Di	didymium, a mixture of all	F	farad (capacitance)
	lanthanons	F	faraday (96,487 C)
dia	diameter	f	femto (prefix for 10^{-15})
dil	dilute		

FAO	Food and Agriculture Organization (United Nations)	hyd	hydrated, hydrous
		hyg	hygroscopic
		Hz	hertz
fcc	face-centered cubic	i (eg, Pr^i)	iso (eg, isopropyl)
FDA	Food and Drug Administration	i-	inactive (eg, i-methionine)
		IACS	International Annealed Copper Standard
FEA	Federal Energy Administration	ibp	initial boiling point
FHSA	Federal Hazardous Substances Act	IC	integrated circuit
		ICC	Interstate Commerce Commission
fob	free on board	ICT	International Critical Table
fp	freezing point		
FPC	Federal Power Commission	ID	inside diameter; infective dose
FRB	Federal Reserve Board	ip	intraperitoneal
frz	freezing	IPS	iron pipe size
G	giga (prefix for 10^9)	ir	infrared
G	gravitational constant = 6.67×10^{11} N·m^2/kg^2	IRLG	Interagency Regulatory Liaison Group
g	gram	ISO	International Organization Standardization
(g)	gas, only as in H$_2$O(g)		
g	gravitational acceleration	ITS-90	International Temperature Scale (NIST)
gc	gas chromatography		
gem-	geminal	IU	International Unit
glc	gas–liquid chromatography	IUPAC	International Union of Pure and Applied Chemistry
g-mol wt; gmw	gram-molecular weight		
GNP	gross national product	IV	iodine value
gpc	gel-permeation chromatography	iv	intravenous
		J	joule
GRAS	Generally Recognized as Safe	K	kelvin
		k	kilo (prefix for 10^3)
grd	ground	kg	kilogram
Gy	gray	L	denoting configurational relationship
H	henry		
h	hour; hecto (prefix for 10^2)	L	liter (for fluids only) (5)
ha	hectare	l-	$levo$-, levorotatory
HB	Brinell hardness number	(l)	liquid, only as in NH$_3$(l)
Hb	hemoglobin	LC$_{50}$	conc lethal to 50% of the animals tested
hcp	hexagonal close-packed		
hex	hexagonal	LCAO	linear combination of atomic orbitals
HK	Knoop hardness number		
hplc	high performance liquid chromatography	lc	liquid chromatography
		LCD	liquid crystal display
HRC	Rockwell hardness (C scale)	lcl	less than carload lots
		LD$_{50}$	dose lethal to 50% of the animals tested
HV	Vickers hardness number		

LED	light-emitting diode	$N\text{-}$	denoting attachment to nitrogen
liq	liquid		
lm	lumen	n (as n_D^{20})	index of refraction (for 20°C and sodium light)
ln	logarithm (natural)		
LNG	liquefied natural gas	^{n} (as Bu^n),	
log	logarithm (common)	$n\text{-}$	normal (straight-chain structure)
LOI	limiting oxygen index		
LPG	liquefied petroleum gas	n	neutron
ltl	less than truckload lots	n	nano (prefix for 10^9)
lx	lux	na	not available
M	mega (prefix for 10^6); metal (as in MA)	NAS	National Academy of Sciences
M	molar; actual mass	NASA	National Aeronautics and Space Administration
$\overline{M}_w$	weight-average mol wt		
$\overline{M}_n$	number-average mol wt	nat	natural
m	meter; milli (prefix for 10^{-3})	ndt	nondestructive testing
		neg	negative
m	molal	NF	*National Formulary*
$m\text{-}$	meta	NIH	National Institutes of Health
max	maximum		
MCA	Chemical Manufacturers' Association (was Manufacturing Chemists Association)	NIOSH	National Institute of Occupational Safety and Health
		NIST	National Institute of Standards and Technology (formerly National Bureau of Standards)
MEK	methyl ethyl ketone		
meq	milliequivalent		
mfd	manufactured		
mfg	manufacturing		
mfr	manufacturer	nmr	nuclear magnetic resonance
MIBC	methyl isobutyl carbinol		
MIBK	methyl isobutyl ketone	NND	New and Nonofficial Drugs (AMA)
MIC	minimum inhibiting concentration		
		no.	number
min	minute; minimum	NOI-(BN)	not otherwise indexed (by name)
mL	milliliter		
MLD	minimum lethal dose	NOS	not otherwise specified
MO	molecular orbital	nqr	nuclear quadruple resonance
mo	month		
mol	mole	NRC	Nuclear Regulatory Commission; National Research Council
mol wt	molecular weight		
mp	melting point		
MR	molar refraction	NRI	New Ring Index
ms	mass spectrometry	NSF	National Science Foundation
MSDS	material safety data sheet		
mxt	mixture	NTA	nitrilotriacetic acid
μ	micro (prefix for 10^{-6})	NTP	normal temperature and pressure (25°C and 101.3 kPa or 1 atm)
N	newton (force)		
N	normal (concentration); neutron number		

NTSB	National Transportation Safety Board	qv	quod vide (which see)
O-	denoting attachment to oxygen	R	univalent hydrocarbon radical
o-	ortho	(*R*)-	rectus (clockwise configuration)
OD	outside diameter	*r*	precision of data
OPEC	Organization of Petroleum Exporting Countries	rad	radian; radius
o-phen	*o*-phenanthridine	RCRA	Resource Conservation and Recovery Act
OSHA	Occupational Safety and Health Administration	rds	rate-determining step
		ref.	reference
owf	on weight of fiber	rf	radio frequency, *n.*
Ω	ohm	r-f	radio frequency, *adj.*
P	peta (prefix for 10^{15})	rh	relative humidity
p	pico (prefix for 10^{-12})	RI	Ring Index
p-	para	rms	root-mean square
p	proton	rpm	rotations per minute
p.	page	rps	revolutions per second
Pa	pascal (pressure)	RT	room temperature
PEL	personal exposure limit based on an 8-h exposure	RTECS	Registry of Toxic Effects of Chemical Substances
		ˢ (eg, Buˢ);	
pd	potential difference	*sec*-	secondary (eg, secondary butyl)
pH	negative logarithm of the effective hydrogen ion concentration	S	siemens
		(*S*)-	sinister (counterclockwise configuration)
phr	parts per hundred of resin (rubber)	S-	denoting attachment to sulfur
p-i-n	positive-intrinsic-negative	*s*-	symmetric(al)
pmr	proton magnetic resonance	s	second
p-n	positive-negative	(s)	solid, only as in $H_2O(s)$
po	per os (oral)	SAE	Society of Automotive Engineers
POP	polyoxypropylene		
pos	positive	SAN	styrene-acrylonitrile
pp.	pages	sat(d)	saturate(d)
ppb	parts per billion (10^9)	satn	saturation
ppm	parts per million (10^6)	SBS	styrene–butadiene–styrene
ppmv	parts per million by volume	sc	subcutaneous
ppmwt	parts per million by weight	SCF	self-consistent field; standard cubic feet
PPO	poly(phenyl oxide)		
ppt(d)	precipitate(d)	Sch	Schultz number
pptn	precipitation	sem	scanning electron microscope(y)
Pr (no.)	foreign prototype (number)		
pt	point; part		
PVC	poly(vinyl chloride)	SFs	Saybolt Furol seconds
pwd	powder	sl sol	slightly soluble
py	pyridine	sol	soluble

soln	solution	*trans-*	isomer in which
soly	solubility		substituted groups are
sp	specific; species		on opposite sides of
sp gr	specific gravity		double bond between C
sr	steradian		atoms
std	standard	TSCA	Toxic Substances Control
STP	standard temperature and		Act
	pressure (0°C and 101.3	TWA	time-weighted average
	kPa)	Twad	Twaddell
sub	sublime(s)	UL	Underwriters' Laboratory
SUs	Saybolt Universal seconds	USDA	United States Department
syn	synthetic		of Agriculture
^{t} (eg, Bu^t),		USP	*United States*
t-, tert-	tertiary (eg, tertiary		*Pharmacopeia*
	butyl)	uv	ultraviolet
T	tera (prefix for 10^{12}); tesla	V	volt (emf)
	(magnetic flux density)	var	variable
t	metric ton (tonne)	*vic-*	vicinal
t	temperature	vol	volume (not volatile)
TAPPI	Technical Association of	vs	versus
	the Pulp and Paper	v sol	very soluble
	Industry	W	watt
TCC	Tagliabue closed cup	Wb	weber
tex	tex (linear density)	Wh	watt hour
T_g	glass-transition	WHO	World Health
	temperature		Organization (United
tga	thermogravimetric		Nations)
	analysis	wk	week
THF	tetrahydrofuran	yr	year
tlc	thin layer chromatography	(Z)-	zusammen; together;
TLV	threshold limit value		atomic number

Non-SI (Unacceptable and Obsolete) Units		Use
Å	angstrom	nm
at	atmosphere, technical	Pa
atm	atmosphere, standard	Pa
b	barn	cm^2
bar†	bar	Pa
bbl	barrel	m^3
bhp	brake horsepower	W
Btu	British thermal unit	J
bu	bushel	m^3; L
cal	calorie	J
cfm	cubic foot per minute	m^3/s
Ci	curie	Bq
cSt	centistokes	mm^2/s
c/s	cycle per second	Hz

†Do not use bar (10^5 Pa) or millibar (10^2 Pa) because they are not SI units, and are accepted internationally only for a limited time in special fields because of existing usage.

Non-SI (Unacceptable and Obsolete) Units		Use
cu	cubic	exponential form
D	debye	C·m
den	denier	tex
dr	dram	kg
dyn	dyne	N
dyn/cm	dyne per centimeter	mN/m
erg	erg	J
eu	entropy unit	J/K
°F	degree Fahrenheit	°C; K
fc	footcandle	lx
fl	footlambert	lx
fl oz	fluid ounce	m^3; L
ft	foot	m
ft·lbf	foot pound-force	J
gf den	gram-force per denier	N/tex
G	gauss	T
Gal	gal	m/s^2
gal	gallon	m^3; L
Gb	gilbert	A
gpm	gallon per minute	(m^3/s); (m^3/h)
gr	grain	kg
hp	horsepower	W
ihp	indicated horsepower	W
in.	inch	m
in. Hg	inch of mercury	Pa
in. H_2O	inch of water	Pa
in.-lbf	inch pound-force	J
kcal	kilo-calorie	J
kgf	kilogram-force	N
kilo	for kilogram	kg
L	lambert	lx
lb	pound	kg
lbf	pound-force	N
mho	mho	S
mi	mile	m
MM	million	M
mm Hg	millimeter of mercury	Pa
mμ	millimicron	nm
mph	miles per hour	km/h
μ	micron	μm
Oe	oersted	A/m
oz	ounce	kg
ozf	ounce-force	N
η	poise	Pa·s
P	poise	Pa·s
ph	phot	lx
psi	pounds-force per square inch	Pa
psia	pounds-force per square inch absolute	Pa
psig	pounds-force per square inch gage	Pa
qt	quart	m^3; L
°R	degree Rankine	K
rd	rad	Gy
sb	stilb	lx
SCF	standard cubic foot	m^3
sq	square	exponential form
thm	therm	J
yd	yard	m

BIBLIOGRAPHY

1. The International Bureau of Weights and Measures, BIPM (Parc de Saint-Cloud, France) is described in Appendix X2 of Ref. 4. This bureau operates under the exclusive supervision of the International Committee for Weights and Measures (CIPM).
2. *Metric Editorial Guide (ANMC-78-1)*, latest ed., American National Metric Council, 5410 Grosvenor Lane, Bethesda, Md. 20814, 1981.
3. *SI Units and Recommendations for the Use of Their Multiples and of Certain Other Units (ISO 1000-1981)*, American National Standards Institute, 1430 Broadway, New York, 10018, 1981.
4. Based on *ASTM E380-89a (Standard Practice for Use of the International System of Units (SI))*, American Society for Testing and Materials, 1916 Race Street, Philadelphia, Pa. 19103, 1989.
5. *Fed. Reg.*, Dec. 10, 1976 (41 FR 36414).
6. For ANSI address, see Ref. 3.

R. P. Lukens
ASTM Committee E-43 on SI Practice

Continued

SILICON COMPOUNDS

SYNTHETIC INORGANIC SILICATES

Commercial soluble silicates have the general formula

$$M_2O \cdot mSiO_2 \cdot nH_2O$$

where M is an alkali metal and m, the modulus, and n are the number of moles of SiO_2 and H_2O, respectively, per mole of M_2O. The composition of commercial alkali silicates is typically described by the weight ratio of SiO_2 to M_2O. Sodium silicates [1344-09-8] are the most common; potassium silicate [1312-76-1] and lithium silicates [12627-14-4] are manufactured to a limited extent for special applications. These materials are usually manufactured as glasses that dissolve in water to form viscous, alkaline solutions. The modulus or ratio, m, in commercial sodium silicate products typically varies from 0.5 to 4.0. The most common form of soluble silicate has an m value of 3.3.

The soluble alkali metal silicates have many uses, the largest and most rapidly growing of which arises from the ability to serve as a source of reactive primary silica (qv) species. Knowledge of soluble glass has been traced to antiquity, and Goethe is known to have experimented with it in 1768. Industrial development, however, began in Germany in the early nineteenth century (1). The common form was called wasserglass. Three principal uses in Europe

1

during that time for waterglass were in treating curtains to decrease flammability, as a dung replacement in textile manufacture, and in making bar soaps. By the middle of the nineteenth century silicates were being produced in Germany by Wacker, in France by Kuhlmann, and in Britain by Gossage. In the United States, imported silicates were also being used as a corrosion-inhibiting coating for cannons and for treating wooden docks. Commercial production began in North America during the Civil War. Silicates were used in laundry soaps as a replacement for rosin, which was scarce because of the war. Sodium carbonate [497-19-8] (soda ash) and sand [14808-60-7] fused in an open-hearth furnace produced a glass that was cooled, crushed, and dissolved in water (see SOAP).

These glasses, amorphous polymeric salts of a strong base and a weak acid, give highly alkaline solutions that made them an economical builder for bar soaps and later for detergents based mostly on their solution chemistry. Commercial availability of these solutions also led to other uses, based primarily on their physical properties, ie, the glassy nature of highly concentrated sodium silicate solutions made them effective adhesives and binders. The third broad market for soluble silicates was provided later by the manufacture of synthetic pigments (qv), fillers, gels (qv), clays (qv), and zeolites (qv), where the silicates might be regarded as value-added sand.

As of the mid-1990s, soluble silicates are used primarily as sources of reactive silica (57%), in detergency (qv) (23%), in pulp (qv) and paper (qv) production (7%), for adhesives and binders (5%), and in other applications (8%). The structure and chemistry of solutions containing polymeric silicate species have been characterized using modern analytical techniques. This improved understanding of silicate speciation contributes to the development of new markets. Thus, the sodium silicates constitute a versatile, stable, and growing commodity and are ranked among the top 50 commodity chemicals.

Structure

The crystalline mineral silicates have been well characterized and their diversity of structure thoroughly presented (2). The structures of silicate glasses and solutions can be investigated through potentiometric and dye adsorption studies, chemical derivatization and gas chromatography, and laser Raman, infrared (ftir), and ^{29}Si Fourier transform nuclear magnetic resonance (^{29}Si ft-nmr) spectroscopy. References 3–6 contain reviews of the general chemical and physical properties of silicate materials.

Silicate Glasses. Synthetic silicates and silica are composed of oligomers of the orthosilicate ion, SiO_4^{4-}. Orthosilicate monomers have a tetrahedral structure. Silicate tetrahedra can be used to construct more complex structures according to Pauling's rules, ie, Si–O–Si bonds are permitted only at polygon corners. As shown in Table 1, structural categories are possible. The Q^s notation refers to the connectivity of silicon atoms (7). The value of the superscript, s, defines the number of nearest neighbor silicon atoms to a given Si (Fig. 1). Many framework structural modifications are available within these categories. Bonding the monomers at all vertices yields the completely condensed structure of quartz, $(SiO_2)_n$. Quartz does not exhibit a close-packed structure because of constraints imposed by Pauling's rules. Therefore, alternate vacancies are present

Table 1. Silicate Structural Categories

Silicate type	Unit structure	Mineral examples		Formula	Q^s structures[a,b]
		Name	CAS Registry Number		
oligosilicates, discreet					
orthosilicate	SiO_4^{4-}	zircon	[14940-68-2]	$Zr[SiO_4]$	Q^0
pyrosilicate	$Si_2O_7^{6-}$	thortveitite	[17442-06-7]	$Sc_2[Si_2O_7]$	Q^1Q^1
cyclosilicates, discrete cyclic					
trimer	$Si_3O_9^{6-}$	benitoite	[15491-35-7]	$BaTi[Si_3O_9]$	$(Q^2)_3$
tetramer	$Si_4O_{12}^{8-}$	papagoite	[12355-62-3]	$Ca_2Cu_2Al_2[Si_4O_{12}](OH)_6$	$(Q^2)_4$
hexamer	$Si_6O_{18}^{12-}$	dioptase	[15606-25-4]	$Cu_{12}[Si_6O_{18}]\cdot6H_2O$	$(Q^2)_6$
polysilicates, chains					
pyroxenes	$(SiO_3^{2-})_n$	wollastonite	[14567-51-2]	$Ca[SiO_3]$	$(Q^2)_n$
amphiboles	$(Si_8O_{22}^{12-})_n$	tremolite	[14567-73-8]	$Ca_2Mg_5[Si_8O_{22}](OH,F)_2$	$(Q^3Q^2)_n$
phyllosilicates, sheets	$(Si_4O_{10}^{2-})_n$	talc	[14807-96-6]	$Mg_3[Si_4O_{10}](OH)_2$	$(Q^3)_n$
tectosilicates, frameworks	$(SiO_2)_n$	silica, quartz	[7631-86-9]		$(Q^4)_n$

[a]Refers to the connectivity of silicon atoms. See Fig. 1.
[b]End groups or surface silicons for more condensed structures are ignored.

3

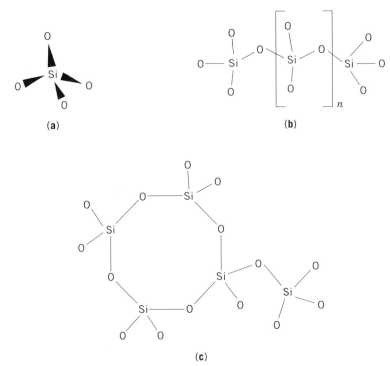

Fig. 1. Examples of Q^s structures for silicates: (**a**) monomer where Q^0 is present; (**b**) metasilicates where $Q^1[Q^2]_n Q^1$ is present; and (**c**) branched-ring structure where $Q^2 Q^2 Q^2 Q^3 Q^1$ is present.

in half of the tetrahedral positions. In the formation of soluble sodium silicate glasses, Na^+ and O^{2-} ions are introduced into the quartz network, breaking siloxane bonds, Si–O–Si, to form SiO^- sites, ie, nonbridging oxygen atoms. Sodium ions are thought to be distributed nonuniformly in the interstices of the disordered silica network. These produce regions rich in SiO_2 polymers and other areas enriched in cations (8). Mineral silicate structures are expected to exist in regions of short-range order as described by a random polymer order–disorder equilibrium model that is similar to the flickering cluster model of water structure. Thus, the glass state is envisioned as a complex mixture of silicate anions that contains alkali cations distributed in the interstitial voids, bonded, on average, by a number of oxygen atoms equal to the cation coordination number.

The physical and chemical properties of silicate glasses depend on the composition of the material, ion size, and cation coordination number (9). A melt or glass having a SiO_2/Na_2O ratio of 1, ie, sodium metasilicate [*1344-09-8*], is expected to possess a high proportion of $(SiO_2^{3-})_n$ chains. At a ratio of 2, sheets might predominate. However, little direct evidence has been shown for a clear predominance of any of these structures. The potential structures of silicate melts of different ratios are discussed in detail elsewhere (10–12).

Crystalline Alkali Silicates. The most common crystalline soluble silicates belong to the metasilicate family, $Na_2O \cdot SiO_2 \cdot nH_2O$, in commercial notation. Anhydrous sodium monopolysilicate [6834-92-0], Na_2SiO_3, contains SiO_2 chains, whereas the hydrates $Na_2H_2SiO_4 \cdot xH_2O$, where $x = n - 1$, contain only the dihydrogen monosilicate ion (13). The structures of the series sodium dihydrogen monosilicate tetrahydrate [10213-79-3], sodium dihydrogen monosilicate pentahydrate [35064-64-3], sodium dihydrogen monosilicate heptahydrate [27121-04-6], and sodium dihydrogen monosilicate octahydrate [13517-24-3] vary primarily by the order and coordination of the hydrated sodium ion. However, some differences in the symmetry and interatomic bonding of the $[Si(OH)_2O_2]^{2-}$ groups are observed. Only the anhydrous and tetrahydrate forms, ie, the so-called pentahydrate, are of commercial importance. However, solutions of silicates having values of $m < 1.6$ and solids $\geq 20\%$ are unstable with respect to the octahydrate.

Other known mineral structures of sodium silicates include natrosilite [56941-93-6], $Na_2Si_2O_5$; magadiite [12285-88-0], $Na_2Si_{14}O_{29} \cdot 11H_2O$; kenyaite [12285-95-9], $Na_2Si_{22}O_{45} \cdot 10H_2O$; makatite [27788-50-7], $Na_2Si_4O_9 \cdot 5H_2O$; and kanemite [38785-33-0], $NaHSi_2O_5 \cdot 3H_2O$, in addition to the aforementioned synthetic metasilicates (14). Kenyaite and magadiite (15) are found associated with other siliceous mineral deposits formed in sodium carbonate-rich alkaline waters. Materials having similar chemical compositions have formed in sodium silicate solutions during glass dissolution, particularly when the ratio exceeds about 2.5. Silicate producers over the years have attempted to find markets for these nuisance by-products of sodium silicate manufacture. The preparation of magadiite and some of its intercalation compounds were first described (16) in 1972. The synthesis of kenyaite was reported later (17). As of the mid-1990s these so-called tetrasilicates are receiving renewed attention in the detergent and catalysts markets owing to their layered structure.

Dissolution

The dissolution of soluble silicates is of considerable commercial importance. Its rate depends on the glass ratio, solids concentration, temperature, pressure, and glass particle size. Commercially, glasses are dissolved in either batch atmospheric or pressure dissolvers or continuous atmospheric processes. Dissolution of sodium silicate glass proceeds through a two-step mechanism that involves ion exchange (qv) and network breakdown (18).

Ion exchange $\equiv Si-ONa + H_2O \rightleftharpoons \equiv Si-OH + Na^+ + OH^-$ (1)

Network breakdown $\equiv Si-O-Si\equiv + HO^- \rightleftharpoons \equiv Si-O^- + HO-Si\equiv$ (2)

Thus, silica is removed from the glass following leaching of the alkali cations. The rate of SiO_2 extraction increases as the pH of the solution rises above

9. Extraction of alkali occurs more readily at pH values below the pK_a of the glass surface silanol functionalities. If sufficient hydroxide activity is not generated in the ion-exchange step, the silica network does not decompose and glass leaching results. The transition state in the decomposition process may involve pentacoordinate silicon atoms. The presence of Na^+ in the glass network retards the overall dissolution rate by hindering nucleophilic attack by hydroxide. The rate of SiO_2 removal from a potassium silicate glass depends on the rate of alkali leaching and not on the quantity of alkali removed (19). Smaller cations having higher charge densities produce less soluble silicate glasses as $Li^+ < Na^+ < K^+$. The presence of multivalent metal ion impurities, eg, Al^{3+}, Ca^{2+}, or Fe^{3+}, also significantly reduces glass solubility. A correlation exists between the ratio of free and total silanol groups and glass solubility (20). This ratio varies linearly as different metal ions are added to Na-, K-, or Li-based silicate glasses (4.0 mol ratio). The addition of water depresses this ratio significantly.

A tower operated at atmospheric pressure for dissolving glass has been described (21). The dissolution rate is independent of liquor concentration and circulation rate. The principal factors are temperature, glass composition, and particle surface area. The glass must be sized to avoid a phenomenon referred to as sticker, which occurs when a dissolving glass mass solidifies. Studies of the dissolution rate of a 2.0 ratio sodium silicate glass into concentrated solutions indicate that the rate of dissolution, expressed as kg dissolved per hour per kg of glass, is independent of the initial particle size (22). In addition, only linear increases in the solution concentration as a function of time were observed under conditions in which Na^+ ion exchange was suppressed by an increase of Na^+ activity of the dissolving liquor. The rate of increase in solution concentration appears to be related inversely to the sodium ion activity. It is also expected that the dissolution process would be sensitive to the amount of CO_3^{2-} ion either in the glass owing to incomplete fusion or on the surface owing to interaction of stored glass with air.

Silicates in Solutions. The distribution of silicate species in aqueous sodium silicate solutions has long been of interest because of the wide variations in properties that these solutions exhibit with different moduli (23–25). Early work led to a dual-nature description of silicates as solutions composed of hydroxide ions, sodium ions, colloidal silicic acid, and so-called crystalloidal silica (26). Crystalloidal silica was assumed to be analogous to the simple species then thought to be the components of crystalline silicate compounds. These include charged aggregates of unit silicate structures and silica (ionic micelles), and well-defined silicate anions.

Subsequent research using light-scattering techniques showed that stable silicate solutions did not contain large particles (27). Aggregation of particles was detected, however, in highly concentrated, high ratio silicate solutions. Well characterized sodium silicate solutions free of trace metal impurities and visible colloidal particles must be used when studying the properties of solutions. Conclusions from studies which do not control for ion impurities are suspect.

Polymers in solutions of a variety of alkali silicates have been studied by preparing, separating, and identifying trimethylsilyl derivatives using gas chromatography (28–31). This work gives strong evidence for the presence of a variety of Si_1–Si_8 silicate structures, even in highly alkaline, dilute silicate solutions. Highly polymerized Q^1, Q^2, Q^3, and Q^4 species were detected, especially in solutions having compositions close to the solubility limit of amorphous silica. Conclusions concerning the quantitative speciation of silicate solutions are weakened, however, because the derivatization method induces polymerization and rearrangement reactions (32). Gel chromatography (33) indicates that the solutions contain a complex mixture of silicate anions in dynamic equilibrium.

Laser Raman spectroscopy and ^{29}Si ft-nmr spectroscopy have been used to examine directly the structure of silicate species in solution (34–41). Early laser Raman spectroscopic studies concluded that dilute, 0.3 M sodium silicate solutions of varying ratios (m = 0.33–3.3) contain an equilibrium mixture of monomeric silicate, $SiO(OH)_3^-$ and $SiO_2(OH)_2^{2-}$, and polymeric anions, regardless of their histories and methods of preparation. Subsequent research showed that monomeric and dimeric silicate species polymerize to form cyclic polysilicate anions, especially under alkaline conditions (pH 11), and that the distribution of ionic species changes upon aging of the solution (34–36). The ^{29}Si nmr spectra of alkali silicate solutions of varying ratios have been measured and the types of Si-containing species and their relative concentrations have been estimated (Fig. 2) (37). Other researchers have employed more specialized techniques, such as two-dimensional nmr, to identify tentatively 18 individual ionic species in potassium silicate solutions (38–41). Cage-like structures based on three-, four-, and five-membered rings were detected, especially in weakly alkaline solutions. Model-building studies showed that structures based on cyclic Si_4 and Si_5 species were compatible with the observed properties of larger solution-phase silicate ions. The proportion of four- and five-membered rings increased as the silicate particles grew (5,42). Structures having fivefold symmetry are postulated to account for the production of spherical particles and for the tendency of silicate particles to avoid crystallizing (5).

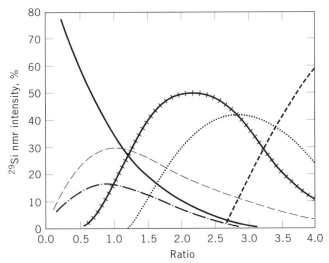

Fig. 2. Distribution of silicon centers in soluble silicate solutions from ^{29}Si Fourier-transform nmr spectroscopy (7,37), where (——) represents Q^0; (— —), Q^1; (—·—); Q^2 (ring); (⊢⊢), Q^2 (chain); (·····), Q^3; and (– – –), Q^4. See Table 1. Values are estimated.

Among the techniques for species determination in soluble silicates, ^{29}Si nmr spectroscopy gives the most information about equilibrium silicate solutions, but trimethylsilylation provides the best means for studying the dynamics of nonequilibrium systems (29,42). An equilibrium state is attained rapidly in relatively pure systems under alkaline conditions, ie, pH $>$ 10. These equilibrium states and the time needed to achieve them appear to be influenced by impurities, their concentration, and state of distribution, eg, in colloidal silicate particles (23). These impurities, eg, CO_3^{2-}, Cl^-, SO_4^{2-}, Al^{3+}, Fe^{3+}, Ca^{2+}, Mg^{2+}, Zn^{2+}, Mn^{2+}, and others, roughly in order of prevalence, come from the sodium carbonate, sand, water, refractories, and other sources and can range from $>$2000 to $<$1 ppm. For example, chloride has been analyzed from 130 to 1900 ppm and Mn^{2+} from 100 ppb to 1.8 ppm. They can vary with processing conditions and locations of manufacturing facilities.

In acidic solutions, equilibrium is achieved more slowly. Polymerization of smaller species appears to occur sequentially; a given polymer species first increases in size and then disappears, presumably because of its inclusion in higher order polymers. Depolymerization of silicate species appears to be rapid, because crystalline Na_2SiO_3 and $Na_2H_2SiO_4 \cdot 8H_2O$ yield equivalent distributions of silicate species in water upon dissolution.

Sjöberg and co-workers combined emf and nmr data to develop a model to describe polymerization and protonation of silicate species in sodium silicate solutions (43,44). Thirteen polynuclear species are described. This model yields useful insights into the changes in polymerization of sodium silicate solutions as $[H^+]$, $[Na^+]$, and $[SiO_2]$ change. Higher order polynuclear species predominate as $[SiO_2]$ increases or pH drops under alkaline conditions. The results suggest that polymers near the solubility limits of amorphous silica contain up to 50 Si atoms and the average charge-to-Si ratio is roughly 0.5. It is expected that these large particles possess a significant portion of their silanol sites trapped internally or on the surface of the particle and that the average connectivity value, s, approaches a limiting value of 3.5 (29).

Polymerization in Solution

Polymerization and depolymerization of silicate anions and their interactions with other ions and complexing agents are of great interest in sol–gel and catalyst manufacture, detergency, oil and gas production, waste management, and limnology (45–50). The complex silanol condensation process may be represented empirically by

$$\equiv Si-OH \overset{K_a}{\rightleftharpoons} \equiv Si-O^- + H^+ \qquad (3)$$

$$\equiv Si-O^- + HO-Si\equiv \underset{k_{-2}}{\overset{k_2}{\rightleftharpoons}} \equiv Si-O-Si\equiv + HO^- \qquad (4)$$

Condensation occurs most readily at a pH value equal to the pK_a of the participating silanol group. This representation becomes less valid at pH values above 10, where the rate constant of the depolymerization reaction (k_{-2}) becomes significant and at very low pH values where acids exert a catalytic influence on polymerization. The pK_a of monosilicic acid is 9.91 ± 0.04 (51). The pK_a value of Si−OH decreases to 6.5 in higher order silicate polymers (52), which is consistent with pK_a values of 6.8 ± 0.2 reported for the surface silanol groups of silica gel (53). Thus, the acidity of silanol functionalities increases as the degree of polymerization of the anion increases. However, the exact relationship between the connectivity of the silanol silicon and SiOH acidity is not known.

The state of ionization of the silica particle surface controls the rate of polymerization following homogeneous nucleation. The rate of reaction of dissolved silicate at the surface of amorphous SiO_2 is proportional to the density of ionized silanol groups. The degree of surface ionization also influences the value of the surface tension and hence the rate of homogeneous nucleation. Addition of salts to increase ionic strength accelerates homogeneous nucleation and deposition of silicate by increasing the extent of SiOH ionization and decreasing SiO_2 solubility. Incorporation of these concepts into a quantitative theoretical model of homogeneous nucleation accurately predicts the rate of silica particle growth under well-defined laboratory conditions (50).

Silicate polymerization in dilute solutions at pH values up to ca 10 is sensitive to pH and other factors that generally influence colloidal systems, eg, ionic strength, dielectric constant, and temperature. Larger particles grow at the expense of smaller particles (Ostwald ripening), especially in strongly alkaline solutions where the latter dissolve more readily. This results from the tendency of the smaller particles to condense at the surfaces of the larger particles. For example, polymerization occurs primarily on particle surfaces when colloidal silica particles are dispersed in a soluble silicate solution. Lower pH values and higher ionic strength conditions promote the growth of smaller particles.

Extensive research on sol−gel processing of silicic acid esters, eg, tetraethoxysilane (TEOS), $Si(OC_2H_5)_4$, in alcohol−water mixtures has elucidated silica polymerization in nonaqueous solvents (54,55). The relative rates of hydrolysis and condensation depend on the substrate, water, and catalyst (acid or base) concentrations; these rates determine the polymer structure. For example, acid-catalyzed hydrolysis of TEOS at low water concentrations produces linear polymers. These solutions yield fibers upon spinning. Conversely, high H_2O/TEOS ratios favor the formation of highly cross-linked polymers under acidic conditions. Microspherical silica particles form in the presence of ammonia (56). Catalyst and substrate concentrations have similar effects on alkali silicate polymerization in aqueous solution. Silica polymers produced by acid hydrolysis of TEOS and acidification of sodium silicate exhibit a ladder framework structure of three SiO units in width through the linear condensation of oligomers. This solution forms fibers upon unidirectional freezing (48). Numerous investigators have attempted to produce polysiloxane (silicone) fluids by the esterification of alkali and mineral silicates. However, the process yields polymers of insufficient molecular weight for practical utility. If the concentration of SiO_2 is sufficiently high, ca 1 wt %, interparticle aggregation and ultimately network formation (gelation) occur, yielding a continuous structure throughout the medium. The gel structure initially encompasses the entire system and appears to be uniform, but subsequent condensation within the gel causes shrinkage and release of water, ie, syneresis. In more concentrated solutions, network formation occurs to a sufficient extent to produce a gel point at a lower total conversion of low to high molecular weight polymers (57). Therefore, the pH values of the maximum gelation rate of dilute silicates are lower than those for more concentrated solutions, given a constant initial degree of polymerization. Furthermore, a second-order reaction mechanism predicts the pH changes that are observed during polymerization; this prediction is based on the initial pH value of the solution and the acid-dissociation constants of the silanol groups. A quantitative description of the polymerization–depolymerization kinetics requires a more complete representation of the specific anions formed in solution, an understanding of the influence of trace impurities for commercial systems, and greater knowledge of the activity variations of the potential condensation sites present on the silica in solution.

Silicate solutions of equivalent composition may exhibit different physical properties and chemical reactivities because of differences in the distributions of polymer silicate species. This effect is keenly observed in commercial alkali silicate solutions with compositions that lie in the metastable region near the solubility limit of amorphous silica. Experimental studies have shown that the precipitation boundaries of sodium silicate solutions expand as a function of time, depending on the concentration of metal salts (29,58). Apparently, the high viscosity of concentrated alkali silicate solutions contributes to the slow approach to equilibrium.

Chemical Activity

Silica Polymer–Metal Ion Interactions in Solution. The reaction of metal ions with polymeric silicate species in solution may be viewed as an ion-exchange process. Consequently, it might be expected that silicate species acting as ligands would exhibit a range of reactivities toward cations in solution (59). Silica gel forms complexes with multivalent metal ions in a manner that indicates a correlation between the ligand properties of the surface Si–OH groups and metal ion hydrolysis (60,61). For Cu^{2+}, Fe^{3+}, Cd^{2+}, and Pb^{2+},

$$pK_1^s = 0.09 + 0.62p^*K_1 \tag{5}$$

where K_1^s and $^*K_1$ are the stability constants of the surface complex and metal ion hydrolysis complex, respectively. Metal ion adsorption on silica gel may be initiated at a pH value corresponding to surface nucleation. This seems to relate to a reduction of cation–solvent interactions leading to conditions favorable for adsorption of hydrated metal ions from solution. Metal ion hydrolysis may be required before adsorption (62); direct participation by unhydrated ions has also been proposed (61). Other studies suggest that cations adsorb to silica gel surfaces as a result of hydroxyl ion adsorption, which carries an equivalent amount of cations to the surface (63).

At a given pH value, the solution activities of Ca^{2+}, Mg^{2+}, and Cu^{2+} decrease to a greater extent in the presence of SiO_2 derived from 2.0- and

3.8-ratio silicates than they do in solutions prepared from sodium orthosilicate (64). Thus, highly polymerized silicate anions appear to interact with metal ions in solution in a manner analogous to silica gel, and the interaction decreases as the degree of silicate polymerization decreases. This is consistent with the observation that silica suspended in solutions of polyvalent metal salts begins to adsorb metal ions when the pH value is raised to within 1–2 pH units of the OH^- concentration at which the corresponding metal hydroxides precipitate (65).

The increased acidity of the larger polymers most likely leads to this reduction in metal ion activity through easier development of active bonding sites in silicate polymers. Thus, it could be expected that interaction constants between metal ions and polymer silanol sites vary as a function of time and the silicate polymer size. The interaction of cations with a silicate anion leads to a reduction in pH. This produces larger silicate anions, which in turn increases the complexation of metal ions. Therefore, the metal ion distribution in an amorphous metal silicate particle is expected to be nonhomogeneous. It is not known whether this occurs, but it is clear that metal ions and silicates react in a complex process that is comparable to metal ion hydrolysis. The products of the reactions of soluble silicates with metal salts in concentrated solutions at ambient temperature are considered to be complex mixtures of metal ions and/or metal hydroxides, coagulated colloidal size silica species, and silica gels.

Effect on Oxide–Water Interfaces. The adsorption (qv) of ions at clay mineral and rock surfaces is an important step in natural and industrial processes. Silicates are adsorbed on oxides to a far greater extent than would be predicted from their concentrations (66). This adsorption maximum at a given pH value is independent of ionic strength, and maximum adsorption occurs at a pH value near the pK_a of orthosilicate. The pH values of maximum adsorption of weak acid anions and the pK_a values of their conjugate acids are correlated. This indicates that the presence of both the acid and its conjugate base is required for adsorption. The adsorption of silicate species is far greater at lower pH than simple acid–base equilibria would predict. This may be due to ion–oxide surface interactions or to ions already present on the surface. Alternatively, this deviation may reflect varying silicate polymer acidity. Similar behavior has been observed for the adsorption of aqueous silica to the surface of γ-Al_2O_3 (67). Divalent metal ions tend to reduce silicate adsorption.

The addition of polymeric silicate anions to oxide mineral suspensions increases the magnitude of the negative surface charge of the mineral particles (68). When silicate solutions are used instead of NaOH to adjust the pH of suspensions of ground quartz in 0.1 mM $PbCl_2$, the pH of maximum positive particle surface charge shifts to lower values and the pH range of positive charge narrows. These effects are more pronounced for polymeric silicates. A similar reduction of the influence of multivalent cations on quartz surface properties occurs in 0.1 mM $FeCl_3$ and mixtures of $PbCl_2$ and $FeCl_3$ regardless of Ca or Mg hardness. The influence of silicate polymers on iron oxide sol surfaces has also been studied (69). The effectiveness of soluble silica in discharging and recharging the sol surface increases with the modulus of the original sodium silicate solution. Thus, soluble silicates adsorb specifically to oxide surfaces and play a significant role in maintaining a negative surface charge on oxide particles in the presence of cations that could otherwise reverse the surface charge.

Characteristics

The characteristics of soluble silicates relevant to various uses include the pH behavior of solutions, the rate of water loss from films, and dried film strength. The pH values of silicate solutions are a function of composition and concentration. These solutions are alkaline, being composed of a salt of a strong base and a weak acid. The solutions exhibit up to twice the buffering action of other alkaline chemicals, eg, phosphate. An approximately linear empirical relationship exists between the modulus of sodium silicate and the maximum solution pH for ratios of 2.0 to 4.0.

$$pH = 13.4 - 0.69m \qquad (6)$$

This relationship permits the ratio of a pure, concentrated sodium silicate solution to be estimated. The rate of water loss from silicate solutions and films increases at higher ratios. If films or solutions are dried at a given temperature and humidity, a change in condition initiates further drying or rehydration, depending on whether drying is sufficient to insolubilize the silicate. Hydrated glass films made from silicates with higher ratios or from those containing metal ions that reduce solubility rehydrate less rapidly. Finally, the dried-film characteristics depend on the glass ratio. Silicates having higher ratios produce more brittle films.

Manufacture and Processing

Soluble silicate glasses are usually manufactured in oil- or gas-fired open-hearth regenerative furnaces. Newer continuous-flow glass melters are equipped with high intensity gas burners or plasmas (70,71). These latter technologies offer significant advantages over the conventional batch-melting processes. Glass composition and shutdown and start-up procedures can be changed rapidly because no large molten glass reservoir is maintained (70). In conventional processes, the glass is made by the reaction of sand and sodium carbonate (soda ash) at 1100–1400°C, a temperature sufficiently high to provide a reasonable rate of quartz dissolution in the molten batch and manageable melt viscosity. Furnace life is dependent on the quality of the refractories, corrosivity of the melt, and manufacturing technique. Generally, three years is regarded as normal between rebuilds. The liquidus diagrams for sodium silicate glasses and lines of constant melt viscosity are shown in Figure 3. The rate of reaction of quartz with Na_2CO_3 is controlled by silica diffusion and varies inversely with the square of the radius of the quartz particles (72). As the Na_2CO_3 melt envelops the sand grains, the silica network breaks down and diffuses slowly into the melt. Thermogravimetric analysis indicates that the kinetics of the reaction of sodium carbonate and SiO_2 may be adapted to a modified Ginstling and Brounsthein model (73,74). The glass melts are highly corrosive toward refractory materials compared with the common bottle glasses and the furnace must be carefully designed. Electric melting furnaces can be used satisfactorily where low cost electric power is available (75) and offer the advantage of eliminating contact with products of fuel combustion, eg SO_x, which could affect the glass.

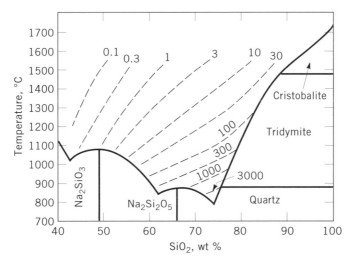

Fig. 3. The viscosity and liquidus curves for molten sodium silicates, where the numbers on the dashed lines are viscosity in Pa·s (1). Courtesy of The PQ Corp. To convert Pa·s to poise, divide by 10.

The glass product can be drawn and formed into solid lumps or drawn directly into a rotary dissolver. The glass lumps can also be dissolved in a pressure apparatus at pressures up to 690 MPa (100 psi). The concentrations and ratios of the solutions are monitored during manufacture using alkali–gravity–viscosity (AGV) charts (Fig. 4). These charts and a knowledge of the density and Na_2O content of the solution may be used to determine the approximate ratio and

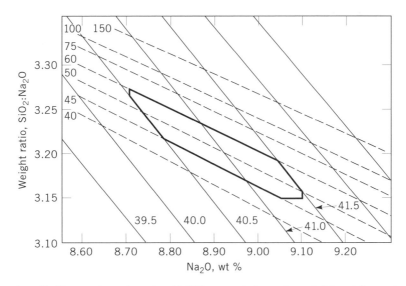

Fig. 4. An alkali–gravity–viscosity (AGV) control chart for a 3.22-weight ratio sodium silicate solution, where (——) and (— — —) represent lines of constant density and viscosity, respectively. The area enclosed within the emboldened box represents the range of properties for a typical product. Courtesy of The PQ Corp.

solids concentration of a product. Lump glass is sold as such to processors having dissolving capability or ground to powders of various particle sizes and size distributions. Silicate solutions up to a ratio of ca 2.65 can be manufactured in an autoclave at ca 160°C by dissolving finely ground sand in a NaOH solution. Higher ratio silicates can be prepared from amorphous silica. Caustic soda can be added to reduce the ratio. The choice of the process to obtain a certain ratio is usually based on economics. The solids concentration of the commercial solutions depends on the maximum viscosity that can be tolerated. Figure 5 illustrates the variation of the viscosity of the solution with concentration. The more siliceous products show abrupt increases in viscosity with solids content. This property is valuable in certain adhesive applications. As a high ratio solution is evaporated, the viscosity increases to a point where solid solutions or highly hydrated glasses form. These materials have commercial significance because the rate of dissolution of soluble silicate solids is a function of particle size, ratio, and water content. These dried silicate solutions, or hydrous silicates, which contain ca 20% water, are stable enough to be handled commercially. They dissolve with rapidity compared to the ground glasses, making them attractive in certain applications. Thus, powdered soluble silicate products having a wide range of ratios and dissolution properties can be manufactured to meet the demands of diverse applications (Table 2). Hydrous silicates are manufactured in drum dryers or spray towers.

Crystalline sodium metasilicates are manufactured by processing highly concentrated solutions of sodium silicate (1.0 ratio) or by direct fusion of sand and soda ash, followed by grinding and sizing. Solutions at a metasilicate ratio are dried in a dryer (rotary moving bed or fluid bed). The metasilicate solution is sprayed onto the bed and the beads formed are screened as they leave the dryer. The fines are recycled to provide seed material for further growth. This process yields a uniformly sized, readily soluble, anhydrous sodium metasilicate (ASM) product. The only commercially available hydrated sodium metasilicate,

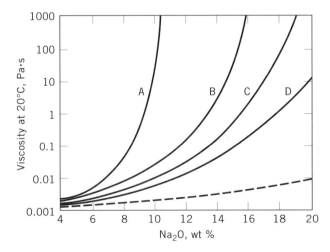

Fig. 5. Viscosities of sodium silicate solutions vs percent solids: A, ratio of 3.22: B, 2.4; C, 2.0; D, 1.6; and (— — —), NaOH. To convert Pa·s to poise, divide by 10. Courtesy of The PQ Corp.

Table 2. Solution Rates of Amorphous Silicate Powders[a]

Silicate, wt % ratio	Particle size, μm (mesh)[b]	Time needed to dissolve, min					
		at 25°C			at 50°C		
		50%	75%	100%	50%[c]	75%	100%
3.22 ratio sodium							
anhydrous glass	230 (65)	3600			30[d]		
hydrated[e]	149 (100)	19	45		0.9	1.3	1.7
2.00 ratio sodium							
anhydrous glass	230 (65)	600	4200		17		60
hydrated[e]	149 (100)	0.45	0.90		0.25	0.37	0.48
2.50 ratio potassium, anhydrous glass	230 (65)	60	450	2900	12	45	

[a]Three parts water to one part silicate powder by wt. [b]Tyler screen. [c]Unless otherwise noted. [d]Value corresponds to 15% dissolved. [e]Has 18.5 wt % H_2O.

$Na_2H_2SiO_4 \cdot 4H_2O$, often called a pentahydrate, is manufactured by preparing a solution of pentahydrate composition at 72.2°C and allowing the mass to cool.

Sodium orthosilicates are produced by blending ASM and NaOH beads or by fusion and grinding as in the direct manufacture of ASM. The relationships of these processes are shown in Figure 6.

Potassium silicates are manufactured in a manner similar to sodium silicates by the reaction of K_2CO_3 and sand. However, crystalline products are not manufactured and the glass is supplied as a flake. A 3.90 mole ratio potassium silicate flake glass dissolves readily in water at ca 88°C without pressure by incremental addition of glass to water. The exothermic heat of dissolution causes the temperature of the solution to rise to the boiling point. Lithium silicate solutions are usually prepared by dissolving silica gel in a LiOH solution or mixing colloidal silica with LiOH.

Commercial Products

The average composition and pertinent properties of commercial soluble silicates are given in Table 3. The largest volume of these materials is sold as liquids, differentiated by the SiO_2/Na_2O ratio and specific gravity. The powdered forms are useful as ingredients in dry-blended products; to control silicate reactivity; for convenience in handling and storage; or to increase silicate solution concentrations above the ranges available in liquid form. Sodium orthosilicate and metasilicates afford high alkalinity. The federal specifications covering commercial silicates are shown below:

Commodity	Specification number
sodium metasilicate	O-S-604D
type I, pentahydrate	
type II, anhydrous	
sodium orthosilicate	P-S-651E
sodium silicate solution	O-S-605D

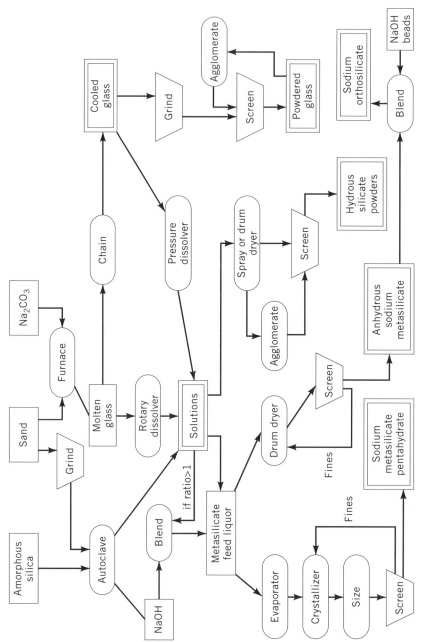

Fig. 6. Manufacturing routes for commercial sodium silicates. Courtesy of The PQ Corp.

16

Table 3. Commercial Sodium and Potassium Silicates

Commercial silicates	SiO₂, wt %	Wt ratioᵃ, SiO₂:M₂O	Modulusᵃ, SiO₂:M₂O	Softening pointᵇ, °C	Flow pointᶜ, °C	H₂O, wt %	°Bauméᵈ at 20°C	d_{20}^{20}, g/cm³	Viscosity, at 20°C, Pa·sᵉ	pH
anhydrous glasses										
sodium silicates	75.7	3.22	3.33	655	840					
	66.0	2.00	2.06	590	760					
potassium silicates	70.7	2.50	3.92	700	905					
hydrated amorphous powders										
sodium silicates	61.8	3.22	3.33			18.5				
	64.0	2.00	2.06			18.5				
solutions										
sodium silicates	31.5	1.60	1.65				58.5	1.68	7.00	12.8
	36.0	2.00	2.06				59.3	1.69	70.00	12.2
	26.5	2.50	2.58				42.0	1.41	0.06	11.7
	31.7	2.88	2.97				47.0	1.49	0.96	11.5
	28.7	3.22	3.32				41.0	1.39	0.18	11.3
	25.3	3.75	3.86				35.0	1.32	0.22	10.8
potassium silicates	20.8	2.50	3.93				29.8	1.259	0.04	11.30
	19.9	2.20	3.45				30.0	1.261	0.01	11.55
	26.3	2.10	3.30				40.0	1.381	1.05	11.70
	29.5	1.80	2.83				47.7	1.490	1.30	12.15
crystalline solids										
sodium orthosilicate	28.8	60.8ᶠ	0.50			9.5				
anhydrous sodium metasilicate	47.1	51.0ᶠ	1.00			2.0				
sodium metasilicate pentahydrate	26.4	29.3ᶠ	1.00			42.0				
sodium sesquisilicate [1344-09-8]	24.1	36.7ᶠ	0.67			38.1				

ᵃM represents Na or K. ᵇViscosity reaches 4 MPa·s (4×10^7 P). ᶜViscosity reaches 10 kPa·s (10^5 P). ᵈTo convert Be° to sp gr, divide 145 by (145 − Be°). ᵉTo convert Pa·s to P, divide by 10. ᶠValue is wt % of M₂O.

17

Handling and Storage.   Liquid silicates are available in drums or bulk quantities. Bulk shipments are transported by truck, rail, or tanker; large volumes are transferred by gravity or pumping. Closed, vented, or carbon-steel tanks with capacities of 45–57 m^3 are recommended for rail cars and not less than 23 m^3 for tank trucks. Powdered silicates are available in 50- or 100 lb (23 or 45-kg) bags, 300 or 350-lb (136 or 159-kg) fiber drums, or 2000-lb (907-kg) semibulk bags. A pressure-differential bulk unloading system is recommended for transfer. A minimum storage capacity of 28 metric tons is advisable for bulk. The storage bins should be controlled for humidity because of the hygroscopic nature of powdered silicates and to minimize uptake of CO_2 (76). Although the packaging of most bagged products is designed to provide humidity protection for reasonable periods of time, additional protection may be required if unusually high humidity or prolonged storage is anticipated.

Regulatory Status

Additives to Food and Potable Water.   Sodium silicate is generally recognized as safe (GRAS) by the FDA in fabrics and when exposed to food (77). It is also recognized as a secondary direct food additive when used in boiler water for food-contact steam (78). In addition, it has unpublished GRAS status as a corrosion inhibitor in potable water. The safety of sodium silicate in these and other food-related applications has been extensively reviewed (79). Sodium metasilicate has been affirmed GRAS for washing and lye peeling of fruits, vegetables, and nuts (80); in tripe denuding and hog scaling; and as a corrosion inhibitor in canned and bottled drinking water (81). The use of sodium silicate in canned emergency drinking water at concentrations up to 100 ppm is described in a U.S. federal military specification (82). Sodium silicate and sodium metasilicate are authorized by EPA and many state agencies as drinking water additives for clarification of potable water or as a corrosion inhibitor in piped water systems (83). A specification for sodium silicate appears in the Water Chemicals Codex (84) and the American Waterworks Association Standards (85).

Transportation and Disposal.   Only highly alkaline forms of soluble silicates are regulated by the U.S. Department of Transportation (DOT) as hazardous materials for transportation. When discarded, these are classified as hazardous waste under the Resource Conservation and Recovery Act (RCRA). Typical members of this class are sodium silicate solutions having silica-to-alkali ratios of less than 1.6 and sodium silicate powders with ratios of less than 1.0. In the recommended treatment and disposal method, the soluble silicates are neutralized with aqueous acid (6 M H^+ or equivalent), and the resulting silica gel is disposed of according to local, state, and federal regulations. The neutral liquid, a salt solution, can be flushed into sewer systems (86).

Consumer Products.   The Consumer Product Safety Commission's (CPSC) labeling criteria under the Federal Hazardous Substances Act are based on biological testing. In the absence of specific data, CPSC requires specific cautionary statements for consumer products containing certain types and amounts of sodium silicates (87).

Occupational Safety and Health.   OSHA has set no specific limits for sodium and potassium silicates (88). A prudent industrial exposure standard could range from the permissible exposure limit (PEL) for inert or nuisance particulates to the PEL for sodium hydroxide, depending on the rate of dissolution and the concentration of airborne material. Material safety data sheets issued by silicate producers should be consulted for specific handling precautions, recommended personal protective equipment, and other important safety information.

Production and Economic Aspects

Worldwide silicate production is estimated to be approximately 4×10^6 t/yr. The 1995 prices for typical products are given in Table 4. The price for a liquid product, normalized to a 100% solids basis, is usually close to the price of the corresponding bulk glass. International trade of silicates, especially liquid products, is limited by the cost of transportation rather than production costs. Production of silicates has increased in the 1990s (Table 5). This growth is primarily attributed to an increase in the use of silicate to make derivatives and also to use in peroxide bleaching.

The main merchant producers of soluble silicates in the United States are The PQ Corporation, OxyChem, Crosfields Chemicals, Power Silicates, and Chemical Products. Several companies including PPG Industries, W. R. Grace, J. M. Huber, DuPont, and Albemarle Corporation produce silicates for captive use. There were more than 30 soluble silicate plants in the United States in 1996 and production capabilities were more than adequate for demands. A key producer worldwide is The PQ Corporation, which has plants in both North and South America, Europe, Asia, and Australia. In Japan, Toso Sangyo, Fuji Chemical, and Nippon Chemical are the largest producers. Henkel, Rhône-Poulenc, Foret, and Crosfield Chemicals are principal producers in Europe (90).

Table 4. 1995 Prices of Silicate Products

Product	Truckload prices fob plant, $/t		
	Bulk	Bags	Drums
sodium silicate lump glass			
3.22 ratio		480	
2.00 ratio		620	
anhydrous sodium metasilicate	670	690	760
sodium metasilicate pentahydrate	440	450	520
hydrous sodium silicate			
2.00 ratio, 18 wt % water		840	960
3.22 ratio, 18.5 wt % water		730	850
potassium silicate flake glass, 2.50 ratio	1200		

Analysis and Test Methods

Classical wet chemical methods and instrumental techniques are used for the analysis of soluble silicates (91). Silica is determined by gravimetric techniques or by the fluorosilicate volumetric procedure.

$$Si(OH)_4 + 6\,NaF \longrightarrow Na_2SiF_6 + 4\,NaOH \qquad (7)$$

Table 5. Production and Shipment of Sodium Silicates, 1955–1995,[a] t × 10³

Year	Sodium silicate solution[b]		Sodium metasilicate[c]		Sodium orthosilicate production[d]
	Production	Shipment	Production	Shipment	
1955	570	422	142	117	34
1965	533	414	228	186	38
1970	570	423	204	162	31
1975	656	501	190	165	
1980	730	493	152	132	11
1985	627	422	97	92	
1986	717	465	88	97	
1987	863	596	97	97	
1988	736	492	97	101	
1989	755	518	93	97	
1990	740	494	98	93	
1991	789	497	104	90	
1992	854	496	105	88	
1993	905		93		
1994	941		88		
1995	1020		82		

[a]Ref. 89.
[b]Soluble sodium silicate glass solid and liquid (anhydrous). Excludes quantities consumed in the manufacture of meta-, ortho-, and sesquisilicates. Includes quantities consumed in the manufacture of glass powder, hydrated glasses, and precipitated products. Shipment figures include unspecified amounts shipped to other plants for the manufacture of meta-, ortho-, or sesquisilicates.
[c]All sodium metasilicate products (SIC No. 2819743) are given on a 100 wt % sodium metasilicate pentahydrate basis.
[d]Estimated production data for 100 wt % sodium orthosilicate (SIC No. 2819745).

The sodium hydroxide is titrated with HCl. In a thermometric titration (92), the silicate solution is treated first with hydrochloric acid to measure Na_2O and then with hydrofluoric acid to determine precipitated SiO_2. Lower silica concentrations are measured with the silicomolybdate colorimetric method or instrumental techniques. X-ray fluorescence, atomic absorption and plasma emission spectroscopies, ion-selective electrodes, and ion chromatography are utilized to detect principal components as well as trace cationic and anionic impurities. Fourier transform infrared, ft-nmr, laser Raman, and x-ray photoelectron spectroscopic techniques allow identification of the molecular structure of silicates in solid materials and solutions.

Health, Safety, and Environmental Aspects

The primary hazard of commercial soluble silicates is their moderate-to-strong alkalinity. Contact–exposure effects can range from irritation to corrosion, depending on the concentration of the silicate solution, the silica-to-alkali ratio, the sensitivity of the tissue exposed, and the duration of exposure. Sodium silicate solutions of commercial concentrations having a $m < 1.8$ and silicate powders of $m < 2.4$ attack the skin when tested according to Federal Hazardous Substances Act (FHSA) protocols (93). Soluble silicates are rapidly absorbed and eliminated if ingested or inhaled (94,95). Trace quantities of silicon are essential in nutrition, possibly as a metal ion bioavailability attenuator (96), but siliceous urinary calculi may result if normal dietary amounts are greatly exceeded (79).

Compounds of silicon and oxygen are the primary constituents of the earth's land mass. Dissolved silica is a minor yet ubiquitous constituent of the hydrosphere. The earth's biomass contains dissolved silica, amorphous solid-phase SiO_2, and silica that is bound to organic matter. Dissolved SiO_2 is supplied to the environment by chemical and biochemical weathering processes at a rate of ca 6 teramoles Si per year. These inputs are roughly balanced by net outputs of biogenic silica (97,98). The immense magnitude of the flux of the natural silica cycle can be appreciated by considering the average silica weathering rate for watersheds: ca 2000 kg/(km^2·yr) (99). Dissolved silica may become depleted in inland waters when algal growth occurs rapidly, resulting in the replacement of diatoms by species that accelerate eutrophication (100). Commercial soluble silicates have a higher degree of polymerization than natural dissolved silica because the former contain higher SiO_2 concentrations. However, the commercial silicates are thought to depolymerize rapidly to molecular species that are indistinguishable from natural material upon dilution (101).

Uses

Alkali silicates are used as components, rather than reactants, in many applications. In many cases they only contribute partially to overall performance. Utility factors are generally not as easy to identify. Their benefit usually depends on the surface and solution chemical properties of the wide range of highly hydrophilic polymeric silicate ions deliverable from soluble silicate products or their proprietary modifications. In most cases, however, one or two of the many possible influences of these complex anions clearly express themselves in final product performance at a level sufficient to justify their use (102). Estimates of the 1995 U.S. consumption of sodium silicates are shown in Table 6.

Detergency. If the volume of sodium silicate used to make zeolites (qv) is included in this market classification, then the largest single use for soluble silicates is in soaps and detergents (ca 42%) (see DETERGENCY). The soluble silicates help provide constant pH values in an effective range and aid in the saponification of oils and fats by means of their alkalinity and buffering capacity. They confer crispness to spray-dried detergent granules because of their glassy nature upon dehydration. In addition, silicates enhance the effectiveness of surfactants (qv) by lowering the activity of certain metal ions; help disperse soil particles; and prevent corrosion of soft metal, eg, tin, aluminum, copper, and brass, owing to complex formation. In dishwashing detergents, silicates protect chinaware and metal utensils from the problem of chelate blush. In addition, silicates reduce the activity of Mg^{2+} hardness and work well with soda ash and zeolites which selectively reduce Ca^{2+} hardness (103).

Water Treatment. Silicates are used in water treatment to prepare activated silica sol, a stable, acid-polymerized silicate suspension that functions as an aid in the alum coagulation of matter suspended in raw and wastewater streams; to inhibit corrosion of metal surfaces in contact with water; to stabilize reduced iron and manganese in water supplies; and in boiler water (104–106).

Table 6. Sodium Silicate Consumption

Application	U.S. market, %		
	1975	1985	1995
detergents	30	26	23
silica source			
zeolites–detergent	1	5	18
silicate-based pigments	20	21	16
cracking catalysts	10	12	15
silica gels	3	2	4
silica sols	1	2	3
titania pigments/other	2	3	1
Total	*37*	*45*	*57*
other			
adhesives	4	5	2
cements/ceramics	4	3	2
roofing granules	3	4	4
ore flotation	2	1	1
water treatment	2	2	2
pulp/textile bleaching	6	5	7
foundry binder	1	1	1
welding rods	1	1	1
miscellaneous	10	7	
Total	*33*	*29*	*20*

Similar applications are found in other areas where large quantities of water are consumed, such as detergency, papermaking, bleaching, and oil recovery. These applications are based on the alkalinity of silicates and the ability of the polymeric silicate anion to form complexes with metal ions in solution and adsorb at charged interfaces. Silicates are effective in high hardness systems where ions other than Mg^{2+} and Ca^{2+} are present. Compared with other inorganic sequestration agents, silicates are insensitive to Ca^{2+}.

Corrosion Control. Silica in water exposed to various metals leads to the formation of a surface less susceptible to corrosion. A likely explanation is the formation of metallosilicate complexes at the metal–water interface after an initial disruption of the metal oxide layer and formation of an active site. This modified surface is expected to be more resistant to subsequent corrosive action via lowered surface activity or reduced diffusion.

Waste Treatment. Soluble silicates, when effectively used in the formation of cement-solidified waste forms, cause a reduction in leaching rate (107,108). This effect could be described by assuming that the rate of leaching was attenuated by changes in reactant or product diffusion through the solid. The concept of a permeability reduction, in addition to any specific solubility reduction owing to the formation of a new silicate species, explains why these waste forms often show higher levels of leachability when mechanically degraded. It also has been used to explain why substances which would not be expected to react with soluble silica can show reduced leachability (109).

Mineral Beneficiation. Sodium silicates are used in froth flotation as strong and selective settling agents, where they increase the hydrophilicity of the mineral particles. Small amounts of silicates may activate the flotation of some calcium-containing minerals. Soluble silicates are excellent dispersants of slimes formed by finely ground minerals (110). Silicates are adsorbed selectively at mineral surfaces in low concentrations and promote separation of particles of different sizes by causing different rates of settling. However, excess silicate settles all minerals. Soluble silicates deactivate certain mineral surfaces to flotation by preventing absorption by a collector, such as oleic acid. High ratio silicates are more active and appear to derive their values from their ability to attenuate particle separation at low cost by the development of appropriate hydrophobic and hydrophilic surface properties. The latter effect results from simple anion adsorption; the former is inherent to the mineral or a result of collector adsorption. Silicates are inexpensive additives to the wet ore grinding process where they reduce wear on steel grinding balls and rods (111); 85% of those steel losses are a result of corrosion (112).

Adhesives and Binders. Silicates are often used to bind sand in foundries, or to bind other minerals either by means of an acid added directly, as in the CO_2 process, or indirectly, using an organic ester that hydrolyzes to a carboxylic acid and gels the silicate sand or mineral mass (113–115). Silicates are used extensively as adhesives in spiral tube winding, fiber drums, end sealing, laminating metal foil to paper, and in corrugated boxes. They are also employed in the manufacture of refractory and acid-resistant mortars and cements. Potassium silicates are preferred when high refractory qualities and a nonblooming surface are required. In gunning applications, sodium silicates might be preferred because of their tackiness. Soluble potassium silicates are employed as binders of phosphors to black-and-white television screems. Silicates are used in pelletizing, granulating, and briquetting of fine particles and as a vehicle for water-based coatings. If a water- or humidity-resistant system is not required in any of the preceding applications, simple air drying or heating is sufficient (116,117). However, a setting agent, ie, a reactive pigment or heavy metal salt; post-treatment, ie, baking or curing with an acid wash; or glyoxal curing is necessary where water resistance is required (118). Lithium enhances the water resistance of silicate–sand foundry binders (119). The degree of water resistance depends on the setting agents and the method and temperature of curing. Water resistance is essential for welding-rod coatings, roofing granules, and zinc-rich paints for iron and steel.

Enhanced Oil Recovery. Silicates are employed in enhanced oil recovery for reasons that are related to their function in detergency and ore beneficiation. The presence of silicate enhances surfactant effectiveness, especially in hard water reservoir brines (120). The addition of silicates, both before and during direct surfactant floods, eg, micellar-polymer preflushes and low tension waterflooding with silicate, indicates that they retard surfactant adsorption by the substrate and improve sweep efficiency. The latter phenomenon may involve selective permeability reduction (121–123). In this manner silicates improve water wetting in the reservoir. The dissolved silica in silicates enables the flooding solutions to function as a buffering fluid by preventing undesirable mineral–fluid reactions that slow chemical propagation. Silicate polymerization and gelation has been used in subterranean formations to plug large cavities and reduce unwanted fluid flows. The controlled placement and *in situ* gelation of silicates can solve many problems encountered in the drilling and completion of wells and the production of oil and gas through water flooding and enhanced treatment technologies (124,125).

Bleaching. Silicates are utilized in combination with hydrogen peroxide to improve pulp and textile bleaching efficiency (126,127). Approximately 40% of the improvement in bleaching efficiency, as measured by brightness, can be attributed in equal measure to increased alkalinity and ionic strength of the bleaching bath (126). Other mechanisms proposed for enhancement include the formation of highly reactive peroxysilicates in solution, adsorption of free-radical intermediates from H_2O_2 disproportionation and products of lignin darkening reactions, and modification of peroxide equilibrium. A plausible explanation is that the silicate inactivates iron and manganese species which catalyze peroxide decomposition (128). However, it is a concern in the industry that the benefits of silicate might be partially offset by the formation of siliceous deposits on bleaching and papermaking equipment and on pulp fibers (129).

Deflocculation and Slurry Thinning. Silicates are used as deflocculants, ie, agents that maintain high solids slurry viscosities at increased solids concentrations. Soluble silicates suppress the formation of ordered structures within clay slurries that creates resistance to viscous flow within the various sytems. Laboratory trials are necessary, because the complexity of the systems precludes the use of a universal deflocculant. Silicates are employed in thinning of lime-stone or clay slurries used in the wet-process manufacture of cements and bricks, clay refining, and petroleum drilling muds (see also FLOCCULATING AGENTS).

Soil Stabilization/Grouting. Sodium silicates have been used in soil stabilization for most of the twentieth century. This usage represents the largest market in Japan in the mid-1990s. Sodium silicates can be used for water control and soil stabilization in tunneling and excavation projects. These grouts are strong, reliable, and environmentally safe. There are many setting (gelation) agents for these systems. However, inorganic agents are generally used when seeking high strength and organic agents, eg, esters, when seeking greater control over the gel time. The latter advantage is negated somewhat by increased cost and reduced strength. This controlled placement and gelation technology is also used in oil-field applications and is being adapted for *in situ* stabilization of soils contaminated with hazardous wastes.

Derivatives

In the chemical processing industry (CPI), alkali silicates are valued as a reactive source of $(SiO_2)_n$ structural units. They may be viewed (130) as silica dissolved or dispersed in a hydroxide ion-rich aqueous system. Soluble silica as an intermediate can react with acids and bases to form a wide range of final products from seemingly simple condensed forms of relatively pure noncrystalline silica, precipitates, gels, and sols, to highly complex crystalline metallosilicates like those found in the broad class of aluminosilicates, zeolites. The key properties of an intermediate are likely to be silica concentration, ratio, supporting cation, type and level of impurities, and consistency in these factors. These factors are particularly important when manufacturing catalysts, highly selective sorbents, and other high performance materials where trace metals could affect final product performance.

Silica Sols, Precipitated Silicas, and Silica Gels. Solutions of soluble silicates may be modified by ion exchange, acid addition, and post-treatment. These products are distinguished primarily by their SiO_2 and H_2O content, pore volume and surface area, particle size and morphology, and content of residual salts. Silica sols (131) are manufactured from diluted 3.3 ratio sodium silicate solutions by H^+/Na^+ ion exchange and concentration. This process yields a dispersion of colloidal silica particles used in antislip agents, castings, binders, and polishing solutions for silicon wafers.

Precipitated silicas (132) are made by counteracting the interparticle forces that hold large polymeric silicate anions in solution by the addition of an organic compound or a sodium salt and mineral acid. The anions grow and coagulate into clusters which are filtered, washed, dried, classified, and sometimes surface treated. Silica gels (133) are prepared by acidifying concentrated 3.3 ratio sodium silicate solutions. Acidification rapidly produces a gel network, called a hydrogel, after it passes through the sol stage. Xerogels and aerogels (qv) are manufactured by gelation, followed by milling, washing, and drying. These powdered products have a wide range of properties (134) and are used as reinforcing agents or fillers in rubber (white carbon black), plastic films and paints, carriers, anticaking and flow-control agents, defoamers, desiccants, toothpaste thickeners and abrasives, beer clarifiers, and as catalyst and chromatography supports.

Synthetic Insoluble Silicates. Insoluble crystalline silicates, ie, mineral-type compounds, are synthesized from soluble silicates by precipitation, gelation, ion exchange, and hydrothermal techniques. Hydrothermal treatment of partially neutralized, high mole ratio ($m = 12-50$), sodium silicate solutions yields neutral alkali polysilicates that exhibit a layered structure and high ion-exchange capacity (135,136). These and other lamellar silicates can be utilized either alone or modified via pillaring (137) as adsorbents and catalysts.

The zeolites $M_{x/n}[(AlO_2)_x(SiO_2)_y] \cdot zH_2O$, where M is usually an alkali or alkaline-earth metal ion, are of great commercial importance. These materials, which serve as a model system for the advanced inorganic materials revolution (138), are made by hydrothermal methods and their architectural variations seem limitless. Synthetic zeolites are generally purer and less expensive than their natural analogues and almost all commercial applications utilize the former (139). These crystalline materials contain SiO_4 tetrahedra in which some of the Si atoms are replaced by Al (or other metals) in the basic structural network. Zeolites have an open framework structure, which varies as the Si/Al ratio and mobile charge-neutralizing cation, M, vary (140). These microporous materials, referred to as molecular sieves (qv), are used as ion exchangers, selective adsorbents, and catalysts (141). Other applications include radioactive waste treatment, plastics additives, and controlled release of chemicals.

The nature and concentration of the modifier of the SiO_4 framework, eg, Al, influence the structure and ion-exchange properties of the zeolite. Gallo-, ferro-, and titanosilicate zeolites have been synthesized (142–144). Solvent properties may control zeolite structure formation. For example, silica–sodalite prepared in ethylene glycol is composed of four- and six-membered ring structures, whereas the structure of highly siliceous zeolites obtained from aqueous reaction systems and organic bases contains five-membered rings (145,146). Framework structure and acidity influence the function of zeolites as catalysts. Solid-state ^{29}Si

OLSON LIBRARY
NORTHERN MICHIGAN UNIVERSITY
MARQUETTE, MICHIGAN 49855

nmr has become a powerful tool in the investigation of the synthetic insoluble silicates (147).

Another insoluble silicate of commercial interest is the magnesium silicate smectic clay, hectorite, which may be prepared from 3.2 ratio sodium silicate, LiF, $MgSO_4$, and Na_2CO_3 (148). This inorganic thickening agent produces translucent, thixotropic gels that are not tacky, gummy, or stringy. The gels are used in antiperspirants, gel toothpastes, shampoos, cosmetics (qv), paints (qv), and cleaning products. Although numerous other crystalline silicates can be prepared at high temperatures and pressures, eg, asbestiform minerals and calcium silicate, the commercial applications and volumes are not significant in comparison with those of hectorite and zeolites. In many cases, these other crystalline silicates are of interest in understanding the chemistry of cement formation and mineral development. However, there are several significant products and markets for amorphous silicates made by reaction of soluble salts of magnesium, aluminum, calcium, and other metals with soluble alkali silicates. The precipitation of this class of materials has been reviewed (149). These materials are useful as sorbents, ion exchangers, rubber reinforcers, paint hardeners, fillers, thixotropes, and pigments.

BIBLIOGRAPHY

"Soluble Silicates and Synthetic Insoluble Silicates" in *ECT* 1st ed., Vol. 12, pp. 303–330 and "Synthetic Inorganic Silicates" under "Silicon Compounds" in *ECT* 2nd ed., Vol. 18, pp. 134–165, by J. H. Wills, Philadelphia Quartz Co.; in *ECT* 3rd ed., Vol. 20, pp. 855–880, by J. S. Falcone, Jr., The PQ Corp.

1. D. Barby and co-workers, in R. Thomson, ed., *The Modern Inorganic Chemical Industry*, Chemical Society, London, 1977, p. 320.
2. F. Liebau, *Structural Chemistry of Silicates*, Springer-Verlag, Berlin, 1985.
3. J. G. Vail, *Soluble Silicates*, Reinhold Publishing Corp., New York, 1952.
4. H. H. Weldes and K. R. Lange, *Ind. Eng. Chem.* **61**, 29 (1969).
5. L. S. Dent Glasser, *Chem. Brit.* **18**, 33 (Jan. 1982).
6. J. S. Falcone, Jr., ed., *Soluble Silicates*, ACS Symposium Series 194, American Chemical Society, Washington, D.C., 1982.
7. G. Von Engelhardt and co-workers, *Z. Anorg. Allg. Chem.* **418**, 17 (1975).
8. B. D. Mosel and co-workers, *Phys. Chem. Glasses* **15**(6), 154 (1974).
9. S. P. Zhdanov, *Fiz. Khim. Stekla* **4**(5), 515 (1978).
10. H. P. Calhoun, C. R. Masson, and J. M. Jansen, *J. Chem. Soc. Chem. Commun.*, 576 (1980).
11. S. A. Brawer and W. B. White, *J. Non-Cryst. Solids* **23**, 261 (1977).
12. S. Urnes and co-workers, *J. Non-Cryst. Solids* **29**, 1 (1978).
13. L. S. Dent Glasser and P. B. Jamieson, *Acta Crystallogr.* **B32**(3), 705 (1976).
14. K. Beneke and G. Lagaly, *Am. Mineral.* **62**, 763 (1977).
15. O. P. Bricker, *Am. Mineralogist* **54**, 1026 (1969).
16. G. Lagaly, K. Beneke, and A. Weiss, in J. M. Serratoga, ed., *Proc. Int. Clay Conf. Madrid, 1972*, Div. Ciencias CSIC, Madrid, 1973, p. 662.
17. K. Beneke and G. Lagaly, *Am. Mineralogist* **68**, 818 (1983).
18. T. M. El-Shamy, J. Lewins, and R. W. Douglas, *Glass Technol.* **13**(3), 81 (1972).
19. R. W. Douglas and T. M. El-Shamy, *J. Am. Ceram. Soc.* **50**(1), 1 (1967).
20. C. Wu, *J. Am. Ceram. Soc.* **63**(7–8), 453 (1980).

21. F. R. Jorgensen, *J. Appl. Chem. Biotechnol.* **24**, 303 (1977).
22. R. W. Spencer and J. S. Falcone, Jr., technical data, The PQ Corp., Conshohocken, Pa., 1980.
23. D. Hoebbel, R. Ebert, *Z. für Chem.* **28**(2), 41–51 (1988).
24. D. Hoebbel and co-workers, *Z. Anorg. Allg. Chem.* **558**, 171–188 (1988).
25. J. S. Falcone, Jr., in P. W. Brown, ed., *Cements Research Progress—1988*, American Ceramic Society, Westerville, Ohio, 1989, p. 277.
26. R. W. Harman, *J. Phys. Chem.* **32**, 44 (1928).
27. R. V. Nauman and P. Debye, *J. Phys. Chem.* **55**, 1 (1951); 65, 5 (1961).
28. C. W. Lentz, *Inorg. Chem.* **3**(4), 574 (1964).
29. L. S. Dent Glasser and E. E. Lachowski, *J. Chem. Soc. Dalton Trans.*, 393 (1980).
30. B. R. Currell and J. R. Parsonage in G. E. Ham, ed., *Advances in Organometallic and Inorganic Polymer Science*, Marcel Dekker, Inc., New York, 1982.
31. N. H. Ray and R. J. Plaisted, *J. Chem. Soc. Dalton Trans.*, 475 (1983).
32. L. S. Dent Glasser and S. K. Sharma, *Br. Polym. J.* **6**, 283 (1974).
33. T. Tarutani, *J. Chromatogr.* **313**, 33 (1984).
34. E. Freund, *Bull. Soc. Chim. Fr.* (7–8), 2238 (1973).
35. A. Marinangeli and co-workers, *Can. J. Spectrosc.* **23**(6), 173 (1978).
36. R. Alvarez and D. L. Sparks, *Nature (London)* **318**(6047), 649 (1985).
37. H. C. Marsmann, *Z. Naturforsch.* **29B**(7–8), 495 (1974).
38. R. K. Harris and C. T. G. Knight, *J. Chem. Soc. Faraday Trans. 2* **79**, 1525 (1983).
39. *Ibid.*, 1535.
40. R. K. Harris and co-workers, *J. Magn. Reson.* **57**, 115 (1984).
41. R. K. Harris and co-workers, *J. Mol. Liq.* **29**, 63 (1984).
42. L. S. Dent Glasser and E. E. Lachowski, *J. Chem. Soc. Dalton Trans.*, 399 (1980).
43. S. Sjöberg, L. Öhman, and N. Ingri, *Acta. Chim. Scand.* **A39**, 93 (1985).
44. I. Svensson, S. Sjöberg, and L. Öhman, *J. Chem. Soc. FT1* **82**, 3635 (1986).
45. R. K. Iler, *The Chemistry of Silica*, John Wiley & Sons, Inc., New York, 1979, p. 172.
46. A. R. Marsh, G. Klein, and T. Vermeulen, *Polymerization Kinetics and Equilibria of Silicic Acid in Aqueous Systems*, No. LBL 4415, National Technical Information Service, Springfield, Va., 1975.
47. K. R. Anderson, L. S. Dent Glasser, and D. N. Smith in Ref. 6, p. 115.
48. M. F. Bechtold, W. Mahler, and R. A. Schum, *J. Polym. Sci. Polym. Chem. Ed.* **18**, 2823 (1980).
49. R. K. Iler, *J. Colloid Chem. Interface Sci.* **75**(1), 138 (1980).
50. O. Weres, *J. Colloid Chem. Interface Sci.* **84**(2), 379 (1981).
51. R. Schwartz and W. D. Muller, *Z. Anorg. Allg. Chem.* **296**, 273 (1958).
52. V. N. Belyakov and co-workers, *Ukr. Khim. Zh. Russ. Ed.* **40**(3), 236 (1974).
53. P. W. Schindler and H. R. Kamber, *Helv. Chim. Acta* **51**, 1781 (1968).
54. L. C. Klein, *Ann. Rev. Mater. Sci.* **15**, 227 (1985).
55. C. J. Brinker, D. E. Clark, and D. R. Ulrich, eds., *Better Ceramics through Chemistry*, Elsevier-North Holland, New York, 1984.
56. W. Stöber, A. Fink, and E. Bohn, *J. Colloid Interface Sci.* **26**, 62 (1968).
57. K. R. Lange and R. W. Spencer, *Environ. Sci. Technol.* **2**(3), 212 (1969).
58. E. P. Katsanis, P. H. Krumrine, and J. S. Falcone, Jr., "Chemistry of Precipitation and Scale Formation in Geological Systems," SPE preprint 11802, *National Symposium on Oil Field and Geothermal Chemistry, Denver, Colo., June 1, 1983*, Society of Petroleum Engineers, 1983.
59. L. N. Allen, E. Matijevic, and L. Meites, *J. Inorg. Nucl. Chem.* **33**, 1293 (1971).
60. R. W. Maatman and co-workers, *J. Phys. Chem.* **68**(4), 757 (1964).
61. P. W. Schindler and co-workers, *J. Colloid Interface Sci.* **55**, 469 (1976).
62. R. O. James and T. W. Healy, *J. Colloid Interface Sci.* **40**, 42, 53, 65 (1972).

63. I. M. Kolthoff and V. A. Stenger, *J. Phys. Chem.* **36**, 2113 (1932); **38**, 249, 475 (1934).
64. J. S. Falcone, Jr., in Ref. 6, p. 133.
65. Ref. 45, p. 667.
66. F. J. Hingston, R. J. Atkinson, A. M. Posner, and J. P. Quirk, *Nature* **215**, 1459 (1967).
67. C. P. Huang, *Earth Plant. Sci. Lett.* **27**, 265 (1975).
68. F. Tsai and J. S. Falcone, paper presented at the *ACS/CSJ Chemical Congress*, Honolulu, Hawaii, Apr. 6, 1979.
69. F. J. Hazel, *J. Phys. Chem.* **49**, 520 (1945).
70. U.S. Pat. 4,545,798 (Oct. 8, 1985), J. M. Matesa.
71. *Ceram. Bull.* **65**(7), 1027 (1986).
72. C. Kröger, *Glass Ind.* **37**(3), 133 (1956).
73. W. R. Ott, *Ceramurgia* **5**(1), 37 (1979).
74. A. M. Ginstling and B. I. Brounsthein, *J. Appl. Chem. USSR* **23**, 1327 (1950).
75. A. G. Pincus and G. M. Dicken, eds., *Electric Melting in the Glass Industry*, Magazine for Industry, Inc., New York, 1976.
76. D. E. Veinot and co-workers, *J. Non-Cryst. Solids* **127**, 221–226 (1991).
77. 21 CFR §182.70–182.90, rev. Apr. 1, 1979.
78. 21 CFR §173.310, rev. Apr. 1, 1979.
79. Select Committee on GRAS Substances, *Evaluation of the Health Aspects of Certain Silicates as Food Ingredients, SCOGS-61, NTIS Pb 301-402/AS*, Federation of American Societies for Experimental Biology, Springfield, Va., 1979, p. 26.
80. 21 CFR §184.1769a, Sept. 25, 1985.
81. 21 CFR §173.315, Apr. 15, 1984.
82. Federal military specification MIL-W-15117D, Amendment 3, Jan. 22, 1975, Washington, D.C.
83. 40 CFR §141.82, *National Primary Drinking Water Regulations-Description of Corrosion Control Treatment Requirements*, rev. July 1, 1991.
84. Committee on Water Treatment Chemicals, Food and Nutrition Board, Assembly of Life Sciences, National Research Council, *Water Chemicals Codex*, National Academy Press, Washington, D.C., 1982, p. 63.
85. *ANSI/AWWA Standard B-404-80*, American Waterworks Association, Denver, Colo., 1980.
86. *Sodium Silicate-Oil and Hazardous Materials—Technical Assistance Data System*, National Institute of Health, Washington, D.C. (an on-line database).
87. *Hazardous Labelling Guide*, 9010.125, App. 1, U.S. Consumer Products Safety Commission, Washington, D.C., 1975, p. 45.
88. CFR §1910.1000, rev. July 1, 1979.
89. U.S. Dept. of Commerce, Bureau of the Census, Washington, D.C., summarized from *MQ28A: Inorganic Chemicals*, ftp.census.gov/pub/industry/mq28a961.TXT.
90. *Chem. Week*, 40 (Mar. 8, 1995).
91. J. L. Bass, in Ref. 6, p. 17.
92. H. Strauss and R. Rutkowski, *Silikattechnik* **29**(11), 339 (1978).
93. W. L. Schleyer and J. L. Blumberg, in Ref. 6, p. 49.
94. G. M. Berke and T. W. Osborne, *Food Cosmet. Toxicol.* **17**, 123 (1979).
95. R. Michon, *C. R. Acad. Sci.* **243**, 2194 (1956).
96. J. D. Birchall, in H. E. Bergna, ed., *The Colloid Chemistry of Silica*, The American Chemical Society, Washington, D.C., 1994, Chapt. 31, p. 601.
97. P. Tréguer and co-workers, *Science*, **268**, 375 (Apr. 21, 1995).
98. J. S. Falcone, Jr., J. G. Blumberg, "Anthropogenic Silicates", in *Handbook of Environmental Chemistry*, Vol. 3, Pt. F, O. Hutzinger, ed., Springer-Verlag, Berlin, 1992.

99. M. A. Soukup, *The Limnology of a Eutrophic Hardwater New England Lake with Major Emphasis on the Biogeochemistry of Dissolved Silica*, No. 75-27-527, University Microfilms, Ann Arbor, Mich., 1975.

100. P. Kilham, *Limnol. Oceanogr.* **16**(1), 10 (1971).

101. J. L. O'Connor, *J. Phys. Chem.* **65**(1), I (1961).

102. J. S. Falcone, Jr., in H. Bergna, ed., *Colloid Chemistry of Silica*, ACS Advances in Chemistry Series No. 234, ACS Books, Washington, D.C., 1994, Chapt. 30, p. 595.

103. R. S. Schreiber, in Ref. 6, p. 271.

104. E. P. Katsanis, W. B. Esmonde, and R. W. Spencer, *Maater. Perform.* **25**, 19 (1986).

105. M. G. Browman, R. B. Robinson, and G. D. Reed, *Environ. Sci. Technol.* **23**, 566 (1989).

106. R. B. Robinson, G. D. Reed, and B. Frazier, *J. AWWA*, **84**(2), 77–82 (1992).

107. E. L. Davis, J. S. Falcone, Jr., S. D. Boyce, and P. H. Krumrine, "Mechanisms for the Fixation of Heavy Metals in Solidified Wastes Using Soluble Silicates", paper presented at *HWHM 1986 Conference*, Atlanta, Ga., Mar. 4, 1986.

108. J. S. Falcone, Jr., R. W. Spencer, R. H. Reifsnyder, and E. P. Katsanis, in L. P. Jackson and co-workers, eds., *Hazardous and Industrial Waste Management and Testing*, ASTM, Philadelphia, Pa., 1984, p. 213.

109. J. R. Connor, *Chemical Fixation and Solidification of Hazardous Wastes*, Van Nostrand Reinhold, New York, 1990, in particular Chapt. 11, pp. 376–406.

110. J. S. Falcone, Jr., *Mining Engineering* **34**(10), 1493 (1982).

111. G. R. Hoey, W. Dingley, and A. W. Lui, **59**(5), 36 (1975).

112. J. L. Briggs, *Mater. Perform.* **12**(1), 20 (1974).

113. K. E. Nicholas, *The CO_2–Silicate Process in Foundries*, British Cast Iron Research Association, Alvechurch, U.K., 1972.

114. M. Roberts, **133**(2925), 783 (1972).

115. K. J. D. Mackenzie, I. W. M. Brown, P. Ranchod, and R. H. Meinhold, *J. Mater. Sci.* **26**, 763–768 (1991).

116. L. S. Dent Glasser and C. K. Lee, *J. Appl. Chem. Biotechnol.* **21**(5), 127 (1971).

117. N. R. Horikawa, K. R. Lange, and W. L. Schleyer, *Adhes. Age.* **10**(7), 30 (1967).

118. L. S. Dent Glasser, E. G. Grassick, and E. E. Lachowski, *J. Chem. Biotechnol.* **29**, 283 (1979).

119. I. B. Ailin-Pyzik, R. W. Spencer, and J. S. Falcone, Jr., *Trans. Am. Foundrymen's Soc.* **89**, 543 (1981).

120. J. S. Falcone, Jr., P. H. Krumrine, and G. C. Schweiker, *JAOCS* **59**(10), 826A (1982).

121. P. H. Krumrine, J. S. Falcone, Jr., and W. E. Gittler, "The Evaluation and Use of Alkaline Chemicals for Enhanced Oil Recovery," paper presented at the *3rd European Symposium on Improved Oil Recovery*, Rome, Italy, Apr. 18, 1985.

122. P. H. Krumrine and J. S. Falcone, Jr., *SPE Reservoir Eng.* (2), 62 (1988).

123. P. H. Krumrine, in Ref. 6, p. 187.

124. J. K. Borchardt, *Colloids and Surfaces* **63**, 189 (1992).

125. W. F. Hower and J. Ramos, *J. Pet. Technol.* **9**, 17 (1957).

126. G. W. Kutney, *Pulp Paper* (*Canada*) **86**(12), T402 (1985).

127. M. J. Palin, D. C. Teasdale, and L. Benisek, *J. Soc. Dyers Colour.* **99**, 261 (1983).

128. J. L. Colodette, S. Rothenberg, and C. W. Dence, *J. Pulp and Paper Sci.* **15**, J3 (1989).

129. W. C. Froass, *Interactions of Calcium, Magnesium and Silicate Under Peroxide Bleaching Conditions*, M.S. Dissertation, SUNY College of Environmental Science and Forestry, Syracuse, N.Y., June, 1991.

130. J. R. McLaughlin, "The Properties, Applications and Markets for Alkaline Solutions of Soluble Silicates," paper presented to *CMRA*, New York, May, 1976.

131. C. C. Payne, in Ref. 101, Chapt. 29, p. 581.

132. H. K. Ferch, in Ref. 101, Chapt. 24, p. 481.

133. R. E. Patterson, in Ref. 101, Chapt. 32, p. 617.
134. D. Barby, in G. D. Parfitt and K. S. W. Sing, eds., *Characteristics of Powdered Surfaces*, Academic Press, Ltd., London, 1976, p. 353.
135. U.S. Pat. 5,236,681 (Aug. 17, 1993), P. Chu, G. W. Kirker, S. Krishnamurthy, and J. Vartuli (to Mobile Oil Corp.).
136. W. Schwieger and co-workers, *Z. Chem.* **25**(6), 228 (1985).
137. J. S. Dailey and T. J. Pinnavaia, *Chem. Mater.* **4**, 855 (1992).
138. A. K. Cheetham, *Science*, **264**, 794 (May 6, 1994).
139. P. H. Shimizu, *Soap Cosmet. Chem. Spec.* **53**, 33 (1977).
140. J. M. Newsam, *Science* **231**, 1093 (1986).
141. W. Hölderich, M. Hesse, and F. Näumann, *Angew. Chem. Int. Ed. Engl.* **27**, 226–246 (1988).
142. X. Liu and J. H. Thomas, *J. Chem. Soc., Chem. Commun.* (21), 1544 (1985).
143. R. Szostak, V. Nair, and T. L. Thomas, *J. Chem. Soc. FT1* **83**, 487 (1987).
144. Jpn. Pat. 127,217 (July 6, 1985).
145. D. M. Bibby and M. P. Dale, *Nature* **317**, 157 (1985).
146. C. T. G. Knight and co-workers, *Stud. Surf. Sci. Catal.* **97**, 483 (1995).
147. G. Engelhardt and H. Koller, *NMR Basic Principles and Progress*, Vol. 31, Springer-Verlag, Berlin, 1994.
148. U.S. Pat. 3,586,478 (June 22, 1971), B. S. Neuman (to Laporte Industries, Ltd.).
149. A. Packter, *Cryst. Res. Technol.* **21**, 575–585 (1986).

General References

H. Bergna, ed., *Colloid Chemistry of Silica*, ACS Advances in Chemistry Series Number 234, ACS Books, Washington, D.C., 1994.
G. Engelhardt and H. Koller, *NMR Basic Principles and Progress*, Vol. 31, Springer-Verlag, Berlin, 1994.
J. S. Falcone, Jr., and J. G. Blumberg, in O. Hutzinger, ed., *Handbook of Environmental Chemistry*, Vol. 3, Pt. F, Springer-Verlag, Berlin, 1992.
G. Engelhardt and D. Michel, *High Resolution Solid State NMR of Silicates and Zeolites*, Chichester, John Wiley & Sons, Ltd., 1987.
F. Liebau, *Structural Chemistry of Silicates*, Springer-Verlag, Berlin, 1985.
L. S. Dent Glasser, *Chem. Brit.*, 33, 36, 38, 39 (Jan. 1982).
J. S. Falcone, Jr., ed., *Soluble Silicates, ACS Symposium Series, No. 194*, American Chemical Society, Washington, D.C., 1982.
R. K. Iler, *The Chemistry of Silica*, John Wiley & Sons, Inc., New York, 1979.
N. Ingri, in G. Bendtz and I. Lindquist, eds., *Biochemistry of Silica and Related Compounds*, Plenum Press, New York, 1978, p. 3.
D. Barby and co-workers, in R. Thomson, ed., *The Modern Inorganic Chemical Industry*, Chemical Society, London, 1977, p. 320.
W. Eitel, *Silicate Science*, Vols. 2, 3, and 4, Academic Press, Inc., New York, 1964.
J. G. Vail, *Soluble Silicates*, Reinhold Publishing Corp., New York, 1952.

JAMES S. FALCONE, JR.
West Chester University

SILICON HALIDES

The study of silicon halides began in the early 1800s (1). Since then, essentially all of the monomeric silicon halides have been extensively studied and reported in the literature. These include mixed silicon halides and halohydrides. A large number of halogenated polysilanes have also been reported (1). Despite the extensive research in silicon halides, only two of these chemicals are produced on a large industrial scale (excluding organohalosilanes). These are tetrachlorosilane [10026-04-7], $SiCl_4$, and trichlorosilane [1025-78-2], $HSiCl_3$.

Physical Properties

The physical properties of silicon tetrahalides are listed in Table 1; those of the halohydrides are listed in Table 2. A more complex review of the physical properties of these chemicals is available (2). Detailed lists of properties of the colorless fuming liquids, silicon tetrachloride and trichlorosilane, are given in Table 3. A review of the physical and thermodynamic properties of silicon tetrachloride is given in Reference 3.

Table 1. Properties of Silicon Tetrahalides

Compound	CAS Registry Number	Mp, °C	Bp, °C	Density^a, g/cm^3	Bond energy, kJ/mol^b
SiF_4	[7783-61-1]	−95.0	−90.3	1.66_{-95}	146
$SiCl_4$	[10026-04-7]	−68.8	56.8	1.48_{20}	381
$SiBr_4$	[7789-66-4]	5	155.0	2.81_{29}	310
SiI_4	[13465-84-4]	124	290.0		234

^{a}Subscripted values are temperature in °C.
^{b}To convert J to cal, divide by 4.184.

Table 2. Properties of Silicon Halohydrides

Compound	CAS Registry Number	Mp, °C	Bp, °C	Density^a, g/cm^3
H_3SiF	[13537-33-2]		−99.0	
H_2SiF_2	[13824-36-7]	−122.0	−77.8	
$HSiF_3$	[13465-71-9]	−131.2	−97.5	
H_3SiCl	[13465-78-6]	−118.0	−30.4	1.145_{-113}
H_2SiCl_2	[4109-96-0]	−122.0	8.3	1.42_{-122}
$HSiCl_3$	[1025-78-2]	−128.2	31.8	1.3313_{25}
H_3SiBr	[13465-71-1]	−94.0	1.9	1.531_{20}
H_2SiBr_2	[13768-94-0]	−70.1	66.0	2.17_0
$HSiBr_3$	[7789-57-0]	−73.0	111.8	2.7_{17}
H_3SiI	[13598-42-0]	−57.0	45.4	$2.035_{14.8}$
H_2SiI_2	[13760-02-6]	−1.0	149.5	$2.724_{20.5}$
$HSiI_3$	[13465-72-0]	8.0	111.0^b	3.314_{20}

^{a}Subscripted values are temperature in °C.
^{b}At 2.9 kPa (21.8 mm Hg).

Table 3. Physical Properties of Silicon Tetrachloride and Trichlorosilane[a]

Property	$SiCl_4$	$SiHCl_3$
refractive index	1.4146	1.3983
d_4^{25}, g/cm^3	1.4736	1.3313
viscosity, at 25°C, mm^2/s(=cSt)	0.35	0.23
vapor pressure, kPa[b]		
at −50°C		1.7
−30°C	1.8	6.3
−10°C	6.1	18.1
10°C	16.4	44.2
30°C	38.5	94.7
50°C	79.9	182.9
heat of vaporization, J/g[c]	167.4	195.4
standard heat of formation, at 25°C, kJ/kg[c]	−4075.6	−3861.5
flash point, Cleveland open cup, °C		−28
autoignition temperature, °C		182
specific heat, J/g[c]	0.20	0.96
coefficient of expansion, °C^{-1}	0.0011	0.0019

[a]Both $SiCl_4$ and $SiHCl_3$ are soluble in organic solvents and react with H_2O and alcohol.   [b]To convert kPa to mm Hg, multiply by 7.5.   [c]To convert J to cal, divide by 4.184.

Several of the mixed silicon halides are formed simply by heating a mixture of the tetrahalosilanes at moderate temperature (4), eg,

$$SiCl_4 + SiBr_4 \rightleftharpoons SiBrCl_3 + SiBr_2Cl_2 + SiBr_3Cl$$

Mixed halosilanes can also be prepared by heating a mixture of the appropriate halides of silicon and aluminum (5,6). Some physical properties of these mixed tetrahalosilanes are listed in Table 4. In addition, the properties of several

Table 4. Properties of Mixed Silicon Halides

Compounds	CAS Registry Number	Mp, °C	Bp, °C
$SiBr_3Cl$	[13465-76-4]	−208	128
$SiBr_2Cl_2$	[13465-75-3]	−45.5	104.4
$SiBrCl_3$	[13465-74-2]	−62.0	80
$SiClI_3$	[13932-03-1]	2.0	234–237
$SiCl_2I_2$	[13977-54-3]	<−60.0	172
$SiCl_3I$	[13465-85-5]	<−60.0	113–114
$SiBr_3I$	[13536-76-0]	14.0	192
$SiBr_2I_2$	[13550-39-5]	38.0	230–231
$SiBrI_3$	[13536-68-0]	53.0	255
$SiBr_3F$	[18356-67-7]	−82.5	83.8
$SiCl_3F$	[14965-52-7]	−120.8	12.2
$SiBr_2F_2$	[14188-35-3]	−66.9	13.7
$SiCl_2F_2$	[18356-71-3]	−139.7	−32.2
$SiClF_3$	[14049-36-6]	−142.0	−70.0
$SiBrF_3$	[14049-39-9]	−70.5	−41.7
$SiBrCl_2F$	[28054-58-2]	−112.3	35.5
$SiBr_2ClF$	[28054-61-7]	−99.2	59.5

halogenated polysilanes are listed in Table 5. A summary of the preparation and properties of the chlorinated polysilanes has been published (7).

Table 5. Properties of Some Halogenated Polysilanes

Compound	CAS Registry Number	Mp, °C	Bp, °C	Density[a], g/cm^3
Si_2F_6	[13830-68-7]	−18.6	−19.1	0.008_0
Si_2Cl_6	[13465-77-5]	−1.0	144.5	1.5624_{15}
Si_3Cl_8	[13596-23-1]	−67.0	216.0	1.61_{15}
Si_4Cl_{10}	[13763-19-4]	93.6	149[b]	
Si_5Cl_{12}	[13596-24-2]	−80.0	190[b]	
Si_6Cl_{14}	[13596-25-3]	319.0 dec		
Si_2Br_6	[13517-13-0]	95.0	265	
Si_3Br_8	[54804-32-9]	133.0		
Si_4Br_{10}	[81626-34-8]	185.0		
Si_2I_6	[13510-43-5]	250.0 dec		

[a]Subscripted values are temperature in °C.
[b]At 0.2 kPa (1.5 mm Hg).

Chemical Properties

Silicon halides are typically tetrahedral compounds. The silicone–halogen bond is very polar; thus the silicon is susceptible to nucleophilic attack, which in part accounts for the broad range of reactivity with various chemicals. Furthermore, reactivity generally increases with the atomic weight of the halogen atom.

Halosilanes are very reactive toward protic chemicals. They generally react violently with water, forming silicon dioxide and the respective hydrohalogens. Other examples include reaction with alcohols and amines as follows:

$$SiCl_4 + 4\ CH_3CH_2OH \longrightarrow Si(OCH_2CH_3)_4 + 4\ HCl$$

$$SiCl_4 + \text{excess } HN(CH_3)_2 \longrightarrow Si(N(CH_3)_2)_4 + 4\ HN(CH_3)_2 \cdot HCl$$

Such substitution reactions typically are reversible and are facilitated by the removal of the hydrohalogen being formed. This is commonly accomplished by adding a tertiary amine or excess amine reactant, which generally precipitates as the amine hydrogen halide. The halosilanes are reduced to silicon with hydrogen at elevated temperatures. Tetrafluorosilane [7783-61-1] reacts with hydrogen only above 2000°C, whereas tetrachlorosilane and tetraiodosilane [13465-84-4] are reduced by hydrogen at 1000–2000°C (1,8–10). The silicon halohydrides and mixed halides are reduced at lower temperatures.

Silicon halides are stable to oxygen at room temperature but react at high temperature to form mixtures of oxyhalosilanes (11,12). Silicon halides are also reduced by a number of metals. One of the earliest methods of producing silicon was the reduction of a silicon halide with sodium or potassium (1). A similar reduction process with zinc was the first commercial process for producing semiconductor-grade silicon from silicon tetrachloride (13,14).

Magnesium does not form stable Grignard reagents with silicon halides, although some silicon halohydrides do react, forming polysilanes (15).

$$2 \text{ HSiBr}_3 + 3 \text{ Mg} \longrightarrow 2/x \, (\text{SiH})_x + 3 \text{ MgBr}_2$$

All silicon halides are readily reduced by hydride ions or complex hydrides (qv) to silicon hydrides (16–18).

$$4 \text{ R}_2\text{AlH} + \text{SiX}_4 \longrightarrow \text{SiH}_4 + 4 \text{ R}_2\text{AlX}$$

Historically, among the most important reactions of silicon halides are those occurring with metal alkyls and metal alkyl halides. The Grignard reaction, for example, was the first commercial process for manufacturing organosilicon compounds, which were later converted to silicones (19).

$$\text{SiCl}_4 + n \text{ RMgX} \longrightarrow \text{R}_n\text{SiCl}_{4-n} + n \text{ MgXCl}$$

The silicon halohydrides are particularly useful intermediate chemicals because of their ability to add to alkenes, as follows (19,20):

$$\text{HSiCl}_3 + \text{RCH}=\text{CH}_2 \longrightarrow \text{RCH}_2\text{CH}_2\text{SiCl}_3$$

This reaction, catalyzed by uv radiation, peroxides, and some metal catalysts, eg, platinum, led to the production of a broad range of alkyl and functional alkyl trihalosilanes. These alkylsilanes have important commercial value as monomers and are also used in the production of silicon fluids and resins. Additional information on the chemistry of silicon halides is available (19,21–24).

Manufacturing Processes

Silicon halides can be easily prepared by the reaction of silicon or silicon alloys and the respective halogens (24). Fluorine and silicon react at room temperature to produce silicon tetrafluoride. Chlorine reacts with silicon exothermally, but the mixture must be heated to several hundred degrees centigrade to initiate the reaction (25). Bromine and iodine react with silicon at red heat.

Hydrogen halides also react freely with elemental silicon at moderate temperatures to yield halosilanes (26–31).

$$2 \text{ Si} + 7 \text{ HCl} \longrightarrow \text{HSiCl}_3 + \text{SiCl}_4 + 3 \text{ H}_2$$

Although a mixture of silicon halides and hydrides is formed, the formation of silicon tetrachloride can be maximized by increasing the temperature (22). The direct reaction of silicon and hydrogen chloride (qv) is also the primary manufacturing procedure for producing trichlorosilane, HSiCl_3. In this case, the reactor-bed temperature is maintained so as to optimize the yield of trichlorosilane. Trichlorosilane can also be produced according to the following reaction in a fluid-bed reactor (30).

$$3\,SiCl_4 + 2\,H_2 + Si \underset{500°C}{\overset{Cu}{\rightleftharpoons}} 4\,HSiCl_3$$

At equilibrium the vapors are predominantly hydrogen and silicon tetrachlorides. However, these can be easily removed from the trichlorosilane and recycled. A once-common commercial manufacturing procedure for silicon tetrachloride was the reaction of chlorine gas with silicon carbide.

$$SiC + 2\,Cl_2 \longrightarrow SiCl_4 + C$$

This reaction is highly exothermic; thus, it is difficult to control the reaction temperature (31). The oldest method for producing $SiCl_4$ is the direct reaction of silica and chlorine in the presence of carbon as a reducing agent (24).

$$SiO_2 + 2\,C + 2\,Cl_2 \longrightarrow SiCl_4 + 2\,CO$$

In one modification of this procedure, the starting material is pyrolyzed rice hulls in place of more conventional forms of silicon dioxide (31). Another unique process involves chlorination of a combination of SiC and SiO_2 with carbon in a fluid-bed reactor (32). The advantages of this process are that it is less energy-intensive and substantially free of lower silicon chlorides.

Silicon Tetrachloride.   Most commercially available silicon tetrachloride is made as a by-product of the production of alkylchlorosilanes and trichlorosilane and from the production of semiconductor-grade silicon by thermal reduction of trichlorosilane.

$$2\,HSiCl_3 \xrightarrow{>1000°C} Si + SiCl_4 + 2\,HCl$$

The United States is the largest regional producer of silicon tetrachloride. The leading U.S. producers are Dow Corning and General Electric. The next-leading global producing area is Japan, led by Shin-Etsu, Tokuyama Soda, and Mitsubishi Materials. The remaining silicon tetrachloride is produced in Europe by Wacher and Hüls.

Trichlorosilane.   The primary production process for trichlorosilane is the direct reaction of hydrogen chloride gas and silicon metal in a fluid-bed reactor. Although this process produces both trichlorosilane and silicon tetrachloride, production of the latter can be minimized by proper control of the reaction temperature (22). A significant amount of trichlorosilane is also produced by thermal rearrangement of silicon tetrachloride in the presence of hydrogen gas and silicon.

The two largest global producers of trichlorosilane are Wacher in Europe and Dow Corning in the United States. In addition, there are three primary producers in Japan: Tokuyama Soda, Mitsubishi Materials, and Osaka Titanium.

Health and Safety Factors

Halosilane vapors, except for fluorosilanes, react with moist air to produce the respective hydrohalogen acid mists. Federal standards have not been set for exposure to halosilanes, but it is generally believed that there is no serious risk if vapor concentrations are maintained below a level that produces an irritating concentration of acid mist. The exposure threshold limit value (TLV) for HCl is 5 ppm, expressed as a ceiling limit, which means that no exposure above 5 ppm should be permitted. Because most people experience odor and some irritation at or below 5 ppm, HCl is considered to have good warning properties. Liquid halosilanes react violently with water to produce the respective hydrohalogens. Contact with skin and eyes produces severe burns. Eye contact is particularly serious and may result in loss of sight.

Halosilanes should only be handled in areas that are equipped with adequate ventilation, eye-wash facilities, and safety showers. It is recommended that personnel handling halosilanes wear rubber aprons and gloves and chemical safety goggles. Furthermore, all personnel handling halosilanes should be thoroughly trained in safe handling procedures, the hazardous characteristics of halosilanes, and emergency procedures for all foreseeable emergencies.

Uses

Silicon Tetrachloride. Although the range of industrial applications is broad, the vast majority (ca 95%) of $SiCl_4$ use is in the manufacture of fumed silica. This production process is carried out by burning silicon tetrachloride in a mixture of hydrogen and oxygen (33). High purity grades of silicon tetrachloride are also used to prepare silica, which is used in the manufacture of optical waveguides, ie, fiber optics (qv). Some silicon tetrachloride is also used to produce polycrystalline silicon for the semiconductor (qv) industry.

Silicon tetrachloride is also used to prepare silicate esters, eg, tetraethylorthosilicate:

$$SiCl_4 + 4\ CH_3CH_2OH \longrightarrow Si(OCH_2CH_3)_4 + HCl$$

Silicate esters are used in the production of coating and refractories and in some semiconductor manufacturing operations. A broad range of purity grades of silicon tetrachloride are available to meet the requirements of these different applications.

Trichlorosilane. There are essentially only two large industrial applications for trichlorosilane. These are the synthesis of organotrichlorosilanes and the production of semiconductor-grade silicon metal (see SILICON AND SILICON ALLOYS, PURE SILICON). In the production of semiconductor-grade silicon metal, the purified trichlorosilane is reduced in the presence of hydrogen at temperatures greater than 1000°C. Although a large number of silicon halides and silicon halohydrides can be reduced to semiconductor-grade silicone, trichlorosilane is most commonly used because of its favorable balance of manufacturing, purification, handling, and chemical reduction properties (1,12,30).

Dichlorosilane. Dichlorosilane [4109-96-0] is produced in relatively modest commercial quantities compared to the above chlorosilanes. This silane is generally recovered as a by-product of the production of other silanes. It is used exclusively in the semiconductor industry to produce a range of inorganic films.

Although consumption of dichlorosilane is modest, it is a very high value material, generally selling in the range of hundreds of dollars per kilogram, depending on purity grade. Dichlorosilane is sold in cylinders by gas distributors such as Solkatronic, Air Products, and Matheson.

BIBLIOGRAPHY

"Silicon Halides" under "Silicon Compounds," in *ECT* 1st ed., Vol. 12, pp. 368–370 by E. G. Rochow, Harvard University; in *ECT* 2nd ed., Vol. 18, pp. 166–172, by A. R. Anderson, Anderson Development Co.; in *ECT* 3rd ed., Vol. 20, pp. 881–887, by W. Collins, Dow Corning Corp.

1. W. R. Runyan, *Silicon Semiconductor Technology*, Texas Instrument Electronics Series, McGraw-Hill Book Co., Inc., New York, 1965.
2. E. A. Ebsworth, *Volatile Silicon Compounds*, Pergamon Press Ltd., Oxford, U.K., 1963.
3. C. L. Yaws, G. Hsu, P. N. Shah, P. Lubwak, and P. M. Patel, *Solid State Technol.* **22**(2), 65 (1979).
4. H. H. Anderson, *J. Am. Chem. Soc.* **67**, 859 (1945).
5. M. Schmeisser and H. Jenkner, *Z. Naturforsch.* **7B**, 191 (1952).
6. K. Moedritzer, *Organomet. Chem. Rev.* **1**, 179 (1966).
7. E. F. Hengge, *Rev. Inorg. Chem.* **2**(2), 139 (1980).
8. N. C. Cook, J. K. Walfe, and J. D. Cobine, paper presented at *The 128th Meeting American Chemical Society*, Minneapolis, Minn., 1955.
9. R. Schwarz and H. Merckback, *Z. Anorg. Allgem. Chem.* **232**, 241 (1937).
10. R. B. Litton and H. C. Anderson, *J. Electrochem. Soc.* **101**, 287 (1954).
11. A. D. Gaunt, H. Mackle, and L. E. Sutton, *Trans. Faraday Soc.* **47**, 943 (1954).
12. D. W. S. Chambers and C. J. Wilkins, *J. Chem. Soc.*, 5088 (1960).
13. D. W. Lyon, C. M. Olson, and E. D. Lewis, *J. Electrochem. Soc.* **96**, 359 (1949).
14. U.S. Pat. 2,773,745 (Dec. 11, 1956), K. H. Butler and C. M. Olson (to E. I. du Pont de Nemours & Co., Inc.).
15. G. Schott, W. Herman, and R. Hirschmann, *Angew. Chem.* **68**, 213 (1956).
16. A. E. Finholt, A. G. Bond, K. E. Wilzback, and H. I. Schlesinger, *J. Am. Chem. Soc.* **69**, 2692 (1947).
17. J. E. Baines and C. Eaborn, *J. Chem. Soc.*, 1436 (1956).
18. E. G. Rochow, *J. Am. Chem. Soc.* **67**, 963 (1945).
19. W. Noll, *Chemistry and Technology of Silicones*, Academic Press, Inc., New York, 1968.
20. U.S. Pat. 2,823,218 (May 12, 1955), J. L. Speier and D. E. Hook (the Dow Corning Corp.).
21. A. G. MacDiarmid, *Organometallic Compounds of the Group IV Elements*, Vol. 2, *The Bond to Halogens and Halogenoids*, Marcel Dekker, Inc., New York, 1972.
22. V. Bazant, V. Choalovsky, and J. Rathousky, *Organosilicon Compounds*, Vol. 1, *Chemistry of Organosilicon Compounds*, Academic Press, Inc., New York, 1965.
23. W. Noll, *Chemistry and Technology of Silicone*, Academic Press, Inc., New York, 1968.
24. E. Hengge in V. Gutmann, ed., *Inorganic Silicon Halides in Halogen Chemistry*, Vol. 2, Academic Press, Inc., New York, 1967.
25. K. A. Andrianov, *Dokl. Akad. Nauk. SSSR* **28**, 66 (1940).
26. H. Buffard, F. Wohler, *Am. Chem.* **104**, 94 (1857).
27. S. Friedel and J. Crafts, *Am. Chem.* **147**, 355 (1863).
28. L. Gatterman, *Chem. Ber.* **22**, 186 (1889).

29. *Inorg. Synth.* **1**, 38 (1939).

30. *Feasibility of the Silane Process for Producing Semiconductor Grade Silicon*, Jet Propulsion Laboratory Contract 954334, June 1979.

31. P. K. Basu, Ph.D. dissertation, *Development of a Process for the Manufacture of Silicone Tetrachloride from Rice Hulls*, University of California, Berkeley, 1972.

32. U.S. Pat. 3,173,758 (Mar. 16, 1965), R. N. Secord (to Cabot Corp.).

33. L. J. White and G. L. Duffy, *Ind. Eng. Chem.* **3**, 235 (1959).

WARD COLLINS
Dow Corning Corporation

SILANES

The properties and applications of commercially important hydride functional silanes, ie, compounds having a Si–H bond; halosilanes, ie, compounds having a Si–X bond; and organosilanes, ie, compounds having a Si–C bond, are discussed herein. Compounds having Si–OSi bonds are called siloxanes or silicones. Those having a Si–OR bond are called silicon esters. Siloxanes and silicon esters are discussed elsewhere in the *Encyclopedia* (see SILICON COMPOUNDS, SILICON ESTERS; SILICON COMPOUNDS, SILICONES).

Silane [7803-62-5], SiH_4, is the simplest silicon compound and provides the basis of nomenclature for all silicon chemistry (1). Compounds are named as derivatives of silane. The substituents, whether inorganic or organic, are prefixed. Examples are trichlorosilane [10025-78-2], $HSiCl_3$; disilane [1590-87-0], H_3SiSiH_3; methyldichlorosilane [75-54-7], CH_3SiHCl_2; methylsilane [992-94-9], CH_3SiH_3; diethylsilane [542-91-6], $(C_2H_5)_2SiH_2$; and triethylsilane [617-86-7], $(C_2H_5)_3SiH$. Two or more substituents are listed alphabetically, adhering to the following rules: substituted organic moieties are named first, followed by simple organic fragments; alkoxy substituents are named next, followed by acyloxy, halogen, and pseudohalogen groups. For example, ethylmethylethoxysilane [68414-52-8], $C_2H_5(CH_3)SiH(OC_2H_5)$, and 3-chloropropylmethylchlorosilane [33687-63-7], $ClCH_2CH_2CH_2SiH(CH_3)Cl$, are correct. The complete rules for nomenclature are available (2).

Inorganic Hydride Functional Silanes

Hydride functional silanes are sometimes simply referred to as silanes. The classic work in the field was completed in the early 1900s and involved the study of silane and higher binary silanes, Si_nH_{2n+2}, by means of precision vacuum techniques (3). Only a few of the thousands of hydride functional silanes reported have any commercial significance. These include inorganic silanes, organic silanes, and polymeric siloxanes. Despite the small number, a wide range of applications has developed for such compounds, eg, in the manufacture of high purity and electronic-grade silicon metal (see SILICON AND SILICON ALLOYS, PURE SILICON) and in epitaxial silicon deposition (see ELECTRONIC MATERIALS; INTEGRATED CIRCUITS; SEMICONDUCTORS); as selective reducing agents; as monomers; and as elastomer intermediates (see ELASTOMERS, SYNTHETIC). Not least is the use of these materials as intermediates for production of other silanes and silicones.

The inorganic silanes of commercial importance include silane, dichlorosilane, and trichlorosilane. The last, trichlorosilane, is preponderant. It is not only the preferred intermediate for the first two, but it is also used in the production of high purity silicon metal and as an intermediate for silane adhesion promoters, coupling agents, silicone resin intermediates, and surface treatments. Other silanes that appear to have potential in solar electronics are monochlorosilane, disilane, and some silylmetal hydrides. Silicon-based thin films (qv) employed in microelectronic applications are described as hydrogenated amorphous silicon (a-Si–H). An understanding of the physical and chemical behavior of simple silicon compounds, both as precursors and models, is essential for microelectronic device fabrication. Analogously, hydrogenated amorphous silicon carbide is of interest as a wide-gap solar cell material. Siloxene [27233-73-4], $(H_6Si_6O_3)_x$, an inorganic polymer containing silicon hydride bonds, is of interest as a catalyst and as a model for luminescent porous silicon (4) (see METAL-CONTAINING POLYMERS (SUPPLEMENT); SOLAR ENERGY).

Physical Properties. Silanes and chlorosilanes have boiling points, melting points, and dipole moments comparable to those of simple hydrocarbons (qv) and chlorinated hydrocarbons. Moreover, both silanes and hydrocarbons are colorless gases or liquids at room temperature. The similarity ends, however, with these simple physical characteristics (5). Silane, chlorosilane, disilane, and trisilylamine are pyrophoric, igniting immediately on contact with air. The chlorosilanes react with moist air, liberating hydrogen chloride. Dichlorosilane hydrolyzes to a polymeric material that may ignite spontaneously. Even trichlorosilane is highly flammable. The ability of chlorosilanes to permeate or solvate materials of construction, coupled with their hydrolysis to corrosive hydrogen chloride, formation of abrasive silica, and ability to act as reducing agents, makes these compounds difficult to handle.

The simple inorganic silanes are similar to carbon in that each forms stable, covalent, single bonds. Double bonds involving silicon and silicon carbon are relatively unstable. Whereas examples of isolable silicon double-bond-containing materials have been reported (6,7), these do not constitute a part of industrial silane technology. Most silane materials have a tetrahedral bonding geometry consistent with formation of sp^3 hybrid orbitals. Although in some cases participation of $3d$ orbitals in five- or six-coordinate silicon compounds, eg, SiF_6^{-2}, has been invoked, the degree of participation is debated. Silicon is more electropositive than hydrogen and carbon, generally leading to a more polar bond structure than that which occurs in the carbon analogues. The expected inductive release of electrons from R_3Si, however, does not occur. Disiloxanes, for example, are less basic than ethers. A factor which may contribute to the greater reactivity of silicon compared to that of carbon is silicon's greater size.

The physical and thermodynamic properties of silane in the context of semiconductor applications have been reviewed in detail (8). Tabulations of properties of various silanes in the context of inorganic chemistry have also been published (9). Table 1 contains selected physical properties of inorganic silanes.

Thermal Properties. Silanes have less thermal stability than hydrocarbon analogues. The C–H bond energy in methane is 414 kJ/mol (98.9 kcal/mol); the

Table 1. Properties of Inorganic Silanes

Parameter	SiH_4	H_3SiCl	H_2SiCl_2	$HSiCl_3$	H_3SiSiH_3	$H_3SiOSiH_3$	$(H_3Si)_3N$
CAS Registry Number	[7803-62-5]	[13465-78-6]	[4109-96-0]	[10025-78-2]	[1590-87-0]	[13862-73-4]	[13862-16-3]
mp, °C	−185	−118	−122	−126.5	−132.5	−144	−105.6
bp, °C	−111.9	−30.4	8.2	31.9	−14.5	−15.2	52
vapor pressure,[a] Pa[b]	71_{-118}		13_{-34}	$53_{14.5}$			14.5_0
ΔH_{vap}, kJ/mol[c]	12.5	20	25.2	26.6	21.2	21.6	
ΔH_{fus}, kJ/mol[c]	0.67						
critical temperature, °C	−3.5		176	234	109		
critical pressure, MPa[d]	472		455	365			
ΔH_f, kJ/mol[c]	32.6		−314	−482			
dipole moment, $C \cdot m \times 10^{-30}$		4.347	3.913	3.24		0.8	0
density,[a] g/cm³	0.68_{-185}	1.145_{-113}	1.22	1.34	0.69_{-15}	0.881_{-15}	0.895_{-106}
autoignition temperature, °C	f	f	55	215	f	<50	f

[a]Subscripted numbers are temperature in °C.
[b]To convert Pa to mm Hg, multiply by 7.5.
[c]To convert J to cal, divide by 4.184.
[d]To convert MPa to psi, multiply by 145.
[e]To convert C·m to debye, divide by 3.336×10^{-30}.
[f]Pyrophoric.

40

Si–H bond energy in silane is 378 kJ/mol (90.3 kcal/mol) (10). Silane, however, is one of the most thermally stable inorganic silanes. Decomposition occurs at 500°C in the absence of catalytic surfaces, at 300°C in glass vessels, and at 180°C in the presence of charcoal (11). Disilanes and other members of the binary series are less stable. Halogen-substituted silanes are subject to disproportionation reactions at higher temperatures (12).

The thermal decomposition of silanes in the presence of hydrogen into silicon for production of ultrapure, semiconductor-grade silicon has become an important art, known as the Siemens process (13). A variety of process parameters, which usually include the introduction of hydrogen, have been studied. Silane can be used to deposit silicon at temperatures below 1000°C (14). Dichlorosilane deposits silicon at 1000–1150°C (15,16). Trichlorosilane has been reported as a source for silicon deposition at >1150°C (17). Tribromosilane is ordinarily a source for silicon deposition at 600–800°C (18). Thin-film deposition of silicon metal from silane and disilane takes place at temperatures as low as 640°C, but results in amorphous hydrogenated silicon (19).

Chemical Properties. *Oxidation.* All inorganic silicon hydrides are readily oxidized. Silane and disilane are pyrophoric in air and form silicon dioxide and water as combustion products; thus, the soot from these materials is white. The activation energies of the reaction of silane with molecular and atomic oxygen have been reported (20,21). The oxidation reaction of dichlorosilane under low pressure has been used for the vapor deposition of silicon dioxide (22).

Water and Alcohols. Silanes do not react with pure water or slightly acidified water under normal conditions. A rapid reaction occurs, however, in basic solution with quantitative evolution of hydrogen (3). Alkali leached from glass is sufficient to lead to the hydrolysis of silanes.

$$SiH_4 + 2\,KOH + H_2O \longrightarrow K_2SiO_3 + 4\,H_2$$

$$Si_2H_6 + 4\,KOH + 2\,H_2O \longrightarrow 2\,K_2SiO_3 + 7\,H_2$$

Complete basic hydrolysis, followed by the quantitative measurement of hydrogen formed, can be used to determine the number of Si–H and Si–Si bonds present in a particular compound. One molecule of H_2 is liberated for each Si–H and Si–Si bond present. The total silicon content can be obtained from analysis of the resulting silicate solution.

Silane reacts with methanol at room temperature to produce methoxy-monosilanes such as $Si(OCH_3)_4$ [78-10-4], $HSi(OCH_3)_3$, and $H_2Si(OCH_3)_2$ [5314-52-3], but not H_3SiOCH_3 [2171-96-2] (23). The reaction is catalyzed by copper metal. In the presence of alkoxide ions, SiH_4 reacts with various alcohols, except CH_3OH, to produce tetraalkoxysilanes and hydrogen (24).

Halogens, Hydrogen Halides, and Other Covalent Halides. Most compounds containing Si–H bonds react very rapidly with the free halogens. An explosive reaction takes place when chlorine or bromine is allowed to react with SiH_4 at room temperature, presumably forming halogenated silane derivatives (3). At lower temperatures, the reactions are moderated considerably, for example,

$$SiH_4 + Br_2 \xrightarrow[-80°C]{} SiH_3Br\ (+SiH_2Br_2) + HBr$$

Halogen derivatives also form when the silanes are allowed to react with anhydrous hydrogen halides, ie, HCl, HBr, or HI, in the presence of an appropriate aluminum halide catalyst (25,26). The reactions are generally quite moderate and can be carried out at room temperature or slightly above, ie, 80–100°C:

$$SiH_4 + 3\ HX \xrightarrow{Al_2X_6} SiHX_3 + 3\ H_2$$

where X is Cl, Br, or I. Hydrogen bromide reacts more readily than HCl or HI. Increasing the temperature or the duration of reactions generally leads to the formation of more fully halogenated derivatives.

Metals and Metal Derivatives.   Silane reacts with alkali metals dissolved in various solvents, forming as the chief product the silyl derivative of the metal, potassium being the most commonly studied, eg, $KSiH_3$ [13812-63-0] (27–30). When 1,2-dimethoxyethane or bis(2-methoxyethyl)ether are used as solvents, two competing reactions occur, where M is an alkali metal.

$$SiH_4 + 2\ M^0 \longrightarrow SiH_3M + MH$$
$$SiH_4 + M^0 \longrightarrow SiH_3M + 1/2\ H_2$$

Using hexamethylphosphoramide as the solvent, only the second reaction occurs. Disilane also reacts with potassium in 1,2-dimethoxyethane to form $KSiH_3$, although SiH_4 and nonvolatile polysilanes are also produced (28,31). Pure crystalline $KSiH_3$ prepared from SiH_4 and potassium in 1,2-dimethoxyethane has been obtained by slow evaporation of the solvent. When liquid ammonia is used as the solvent, only a small fraction of SiH_4 is converted into metal salt; most of the SiH_4 undergoes ammonolysis (32).

Disilane undergoes disproportionation in 1,2-dimethoxyethane, forming SiH_4 and a solid material, $+(SiH_2)_x$, when an alkali metal salt, eg, KH or LiCl, is present (33,34).

Silanes react with alkyllithium compounds, forming various alkylsilanes. Complete substitution is generally favored; however, less substituted products can be isolated by proper choice of solvent. All four methylsilanes, vinylsilane [7291-09-1], and divinylsilane [18142-56-8] have been isolated from the reaction of SiH_4 and the appropriate alkyllithium compound with propyl ether as the solvent (35). Methylsilane and ethyldisilane [7528-37-2] have been obtained in a similar reaction (36).

Electrical Discharge, Irradiation, and Photolysis.   Early reports of the decomposition of SiH_4 in an electrical discharge indicated that the main products were hydrogen, solid silicon subhydrides of composition $SiH_{1.2-1.7}$, and small quantities of higher silanes (37). However, more recent studies indicate that under certain conditions reasonably large quantities of higher silanes up to *n*-Si_4H_{10} [7783-29-1] and *iso*-Si_4H_{10} [13597-87-0] plus smaller amounts of various isomers of higher silanes up to Si_8H_{18} can be produced by this method (38–40). In addition, mixed-hydride derivatives can be prepared by subjecting mixtures of SiH_4 and certain other volatile hydrides to such a discharge. Thus, SiH_3GeH_3 [13768-63-3], SiH_3PH_2 [14616-47-8], and SiH_3AsH_2 [15455-99-9] have been prepared from SiH_4-GeH_4, SiH_4-PH_3, and SiH_4-AsH_3 mixtures, respectively (41,42). Although both disilylphosphine [*14616-42-3*], $(SiH_3)_2PH$, and disilanylphosphine, $SiH_3SiH_2PH_2$, are obtained in the SiH_4-PH_3 system, these can be obtained free of each other in the discharge of $SiH_4-SiH_3PH_2$ and $Si_2H_6-PH_3$ (43). Methyldisilane [*13498-43-6*] can be isolated from the products of a $SiH_4-CH_3SiH_3$ or a $SiH_4-(CH_3)_2O$ discharge system (33,44). Hexachlorodisilane [*13465-77-5*] and hexafluorodisilane [*13830-68-7*] can be prepared from trichlorosilane and trifluorosilane [*13465-71-9*], respectively (45). Glow discharge also provides a method for a high deposition rate of silicon from disilane (46). Silicon has been produced from silane by Penning discharge (47). Similarly, amorphous hydrogenated silicon nitride has been produced from silane–ammonia–nitrogen mixtures (48).

Photolysis studies of SiH_4 have involved a sensitizer, eg, mercury vapor, because SiH_4 does not absorb uv-radiation at wavelengths above 185 nm (49). The mercury-sensitized photolysis of SiH_4 leads to the formation of H_2, Si_2H_6, Si_3H_8 [*7783-26-8*], and polymeric solid silanes; the quantum yields of H_2 and Si_2H_6 are ca 1.6 and 0.6, respectively (50,51). The photolysis of SiH_4 in the presence of GeH_4 or CH_3I produces SiH_3GeH_3 or CH_3SiH_3, respectively (52).

Manufacture and Processing. There are four methods of production of compounds containing a Si–H bond that are noteworthy. Silicides of magnesium, aluminum, lithium, iron, and other metals react with acids or their ammonium salts to produce silane and higher binary silanes. This method, uniquely applicable to the inorganic silanes, is the historic method for production of the silicon hydrides. Under optimum conditions, ie, the addition of magnesium silicide to dilute phosphoric acid, a 23% conversion to volatile silanes is possible (3). The composition of the silane mixture is 40 wt % SiH_4, 30 wt % Si_2H_6, 15 wt % Si_3H_8, and 15 wt % higher silanes. The highest reported yields of silicon hydrides have been achieved by treatment of magnesium silicide with NH_4Br in liquid ammonia at $-33°C$ and $N_2H_4 \cdot 2HCl$ in hydrazine (53–55).

The formation of silane from magnesium silicide by acidic hydrolysis is thought to involve one of two paths after formation of a dihydroxymagnesiosilane intermediate (56).

$$Mg_2Si + 2\,H_2O \xrightarrow{\text{HCl}} \begin{bmatrix} HO-Mg \\ \diagdown SiH_2 \\ HO-Mg \diagup \end{bmatrix}$$

$$[SiH_2O] + Mg(OH)_2 + 2\,H_2 + 2\,H_2O + MgCl_2 \qquad SiH_4 + 2\,Mg(OH)_2$$

The relatively low yields of silane provided by this method together with the difficulties in subsequently purifying the materials led to the discontinuation of commercial processes based on this technology in the United States in the late 1960s and early 1970s. This method was generally abandoned in favor of methods involving reduction of silicon halides. Interest in disilane has been sufficient to maintain magnesium silicide-based production in Japan. Treatment of calcium silicide with HCl–ethanol or glacial acetic acid yields the complex polymer called siloxene (57,58).

The reductions of chlorosilanes by lithium aluminum hydride, lithium hydride, and other metal hydrides, MH, offers the advantages of higher yield and purity as well as flexibility in producing a range of silicon hydrides comparable to the range of silicon halides (59). The general reaction is as follows:

$$R_n SiX_{4-n} + (4-n) MH \longrightarrow R_n SiH_4 + (4-n) MX$$

where X is a halogen. The most versatile reagent is lithium aluminum hydride; the most convenient solvent is tetrahydrofuran, followed by ethyl ether and bis(methoxyethyl)ether, although tertiary amines in hydrocarbons have also been reported as reaction media (60). The reduction occurs at room temperature. This approach is employed in the commercial production of silane from tetrafluorosilane utilizing sodium aluminum hydride (61,62). Limitations of the lithium aluminum hydride reagent system include the inability to produce partially reduced materials, eg, $SiCl_4$ is reduced completely to SiH_4 without formation of SiH_2Cl_2; and the catalytic rearrangement of susceptible bonds because of the formation of by-product aluminum chloride.

Lithium hydride is perhaps the most useful of the other metal hydrides. The principal limitation is poor solubility, which essentially limits reaction media to such solvents as dioxane and dibutyl ether. Sodium hydride, which is too insoluble to function efficiently in solvents, is an effective reducing agent for the production of silane when dissolved in a LiCl–KCl eutectic at 348°C (63–65). Magnesium hydride has also been shown to be effective in the reduction of chloro- and fluorosilanes in solvent systems (66) and eutectic melts (67).

Methods of direct reduction of chlorosilanes using hydrogen at high temperatures have historically been inefficient processes (68–70). Significant process innovations, involving the hydrogenation of silicon tetrachloride over Si–Cu at less than 2.45 MPa (500 psi), proceed in good conversion (71,72) and allow commercial processes.

$$3\ SiCl_4 + 2\ H_2 + Si^0 \xrightarrow{\ 550\ K\ } 4\ HSiCl_3$$

At atmospheric pressure, the conversion to trichlorosilane is limited to about 16%. The conversion of $SiCl_4$ to $HSiCl_3$ was found to be at equilibrium. If contact time was greater than 45 s and the mole ratio of hydrogen to silicon tetrachloride 1:1, then at 14 kPa (2 psi) and 550°C, the $HSiCl_3$ mole fraction reached 0.7 but substantial formation of H_2SiCl_2 occurred (62). Enhancements in yield have been reported through preactivating the silicon mass by removal of oxides (73) and the rapid thermal quench of the effluent gas stream (74). The reduction of silicon tetrachloride in a plasma has also been reported (75).

Disproportionation reactions of silicon hydrides occur readily in the presence of a variety of catalysts. For example:

$$4\ HSiCl_3 \xrightarrow{\ catalyst\ } SiH_4 + 3\ SiCl_4$$

The most common catalysts in order of decreasing reactivity are halides of aluminum, boron, zinc, and iron (76). Alkali metals and their alcoholates, amines, nitriles, and tetraalkylureas have been used (77–80). The largest commercial processes use a resin–catalyst system (81). Trichlorosilane refluxes in a bed of anion-exchange resin containing tertiary amino or quaternary ammonium groups. Contact time can be used to control disproportionation to dichlorosilane, monochlorosilane, or silane.

Direct synthesis is the preparative method that ultimately accounts for most of the commercial silicon hydride production. This is the synthesis of halosilanes by the direct reaction of a halogen or halide with silicon metal, silicon dioxide, silicon carbide, or metal silicide without an intervening chemical step or reagent. Trichlorosilane is produced by the reaction of hydrogen chloride and silicon, ferrosilicon, or calcium silicide with or without a copper catalyst (82,83). Standard purity is produced in a static bed at 400–900°C.

$$2\ Si^0 + 7\ HCl \longrightarrow HSiCl_3 + SiCl_4 + 3\ H_2$$

Despite the apparent stoichiometry, this technology generates only 10–20% trichlorosilane. The balance is tetrachlorosilane. Trichlorosilane is in essence a recoverable by-product in a feedstock process for the production of fumed silica. High purity trichlorosilane is usually produced in a fluidized bed at 300–425°C. Hydrogen chloride is typically employed, although separate feeds of hydrogen and chlorine can be combusted in the reactor. Metallurgical silicon feedstock typically has a median particle size of 40–60 mesh (ca 420–250 μm); principal impurities are iron and aluminum as a result of primary silicon production and the milling employed in size reduction (qv) (see SILICON AND SILICON ALLOYS, CHEMICAL AND METALLURGICAL). A gas flow rate 4–20 times the minimum fluidization velocity is employed (84). A typical reactor is constructed of carbon steel, has a diameter of 0.5–0.8 m, operates at 2–5 kPa (0.3–0.7 psi), and has an output of 600–1000 kg/h. The heat of reaction, -141.8 kJ/mol (-33.9 kcal/mol) at 298°C (85), energetically propagates the initiated reaction and cooling must be provided. The conversion to trichlorosilane under these conditions is typically 80–88%; the remainder is silicon tetrachloride, 1–4% higher silicon halides consisting mainly of hexachlorodisilane and pentachlorodisilane, and 1–2% dichlorosilane. Pilot-scale technology for trichlorosilane developed independently from commercial technology has been reported in detail (86). Similarly, laboratory-scale production of tribromosilane has been reported with a fixed bed at 360–400°C (87). Additionally, inorganic silanes can be formed by cleavage reactions of aromatic organosilanes, eg, (88,89):

$$C_6H_5SiH_3 + HI \longrightarrow ISiH_3 + C_6H_6$$

Other, less satisfactory methods for forming silicon hydrogen bonds are shown below (90–94), where diphos = $(C_6H_5)_2PCH_2CH_2P(C_6H_5)_2$.

$$Cl_3SiSiCl_3 + HCl \longrightarrow HSiCl_3 + SiCl_4$$

$$Cl_3SiSiCl_3 + NH_3 \longrightarrow HSiCl_3 + SiCl_3NH_2$$

$$SiCl_4 + HCHO \longrightarrow H_3SiCl + H_2SiCl_2 + \text{other products}$$

$$3\ SiCl_4 + 4\ Al^0 + 6\ H_2 \longrightarrow 3\ SiH_4 + 4\ AlCl_3$$

$$HSiCl_3 + (diphos)_2CoH \xrightarrow{175°C} H_2SiCl_2 + (diphos)_2CoCl$$

Economic Aspects. Trichlorosilane is the only inorganic silicon hydride produced in large scale. The annual U.S. production is ca 30,000–35,000 metric tons. Substantial quantities of this material are also generated in the production of silicon tetrachloride streams used in fumed silica (qv) production, but these are never isolated. The merchant cost of trichlorosilane is $2–$10/kg, depending on grade and container. Trichlorosilane is produced in the United States by Dow Corning, Degussa, General Electric, and Texas Instruments. Other worldwide producers are Hüls, Wacker, Chisso, Mitsubishi, Shin-etsu, and Tokoyama. Most (60–65%) trichlorosilane is used in the production of electronic-grade silicon (Fig. 1). Polycrystalline silicon processes consume most of the material, but some is used for epitaxial deposition. Hemlock Semiconductor, a joint venture

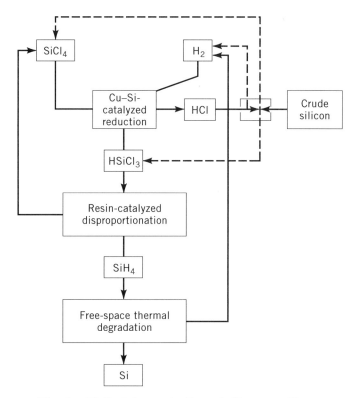

Fig. 1. Method for production of ultrapure silicon.

dominated by Dow Corning, produces polycrystalline silicon from trichlorosilane by a Siemens-type process. Hemlock Semiconductor had a nameplate capacity of 3500 t in 1995 and was scheduled to expand to 5000 t in 1997. Other producers of polycrystalline silicon from trichlorosilane by similar processes include Komatsu, Tokoyama, Mitsubishi, and Shin-etzu. Other uses for trichlorosilane include disproportionation to silicon hydrides and the conversion to organosilanes, eg, adhesion promoters.

In the United States, most of the consumption of silane is internal. Silane is consumed in a free-space polycrystalline silicon production facility (3000 t/yr) at Moses Lake, Washington. This facility was built by Union Carbide but as of 1996 was operated by Advanced Silicon Materials Inc., a joint venture controlled by Komatsu. Silane is also consumed in a fluidized-bed polycrystalline silicon production facility (1250 t/yr) in Pasadena, Texas, built by Ethyl Corporation and operated as of 1996 by MEMC, a publicly traded company 51% owned by Hüls. The world merchant market for silane is about 250 t/yr. The bulk price for silane is $80–$120/kg; much higher pricing is placed on small packages. The primary use of silane is in microelectronics. Minor applications exist in reprography, ie, in photocopiers, and in specialty glass technology.

Other specialty silanes used in microelectronic applications include dichlorosilane and disilane. Tribromosilane [7789-57-3], iodosilanes, and trisilylamine [13862-16-3] are of interest for microelectronics in low temperature deposition technologies.

Health and Safety Factors, Toxicology. The acute hazards of silicon hydrides are overwhelmingly important in considering worker safety. Silane is pyrophoric. Dichlorosilane, methylsilane, and dimethylsilane have demonstrated flow dynamic ignition. High levels of explosion severity have been reported under certain conditions (95). Chlorine-containing compounds generate hydrogen chloride on contact with water and other protic materials. At low concentrations, chlorosilanes affect nasal and pulmonary membranes. Only a minimal amount of toxicity information is available on these materials. The LD_{50} for trichlorosilane is 1050 mg/kg (96).

Organic Hydride Functional Silanes

The organohydrosilane of greatest commercial importance is methyldichlorosilane [75-54-7]. Careful hydrolysis of this material with water affords polymethylhydrosiloxanes, primarily used in the textile industry to waterproof and improve the wear resistance of fabrics (see WATERPROOFING AND WATER/OIL REPELLANCY). This polymer is also used as a waterproofing agent in the leather (qv) industry, the paper (qv) industry for sizing, electronic applications, and in construction for enhancing the water resistance of gypsum board (97–99). Similar fluids based on ethyldichlorosilane [1789-58-8] have been developed for commercial application in the CIS (former USSR), but as of 1996 these were not readily available elsewhere. Methyldichlorosilane is also used captively by silane and silicone producers in thermal condensation reactions to produce vinyl, phenyl, and cyanoalkyl precursors to silicone fluids. The addition reaction of methyldichlorosilane to fluorocarbon alkenes has enabled production of methyltrifluoropropyl silicone fluids, gums (qv), and rubbers. Trialkoxysilanes, eg, triethoxysilane [*998-30-1*] and trimethoxysilane [*2487-90-3*], are being used to prepare a number of organic coupling agents utilized by the plastics industry as adhesion promoters (see ADHESIVES). Organosilanes containing one or more Si–H bonds have excellent reducing capabilities (100).

Physical Properties. The physical properties of organosilanes are determined largely by the properties of the silicon atom (Table 2). Because silicon is larger and less electronegative than either carbon or hydrogen, the polarity of the Si–H bond is opposite to that of the C–H bond (Table 3). This difference in polarity imparts hydride character to the Si–H bonds of organosilanes. This difference in electronegativities is not as great as in ionic hydrides (qv), eg, LiH, NaH, and CaH$_2$; thus, the Si–H bond is still largely (98%) covalent. The size of the silicon atom is greater than the carbon atom, and this increase in atomic volume enables nucleophilic attack on the silicon to occur more readily than on carbon. Electrophilic attack on hydrogen bonded to silicon is also facilitated by the small steric constraints of hydrogen and the increased Si–H bond length. Additionally, the Si–H bond energy is somewhat lower than that of C–H, as is reflected in the thermal stabilities of such bonds. Organohydrosilanes begin to decompose at 440–460°C through homolytic cleavage of the Si–H bond and subsequent radical formation. The length and strength of silane bonds are also manifest in the vibrational spectra. Vibration frequencies of silanes, 2100–2260 cm^{-1}, are lower than those of C–H, 2700–3000 cm^{-1}. The presence of inductive groups, eg, halogen, vinyl, phenyl, or fluoroalkyl, on silicon increases the vibrational frequency of the Si–H bond. Conversely, the presence of alkyl substituents generally decreases the vibrational frequency because of the donation of electron density. A comparison of some physical properties of carbon and silicon and their bonds to hydrogen is given in Table 3.

Unlike carbon, the silicon atom may utilize vacant 3*d* orbitals to expand its valence beyond four, to five or six, forming additional bonds with electron donors. This is shown by isolated amine complexes. The stability of the organosilane amine complexes varies over a wide range and depends on the nature of the donor and acceptor (2).

Chemical Properties. Organohydrosilanes undergo a wide variety of chemical conversions. The Si–H bond of organohydrosilanes reacts with elements of most groups of the Periodic System, especially Groups 16(VIA) and 17(VIIA). There are no known reactions if the Si–H bond is replaced by stable bonds of silicon with elements of Groups 2(IIA), 13(IIIA), and 8–10(VIII).

Oxidation. Whereas the majority of inorganic silanes are oxidized spontaneously on contact with oxygen and air, the simplest organosilane, ie, methylsilane, is stable to air and does not spontaneously inflame, although flow-dynamic ignition has been reported. Methylsilane has been reported to explode in the presence of mercury and oxygen (102). The larger-chain alkylsilanes are more stable but ignite spontaneously when vaporized in oxygen under pressure. Phenyl and cyclohexylsilane can be distilled open to the atmosphere. Trialkyl- and triarylsilanes are more stable and have been distilled at as high as 325°C without decomposition (103). Differences between organosilanes and inorganic silanes are attributed to the electronic and steric effects of the organic substituents. The oxidizability of the Si–H bonds is much greater than that of C–H bonds. This difference is manifested by the ease

Table 2. Properties of Hydride Functional Organosilanes, Siloxanes, and Silazanes[a]

Compound	CAS Registry Number	Mol wt	Bp,[b] °C	Mp, °C	d_4^{20}	n_D^{20}
CH_3SiH_3	[992-94-9]	46.1	−57	−157	0.6277[c]	
$(CH_3)_3SiSiH_3$	[18365-32-7]	104.3	80−81			
$C_2H_5SiH_3$	[2814-79-1]	60.2	−14	−180	0.6396[d]	
$C_4H_9SiH_3$	[1600-29-9]	88.2	56	−138	0.6764	1.3922
$C_6H_5SiH_3$	[694-53-1]	108.2	120	<−60	0.8681	1.5125
$C_6H_{13}SiH_3$	[1072-14-6]	116.3	114−115	−98	0.7182	1.4129
$C_{18}H_{37}SiH_3$	[18623-11-5]	284.6	$195−196_2$	29		
$(CH_3)_2SiH_2$	[1111-74-6]	60.2	−20	−150	0.6377[e]	
$CH_3(C_6H_5)SiH_2$	[766-08-5]	122.2	139		0.889	1.506
$(C_2H_5)_2SiH_2$	[542-91-6]	88.2	56	−134	0.6832	1.3920
$(C_2H_5O)_2SiH_2$	[18165-68-9]	120.2	90−92			
$(C_6H_5)_2SiH_2$	[775-12-2]	184.3	$100−101_4$		0.9964	1.5756
$CH_3SiH(Cl)_2$	[75-54-7]	115.0	41−42	−93	1.105	1.422
$CH_3SiH(OCH_3)_2$	[16881-77-9]	106.2	61	−136	0.861	1.360
$CH_3SiH(OC_2H_5)_2$	[2031-62-1]	134.3	94.5		0.829[f]	1.372[f]
$CH_3SiH(Cl)N(CH_3)_2$	[18209-60-4]	123.7	85−87			
$CH_3SiH[N(CH_3)_2]_2$	[3768-57-8]	176.4	112−113			
$CH_3SiH(OOCCH_3)_2$	[3435-15-2]	162.2	$83−84_6$		1.1081	1.4022
$CH_3SiH(C_2H_5)Cl$	[6374-21-6]	108.6	67−68		0.8816	1.4020
$(CH_3)_2SiHCl$	[1066-35-9]	94.6	36	−111	0.851	1.3827
$(CH_3)_2SiHC_2H_5$	[758-21-4]	88.2	46		0.6681	1.3783
$(CH_3)_2SiHOC_2H_5$	[14857-34-2]	104.2	54		0.7572	1.3683
$(CH_3)_2SiHN(C_2H_5)_2$	[13686-66-3]	131.3	109−110			
$(CH_3)_3SiH$	[993-07-7]	74.2	67	−136	0.6375	
$(CH_3O)_3SiH$	[2487-90-3]	122.2	86−87	−114	0.860	1.3687
$C_2H_5SiH(Cl_2)$	[1789-58-8]	129.1	75	−107	1.0926	1.4148
$(C_2H_5)_2SiH(Cl)$	[1609-19-4]	122.7	100		0.889	1.4152
$(C_2H_5)_2SiH(CH_3)$	[760-32-7]	102.3	77−78		0.7054	1.3984
$(C_2H_5)_3SiH$	[617-86-7]	116.3	107−108	−157	0.7318	1.4119
$(C_2H_5O)_3SiH$	[998-30-1]	164.3	131.5	−170	0.875	1.337
$(2-ClC_2H_4O)_3SiH$	[10138-79-1]	267.6	$117−118_{0.3}$		1.2886	1.4577
$(C_3H_7)_3SiH$	[998-29-8]	158.4	173		0.7580	1.4272
$(C_6H_5)_2SiH(Cl)$	[1631-83-0]	218.8	$143_{1.3}$		1.118	1.581
$C_6H_5SiH(Cl_2)$	[1631-84-1]	177.1	$65−66_{1.3}$		1.212	1.526
$C_6H_5SiH(CH_3)Cl$	[1631-82-9]	156.7	113_{13}		1.054	1.571
$C_6H_5SiH(CH_3)_2$	[766-77-8]	136.3	156−157		0.8891	1.4995
$C_6H_5(H_2C=CH)-$ $SiH(CH_3)$	[17878-39-6]	148.3	$56−57_{0.9}$		0.891	1.5115
$(C_6H_5)_2SiH(CH_3)$	[776-76-1]	198.3	266−267		0.9945	1.5717
$(C_6H_5)_3SiH$	[789-25-3]	260.4	$160−165_{0.4}$	42.4		
$(C_6H_{13})_3SiH$	[2929-52-4]	284.6	160−161		0.7992	1.448
$(C_8H_{17})_3SiH$	[18765-09-8]	368.8	$163−165_{0.02}$		0.821	1.454
$[(CH_3)_2SiH]_2O$	[3277-26-7]	134.3	70−71		0.757	1.370
$[(CH_3)_2SiH]_2NH$	[15933-59-2]	133.3	99−100		0.766	1.4044
$(CH_3)_3SiOSiH(CH_3)_2$	[1438-82-0]	148.4	85−86		0.7578	1.3740
$[(CH_3)_3SiO]_2SiH(CH_3)$	[1873-88-7]	222.5	141−142		0.8136	1.3815
$[CH_3SiHO]_4$	[2370-88-9]	240.5	134−135	−69	0.9912	1.3870
$[SiH(CH_3)_2]_2$	[814-98-2]	118.3	118.3	84−86	0.7202	1.4290

[a]Refs. 2 and 101. [b]Subscripted numbers are pressures in kPa. To convert kPa to mm Hg, multiply by 7.5. [c]At −58°C. [d]At −14°C. [e]At −20°C. [f]At 25°C.

Table 3. Physical Properties of Carbon and Silicon and Their Bonds to Hydrogen[a]

Element	Electronegativity	Covalent radius, pm	Usual coordination number	Bond with hydrogen	
				Bond length, pm	Bond energy, kJ/mol[b]
silicon	1.8	117	4	149	369–378
hydrogen	2.1	30		74	432
carbon	2.5	77	4	106–110	368–506

[a]Ref. 10, 59.
[b]To convert J to cal, divide by 4.184.

of oxidation of organohydrosilanes with metal oxides. The most noteworthy of these is mercuric oxide. Reactions of alkylsilanes with mercuric oxide in inert solvents, eg, toluene, afford nearly quantitative yields of silanols or silanediols.

$$R_2SiH_2 + 2\,HgO \xrightarrow[-80-20°C]{toluene} R_2Si(OH)_2 + 2\,Hg^0$$

Oxidation has also been cited as occurring in the cure of polymethylhydrosiloxane [9004-73-3] (PMHS) on cellulose acetate fibers. Investigation of the cured, cross-linked silicone shows no evidence of the Si–H bond. The same compound under an atmosphere of nitrogen does not cure and retains the Si–H bonds (99).

Hydrolysis. As for inorganic silanes, no reaction occurs between organohydrosilanes and water. The presence of acidic or alkaline catalysts, however, brings about the reaction according to the following scheme:

$$\equiv Si-H + H_2O \xrightarrow{catalyst} \equiv Si-OH + H_2$$

The ease of hydrolysis depends on the pH and is more rapid under alkaline than acidic conditions. Organosilanes are most stable in weakly acidic media. The rate and extent of hydrolysis reactions are conveniently monitored by measurement of the rate and amount of hydrogen evolved. Alkaline hydrolysis is more widely used and has been more extensively studied. Alkaline catalysts may be organic, eg, triethylamine, pyridine, etc, or inorganic, eg, KOH, NaOH, NH$_4$OH, etc. The rate and completeness of these reactions depend on the nature and amount of catalyst, solvent, reagent concentration, temperature, and organic substituents of the organosilane. The substituents influence rate through steric and inductive effects. Such effects determine the ease of formation of the pentacoordinate silicon complex believed to be the transition state (104,105). The mechanism is thought to proceed through rapid attack of the base on the organohydrosilane with formation of a pentacoordinate silicon via bonding through 3d orbitals. Slow decomposition of this complex through interaction with a solvent proton affords the product.

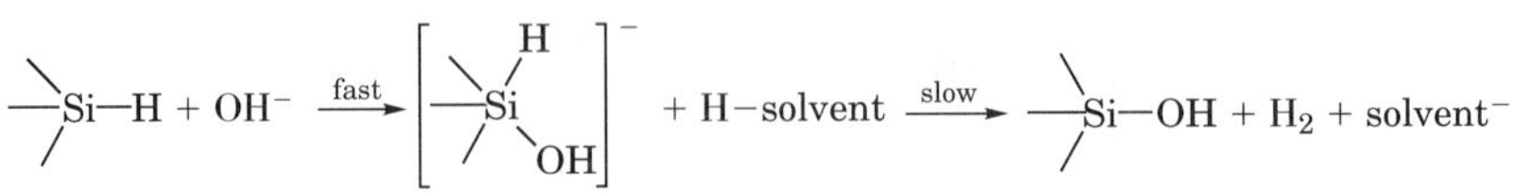

Steric and inductive effects determine the rate of formation of the pentacovalent silicon reaction complex. In alkaline hydrolysis, replacement of a hydrogen by alkyl groups, which have lower electronegativity and greater steric requirements, leads to slower hydrolysis rates. Replacement of alkyl groups with bulkier alkyl substituents has the same effect. Reaction rates decrease according to:

$$C_4H_9SiH_3 > CH_3(i\text{-}C_3H_7)SiH_2 > (C_2H_5)_2SiH_2 > (C_2H_5)_3SiH > (C_3H_7)_3SiH$$

Mechanistically the rate-determining step is nucleophilic attack involving the hydroxide ion and the more positive silicon atom in the Si–H bond. This attack has been related to the Lewis acid strength of the corresponding silane, ie, to the ability to act as an acceptor for a given attacking base. Similar inductive and steric effects apply for acid hydrolysis of organosilanes (106).

Alcohols, Phenols, Silanols, and Carboxylic Acids. The catalyzed reaction of organosilanes with hydroxyl-containing organic compounds affords organoalkoxy and organoaryloxysilanes, usually in high yields (107). Alkali–metal oxides, hydrogen halides, and metal halides are most often used as catalysts. Of the metal halides, $ZnCl_2$ and $SnCl_2$ are the most commonly used:

$$\equiv Si\text{-}H + HOR \xrightarrow{\text{catalyst}} \equiv Si\text{-}OR + H_2$$

where R is an alkyl or aryl. The reaction of organosilanes with alkali metal alkoxides proceeds vigorously and in high yields. Aryloxy derivatives of silanes can be produced from phenols in yields up to 90% by catalysis with colloidal nickel and less effectively with other metals, eg, copper, cobalt, chromium, zinc, and tin (108). Reactivity increases for selected alcohols in the order $CH_3OH <$ $C_2H_5OH < n\text{-}C_3H_7OH < n\text{-}C_4H_9OH$ (109), suggesting that the alkoxide is the active nucleophile.

$$C_6H_5SiH_3 + 3\ ROH \longrightarrow C_6H_5Si(OR)_3 + 3\ H_2$$

Organic amines, eg, pyridine and piperidine, have also been used successfully as catalysts in the reactions of organosilanes with alcohols and silanols. The reactions of organosilanes with organosilanols lead to formation of siloxane bonds. Nickel, zinc, and tin also exhibit a catalytic effect.

$$\equiv Si\text{-}H + HOSi\equiv \xrightarrow{\text{catalyst}} \equiv Si\text{-}OSi\equiv + H_2$$

The reaction is of practical importance in the vulcanization of silicone rubbers (see RUBBER COMPOUNDING). Linear hydroxy-terminated polydimethylsiloxanes are conveniently cross-linked by reaction with methyldiethoxysilane or triethoxysilane [998-30-1]. Catalysts are amines, carboxylic acid salts of divalent metals such as Zn, Sn, Pb, Fe, Ba, and Ca, and organotin compounds. Hydroxy-terminated polysiloxanes react with Si–H-containing polysiloxanes to produce cross-linked materials. Metal salts of carboxylic acids are used as catalysts.

The liberated hydrogen can act as a blowing agent to produce silicone foam rubber (110).

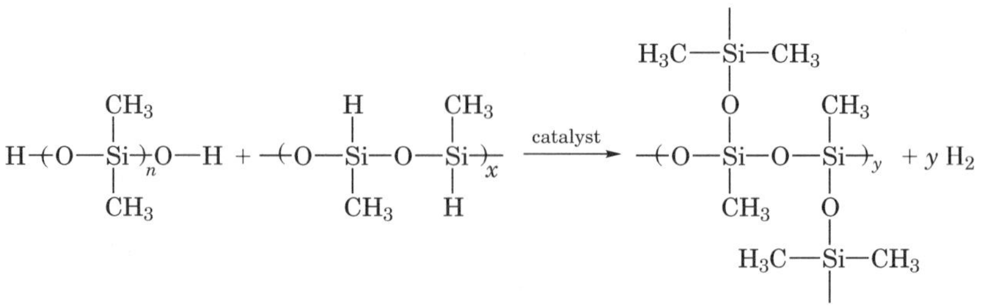

$$
\text{H} \underset{\substack{| \\ CH_3}}{\overset{\substack{CH_3 \\ |}}{\left(O-Si\right)_n O}} \text{H} + \underset{\substack{| \\ CH_3}}{\overset{\substack{H \\ |}}{\left(O-Si\right.}} \underset{\substack{| \\ H}}{\overset{\substack{CH_3 \\ |}}{-O-Si)_x}} \xrightarrow{\text{catalyst}} \quad +y\,H_2
$$

Whereas metal salts of carboxylic acids catalyze the above reactions, these are not sufficiently basic to cleave Si–H bonds. Mercury salts of organic acids in the presence of silver perchlorate, however, do react to produce organoacyloxysilanes (111).

$$
R_3'Si\text{-}H \xrightarrow{\overset{O}{\overset{\|}{Hg(OCR)_2}}} R_3'SiO\overset{O}{\overset{\|}{C}}R + R\overset{O}{\overset{\|}{C}}OH + Hg^0
$$

Organoacyloxysilanes are also produced by reaction of organosilanes with carboxylic acids in the presence of strong mineral acids, eg, sulfuric and hydroiodic acids. Trialkylacyloxysilanes have been obtained in 81–87% yield from monocarboxylic acids in the presence of aluminum and iodine.

$$
R_3SiH + HOCR' \longrightarrow R_3SiOCR' + H_2
$$

Peracylation of polymethylhydrosiloxane to produce a cross-linked or cross-linkable material is achieved by reaction with acetic acid and is catalyzed by anhydrous zinc chloride (112). This reaction can be extended to monomeric organosilanes under similar conditions.

Amines and Phosphines. As in reactions of alcohols and acids with organosilanes, reaction of the Si–H bond with amines and phosphines proceeds only under catalysis. Alkali metal amides or phosphines are the catalysts of choice and effect replacement of the Si–H bond with Si–N or Si–P bonds, respectively. Catalytic activity of the alkali metals for these reactions is K > Na > Li (113). Reactions of triorganosilanes with ammonia (qv) and amines have been the most widely studied. Ammonia affords the disubstituted nitrogen compounds or disilazanes with no trisubstitution being observed. This presumably occurs because of steric crowding at nitrogen. For example, reactions for the formation of triethyldisilazane [2117-18-2] and triphenyldisilazane [4158-64-9] are as follows:

$$
2\,(C_2H_5)_3SiH + NH_3 \xrightarrow{NaNH_2} ((C_2H_5)_3Si)_2NH + H_2
$$

$$
2\,(C_6H_5)_3SiH + NH_3 \xrightarrow{NaNH_2} ((C_6H_5)_3Si)_2NH + H_2
$$

Primary organic amines react with triethylsilane in the presence of the appropriate potassium amides to produce organoaminotriethylsilanes with yields of 82–92%.

$$(C_2H_5)_3SiH + RNH_2 \xrightarrow{\text{KNHR}} (C_2H_5)_3SiNHR + H_2$$

where R is C_2H_5 [3294-33-5], i-C_3H_7 [5277-20-3], or C_6H_5 [18106-48-4]. No reaction occurs using secondary or tertiary amines under similar reaction conditions. Steric factors are also important, as branched primary amines require higher temperatures.

In the presence of hydrogen and certain precious metal and acidic catalysts, dihydridosilanes react with ammonia to form silazane oligomers (114).

$$n\,(C_2H_5)_2SiH_2 + n\,NH_3 \xrightarrow{\text{Ru}_3(\text{CO})_{12},\,H_2} ((C_2H_5)_2SiNH)_n + n\,H_2$$

The same reaction is exploited on a commercial scale to convert higher silazanes to cyclic silazanes, but the hydride intermediate is never isolated (115).

$$(CH_3)_2SiNH)_n \longrightarrow [H(CH_3)_2SiNH_2] \xrightarrow{\text{Pt/C, }H_2} c\text{-}((CH_3)_2SiNH)_3$$

Halogens and Halogen Compounds. The reaction of organosilanes with halogens and halogen compounds usually proceeds in good yield through cleavage of the Si–H bond and formation of the Si–X bond. This reaction can be achieved by direct action of halogen on the organosilane or by interaction with halogen-containing organic and inorganic compounds. Reaction with fluorine, however, does not proceed satisfactorily. Cleavage not only of the Si–H bond, but also of the C–Si and C–H bonds, occurs. Fluorination of organosilanes has been achieved by reaction with inorganic fluorides, eg, AgF and SbF_3. Direct halogenation with chlorine, bromine, and iodine proceeds smoothly, however. Reactions are normally carried out in an inert solvent, eg, carbon tetrachloride, hexane, ethyl bromide, benzene, or chloroform. The reaction rate is further controlled by maintaining low reaction temperatures and controlling the rate of halogen addition. Steric requirements are not a significant factor, as evidenced by the facile iodination of tricyclohexylsilane [1629-47-6]. These reactions are fast and convenient and allow partial replacement of Si–H bonds (103):

$$R_3SiH + X_2 \longrightarrow R_3SiX + HX$$
$$C_6H_5SiH_3 + Br_2 \longrightarrow C_6H_5SiH_2Br + HBr$$

where R is an alkyl, a cycloalkyl, or aryl, and X_2 is Cl_2, Br_2, or I_2.

Organosilanes may react with hydrogen halides to afford organohalosilanes and hydrogen. Hydrogen fluoride is not generally used for conversion of organosilanes to organofluorosilanes. The action of hydrogen fluoride readily cleaves Si–H and Si–C bonds, forming Si–F bonds. In the presence of copper salts, however, HF fluorinates the Si–H bond without splitting the Si–C bond (100).

Using hydrogen chloride and hydrogen bromide, reactions are catalyzed by a Lewis acid, eg, $AlCl_3$, ie, where X = Cl or Br:

$$\equiv Si-H + HX \xrightarrow{AlX_3} \equiv Si-X + H_2$$

Hydrogen iodide iodinates trialkylsilanes in good yield in boiling carbon tetrachloride with no aluminum halide present (116). This can perhaps be explained on the basis that some free iodine is always present in equilibrium with hydrogen iodide.

A convenient synthesis of organochlorosilanes from organosilanes is achieved by reaction with inorganic chlorides of Hg, Pt, V, Cr, Mo, Pd, Se, Bi, Fe, Sn, Cu, and even C. The last compounds, tin tetrachloride, copper(II) chloride, and, under catalytic conditions, carbon tetrachloride (117,118), are most widely used.

$$SiH_4 + SnCl_4 \longrightarrow HSiCl_3 + SnCl_2 + HCl$$

Tin tetrachloride has been used to prepare the sterically hindered triisopropyl-chlorosilane [13154-24-0] (119). Organobromosilanes are obtained under similar conditions through reaction with cupric and mercuric bromide. These reactions are most suitable for stepwise displacement of hydrogen to form mixed hydridochlorosilanes or in systems sensitive to halogen (120). Hydrides have also been displaced using organic bromides. Heating triethylsilane and allyl bromide, propyl bromide, and methyl β-bromopropionate produces triethylbromosilane [1112-48-7] in 10–100% yields (121). Organochlorosilanes have been obtained from alkyl and alkenyl chlorides, eg, allyl, neopentyl, hexyl, and propyl chlorides. Unlike bromo compounds, these reactions proceed when catalyzed by small amounts of $AlCl_3$. Such reactions show potential for reduction of alkyl halides (122).

$$C_6H_{11}Cl + (C_2H_5)_3SiH \xrightarrow{AlCl_3, \, 0°C} C_6H_{12} + (C_2H_5)_3SiCl$$

In the context of commercial silane production, these frequently lead to by-product reactions, as for example the production of allyldimethylsilane by a Grignard route (see GRIGNARD REACTIONS).

$$3\,CH_2\!=\!CHCH_2Cl + 2\,H(CH_3)_2SiCl \xrightarrow{Mg} H(CH_3)_2SiCH_2CH\!=\!CH_2 + HCl + 2\,MgCl_2$$
$$+ CH_2\!=\!CHCH_2(CH_3)_2SiCH_2CH\!=\!CH_2$$

Even more importantly, these lead to reduction by-products in the commercial hydrosilylation reaction with allyl chloride.

$$2\,CH_2\!=\!CHCH_2Cl + 2\,H(CH_3)_2SiCl \xrightarrow{Pt} Cl(CH_3)_2SiCH_2CH_2CH_2Cl + CH_3CH\!=\!CH_2$$
$$+ (CH_3)_2SiCl_2$$

Geminal polyhalides also react with organosilanes under peroxide catalysis. For example, triethylsilane affords triethylchlorosilane in good yield upon reaction with carbon tetrachloride in the presence of benzoyl peroxide (bpo) at 80°C (94,100,102).

$$(C_2H_5)_3SiH + CCl_4 \xrightarrow{\text{bpo}} (C_2H_5)_3SiCl + HCCl_3$$

Acid halides, eg, benzoyl chloride, acetyl chloride, and benzoyl bromide, have been used to prepare Si–Cl and Si–Br compounds from organosilanes. Acetyl chloride proceeds to higher yield when catalyzed by aluminum chloride.

Metals and Organometallic Compounds. There are no reports of the direct reaction of the Si–H bond in organosilanes with magnesium, zinc, mercury, aluminum, and other elements of the Groups 2(IIA), 12(IIB), and 13(IIIA) metals. The alkali metals, ie, sodium, potassium, and their alloys, react with arylsilanes in amines or ammonia to produce the arylsilyl derivatives of these metals. However, these reactions should be regarded as indirect examples of hydrogen replacement on silicon because they probably go through an amide intermediate. Direct reaction between arylsilanes and alkali metals occurs when alloys of potassium and sodium are used. The bond between the triarylsilicon moiety and the alkali metal is usually quite stable and reacts in a variety of ways (123,124). Formation of potassium triphenylsilicide [*15487-82-8*], di(triphenylsilane) [*1450-23-3*], and triphenylorganosilane occur as shown:

$$(C_6H_5)_3SiH \xrightarrow{\text{Na/K}} (C_6H_5)_3SiK + NaH$$

$$(C_6H_5)_3SiK + (C_6H_5)_3SiH \longrightarrow (C_6H_5)_3SiSi(C_6H_5)_3$$

$$(C_6H_5)_3SiK + RX \longrightarrow (C_6H_5)_3SiR + KX$$

where R is CH$_3$ [*791-29-7*], C$_6$H$_5$ [*1048-08-4*], or (CH$_3$)$_3$Si [*1450-18-6*], and X is Cl or Br.

The preparation of cyclic polysilanes from dialkylchlorosilanes and lithium metal in tetrahydrofuran presumably involves a silicon–lithium intermediate (125).

$$5\text{-}7 \ (CH_3)_2SiHCl \xrightarrow{\text{Li}} [(CH_3)_2Si]_{5\text{-}7} + 5\text{-}7 \ HCl$$

Alkylation and arylation of organosilanes occur readily with alkyl and aryl alkali metal compounds. Yields from these reactions are good but are influenced by steric requirements on both silane and metal compounds. There is little inductive effect by the organic groups attached to silicon, as measured by the yield of products (126,127). These reactions proceed more readily in tetrahydrofuran and ethyl ether than in ligroin or petroleum ether, where R and R$'$ are alkyl or aryl and M is Li, Na, or K.

$$R_nSiH_{(4-n)} + (4-n) \ R'M \longrightarrow R_nSiR'_{4-n} + (4-n) \ MH$$

Dehydrogenative Coupling of Hydride Functional Silanes. The autocoupling of dihydridosilanes was first observed using Wilkinson's catalyst (128). A considerable effort has been undertaken to enhance catalyst turnover and increase the molecular weight of polysilane products (129) because the materials have commercial potential in ceramic, photoresist, and conductive polymer technology.

$$n\ RSiH_3 \xrightarrow{\text{catalyst}} +\!\!\!\!\!\begin{array}{c} H \\ | \\ Si \\ | \\ R \end{array}\!\!\!\!\!\Big)_{\!\!n} + n\ H_2$$

There appear to be two general classes of catalysts. Cyclopentadienyl (Cp) transition-metal catalysts of the general structure $Cp_x MR_y$ were introduced first (130) and then highly elaborated (131). A second class of catalyst has the general structure $Cl_x ML_y$, where M is an electron-rich metal. A typical example is $ClRh(P(C_6H_5)_3)_3$. The latter class also leads to disproportionation reactions of silanes.

Organosilanes as Reducing Agents. The two principal categories of reductive chemistry of hydridosilanes are hydrosilylation and ionic reduction. Hydrosilylation is the catalyzed addition of a hydridosilane to a multiply bonded system. This chemistry is a principal technology in silicon–carbon bond formation. Ionic reduction by silanes, a class of chemistry more properly considered within the context of organic synthesis, is the subject of detailed reviews (132).

The hybridic nature of the Si–H bond is utilized to generate C–H bonds by ionic hydrogenation according to the following general mechanism, in which a hydride is transferred to a carbocation.

$$\begin{array}{c} \diagdown \\ C=C \\ \diagup \end{array} \xrightarrow{H^+} \begin{array}{c} | \\ -CH-C^+\!\!- \\ | \end{array} \xrightarrow{H^-} \begin{array}{c} | \\ -CH-CH- \\ | \quad | \end{array}$$

A catalyst, usually acid, is required to promote chemoselective and regioselective reduction under mild conditions. A variety of organosilanes can be used, but triethylsilane in the presence of trifluoroacetic acid is the most frequently reported. Use of this reagent enables reduction of alkenes to alkanes. Branched alkenes are reduced more readily than unbranched ones. Selective hydrogenation of branched dienes is also possible.

Carboxylic acids, esters, amides, nitriles, nitro groups and most aromatic nuclei are not reduced under ionic hydrogenation conditions (133). An organosiloxane, polymethylhydrosiloxane [9004-73-3] (PMHS), is most economically favored for large-scale reductions. Polymethylhydrosiloxane is a versatile low cost hydride transfer reagent having a hydride equivalent weight of 60. Reactions are catalyzed by Pd[0] or dibutyltinoxide. The choice of reaction conditions leads to chemoselective reduction, eg, allyl reductions in the presence of ketones and aldehydes (134–136). Esters are reduced to primary alcohols in the presence of titanium isopropoxide (137).

An incomplete summary of silanes as reducing agents is provided in Table 4.

Silicon–Carbon Bond-Forming Reactions. After the Rochow-Müller direct process, the hydrosilylation reaction (139),

$$RCH{=}CH_2 + HSiCl_3 \xrightarrow{Pt} RCH_2CH_2SiCl_3$$

and the Barry arylation (140,141),

account for the most significant commercial production of organosilanes.

Manufacture and Processing. *Direct Process.* The preparation of organosilanes by the direct process, first reported in 1945, is the primary

Table 4. Organosilanes as Reducing Agents[a,b]

Substrate[c]	$(C_2H_5)_3SiH$	$(C_6H_5)_3SiH$	$(C_6H_5)_2SiH_2$	PMHS[d]	Products[c]
C=C	++				CHCH
CX	+	+			CH
C=O	++	+	+	+	CH_2, CHOH, COR
ROH	+	++	+		RH
RC(=O)Cl	++				RCHO
RCOOR'	++				RCH_2OR'
RCONR$'_2$	+				$RCH_2NR'_2$
RCN	++				RCHO
RNCO	+				RNHCHO
ArNO$_2$				+	$ArNH_2$
RCH=NR'	+		+		RCH_2NHR'
P=O, PCl		+	++	+	PH

[a]Ref. 138.
[b]++ = recommended; + = reported.
[c]X = halogen; Ar = aryl.
[d]PMHS = polymethylhydrosiloxane.

method used commercially (142,143). Organosilanes in the United States, France, Germany, Japan, and the CIS are prepared by this method, including CH_3SiHCl_2, $(CH_3)_2SiHCl$, and $C_2H_5SiHCl_2$. Those materials are utilized as polymers and reactive intermediates. The synthesis involves the reaction of alkyl halides, eg, methyl and ethyl chloride, with silicon metal or silicon alloys in a fluidized bed at 250–450°C:

$$x\ RX + y\ Si^0 \xrightarrow{250-450°C} R_3SiX + R_2SiX_2 + RSiX_3 + RSiHX_2 + R_2SiHX + R_4Si$$

where R is CH_3 or C_2H_5. The yields of the desired products can be maximized by adjusting temperature, contact time, catalyst (usually copper or copper salts), and catalyst content. Methyldichlorosilane yields can also be increased by the addition of hydrogen to the methyl chloride (144,145). The use of alloys of silicon with copper or cobalt activated with copper chloride permits higher yields of $C_2H_5SiH(Cl_2)$ and $(C_2H_5)_2SiH(Cl)$ (146). Direct process still residues, which contain large amounts of methyldisilanes, can be converted to methylchlorosilanes by reaction with HCl. Yields are low, however, and the process is of little preparative value. The direct process has been extended to the production of alkoxysilanes. Reaction of silicon with ethanol or methanol containing less than 2000 ppm water produces $(C_2H_5O)_3SiH$ and $(CH_3O)_3SiH$ using cuprous chloride as the catalyst. Yields of trialkoxysilane in the instances where methyl and ethyl alcohols are employed are 86 and 91%, respectively (147,148).

$$3\ ROH + Si^0 \xrightarrow{CuCl,\ 220°C} (RO)_3SiH + H_2$$

Dialkylamino-substituted silanes have also been obtained by a similar process (149–151).

$$3\ (CH_3)_2NH + Si^0 \xrightarrow{250°C} ((CH_3)_2N)_3SiH + H_2$$

Reduction with Metal Hydrides. Organosilanes can be synthesized most conveniently on pilot, bench, and laboratory scale by reduction of organic-substituted halo- and alkoxysilanes, using metal hydrides. As for inorganic silanes, the most effective reducing agent is lithium aluminum hydride, which offers advantages of reduction under mild conditions, high yield, and good purity of reaction products. Commonly employed solvents are ether and tetrahydrofuran. Mechanistically, the reaction is

$$4\ R_nSiX_{(4-n)} + (4-n)\ LiAlH_4 \longrightarrow 4\ R_nSiH_{(4-n)} + (4-n)\ LiX + (4-n)\ AlX_3$$

where X = F, Cl, Br, I, OR, OC_6H_5SR, or NR_2.

The versatility of lithium aluminum hydride permits synthesis of alkyl, alkenyl, and arylsilanes. Silanes containing functional groups, such as chloro, amino, and alkoxyl in the organic substituents, can also be prepared. Mixed compounds containing both SiCl and SiH cannot be prepared from organopolyhalosilanes using lithium aluminum hydride. Reduction is invariably complete.

Other reducing agents that have been used to prepare organosilanes are lithium hydride, sodium hydride, and the lithium, sodium, potassium, and aluminum borohydrides. Of these reagents, lithium hydride has been most widely used. This compound is a weaker reducing agent than lithium aluminum hydride and lower yields are generally obtained. Satisfactory reductions generally require higher temperatures, prolonged reaction times, and excess LiH. Reductions are run in high boiling solvents, eg, dioxane and dibutyl ether. This reducing agent is usually employed where there is a danger of side reactions with the aluminum halides formed as by-products from lithium aluminum hydride reductions. Similarly, magnesium hydride [7693-27-8], which has been shown to reduce fluoro- and chlorosilanes, appears to have significant potential where catalytically inactive by-products are required (152). Yields of organohalosilanes and organoalkoxysilanes are listed in Table 5.

Organosilanes are also produced by reaction of organohalosilanes and organoalkoxysilanes with organometallic compounds. Sterically hindered Grignard reagents having a proton in the beta position relative to MgX are the most effective for this type of reduction (153,154). Although it was used to prepare the first isolated organosilanes, this method is of limited value. Organolithium reagents, eg, t-butyllithium, have also produced organohydrosilanes on reaction with organochlorosilanes and tetrahalosilanes. Tri-t-butylsilane [18159-55-2], which cannot be prepared by other reaction methods, is synthesized successfully by this route beginning with t-butyltrichlorosilane [18171-74-9] (155). Success of this reaction is attributed to the t-butyllithium acting as a reducing agent.

$$t\text{-}C_4H_9SiCl_3 + 3\ t\text{-}C_4H_9Li \longrightarrow (t\text{-}C_4H_9)_3SiH + (CH_3)_2C{=}CH_2 + 3\ LiCl$$
$$10\%$$

$$SiF_4 + 4\ t\text{-}C_4H_9Li \longrightarrow (t\text{-}C_4H_9)_3SiH + (CH_3)_2C{=}CH_2 + 4\ LiF$$
$$80\%$$

Table 5. Yields of Organosilanes via Reduction with LiAlH and LiH[a]

Compound	CAS Registry Number	Yield, % LiAlH$_4$	LiH
CH$_3$SiH$_3$	[992-94-9]	80–90	
ClCH$_2$SiH$_3$	[10112-09-1]	80	
C$_2$H$_5$SiH$_3$	[2814-79-1]	80–90	
(C$_2$H$_5$)$_2$SiH$_2$	[542-91-6]	80–90	66
(C$_2$H$_5$)$_3$SiH	[617-86-7]	80–90	20
(C$_2$H$_3$)$_3$SiH	[2372-31-8]	52	39
C$_2$H$_5$(CH$_2$=CH)SiH$_2$	[18243-33-9]		24
C$_4$H$_9$SiH$_3$	[1600-29-9]	80–90	
CH$_2$(CH$_2$)$_3$SiH$_2$	[288-06-2]	47	27
C$_3$H$_7$SiH$_3$	[13154-66-0]	80	
C$_6$H$_5$SiH$_3$	[694-53-1]	86	
(C$_6$H$_5$)$_2$SiH$_2$	[775-12-2]	78	
C$_6$H$_{13}$SiH$_3$	[1072-14-6]		65
(C$_6$H$_{13}$)$_3$SiH	[2929-52-4]	94	

[a]Ref. 59.

Disproportionation. Disproportionation reactions have also been used to prepare organosilanes. These reactions involve interaction of organosilanes and other silicon compounds containing organic, alkoxy, and halogen groups bound to silicon. Reactions are catalyzed by a variety of materials, including alkali metals, alkali metal alcoholates, and Lewis acids, eg, aluminum, zinc, iron, and boron halides (156). Aluminum chloride is the most active and widely used catalyst. It enables facile preparation of various alkyl and dialkylsilanes according to the following scheme:

$$R_nSiX_{(4-n)} + (4 - n) R_3^\bullet SiH \longrightarrow R_nSiH_{(4-n)} + (4 - n) R_3^\bullet SiX$$

where R and R$'$ are alkyl, X is Cl or Br, and $n = 1$ or 2.

Organochlorosilanes containing Si–H disproportionate in the presence of aluminum chloride without addition of more organosilane. Organic groups can be replaced by hydrogen (157). For example, tetraphenylsilane [*1048-08-4*] can be made from phenylmethylsilane [*766-08-5*].

$$C_6H_5(CH_3)SiH_2 \xrightarrow{\text{AlCl}_3} (C_6H_5)_4Si + CH_3SiH_3 + (CH_3)_2SiH_2$$

Similar disproportionation reactions are catalyzed by organic catalysts, eg, adiponitrile, pyridine, and dimethylacetamide. Methods for the redistribution of methylhydridosilane mixtures from the direct process have been developed to enhance the yield of dimethylchlorosilane (158).

Grignard Reagents. Grignard reagents are utilized to transfer organic groups to silicon. In contrast to more basic organometallic compounds such as alkyl lithium reagents, cleavage of the Si–H bond does not occur. However, in tetrahydrofuran at elevated temperature and in the case of polyhydridosilanes in diethyl ether, the hydride can be displaced (159).

$$C_6H_5SiH_3 + C_6H_5MgBr \longrightarrow (C_6H_5)_2SiH_2$$

In general, Grignard reagents are useful in the synthesis of mixed hydridochlorosilanes because these reagents can effect stepwise substitution of the halogen, eg,

$$HSiCl_3 + 2\ i\text{-}C_3H_7MgCl \longrightarrow (i\text{-}C_3H_7)_2SiHCl + 2\ MgCl_2$$

Addition to Olefins. Organohydrosilanes can also be prepared by addition of halosilanes and organosilanes containing multiple Si–H bonds to olefins. These reactions are catalyzed by platinum, platinum salts, peroxides, ultraviolet light, or ionizing radiation.

$$H_2SiCl_2 + RCH{=\!=}CH_2 \xrightarrow{\text{H}_2\text{PtCl}_6} RCH_2CH_2SiHCl_2$$

$$RSiH_3 + R'CH{=\!=}CH_2 \longrightarrow R'CH_2CH_2SiH_2(R)$$

Economic Aspects. The only organic silane that is produced on a large scale is methyldichlorosilane. It is a by-product in the direct process for methylchlorosilane production, representing 0.5–3.0% of the overall output. An estimate of worldwide methyldichlorosilane production for 1995 is in the 3,000–20,000 t/yr range. Because there is no distinct process technology operated for the production of methyldichlorosilane, its availability varies in accord with both overall methylchlorosilane production and its demand in silicon production. Domestic producers of methyldichlorosilane are General Electric and Dow Corning. Pricing is $4–$6/kg. Polymethylhydrosiloxane, the polymeric derivative of methyldichlorosilane, is also available from these suppliers. Prices are $7–$10/kg. Other specialty organic silicon hydrides bear much higher pricing. The most significant products are dimethylchlorosilane, tetramethyldisiloxane, tetramethylcyclotetrasiloxane, triethylsilane, diethylmethylsilane, methylsilane, diphenylsilane, and various hydride-containing silicone copolymers. The cost of these materials varies between $35–$400/kg. Producers of these materials are Dow Corning, General Electric, Gelest, Harris Specialty Chemicals, and Hüls.

Halosilanes

Only three inorganic halosilanes are produced on a large industrial scale, ie, tetrachlorosilane [10026-04-7], tetrafluorosilane [7783-61-1], and trichlorosilane.

Physical Properties.   The bonds between silicon and halogen are predominantly covalent in character, exhibiting bond energies consistent with good thermal stability. The bond energy of Si–F is >550 kJ/mol (131 kcal/mol) making Si–F one of the strongest covalent bonds. The extreme polarization of the bonds, however, results in materials that are extremely sensitive to nucleophilic attack. All halosilanes, for example, fume in air from the presence of moisture liberating the hydrogen halide. These compounds are thus extremely corrosive materials in open environments. The parent halogen is soluble in tetrahalosilanes. For example, chlorine is soluble to the extent of 1% in tetrachlorosilane. Physical properties for halosilanes of commercial significance are given in Table 6.

Chemical Properties.   *Oxidation.*   Silicon halides are stable to oxygen at room temperature, but react at elevated temperatures to form, in the case of

Table 6. Properties of Halosilanes

Property	SiF_4	$SiCl_4$	$SiBr_4$	SiI_4
CAS Registry Number	[7783-61-1]	[10026-04-7]	[7789-66-4]	[13465-84-4]
mp, °C	−90.3	−70.4	5.4	120.5
bp, °C	−95.7 sub	57	155	287.5
vapor pressure,[a] kPa[b]	68.6_{-100}	25.9_{20}	0.24_0	
H_{vap}, kJ/mol[c]	18.7	28.7	37.9	26.6
bond energy, kJ/mol[c]	565	381	310	244
critical temperature, °C	−14.15	233.6	383	
critical pressure, kPa[d]	37.3	36.8		
density,[a] g/cm^3	1.598_{-80}	1.4707	2.771	4.2

[a]The subscripted numbers are temperature in °C.
[b]To convert kPa to mm Hg, multiply by 7.5.
[c]To convert J to cal, divide by 4.184.
[d]To convert kPa to psi, multiply by 0.145.

chlorides, oxychlorosilanes. Tetrachlorosilane reacts at 950–970°C to form hexa-chlorodisiloxane (160,161).

$$2 \, SiCl_4 + 1/2 \, O_2 \longrightarrow Cl_3SiOSiCl_3 + Cl_2$$

Tetrabromosilane reacts with oxygen at 670–695°C to form polybromosiloxanes and at 900°C to form silicon dioxide.

Reaction with Silicon. At elevated temperature (1000°C) tetrachlorosilane attacks pure crystalline silicon to form a mixture of higher chlorosilanes including Si_2Cl_6, Si_3Cl_8, and Si_4Cl_{10}. Under vacuum at 1000°C, silicon tetrafluoride reacts with silicon to form difluorosilene (162), which, using a rapid thermal quench below −60°C, forms polydifluorosilane, a wax-like polymer. The polydi-fluorosilane reverts to silicon and silicon tetrafluoride at moderate temperature.

$$Si^0 + SiF_4 \xrightarrow{1000°C} SiF_2 \xrightarrow{-76°C} (SiF_2)_n \xrightarrow{0°C} n/2 \, Si^0 + n/2 \, SiF_4$$

In principle, this technology represents a possible alternative for closed-loop purification of silicon, but transport and purification efficiencies are not sufficient to be enabling (163).

Protic Materials. The hydrolytic behavior of tetrachlorosilane was described by Mendeleyev (164), and is summarized herein. Orthosilicic acid formed by the action of water on tetrachlorosilane:

> does not remain in that form, but loses part of its water with extraordinary ease...
> The hydrate formed is not actually obtained with as high a water content as
> corresponds to $Si(OH)_4$. ..In the hydrates $nSiO_2 \cdot mH_2O$, m becomes smaller than
> n... This loss of water proceeds, in the natural hydrates, in perfect sequence, and
> so to speak, imperceptibly, until n becomes incomparably greater than m.... The
> structure of silica is polymeric, complex, instead of simple, as it is expressed by its
> empirical formula.

The heat of reaction of tetrachlorosilane with an excess of water is 290.0 kJ/mol (69.3 kcal/mol). The reaction of tetrafluorosilane with excess water contrasts with the other halosilanes, because it leads to formation of hexafluo-rosilicic acid and a hydrous silica.

$$2 \, SiF_4 + 2 \, H_2O \longrightarrow H_2SiF_6 + SiO_2 + 2 \, HF$$

The reaction of halosilanes with alcohols proceeds analogously. Difunctional amines react with tetrahalosilanes to yield tetrakis(dialkylamino)silanes.

$$8 \, (CH_3)_2NH + SiCl_4 \longrightarrow Si(N(CH_3)_2)_4 + 4 \, (CH_3)_2NH_2^+ + 4 \, Cl^-$$

The analogous reaction with ammonia leads ultimately to silicon nitride. In the past, hydrocarbon soluble fractions of the ammonolysis were incorrectly referred to as silicon diimide. This improper designation occasionally persists as of the mid-1990s.

Reduction/Reaction with Hydrogen. Tetrafluorosilane reacts with hydrogen only above 2000°C. Tetrachlorosilane can be reduced by hydrogen at 1200°C. Tetraiodosilane can be reduced to silicon at 1000°C (165). Reduction of tetrafluorosilane with potassium metal to silicon was the first method used to prepare silicon (see SILICON AND SILICON ALLOYS). The reduction of silicon tetrachloride by zinc metal led to the first semiconductor-grade silicon (166,167).

$$K_2SiF_6 + 4\,K \longrightarrow 6\,KF + Si$$
$$SiCl_4 + 2\,Zn \longrightarrow 2\,ZnCl_2 + Si$$

Sodium and magnesium do not react with tetrachlorosilane at room temperature, but do so at elevated temperatures and in the presence of polar aprotic solvents at moderately elevated temperatures. The Wurtz-Fittig coupling of organosilanes to form disilanes (168) and polysilanes (169) is usually accomplished using molten sodium in toluene or xylene.

$$2\,R_3SiCl + 2\,Na \longrightarrow R_3SiSiR_3 + 2\,NaCl$$
$$n\,R_2SiCl_2 + 2n\,Na \longrightarrow (R_2Si)_n + 2n\,NaCl$$

Reduction of halosilanes with hydrides leads to the formation of hydride functional silanes and is considered in that section.

Reaction With Hydrogen and Oxygen. The combustion of silicon tetrachloride in a hydrogen oxygen (air) flame is an important technology leading to fumed or pyrogenic silica.

$$SiCl_4 + 2\,H_2 + O_2 \longrightarrow SiO_2 + 4\,HCl$$

The mechanism for the formation of silica is complex because oxidation, reduction, and hydrolysis pathways are all possible.

Organometallics. Halosilanes undergo substitution reactions with alkali metal organics, Grignard reagents, and alkylaluminums. These reactions lead to carbon–silicon bond formation.

$$RMgCl + R'SiCl_3 \longrightarrow RR'SiCl_2 + 2\,MgCl_2$$
$$4\,RLi + SiCl_4 \longrightarrow R_4Si + 4\,LiCl$$
$$AlR_3 + 3\,R_3'SiCl \longrightarrow 3\,RR_3'Si + AlCl_3$$

Manufacturing. Tetrachlorosilane can be manufactured directly by the reaction of chlorine on silicon metal or ferrosilicon at 500°C or silicon carbide (170).

$$Si + 2\,Cl_2 \longrightarrow SiCl_4$$
$$SiC + 2\,Cl_2 \longrightarrow SiCl_4 + C$$

The production of silicon tetrachloride by these methods was abandoned worldwide in the early 1980s. Industrial tetrachlorosilane derives from two processes associated with trichlorosilane, the direct reaction of hydrogen chloride on silicon primarily produced as an intermediate for fumed silica production, and as a by-product in the disproportionation reaction of trichlorosilane to silane utilized in microelectronics. Substantial quantities of tetrachlorosilane are produced as a by-product in the production of zirconium tetrachloride, but this source has decreased in the 1990s owing to reduction in demand for zirconium in nuclear facilities (see NUCLEAR REACTORS). The price of tetrachlorosilane varies between $1/kg and $25/kg, depending on grade and container.

Tetrabromosilane and tetraiodosilane are produced by the direct reaction between silicon and bromine at 500°C and silicon and iodine at 600°C. There is no commercial production of these materials as of mid-1996.

Although tetrafluorosilane can be readily produced by the action of hydrogen fluoride on silica, its production is a by-product of HF production by the reaction of fluorospar and sulfuric acid and as a by-product from phosphate fertilizer production by the treatment of fluoroapatite with sulfuric acid (171). The most significant U.S. production is by IMC-Agrico at Uncle Sam, Louisiana.

$$4 \ Ca_5(PO_4)_3F + 4 \ H_2SO_4 + SiO_2 \longrightarrow 4 \ Ca_5(PO_4)_3HSO_4 + 2 \ H_2O + SiF_4$$

A small by-product stream is also realized in Europe from glass-etching by HF. Laboratory-scale production is readily accomplished by exchange between metal fluorides and chlorosilanes (172).

$$SiCl_4 + 2 \ PbF_2 \longrightarrow SiF_4 + PbCl_2$$

$$n\text{-}(C_4H_9)_3SnF + (CH_3)_3SiCl \longrightarrow (CH_3)_3SiF + n\text{-}(C_4H_9)_3SnCl$$

Halosilanes can also be produced by displacement of amines with a hydrogen halide (173).

$$SiH(NHC_6H_5)_3 + 6 \ HI \longrightarrow SiHI_3 + 3 \ (C_6H_5NH_3)I$$

Mixed halosilanes are produced by warming a mixture of tetrahydrosilanes.

$$SiCl_4 + SiBr_4 \longrightarrow 2 \ SiCl_2Br_2$$

Alternatively, they can be prepared by the reaction of hydridohalosilanes with halogen (174).

Health and Safety. Halosilane vapors react with moist air to produce the respective hydrohalogen acid mist. Federal standards have not set exposure to halosilanes, but it is generally believed that there is no serious risk if vapor concentrations are maintained below a level that produces an irritating concentration of acid mist. The exposure threshold limit value (TLV) for HCl is 5 ppm, expressed as a ceiling limit. Because most people experience odor and irritation at or below 5 ppm, HCl is considered to have good warning properties.

Uses. The overwhelming use for tetrachlorosilane is in the production of fumed silica. The tetrachlorosilane used as a raw material for fumed silica is of generally low purity (80–95%) and typically contains some trichlorosilane as well as methyltrichlorosilane (introduced from other manufacturing sources). Tetrachlorosilane utilized as an intermediate for tetraalkoxysilanes is typically >98% pure. Extremely high purity grades of >99.99% having controlled metal contents below 1 ppm are utilized in the production of silica fiber optic waveguides. Tetrafluorosilane is primarily an intermediate for the production of fluorosilicic acid, used as a sterilant for glass bottles, and in electroplating applications (see FLUORINE COMPOUNDS, INORGANIC). Commercial production of silane from tetrafluorosilane accounts for about 5000 t/yr at the MEMC facility located at Pasadena, Texas.

BIBLIOGRAPHY

"Silicon Compounds" in *ECT* 1st ed., Vol. 12, pp. 365–392, by E. G. Rochow, Harvard University; "Silicon Compounds, Silanes" in *ECT* 2nd ed., pp. 172–215, by C. H. Van Dyke, Carnegie-Mellon University; in *ECT* 3rd ed., Vol. 20, pp. 887–911, by B. Arkles and W. R. Peterson, Jr., Petrarch Systems, Inc.

1. P. E. Verkade, *Chem. Weekbl.* **47**, 309 (1951).
2. V. Bazant, V. Chvalovsky, and J. Rathovsky, *Organosilicon Compounds*, Vol. 1, Academic Press, Inc., New York, 1965.
3. A. Stock, *Hydrides of Boron and Silicon*, Cornell University Press, Ithaca, N.Y., 1933.
4. Y. Yamaguchi, *Mol. Eng.* **3**, 301–309 (1994).
5. R. West and T. Barton, *Chem. Ed.* **57**, 165, 364 (1980).
6. R. West and J. Michl, *Science* **214**, 1344 (1980).
7. A. G. Brook, F. Abdesaken, and B. Gutekunst, in *Proceedings of the XV Silicon Symposium*, Mar. 1981, Duke University, Durham, N.C.; A. G. Brook and co-workers, *J. Am. Chem. Soc.* **104**, 5667 (1982).
8. R. Borreson and co-workers, *Solid State Technol.* **21**, 43 (1978).
9. E. Rochow in *Comprehensive Inorganic Chemistry*, Vol. 1, Pergamon Press, Inc., Elmsford, N.Y., 1973, p. 1353.
10. R. Walsh, *Accounts Chem. Res.* **14**, 246 (1981).
11. D. G. White and E. G. Rochow, *J. Am. Chem. Soc.* **76**, 3897 (1954).
12. S. N. Borisov, M. G. Voronkov, and B. N. Dolgov, *Dokl. Akad. Nauk SSSR* **93**, 114 (1957).
13. Ger. Offen. 1,102,117 (May 18, 1954) (to Siemens-Haska AG).
14. W. Claassen and J. Bloehm, *Phillips J. Res.* **36**(2), 122 (1981).
15. D. J. DeLong, *Solid State Technol.* **15**(10), 43 (Aug. 19, 1972).
16. U.S. Pat. 3,400,660 (Aug. 19, 1975), H. Bradley (to Union Carbide).
17. W. R. Runyan, *Silicon Semiconductor Technology*, McGraw-Hill Book Co., Inc., New York, 1965.
18. U.S. Pat. 4,318,942, L. M. Woerner and E. B. Moore (to J. C. Schumacher Co.).
19. B. A. Scott, D. R. Estes, and D. B. Beach in J. Y. Corey, E. R. Corey, and P. P. Gaspar, eds., *Silicon Chemistry*, Ellis Horwood/Wiley, Chichester, U.K., 1988, p. 369.
20. S. A. Arutyunyan and E. N. Sarkisyan, *Arm. Khim. Zh.* **34**(1), 3 (1981).
21. T. G. Mkryan, E. N. Sarkisyan, and S. A. Arutyunyan, *Arm. Khim. Zh.* **34**(1), 3 (1981).
22. U.S. Pat. 4,239,811 (Dec. 16, 1980), B. M. Kiemlage (to IBM).
23. B. Sternbach and A. G. MacDiarmid, *J. Am. Chem. Soc.* **81**, 5109 (1959).
24. J. S. Peake and W. H. Nebergall, Y. T. Chen, *J. Am. Chem. Soc.* **74**, 1526 (1952).
25. L. G. L. Ward and A. G. MacDiarmid, *J. Am. Chem. Soc.* **82**, 2151 (1960).

26. M. Aberdini, C. H. VanDyke, and A. G. MacDiarmid, *J. Inorg. Nuc. Chem.* **28**, 1373 (1966).
27. S. Cradock, G. A. Gibbon, and C. H. Van Dyke, *Inorg. Chem.*, **6**, 1751 (1967).
28. E. Amberger and E. Mulhofer, *J. Organomet. Chem.*, **12**, 557 (1968).
29. E. Amberger, R. Romer, and A. Layer, *J. Organomet. Chem.* **12**, 417 (1968).
30. M. A. Ring and D. M. Ritter, *J. Am. Chem. Soc.* **83**, 802 (1961).
31. S. P. Garrity and M. A. Ring, *Inorg. Nucl. Chem. Lett.* **4**, 77 (1968).
32. D. S. Rustad and W. L. Jolly, *Inorg. Chem.* **6**, 1986 (1967).
33. R. C. Kennedy, L. P. Freeman, A. P. Fox, M. A. Ring, *J. Inorg. Nucl. Chem.* **28**, 1373 (1966).
34. J. A. Morrison and M. A. Ring, *Inorg. Chem.* **6**, 100 (1967).
35. E. A. Groschwitz, W. M. Ingle, and M. A. Ring, *J. Organomet. Chem.* **9**, 1373 (1966).
36. W. J. Boldue and M. A. Ring, *J. Organomet. Chem.* **9**, 421 (1967).
37. R. Schwarz and F. Heinrich, *Z. Anorg. Chem.* **221**, 227 (1935).
38. S. D. Gokhale and W. L. Jolly, *Inorg. Chem.* **3**, 946 (1964).
39. E. J. Spanier and A. G. MacDiarmid, *Inorg. Chem.*, **1**, 432 (1962).
40. S. D. Gokhale, J. E. Drake, and W. L. Jolly, *J. Inorg. Nuc. Chem.* **27**, 1911 (1965).
41. E. J. Spanier and A. G. MacDiarmid, *Inorg. Chem.* **2**, 215 (1963).
42. S. D. Gokhale and W. L. Jolly, *Inorg. Chem.* **3**, 1141 (1964).
43. S. D. Gokhale and W. L. Jolly, *Inorg. Chem.* **4**, 596 (1965).
44. M. Abedini and A. G. MacDiarmid, *Inorg. Chem.* **5**, 2040 (1966).
45. A. A. Kirpichinikova and co-workers, *Zh. Vses. Khim. Obshchest.* **22**, 465 (1977).
46. B. Scott and M. Brodsky, *Bull. Am. Phys. Soc.* **25**, 299 (1980).
47. T. Hiiao and co-workers, *J. Appl. Phys.* **52**(12), 7453 (1981).
48. M. J. Loboda and J. A. Seifferly, *J. Mater. Res.* **11**, 391 (1996).
49. H. J. Emeleus and K. Stewart, *Trans. Farady Soc.* **32**, 1677 (1936).
50. N. Nihi and G. J. Mains, *J. Phys. Chem.* **68**, 304 (1964).
51. B. Reimann and R. Potzinger, *Ber. Bunsenges. Phys. Chem.* **80**, 565 (1976).
52. G. A. Gobbon, T. Rosseau, C. H. VanDyke, and G. J. Mains, *Inorg. Chem.* **5**, 114 (1966).
53. W. C. Johnson and J. R. Hogness, *J. Am. Chem. Soc.* **56**, 1252 (1934).
54. W. C. Johnson and S. Isenberg, *J. Am. Chem. Soc.* **57**, 1359 (1935).
55. F. Feher and W. Tromm, *Z. Anorg. Allgem., Chem.* **282**, 29 (1955).
56. R. Schwarz and E. Konrad, *Ber.* **55**, 3242 (1952); *Z. Anorg. Allg. Chem.* **215**, 288 (1938).
57. G. Scott and D. Naumann, *Z. Anorg. Allg. Chem.* **291**, 103, 112 (1957).
58. Y. Ono, Y. Sndoda, and T. Keii, *J. Am. Chem. Soc.* **97**, 5284 (1975).
59. A. D. Petrov, B. F. Mironov, V. A. Ponomorenko, and R. A. Cherpyshev, *Synthesis of Organosilicon Monomers*, Consultants Bureau, New York, 1964.
60. U.S. Pat. 4,927,616, E. M. Marlett (to Ethyl Corp.).
61. Eur. Par. Appl. EP 87-116243, E. M. Marlett and F. W. Frey (to Ethyl).
62. Eur. Pat. Appl. EP 89-14115, J. E. Boone, D. M. Richards, and J. Bossier (to Ethyl Corp.).
63. W. Sundermeyer and O. Glemser, *Angew. Chem.* **70**, 625 (1958).
64. O. Glemser and W. Lohman, *Z. Anorg. Allg. Chem.* **275**, 260 (1964).
65. U.S. Pat. 4,051,136 (Aug. 9, 1977), R. E. Franklin, W. A. Francis, and G. Tanaron (to Union Carbide).
66. Can. Pat. 2,121,931 (1994), K. Klein and co-workers; Germ. Offen. 43 13 130.1-44 (1993).
67. Ger. Offen. DE 3409172 W. Porcham (to Swarovski).
68. D. T. Hurd, *J. Am. Chem. Soc.*, **67**, 1545 (1945).
69. U.S. Pat. 2,458,703 (Jan. 11, 1949), D. B. Hatcher (to Libby-Owens-Ford).
70. D. L. Bailey, personal communication. This work was performed by E. O. Brimm at Union Carbide in the 1940s.

71. J. Y. P. Mui, Low Cost Solar Array Project, NASA/JPL Contract 954334, Union Carbide Corp., June 1979.
72. J. Y. P. Mui and D. Seyferth, *Investigation of the Hydrogenation of Silicon Tetrachloride*, DOE/JPL/955382-79/8, Department of Energy, Washington, D.C., 1981.
73. Ger. Offen. DE 3341340 A1 (May 23, 1985), A. Schnegg and R. Rurlaender.
74. Ger. DE3809784 C1 (July 13, 1988), K. Ruff.
75. PCT Int. Appl. WO 81-03168 (Nov. 12, 1981), K. R. Sarma and M. J. Rice, Jr. (to Motorola).
76. U.S. Pat. 2,627,451 (Feb. 3, 1953); U.S. Pat. 2,735,861, C. E. Erikson and G. H. Wagner (to Union Carbide).
77. U.S. Pat. 2,745,860 (May 15, 1956), D. L. Bailey (to Union Carbide).
78. H. J. Emeleus and N. Miller, *J. Chem. Soc.*, 819 (1939).
79. U.S. Pat. 2,732,282 (Jan. 24, 1956), D. L. Bailey, G. H. Wagner, and P. W. Shafer (to Union Carbide).
80. Ger. Offen. 2,550,076 (1976), G. Marin.
81. U.S. Pat. 3,968,199 (July 6, 1976), C. J. Bakey (to Union Carbide).
82. E. Helfrich and J. Hausen, *Ber.* **57B**, 795 (1924).
83. W. C. Breneman, in K. M. Lewis and D. G. Rethwish, eds., *Catalyzed Direct Reactions of Silicon*, Elsevier Science, Inc., New York, 1993, p. 441.
84. Jpn. Kokai Tokkyo Koho JP 63/170210 A2 [88/170210] (July 14, 1988), H. Yamada, H. Ogawa, T. Hosokawa, and M. Tachikawa.
85. D. R. Stull and H. Prophet, *JANAF Thermochemical Tables*, National Bureau of Standards, Washington, D.C., 1971.
86. J. A. Cervantes and co-workers, in Ref. 83, p. 459.
87. W. C. Schumb, *Inorg. Synth.* **1**, 38 (1939).
88. L. G. L. Ward, *Inorg. Synth.* **11**, 159 (1968).
89. G. Tamizhmani, M. Cocivera, R. T. Oakley, and P. Del Bel Belluz, *Chem. Mat.* **2**, 473 (1990).
90. C. J. Wilkins, *J. Chem. Soc.*, 3409 (1953).
91. H. Brederman, T. J. Thor, and H. J. Waterman, *Research* **7**, 829 (1959).
92. D. J. Hurd, *J. Am. Chem. Soc.* **67**, 1545 (1945).
93. H. L. Jackson, F. D. Marsh, and E. L. Muetterties, *Inorg. Chem.* **2**, 43 (1963).
94. N. J. Archer, R. N. Hazeldine, and R. V. Parish, *J. Orgmet. Chem.* **81**, 335 (1974).
95. K. Sharp, A. Arvidson, and T. C. Elvey, *J. Electrochem. Soc.* **129**, 2346 (1982).
96. *Registry of Toxic Effects of Chemical Substances*, National Institute of Occupational Safety and Health, (NIOSH), Washington, D.C., 1976.
97. F. Fortes, *Ind. Eng. Chem.* **46**, 2325 (1954).
98. R. R. McGregor, *Silicones and Their Uses*, McGraw-Hill Book Co., Inc., New York, 1954.
99. W. Noll, *Chemistry and Technology of Silicones*, Academic Press, Inc., New York, 1968.
100. Y. Nagai, *Org. Prep. Proc. Int.* **12**, 15 (1980).
101. B. Arkles, *Silicon, Germanium, Tin, and Lead Compounds: A Survey of Properties and Chemistry*, Gelest Inc., Tullytown, Pa., 1995.
102. H. J. Emeleus and K. Stewart, *Nature*, 397 (1935).
103. C. Eaborn, *Organosilicon Compounds*, Buttersworth Scientific Publications, Lander, U.K., 1960.
104. J. E. Baines and C. Eaborn, *J. Chem. Soc.*, 4023 (1955).
105. L. H. Sommer, O. Bennet, P. G. Campbell, and D. R. Weyenberg, *J. Am. Chem. Soc.* **79**, 3295 (1957).
106. J. E. Baines and C. Eaborn, *J. Chem. Soc.*, 7436 (1956).
107. E. Lukevics and M. Dzintara, *J. Organomet. Chem.* **295**, 265 (1985).
108. W. S. Miller, J. S. Peake, and W. H. Nebergall, *J. Am. Chem. Soc.* **79**, 5604 (1957).

109. L. H. Sommer and C. L. Frye, *J. Am. Chem. Soc.* **79**, 3295 (1957).
110. C. L. Lee, R. G. Niemi, and K. M. Kelley, *J. Cell. Plast.* **13**, 62 (1977).
111. C. Eaborn, *J. Chem. Soc.*, 23517 (1955).
112. U.S. Pat. 2,658,908 (Nov. 10, 1953), S. Nitzsche and E. Pirson (to Wacker).
113. B. N. Dolgov, N. P. Kharitonov, and M. G. Voronkov, *Zh. Obsch. Khim.* **24**, 678 (1954).
114. U.S. Pat. 4,612,383 (1986); U.S. Pat. 4,788,309 (1988), R. Laine and Y. Blum (to SRI).
115. U.S. Pat. 4,577,039 (Mar. 18, 1986), B. Arkles and B. Hamon (to Petrarch Systems).
116. M. G. Voronkov and Yu I. Khudobin, *Izv. Akad. Nauk SSSR*, 805 (1956).
117. Y. Nagai, H. Matsumoto, T. Yagihara, and K. Moxishita, *Kogyo Kagaku Zasshi* **71**, 1112 (1968).
118. R. Boukherroub, C. Chatgilialoglu, and G. Manuel, *Organometallics* **15**, 1508 (1996).
119. R. G. Cunico and L. Bedell, *J. Org. Chem.* **45**, 4797 (1980).
120. H. H. Anderson, *J. Am. Chem. Soc.* **69**, 2600 (1958).
121. H. Westermark, *Acta Chem. Scand.* **8**, 1086 (1954).
122. M. P. Doyle and C. T. West, *J. Org. Chem.* **41**, 1393 (1976).
123. R. A. Benkeser, H. Landesman, and D. J. Foster, *J. Am. Chem. Soc.* **74**, 648 (1952).
124. R. A. Benkeser and D. J. Foster, *J. Am. Chem. Soc.* **74**, 4200 (1952).
125. U.S. Pat. 4,276,424 (June 30, 1981), B. Arkles and W. R. Peterson, Jr. (to Petrarch Systems).
126. R. A. Benkeser and F. J. Riel, *J. Am. Chem. Soc.* **73**, 3472 (1951).
127. H. Gilman and J. J. Goodman, *J. Org. Chem.* **22**, 45 (1957).
128. I. Ojima, S.-I. Inabe, T. Kogure, and Y. Nagai, *J. Organomet. Chem.* **55**, C7 (1973).
129. J. Y. Corey, *Adv. Silicon Chem.* **1**, 327 (1991).
130. C. Aitken, J. Harrod, and E. Samuel, *J. Am. Chem. Soc.* **108**, 4059 (1986).
131. T. D. Tilley, *Acc. Chem. Res.* **26**, 22 (1993).
132. W. P. Weber, *Silicon Reagents for Organic Synthesis*, Springer-Verlag, New York, 1983, p. 273.
133. D. N. Kursanov, Z. N. Parnes, and N. M. Loim, *Synthesis*, 633 (1974).
134. J. Lipowitz and co-workers, *J. Org. Chem.* **38**, 162 (1973).
135. E. Keinan and co-workers, *Israel. J. Chem.* **24**, 82 (1984); *J. Org. Chem.* **48**, 3545 (1983).
136. T. Mukaiyama and co-workers, *Chem. Lett.*, 1727 (1983).
137. M. Reding and co-workers, *J. Org. Chem.* **60**, 7884 (1995).
138. *Hydrosilanes as Reducing Agents*, Technical Bulletin, Chisso Corp., Tokyo, Japan, 1980.
139. B. Marciniec, *Comprehensive Handbook of Hydrosilylation*, Pergamon Press, New York, 1992.
140. U.S. Pat. 2,494,560 (1950); U.S. Pat. 2,626,929 (1953), A. J. Barry (to Dow Corning).
141. A. J. Barry, J. W. Gilkey, D. E. Hook, *Metal-Organic Compounds*, ACS Advances in Chemistry, Vol. 23, American Chemical Society, Washington, D.C., 1959, p. 246.
142. E. G. Rochow, *J. Am. Chem. Soc.* **67**, 693 (1945).
143. U.S. Pat. 2,380,995 (Aug. 7, 1945), E. G. Rochow (to General Electric).
144. U.S. Pat. 2,380,998 (Aug. 7, 1945), M. M. Sprung and W. F. Gilliam (to General Electric).
145. U.S. Pat. 4,973,725 (Nov. 27, 1990), K. Lewis, J. Larnard, and B. Kanner (to Union Carbide).
146. Brit. Pat. 681,387 (1952) R. Decker and H. Holz.
147. Jpn. Kokai Tokkyo Koho 80 28,928 (Apr. 6, 1980), S. Suzuki, T. Imaki, and T. Yamamura (to Mitsubishi).
148. Jpn. Kokai Tokkyo Koho 80 28,929 (Apr. 6, 1980), S. Suzuki, T. Imaki, and T. Yamamura (to Mitsubishi).
149. B. Kanner, W. B. Herdle, J. M. Quirk, in J. Y. Corey, E. R. Corey, and P. P. Gaspar, eds., *Silicon Chemistry*, Ellis Horwood/Wiley, Chichester, U.K., 1988, p. 123.

150. U.S. Pat. 4,255,348 (Mar. 10, 1981), B. Kanner and W. B. Herdle (to Union Carbide).
151. K. M. Lewis, B. K. C. Tan, T. E. Childress, and D. McLeod, in K. M. Lewis and D. G. Rethwisc, eds., *Catalyzed Direct Reactions of Silicon*, Elsevier Science, Inc., New York, 1993, p. 559.
152. Can. Pat. 2,121,931 (1994); Ger. Offen. P 43 13 130.1–44 (1993), K. Klein and co-workers.
153. M. C. Harvey, W. H. Nebergall, and J. S. Peake, *J. Am. Chem. Soc.* **79**, 7762 (1957).
154. M. B. Lacout-Loustalet, J. P. Dupin, F. Metras, and J. Valade, *J. Organomet. Chem.* **31**, 337 (1971).
155. E. M. Dexheimer and L. Spialter, *Tetrahedron Lett.*, 1771 (1975).
156. U.S. Pat. 2,745,860 (May 15, 1956), D. L. Bailey (to Union Carbide).
157. J. L. Speier and R. E. Zimmerman, *J. Am. Chem. Soc.* **77**, 6395 (1955).
158. U.S. Pat. 5,493,043 (Feb. 20, 1996), O. W. Marko (to Dow Corning).
159. H. Gilman and E. A. Zeuch, *J. Am. Chem. Soc.* **79**, 4560 (1957).
160. A. D. Gaunt, H. Mackle, and L. E. Sutton, *Trans. Faraday Soc.* **47**, 943 (1954).
161. B. A. Grigor and C. J. Wilkins, *Inorg. Syn.* **7**, 23 (1963).
162. P. L. Timms, R. A. Kent, T. C. Ehlert, and J. L. Margrave, *J. Am. Chem. Soc.* **87**, 2824 (1965).
163. W. Ingle, R. Rosler, S. Thompson, and R. Chaney, *Semiconductor-Grade Solar Silicon Purification Project*, NASA-CR-158868, Rept.-2257/12, Cape Canavral, Fla., 1979.
164. D. I. Mendeleyev, *Principles of Chemistry*, St. Petersburg, CIS, 1871.
165. U.S. Pat. 3,006,737 (Oct. 31, 1961), G. H. Moats, B. Rubin, and W. B. Jackson.
166. D. W. Lyon, C. M. Olso, and E. D. Lewis, *J. Electrochem. Soc.* **96**, 359 (1949).
167. U.S. Pat. 2,773,745 (Dec. 11, 1956), K. Butler and C. M. Olson (to DuPont).
168. H. Gilman and T. C. Wu, *J. Am. Chem. Soc.* **75**, 3762 (1953).
169. R. D. Miller and co-workers, in L. F. Tompson and C. G. Wilson, eds., *Materials for Microlithography*, ACS Symposium Series 266, American Chemical Society, Washington, D.C., 1984, p. 293.
170. K. A. Andrianov, *Dokl. Akad. Nauk SSSR* **28**, 66 (1940).
171. U.S. Pat. 2,833,628 (1958), Molstad (to W. R. Grace).
172. D. K. Padma, B. S. Suresh, and A. R. Vasudevamurthy, *J. Fluoro. Chem.* **14**, 327 (1979).
173. E. Hengge and F. Höfler, *Z. Naturforsch.* **26a**, 768 (1971).
174. R. West and E. G. Rochow, *Inorg. Syn.* **4**, 41 (1953).

BARRY ARKLES
Gelest, Inc.

SILICON ESTERS

Silicon esters are silicon compounds that contain an oxygen bridge from silicon to an organic group, ie, Si–OR. The earliest reported organic silicon compounds contain four oxygen bridges and are often named as derivatives of orthosilicic acid, $Si(OH)_4$. The most conspicuous material is tetraethyl orthosilicate [*78-10-4*], $Si(OC_2H_5)_4$. The advent of organosilanes that contain silicon–carbon bonds, Si–C, initiated an organic nomenclature by which compounds are named as alkoxy derivatives. For example $Si(OC_2H_5)_4$ becomes tetraethoxysilane. The compound $CH_3Si(OCH_3)_3$ is named methyltrimethoxysilane. Whereas the latter usage is preferred, the literature even in the mid-1990s, particularly in ceramics (qv) technology, contains the older terms. Acyloxysilanes, eg, tetraacetoxysilane, $Si(OOCCH_3)_4$, are also members of this class. The chemistry and applications of acyloxysilanes are significantly different from those of the alkoxysilanes.

Applications for tetraalkoxysilanes cover a broad range. These compounds are classified roughly according to whether the Si−OR bond is expected to remain intact or to be hydrolyzed in the final application. Applications in which the Si−OR bond is hydrolyzed include binders for foundry-mold sands used in investment and thin-shell castings, binders for refractories (qv), resins, coatings (qv), sol−gel glasses, cross-linking agents, and adhesion promoters. Applications in which the Si−OR bond remains intact include heat-transfer and hydraulic fluids (qv). In general, the lower molecular weight compounds, eg, tetraethoxysilane and tetramethoxysilane, are used in reactive applications, whereas compounds such as tetra-2-ethylhexoxysilane are associated with mechanical applications. Methyl- and phenyltrialkoxysilanes are primarily used in the production of silicone resins and coatings. Longer-chain materials, eg, propyl-, isobutyl-, and octyltrialkoxysilanes, are used in hydrophobic coatings, primarily for masonry and concrete. The hydridoalkoxysilanes, triethoxysilane and trimethoxysilane, are intermediates for the production of organofunctional silanes. Organosilane esters in which there is a functional or reactive substitution of the organic radical are used as coupling agents. Tetraethoxysilane and its polymeric derivatives account for >75% of all production of nonfunctional silane esters.

Properties

The tetraalkoxysilanes possess excellent thermal stability and liquid behavior over a broad temperature range that widens with length and branching of the substituents. The physical properties of the silane esters, particularly the polymeric esters containing siloxane bonds, ie, Si−O−Si, are often likened to the silicone oils. These have low pourpoints and similar temperature−viscosity relationships. The alkoxysilanes generally have sweet, fruity odors that become less apparent as molecular weight increases. With the exception of tetramethoxysilane, trimethoxysilane, triethoxysilane, and a few closely related compounds that can be absorbed into corneal tissue, causing eye damage, the alkoxysilanes generally exhibit low levels of toxicity.

The physical properties of commercial alkoxysilanes are provided in Table 1. Two classes of silane esters have very distinct properties and are generally considered apart from alkoxysilanes. Silatranes are compounds derived from trialkanolamines and have silicon−nitrogen coordination. These are generally hydrolytically stable and have unique physiological properties (3). A second special class of monomeric esters are cyclic diesters of polyethyleneoxide glycols designated sila-crowns, which have application as catalysts (4). Neither silatranes nor sila-crowns are considered herein.

Aryloxy- and acyloxysilanes are often solids. The aryloxysilanes have excellent thermal stability. Acyloxy and mixed acyloxyalkoxysilanes have poor thermal stability. Thermal decomposition has been noted at temperatures as low as 110°C and is generally observed by 170°C.

The most significant difference between the alkoxysilanes and silicones is the susceptibility of the Si−OR bond to hydrolysis (see SILICON COMPOUNDS,

SILICONES). The simple alkoxysilanes are often operationally viewed as liquid sources of silicon dioxide (see SILICA). The hydrolysis reaction, which yields polymers of silicic acid that can be dehydrated to silicon dioxide, is of considerable commercial importance. The stoichiometry for hydrolysis for tetraethoxysilane is

$$\text{Si(OC}_2\text{H}_5)_4 + 2\,\text{H}_2\text{O} \xrightarrow{\text{acid or base}} \text{SiO}_2 + 4\,\text{C}_2\text{H}_5\text{OH}$$

Silicon dioxide never forms directly during hydrolysis. Intermediate ethoxy derivatives of silicic acid and polysilicates form as hydrolysis progresses. The polysilicates grow in molecular weight until most or all of the ethoxy groups are removed and a nonlinear network of Si–O–Si remains. The development of cyclic and cube structures containing 3–8 silicon atoms also occurs (5–7). A numerical modeling system for the hydrolysis of tetraethoxysilane has been developed (8). The viscosity of the solution increases until gelation or precipitation. Partially hydrolyzed materials of this type often contain more than enough silanols, Si–OH, to displace most of the remaining ethoxy groups in an acid- or base-catalyzed condensation. The stoichiometric equation for partial hydrolysis is

$$n\,\text{Si(OC}_2\text{H}_5)_4 + 2x\,\text{H}_2\text{O} \xrightarrow{\text{acid or base}} \left(\!\!-\text{Si(OC}_2\text{H}_5)_{4(1-x)}\text{O}_{2x}\!\!-\right)_{\overline{y}} + 4x\,\text{ROH}$$

where x is the mol % partial hydrolysis. If the alkoxysilane is an organoalkoxysilane, eg, methyltriethoxysilane or phenyltriethoxysilane, the hydrolysis proceeds analogously to give the organosilsesquioxanes, $(\text{RSiO}_{1.5})_n$, instead of the dioxides, $(\text{SiO}_2)_n$. Likewise diorganodialkoxysilanes yield silanediols upon hydrolysis. Whereas the hydrolysis process is not usually considered to be an equilibrium reaction, the equilibrium constant for the reaction is $\sim\!2 \times 10^{-3}$. The reversibility of the reaction only plays a significant role when the hydrolysis products are soluble (9).

Redistilled tetraethoxysilane containing less than 1 ppm chloride added to neutral $18 \times 10^6\,\Omega$ water purged with nitrogen in fluorocarbon bottles does not hydrolyze to a gel for over six months (10). Without special precautions, tetraethoxysilane hydrolyzes to a gel in $\sim\!10$ d; tetramethoxysilane hydrolyzes in $\sim\!2$ d; and tetra-n-butoxysilane in $\sim\!25$ d. The hydrolysis reaction is catalyzed by acid or base. Acid-catalyzed hydrolysis generally proceeds more rapidly than base hydrolysis, and leads to more linear polymers than base hydrolysis. In contrast to base-catalyzed hydrolysis, there is a significant rate difference between the rate of hydrolysis of the first and second alkoxy groups in the presence of acid (see Table 1).

For binder preparation, dilute hydrochloric or acetic acids are preferred, because these facilitate formation of stable silanol condensation products. When more complete condensation or gelation is preferred, a wider range of catalysts, including moderately basic ones, is employed. These materials, which are often called hardeners or accelerators, include aqueous ammonia, ammonium carbonate, triethanolamine, calcium hydroxide, magnesium oxide, dicyclohexylamine, alcoholic ammonium acetate, and tributyltin oxide (11,12).

Table 1. Physical Properties of Silane Esters[a]

Compound	CAS Registry Number	Formula	Boiling point[b], °C
Monoorganoalkoxysilanes			
methyltrimethoxysilane	[1185-55-3]	$CH_3Si(OCH_3)_3$	102–103
methyltriethoxysilane	[2031-67-6]	$CH_3Si(OC_2H_5)_3$	141–143
ethyltrimethoxysilane	[5314-55-6]	$C_2H_5Si(OCH_3)_3$	124–125
ethyltriethoxysilane	[78-07-8]	$C_2H_5Si(OC_2H_5)_3$	158–159
propyltrimethoxysilane	[1067-25-0]	$C_3H_7Si(OCH_3)_3$	142
propyltriethoxysilane	[141-57-1]	$C_3H_7Si(OC_2H_5)_3$	179–180
isobutyltrimethoxysilane	[18395-30-7]	$i\text{-}C_4H_9Si(OCH_3)_3$	154
pentyltriethoxysilane	[2761-24-2]	$C_5H_{11}Si(OC_2H_5)_3$	$95\text{--}6_{1.3}$
octyltriethoxysilane	[2943-75-1]	$C_8H_{17}Si(OC_2H_5)_3$	$98\text{--}99_{0.27}$
octadecyltrimethoxysilane	[3069-42-9]	$C_{18}H_{37}Si(OCH_3)_3$	$170_{0.013}$
octadecyltriethoxysilane	[112-04-9]	$C_{18}H_{37}Si(OC_2H_5)_3$	$165\text{--}169_{0.27}$
phenyltriethoxysilane	[780-69-8]	$C_6H_5Si(OC_2H_5)_3$	$112\text{--}113_{1.3}$
Tetraorganoxysilanes and polyorganoxysiloxanes			
tetramethoxysilane	[681-84-5]	$Si(OCH_3)_4$	121–122
tetraethoxysilane	[78-10-4]	$Si(OC_2H_5)_4$	169
tetrapropoxysilane	[682-01-9]	$Si(O\text{-}n\text{-}C_3H_7)_4$	224–225
tetraisopropoxysilane[e]	[1992-48-9]	$Si(O\text{-}i\text{-}C_3H_7)_4$	185–186
tetrabutoxysilane	[4766-57-8]	$Si(O\text{-}n\text{-}C_4H_9)_4$	$115_{0.4}$
tetrakis(s-butoxy)silane	[5089-76-9]	$Si(O\text{-}sec\text{-}C_4H_9)_4$	$87_{0.27}$
tetrakis(2-ethyl-butoxy)silane	[78-13-7]	$Si(OCH_2CH(C_2H_5)_2)_4$	$166\text{--}172_{0.27}$
tetrakis(2-ethyl-hexoxy)silane	[115-82-2]	$Si(OCH_2CH(C_2H_5)(C_4H_9))_4$	$194_{0.13}$
tetrakis(2-methoxy-ethoxy)silane	[2157-45-1]	$Si(OCH_2CH_2OCH_3)_4$	$179\text{--}182_{14.7}$
tetraphenoxysilane[e]	[1174-72-7]	$Si(OC_6H_5)_4$	$236\text{--}237_{0.13}$
hexaethoxydisiloxane[e]	[2157-42-8]	$(C_2H_5O)_3SiOSi(OC_2H_5)_3$	230–232
ethylsilicate 40[i]	[18954-71-7]	ca $(OSi(OC_2H_5)_2)_{4\text{--}5}$	290–310
Acyloxysilanes			
tetracetoxysilane	[5623-90-3]	$Si(OOCCH_3)_4$	$148_{0.8}$
methyltriacetoxysilane	[4253-34-3]	$CH_3Si(OOCCH_3)_3$	$87\text{--}88_{0.4}$
ethyltriacetoxysilane	[17689-77-9]	$C_2H_5Si(OOCCH_3)_3$	107_1
di-t-butoxydiacetoxysilane	[13170-23-5]	$(t\text{-}C_4H_9O)_2Si(OOCCH_3)_2$	$102_{0.7}$

[a]Ref. 1. [b]Subscript denotes pressure, other than atmospheric, in kPa. To convert kPa to psi, multiply by 0.145. [c]To convert kJ to kcal, divide by 4.184. [d]Ref. 2. Value may be questionable. [e]Model compound; not of commercial significance. [f]At 38°C. [g]At 60°C. [h]At 55°C. [i]Nominal values; commercial values may vary. Properties given are for the average compound containing 40 wt % silicon dioxide.

Sol–Gel Process Technology and Chemistry

The complete hydrolysis of tetraalkoxysilanes under highly controlled conditions, usually without the presence of fillers, is associated with sol–gel technology (qv). Sol–gel is a method for preparing specialty metal oxide glasses and ceramics by

Table 1. (*Continued*)

Melting point, °C	Density, g/cm^3	Refractive index, n_D	ΔH_{vap}, kJ/mol^c	Viscosity, mm^2/s(=cP)	Flash-point, °C	LD$_{50}$ (oral, rat), mg/kg
			Monoorganoalkoxysilanes			
	0.955	1.3646		0.5	8	
	0.895	1.3832		0.6	23	12,500^d
	0.949	1.3838		0.5	27	
	0.896	1.3955	32.6	0.7	40	13,720
	0.939	1.3880			34	7,420
	0.892	1.3956			57	
	0.933	1.3960			42	>2,000
	0.895	1.4059			68	
<−40	0.875	1.4160		1.9	100	
13−17	0.885	1.4391			140	
10−12	0.870	1.4386				>5,000
	0.996	1.4718	47.7		96	2,830
			Tetraorganoxysilanes and polyorganoxysiloxanes			
2	1.032	1.3668	46.8	0.5	20	700
−85	0.934	1.3838	46.0	0.7	46	6,270
<−80	0.916	1.4012		1.7	95	
<−22	0.887	1.3845	46.8	1.2	60	
<−80	0.899	1.4128	61.9	2.3	110	
	0.885	1.4000		2.1^f	104	
<−70	0.892	1.4309		4.4^f	116	22,130
<−80	0.88	1.4388	70.6	6.8^f	188	>22,000
<−70	1.079	1.4219		4.4	140	
48−49	1.141	1.554^g		6.6^h		
	0.998	1.3914				
−90	1.05−1.06	1.3914		4−5	43	
			Acyloxysilanes			
110 sub	1.06	1.4220				
40	1.175	1.4083			85	
7−9	1.143	1.4123			106	
−4	1.0196	1.4040			95	

hydrolyzing a chemical precursor or mixture of chemical precursors that pass sequentially through a solution state and a gel state before being dehydrated to a glass or ceramic. The use of sol–gel technology has increased dramatically since 1980. A variety of techniques have been developed to prepare fibers, microspheres, thin films (qv), fine powders, and monoliths. Applications for this technology include protective coatings, catalysts, piezoelectric devices, waveguides, lenses, high strength ceramics, superconductors, insulating materials, and nuclear waste encapsulation. The flexibility of sol–gel technology allows unique access to multicomponent oxide systems and low temperature process regimes. An excellent review of sol–gel chemistry is available (13).

Preparation of metal oxides by the sol–gel route proceeds through three basic steps: (1) partial hydrolysis of metal alkoxides to form reactive monomers; (2) the polycondensation of these monomers to form colloid-like oligomers (sol formation); and (3) additional hydrolysis to promote polymerization and cross-linking leading to a three-dimensional matrix (gel formation). Although presented herein sequentially, these reactions occur simultaneously after the initial processing stage.

Monomer formation or partial hydrolysis

$$Si(OC_2H_5)_4 + H_2O \longrightarrow (C_2H_5O)_3SiOH + C_2H_5OH$$

Sol formation or polycondensation

$$y\,(C_2H_5O)_3SiOH \xrightarrow[\text{dimers}]{\text{monomers}} (C_2H_5O)_3Si(OSi(OC_2H_5)_2)_nOH + n\,C_2H_5OH$$

Gelation or cross-linking

$$(C_2H_5O)_3Si(OSi(OC_2H_5)_2)_nOH \xrightarrow{n\,H_2O} (C_2H_5O)_3Si(OSi(OC_2H_5)OH)_nOH + n\,C_2H_5OH \xrightarrow{-H_2O}$$

where R is C_2H_5.

As polymerization and cross-linking progress, the viscosity of the sol gradually increases until the sol–gel transition point is reached. At this point the viscosity abruptly increases and gelation occurs. Further increases in cross-linking are promoted by drying and other dehydration methods. Maximum density is achieved in a process called densification in which the isolated gel is heated above its glass-transition temperature. The densification rate and transition (sintering) temperature are influenced primarily by the morphology and composition of the gel.

Nonhydrolytic methods for the formation of silicon dioxide from tetraalkoxysilanes have been reported (14–15). Others have been reinvestigated (16):

$$2\,HCOOH + Si(OC_2H_5)_4 \longrightarrow SiO_2 + 2\,C_2H_5OH + 2\,HCOOC_2H_5$$

$$Si(OCH_2C_6H_5)_4 + SiCl_4 \longrightarrow SiO_2 + 4\,C_6H_5CH_2Cl$$

The Si–OR bond undergoes a variety of reactions apart from hydrolysis and condensation. In one of the more important aspects of reactivity, it is associated with the production of silicone intermediates and with cross-linking reactions for silicone room temperature vulcanizing materials (RTVs) (17). The reactivity of the Si–OR bond is in many cases analogous to the Si–Cl bond, except that the reactions are more sluggish. These reactions become increasingly more sluggish with greater bulk and steric screening of the alkoxy group. Reactions that have been reviewed (18–20) include the following:

$$— SiOR + R'MgCl \longrightarrow — SiR' + Mg(OR)Cl$$

$$— SiOR + HOSi \longrightarrow — Si — O — Si — + ROH$$

$$— SiOR + HOB \longrightarrow — SiOB — + ROH$$

$$— SiOR + (R'\overset{O}{\overset{\|}{C}})_2O \longrightarrow — Si O\overset{O}{\overset{\|}{C}}R' + R'\overset{O}{\overset{\|}{C}}OR$$

$$2 — SiOR + R'CHO \xrightarrow{H_2SO_4} — Si — O — Si — + R'CH(OR)_2$$

$$— SiOR + ClSi \xrightarrow{catalyst} — Si — O — Si — + RCl$$

$$— SiOR + HON = R' \xrightarrow{catalyst} — Si — ON = R' + ROH$$

$$— SiOR + R'COCl \longrightarrow — SiCl + R'COOR$$

In comparison to the Si–OR bond, the Si–C bond can be considered essentially unreactive if the organic moiety is a simple unsubstituted hydrocarbon. If the organic moiety is substituted as in the case of a trialkoxysilane, the chemistry is more appropriately considered elsewhere (see SILICON COMPOUNDS, SILANES; SILICON COMPOUNDS, SILYLATING AGENTS).

Simple alkyl- and aryltrialkoxysilanes have three rather than four matrix coordinations in the polymeric hydrolysates, leading to less rigid structures than those derived from tetraalkoxysilanes. These and other changes in physical characteristics, eg, wetting and partition properties, make these materials more appropriate in a variety of coating applications, where tetraalkoxysilanes are not acceptable. These materials are variously referred to as T-resins, organosilsesquioxanes, and ormosils, from the term organic modified silicas. Methylsilsesquioxanes are stable to 400°C. Phenylsilsesquioxanes are stable to 475°C.

Preparation

The principal method of silicon ester production is described by Von Ebelman's 1846 synthesis (21):

$$SiCl_4 + 4 C_2H_5OH \longrightarrow Si(OC_2H_5)_4 + 4 HCl$$

The reaction is generalized to

$$R_{(4-n)}SiCl_n + n R'OH \longrightarrow R_{(4-n)}Si(OR')_n + n HCl$$

Process considerations must not only take into account characteristics of the particular alcohol or phenol to be esterified, but also the self-propagating by-product reaction, which results in polymer formation.

$$ROH + HCl \longrightarrow RCl + H_2O$$

$$H_2O + SiCl_4 \longrightarrow [SiCl_2O] + 2\ HCl$$

Methods used to remove hydrogen chloride include the use of refluxing solvents or reaction mixtures, sparging dry air or nitrogen through the reaction mixture, and conducting the reaction in vapor phase or under applied vacuum. Amines can be employed as base-acceptors, but generally this is not practical commercially. In batch processes, the alcohol is always added to the chlorosilane. Continuous processes involve (1) pumping the alcohol and chlorosilane together in a mixing section, (2) introducing the chlorosilane vapor countercurrent to liquid alcohol, or (3) introducing chlorosilane vapor in a two-column reaction distillation scheme in which substoichiometric alcohol is introduced center-column to a chlorosilane, removing HCl and unreacted alcohol in the overhead, and in the second stage a slight excess of the alcohol is introduced center-column, recovering the excess overhead and removing product from the bottom. All processes provide a method for removal of by-product hydrogen chloride. The energy of activation for the reaction of ethanol with silicon tetrachloride in the vapor phase is 64.9 kJ/mol (15.5 kcal/mol) (22). The initial stages of the esterification processes are endothermic because the heat of evaporation of HCl cools the reaction mixture. In the last stages of esterification, the mixtures are usually heated during the final addition of alcohol. Tertiary alkoxides cannot be formed in this manner.

In the batch production of tetraethoxysilane from silicon tetrachloride, the initial reaction product contains at least 90 wt % tetraethoxysilane with a 28 wt % SiO_2 content. Distillation removes alcohol and high boiling impurities, and the distilled product contains at least 98% tetraethoxysilane and is called pure ethyl silicate. Partially hydrolyzed or polymeric versions where substantial portions have an average of 4–5 silicon atoms and 40 wt % SiO_2 or have an average of 5–8 silicon atoms and a 50 wt % SiO_2 content are referred to as ethylsilicate 40 and ethylsilicate 50, respectively. Ethylsilicate 50 has a branched structure with approximately 30–35% of all silicon atoms bonded to two others by oxygen bridges; 35–38% are bonded to three silicon atoms; and 12–16% are bonded to four silicon atoms. Model systems for hydrolyzed tetraethoxysilane which include cube as well as cyclic structures have been prepared (4).

Although known since the 1940s and 1950s (23,24), catalyzed direct reactions of alcohols using silicon metal have become important commercial technology in the 1990s for production of lower esters. Patents have reported the reactions of methanol and ethanol with silicon in high boiling solvents, or in contained reaction products to give high yields of trialkoxysilanes and tetraalkoxysilanes (25,26). It has been demonstrated that in the presence of a methoxy compound, where M is a metal, and under moderate pressure, substantial improvements in yield can be achieved (27).

$$Si + 4\ ROH \xrightarrow[\text{Si(OR)}_4]{\text{Fe, CH}_3\text{OM}} Si(OR)_4 + 2\ H_2$$

The synthesis of triethoxysilane (28) and trimethoxysilane (29) has also been achieved by direct process. In 1980 there were no direct processes for the production of alkoxysilanes. In 1995 Silbond in Weston, Michigan, and Carboline in St. Louis, Missouri, operated processes for the production of tetraethoxysilane in the United States, and OSi/Witco announced start-up of a process to produce triethoxysilane and tetraethoxysilane in Termoli, Italy.

Apart from the direct action of an alcohol on a chlorosilane or silicon, the only other commercial method used to prepare alkoxysilanes is transesterification.

$$Si(OR)_4 + 4 R'OH \underset{}{\overset{catalyst}{\rightleftharpoons}} Si(OR')_4 + 4 ROH$$

Transesterification, an equilibrium reaction, is practical only when the alcohol to be esterified has a high boiling point and the leaving alcohol can be removed by distillation. The most widely used catalysts are sodium alcoholates and organic titanates, although amines are also used (30,31).

A provocative reaction of ethylene glycol directly with silicon dioxide that leads to a complex mixture of oligomeric and cyclic ester species has been reported (32). This reaction proceeds in the presence of sodium hydroxide or in the presence of high boiling tertiary amines (33).

Other preparative methods for alkoxysilanes, in approximate order of declining utility, are given by the following equations (34–40):

$$\equiv SiCl + (RO)_3CH \longrightarrow \equiv SiOR + RCl + RO\overset{\overset{\displaystyle O}{\displaystyle \|}}{C}H$$

$$\equiv SiCl + NaOR \longrightarrow \equiv SiOR + NaCl$$

$$\equiv SiH + HOR \overset{catalyst}{\longrightarrow} \equiv SiOR + H_2$$

$$\equiv SiCl + RCH\overset{\overset{\displaystyle O}{\diagdown}}{-}CH_2 \longrightarrow \equiv SiOCH_2CH(Cl)R$$

$$\equiv SiOH + HOR \longrightarrow \equiv SiOR + H_2O$$

$$\equiv SiCl + CH_3NO_2 \longrightarrow \equiv SiOCH_3 + NO_2Cl$$

$$\equiv SiSR + HOR \longrightarrow \equiv SiOR + H_2S$$

The acyloxysilanes are produced by the reaction of an anhydride and a chlorosilane.

$$SiCl_4 + 4 (R\overset{\overset{\displaystyle O}{\displaystyle \|}}{C})_2O \longrightarrow Si(O\overset{\overset{\displaystyle O}{\displaystyle \|}}{C}R)_4 + 4 R\overset{\overset{\displaystyle O}{\displaystyle \|}}{C}Cl$$

The analogous reaction between anhydrides and alkoxysilanes also produces acyloxysilanes. The direct reaction of acids with chlorosilanes does not cleanly lead to full substitution. Commercial production of methyltriacetoxysilane directly from methyltrichlorosilane and acetic acid has been made possible by the addition of small amounts of acetic anhydride or EDTA, or acceptance of dimethyltetraacetoxydisiloxane in the final room temperature vulcanizing (RTV) application (41–43). A reaction which leads to the formation of acyloxysilanes is the interaction of acid chlorides with silylamides.

Economic Aspects

Tetraethoxysilane and its polymeric derivatives account for >90% of the dollar value of nonaryl- or alkyl-substituted esters. The leading U.S. suppliers are Silbond and Hüls America. Silbond manufacture is limited to ethylsilicates. Hüls imports ethylsilicate produced at its Rheinfelden and Dresden, Germany, facilities. Eagle-Picher announced start-up of an electronic-grade tetraethoxysilane unit at Miami, Oklahoma, in September 1995. Yamanaka operates a similar facility in Japan. The U.S. market for the ethylsilicates is estimated at 3000–3500 metric tons. Pricing ranges from $2.40 to $6.00/kg, depending on grade and quantity.

Nonethyl ester prices are $3–$100/kg. Hüls, Gelest, and Harris Specialty Chemical are U.S. producers. Alkyltrialkoxysilanes are produced by Hüls at Mobile, Alabama, Harris Specialty Chemical at Gainesville, Florida, and OSi/Witco in Sistersville, West Virginia. In 1996, AlliedSignal announced the sale of its Jayhawk facility, which produced alkylalkoxysilanes, to Allco/Inspec, and the future production of alkylalkoxysilanes at the facility was not certain. Gelest manufactures specialty alkylsilane esters at its Tullytown, Pennsylvania, facility. Shin-etsu and Dai-Hachi produce alkylalkoxysilanes in Japan.

The bulk of acyloxysilanes is produced and used by the principal U.S. silicone rubber producers, ie, Dow Corning and General Electric (see SILICON COMPOUNDS, SILICONES).

Toxicity

The alkoxysilanes generally have a low order of toxicity, which may be associated with their alcoholic products of hydrolysis. Notable exceptions are tetramethoxysilane and two hydridosilanes, trimethoxysilane and triethoxysilane. Triethoxymethoxysilane is the lowest member of the series in which the hazard is substantially reduced. Vapors of these materials may be absorbed directly into corneal tissue, causing blindness (44). The onset of corneal damage is noted by a scratchy feeling in the eyes, usually 2–4 h after exposure. The effects of exposure to the methoxysilanes are rarely reversible. This is a significant consideration for worker safety. Especially because of the pleasant minty or fruity fragrance, exposure to silicon esters is frequently ignored.

Uses

Precision Casting. The ethoxysilanes are used as binders in precision casting for investment and thin-shell processing (45–47). Ethylsilicate 40 and its partial hydrolysates are preponderant. In the investment process, 3–10% excess water is added to a prehydrolyzed silicate binder. This is mixed with refractory material. If the refractory material contains magnesium oxide or calcium hydroxide, gelation occurs in 40–60 min. If these additives are not present or an accelerated cure is required, catalysts are added to the binder prior to mixing with the refractory. In the thin-shell process, fusible patterns are dipped into slurries made of a refractory and ethylsilicate binder. Curing is accomplished by air drying or exposure to ammonia vapor. Ethanol from hydrolysis is either allowed to evaporate or is burned off prior to firing. The utilization of silicate esters has diminished since the 1980s, owing to their partial replacement by colloidal silicas.

Cements and Ceramics. Refractory cements and ceramics (qv) are prepared from slurries of silica, zirconia, alumina, or magnesia and a prehydrolyzed silicate (see CEMENT). Calcining at 1000°C yields cured refractory shapes (48,49).

Glass Frosting. Deposition of silicon dioxide is used to impart a translucent coating on glass (qv) (50). The surfaces are either exposed to tetraethoxysilane or tetramethoxysilane under high moisture conditions, or the alkoxysilanes are ignited and the resulting powder is applied to the surface.

Paints and Coatings. Ethoxysilanes are used in high temperature, zinc-rich paints (see PAINT) (51,52). Methyl- and phenyltrialkoxysilanes are used to prepare abrasion-resistant coatings for plastics (53), particularly polycarbonate (54,55), and dielectric coatings and seals for high voltage electrical components, including television tubes.

Sol–Gel Glasses and Ceramics. Although sol–gel is actually a process, not a product, several classes of materials are associated with the sol–gel process. Sol–gel-derived materials include fine powders, coatings and monoliths, and aerogels (qv). Sol–gel-derived powders are produced by the Stöber process (56) and variations of it. The products are used in catalysis and chromatography. Unmodified silica coatings are employed in the preparation of ion-free coatings on glass used in liquid crystal displays (LCDs), coatings for eyewear, and industrial and automotive plastic glazing. Small-diameter lenses and gradient index (GRIN) optics are in commercial development, as of 1996. Aerogels derived from tetraethoxysilane have extremely great potential as insulating materials. BASF has introduced a product based on this technology called Basogel.

Water Repellents. Protective and consolidating coatings for masonry and other applications are produced from methyl-, propyl-, isobutyl-, and octyltrialkoxysilanes (57). Applications for these materials are in two principal markets: vertical, ie, buildings; and horizontal, ie, bridge decks, parking garages, etc. Performance characteristics such as substrate penetration often make use of lower alkyltrialkoxysilanes, important where salt penetration is a concern. A water repellent based on isobutyltrimethoxysilane is marketed under the trade name of Chem-Trete by Hüls. The ability of longer alkyls to form stable emulsions makes them preferred for cost and safety (58). A product based on octyltriethoxysilane is marketed under the trade name Enviroseal by Harris Specialty Chemical.

Bonded Phases. Substrate-bond hydrocarbon coatings for high pressure liquid chromatography (hplc) and flash chromatography are prepared from octyltrialkoxysilanes and other long-chain alkyltrialkoxysilanes (see CHROMATOGRAPHY).

Hydraulic and Heat-Transfer Fluids. Hydraulic fluids (qv) for high altitude supersonic aircraft and thermal exchange applications including solar panels employ fluids such as tetrakis(2-ethylhexoxy)silane. These products have been marketed under the trade name Coolanol by Monsanto (see HEAT-EXCHANGE TECHNOLOGY).

Silicone Room Temperature Vulcanizing Cross-Linking. Condensation-cured polydimethylsiloxanes contain terminal silanol groups which condense with the silanols produced by ambient moisture hydrolysis of acyloxysilanes.

Methyltriacetoxysilane, ethyltriacetoxysilane, and tetraacetoxysilane are the most commonly used cross-linking agents.

Spin-On Glass. In microelectronic applications, films of silicon dioxide are deposited on silicon substrates by the application of a partially hydrolyzed solution of tetraethoxysilane or methyltriethoxysilane (59,60). A product based on this technology is marketed under the name Accuspin by AlliedSignal.

Chemical Vapor Deposition. Chemical vapor deposition (CVD) of silicon dioxide from tetraethoxysilane assisted by the presence of oxygen and a plasma is an important technology for the deposition of pure and modified dielectrics for microelectronics (61). An alternative method for the deposition of silicon dioxide utilizes di-*t*-butoxydiacetoxysilane (62).

BIBLIOGRAPHY

"Silicon Esters and Ethers" under "Silicon Compounds" in *ECT* 1st ed., Vol. 12, pp. 371–372, by E. G. Rochow, Harvard University; in *ECT* 2nd ed., Vol. 18, pp. 216–221, by A. R. Anderson, Anderson Development Co.; in *ECT* 3rd ed., Vol. 20, pp. 912–921, by B. Arkles, Petrarch Systems, Inc.

1. B. Arkles, *Silicon, Germanium, Tin, and Lead Compounds: A Survey of Properties and Chemistry*, Gelest, Inc., Tullytown, Pa., 1995.
2. N. Sax and R. Lewis, *Dangerous Properties of Industrial Materials*, Van Nostrand Reinhold Co., Inc., New York, 1989.
3. M. G. Voronkov, in G. Bendz and I. Lindqvist, eds., *Biochemistry of Silicon and Related Problems*, Plenum Publishing Corp., New York, 1977, p. 395.
4. B. Arkles, R. Anderson, and K. King, *Organometallics* **2**, 454, 1983.
5. J. F. Brown, Jr., *J. Am. Chem. Soc.* **87**, 4317, 1965.
6. P. Cagle and co-workers, in B. Zelinski and co-workers, eds., *Better Ceramics Through Chemistry IV, MRS Proc. 180*, 1990, p. 961.
7. F. Feher, D. Newman, and J. Walzer, *J. Am. Chem. Soc.* **111**, 1741, 1989.
8. B. D. Kay and R. A. Assink, *J. Non Cryst. Sol.* **104**, 112, 1988.
9. B. Arkles, J. Steinmetz, J. Zazyczny, and P. Mehta in K. L. Mittal, ed., *Silanes & Other Coupling Agents*, VSP, 1993, p. 91.
10. Technical data, Gelest, Inc., Tullytown, Pa., 1992.
11. U.S. Pat. 2,550,923 (May 1, 1951), C. Shaw, J. E. Hocksford, and W. E. Smith (to Shaw and Langish-Smith); U.S. Pat. 2,795,022 (June 11, 1957), N. Shaw (to Shaw Process).
12. H. G. Emblem and T. R. Turger, *Trans. Br. Ceram. Soc.* **78**(5), 1979.
13. C. J. Brinker, G. W. Scherer, *Sol-Gel Science*, Academic Press, Orlando, Fla., 1990.
14. U.S. Pat. 4,950,779 (Aug. 21, 1990), J. H. Wengrovius and V. M. VanValkenburgh (to General Electric).
15. U.S. Pat. 5,441,718 (Aug. 15, 1995), K. Sharp (to E. I. du Pont de Nemours & Co., Inc.).
16. S. Acosta and co-workers, in *Better Ceramics Through Chemistry VI, MRS Proc. 346*, 1994, p. 43.
17. E. L. Warrick, O. R. Pierce, K. E. Polmanteer, and J. C. Saam, *Rubber Chem. Tech. Revs.* **52**, 437 (1979).
18. B. Arkles, in G. Silverman and P. Rackita, eds., *Grignard Reagents*, Marcel Dekker, New York, 1996, p. 667.
19. R. C. Mehrotra, V. D. Gupta, and G. Srivastiva, *Rev. Silicon, Germanium, Tin Lead Comp.* **1**, 299 (1975).
20. M. G. Voronkov, V. P. Mileshevich, and Yu. A. Yuzhelevski, *The Siloxane Bond*, Plenum Publishing Corp., New York, 1978.

21. J. Von Ebelman, *Ann. Chem.* **57**, 319 (1846).
22. V. G. Ukhtomshii, *Izv. Vyssh. Ucheb. Zaved. Khim. Khim. Tekhnol.*, **19**(7), 146 (1976).
23. U.S. Pat. 2,473,260 (June 26, 1946), E. G. Rochow (to General Electric).
24. U.S. Pat. 3,072,700 (Jan. 8, 1963), N. de Wit (to Union Carbide).
25. B. Kanner and K. M. Lewis, in K. M. Lewis and D. G. Rethwisch, eds., *Catalyzed Direct Reactions of Silicon*, Elsevier Science, Inc., New York, 1993, p. 39.
26. U.S. Pat. 4,323,690 (Apr. 6, 1982), J. F. Montle, H. J. Markowsi, P. D. Lodewyck, and D. F. Schneider (to Carboline).
27. U.S. Pat. 4,113,761 (Sept. 12, 1978), G. Kreuzberg, A. Lenz, and W. Rogher (to Dynamit Nobel).
28. Jpn. Kokai Tokkyo Koho 80 28,929 (Feb. 29, 1980), S. Suzuki, T. Imaki, and T. Yamamura (to Mitsubishi).
29. U.S. Pat. 5,084,590 (Jan. 28, 1992), J. S. Ritscher and T. E. Childress (to Union Carbide).
30. P. D. George and J. R. Ladd, *J. Am. Chem. Soc.* **75**, 987 (1953).
31. H. Steimann, G. Tschernko, and H. Hamann, *Z. Chem.* **17**, 89 (1977).
32. R. M. Laine, *Nature* **353**, 640 (1991).
33. R. M. Laine, *Chem. Mat.* **6**, 2177 (1994).
34. L. M. Shore, *J. Am. Chem. Soc.* **76**, 1390 (1959).
35. U.S. Pat. 2,381,137 (May 14, 1942), W. I. Patnode and R. O. Sauer (to General Electric).
36. D. Seyferth and E. G. Rochow, *J. Org. Chem.* **20**, 250 (1955).
37. V. Day, W. Klemperer, V. Mainz, and D. Millar, *J. Am. Chem. Soc.* **107**, 8262 (1985).
38. A. Weiss and G. Reiff, *Z. Anorg. Allg. Chem.* **311**, 151 (1961).
39. M. E. Havill, I. Jofee, and H. W. Post, *J. Org. Chem.* **13**, 280 (1948).
40. U.S. Pats. 2,569,455 and 2,459,746 (1949), J. B. Culbertson, H. D. Erasmus, and F. M. Fowler (to Union Carbide).
41. U.S. Pat. 3,9784,198 (Aug. 10, 1976), B. Ashby (to General Electric).
42. U.S. Pat. 4,329,484, L. P. Peterson (to General Electric).
43. U.S. Pat. 4,332,956 (June 1, 1982), L. A. Tolentino (to General Electric).
44. H. F. Smith, *The Effect of Tetramethylorthosilicate on the Eyes*, Carbide and Carbon Chemicals Industrial Fellowship 274-1, Mellon Institute, Pittsburgh, Pa.
45. A. Dunlop, *Foundry Trade J.* **75**, 107 (1945).
46. A. E. Focke, *Met. Prog.* **49**, 489 (1945).
47. U.S. Pat. 2,678,282 (May 11, 1954), C. Jones (to Pilkington Bros. Ltd.).
48. H. G. Emblem and I. R. Walters, *J. Appl. Chem., Biotechnol.* **27**, 618 (1977).
49. U.S. Pat. 2,678,282 (May 11, 1954), C. Jones (to Pilkington Bros. Ltd.).
50. U.S. Pat. 2,596,896 (Mar. 20, 1951), M. Pipken (to General Electric).
51. U.S. Pat. 3,056,684 (Oct. 11, 1962), S. L. Lopata (to Carboline).
52. D. M. Berger, *Met. Finishing* **72**(4), 27 (1979).
53. U.S. Pat. 4,197,230 (Apr. 8, 1980), R. H. Baney and L. A. Harris (to Dow Corning).
54. U.S. Pat. 4,495,360 (Jan. 22, 1985), B. T. Anthony (to General Electric).
55. U.S. Pat. 4,491,508 (Jan. 1, 1985), D. R. Olson and K. K. Webb (to General Electric).
56. W. Stöber and co-workers, *J. Coll. Interface Sci.* **26**, 62 (1968).
57. U.S. Pat. 2,916,461 (Dec. 8, 1959), K. W. Krantz (to General Electric).
58. U.S. Pat. 4,648,904 (Mar. 10, 1987), R. Depasquale and M. Wilson (to SCM).
59. U.S. Pat. 3,915,766 (Oct. 28, 1975), G. F. Pollack and J. G. Fish (to Texas Instruments).
60. U.S. Pat. 4,103,065 (July 25, 1978), D. Gagnon (to Owens-Illinois).
61. S. Fisher and co-workers, *Solid State Technol.*, 55 (Sept. 1993).
62. G. Smolinsky and R. Dean, *Mat. Lett.* **4**, 256 (1986).

BARRY ARKLES
Gelest, Inc.

SILICONES

Silicones are a class of polymers having the formula $(R_m Si(O)_{4-m/2}))_n$, where $m = 1-3$ and $n \geq 2$. The most common are the polydimethylsiloxanes (PDMS).

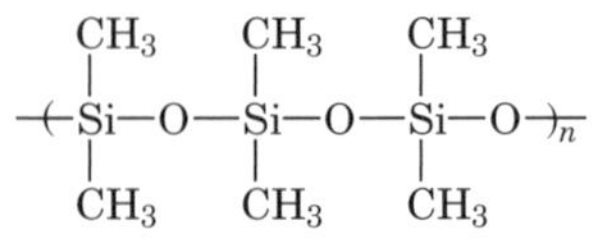

$$\left[\begin{array}{ccc} CH_3 & CH_3 & CH_3 \\ | & | & | \\ Si-O-Si-O-Si-O \\ | & | & | \\ CH_3 & CH_3 & CH_3 \end{array}\right]_n$$

Silicones are the subject of many reviews (1–8). Commercial products include fluids, filled fluids and gums, greases, resins and rubber (1,2). Various forms of silicones and examples of applications are listed in Table 1.

The designations M, D, T, and Q are used, respectively, for mono-, di-, tri-, and quaternary coordination of oxygen around silicon in silicones. A T group can also be written as $CH_3SiO_{3/2}$; a Q group as SiO_2. When groups other than methyl are present, these groups are indicated with a superscript; eg, D^{Vi} represents a methyl vinyl siloxy group, $(CH_3)(CH_2{=}CH)SiO$. Resins are often composed as M_xQ, M_xT, $M_xD_yT_zQ$, etc. This common shorthand notation for silicones is shown in Figure 1.

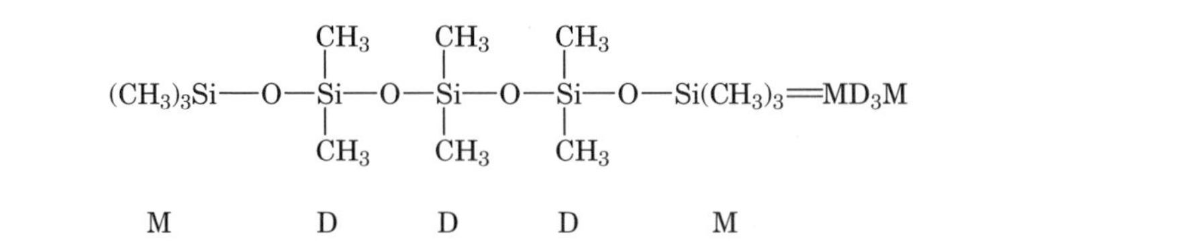

M group   D group   T group   Q group

For example,

$$(CH_3)_3Si-O-\underset{CH_3}{\overset{CH_3}{Si}}-O-\underset{CH_3}{\overset{CH_3}{Si}}-O-\underset{CH_3}{\overset{CH_3}{Si}}-O-Si(CH_3)_3 {=} MD_3M$$

M   D   D   D   M

Fig. 1. Widely accepted abbreviations used for silicone groups.

History of Silicones

Although silicon is quite abundant in nature, there are no natural forms of organosilicon (2). The first covalent compound of silicon, SiF_4, was made in 1771 (9). Berzelius, in 1823, was the first to make amorphous silicon by potassium reduction of SiF_4; he also found that the combination of SiO_2, CuO, and C had greater hardness than SiO_2 alone. In 1896 Morrison made silicon metal

Table 1. Silicone Products and Their Uses

Commercial product	Use
fluids: heat-stable liquids	lubricants, water repellents, defoamers, release agents, surfactants
filled fluids and gums	valve lubricants, moistureproof sealants for electrical connectors, pressure-sensitive adhesives, personal care products
grease: fluid and carbon black or soap	nonflow lubricants, polishes
resins: cross-linked materials	electrical insulation, lubricant and paint additives, release formulations, water repellents
rubbers: fluids or gums and surface-treated fillers; elastic with good tensile strength	electrical insulation, medical devices, seals, textile coatings, foams

by the reaction of SiO_2 and C in an electric arc furnace, a process by which metallurgical-grade silicon is produced commercially in the 1990s. Ironically, in 1900 Oswalt announced that "elementary silicon has no application or any interesting properties" (10). Six years later, in what would prove to be an important discovery, Vigreaux treated SiF_6 with copper at 1000°C to give Cu_xSi. Alkoxy silanes were first prepared in the mid-1800s (11,12).

Kipping is considered the father of organosilicon chemistry. He employed Grignard reagents to make compounds of the general class $R_{4-x}SiCl_x$ (13,14). Kipping reportedly remarked that the polymers obtained from the hydrolysis of chlorosilanes were "omnipresent nuisances" (15). In the 1930s Stock made significant contributions by carrying out gas-phase reactions to prepare silicon hydrides and compounds having Si–Si bonds (16).

The beginning of the silicone industry has been well documented (15,17). In 1930, Corning Glass made the first deliberate effort to find a use for silicones, and in 1939 the company began the etherless methyl Grignard process as a method for making methyl silicones. One of the first commercial products, DC 4, was composed of silica in PDMS fluid and used in high altitude aircraft applications. Another early product was DC 990A resin, $C_6H_5(C_2H_5)Si(OH)_2$, which when cross-linked at high temperature had good dielectric strength, low dielectric loss, and good thermal stability.

In 1940 Rochow discovered the direct process, also called the methyl-chlorosilane (MCS) process, in which methyl chloride is passed over a bed of silicon and copper to produce a variety of methylchlorosilanes, including dimethyldichlorosilane [75-78-5], $(CH_3)_2SiCl_2$. Working independently, Müller made a similar discovery in Germany. Consequently, the process is frequently called the Rochow process and sometimes the Rochow-Müller reaction. These discoveries were followed by two key publications describing the work that marked the beginning of the commercial silicone industry (18,19). Production increased rapidly with the need for silicones in World War II. In 1943, the Dow Corning Corp. was formed in Midland, Michigan as a joint venture between Corning Glass and Dow Chemical. In 1947 GE opened a plant in Waterford, New York for manufacture of silicones, and in 1949 Union Carbide opened a silicone manufacturing plant in Tonawanda, New York.

Chemistry

The chemistry of silicones has been described in many textbooks (1,20,21).

Direct Process.   Passing methyl chloride through a fluidized bed of copper and silicon yields a mixture of chlorosilanes. The rate of methylchlorosilane (MCS) production and chemical selectivity, as determined by the ratio of dimethydichlorosilane to the other compounds formed, are significantly affected by trace elements in the catalyst bed; very pure copper and silicon gives poor yield and selectivity (22).

Much of the early work on the direct process was devoted to understanding the structure of the actual catalyst. Many reports suggest that Cu_3Si, or eta-phase, is a key catalytic species in the MCS reaction (9,23–26). Reactions employing Cu_3Si and Cu_5Si have been shown in the 1990s to be good models for the direct process (27). In a single-crystal study of the surface interaction of Si(100) with various forms of copper as models of the direct process, the highest selectivity obtained was for a mixture of 82% Cu and 18% Cu_2O, similar to the selectivity obtained in a fluidized-bed reactor (28).

In one of the first comprehensive studies on the effect of trace elements (28), the importance of zinc and tin, among others, as promoters was recognized. The effects of three promoters and their synergy have been investigated empirically (22). The results show that the reaction rate of the direct process decreases as a function of catalyst bed composition: SiCuZnSn > aged SiCu > SiCuAlSn ≅ SiCuZn > SiCuAl > fresh SiCu > SiCuZn. The selectivity for the formation of $(CH_3)_2SiCl_2$ is also dependent on catalyst composition and decreases in the following way: SiCuZnSn > SiCuAlSn ≅ SiCuZn > aged SiCu > SiCuSn > SiCuAl > fresh SiCu. This study concluded that zinc functions as a methylating agent, and that aluminum is an impurity in silicon and therefore contradictory results were obtained. Nevertheless, aluminum is a synergistic promoter with tin for rate and selectivity, and it appears that aluminum modifies the alloy properties of CuZnAlSi. Finally, tin is synergistic with zinc. The effect of aluminum on alloying properties has been discussed (29). However, more fundamental work remains to be done to understand the direct process at the molecular level. Zinc and tin surface enrichment appears to be essential for high activity and reactivity. Additives that impede silicon and/or zinc surface enrichment have a negative impact on rate, lead is a poison for this reason (30).

Alternatives to the methyl chloride direct process have been reviewed (31). Processes to make phenyl and ethyl silicones have employed direct-process chemistry. Phenyl chloride has been used in place of methyl chloride to make phenylchlorosilanes (15). In addition, phenylchlorosilanes are produced by the reaction of benzene, $HSiCl_3$, and BCl_3 (17,31). Ethylsilicones have been made primarily in the CIS, where the direct process is carried out with ethyl chloride in place of methyl chloride (32). Vinyl chloride can also be used in the direct process to produce vinylchlorosilanes (31). Alternative methods for making vinylchlorosilanes include reaction of vinyl chloride with $HSiCl_3$ or the platinum-catalyzed hydrosilylation of acetylene with $HSiCl_3$.

Some new direct-process-like reactions have been described. The reaction between methanol and Cu/Si gives $(CH_3O)_3SiH$ (eq. 1), and the reaction of methanol over Si/B gives $(CH_3O)_4Si$ (33,34).

$$3\ CH_3OH + Si \xrightarrow[\text{promoters}]{\text{Cu, fluidized-bed}} (CH_3O)_3SiH + H_2 \qquad (1)$$

The reaction of alcohols other than methanol has been examined under direct-process conditions as a method to make alkyl silicates (35). The use of HCl in the MCS reaction gas stream increases CH_3Cl_2SiH without decreasing the total combined yield of $(CH_3)_2SiCl_2$ and CH_3Cl_2SiH; no significant undesirable $SiCl_4$ or $HSiCl_3$ forms (36,37). Similar results are obtained by using hydrogen in place of HCl (38). Hydrogen in the direct-process stream enriches SiH monomers, but promoters normally used in the direct process usually retard SiH monomer formation. Zinc is particularly deleterious to CH_3HSiCl_2 formation (39,40). Other forms of direct siloxane processes include the reaction of dimethyl ether and methyl halide over Si/Cu to give $(CH_3)_2Si(OCH_3)_2$ (41,42). More recently the reaction of $(CH_3)_2NH$ and silicon to give $((CH_3)_2N)_3SiH$ has been reported (43).

Methylchlorodisilanes are by-products of the direct-process residue, commonly called high boiling point residue or simply residue, and are formed in about 4% of the total $(CH_3)_2SiCl_2$ produced, which in 1994 was about 30,000 tons per year. Disilanes are key constituents of the residue, and novel reactions forming Si–Cl bonds have been described (44,45). Some chemical reactions of direct-process disilanes are shown in Figure 2. Cleavage chemistry of Si–Si compounds with HCl practiced industrially has also been described (47).

Cleavage reactions:

$$Cl_2CH_3Si-SiCl(CH_3)_2 + HCl \xrightarrow{(C_4H_9)_3NHCl} Cl_2CH_3SiH + Cl_2Si(CH_3)_2 \qquad (i)$$

Disproportionation:

$$n\ Cl_2CH_3Si-SiCl(CH_3)_2 \xrightarrow{\text{base catalyst}} n\ CH_3SiCl_3 + \underset{Cl}{\overset{CH_3}{-(\!-Si\!-\!)_n}} \qquad (ii)$$

Si–C-bond-forming reactions:

$$Cl_2CH_3Si-SiCl(CH_3)_2 + R-C\equiv C-R \xrightarrow{Pt} Cl_2CH_3Si\underset{CR=CR}{\diagdown\diagup}SiCl(CH_3)_2 \qquad (iii)$$

$$Cl(CH_3)_2Si-Si(CH_3)_2Cl + CH_2=CH_2 \longrightarrow Cl(CH_3)_2SiCH_2CH_2Si(CH_3)_2Cl \qquad (iv)$$

$$ArCOCl + Cl_2CH_3Si-SiCl(CH_3)_2 \xrightarrow{Pd} ArSiCl(CH_3)_2 + Cl_3SiCH_3 + CO \qquad (v)$$

Fig. 2. Chemistry of direct-process residue methylchlorodisilanes. The reactions are discussed in detail in the following References: (i), 46; (ii), 44,45; (iii), 47; (iv), 48; and (v), 49,50.

A significant use of direct-process waste is realized by C_6H_5-Si bond formation via silylative decarbonylation (Fig. 2) (49,50). A novel route to C_6H_5-Si bond formation has also been described (eq. 2) (51).

Synthesis of Silicone Monomers and Intermediates.   Another important reaction for the formation of $Si-C$ bonds, in addition to the direct process and the Grignard reaction, is hydrosilylation (eq. 3), which is used for the formation of monomers for producing a wide range of organomodified silicones and for cross-linking silicone polymers (8,52–58). Formation of ether and ester bonds at silicon is important for the manufacture of curable silicone materials. Alcoholysis of the $Si-Cl$ bond (eq. 4) is a method for forming silyl ethers. HCl removal is typically accomplished by the addition of tertiary amines or by using $NaOR'$ in place of $R'OH$ to form NaCl.

$$R_3SiH + R'CH = CH_2 \xrightarrow{\text{catalyst}} R_3SiCH_2CH_2R' \qquad (3)$$

$$RSiCl_3 + 3\ R'OH \longrightarrow RSi(OR')_3 + 3\ HCl \qquad (4)$$

An alternative to alcohol or metal alkoxide is the use of orthoformate (eq. 5). Generally, higher alkoxy groups hydrolyze more slowly than lower alkoxy groups. Alkoxy exchange, ie, replacing R' with R'' in R_3SiOR', requires a catalyst such as mineral acid, Lewis acid (alkyl titanate), or strong base (eq. 6).

$$R_3SiCl + HC(OR')_3 \xrightarrow{\text{ROH}} R_3SiOR' + R'Cl + HCOOR' \qquad (5)$$

$$H_2N(CH_2)_3Si(OCH_3)_3 + 3\ ROH \xrightarrow{\text{NaOR}} H_2N(CH_2)_3Si(OR)_3 + 3\ CH_3OH \qquad (6)$$

An important end group in silicone chemistry is the acetoxy group; the familiar silicone sealants release acetic acid during moisture cure of these acetoxy-stopped polymers. Acetoxysilanes hydrolyze more readily than alkoxy groups. Acylation of a chlorosilane can be accomplished by the addition of sodium acetate or by reaction with acetic anhydride. Other reactions that permit formation of organofunctional silicones are shown in Figure 3.

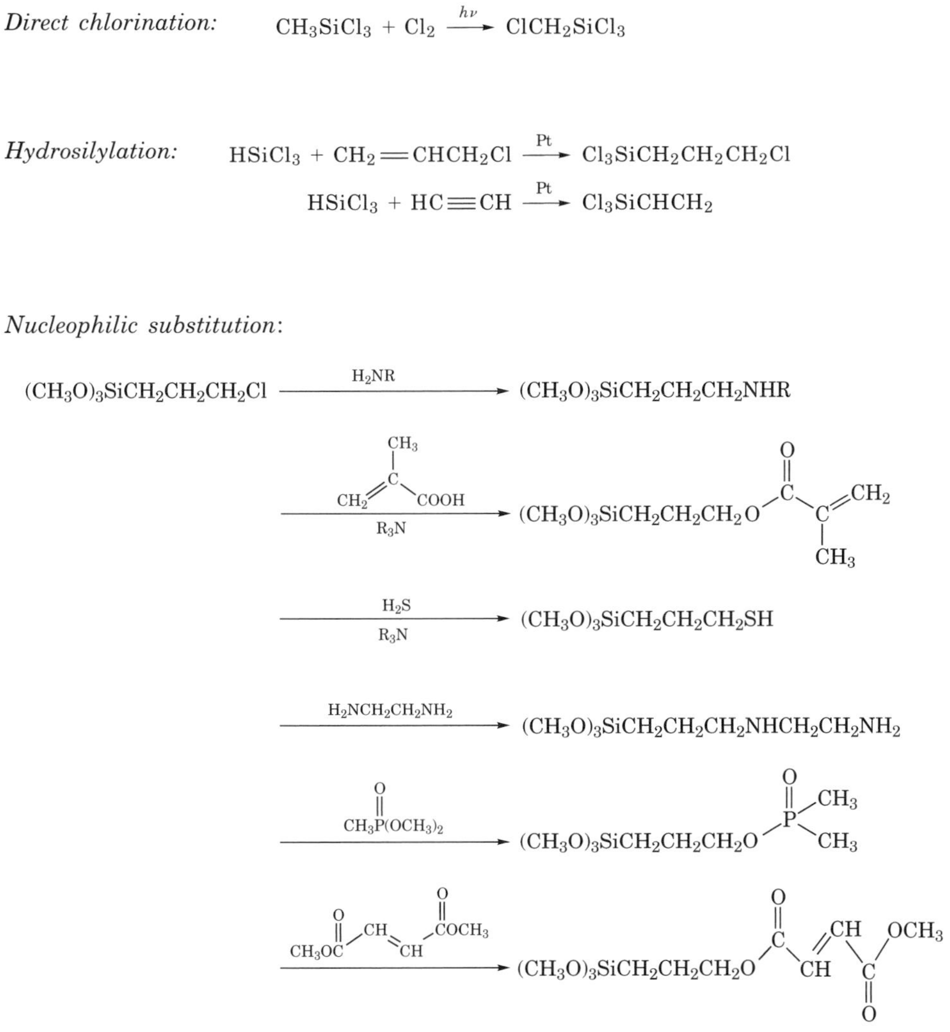

Direct chlorination: $CH_3SiCl_3 + Cl_2 \xrightarrow{h\nu} ClCH_2SiCl_3$

Hydrosilylation: $HSiCl_3 + CH_2{=}CHCH_2Cl \xrightarrow{Pt} Cl_3SiCH_2CH_2CH_2Cl$

$HSiCl_3 + HC{\equiv}CH \xrightarrow{Pt} Cl_3SiCHCH_2$

Nucleophilic substitution:

Fig. 3. Synthetic routes to some common organofunctional silanes.

Polymerization

The manufacture of polydimethylsiloxane polymers is a multistep process. The hydrolysis of the chlorosilanes obtained from the direction process yields a mixture of cyclic and linear silanol-stopped oligomers, called hydrolysate (eq. 7) (21). In some cases, chloro-stopped polymers can also be obtained (59).

$$\underset{\overset{|}{CH_3}}{\overset{\overset{CH_3}{|}}{Cl-Si-Cl}} \xrightarrow[-HCl]{+H_2O} \left[\underset{\overset{|}{CH_3}}{\overset{\overset{CH_3}{|}}{HO-Si-OH}}\right] \xrightarrow{-H_2O} \underset{\overset{|}{CH_3}}{\overset{\overset{CH_3}{|}}{\left(\!\!-Si-O-\right)_{\!\!m}}} + \underset{\overset{|}{CH_3}}{\overset{\overset{CH_3}{|}}{HO-Si-O}}\underset{\overset{|}{CH_3}}{\overset{\overset{CH_3}{|}}{\left(\!\!-Si-O-\right)_{\!\!n}}}\underset{\overset{|}{CH_3}}{\overset{\overset{CH_3}{|}}{Si-OH}}$$

$$(7)$$

The ratio of cyclic to linear oligomers, as well as the chain length of the linear siloxanes, is controlled by the conditions of hydrolysis, such as the ratio of chlorosilane to water, temperature, contact time, and solvents (60,61). Commercially, hydrolysis of dimethyldichlorosilane is performed by either batch or a continuous process (62). In the typical industrial operation, the dimethyldichlorosilane is mixed with 22% azeotropic aqueous hydrochloric acid in a continuous reactor. The mixture of hydrolysate and 32% concentrated acid is separated in a decanter. After separation, the anhydrous hydrogen chloride is converted to methyl chloride, which is then reused in the direct process. The hydrolysate is washed for removal of residual acid, neutralized, dried, and filtered (63). The typical yield of cyclic oligomers is between 35 and 50%. The mixture of cyclic oligomers consists mainly of tetramer and pentamer. Only a small amount of cyclic trimer is formed.

The complete conversion of the dimethyldichlorosilane to only linear oligomers is also possible in the continuous hydrolysis operation. In this operation, the cyclics are separated from linear oligomers by a stripping process and are mixed again with dimethyldichlorosilane. This mixture undergoes equilibration to chloro-terminated oligomers and is subsequently hydrolyzed. The silanol-stopped linear oligomers are directly used in the manufacture of silicone polymers.

Dimethyldichlorosilane can also be converted into siloxane silanol-stopped oligomers by a methanolysis process (eq. 8).

$$\underset{\overset{|}{CH_3}}{\overset{\overset{CH_3}{|}}{Cl-Si-Cl}} \xrightarrow[-CH_3Cl]{+CH_3OH} \underset{\overset{|}{CH_3}}{\overset{\overset{CH_3}{|}}{\left(\!\!-Si-O-\right)_{\!\!m}}} + \underset{\overset{|}{CH_3}}{\overset{\overset{CH_3}{|}}{HO-Si-O}}\underset{\overset{|}{CH_3}}{\overset{\overset{CH_3}{|}}{\left(\!\!-Si-O-\right)_{\!\!n}}}\underset{\overset{|}{CH_3}}{\overset{\overset{CH_3}{|}}{Si-OH}} \qquad (8)$$

In contrast to the hydrolysis technology, the methanolysis process allows for the one-step synthesis of organosiloxane oligomers and methyl chloride without formation of hydrochloric acid (64,65). The continuous methanolysis can also yield quantitatively linear silanol-stopped oligomers by recycle of the cyclic fraction into the hydrolysis loop.

If the linear fraction of siloxane oligomers is used directly in the manufacture of silicone polymers, extremely pure (greater than 99.99%) dimethyldichlorosilane is required. A higher content of methyltrichlorosilane can produce significant amounts of trifunctional units and considerably affect the physical properties of the final products. If such high purity dimethyldichlorosilane is not achieved, an additional step, called cracking, must be included in the production scheme (66).

During the cracking process, the hydrolyzate is depolymerized in the presence of strong base or acids to yield cyclic monomers, primarily octamethylcyclotetrasiloxane (D_4) and decamethylcyclopentasiloxane (D_5), which are distilled from the reaction mixture. The trifunctional by-products remain in the pot and are periodically removed.

Polycondensation. The linear fraction of hydrolysate, ie, oligosiloxane-α,ω-diols whose viscosity is from 10 to 100 mPa (=cP), is converted further to silicone fluids and high molecular weight gums by polycondensation of the silanol end groups (eq. 9). Polycondensation is an equilibrium process. In spite of the relatively high equilibrium constant for this reaction (K at 35°C = 860 ± 90), water removal is required to obtain high molecular weight siloxane polymers (67).

$$\underset{\substack{\text{CH}_3 \\ | \\ \text{H}_3\text{C}-\overset{|}{\underset{|}{\text{Si}}}-\text{OH} \\ | \\ \text{CH}_3}}{} + \underset{\substack{\text{CH}_3 \\ | \\ \text{HO}-\overset{|}{\underset{|}{\text{Si}}}-\text{CH}_3 \\ | \\ \text{CH}_3}}{} \underset{\text{catalyst}}{\overset{K_{eq}}{\rightleftharpoons}} \underset{\substack{\text{CH}_3 \quad\ \text{CH}_3 \\ | \qquad\ | \\ \text{H}_3\text{C}-\overset{|}{\underset{|}{\text{Si}}}-\text{O}-\overset{|}{\underset{|}{\text{Si}}}-\text{CH}_3 \\ | \qquad\ | \\ \text{CH}_3 \quad\ \text{CH}_3}}{} + \text{H}_2\text{O} \qquad (9)$$

Although the low molecular weight silanols such as trimethylsilanol or dimethylsilanediol undergo condensation thermally, the higher molecular weight oligomers are much more stable and their polycondensation must be catalyzed. Many catalytic systems capable of promoting polycondensation of siloxanediols have been described in patents and open literature (59,68,69). These catalysts include strong acids such as HCl, HBr, H_2SO_4, $HClO_4$, and CF_3SO_3H; phosphonitrilic chlorides such as $(Cl_3PN(PCl_2N)_nPCl_3)^+PCl_6^-$; strong bases such as KOH, NaOH, and $(CH_3)_4NOH$; as well as amines, amine salts of carboxylic acids, ion exchange resin, and clays activated with mineral acids. One of the most efficient and extensively used in industry is the catalytic system based on phosphonitrilic chlorides (70–72).

The first mechanistic studies of silanol polycondensation on the monomer level were performed in the 1950s (73–75). The condensation of dimethylsiloxanediol in dioxane exhibits second-order kinetics with respect to diol and first-order kinetics with respect to acid. The proposed mechanism involves the protonation of the silanol group and subsequent nucleophilic substitution at the silicone (eqs. 10 and 11).

$$\underset{\substack{\text{CH}_3 \\ | \\ \text{H}_3\text{C}-\overset{|}{\underset{|}{\text{Si}}}-\text{OH} \\ | \\ \text{CH}_3}}{} + \text{HA} \rightleftharpoons \underset{\substack{\text{CH}_3 \\ | \\ \text{H}_3\text{C}-\overset{|}{\underset{|}{\text{Si}}}-\text{OH}_2^+\text{A}^- \\ | \\ \text{CH}_3}}{} \qquad (10)$$

$$H_3C-\underset{\underset{\displaystyle CH_3}{|}}{\overset{\overset{\displaystyle CH_3}{|}}{Si}}-OH_2{}^+ \ A^- \ + HO-\underset{\underset{\displaystyle CH_3}{|}}{\overset{\overset{\displaystyle CH_3}{|}}{Si}}-CH_3 \ \rightleftharpoons \ H_3C-\underset{\underset{\displaystyle CH_3}{|}}{\overset{\overset{\displaystyle CH_3}{|}}{Si}}-O-\underset{\underset{\displaystyle CH_3}{|}}{\overset{\overset{\displaystyle CH_3}{|}}{Si}}-CH_3 \ +H_2O \ + HA \quad (11)$$

The condensation catalyzed by a strong base is first order with respect to substrate and catalyst (74,75). Because of the high acidity of silanol, all the alkali metal base (MtOH) is usually transformed into the silanolate anion. In the rate-determining step, the silanolate anion attacks the silicon atom in the silanol end group (eq. 12 and 13).

$$H_3C-\underset{\underset{\displaystyle CH_3}{|}}{\overset{\overset{\displaystyle CH_3}{|}}{Si}}-OH \ + MtOH \ \rightleftharpoons \ H_3C-\underset{\underset{\displaystyle CH_3}{|}}{\overset{\overset{\displaystyle CH_3}{|}}{Si}}-O^-Mt^+ \ + H_2O \quad (12)$$

$$H_3C-\underset{\underset{\displaystyle CH_3}{|}}{\overset{\overset{\displaystyle CH_3}{|}}{Si}}-O^-Mt^+ + HO-\underset{\underset{\displaystyle CH_3}{|}}{\overset{\overset{\displaystyle CH_3}{|}}{Si}}-CH_3 \ \rightleftharpoons \ H_3C-\underset{\underset{\displaystyle CH_3}{|}}{\overset{\overset{\displaystyle CH_3}{|}}{Si}}-O-\underset{\underset{\displaystyle CH_3}{|}}{\overset{\overset{\displaystyle CH_3}{|}}{Si}}-CH_3 \ + MtOH \quad (13)$$

Early studies of the condensation reaction on the monomer level did not give the full picture of this process and only in the 1980s was polycondensation of siloxanols studied by using oligomeric model compounds (76,77). These studies revealed that in the presence of strong protic acids three processes must be considered: linear condensation (eq. 14), cyclization (eq. 15), and disproportionation (eq. 16).

$$HO\!-\!\!\left(\!\underset{\underset{\displaystyle CH_3}{|}}{\overset{\overset{\displaystyle CH_3}{|}}{Si}}\!-\!O\!\right)_{\!\!n}\!\!H \ + H_2O \quad (14)$$

linear condensation

$$HO\!-\!\!\left(\!\underset{\underset{\displaystyle CH_3}{|}}{\overset{\overset{\displaystyle CH_3}{|}}{Si}}\!-\!O\!\right)_{\!\!n}\!\!H \ \xrightarrow{\text{catalyst}}$$

cyclization

$$\left[\!\underset{\underset{\displaystyle CH_3}{|}}{\overset{\overset{\displaystyle CH_3}{|}}{Si}}\!-\!O\!\right]_{\!\!n} \ + H_2O \quad (15)$$

disproportionation

$$HO\!-\!\!\left(\!\underset{\underset{\displaystyle CH_3}{|}}{\overset{\overset{\displaystyle CH_3}{|}}{Si}}\!-\!O\!\right)_{\!\!n-1}\!\!H \ + HO\!-\!\!\left(\!\underset{\underset{\displaystyle CH_3}{|}}{\overset{\overset{\displaystyle CH_3}{|}}{Si}}\!-\!O\!\right)_{\!\!n+1}\!\!H \quad (16)$$

The relative contributions from these processes strongly depend on the re-action conditions, such as type of solvent, substrate and water concentration, and acidity of catalyst (78,79). It was also discovered that in acid–base inert solvents, such as methylene chloride, the basic assistance required for the condensation process is provided by another silanol group. This phenomena, called intra–inter catalysis, controls the linear-to-cyclic products ratio, which is constant at a wide range of substrate concentrations.

The behavior of oligosiloxanediols in the presence of strong bases is differ-ent. The contribution to the overall process of the disproportionation reaction, involving a migration of the ultimate siloxane unit between siloxane molecules, is much greater and may even completely dominate the polycondensation reaction (80). The reactivity enhancement of the siloxane bond adjacent to the silanolate anion can be understood in terms of $n(0) \rightarrow \sigma^*(SiO)$ conjugation.

Ring-Opening Polymerization.   Ring-opening polymerization of cyclic oligosiloxanes is an alternative to the polycondensation method of manufac-turing siloxane polymers. Commercially, the polymerization of unstrained oc-tamethyltetracyclosiloxane (D_4) is the most important (59,68). In the presence of catalysts such as strong acids or bases, D_4 undergoes equilibrium polymer-ization, which results in a mixture of high molecular weight polymer and low molecular weight cyclic oligomers.

The position of the equilibrium depends on a number of factors, such as concentration of siloxane units and the nature of substituents on the silicon, but is independent of the starting siloxane composition and the polymerization con-ditions (81,82). For a bulk polymerization of dimethylsiloxane, the equilibrium concentration of cyclic oligomers is approximately 18 wt % (83). The equilibrium mixture of cyclosiloxanes is composed of a continuous population to at least D_{400}, but D_4, D_5, and D_6 make over 95 wt % of the total cyclic fraction (84).

The ring-opening polymerization of D_4 is controlled by entropy, because thermodynamically all bonds in the monomer and polymer are approximately the same (21). The molar cyclization equilibrium constants of dimethylsiloxane rings have been predicted by the Jacobson-Stockmayer theory (85). The ring–chain equilibrium for siloxane polymers has been studied in detail and is the subject of several reviews (82,83,86–89). The equilibrium constant of the formation of each cyclic is approximately equal to the equilibrium concentration of this cyclic, $K_n \approx [(SiR_2O)_n]_{eq}$. Thus the total concentration of cyclic oligomers in the equilibrium is independent of the initial monomer concentration. As a consequence, the amount of linear polymer decreases until the critical dilution point is reached, at which point only cyclic products are formed.

Anionic Polymerization of Cyclic Siloxanes.   The anionic polymerization of cyclosiloxanes can be performed in the presence of a wide variety of strong bases such as hydroxides, alcoholates, or silanolates of alkali metals (59,68).

Commercially, the most important catalyst is potassium silanolate. The activity of the alkali metal hydroxides increases in the following sequence: LiOH < NaOH < KOH < CsOH, which is also the order in which the degree of ionization of their hydroxides increases (90). Another important class of catalysts is tetraalkyl ammonium or phosphonium hydroxides or silanolates (91–93). These catalysts undergo thermal degradation when the polymer is heated above the temperature required (typically >150°C) to decompose the catalyst, giving volatile products and the neutral, thermally stable polymer.

The manufacture of silicone polymers via anionic polymerization is widely used in the silicone industry. The anionic polymerization of cyclic siloxanes can be conducted in a single-batch reactor or in a continuously stirred reactor (94,95). The viscosity of the polymer and type of end groups are easily controlled by the amount of added water or triorganosilyl chain-terminating groups.

The mechanism of anionic polymerization of cyclosiloxanes has been the subject of several studies (96,97). The first kinetic analysis in this area was carried out in the early 1950s (98). In the general scheme of this process, the propagation/depropagation step involves the nucleophilic attack of the silanolate anion on the silicon, which results in the cleavage of the siloxane bond and formation of the new silanolate active center (eq. 17).

$$
\begin{array}{c}
\text{CH}_3 \\
| \\
\text{\textasciitilde Si{-}O}^-\text{Mt}^+ + \\
| \\
\text{CH}_3
\end{array}
\quad
\begin{array}{c}
\text{H}_3\text{C} \quad \text{CH}_3 \\
\diagdown \diagup \\
\text{Si} \\
\text{O} \; (\\
\diagdown \\
\text{Si} \\
\diagup \diagdown \\
\text{H}_3\text{C} \quad \text{CH}_3
\end{array}
\quad \rightleftharpoons \quad
\begin{array}{c}
\text{CH}_3 \quad \text{CH}_3 \; \text{CH}_3 \\
| \quad\quad | \quad\quad | \\
\text{\textasciitilde Si{-}O{-}Si{-}Si{-}O}^-\text{Mt}^+ \\
| \quad\quad | \quad\quad | \\
\text{CH}_3 \quad \text{CH}_3 \; \text{CH}_3
\end{array}
\qquad (17)
$$

The kinetics of this process is strongly affected by an association phenomenon. It has been known that the active center is the silanolate ion pair, which is in equilibrium with dormant ion pair complexes (99,100). The polymerization of cyclosiloxanes in the presence of potassium silanolate shows the kinetic order 0.5 with respect to the initiator, which suggests the principal role of dimer complexes (101).

The search for new, high performance materials requires the synthesis of well-defined, narrow molecular weight distribution, cyclic-free, homo- and copolymers. Synthesis of these polymers can be accomplished by the kinetically controlled polymerization of the strained monomer, hexaalkylcyclotrisiloxanes. In the presence of the proper initiator and under the right reaction conditions, the polymerization of hexamethylcyclotrisiloxane (D$_3$) can proceed as a classical living polymerization. The most frequently used initiator that ensures fast and quantitative initiation and propagation, free of depolymerization and chain-scrambling processes, is lithium silanolate (102,103). The rate of D$_3$ polymerization in the presence of lithium silanolate is slow. To accelerate the polymerization process, a cation-interacting solvent such as THF is commonly employed (104,105). Instead of tetrahydrofuran (THF), some nucleophilic additives such as hexamethyl phosphoramide (HMPT) or dimethyl sulfoxide (DMSO), as well as chelating agents such as cryptates, can also be used (106–111). These additives interact strongly with the lithium cation and convert the intimate ion pair into a more reactive, separated one.

Cationic Polymerization of Cyclic Siloxanes. The cationic polymerization of cyclic siloxanes is often a preferred method for the synthesis of siloxane polymers. This process, which can be performed at relatively low temperature, can be applied to the synthesis of polysiloxanes having base-sensitive substituent such as SiH, and the catalyst can be easily deactivated. The first high molecular weight siloxane polymer was prepared by the cationic ring-opening polymerization of D_4 in the presence of sulfuric acid (112). Since that time many catalytic systems have been reported in patents and open literature, including strong protic acids such as CF_3SO_3H (triflic acid), $HClO_4$, H_2SO_4, aryl- and alkylsulfonic acids, heterogeneous catalysts such as ion-exchange resins, acid-treated graphite, and acid-treated clays, as well as some Lewis acids such as $SnCl_4$ (113–125). Polymerization in the presence of Lewis acids is a subject of controversy. Strong protic acids such as $HSnCl_5$, the product of the reaction of Lewis acid with water or other protic impurities, are proposed in most cases as the true catalyst (126). However, it has been reported that some nonprotic systems are able to initiate polymerization of cyclosiloxanes (127).

Despite a long history of commercial application, the mechanism of cationic polymerization of cyclic siloxanes is not as well understood as the anionic process. The polymerization of D_3 and D_4 in the presence of CF_3SO_3H has been a subject of intensive studies (113,128–132). Several unusual kinetic results have been observed, such as the apparent negative order in the monomer, negative activation energy, or a strong effect of water on the kinetics of polymerization (131–135). Although still subject to controversy as of this writing (1996), a complex mechanism of acid-catalyzed polymerization has been proposed (97,135,136). The mechanism consists of four processes: initiation-ring opening, step growth by homo- and heterofunctional condensation, chain propagation by direct monomer addition, and end group interconversion. Hydrogen bond association plays a crucial role in this process (97,135,136). In spite of these extensive studies, many unanswered questions remain, such as the structure of the active propagation center.

Emulsion Polymerization. Even though siloxane bond formation is an equilibrium process, it is possible to form siloxane polymers by polycondensation or ring-opening polymerization in aqueous emulsions (137–139). D_4 can be converted into high molecular weight polymer by emulsion polymerization in the presence of dodecylbenzenesulfonic acid [27176-87-0] (DBSA), which acts as both emulsifying surfactant and catalyst (140). It is also possible to obtain high molecular weight polymers by polycondensation of α,ω-dihydroxy-stopped oligosiloxanes in an aqueous emulsion (141,142) employing DBSA as the surface-active catalyst. The polycondensation involves the reaction of a complex of silanol with two molecules of DBSA with another silanol end group (143). The anionic emulsion polymerization of D_4 in the presence of benzyldimethyldodecylammonium hydroxide as a surface-active catalyst has been reported (144) and a reaction scheme proposed, which involves a quaternary ammonium silanolate as an active species.

Radiation-Induced Polymerization. In 1956 it was discovered that D_3 can be polymerized in the solid state by γ-irradiation (145). Since that time a number of papers have reported radiation-induced polymerization of D_3 and D_4 in the solid state (146,147). The first successful polymerization of cyclic siloxanes in the liquid state (148) and later work (149) showed that the polymerization of cyclic siloxanes induced by γ-irradiation has a cationic nature. The polymerization is initiated by a cleavage of Si–C bond and formation of silylenium cation.

The γ-radiation-induced polymerization requires an extremely high purity reaction system. Trace amounts of water can terminate a cationic reaction and inhibit polymerization. Organic bases such as ammonia and trimethylamine also inhibit polymerization. The γ-radiation-induced polymerization of a rigorously dried D_4 obeys the Hayashi-Williams equation for completely pure systems (150).

Plasma Polymerization.   A need for well-defined, thin polymer films for applications in optics, electronics, or biomedicine stimulated the development of plasma-induced polymerization during the 1980s and early 1990s. The special interest in organosilicon monomers may be because many of these monomers are sufficiently volatile, nontoxic, nonflammable, and relatively inexpensive. Also, plasma-polymerized organosilicone films have the natural chemical affinity of a single-crystalline silicone, and the properties of these films can be varied widely by the choice of monomer and polymerization parameters (151). The mechanism of plasma polymerization is still not well understood as of this writing and differs substantially from the conventional ring-opening polymerization of cyclic monomers. The dissociation and ionization of organosilicone monomers by low energy electron impact is the dominant source of radical fragments and a variety of ion–molecules and ion–radicals. In the case of hexamethyldisiloxane, extension of the monomer molecule by one dimethylsiloxane unit and elimination of the trimethylsilenium ion was proposed as a crucial step in the polymer film formation. The building of Si–C–Si links also plays an important role. Cyclic monomers such as D_4 polymerize via formation of a network of siloxane chains and rings. The resulting films are usually hard and scratch-resistant but often show considerable stress (152). Such films can be used as corrosion-resistant coatings on metals or as membranes in gas sensor devices (153,154)

Silicone Network Formation

Silicone rubber has a three-dimensional network structure caused by cross-linking of polydimethylsiloxane chains. Three reaction types are predominantly employed for the formation of silicone networks (155): peroxide-induced free-radical processes, hydrosilylation addition cure, and condensation cure. Silicones have also been cross-linked using radiation to produce free radicals or to induce photoinitiated reactions.

Peroxide Cure. The use of peroxide catalysts at elevated temperatures is one of the most common methods for the preparation of silicone networks (156). Typical peroxides include the aroyl chlorides such as dibenzoyl (157), bis-*p*-chlorobenzoyl, and bis-2,4-dichlorobenzoyl peroxides; alkylaroyl peroxides such as dicumyl peroxide; and dialkyl peroxides such as di-*t*-butyl peroxide and 2,5-dimethyl-2,5-di-*t*-butylperoxyhexane (see PEROXY COMPOUNDS, ORGANIC). The amount and type of peroxide used determine the cure temperature of the composition as well as the ultimate properties. Diaroyl peroxides can be used with either polydimethysiloxanes or with vinylmethylpolysiloxanes. Bis-2,4-dichlorobenzoyl peroxide has the highest rate of decomposition and the lowest decomposition temperature. Cross-linking by peroxides occurs by the generation of free radicals via homolytic cleavage of the peroxide at elevated temperatures. The peroxy radicals can abstract hydrogen atoms from methyl groups, forming ethylenic linkages between siloxane chains (158). Model studies have confirmed the formation of $SiCH_2CH_2Si$ as the predominant cross-linking reaction (159).

$$ArOOAr \longrightarrow 2\ ArO\cdot$$

$$ArO\cdot + CH_3Si \longrightarrow ArOH + \cdot CH_2Si$$

$$\cdot CH_2Si + \cdot CH_2Si \longrightarrow SiCH_2CH_2Si$$

Polymerization of methylvinylpolysiloxanes can be activated using alkylaroyl or dialkyl peroxides (59,160). A radical can be formed at either the methyl or vinyl site; it is generally agreed that the reaction through the vinyl group is energetically favored. Several mechanisms have been proposed to explain the reaction. One suggestion is that the radical is first formed on a methyl group which can then attack either a methyl group or a vinyl group (161). Another proposal is that the cross-link occurs primarily through the vinyl group (162). The first step is addition of the peroxy radical to the vinyl moiety (163). The resultant radical can then attack either another vinyl group or a methyl group. Termination can occur by coupling or by hydrogen abstraction from adjacent peroxide molecules. Rheometric studies indicate that methylvinylsilicone resins do undergo cross-linking via the methyl groups but at a slower rate than through the vinyl groups (164).

Hydrosilylation or Addition Cure. Hydrosilylation, the reaction between a silicon hydride and a vinyl group to form an ethylenic linkage (eq. 3), is extremely important for the formation of silicone networks (52,58,165,166). Cross-linking occurs between polymers containing multifunctional hydride and vinyl groups. No by-products are produced by this cure mechanism. Addition to the olefin may be either Markownikoff or anti-Markownikoff, depending on the olefin. Hydrosilylation is generally catalyzed by Group 8–10 (VIII) metal complexes or supported metal catalysts. Platinum compounds are the most widely used (52,58). Soluble platinum olefin catalysts such as those typically used in industry have high turnover rates and are useful in concentrations as low as 1–2 ppm (167). Rhodium(I) complexes have been reported to have similar reactivity to soluble platinum systems (168).

Several side reactions or post-curing reactions are possible. Disproportionation reactions involving terminal hydride groups have been reported (169). Excess SiH may undergo hydrolysis and further reaction between silanols can occur

(170–172). Isomerization of a terminal olefin to a less reactive internal olefin has been noted (169). Vinylsilane/hydride interchange reactions have been observed (165).

Several mechanisms have been proposed for the homogeneous hydrosilylation reaction. The Chalk-Harrod mechanism (173) contains elementary steps reminiscent of homogeneous hydrogenation with successive oxidative addition and reductive elimination steps. A mechanism has been proposed based on the intermediacy of colloidal species (174,175). Colloids were detected by transmission electron microscopy after evaporation of actual catalytic reactions. Extended x-ray absorption fine-structure (EXAFS) studies have shown two platinum-containing intermediates present during hydrosilylation (176). Under conditions of high vinyl concentrations, a mononuclear platinum species containing six Pt–C bonds, such as platinum coordinated to three vinyl groups, is present, but in the presence of high concentrations of silicon–hydride or with poorly coordinating olefins such as hexenes, a multinuclear platinum species with silicon ligands is the predominant species. The two platinum species can interconvert. Colloids were found in a post-mortem analysis upon evaporation or destabilization of the reactant solutions. A mechanism has been proposed that is similar to the Chalk-Harrod mechanism with the addition of several competing equilibriums and colloid formation at the end of the reaction.

$$R_3SiH + R'H_2C\!=\!CHR' \xrightarrow{\text{Pt}} \begin{array}{c} \xrightarrow{\text{high vinyl}} Pt(H_2C\!=\!CHR')_x \\ \updownarrow \\ \xrightarrow{\text{high SiH}} Pt_x(R_3Si)_y \end{array}$$

Inhibitors are often included in formulations to increase the pot life and cure temperature so that coatings or moldings can be conveniently prepared. An ideal silicone addition cure may combine instant cure at elevated temperature with infinite pot life at ambient conditions. Unfortunately, real systems always deviate from this ideal situation. A proposed mechanism for inhibitor (I) function is an equilibrium involving the inhibitor, catalyst ligands (L), the silicone–hydride groups, and the silicone vinyl groups (177).

$$PtL_x + n\,I \rightleftharpoons PtL_{x-n}I_n$$

R_3SiH and $CH_2\!=\!CHR'$ interact with both PtL_x and $PtL_{x-n}I_n$. Complexing or chelating ligands such as phosphines and sulfur complexes are excellent inhibitors, but often form such stable complexes that they act as poisons and prevent cure even at elevated temperatures. Unsaturated organic compounds are preferred, such as acetylenic alcohols, acetylene dicarboxylates, maleates, fumarates, eneynes, and azo compounds (178–189). An alternative concept has been the encapsulation of the platinum catalysts with either cyclodextrin or in thermoplastics or silicones (190–192).

Condensation Cure. The condensation of silanol groups to form siloxanes is an extremely important industrial reaction and may be represented in its simplest form as follows:

$$SiOH + SiOH \longrightarrow SiOSi + H_2O$$

Unfortunately, because self-condensation of silanols on the same silicone can occur almost spontaneously, the reaction of disilanol or trisilanol compounds with telechelic silanol polymers to form a three-dimensional network is not feasible. Instead, the telechelic polymers react with cross-linkers containing reactive groups such as alkoxysilanes, acyloxysilanes, silicon hydrides, or methylethyloximesilanes, as in the reactions in equations 18–21 (155).

$$SiOH + Si(OCH_3) \longrightarrow SiOSi + CH_3OH \tag{18}$$

$$SiOH + Si(OOCCH_3) \longrightarrow SiOSi + CH_3COOH \tag{19}$$

$$SiOH + SiH \longrightarrow SiOSi + H_2 \tag{20}$$

$$SiOH + Si(ON\!=\!\underset{\underset{CH_3}{|}}{C}CH_2CH_3) \longrightarrow SiOSi + HON\!=\!\underset{\underset{CH_3}{|}}{C}CH_2CH_3 \tag{21}$$

Condensation catalysts include both acids and bases, as well as organic compounds of metals. Both tin(II) and tin(IV) complexes with carboxylic acids are extremely useful. It has been suggested that the tin catalyst is converted to its active form by partial hydrolysis followed by reaction with the hydrolyzable silane to yield a tin–silanolate species (eqs. 22 and 23) (193,194).

$$R_2Sn(OCOR')_2 + H_2O \longrightarrow R_2Sn(OCOR')OH + R'COOH \tag{22}$$

$$R_2Sn(OCOR')OH + Si(OR)_4 \longrightarrow R_2Sn(OCOR')(OSi(OR)_3 + ROH \tag{23}$$

The organotin silanolate can then react with the polydimethylsiloxane diol by either attack on the SiOC bond or by silanolysis of the SnOC bond (193,194). Other metal catalysts include chelated salts of titanium and tetraalkoxytitanates. Formation of a cross-linked matrix involves a combination of the three steps in equations 24–26.

$$Si(OR) + H_2O \longrightarrow SiOH + ROH \tag{24}$$

$$SiOH + SiOH \longrightarrow SiOSi + H_2O \tag{25}$$

$$Si(OR) + SiOH \longrightarrow SiOSi + ROH \tag{26}$$

Relative hydrolysis and condensation rate studies of multifunctional silanes, $Si(OR)_x$, under acidic and basic catalysis showed that the first (OR) group hydrolyzes much more readily than subsequent groups (195). Silanol–silanol condensation is much slower than silanol–alkoxysilane condensation, even if the alkoxysilane is monofunctional, thus suggesting that chain extension is insignificant in the presence of a cross-linker (196–199).

Condensation cure can also be carried out in emulsions (200–209). In this case, the cross-linker and polydimethylsiloxanediol are emulsified using anionic, cationic, or nonionic surfactants in water, and a condensation catalyst such as dibutyltin dilaurate is added. The polymer can then undergo cross-linking, forming a continuous film when the water is evaporated.

High Energy Radiation Cure.   Silicones can be vulcanized by irradiation. The energy sources include radioactive cobalt (^{60}Co), Van de Graff generators, and resonance transformers (210). The properties of vulcanized silicones cross-linked by high energy irradiation are indistinguishable from those catalyzed with peroxides in the unaged state (211). But silicones cross-linked by irradiation have two advantages: better resistance to aging in humid environments and hydrolytic depolymerization at elevated temperatures because the compositions are catalyst-free (212).

Radiation cure of methylvinyl silicone rubber results in a random cross-linking process with no differentiation between methyl and vinyl groups (213). Substitution of phenyl for methyl groups on the siloxane increases the radiation resistance. Silicone hydride groups are highly sensitive to radiation (214).

Photoinitiated Radiation Cure.   Attachment of photopolymerizable organic groups to a polysiloxane backbone provides a convenient means to utilize uv radiation to promote cure (215). Examples of organic groups include mercaptans, acrylates, and oxiranes. Thiols can be added across an olefin using an aromatic ketone as the initiator (216,217). Unfortunately, the odors associated with thiols have limited the commercial acceptance of this technology. Silicones containing pendant acrylate groups have been prepared by reaction of acrylic acid with an epoxy-functional siloxane or by reaction of a halo-organo-functionalized siloxane with acrylic acid (218,219). Cross-linking occurs by a free-radical mechanism and the reaction conditions must be inert. Vinyl ethers and oxiranes can be cationically photopolymerized in the presence of diaryliodonium or sulfonium salts (220–226). This chemistry can be carried out in the presence of oxygen.

Hydrosilylation can also be initiated by a free-radical mechanism (227–229). A photochemical route uses photosensitizers such as peresters to generate radicals in the system. Unfortunately, the reaction is quite sluggish. In several applications, radiation is used in combination with platinum and an inhibitor to cure via hydrosilylation (230–232). The inhibitor is either destroyed or deactivated by uv radiation.

Characterization of Silicone Networks.   The cross-linking of silicones as a function of time can be monitored using a variety of techniques such as infrared spectroscopy, dynamic mechanical analysis, dielectric spectroscopy, ultrasound, differential scanning calorimetry, and thermomechanical analysis (233–239). Infrared spectroscopy is especially useful with addition cure systems because the disappearance of the SiH peak in the starting silicone hydride can be monitored. Dynamic mechanical analysis has been used to study the critical time for gelation of a polydimethylsiloxane network (233). In a study of the isothermal dielectric properties of condensation-cured networks at room temperature as a function of catalyst concentration during cure, the effect of catalyst concentration was reflected in the value of the loss factor (234). The gel point of an addition-cured network can be determined using ultrasound techniques (236). A step-like increase in the longitudinal wave velocity is exhibited at the sol–gel transition point. Differential scanning calorimetry has been used with hydrosilylation-cured systems to determine rates of conversion and to develop a kinetic model for the rubber injection-molding process (238).

In addition to the above techniques, inverse gas chromatography, swelling experiments, tensile tests, mechanical analyses, and small-angle neutron scattering have been used to determine the cross-link density of cured networks (240–245). ^{29}Si solid-state nmr and chemical degradation methods have been used to characterize cured networks structurally (246). ^{1}H- and ^{2}H-nmr and spin echo experiments have been used to study the dynamics of cured silicone networks (247–250).

Model Networks. Construction of model networks allows development of quantitative structure property relationships and provide the ability to test the accuracy of the theories of rubber elasticity (251–254). By definition, model networks have controlled molecular weight between cross-links, controlled cross-link functionality, and controlled molecular weight distribution of cross-linked chains. Silicones cross-linked by either condensation or addition reactions are ideally suited for these studies because all of the above parameters can be controlled. A typical condensation-cure model network consists of an α, ω-polydimethylsiloxanediol, tetraethoxysilane (or alkyltrimethoxysilane), and a tin-cure catalyst (255). A typical addition-cure model is composed of α, ω-vinylpolydimethylsiloxane, tetrakis(dimethylsiloxy)silane, and a platinum-cure catalyst (256–258).

Using both condensation-cured and addition-cured model systems, it has been shown that the modulus depends on the molecular weight of the polymer and that the modulus at rupture increases with increased junction functionality (259). However, if a bimodal distribution of chain lengths is employed, an anomalously high modulus at high extensions is observed. Finite extensibility of the short chains has been proposed as the origin of this upturn in the stress–strain curve.

Monodisperse model networks are prepared using prepolymers with polydispersivity close to 1.0. Equilibrium tensile measurements on networks prepared from fractionated polydimethylsiloxane show single-curve elastic moduli of all the networks (260,261). Dynamic mechanical measurements have been reported for networks prepared by telechlelic condensation of monodisperse polydimethylsiloxane with tetraethylorthosilicate (262). The tensile and swelling behaviors of polydimethylsiloxane networks of high junction functionality have been determined (263). In an examination of the mechanical behavior and swelling properties of networks prepared from cross-linking fractionated vinylmethylpolydimethylsiloxane with pentamethylcyclopentasiloxane (264), the dynamic mechanical properties were measured, which were in good agreement with the swelling rates. It has been found that a lowering of the elastic modulus increases broadness of the molecular weight distribution of networks (265). There is a good correlation between theoretical predictions and the experimental value of the tensile modulus for monodisperse networks prepared from vinyl-terminated silicones and tetrakis(dimethylsiloxy)silane (246).

Filled Silicone Networks. Few applications use silicone elastomers in the unfilled state. The addition of fillers (qv) results in a several-fold improvement in properties, and fillers can be broadly categorized as reinforcing and nonreinforcing (or semireinforcing). Reinforcing fillers increase tensile strength, tear strength, and abrasion resistance, whereas nonreinforcing fillers are used as additives for reducing cost, improving heat stability, imparting color, and increasing electrical conductivity (266).

Nonreinforcing fillers include calcium carbonate, clays, silicates, and aluminates that are primarily used as extending fillers; they serve the purpose of reducing cost per unit volume. Pigment-grade oxides, especially ferric oxides, are used as fillers for high temperature compounds in oxidizing environments. Other fillers such as fumed titania, alumina, and zirconia find applications for extended heat stability, for imparting color, and for improving the electrical conductivity of the formulation. Although carbon black is the most common reinforcing filler for other vulcanized rubber/elastomeric applications, it is not commonly used as a reinforcing filler in silicones. Instead, it finds applications in improving electrical conductivity and pigmentation (267) (see CARBON, CARBON BLACK).

Fillers that are used for providing reinforcement include finely divided silicas prepared by vapor-phase hydrolysis or oxidation of chlorosilanes, dehydrated silica gels, precipitated silicas, diatomaceous silicas, and finely ground high assay natural silicas (268–272). The size, structure, and surface chemistry of the filler all play important roles in determining the final degree of reinforcement (273,274). The most important criteria for reinforcement are the availability of sufficient surface area and a strong van der Waals or hydrogen bonding interaction between the polymer and the filler; however, a covalent linkage is not essential (275,276). Fillers made by the fumed process have typical surface areas around 200 m^2/g and provide the maximum reinforcement. Several comparative studies of the effectiveness of fumed vs precipitated silica can be found in the literature (271,277,278).

Several properties of the filler are important to the compounder (279). Properties that are frequently reported by fumed silica manufacturers include the acidity of the filler, nitrogen adsorption, oil absorption, and particle size distribution (280,281). The adsorption techniques provide a measure of the surface area of the filler, whereas oil absorption is an indication of the structure of the filler (282). Measurement of the silanol concentration is critical, and some techniques that are commonly used in the industry to estimate this parameter are the methyl red absorption and methanol wettability (273,274,277) tests. Other techniques include various spectroscopies, such as diffuse reflectance infrared spectroscopy (drift), inverse gas chromatography (igc), photoacoustic ir, nmr, Raman, and surface forces apparatus (277,283–290).

Structuring refers to the formation of an elastic mass before cure that impedes normal processing operations such as molding and extrusion. Intensive working may be required to restore plasticity. Plasticity and process aids are therefore incorporated as monomeric or oligomeric organosilicon compounds. Alternatively, the silanol concentration on the silica surface can be controlled by filler treatment to prevent structuring (291,292). Reaction of the silica particles with hot vapors of low molecular weight cyclic siloxanes and hexamethyldisilazane (HMDZ) are some of the commonly used filler treatments in the industry (267,293).

The final mechanical properties of the compound are a function of the concentration of the reinforcing filler in the formulation. In the small strain limit,

Einstein relationships have been proposed for changes in viscosity and modulus as a function of volume fraction of the filler (294,295). Although these relationships work well for spherical and low structure fillers, they do not correctly predict the final properties of systems filled with high structure nonspherical particles. The high structure of the filler leads to a system that is not free draining, resulting in an augmentation of the filler concentration caused by the rubber trapped in the internal void space (occluded rubber) (296–299). An empirical modification to the Einstein relationship for modulus has been proposed (300). However, these relationships do not account for the secondary agglomeration that results in highly non-Newtonian and thixotropic behavior (301–308). The reasons for improvement in the failure properties of these systems are not well understood (307,309,310). Some mechanisms that have been proposed include the ability of a filled system to increase energy dissipation at the tip of the crack through viscoelastic processes, and the ability of the dispersed particles in cooperatively arresting or deflecting the growth of cracks, thereby delaying the onset of catastrophic failure. A phenomenon often discussed in connection with filler reinforcement is stress softening, which probably arises because of the progressive detachment, or breaking of network chains attached to filler particles (311).

A good dispersion of the filler particles is essential for ensuring satisfactory ultimate properties (312–318). The first processing step involves the initial incorporation of the filler in the polymer matrix, and is limited primarily by the wettability of the filler. The next step involves a fracture of the aggregates and their uniform dispersion under shear. Heat treatment is often used to promote wetting and improve polymer–filler interactions (319). Dispersion is measured in terms of the remaining agglomerates, which cause premature failure in tensile testing and other ultimate properties. Trends in obtaining uniform dispersions include *in situ* precipitation of silica by catalytic hydrolysis of tetraethyl orthosilicate (TEOS) in a preformed silicone matrix using sol–gel techniques (320).

Properties and Uses

Silicone properties and uses have been discussed extensively (59,321–324).

Silicone Fluids. Silicone fluids are used in a wide variety of applications, including damping fluids, dielectric fluids, polishes, cosmetic and personal care additives, textile finishes, hydraulic fluids, paint additives, and heat-transfer oils. Polydimethylsiloxane oils are manufactured by the equilibrium polymerization of cyclic or linear dimethylsilicone precursors. Trifunctional organosilane end groups, typically trimethylsilyl (M), are used, and the ratio of end group to chain units (D), ie, M/D, controls the ultimate average molecular weight and viscosity (112). Low viscosity fluids, $<10^5$ mm^2/s(=cSt), are generally prepared by acid-catalyzed equilibration. The reaction can be run continuously or in a glass-lined batch reactor at temperatures up to 180°C. Solid acidified montmorillonite (Filtrol) or sulfuric acid on carbon catalysts are removed by filtration, and the end products are stripped under high vacuum and temperature to remove residual low molecular weight cyclic or linear siloxanes (325).

High molecular weight ($>10^6$ mm^2/s(=cSt)) silicone oils and gums are prepared by base-catalyzed, ring-opening polymerization of D$_3$ or D$_4$, or by condensation polymerization of silanol-terminated PDMS. Both methods are practiced commercially. Potassium silanoate, prepared from HO–(D_n)–OH and KOH, or transient catalysts such as tetramethylammonium or tetrabutylphosphonium hydroxide, are used in the ring-opening polymerization method (98,326,327). The potassium silanoate is quenched at the end of reaction, eg, with phosphoric acid. The transient catalysts can be decomposed at temperatures >150°C. Condensation polymerization also requires a catalyst, eg, linear phosphornitrilic chlorides (LPNC) (70–72). Ultimate molecular weight is controlled by the removal of water. LPNC catalysts can be deactivated thermally or by the addition of bases such as hexamethyldisilazane.

The physical properties of polydimethylsiloxane fluids have been extensively studied (59,61,328,329). The properties of a typical 50-mm^2/s(=cSt) silicone oil are shown in Table 2. Linear silicone fluids have low melting points and second-order (glass) transition temperatures. High molecular weight PDMS has a T_m of −51°C and a T_g of −86°C (330). The low intermolecular forces in silicones also lead to low boiling points, low activation energies for viscous flow, high compressibility, and generally Newtonian flow behavior. High molecular weight polymers ($M_w > 35,000$) exhibit non-Newtonian flow, especially at high shear rates. This same phenomenon also results in a material having relatively poor physical properties at high molecular weight.

Unlike the linear analogues, cyclic dimethylsiloxanes are low melting solids and the T_m generally decreases with increasing ring size. Physical properties for a series of M-stopped oligomers, MD$_x$M, and cyclic compounds, D$_x$, are listed in Table 3.

Many of the applications for silicone oils are derived from the wide temperature range over which they can be used. Addition of a small amount of

Table 2. Typical Silicone Fluid Properties[a]

Property	Value
Physical	
η at 25°C, mm^2/s(=cSt)	50
viscosity–temperature coefficient	0.59
density, g/mL	0.963
refractive index	1.402
pour point, °C	−55
surface tension at 25°C, mN/m(=dyn/cm)	20.8
thermal expansion, cm^3/cm^3/°C	0.00106
max. volatiles at 150°C, %	0.5
specific heat	0.36
Electrical	
dielectric strength, V/min	35,000
ϵ at 60 Hz	2.72
dissipation factor	0.0001
volume resistivity, Ω·cm	1×10^{14}
service temperature, continuous, °C	150
max. temperature in air, °C	200
max, temperature in N$_2$, °C	300

[a]Refs. 59 and 328.

Table 3. Properties of MD$_x$M and D$_x$ [a]

Compound	CAS Registry Number	MDT formula	Melting point, °C	Boiling point, °C	Density, d^{20}, g/mL	Refractive index, n_D^{20}	Viscosity (η) at 25°C, mm^2/s (=cSt)	Flash point, °C
hexamethyldisiloxane	[107-46-0]	MM	−67	99.5	0.7636	1.3774	0.65	−9
octamethyltrisiloxane	[107-51-7]	MDM	−80	153	0.8200	1.3840	1.04	37
decamethyltetrasiloxane	[141-62-8]	MD$_2$M	−76	194	0.8536	1.3895	1.53	70
dodecamethylpentasiloxane	[141-63-9]	MD$_3$M	−80	229	0.8755	1.3925	2.06	94
tetradecamethylhexasiloxane	[107-52-8]	MD$_4$M	−59	245	0.8910	1.3948	2.63	118
hexadecamethylheptasiloxane	[541-01-5]	MD$_5$M	−78	270	0.9012	1.3965	3.24	133
octadecamethyloctasiloxane	[556-69-4]	MD$_6$M	−63	290	0.9099	1.3970	3.88	144
eicosamethylnonasiloxane	[2652-13-3]	MD$_7$M		307.5	0.9180	1.3980	4.58	159
hexamethylcyclotrisiloxane	[541-05-9]	D$_3$	64.5	134				
octamethylcyclotetrasiloxane	[556-67-2]	D$_4$	17.5	175.8	0.9561	1.3968	2.30	69
decamethylcyclopentasiloxane	[541-02-6]	D$_5$	−44	210	0.9593	1.3982	3.87	
dodecamethylcyclohexasiloxane	[540-97-6]	D$_6$	−3	245	0.9672	1.4015	6.62	
1,3,5,7,9,11,13-heptaethyl-cycloheptasiloxane	[17909-36-3]	D$_7$	−32	154[b]	0.9730	1.4040	9.57	
hexadecamethylcyclooctasiloxane	[556-68-3]	D$_8$	31.5	290	1.1770[c]	1.4060	13.23	

[a] Ref. 328.
[b] At 2.7 kPa (20 mm Hg).
[c] Crystals.

103

phenylsiloxane in the form of diphenyl or phenymethylsiloxy groups, and/or incorporation of a slight amount of branching, suppresses the T_m of the dimethyl-silicones and can reduce the pour points to as low as $-100\,°C$ (331). Methylsilicone oils are stable in air at $150\,°C$ for extremely long periods of time, and undergo only slow degradation up to $200\,°C$. In vacuum or inert atmospheres, dimethylsilicone oil can be heated to $300\,°C$ for limited amounts of time (332). Heat-resistant silicone fluids have increasing amounts of phenyl-containing substituents. High molecular weight methylphenylsilicones can be used in air at temperatures up to $250\,°C$ for several hundred hours and can withstand brief exposures at up to $450\,°C$ in the presence of inert gases (333). Table 4 compares silicone polymers with equivalent petroleum-based materials, and demonstrates the wide temperature use of silicone materials.

Dimethylsilicone fluids decompose via two principal mechanisms: retrocyclization to low volatile cyclic siloxanes such as D_3 and D_4, and thermal oxidation of the alkyl side chains to give formaldehyde, CO_2, water, and T groups (328). The retrocyclization process is catalyzed by acids or bases and can occur at temperatures above $140\,°C$. Catalytic acidic or basic sites on glassware and metallic containers are often the source of degradation of PDMS fluids. Heating in air to $200-250\,°C$ causes thermal oxidation of PDMS, and the cross-linking from T and Q group formation ultimately results in gellation. Stabilizers such as p-aminophenol, naphthols, metal acetonylacetonates, and iron octoate are commonly used to improve thermal stability (334–338).

For higher molecular weight polydimethylsiloxanes ($M_n > 2500$), the number-average molecular weight is related to the bulk viscosity by the following formula, where the viscosity units are $mm^2/s(=cSt)$.

$$\log(\text{viscosity}) = 1.00 + 0.0123\ (M_n)^{1/2}$$

The number-average molecular weight of dimethylsiloxane can also be determined from the intrinsic viscosity [η, dL/g] (extrapolated to zero viscosity) in toluene or methyl ethyl ketone according to the following equation (339,340):

$$[\eta] = 2 \times 10^{-4}\ (M_n)^{0.66} \qquad [\eta] = 8 \times 10^{-4}\ (M_n)^{0.5}$$

toluene   methyl ethyl ketone

Compared with petroleum-based fluids, silicone oils show relatively small changes in viscosity as a result of temperature change (59,328). A common

Table 4. Viscosity of Silicone Oil vs Petroleum Oil

Temperature, °C	Silicone η, $mm^2/s(=cSt)$	Petroleum η, $mm^2/s(=cSt)$
100	40	11
38	100	100
-18	350	11,000
-37	660	230,000
-57	1560	solid

measure of the viscosity change with temperature is the viscosity–temperature coefficient (VTC) (341). Typical dimethylsilicone VTC is 0.6 or less. Phenylsilicones are slightly higher. Organic oils are typically 0.8 or greater. The viscosity–temperature curves follow equations 27 and 28, where a, b, and c are constants.

$$\log(\eta + b) = a \log T + c \tag{27}$$

$$VTC = \frac{\eta_{38C} - \eta_{99C}}{\eta_{38C}} \tag{28}$$

Figure 4 shows viscosity–temperature curves for a series of silicone oils from 20 to 10,000 mm^2/s(=cSt), an SAE 10 petroleum oil, and a diester-based oil (274). Silicones with phenyl, trifluoropropyl, or larger alkyl groups show larger viscosity changes with increasing temperature than dimethysilicone oils. Low molecular weight silicone oils (<1000 mm^2/s) exhibit Newtonian flow behavior and undergo only small changes in viscosity under shearing conditions (342). Higher viscosity PDMS fluids shear thin at higher shear rates. The fluids are stable to shear stress, however, and recover to their original viscosity after being passed through small capillaries under pressure. Figure 5 shows the effect of shear rate on viscosity for a series of PDMS fluids of different molecular weights.

Many of the unique properties of silicone oils are associated with the surface effects of dimethylsiloxanes, eg, imparting water repellency to fabrics, antifoaming agents, release liners for adhesive labels, and a variety of polishes and waxes (343). Dimethylsilicone oils can spread onto many solid and liquid

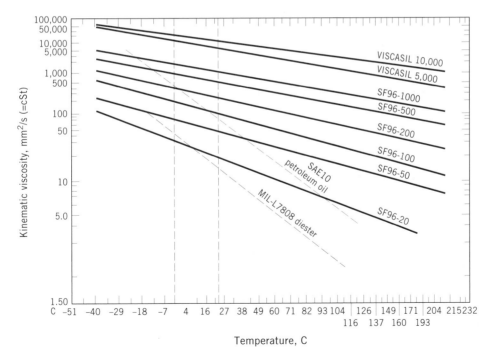

Fig. 4.   Kinematic viscosity–temperature relationship of dimethylsilicone fluids.

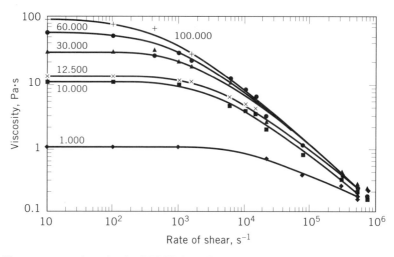

Fig. 5. Shear rate vs viscosity for PDMS. Numbers on curves indicate molecular weights. To convert Pa·s to poise, multiply by 10.

surfaces to form films of molecular dimensions (344,345). This phenomenon is greatly affected by even small changes in the chemical structure of siloxane in the siloxane polymer. Increasing the size of the alkyl substituent from methyl to ethyl dramatically reduces the film-forming ability of the polymer (346). The phenyl-substituted silicones are spread onto water or solid surfaces more slowly than PDMS (347).

Dimethylsilicone polymers are often described as having a combination of silicate and paraffin structures, and the orientation of the polymer chains onto surfaces, physically, by chemical affinity, or bonding, can contribute to the observed surface properties. The surface tension of hexamethyldisiloxane, MM, is 15.7 mN/m(=(dyn/cm)). Increasing incorporation of MD_xM increases the value to about 20 dyn/cm for higher viscosity fluids. Silicone fluids are characterized by their hydrophobicity (resistance to water), and the contact angle between PDMS and water is between 100 and 110°. Carefully deposited dimethysilicone on soda lime glass has a contact angle of about 60° at room temperature; the angle increases with increasing temperature to about 100° at 390°C and then decreases at the decomposition temperature (59).

Gases are soluble in dimethylsilicone polymers and PDMS is permeable to water vapor. The solubility of air, nitrogen, and carbon dioxide at 101.3 kPa (1 atm) is 0.17, 0.19, and 1.00 cm^3/cm^3, respectively (348). About 250 to 300 ppb of water can dissolve in dimethylsiloxane at 25°C and 95% relative humidity (349). The water solubility of silanol-terminated oligomers increases with decreasing molecular weight, and the lowest monomer, dimethylsilanediol, is water-miscible. Silicone oils are soluble in nonpolar organic solvents such as benzene, toluene, dimethylether, chloroform, methyl ethyl ketone, methylene chloride, or kerosene. Incorporation of phenyl and especially trifluoropropyl or β-cyanoethyl functionality onto the polymer decreases the solubility of siloxane to organic solvents (342). They are slightly soluble in acetone, ethanol, and butanol, and relatively immiscible in methanol, ethylene glycol, and water. Exposure to strong acids or bases reequilibrates PDMS.

Silicone fluids have good dielectric properties, loss factor, specific resistance, and dielectric strength at normal operating conditions, and the properties vary only slightly with temperature (59,328,350). The properties in combination with relatively low flammability have led to the use of silicones in transformers and other large electrical applications (351). The dielectric constant of a 1000-cSt oil is 2.8 at 30°C and 2.6 at 100°C. The loss factor is low, 1.2×10^{-4} at 20°C, and behaves irregularly with frequency. Specific resistance is 4×10^{15} ohm-cm at 20°C and decreases with increasing temperature or with exposure to moisture. The dielectric strength is 120 kV/cm (20°C, 50 cP, 6 kV/s).

Silicone oils are good hydrodynamic lubricants but have generally poor frictional lubricating properties (352–354). The latter can be improved by incorporating chlorophenyl groups into the polymer side chains (355). For steel on steel, the coefficient of friction is about 0.3–0.5. The load-bearing capacity of PDMS (Almen-Wieland machine) is only 50–150 kg, compared with ~1000 kg for polychlorophenylmethylsiloxane and up to 2000 kg for mineral oil.

The thermal conductivity of dimethylsilicone with viscosities >100 mm^2/s(=cSt) is 15.5 J/(s·cm·°C) (3.7×10^{-4} cal/(s·cm·°C)), and is roughly constant with increasing viscosity (356). The specific heat is 1.55–1.70 J/(g·°C) (0.37–0.41 cal/(g·°C)) over the 20–200°C range and is practically independent of viscosity (357). Thermal expansion of PDMS is ~0.1%/°C (356).

Liquid silicone oils are highly compressible and remain liquid over pressure ranges where normal paraffin oils have already solidified (358,359). This property, combined with a wide temperature use range, is the reason for silicone use in a large number of hydraulic applications. The adiabatic compressibility of 1000-cSt dimethylsilicone oils is ~1000 Pa^{-1} (100 cm^2/dyn) and decreases slightly with increasing viscosity, similar to paraffin oils. PDMS oils can withstand pressures greater than 3.4 GPa (35,000 kg/cm^2) even though they are compressed >30%.

Dimethylsilicone fluids are transparent to visible light and microwave but absorb ultraviolet radiation at wavelengths below 280 nm (360). Siloxanes absorb strongly in the infrared between 1000 and 1100 cm^{-1} (Si–O stretching), 1300–1350 cm^{-1} (Si–C stretching), and 2950–3000 cm^{-1} (C–H stretching) (361). The speed of sound in PDMS is 987.3 m/s for a 1000–cSt fluid at 30°C and decreases with increasing temperature (362). Methylsilicone fluids are cross-linked when exposed to gamma ray or electron beam irradiation (363). Polymers containing aromatic substituents are more resistant than dimethylpolymers.

Silicone Heat-Cured Rubber. Silicone elastomers are made by vulcanizing high molecular weight (>5 × 10^5 mol wt) linear polydimethylsiloxane polymer, often called gum. Fillers are used in these formulations to increase strength through reinforcement. Extending fillers and various additives, eg, antioxidants, adhesion promoters, and pigments, can be used to obtain certain properties (59,357,364).

Peroxides are typical vulcanizing agents and the mechanism of cure involves free-radical abstraction of a silicon methyl group proton and subsequent dimerization of the methyl radicals to form ethylene cross-links (eqs. 29–31) (365,366). Vinyl-containing polymers are often used to control the cross-linking reaction. Commonly used peroxides include di-*t*-butyl peroxide [*110-05-4*], benzoyl peroxide [*94-36-0*], di(*p*-cumyl) peroxide, and di(*p*-chlorophenyl) peroxide. The choice of peroxide is made based on the desired cure temperature and rate. Table 5 lists some common peroxide curing agents, typical cure temperatures, and some recommended processing conditions.

$$(C_6H_5COO)_2 \longrightarrow 2\ C_6H_5COO\cdot \tag{29}$$

$$C_6H_5COO\cdot + (CH_3)_2SiO \longrightarrow \cdot CH_2(CH_3)SiO + C_6H_5COOH \tag{30}$$

$$2\cdot CH_2(CH_3)SiO \longrightarrow OSi(CH_3)CH_2CH_2(CH_3)SiO \tag{31}$$

Unlike natural rubber, silicone rubber does not stress-crystallize when elongated, which leads to relatively poor physical properties. Unfilled silicone rubber has only a 0.35-MPa (50-psi) tensile stress at break. To overcome this, silicone rubber is compounded with 10 to 25 wt % reinforcing fillers, typically fumed silica, to improve the final rubber product properties (268,270,271,367). Other common fillers include precipitated silica, titanium dioxide, calcium carbonate, magnesium oxide, and ferric oxide. Pigments and colorants are also used. The addition of fillers to the gum stock can result in structuring when stored, which decreases the workability of the material. To prevent structuring, the filler is often treated with agents such as hexamethyldisilazane to reduce surface hydroxyl functionality, or with the addition of processing aids such as silicone oils, diphenylsilanediol, or dimethylpinacoxysilane (267,368).

The processing methods for silicone rubber are similar to those used in the natural rubber industry (59,369–371). Polymer gum stock and fillers are compounded in a dough or Banbury-type mixer. Catalysts are added and additional compounding is completed on water-cooled roll mills. For small batches, the entire process can be carried out on a two-roll mill. Heat-cured silicone rubber is

Table 5. Cure Agents for Silicone Rubber[a]

Curing agent	Commercial name	Temperature, °C	Property
Active paste[b] *composition*			
di(2,4-dichlorobenzoyl) peroxide	Cadox TS-50 or Luperco CST	104–132	hot-air vulcanizing
benzoyl peroxide	Cadox BS or Luperco AST	116–138	molding, steam curing
Active powder composition			
dicumyl peroxide[c]	DI-CUP 40C	154–177	molding thick sections, bonding, steam curing
2,5-di(*t*-butylperoxy)-2,5-dimethylhexane[d]	Varox, Luperco 101XL, or Luperco 101[e]	166–182	molding thick sections, bonding, steam curing

[a]Ref. 357.
[b]50% active paste composition.
[c]40% active powder composition.
[d]50% active powder composition.
[e]100% active liquid.

commercially available as gum stock, reinforced gum, partially filled gum, un-catalyzed compounds, dispersions, and catalyzed compounds. The latter is ready for use without additional processing. Before being used, silicone rubber is often freshened, ie, the compound is freshly worked on a rubber mill until it is a smooth continuous sheet. The freshening process eliminates the structuring problems associated with polymer–filler interactions.

It is common practice in the silicone rubber industry to prepare specific or custom mixtures of polymer, fillers, and cure catalysts for particular applications. The number of potential combinations is enormous. In general, the mixture is selected to achieve some special operating or processing requirement, and the formulations are classified accordingly. Table 6 lists some of the commercially important types.

Silicone rubber is most commonly fabricated by compression-molding-catalyzed gum stock at 100–180°C under 5.5–10.3 MPa (800–1500 psi) pressure. Mold release compounds are usually employed. Under these conditions the rubber is cured in a few minutes. Extrusion processing is used in the manufacture of tubes, rods, wire and cable insulation, and continuous profiles. Initial properties, so-called green strength, are obtained by curing in hot air or steam tunnels from 276 to 690 kPa (40–100 psi) at 300–450°C for several minutes. Final physical properties are achieved by post-curing in air or steam, typically for 30–90 minutes. When silicone rubber must be bonded to other surfaces, eg, metals, plastics, or ceramics, primers are used. Silicate or titanate esters from the hydrolysis of tetraethylorthosilicate or tetraethyltitanate are often used as primers. Silicone rubber-coated textiles and glass cloth are made by initially dissolving gum stock in solvent and applying the rubber by dip coating. The fabric is then dried and the rubber cured in heated towers. Tubes and hose can

Table 6. Properties of Different Classes of Silicone Rubbers[a]

Class	Hardness, durometer	Tensile strength, MPa[b]	Elongation, %	Compression set[c], %	Useful temperature range, °C Min	Max	Tear strength, J/cm²[d]
general purpose	40–80	4.8–7.0	100–400	15–50	−60	260	0.9
low compression set	50–80	4.8–7.0	80–400	10–15	−60	260	0.9
extreme low temperature	25–80	5.5–10.3	150–600	20–50	−100	260	3.1
extreme high temperature	40–80	4.8–7.6	200–500	10–40	−60	315	
wire and cable	50–80	4.1–10.3	100–500	20–50	−100	260	
solvent-resistant	50–60	5.8–7.0	170–225	20–30	−68	232	1.3
high strength flame retardant	40–50	9.6–11.0	500–700				2.8–3.8

[a]Refs. 372–376.
[b]To convert MPa to psi, multiply by 145.
[c]At 150°C, 22 h.
[d]To convert J/cm² to lbf/in., multiply by 57.1.

be formed on mandrels from this fabric into complex shapes. Foamed or sponge silicone rubber can be made by incorporating chemical blowing agents into the rubber stock, which eliminates typically nitrogen or carbon dioxide under the thermal curing conditions. Sponge silicone rubber made in this way has a closed cell structure and densities of $0.4–1.0$ g/cm^3.

Vulcanized silicone rubber is characterized by its wide temperature use range (-50 to $>200°C$), excellent electrical properties, and resistance to air oxidation and weathering conditions. Silicone rubber is also extremely permeable to gases and water vapor. The mechanical properties of silicone rubber are generally inferior to most organic (butyl) rubbers at room temperature. Tables 6, 7, and 8 list some typical physical properties for vulcanized silicone rubber. Silicone rubber can be made with varying degrees of hardness. Shore A values from 20 to 90 can be made; the best physical properties are obtained from 50 to 70. The properties of silicone change with temperature (376). The Young's modulus drops from 10^5 to 2×10^2 MPa (14.5×10^4 to 2.9×10^4 psi) from $-50°C$ to room temperature and then is fairly constant to $260°C$. Tensile strength at break decreases from 6.9 MPa (1000 psi) at $0°C$ to 2.1 MPa (300 psi) at $300°C$. Typical elongation at break values range from 300 to 700%, depending on the composition.

The compression set of silicone rubber is similar to organic types of rubber at low ($0–50°C$) temperatures, ranging from 5 to 15% (380). Above $50°C$, silicone

Table 7. Properties of Silicone Gums[a]

Type	CAS Registry Number	Density, d^{25}, g/cm^3	T_g, °C	ASTM D926, Williams plasticity
(CH$_3$)$_2$SiO	[9016-00-6]	0.98	-123	95–125
CH$_3$(C$_6$H$_5$)SiO	[9005-12-3]	0.98	-113	135–180
CH$_3$(CF$_3$CH$_2$CH$_2$)SiO	[25791-89-3]	1.25	-65	

[a]Refs. 331, 373–375.

Table 8. Permeability of Silicone Elastomers[a]

Type	CAS Registry Number	Temperature, °C	Gas	Permeability, nmol/(m·s·kPa)[b]
dimethyl silicone	[63148-62-9]	25	CO$_2$	180
		25	O$_2$	35
		25	air	19.6
		30	butane	0.89
		70	butane	0.76
fluorosilicones[c]	[63148-56-1]	26	CO$_2$	35
		26	O$_2$	5.8
nitrile silicone	[70775-91-6]	31	CO$_2$	105
		31	O$_2$	17.8

[a]Refs. 377–379.
[b]To convert nmol/(m·s·kPa) to (cm^3·cm)/(s·cm^2·mm Hg), multiply by 3×10^{-8}.
[c]Methyltrifluoropropyl silicones.

rubber is superior, but compression set increases with time and temperature. Silicone rubber is more tear-sensitive than butyl rubber, and the degree of sensitivity is a function of filler size and dispersion, cross-link density, and curing conditions. The electrical properties of silicone rubber are generally superior to organic rubbers and are retained over a temperature range from -50 to $250°C$ (51). Typical electrical values for a heat-cured silicone rubber are shown in Table 9.

Table 9. Electrical Properties of Typical Silicone Rubber[a]

Property	Value
volume resistivity[b], $\Omega \cdot cm$	$1 \times 10^{14} - 1 \times 10^{16}$
electric strength, V/25.4 μm (=V/mil)	$400 - 700$
dielectric constant, 60 Hz	$2.95 - 4.00$
power factor, 60 Hz	$0.001 - 0.01$
surface resistance, Ω	$3.0 \times 10^{13} - 4.5 \times 10^{14}$
dielectric loss factor, tan δ	$5 \times 10^{-4} - 4 \times 10^{-3}$

[a]Ref. 35.
[b]Ref. 56.

Silicone rubber film is 10 to 20 times more permeable to gases and water vapor than organic rubber (380). The water permeability of silicone rubber is $\sim 14 \times 10^6$ mol/(m·s·Pa) (1.2×10^6 g/(h·cm·torr)), which means that silicone rubber can absorb about 35 mg of water per square centimeter of surface area after seven days exposure (377). Table 8 shows the permeability of silicone to common gases. Organic solvents can diffuse and swell into silicone rubber, significantly decreasing the physical properties of the material. The degree of swelling depends on the solubility parameters of the solvent and the rubber, as illustrated in Figure 6.

Solvent-resistant rubber based on either trifluoropropylmethylsiloxane or β-cyanoethylmethylsiloxane has been developed for applications, eg, as fuel tank sealants, where the material will be exposed to aggressive solvents. Those based on trifluoropropylmethylsiloxane are more important commercially. Pure water has little effect on silicone; however, long exposures in the presence of acid or

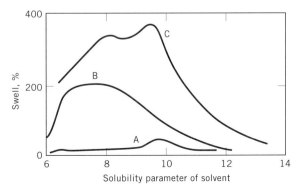

Fig. 6. Swelling of silicone rubber by solvents (381). The fluorosilicones (A) are methyl trifluoropropyl silicones. B = dimethylsilicone; C = methyl(phenyl)silicone.

base catalysts causes degradation and reversion of the rubber to a sticky gum (376,382). Prolonged exposure to temperatures above 300°C in air causes the rubber to stiffen and ultimately to become brittle (376). As with silicone oil, the principal chemical pathways of degradation are retrocyclization and oxidation of the polymer alkyl groups. Heating in air at 125°C causes a decrease in elongation and an increase in hardness, but no important changes in tensile properties are observed.

Silicone rubber burns with a high char yield, and the residual material is nonconducting silicon dioxide. The limiting oxygen index for a typical rubber formulation is about 20. Flame-retardant versions are available with oxygen index values as great as 40–50 (372). In general, silicone rubber is not resistant to gamma or electron beam radiation and undergoes cross-linking, which leads to embrittlement (383). Increasing levels of phenyl-containing silicone in the rubber improves the radiation resistance (384).

Silicone Liquid-Injection-Molding Rubber. An increasingly important processing technique for silicone rubber is liquid injection molding. Unlike heat-cured rubber, which is typically compression-molded from high viscosity gum stock, liquid-injection-molded (LIM) rubber is made from low viscosity starting materials, 1000–2000 mPa·s(=cP), and is cured in molds similar to those used for plastic injection molding. The principal advantages of LIM include rapid cycle times and the ability to fill complex mold shapes because of the low viscosity of the inputs. Rubber parts can be cured in 10–40 s using low molding pressures, ca 2–20 MPa (300–3000 psi), and low curing temperatures, typically 150–260°C (385). LIM processing is being increasingly used for applications such as electrical connectors, O-ring seals, valves, electrical components, health care products, and sporting equipment such as goggles and scuba masks.

Silicone LIM rubber is made from a two-component polymer system. One part (Part B) contains a linear polydimethysiloxane polymer with pendent Si–H functionality, reinforcing fillers such as fumed silica, extending fillers, pigments, and stabilizers. The second part (Part A) contains linear polydimethylsiloxane with terminal and pendent vinyl groups; reinforcing and extending fillers; a platinum hydrosilylation catalyst; and a catalyst inhibitor, commonly olefin, amine, or phosphine ligands. After mixing and heating, the catalyst initiates the cross-linking reaction by addition of the Si–H group to the double bond (eq. 32). Latent cure catalysts have been developed that allow the formulation of one-component products (386). These systems work by incorporation of platinum ligands that deactivate the hydrosilylation catalysts at room temperature; however, when heated to temperatures above 100°C, these catalysts become active.

(32)

Part A   Part B

Cured silicone LIM rubber can be fabricated with physical properties equivalent to heat-cured rubber (385). Shore A hardness can range from 30 to 70, depending on formulations. Typical physical properties include tensile strengths as high as 9.7 MPa (1400 psi), 500–775% elongation at break, and tear strength of >30 N/mm (180 lb/in.). Compression sets of less than 10% can be achieved if the material is baked after processing.

Foam Rubber. Flexible foamed silicone rubber can be fabricated with a flame retardancy greatly superior to that of the urethane-type foam. A self-blowing, low to medium density ($80–240$-kg/cm^3($5–15$-lb/ft^3)) silicone foamed rubber can be prepared using polymers similar to those used in LIM products (387–390). Two components are mixed at room temperature: one part is pendent SiH containing polydimethylsiloxane and fillers, and the other consists of terminal and pendant vinyl containing PDMS, reinforcing and extending fillers, a platinum hydrosilylation catalyst and inhibitor, water, alcohol, and an emulsifying agent. Typical time for foam formation is 20 min.

Two chemical reactions occur simultaneously when mixed. One is the platinum-catalyzed reaction of hydroxyl groups from water or alcohol with polymer SiH to give hydrogen gas (eq. 33). This reaction is the source of the blowing agent that forms the foam. The second reaction is the cross-linking of SiH and Si–vinyl, which increases the polymer viscosity and ultimately gels and cures to give an elastomer. This is the same cross-linking reaction depicted in equation 32. Proper control of the kinetics of these two reactions is critical to achieving a foam having good physical properties.

$$
\underset{\substack{| \\ CH_3}}{\overset{\substack{CH_3 \\ |}}{{+}\!{(\!-}Si{-}O{-}\!)}}\ \underset{\substack{| \\ H}}{\overset{\substack{CH_3 \\ |}}{Si{-}O{+}_x}} + H_2O \longrightarrow \underset{\substack{| \\ CH_3}}{\overset{\substack{CH_3 \\ |}}{{+}\!{(\!-}Si{-}O{-}\!)}}\ \underset{\substack{| \\ OH}}{\overset{\substack{CH_3 \\ |}}{Si{-}O{+}_x}} + H_2 \qquad (33)
$$

Silicone foam thus formed has an open cell structure and is a relatively poor insulating material. Cell size can be controlled by the selection of fillers, which serve as bubble nucleating sites. The addition of quartz as a filler greatly improves the flame retardancy of the foam; char yields of >65% can be achieved. Because of its excellent flammability characteristics, silicone foam is used in building and construction fire-stop systems and as pipe insulation in power plants. Typical physical properties of silicone foam are listed in Table 10.

Silicone Resins. Silicone resins are an unusual class of organosiloxane polymers. Unlike linear poly(siloxanes), the typical silicone resin has a highly branched molecular structure. The most unique, and perhaps most useful, characteristics of these materials are their solubility in organic solvents and apparent miscibility in other polymers, including silicones. The incongruity between solubility and three-dimensional structure is caused by low molecular weight ($M_n < 10{,}000$ g/mol) and broad polydispersivity of most silicone resins.

A wide variety of organosilicone resins containing a combination of M, D, T, and/or Q groups have been prepared and many are commercially manufactured. In addition, resins containing hydrosilation-reactive SiH and SiVi groups or other functionalities, including OH and phenyl groups, are known. Two classes of silicone resins are most widely used in the silicone industry: MQ and TD resins.

Table 10. Physical and Flammability Properties of Silicone Foam Rubber[a]

Property	Value
density, kg/m^3	80–240
tensile strength, MPa[b]	0.52
elongation, %	90
thermal conductivity, W/m·K	0.06
operating temperature range, °C	−60 to 205
limiting oxygen index	30
UL-94 flammability (3.2-mm thick)	V-0
flame spread index (ASTM E162)	16
smoke density flaming (ASTM E662)	18
smoke density smoldering (ASTM E662)	9
verticle burn (FAA 25.853)	compliant

[a]Ref. 391.
[b]To convert MPa to psi, multiply by 145.

MQ Resins. These resins are composed of clusters of quadrafunctional silicate Q groups end-capped with monofunctional trimethylsiloxy M groups. The structure of an MQ resin molecule is defined by three characterization parameters: M/Q ratio, molecular weight, and % OH. Standard analytical techniques have been used to quantify these parameters, including ^{29}Si-nmr to determine M/Q ratio, gpc for molecular weight, and ftir for % OH (392,393). Most commercially useful MQ resins have an M/Q ratio between 0.6 and 1. Ratios lower than 0.6 result in insoluble solids, whereas those greater than 1 produce liquids (394). Molecular weight is related to M/Q ratio; higher molecular weights is associated with lower M/Q ratios. Typical molecular weights are M_n = 10,000 g/mol for a resin with an M/Q of 0.6, and M_n = 2000 g/mol for an M/Q of (393,394). The silanol content (% OH by weight) of these resins, typically ranging between 0 and 3%, is process-dependent (Fig. 7). An ir study indicates that OH groups are intramolecularly associated through hydrogen bonding (394). A computer-generated molecular structure of a common commercially prepared MQ resin

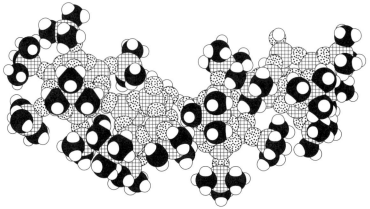

Fig. 7. Molecular structure of a typical MQ resin, $[M_{0.62}Q^{OH}_{0.17}Q_{0.83}]_{32}$, where $\bigcirc$ = H, $\bullet$ = C, $\oplus$ = Si, and $\circledcirc$ = O.

has been reported (393). This structure, shown in Figure 7, consists of a chain of silicate clusters end-capped with trimethylsiloxy groups.

MQ resins are commercially manufactured by one of two processes: the ethyl silicate or the sodium silicate process. In the ethyl silicate process, these resins were first prepared by cohydrolysis of tetraethoxysilane and trimethylchlorosilane in the presence of an aromatic solvent (eq. 34). This process is versatile and reproducible; it can be used to prepare soluble MQ resins with M/Q ratios ranging between 0.6 and 4. The products of these reactions typically contain high levels of residual alkoxysilane groups.

$$\text{Si(OCH}_2\text{CH}_3)_4 + \text{(CH}_3)_3\text{SiCl} \xrightarrow[\text{2. H}_2\text{O}]{\text{1. solvent}} [\text{M}_x\text{Q}]_y \qquad (34)$$

A more economical route to MQ resin uses low cost sodium silicate and trimethylchlorosilane as inputs (eq. 35) (395). The sodium silicate process is initiated by acidifying an aqueous sodium silicate solution to a pH of 2. The resulting hydrosol quickly builds molecular weight. The rate of this increase is moderated by the addition of an alcohol such as 2-propanol. The hydrosol is subsequently silylated by the addition of trimethylchlorosilane. This process, which is kinetically sensitive and limited to synthesizing M/Q ratios of 1 or less, is preferred when MQ resins having high (>1%) OH content are required (395).

$$[(\text{SiO}_2)(\text{Na}_2\text{O})_{0.3}]_{5.5} \text{ (aq) + HCl (aq)} \longrightarrow \text{silicate hydrosol} \xrightarrow{\text{(CH}_3)_2\text{CHOH/solvent}} \text{stabilized silicate}$$

$$\text{hydrosol} \xrightarrow[\text{2. strip/filter}]{\text{1. (CH}_3)_3\text{SiCl}} \text{M}_{0.7}\text{Q resin} \qquad (35)$$

Both the ethyl and sodium silicate processes can be modified by substituting SiH- or SiVi-functional chlorosilanes or combinations of chlorosilanes to produce hydrosilation-reactive MQ resins (396,397).

The most prominent use of MQ resins is as the tackifying agent for silicone pressure-sensitive adhesives (PSA) (398,399). The other main component of silicone PSA is a silicone gum. This mixture of MQ resin and silicone gum is applied to a tape backing such as polytetrafluoroethylene (PTFE) or poly(ethylene terephthalate) (PET) and cured in the presence of a peroxide catalyst to provide an adhesive tape. MQ resins are also commonly used as control-release additives for silicone paper release products, as reinforcing fillers for liquid-injection-moldable silicones, in masonry sealants, and in leather/textile water-repellent coatings (398,400–404). Additional uses of MQ resins are as surfactants. MQ resins are commercially attractive for use in defoaming applications as well as for stabilizing frothed urethane for high density carpet backing. Virtually every use for MQ resins is as a blend with a poly(dimethylsiloxane). These blends have been characterized as interpenetrating networks (402,405,406). MQ resin/PDMS blends are microphase-separated. One phase is PDMS-rich and has a glass-transition temperature of $-107°C$. This T_g does not change with changes in the blend ratio. The other phase is an MQ resin/PDMS-miscible phase whose T_g varies linearly from -100 to $200°C$ as the composition of the blend is changed from 30 to 90% MQ resin (394). These unique viscoelastic properties contribute to the versatility of silicone products containing MQ resin.

TD Resins.   The other important class of silicone resins is TD resins. These materials are simply prepared by cohydrolyzing mixtures of chlorosilanes in organic solvents (eq. 36), where $R = CH_3$ or C_6H_5 (407).

$$x \ RSiCl_3 + y \ R_2SiCl_2 \xrightarrow[\text{solvent}]{H_2O} T_x D_y \ \text{resin} \tag{36}$$

A variety of liquid and solid resins can be prepared by varying the T/D ratio. Commercial TD resins are available, containing from 5 to 95% T groups. Mixtures of methyl and phenylchlorosilanes are also frequently used. Some TD resins are modified by the addition of catalysts to reduce the silanol content, thereby increasing the molecular weight. Contrary to MQ resins, which are very stable, TD resins are unstable toward silanol condensation reactions.

TD resins are used as protective coatings, electrical coatings, saturants, laminates, and water repellents (59). TD resins are also useful in high performance paints. Compositions high in silanol content are utilized in reactive formulations. Low silanol TD resins are used as a nonreactive additive to alkyd paint formulations. Attractive features of TD resin-based protective coatings include superior, uv-resistant weatherability and excellent high and low temperature properties. Silicone electrical coatings are preferred when good dielectric insulation is required over a broad temperature range. TD resins are also used to saturate glass or Kevlar fabrics. After curing, these flexible composites are used in a variety of applications. TD resin-based water repellents are used chiefly with siliceous substrates. Some typical TD silicone resin properties are as follows: dielectric strength = 68,000 V/mm; dielectric constant = 2.9 at 60 Hz and 25°C; dissipation factor = 0.006 at 60 Hz and 25°C; surface resistivity = 1×10^{14} Ω/in.^{2}; and volume resistivity = 1×10^{14} $\Omega \cdot$cm.

Organosilicone Coating Products.   Silicone products are used in a large variety of coatings applications; most prominent among these are silicone pressure-sensitive adhesives (PSA), plastic hardcoats, and paper release coatings (398,399).

Pressure-Sensitive Adhesives.   Silicone PSAs are used primarily in specialty tape applications that require the superior properties of silicones, including resistance to harsh chemical environments and temperature extremes (398,399). Silicone PSAs are also used in applications requiring long service life, electrical insulation, and protection from moisture. Another distinctive advantage of silicone PSAs is their ability to wet low surface energy tape substrates such as PTFE.

Silicone PSAs are blends or interpenetrating networks (IPNs) composed of a tackifying MQ resin cured in a cross-linked poly(siloxane) network. The synthesis and structure of MQ resins have been described above. The poly(siloxane) network is traditionally derived by free-radical cross-linking of a high molecular weight PDMS polymer or gum using a peroxide catalyst, such as benzoyl peroxide or 2,4-dichlorobenzoyl peroxide. The curing reaction is performed immediately after the PSA has been coated onto a tape substrate, such as PET, PTFE, or Kapton. Uncured PSAs are supplied as a solution in an organic solvent. Some silicone PSAs also incorporate phenyl groups onto the gum portion of the adhesive to increase the use temperature.

Another important aspect of the chemistry of silicone PSAs is the molecular interaction between the tackifying MQ resin and the PDMS network. For optimal adhesive properties it is important that some covalent bonding exists between the resin and the PDMS. This bonding is usually achieved by promoting a silanol condensation reaction between residual OH groups on the resin and on the end caps of the PDMS. The condensation reaction can occur either during blending of the PSA components or during processing of the tape. Studies of the viscoelastic properties of cured silicone PSAs indicate that these materials are microphase-separated (394,405,406). The adhesive properties of PSAs are a function of the unique rheological properties of the MQ resin/PDMS blends.

The key adhesive properties of a silicone PSA are tack, peel adhesion, and cohesive strength (lap shear). The required balance of these properties is controlled by several factors. A PSA composition rich in MQ resin usually favors low tack and high peel adhesion and cohesive strength values. A high cross-link density or a larger number of covalent bonds between the MQ resin and the PDMS network favors high cohesive strength at the expense of tack and peel adhesion. Low peel adhesion is observed when uncured cyclic or linear silicones are present. Other PSA properties important to the tape manufacturer, ie, viscosity, solvent content, cure time, and temperature, are typically controlled by varying the type and amount of organic solvent and peroxide cure catalyst. Primers are also frequently used to promote adhesion of the PSA to the tape backing.

One advance in silicone PSA chemistry is the use of Pt-catalyzed hydrosilation reactions as the method of cure (408). Advantages of hydrosilation-cured silicone PSAs include lowering the level of solvent, lowering cure temperatures, increasing line speeds, as well as the ability to be used with a broader range of tape backings. In this technology, lower molecular weight polymers containing hydrosilation-reactive SiH and SiVi groups are substituted for the traditional OH-end-capped PDMS gum. A Pt catalyst and inhibitor are used instead of a peroxide.

There are several important, specialized applications for silicone PSA tapes. Platers tapes mask selected areas of parts during etching or plating operations. Masking tapes are also used as protective coatings against high temperatures, radiation, harsh chemical environments, or moisture. These tapes are used frequently in the manufacture of printed circuit boards (409). Splicing tapes are used to join plastic films. Silicone PSAs are often used to splice low surface energy materials or to provide high cohesive strengths at temperature extremes. Plasma or flame spray tapes are used to protect selected metal surfaces during sandblasting or flame spraying operations. Silicone PSAs are particularly useful as the adhesive for electrical insulating tapes. A common application is as a wire wrap in motor coils. Silicone PSAs are also used in medical applications, notably as an adhesive for bandages and transdermal drug delivery systems (410).

Silicone Hardcoats. Silicone hardcoat technology evolved from the need to develop thin film coatings to impart abrasion and chemical resistance to plastic substrates. The first commercial silicone hardcoat products were developed in the 1970s (411–413). The basic chemistry involves first hydrolyzing a trialkoxysilane in the presence of an aqueous colloidal silica solution. The resulting solution is then diluted with alcohols such as isopropanol or *n*-butanol and formulated with various additives to give the final coating product. The coating and curing chemistry of silicone hardcoats is directly related to the widely studied field of sol–gel chemistry (414). The early silicone hardcoats were thermally cured to give thin, glass-like films having excellent adhesion to a wide variety of substrates. These coatings were used in applications such as safety lenses in eyeglasses, acrylic sheets for architectural glazing, and in automotive lighting.

Advances in silicone hardcoat technology include the development of weathereable and uv-curable hardcoats. Weatherable hardcoats are used in exterior applications such as polycarbonate windows and automotive lighting. These thermally cured coatings contain copolymerizable uv stabilizers which provide protection from solar radiation to photosensitive polymeric substrates (415). Uv-curable silicone hardcoats are prepared using functionalized trialkoxysilanes, which are hydrolyzed in the presence of aqueous colloidal silica. The resulting resins are diluted in photocurable monomers such as acrylates, epoxides, or vinyl ethers, and the alcohol/water solvent is then removed to yield solvent-free coating compositions (416). These can be cured using uv or electron beam radiation to give transparent, abrasion- and chemical-resistant coatings on glass, metal, or a variety of plastic substrates. These silicone hardcoats are replacing the traditional solvent-borne, thermally cured coatings in those applications requiring low temperature cure and reduced solvent emissions.

Paper Release Coatings. Paper release coatings are a fast-growing segment of silicone technology. These materials are used in label systems in which the silicone coating is part of the disposable paper liner. The role of the silicone is to provide a low surface energy interface between the paper liner and the adhesive label. Thermally cured paper release coatings are usually solventless mixtures of an SiVi-terminated PDMS, an SiH-containing cross-linker, a Pt hydrosilation catalyst, and a cure inhibitor. Having reactive SiVi groups on the ends of the polymer chains rather than in the middle of the chain is preferred for faster cure (417). Typical hydrosilation inhibitors are alkynols or alkyl maleates. Control release additives are frequently used in paper release coatings to increase the peel strength required to remove the linear from the adhesive label. These additives are usually MQ resins that function by changing the viscoelastic properties of the coating. It is also preferred that these resins contain reactive SiVi groups that participate in the curing reactions.

Technology of silicone paper release coatings is being driven by the need for faster line speeds for coating paper. Advances include faster thermal curing and radiation-curable paper coatings. Substitution of 5-hexenyl groups for vinyl groups has been reported to increase the cure rate of thermal paper release coatings (418–420). However, the industry is rapidly evolving toward the use of uv-curable products. These coatings use silicone polymers and control release resin additives containing photocurable groups such as epoxides, acrylates, or vinyl ethers (421).

Typical substrates for silicone release coatings are supercalendered kraft paper, glassines, and thermally sensitive films such as polyethylene and polypropylene. Ideal curing conditions are 150°C or lower, and line speeds are as fast as 460 m/min. Key properties for release coatings are cure speed, integrity of cure, and stable release values.

Room-Temperature Vulcanizable Silicones. Moisture-curable, room-temperature vulcanizable (RTV) silicones represent one of the largest-volume and commercially most successful silicone technologies. When exposed to atmospheric moisture, RTV silicones undergo hydrolysis and condensation reactions and cure into high strength elastomers. The rate of cure is dependent on temperature and humidity. In the uncured state, these one-part products have a shelf life of six months to several years. The cured elastomers have excellent primerless adhesion to substrates as varied as glass, metals, wood, masonry, and plastics. Additional benefits include excellent weatherability, durability, electrical insulation, chemical resistance, stability at high temperatures, and flexibility at low temperatures. RTV products that have a wide range of viscosities, from low viscosity self-leveling potting compositions to high viscosity thixotropic pastes, are commercially available. One of the few disadvantages is a slow rate of cure, often requiring more than seven days in low humidities to achieve optimum properties. The cure speed can be increased by raising the temperature or by using a two-part formulation that includes water in one of the parts. Sealants and adhesives are the chief uses of RTV silicones.

RTV Silicone Chemistry. There are two basic cure chemistries used by RTV silicones: the acetoxy-based and the alkoxy-based cure systems. Acetoxy-based RTV silicones were first commercialized in the early 1960s (422,423). The general chemical reactions of these first-generation products are shown in Figure 8.

In the first reaction, a multifunctional acetoxysilane cross-linker, such as methyltriacetoxysilane, reacts with a silanol-terminated PDMS, which releases acetic acid and forms a polymer containing multiple, hydrolytically unstable acetoxysilane end caps. When exposed to atmospheric moisture, this functionalized polymer hydrolyzes, releasing more acetic acid and producing diol-functionalized polymer end caps. This intermediate then undergoes self-condensation reactions, resulting in a silicone rubber. Tin(IV)-derived catalysts are frequently used to accelerate the cure of these products. Acetoxy-based RTV silicones have the advantage of having a long shelf life in the uncured state; however, their commercial applications are limited by the odor and corrosive nature of acetic acid.

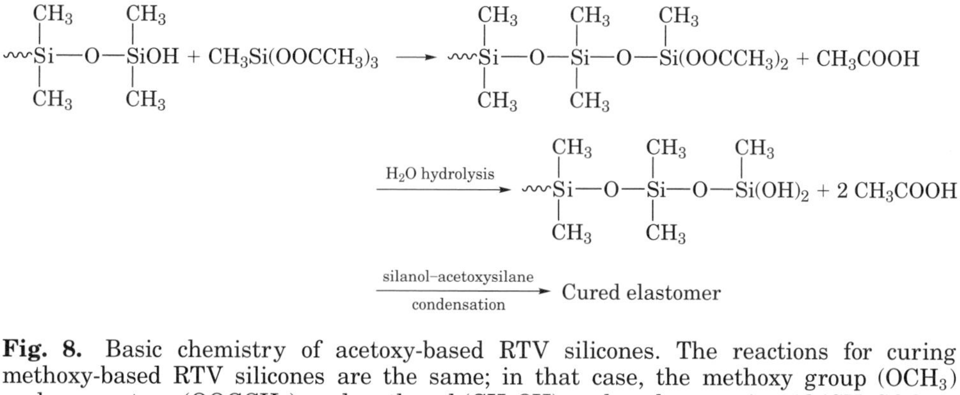

Fig. 8. Basic chemistry of acetoxy-based RTV silicones. The reactions for curing methoxy-based RTV silicones are the same; in that case, the methoxy group (OCH_3) replaces acetoxy ($OOCCH_3$), and methanol (CH_3OH), rather than acetic acid (CH_3COOH), is formed.

Methoxy-based RTV silicones were commercialized in the late 1960s to overcome the deficiencies of their acetoxy predecessors (424). As described in Figure 8, the general chemistry of methoxy-based RTV silicones is similar to their acetoxy analogues.

The principal differences are that methoxy-based cross-linkers such as methyltrimethoxysilane do not spontaneously react with silanol-end-capped PDMS, and methoxysilane-terminated PDMS does not react with atmospheric moisture under pH-neutral conditions. Catalysts are required for both of these reactions. Amines, Ti complexes, Al complexes, and ammonium carboxylate salts have been reported to be end-capping catalysts (425–428). Tin- and titanium-based catalysts are chiefly used as moisture cure catalysts (425,426,429). Unfortunately, these catalysts also promote undesirable siloxane redistribution reactions, which lead to instability of these RTV silicones in the uncured state (430). Two strategies are used to prepare commercially useful methoxy-based RTV silicones. In the first, some Ti- and Sn-based cure catalysts containing chelated ligands, and an end-capping catalyst based on ammonium formate salts, have been found to be poor siloxane redistribution catalysts, and therefore are useful catalysts in RTV silicones having shelf lives of six months or greater (425,426,431–443). In the second, one-part RTV silicones with a higher degree of stability are achieved by removing methanol either physically or with a methanol scavenger such as hexamethyldisilazane (429). Removal of methanol has been reported to block destabilizing siloxane redistribution reactions (430).

RTV Silicone Compositions, Properties, and Applications. A typical commercial RTV silicone is a blend of a PDMS, end-capped with OH groups or alkoxysilanes. The compositions typically contain excess cross-linking silane, reinforcing fillers, plasticizing oils, adhesion-promoting silanes, a cure catalyst, and other optional additives. The key properties are the tensile strength and adhesion of the cured elastomer, the shelf life and thixotropic behavior of the uncured product, and the cure rate. The tensile strength of an RTV silicone is a function of the cross-link density of the silicone elastomer and the amount and type of reinforcing filler. Useful fillers include high surface area, silane-treated fumed silica, and calcium carbonate. Plasticizing fluids are usually trimethylsiloxy-end-capped PDMS fluids, which are used to improve thixotropy and adhesion. The outstanding adhesive properties of RTV silicones are affected by almost every ingredient of the composition. Adhesion promoters are frequently used to enhance adhesion to difficult substrates such as glass, metals, and plastics. A wide variety of trialkoxy- or dialkoxy-substituted functional silanes are used as adhesion promoters. Examples include γ-aminopropyltrimethoxysilane, cyanoethyltriethoxysilane, and glycidoxypropyltrimethoxysilane. Other additives used in RTV silicones are iron and rare-earth oxides for improved high temperature stability, platinum for flame retardancy, fungicides for high humidity applications, and a variety of pigments. One-part silicone RTV silicones are sold in three primary markets: household consumer products, construction products, and industrial adhesives. Noncorrosive RTV silicones using methoxy-based cure chemistry are useful as electronic potting compositions. Typical RTV silicon properties are given in Table 11.

Table 11. Typical RTV Silicone Properties

Property	Value
density, g/cm^3	1.07
hardness, Shore A	30–35
tensile strength, MPa[a]	2.55–3.24
elongation, %	375–450
service temperature, continuous, °C	200–260
max temperature, short periods, °C	260–315
dielectric strength, kV/mm	19.7

[a]To convert MPa to kgf/cm^2, multiply by 10.2.

Analysis and Testing of Silicones

The unique chemical, physical, and spectroscopic properties of organosilicon compounds are reflected in the analytical methodology used for the detection, quantification, and characterization of these compounds. Several thorough, up-to-date reviews dealing with analytical methods applied to silicones have been published (434–436).

Chemical Analysis. The presence of silicones in a sample can be ascertained qualitatively by burning a small amount of the sample on the tip of a spatula. Silicones burn with a characteristic sparkly flame and emit a white sooty smoke on combustion. A white ashen residue is often deposited as well. If this residue dissolves and becomes volatile when heated with hydrofluoric acid, it is most likely a siliceous residue (437). Quantitative measurement of total silicon in a sample is often accomplished indirectly, by converting the species to silica or silicate, followed by determination of the heteropoly blue silicomolybdate, which absorbs at 800 nm, using atomic spectroscopy or uv spectroscopy (438–443). Pyrolysis gc followed by mass spectroscopic detection of the pyrolysate is a particularly sensitive tool for identifying silicones (442,443). This technique relies on the pyrolytic conversion of silicones to cyclics, predominantly to D_3 [541-05-9], which is readily detected and quantified (eq. 37).

$$\text{~~}(CH_3)_2SiO\text{~~}_n \xrightarrow{\Delta} \begin{array}{c} H_3C \quad CH_3 \\ Si \\ O \quad O \\ H_3C-Si \quad Si-CH_3 \\ H_3C \quad O \quad CH_3 \end{array} + [(CH_3)_2SiO]_x \tag{37}$$

$$D_3$$

Functional groups in silicone monomers and polymers are characterized by chemical reactivity. For example, chlorosilanes are hydrolyzed, and the halogen is quantified by titration with alkali or silver nitrate. Other types of halogen substitution may require more drastic methods of decomposition. Silicon hydride is assayed by determining the hydrogen evolved after base-catalyzed hydrolysis or alcoholysis. Silanol concentration can be calculated by measuring the methane evolved with methyl Grignard reagent. Water is corrected for by reaction with calcium hydride, which, unless specially prepared, does not react with silanol. Water and silanol levels can also be measured separately by ir techniques.

The structural architecture of silicone polymers, such as the number of D, T, and Q sites and the number and type of cross-link sites, can be determined by a degradative analysis technique in which the polymer is allowed to react with a large excess of a capping agent, such as hexamethyldisiloxane, in the presence of a suitable equilibration catalyst (eq. 38). Triflic acid is often used as a catalyst because it promotes the depolymerization process at ambient temperature (444). A related process employs the KOH- or KOC_2H_5-catalyzed reaction of silicones with excess $Si(OC_2H_5)_4$ (eq. 39) to produce ethoxylated methylsilicon species, which are quantitatively determined by gc (445).

$$[(CH_3)_3SiO]_m\,[(CH_3)_2SiO]_n\,[CH_3SiO_3]_o\,[SiO_4]_p \; + \; (CH_3)_3SiOSi(CH_3)_3 \longrightarrow$$

$$\quad\quad M \quad\quad\quad\quad D \quad\quad\quad\quad T \quad\quad\quad Q$$

$$n\,(CH_3)_2Si[OSi(CH_3)_3]_2 \; + \; o\,CH_3Si[OSi(CH_3)_3]_3 \; + \; p\,Si[OSi(CH_3)_3]_4 \quad (38)$$

$$[(CH_3)_3SiO]_m\,[(CH_3)_2SiO]_n\,[CH_3SiO_3]_o\,[SiO_4]_p \; + \; Si(OC_2H_5)_4 \; \xrightarrow[\text{or } KOC_2H_5]{\text{KOH}}$$

$$\quad\quad M \quad\quad\quad\quad D \quad\quad\quad\quad T \quad\quad\quad Q$$

$$m\,(CH_3)_3SiOC_2H_5 \; + \; n\,(CH_3)_2Si(OC_2H_5)_2 \; + \; o\,CH_3Si(OC_2H_5)_3 \quad (39)$$

Instrumental Methods. A variety of spectroscopic techniques are available for the characterization of silicones. Descriptions of these techniques and literature references relevant to silicone analysis are summarized in Table 12.

A number of techniques have been developed for the trace analysis of silicones in environmental samples. In these analyses, care must be taken to avoid contamination of the samples because of the ubiquitous presence of silicones, particularly in a laboratory environment. Depending on the method of detection, interference from inorganic silicate can also be problematic, hence nonsilica-based vessels are often used in these determinations. Silicones have been extracted from environmental samples with solvents such as hexane, diethyl ether, methyl isobutylketone, ethyl acetate, and tetrahydrofuran (THF) using either sequential or Soxhlet techniques (470–475). Silicones of a wide range of molecular weights and polarities are soluble in THF. This feature, coupled with its volatility and miscibility with water, makes THF an excellent solvent for the extraction of silicones from wet samples, ie, soils and sediments. Trace levels of silicones extracted from environmental samples have been measured by a number of techniques, including atomic absorption spectroscopy (aas), inductively coupled plasma-atomic emission spectroscopy (ICP-aes), pyrolysis gc-ms, as well as ^{1}H- and ^{29}Si-nmr spectroscopy (459,468,472,476–481). Low molecular weight, volatile silicones can be determined by gc or gc-ms techniques, either directly or after derivatization with a suitable silylating agent. Water-soluble silicones can be determined by HCl digestion, followed by extractive derivatization with hexamethyldisiloxane and gc analysis of the trimethylsilyl-capped monomeric species (482).

Table 12. Spectroscopic Methods for Analysis of Silicones

Technique	Information provided	Reference
mass spectrometry	identification, structural information, molecular weight information, and quantification; information is dependant on the mode of ionization; particularly useful when combined with a chromatographic separation technique, such as in gc-ms, hplc-ms, and pyrolysis gc-ms	442, 443, 446–448
ir/uv spectroscopy	identification, structural information, quantification and functional group identification; can be employed as detectors in chromatographic separations	449–457
nmr[a]	detailed structural information; molecular weights can be determined by end group analysis; solid-state cp-mas techniques can be used to characterize solid samples and cross-linked elastomers	458–465
atomic absorption	quantitatively detects total silicon in the sample; sensitivities in the ppb range are possible; particularly useful when used in combination with chromatographic separation devices; can be used to detect trace elemental contaminants in silicones	466–469

[a]Of particular interest are ^{1}H, ^{13}C, and ^{29}Si nuclei.

The use of separation techniques, such as gel permeation and high pressure liquid chromatography interfaced with sensitive, silicon-specific aas or ICP detectors, has been particularly advantageous for the analysis of silicones in environmental extracts (469,483–486). Supercritical fluid chromatography coupled with various detection devices is effective for the separation of silicone oligomers that have molecular weights less than 3000 Da. Time-of-flight secondary ion mass spectrometry (TOF-sims) is applicable up to 10,000 Da (487).

Determination of Physical Properties. Common properties of silicone polymers, such as refractive index, density, and viscosity, are measured using conventional techniques (488). Molecular weights can be determined by standard methods, provided suitable reference standards are available. Empirical viscosity/molecular weight correlations are also useful as a simple means of obtaining approximate molecular weights (489). End group analysis by suitable spectroscopic or chemical methods can give reliable number-average molecular weight data. Long-chain branching in silicone polymers can be ascertained chemically, by determining the ratio of branch sites to end groups using degradation methods (444,445), or instrumentally, by measuring molecular size–viscosity relationships, for example, using a gpc instrument coupled to a differential viscometry or light-scattering detector (490,491). The level of cross-linking in silicone polymers, quantified in terms of the average molecular weight of strands between crosslinks, is often measured by solvent swelling or modulus measurements, which are related to the combined effects of physical and chemical cross-links (492,493). Chemical cross-links can be quantified by nmr techniques or be degradative analysis to determine actual concentrations of T and Q cross-link sites.

Thermal stability, highly significant in silicone evaluation, is monitored by thermogravimetric analysis (tga), differential scanning calorimetry (dsc), or dynamic mechanical analysis (dma). Because environmental factors and chemical neutrality affect performance at high temperatures, oven-aging tests on finished products are frequently required. Measurement of weight loss on heat aging is frequently more sensitive than direct chemical methods for detecting the presence of trace amounts of strong acid or base, which reduce polymer stability.

A great variety of analytical techniques have been applied to silicone processing and quality control. In the early days of commercial production, the composition of chlorosilane intermediate mixtures was determined by analytical distillation and weight determination of fractions, supplemented by wet analysis of hydrolyzable chlorine and specific gravity measurements. These methods have largely been supplanted by gc and gc-ms methods. Gas chromatography is also used to monitor alkoxysilanes, acyloxysilanes, and the lower boiling point linear and cyclic siloxanes. On-line (in-process) monitoring of continuous and semicontinuous processes employs these and other relevant techniques.

Quality control testing of silicones utilizes a combination of physical and chemical measurements to ensure satisfactory product performance and processibility. For example, in addition to the usual physical properties of cured elastomers, the plasticity of heat-cured rubber and the extrusion rate of TVR elastomers under standard conditions are important to the customer. Where the silicone application involves surface activity, a use test is frequently the only reliable indicator of performance. For example, the performance of an antifoaming agent can be tested by measuring the foam reduction when the silicone emulsion is added to an agitated standard detergent solution. The product data sheets and technical bulletins from commercial silicone producers can be consulted for more information.

Properties of silicone fluids can be determined by a number of ASTM tests that were developed for the petroleum industry. ASTM D2225 describes methods of testing silicone fluids. ASTM tests typically used for determining standard physical properties are as follows: flash point, D92 and D93; pour point, D97; specific gravity, D1298; viscosity, D445; refractive index, D1807; and volume resistivity, D1169.

Health and Environmental Aspects

Few materials of commercial importance possess a range of applications as diverse as silicones. The widespread acceptance of silicones has given rise to remarkable growth of these materials since their first commercial production in the mid-1940s. In 1950, approximately 10^3 metric tons of silicones were produced. By 1965, production amounted to 1.4×10^4 metric tons and in 1977, world production reached 1.5×10^5 tons (494). By comparison, 5×10^5 tons of phthalate plasticizers were produced in 1977. More recently, silicones have found increasing use in personal care products, such as deodorants, shampoos, and cosmetics. As a result of their widespread use and remarkable inertness toward light, heat, and chemical agents, an understanding of the environmental fate and distribution of silicones is important. Several reviews dealing with this subject are available (495–498).

Organosilicones have been detected in terrestrial, aquatic, and atmospheric environments (Table 13). The highest concentrations of environmental silicones are found in the sludges of wastewater treatment plants and upstream of process plants. Much lower levels are found in water or sediment samples except at certain point sources, such as in effluents of dyeing factories and other industries that employ silicones in their processes. In general, levels of organosilicones in sludges and sediments have been found to correlate with the total organic content of these media as determined by pyrolysis weight losses. This implies a migratory aptitude similar to that of other organic pollutants. For example, analysis of undisturbed, ^{210}Pb-dated sediment taken from Puget Sound in Washington indicates the presence of organic silicon in a 15-yr-old sludge but not in a 60-year-old sample (504,505). This result suggests that silicones do not migrate appreciably of their own accord but follow sediment movement.

Silicones have been detected in water from various rivers and in effluent from wastewater treatment facilities. They are barely detectable in water from municipal treatment facilities. Water solubility (S) of silicones has been found to correlate with the octanol–water partition coefficients (K_{ow}) for a series of trimethylsilyl-capped siloxane oligomers of various molecular weights according to equation 40 (506).

$$\log K_{ow} = 3.047 - 1.048 \log S \qquad (40)$$

This expression can be used to predict solubilities from the octanol-water partition coefficients. Solubility and K_{ow} data for four oligomeric siloxanes are listed in Table 14.

Table 13. Silicones in Various Environmental Samples

Location	Silicone, ppm[a]	Organic silicone, pp[b]	Reference
Nagara River[c] sediment	0.3–5.8	2–54.2	477
dye factory	41–6290	0–1150	477
sewage plant sludge	144	10.2	477
night soil plant	34.3		477
domestic wastewater treatment plant		2.4–4.9	477
Blue Plains[d] wastewater treatment plant		4	499
sludge	89–104		499, 500
filter cake	26–46		499
Potomac River[d]	0.46–3.07		500
Delaware Bay	0.1–1.56	34	500
Chesapeake Bay	0–36.1	30	500, 501
Bight[e] wastewater treatment plant	0.48		502
Hudson River[e] sediment	10–350		503

[a]Wt Si/dry wt.
[b]Aquatic samples.
[c]Japan.
[d]Washington, D.C.
[e]New York.

**Table 14. Effect of Siloxane Molecular Weight on Water
Solubility and Octanol-Water Partition Coefficient**

Molecular weight	Solubility in H_2O, ppm	$\log K_{ow}$
1,200	1.60	2.86
6,000	0.56	3.26
25,000	0.17	3.83
56,000	0.076	4.25

The rapid drop in solubility and increase in K_{ow} with increasing molecular weight are evident from these data. Table 13 lists typical levels of silicones reported for aquatic samples, as well as terrestrial samples.

Because of their hydrophobic nature, silicones entering the aquatic environment should be significantly absorbed by sediment or migrate to the air–water interface. Silicones have been measured in the aqueous surface microlayer at two estuarian locations and found to be comparable to levels measured in bulk (505). Volatile surface siloxanes become airborne by evaporation, and higher molecular weight species are dispersed as aerosols.

Silicone contamination has been implicated as a cause of failure in telephone switching systems and other devices that contain relay switch contacts (507). Analysis of airborne particulates near telephone switching stations showed the presence of silicones at these locations. Where the indoor use of silicones is intentionally minimized, outdoor levels were found to be higher than inside concentrations (508). Samples of particulates taken at two New Jersey office buildings revealed silicone levels that were considerably higher indoors than outdoors. In these cases, indoor silicone aerosols are believed to be generated primarily by photocopiers, which use silicone fuser oils.

Airborne organic silicon has been detected in samples collected at Barrow, Alaska (509). Organic silicon levels corresponding to an airborne concentration of 8 ng/m^3 were detected. As a comparison, these samples were determined to hold approximately 20 ng/m^3 of phthalate-based plasticizers.

Knowledge of the transformation of silicones under various environmental conditions is key to understanding the fate of these materials. Model studies predict that a large fraction of silicones entering the environment through wastewater treatment systems, ie, municipal treatment plants or septic systems, will ultimately be deposited on soils as a result of absorption or of sludge amendment (510). The chemistry of silicones on soils is thus an important factor in assessing the overall environmental impact of these materials. Silicone fluids degrade when exposed to soils (474,511,512). The rate of degradation depends on soil type and moisture content. Degradation is slow but observable on moist soils and becomes quite rapid when the available moisture content of the soil drops below 5%. The products of soil-induced abiotic degradation are typically silanol-terminated monomers and oligomers. Dimethylsilane-1,1-diol (**1**), the simplest monomeric unit derived from silicone hydrolysis, is an important degradation product of abiotic silicone degradation. Diol (**1**) is produced when silicones are exposed to soils and sediments even when the overall extent of polymer degradation is small, indicating a possible kinetic preference for its formation. As a polar, water-soluble species, (**1**) exhibits environmental distribution properties drastically different from the silicones from which it is derived (513). Differences in the chemistry of (**1**) and the parent silicones are also expected. In organic solvents, (**1**) undergoes rapid self-condensation and forms oligomeric products, whereas in dilute (<400 ppm) aqueous solution (eq. 41), equilibrium among (**1**) and water-soluble oligomeric silanols is dominated by (**1**) (514). Therefore, (**1**) is expected to be the principal waterborne silicone species found in most environmental samples.

$$(41)$$

(**1**)

Irradiation of water-soluble silicon species such as (**1**) with artificial sunlight (xenon arc lamp) in the presence of environmentally significant levels of nitrate led to their rapid disappearance ($t_{1/2}$ = 4–9 d) (515). Silicate was determined to be the ultimate photodegradation product. Transformation of the soluble silicones to silicate under these conditions was complete in about 35 days. Irradiation of an emulsified high molecular weight silicone fluid (degree of polymerization (DP) = 50) resulted in negligible (1.7%) conversion to silicate in the same time period. Degradation did not occur in the dark or in the absence of nitrate. Nitrate is thought to act in this system by providing a source of photochemically generated hydroxyl radicals (eq. 42). Oxidative attack of these reactive species on the silicon–carbon bonds leads eventually to silicate and CO_2.

$$NO_3^- + H_2O \xrightarrow{h\nu} NO_2 + HO^\cdot + HO^- \qquad (42)$$

Water-soluble silanols such as (**1**) were found to undergo successive oxidative demethylations with tropospheric ultraviolet irradiation in the presence of suitable chromophores, such as nitrogen oxides (516). The water-soluble methylated silicones did not promote diatom (*Navicula pelliculosa*) growth but the demethylated photo products did. The sequence of soil-induced degradation of silicones to water-soluble species such as (**1**), followed by light-induced conversion to silicate, suggests a pathway, conceptually at least, for the mineralization of silicones.

Polydimethylsiloxane does not biodegrade at an observable rate under normal conditions of environmental exposure. For example, a ^{14}C-labeled 300-cSt PDMS fluid showed no evidence of biodegradation when exposed to activated sludge over a period of 70 days (517). In a similar microbiological screening study, silicones were found to be inert to all applied cultural conditions (518). However, when ^{14}C-labeled PDMS was applied to soil and allowed to degrade to low molecular weight species, the degradation products were further degraded to silicate and [14]CO_2, presumably via a biotic pathway (519). ^{14}C-dimethylsilanediol is biodegraded on soils at a rate of $\approx$1% per month (520). The rate of [14]CO_2 production can be increased an order of magnitude by the addition of certain carbon sources, such as 2-propanol and methylsulfone, to the medium. Thus, by a combination of abiotic, soil-mediated degradation and biotic or light-induced conversion of the degradation products, silicones can undergo complete mineralization in the environment. The rates at which this mineralization occurs depend on the conditions of environmental exposure.

Economic Aspects

Silicones, an important item of commerce, are widely available commercially (9,494). The principal manufacturers of silicone operate direct-process reactors to produce dimethyldichlorosilane and, ultimately, polydimethylsiloxane. Typical plants produce more than 450 t per year. The silicone industry is a global enterprise in the 1990s, with principal producers in the United States (Dow Corning, GE, and OSi), Europe (Wacker Chemie, Hüls, Rhône-Poulenc, and Bayer), and Southeast Asia (Shin-Etsu, Toshiba Silicones, and Dow Corning, Japan). Table 15 lists the approximate sales of the principal producers for 1991.

In addition to the basic silicone producers, many other companies have developed product lines as specialty suppliers or users of silicones. This trend of diversification started in the 1970s and has continued through the 1990s.

Table 15. Approximate Sales for Principal Silicone Producers in 1991[a]

Company	Sales, $ millions
Dow Corning	1850
General Electric	750
Shin-Etsu	750
Wacker Chemie	675
Rhône-Poulenc	460
Union Carbide[b]	300
others[c]	465

[a]Data from U.S. Dept. of Commerce, Springfield, Virginia.
[b]Now OSi.
[c]Hüls divested its silicones operation, Petrarch, to UTI. Other silicone producers in 1995 are PCR, Gelest, Th. Goldschmidt, Bayer, Dow-Toray, Toshiba Silicones, and Nippon Unicar. Silicones are also produced in the Ukraine, CIS, Czech Republic, and China.

BIBLIOGRAPHY

"Silicones" in *ECT* 1st ed., Vol. 12, pp. 393–413, by R. R. McGregor, Mellon Institute of Industrial Research; in *ECT* 2nd ed., Vol. 18, pp. 221–257, by R. Meals, General Electric Co.; "Silicon Compounds (Silicones)" in *ECT* 3rd ed., Vol. 20, pp. 922–962, by B. B. Hardman and A. Torkelson, General Electric Co.

1. E. P. Plueddemann, *Silane Coupling Agents*, Plenum Press, New York, 1983.
2. E. G. Rochow, *Silicon & Silicones*, Springer-Verlag, Berlin, 1987.
3. D. Hunter, *Chem. Week*, 24 (Feb. 19, 1992).
4. A. I. Gorbunfov, A. P. Belyi, and G. G. Filippov, *Rus. Chem. Rev.* **43**, 291 (1974).

5. A. D. Petrov, B. F. Mironov, V. A. Ponomarenko, and E. A. Chernyshev, *Synthesis of Organosilicon Monomers*, Consultants Bureau, New York, 1964, p. 36.
6. J. M. Zeigler and F. W. G. Fearon, eds., *Silicon-Based Polymer Science*, ACS Advances in Chemistry Series 224, American Chemical Society, Washington, D.C., 1990.
7. F. O. Stark, J. R. Falender, and A. P. Wright, in G. Wilkinson, F. G. A. Stone, and E. W. Abel, eds., *Comprehensive Organometallic Chemistry*, Pergamon, Oxford, U.K., 1982, Chapt. 9.3, p. 305.
8. C. Eaborn and R. W. Bott, in A. G. MacDiarmid, ed., *The Bond to Carbon*, Vol. I, Part 1, Marcel Dekker, New York, 1968, p. 105.
9. R. J. H. Voorhoeve, *Organohalosilanes: Precursors to Silicones*, Elsevier, New York, 1967.
10. R. R. McGregor, *Silicones & Their Uses*, McGraw-Hill Book Co., Inc., New York, 1954.
11. A. Ladenburg, *Ann. Chem.* **164**, 300 (1872).
12. J. J. Ebelman, *Compt. Rend.* **19**, 398 (1844).
13. F. S. Kipping, *Proc. Chem. Soc.* **20**, 15 (1904).
14. W. Dilthey, *Chem. Ber.* **37**, 1139 (1904).
15. H. A. Liebhafsky, *Silicones Under the Monogram*, John Wiley and Sons, Inc., New York, 1978.
16. A. Stock, *Hydrides of Boron and Silicon*, Cornell University Press, Ithaca, N.Y., 1933, p. 20.
17. E. L. Warrick, *Forty Years of Firsts*, McGraw Hill Book Co., Inc., New York, 1990.
18. E. G. Rochow and W. Gilliam, *J. Am. Chem. Soc.* **63**, 798 (1941).
19. J. F. Hyde and R. DeLong, *J. Am. Chem. Soc.* **63**, 1194 (1941).
20. V. Chvalovsky, in B. J. Aylett, ed., *Organometallic Derivatives of the Main Group Elements*, Butterworths, London, 1975.
21. T. C. Kendrick, B. Parbhoo, and J. W. White, in S. Patai and Z. Rappoport, eds., *The Silicon-Heteroatom Bond*, John Wiley & Sons, Ltd., Chichester, U.K., 1989, Chapt. 3.
22. L. D. Gasper-Galvin, D. G. Rethwisch, D. M. Sevenich, and H. B. Fridrich, in K. M. Lewis and D. G. Rethwisch, eds., *Catalyzed Direct Reactions of Silicon*, Elsevier, Amsterdam, the Netherlands, 1993, p. 279.
23. B. Imelik and P. Trambonze, *Bull. Soc. Chim. France*, **238**, 680 (1954).
24. V. S. Fikhtengol'ts and A. I. Klebansky, *J. Gen. Chem.* (*USSR*) **27**, 2535 (1957).
25. A. I. Klebansky and V. S. Fikhtengol'ts, *J. Gen. Chem.* (*USSR*) **26**, 2795 (1956).
26. T. C. Frank, K. B. Kester, and J. L. Falconer, *J. Catal.* **91**, 44 (1985).
27. N. Floquet, S. Yilmaz, and J. L. Falconer, *J. Catal.* **148**, 348 (1994).
28. W. J. Ward, A. Ritzer, K. M. Carroll, and J. W. Flock, *J. Catal.* **100**, 240 (1986).
29. G. Laroze, J. L. Plaque, G. Weber, and B. Gillot, in H. A. Øye and H. Rong, eds., *Silicon for Chemical Industry, Geironger, Norway, June 1992*, Institute of Inorganic Chemistry, NTH, Trondheim, Norway, p. 151.
30. K. M. Lewis, D. McLeod, and B. Kanner, in J. W. Ward, ed., *Catalysis*, Elsevier, Amsterdam, the Netherlands, 1988, p. 415.
31. B. Kanner and K. M. Lewis, in Ref. 22, p. 1.
32. V. M. Mikhaylov and V. N. Penskii, *Production of Monomer and Polymer Organosilicon Compounds*, English transl., No. AD-779129, NTIS, U.S. Dept. of Commerce, Springfield, Va., p. 18.
33. U.S. Pat. 4,727,173 (Feb. 23, 1988), F. D. Mendicino (to Union Carbide), U.S. Pat. 2,473,260 (June 14, 1949), E. G. Rochow (to General Electric Co.).
34. U.S. Pat. 4,778,910 (Oct. 18, 1988), J. O. Stoffer, J. F. Montle, and N. L. D. Somasivi.
35. R. J. Ayen and J. H. Burk, in C. J. Brinker, D. E. Clark, and D. R. Ulrich, eds., *Mater. Res. Soc., Symp. Proc.* **73**, 801 (1986).
36. U.S. Pat. 4,966,986 (Oct. 30, 1990), R. L. Halm and R. H. Zapp (to Dow Corning).
37. U.S. Pat. 4,962,220 (Oct. 9, 1990), R. L. Halm and R. H. Zapp (to Dow Corning).

38. U.S. Pat. 4,965,388 (Oct. 23, 1990), R. L. Halm, R. H. Zapp, and R. D. Streu (to Dow Corning).
39. U.S. Pat. 4,973,725 (Nov. 27, 1990), K. Lewis (to Union Carbide).
40. K. M. Lewis, R. A. Cameron, J. M. Larnerd, and B. Kanner, *Xth International Symposium on Organosilicon Chemistry*, Aug. 15–20, 1993, Poznan, Poland, p. 78.
41. U.S. Pat. 4,088,669 (May 9, 1978), J. R. Malek, J. L. Speier, and A. P. Wright (to Dow Corning).
42. U.S. Pat. 4,593,114 (June 3, 1986), K. M. Lewis and B. Kanner (to Union Carbide).
43. U.S. Pat. 4,255,348 (Mar. 10, 1981), B. Kanner and W. B. Herdle (to Union Carbide).
44. M. Kumada and K. Tamao, *Adv. Organomet. Chem.* **6**, 19 (1968).
45. E. Hengge, *Topics Curr. Chem.* **51**, 1 (1974).
46. R. Calas, J. Dunogues, G. Deleris, and N. Duffaut, *J. Organomet. Chem.* **225**, 117 (1982).
47. K. Tamao, T. Hayashi, and M. Kumada, *J. Organomet. Chem.* **114**, C9 (1976).
48. T. Hayashi, T. Kobayaski, A. Kawamoto, H. Tamashira, and M. Tanaka, *Organometallics*, **9**, 280 (1990).
49. J. D. Rich, *J. Am. Chem. Soc.* **111**, 5886 (1989).
50. J. D. Rich, *Organometallics*, **8**, 2609 (1989).
51. J. Stein, K. X. Lettko, J. A. King, and R. E. Colborn, *J. Appl. Polym. Sci.* **51**, 815 (1994).
52. I. Ojima, in S. Patai and Z. Rappoport, eds., *The Chemistry of Organic Silicon Compounds*, John Wiley & Sons, Inc., New York, 1989.
53. D. A. Armitage, in Ref. 7, Vol. 2, p. 117.
54. J. L. Speier, *Adv. Organomet. Chem.* **17**, 407 (1979).
55. J. F. Harrod and A. J. Chalk, in I. Wender and P. Pino, eds., *Organic Synthesis Via Metal Carbonyls*, John Wiley & Sons, Inc., New York, 1977, p. 673.
56. E. Lukevics, Z. V. Belyakov, M. G. Pomerantseva, and M. G. Voronkov, in D. Seyferth, ed., *J. Organomet. Chem. Library*, **5** 1–179 (1977).
57. A. J. Chalk and J. F. Harrod, *J. Am. Chem. Soc.* **87**, 16 (1965).
58. B. Marciniec, *Comprehensive Handbook on Hydrosilylation*, Pergamon Press, U.K., 1992.
59. W. Noll, *Chemistry and Technology of Silicones*, Academic Press, Inc., New York, 1978.
60. J. A. Semlyen, *Adv. Polym. Sci.* **21**, 41 (1976).
61. J. E. McGrath and J. S. Riffle, *Silicon Compounds, Register and Review*, No. S-5, Petrarch Systems, Inc., Bristol, Pa., 1982, p. 62.
62. R. Gutoff, *Ind. Eng. Chem.* **49**, 1807 (1957).
63. U.S. Pat. 4,609,751 (Sept. 2, 1986), A. L. Hajjar (to General Electric Co.).
64. U.S. Pat. 4,108,882 (Aug. 22, 1978), L. G. Mahone (to Dow Corning).
65. U.S. Pat. 4,366,324 (Dec. 28, 1982), K. Habata, K. Ichikawa, and M. Shimizu (to Shin-Etsu Chemical Co.).
66. Brit. Pat. 843,273 (1960), A. N. Pines (to Union Carbide).
67. L. Wilczek and J. Chojnowski, *Makromol. Chem.* **184**, 77 (1983).
68. M. G. Voronkov, V. P. Mileshkevich, and Yu. A. Yuzhelevskii, *The Siloxane Bond*, Consultants Bureau, New York, 1970.
69. J. J. Lebrun and H. Porte, *Polysiloxanes*, in G. Allen, J. C. Bevington, G. C. Eastmond, A. Ledwith, S. Russ, and P. Sigwalt, eds., *Comprehensive Polymer Science*, Vol. 5, Pergamon, Oxford, U.K., 1989.
70. U.S. Pat. 3,839,388 (Oct. 1, 1974), W. Hechtl, S. Nitzsche, and E. Wohlfarth (to Wacker-Chemie GmbH).
71. U.S. Pat. 4,739,026 (Apr. 19, 1988), M. Riederer and M. Piechler (to Wacker-Chemie GmbH).

72. U.S. Pat. 5,210,131 (May 11, 1993), J. Gilson and J. de la Crol Habimana (to Dow Corning).

73. W. T. Grubb, *J. Am. Chem. Soc.* **76**, 3408 (1954).

74. Z. Lasocki and S. Chrzczonowicz, *J. Polym. Sci.* **59**, 259 (1962).

75. Z. Lasocki and S. Chrzczonowicz, *Bull. Acad. Polon. Sci., Ser. Sci. Chim.* **9**, 519 (1961).

76. J. Chojnowski, S. Rubinsztajn, and L. Wilczek, *J. Chem. Soc., Chem. Commun.* **69** (1984).

77. J. Chojnowski, S. Rubinsztajn, and L. Wilczek, *Macromolecules*, **20**, 2345 (1987).

78. M. Cypryk, S. Rubinsztajn, and J. Chojnowski, *J. Organometal. Chem.* **446**, 91 (1993).

79. S. Rubinsztajn, M. Cypryk, and J. Chojnowski, *Macromolecules*, **26**, 5389 (1993).

80. J. Chojnowski, K. Kazimierski, S. Rubinsztajn, and W. Stanczyk, *Makromol. Chem.* **187**, 2345 (1986).

81. P. V. Wright and J. Am Semlyen, *Polymer*, **10**, 462 (1969).

82. J. A. Semlyen, in J. A. Semlyen, ed., *Cyclic Polymers*, Elsevier, Applied Science Publisher, London, 1986.

83. J. B. Carmichael, *Rubber Chem. Technol.* **40**, 1084 (1967).

84. J. F. Brown and G. M. F. Slusarczuk, *J. Am. Chem. Soc.* **87**, 931 (1965).

85. H. Jacobsen and W. H. Stockmayer, *J. Phys. Chem.* **18**, 1600 (1950).

86. J. Carmichael and J. Heffel, *J. Phys. Chem.* **69**, 2218 (1965).

87. J. Carmichael and R. Winger, *J. Polym. Sci. A*, **3**, 971 (1965).

88. J. Semlyen and P. Winger, *Polymer*, **10**, 543 (1969).

89. J. A. Semlyen, in S. J. Clarson and J. A. Semlyen, eds., *Siloxane Polymers*, Ellis Horwood and Prentice Hall, Englewood Cliffs, N.J., 1993.

90. D. T. Hurd and R. C. Osthoff, *J. Am. Chem. Soc.* **76**, 249 (1954).

91. U.S. Pat. 3,041,362 (June 26, 1962), R. L. Merker and W. A. Piccoli (to Dow Corning).

92. A. K. Gilbert and S. W. Kantor, *J. Polym. Sci.* **40**, 35 (1959).

93. U.S. Pat. 5,089,450 (Feb. 18, 1992), I. Watanuki, N. Kodana, and M. Sato (to Shin-Etsu Chemical Co.).

94. U.S. Pat. 4,250,290 (Feb. 10, 1981), L. P. Petersen (to General Electric Co.).

95. U.S. Pat. 4,128,568 (Dec. 5, 1978), W. Buchner, B. Degan, L. Fries, J. Helmut, R. Mundil, and K. H. Rudolph (to Bayer Actiengesellschaft).

96. P. V. Wright, in K. J. Ivin and T. Seagusa, eds., *Ring Opening Polymerization*, Elsevier, London, 1984.

97. J. Chojnowski, in Ref. 89.

98. W. T. Grubb and R. C. Osthoff, *J. Am. Chem. Soc.* **77**, 1405 (1955).

99. J. Chojnowski and M. Mazurek, *Makromol. Chem.* **176**, 2999 (1975).

100. M. Mazurek and J. Chojnowski, *Macromolecules*, **11**, 347 (1978).

101. M. Shinohara, *Am. Chem. Soc.* **14**, 1209 (1973).

102. M. Szwarc, *Carboanions Living Polymers and Electron Transfer Processes*, John Wiley & Sons, Inc., New York, 1968.

103. U.S. Pat. 3,337,497 (Aug. 22, 1967), E. E. Bostick (to General Electric Co.).

104. U. Maschke and T. Wagner, *Macromol. Chem.* **193**, 2453 (1992).

105. J. G. Zilliox, J. E. Roovers, and S. Bywater, *Macromolecules*, **8**, 573 (1975).

106. Y. Gnanou and P. Rempp, *Makromol. Chem.* **189**, 1997 (1988).

107. Yu. A. Yuzhelevskii, E. G. Kagan, and N. N. Fedoseeva, *Dokl. Akad. Nauk SSSR*, **190**, 647 (1970).

108. W. H. Dickstein and C. P. Lillya, *Macromolecules*, **22**, 3882 (1989).

109. L. C. Lee and O. K. Johannson, *J. Polym. Sci. Polym. Chem. Ed.* **14**, 729 (1976).

110. T. Suzuki, *Polymer*, **30**, 333 (1989).

111. S. Boileau, in J. E. McGrath, ed., *Anionic Polymerization*, ACS Symposium Series 166, American Chemical Society, Washington, D.C., 1981.

112. W. Patnode and D. F. Wilcock, *J. Am. Chem. Soc.* **68**, 358 (1946).
113. J. Chojnowski, M. Mazurek, M. Scibiorek, and L. Wilczek, *Makromol. Chem.* **175**, 3299 (1974).
114. U.S. Pat. 5,241,032 (Aug. 31, 1993), K. Kobayashi and Y. Yamamoto (to Dow Corning Toray Silicone Co., Ltd).
115. K. A. Andrianov, M. I. Shkolnik, V. M. Kopylow, and N. N. Bravina, *Vysokomol. Soed.* **B16**, 893 (1974).
116. U.S. Pat. 5,206,330 (Apr. 27, 1993), K. Kobayashi, N. Ida, and S. Mori (to Shin-Etsu Chemical Co.).
117. U.S. Pat. 3,297,725 (Jan. 10, 1967), G. Dieckelmann and W. Gundel (to Deutsche Hydrierwerke GmbH).
118. V. N. Gruber and L. S. Mukhina, *Vysokomol. Soed.* **3**, 174 (1961).
119. M. T. Bryk, N. N. Bagley, and O. D. Kurilenko, *Vysokomol. Soed.* **A17**, 1034 (1975).
120. U.S. Pat. 3,853,933 (Dec. 10, 1974), G. R. Siciliano (to General Electric Co.).
121. U.S. Pat. 3,853,934 (Dec. 10, 1974), G. R. Siciliano (to General Electric Co.).
122. U.S. Pat. 2,389,477 (Nov. 20, 1945), J. G. E. Wright and J. Marsden (to General Electric Co.).
123. L. N. Lewis, D. S. Johnson, and J. D. Rich, *J. Organomet. Chem.* **448**, 15 (1993).
124. P-S. Chung and M. A. Buese, *J. Am. Chem. Soc.* **115**, 11475 (1993).
125. D. Poczynok and M. A. Buese, *J. Polym. Sci. Polym. Chem.* **31**, 893 (1993).
126. T. C. Kendrick, *J. Chem. Soc.*, 2027 (1965).
127. E. Jordan, L. Lestel, S. Boileau, H. Charadame, and A. Gandini, *Makromol. Chem.* **190**, 267 (1989).
128. J. J. Lebrun, G. Sauvet, and P. Sigwalt, *Makromol. Chem., Rapid Comm.* **3**, 757 (1982).
129. J. Chojnowski, S. Rubinsztajn, and L. Wilczek, *L'Actualite Chimique*, **3**, 56 (1986).
130. J. Chojnowski, M. Scibiorek, and J. Kowalski, *Makromol. Chem.* **178**, 1351 (1976).
131. L. Wilczek and J. Chojnowski, *Macromolecules*, **14**, 9 (1981).
132. G. Sauvet, J. J. Lebrun, and P. Sigwalt, in E. J. Goethals, ed., *Cationic Polymerization and Related Processes*, Academic Press, London, 1984.
133. J. Chojnowski, L. Wilczek, and S. Rubinsztajn, in Ref. 135.
134. P. Sigwalt, C. Gobin, P. Nicol, M. Moreau, and M. Masure, *Macromol. Chem., Macromol. Symp.* **42/43**, 229 (1991).
135. P. Sigwalt, M. Masure, M. Moreau, and R. Bischoff, *Macromol. Chem., Macromol. Symp.* **73**, 147 (1993).
136. L. Wilczek and J. Chojnowski, *Makromol. Chem.* **180**, 117 (1979).
137. U.S. Pat. 3,294,725 (Dec. 27, 1966), D. E. Findlay and D. R. Weyenberg (to Dow Corning).
138. U.S. Pat. 4,288,356 (Sept. 8, 1981), D. J. Huebner and D. R. Weyenberg (to Dow Corning).
139. U.S. Pat. 4,999,398 (Mar. 12, 1991), D. Graiver and O. Tanaka (to Dow Corning).
140. D. E. Findlay, D. R. Weyenberg, and J. Cekada Jr., *J. Polym. Sci.* **C27**, 27 (1969).
141. U.S. Pat. 4,244,849 (Jan. 13, 1981), J. C. Saam (to Dow Corning).
142. Eur. Pat. 166,397 (1986), D. J. Huebner and J. C. Saam (to Dow Corning).
143. J. C. Saam and D. J. Huebner, *J. Polym. Sci., Polym. Chem. Ed.* **20**, 3351 (1982).
144. A. De Gunzbourg, J. C. Favier, and P. Hemery, *Polym. International*, **35**, 179 (1994).
145. E. J. Lawton, W. T. Grubb, and J. S. Baldwin, *J. Polym. Sci.* **19**, 455 (1956).
146. C. J. Wolf and A. C. Stewart, *J. Phys. Chem.* **66**, 1119 (1962).
147. A. S. Chawla and L. E. St. Pierre, *Adv. Chem. Ser.* **91**, 229 (1969).
148. A. S. Chawla and L. E. St. Pierre, *J. Appl. Polym. Sci.* **16**, 1887 (1972).
149. D. M. Naylor, V. T. Stannett, A. Deffieux, and P. Sigwalt, *J. Polym. Sci., Polym. Lett.* **24**, 319 (1986).

150. D. M. Naylor, V. T. Stannett, A. Deffieux, and P. Sigwalt, *Polymer*, **31**, 954 (1990).
151. A. M. Wrobel and M. R. Wertheimer, in R. d'Agostino, ed., *Plasma Deposition, Treatment and Etching of Polymers*, Academic Press, Inc., San Diego, Calif., 1990.
152. C. Rau and W. Kulisch, *Thin Solid Films*, **249**, 28 (1994).
153. D. L. Cho and H. Yasuda, *J. Appl. Polym. Sci., Appl. Polym. Symp.* **42**, 327 (1988).
154. N. Inagaki and Y. Hashimoto, *J. Appl. Polym. Sci., Appl. Polym. Symp.* **42**, 221 (1988).
155. D. R. Thomas, in S. J. Clarson and J. A. Semlyen, *PTR*, Prentice Hall, Englewood Cliffs, N.J., 1993.
156. "Organic Peroxides" and "Free Radical Initiators," *Modern Plastics Encyclopedia*, Vol. 62 (13), McGraw-Hill, Book Co., Inc., New York, 1985, p. 166.
157. U.S. Pat. 2,448,565 (Sept. 7, 1948), G. Wright and C. S. Oliver (to General Electric Co.).
158. M. L. Dunham, D. L. Bailey, and R. M. Miner, *Ind. Eng. Chem.* **49**(9), 1373 (1957).
159. S. W. Kantor, *ACS 130th Meeting*, American Chemical Society, Washington, D.C., Sept. 1956.
160. U.S. Pat. 2,445,794 (July 27, 1948), J. Marsden (to General Electric Co.).
161. R. J. Cush and H. W. Winnan, in A. Whelan and K. S. Lee, eds., *Developments in Rubber Technology*, Vol. 2, Applied Science Publishers, Ltd., London, 1981.
162. D. Wrobel, in G. Koerner, M. Schulze, and J. Weis, eds., *Silicones, Chemistry and Technology*, Vulkan-Verlag, Germany, 1991.
163. K. E. Polmanteer, *Rubber Chem. Technol.* **61**, 489 (1988).
164. J. B. Class and R. P. Grasso, *Rubber Chem. Technol.* **66**, 605 (1993).
165. J. L. Speier, *Adv. Organometal. Chem.* **17**, 407 (1979).
166. J. P. Collman and L. S. Hegedus, *Principles and Application of Organotransition Metal Chemistry*, University Science Books, Calif., 1980.
167. U.S. Pat. 3,775,452 (Nov. 27, 1973), B. D. Karstedt (to General Electric Co.).
168. M. Heidingsfeldova, M. Schatz, M. Czakoova, and M. Capka, *J. Appl. Poly. Sci.* **42**, 179 (1991).
169. C. W. Macosko and J. C. Saam, *Polymer Bull.* **18**, 463 (1987).
170. A. V. Gorshov, Y. M. Kopylov, A. A. Dontsov, and L. Z. Khazen, *Int. Poly. Sci. Tech.* **13**, T/26 (1986).
171. X. Quan, *Poly. Eng. Sci.* **29**, 1419 (1989).
172. A. M. Podoba, E. A. Goldovski, and A. A. Dortsov, *Int. Poly. Sci. Tech.* **14**, T/42 (1987).
173. A. J. Chalk and J. F. Harrod, *J. Am. Chem. Soc.* **87**, 16 (1965).
174. L. N. Lewis and N. J. Lewis, *J. Am. Chem. Soc.* **108**, 7228 (1986).
175. L. N. Lewis, *J. Am. Chem. Soc.* **112**, 5998 (1990).
176. L. N. Lewis, J. Stein, and Y. Gao, unpublished results.
177. U.S. Pat. 5,331,075 (July 19, 1994), C. A. Sumpter, L. N. Lewis, and W. B. Lawrence.
178. U.S. Pat. 4,603,168 (July 29, 1986), S. Sasaki and Y. Hamada (to Toray Silicones).
179. U.S. Pat. 4,490,488 (Dec. 25, 1984), R. J. Cush (to Dow Corning).
180. U.S. Pat. 3,445,420 (May 20, 1969), G. J. Kookootsedes and E. P. Plueddemann (to Dow Corning).
181. U.S. Pat. 4,336,364 (June 22, 1982), M. T. Maxson (to Dow Corning).
182. U.S. Pat. 4,347,346 (Aug. 31, 1982), R. P. Eckberg (to General Electric Co.).
183. U.S. Pat. 4,256,870 (Mar. 17, 1981), R. P. Eckberg (to General Electric Co.).
184. U.S. Pat. 4,783,552 (Nov. 8, 1988), P. Y. K. Lo, L. E. Thayer, and A. P. Wright (to Dow Corning).
185. U.S. Pat. 4,774,111 (Sept. 27, 1988), P. Y. K. Lo (to Dow Corning).
186. U.S. Pat. 4,465,818 (Aug. 14, 1984), A. Shirahata and S. Shosaku (to Toray Silicones).
187. U.S. Pat. 4,472,563 (Sept. 18, 1984), G. Chandra, P. Y. K. Lo, and Y. A. Peters (to Dow Corning).

188. U.S. Pat. 5,122,585 (June 16, 1992), C. A. Sumpter and L. N. Lewis (to General Electric Co.).
189. U.S. Pat. 5,206,329 (Apr. 27, 1993), C. A. Sumpter, L. N. Lewis, and S. J. Danishefsky (to General Electric Co.).
190. U.S. Pat. 5,025,073 (June 18, 1991), L. N. Lewis and C. A. Sumpter (to General Electric Co.).
191. U.S. Pat. 5,015,691 (May 14, 1991), L. N. Lewis and T. C. Chang (to General Electric Co.).
192. U.S. Pat. 5,254,656 (Oct. 19, 1993), C. J. Bilgrien and B. A. Witucki (to Dow Corning).
193. V. V. Severnyi, R. M. Minas'yan, I. A. Makarenko, and N. M. Bizyuakova, *Vysokomoh Soedin. Ser A (Eng. Ed.)*, **18**, 1464 (1976).
194. F. W. Van der Weij, *Makromol. Chem.* **181**, 2541 (1980).
195. K. A. Smith, *J. Org. Chem.* **51**, 3827 (1986).
196. X. W. He, J. M. Widmaier, J. E. Herz, and G. C. Meyer, *Eur. Polym. J.* **24**, 1145 (1988).
197. M. A. Sharef and J. E. Mark, *Makromol. Chem.* **190**, 495 (1989).
198. S. J. Clarson, Z. Wang, and J. E. Mark, *Eur. Polym. J.* **26**, 621 (1990).
199. M. Guibergia-Pierron and G. Sauvet, *Eur. Polym. J.* **28**, 29 (1992).
200. J. C. Saam, D. Graiver, and M. Baile, *Rubber Chem. Technol.* **54**, 976 (1981).
201. D. Graiver, D. J. Huebner, and J. C. Saam, *Rubber Chem. Technol.* **56**, 918 (1983).
202. S. B. Smith and J. Waterborne, *Coatings*, **10**, 13 (1983).
203. D. T. Liles and H. V. Lefler, *Proc. Waterborne, Higher Solids and Powder Coatings Symp.* **18**, 161 (1991).
204. U.S. Pat. 4,608,412 (Aug. 26, 1986), A. L. Frieberg (to Dow Corning).
205. J. Stein, T. M. Leonard, and J. F. Smith, *J. Appl. Polym. Sci.* **47**, 667 (1993).
206. U.S. Pat. 5,034,455 (July 23, 1991), J. Stein and T. M. Leonard (to General Electric Co.).
207. U.S. Pat. 4,814,368 (Mar. 21, 1989), J. Stein, T. M. Leonard, and S. L. Pratt (to General Electric Co.).
208. U.S. Pat. 4,877,828 (Oct. 31, 1989), J. Stein and T. M. Leonard (to General Electric Co.).
209. U.S. Pat. 5,321,075 (June 14, 1994), D. T. Liles (to Dow Corning).
210. L. M. Epstein and N. S. Marans, *Rubber Age*, **82**, 825 (1958).
211. A. Vokal, P. Kourim, J. Sussmilchova, M. Heidingsfeldova, and B. Kopecky, *Radiat. Phys. Chem.* **28**, 497 (1986).
212. E. L. Warrick, *Ind. Eng. Chem.* **47**, 2388 (1955).
213. W. Ming-Jun, D. Cong, M. Zue-Teh, and D. Lianj-Chang, paper presented at the *International Conference on Radiation Processing* for *Plastics and Rubber*, Canterbury, U.K., Mar. 28–30, 1984.
214. J. F. Zack, E. L. Warrick, and G. Knoll, *J. Chem. Eng. Data* **6**, 279 (1961).
215. R. P. Eckberg, "Chemistry and Technology of Radiation Curable Silicone Release Coatings," in D. Satas, ed., *Advances in Pressure Sensitive Adhesives*, Satas & Associates, Warwick, R.I., 1992.
216. U.S. Pat. 4,052,529 (Oct. 4, 1977), G. N. Bokerman, J. A. Calquhoun, and D. J. Gordon (to Dow Corning).
217. U.S. Pat. 4,303,484 (Dec. 1, 1981), M. Takamizawa, F. Annaka, Y. Takasaki, and H. Aoki (to Shin-Etsu).
218. U.S. Pat. 4,348,454 (Sept. 7, 1982), R. P. Eckberg (to General Electric Co.).
219. U.S. Pat. 4,293,678 (Oct. 6, 1981), R. G. Carter (to Union Carbide).
220. R. P. Eckberg, *UV Cure of Epoxysilicones, Radiation Curing in Polymer Science and Technology*, Vol. IV, Elsevier Publishers, Ltd., U.K., 1993.
221. J. V. Crivello and J. H. W. Lam, *J. Poly. Sci.* **17**, 977 (1979).

222. J. V. Crivello, J. L. Lee, and D. A. Conlan, *Proc. Radcure VI Conference*, Dearborn, Mich., 1982.
223. J. V. Crivello, *Chemtech.* **10**, 624 (1980).
224. U.S. Pat. 4,279,717 (July 21, 1981), R. P. Eckberg (to General Electric Co.).
225. U.S. Pat. 4,421,904 (Dec. 20, 1983), R. P. Eckberg and R. W. LaRochelle (to General Electric Co.).
226. R. P. Eckberg, K. D. Riding, M. E. Grenoble, and S. M. John, *Adhesive Age*, 24 (Apr. 1989).
227. U.S. Pat. 4,064,027 (Dec. 20, 1977), G. A. L. Gant (to Dow Corning).
228. U.S. Pat. 4,335,085 (June 15, 1982), H. Hatanka.
229. U.S. Pat. 4,435,259 (Mar. 6, 1984), M. Chang (to Pitney Bowes).
230. U.S. Pat. 4,510,094 (Apr. 9, 1985), P. J. Drahnak (to 3M).
231. U.S. Pat. 4,600,484 (July 15, 1986), P. J. Drahnak (to 3M).
232. L. D. Boardman, *Organometallics*, **11**, 4194 (1992).
233. S. K. Venkataraman, L. Coyne, F. Chambon, M. Gottlieb, and H. H. Winter, *Polymer*, **30**, 2222 (1989).
234. X. Xu and V. Galiatsatos, *Makrol. Chem., Macromol Symp.* **76**, 137 (1993).
235. C. P. Wong, *J. Mater. Res.* **5**, 795 (1990).
236. A. Shefer, M. Gottlieb, and G. Gorodetsky, *Isr. J. Tech.* **24**, 673 (1988).
237. A. Shefer, G. Gorodetsky, and M. Gottlieb, *Mat. Res. Soc. Symp. Proc.* **177**, 31 (1990).
238. G. L. Batch, C. W. Macosko, and D. N. Kemp, *Rubber Chem. Tech.* **64**, 218 (1990).
239. E. M. Barrall, M. A. Flandera, and J. A. Logan, *Thermochim. Acta*, **5**, 415 (1973).
240. Z. Tan, R. Jaeger, and G. J. Vancso, *Polymer*, **35**, 3230 (1994).
241. V. K. Soni and R. S. Stein, *Macromolecules*, **23**, 5257 (1990).
242. A. N. Gent, G. L. Lium, and M. Mazurek, *J. Polym. Sci., Part B: Polym. Phys.* **32**, 271 (1994).
243. A. L. Andrady, M. A. Lloente, and J. E. Mark, *Polym. Bull.* **26**, 357 (1991).
244. S. Venkatraman, *J. Appl. Poly. Sci.* **48**, 1383 (1993).
245. A. N. Falcao, J. Skov Pederson, and K. Mortensen, *Macromolecules*, **26**, 5350 (1993).
246. S. L. Bontems, J. Stein, and M. A. Zumbrum, *J. Polym. Sci. Part A: Polym. Chem.* **31**, 2697 (1993).
247. G. Simon and H. Schneider, *Makrol. Chem., Macromol. Symp.* **52**, 233 (1991).
248. M. G. Brereton, *Macromolecules*, **24**, 6160 (1991).
249. P. Sotta and B. Deloche, *J. Chem. Phys.* **100**, 4591 (1994).
250. A. Lapp, M. Daoud, G. Jannink, and B. Farago, *J. Noncryst. Sol.* **172–174**, 862 (1994).
251. V. Galiatsatos and J. E. Mark, in J. D. Zeigler and F. G. W. Fearon, eds., *Advances in Silicon-Based Polymer Science*, ACS Publishing, Washington, D.C., 1990.
252. V. Galiatsatos and J. E. Mark, *Macromolecules*, **20**, 2631 (1987).
253. S. J. Clarson, V. Galiatsatos, and J. E. Mark, *Macromolecules*, **23**, 1504 (1990).
254. J. E. Mark, *Adv. Polym. Sci.* **44**, 1 (1982).
255. J. E. Mark and M. A. Llorente, *J. Am. Chem. Soc.* **102**, 632 (1980).
256. M. V. Valles and C. W. Macosko, *Macromolecules*, **12**, 673 (1979).
257. M. A. Llorente and J. E. Mark, *Macromolecules*, **13**, 681 (1980).
258. K. O. Meyers and E. W. Merrill, *Elastom. Rubber Elastic.* **193**, 329 (1982).
259. J. E. Mark, *Acc. Chem. Res.* **27**, 271 (1994).
260. J. E. Mark and J. L. Sullivan, *J. Chem. Phys.* **66**, 1006 (1977).
261. J. E. Mark, *Makrol. Chem. Suppl.* **2**, 87 (1979).
262. R. Kosfeld, M. Hess, and D. Hansen, *Poly. Bull.* **3**, 603 (1980).
263. K. O. Meyers, M. L. Bye, and E. W. Merrill, *Macromolecules*, **13**, 1045 (1980).
264. N. Rennar. Kaust. + *Gummi.* **6**, 480 (1989).
265. K. H. Schimmel and G. Heinrich, *Coll. Polym. Sci.* **269**, 1003 (1991).

266. A. I. Medalia and G. Kraus, in J. E. Mark, B. Erman, and F. R. Eirich, eds., *Reinforcement of Elastomers by Particulate Fillers in Science and Technology of Rubber*, Academic Press, Inc., San Diego, Calif., 1994.

267. U.S. Pat. 2,938,009 (May 24, 1960), G. R. Lucas (to General Electric Co.).

268. B. B. Boonstra, H. Cochrane, and E. M. Dannenberg, *Rubber Chem. Technol.* **48**, 448 (1975).

269. M. P. Wagner, *Rubber Chem. Technol.* **49**, 703 (1976).

270. E. L. Warrick, O. R. Pierce, and K. E. Polmanteer, *Rubber Chem. Technol.* **52**, 437 (1979).

271. G. Berrod, A. Vidal, E. Papirer, and J. B. Donnet, *J. Appl. Polym. Sci.* **26**, 833 (1981).

272. H. L. Chapman, M. A. Lutz, and K. E. Polmanteer, *Rubber Chem. Technol.* **58**(5), 953 (1985).

273. R. K. Iler, *The Chemistry of Silica*, John Wiley & Sons, Inc., New York, 1979.

274. H. E. Bergna, ed., *Adv. Chem. Ser.*, 234 (1994); C. Macosko and co-workers, *Rubber Chem. Technol.* **67**, 820 (1994).

275. M. Morton, J. C. Healy, and R. L. Denecour, *Proc. Int. Rubber Conf., 5th*, 175 (1967).

276. M. Morton, *Adv. Chem. Ser.* **99**, 490 (1971).

277. A. Burneau and O. Barres, *Langmuir*, **6**, 1364 (1990).

278. M. Zaborski, A. Vidal, G. Ligner, H. Balard, E. Papirer, and A. Burneau, *Langmuir*, **5**, 447 (1989).

279. H. Cochrane and C. S. Lin, *Rubber Chem. Technol.* **66**, 48 (1993).

280. J. Janzen and G. Krauss, *Rubber Chem. Technol.* **44**, 1287 (1971).

281. J. Janzen, *Rubber Chem. Technol.* **55**, 669 (1982).

282. A. I. Medallia, R. R. Juengel, and J. M. Collins, in A. Whelan and K. S. Lee, eds., *Improving the Performance of Rubber Products*, Applied Science Publishers, Barking, Essex, U.K., 1979, Chapt. 5.

283. M. I. Aranguren, C. W. Macosoko, B. Thakkar, and M. Tirrell, *Mat. Res. Soc. Symp. Proc.* **170**, 303 (1990).

284. M. L. Hair, in A. T. Bell and M. L. Hair, eds., *ACS Symp. Ser.* **137**, 1 (1980).

285. A. M. Vidal and E. Papirer, in H. E. Bergna, ed., *The Colloid Chemistry of Silica Advances in Chemistry Series*, **234** (1994).

286. M. Zumbrum, *J. Adhesion*, **46**, 181 (1994).

287. G. E. Maciel, C. E. Bronnimann, R. C. Zeigler, I-Ssuer Chuang, D. R. Kinney, and E. A. Keiter, in Ref. 289, p. 269.

288. L. Garrido, J. L. Ackerman, and J. E. Mark, *Mat. Res. Soc. Symp. Proc.* **171**, 65 (1990).

289. A. Burneau, B. Humbert, O. Barres, J. P. Gallas, and J. C. Lavalley, in Ref. 289, p. 269.

290. G. Vigil, Z. Xu, S. Steinberg, and J. Israelachvili, *J. Collo. Int. Sci.* **165**, 367 (1994).

291. E. M. Dannenburg, *Rubber Chem. Technol.* **48**, 410 (1975).

292. S. K. Mandal and D. K. Basu, *Rubber Chem. Technol.* **67**, 672 (1994).

293. M. T. Maxson and C. L. Lee, *Rubber Chem. Technol.* **55**, 233 (1982).

294. E. Guth, R. Simha, and O. Gold, *Kolloid-Z.* **74**, 266 (1936).

295. E. Guth and O. Gold, *Phys. Rev.* **53**, 322 (1938).

296. F. Bueche, *J. Appl. Polym. Sci.* **5**, 271 (1961).

297. L. Mullins and N. R. Tobin, *J. Appl. Polym. Sci.* **9**, 2993 (1965).

298. A. I. Medalia, *J. Colloid Interface Sci.* **32**, 115 (1970).

299. G. Krauss, *J. Polym. Sci. B*, **8**, 601 (1970); *Rubber Chem. Technol.* **44**, 199 (1971).

300. S. Wolff and J. B. Donnet, *Rubber Chem. Technol.* **63**, 32 (1990).

301. A. R. Payne, *J. Appl. Polym. Sci.* **7**, 873 (1965).

302. A. R. Payne, in G. Krauss, ed., *Reinforcement of Elastomers*, Wiley-Interscience, New York, 1965.

303. A. I. Medalia, *Rubber Chem. Technol.* **46**, 877 (1973).
304. G. Krauss, *J. Appl. Polym. Sci., Appl. Polym. Symp.* **39**, 75 (1984).
305. N. Nagata, T. Kobatake, H. Watanabe, A. Ueda, and A. Yoshioka, *Rubber Chem. Technol.* **60**, 837 (1987).
306. F. T. Tsutsumi, M. Sakakibara, and N. Oshima, *Rubber Chem. Technol.* **63**, 8 (1990).
307. A. I. Medallia, *Rubber Chem. Technol.* **64**, 481 (1991).
308. M. I. Aranguren, E. Mora, J. V. DeGroot, and J. C. W. Macosko, *J. Rheol.* **36**, 1165 (1992).
309. B. B. Boonstra, H. Cochrane, and E. M. Dannenberg, *Rubber Chem. Technol.* **48**, 558 (1975).
310. A. Voet, J. C. Morawski, and J. B. Donnet, *Rubber Chem. Technol.* **50**, 342 (1977).
311. L. Mullins, *Rubber Chem. Technol.* **42**, 339 (1969).
312. C. R. Herd, G. C. McDonald, and W. M. Hess, *Rubber Chem. Technol.* **65**, 107 (1992).
313. G. C. McDonald and W. M. Hess, *Rubber Chem. Technol.* **50**, 842 (1977).
314. B. B. Boonstra and A. I. Medalia, *Rubber Age (N.Y.)* **92**, 892 (1963); **93**, 82 (1963).
315. S. Shiga and M. Furuta, *Rubber Chem. Technol.* **58**, 1 (1985).
316. W. H. Hess, *Rubber Chem. Technol.* **64**, 386 (1991).
317. F. Bohin and I. Manas-Zloczower, *Rubber Chem. Technol.* **67**, 602 (1994).
318. Y. Li, M. J. Wang, T. Zhang, F. Zhang, and X. Fu, *Rubber Chem. Technol.* **67**, 693 (1994).
319. A. M. Gessler, *Rubber Age*, **101**, 54 (1969).
320. J. E. Mark, S. Wang, P. Xu, and J. Wen, *Mat. Res. Symp. Proc.* **274**, 77 (1992).
321. H. W. Post, *Silicones and Other Organosilicon Compounds*, Reinhold, New York, 1949.
322. M. G. Voronkov, V. P. Mileshkevich, and Y. A. Yufhelevskii, *The Siloxane Bond*, Consultants Bureau, New York, 1978.
323. O. K. Johannson and C. L. Lee, in K. C. Frisch, ed., *Cyclic Monomers*, Wiley-Interscience, New York, 1972, p. 459.
324. S. J. Clarson, *Siloxane Polymers*, Prentice Hall, Engelwood Cliffs, N.J., 1993.
325. U.S. Pat. 2,504,388 (Apr. 18, 1949), O. A. Braley and C. L. Moyle (to Dow Corning).
326. Ger. Pat. 872,087, W. H. Daudt and J. F. Hyde (to Dow Corning).
327. U.S. Pat. 2,766,220 (Oct. 9, 1956), S. W. Kantor (to General Electric Co.).
328. R. N. Meals and F. M. Lewis, *Silicones*, Reinhold Publishing Co., New York, 1959.
329. S. J. Clarson, *New J. Chem.* **17**, 711 (1993).
330. M. Varma-Nair, J. P. Wesson, and B. Wunderlich, *J. Thermal Anal.* **35**, 1919 (1989).
331. K. E. Polmanteer and M. J. Hunter, *J. Appl. Polym. Sci.* **1**, 3 (1959).
332. C. M. Murphy and D. C. Smith, *Ind. Engng. Chem.* **42**, 2462 (1950).
333. M. V. Sobolevskii and co-workers, *Plast. Massy*, **3**, 13 (1962).
334. U.S. Pat. 2,389,802 (Nov. 27, 1945), R. R. McGregor and E. L. Warrick (to Dow Corning).
335. U.S. Pats. 2,389,804 and 2,389,806 (Nov. 27, 1945), R. R. McGregor and E. L. Warrick (to Dow Corning).
336. U.S. Pat. 2,465,296 (Mar. 22, 1949), J. Swiss (to Westinghouse).
337. H. R. Baker, R. E. Kagarise, J. G. O'Rear, and P. J. Sneigoski, *J. Chem. Eng. Data*, **11**, 110 (1966).
338. U.S. Pat. 3,865,784 (Feb. 11, 1975), R. S. Neale and A. N. Oines (to Union Carbide).
339. A. J. Barry, *J. Appl. Phys.* **17**, 1020 (1946).
340. P. J. Flory, L. Mandelkern, J. B. Kisinger, and W. B. Schultz, *J. Am. Chem. Soc.* **74**, 3364 (1952).
341. *Fluids Handbook*, GE Silicones, Waterford, N.Y., Apr. 1993.
342. A. J. Barry and H. N. Heck, in F.G.A. Stone and W.A.G. Grahan, eds., *Inorganic Polymers*, Academic Press, Inc., New York, 1962.

343. G. Schmaucks, R. Wagner, and R. Wersig, *J. Organomet. Chem.* **446**, 9 (1993).
344. W. Noll, *Kolloid-Z.* **211**, 98 (1966).
345. H. W. Fox, P. W. Taylor, and W. A. Zisman, *Ind. Eng. Chem.* **38**, 1401 (1947).
346. W. Noll, *IUPAC Symposium for Organosilicon Chemistry*, Prague, Czechoslovakia, 1965.
347. H. Ruether, C. Maass, and G. Reichel, *Plaste Kautschuk*, **7**, 171 (1960).
348. P. Cannon, L. E. St. Pierre, and A. A. Miller, *J. Chem. Eng. Data*, **5**, 236 (1960).
349. G. E. Vogel and F. O. Stark, *J. Chem. Eng. Data*, **9**, 599 (1964).
350. E. B. Baker, A. J. Barry, and M. J. Hunter, *Ind. Eng. Chem.* **38**, 1117 (1946).
351. R. R. Buch, *J. Fire Saf.* **17**, 1 (1991).
352. C. C. Currie and M. C. Hommel, *Ind. Engng. Chem.* **42**, 2452 (1950).
353. G. Grand and C. C. Currie, *Mechan. Engng.* **73**, 311 (1951).
354. D. H. Demby, S. J. Stoklosa, and A. Gross, *Chem. Ind.* **48**, 183 (1993).
355. U.S. Pat. 2,599,984 (June 10, 1952), H. J. Fletcher and M. J. Hunter (to Dow Corning).
356. R. R. McGregor, *Silicones*, McGraw-Hill Book Co., Inc., New York, 1954.
357. *Rubber Fabricator Handbook*, GE Silicones, Waterford, N.Y., Dec. 1993.
358. P. W. Bridgeman, *Proc. Am. Acad. Arts Sci.* **77**, 129 (1947).
359. *Ibid.*, p. 115.
360. Y. Isreli, J. Lacoste, and J. Cavezzan, *Poly. Deg. Stabil.* **37**, 201 (1992).
361. E. D. Lipp, *Appl. Spectrosc.* **45**, 477 (1991).
362. A. Weissler, *J. Am. Chem. Soc.* **71**, 93 (1949).
363. F. Clark, *Insulating Materials for Design and Engineering Practice*, John Wiley & Sons, Inc., New York, 1962.
364. S. Fordham, *Silicones*, Philosophical Library, New York, 1960.
365. A. M. Bueche, *J. Poly. Sci.* **15**, 105 (1955).
366. M. L. Dunham, D. L. Bailey, and R. Y. Mixer, *Ind. Engng. Chem.* **49**, 1373 (1957).
367. A. M. Bueche, *J. Polym. Sci.* **25**, 139 (1957).
368. U.S. Pat. 3,635,743 (Jan. 18, 1972), A. H. Smith (to General Electric Co.).
369. U.S. Pat. 2,448,756 (Sept. 7, 1948), M. C. Agens (to General Electric Co.).
370. R. A. Labine, *Chem. Eng.* **67**, 102 (1960).
371. W. J. Bobear, in M. Morton, ed., *Rubber Technology*, 2nd ed., D. Van Nostrand Co., New York, 1973.
372. W. Lynch, *Handbook of Silicone Rubber Fabrication*, D. Van Nostrand Co., New York, 1978.
373. *Silastic Compounding System*, Dow Corning, Midland, Mich., 1977.
374. *Silicone Rubber Fabricators Handbook*, General Electric Co., Waterford, N.Y., 1980.
375. J. Brandrup and E. H. Immergut, eds., *Polymer Handbook*, Wiley-Interscience, New York, 1967.
376. F. M. Lewis, *Rubber Chem. Technol.* **35**, 1222 (1962).
377. J. A. Barrie and B. Platt, *Polymer*, **4**, 303 (1963).
378. R. M. Barrier, J. A. Barrie, and N. K. Raman, *Rubber Chem. Technol.* **36**, 642, 651 (1963).
379. C. J. Major and K. Kammermeyer, *Mod. Plast.* **39**, 135 (1962).
380. C. W. Roush and S. A. Bailey, Jr., *Rubber Age*, **84**, 75 (1958).
381. K. B. Yerrick and H. N. Beck, *Rubber Chem. Technol.* **37**, 261 (1964).
382. R. Harrington, *Rubber Age*, **84**, 798 (1959).
383. D. J. Fischer, R. G. Chaffee, and V. Flegel, *Rubber Age*, **87**, 59 (1960).
384. S. D. Gehman and G. C. Gregson, *Rubber Rev.* **33**, 1429 (1960).
385. *LIM Product Data Sheet*, GE Silicones, Waterford, N.Y., May, 1994.
386. U.S. Pat. 4,418,157 (Nov. 29, 1983), F. J. Modic (to General Electric Co.).
387. U.S. Pat. 4,026,843 (May 31, 1977), R. E. Kittle (to Dow Corning).

388. U.S. Pat. 4,851,452 (July 25, 1989), D. C Gross, L. N. Lewis, and C. L. Haig (to General Electric Co.).
389. U.S. Pat. 4,879,317 (Nov. 7, 1989), K. A. Smith and C. L. Haig (to General Electric Co.).
390. U.S. Pat. 4,550,125 (Oct. 29, 1985), C-L. Lee, M. T. Maxson, and J. A. Rabe (to Dow Corning); U.S. Pat. 4,555,529 (Nov. 26, 1985), C-L. Lee, M. T. Maxson, and J. A. Rabe (to Dow Corning).
391. *Foam Products Data Sheets*, GE Silicones, Waterford, N.Y., Dec. 1992; R. D. Steinmeyer, M. A. Becker, E. D. Lipp, and R. B. Taylor, in A. L. Smith, ed., *The Analytical Chemistry of Silicones*, John Wiley & Sons, Inc., New York, 1991, pp. 294–371.
392. R. C. Smith and co-workers, in A. C. Smith, ed., *Analysis of Silicones*, John Wiley & Sons, Inc., New York, 1974, p. 113.
393. J. H. Wengrovius, T. B. Burnell, M. A. Zumbrum, and J. T. Bendler, presentation at *XXIIth Organosilicon Symposium*, Troy, N.Y., Mar. 1994.
394. J. H. Wengrovius, T. B. Burnell, and M. A. Zumbrum, presentation at *Xth International Symposium on Organosilicon Chemistry*, Poznan, Poland, Aug. 1993.
395. U.S. Pat. 2,676,182 (1954), W. H. Daudt and L. J. Tyler (to Dow Corning).
396. U.S. Pat. 4,707,531 (1987), A. Shirahata (to Toray Silicones).
397. U.S. Pat. 5,124,212 (1992), C. Lee and M. A. Lutz (to Dow Corning).
398. L. A. Sobieski and T. J. Tangney, in D. Satas, ed., *Handbook of Pressure Sensitive Adhesive Technology*, 2nd ed., Van Nostrand Reinhold, New York, 1989, p. 508.
399. D. F. Merrill, *SAMPE Quarterly*, **16**(4), 40 (1985).
400. U.S. Pat. 3,527,659 (Sept. 8, 1970), J. W. Keil (to Dow Corning).
401. U.S. Pat. 4,611,042 (Sept. 9, 1986), S. A. Rivers-Farrell and A. P. Wright (to Dow Corning).
402. U.S. Pat. 4,529,758 (July 16, 1985), F. J. Traver (to General Electric Co.).
403. U.S. Pat. 2,672,455 (Mar. 16, 1954), C. C. Currie (to Dow Corning).
404. U.S. Pat. 2,678,893 (May 18, 1954), T. A. Kauppi (to Dow Corning).
405. B. C. Copley, *Org. Coatings Appl. Poly. Sci. Proc.* **48**, 121 (1983).
406. B. C. Copley, in L. H. Lee, ed., *Polymer Science and Technology*, Plenum Press, New York, 1984, p. 257.
407. U.S. Pat. 2,842,522 (July 8, 1958), C. L. Frye (to Dow Corning).
408. S. B. Lin, *Int. J. Adhesion Adhesives*, **14**(3), 185 (1994).
409. R. Clark, *Handbook of Printed Circuit Manufacturing*, Van Nostrand Reinhold, New York, 1985.
410. S. Huie, P. F. Schmit, and J. S. Warren, *Adhesives Age*, **28**(8), 30 (1985).
411. U.S. Pat. 3,861,939 (Jan. 21, 1975), D. F. Merrill and P. J. Lavan (to General Electric Co.).
412. U.S. Pat. 4,027,073 (May 31, 1977), H. A. Clark (to Dow Corning).
413. U.S. Pat. 4,177,315 (Dec. 4, 1979), R. W. Ubersax (to Du Pont).
414. C. J. Brinker and G. W. Scherer, *Sol Gel Science: The Physics and Chemistry of Sol Gel Processing*, Academic Press, Inc., San Diego, Calif., 1990.
415. U.S. Pat. 4,495,360 (Jan. 22, 1985), B. T. Anthony (to General Electric Co.).
416. U.S. Pat. 4,491,508 (Jan. 1, 1985), D. R. Olson and K. K. Webb (to General Electric Co.).
417. R. P. Eckberg, *TAPPI Proceedings–1987 Polymers, Laminations and Coatings Conference*, Tappi Press, Atlanta, Ga., 1987, p. 21.
418. J. D. Jones, *Release Liner Markets and Technology—Conference Proceedings*, Chicago, Ill., 1992.
419. U.S. Pat. 4,609,574 (Sept. 2, 1986), J. R. Keryk, P. Y. K. Lo, and L. E. Thayer (to Dow Corning).
420. Eur. Pat. Appl. 523,660 (1993), C. Herzig, B. Deubzer, and D. Huettner.

421. R. P. Eckberg, in D. Satas, ed., *Advances in Pressure Sensitive Adhesive Technology-1*, Satas & Associates, Warwick, R.I., 1992, p. 50.

422. U.S. Pat. 3,133,891 (May 19, 1964), L. Ceyzeriat (to Rhône-Poulenc SA).

423. U.S. Pat. 3,035,016 (May 15, 1962), L. B. Bruner (to Dow Corning).

424. U.S. Pat. 3,127,363 (Mar. 31, 1964), S. Nitzsche and W. Manfred (to Wacker-Chemie GmbH).

425. U.S. Pat. 3,334,067 (Aug. 8, 1967), D. R. Weyenberg (to Dow Corning).

426. U.S. Pat. 3,542,901 (Nov. 24, 1970), K. G. Cooper and P. R. A. Hansen (to Midland Silicones, Ltd.).

427. U.S. Pat. 4,489,199 (Dec. 18, 1994), J. H. Wengrovius (to General Electric Co.).

428. U.S. Pat. 4,515,932 (May 7, 1985), R. K. Chung (to General Electric Co.).

429. U.S. Pat. 4,395,526 (July 26, 1983), M. A. White, R. A. Smith, M. D. Beers, R. T. Swiger, and G. M. Lucas (to General Electric Co.).

430. J. H. Wengrovius, V. M. VanValkenburgh, and J. F. Smith, presentation at *XXV Organosilicon Symposium*, Los Angeles, Calif., Apr. 1992.

431. U.S. Pat. 4,667,007 (May 19, 1987), J. H. Wengrovius and T. P. Lockhart (to General Electric Co.).

432. U.S. Pat. 4,788,170 (Nov. 29, 1988), J. H. Wengrovius (to General Electric Co.).

433. U.S. Pat. 4,863,992 (Sept. 5, 1989), J. H. Wengrovius, J. E. Hallgren, J. Stein, and G. M. Lucas (to General Electric Co.).

434. J. D. Winefordner and I. M. Kolthoff, ed., "Chemical Analysis," in A. Lee Smith, ed., *The Analytical Chemistry of Silicones*, Vol. 12, John Wiley & Sons, Inc., New York, 1991.

435. T. V. Kirillova, O. P. Trokhachenova, R. R. Tarasyants, and S. M. Chernykh, *Zh. Anal. Khim.* **47**, 159 (1992).

436. B. B. Hardman and A. Torkelson, in J. I. Kroschwitz, ed., *Encyclopedia of Polymer Science and Engineering*, 2nd ed., Vol. 15, John Wiley & Sons, Inc., New York, 1989, p. 291.

437. M. C. Angelotti, in Ref. 443, Chapt. 3, p. 48.

438. A. L. Smith and R. D. Parker, in Ref. 446.

439. H. J. Horner, J. E. Weiler, and N. C. Angelotti, *Anal. Chem.* **32**, 858 (1960).

440. W. G. Doeden, E. M. Kushibab, and A. C. Ingala, *J. Am. Oil Chem. Soc.* **57**, 73 (1980).

441. D. A. McCamey, D. P. Ianelli, L. J. Bryson, and T. M. Thorpe, *Anal. Chim. Acta*, **188**, 119 (1986).

442. J. A. Moore, in Ref. 446, p. 427.

443. S. Fujimoto, H. Ontani, and S. Buge, *Z. Anal. Chem.* **331**, 342 (1988).

444. M. A. Buese, F. L. Keohan, and S. A. Swint, *J. Polym. Sci., Part A: Polym. Chem. Ed.* **29**, 303 (1991).

445. P. J. Garner and R. C. Smith, *Pittsburgh Conference and Exposition*, Abstract No. 969, New Orleans, La., 1985.

446. J. D. Pinkston, G. Owens, L. Burkes, T. Delaney, D. Millington, and D. Malyby, *Anal. Chem.* **60**, 962 (1988).

447. W. VanderHeuvel, J. Smith, R. Firestone, and J. Beck, *Anal. Lett.* **5**, 285 (1972).

448. M. Vincenti, E. Pellizetti, A. Guarino, and S. Costanzi, *Anal. Chem.* **64**, 1879 (1992).

449. E. D. Lipp and A. Lee Smith, in Ref. 443, Chapt. 11, p. 305.

450. P. R. Griffiths and J. A. Haseth, *Fourier Transform Infrared Spectroscopy*, John Wiley & Sons, Inc., New York, 1986.

451. A. L. Smith, *Applied Infrared Spectroscopy*, John Wiley & Sons, Inc., New York, 1979.

452. J. D. Ingle, Jr. and S. R. Crouch, *Spectrochemical Analysis*, Prentice Hall, Old Tappan, N.H., 1988.

453. P. R. Griffiths, S. L. Pentony, Jr., A. Giorgetti, and K. H. Shafer, *Anal. Chem.* **58**, 1349A (1986).

454. A. L. Smith, in J. R. Durig, ed., *Chemical, Biological and Industrial Applications of Infrared Spectroscopy*, John Wiley & Sons, Inc., New York, 1985.

455. E. Kohn and M. E. Chisum, *ACS Symp. Ser.* **352**, 169 (1987).

456. P. J. Madek and E. Marechal, *J. Polym. Sci.* **18**, 2417 (1980).

457. E. D. Lipp, *Appl. Spectrosc. Rev.* **27**, 385 (1992).

458. R. B. Taylor, B. Parbhoo, and D. M. Fillmore, in Ref. 443, Chapt. 12, p. 347.

459. E. A. Williams, *Polym. Prepr., Am. Chem. Soc., Div. Polym. Chem.* **31**, 119 (1990).

460. P. Fux, *Analyst*, **115**, 179 (1990).

461. J. F. Hampton, C. W. Lacefield, and J. F. Hyde, *Inorg. Chem.* **4**, 1659 (1965).

462. E. A. Williams, in *The Chemistry of Organic Silicon Compounds*, S. Patai and Z. Rappoport, eds., Wiley, London, 1989, Ch. 8.

463. P. J. Kanyha and S. Brey, *22nd Congr. AMPERE Magn. Reson. Relat. Phenom. Proc.*, 341 (1984).

464. E. A. Williams, in G. A. Webb, ed., *Annual Reports on NMR Spectroscopy*, Vol. 15, Academic Press, London, 1983, p. 235.

465. K. Beshah, J. E. Mark, and J. L. Ackerman, *J. Polym. Sci., Part B, Polym. Phys.* **24**, 1207 (1986).

466. K. Beshah, J. E. Mark, and J. L. Ackerman, *Macromolecules*, **19**, 2194 (1986).

467. N. W. Lytle, in Ref. 443, Chapt. 14.

468. R. Pellenbarg, A. C. Siglio, and A. Hattori, *Marine Estuarine Geochem., Proc. Symp.*, 121 (1985).

469. N. Watanabe, H. Nagase, and Y. Ose, *Sci. Tot. Environ.* **73**, 1 (1988).

470. R. B. Annelin and C. L. Frye, *Sci. Tot. Environ.* **83**, 1 (1989).

471. G. E. Bately and J. W. Hayes, *Aust. J. Mar. Freshwater Res.* **42**, 287 (1991).

472. R. Pellenbarg, *Env. Sci. Technol.* **13**, 565 (1979).

473. R. R. Buch and D. N. Ingebrigtson, *Env. Sci. Technol.* **13**, 676 (1979).

474. J. C. Carpenter, J. A. Cella, and S. B. Dorn, *Env. Sci. Technol.* **29**(4), 864–868 (1995).

475. R. G. Lehmann, *Env. Toxicol. Chem.* **12**, 1851 (1993).

476. C. Anderson, K. Hochgeshwender, and W. R. Weidmann, *Chemosphere*, **16**, 10 (1987).

477. N. Watanabe, H. Nagase, T. Nakamura, E. Watanabe, and T. Ose, *Sci. Tot. Environ.* **35**, 91 (1984).

478. C. J. Weschler, *Atmospher. Environ.* **15**, 1365 (1981).

479. C. J. Weschler, *Sci. Tot. Environ.* **73**, 53 (1988).

480. Centre European Des Silicones (CES), *Proceedings of the Joint American European Meeting Anal. Quest. Concern. Si-Organ. Environ.*, Washington, D.C., 1993.

481. R. E. Pellenbarg, *Appl. Organometall. Chem.* **5**, 107 (1991).

482. L. G. Mahone and co-workers, *Env. Sci. Technol.* **2**, 307 (1983).

483. R. M. Cassidy, M. T. Hurteau, J. P. Mislan, and R. W. Ashley, *J. Chrom. Sci.* **14**, 444 (1976).

484. W. R. Biggs and J. C. Fetzer, *Anal. Chem.* **61**, 236 (1989).

485. K. H. Forbes, J. F. Vecchiarelli, P. C. Uden, and R. M. Barnes, *Anal. Chem.* **62**, 2033 (1990).

486. S. B. Dorn and E. M. Skelley-Frame, *Analyst*, **119**, 1687 (1994).

487. B. Hagenhoff, A. Benninghoven, H. Barthel, and W. Zoller, *Anal. Chem.* **63**, 2466 (1991).

488. O. L. Flannagan and N. R. Langley, in Ref. 443, Chapt. 4.

489. M. Kurata and Y. Tsunashima, in J. Brandrup and E. H. Immergut, ed., *Polymer Handbook*, 3rd ed., John Wiley & Sons, Inc., New York, 1989.

490. E. E. Drott and R. A. Mendelson, *J. Polym. Sci., Part A-2*, **8**, 1361, 1373 (1970).

491. F. M. Mirabella and L. Wild, *Polym. Mater. Sci. Eng.* **59**, 7 (1988).

492. N. R. Langley and J. D. Ferry, *Macromolecules*, **1**, 353 (1968).

493. K. O. Meyers, M. L. Bye, and E. W. Merrill, *Macromolecules*, **13**, 1045 (1980).

494. W. J. Storck, *Chem. Eng. News*, **56**, 8 (1978).

495. C. L. Frye, *Sci. Total Environ.* **73**, 17 (1988).

496. C. W. Lentz, *Ind. Res. Dev.* **22**, 139 (1980).

497. A. Opperhuizen, G. M. Asyee, and J. R. Parson, *CEC 5th Eur. Symp. (Rome)*, **176**, (1988).

498. A. Spacie, *No. PB247778*, Syracuse Research Corp. Rept. Syracuse, N.Y., 1984.

499. R. Pellenbarg, A. C. Siglio, and A. Hattori, *Marine Estuarian Geochem.* **121** (1985).

500. R. Pellenbarg, *Env. Sci. Technol.* **13**, 565 (1979).

501. R. Pellenbarg, *Mar. Poll. Bull.* **13**, 427 (1982).

502. R. Pellenbarg, *Mar. Poll. Bull.* **10**, 267 (1970).

503. W. Balz, J. A. Cella, J. C. Carpenter, S. B. Dorn, and E. Skelly-Frame, *Abstracts of the 205th ACS Meeting, Denver, Colo., 1992*, American Chemical Society, Washington, D.C., pp. 17, 18.

504. R. Pellenbarg, *Sci. Total Environ.* **73**, 11 (1988).

505. R. E. Pellenbarg and D. E. Tevault, *Env. Sci. Technol.* **20**, 743 (1986).

506. N. Watanabe, T. Nakamura, and E. Watanabe, *Sci. Total Environ.* **38**, 167 (1984).

507. N. M. Kitchen and C. A. Russell, *Proceedings of the 21st Annual Holm Seminar on Electrical Contacts*, Illinois Institute of Technology, Oct. 14–16, 1975.

508. C. J. Weschler, *Sci. Total Environ.* **73**, 53 (1988).

509. C. J. Weschler, *Atmos. Environ.* **15**, 1365 (1981).

510. *Fed. Reg.* **58**, 9257 (1993).

511. R. R. Buch and D. N. Ingebrigtson, *Env. Sci. Technol.* **13**, 676 (1979).

512. R. G. Lehmann, S. Varaprath, and C. L. Frye, *Environ. Toxicol. Chem.* **13**, 1061 (1994).

513. J. A. Cella and J. C. Carpenter, *XXVII Organosilicon Symposium*, Abstract No. B-8, Troy, N.Y., 1994.

514. J. Spivack, *XXVII Organosilicon Symposium*, Troy, N.Y., 1994.

515. C. Anderson, K. Hochgeshwender, H. Weidemann, and R. Wilmes, *Chemosphere*, **16**, 2567 (1987).

516. R. R. Buch, T. H. Lane, R. B. Annelin, and C. L. Frye, *Environ. Toxicol. Chem.* **3**, 215 (1984).

517. E. J. Hobbs, M. L. Keplinger, and J. C. Calandra, *Env. Res.* **15**, 229 (1975).

518. R. F. Sharp and H. O. W. Eggins, *Int. Biodeter. Bull.* **6**, 19 (1970).

519. R. G. Lehmann, S. Varaprath, and C. L. Frye, *Environ. Toxicol. Chem.* **13**, 1753 (1994).

520. J. Spivack and J. C. Carpenter, personal communication.

General Reference

D. Scott, *J. Am. Chem. Soc.* **68**, 2294 (1946).

JONATHAN RICH
JAMES CELLA
LARRY LEWIS
JUDITH STEIN
NAVJOT SINGH
GE Corporate Research and Development

SLAWOMIR RUBINSZTAJN
JEFF WENGROVIUS
GE Silicones

SILYLATING AGENTS

Silylation of Organic Compounds

Silylation is the replacement of one or more active hydrogens from an organic molecule by a trisubstituted silyl, R_3Si-, group. The active hydrogen is usually an alcohol, carboxylic acid, or phenol, ie, $-OH$; an amine, amide, or urea, $-NH$; or a thiol, $-SH$, and the silylating agent is usually a trimethylsilyl halide, dimethylsilyl dihalide, or a trimethylsilyl nitrogen-functional compound. Newer, more reactive silylating agents cleave esters and ethers. A mixture of silylating agents may be used. A mixture of trimethylchlorosilane and hexamethyldisilazane is more reactive than either reagent alone, and the by-products combine to form ammonium chloride.

$$(CH_3)_3SiNHSi(CH_3)_3 + (CH_3)_3SiCl + 3\ ROH \longrightarrow 3(CH_3)_3SiOR + NH_4Cl$$

Derivatizing an organic compound for analysis may require only a few drops of reagent selected from silylating kits supplied by laboratory supply houses. Commercial synthesis of penicillins requires silylating agents purchased in tank cars from the manufacturer (see ANTIBIOTICS, β-LACTAMS–PENICILLINS AND OTHERS).

Typical commercial silylating agents are listed in Table 1. The first three silylating agents in the table are available in bulk quantities and are most suitable for large-scale commercial silylation. The chlorosilanes are generally used in combination with an acid acceptor, eg, triethylamine. The nitrogen-functional silanes each have certain advantages for particular applications. Fluorinated silylating agents give enhanced rates of reaction and more volatile by-products.

The chlorosilanes are clear liquids that should be treated as strong acids. They react readily with water to form corrosive HCl gas and liquid. Liquid chlorosilanes and their vapors are corrosive to the skin and extremely irritating to the mucous membranes of the eyes, nose, and throat. The nitrogen-functional silanes react with water to form ammonia, amines, or amides. Because ammonia and amines are moderately corrosive to the skin and very irritating to the eyes, nose, and throat, silylamines should be handled like organic amines. Trimethylsilyl trifluoromethanesulfonate and trimethylsilyl iodide form very corrosive acidic products.

The techniques of silylation and their application in analysis have been reviewed (1,2), as have the intermediate steps in organic synthesis (3). Summaries of silylation applications are available (4,5).

Derivatization for Analysis. Silylation of organic materials has been an invaluable tool in analytical chemistry to allow ready analysis by gas–liquid chromatography (glc), mass spectrometry (qv), and thin-layer chromatography (tlc) (see CHROMATOGRAPHY). There are four main reasons to derivatize a compound for analysis: to increase volatility, to increase thermal stability, to enhance detectability, and to improve separation. Silylating kits offered by laboratory supply houses may contain mixtures of chlorine- and nitrogen-functional silylating agents and activating solvents or catalysts. Methodology for derivatization

Table 1. Methyl Silylating Agents

Chemical name	CAS Registry Number	Common abbreviation	Formula
trimethylchlorosilane	[75-77-4]	TMCS	$(CH_3)_3SiCl$
dimethyldichlorosilane	[75-78-5]	DMCS	$(CH_3)_2SiCl_2$
hexamethyldisilazane	[999-97-3]	HMDZ	$(CH_3)_3SiNHSi(CH_3)_3$
chloromethyldimethyl-chlorosilane	[1719-57-9]	CMDMS	$ClCH_2(CH_3)_2SiCl$
N,N'-bis(trimethylsilyl)-urea	[18297-63-7]	BSU	$[(CH_3)_3SiNH]_2CO$
N-trimethylsilyldiethyl-amine	[996-50-9]	TMSDEA	$(CH_3)_3SiNH(C_2H_5)_2$
N-trimethylsilylimid-azole	[18156-74-6]	TSIM	$(CH_3)_3SiN\underset{=N}{\boxed{}}$
N,O-bis(trimethylsilyl)-acetamide	[10416-59-8]	BSA	$(CH_3)_3SiN{=}C(CH_3)OSi(CH_3)_3$
N,O-bis(trimethylsilyl)-trifluoroacetamide	[21149-38-2]	BSTFA	$(CH_3)_3SiN{=}C(CF_3)OSi(CH_3)_3$
N-methyl-N-trimethyl-silyltrifluoroacetamide	[24589-78-4]	MSTFA	$(CH_3)_3SiN(CH_3)COCF_3$
t-butyldimethylsilylimid-azole	[54925-64-3]	TBDMIM	$t\text{-}C_4H_9(CH_3)_2SiN\underset{=N}{\boxed{}}$
N-trimethylsilylacet-amide	[13435-12-6]	MTSA	$(CH_3)_3SiNHCOCH_3$
trimethylsilyl trifluoro-methanesulfonate	[27607-77-8]	TMS triflate	$(CH_3)_3SiOSO_2CF_3$
trimethylsilyl iodide	[16029-98-4]	TMSI	$(CH_3)_3SiI$

and recommended specific reagents for the various applications may be found in the catalogs of suppliers.

N-Trimethylsilyldiethylamine (TMSDEA) is a strongly basic silylating reagent and is particularly useful for derivatizing low molecular weight acids. The reaction by-product, diethylamine, is volatile enough to be easily removed from the reaction medium.

N-Trimethylsilylimidazole (TSIM) is the strongest hydroxy silylator available and is the reagent of choice for carbohydrates and steroids. This reagent is unique in that it reacts quickly and smoothly with hydroxyls and carboxyl groups but not with amines. This characteristic makes TSIM particularly useful in multiderivatization schemes for compounds containing both hydroxyl and amine groups.

N,O-Bis(trimethylsilyl)acetamide (BSA) reacts quantitatively under relatively mild conditions with a wide variety of compounds to form volatile, stable trimethylsilane (TMS) derivatives for glc analysis. It has been used extensively for derivatizing alcohols, amines, carboxylic acids, phenols, steroids, biogenic amines, alkaloids (qv), etc. It is not recommended for use with carbohydrates (qv) or very low molecular weight compounds. Reactions are generally fast and the reagent is usually used in conjunction with a solvent, eg, pyridine or dimethylformamide (DMF). When used with DMF, it is the reagent of choice for derivatizing phenols (see PHENOL).

N,O-Bis(trimethylsilyl)trifluoroacetamide (BSTFA) is a powerful trimethylsilyl donor having approximately the same donor strength as the unfluorinated analogue BSA. Reactions of BSTFA are similar to those of BSA. The main advantage of BSTFA over BSA is the greater volatility of the former's reaction by-products, ie, monotrimethylsilyltrifluoroacetamide and trifluoroacetamide. This physical characteristic is particularly useful in the gas chromatography (gc) of some of the lower boiling TMS-amino acids and TMS Krebs cycle acids where the by-products of BSA may have similar retention characteristics and thus obscure these derivatives on the chromatogram. The by-products of BSTFA usually elute with the solvent front.

For the derivatization of fatty acid amides, slightly hindered hydroxyls, and other difficultly silylatable compounds, BSTFA containing 1 wt % trimethylchlorosilane is used. This catalyzed formulation is stronger than BSTFA alone. When silylation reagents are consumed in the hydrogen flame, silicon dioxide, SiO_2, is formed. *N,O*-Bis(trimethylsilyl)trifluoroacetamide contains three fluorine atoms which form HF in the detector flame and react with the SiO_2 to form volatile products. This removal of SiO_2 provides a decrease in detector fouling and background noise.

N-Methyl-*N*-trimethylsilyltrifluoroacetamide (MSTFA) is the most volatile TMS-amide available; it is more volatile than BSTFA or BSA. Its by-product, *N*-methyltrifluoroacetamide, has an even lower retention time in glc than MSTFA. This is of considerable value in glc determinations where the reagent or by-products obscure the derivative on the chromatogram. Silylation of steroids (qv) shows MSTFA to be significantly stronger in donor strength than BSTFA or BSA. *N*-Methyl-*N*-trimethylsilyltrifluoroacetamide silylates hydrochloride salts of amines directly.

The *t*-butyldimethylsilyl group introduced by TBDMIM has a number of advantages in protecting alcohols (6). The silylated alcohol hydrolyzes more slowly than an alcohol silylated with TMS by a factor of 10^4. The silyl ether is also stable to powerful oxidizing and reducing agents, but it can easily be removed by aqueous acetic acid or tetrabutylammonium fluoride in tetrahydrofuran.

TMS triflate [27607-77-8] is an extremely powerful silylating agent for most active hydrogens. It surpasses the silylating potential of TMCS by a factor of nearly 10^9. It readily converts 1,2- and 1,3-diketones into disilylated dienes (7).

Trimethylsilyl iodide [16029-98-4] (TMSI) is an effective reagent for cleaving esters and ethers. The reaction of hexamethyldisilane [1450-14-2] with iodine gives quantitative conversion to TMSI. A simple mixture of trimethylchlorosilane and sodium iodide can be used in a similar way to cleave esters and ethers (8), giving silylated acids or alcohols that can be liberated by reaction with water.

$$RC\overset{\overset{\displaystyle O}{\|}}{O}R' + (CH_3)_3SiI \longrightarrow RC\overset{\overset{\displaystyle O}{\|}}{O}Si(CH_3)_3 + R'I$$

$$ROR' + (CH_3)_3SiCl + NaI \longrightarrow ROSi(CH_3)_3 + R'I + NaCl$$

It is possible to use halogen-sensitive detectors in glc analysis of active hydrogen compounds by silylating them with a halogenated silylating agent, eg, CMDMS (9).

Silylation in Organic Synthesis.   Silyl blocking agents are used in organic synthesis to protect sensitive functional groups, to alter reactivity and solubility, and to increase stability of intermediates. Silylation applications in pharmaceutical synthesis have been used to protect a wide range of OH groups, eg, alcohols in prostaglandins (qv) and steroid synthesis, enols in the synthesis of nucleosides and steroids (qv), and carboxylic acids and sulfenic acids in the synthesis of penicillins and cephalosporins (6) (see ANTIBIOTICS). Silylation has its broadest use in the commercial synthesis of penicillins. The blocking effect of trimethylsilyl and dimethylsilyl groups on 6-aminopenicillanic acid (6-APA) (**1**) has played an important role in the total synthetic production of semisynthetic penicillins.

$$H_2N \quad \overset{H}{\underset{}{\diagup}} \quad S \quad \diagdown CH_3$$

(**1**)

Protection of carboxylic acids and sulfenic acids requires efficient silyl donors, eg, BSA, MTSA, and bis(trimethylsilyl)urea [18297-63-7] (BSU). BSU is often prepared *in situ* from hexamethyldisilazane and urea to yield over 90% of the silylated derivative in synthesis of cephalosporins (5).

It is possible to synthesize 1,2,3-triazoles from acetylenes and hydrazoic acid, but the instability of hydrazoic acid has limited this application. Sodium azide is silylated readily with trimethylchlorosilane to produce trimethylsilyl azide [4648-54-8], $(CH_3)_3SiN_3$, which reacts with acetylenes to produce high yields of 1,2,3-triazoles (10).

$$(CH_3)_3SiN_3 + C_6H_5C{\equiv}CH \longrightarrow (CH_3)_3Si{-}N{-}N \overset{H_2O}{\longrightarrow} HN{-}N + ((CH_3)_3Si)_2O$$

In this case, the substitution of the TMS group for hydrogen in HN_3 imparts a degree of thermal stability to the otherwise unstable azide and also acts as a blocking agent in allowing the direct synthesis of the triazole. Trimethylsilyl azide can be distilled at atmospheric pressure without decomposition (bp 95°C). Similarly, diazomethane, potentially very explosive, is thermally stable when trimethylsilated, $(CH_3)_3SiCHN_2$ [18107-18-1].

Two techniques have been described for producing trimethylsilyl enol ethers from aldehydes or ketones (10): reaction of $(CH_3)_3SiCl$ and $(C_2H_5)_3N$ in DMF; and reaction of $LiN(C_2H_5)_2$, which generates enolate ions in the presence of

$(CH_3)_3SiCl$. The resulting enol ethers can undergo a wide variety of reactions at the double bond, making this type of reaction important in hormone synthesis (11) (see HORMONES).

Silylation of Inorganic Compounds

Silicate Modifications. A method has been described in which silicate minerals are simultaneously acid-leached and trimethylsilyl end-blocked to yield specific trimethylsilyl silicates having the same silicate structure as the mineral from which these were derived (12). Olivine, hemimorphite, sodalite, natrolite, laumontite, and sodium silicates are converted to TMS derivatives of orthosilicates, pyrosilicates, cyclic polysilicates, etc, making it possible to classify the minerals according to their silicate structure. The same technique is used to analyze the siloxanol structure of aqueous solutions of vinyltrimethoxysilane (13). Certain anionic siliconates stabilize solutions of alkali silicates to give stable solutions in water or alcohols at any pH (14). Such silicate–siliconate mixtures are used as corrosion inhibitors in glycol antifreeze (15) (see ANTIFREEZES AND DEICING FLUIDS; CORROSION AND CORROSION CONTROL).

Ziegler-Natta Polymerization. The polymerization of propylene with Ziegler-Natta catalysts, ie, complexes of $TiCl_3$–$(C_2H_5)_3Al$ on $MgCl_2$ supports, is significantly affected by external addition to the reactor of organo(alkoxy)-silanes with the propylene feed. The nature of the organic group(s) and alkoxy group(s) affects the catalyst activity and the microstructure of the polymer (16). Silane donors such as phenyltriethoxysilane [*780-69-8*], cyclohexylmethyl-dimethoxysilane [*17865-32-6*], and dicyclopentyl dimethoxysilane [*126990-35-0*] are used commercially. The silane coordinates with the Ziegler-Natta catalyst to increase the isotacticity of polypropylene to 98% or higher and increase the polymer yield per amount of catalyst.

Silylation of Inorganic Surfaces

Alkyl Silylating Agents. Alkyl silylating agents convert mineral surfaces to water-repellent, low energy surfaces useful in water-resistant treatments for masonry, electrical insulators, packings for chromatography, and noncaking fire extinguishers. Methylchlorosilanes react with water or hydroxyl groups at the surface to liberate HCl and deposit a thin film of methylpolysiloxane, which has a low critical surface tension and is therefore not wetted by water (17). Ceramic insulators can be treated with methylchlorosilane vapors or solutions in inert solvents to maintain high electrical resistivity under humid conditions (18). The corrosive action of the evolved HCl can be avoided by prehydrolyzing the chlorosilanes in organic solvent and applying them as organic solutions of organopolysiloxanols. Hydrolyzed methylchlorosilanes also dissolve in aqueous alkali and are then applied as aqueous solutions of sodium methylsiliconates.

The siliconates are neutralized by carbon dioxide in the air to form an insoluble, water-resistant methylpolysiloxane film within 24 hours. Treatment of brick, mortar, sandstone, concrete, and other masonry protects the surface from spalling, cracking, efflorescence, and other types of damage caused by water (see SURFACE MODIFICATION (SUPPLEMENT); WATERPROOFING AND WATER/OIL REPELLENCY).

Silanes can alter the critical surface tension of a substrate in a well-defined manner. Critical surface tension is associated with the wettability or release qualities of a substrate. Liquids having a surface tension below the critical surface tension, γ_c, of a substrate wet the surface. Critical surface tensions of a number of typical surfaces are compared to γ_c of silane-treated surfaces in Table 2 (19).

Celite or firebrick packing for glc columns is often treated with TMCS, DMCS, or other volatile silylating agents (see Table 1) to reduce tailing by polar organic compounds. A chemically bonded methyl silicone support is stable for temperature programming to 390°C and allows elution of hydrocarbons up to C_{50} (20).

Table 2. Critical Surface Tensions, γ_c, mN/m(=dyn/cm)[a,b]

Surface	γ_c
polytetrafluoroethylene	18.5
methyltrimethoxysilane	22.5
vinyltriethoxysilane	25.0
paraffin wax	25.5
ethyltrimethoxysilane	27.0
propyltrimethoxysilane	28.5
glass, soda–lime (wet)	30.0
polychlorotrifluoroethylene	31.0
polypropylene	31.0
polyethylene	33.0
(3,3,3-trifluoropropyl)trimethoxysilane	33.5
(3-(2-aminoethyl)aminopropyl)trimethoxysilane	33.5
polystyrene	34.0
cyanoethyltrimethoxysilane	34.0
aminopropyltriethoxysilane	35.0
poly(vinyl chloride)	39.0
phenyltrimethoxysilane	40.0
(3-chloropropyl)trimethoxysilane	40.5
(3-mercaptopropyl)trimethoxysilane	41.0
(3-glycidoxypropyl)trimethoxysilane	42.5
poly(ethylene terephthalate)	43.0
copper (dry)	44.0
aluminum (dry)	45.0
iron (dry)	46.0
nylon-6,6	46.0
glass, soda–lime (dry)	47.0
silica, fused	78.0

[a]Ref. 19.
[b]Critical surface tensions for silanes refer to treated surfaces.

High performance liquid chromatography (hplc) combines the bonded solid phase of gas chromatography with the methodology of column liquid chromatography. The most popular type of hplc involves low polarity silicone-bonded surfaces having more highly polar liquids in a process termed reversed-phase chromatography (21). A favorite stationary phase is fine particle silica treated with octadecyltrichlorosilane [112-04-9]. In some instances, octyl-, phenyl-, cyclohexyl-, or ethylsilanes can be used to obtain improved selectivity. Metal ions on the bonded phase can be used to enhance separation of polar molecules through liqand-exchange chromatography (lec) (22). Metal ions, eg, copper, are retained on silica gel that has been treated with chelating functional silanes (see CHELATING AGENTS).

Organofunctional Silylating Agents

Whereas alkylsilylating agents provide low energy surfaces designed for release, a series of organofunctional silylating agents is offered commercially as adhesion promoters. Principal applications have been as coupling agents in glass- or mineral-reinforced organic resin composites and as adhesion promoters for paints, inks, coatings, and adhesives. Organofunctional silanes are also used to control orientation of liquid crystals, bind heavy-metal ions, immobilize enzymes and cell organelles, modify metal oxide electrodes, and to bind antimicrobial agents to surfaces. These and other applications have been described (23). Representative commercial silane coupling agents are listed in Table 3. These compounds can also be used as chemical intermediates for preparing other more specialized organofunctional silanes (24).

Liquid Crystals. In liquid crystal displays, clarity and permanence of image is enhanced if the display can be oriented parallel or perpendicular to the substrate. Oxide surfaces treated with dimethyloctadecyl-3-trimethoxysilylpropylammonium chloride [27668-52-6], $C_{18}H_{37}N^+(CH_3)_2CH_2CH_2$-$CH_2Si(OCH_3)_3Cl^-$, tend to orient liquid crystals perpendicular to the surface (see LIQUID CRYSTALLINE MATERIALS); parallel orientation is obtained on surfaces treated with N-methylaminopropyltrimethoxysilane [3069-25-8], CH_3NH-$CH_2CH_2CH_2Si(OCH_3)_3$ (25).

Ion Removal and Metal Oxide Electrodes. The ethylenediamine (en)-functional silane, shown in Table 3 (No. 5), has been studied extensively as a silylating agent on silica gel to preconcentrate polyvalent anions and cations from dilute aqueous solutions (26,27). Numerous other chelate-functional silanes have been immobilized on silica gel, controlled-pore glass, and fiber glass for removal of metal ions from solution (28,29).

Metal oxide electrodes have been coated with a monolayer of this same diaminosilane (see Table 3, No. 5) by contacting the electrodes with a benzene solution of the silane at room temperature (30). Electroactive moieties attached to such silane-treated electrodes undergo electron-transfer reactions with the underlying metal oxide (31). Dye molecules attached to silylated electrodes absorb light coincident with the absorption spectrum of the dye, which is a first step toward simple production of photoelectrochemical devices (32) (see PHOTOVOLTAIC CELLS).

Table 3. Commercial Silane Coupling Agents

No.	Silane coupling agent	CAS Registry Number	Formula	Application in plastics
1	vinyltrimethoxysilane	[2768-02-7]	$CH_2{=}CHSi(OCH_3)_3$	polyethylene, unsaturated polymers
2	3-methacryloxypropyltrimethoxysilane	[2530-85-0]	$CH_2{=}C(CH_3)\overset{\overset{O}{\|}}{C}O(CH_2)_3Si(OCH_3)_3$	unsaturated polymers
3	vinylbenzyl cationic silane	[34937-00-3]	$CH_2{=}CHC_6H_4CH_2NHCH_2{-}$ $CH_2NH(CH_2)_3Si(OCH_3)_3{\cdot}HCl$	all polymers
4	3-aminopropyltriethoxysilane	[919-30-2]	$H_2NCH_2CH_2CH_2Si(OC_2H_5)_3$	epoxies, phenolics, nylon
5	N-(2-aminoethyl)-3-aminopropyltrimethoxysilane	[1760-24-3]	$H_2NCH_2CH_2NH(CH_2)_3{-}$ $Si(OCH_3)_3$	epoxies, phenolics, nylon
6	3-glycidoxypropyltrimethoxysilane	[25704-87-4]	$\overset{O}{\overset{\triangle}{CH_2CHCH_2}}O(CH_2)_3Si(OCH_3)_3$	most thermosetting resins
7	bis(3-triethoxysilylpropyl)-tetrasulfide	[40372-72-3]	$((CH_3CH_2O)_3SiCH_2CH_2{-}$ $CH_2SS)_2$	rubber, polysulfides
8	3-mercaptopropyltrimethoxysilane	[4420-74-0]	$HSCH_2CH_2CH_2Si(OCH_3)_3$	rubber, epoxies, polysulfides
9	3-chloropropyltrimethoxysilane	[25512-39-4]	$ClCH_2CH_2CH_2Si(OCH_3)_3$	epoxies

Antimicrobials. Surface-bonded organosilicon quaternary ammonium chlorides have enhanced antimicrobial and algicidal activity (33). Thus, the hydroysis product of dimethyloctadecyl-3-trimethoxysilylpropylammonium chloride [27668-52-6] exhibits antimicrobial activity against a broad range of microorganisms while chemically bonded to a variety of surfaces. The chemical is not removed from surfaces by repeated washing with water, and its antimicrobial activity is not attributed to a slow release of the chemical but rather to the surface-bonded chemical.

Polypeptide Synthesis and Analysis. Silica or controlled-pore glass supports treated with (chloromethyl)phenylethyltrimethoxysilane [68128-25-6] or its derivatives are replacing chloromethylated styrene–divinylbenzene (Merrifield resin) as supports in polypeptide synthesis. The silylated support reacts with the triethylammonium salt of a protected amino acid. Once the initial amino acid residue has been coupled to the support, a variety of peptide synthesis methods can be used (34). At the completion of synthesis, the anchored peptide is separated from the support with hydrogen bromide in acetic acid (see PROTEIN ENGINEERING; PROTEINS).

Edman degradations can be accomplished by treating aminopropylsilylated supports (a silica of 1.0–7.5-nm pore size yields aminopropylsilane in concentrations of ca 2×10^{-7} mol/g of glass) with the peptide to be analyzed in the presence of dicyclohexylcarbodiimide (35). The carboxyl end of the peptide bonds to the amino group of the silane through an amide group. The bound peptide is then treated with phenylthiocyanate in the presence of base to yield an N-terminal phenylthiocarbamyl derivative which, on treatment with acid, cyclizes to a phenylthiohydantoin and cleaves. The hydantoin is analyzed and the process is repeated with the bound peptide residue.

Immobilized Enzymes and Metal-Complex Catalysts. Use of enzymes to catalyze reactions in cell-free systems has been limited by the difficulty of enzyme isolation, lability of the enzymes, and difficulty in effecting clean separations of enzymes from reaction mixtures. An approach that has circumvented some of these problems is to attach enzymes to solid support materials (36). The most frequently used technique for immobilizing enzymes on a solid support involves reducing N-(3-triethoxysilylpropyl)-p-nitrobenzamide [60871-86-5] groups after attachment to silica or controlled-pore glass to give aniline derivatives, then converting them to diazonium salt, and effecting coupling through azo linkage to the tyrosine of the proteins (see ENZYMES IN ORGANIC SYNTHESIS; ENZYME APPLICATIONS, INDUSTRIAL).

Two general methods are available for immobilizing metal-complex catalysts on metal oxide surfaces by the use of ligand–silane coupling reagents of the type X_3SiL, where X is a hydrolyzable group and L is a ligand (37). In Method A, the ligand silane reacts with surface hydroxyl groups to form a ligand metal oxide, usually a ligand silica, and then a metal-complex precursor, M, reacts with the functionalized surface. In Method B, a metal–liquid silane complex is first formed in solution and then reacts with surface hydroxyls of the support. Both methods give the same metal–silane complex on the metal surface: {metal}–O–SiX_2LM, eg, (($\equiv$SiOSi$(CH_2)_2$P$(C_6H_5)_2)_2$RhCl(CO). Supported metal-complex catalysts have been used in hydrogenation, hydroformylation, hydrosilylation, isomerization, and other chemical reactions (see CATALYSTS, SUPPORTED).

Reinforced Composites. Silane coupling agents modify the interface between inorganic surfaces and organic resins to improve the adhesion between resin and surface, thus improving physical properties and water resistance of reinforced plastics. Suitable coupling agents are available for any of the common plastics and metal, glass, or many other inorganic reinforcements. Principal applications for these coupling agents are in reinforced plastics for boats, storage tanks, pipes, automobiles, and architectural structures (see LAMINATED AND REINFORCED PLASTICS). Other applications are in the treatment of mineral fillers and pigments for paint and rubber, in primers to improve the adhesion of paints, inks, coatings, and adhesives to metals and other inorganic surfaces, and in tarnish and corrosion inhibitors for silver, copper, aluminum, and steel (see FILLERS; CORROSION AND CORROSION CONTROL).

Commercial silane coupling agents are soluble in acidified water or become soluble as the alkoxy groups hydrolyze from silicon. The resulting aqueous solutions are stable in water for at least several hours, but they may become insoluble as the silanols condense to siloxanes. Freshly prepared aqueous solutions of the silane coupling agents are therefore applied to glass filaments by the fiber manufacturer with a polymeric substance, a lubricant, and an antistatic agent as a complete size for glass mat or woven roving. Glass cloth woven from glass fiber with a starch–oil size can be heat-cleaned to burn off the organic size and treated with a dilute solution of the desired silane coupling agents. This operation is accomplished by the glass weaver for electronic applications (38). The total amount of silane coupling agent applied is generally 0.1–0.5% of the weight of the glass. The improvement in laminate properties imparted by silane coupling agents in typical glass-cloth laminates is summarized in Table 4 (39). All results are based on compression-molded test samples containing 60–70 wt % glass in the laminate (see LAMINATED MATERIALS, GLASS).

The nature of adhesion through silane coupling agents has been studied extensively by advanced analytical techniques (38,39). It is fairly well established that silane coupling agents form M–O–Si bonds on mineral surfaces where M = Si, Ti, Al, Fe, etc. Condensation and hydrolysis of coupling agent silanols on a silica surface are in true equilibrium in water (40). Equilibrium constants suggest that bonding of a typical silane coupling agent, $RSi(OH)_3$, to silica has a thousandfold advantage in water resistance over a simple alkoxy bond between a hydroxyl-functional polymer and silica. Mechanical properties of filled polymer castings are improved by treatment of a wide range of mineral fillers with appropriate silane coupling agents (41).

Oxane bonds, M–O–Si, are hydrolyzed during prolonged exposure to water but reform when dried. Adhesion in composites is maintained by controlling conditions favorable for equilibrium oxane formation, ie, maximum initial oxane bonding, minimum penetration of water to the interface, and optimum morphology for retention of silanols at the interface. The inclusion of a hydrophobic

Table 4. Silane Performance in Glass-Cloth-Reinforced Laminates[a]

Resin	Silane coupling agent, No.[b]	Flexural strength improvement, %	
		Dry	Wet[c]
epoxy			
anhydride	9	20	300
aromatic amine	3,9	50	140
dicyandiamide	3,6	30	70
polyester	2,3	60	140
vinyl ester	2,3	40	65
phenolic	3–6	40	120
melamine	3,6	100	250
nylon	3–6	65	130
polycarbonate	3–6	65	230
polyterephthalate	3	50	50
polystyrene	3	40	110
acrylonitrile–butadiene–styrene terpolymer	3,6	30	50
styrene–acrylonitrile copolymer	3	40	70
poly(vinyl chloride)	3–5	60	80
polyethylene	3	130	130
polypropylene	3	90	90

[a]Ref. 38. [b]See Table 3. [c]Wet = boiling for 72 h in water for epoxies, 2 h for other resins.

silane, such as phenyltrimethoxysilane [2996-92-1], with the organofunctional silane increases thermal stability of the silane and make the bond more water resistant (42).

Although simplified representations of coupling through organofunctional silanes often show a well-aligned monolayer of silane forming a covalent bridge between polymer and filler, the actual situation is much more complex. Coverage by hydrolyzed silane is more likely to be equivalent to several monolayers. The hydrolyzed silane condenses to oligomeric siloxanols that initially are soluble and fusible but ultimately can condense to rigid cross-linked structures. Contact of a treated surface with polymer matrix is made while the siloxanols still have some degree of solubility. Bonding with the matrix resin then can take several forms. The oligomeric siloxanol layer may be compatible in the liquid-matrix resin and then form a true copolymer during resin cure. Partial solution compatibility is also possible and an interpenetrating polymer network can form as the siloxanols and matrix resin cure separately with a limited amount of copolymerization. Probably all thermosetting resins are coupled to silane-treated fillers by some modification of these two extremes.

Interdiffusion of silane primer segments with matrix molecules having no cross-linking may become a factor in bonding of thermoplastic polymers (43). This is the suggested mechanism when a silane–thermoplastic copolymer is used as a primer or coupling agent for the corresponding unmodified thermoplastic. A siloxanol layer may also diffuse into a nonreactive thermoplastic layer and then cross-link at the fabrication temperature. Structures, in which only one of the interpenetrating phases cross-links, have been designated pseudointerpenetrating networks. Amine-functional silanes (Table 3, No. 3, 4, and 5) probably function in this manner in coupling to polyolefins and possibly to other thermoplastics. Layers of amine-functional siloxanols in the absence of matrix resins cure at 150°C to very hard, tough films.

Peroxides or other additives, eg, chlorinated paraffin, may also cause the thermoplastic resin to cross-link with the siloxanols. In this case, a true interpenetrating polymer network forms, in which both phases are cross-linked.

Performance of coupling agents in reinforced composites may depend as much on physical properties resulting from the method of application as on the chemistry of the organofunctional silane. Physical solubility or compatibility of a siloxanol layer is determined by the nature and degree of siloxane condensation on a mineral surface.

BIBLIOGRAPHY

"Silylating Agents" under "Silicon Compounds," in ECT 2nd ed., Vol. 18, pp. 260–268, by E. P. Plueddemann, Dow Corning Corp.; in ECT 3rd ed., Vol. 20, pp. 962–973, by E. Plueddemann, Dow Corning Corp.

1. A. E. Pierce, Silylation of Organic Compounds, Pierce Chemical Co., Rockford, Ill., 1968.
2. K. Balu and G. Kind, eds., Handbook of Derivatives for Chromatography, Heyden, London, 1977.
3. J. F. Klebe, in E. Taylor, ed., Advances in Organic Chemistry, Wiley-Interscience, New York, 1972, p. 97.
4. C. A. Roth, Ind. Eng. Chem. Prod. Res. Develop. 11, 134 (1972).

5. B. E. Cooper, *Process Biochem.* **15**, 9 (Jan. 1980).
6. E. J. Corey and T. Ravindranathan, *J. Am. Chem. Soc.* **94**, 4013 (1972).
7. G. Simchen, in G. L. Larson, ed., *Advances in Silicon Chemistry*, Vol. 1, JAI Press, Greenwich, Conn., 1991, pp. 196, 219.
8. T. Morita, Y. Okamoto, and H. Sakurai, *J. Chem. Soc., Chem. Comm.* **20**, 874 (1978).
9. C. A. Bache, L. E. St. John, Jr., and D. J. Lisk, *Anal. Chem.* **40**, 1241 (1968).
10. L. Birkofer, A. Ritter, and H. Uhlenbrauck, *Chem. Ber.* **96**, 2750, 3280 (1963).
11. Y. Horiguchi, E. Nakamura, and I. Kuwajima, *Tetrahedron Lett.* **30**, 3523 (1989).
12. C. W. Lentz, *Inorg. Chem.* **3**, 574 (1964).
13. E. P. Plueddemann, *Proceedings of the 24th Annual Technological Conference, SPI, Reinforced Plastics/Composite Division*, Washington, D.C., 1969, Paper 19-A.
14. E. P. Plueddemann, *Silane Coupling Agents*, 2nd ed., Plenum Press, New York, 1991, Chapt. 3.
15. U.S. Pat. 4,370,255 (Jan. 25, 1983), E. P. Plueddemann (to Dow Corning Corp.).
16. M. Harkonen, J. V. Seppala, and T. Vaananen, *Makromol. Chem.* **192**, 721 (1991).
17. E. G. Shafrin and W. A. Zisman, *Contact Angle, Wettability, and Adhesion, Advances in Chemistry Series No. 43*, American Chemical Society, Washington, D.C., 1964, p. 145.
18. O. K. Johannson and J. J. Torok, *Proc. Inst. Radio Electron. Eng.* **34**, 296 (1946).
19. B. Arkles, *Chemtech*, 768 (Dec. 1977).
20. T. J. Nestrick, L. L. Lamparski, and R. H. Stehl, *Anal. Chem.* **51**, 2273 (1979).
21. S. A. Wise, L. C. Sander, and W. E. May, in D. E. Leyden, ed., *Silanes, Surfaces, and Interfaces*, Gordon and Breach, New York, 1985, pp. 349–370.
22. F. K. Chow and E. Grushka, *Anal. Chem.* **50**, 1346 (1978).
23. D. E. Leyden and W. Collins, eds., *Silylated Surfaces*, Gordon and Breach Science Publishers, London, 1980.
24. Ref. 14, Chapt. 2.
25. F. J. Kahn, G. N. Taylor, and H. Schonborn, *Proc. IEEE* **61**, 823 (1973).
26. D. E. Leyden and G. H. Luttrell, *Anal. Chem.* **47**, 1612 (1975).
27. D. E. Leyden, in Ref. 23, pp. 321–331.
28. U.S. Pat. 4,071,546 (Jan. 31, 1978), E. P. Plueddemann (to Dow Corning Corp.).
29. T. G. Waddell, D. E. Leyden, and D. M. Hercules, in Ref. 23, pp. 55–72.
30. M. Murray, in Ref. 23, pp. 125–134.
31. P. R. Moses and R. W. Murray, *J. Am. Chem. Soc.* **98**, 7435 (1976).
32. N. R. Armstrong, in Ref. 23, pp. 159–170.
33. A. J. Isquith, E. A. Abbott, and P. A. Walters, *Applied Microbiol.* **23**, 859 (1973).
34. W. Parr and K. Grohmann, *Tetrahedron Lett.* **28**, 2633 (1971).
35. W. Machleidt, *Proc. Int. Conf. Solid Phase Methods in Protein Sequence Anal.*, 17 (1975).
36. M. Lynn, in H. W. Weetall, ed., *Immobilized Enzymes, Antigens, and Peptides*, Marcel Dekker, New York, 1975; W. H. Scouten, in Ref. 21, pp. 59–72.
37. T. J. Pinnavaia, J. G-S. Lee, and M. Abeduri, in Ref. 23, Chapt. 16.
38. E. P. Plueddemann, *Silane Coupling Agents*, 1st ed., Pergamon Press, New York, 1982, Chapt. 4.
39. Ref. 38, Chapt. 5.
40. E. R. Pohl and F. D. Osterholtz, *J. Adhesion Sci. Technol.* **6**, 127 (1992).
41. S. Sterman and J. G. Marsden, *Plastics Technol.* **9**, 39 (May 1963).
42. P. G. Pape and E. P. Plueddemann, *J. Adhesion Sci. Technol.* **5**, 831 (1991).
43. M. K. Chaudhury, T. M. Gentle, and E. P. Plueddemann, *J. Adhesion Sci. Technol.* **1**, 29 (1987).

Peter G. Pape
Dow Corning Corporation

SILK

Silks can be defined as externally spun fibrous protein secretions. Of all the natural fibers, silks represent the only ones that are spun. Silk fibers have been used in textiles for thousands of years owing to their unique visual luster, tactile properties, and durability. These fibers are remarkable materials displaying unusual mechanical properties. Strong, extensible, and compressible, silks display interesting electromagnetic responses, particularly in the uv range for insect entrapment; form liquid crystalline phases related to processing; and exhibit piezoelectric properties (1). Silks were used in optical instruments as late as the mid-1900s because of their fine diameter and high strength and stability over a range of temperatures. Spider silks are reportedly used in the South Pacific for gill nets, dip nets, fishing lures, as well as in weaving and ceremonial dress (1).

Types of Silk

Silks are synthesized by a variety of organisms, including silkworms, spiders, scorpions, mites, and flies. Few of these silks have been characterized. Silks differ in properties, composition, and morphology, depending on the source. Silkworm cocoon silk from *Bombyx mori* is the most well characterized owing to the extensive use of these fibers in the textile industry for over 5000 years in a practice originating in China. The dragline silk from the orb-weaving spider, *Nephila clavipes*, is the most well characterized of the different spider silks. Spider silk has not been domesticated for textile applications because spiders are more difficult to raise in large numbers on account of their solitary and predatory nature. Unlike the cocoon silk from the silkworm, orb webs are not reelable as a single fiber.

Silkworm Cocoon Silk. The cocoon silk from *B. mori* contains two structural fibroin filaments coated with a family of glue-like sericin proteins, resulting in a single thread having a diameter of 10 to 25 μm. Wild silkworms generally have larger-diameter threads, ranging up to 65 μm and displaying a variety of different cross-sectional morphologies. The life cycle of *B. mori* runs for 55 to 60 days and the organism passes through a series of developmental stages or molts. Silk production occurs during cocoon formation around day 26 in the cycle during the fifth larval instar just before molt to the pupa. The silkworm passes through four different metamorphosizing phases: egg or embryo, larva, pupa, and moth (adult). Smaller quantities of silk are produced at all larval stages except during molts.

Spider Silk. Spider silks function in prey capture, reproduction, and as vibration receptors, safety lines, and dispersion tools. Spider silks are synthesized in glands located in the abdomen and spun through a series of orifices (spinnerets). The types and nature of the various silks are diverse and dependent on the type of spider (2). Some general categories of silks and the glands responsible for their production are listed in Table 1.

Table 1. Function and Location of Spider Silk Glands[a]

Silk	Gland	Function
dragline	major ampullate	orb web frame and radii construction, safety line
viscid	flagelliform	prey capture
glue-like	aggregate	prey capture, attachment
minor ampullate	minor ampullate	orb web frame construction
cocoon	cylindrical	reproduction
wrapping	aciniform	wrapping captured prey
attachment	piriform	attachment to environmental substrates

[a]Ref. 3.

Structure

Composition.   Table 2 summarizes the amino acid composition of various silk proteins (see AMINO ACIDS; PROTEINS). The silkworm cocoon silk contains two structural proteins, the fibroin heavy chain (mol wt ca 325,000) and fibroin light chain (mol wt ca 25,000), plus the family of sericin proteins (mol wt 20,000–310,000) to hold the fibroin chains together in the final cocoon fiber. Other silks, such as the caddis fly and aquatic midge, which spin silks underwater to form sheltered tubes, have also been characterized and consist of a family of proteins having high cysteine content and running from low to very high ($>10^6$) molecular weights (4). Some silkworm silks from wild strains have been reported to contain up to 95% alanine (1).

The consensus crystalline amino acid repeat in the *B. mori* silkworm cocoon silk fibroin heavy chain is the 59mer: GAGAGSGAAG[SGAGAG]$_8$Y. More

Table 2. Amino Acid Composition of Silks

Amino acid	Abbreviation	*B. mori*		*P. c. ricini*[a]	*N. clavipes*
		Fibroin	Sericin	Fibroin	Dragline
glycine	G	42.9	13.5	33.2	37.1
alanine	A	30.0	5.8	48.4	21.1
serine	S	12.2	34.0	5.5	4.5
tyrosine	Y	4.8	3.6	4.5	
aspartic acid–asparagine	D, N	1.9	14.6	2.7	2.5
arginine	R	0.5	3.1	1.7	7.6
histidine	H	0.2	1.4	1.7	0.5
glutamic acid–glutamine	E, Q	1.4	6.2	0.7	9.2
lysine	K	0.4	3.5	0.2	0.5
valine	V	2.5	2.9	0.4	1.8
leucine	L	0.6	0.7	0.3	3.8
isoleucine	I	0.6	0.7	0.4	0.9
phenylalanine	F	0.7	0.4	0.2	0.7
proline	P	0.5	0.6	0.4	4.3
threonine	T	0.9	8.8	0.5	1.7
methionine	M	0.1	0.1	trace	0.4
cysteine	C	trace	0.1	trace	0.3
tryptophan	W			0.6	2.9

[a]*Philosamia cynthia ricini* = wild-type silkworm silk.

detailed analysis of these repeats indicates that the fibroin contains alternate arrays of repeating GAGAGS and GAGAGY. Valine or tyrosine replacements for alanine exist in the second repeat (5). These core repeats are surrounded by homogenous nonrepetitive amorphous domains. The repetitive structures in the protein are thought to be the result of genetic level continuous unequal crossovers or genetic recombination events during evolution.

The spider dragline silk from the principal ampullate gland contains at least one protein, called MaSp1, for major ampullate silk protein, previously termed spidroin 1; mol wt is around 275,000 (6). It remains unclear whether additional proteins play a significant role in the dragline silk fiber. There is no sericin or glue-like protein associated with the dragline fiber. MaSp1 contains amino acid repeats considerably shorter than those found in the silkworm fibroin and not as highly conserved. The repeats contain polyalanine domains consisting of from six to nine residues, and a 15-amino acid region showing a GGX repeat motif, where X = alanine, tyrosine, leucine, or glutamine (6,7).

Secondary Structure. The silkworm cocoon and spider dragline silks are characterized as an antiparallel β-pleated sheet wherein the polymer chain axis is parallel to the fiber axis. Other silks are known to form α-helical (bees, wasps, ants) or cross-β-sheet (many insects) structures. The cross-β-sheets are characterized by a polymer chain axis perpendicular to the fiber axis and a higher serine content. Most silks assume a range of different secondary structures during processing from soluble protein in the glands to insoluble spun fibers.

The crystalline structure of silk was first described in the 1950s as an antiparallel, hydrogen-bonded β-sheet based on the characterization of *B. mori* fibroin (8), and further modifications to this early model have been made over the years (9,10). Two crystalline forms for silk have been characterized based on x-ray diffraction and ^{13}C-cross-polarization magic angle spinning (cp/mas) nmr spectroscopy. The random coil or silk I, ie, the prespun pseudocrystalline form of silk present in the gland in a water-soluble state, is predominant in the gland; silk II, ie, the spun form of silk which is insoluble in water, becomes predominant once the protein is spun into fiber (9,11). The unit cell parameters in the silk II structure are 940 pm (a, interchain), 697 pm (b, fiber axis), and 920 pm (c, intersheet). The chains run antiparallel with interchain hydrogen bonds roughly perpendicular to the chain axis between carbonyl and amine groups. Van der Waal forces stabilize intersheet interactions owing to the predominance of short side-chain amino acids (glycine, alanine, serine) in these crystalline regions. Solid-state ^{13}C-nmr studies of *N. clavipes* dragline silk have concluded that the polyalanine runs in the MaSp1 are in a β-sheet conformation (12).

The structure of silk I remains incompletely understood. This structure is unstable and, upon shearing, drawing, heating, spinning, or exposure in an electric field or exposure to polar solvents such as methanol or acetone, converts to silk II. The change in unit cell dimensions during the transition from silk I to silk II during fiber spinning is most significant in the intersheet plane, where an 18.3% decrease in distance occurs between overlying sheets (13). This change results in the exclusion of water, thus reducing solubility of the protein. Silk II is more energetically stable than silk I and the energy barrier for the transition is low, whereas the return barrier is high and considered essentially irreversible (1,14). Because the instability or metastable nature of the silk I conformation leads to difficulty in obtaining an orientated sample for fiber diffraction analysis and detailed structural characterization, different models have been proposed to describe the silk I structure (14,15).

Crystallinity. Generally, spider dragline and silkworm cocoon silks are considered semicrystalline materials having amorphous flexible chains reinforced by strong stiff crystals (3). The orb web fibers are composite materials (qv) in the sense that they are composed of crystalline regions immersed in less crystalline regions, which have estimates of 30–50% crystallinity (3,16). Earlier studies by x-ray diffraction analysis indicated approximately 62–65% crystallinity in cocoon silk fibroin from the silkworm, approximately 50–63% in wild-type silkworm cocoons, and lesser amounts in spider silk (17).

Structure of the Spider Orb Web. The construction of the orb web is a feat of engineering involving material tailoring, optimization of material interfaces, and conservation of resources to promote survival of the spider (2,3). In addition, the web absorbs water from the atmosphere and ingestion by the spider may provide a significant contribution to water intake needs (18). Around 70% of the energy is dissipated through viscoelastic processes upon impact by a flying insect into the web (3). Thus the web balances stiffness and strength against extensibility, both to keep the web from breaking and to keep the insect from being ejected from the web by elastic recoil (3). The ability to dissipate the kinetic energy of a flying insect impacting the web is based on the hysteresis of radial threads and also aerodynamic damping by the web (18,19). Some orb webs appear to be at least in part recycled by ingestion as a conservation tool, and some of the amino acids are reused in new webs.

Processing

In Vivo **Processing.** Silks are synthesized in specialized glands within the organism. Initially, some degree of self-organization or assembly occurs as a result of protein–protein interactions among the crystalline repeats in the protein chains (13). In the spider gland, changes in physiological conditions such as pH and salt concentrations accompany the processing and, presumably, help maintain solubility, despite increasing protein concentration during passage through the various regions of the gland (1). In silkworm there are three distinct regions to the glands and two sets of these organs feeding into one final thread (20,21). The fibroin is synthesized in the posterior region of the gland and the protein moves by peristalsis to the middle region of the gland where it is stored as a viscous aqueous solution until needed for spinning. The protein concentration is 12–15% in the posterior region of the gland where fibroin chain synthesis occurs, increases to around 20–30% in the middle region of the gland where the fibroin is stored and sericin is synthesized, and is significantly higher in the anterior region of the gland where spinning is initiated (20). The two lobes of the gland join just before the spinnerets in the anterior region and the fiber is spun into air. Aside from binding together the fibroin chains in the final spun fiber, sericin in this process may function in plasticization to ease the flow through the spinneret, as a reservoir for divalent cations, or as a water-holding medium to promote plasticization of the fiber after spinning.

Rheological experiments indicate that crystallinity in the fiber correlates positively with shear and draw rates, and an extrusion rate of around 50 cm/min was found to be a minimum threshold for the appearance of birefringence and the conversion of the soluble silk solution in the gland to the β-sheet found in the spun fiber (20,22). In the posterior region of the gland, 0.4–0.8 mm in diameter, the silk solution is optically featureless, a range of secondary structures are present, including random coil and silk I, and the shear rate is low. In the middle region of the gland, the diameter is 1.2–2.5 mm, streaming birefringence is observed, and the shear rate is also low. In the anterior region of the gland, the diameter is narrow, 0.05–0.3 mm, the shear rate is high, water appears to be actively transported out of the gland, the pH decreases, and active ion exchange occurs. Viscosity also increases but presumably decreases prior to spinning as a result of the liquid crystalline phase. At this point the characteristic silk II structure forms. In the pair of major ampullate glands in the spider, which are the location of dragline protein synthesis, a similar process occurs as summarized for the silkworm. This gland is smaller, however, and there is no sericin contribution in the middle region of the gland. A lyotropic nematic liquid crystalline phase of the protein forms prior to spinning in both the spider and the silkworm, as well as in many of the different glands of the spider responsible for the different silks (23).

Commercial and Artificial Processing. Commercially, silkworm cocoons are extracted in hot soapy water to remove the sticky sericin protein. The remaining fibroin or structural silk is reeled onto spools, yielding approximately 300–1200 meters of usable thread per cocoon. These threads can be dyed or modified for textile applications. Production levels of silk textiles in 1992 were 67,000 metric tons worldwide. The highest levels were in China, at 30,000 t, followed by Japan, at 17,000 t, and other Asian and Oceanian countries, at 14,000 t (24). Less than 3000 metric tons are produced annually in each of Eastern Europe, Western Europe, and Latin America; almost no production exists in North America, the Middle East, or Africa. 1993 projections were for a continued worldwide increase in silk textile production to 75,000 metric tons by 1997 and 90,000 metric tons by 2002 (24).

Most solvents used to solubilize globular proteins do not suffice for silks owing to extensive hydrogen bonding and van der Waals interactions, and the exclusion of water from the intersheet regions. Silks are insoluble in water, dilute acids and alkali, and most organic solvents; they are resistant to most proteolytic enzymes (6,25). Silkworm fibroin can be solubilized by first degumming or removing the sericin using boiling soap solution or boiling dilute sodium bicarbonate solution, followed by immersion of the fibroin in high concentration salt solutions such as lithium bromide, lithium thiocyanate, or calcium chloride. These salt solutions can also be used to solubilize spider silk, as can high concentrations of propionic acid–hydrochloric acid mixtures and formic acid (6). After solubilization in these aggressive solvents, dialysis into water or buffers can be used to remove the salts or acids, although premature reprecipitation is a common problem. Ternary-phase diagrams of silk, water, and chaotropic salt for processing windows have been published (26,27) for native silkworm silks and for genetically engineered versions of silkworm silk.

Films or membranes of silkworm silk have been produced by air-drying aqueous solutions prepared from the concentrated salts, followed by dialysis (11,28). The films, which are water soluble, generally contain silk in the silk I conformation with a significant content of random coil. Many different treatments have been used to modify these films to decrease their water solubility by converting silk I to silk II in a process found useful for enzyme entrapment (28). Silk membranes have also been cast from fibroin solutions and characterized for permeation properties. Oxygen and water vapor transmission rates were dependent on the exposure conditions to methanol to facilitate the conversion to silk II (29). Thin monolayer films have been formed from solubilized silkworm silk using Langmuir techniques to facilitate structural characterization of the protein (30). Resolubilized silkworm cocoon silk has been spun into fibers (31), as have recombinant silkworm silks (32).

Properties

Mechanical Properties. The mechanical properties of silks are an intriguing combination of high strength, extensibility, and compressibility (Table 3).

Resistance to axial compressive deformation is another interesting property of the silk fibers. Based on microscopic evaluations of knotted single fibers, no evidence of kink-band failure on the compressive side of a knot curve has been observed (33,35). Synthetic high performance fibers fail by this mode even at relatively low strain levels. This is a principal limitation of synthetic fibers in some structural applications.

Fibers. *B. mori* cocoon silk ranges from 10 to 25 μm in diameter; dragline silk from *N. clavipes* from 2.5 to 4.5 μm in diameter. Web fibers from some

Table 3. Mechanical Properties of Silks and Other Fibers[a]

Fiber	Elongation, %	Modulus, GPa[b]	Strength, GPa[b]	Energy to break, J/kg[c]
		Fibroins		
B. mori	15–35	5	0.6	7×10^4
other silkworms	12–50	2–4	0.1–0.6	$(3-6) \times 10^4$
		Draglines		
N. clavipes				
quasistatic[d]	9–11	22–60	1.1–2.9	$(3.7-12) \times 10^4$
high strain[e]	10	20		
other spiders	10–39	2–24	0.2–1.8	$(1-10) \times 10^4$
		Other fibers		
nylon	18–26	3	0.5	8×10^4
cotton	5–7	6–11	0.3–0.7	$(5-15) \times 10^3$
Kevlar	4	100	4	3×10^4
steel	8	200	2	2×10^3

[a]Refs. 1, 25, 33, and 34.
[b]1 GPa = 10^9 N/m^2. To convert GPa to psi, multiply by 145,000.
[c]To convert J to cal, divide by 4.184.
[d]Instron tensile test rates of 10%/s.
[e]Rates of >500,000%/s.

spiders have diameters as low as 0.01 μm. A silking rate of around 1 cm/s is considered equivalent to natural spinning rates for the spider (20). Some spider silks have been observed to supercontact up to around 50% when unconstrained and exposed to high moisture; other silks such as the silkworm cocoon silk do not contract under similar experimental conditions (36). A skin core has been reported using light microscopy and electron microscopy. However, more recent data using dragline silk from *N. clavipes* refute this finding (37).

Thermal Properties. Spider dragline silk was thermally stable to about 230°C based on thermal gravimetric analysis (tga) (33). Two thermal transitions were observed by dynamic mechanical analysis (dma), one at -75°C, presumed to represent localized mobility in the noncrystalline regions of the silk fiber, and the other at 210°C, indicative of a partial melt or a glass transition. Data from thermal studies on *B. mori* silkworm cocoon silk indicate a glass-transition temperature, T_g, of 175°C and stability to around 250°C (37). The T_g for wild silkworm cocoon silks were slightly higher, from 160 to 210°C.

Genetic Engineering

An understanding of the genetics of silk production in silkworms and spiders should help in developing processes for higher levels of silk expression generated by recombinant deoxyribonucleic acid (DNA) methods. The ability of the silkworm to produce large amounts of protein, around 300 μg of fibroin per epithelial cell lining the silk producing gland, has generated a great amount of interest from molecular biologists and developmental biologists in elucidating the genetic regulation of this system. Genetically engineered or recombinant DNA silkworm and spider silks have been produced using either native genes or synthetic genes (32,38,39). It is interesting to note that studies of variations in native silkworm populations indicate differences in sizes of the crystalline-encoding domains resulting from a high degree of polymorphism in length and organization of the fibroin gene and thus the encoded proteins (40,41). A significant degree of variation in protein size apparently can be tolerated in native populations of silkworms and possibly spiders. Owing to the highly repetitive nature of the genes and the encoded proteins, the deletion or addition of repeats has little impact on secondary structure and functional performance within a certain window of sizes.

Applications

Because of the advent of improved analytical techniques, together with the tools of biotechnology, a new generation of products is envisioned with silk. The ability to tailor polymer structure to a precise degree leads to interesting possibilities in the control of macroscopic functional properties of fibers, membranes, and coatings, as well as improved control of processing windows. Biotechnology offers the tools with which to solve limitations in spider silk production that have not been overcome with traditional domestication and breeding approaches, such as those used successfully with the silkworm. This is of interest because of the variety of silk structures available and the higher modulus and strength as compared to silkworm silk.

Hybrid silk fibers containing synthetic fiber cores having silk coextruded or grafted have been synthesized. Cosmetics (qv) and consumer products such as hair replacements and shampoos containing silk have also been marketed. Sutures, biomaterials for tissue repairs, wound coatings, artificial tendons, bone repair, and related needs may be possible applications, assuming immunological responses to the silks are controllable (see PROSTHETICS AND BIOMEDICAL DEVICES). It is also reasonable to speculate on the use of silk webbing for tissue cell growth, nerve cell growth, and brain repair applications as temporary scaffolding during regrowth and reinfusion after surgery. Cell culture petri plates having genetically engineered silkworm silks containing cell binding or adhesive domains have already been produced and are sold commercially. These recombinant silks are stable during injection molding with polystyrene. The demonstration of fiber spinning from resolubilized silkworm silk provides further opportunities in material fabrication using native and genetically engineered silk proteins.

BIBLIOGRAPHY

"Silk" in *ECT* 1st ed., Vol. 12, pp. 414–452, by A. C. Hayes, North Carolina State College School of Textiles; in *ECT* 2nd ed., Vol. 18, pp. 269–279, by A. C. Hayes, North Carolina State University School of Textiles; in *ECT* 3rd ed., Vol. 20, pp. 973–981, by C. D. Livengood, North Carolina State University.

1. D. L. Kaplan and co-workers, eds., *Silks: Materials Science and Biotechnology*, American Chemical Society, Washington, D.C., 1994.
2. M. W. Denny, in J. F. V. Vincent and J. D. Currey, eds., *Mechanical Properties of Biological Materials*, Cambridge University Press, New York, 1980, p. 247.
3. J. M. Gosline, M. E. DeMont, and M. W. Denny, *Endeavour*, **10**, 37 (1986).
4. S. T. Case and co-workers, in Ref. 1, p. 80.
5. K. Mita and co-workers, *J. Mol. Bio.* **203**, 917 (1988).
6. C. M. Mello and co-workers, in Ref. 1, p. 67.
7. M. Xu and R. V. Lewis, *Proc. Nat. Acad. Sci. USA*, **87**, 7120 (1990).
8. R. E. Marsh, R. B. Corey, and L. Pauling, *Biochimica Biophysica Acta*, **16**, 1 (1955).
9. R. D. B. Fraser and T. P. MacRae, *Conformation in Fibrous Proteins*, Academic Press, Inc., New York (1973).
10. F. Colonna–Cesari, S. Premilat, and B. Lotz, *J. Mol. Bio.* **95**, 71 (1975).
11. T. Asakura and co-workers, *Macromolecules*, **18**, 1841 (1985).
12. A. Simmons, C. Michal, and L. W. Jelinski, *Science*, **271**, 84 (1996).
13. D. L. Kaplan and co-workers, *Proc. Materials Res. Soc. Symp.* **255**, 19 (1992).
14. S. A. Fossey and co-workers, *Biopolymers*, **3**, 1529 (1991).
15. B. Lotz and H. D. Keith, *J. Mol. Bio.* **61**, 201 (1971).
16. B. Thiel and co-workers, *Proc. Materials Res. Soc. Symp.* **330**, 21 (1994).
17. J. O. Warwicker, *J. Mol. Bio.* **2**, 350 (1960).
18. D. T. Edmonds and F. Vollrath, *Proc. Royal Soc. London*, **B248**, 145 (1992).
19. L. H. Lin, D. T. Edmonds, and F. Vollrath, *Nature*, **273**, 146 (1995).
20. J. Magoshi, Y. Magoshi, and S. Nakamura, in Ref. 1, p. 292.
21. E. K. Tillinghast and M. A. Townley, in Ref. 1, p. 29.
22. E. Ilzuka, *J. Appl. Poly. Sci. Japan*, **41**, 163 (1985).
23. K. Kerkam and co-workers, *Nature*, **349**, 596 (1991).
24. Technical data, The Freedonia Group, Inc., New York, 1993.

25. D. L. Kaplan and co-workers, in D. Byrom ed., *Biomaterials: Novel Materials from Biological Sources*, Stockton Press, New York, 1991, p. 1.
26. J. P. Anderson, M. Stephen-Hassard, and D. C. Martin, in Ref. 1, p. 137.
27. C. Viney and co-workers, in Ref. 1, p. 120.
28. M. Demura and T. Asakura, *J. Membrane Sci.* **59**, 39 (1991).
29. N. Minoura, M. Tsukada, and M. Nagura, *Biomaterials*, **11**, 430 (1990).
30. W. S. Muller and co-workers, *Langmuir*, **9**, 1857 (1993).
31. PCT W093/15244 (Aug. 5, 1993), R. Lock (to E. I. du Pont de Nemours & Co., Inc.).
32. J. Cappello and co-workers, *Biotech. Prog.* **6**, 198 (1990).
33. P. M. Cunniff and co-workers, *Polym. Adv. Tech.* **5**, 401 (1994).
34. P. D. Calvert, in *Encyclopedia of Materials Science and Engineering*, Pergamon Press, Oxford, U.K., 1988, p. 334.
35. D. V. Mahoney and co-workers, in Ref. 1, p. 196.
36. R. W. Work, *J. Arachnology*, **9**, 299 (1981).
37. S. Nakamura, J. Magoshi, and Y. Magoshi, in Ref. 1, p. 211.
38. Y. Suzuki and Y. Oshima, *Cold Spring Harbor Symp. Quant. Bio.* **42**, 947 (1977).
39. J. T. Prince and co-workers, *Biochemistry*, **34**, 10879 (1995).
40. K. U. Sprague and co-workers, *Cell*, **17**, 407 (1979).
41. L. P. Gage and R. F. Manning, *J. Bio. Chem.* **255**, 9444 (1980).

DAVID L. KAPLAN
CHARLENE MELLO
STEPHEN FOSSEY
STEVEN ARCIDAICONO
U.S. Army Natick Research,
Development, & Engineering Center

SILLEMANITE. See REFRACTORIES.

SILVER AND SILVER ALLOYS

Silver [7440-22-4], Ag (at no. 47), is a white, lustrous, soft, malleable metal having the highest known electrical and thermal conductivities. It is the most highly reflective of all the metals in the visible spectrum and second only to gold in the long-wave infrared. Along with its colorful neighbors copper and gold in the Periodic Table Group II (IB) metals, silver occurs naturally in metallic form and thus was among the earliest metals used by humans. Silver is widely used for jewelry, tableware, holloware, and coinage. Its advantageous chemical and physical properties find extensive application in photography (qv), electrical and electronic components (see ELECTRICAL CONNECTORS; ELECTRONIC MATERIALS), solders and brazing alloys (qv), catalysts, clinical treatment, mirrors, high performance journal bearings, and sanitation.

Properties

Some properties of silver are summarized in Table 1. The solar energy transmittance and neutron-absorption characteristics of silver are shown in Figures 1 and 2, respectively. Thermal properties are given in Table 2. Other properties are given in References 1, 3, and 4.

In the electromotive force series of the elements, silver is less noble than only Pd, Hg, Pt, and Au. All provide high corrosion resistance. Silver cannot form oxides under ambient conditions. Its highly reactive character, however, results in the formation of black sulfides on exposure to sulfur-containing atmospheres.

Silver has the highest (97%) visual light reflection and is second only to gold in its reflectance of long-wave infrared radiation (Ag 92%, Au 97%). Transparent coatings of silver sputtered onto window glass, ie, low E-glass, are the most reflective coatings for thermal insulating windows (see Fig. 1).

Laser stimulation of a silver surface results in a reflected signal over a million times stronger than that of other metals. Called laser-enhanced Raman spectroscopy, this procedure is useful in catalysis. The large neutron cross section of silver (see Fig. 2), makes this element useful as a thermal neutron flux monitor for reactor surveillance programs (see NUCLEAR REACTORS).

The fatigue resistance of silver in bearings is superior to that of all other materials, which accounts for its use in diesel locomotive crankshafts (see BEARING MATERIALS). A flash lead–tin overplating is applied to trap or collect particles that would otherwise score the hard surface of silver. Silver is widely used to prevent galling in bearings, acting as a dry film lubricant owing to its tendency toward interplanar shear. Silver is also widely used to prevent seizing of bearings experiencing long idle periods. On start-up, the low shear strength of silver permits instant response.

Oxidation States.   The common oxidation state of silver is $+1$, ie, Ag^+, as found in AgCl, which is used with Mg in sea- or freshwater-activated batteries (qv); $AgNO_3$, the initial material for photographic materials, medical compounds, catalysts, etc; and silver oxide, Ag_2O, an electrode in batteries (see SILVER COMPOUNDS). Few Ag^{2+} compounds are known. The aqua ion $[Ag(H_2O)_4]^{2+}$, which has one unpaired electron, is obtained from an $HClO_4$ or HNO_3 solution by oxidation of Ag^+ with ozone or by dissolution of AgO in acid. The potentials for the Ag^{2+}/Ag^+ couple, $+2.00$ V in 4 M $HClO_4$ and $+1.93$ in 4 M HNO_3, show that Ag^{2+} is a powerful oxidizing agent.

Silver in the $+3$ oxidation state, including silver peroxide, ie, black oxide, marketed as AgO, is obtained by the action of the vigorous oxidizing agent $S_2O_8^{2-}$ on Ag_2O or other Ag compounds. X-ray and neutron diffraction analyses show the nominal AgO unit cell to be $Ag_2O \cdot Ag_2O_3$. Both Ag^+ and Ag^{3+} are present. Another compound of potentially important commercial value is Ag_4O_4, which has a unit cell of two Ag^+ and two Ag^{3+} ions. Its preparation is as follows:

$$4\ AgNO_3 + 2\ Na_2S_2O_8 + 8\ NaOH \longrightarrow Ag_4O_4 + 4\ Na_2SO_4 + 4\ NaNO_3 + 4\ H_2O$$

Tetrasilver tetroxide is a powerful oxidizer for sanitizing swimming pools, hot tubs, and industrial cooling system waters (see WATER, TREATMENT OF SWIMMING POOLS, SPAS, AND HOT TUBS). This oxide is slightly soluble and its dissociation into silver ions is enhanced by the addition of the oxidizer $K_2S_2O_8$. Bivalent

Table 1. Properties of Silver

Parameter	Value	Reference
atomic mass, amu	107.8682	1
melting point[a], °C	961.93	1
electron structure	[Kr] $4d^{10}\,5s^1$	1
isotopic abundance[b], %		
106.9051	51.84	1
108.9048	48.16	1
atomic diameter, pm	2883	1
ionic radius, pm	126	
crystal structure	fcc, Fm3m	1
lattice spacing, a_0, at 20°C, pm	0.4078 nm	1
ionization potentials, eV[c]		
$Ag \rightarrow Ag^+ + e^-$	7.574	1
$Ag^+ \rightarrow Ag^{2+} + e^-$	21.960	1
$Ag^{2+} \rightarrow Ag^{3+} + e^-$	36.10	1
electrochemical potential[d], V	0.798	1
density, at 20°C, g/cm^3		
annealed	10.492	1
hard drawn	10.43	3
at 0 K	10.63	1
boiling point, °C	2187	4
recrystallization temperature[e], °C	20–200	
tensile strength, 5-mm dia wire, MPa[f]		
annealed at 600°C	125–186	3
50% cold worked	290	3
elongation, 5-mm dia wire, % in 5.08 cm		
at 20°C, annealed at 600°C	43–50	
50% cold worked	3–5	5
electrical resistivity[g], ρ, Ω/m		
20 K	0.00422×10^{-8}	6
273.15 K (=0°C)	1.467×10^{-8}	6
500 K	2.875×10^{-8}	6
1235.08 K, solid	8.415×10^{-8}	6
1235.08 K, liquid[h]	17.30×10^{-8}	6
vapor pressure, Pa[i]		
684°C	10^{-6}	7
828°C	10^{-4}	7
1028°C	10^{-2}	7
1330°C	1	7
1543°C	10^1	7
1825°C	10^2	7
hardness, Brinell, kg/mm^{2j}	25–30	5
elastic constants, 0°C, GPa[k]		
C_{11}	131.5	1
C_{12}	97.33	1
C_{44}	51.1	1
elastic properties[l]		
Young's modulus at 20°C, GPa[k]	91.3	1
modulus of rigidity at 20°C, GPa[k]	26.9–29.7	5

Table 1. (*Continued*)

Parameter	Value	Reference
Poisson's ratio, at 293 K		
annealed	0.364	1
hard drawn	0.39	
viscosity, molten, at 1043°C, mPa·s(=cP)	3.697	4
thermal neutron capture cross section, m^2	63×10^{-28}	8
Fermi energy, eV	5.52	1
Fermi surface, spherical	necks at $\langle 111 \rangle$	1
work function, electrons		
thermionic, eV	3.09–4.31	4
photoelectric, eV	3.67–4.81	4

[a] A partial pressure of 20 kPa (2.9 psi) O_2 results in a freezing point of ca 950°C.
[b] See also Ref. 2.
[c] Electron volts, 1 eV = 1.602×10^{-19} J.
[d] For the equation, $Ag^+ + e^- \rightarrow Ag$.
[e] Depends on amount of dissolved oxygen and trace elements in solid solution in silver, plus cold work.
[f] To convert MPa to psi, multiply by 145.
[g] Total electrical resistivity of 99.999% pure or purer bulk silver. Impurities increase resistivity.
[h] Liquid 99.95% pure or purer.
[i] To convert Pa to mm Hg, multiply by 0.0075.
[j] At 20°C, annealed at 600°C.
[k] To convert GPa to psi, multiply by 145,000.
[l] 1 Pascal = 10 dyne/cm^2 = 1.45×10^{-4} lbf/$in.^2$.

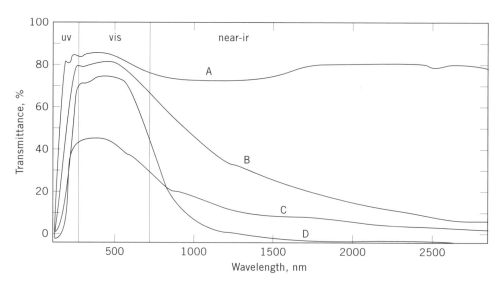

Fig. 1. Solar energy transmittance through glass and glass having sputtered silver coatings, where A corresponds to clear float glass; B, to LoE, one layer of sputtered silver in stack of oxides (prevents tarnishing and eliminates internal reflections) deposited on one interior surface of two layers of glass (an inert gas (argon) is placed between panes); C, to LoE$_1$, same as B except silver and oxide layer thicknesses selected to provide lower solar energy minimum transmission; and D, LoE$_2$, similar to B except that two separated layers of sputtered silver located in stack of oxides (9). Courtesy of Cardinal IG.

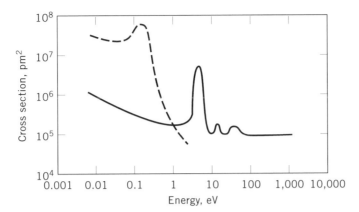

Fig. 2. Total neutron cross sections of silver (——) and cadmium (– – –) (10). To convert pm^2 to barns, multiply by 10^{-4}.

Table 2. Thermal Properties of Silver

Parameter	Value	Reference
thermal conductivity[a] at 0°C, W/(m·K)	418.7	3
linear coefficient of thermal expansion, m/(m·K)		
av 0–100°C	19.68×10^{-6}	
at 500°C	20.61×10^{-6}	3
specific heat, mean, J/(kg·K)[b]		
solid at 0°C	234	3
at 100°C	238	3
at 527°C	282	3
at 961°C	297	3
liquid at 961–2227°C	310	3
latent heat of fusion, J/g[b]	104.2	4
latent heat of vaporization, kJ/g[b]	2.636	
heat capacity, C_p, at 0–800°C, J/(g·°C)[b,c]	$C_p = 0.2320 + 0.06035 \times 10^{-4}\ t - 0.67896 \times 10^{-8}\ t^2$	4
thermal electromotive force vs NBS Pt 27 (NBS Standard), mV		
at 100°C	+0.76	5
300°C	+3.20	5
500°C	+6.50	5
700°C	+10.80	5

[a] 1 W/(m·K) = 2.39×10^{-3} cal/(s·cm·°C) = 0.578 (Btu·in.)/(h·ft^2·°F).
[b] To convert J to cal, divide by 4.184.
[c] Temperature, t, in °C.

and trivalent silver disinfectants have been shown to be from 50 to 200 times more effective as sanitizers than monovalent silver compounds.

Trivalent complexes of biguanide and periodate are also prepared commercially for water sanitation. The trivalent silver periodate, for example, is prepared by the action of potassium periodate and potassium hydroxide on Ag_4O_4:

$$Ag_4O_4 + 6\ KOH + 4\ KIO_4 \longrightarrow 2\ K_5H_2[Ag(IO_6)_2] + Ag_2O + H_2O$$

Oxygen Reactivity. Silver is second only to gold as the element having the weakest interaction with oxygen, providing silver with its superior sparking and combustion resistance. Silver must be oxidized chemically or electrolytically to form Ag_2O. Thermodynamically, silver exists only in the metallic state above 200°C.

Solid silver is more permeable by oxygen than any other metal. Oxygen moves freely within the metallic silver lattice, not leaving the surface until two oxygen atoms connect to form O_2. This occurs at ~300°C. Below this temperature silver is an efficient catalyst for gaseous oxidative chemical reactions. Silver is also an extremely efficient catalyst for aqueous oxidative sanitation.

Oxygen migrates through silver in the atomic state, its rate of diffusion increasing with rising temperature. An instrument using this effect provides a stream of pure atomic oxygen *in vacuo* through a ca 0.25-μm silver membrane, operating at ca 350–450°C with oxygen pressure behind the membrane. A flux of low energy electrons is aimed at the membrane to strike off the atoms of oxygen as they reach the outer surface of the silver.

Molten silver dissolves nearly 10 times its own volume of oxygen, ie, 0.32 wt % above its melting point, and ejects much but not all of the O_2 violently as it solidifies. There appears to be no lower temperature limit at which oxygen does not dissolve in silver.

Ordinary silver bullion processed in air may contain ca 200 ppm oxygen. Commercial silver bullion prepared under nitrogen may contain as little as 25 ppm oxygen. During casting some oxygen may be introduced to convert base metal impurities into oxides. Because these oxides do not enter into the solid solution, they have no effect on the annealing and recrystallization temperature of the silver critical to the silversmith.

Ignition Suppression. The formation of the weak silver oxide produces such little heat (14 J/g (3.3 cal/g)) that silver does not sustain combustion even when ignited. Thus silver finds important use as a spark suppressor in oxygen-rich systems such as in the space shuttle, where solid silver parts are used in the high pressure liquid oxygen pumps and wherever sparks might be generated by rubbing or friction.

Catalysis. The mechanism of hydrogen abstraction from alcohols to form aldehydes (qv) over silver has been elucidated (11). Silver is the principal catalyst for the production of formaldehyde (qv), the U.S. production of which was 4×10^6 metric tons in 1993. The catalytic oxidation of ethylene (qv) to ethylene oxide (qv) is unique to silver. The U.S. production of ethylene oxide was 2.9×10^6 metric tons in 1993. Silver is an effective oxidative catalyst because it breaks oxygen molecules into atoms, adsorbing those atoms lightly on the surface and providing sites for the oxidation of oncoming chemical compounds or bacteria. The silver itself remains chemically inactive.

Sanitation. The power of silver to inactivate bacteria was confirmed in 1893. The exact biological mechanism(s), however, remain to be clearly identified. Considerable research demonstrates the effective sanitation of swimming pools, hot tubs, and cooling towers by electrically generated silver and copper ions (12). When silver is plated on alumina particles in an aqueous medium, the very weak (ca 29–34 J/mol (7–8 kcal/mol)) bond silver forms with oxygen results in powerful oxidation power providing a 99% bacteria kill in a single pass and a 100% kill in ozone-charged water.

Hydrogen and Nitrogen.   Silver dissolves about 0.02 mL of hydrogen per mL of Ag^0 at 600°C, and 0.05 mL at 900°C. Nitrogen, insoluble in silver, is a suitable atmosphere for casting and melting silver to avoid oxygen absorption.

Dissolution of Silver.   Silver is dissolved by oxidizing acids and alkali metal cyanide solutions in the presence of oxygen. The latter method is the principal technique for dissolving silver from ore. Silver has extensive solubility in mercury (qv) and low melting metals such as sodium, potassium, and their mixtures. Cyanide solutions of silver are used for electroplating and electroforming. The silver is deposited at the cathode either as pure crystals or as layers on a mandrel.

A dissolution reaction with KCN is as follows:

$$4\,Ag + 8\,KCN + O_2 + 2\,H_2O \longrightarrow 4\,KAg(CN)_2 + 4\,KOH$$

Nitric acid dissolves silver at all concentrations. This is the principal chemical reaction for the dissolution of silver into the soluble nitrate, which is the chemical intermediate for the production of electroplated ware, catalysts, battery plates, pharmaceuticals, mirrors, and silver halides for photographic materials. Nitric acid removes silver from the residual pellet in the gold fire assay.

Tarnish.   No passivation treatment to prevent silver from tarnishing exists. Sulfides in the atmosphere react to form Ag_2S. The rate of tarnishing increases with the proportion of copper in the alloy, the latter forming a black Cu_2S deposit. Increasing humidity hastens tarnish formation. As the tarnish film thickens, the rate of tarnish formation is reduced. Tarnishing can be reduced in storage by the use of sulfur-free paper or paper impregnated with copper or cadmium acetate which preferentially absorbs H_2S.

A list of a wide variety of compounds and their reaction with silver is available (13).

Occurrence

The American cordillera extending from Alaska to Bolivia has been the most productive source of silver wherever it is associated with Tertiary age intrusive volcanic rocks, mostly concentrated by hydrothermal action. The largest producing mine in the cordillera is at Potosi, Bolivia, where the total silver output since the 1500s is estimated at over 31,000 metric tons.

Mexico, the world's leading producer of silver since the Spanish conquest, obtains virtually its entire silver production from lead–zinc mines in the central cordillera. Mexico retained its dominance in silver production until the discovery of the Comstock Lode in Nevada in 1859. Discoveries in Colorado, Arizona, and Montana placed the United States as the world's top silver producer from 1871 until 1900. As these mines played out, Mexico's vast resources returned it to its former position of dominance.

The most important body of primary silver ore in the United States in the 1990s is located in Silver Valley, the Coeur d'Alene Mining District of Shoshone

County, Idaho, which produces >200 t/yr of silver. The main ore mineral is tetrahedrite [*12054-35-2*], associated with sulfides of lead, copper, iron, and zinc.

In the Commonwealth of Independent States (CIS, formerly the USSR), nearly 50% of the CIS production comes from Kazakhstan. Silver is produced from the lead–zinc mines at Ostkamen, Shymkent, and Leninogorsk (ca 1000 t) and also in Russia's Far East, where it is a by-product of the tin deposits near Khabarovsk, and of the copper and gold deposits in the Ural Mountains.

Canada's most productive district is in the Canadian shield at Cobalt, Ontario. Nearly 97% of the silver values in this pre-Cambrian age ore were found as great slabs of nearly pure silver. One slab was 474 kg. The huge open-pit Kidd Creek Mine has had an annual output of over 80 t of silver.

Some 60 silver minerals are known. The most important economically are argentite [*1332-04-3*], Ag_2S; cerargyrite [*14358-96-4*], $AgCl$; polybasite [*53810-31-4*], $Ag_{16}Sb_2S_{11}$; proustite [*15152-58-4*], Ag_3AsS_3; pyrargyrite [*15123-77-0*], Ag_3SbS_3; stephanite [*1302-12-1*], Ag_5SbS_4; tetrahedrite, $Cu_3(AsSb)S_3$; and the tellurides. Silver is commonly associated with gold (see GOLD AND GOLD COMPOUNDS), copper (qv), lead (qv), and zinc (see ZINC AND ZINC ALLOYS) ores.

Resources.   World resources of silver are estimated to be about half a million tons. However, only about 250,000 metric tons are considered economically recoverable reserves. These are associated with ores of copper, gold, lead, and zinc, and extraction depends on the economic recovery of those metals. Canada and the CIS vie for the greatest reserves of silver in the ground.

Mining and Processing

Silver utensils dating to ca 3000 BC attest to the antiquity of silver production. By ca 1000 BC the argentiferous lead ores at Mount Laurium, Greece, were smelted, ie, melted in a furnace, and followed by cupellation, ie, blowing air across the pool of metal to oxidize the base metals into slag. An estimated 8000 t of silver produced from the mines at Laurium greatly contributed to the wealth of Athens. Studies of Greenland ice and lake sediments throughout northern Europe show lead levels hundreds of times higher than natural, beginning ca 500 BC. This serves as evidence of early large-scale silver smelting by the Greeks and Romans.

The discovery of aqua regia by the Arab alchemist Jabir Ibn Hayyan (720–813 AD) provided a new extraction technology. Amalgamation of silver in ores with mercury was extensively used during the late fifteenth century by the Spaniards in Mexico and Bolivia. In 1861 the complex ores of the Comstock Lode, Nevada, were ground together with mercury, salt, copper sulfate, and sulfuric acid, and then steam-heated to recover the silver.

The chlorination process, introduced in Europe in 1843, roasted ore with chlorides, followed by a hot brine leach and subsequent precipitation of the silver on copper. In 1887 it was discovered that gold and silver can be recovered by sodium cyanide, and this process displaced the dangerous chlorination process. By 1907 the cyanide process, where a cyanide solution is mixed with zinc dust to precipitate the silver, was universally in use.

In the 1980s, zinc precipitation was replaced by a method involving the passing of the solution over activated carbon to adsorb the precious metals, which are then stripped from the charcoal by a hot caustic solution. Electrowinning removes the precious metals from this solution, depositing them on the cathode. One benefit of this process is that the filtration or deareation steps of the zinc precipitation are not required. Thus the environmental hazard of the zinc salts is eliminated.

Heap leaching, ie, spraying of sodium cyanide solution over roughly crushed ores heaped on an impervious pad, has become the most economical way of recovering precious metal values from very low grade ores. Recoveries are lower, however, ca 70%, for heap leaching; ca 85% for milling and cyanidization.

Production and Consumption

Mine Production of Silver. World production of silver by region is given in Table 3. Some 900,000 metric tons are estimated to have been mined since early times. By the year 1500 world mine production was about 50 t/yr. In 1992 world production exceeded 14,900 metric tons. Following the breakup of the Soviet Union, previously undisclosed data showed that the USSR led world silver production during 1979–1980 at about 1550 metric tons. During the early 1990s the production in this region exceeded 2000 t/yr.

U.S. silver production from 1985 to 1994 averaged 1588 t/yr. Less than one quarter of this output comes from silver mine districts, however. About half is as by-product of gold mines; about one quarter comes from copper and lead–zinc mines. The silver production in Mexico from 1985 to 1994 averaged 2256 t/yr, and Peru, at the southern extremity of the cordillera, where silver is a by-product of copper and lead–zinc mines, averaged 1810 t/yr.

Consumption. World consumption of silver in 1994, including the use of scrap, was about 23,300 t. Industrial and decorative uses consumed 8,690 t; jewelry and silverware, 6,477 t; photography, 6,820 t; and official coins, 1,335 t.

The demand for silver is an indicator of technically advanced products production. In 1950, when the population in the free-market countries was 1.7 billion, consumption of silver was 2.9 $t/10^6$ people. By 1992, the population had increased to 4.2 billion and silver consumption reached 4.2 $t/10^6$ people. Fabrication demand by country is given in Table 4 and by category in Table 5.

Table 3. Silver Mine Production, $t^{a,b}$

Region	1985	1990	1994
Africa	496	570	564
Asia	678	524	597
Europe	1,762	1,728	1,594
North America	4,573	5,858	4,443
Central and South America	2,618	3,118	3,168
Oceania	1,088	1,182	1,077
CIS[c]	1,960	2,239	1,499
Total[d]	*13,312*	*15,992*	*13,817*

[a]Ref. 14.
[b]To convert metric tons to millions of troy ounces, divide by 31.1.
[c]Formerly USSR.
[d]Includes China and North Korea.

Table 4. Fabrication Demand by Country, t [a,b]

Country	1985	1990	1994
United States	3,704	3,982	4,190
Belgium	510	597	629
Canada	292	186	70
France	597	765	873
Germany	1,060	1,681	1,616
Hong Kong	100	100	90
India		1,315	2,834
Italy	759	1,426	1,496
Japan	2,258	3,601	3,353
Mexico	292	478	921
the Netherlands	31	86	96
Peru		46	28
South Korea		212	496
Thailand		750	905
Taiwan		140	160
United Kingdom and Ireland	607	801	1,043

[a] Ref. 14.
[b] Includes scrap.

Table 5. U.S. Fabrication Demand of Silver, t [a]

Product category	1985	1990	1994
photography	1801	2084	2109
electrical contacts and conductors	855	709	650
batteries	78	93	106
catalysts	75	93	124
sterlingware	109	109	200
jewelry	180	62	165
silverplate	115	87	93
mirrors	31	37	44
miscellaneous	448	425	444
coinage	12	283	255
Total	*3704*	*3982*	*4190*

[a] Refs. 14 and 15.

Secondary Silver Recovery.   The consumption of silver normally exceeds its mine production; therefore recovery from scrapped products, such as electrical gear, coins, and photographic film and solutions, is critical to its supply. In 1994, the silver gleaned from scrap throughout the world was estimated to exceed 4100 t. The overwhelming proportion of this was from photographic material (see RECYCLING, NONFERROUS METALS) (16).

Standards and Specifications

Commodity exchanges require good delivery silver bullion to be 999 parts per 1000 fine silver. Samples of specifications for silver are given in Table 6. Specifications for silver bullion, brazing alloys, electrical contact alloys, etc, are pub-

Table 6. Specifications for Silver and Silver Alloys, Wt %

Grade	Designation	Ag[a]	Ag + Cu[a]	Fe[b]	Pb[b]	Other[b]
ASTM B413[c]	Grade 99.90 Refined silver	99.90	99.95	0.002	0.025	Bi 0.001, Cu 0.08
MIL-S-13282b[d]	Grade A silver[e]	99.95	99.98	0.005	0.025	Bi 0.001, Cu 0.10, Se 0.001, Te 0.001
SAE/UNS P07931[f],	sterling silver, silversmith's grade	92.10–93.50	6.50–7.90[g]	0.05	0.03	Ca 0.05, Zn 0.06, all others 0.06

[a]Value is minimum. [b]Value is maximum. [c]Ref. 17. [d]Ref. 18. [e]For anodes. [f]Ref. 19. [g]Cu only.

lished by ASTM, the American Welding Society, Japanese Industrial Standards, SAE (Aerospace Materials Specifications), and the U.S. Department of Defense.

Assaying and Analysis

The fire assay, the antecedents of which date to ancient Egypt, remains the most reliable method for the accurate quantitative determination of precious metals in any mixture for concentrations from 5 ppm to 100%. A sample is folded into silver-free lead foil cones, which are placed in bone-ash cupels (cups) and heated to between 1000 and 1200°C to oxidize the nonnoble metals. The oxides are then absorbed into a bone-ash cupel (ca 99%) and a shiny, uniformly metallic-colored bead remains. The bead is brushed clean, rolled flat, and treated with CP grade nitric acid to dissolve the silver. The presence of trace metals in that solution is then determined by instrumental techniques and the purity of the silver determined by difference.

Chemical analysis methods may be used for assay of silver alloys containing no interfering base metals. Nitric acid dissolution of the silver and precipitation as AgCl, or the Gay-Lussac-Volhard titration methods are used interchangeably for the higher concentrations of silver. These procedures have been described (4).

Instrumental methods for quantitative determination of silver purity include (1) atomic absorption, which is the fractional decrease in intensity of radiation per number of atoms per unit area. It is fast, accurate, and economical. Silver detection is highly sensitive and this analysis is widely used for ores and concentrates down to ca 2–3% silver. (2) Emission spectroscopy and (3) mass spectrometry (qv) are used to quantitatively assay trace elements in high purity silver. The purity of the silver is found by subtracting the proportion of the trace elements present. The use of ms assay is a necessity for electronics applications, where 999.99 or five nines fine silver or better is required. (4) The inductively coupled plasma technique functions on the same principle as emission spectroscopy, but uses solutions put through an 8,000–10,000 K plasma of argon. The growing popularity of this method is based on the fast, simple analysis. The usual interpretation of photographic plates is avoided (see PLASMA TECHNOLOGY). Lastly (5), x-ray fluorescence is simple, rapid, and accurate. This technique leads to identification of elements by their characteristic x-ray emission. The composition of the sample is obtained by converting the x-ray intensities into metal concentrations. However, each type of sample must be individually calibrated.

Health and Safety Factors

Silver is one of the very few antimicrobial agents for which epidemiological data extends over millennia of use. Silver metal dissolves in water to the extent of 5 parts per billion (ppb), making the water sufficiently toxic to kill such organisms as *E. coli* and *B. typhosus.* Silver blocks growth of *E. coli* cells as a result of both surface binding and intracellular uptake. On the surface of a bacterium, and on fish gills, silver blocks transmission of oxygen, leading to expiration. Silver is not toxic to humans because it does not react with mammalian cells, and the blood brain barrier blocks the silver from nerve cells of the brain and spinal column, where it would exhibit toxicity.

The U.S. Environmental Protection Agency, under the Safe Drinking Water Act, set the secondary contaminant level for silver in drinking water at 0.1 mg/L (20). Secondary contaminants are not considered to be hazardous to health and thus the limits are not federally enforceable.

Silver Alloys

The atomic radius of silver (144 pm) is within about 15% of many elements, permitting solid solutions with Al, Au, Be, Bi, Cu, Cd, Ge, In, Mn, Pb, Pd, Pt, Sb, Sn, Th, and Zn. These metals form useful brazing, jewelry, and soldering alloys. Copper is the only metal with which silver forms a simple eutectic between two solid solutions (Fig. 3). Silver has extremely limited solubility in B, C, Co, Cr, Fe, Ge, Ir, Ni, Mg, Mo, Se, Si, Te, Ti, and W. Thus these metals may be brazed by silver alloys without serious erosion during welding (qv).

Silver's advantageous physical, chemical, electrical, and thermal conductive properties are used in a variety of alloys. For example, (*1*) a 3.5% Ag, 0.5% Zr copper alloy provides a high strength, thermally conductive, fatigue-resistant liner of the hottest portion of the space shuttle engine main combustion chamber (see ABLATIVE MATERIALS). (*2*) The oldest standard silver jewelry alloy is sterling, 92.5% Ag, 7.5% Cu, which, when heated and rapidly quenched, results in a soft, highly workable, ductile alloy. Annealing results in extended wear life for silver coins and tableware.

Brazes and Solders.   Silver imparts high tensile strength, ductility, thermal conductivity, bactericidal properties, and unusual wettability to most metals for soldering and brazing alloys. Some 1200 t of silver are used worldwide annually for this application. Alloys of Pb, Sb, and Sn provide low temperature solders (143–320°C); Au and Ge provide brazes melting at ca 525°C; Cd, Cu, Mn, Ni, P, Sn, and Zn provide high temperature brazes (660–970°C). A highly ductile 92% Ag, 4% Ti, 1% In alloy provides strong ceramic-to-ceramic joints at 1000°C. A 65% Sn, 25% Ag, and 10% Sn alloy provides unique wetting and resistance to thermal fatigue properties that are critical for the bonding of silicon chips to metallic substrates.

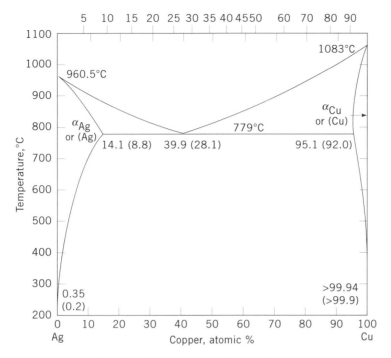

Fig. 3. Silver–copper phase diagram.

Uses

Coinage. As a store of value, silver metal has been traded since prehistoric times. About 700 BC, trade silver was issued in the form of coins. Because silver is soft, the coin silver alloy (90 Ag, 10 Cu) was adopted in the United States in 1837 and also by other national governments. Following World War II, burgeoning industrial demand consumed much of mine production, and with the inflation of national currencies, the decreasing monetary value of silver coins to below that of their bullion value brought an end to circulating silver coinage. Coinage consumption of silver as of the 1990s is split between commemorative coins, made from sterling and coin silver alloys, and bullion coins, made of pure silver, such as the Mexican Onza and the U.S. Eagle.

Photography. The introduction into photographic techniques of tabular silver halide crystals in 1983 markedly improved the efficiency of silver halides for capturing light, decreasing light scattering, and allowing greater adsorption of sensitizing dyes, extending the film's sensitivity to 1400 nm in the infrared. In the best fine-grain color negative film, a 35-mm frame contains some 25 million image elements. The best digital electronic system contains about two million image elements. The reaction of light with halides is one of reduction. The silver ion is reduced to silver, and the halogen atoms react with the emulsion, making the process irreversible (see PHOTOGRAPHY).

Photochromic glass used for sunglasses contains silver, copper, and halogen salts. Sunlight reduces the silver ion to a silver atom, darkening the glass. The released halogen atoms are trapped by the copper, and on removal of sunlight are recovered by the silver, restoring the glass to colorlessness, a readily reversible process. The silver crystallites block up to 97% of harmful uv rays.

Electrical Contacts.   Silver combines the highest electrical and thermal conductivity and freedom from corrosion with low price. Its high resistance to arcing and mechanical wear may be increased by combining it with other materials such as graphite and cadmium oxide, which have superior nonsticking and nonwelding properties, and tin and indium oxides. The low power membrane switch consists of pure silver surfaces on the inside surfaces of two opposing sealed Mylar sheets that make an electrical contact. This switch finds universal use under the keys of millions of computer keyboards and control panels. Above 800 A, contacts are made of silver infiltrated into a tungsten matrix to avoid arc welding. On annealing, sputtered silver contacts can provide a resistivity of 10^{-9} $\Omega \cdot cm^2$. These are useful for high temperature superconductor switching. Tarnish has little effect on silver contacts because the film, though dark, is soft and easily wiped off a contact when a toggle switch is activated.

Silver Thick Films.   About half of the silver consumed in the United States for its electrical properties is used by the electronics industry. Of this amount some 40% is used for the preparation of thick-film pastes in circuit paths and capacitors. These are silk-screened onto ceramic or plastic circuit boards for multilayer circuit sandwich components.

Electroplating.   The use of silver for plating tableware and decorative objects in the United States exceeds 100 t/yr. Silver is normally plated using potassium cyanide solutions, which exhibit the highest plating current densities. The advent of computer-monitored electroplating has allowed consistent production line plating of gold–silver alloys.

Electroless Plating.   The historic three-part mirror silvering system of $AgNO_3$ and ammonia (to retard precipitation of the silver), a solution of NaOH, and a reducing solution of invert sugar (later formaldehyde) (ca 1835), has been replaced by a two-solution silvering system of NaOH and ammoniacal $AgNO_3$, mixed with a glucose solution containing an amine group. This allows electroless plating of mirrors at lower temperatures with better adhesion. Both require glass pretreated with $SnCl_2$. The new system lends itself to automated production of video compact disks, ornaments, thermos bottles, bottle caps, etc. About 40 t/yr of silver are used in the United States for manufacturing mirrors.

Magnetron Sputtered Reflective Coatings.   Silver atoms sputtered *in vacuo* onto glass and polyester films entered the domestic window market in 1979 and are primary contenders for energy savings. The invisible silver coatings on ordinary double-pane windows where argon is used as the insulating gas can approach a 96% reflectivity in the long infrared (see Fig. 1). Multilayer polyester films have achieved an R-value of 10.

Sputtered silver mirrors are used for solar energy (qv) collectors and astronomical telescope mirrors. Approximately 3 t/yr of silver are used in the United States for low emissivity windows.

Dental Amalgam.   Silver–mercury dental amalgams have been universally used since the late 1830s. These have proven to be the most successful tooth filling materials in terms of both bactericidal and physical properties. This includes the ability to be pressed into tooth cavity irregularities and then to expand slightly on solidification to provide a firm fit. The silver alloys incorporated in the amalgam have varied widely in composition: Ag 25–70 wt %, Cu 0–30 wt %, Sn 0–30 wt %, and Zn 0–1 wt %. The mercury-to-alloy ratio has ranged from 43–54% (see DENTAL MATERIALS).

In 1993 the National Institute of Standards and Technology (NIST) announced a mercury-free, silver-coated tin filling material. The silver-coated tin particles are stripped of oxygen and stored in an oxygen-inhibiting solution until they are packed into the tooth cavity. When this solution is expelled during packing in the cavity by the dentist, the particles are freed to cold-weld, forming a true metallurgical bond.

Other Uses. *Bearings.* Silver is plated in all main-shaft bearings of jet engines because the silver provides a low coefficient of friction and superior fatigue and corrosion resistance, and has sufficient lubricity to serve as an emergency lubricant in case of oil failure.

Giant Magnetoresistance. Silver and cobalt, although mutually insoluble, may be ground together and fused to exhibit giant magnetoresistance when a magnetic field is applied. Resistance changes as great as 50% are observed. The Co-rich precipitates in an Ag-rich matrix find use in magnetic-tape read heads, sensing magnetism devices, etc (see MAGNETIC MATERIALS).

Economic Aspects

New York Commodity Market (COMEX) prices from 1982 to 1993 are given in Table 7. The economic aspects of silver have been reviewed (14).

Table 7. Spot Settlement Price for Silver[a], $/kg

Year	High	Low	Average
1982	363.30	154.64	254.95
1985	221.51	176.19	197.41
1990	172.01	126.67	154.97
1994	185.83	146.93	169.76

[a]Ref. 21.

BIBLIOGRAPHY

"Silver and Silver Alloys" in *ECT* 1st ed., Vol. 12, pp. 426–438, by E. H. Konrad and F. A. Meier (Analysis), The American Platinum Works; in *ECT* 2nd ed., Vol. 18, pp. 279–294, by C. D. Coxe, Handy & Harman; in *ECT* 3rd ed., Vol. 21, pp. 1–15, by G. H. Sistare, Consultant.

1. D. R. Smith and F. R. Fickett, *J. Res. Nat. Inst. Stds. Technol.*, (100), 119–171 (1995).
2. D. Strominger, J. M. Hollander, and G. T. Seaborg, *Rev. Mod. Phys.* **30**, 585 (1958).
3. B. A. Rogers and co-workers, *Silver: Its Properties and Industrial Uses*, NBS Circular C412, U.S. Government Printing Office, Washington, D.C., 1936.
4. A. Butts, *Silver Economics, Metallurgy, and Use*, Van Nostrand Reinhold Co., Inc., New York, 1967. A comprehensive study of silver.

5. Technical Bulletin, Vol. 10, No. 3, Engelhard Industries, Iselin, N.J., 1970, p. 80. Data on silver.
6. R. A. Matula, *J. Phys. Chem. Ref. Data* **8**, 4 (1979). American Institute of Physics resistivity tables.
7. N. Nesmayanov, *Vapor Pressure of the Elements*, Academic Press, Inc., New York, 1963.
8. D. J. Hughes and R. B. Schwartz, *Neutron Cross Section*, USAEC report BNL-325, U.S. Atomic Energy Commission, Washington, D.C., 1958.
9. *Glass Products for Windows and Doors*, Cardinal IG, Minnetonka, Minn., 1994, p. 21.
10. E. Bleuler and G. J. Goldsmith, *Experimental Nucleonics*, Holt, Reinhart and Winston, Inc., New York, 1952.
11. H. K. Plummer, Jr., W. L. H. Watkins, and H. S. Gandhi, *Appl. Catal.* **29**, 261–283 (1987).
12. C. P. Gerba and R. B. Thurman, *CRC Crit. Rev. Environ. Control* **18**(4), 295 (1989).
13. L. Addicks, *Silver in Industry*, Reinhold Publishing Co., New York, 1940, pp. 368–393.
14. *World Silver Survey, 1995*, The Silver Institute, Washington, D.C.
15. *World Silver Survey, 1994*, The Silver Institute, Washington, D.C.
16. P. Krause, *1992 World Silver Survey*, The Silver Institute, Washington, D.C., p. 73.
17. *Book of Standards*, Vol. 2.04, American Society for Testing and Materials, Philadelphia, Pa., 1995.
18. *Silver and Silver Alloys*, MIL-S-13282 B, Naval Publications Center, U.S. Department of Defense, Philadelphia, Pa., 1964.
19. *Unified Numbering System*, SAE/HS 1086a, SAE, Inc., Warrendale, Pa.
20. *Fed. Reg.* **56**, 1471 (Jan. 14, 1991).
21. *COMEX*, New York, 1995.

General References

World Silver Survey, The Silver Institute, Washington, D.C., 1992, 1995.
T. P. Mohide, *Silver*, Ontario Ministry of Mineral Resources, Toronto, Canada, 1985.
Mineral Facts and Problems, U.S. Bureau of Mines, Washington, D.C., 1965.
I. E. Wachs and R. J. Madix, *Surface Sci.* **76**, 531 (1978).
M. S. Antleman, in *IPMI Proceedings*, International Precious Metals Institute, Allentown, Pa., 1992, p. 141.
R. J. Araujo, *Contemp. Phys.* **21**(1), 77 (1980).
M. J. Carey and co-workers, *Appl. Phys. Lett.* **61**(24), 14 (Dec. 1992).
J. W. Ekin and co-workers, *Appl. Phys. Lett.* **52**(4), 331 (1988).
J. G. Hust and A. F. Clark, *Cryogenics*, **13**(6), 325 (June 1973).
J. A. Eisele, A. F. Colombo, and G. E. McClelland, *Separ. Sci. Technol.* **18**(12,13), 19081 (1983).
C. L. Fox and B. W. Rappole, *Surg. Gynecol. Obstet.* **128**, 1021 (May 1969).
The Fire Assay for Gold, The Gold Institute, Washington, D.C., 1985.
C. F. Heinig, Jr., *Ozone Sci. Eng.*, 533–546 (Dec. 1993).
F. Habashi, *CIM Bull.* **80**, 108 (Sept. 1987).
C. F. Key, J. G. Austin, and J. W. Bransford, *Flammability of Materials in Gaseous Oxygen Environments*, NASA TMX-64783, National Aeronautics and Space Administration, Washington, D.C., Sept. 1973.
Technology Assessment Conference Statement on Effects and Side Effects of Dental Restorative Materials, National Institute of Dental Research, NIH, Bethesda, Md., Aug. 26–29, 1991.
M. L. Malhotra and K. Asgar, *J. Am. Dental Assn.* (96), 444 (Mar. 1978).
R. A. Outlaw and M. R. Davidson, *Vacuum Sci. Tech. A* (12), 854 (May–June 1994).
K. W. Nageli, *Schweiz. Naturf. Ges.* **33**, 174 (1893).

Personal communication, D. H. Corrigan, VP, Handy & Harman, Oct. 1994.
Personal communication, P. Krause, President (ret.) Ilford Corp., June 1994.
R. M. Levine, *World Silver Survey*, The Silver Institute, Washington, D.C., 1994, p. 117.

SAMUEL F. ETRIS
The Silver Institute

SILVER COMPOUNDS

Silver, a white, lustrous metal, slightly less malleable and ductile than gold (see GOLD AND GOLD COMPOUNDS), has high thermal and electrical conductivity (see SILVER AND SILVER ALLOYS). Most silver compounds are made from silver nitrate [7761-88-8], $AgNO_3$, which is prepared from silver metal.

Some silver metal is found in nature, frequently alloyed with other metals such as copper, lead, or gold. Naturally occurring silver compounds, however, are the primary sources of silver. The most abundant naturally occurring silver compound is silver sulfide [21548-73-2] (argentite), Ag_2S, found alone and combined with iron, copper, and lead sulfides. Other naturally occurring silver compounds are silver sulfonantimonite [15983-65-0] (pyrargyrite), Ag_3SbS_3, silver arsenite [15122-57-3] (proustite), Ag_3AsS_3, silver selenide [1302-09-6], and silver telluride [12653-91-7]. Silver chloride (chlorargyrite) and silver iodide (iodargyrite) have also been found in substantial quantities in the western United States.

The average concentration of silver in the earth's crust has been estimated to be 0.1 mg/kg (1). An examination of 130 sources of natural waters in the United States in the 1960s detected silver in 6.6% of the samples in concentrations from 0.1–38 μg/L (2). The mean concentration was 2.6 μg/L. However, more recent data using ultraclean sampling (qv) and modern analytical techniques indicate that the higher concentrations reported in the earlier study were probably a result of contamination and that the true concentration of silver in natural waters is more on the order of 3–5 ng/L (3–5). Silver is rapidly adsorbed onto particles present in natural waters, and its concentration after a discharge is quickly reduced to background or upstream levels through dilution and sedimentation (qv) (3–5).

Silver belongs to Group II (IB) of the Periodic Table. The metal has a $4d^{10}5s^1$ outer electronic configuration. Silver has been shown to have three possible positive oxidation states, but only silver(I) is stable in aqueous solution. The silver(II) ion, a powerful oxidizing agent, is a transient species in solution. The two silver(II) compounds that have been studied are the oxide and the fluoride. Silver(III) exists only when stabilized through complex formation.

Only three simple silver salts, ie, the fluoride, nitrate, and perchlorate, are soluble to the extent of at least one mole per liter. Silver acetate, chlorate, nitrite, and sulfate are considered to be moderately soluble. All other silver salts are, at most, sparingly soluble; the sulfide is one of the most insoluble salts known. Silver(I) also forms stable complexes with excess ammonia, cyanide, thiosulfate, and the halides. Complex formation often results in the solubilization of otherwise insoluble salts. Silver bromide and iodide are colored, although the respective ions are colorless. This is considered to be evidence of the partially covalent nature of these salts.

Silver compounds, available from commercial suppliers, are expensive. Reagent grades of silver(I) carbonate, cyanide, diethyldithiocarbamate, iodate, nitrate, oxide, phosphate, and sulfate are available. Standardized solutions of silver nitrate are also available for analytical uses. Purified grades of silver(I) acetate, bromide, cyanide, and iodide can be purchased; silver nitrate is also made as a USP XX grade for medicinal uses (6).

Many silver compounds are unstable to light, and are thus shipped in brown glass or opaque plastic bottles. Silver compounds that are oxidants, eg, silver nitrate and iodate, must be so identified according to U.S. Department of Transportation (DOT) regulations. Compounds such as silver cyanide, which is toxic owing to its cyanide content, must carry a poison label. However, most silver compounds are essentially nontoxic.

Silver(I) Compounds

The solubility and solubility product constants, K_{sp}, for many silver compounds are given in Table 1 (7–9).

Silver Acetate. Silver acetate, $H_3CCOOAg$, is prepared from aqueous silver nitrate and acetate ion. Colorless silver acetate crystals and solutions made from this salt are unstable to light.

Silver Azide. Silver azide, AgN_3, is prepared by treating an aqueous solution of silver nitrate with hydrazine (qv) or hydrazoic acid. It is shock-sensitive and decomposes violently when heated.

Silver Acetylide. Silver acetylide [7659-31-6] (silver carbide), Ag_2C_2, is prepared by bubbling acetylene through an ammoniacal solution of silver nitrate. Silver acetylide is sensitive to the point of undergoing detonation on contact.

Silver Bromide. Silver bromide, AgBr, is formed by the addition of bromide ions to an aqueous solution of silver nitrate. The light yellow to green-yellow precipitate is less soluble in ammonia than silver chloride, but it easily dissolves in the presence of other complexing agents, such as thiosulfate ions.

Silver bromide crystals, formed from stoichiometric amounts of silver nitrate and potassium bromide, are characterized by a cubic structure having interionic distances of 0.29 nm. If, however, an excess of either ion is present, octahedral crystals tend to form. The yellow color of silver bromide has been attributed to ionic deformation, an indication of its partially covalent character. Silver bromide melts at 434°C and dissociates when heated above 500°C.

Silver bromide is significantly more photosensitive than silver chloride, resulting in the extensive use of silver bromide in photographic products. The crystal structure of photographic silver bromide is often octahedral.

Table 1. Solubility and Solubility Products of Silver(I) Compounds

Silver(I) compound	CAS Registry Number	Aqueous solubility[a], g/L H_2O	K_{sp}[a]
acetate	[563-63-3]	1.1×10	
azide	[13863-88-2]		2.9×10^{-9}[b]
bromate	[7783-89-3]	1.6[b]	
bromide	[7785-23-1]	1.3×10^{-4}	3.3×10^{-13}
carbonate	[534-16-7]	3.3×10^{-2}	8.2×10^{-12}
chlorate	[7783-92-8]	9.0×10^{0}[b]	
chloride	[7783-90-6]	1.9×10^{-3}	1.8×10^{-10}
chromate	[7784-01-2]	3.6×10^{-2}	1.9×10^{-12}[b]
cyanide	[506-64-9]	2.3×10^{-5}	1.6×10^{-14}
fluoride	[7775-41-9]	1.8×10^{3}	
iodate	[7783-97-3]	4.4×10^{-1}[b]	3.1×10^{-8}[b]
iodide	[7783-96-2]	2.6×10^{-6}	8.5×10^{-17}
nitrate	[7761-88-8]	2.16×10^{2}[c]	
nitrite	[7783-99-5]	4.2	
oxide	[20667-12-3]	2.2×10^{-2}	
selenate	[7784-07-8]	1.2[b]	4×10^{-9}[b]
sulfate	[10294-26-5]	8.3	1.2×10^{-5}
sulfide	[21548-73-2]	1.4×10^{-4}[b]	1.0×10^{-50}[b,d]
			3.8×10^{-52}[e]
			5.9×10^{-52}[f]
sulfite	[13465-98-0]		2×10^{-14}[b]
thiocyanate	[1701-93-5]	1.3×10^{-4}[b]	1.0×10^{-12}[b]

[a]At 25°C unless otherwise indicated.   [b]At 20°C.   [c]Value is per 100 g H_2O.   [d]Ref. 4.   [e]Ref. 5.   [f]Ref. 6.

Silver Carbonate. Silver carbonate, Ag_2CO_3, is produced by the addition of an alkaline carbonate solution to a concentrated solution of silver nitrate. The pH and temperature of the reaction must be carefully controlled to prevent the formation of silver oxide. A suspension of Ag_2CO_3 is slightly basic because of the extensive hydrolysis of the ions present. Heating solid Ag_2CO_3 to 218°C gives Ag_2O and CO_2.

Silver Chloride. Silver chloride, AgCl, is a white precipitate that forms when chloride ion is added to a silver nitrate solution. The order of solubility of the three silver halides is $Cl^- > Br^- > I^-$. Because of the formation of complexes, silver chloride is soluble in solutions containing excess chloride and in solutions of cyanide, thiosulfate, and ammonia. Silver chloride is insoluble in nitric and dilute sulfuric acid. Treatment with concentrated sulfuric acid gives silver sulfate.

Silver chloride crystals are face-centered cubic (fcc), having a distance of 0.28 nm between each ion in the lattice. Silver chloride, the most ionic of the halides, melts at 455°C and boils at 1550°C. Silver chloride is very ductile and can be rolled into large sheets. Individual crystals weighing up to 22 kg have been prepared (10).

The silver ion in silver chloride can be readily reduced by light, and is used to a great extent in photographic print papers. Sufficient light intensity and time leads to silver chloride decomposing completely into silver and chlorine.

Silver Chromate.   Silver chromate, Ag_2CrO_4, is prepared by treating silver nitrate with a solution of chromate salt or by heating a suspension of silver dichromate [7784-02-3].

Silver Cyanide.   Silver cyanide, AgCN, forms as a precipitate when stoichiometric quantities of silver nitrate and a soluble cyanide are mixed. Silver(I) ion readily forms soluble complexes, ie, $Ag(CN)_2^-$ or $Ag(CN)_3^{2-}$, in the presence of excess cyanide ion.

Silver Fluoride.   Silver fluoride, AgF, is prepared by treating a basic silver salt such as silver oxide or silver carbonate, with hydrogen fluoride. Silver fluoride can exist as the anhydrous salt, a dihydrate [72214-21-2] (<42°C), and a tetrahydrate [22424-42-6] (<18°C). The anhydrous salt is colorless, but the dihydrate and tetrahydrate are yellow. Ultraviolet light or electrolysis decomposes silver fluoride to silver subfluoride [1302-01-8], Ag_2F, and fluorine.

Other Silver Halogen-Containing Salts.   All silver halides are reduced to silver by treating an aqueous suspension with more active metals, such as magnesium, zinc, aluminum, copper, iron, or lead. Alternatively, the dry salts are reduced by heating with turnings or powders of these metals. Photolyzed silver halides are also reduced by organic reducing agents or developers, eg, hydroquinone, p-aminophenol, and p-phenylenediamine, during photographic processing (see PHOTOGRAPHY).

Halates.   Silver chlorate, $AgClO_3$, silver bromate, $AgBrO_3$, and silver iodate, $AgIO_3$, have been prepared. The halates may decompose explosively if heated.

Silver Iodide.   Silver iodide, AgI, precipitates as a yellow solid when iodide ion is added to a solution of silver nitrate. It dissolves in the presence of excess iodide ion, forming an AgI_2 complex; however, silver iodide is only slightly soluble in ammonia and dissolves slowly in thiosulfate and cyanide solutions.

Silver iodide exists in one of three crystal structures depending on the temperature, a phenomenon frequently referred to as trimorphism. Below 137°C, silver iodide is in the cold cubic, or γ-form; at 137–145.8°C, it exists in the green-yellow colored hexagonal, or β-form; above 145.8°C, the yellow cubic or α-form of silver iodide is the stable crystal structure. Silver iodide decomposes into its elements at 552°C.

Although silver iodide is the least photosensitive of the three halides, it has the broadest wavelength sensitivity in the visible spectrum. This feature makes silver iodide particularly useful in the photographic industry. It resists reduction by metals, but is reduced quantitatively by zinc and iron in the presence of sulfuric acid.

Silver Nitrate.   Silver nitrate, $AgNO_3$, is the most important commercial silver salt because it serves as the starting material for all other silver compounds. It is prepared by the oxidation of silver metal with hot nitric acid. The by-products are nitrogen oxides, NO and NO_2, which are vented to the atmosphere or scrubbed out of the fumes with an alkaline solution. Heavy metal impurities, such as copper, lead, and iron, are precipitated by increasing the pH of the solution to 5.5–6.5 with silver oxide and then boiling. The solution containing silver nitrate is made slightly acid, heated, evaporated, and then poured into pans to cool and crystallize. The crystals are washed, centrifuged, and dried. They can be further purified by recrystallization from hot water.

The Kestner-Johnson dissolver is widely used for the preparation of silver nitrate (11). In this process, silver bars are dissolved in 45% nitric acid in a pure oxygen atmosphere. Any nitric oxide, NO, produced is oxidized to nitrogen dioxide, NO_2, which in turn reacts with water to form more nitric acid and nitric oxide. The nitric acid is then passed over a bed of granulated silver in the presence of oxygen. Most of the acid reacts. The resulting solution contains silver at ca 840 g/L (12). This solution can be further purified using charcoal (13), alumina (14), and ultraviolet radiation (15).

The manufacture of silver nitrate for the preparation of photographic emulsions requires silver of very high purity. At the Eastman Kodak Company, the principal U.S. producer of silver nitrate, 99.95% pure silver bars are dissolved in 67% nitric acid in three tanks connected in parallel. Excess nitric acid is removed from the resulting solution, which contains 60–65% silver nitrate, and the solution is filtered. This solution is evaporated until its silver nitrate concentration is 84%. It is then cooled to prepare the first crop of crystals. The mother liquor is purified by the addition of silver oxide and returned to the initial stages of the process. The crude silver nitrate is centrifuged and recrystallized from hot, demineralized water. Equipment used in this process is made of ANSI 310 stainless steel (16).

Silver nitrate forms colorless, rhombic crystals. It is dimorphic and changes to the hexagonal rhombohedral form at 159.8°C. It melts at 212°C to a yellowish liquid which solidifies to a white, crystalline mass on cooling. An alchemical name, lunar caustic, is still applied to this fused salt. In the presence of a trace of nitric acid, silver nitrate is stable to 350°C. It decomposes at 440°C to metallic silver, nitrogen, and nitrogen oxides. Solutions of silver nitrate are usually acidic, having a pH of 3.6–4.6. Silver nitrate is soluble in ethanol and acetone.

In the absence of organic matter, silver nitrate is not photosensitive. It is easily reduced to silver metal by glucose, tartaric acid, formaldehyde, hydrazine, and sodium borohydride.

Silver Nitrite. Silver nitrite, $AgNO_2$, is prepared from silver nitrate and a soluble nitrite, or silver sulfate and barium nitrite.

Organic Acid Salts. Slightly soluble or insoluble silver salts are precipitated when mono- and dicarboxylic aliphatic acids or their anions are treated with silver nitrate solutions. Silver behenate [2489-05-6], $C_{22}H_{43}O_3Ag$; silver laurate [18268-45-6], $C_{12}H_{23}O_2Ag$; and silver stearate [3507-99-1], $C_{18}H_{35}O_2Ag$, are used in commercial applications. Silver oxalate [533-51-7], $C_2O_4Ag_2$, decomposes explosively when heated.

Silver Oxide. Silver oxide, Ag_2O, a dark brown-to-black material, is formed when an excess of hydroxide ion is added to a silver nitrate solution. Silver oxide can also be prepared by heating finely divided silver metal in the presence of oxygen. Anhydrous silver oxide is difficult to prepare. Reagent-grade material may contain up to 1 wt % water. Silver oxide is soluble in most silver-complexing agents. However, if a suspension of silver oxide in excess hydroxide is treated with ammonium hydroxide, highly explosive fulminating silver, or Bertholet's silver (17), may be produced.

When heated to 100°C, silver oxide decomposes into its elements, and is completely decomposed above 300°C. Silver oxide and sulfur form silver sulfide. Silver oxide absorbs carbon dioxide from the air, forming silver carbonate.

Silver Permanganate. Silver permanganate [7783-98-4], $AgMnO_4$, is a violet solid formed when a potassium permanganate solution is added to a silver nitrate solution. It decomposes upon heating, exposure to light, or by reaction with alcohol.

Perhalates. Whereas silver perchlorate [7783-93-9], $AgClO_4$, and silver periodate [15606-77-6], $AgIO_4$, are well known, silver perbromate [54494-97-2], $AgBrO_4$, has more recently been described (18). Silver perchlorate is prepared from silver oxide and perchloric acid, or by treating silver sulfate with barium perchlorate. Silver perchlorate is one of the few silver salts that is appreciably soluble in organic solvents such as glycerol, toluene, and chlorobenzene.

Silver Phosphates. Silver phosphate [7784-09-0], or silver orthophosphate, Ag_3PO_4, is a bright yellow material formed by treating silver nitrate with a soluble phosphate salt or phosphoric acid. Silver pyrophosphate [13465-97-9], $Ag_4P_2O_7$, is a white salt prepared by the addition of a soluble pyrophosphate to silver nitrate. Both the phosphate and the pyrophosphate are light-sensitive. Silver pyrophosphate turns red upon exposure to light.

Silver Selenate. Silver selenate, Ag_2SeO_4, is prepared from silver carbonate and sodium selenate (see SELENIUM AND SELENIUM COMPOUNDS).

Silver Sulfate. Silver sulfate, $Ag_8Se_2O_4$, is prepared by treating metallic silver with hot sulfuric acid. Alternatively, a solution of silver nitrate is acidified with sulfuric acid and the nitric acid is evaporated, leaving a solution of silver sulfate. Silver sulfate is more soluble in sulfuric acid than in water because of the formation of silver hydrogen sulfate [19287-89-9], $AgHSO_4$.

Silver sulfate decomposes above 1085°C into silver, sulfur dioxide, and oxygen. This property is utilized in the separation of silver from sulfide ores by direct oxidation. Silver sulfate is reduced to silver metal by hydrogen, carbon, carbon monoxide, zinc, and copper.

Silver Sulfide. Silver sulfide, Ag_2S, forms as a finely divided black precipitate when solutions or suspensions of most silver salts are treated with an alkaline sulfide solution or hydrogen sulfide. Silver sulfide has a dimorphic crystal structure. Transition from the rhombic (acanthite) to the cubic (argentite) form occurs at 175°C. Both crystal structures are found in nature.

Silver sulfide is one of the most insoluble salts known. It is not solubilized by nonoxidizing mineral acids, but it is soluble in concentrated nitric acid, concentrated sulfuric acid, and alkaline cyanide solutions.

Silver and sulfur combine even in the cold to form silver sulfide. The tendency of silver to tarnish is an example of the ease with which silver and sulfur compounds react. Polishes that contain silver complexing agents, such as chloride ion or thiourea, are used to remove silver tarnish.

Silver sulfide is exceptionally stable in air and sunlight, but decomposes when heated to 810°C. Moss silver (filiform silver), consisting of long hair-like growths of pure silver, is formed when silver sulfide is heated for a prolonged period at elevated temperatures below 810°C.

Silver Sulfite. Silver sulfite, Ag_2SO_3, is obtained as a white precipitate when sulfur dioxide is bubbled through a solution of silver nitrate. Silver sulfite is unstable to light and heat, and solutions decompose when boiled.

Silver Tetrafluoroborate. Silver tetrafluoroborate [14104-20-2], $AgBF_4$, is formed from silver borate and sodium borofluoride or bromine trifluoride. It is soluble in organic solvents.

Silver Thiocyanate. Silver thiocyanate, AgSCN, is formed by the reaction of stoichiometric amounts of silver ion and a soluble thiocyanate.

Silver Thiosulfate. Silver thiosulfate [23149-52-2], $Ag_2S_2O_3$, is an insoluble precipitate formed when a soluble thiosulfate reacts with an excess of silver nitrate. In order to minimize the formation of silver sulfide, the silver ion can be complexed by halides before the addition of the thiosulfate solution. In the presence of excess thiosulfate, the very soluble $Ag_2(S_2O_3)_3^{4-}$ and $Ag_2(S_2O_3)_5^{6-}$ complexes form. These soluble thiosulfate complexes, which are very stable, are the basis of photographic fixers. Silver thiosulfate complexes are oxidized to form silver sulfide, sulfate, and elemental sulfur (see THIOSULFATES).

Silver(I) Complexes

Silver ions form a number of complexes with both π-bonding and non-π-bonding ligands. Linear polynuclear complexes are known. The usual species are AgL and AgL_2, but silver complexes up to AgL_4 have been identified. Many of these complexes have commercial application.

Ammonia and Amine Complexes. In the presence of excess ammonia (qv), silver ion forms the complex ions $Ag(NH_3)_2^+$ and $Ag(NH_3)_3^+$. To minimize the formation of fulminating silver, these complexes should not be prepared from strongly basic suspensions of silver oxide. Highly explosive fulminating silver, believed to consist of either silver nitride or silver imide, may detonate spontaneously when silver oxide is heated with ammonia or when alkaline solutions of a silver–amine complex are stored. Addition of appropriate amounts of HCl to a solution of fulminating silver renders it harmless. Stable silver complexes are also formed from many aliphatic and aromatic amines, eg, ethylamine, aniline, and pyridine.

Cyanide Complexes. Insoluble silver cyanide, AgCN, is readily dissolved in an excess of alkali cyanide. The predominant silver species present in such solutions is $Ag(CN)_2^-$, with some $Ag(CN)_3^{2-}$ and $Ag(CN)_4^{3-}$. Virtually all silver salts, including the insoluble silver sulfide, dissolve in the presence of excess cyanide because the dissociation constant for the $Ag(CN)_2^-$ complex is only 4×10^{-19} (see CYANIDES).

Halide Complexes. Silver halides form soluble complex ions, AgX_2^- and AgX_3^{2-}, with excess chloride, bromide, and iodide. The relative stability of these complexes is $I^- > Br^- > Cl^-$. Complex formation affects solubility greatly. The solubility of silver chloride in 1 N HCl is 100 times greater than in pure water.

Olefin Complexes. Silver ion forms complexes with olefins and many aromatic compounds. As a general rule, the stability of olefin complexes decreases as alkyl groups are substituted for the hydrogen bonded to the ethylene carbon atoms (19).

Sulfur Complexes. Silver compounds other than sulfide dissolve in excess thiosulfate. Stable silver complexes are also formed with thiourea. Except for the cyanide complexes, these sulfur complexes of silver are the most stable. In photography, solutions of sodium or ammonium thiosulfate fixers are used to solubilize silver halides present in processed photographic emulsions. When insoluble silver thiosulfate is dissolved in excess thiosulfate, various silver complexes form. At low thiosulfate concentrations, the principal silver species is $Ag_2(S_2O_3)_2^{2-}$; at high thiosulfate concentrations, species such as $Ag_2(S_2O_3)_6^{10-}$

are present. Silver sulfide dissolves in alkaline sulfide solutions to form complex ions such as $Ag(S_4)_2^{3-}$ and $Ag(HS)_4^{2-}$. These ions are found in hot springs and waters highly charged with H_2S (20). Silver forms stable, slightly ionized salts with aliphatic and aromatic thiols.

Other Oxidation States

Silver(II) Compounds. Silver(II) is stabilized by coordination with nitrogen heterocyclic bases, such as pyridine and dipyridyl. These cationic complexes are prepared by the peroxysulfate oxidation of silver(I) solutions in the presence of an excess of the ligand. An extensive review of the higher oxidation states of silver has been published (21).

Silver Difluoride. Silver(II) fluoride [7783-95-1], AgF_2, is a brown-to-black hygroscopic material obtained by the treatment of silver chloride with fluorine gas at 200°C (22) or by the action of fluorine gas on silver metal. It is a strong oxidizing agent and a strong fluorinating agent. When heated above 450°C, it decomposes into AgF and fluorine. It also decomposes in aqueous solutions unless stabilized with concentrated HNO_3.

Silver(II) Oxide. Silver(II) oxide [1301-96-8, 35366-11-1], AgO, is prepared by persulfate oxidation of Ag_2O in basic medium at 90°C or by the anodic oxidation of solutions of silver(I) salts. This black oxide is stable to 100°C, dissolves in acids, and evolves oxygen and gives some Ag^{2+} in solution. It was previously believed to be a peroxide, but chemical evidence has ruled out this possibility. The most likely formula of this diamagnetic material is $Ag(I)Ag(III)O_2$ (19).

Silver(II) oxide is a strong oxidant. Reactions in alkaline medium have been studied extensively (19). It decomposes in aqueous solution unless stabilized with concentrated nitric acid.

Silver(III) Compounds. No simple silver(III) compounds exist. When mixtures of potassium or cesium halides are heated with silver halides in a stream of fluorine gas, yellow $KAgF_4$ [23739-18-6] or $CsAgF_4$ [53585-89-0], respectively, are obtained. These compounds are diamagnetic and extremely sensitive to moisture (21). When Ag_2SO_4 is treated with aqueous potassium persulfate in the presence of ethylenedibiguanidinium sulfate, the relatively stable Ag(III)-ethylenebiguanide complex is formed.

Economic Aspects

The cost of various silver compounds is a function of the silver market price. In 1980, the estimated usage of silver in the United States was 3730 metric tons (120 × 10^6 troy oz) (23). This silver is derived from silver mined within the United States; silver recycled or reclaimed from secondary sources, eg, coinage, flatware, jewelry, and photographic materials; and imported silver. In 1980, Canada, Mexico, and Peru, the principal exporters of silver to the United States, shipped 1670 tons (53.8 × 10^6 troy oz) as silver bullion and silver compounds. U.S. imported 2799 t and exported 964 t in 1988 (23).

New silver accounts for only a portion of the silver used in the United States. Recycled silver makes up the difference. Availability of recycled silver is dependent on market price. As the market price increases, so does the flow of recycled silver (see RECYCLING, NONFERROUS METALS). The New York price reached an all-time high of \$1543/kg (\$48.00/troy oz) on January 21, 1980, primarily as a result of speculation. The price fell to \$347.30/kg (\$10.80/troy oz) four months later as the pressure of speculative activity in the silver market lessened. Comprehensive reviews of the silver market are published yearly; the New York prices between 1985 and 1988 were as follows (23).

Year	High	Low	Average
1985	216	179	197
1986	199	157	176
1987	328	172	225
1988	257	193	210

Analytical Test Methods

Qualitative. The classic method for the qualitative determination of silver in solution is precipitation as silver chloride with dilute nitric acid and chloride ion. The silver chloride can be differentiated from lead or mercurous chlorides, which also may precipitate, by the fact that lead chloride is soluble in hot water but not in ammonium hydroxide, whereas mercurous chloride turns black in ammonium hydroxide. Silver chloride dissolves in ammonium hydroxide because of the formation of soluble silver–ammonia complexes. A number of selective spot tests (24) include reactions with p-dimethylamino-benzlidenerhodanine, ceric ammonium nitrate, or bromopyrogallol red [16574-43-9]. Silver is detected by x-ray fluorescence and arc-emission spectrometry. Two sensitive arc-emission lines for silver occur at 328.1 and 338.3 nm.

Quantitative. Classically, silver concentration in solution has been determined by titration with a standard solution of thiocyanate. Ferric ion is the indicator. The deep red ferric thiocyanate color appears only when the silver is completely titrated. Gravimetrically, silver is determined by precipitation with chloride, sulfide, or 1,2,3-benzotriazole. Silver can be precipitated as the metal by electrodeposition or chemical reducing agents. A colored silver diethyldithiocarbamate complex, extractable by organic solvents, is used for the spectrophotometric determination of silver complexes.

Highly sensitive instrumental techniques, such as x-ray fluorescence, atomic absorption spectrometry, and inductively coupled plasma optical emission spectrometry, have wide application for the analysis of silver in a multitude of materials. In order to minimize the effects of various matrices in which silver may exist, samples are treated with perchloric or nitric acid. Direct-aspiration atomic absorption (25) and inductively coupled plasma (26) have silver detection limits of 10 and 7 μg/L, respectively. The use of a graphic furnace in an atomic absorption spectrograph lowers the silver detection limit to 0.2 μg/L.

Instrumental methods are useful for the determination of the total silver in a sample, but such methods do not differentiate the various species of silver that may be present. A silver ion-selective electrode measures the activity of the silver ions present in a solution. These activity values can be related to the concentration of the free silver ion in the solution. Commercially available silver ion-selective electrodes measure Ag^+ down to 10 μg/L, and special silver ion electrodes can measure free silver ion at 1 ng/L (27) (see ELECTROANALYTICAL TECHNIQUES).

Health and Safety Factors

Silver compounds that generate significant quantities of free silver ion in solution, eg, silver nitrate, can be toxic to bacteria and freshwater aquatic organisms. At one time these silver compounds were also thought to cause adverse health effects. However, in 1989, the medical profession determined that the deposition of silver in internal organs and skin (argyria and argyrosis) did not impair the functions of the affected organs. Thus argyria and argyrosis are considered to be cosmetic effects, not adverse health effects. In response to this, the U.S. Environmental Protection Agency (EPA) removed silver from the list of maximum contaminant levels (MCL) regulated in drinking water in 1990 (28).

Silver compounds having anions that are inherently toxic, eg, silver arsenate and silver cyanide, can cause adverse health effects. The reported rat oral LD values for silver nitrate, silver arsenate [13510-44-6], and silver cyanide are 500–800 (29), 200–400 (29), and 123 mg/kg (30), respectively. Silver compounds or complexes in which the silver ion is not biologically available, eg, silver sulfide and silver thiosulfate complexes, are considered to be without adverse health effects and essentially nontoxic.

Effects from chronic exposure to soluble silver and silver compounds seem to be limited to deposition without evidence of health impairment (29,31,32). Argyria and argyrosis have resulted from therapeutic and occupational exposures to silver and its compounds. These disorders are characterized by either localized or general deposition of a silver–protein complex in parts of the body, and impart a blue-gray discoloration to the areas affected. In generalized argyria, characteristic discoloration may appear on the face, ears, forearms, and under the fingernails (33). Although the exact quantities of silver required for the development of argyria are not known, estimates based on therapeutic exposure suggest that the gradual accumulation of 1–5 g leads to generalized argyria (34). Localized argyria can occur as a result of the prolonged handling of metallic silver, which causes silver particles to be embedded in the skin and subcutaneous tissues via sweat-gland pores, or following the application of silver compounds to abraded skin areas (35). In argyrosis, silver is deposited primarily in the cornea and conjunctiva. This does not, however, appear to cause visual impairment (29,32).

Occupational argyria is uncommon in the 1990s because of effective industrial hygiene practices. In 1980, the American Conference of Governmental Industrial Hygienists (ACGIH) adopted a TLV for airborne silver metal particles of 100 μg/m^3, and proposed a TLV of 10 μg/m^3, as silver, for airborne soluble silver compounds (36). These values were selected to protect against argyria and argyrosis from industrial exposures to silver and silver compounds. Argyria caused by therapeutic use of silver compounds is extremely rare because of the availability of silver-free antiseptics and antibiotics (37).

In 1980, the EPA published ambient water quality criteria for silver. An upper limit of 50 μg/L in natural waters was set to provide adequate protection against adverse health effects (38). In 1992, EPA deleted the human health criteria for silver from the ambient water quality criteria to be consistent with the drinking water standards (39).

Environmental Impact

The impact that a silver compound has in water is a function of the free or weakly complexed silver ion concentration generated by that compound, not the total silver concentration (3–5,27,40–42). In a standardized, acute aquatic bioassay, fathead minnows were exposed to various concentrations of silver compounds for a 96-h period and the concentration of total silver lethal to half of the exposed population (96-h LC_{50}) determined. For silver nitrate, the value obtained was 16 μg/L. For silver sulfide and silver thiosulfate complexes, the values were >240 and >280 mg/L, respectively, the highest concentrations tested (27).

The chronic aquatic effects which relate silver speciation to adverse environmental effects were studied on rainbow trout eggs and fry. The maximum acceptable toxicant concentration (MATC) for silver nitrate, as total silver, was reported to be 90–170 ng/L (43). Using fathead minnow eggs and fry, the MATC, as total silver, for silver thiosulfate complexes was reported as 21–44 mg/L, and for silver sulfide as 11 mg/L, the maximum concentration tested (27).

Free ionic silver readily forms soluble complexes or insoluble materials with dissolved and suspended material present in natural waters, such as sediments and sulfide ions (44). The hardness of water is sometimes used as an indicator of its complex-forming capacity. Because of the direct relationship between the availability of free silver ions and adverse environmental effects, the 1980 ambient freshwater criterion for the protection of aquatic life is expressed as a function of the hardness of the water in question. The maximum recommended concentration of total recoverable silver, $C_{\text{Ag (aq)}}$ in fresh water is thus given by the following expression (45) in μg/L.

$$C_{\text{Ag (aq)}} = \exp^{(1.72\ (\ln(\text{hardness}))-6.52)}$$

For example, at hardnesses of 50, 100, and 200 mg/L, as $CaCO_3$, the concentration of total recoverable silver should not exceed 1.2, 4.1, and 13 μg/L, respectively. The total recoverable concentration of silver in salt water was recommended not to exceed 2.3 μg/L.

As of the 1990s, the EPA recommends that the dissolved form of silver be used as a better estimate of the bioavailable fraction and recommends using 85% of the total recoverable quantity. Thus, in fresh water at hardnesses of 50, 100, and 200 mg/L CaCO, the concentration of dissolved silver should not exceed 1.0, 3.5, and 11 μg/L, respectively. The concentration of dissolved silver in salt water should not exceed 1.9 μg/L (46).

In the manufacture of photographic materials, silver is originally present as a halide. When light-exposed photographic films and papers are processed, the silver halide that has not been affected by light is normally removed by solubilization as a thiosulfate complex using a thiosulfate-containing fixing bath. Before disposing of exhausted fixing baths, most of the silver is recovered, frequently by metallic exchange or by electrolytic reduction. The resulting concentrations of silver thiosulfate complex in the final effluents are 0.1–20 mg/L, as total silver (27).

In secondary wastewater treatment plants receiving silver thiosulfate complexes, microorganisms convert this complex predominately to silver sulfide and some metallic silver (see WASTES, INDUSTRIAL). These silver species are substantially removed from the treatment plant effluent at the settling step (47,48). Any silver entering municipal secondary treatment plants tends to bind quickly to sulfide ions present in the system and precipitate into the treatment plant sludge (49). Thus, silver discharged to secondary wastewater treatment plants or into natural waters is not present as the free silver ion but rather as a complexed or insoluble species.

Uses

Analysis. The ability of silver ion to form sparingly soluble precipitates with many anions has been applied to their quantitative determination. Bromide, chloride, iodide, thiocyanate, and borate are determined by the titration of solutions containing these anions using standardized silver nitrate solutions in the presence of a suitable indicator. These titrations use fluorescein, tartrazine, rhodamine 6-G, and phenosafranine as indicators (50).

Silver diethyldithiocarbamate [1470-61-7] is a reagent commonly used for the spectrophotometric measurement of arsenic in aqueous samples (51) and for the analysis of antimony (52). Silver iodate is used in the determination of chloride in biological samples such as blood (53).

Combination silver–silver salt electrodes have been used in electrochemistry. The potential of the common Ag/AgCl (saturated)–KCl (saturated) reference electrode is +0.199 V. Silver phosphate is suitable for the preparation of a reference electrode for the measurement of aqueous phosphate solutions (54). The silver–silver sulfate–sodium sulfate reference electrode has also been described (55).

Batteries. Primary, ie, nonrechargeable, batteries containing silver compounds have gained in popularity through use in miniaturized electronic devices. The silver oxide–zinc cell has a cathode of Ag_2O or AgO. These cells are characterized by a high energy output per unit weight and a fairly constant voltage, ca 1.5 V, during discharge. Originally used almost exclusively for military applications, satellites, and space probes, silver oxide–zinc batteries are used as of this writing (ca 1996) as power sources for wrist watches, pocket calculators, and hearing aids. Silver batteries have excellent shelf stability. Ninety percent of the original capacity is retained after one year of storage at 21°C (see BATTERIES, PRIMARY CELLS). Silver chromate is one of several oxidizing agents that can be used in lithium primary batteries (56).

Silver sulfide, when pure, conducts electricity like a metal of high specific resistance, yet it has a zero temperature coefficient. This metallic conduction is believed to result from a few silver ions existing in the divalent state, and thus providing free electrons to transport current. The use of silver sulfide as a solid electrolyte in batteries has been described (57).

Catalysts. Silver and silver compounds are widely used in research and industry as catalysts for oxidation, reduction, and polymerization reactions. Silver nitrate has been reported as a catalyst for the preparation of propylene oxide (qv) from propylene (qv) (58), and silver acetate has been reported as being a suitable catalyst for the production of ethylene oxide (qv) from ethylene (qv) (59). The solubility of silver perchlorate in organic solvents makes it a possible catalyst for polymerization reactions, such as the production of butyl acrylate polymers in dimethylformamide (60) or the polymerization of methacrylamide (61). Similarly, the solubility of silver tetrafluoroborate in organic solvents has enhanced its use in the synthesis of 3-pyrrolines by the cyclization of allenic amines (62).

Silver carbonate, alone or on Celite, has been used as a catalyst for the oxidation of methyl esters of D-fructose (63), ethylene (64), propylene (65), trioses (66), and α-diols (67). The mechanism of the catalysis of alcohol oxidation by silver carbonate on Celite has been studied (68).

Silver sulfate has been described as a catalyst for the reduction of aromatic hydrocarbons to cyclohexane derivatives (69). It is also a catalyst for oxidation reactions, and as such has long been recommended for the oxidation of organic materials during the determination of the COD of wastewater samples (70,71) (see WASTES, INDUSTRIAL; WATER, INDUSTRIAL WATER TREATMENT).

Cloud Seeding. In 1947, it was demonstrated that silver iodide could initiate ice crystal formation because, in the β-crystalline form, it is isomorphic with ice crystals. As a result, cloud seeding with silver iodide has been used in weather modifications attempts such as increases and decreases in precipitation (rain or snow) and the dissipation of fog. Optimum conditions for cloud seeding are present when precipitation is possible but the nuclei for the crystallization of water are lacking.

Silver iodide crystals, or smoke, for cloud seeding are produced predominantly by ground-based steady-state generators. Short-term (5–10-min) flares are also used. In one study, ground-based generators produced an average of 255 grams silver iodide crystals per hour. Each generator is designed to cover a 259-km (100-mi) target area. Cloud seeding is reviewed in Reference 72. Studies of high alpine mountain lakes determined that there were no adverse environmental effects from cloud seeding operations (73).

Electroplating. Most silver-plating baths employ alkaline solutions of silver cyanide. The silver cyanide complexes that are obtained in a very low concentration of free silver ion in solution produce a much firmer deposit of silver during electroplating than solutions that contain higher concentrations. An excess of cyanide beyond that needed to form the $Ag(CN)_2^-$ complex is employed to control the Ag^+ concentration. The silver is added to the solution either directly as silver cyanide or by oxidation of a silver-rod electrode. Plating baths frequently contain 40–140 g/L of silver cyanide (74) (see ELECTROPLATING).

Medicinal Preparations. Silver nitrate is used in medicine in the form of a stick, usually containing 1–3% silver chloride, or in solutions of varying concentrations. Uses of silver in medicine as of the 1990s are much reduced from earlier in the twentieth century because of the availability of a broad spectrum of other remedies. However, silver preparations are not likely to cause sensitization. An example of a procedure still in use is the drop of 1% silver nitrate required in many states to be placed in the eyes of newborn infants as a prophylactic against ophthalmia neonatorum. When applied topically, silver sulfadiazine [22199-08-2] has proven effective in preventing infections in burn victims. Aqueous solutions containing 10–20% silver nitrate, or a solid stick, are highly corrosive and can be applied locally to remove warts or cauterize wounds (see DISINFECTANTS AND ANTISEPTICS).

Mirrors.   The use of silver for the production of mirrors results in a highly reflective coating. The mirror is produced by the reaction of two separate solutions on the glass surface to be coated. One contains silver ammonia complex, the other an organic reducing agent, eg, formaldehyde, sodium potassium tartrate, sugar, or hydrazine. After rinsing, the silver coating can be protected by copper plating or silicone coating (75).

Photography.   The largest single use of silver and its compounds is in the photographic industry (76,77). Silver nitrate and a halide salt of an alkali metal or an ammonium halide give a light-sensitive silver halide. The silver halide can account for up to 30–40% of the total emulsion weight. Gelatin is the other primary constituent (see EMULSIONS; PHOTOGRAPHY). Many different silver halide emulsions are manufactured; the ratios of the halides and preparation details are adjusted according to the specific properties and applications desired. For photographic papers and the emulsions of low sensitivity, silver chloride, chlorobromide, or bromide is employed. Emulsions of high sensitivity are primarily bromide plus up to 10 mol % of silver iodide. Pure silver iodide emulsions are not commercially important and silver fluoride has no comparable photographic use.

Heat-processed photographic systems have been described that utilize silver behenate, silver laurate, and silver stearate. These silver salts are coated on paper in the presence of organic reducing agents (78,79).

Newer methods of image recording seek to avoid the high cost of silver. However, continued research has not led to systems that are able to offer the same combination of high sensitivity, high image density, exceptional resolution, permanence, and tricolor recording. In color photography (qv), dyes comprise the finished image, and the emulsion silver is removed during processing. Silver ions present in the photographic fixing solution as a silver thiosulfate complex can be recovered by metallic replacement, electroplating, or chemical precipitation.

Other Uses.   Photochromic glass contains silver chloride (80) and silver molybdate [13765-74-7] (81) (see CHROMOGENIC MATERIALS). An apparatus coated with silver nitrate has been described for the detection of rain or snow (82). Treatment with silver-thiosulfate complex has been reported as dramatically increasing the post-harvest life of cut carnations (83). Silver sulfate has been used in the electrolytic coloring of aluminum (84). Silver sulfate also imparts a yellowish red color to glass bulbs (85).

BIBLIOGRAPHY

"Silver Compounds" in *ECT* 1st ed., Vol. 12, pp. 438–443, by F. A. Meier, The American Platinum Works; in *ECT* 2nd ed., Vol. 18, pp. 295–309, by T. N. Tischer, Eastman Kodak Co.; in *ECT* 3rd ed., Vol. 21, pp. 16–32, by H. B. Lockhart, Jr., Eastman Kodak Co.

1. I. C. Smith and B. L. Carson, *Trace Metals in the Environment*, Vol. 2, *Silver*, Ann Arbor Science, Ann Arbor, Mich., 1977, p. 12.
2. J. F. Kopp and R. C. Kroner, *Trace Metals in Waters of the United States*, U.S. Dept. of the Interior, FWPCA, Cincinnati, Ohio, 1970, p. 8.
3. A. Andren, ed., *Proceedings from the 1st International Conference on Silver: Fate, Transport & Toxicity in the Environment*, University of Wisconsin, Madison, 1994.

4. A. Andren, ed., *Proceedings from the 2nd International Conference on Silver: Fate, Transport & Toxicity in the Environment*, University of Wisconsin, Madison, 1995.

5. A. Andren, ed., *Proceedings from the 3rd International Conference on Silver: Fate, Transport & Toxicity in the Environment*, University of Wisconsin, Madison, 1996.

6. *The United States Pharmacopeia XX (USP XX–NFXV)*, The United States Pharmacopeial Convention, Inc., Rockville, Md., 1980, p. 723.

7. T. R. Hogness and W. C. Johnson, *Qualitative Analysis and Chemical Equilibrium*, 4th ed., Holt, Rinehart & Winston, Inc., New York, 1954, p. 565.

8. S. F. Ravitz, *J. Phys. Chem.* **40**, 61 (1936).

9. A. F. Kapistinski and I. A. Korschunov, *J. Phys. Chem. (USSR)* **14**, 134 (1940).

10. N. R. Nail, F. Moser, P. E. Goddard, and F. Urback, *Rev. Sci. Instrum.* **28**, 275 (1957).

11. U.S. Pats. 2,581,518 and 2,581,519 (Feb. 17, 1948), T. Critchley (to Johnson and Sons' Smelting Works).

12. Ref. 1, p. 203.

13. U.S. Pat. 2,543,792 (Nov. 2, 1949), M. Marasco and J. A. Moede (to E. I. du Pont de Nemours & Co., Inc.).

14. U.S. Pat. 2,614,029 (Feb. 21, 1951), J. A. Moede (to E. I. du Pont de Nemours & Co., Inc.).

15. U.S. Pat. 2,940,828 (Oct. 29, 1957), J. A. Moede (to E. I. du Pont de Nemours & Co., Inc.).

16. *Chem. Eng.* **59**, 217 (Oct. 1952).

17. F. Raschig, *Ann. Chem.* **233**, 93 (1886).

18. U.S. Pat. 4,022,811 (May 5, 1977), K. Baum, C. D. Beard, and V. Grakaukas (to U.S. Dept. of the Navy).

19. F. A. Cotton and G. Wilkinson, *Advanced Inorganic Chemistry*, Wiley-Interscience, New York, 1962, p. 642.

20. Ref. 1, p. 448.

21. J. McMillan, *Chem. Rev.* **62**, 65 (1962).

22. H. F. Priest, *Inorg. Synth.* **3**, 176 (1950).

23. *The Silver Market—1988*, Handy and Harman, New York, 1989.

24. F. Feigl, *Spot Tests in Inorganic Analysis*, 5th ed., Elsevier Publishing Co., New York, 1958, pp. 58–64.

25. *Methods for Chemical Analysis of Water and Wastes*, EPA No. 600/4-79-020, U.S. Environmental Protection Agency, Cincinnati, Ohio, 1979, pp. 272.1-1 and 272.2-1.

26. *Fed. Reg.* **44**(233), 69559 (1979).

27. H. B. Lockhart, Jr., *The Environmental Fate of Silver Discharged to the Environment by the Photographic Industry*, Eastman Kodak Co., Rochester, N.Y., 1980.

28. *Fed. Reg.* **56**(20), 3573–3575 (1991).

29. Technical data, Health, Safety and Human Factors Laboratory, Eastman Kodak Co., Rochester, N.Y., 1980.

30. *Registry of Toxic Effects of Chemical Substances*, National Institute of Occupational Safety and Health (NIOSH), Cincinnati, Ohio, 1978.

31. K. D. Rosenman, A. Noss, and S. Kon, *J. Occup. Med.* **21**, 430 (1979).

32. A. P. Moss, A. Sugar, and N. A. Hargett, *Arch. Ophthalmol.* **97**, 905 (1979).

33. E. Browning, *Toxicity of Industrial Metals*, Butterworths, London, 1961, pp. 262–267.

34. W. R. Hill and D. M. Pillsbury, *Argyria in the Pharmacology of Silver*, Williams & Wilkins Co., Baltimore, Md., 1939.

35. W. R. Buckley, *Arch. Dermatol.* **88**, 99 (1963).

36. *Threshold Limit Values for Chemical Substances and Physical Agents in the Workroom Environment with Intended Changes for 1980*, American Conference of Governmental Industrial Hygienists, Cincinnati, Ohio, 1980, p. 28.

37. Ref. 1, p. 264.

38. U.S. Environmental Protection Agency, *Ambient Water Quality Criteria for Silver*, PB-81-117822, Washington, D.C., 1980, pp. 17-1–17-11.

39. *Fed. Reg.* **57**(246), 60910 (Dec. 22, 1992).

40. C. J. Tehaar, W. S. Ewell, S. P. Dziuba, and D. W. Fassett, *Photogr. Sci. Eng.* **16**, 370 (1972).

41. C. J. Terhaar, W. S. Ewell, S. P. Dziuba, W. W. White, and P. J. Murphy, *Water Res.* **11**, 101 (1977).

42. C. F. Cooper and W. C. Jolly, *Water Resour. Res.* **6**, 88 (1970).

43. P. H. Davis, J. P. Goettl, Jr., and J. R. Sinley, *Water Res.* **12**, 113 (1978).

44. M. A. Callahan and co-workers, *Water-Related Environmental Fate of 129 Priority Pollutants*, Vol. 1, EPA-440/4-79/029a, U.S. Environmental Protection Agency, Washington, D.C., 1980, pp. 17-1–17-11.

45. Ref. 38, p. B-13.

46. *Fed. Reg.* **60**(86), 22228 (May 4, 1995).

47. C. C. Bard, J. J. Murphy, D. L. Stone, and C. J. Terhaar, *J. Water Poll. Contr. Fed.* **48**, 389 (1976).

48. J. B. F. Scientific Corp., *Pathways of Photoprocessing Chemicals in Publicly Owned Treatment Works*, National Association of Photographic Manufacturers, Inc., Harrison, N.Y., 1977.

49. H. D. Feiler, P. J. Storch, and A. Shattuck, *Treatment and Removal of Priority Industrial Pollutants at Publicly Owned Treatment Works*, EPA-400/1-79-300, U.S. Environmental Protection Agency, Washington, D.C., 1979.

50. J. Gassett, R. C. Denney, G. H. Jeffery, and J. Mendham, eds., *Vogel's Textbook of Quantitative Inorganic Analysis*, 4th ed., Longman, Inc., New York, 1978, pp. 279–288.

51. E. P. Welsh, *Geol. Surv. Open File Rep. (U.S.)*, 79 (1979).

52. Y. Yamamoto, M. Kanke, and Y. Mizukami, *Chem. Lett.* **7**, 535 (1972).

53. W. Sendroy, *J. Biol. Chem.* **120**, 335 (1937).

54. H. Tischner, E. Wendler-Kalsch, and H. Kaesche, *Corrosion (Houston)* **39**, 510 (1980).

55. D. A. Shores and R. C. John, *J. Appl. Electrochem.* **10**, 275 (1980).

56. P. Cignini, M. Scovi, S. Panero, and G. Pistoia, *J. Power Sources* **3**, 347 (1978).

57. Jpn. Pat. 80 119,366 (Mar. 7, 1979) (to Citizen Watch Co., Ltd.).

58. Jpn. Pat. 80 85,574 (June 27, 1980) (to Showa Denko KK).

59. Fr. Pat. 5388 (Apr. 28, 1978), J. M. Cognion and J. Kervennal (to Produits Chimiques Ugine Kuhlman).

60. M. Sahan and C. Senvar, *Chim. Acta Turc.* **8**, 55 (1980).

61. S. P. Manikam, N. R. Subbaratnam, and K. Venkatardo, *J. Polym. Sci. Polym. Chem. Ed.* **18**, 1679 (1980).

62. A. Claesson, C. Sahlberg, and K. Luthman, *Acta Chem. Scand. Ser. B* **B33**, 309 (1979).

63. H. Hammer and S. Morgenlie, *Acta Chem. Scand. Ser. B* **B32**, 343 (1978).

64. U.S. Pat. 4,102,820 (July 25, 1978), S. B. Cavitt (to Texaco Development Corp.).

65. Ger. Pat. 2,312,429 (Sept. 27, 1973), C. Piccinini, M. Morello, and P. Rebora (Snam Progetti SpA).

66. S. Morgenlie, *Acta Chem. Scand.* **27**, 3009 (1973).

67. J. Bastard, M. Fetizon, and J. C. Gromain, *Tetrahedron* **29**, 2867 (1973).

68. F. J. Kakis, M. Fetizon, N. Douchkine, M. Golfier, P. Morgues, and T. Prange, *J. Org. Chem.* **29**, 523 (1974).

69. U.S. Pat. 4,067,915 (July 17, 1980), S. Matsuhira, M. Nishino, and Y. Yasuhara (to Toray Industries, Inc.).

70. M. C. Rand, A. E. Greenberg, and M. J. Taras, eds., *Standard Methods for the Examination of Water and Wastewater*, 14th ed., American Public Health Association, Washington, D.C., 1976, pp. 550–554.

71. R. Wilson, *A Study of New Catalytic Agents to Determine Chemical Oxygen Demand*, PB-270965, National Technical Information Service (NTIS), Washington, D.C., 1977.

72. Ref. 1, pp. 220–225.

73. P. H. Davies and J. P. Goettl, Jr., in D. A. Klein, ed., *Environmental Impacts of Artificial Ice Nucleating Agents*, Dowden, Hutchinson and Ross, Inc., Stroudsburg, Pa., 1978, pp. 149–161.

74. M. A. Orr in A. Butts and C. D. Coxe, eds., *Silver—Economics, Metallurgy and Use*, D. Van Nostrand Co., Inc., Princeton, N.J., 1967, p. 185.

75. R. D. Pohl, in Ref. 74, p. 193.

76. B. H. Carroll, G. C. Higgins, and T. H. James, *Introduction to Photographic Theory, The Silver Halide Process*, John Wiley & Sons, Inc., New York, 1980.

77. T. H. James, ed., *Theory of the Photographic Process*, Macmillan, Inc., New York, 1977.

78. Ger. Pat. 2,855,932 (July 5, 1979), K. Akashi, M. Akiyama, T. Shiga, T. Matsui, Y. Hayashi, and T. Kimura (to Asahi Chemical Industry Co., Ltd.).

79. Jpn. Pat. 79 76926 (May 25, 1979) (to Fuji Photo Film Co., Ltd.).

80. U.S. Pat. 3,252,374 (Feb. 15, 1962), S. D. Stookey (to Corning Glass Works).

81. Belg. Pat. 644,989 (Sept. 10, 1964), L. C. Sawchuk and S. D. Stookey (to Corning Glass Works).

82. Jpn. Pat. 80 19,375 (May 26, 1980) (to Mitsui Toastsu Chemicals, Inc.).

83. G. G. Dimallo and J. Van Staden, *Z. Pflanzenphysiol.* **99**, 9 (1980).

84. Jpn. Pat. 78 04,504 (Feb. 17, 1978), T. Abe, Y. Uchiyama, and T. Ohtsuka (to Showa Aluminum KK).

85. Jpn. Pat. 78 03,412 (Jan. 13, 1978), M. Sangen (to Matsushita Electronics Corp.).

C. ROBERT CAPPEL
Eastman Kodak Company

SIMULTANEOUS HEAT AND MASS TRANSFER

Heat transfer and mass transfer occur simultaneously whenever a transfer operation involves a change in phase or a chemical reaction. Of these two situations, only the first is considered herein because in reacting systems the complications of chemical reaction mechanisms and pathways are usually primary (see HEAT-EXCHANGE TECHNOLOGY). Even in processes involving phase changes, design is frequently based on the heat-transfer process alone; mass transfer is presumed to add no complications. But in fact mass transfer (qv) effects do influence and can even limit the process rate.

In processes where a condensing vapor or vapor from a liquid phase moves through an inert gas, eg, condensation in the presence of air, drying, humidification, crystallization (qv), and boiling of a multicomponent liquid, mass-transfer as well as heat-transfer effects are important (see AIR CONDITIONING; DISTILLATION; EVAPORATION). Such processes are discussed elsewhere in the *Encyclopedia*, but the primary emphasis is on either the heat transfer or the mass transfer taking place. Herein the interactions between heat and mass transfer in such processes are discussed, and applications to humidification, dehumidification, and water cooling are developed. These same principles are applicable to other operations.

Condensation and Vaporization as Effected by Simultaneous Heat and Mass Transfer

Consider the interphase transfer that occurs when one or more components change phase in the presence of inert or less active components. The transferring component must be transported through its original phase to the boundary, and must then escape into the second phase. The phase change involves a heat effect. Energy is transported to or from the boundary to balance the phase-change heat effect. The boundary temperature is influenced by the rate of heat transfer, and this determines the fugacity of the diffusing component at the boundary. Thus, to describe the process, rate equations for heat transfer and mass transfer must be written along with material balances for the components present and an energy balance. Appropriate boundary conditions must be applied, and the resulting set of differential and algebraic equations solved. The rate equations for heat and mass transfer express the rate of transport in terms of the driving force divided by the resistance across the transfer path. For heat transfer the driving force is expressed as a temperature difference, whereas the resistance is the reciprocal of the transport area times a coefficient. Here the concentration driving force should be a fugacity or activity difference with the coefficient in consistent units. However, these properties are not directly measurable so the driving force is expressed in terms of mole fractions, partial pressures, or mole ratios. The use of these terms requires the use of consistent coefficient values and limits the usefulness of the equations to systems that obey Raoult's law, or requires the use of empirical nonideality coefficients. Herein it is assumed that Raoult's law holds.

This process has been used for various situations (1–14). For the condensation of a single component from a binary gas mixture, the gas-stream sensible heat and mass-transfer equations for a differential condenser section take the following forms:

$$G \cdot C_p \frac{dT_G}{dA} = -h_g \cdot (T_g - T_s) \frac{\epsilon}{e^\epsilon - 1} \tag{1}$$

$$\frac{dV}{dA} = -k_g \cdot (P_g - P_s) \tag{2}$$

No condensation is taking place here in the bulk gas phase. If condensation does take place so that fogging occurs, these equations become

$$G \cdot C_p \frac{dT_G}{dA} = -h_g \cdot (T_g - T_s) \frac{\epsilon}{e^\epsilon - 1} + \lambda \frac{dF}{dA} \tag{3}$$

$$\frac{dV}{dA} = -k_g \cdot (P_g - P_s) - \frac{dF}{dA} \tag{4}$$

The term $\epsilon/(e^\epsilon - 1)$, which appears in equations 1 and 3, was first developed to account for the sensible heat transferred by the diffusing vapor (1). The quantity ϵ represents the group $M_i \cdot C_{pi}/h_g$, the ratio of total transported energy to convective heat transfer. Thus it may be thought of as the fractional influence of mass transfer on the heat-transfer process. The last term of equation 3 is the latent heat contributed to the gas phase by the fog formation. The vapor loss from the gas phase through both surface and gas-phase condensation can be related to the partial pressure of the condensing vapor by using Dalton's law and a differential material balance.

The effect on the coolant temperature of latent and sensible heat transferred to the surface from the condensing vapor is as shown in equation 5:

$$L \cdot M_L \cdot C_w \frac{dT_w}{dA} = \pm h_o \cdot (T_s - T_w) \tag{5}$$

where the $\pm$ sign is negative for countercurrent flow.

Assuming a linear relation between surface temperature and corresponding vapor pressure of the condensable component allows a heat balance to be written from gas phase to the surface:

$$h_o \cdot (T_s - T_w) = h_g \cdot (T_g - T_s) \frac{\epsilon}{1 - e^\epsilon} + k_g \cdot \lambda (P_g - P_s) \tag{6}$$

Combining equation 6 with the heat- and mass-transfer rate expressions gives

$$w \cdot C_w \frac{dT_w}{dA} = e^\epsilon \cdot G \cdot C_p \frac{dT_g}{dA} + \frac{V' \cdot \lambda P}{(P_t - P_{go})(P_t - P_{gf})} \cdot \frac{dP_g}{dA} \tag{7}$$

Equations 6 and 7 are not affected by fogging because the latent heat thus obtained is retained as sensible heat in the gas phase.

These basic relations have been solved for a wide range of cooler–condenser conditions and for different complexities of systems. A design procedure based on the assumption that the mixture is saturated throughout the condensation process has been developed (2). This assumption was later shown to depend on the rate of diffusion of the condensing component: some cases having rapidly diffusing components tend to superheat, and others having slowly diffusing vapors tend to subcool. The same approach extended to superheated mixtures has been used to develop the following equation for calculating T and partial pressures, P, during condensation (3):

$$\frac{dP_g}{dT_g} = \frac{P_t - P_g}{(Le)^{2/3} \cdot P_{BM}} \cdot \frac{P_g - P_s}{T_g - T_s} \cdot \frac{e^\epsilon - 1}{\epsilon} \tag{8}$$

This relation was tested experimentally for water condensation in various gases and found to be acceptable (4). It has also been solved via analog computers (5), and in another instance, for a set of conditions ranging from superheating to fogging (6). The use of standard j-factor correlations (see also HEAT-EXCHANGE

TECHNOLOGY, HEAT TRANSFER; MASS TRANSFER) for coefficients of heat and mass transfer have been incorporated into the solution method (4,7,8). Experimental verification has been supplied by workers in the field of absorption (qv) (9–11), condensation (12,13), liquid–liquid extraction (see EXTRACTION, LIQUID–LIQUID) (8,14), and distillation (qv) (15), and in laboratory experiments where free convection played a significant role (15,16).

In considering the effect of mass transfer on the boiling of a multicomponent mixture, both the boiling mechanism and the driving force for transport must be examined (17–20). Moreover, the process is strongly influenced by the effects of convective flow on the boundary layer. In Reference 20 both effects have been taken into consideration to obtain a general correlation based on mechanistic reasoning that fits all available data within ±15%.

The boiling mechanism can conveniently be divided into macroscopic and microscopic mechanisms. The macroscopic mechanism is associated with the heat transfer affected by the bulk movement of the vapor and liquid. The microscopic mechanism is that involved in the nucleation, growth, and departure of gas bubbles from the vaporization site. Both of these mechanistic steps are affected by mass transfer.

The final correlation for the overall boiling heat-transfer coefficient in pipes or channels (20) is a direct addition of the macroscopic (mac) and microscopic (mic) contributions to the coefficient:

$$h_T = h_{\text{mac}} + h_{\text{mic}}$$

$$= \left[0.023(Re_{L\text{-only}})^{0.8}(Pr_L)^{0.4}\frac{k_L}{D_i} \right] \left[\frac{\left(\dfrac{dP}{dz}\right)_{2\phi}}{\left(\dfrac{dP}{dz}\right)_{L\text{-only}}} \right]^{0.444} f(Pr_L)\left[\frac{\Delta\tilde{T}}{\Delta T_s} \right]_{\text{mac}}$$

$$+\, 0.00122\, \frac{k_L^{0.79}\cdot C_{PL}^{0.45}\cdot \rho_L^{0.49}\cdot g_c^{0.25}}{\sigma^{0.5}\mu_L^{0.29}\lambda^{0.24}\rho_V^{0.24}}\, \Delta T_s^{0.24}\cdot\Delta P_s^{0.75}\cdot S_{\text{binary}}\cdot Re_{2\phi} \qquad (9)$$

where

$$f(Pr_L) = \left[\frac{Pr_L + 1}{2} \right]^{0.444} \qquad (10)$$

$$\left(\frac{\Delta\tilde{T}}{\Delta T_s} \right)_{\text{mac}} = 1 - \frac{(1 - y_j)q}{\rho_{\text{avg}}\lambda k_Y \Delta T_s}\cdot\frac{\Delta T_s}{\Delta x_{jP_B}} \qquad (11)$$

$$k_Y = 0.023(Re_{2\phi})^{0.8}\cdot(S_c)^{0.4}\frac{k_L}{D}$$

$$Re_{2\phi} = Re_{L\text{-only}}\left[f(Pr_L)\left[\frac{\left(\dfrac{dP}{dz}\right)_{2\phi}}{\left(\dfrac{dP}{dz}\right)_{L\text{-only}}} \right]^{0.444} \right]^{1.25} \qquad (12)$$

$$S_{\text{binary}}\cdot Re_{2\phi} = \frac{1}{1 - \dfrac{C_{PL}(y_j - x_j)}{\lambda}\cdot\dfrac{\partial T}{\partial x_j}\left[\dfrac{\alpha}{D} \right]^{1/2}}S\cdot Re_{2\phi} \qquad (13)$$

In the macroscopic heat-transfer term of equation 9, the first group in brackets represents the usual Dittus-Boelter equation for heat-transfer coefficients. The second bracket is the ratio of frictional pressure drop per unit length for two-phase flow to that for liquid phase alone. The Prandtl-number function is an empirical correction term. The final bracket is the ratio of the binary macroscopic heat-transfer coefficient to the heat-transfer coefficient that would be calculated for a pure fluid with properties identical to those of the fluid mixture. This term is built on the postulate that mass transfer does not affect the boiling mechanism itself but does affect the driving force.

Likewise, the microscopic heat-transfer term takes accepted empirical correlations for pure-component pool boiling and adds corrections for mass-transfer and convection effects on the driving forces present in pool boiling. In addition to dependence on the usual physical properties, the extent of superheat, the saturation pressure change related to the superheat, and a suppression factor relating mixture behavior to equivalent pure-component heat-transfer coefficients are correlating functions.

Description of Gas–Vapor Systems

In engineering applications, the transport processes involving heat and mass transfer usually occur in process equipment involving vapor–gas mixtures where the vapor undergoes a phase transformation, such as condensation to or evaporation from a liquid phase. In the simplest case, the liquid phase is pure, consisting of the vapor component alone.

The system of primary interest, then, is that of a condensable vapor moving between a liquid phase, usually pure, and a vapor phase in which other components are present. Some of the gas-phase components may be noncondensable. A simple example would be water vapor moving through air to condense on a cold surface. Here the condensed phase, characterized by T and P, exists pure. The vapor-phase description requires y, the mole fraction, as well as T and P. The nomenclature used in the description of vapor-inert gas systems is given in Table 1.

The humidity term and such derivatives as relative humidity and molal humid volume were developed for the air–water system. Use is generally restricted to that system. These terms have also been used for other vapor–noncondensable gas phases.

For the air–water system, the humidity is easily measured by using a wet-bulb thermometer. Air passing the wet wick surrounding the thermometer bulb causes evaporation of moisture from the wick. The balance between heat transfer to the wick and energy required by the latent heat of the mass transfer from the wick gives, at steady state,

$$-k_Y \cdot A \cdot (Y_1 - Y_w) \cdot \lambda_w = (h_g + h_r) \cdot A \cdot (T_1 - T_w) \tag{14}$$

$$T_1 - T_w = \frac{k_Y \lambda_w}{(h_g + h_r)} (Y_w - Y_1) \tag{15}$$

Table 1. Definitions of Humidity Terms

Term	Meaning	Units	Symbol[a]
humidity	vapor content of a gas	mass vapor per mass noncondensable gas	$Y' = Y\frac{M_a}{M_b}$
molal humidity	vapor content of a gas	moles vapor per mole noncondensable gas	Y
relative saturation or relative humidity	ratio of partial pressure of vapor to partial pressure of vapor at saturation	kPa/kPa, or mole fraction per mole fraction, expressed as %	$\frac{y}{y_s} \times 100$
percent saturation or percent humidity	ratio of concentration of vapor to the concentration of vapor at saturation with concentrations expressed as mole ratios	mole ratio per mole ratio, expressed as %	$\frac{Y}{Y_s} \times 100$
molal humid volume	volume of 1 mol of dry gas plus its associated vapor	$m^3/mol^{a,b}$	$V_h = (1 + Y)$ $\times 0.0224\frac{T}{273}$ $\times 1.013\, P^{-1}$
molal humid heat	heat required to raise the temperature of 1 mol of dry gas plus its associated vapor 1°C	$J/(mol \cdot °C)^a$	$c_h = c_b + Y c_a$
adiabatic saturation temperature	temperature that would be attained if the gas were saturated in an adiabatic process	°C or K	T_{sa}
wet-bulb temperature	steady-state temperature attained by a wet-bulb thermometer under standardized conditions	°C or K	T_w
dew-point temperature	temperature at which vapor begins to condense when the gas phase is cooled at constant pressure	°C or K	T_d

[a]Concentration is on the basis of dry gas.
[b]When T is in K and P is in Pa.

If radiant energy transfer can be prevented, the following equation is used:

$$T_1 - T_w = \frac{k_Y \cdot \lambda_w}{h_g}(Y_w - Y_1) \qquad (16)$$

Thus, a measurement of the wet-bulb temperature, T_w, and the temperature T_1, allows the molal humidity, Y_1, to be calculated because Y_w is known. The use of molal humidity as the mass-transfer driving force is conventional and convenient because of the development of humidity data for, especially, the air–water system. The mass-transfer coefficient must be expressed in consistent units.

Another relationship between temperature and humidity results from considering the path of T and Y during an adiabatic saturation process. If a countercurrent packed column exists such that no heat flows from or to the surroundings and the liquids (water) stream is recycled, the gas stream (air) passes once through the unit. In this case the liquid reaches a steady-state temperature. If the column is very tall, the gas exit temperature would reach the temperature of the recycled liquid. Figure 1 shows the physical arrangement and the nomenclature. Material and energy balances written for an envelope encircling the exit and entrance streams from this column using enthalpies in terms of molal humid heats, latent heats, and liquid heat capacities yield the following:

$$C_{h_1}\cdot(T_{\mathrm{sa}} - T_1) = \lambda_{\mathrm{sa}}\cdot(Y_1 - Y_{\mathrm{sa}}) \qquad (17)$$

where sa refers to the adiabatic saturation condition and point 1 is any initial condition. This is the adiabatic saturation equation that traces the path of a moist gas stream as it is humidified under adiabatic conditions.

For the air–water system, Lewis recognized that $C_h = h_g/k_Y$, based on empirical evidence. Thus, the adiabatic saturation equation is identical to the wet-bulb temperature line. In general, again based on empirical evidence (21),

$$\frac{h_g}{k_Y} = C_h\left(\frac{Sc}{Pr}\right)^{0.56} \qquad (18)$$

whereas, normally, $6.83 < h_g/k_Y < 7.82$, for air $Sc \cong Pr \cong 0.70$, and $h_g/k_Y = C_h = 6.94$.

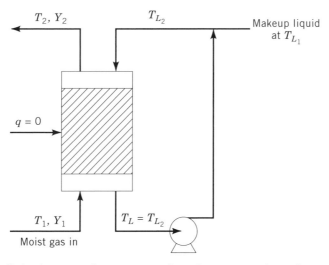

Fig. 1. The adiabatic saturation process, where for a saturation column, $T_2 = T_{\mathrm{sa}}$, and $Y_2 = Y_{\mathrm{sa}}$.

A closer look at the Lewis relation requires an examination of the heat- and mass-transfer mechanisms active in the entire path from the liquid–vapor interface into the bulk of the vapor phase. Such an examination yields the conclusion that, in order for the Lewis relation to hold, eddy diffusivities for heat- and mass-transfer must be equal, as must the thermal and mass diffusivities themselves. This equality may be expected for simple monatomic and diatomic gases and vapors. Air having small concentrations of water vapor fits these criteria closely.

The thermodynamic properties of a vapor–gas mixture, ie, two components, one of which is condensable, are usually presented on a humidity diagram. Figure 2 is the humidity diagram for the air–water vapor phase at normal atmospheric pressure, where humidity is plotted against temperature. Curves are given for saturated vapor and for constant values of relative humidity. Also plotted are lines of constant wet-bulb temperature, or adiabatic saturation lines. These are nearly straight lines having negative slopes slightly less than 30°. These lines are also lines of nearly constant enthalpy, as can be seen by rearranging equation 17. The deviation is in the variations of humidity heat and latent heat along the path. The chart shows values of the enthalpy at saturation as well as lines of constant enthalpy deviation. Thus, the enthalpy can be found by adding the enthalpy deviation to the enthalpy of the saturated gas phase at the wet-bulb temperature. A copy of this diagram covering a greater temperature span is available (see DRYING).

Figure 3 is the humidity chart diagram in molar quantities where enthalpy deviations are not given. Enthalpy may also be calculated from the enthalpy of saturated air and of dry air using % saturation:

$$H = H_{\text{dry}} + (H_{\text{sat}} - H_{\text{dry}}) \cdot (\% \text{ saturation}) \tag{19}$$

Figure 3 gives the % humidity as the measure of vapor concentration, whereas Figure 2 gives relative humidity in %.

For systems other than air–water vapor or for total system pressures different from 101.3 kPa (1 atm), humidity diagrams can be constructed if basic phase-equilibria data are available. The simplest of these relations is Raoult's law, applicable at small solute concentrations:

$$P \cdot y_s = p_i^{\circ} \cdot x_i \tag{20}$$

For a two-component system in which one component exists only in the vapor phase, equation 20 is reduced to the following:

$$y_s = \frac{H_s}{1 + H_s} = \frac{p_i^{\circ}}{P} \tag{21}$$

Calculations for Humidification and Dehumidification Processes

Figure 4 shows the general arrangement and nomenclature for a humidification or dehumidification process, where the subscript 1 refers to the bottom of the column, and subscript 2 to the top. Steady state is assumed. Flow rates and compositions are given in molar terms because this simplifies the results.

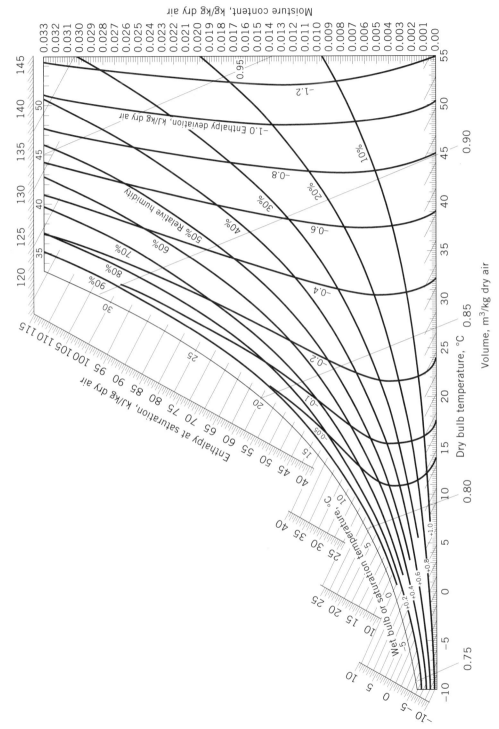

Fig. 2. Psychrometric chart. Below 0°C properties and enthalpy deviation lines are for ice. Courtesy of Carrier Corp. To convert kJ to kcal, divide by 4.184.

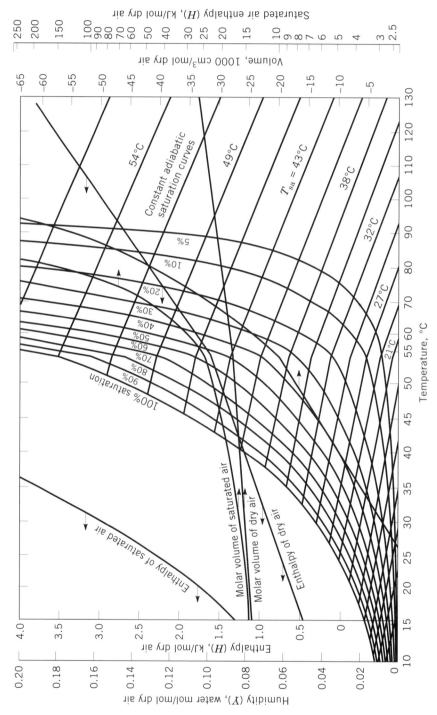

Fig. 3. Humidity chart for the air–water system, molal quantities. To convert kJ to Btu, divide by 1.054; to convert cm³ to ft³, multiply by 35.31 × 10⁻⁶.

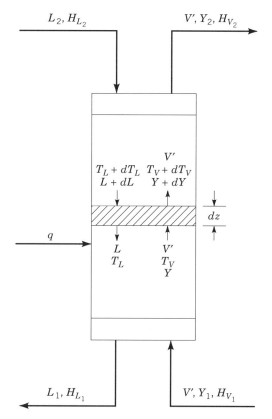

Fig. 4. Arrangement and nomenclature for general humidification–dehumidification process.

Total material, condensable component, and energy balances can be written for the entire column:

$$L_1 - L_2 = V_1 - V_2 \tag{22}$$

$$V' \cdot (Y_2 - Y_1) = L_2 - L_1 \tag{23}$$

$$L_2 \cdot H_{L_2} + V' \cdot H_{V_1} + q = L_1 \cdot H_{L_1} + V' \cdot H_{V_2} \tag{24}$$

Generally, q is small because the outside area is not large in comparison to the amount of heat being transferred, and the energy balance can be simplified. In these conditions it is also convenient to write balances over a differential section of the column. These balances yield the following:

$$V' \cdot dY = dL \tag{25}$$

$$V' \cdot dH_V = d(L \cdot H_L) \tag{26}$$

If the amount of evaporation is small, the change in enthalpy in the liquid phase can be taken as a result of temperature change alone. Using an average (av) liquid flow rate, the following is derived:

$$V' \cdot dH = L_{\text{av(g)}} \cdot C_L \cdot dT_L \tag{27}$$

Similarly, the vapor enthalpy can be expressed in terms of humid heat and latent heat in relation to a base condition:

$$V' \cdot dH = V' \cdot d[C_h \cdot (T_V - T_o) + Y \cdot \lambda_o] = V' \cdot C_h \cdot dT_V + V' \cdot \lambda_o \cdot dY \tag{28}$$

The energy transferred on both sides of the interface in equation 28 can also be written in terms of the appropriate rate expressions. For the liquid phase, it is

$$\frac{L_{\text{av(g)}}}{S} C_L \cdot dT_L = h_L \cdot a (T_L - T_i) \cdot dz \tag{29}$$

For the gas phase, energy transfers both as a result of a thermal driving force and as a by-product of vaporization. Thus,

$$\frac{V'}{S} C_h \cdot dT_V = h_g \cdot a (T_i - T_V) \cdot dz \tag{30}$$

and

$$\frac{V'}{S} \lambda_o \cdot dY = \lambda_o \cdot k_Y \cdot a (Y_i - Y) \cdot dz \tag{31}$$

Combining these two mechanisms for gas-phase transfer, as done in equation 28, yields

$$\frac{V'}{S} dH_V = h_g \cdot a (T_i - T_V) \cdot dz + \lambda_o \cdot k_Y \cdot a (Y_i - Y) \cdot dz \tag{32}$$

Rearranging equation 32 and defining the ratio $h_c \cdot a / (k_Y \cdot a \cdot C_h)$ as r, the psychrometric ratio, give

$$\frac{V'}{S} dH_V = k_Y \cdot a [(C_h \cdot r \cdot T_i + \lambda_o \cdot Y_i) - (C_h \cdot r \cdot T_V + \lambda_o \cdot Y)] \cdot dz \tag{33}$$

For the air–water system, the Lewis relation shows that $r = 1$. Under these conditions, the two parenthetical terms on the right-hand side of equation 33 are enthalpies, and equation 33 becomes the design equation for humidification operations:

$$\frac{V'}{S} dH_V = k_Y \cdot a (H_i - H_V) \cdot dz \tag{34}$$

or

$$\int_{H_{V_1}}^{H_{V_2}} \frac{V' \cdot dH_V}{k_Y \cdot a S (H_i - H_V)} = \int_0^z dz = z \tag{35}$$

The simplification of equation 33 to equation 34 is possible only if $r = 1$; that is, for simple monoatomic and diatomic gases. For other systems the design equation can be obtained by a direct rearrangement of equation 33.

Although equation 35 is a simple expression, it tends to be confusing. In this equation the enthalpy difference appears as driving force in a mass-transfer expression. Enthalpy is not a potential, but rather an extensive thermodynamic function. In equation 35, it is used as enthalpy per mole and is a kind of shorthand for a combination of temperature and mass concentration terms.

The integration of equation 35 requires a knowledge of the mass-transfer coefficient, $k_Y \cdot a$, and also of the interface conditions from which H_i could be obtained. Combining equations 27, 28, and 34 gives a relation balancing transfer rate on both sides of the interface:

$$\frac{V'}{S} dH_V = h_L \cdot a(T_L - T_i) \cdot dz = k_Y \cdot a(H_i - H_V) \cdot dz \tag{36}$$

or

$$\frac{-h_L \cdot a}{k_Y \cdot a} = \frac{H_V - H_i}{T_L - T_i} \tag{37}$$

Thus, the enthalpy and temperature of the vapor–liquid interface are related to the liquid temperature and gas enthalpy at any point in the column through a ratio of heat- and mass-transfer coefficients.

The integration can be carried out graphically or numerically using a computer. For illustrative purposes the graphical procedure is shown in Figure 5. In this plot of vapor enthalpy (H_V or H_i) vs liquid temperature (T_L or T_i), the curved line is the equilibrium curve for the system. For the air–water system, it is the 100% saturation line taken directly from the humidity diagram (see Fig. 3).

The locus of corresponding T_L and H_V points, the operating line for the column, can be obtained by assuming that V' and $L_{av(g)}$ change little and by integrating equation 27 along the length of the column.

$$\int_{H_{V_1}}^{H_{V_2}} V' \cdot dH = \int_{T_{L_1}}^{T_{L_2}} L_{av(g)} \cdot C_L \cdot dT \tag{38}$$

$$V' \cdot (H_{V_2} - H_{V_1}) = L_{av(g)} \cdot C_L \cdot (T_{L_2} - T_{L_1}) \tag{39}$$

or

$$\frac{H_{V_2} - H_{V_1}}{T_{L_2} - T_{L_1}} = \frac{L_{av(g)} \cdot C_L}{V'} \tag{40}$$

Thus, the locus is a straight line, assuming that the ratio on the right side of equation 40 is constant. The actual location of the line can be obtained if both end points are known or if one end point and the slope ($L_{av(g)} \cdot C_L / V'$) can be determined.

On Figure 5 the locus of interface points passes through points F and E. The operating line goes from B to C. In addition to specifying two points on the

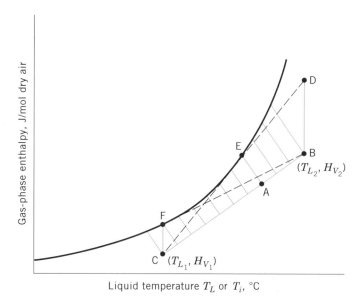

Fig. 5. Adiabatic gas–liquid contacting, graphical representation where point A is an arbitrary point along the column; line CAB is the operating line having slope of $L_{av(g)} \cdot C_L / V'$; point E represents the interface conditions corresponding to point A; and the tie line AE has slope of $h_L \cdot a / (k_Y \cdot a)$. The bold line defines the equilibrium curve, H_i vs T_i. Conditions shown are those of a water-cooling process. To convert J to Btu, divide by 1054.

line itself, B and C, or the slope and one of these points, the column could be required to operate at some convenient gas flow rate greater than minimum. Here the minimum gas flow rate required to cool the liquid from 60 to 30°C is given by line CD. Another possible limiting condition would be the flow of gas with the largest wet-bulb temperature possible to allow water to cool from 60 to 30°C, no matter how large the column (line BF). A design and operating condition can be determined as an acceptable approach to either of these.

Once the operating line is set, interface conditions corresponding to any point on the operating line can be found if heat- and mass-transfer coefficients are available. Then a line of slope $-h_L \cdot a / (k_Y \cdot a)$ connects a point on the operating line, eg, point A, with its corresponding interface condition, point E. This information allows the integration of equation 35 to give the column height. The method is shown graphically in Figure 6, although again a numerical solution is possible.

Determination of the Gas-Phase Temperature. The development given above is in terms of interface conditions, bulk liquid temperature, and bulk gas enthalpy. Often the temperature of the vapor phase is important to the designer, either as one of the variables specified or as an important indicator of fogging conditions in the column. Such a condition would occur if the gas temperature equaled the saturation temperature, that is, the interface temperature. When fogging does occur, the column can no longer be expected to operate according to the relations presented herein but is basically out of control.

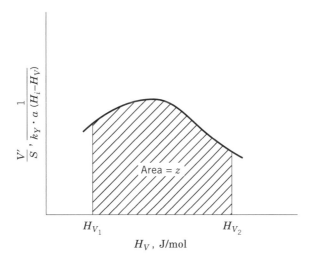

Fig. 6. Integration of the design equation (eq. 35).

Gas-phase temperatures have been obtained by an extension of the graphical method illustrated (22). When equation 30 is divided by equation 34, the result is

$$\frac{V' \cdot C_h \cdot dT_V}{V' \cdot dH_V} = \frac{h_g \cdot a \cdot (T_i - T_V) \cdot dz}{k_Y \cdot a \cdot (H_i - H_V) \cdot dz} \tag{41}$$

or

$$\frac{dT_V}{dH_V} = \frac{h_g \cdot a}{C_h \cdot k_Y \cdot a} \cdot \frac{(T_i - T_V)}{(H_i - H_V)} = \frac{T_i - T_V}{H_i - H_V} \tag{42}$$

The last expression of equation 42 is obtained by applying the Lewis relation, $C_h = h_c/k_Y$. If the differentials of equation 42 are replaced by finite differences, the following obtains:

$$\frac{\Delta T_V}{\Delta H_V} \approx \frac{T_i - T_V}{H_i - H_V} \tag{43}$$

In effect, equation 43 states that the temperature and enthalpy values of the bulk gas phase continuously approach the interface condition at the same point in the column as that for which the gas-phase conditions apply. The graphical application is illustrated in Figure 7. The gas-phase temperature and enthalpy at the bottom of the column are usually known and are plotted at point G_0. The interface conditions at the bottom of the column are given at point H. Then by equation 43 the line GH is the path followed by the gas-phase temperature. This path is followed until the interface condition shifts noticeably, perhaps to point J corresponding to a bulk liquid temperature at point I. Then the gas temperature line approaches point J. The interface conditions shift toward point Z, continuously changing the direction of the gas temperature line. The points at

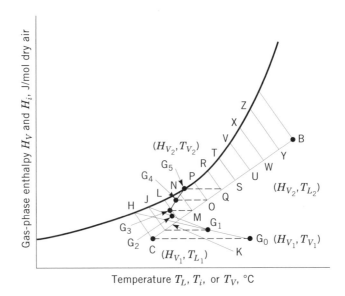

Fig. 7. Determination of the bulk gas-phase temperature path.

which the line changes slope depend on the intervals chosen along the operating line. Here G_1 corresponds to point I, G_2 to point K, G_3 to point M, G_4 to point O, etc.

In the example developed herein fogging occurs at about the time the gas reaches the top of the column. That is far from inevitable and would not have occurred if the operating line had terminated at point U.

The method thus outlined allows the development of a conceptual under-standing of the limits of operation of a humidification column. For actual design, the simplifications used herein may be avoided by handling the fundamental equations numerically by computer.

Transfer Coefficient. The design method described depends for its utility on the availability of mass- and heat-transfer coefficients. Typically, $k_Y \cdot a$ and $h_L \cdot a$ are needed. These must be obtained from the standard correlations for mass and heat transfer, from data reported in the literature (23–30), or from data presented by equipment makers for particular packing (31–33). When this type of information is not available, it is possible to determine heat- and mass-transfer coefficients by a single test using the packing material of interest in a pilot-sized tower. If a steady state is obtained, measurement of air- and water-inlet and -outlet temperatures, and air-inlet and -outlet wet- and dry-bulb temperatures comprises all the necessary information. Interface and operating lines on a T_L–H_V diagram, such as Figure 5, are directly obtained. Because column heights are known, the value of $-h_L \cdot a/(k_Y \cdot a)$ can be obtained by trial and error with the integration demonstrated in Figure 6 and adjustment of the slope of the operating interface condition line until z_{calc} equals the actual column height.

Overall Coefficients. Often overall coefficients of heat and mass trans-fer are available, rather than the film coefficients used earlier. In that case equation 35 can be rewritten as

$$\int_{H_{V_1}}^{H_{V_2}} \frac{V' \cdot dH_V}{k_Y \cdot a \cdot S \cdot (H^* - H_V)} = \int_{o}^{z} dz = z \tag{44}$$

If $k_Y \cdot a$ is constant, this can be written as

$$\frac{V'}{k_Y \cdot a \cdot S} \int_{H_{V_1}}^{H_{V_2}} \frac{dH_V}{(H^* - H_V)} = \text{HTU} \cdot \text{NTU} = z \tag{45}$$

Humidification and Dehumidification Equipment

The addition or removal of a condensable component to or from a noncondensable gas can be accomplished by direct contact between the vapor and the gas. This may be done in a countercurrent tower, usually packed as described elsewhere (see ADSORPTION; DISTILLATION; MASS TRANSFER). The direction of transfer depends on the temperatures of the two streams. Such operations can also be done using spray ponds in which a grid of nozzles sprays liquid, usually water, into the gas phase, usually air. If the air is relatively dry, liquid evaporates into it, both humidifying the air and cooling the liquid. If a large surface of water is available, the same process may be carried out through evaporation from the surface of lake or pond. Usually the purpose is the cooling of process water. As hot water is discharged into the pond, the surface temperature of the pond rises until evaporation (qv) balances the incoming thermal load. A large enough pond surface must be supplied to allow evaporation to balance the thermal load at a manageable temperature rise. This area requirement may exceed the availability of land in the plant site region.

Humidification processes also occur in spray contactors often used to scrub minor components from a gas stream. Here the gas passes through successive sprays of liquid. The liquid is often water but may be specially compounded to enhance absorption of the component to be removed.

Water-Cooling Towers. By far the most common and large-scale mode of humidification processing is in water-cooling towers. As supplies of cooling water become more strained, and as discharge water temperatures are more closely controlled, water cooling and recirculation rather than once-through water use become more common. Two general types of direct-contact cooling towers are in use. The forced-draft tower depends on fans to move the air through the tower. Typically, the tower consists of a set of louvres and baffles over which the water falls, breaking into films and droplets. Air flow may be across this cascading liquid or countercurrent of it. Often both flow arrangements exist in the same tower. Figure 8 shows a cross-sectional view of a cooling tower. Here air flows across the cascading liquid, drawn by a fan located in the outlet duct. In other arrangements the fan can be placed to push the air through the tower.

There are several internal gridwork arrangements, all designed to enhance splashing and film formation in order to give a large water–air interface and

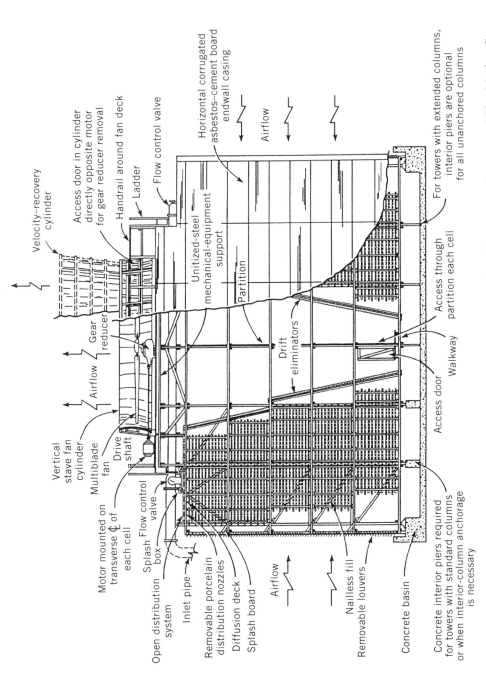

Velocity-recovery cylinder

Access door in cylinder directly opposite motor for gear reducer removal

Handrail around fan deck

Ladder

Flow control valve

Horizontal corrugated asbestos–cement board endwall casing

Airflow

Unitized-steel mechanical-equipment support

Partition

Gear reducer

Airflow

Drift eliminators

Access through partition each cell

Vertical stave fan cylinder

Multiblade fan

Drive shaft

Walkway

For towers with extended columns, interior piers are optional for all unanchored columns

Motor mounted on transverse ₵ of each cell

Splash Flow control box valve

Access door

Open distribution system

Inlet pipe

Removable porcelain distribution nozzles

Diffusion deck

Splash board

Airflow

Nailless fill

Removable louvers

Concrete basin

Concrete interior piers required for towers with standard columns or when interior-column anchorage is necessary

Fig. 8. Transverse cross-sectional view of double-flow induced-draft cooling tower. Courtesy of The Marley Co.

allow a low pressure drop in the air stream passing through. The lattice members were traditionally wood-treated to prevent biological and corrosion attack. More recently, different materials such as transite, various plastic laminates, and ceramics have been used. Packing design has also become more and more specialized, and proprietary designs are offered by most cooling-tower makers.

The thermal design of cooling towers follows the same general procedures already presented. Integration of equation 35 is usually done numerically using the appropriate software, mass-transfer coefficients, saturation enthalpies, etc. In mechanical-draft towers the air and water flows are both supplied by machines, and hence flow rates are fixed. Under these conditions the design procedure is straightforward.

Natural-Draft Cooling Towers. In a natural-draft cooling tower (Fig. 9) the driving force for the air is provided by the buoyancy of the air column in a very tall stack. Stack heights of 100 m are common, and as power-plant sizes increase, the size of single towers is likely to increase also. In the absence of a fan, the air flow rate, G, is no longer an independent variable, but is dependent on the design and operating conditions of the tower. The governing equation for air flow becomes

$$z_t \cdot \Delta\rho = N \frac{G^2}{\rho g_c} \tag{46}$$

where $\Delta\rho$ is the average density difference between the outside air and the air in the stack; z_t is the height of the tower, and N, the resistance to air flow through the tower in velocity heads, is specific for a given tower and can usually be expressed as a constant (34).

For a natural-draft tower, equations 46 and 44 must be solved simultaneously, introducing an expression for $\Delta\rho$ as a function of conditions inside and outside the tower. Up until 1955, this was a cumbersome procedure, and a number of approximate methods were devised to simplify the calculation. Since then, the whole calculation has been done using computer software. For a tower of given dimensions, the air flow can be guessed and the thermal performance and pressure loss calculated. From the thermal performance, the density difference, $\Delta\rho$, can be calculated and the left side of equation 46 compared to the other side. When the correct air flow has been guessed, the equation is satisfied. Iterative procedures using rapid convergence can easily be devised (see COMPUTER-AIDED DESIGN AND MANUFACTURING (CAD/CAM); COMPUTER-AIDED ENGINEERING (CAE)) (35). Cooling-tower manufacturers express confidence in their ability to design cooling towers that meet guaranteed performance and to predict off-design behavior.

Approximate Methods for Predicting Natural-Draft Cooling-Tower Performance. Approximate methods, no longer needed for design work, are useful for rapid estimates of the effects of changing conditions on performance. In addition, a good grasp of the approximate theories leads to a better physical understanding of tower behavior.

An approximate method for integrating equation 44, ie, Merkel's approximation (34,36), leads to the following:

$$\frac{H_m^* - H_2}{T_{L_1} - T_{L_2}} = \frac{L}{2G} + \frac{L}{Kaz} \tag{47}$$

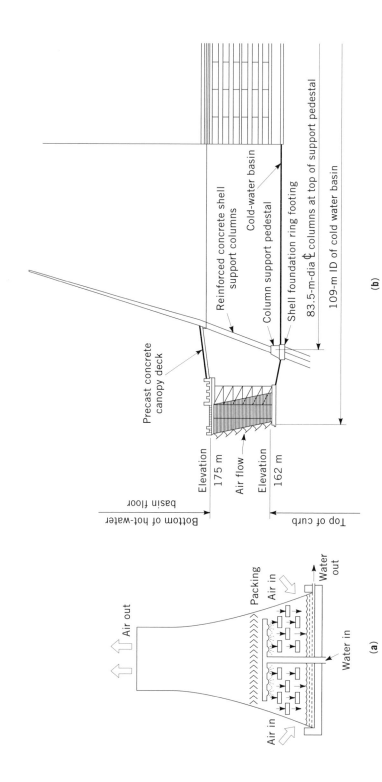

Fig. 9. Natural-draft cooling tower: (**a**) general tower drawing for countercurrent air–water flow arrangement; (**b**) sectional drawing showing arrangement for cross flow of air–water.

214

Using Merkel's approximation and knowing the desired thermal performance, the flow rates, and transfer coefficient, z, can quickly be calculated. The difficulty with this method is that errors of $\geq 10\%$ in z can arise if the cooling range $T_1 - T_2$ is larger than a few degrees.

Equations 46 and 47 have been combined to obtain rapid approximate methods for predicting cooling tower performance (34,37). The most interesting result is obtained from a rather simple analysis (38). If A is the cross-sectional area of the packing, then the liquid total flow rate $W_L = AL$, and the air flow rate $W_G = AG$. Substituting for G in equation 46 gives, with some rearrangement, the following:

$$\frac{A(z_t)^{1/2}}{(N/2)^{1/2}} = \frac{W_G}{(\Delta\rho)^{1/2}\cdot(\rho)^{1/2}} \tag{48}$$

Equation 48 equals D, the duty coefficient of the tower. Let $(N/2)^{1/2} = C^{3/2}$, then

$$\frac{A(z_t)^{1/2}}{C^{3/2}} = -\frac{W_G}{(\Delta\rho)^{1/2}\cdot(\rho)^{1/2}} = D \tag{49}$$

Reference 34 shows that

$$\Delta\rho = 13.465 \times 10^{-5}(\Delta T_V + 0.3124\,\Delta H) \tag{50}$$

Rearranging equation 50, applying the energy balance, and assuming air at standard conditions enters the tower yield (38):

$$\frac{W_L}{D} = 90.59\,\frac{\Delta H}{\Delta T}(\Delta T + 0.3124\,\Delta H)^{1/2} \tag{51}$$

For most cooling towers in the United Kingdom, the exit air is saturated at a temperature close to the mean water temperature in the tower. Hence, if the water temperatures and the air inlet conditions are known, ΔH, ΔT_L, and ΔT_V can all be calculated, and W_L/D can be determined. It was found that the quantity C was approximately constant for these towers, ca 0.4–0.5 (34). If the value of C is known for a given tower, then the left side of equation 49 can be computed and, setting this equal to D, the allowable liquid flow rate can be found. Alternatively, when W_L, T_{L_1}, and air-inlet conditions are given, the equations can be used to find T_{L_2}. A rapid estimate of the effects of off-design conditions can thus be made. Reference 38 presents a nomogram of these equations to facilitate the calculation (see ENGINEERING, CHEMICAL DATA CORRELATION).

Natural-draft cooling towers are extremely sensitive to air-inlet conditions owing to the effects on draft. It can rapidly be established from these approximate equations that as the air-inlet temperature approaches the water-inlet temperature, the allowable heat load decreases rapidly. For this reason, natural-draft towers are unsuitable in many regions of the United States. Figure 10

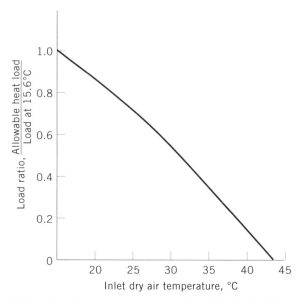

Fig. 10. Effect of inlet dry air temperature on allowable load, where the inlet relative humidity is 50%; water-inlet temperature is 43.3°C; water exit temperature is 32.2°C.

shows the effect of air-inlet temperature on the allowable heat load of a natural-draft tower for some arbitrary numerical values and inlet rh of 50%. The trend is typical.

Trends in Cooling-Tower Use and Development. Natural-draft cooling towers had been rare in the United States because ample water supplies were available for power-plant cooling, and natural-draft towers are best suited for large heat loads. Mechanical-draft towers were used in large numbers for industrial applications and occasionally for power plants. A dramatic change in this situation has occurred. Almost all large post-1980 power plants require cooling towers. Limitations on cooling water use and the return of warm water to rivers has forced the use of all cooling towers. Towers allow the heat load to be dissipated to the air rather than into natural water. The treating of recycled cooling water may pose hazards to aquatic animals, crops, etc, near the tower. In the southern United States, mechanical-draft towers dominate the field; in the northern United States, the situation is not as clear. Because of earlier over-building and public attitudes toward cost recovery, few power plants were built in the 1980s or early 1990s. Many existing plants were, however, upgraded. The issues involved in a decision have been discussed (39,40).

The initial cost of a mechanical-draft cooling tower for a power plant is relatively low, ca $10.00/kW more than a direct-stream cooling system. However, the power required to run the fan is significant, and maintenance must also be considered. In a large power plant, many mechanical-draft towers are required, covering a large area of ground, and problems of water and power distribution become acute. In addition, the plume is discharged close to the ground and can be a source of fog.

The final cost of a natural-draft tower is substantially higher, ca $20.00/kW. However, the maintenance costs are much lower because there are no fans or electrical drives. The natural-draft tower occupies less ground space than the corresponding group of mechanical-draft towers. Additionally, because the plume is discharged ≥ 100 m from the ground, it is much less likely to cause local ground fog. Thus the natural-draft tower is frequently more attractive, and may be chosen even if there is a slight overall cost disadvantage.

The economics of cooling towers has been discussed (38,41). A worthwhile evaluation of the optimum configuration must take into account the interaction between tower performance and plant performance. For example, in considering the additional expense of a cooling tower vs a direct-stream cooling system, it is desirable to optimize the whole plant for each system rather than add the cooling tower to a system optimized for run-of-the-river cooling. Consequently, it is not possible to produce general-cost curves for cooling towers. Each installation must be evaluated separately.

As of the mid-1990s, a large natural-draft cooling tower could cost approximately $12,000,000, according to general cost analysis calculations and annual construction cost ratios. That cost was divided almost equally between the foundation, the packing and water-distribution system, and the shell. Therefore, large cost reductions in any of these items can have a significant effect. Newer, light packings are being developed in a variety of materials, leading to some cost improvements. Packings may be built of fiber-reinforced polymers, chosen for resistance to the heated water to which they are to be exposed, as well as for strength, weight, and cost considerations. The tower shell, made of reinforced concrete, has a hyperbolic shape chosen mainly for structural rather than aerodynamic reasons. A hyperbolic shell requires less concrete than an equally strong cylindrical shell. It seems unlikely that important cost reductions in the shell or foundation can be made.

When all the expenses involved in using wet cooling towers on a power plant are considered, it appears that a mechanical-draft cooling tower system may raise the cost of generating electricity by ca 3%, and a natural-draft tower by ca 6%, over the generating cost of a direct river-cooled power plant. These figures are approximate, but show the effects of a thermal pollution regulatory program on the cost of generating electricity (see THERMAL POLLUTION).

Cooling-Tower Plumes. An important consideration in the acceptability of either a mechanical-draft or a natural-draft tower cooling system is the effect on the environment. The plume emitted by a cooling tower is seen by the surrounding community and can lead to trouble if it is a source of severe ground fog under some atmospheric conditions. The natural-draft tower is much less likely to produce fogging than is the mechanical-draft tower. Nonetheless, it is desirable to devise techniques for predicting plume trajectory and attenuation.

Not only may the cooling-tower plume be a source of fog, which in some weather conditions can ice roadways, but the plume also carries salts from the cooling water itself. These salts may come from salinity in the water, or may be added by the cooling-tower operator to prevent corrosion and biological attack in the column.

Efforts to combat bacteria and corrosion in cooling towers have gone through a long development and are both complex and specific to different waters. Cathodic and anodic inhibitors are used as well as biocides. Systems may include chromates, nitrites, orthophosphates, and ferrocyanides as cathodic inhibitors; zinc, nickel, lead, tin, copper, and silicates as anodic inhibitors; and possibly added materials as biocides and pH controllers. These chemicals can also be carried onto fields surrounding the cooling tower, seriously affecting crop yield or ornamental plantings (see WATER, INDUSTRIAL WATER TREATMENT).

Much work has been done on the modeling of wet cooling-tower plumes; the ultimate aim was to determine their effect on the environment (42–44). The basic approach involves writing the equations of continuity, conservation of energy and momentum, and the equations of motion for the conditions of the plume. These are then solved simultaneously using iterative and numerical methods on large computers. The accuracy of the results depends on whether the boundary conditions and simplifying assumptions are realistic. These are difficult to accomplish because conditions change rapidly, ground configurations are seldom simple, and plume behavior is influenced by a host of casual, nonrepeated situations such as the passing of an airplane or the presence of cloud cover. Modeling owes much to meteorology and especially to the theory of cumulus clouds (see ATMOSPHERIC MODELING).

The recirculation of cooling water via a cooling tower ultimately removes process heat by evaporating water rather than by warming it, as would be the case with once-through systems. When water is especially scarce, it may be necessary to cool process water by transferring the heat to air through indirect heat transfer. This requires dry cooling towers, which have been built in a few dry regions of the United States. These usually take the form of natural-draft cooling towers in which high surface heat-exchange areas replace the usual gridwork. The air-to-circulating-water heat-transfer process passes heat through a solid surface, thus heat-transfer coefficients are low and enormous heat-transfer areas are required. The cost of such towers may be 10-fold that of wet towers, and the availability of tubing for heat transfer becomes a serious problem.

Trends

Work in the area of simultaneous heat and mass transfer has centered on the solution of equations such as 1–18 for cases where the structure and properties of a solid phase must also be considered, as in drying (qv) or adsorption (qv), or where a chemical reaction takes place. Drying simulation (45–47) and drying of foods (48,49) have been particularly active subjects. In the adsorption area the separation of multicomponent fluid mixtures is influenced by comparative rates of diffusion and by interface temperatures (50,51). In the area of reactor studies there has been much interest in monolithic and honeycomb catalytic reactions (52,53) (see EXHAUST CONTROL, INDUSTRIAL). For these kinds of applications psychrometric charts for systems other than air–water would be useful. The construction of such has been considered (54).

Cooling water is a necessity for temperature control. Most industrially generated heat must be dissipated; water is an obvious receptor because heat transfer is relatively rapid and compact. When the heat load is large it is usually necessary to cool and reuse the water. Thus cooling towers are integral parts of power plants, chemical processing operations, and compression steps, and proper design and operation are critical to the entire process. Relative costs vary widely for different process situations. In power plants, cooling towers might represent 10% of the total capital cost.

Simultaneous heat and mass transfer also occurs in drying processes, chemical reaction steps, evaporation, crystallization, and distillation. In all of these operations transfer rates are usually fixed empirically. The process can be evaluated using either the heat- or mass-transfer equations. However, if the process mechanism is to be fully understood, both the heat and mass transfer must be described. Where that has been done, improvements in the engineering of the process usually result (see PROCESS ENERGY CONSERVATION).

Nomenclature

Symbol	Definition	Units
A	interfacial area	m^2
a	interfacial area per unit column volume	m^{-1}
C	heat capacity	$J/(g \cdot K)$
D	molecular diffusivity for mass transfer	m^2/h
F	rate of fog formation	mol/h
G	mass flow rate of gas phase	$kg/(h \cdot m^2)$
g_c	force–mass conversion constant	
H	enthalpy	
HTU	height of a transfer unit	
h_c	convective heat-transfer coefficient	
h_r	coefficient for heat transfer by radiative mechanism	
K_a	overall mass-transfer coefficient per volume of contacting column	
k_g	gas-phase mass-transfer coefficient in partial pressure driving force units	
k_L	liquid-phase thermal conductivity	
k_Y	mass-transfer coefficient in gas-phase mole ratio units	
L	liquid stream molar flow rate	
Le	Lewis number	
M	molecular weight	
N	resistance to air flow in velocity heads	
NTU	number of transfer units	
P	total pressure	
Pr	Prandtl number	
p°	vapor pressure	
q	heat flux	
Re	Reynolds number	
r	psychrometric ratio	
S	suppression factor $(\Delta Te/\Delta T)^{0.99}$	
Sc	Schmidt number	
T	temperature	
$\Delta \tilde{T}$	driving force for the binary macroscopic heat transfer	

V	specific volume	
V	gas-phase molar flow rate	
V'	molar flow rate of noncondensable component	
w	total flow rate	mol/time
x	mole fraction in liquid phase	
Y	mole ratio	
Y'	mass ratio	
y	mole fraction in gas phase	
z	height of column	
ϵ	Ackerman correction term, $= m_i \cdot c_{pi}/h_g$	
λ	latent heat of vaporization	
μ	viscosity	
ρ	density	
σ	surface tension	

Subscripts

BM	mean value for noncondensing component
e	effective
g	gas phase
h	humid value, including gas and vapor
i	interface condition
j	for the jth component (usually less volatile)
L	liquid phase
L-only	for the liquid-phase flow alone
mac	macroscopic contribution
mic	microscopic contribution
o	at reference condition
p	at constant pressure
s	at saturation
sa	at adiabatic saturation condition
V	in the vapor phase
w	for water, or at wet-bulb temperature
Y	the mole ratio driving force
1,2	end points in the process
2ϕ	for two-phase flow

Superscripts

*	bulk concentration in liquid phase but in gas-phase units, or vice versa
'	mass rather than mole basis

BIBLIOGRAPHY

"Simultaneous Heat and Mass Transfer" in *ECT* 3rd ed., Vol. 21, pp. 54–76, by L. A. Wenzel, Lehigh University.

1. G. Ackerman, *Verh. Dtsch. Ing. Forschungsh.* **382**, 1 (1937).
2. A. P. Colburn and O. A. Hougen, *Ind. Eng. Chem.* **26**, 1178 (1934).
3. A. P. Colburn and T. B. Drew, *Trans. Am. Inst. Chem. Eng.* **33**, 197 (1937).
4. F. Stern and F. Votta, Jr., *AIChE J.* **14**, 928 (1968).

5. D. R. Coughanowr and E. O. Stensholt, *Ind. Eng. Chem. Proc. Des. Dev.* **3**, 369 (1964).
6. J. T. Schrodt, *Ind. Eng. Chem. Proc. Des. Dev.* **11**, 20 (1972).
7. J. T. Schrodt, *AIChE J.* **19**, 753 (1973).
8. R. L. Von Berg, *Recent Advances in Liquid–Liquid Extraction*, Pergamon Press, Oxford, U.K., 1971.
9. M. M. Dribika and O. C. Sandall, *Chem. Eng. Sci.* **34**, 733 (1979).
10. H. Hikita and K. Ishimi, *Chem. Eng. Com.* **3**, 547 (1979).
11. ABM Abdul Hye, *Simultaneous Heat and Mass Transfer from a Vertical, Isothermal Surface*, Ph.D. dissertation, University of Windsor, Canada, 1979.
12. T. Mizushina, M. Nakajima, and T. Oshima, *Chem. Eng. Sci.* **13**, 7 (1960).
13. J. T. Schrodt and E. R. Gerhard, *Ind. Eng. Chem. Fund.* **7**, 281 (1968).
14. L. W. Florschuetz and A. R. Khan, *Fourth International Heat Transfer Conference*, Paris, France, 1970.
15. O. C. Sandall and M. M. Dribika, *Inst. Chem. Eng. Sym. Ser.* **56**, 2.5/1 (1979).
16. H. J. Barton and O. Trass, *Can. J. Chem. Eng.* **47**, 20 (1969).
17. L. E. Scriven, *Chem. Eng. Sci.* **10**, 1 (1959).
18. F. Marshall and L. L. Moresco, *Int. J. Heat Mass Transfer*, **20**, 1013 (1977).
19. R. A. W. Schock, *Int. J. Heat Mass Transfer*, **20**, 701 (1977).
20. D. L. Bennett and J. C. Chen, *AIChE J.* **26**, 454 (1980).
21. C. H. Bedingfield and T. B. Drew, *Ind. Eng. Chem.* **42**, 1164 (1950).
22. H. S. Mickley, *Chem. Eng. Prog.* **45**, 739 (1949).
23. S. L. Hensel and R. E. Treybal, *Chem. Eng. Prog.* **48**, 362 (1952).
24. J. Lichtenstein, *Trans. ASME*, **65**, 779 (1943).
25. R. L. Pigford and C. Pyle, *Ind. Eng. Chem.* **43**, 1649 (1951).
26. W. M. Simpson and T. K. Sherwood, *Refrig. Eng.* **52**, 535 (1946).
27. A. E. Surosky and B. F. Dodge, *Ind. Eng. Chem.* **42**, 1112 (1950).
28. F. P. West, W. D. Gilbert, and T. Shimizu, *Ind. Eng. Chem.* **44**, 2470 (1952).
29. J. Weisman and E. F. Bonilla, *Ind. Eng. Chem.* **42**, 1099 (1950).
30. F. Yoshida and T. Tanaka, *Ind. Eng. Chem.* **43**, 1467 (1951).
31. *Countercurrent Cooling Tower Performance*, J. F. Prichard Co., Kansas City, Mo., 1957.
32. *Technical Bulletins R-54-P-5, R-58-P-5*, Marley Co., Kansas City, Mo., 1957.
33. *Performance Curves*, Cooling Tower Institute, Houston, Tex., 1967.
34. H. Chilton, *Proc. IEE, Supply Sect.* **99**, 440 (1952).
35. J. R. Singham, *The Thermal Performance of Natural Draft Cooling Towers*, Imperial College of Science and Technology, Department of Mechanical Engineering, London, 1967.
36. H. B. Nottage, *ASHRAE Trans.* **47**, 429 (1941).
37. I. A. Furzer, *Ind. Eng. Chem. Proc. Des. Dev.* **7**, 555 (1968).
38. S. M. Dalton, D. V. Giovanni, J. S. Maulhetsch, G. T. Preston, and K. E. Yeager, in R. A. Meyers, ed., *Handbook of Energy Technology and Economics*, John Wiley and Sons, Inc., New York, 1983.
39. C. Waselkow, *National Conference on Thermal Pollution*, Federal Water Pollution Control Administration and Vanderbilt University, Nashville, Tenn., 1968.
40. W. R. Shade and A. F. Smith, in Ref. 39.
41. B. Berg and T. E. Larson, *Am. Power Conf.* **35**, 678 (1963).
42. K. G. Baker, *Chem. Proc. Eng.* **56** (Jan. 1967).
43. C. H. Hosler, in *Cooling Towers*, CEP Technical Manual, American Institute of Chemical Engineers, New York, 1972, p. 27.
44. D. B. Hoult, J. A. Fay, and L. J. Forney, *A Theory of Plume Rise Compared with Field Observations*, Fluid Mechanics Laboratory Publication No. 68-2, Massachusetts Institute of Technology, Department of Mechanical Engineering, Cambridge, Mass., 1968.

45. C. T. Karanoudis, Z. B. Maroulis, and D. Marinos-Kouris, *Chem. Eng. Res. Dev.* **72**, 307 (1994).
46. S. Simal, C. Rossello, and A. Berna, *Chem. Eng. Sci.* **49**(22), 3739 (1994).
47. M. C. Robbins and M. N. Ozisik, *Can. J. Chem. Eng.* **69**, 1262 (1991).
48. P. L. Douglas, J. A. T. Jones, and S. K. Mullick, *Chem. Eng. Res. Dev.* **72**, 325, 332, 341 (1994).
49. J. Chirife, *Adv. Drying* **2**, 73 (1983).
50. M. Mazzotti, G. Storti, and M. Morbidelli, *AIChE J.* **40**, 1825 (1994).
51. S. Joshi and J. R. Fair, *Ind. Eng. Chem. Res.* **27**, 2078 (1988).
52. B. A. Finlayson and L. C. Young, *AIChE J.* **25**, 192 (1979).
53. K. Zygourakis and R. Aris, *Chem. Eng. Sci.* **38**, 733 (1983).
54. D. C. Shallcross and S. L. Low, *Chem. Eng. Res. Dev.* **72**, 763 (1994).

General References

Cooling Towers, CEP Technical Manual, American Institute of Chemical Engineering, New York, 1972.
R. E. Treybal, *Mass Transfer Operations*, 2nd ed., McGraw-Hill Book Co., Inc., New York, 1968, pp. 176–220.
A. S. Foust and co-workers, *Principles of Unit Operations*, 2nd ed., John Wiley & Sons, Inc., New York, 1980, Chapt. 17, pp. 420–453.
D. W. Green and J. O. Maloney, *Perry's Chemical Engineers Handbook*, 6th ed., McGraw Hill Book Co., Inc., New York, 1984, pp. 12–24.
E. J. Hoffman, *AIChE J.* **17**, 741 (1971).
A. S. H. Jernqvist, *Br. Chem. Eng.* **11**, 1205 (1966).
F. Kayihan, O. C. Sandall, and D. A. Mellichamp, *Chem. Eng. Sci.* **30**, 1333 (1975).
W. R. Lindberg and R. D. Haberstroh, *AIChE J.* **18**, 243 (1972).
J. L. Manganaro and O. T. Hanna, *AIChE J.* **16**, 204 (1970).
G. L. Standart, R. Taylor, and R. Krishna, *Chem. Eng. Commun.* **3**, 277 (1979).

LEONARD A. WENZEL
Lehigh University

SINGLE CRYSTALS, GROWTH.   See ZONE REFINING.

SIZE ENLARGEMENT

Size enlargement concerns those processes that bring together fine powders into larger masses in order to improve the powder properties. Usually, they produce relatively permanent entities in which the original particles can still be identified, but this is not always necessary, as demonstrated in the formation of amorphous agglomerates by the cooling of melts, or by the production of weak and transient "instant" food agglomerates in which product strength need only be sufficient to withstand downstream handling, packaging, and transportation.

Size enlargement methods have been known for hundreds of years. The roots of the processes can be traced to such ancient techniques as the formation of clay bricks and other building materials, the hammering of implements from sponge iron, the manufacture of various items from precious metal powder, and the preparation of solid molded forms of medicinal agents (1). Agglomeration by heating and roasting techniques became established as a practical operation during the nineteenth century with the need to beneficiate and process fine coals and ores. In this century use of size enlargement has grown rapidly for a number of reasons. The application of high analysis nitrogen fertilizers in intensive agriculture led to development of noncaking and granulated, rather than powdered, products (2). The lower quality of available iron ore resulted in the need to upgrade the resource by grinding and rejection of the liberated impurities, followed by pelletization of the resulting fines into an acceptably coarse product (3,4). Environmental considerations have led to the recovery of many dusts and fine waste powders that can be recycled after size enlargement (5). In addition, modern high volume processing requires consistent feeds with good flow properties, requirements that for powders can often only be met through some form of agglomeration.

Many diverse industries benefit from the use of size enlargement, ranging from the high value, relatively low volume requirements of pharmaceutical manufacturers to the tonnage requirements of the fertilizer and minerals processing industries. Benefits gained from size enlargement are as diverse as the industries in which the operation is used (1,6–8). Dusting losses and caking and lump formation are avoided with improved product appearance through the granulation of fertilizers. The wet granulation of pharmaceutical powders produces nonsegregating powder blends with consistent flow properties. The granulation then feeds high capacity tableting devices which yield tablets of a defined and consistent dosage. Objects with useful structural forms and shapes are produced in powder metallurgy and ceramic forming. The pelleting of carbon black increases bulk density and improves handling qualities. Control of powder properties such as solubility, porosity, and heat-transfer capability can be attained through agglomeration procedures. Agglomeration in liquid suspension, such as the selective oil agglomeration of coal in water suspension, not only recovers the fine particles from the liquid, but does so selectively in that unagglomerated impurity particles remain in suspension (see CARBON, CARBON BLACK; CERAMICS; COAL CONVERSION PROCESSES, CLEANING AND DESULFURIZATION; COFFEE; DRYING; FERTILIZERS; IRON; METALLURGY, POWDER METALLURGY; PHARMACEUTICALS; PLASTICS PROCESSING).

Particle-Bonding Mechanisms

The mechanisms by which particles bond together and grow into agglomerates are affected by the specific size enlargement method. Nevertheless there are certain aspects of the bonding process that are essentially independent of the equipment and method. These aspects are described herein in general terms, followed by more detailed examination of the various size enlargement processes (see COATING PROCESSES, POWDER TECHNOLOGY).

A classification of bonding mechanisms based on the fundamental nature of the interparticle bonds has been widely adopted in the literature and will be used here, together with the theoretical model used to estimate agglomerate strength (9). Rumpf's classification into five categories of particle–particle bridging is summarized in Table 1 together with references from the literature to examples in each category. More than one bonding mechanism is likely to act in a given process. For example, in bonding by tar deposited through solvent evaporation, it is likely that oxidative hardening will also occur. In sintering ores,

Table 1. Classification of Binding Mechanisms According to Rumpf[a]

Class	Mechanism	Representative examples	Refs.
solid bridges	sintering, heat hardening	induration of iron-ore pellets	3,4
		sintering of compacts in powder metallurgy	10
	chemical reaction, hardening binders, "curing"	cement binder for flue-dust pellets	11
		ammoniation–granulation of mixed fertilizers	12
		oxidation of tar binders	13
	incipient melting owing to pressure, friction	briquetting of metals, plastics	14,15
	deposition through drying	crystallization of salts in fertilizer granulation	16
		deposition of colloidal bentonite in dry iron-ore balls	3,4
immobile liquids	viscous binders, adhesives	sugars, glues, gums in pharmaceutical tablets	17
	adsorption layers	instantizing food powders by steam condensation	18
		humidity effects in flow of fine powders	
mobile liquids	liquid bridges (pendular state)	flocculation of fine particles in liquid suspension by immiscible liquid wetting	8
		moistening–mixing of iron-ore sinter mix	3,4
	void space filled or partly filled with liquid (capillary and funicular states)	balling (wet pelletization of ores)	3,4
		soft plastic forming of ceramic powders	
intermolecular and long-range forces	van der Waals, electrostatic, and magnetic	adhesion of fine powders during storage, flow, and handling	19
		spontaneous dry pelletization of fine powders, eg, carbon black, zinc oxide	
mechanical interlocking	shape-related bonding	fracturing and deformation of particles under pressure	20
		fibrous particles, eg, peat moss	21

[a]Refs. 8 and 9.

bonding through chemical reaction also contributes to strength. With very fine powders it is difficult to determine whether bonding through long-range forces or vapor adsorption predominates. Although mechanical interlocking of particles influences agglomerate strength, its contribution is generally considered small in comparison to other mechanisms.

Theoretical Strength of Agglomerates. Based on statistical-geometrical considerations, Rumpf developed the following equation for the mean tensile strength of an agglomerate in which bonds are localized at the points of particle contact (9):

$$\sigma_T \approx \frac{9}{8} \left(\frac{1 - \epsilon}{\epsilon} \right) \frac{H}{d^2} \tag{1}$$

where σ_T is the mean tensile strength per unit section area, Pa; ϵ is the void fraction of the agglomerate; d is the diameter of the (assumed) monosized spherical particles in m; and H is the tensile strength in N, of a single particle–particle bond. To convert Pa to psi, multiply by 0.145×10^{-3}.

In a second main class of agglomerate bonding, particles are embedded in or surrounded by an essentially continuous matrix of binding material rather than having bonding localized at points of particle contact. An important example of this second type of binding is the case in which particles are held together by mobile liquids where adhesion occurs as a result of interfacial tension at the liquid surface and the pressure deficiency (capillary suction) created within the liquid phase by curvature at the liquid surface (22). The various regimes of low viscosity liquid which can exist in an agglomerate are shown in Figure 1.

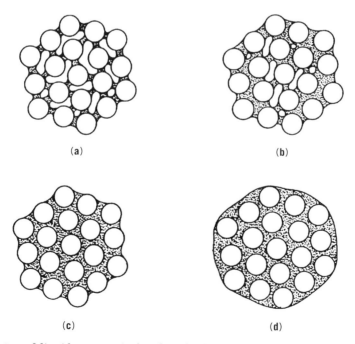

Fig. 1. States of liquid content in bonding by low viscosity liquids: (**a**) pendular, (**b**) funicular, (**c**) capillary, and (**d**) particles surrounded by liquid droplet (8).

In the pendular state, shown in Figure 1a, particles are held together by discrete lens-shaped rings at the points of contact or near-contact. For two uniformly sized spherical particles, the adhesive force in the pendular state for a wetting liquid (contact angle zero degrees) can be calculated (19,23) and substituted for H in equation 1 to yield the following, where γ is the liquid surface tension in N/m.

$$\sigma_T = \frac{9}{4}\left(\frac{1-\epsilon}{\epsilon}\right)\frac{\gamma}{d} \tag{2}$$

When the void space in an agglomerate is completely filled with a liquid (Fig. 1c), the capillary state of wetting is reached, and the tensile strength of the wet particle matrix arises from the pressure deficiency in the liquid network owing to the concave liquid interfaces at the agglomerate surface. This pressure deficiency can be calculated from the Laplace equation for circular capillaries to yield, for liquids which completely wet the particles:

$$\sigma_T = C\left(\frac{1-\epsilon}{\epsilon}\right)\frac{\gamma}{d} \tag{3}$$

where the parameter C has a theoretical value of 6 for uniform spherical particles and ranges between 6.5–8 for nonuniform particles (22,24).

It is evident, by comparing equations 2 and 3, that tensile strength in the pendular state is about one-third that in the capillary state. Intermediate liquid contents in the funicular state (Fig. 1b) yield intermediate values that can be approximated as follows:

$$\sigma_T = sC\left(\frac{1-\epsilon}{\epsilon}\right)\frac{\gamma}{d} \tag{4}$$

where s is the fractional filling of the agglomerate voids. This classical approach to predicting the strength of wet agglomerates has been reviewed (25) and some systems for which this approach needs to be modified are provided.

This discussion of agglomerate strength applies to established and static agglomerates. It has been demonstrated (26) that the viscosity of the binder liquid plays a key role during agglomerate formation, where it provides strength under dynamic strain, such as would occur with intensive mixing. It has been found that under industrially relevant conditions, the strength of the dynamic bridge exceeds the static strength by more than an order of magnitude.

Measuring and Correlating Agglomerate Strength. In practice, simple and quick test methods are most often used to assess the quality of bonding and other desirable properties of product agglomerates. Compression, impact, abrasion, and other types of tests are widely accepted in industry to characterize agglomerate quality in relation to subsequent handling and processing. The mode of agglomerate failure is complex in these tests, making it difficult to relate the results directly to theory.

Compression tests, in which agglomerates are crushed between parallel platens, are probably most universal. To obtain reproducible and accurate results, the rate of loading and method of load application must be strictly controlled. A variety of commercial testers are available to allow this needed control over the compression process. Several means of distributing the load uniformly at the point of contact are used, including covering the platen surface with compressible board. With pliable agglomerates especially, the fracture load is highly dependent on the rate of loading (27,28).

During compression of spherical agglomerates, flattening takes place at the points of contact as particle–particle bonds fail locally and particles are driven into adjoining voids. Small, dense, wedge-like elements form in the regions adjacent to the platens and the agglomerate fails in tension along a circumferential crack joining the poles of loading (double-cone failure) (28,29). Internal frictional effects must be overcome in this compression process, in addition to the tensile strength of the particle matrix given by equations 1–4. For corresponding tensile and compression measurements on wet limestone pellets in the range for which the mean tensile strength is $\sigma_T = 20–83$ kPa (3–12 psi), Rumpf found the ratio of tensile to compressive strength to be $\sigma_T/\sigma_C \approx 0.5–0.77$ (9).

For approximately spherical agglomerates, compression strength is calculated as follows:

$$\sigma_C = L/(\pi D^2/4) \tag{5}$$

where L is the compression force at failure in N, and D is the agglomerate diameter in m. Some typical compressive strengths of various agglomerates are indicated in Figure 2.

According to equation 5, compression test data can usually be correlated through a log–log plot of load at failure against pellet diameter to yield a straight line whose slope is often, but not always, equal to 2 (30,31). The intercept of such a plot is related to the compressive strength, σ_C. Because ultimate failure occurs in tension as explained above for the double-cone failure mechanism, this compressive strength factor can subsequently be correlated with a bonding mechanism through equations 1–4 to account for the effect of such parameters as particle diameter, agglomerate porosity, and the strength of particle–particle bonding (22,32). Alternatively, empirical relationships can be developed to account for the effects of these parameters describing the particle matrix (33,34).

Several other tests of agglomerate quality are done routinely, the details depending on the practice accepted in a specific industry. For iron ore, a drop test is used on wet agglomerates as a measure of their ability to withstand handling up to the point in the process at which they are dried and fired (3). A test might consist of dropping a number of agglomerates from a height of 300 or 450 mm onto a steel plate. The average number of drops required to cause fracture is the drop number. In the ASTM tumbler test for iron-ore pellets (E279) an 11.3-kg sample in the size range 6.4–38.1 mm is placed in a tumbler drum ca 910 mm diameter by 460 mm long and rotated at 24 rpm for a total of 200 revolutions. The drum is equipped with two equally spaced

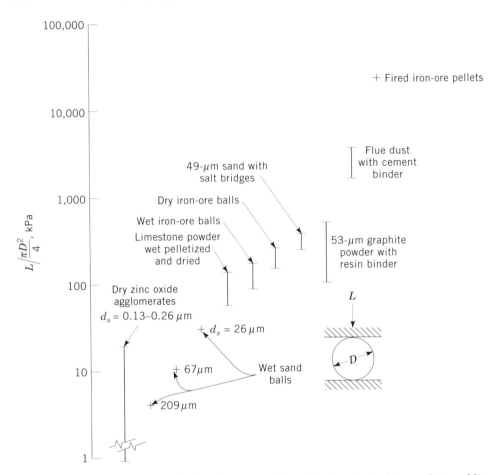

Fig. 2. Compression strength of agglomerates formed into spherical shape by tumbling where d_s = mean disperse particle diameter (8). To convert kPa to psi, multiply by 0.145.

lifters 51 mm high. The abrasion index is given by the weight percentage of plus 6.4 mm material surviving the test and the dust index by the yield of minus 0.6 mm (30 mesh) material. In addition to the tumble drum attrition test, there is wide industry use of screen abrasion tests (35). Preweighed samples of rock salt or potassium granules, for example, are shaken on sieve shakers, with the abraded fines recorded as abrasion index. This can be done with or without the addition of steel ball grinding media in the sieve shaker. Other strength tests such as the Linder rotating-furnace procedure are used with iron-ore pellets in an attempt to determine their reducibility and breakdown under reducing conditions simulating those of a blast furnace (36).

Laboratory tests to assess the caking tendency of fertilizer granules consist of two parts (37). The first entails cake formation in a compression chamber under controlled conditions of air flow, humidity, temperature, etc. In the second part the cake is removed and its crushing strength determined as a measure of the degree of caking. In the tablet disintegration test outlined in USP XX, pharmaceutical tablets are contained in a basket-rack assembly which is immersed in a suitable dissolution fluid in a container and raised and lowered to agitate the tablets at a constant rate of 29–32 cycles per minute for a specified period of time (38). Acceptable tablets disintegrate completely by the end of the test period.

Types of Size Enlargement

Classification of size enlargement methods reveals two distinct categories (8,39). The first is forming-type processes in which the shape, dimensions, composition, and density of the individual larger pieces formed from finely divided materials are of importance. The second is those processes in which creation of a coarse granular material from fines is the objective, and the characteristics of the individual agglomerates are important only in their effect on the properties of the bulk granular product.

    Four principal mechanisms are used to bring fine particles together into larger agglomerates: agglomeration by tumbling and other agitation methods; pressure compaction and extrusion methods; heat reaction, fusion, and drying methods; and agglomeration from liquid suspensions.

Agglomeration by Tumbling and Other Agitation Methods

When fine particles, usually in a moist state, are brought into intimate contact through agitation, binding forces come into action to hold the particles together as an agglomerate. Capillary binding forces caused by wetting with water or aqueous solutions is the most common binding mechanism, but others such as intermolecular forces developed in extremely fine dry powders may also be used to form the agglomerates. Several different forms of agitation may be used. The rolling cascading action of disk and drum devices produces rounded or roughly spherical agglomerates; other types of mixers, described below, generally yield more irregular shapes.

    Agglomerate growth can occur through a number of mechanisms, such as coalescence, crushing and layering, layering of fines, and abrasion transfer (40–42). More than one mechanism may occur simultaneously in a given process. Agglomerates of the order of 1–3 mm diameter formed by coalescence of fine feed particles, or recycled undersize product agglomerates act as seed nuclei for the process. The nuclei grow to larger sizes by the addition of layers of fines supplied continuously to the process.

    To survive and grow in an agitated system, agglomerates must be able to withstand the destructive forces generated by the moving charge of powder. Equation 3 indicates that for given agitation conditions there is a maximum particle size that can produce sufficient tensile strength in the particle matrix to form agglomerates satisfactorily. The addition of fines to a given size distribution of feed material usually improves agglomeration not only because it reduces the mean particle size, but also because it improves packing and reduces voids in the system. Similarly, more intensive agitation conditions generally reduce the size of agglomerates produced from a given material. For wet agglomeration on disk pelletizers top feed size is usually 300–600 μm (ca 30–50 mesh) and at least 25% of the feed powder should be finer than 75 μm (ca 200 mesh) (43). In iron-ore balling, a grind with 40–80% of the material below 44 μm (325 mesh) is normally used.

Agglomerates formed by wet pelletization in balling drums and disks are generally considered to have their internal pores saturated with binding liquid, that is, to be in the capillary state of wetting (see Fig. 1c) (41,44). The weight fraction for the theoretical liquid content of such agglomerates is given by equation 6 where W is the weight fraction of liquid (wet basis); ϵ is the void

$$W = \frac{\epsilon \rho_L}{\epsilon \rho_L + (1 - \epsilon)\rho_S} \tag{6}$$

fraction in the agglomerates; and ρ_L, ρ_S are liquid and particle densities, respectively.

Equation 6 has been fitted to a wide variety of literature data in which solid density ranged from about 1–6 g/cm^3 and liquid densities were generally close to 1 g/cm^3 (44). The following relationships were found.

For average feed-particle diameters <30 μm:

$$W = \frac{1}{1 + 1.85\frac{\rho_S}{\rho_L}} \tag{7}$$

For average feed-particle diameters >30 μm:

$$W = \frac{1}{1 + 2.17\frac{\rho_S}{\rho_L}} \tag{8}$$

These relationships predict the binding liquid content for wet agglomeration with an accuracy of only ca 30%. The liquid content required to agglomerate a particular feed material depends, for example, on the interfacial properties of the system (45). Typical values of moisture content required for balling a variety of materials are listed in Table 2. Very accurate information on the optimum liquid content to agglomerate a particular feed material must be obtained from experimental tests.

Drum and Inclined-Disk Agglomerators.   Although a wide variety of agitation equipment is used industrially to produce agglomerates, rotary drums or cylinders and inclined disks or pans are the most important equipment in terms of tonnages produced. Fertilizer granulation and iron-ore balling represent two of the largest applications (see FERTILIZERS; IRON).

Drum agglomerators (Fig. 3) consist of an inclined rotary cylinder powered by a fixed- or variable-speed drive. Feed material, containing the correct amount of liquid phase, agglomerates under the rolling, tumbling action of the rotating drum. The pitch of the drum (up to 10° from the horizontal) assists material transport down the length of the cylinder. A retaining ring is often fitted at the feed end to prevent spillback of feed. A dam ring may also be used at the exit to increase the depth of material and residence time in the drum. Liquid phase may be introduced either before or immediately after the solids enter the cylinder. With iron ores, the feed is usually premoistened wet filter cake, but

Table 2. Moisture Requirements for Balling a Variety of Materials[a]

Raw material	Approximate size analysis of raw material, less than indicated μm (mesh)[b]	Moisture content of balled product, wt % H_2O
precipitated calcium carbonate	75 (200)	29.5–32.1
hydrated lime	45 (325)	25.7–26.6
pulverized coal	300 (48)	20.8–22.1
calcined ammonium metavandate	75 (200)	20.9–21.8
lead–zinc concentrate	850 (20)	6.9–7.2
iron pyrite calcine	150 (100)	12.2–12.8
specular hematite concentrate	106 (150)	9.4–9.9
taconite concentrate	106 (150)	9.2–10.1
magnetic concentrate	45 (325)	9.8–10.2
direct shipping open-pit ores	1680 (10)	10.3–10.9
underground iron ore	6.4 mm	10.4–10.7
basic oxygen-converter fume	1	9.2–9.6
raw cement meal	106 (150)	13.0–13.9
utilities–fly ash	106 (150)	24.9–25.8
fly ash–sewage sludge composite	106 (150)	25.7–27.1
fly ash–clay slurry composite	106 (150)	22.4–24.9
coal–limestone composite	150 (100)	21.3–22.8
coal–iron-ore composite	300 (48)	12.8–13.9
iron-ore–limestone composite	150 (100)	9.7–10.9
coal–iron-ore–limestone composite	1180 (14)	13.3–14.8

[a]Courtesy of Dravo Engineers and Constructors.
[b]Tyler equivalent scale.

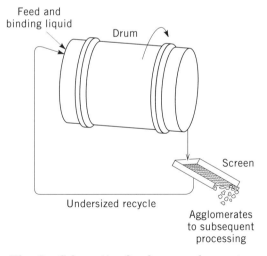

Fig. 3. Schematic of a drum agglomerator.

water sprays may also be located inside the drum for moisture addition to aid control. In mixed fertilizer granulation, various solutions and slurries may be used, but water is the usual wetting agent. Various types of internal scrapers are in use to limit buildup of material on the inside surface and to provide a uniform layer to promote the correct rolling, tumbling action in the drum. Rubber flaps and liners as well as external knockers are used to limit buildup in fertilizer processing.

The drums used for fertilizer granulation range from 1.5 m in diameter by 3 m long, with installed power of 11 kW and rotation rates of 10–17 rpm that process 8 t/h, up to 3 m diameter by 6 m long, with installed power of 112 kW, rotation rates of 7–12 rpm, and a capacity of 51 t/h. These capacities exclude recycle so that the actual throughput of the drums may be much higher. Capacity depends primarily on the quantity of undersize and crushed oversize material recycled to the drum during continuous operation. The recycle ratio (amount recycled/output) varies from <1 (>50% output) to 5 or 6 for hard-to-granulate grades of fertilizer. The grade being granulated, differing formulations for the same grade, differing plant operations, ambient temperatures, and skill all influence capacity in fertilizer granulation (46).

A typical drum used to ball iron ore is 3 m in diameter by 9.5 m long, is driven at 12–14 rpm by a 45-kW drive, and produces 60 t/h product (3,4,47).

An inclined-disk agglomerator consists of a tilted rotating plate equipped with a rim to contain the agglomerating charge (Fig. 4). Solids are fed continuously from above or from the front onto the central part of the disk, and product agglomerates discharge over the rim. Moisture or other binding agents can be sprayed on at various locations on the plate surface. Adjustable scrapers and plows maintain a uniform protective layer of product over the disk surface and also control the flow pattern of material on the disk. Plate angle can be adjusted from 40–70° to the horizontal to obtain the best results, and both constant-speed and variable-speed motors are available as disk drives. Dust covers can be fitted when required. Characteristics of some of the range of inclined disks offered by one manufacturer are given in Table 3.

Operation and control of tumbling agglomerators is affected primarily by the character of the feed powder (size distribution, solubility), by its optimum liquid content for agglomeration, and by retention time in the device (3,48,49). An important parameter for drum agglomeration is the amount of undersize product

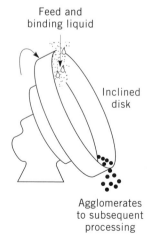

Feed and
binding liquid

Inclined
disk

Agglomerates
to subsequent
processing

Fig. 4.   Schematic of an inclined-disk agglomerator.

Table 3. Characteristics of a Range of Inclined Disks[a]

Diameter, m	Depth, cm	Motor, kW	Speed, rpm	Approximate capacity[b], t/h
0.41	8.9	0.19	12–36 variable	
0.91	20.3	0.75	9–27	0.3
1.40	22.9	2.2	6.7–20.2	0.9
1.83	27.9	3.7	8.1–16.2	1.8
2.44	33.0	11.2	7.5–15	4.1
3.05	39.4	18.6	12.8 fixed	6.4
3.66	44.5	29.8	11.9	11
4.27	49.5	44.7	11.3	15
4.88	55.9	55.9	10.7	24
5.49	61.0	74.6	10.4	32
6.10	66.0	93.3	10.0	40
7.01	76.2	111.9	8.0	53
7.62	76.2	149.2	6.0	63

[a]Courtesy of Feeco International, Inc., 1980.
[b]Based on dry dust at 961 kg/m^3. Capacity depends on type of material and desired product (rates are average for nominal 1.3-cm pellets).

that must be recycled. A disk agglomerator requires experimental adjustment in angle of inclination, speed of rotation, and position of spray(s) and scraper(s). Temperature of operation can be a significant variable for soluble ingredients.

An important feature of the inclined-disk agglomerator is its size separating ability. Feed particles and the smaller agglomerates sift down to the bottom of the tumbling load where, because of their high coefficient of friction, they are carried to the highest part of the disk before rolling downward in an even stream. Larger agglomerates remain closer to the top of the bed, where they travel shorter paths. In continuous operation the largest agglomerates are discharged from the top of the bed over the rim, while smaller ones and feed fines are retained for further growth. Because rotary drums do not possess the inherent classifying action of the inclined disk, agglomerates of a wide size distribution are discharged. Drums are therefore operated in closed circuit with screens to recycle the undersized (and crushed oversized, if present) material.

The inherent classifying action of inclined disks offers an advantage in applications that require accurate agglomerate sizing. Other advantages claimed for the inclined disk include less space requirements and lower cost than drums, as well as sensitivity to operating controls and easy observation of the agglomeration process. These latter features lend versatility in agglomerating a wide variety of materials of different degrees of ease of agglomeration.

Advantages claimed for the drum compared with the disk agglomerator are greater capacity, longer residence time for difficult materials, and less sensitivity to upsets owing to the damping effect of a larger recirculating load. Dusty materials and simultaneous processing steps (chemical reaction or drying during agglomeration) can be handled more easily in a drum.

Many variations on the design of the basic inclined-disk and drum agglomerators are in use. A well-known addition to the inclined disk is a separate reroll ring beyond the rim of the main disk. Normally disks are relatively shallow in depth with a rim height ca 20% of the disk diameter, but deeper configurations are also in use, including those with multistepped sidewalls, deep pan or deep drum designs and a cone pelletizer. Drum agglomerators may incorporate internal baffles (50), lifting blades, or independent paddle shafts in certain applications (51).

Mixer Agglomerators. Mixers are used by various industries in which, by contrast with drum or disk equipment, internal agitators of several designs provide a positive rubbing and shearing action to accomplish both mixing and size enlargement. Horizontal pan mixers, pugmills, and other types of intensive agitation devices are used. The positive cutting-out action of such equipment can handle plastic and sticky powder feeds and its kneading action is claimed to produce denser and stronger granules than the tumbling methods although agglomerates of more irregular shape usually result (see MIXING AND BLENDING).

Pug mixers (blungers, pugmills, paddle mixers) such as that shown in Figure 5 have been widely used in the granulation of fertilizer materials and for the mixing, moistening, and microagglomeration of sinter-strand feed in both the ferrous and nonferrous metallurgical industries. These mixers consist of a horizontal trough with a rotating shaft to which mixing blades or paddles of various designs may be attached. The vessel may be of a single-trough design although a double-trough arrangement is most popular. Twin shafts rotate in opposite directions throwing the materials forward and to the center as the pitched blades on the shaft pass through the charge. Construction is robust, with the body of heavy plate (6.4 or 9.5 mm thick) and hardened agitators or tip inserts. Operational features include fume hoods, spray systems, and stainless steel construction. Provision can be made to feed materials at different points

Fig. 5. Double-shaft mixer used in fertilizer granulation. Courtesy of Renneburg Division of Heyl & Patterson.

along the mixer as well as at the end to ensure that the entire mixing length is used and to add processing versatility. Capacities of these mixers cover a broad range, from typical levels of 20–30 up to several hundred tons per hour.

An intensive countercurrent pan mixer can be used to homogenize feed powders while adding binding liquid to help pregranulate extreme fines before pelletizing. This mixer consists of a rotating mixing pan and two eccentrically mounted mixing stars which operate in the direction opposite to the pan rotation. This equipment operates in a batch mode to handle typically about 30 t/h of fertilizer materials.

Shaft mixers operating at very high rotational speeds are also used to granulate extreme fines, such as clays and carbon black, which may be highly aerated when dry, and plastic or sticky when wet. These machines are generally single-shaft devices in which the paddles are replaced by a series of pins, pegs, or blades. The peg granulator (52) used to agglomerate ceramic clays in the china clay industry and the pinmixer (53) used in the wet pelleting of carbon black and pharmaceutical powders (54) are representative examples. As shown in Figure 6, these mixers consist of a cylindrical shell within which rotates a shaft carrying a multitude of cylindrical rods (pegs or pins) arranged in a helix. The shaft rotates at speeds critical to machine performance. Wet feed or dry feed that is immediately moistened enters the machine at one end and emerges as pellets at the opposite end.

For intensive mixing and agglomerating at high speeds and very short retention times, the vertical plow mixer, also known as the Schugi mixer, can be used (55,56). These high speed mechanical vertical plow-type agglomerators produce agglomerates in the 200 and 2000 μm size range, at typical retention times of 1–2 s.

Powder Clustering. Many applications of size enlargement require only relatively weak, small, cluster-type agglomerates to improve behavior of the

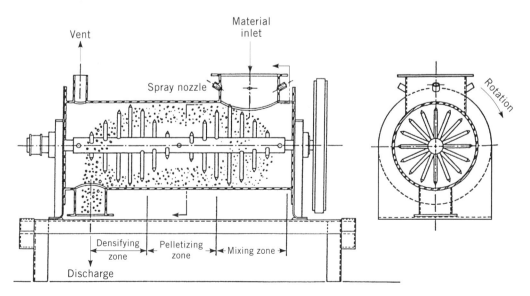

Fig. 6. Pinmixer used in wet-pelleting carbon black (53).

original powder in flow, wetting, dispersion, or dissolution. Tableting feeds in pharmaceutical manufacture, detergent powders, and "instant" food products are examples. In these cases, agglomeration is accomplished by superficially wetting the feed powder, often with less than 5% of bridging liquid in the form of a spray, steam, mist, etc. The wetting is carried out in a relatively dry state in standard or specialized powder mixers in which the mass becomes moist rather than wet or pasty. Equipment used (57) includes sigma-blade and heavy-duty planetary mixers (33), horizontal cylindrical vessels containing mixing and chopping tools (58), and rotary drums of special design (59) to form a constant-density falling curtain along the drum axis into which dispersed binding liquid is directed to form small agglomerates.

Continuous-flow mixing systems are commonly used in the agglomeration of powdered food products. The instantizer agglomerator in Figure 7 is representative of these systems. In Figure 7 the feed powder is introduced to the moistening zone by a pneumatic conveyor and rotary valve. The dry powder falls in a narrow stream between two jet tubes which inject the agglomerating fluid in a highly dispersed state. Steam, water, solvents, or a combination of these are used. Air at ambient temperature is also introduced through radial wall slots in the moistening chamber to induce a vortex motion. Control of this air flow controls the flow pattern and particle temperature. The reduced temperature serves to condense fluid onto the particles while the vortex motion induces particle collisions. Clustered material then drops through an air-heated chamber onto a conditioning conveyor where sufficient time is allowed to reach a uniform moisture distribution. The material then passes to an afterdryer, cooler, and sifter, and is finally bagged (see FOOD PROCESSING).

Pressure Compaction and Extrusion Methods

The compression techniques of size enlargement produce agglomeration by application of suitable forces to particulates held in a confined space. The various methods in use differ in both the means of pressure application and the method used to confine the powder. Unidirectional compaction in punch and die assemblies and in molding presses makes use of a reciprocating punch or ram acting on the particulates in a closed die or mold. In roll-pressing equipment, the particulate material is compacted by squeezing as it is carried into the gap between two rotating rolls. In extrusion systems the particulates undergo shearing and mixing as they are consolidated while being forced through a die or orifice under the action of a screw or roller.

Tableting, pressing, molding, and extrusion operations are commonly used to produce agglomerates of well-defined shape, dimensions, and uniformity in which the properties of each item are important and output is measured in pieces per hour (see CERAMICS, CERAMICS PROCESSING; PHARMACEUTICALS; METALLURGY, POWDER METALLURGY; PLASTICS PROCESSING).

The compaction process of void reduction may be considered to occur by two essentially independent mechanisms (60). The first is the filling of the holes of the same order of size as the original particles. This occurs as particles slide past one another; it is distinguished by the voids being filled with original particles that have undergone only slight size modification by fracture or by plastic

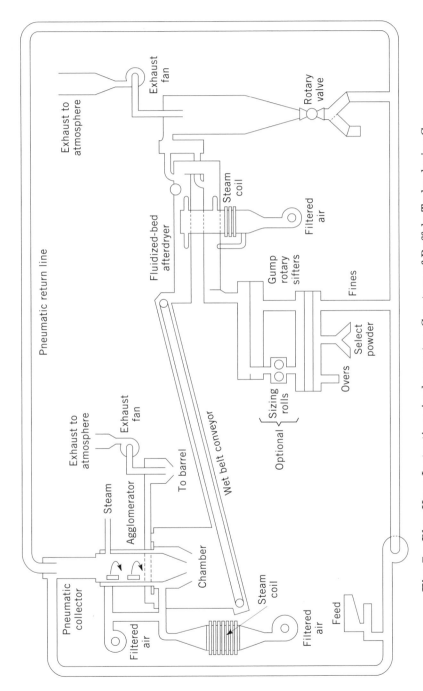

Fig. 7. Blaw-Knox Instantizer-Agglomerator. Courtesy of Buffalo Technologies Corp.

deformation. The second process consists of the filling of voids that are substantially smaller than the original particles by plastic flow or by fragmentation. Many quantitative relationships have been suggested to represent this compaction process, usually in cylindrical die cavities (61–65). A successful theoretical design analysis for roll presses has been developed using small-scale laboratory measurements of the flow and compression properties of the feed powder (66–69).

The success of the compaction operation depends partly on the effective utilization and transmission of applied forces and partly on the physical properties and condition of the mixture being compressed. Friction at the die surface opposes the transmission of the applied pressure in this region, results in unequal distribution of forces within the compact, and hence leads to density and strength maldistribution within the agglomerate (70). Lubricants, both external ones applied to the mold surfaces and internal ones mixed with the powder, are often used to reduce undesirable friction effects (71). For strong compacts, external lubricants are preferable as they do not interfere with the optimum cohesion of clean particulate surfaces. Binder materials may be used to improve strength and also to act as lubricants.

Compacting Presses. In the automotive industry and other metalworking industries, coarse scrap-metal particulates are compressed and recycled to melting operations through piston-type briquetting presses (72,73). Feed materials are typically cast-iron and steel borings or turnings, which tend to bond under pressure at least partially by mechanical interlocking. A reciprocating hydraulic press working on such materials might use a 56-kW hydraulics actuating pump with a rating of 318 t on a 12.7-cm diameter die. Briquettes of cylindrical shape 7.6 cm in length are produced at ca 3–4 t/h. Such compacting presses are not suited to larger tonnages when a small briquette is required. Their reciprocating nature is a disadvantage since this produces nonuniform loads on the drive motors.

Roll Briquetting and Compacting Machines. In roll presses, particulate material is compacted by squeezing as it is carried into the gap between two rolls rotating at equal speed. This is probably the most versatile method of size enlargement because most materials can be agglomerated by this technique with the aid of binders, heat, and/or very high pressures if needed (7,74). The method generally requires less binder and therefore there is little or no requirement for drying the agglomerates. Simplified briquetting and compaction flow diagrams are shown in Figure 8. In briquetting machines, pillow shapes are formed by matching indentations in the rolls. Special shapes, such as the well-known fertilizer spikes, are also made by these machines. Precise design of these pockets based on practical experience is important to ensure optimum briquette density, minimum incidental feather (fines) production, and dependable pocket release of finished briquettes. In compaction machines the agglomerated product is in sheet form as produced by smooth or corrugated rolls. The compacted product can remain in sheet form or can be granulated into the desired particle size on conventional size reduction equipment. Polishing of the particles with or without additional moisture can be done in a rotary drum to round the particles and prevent additional dusting during handling.

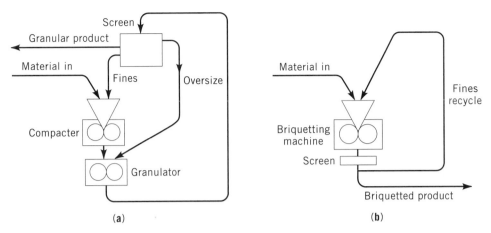

Fig. 8. Simplified compacting and briquetting systems: (**a**) compaction–granulation flow diagram and (**b**) briquetting flow diagram. Courtesy of Hosokawa Bepex Corp.

Roll presses consist of the frame, the two rolls that do the pressing, and the associated bearings, reduction gear, and fixed or variable-speed drive (Fig. 9). Spacers between bearing housings prevent roll contact and allow adjustment of roll spacing. The frame of the press is designed so that all forces are absorbed internally. The rolls are forced together by a hydraulic system which may incorporate a safety valve to prevent overpressure if foreign material intrudes between the roll faces. The rolls consist of a continuous roll shaft, the roll body, and attached molding equipment. The molding surface may be either solid or divided into segments. Segmented rolls are preferred for hot briquetting, as the thermal expansion of the equipment can be controlled more easily. Segmented rolls can be made from harder materials more resistant to wear than can one-piece rolls.

For fine powders that tend to bridge or stick and are of low bulk density, some form of forced feed, such as the tapered screw feeder shown in Figure 9, must be used to deaerate, precompact, and pressurize the feed into the nip. Large machines are available with up to five screw feeders to spread the flow across the rolls, and vacuum hoppers are also used to remove air when densifying low density feeds.

Capacity data for a variety of materials using a range of roll-press sizes are given in Table 4.

Pellet Mills. Pellet mills differ from roll briquetting and compacting machines in that the particulates are compressed and formed into agglomerates by extrusion through a die rather than by squeezing as they are carried into the nip between two rolls. Several types of equipment that use the extrusion principle are available. The die may be a horizontal perforated plate with rollers acting on its upper surface to press material through the plate. Rolls may be either side-by-side with material extruded through one or both of the rolls, or one or more small rolls may be fitted inside a larger die roll. In yet another design, two intermeshed gears are used and material is extruded through die holes located in the gear root.

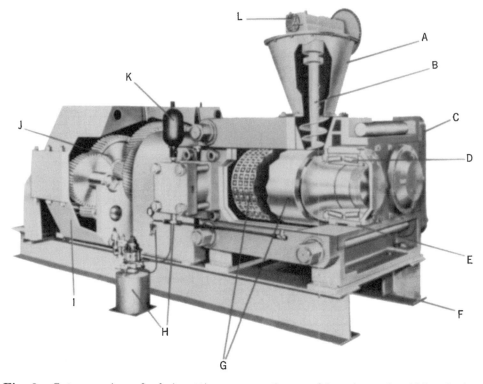

Fig. 9. Cutaway view of a briquetting–compacting machine. A, predensifying feeder; B, feeder screw; C, machine housing; D, antifriction bearing; E, machined bearing block; F, base frame; G, pocketed or corrugated rolls; H, hydraulic system; I, speed reducer; J, gears; K, hydraulic accumulator; and L, feeder drive. Courtesy of Hosokawa Bepex Corp.

The action of the roller and die assembly to produce a shearing and mixing action yields a plastic mix to be pushed through the die. Binders, plasticizing agents, and lubricants may be used to facilitate the process. Probably the most popular design of pellet mills in use is that shown in Figure 10, which utilizes a ring-type die and two or three rollers mounted in a vertical plane. Power is applied to the die to rotate it around the roller assembly, which has a fixed axis. Capacities range from <1 to ≤75 t/h operating from 75–300 rpm with a drive power of 7.5–592 kW. Dies with hole sizes from 2 or 3 to 30 mm are used. Capacities vary greatly depending on speed, hole and die size, moisture content, and other characteristics of the material being pelleted. Scores of materials can be pelleted, from catalysts and carbon materials to rubber crumb, wood pulp and bark, compound animal feeds, and many chemicals.

Heat Reaction, Fusion, and Drying Methods

These methods of size enlargement depend on heat transfer to accomplish particle bonding. Heat may be transferred to the particle agglomerates, as in the drying of a concentrated slurry or paste, the fusion of a mass of fines, or chemical

Table 4. Capacities for Roll Presses, t/h[a]

Roll diameter, cm	25.4	40.6	30.5	26.2	33.0	52.1	71.1	91.4	110
Maximum roll face width, cm	8.3	15.2	10.2	15.2	20.3	34.3	68.6	25.4	137
Roll separating force, t	23	45	36	45	68	136	272	327	600
Carbon									
coal, coke		1.8	0.91		2.7	5.4	23		70
charcoal			7.3			12			
activated					2.7	15			
Metals and ores									
alumina					4.5	9.1	25		
aluminum				1.8	3.6	7.3	18		
brass, copper	0.5			1.4	2.7	5.4	15		
steel-mill waste					4.5	9.1			
iron				2.7	5.4	14	33		
nickel powder					2.3	4.5			
nickel ore						18	33		
stainless steel				1.8	4.5	9.1			
steel								23	
bauxite		1.4				9.1	18		
ferro-metals						9.1			50
Chemicals									
copper sulfate	0.5	1.4		0.91	2.7	5.4	14		
potassium hydroxide				0.9	3.6	7.3			
soda ash	0.5				2.7	5.4	14		
urea	0.23					9.1			
dimethyl terephthalate	0.23				1.8	5.4			
Minerals									
potash						18	73		150
salt				1.8	4.5	8.2			
lime					3.6	7.3	14		
calcium sulfate							12	33	
fluorspar						4.5	9.1	25	
magnesium oxide						1.4	4.5		
asbestos						1.4	2.7		
cement						4.5			
glass batch						4.5	11		

[a]Courtesy of Hosokawa Bepex Corp., 1995.

reaction between particles at elevated temperatures. Alternatively, heat may be removed from the material to cause agglomeration by chilling, as in the solidification–crystallization of a melt or concentrated suspension. Heat transfer may be direct from a heat-transfer fluid or indirect across a heat-transfer surface. As a consequence, heat transfer and drying equipment used is quite varied and includes packed-bed systems of particulates and aggregates, dispersed particle–fluid systems, and heating and cooling on moving surfaces. An equally wide variety of preagglomeration equipment is used to preform powders and pastes

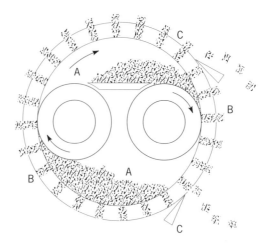

Fig. 10. Operating principle of a common design of pellet mill: A, loose material is fed into pelleting chamber; B, rotation of die and roller pressure forces material through die, compressing it into pellets; and C, adjustable knives cut pellets to desired length. Courtesy of California Pellet Mill Co.

into agglomerates suitable for drying, firing, or chilling. Included are balling devices and pellet mills as described above and extruders using rollers, bars, or wiper blades to force plastic pastes through perforated plates or grids to preform the paste into rods and other shapes (75).

Sintering and Pelletizing. In extractive metallurgy (qv), sintering and pelletizing processes have been developed to allow processing of fine ores, concentrates, and recyclable dusts (3,4). In this connection, sintering refers to a process in which fuel (5–10% coke fines) is mixed with an ore and burned on a grate. A cake of hardened porous material is the resulting agglomerate. In the pelletizing process as applied on a large scale to iron ores, discrete "green" balls produced in balling drums, disks, or cones are dried and hardened by passing combustion gases through a bed of the agglomerates, without fusing the agglomerates as in sintering.

Four separate sequential processes take place during these high temperature operations: drying, preheating, firing or high temperature reaction, and cooling. Ceramic bond formation and grain growth by diffusion are the two important mechanisms for bonding at the high temperatures (1093–1371°C) employed. Simultaneous useful processes may also occur, such as the elimination of sulfur and the decomposition of carbonates and sulfates. The highest tonnage applications are in the beneficiation of iron ores, but nonferrous ores may also be treated (76).

A traveling-grate sintering machine is depicted in Figure 11. The machine consists of a strong frame of structural steel supporting two pallet driving gears and steel tracks or guides. Traveling on the track is an endless line of pallets with perforated bottoms or grates. The train of pallets passes first under a feeding mechanism which lays down a uniform layer of charge, then under an igniter, and finally over a series of wind boxes exhausted by fans. Each wind box is approximately equal in length and width. The speed of the train

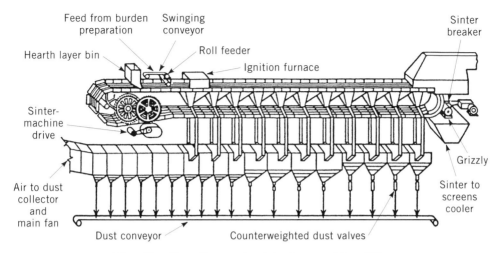

Fig. 11. Traveling-grate sintering machine (77).

of pallets and the volume of air drawn through the charge are controlled so that the combustion layer reaches the grate just as the pallet passes off the machine. The sintering machine is a relatively small part of the equipment needed for a complex sintering plant. Auxiliary devices include conveying and storage equipment, mixing, nodulizing and proportioning equipment, fans, dust collectors, etc.

The capacity of a sintering strand is related directly to the rate at which the burning zone moves downward through the bed (78). This rate is controlled by the air rate through the bed, with the air acting in its function as the heat-transfer medium. The permeability of the sinter bed is thus very important in determining these rates and must be kept high for high throughput. Bed permeability is improved by preagglomeration of the feed material, proper feeding of a material to the sinter strand, and the use of 20–30% of recycled product material to support the needed coarse structure of the bed. A modern sinter strand might be 4 m wide by 61 m long with a capacity of 7200 t/d of ferrous sintered product.

In the pelletizing process for iron ore, wet balls 9.5–13 mm in diameter are dried and heat-hardened on traveling-grate machines adapted from the downdraft sintering grate. Installations differ in air-flow arrangements used to accomplish the process steps of drying, firing, and cooling (3). Pelletizing grates differ from sintering grates in a number of other ways. Pelletizing grates possess a multiplicity of windboxes divided into large groupings to allow recovery of sensible heat (eg, air from the cooling section may be used for drying or combustion purposes). This improves fuel efficiency and adds flexibility to the processing steps. In pelletizing, pellets are held for a long period, relative to that used in sintering, at closely controlled temperatures to effect hardening. In addition, pellet cooling is usually done on the same machine used for heat hardening, whereas sinter is cooled in separate equipment.

In addition to the common straight-grate machine, a circular-grate machine is also in use (79). The drying and firing of pellets may also be done in shaft furnaces or in a combined system using a traveling grate, followed by final firing in a kiln (80).

Drying and Solidification on Surfaces.   In this type of equipment, granular products are formed directly from fluid pastes and melts, without intermediate preforms, by drying or solidification on solid surfaces. Surfaces formed by single or double drums are common. Drum dryers consist of one or more heated metal rolls on which solutions, slurries, or pastes are dried in a thin film. Drying takes place in less than one revolution of the slowly revolving rolls, and a doctor blade scrapes the product off in flake, chip, or granular form. A wide variety of products, such as cereals, fruits, starches, vegetables, meat and fish products, inorganic and organic chemicals, and pharmaceuticals are beneficiated in this way. In drum flakers, a thin film of molten feed is applied to the polished external surface of a revolving, internally cooled drum. Virtually any molten material that solidifies rapidly with cooling can be treated by this method. Although ambient water is normally the cooling liquid, chilled water or other coolants may be used if lower temperatures are needed. The cooled solid is scraped from the drum as a flaked or granular product. Many organic and inorganic food and chemical products are flaked in this way, including caustics, resins, detergents, sugars, waxes, pharmaceuticals, and explosives.

Molten materials can also be cooled to solid products on endless-belt systems, as shown in Figure 12. Some typical materials treated, product and feed characteristics, and capacities of belt cooling systems are given in Table 5.

As illustrated in Figure 12, a number of different feeding and discharge arrangements as well as surface sizes and speeds lend versatility to these methods based on drying and solidification on surfaces. For mobile slurries and fusible and/or soluble particles, these methods offer alternatives to more traditional size enlargement techniques.

Suspended Particle Techniques.   In these methods of size enlargement, granular solids are produced directly from a liquid or semiliquid phase by dispersion in a gas to allow solidification through heat and/or mass transfer. The feed liquid, which may be a solution, gel, paste, emulsion, slurry, or melt, must be pumpable and dispersible. Equipment used includes spray dryers, prilling towers, spouted and fluidized beds, and pneumatic conveying dryers, all of which are amenable to continuous, automated, large-scale operation. Because attrition and fines carryover are common problems with this technique, provision must be made for recovery and recycling.

In the suspension methods, agglomerate formation occurs by hardening of feed droplets into solid particles, by layering of solids deposited from the feed onto existing nuclei, and by adhesion of small particles into aggregates as binding solids from the dispersed feed are deposited. The product size achievable in these methods is usually limited to ca 5 mm and is often much smaller (see DRYING).

In spray drying, the largest particles produced are normally ca 1 mm diameter. Through design and operation of the dryer, however, larger agglomerated particles can be produced, and this technique is used to produce granular dried products in the pharmaceuticals and ceramics industries as well as coarse food powders with instant properties. Although the variables of spray dryer design and operation all interact to influence product characteristics, a number of these have important effects on product size and size distribution (81). In general,

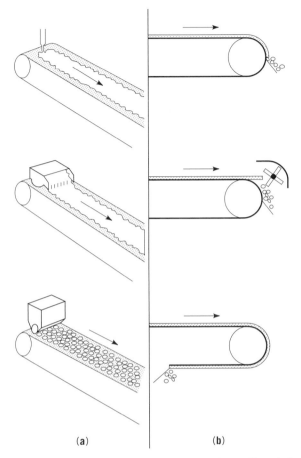

Fig. 12.   (**a**) Typical-feeding and (**b**) discharge systems for endless-belt cooling of molten materials. Courtesy of Sandvik Process Systems Canada Ltd.

the product size is increased by decreasing the intensity of atomization and of spray–air contact, and by lowering the exit temperatures from the dryer. Higher liquid-feed viscosity and feed rate as well as the presence of natural or added binders also favor larger product size.

The flow sheet in Figure 13 shows one example of a system designed to yield agglomerated products. Coarse spray-dried instant food powders are produced directly from liquids in this system. Two stages of agglomeration take place. The initial stage occurs in the atomization zone of the spray dryer, in which relatively cool air is used to retard the evaporation rate and enhance the agglomeration of fines. Further agglomeration is achieved by operating the spray dryer so that the powder is still moist when it leaves the dryer chamber. The agglomerated powder passes out of the bottom of the drying chamber to a vibrating fluid bed where drying is completed, and finally into a second fluid bed for cooling.

Spray cooling or solidification, more commonly known as prilling, is similar to spray drying in that liquid feed is dispersed into droplets at the top of a chamber. These congeal into a solid granular product as they travel down the chamber. The method differs from spray drying in that the liquid droplets are

Table 5. Product Characteristics and Capacity Data for Some Materials Treated in Belt Cooling Systems[a]

Product	Thickness, mm	Feed temp, °C	Discharge temp, °C	Capacity, kg/(h·m²)
resins				
phenolic	1.6	135	43	225
phenolic	1.2–1.3	138	33	277
sulfur	6.4	143	66	269
asphalt	3.2	218	52	90
urea	2.4	191	60	190
ammonium nitrate	1.6	204	71	439
chlorinated wax	1.6	149	38	303
sodium acetate	3.2	82	38	183
hot-melt adhesive	11.1	166	39	70
wax blend	0.6	132	29	129
epoxy resin	1.0	177	38	195

[a]Courtesy of Sandvik Process Systems Canada Ltd., 1995.

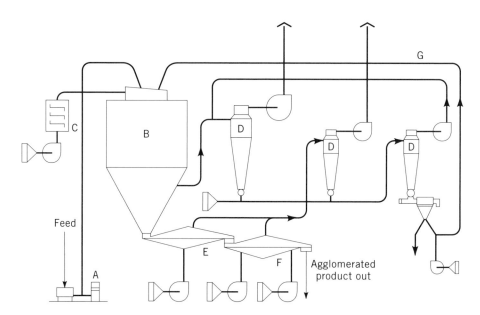

Fig. 13. Spray-dryer system designed for production of agglomerated food powders with instant properties (82): A, liquid-feed system; B, spray-dryer chamber; C, drying air heater; D, cyclones for fines recovery; E, vibrofluidizer as afterdryer; F, vibrofluidizer as aftercooler; and G, fines return to drying chamber.

produced from a melt that solidifies primarily by cooling in the chamber with little, if any, drying. Product size is also generally larger (up to 3 mm dia) than in spray-dried materials. As a result of this relatively large prill size, the process is generally carried out in narrow but very tall towers to ensure that the prills are sufficiently solid when they reach the bottom. Because of the melt feed requirement, prilling is normally limited to materials of low melting point that do not decompose on fusion. Urea and ammonium nitrate fertilizers are traditionally treated by prilling (83,84) (see FERTILIZERS).

Fluid-bed (85–87) and spouted-bed granulation (88,89) both accomplish size enlargement and drying simultaneously by spraying feed liquids (solution, slurry, paste, melt) onto suspended layers of essentially dry particles. The seed-bed particles grow either by coalescence of two or more particles held together by a deposited binder material, or by layering of solids onto the surface of individual particles. Because multiple layers can be deposited, these granulation systems can produce larger granules than spray dryers. The two methods differ in the way the growing bed particles are agitated (Fig. 14). In a fluid bed, a suitable distributor such as a perforated plate passes hot gas uniformly into the base of the particle bed to suspend the particles. In the spouted-bed, hot gas is injected as a single jet into the conical base of the granulation unit causing the bed material to circulate in a fountain-like fashion. Spouted beds were originally developed as an alternative to fluidized beds as a means of contacting gas with solids. The operation of fluidized beds becomes less effective when particles greater than ca 1 mm diameter are to be treated. The gas–solids contacting efficiency is impaired at larger particle sizes, as more and more gas bypasses in the form of large bubbles. In contrast, ca 1 mm is the minimum particle size for which spouting appears to be practical. Thus, spouted-bed granulators allow larger granules to be formed than do fluid-bed granulators (see FLUIDIZATION).

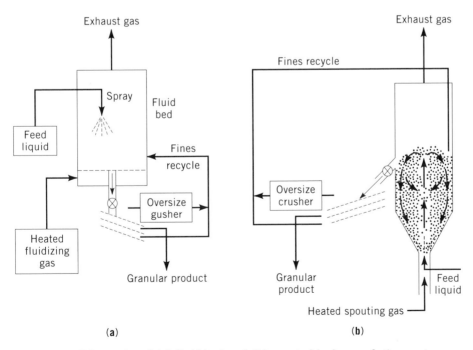

Fig. 14. Schematics of (**a**) fluid-bed and (**b**) spouted-bed granulation systems.

Agglomeration from Liquid Suspensions

Size enlargement of particles contained in liquids is a frequent aid to other operations such as filtration, dewatering, settling, etc. Flocculation procedures are the traditional means used to promote such size enlargement (see FLOCCULATING AGENTS); the product is usually in the form of loose aggregates of an open network structure. Herein, attention is focused on less conventional techniques in which stronger bonding and specialized equipment are used to form large and more permanent agglomerates in liquid suspensions. Table 6 provides examples of such agglomeration processes, in which the objectives include not only the removal or capture of particles in suspension but also the production of granular, including spherical, material, displacement of as much suspending liquid as possible from the product, and the selective agglomeration of one or more components of a multiparticle mixture.

Agglomeration by Competitive Wetting.   Fine particles in liquid suspension can readily be formed into large dense agglomerates of considerable integrity by adding a second or bridging liquid under suitable agitation conditions (103). This second liquid should be effectively immiscible with the suspending liquid and must preferentially wet the solid particles to be agglomerated. A simple example is the addition of oil to an aqueous suspension of fine coal. The oil readily adsorbs preferentially on the carbon particles and forms liquid bridges between these particles by coalescence during the collisions produced by agita-

Table 6. Important Agglomeration Processes Carried Out in Liquid Systems

Process objective	Material treated and process used	References
sphere formation and production of coarse granular products	nuclear fuel and metal-powder production by sol–gel processes	90,91
	manufacturing of small spheres from refractory and high melting point solids (eg, tungsten carbide) by immiscible liquid wetting	92,93
	spherical crystallization: direct agglomeration of crystals during crystallization for drug delivery systems	94
removal and recovery of fine solids from liquid	removal of soot from refinery waters by wetting with oil	95
	recovery of fine coal from preparation plant streams to allow recycling of water	96,97
displacement of suspending liquid	dewatering of various sludges by flocculation, followed by mechanical drainage on filter belts, in revolving drums, etc	98,99
	displacement of moisture from fine coal by wetting with oil	96,97
selective separation of some components in a mixture of particles	removal of ash-forming impurities from coal and tar sands by selective agglomeration	96,100
	coal–gold agglomeration to recover very low concentrations of values in gold ore	101
	solvent extraction and simultaneous soil agglomeration to remediate oil-contaminated soil	102

tion. Inorganic impurity (ash) particles are not wetted by the oil and remain in unagglomerated form in the aqueous slurry (see COAL CONVERSION PROCESSES, CLEANING AND DESULFURIZATION).

The agglomeration phenomena that occur as progressively large amounts of bridging liquid are added to a solids suspension are depicted in Figure 15. The general relationships shown are not specific to a given system, but relate equally well to siliceous particles suspended in oil and collected with water, or to coal particles suspended in water and agglomerated with oil. However, given the need for separation of valuable particles from associated gangue particles, the colloid and surface chemistry involved are usually quite complex. As in the flotation process, selective agglomeration by immiscible liquids depends on the relative wettability of surfaces, and the same fundamentals of surface chemistry apply to the conditioning of particles to yield the required affinity for the wetting liquid (see FLOTATION). A number of examples of ores, fossil fuels, and other particle mixtures in which particle conditioning and selective agglomeration are used to effect separation are given in Reference 103. Mineral recovery applications of the process have been summarized (101).

A most useful feature of the agglomeration technique is its ability to work with extreme fines. Even particles of less than nanometer size (ca 10^{-10} m) can be treated, if appropriate, so that ultrafine grinding can be applied to materials with extreme impurity dissemination to allow recovery of agglomerates of higher purity. A number of applications of liquid-phase agglomeration have reached either the commercial or semicommercial pilot scale of operation.

Oil Agglomeration for Fine Coal Cleaning. In the coal industry, increasingly greater amounts of fine coal must be processed owing to the need to crush to finer sizes for purposes of impurity liberation in low cost physical cleaning

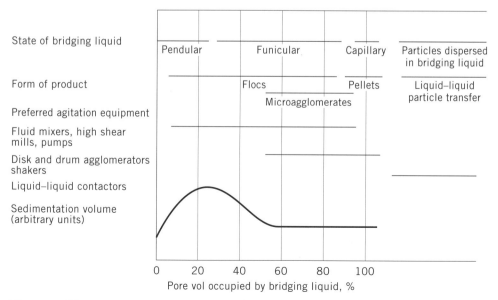

Fig. 15. Phenomena that take place as increasing amounts of an immiscible wetting liquid are added to a suspension of fine particles (103).

processes. Fine coal processing is a significant problem for the coal industry, owing both to the high cost of disposal of reject fines and to the mining cost of lost coal values.

The principal factors affecting coal agglomeration are similar to those affecting flotation, eg, amount and type of collecting oil, degree and type of agitation, pulp density, size consist, and wetting properties of the coal. The most significant operating parameter, however, is the oil concentration. As progressively larger amounts of bridging liquid are added to a suspension of fine particles, a variety of agglomerated products can result. Economic considerations require however, that the smallest quantity possible of oil be used, with the consequent production of very small microagglomerates. This approach requires both an efficient mixer to disperse the oil agglomerant, and centrifugal drying to reach product moisture requirements. A skimming or bubble flotation step has also been added to capture very small agglomerates lost in the screening operation. These process steps are seen in the flow sheet shown in Figure 16, and were used on commercial plants operated in the eastern United States (96) to produce 20–30 t/h of clean coal product agglomerates from 50% minus 325 mesh coal particles contained in a thickener underflow, material which was previously discarded as waste.

Liquid Waste Treatment. Persistent crude petroleum–water emulsions are produced in the recovery and extraction of heavier oils, eg, in the hot water processing of surface-mined oil sands and during *in situ* methods such as steam or water flooding. Oily sludges, formed from water contaminated by fine solids or soil and crude petroleum or bitumens, are an undesirable by-product of most oil recovery or refinery operations. Land farming as a disposal method is increasingly unacceptable, and in any case this and similar approaches ignore the opportunity to recover and reuse valuable aqueous and organic components of the oily wastes.

Liquid-phase agglomeration can play a significant role in oily waste treatment. In the examples discussed for coal, oil is added as an agglomerant to recover and beneficiate the solids. In liquid waste treatment, the waste stream provides the immiscible liquids, and it is then necessary to provide the solid adsorbent for the oil. The commercial fine coal agglomeration facility described in Figure 16 can be thought of as a sizable water treatment facility that recovers clean water containing only a few ppm of organics from the thickener at rates in excess of 2000 L/min. Dispersed, emulsified, and dissolved oils are all susceptible to removal in various degrees by this method.

This approach to liquid waste treatment requires synergy between different process stages or between various industries (industrial ecology). Suitably matched solid adsorbents and waste liquids must be available so that selective wetting and agglomeration take place. Combinations of acceptable feed streams are specific to a given industry or geographic location, but some examples are obvious, such as the emulsions from oil production operations scavenged by thermal coal, eg, in Alberta, Canada; oily sludges from coking/steelmaking operations treated by coal or coke; in chemical/petrochemical complexes, reject or off-spec solvents and other oily wastes treated by solid wastes such as rubber crumb or shredded plastics; and purification of oily wastewater by soot pelletization in an oil gasification plant, used commercially by Shell (95).

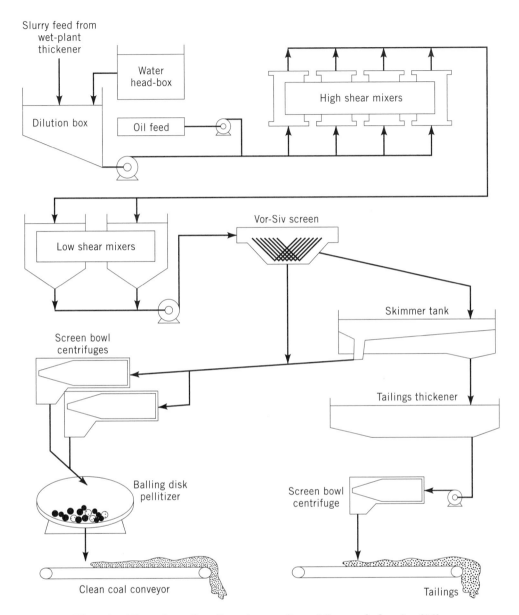

Fig. 16. Flow sheet for oil agglomeration of fine-coal slurries (96).

Spherical Oxide Fuel Particles.   The sol–gel process is a related technique which has been actively developed for the preparation of spherical oxide fuel particles, up to ca 1 mm diameter, for nuclear reactors (90,91). In agglomeration by immiscible liquid wetting, small amounts of a bridging phase adsorbed on the particles coalesce to draw the particles into larger entities. In the sol–gel process, fine particles are initially suspended in an excess of a bridging phase, the suspension is formed into spherical droplets and the excess bridging phase is removed to solidify the droplets into a particulate product. For example, an aqueous sol of colloidal particles such as thoria can be dispersed into droplets in a stream of immiscible water-extracting fluid, such as 2-ethyl-1-hexanol. As water is removed from the sol droplets, the gel formed solidifies and densifies the spherical agglomerates of the colloidal particles. Drying, calcining, and sintering of the recovered microspheres complete the process.

Soil Remediation.   There are many sources of oil-contaminated soils as a result of petrochemical spills and industrial activity in general. For high levels of oil loading (eg, greater than a few weight percent), solvent extraction methods provide important cleaning process options. In general, these processes work well with coarse soils, but have difficulty in handling fine, clay-containing soils because of difficulties in separating extreme fines from the extraction solvent.

Liquid-phase agglomeration can overcome the fines separation problem by forming aggregates of controlled size tailored to the chosen downstream separation method. A size range of 0.5–2 mm has been recommended (102) for soil remediation. This provides optimal drainage and aeration for subsequent bioremediation, which may be needed as a final polishing step. In addition, this size mimics the natural soil size distribution which is important in returning the clean aggregates to the remediated site for fertile agricultural use.

In this process (102), the suspending phase for the soil is selected to be a good sorbent for the oily contaminant. Naphtha is often used. Agglomeration is effected by either water alone if the particles are sufficiently hydrophilic, or by water with added wetting agents to help displace remaining organic contaminants. Process equipment consists of standard conveying, milling, mixing, screening, settling, and filtration modules. In treating soil with ~30% clay in one example, the process removed 95% of a heavy oil contaminant, leaving a residue of 0.3% on the soil, down from about 6% in the feed material. A treatment cost of about $50/t has been estimated for a 240 t/d plant, based on extensive bench and pilot-scale testing of bitumen extraction from oil sands.

NOMENCLATURE

Symbol	Definition	Units
C	parameter in eq. 3	
D	agglomerate diameter	m
d	diameter of monosized spherical particles	m
H	tensile strength of a single particle–particle bond	N
L	compression force at failure	N
s	fractional filling of agglomerate voids	
W	weight fraction of liquid (wet basis)	
γ	liquid surface tension	N/m
ϵ	void fraction of agglomerate	
ρ_L	liquid density	g/cm^3
ρ_S	particle density	g/cm^3
σ_C	compressive strength	Pa
σ_T	mean tensile strength per unit section area	Pa

BIBLIOGRAPHY

"Size Enlargement" in *ECT* 3rd ed., Vol. 21, pp. 77–105, by C. E. Capes, National Research Council of Canada.

1. E. Swartzman, *Trans. Can. Inst. Min. Metall.* **57**, 198 (1954).
2. J. O. Hardesty, *Superphosphate: Its History, Chemistry and Manufacture*, U.S. Dept. of Agriculture, Washington, D.C., 1964, Chapt. 11.
3. D. F. Ball, J. Dartnell, J. Davison, A. Grieve, and R. Wild, *Agglomeration of Iron Ores*, Heinemann, London, 1973.
4. K. Meyer, *Pelletizing of Iron Ores*, Springer-Verlag, Berlin, 1980.
5. W. Pietsch, in *Proceedings of the Sixth International Symposium on Agglomeration*, Society of Powder Technology, Japan, 1993, pp. 837–847.
6. J. E. Browning, *Chem. Eng.* **74**(25), 147 (1967).
7. W. Pietsch, *Size Enlargement by Agglomeration*, John Wiley & Sons, Inc., New York, 1991.
8. C. E. Capes, *Particle Size Enlargement*, Elsevier, Amsterdam, the Netherlands, 1980.
9. H. Rumpf, in W. A. Knepper, ed., *Agglomeration*, Interscience Publishers, a division of John Wiley & Sons, Inc., New York, 1962, pp. 379–414.
10. H. H. Hausner, in Ref. 9, pp. 55–91.
11. D. S. Cahn, *Trans. Soc. Mining Eng. AIME* **250**, 173 (1971).
12. T. P. Hignett, in V. Sauchelli, ed., *Chemistry and Technology of Fertilizers*, Reinhold, New York, 1960, Chapt. 11, pp. 269–298.
13. B. K. Mazumdar, J. M. Sanyal, B. N. Bose, and A. Lahiri, *Indian J. Technol.* **7**, 212 (July 1969).
14. P. L. Waters, *Technical Communication No. 51*, CSIRO, Division of Mineral Chemistry, N.S.W., Australia, May 1969.
15. W. Pietsch, *Chem. Eng. Prog.* **66**, 31 (1970).
16. A. C. Herd, *N. Z. J. Sci.* **17**, 161 (1974).
17. E. Rudnic and J. B. Schwartz, *Remington's Pharmaceutical Sciences*, 18th ed., Mack Publishing Co., Easton, Pa., 1990, pp. 1633–1665.
18. J. D. Jensen, *Food Technol.*, 60 (June 1975).
19. W. B. Pietsch, *J. Eng. Ind.* **91B**(2), 435 (May 1969).
20. A. R. Cooper, Jr. and L. E. Eaton, *J. Am. Ceram. Soc.* **45**(3), 97 (1962).
21. U.S. Pat. 3,844,759 (Oct. 29, 1974), M. M. Ruel and A. F. Sirianni (to Canadian Patents and Development Limited).
22. D. M. Newitt and J. M. Conway-Jones, *Trans. Inst. Chem. Eng.* **36**, 422 (1958).
23. R. A. Fisher, *J. Agric. Sci.* **16**, 492 (1926).
24. P. C. Carman, *Soil Sci.* **52**, 1 (1941).
25. J. R. Wynnyckyj, in Ref. 5, pp. 143–159.
26. B. J. Ennis, J. Li, G. I. Tardos, and R. Pfeffer, *Chem. Eng. Sci.* **45**(10), 3071 (1990).
27. D. S. Kahn and J. M. Karpinski, *Trans. Am. Inst. Min. Eng.* **241**, 475 (1968).
28. P. C. Kapur and D. W. Fuerstenau, *J. Am. Chem. Soc.* **50**, 14 (1967).
29. K. Shinohara and C. E. Capes, *Powder Technol.* **24**, 179 (1979).
30. C. E. Capes, *Powder Technol.* **5**, 119 (1971–1972).
31. C. E. Capes and R. D. Coleman, *Metall. Trans.* **5**, 2604 (1974).
32. C. E. Capes, *Powder Technol.* **4**, 77 (1970–1971).
33. L. Lachman, H. A. Lieberman, and J. L. Kanig, eds., *The Theory and Practice of Industrial Pharmacy*, 3rd ed., Lea and Febiger, Philadelphia, Pa., 1986.
34. B. Hassler and P. G. Kihlstedt, *Cold Bonding Agglomeration*, private communication, 1977.
35. H. Rieschel and K. Zech, *Aufbereitungs-Technik*, **22**(9), 475 (1981). English trans., *Phosphorus and Potassium*, British Sulphur Corp. Ltd., London, Sept.–Oct. 1981.

36. R. Linder, *J. Iron Steel Inst.* **189**, 233 (1958).
37. J. B. Bookey and B. Raistrick, in Ref. 12, Chapt. 18.
38. *United States Pharmacopeia*, XX (USP XX–NF XV), The United States Pharmacopeial Convention, Inc., Rockville, Md., 1980, pp. 958–959.
39. N. G. Stanley-Wood, in M. J. Rhodes, ed., *Principles of Powder Technology*, John Wiley & Sons, Inc., New York, 1990, pp. 193–226.
40. A. A. Adetayo, J. D. Lister, and M. Desai, *Chem. Eng. Sci.* **48**(23), 3951 (1993).
41. C. E. Capes and P. V. Danckwerts, *Trans. Inst. Chem. Eng.* **43**, 116 (1965).
42. K. V. S. Sastry and D. W. Fuerstenau, in K. V. S. Sastry, ed., *Agglomeration 77*, AIME, New York, 1977, pp. 381–402.
43. W. G. Engelleitner, *Ceram. Age* **82**(12), 24, 25, 44, 45 (1966).
44. C. E. Capes, R. L. Germain, and R. D. Coleman, *Ind. Eng. Chem., Process Des. Dev.* **16**, 517 (1977).
45. K. Darcovich, C. E. Capes, and F. D. F. Talbot, *Energy Fuels*, **3**, 64 (1989).
46. G. M. Hebbard, The A. J. Sackett and Sons Co., private communication, July 14, 1977.
47. R. A. Koski, in Ref. 41, pp. 46–73.
48. G. C. Carter and F. Wright, *Inst. Mining Met. Soc. Proc. Adv. in Extractive Metall.*, 89 (1967).
49. W. B. Pietsch, *Aufbereit. Tech.* **7**(4), 144 (1966).
50. H. T. Stirling, in Ref. 9, pp. 177–207.
51. G. R. Hornke and R. E. Powers, paper presented at *The AIME Blast Furnace Coke Oven and Raw Materials Conference*, St. Louis, Mo., Apr. 6, 1959.
52. R. E. Brociner, *Chem. Eng. (London)* **220**, CE 227 (1968).
53. J. A. Frye, W. C. Newton, and W. C. Engelleitner, *Proc. Inst. Briquet. Agglom. Bien. Conf.* **14**, 207 (1975).
54. J. B. Schwartz and co-workers, *Proc. Inst. Briquet. Agglom. Bien. Conf.* **21**, 117 (1989).
55. R. W. Weggel, *Proc. Inst. Briquet. Agglom. Bien. Conf.* **15**, 176 (1977).
56. P. Koenig, *Powder Bulk Eng.* **10**(2), 67 (1976).
57. H. Schubert, *Int. Chem. Eng.* **33**(1), 28 (1993).
58. *Ceram. Age* **86**(3), 15, 16, 18 (1970).
59. C. A. Sumner, *Soap Chem. Spec.* **51**, 29 (July, 1975).
60. A. R. Cooper, Jr. and L. E. Eaton, *J. Am. Ceram. Soc.* **45**(3), 97 (1962).
61. C. L. Huffine, Ph.D. dissertation, Columbia University, New York, 1953.
62. R. S. Spencer, G. D. Gilmore, and R. M. Wiley, *J. Appl. Phys.* **21**, 527 (1950).
63. R. W. Heckel, *Trans. AIME* **221**, 671 (1961).
64. M. J. Donachie, Jr. and M. F. Burr, *J. Met.* **15**, 849 (1963).
65. W. D. Jones, *Fundamental Principles of Powder Metallurgy*, E. Arnold, London, 1960.
66. J. R. Johanson, *Proc. Inst. Briquet. Agglom. Bien. Conf.* **9**, 17 (1965).
67. J. R. Johanson, *Trans. Am. Inst. Mech. Eng. Ser. E. J. Appl. Mech.* **32**, 842 (Dec. 1965).
68. J. R. Johanson, *Proc. Inst. Briquet. Agglom. Bien. Conf.* **11**, 135 (1969).
69. *Ibid.*, **13**, 89 (1973).
70. D. Train, *Trans. Inst. Chem. Eng.* **35**, 258 (1957).
71. K. R. Komarek, *Chem. Eng.* **74**(25), 154 (1967).
72. W. W. Eichenberger, in Ref. 63, p. 52.
73. W. F. Bohm, in Ref. 52, p. 219.
74. W. Pietsch, *Roll Pressing*, 2nd ed., Heyden, London, 1987.
75. I. H. Gibson, in A. S. Goldberg, ed., *Powtech '75*, Heyden, London, 1976, pp. 43–49.
76. T. E. Ban, C. A. Czako, C. D. Thompson, and D. C. Voletta, in Ref. 9, pp. 511–540.
77. R. H. Perry, D. W. Green, and J. O. Maloney, *Chemical Engineers Handbook*, 6th ed., McGraw-Hill, New York, 1984, Sect. 8.

78. R. L. Bennett and R. D. Lopez, *Chem. Eng. Prog. Symp. Ser.* **59**(43), 40 (1963).

79. N. R. Iammartino, *Chem. Eng.*, 76 (May 26, 1975).

80. A. A. Dor, A. English, R. D. Frans, and J. S. Wakeman, paper presented at *The 9th International Mineral Processing Congress*, Prague, Czechoslovakia, 1970, pp. 173–238.

81. K. Masters, *Spray Drying Handbook*, 4th ed., George Goodwin, London, and Halsted Press, a division of John Wiley & Sons, Inc., New York, 1985.

82. K. Masters and A. Stoltze, *Food Eng.* **64** (Feb. 1973).

83. *Can. Chem. Process Ind.* **28**, 299 (May 1944).

84. *Horton Thermaprill Process*, Bull. No. 8305P, HPD Incorporated, Naperville, Ill., 1978.

85. J. W. D. Pictor, *Process Eng.*, 66 (June 1974).

86. W. L. Davies and W. L. Goor, *J. Pharm. Sci.* **60**, 1869 (1971); **61**, 618 (1972).

87. S. Mortensen and S. Hovmand, in D. L. Keairns, ed., *Fluidization Technology*, Vol. II, Hemisphere Publishing Corp., Washington, D.C., 1976, pp. 519–544.

88. K. B. Mathur and N. Epstein, *Spouted Beds*, Academic Press, Inc., New York, 1974.

89. U.S. Pat. 3,231,413 (Jan. 25, 1966), Y. F. Berquin (to Potasse et Engrais Chimiques).

90. J. L. Kelly, A. T. Kleinsteuber, S. D. Clinton, and O. C. Dean, *Ind. Eng. Chem. Process Des. Dev.* **4**, 212 (1965).

91. M. E. A. Hermans, *Powder Metall. Int.* **5**(3), 137 (1973).

92. U.S. Pat. 3,368,004 (Feb. 6, 1968), A. F. Sirianni and I. E. Puddington (to Canadian Patents and Development Limited).

93. C. E. Capes, R. D. Coleman, and W. L. Thayer, paper presented at *The 1st International Conference on the Compaction and Consolidation of Particulate Matter*, London, 1972.

94. Y. Kawashima, F. Cui, H. Takeuchi, T. Niwa, T. Hino, and K. Kiuchi, *Powder Technol.* **78**, 151 (1994).

95. F. J. Zuiderweg and N. van Lookeren Campagne, *Chem. Eng.* (*London*) (220), CE223 (1968).

96. C. E. Capes, in J. W. Leonard, ed., *Coal Preparation*, SME, Littleton, Colo., 1991, pp. 1020–1041.

97. Praxis Engineers, Inc., *Engineering Development of Selective Agglomeration*, Report No. DOE/PC/88879-T6, U.S. DOE, Washington, D.C., Apr. 1993.

98. M. Yusa, H. Suzuki, and S. Tanaka, *J. Am. Water Works Assoc.* **67**, 397 (1975).

99. *Flocpress*, Bull. DB845, Infilco Degremont Inc., Richmond, Va., Sept. 1976.

100. B. D. Sparks and F. W. Meadus, *Energy Processing* **72**, 55 (1979).

101. C. I. House and C. J. Veal, in R. A. Williams, ed., *Colloid and Surface Engineering*, Butterworth-Heinemann, Oxford, U.K., 1992, pp. 188–212.

102. B. D. Sparks, F. W. Meadus, and D. McNabb, paper presented at *AMSE ECO '92 World Conference and Exhibition*, Washington, D.C., 1992.

103. C. E. Capes, A. E. McIlhinney, and A. F. Sirianni, in Ref. 41, pp. 910–930.

General References

References 7, 8, and 38 are also general references.

This article has been issued as NRCC No. 37,589.

C. Edward Capes
K. Darcovich
National Research Council of Canada

SIZE MEASUREMENT
OF PARTICLES

The size distribution of particles in a wide variety of particulate systems is of paramount importance in the chemical processing industries. For example, the compacting and sintering behavior of metallurgical powders, the flow characteristics of granular material, the hiding power of paint (qv) pigments (qv), and the combustion efficiency of powdered coal (qv) and sprayed fluids, are all heavily influenced by the size of the constituent particles (see FLOW MEASUREMENTS; METALLURGY; POWDERS, HANDLING).

A particle is a single unit of material having discrete physical boundaries which define its size, usually in micrometers, μm (1 μm (10^4 Å) = 1×10^{-4} cm = 1×10^{-6} m). The size of a particle is usually expressed by the dimension of its diameter. Typically, particle science is limited to particulate systems within a size range from 10^4 to 10^{-2} μm.

A limitation of the linear dimensional size descriptor is that only particles having simple or defined shapes, such as spheres or cubes, can be uniquely defined by a linear dimension. The common solution to this problem is to describe a nonspherical particle to be equivalent in diameter to a sphere having the same mass, volume, surface area, or settling speed (uniquely defined parameters) as the particle in question (Fig. 1). Therefore, a particle can be described as behaving as a sphere of diameter d. Although this approach makes unique nonspherical particle size characterization possible, it does not come without important consequences because the reported size of a particle is dependent on the physical parameter used in the measurement. A flaky particle falling through a liquid under the influence of gravity is expected to behave as a sphere having a somewhat smaller diameter than that of the same particle measured on the basis of volume equivalence. In reporting particle size data it is therefore necessary to specify the method by which the data were generated. Shape is a parameter which usually influences equivalent sizes, but is not taken into account in most measurement techniques. The variations in diameter equivalence for any specific nonspherical particle can generally be attributed to its shape. Furthermore, it is reasonable to expect variations in particle shape to cause apparent size variations within a particle population, thereby causing artifacts such as widening of the measured size distribution of the population.

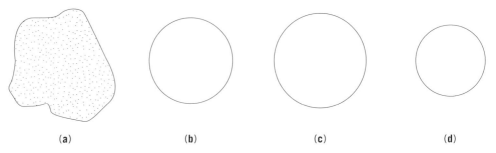

| | | | |
| (a) | (b) | (c) | (d) |

Fig. 1. (a) Particle and examples of equivalent spherical diameters: (b) volume, (c) surface, and (d) settling.

Because of this shape dependence, a limited amount of shape information can be inferred from ratios of spherical equivalence, referred to as shape factors, as obtained by different methods.

The choice of parameter used in the determination of size distribution should include consideration of the information needed in the interpretation of the data. For example, in the case of a manufacturer of paint pigment, the size parameter that best describes the hiding power (performance of the pigment) is the projected area of particles. A powdered catalyst manufacturer is primarily concerned with surface-area equivalence.

Data Representation

Particulate systems composed of identical particles are extremely rare. It is therefore useful to represent a polydispersion of particles as sets of successive size intervals, containing information on the number of particle, length, surface area, or mass. The entire size range, which can span up to several orders of magnitude, can be covered with a relatively small number of intervals. This data set is usually tabulated and transformed into a graphical representation. Size distributions can also be reduced to a single average diameter, such as the mean, median, or mode.

Distribution Averages.   The most commonly used quantities for describing the average diameter of a particle population are the mean, mode, median, and geometric mean. The mean diameter, d, is statistically calculated and in one form or another represents the size of a particle population. It is useful for comparing various populations of particles.

The simplest calculation of the mean, referred to as arithmetic mean (count mean diameter) for data grouped in intervals, consists of the summation of all diameters forming a population, divided by the total number of particles. It can be expressed mathematically by equation 1:

$$\overline{d} = \frac{\sum n_i d_i}{\sum n_i} \tag{1}$$

where n_i is the number of particles in group i having midpoint diameter d_i. Several definitions of the mean are commonly used for various types of comparisons. The most appropriate definition of the mean diameter to be used in any specific application should be the one corresponding most closely to the relevant property of the particle system under study.

The mean volume (mass diameter) is the arithmetic mean diameter of all the particle volumes or masses forming the entire population and, for spherical particles, can be expressed as in equation 2:

$$\overline{d}_m = \overline{d}_v = \left(\sum n_i d_i^3 \Big/ \sum n_i\right)^{1/3} \tag{2}$$

and used when the contribution of particles of various sizes to the total mass of the particles is of importance.

Similarly, the diameter of average surface area (mean surface area diameter), assuming sphericity, can be expressed as in equation 3 and is used when the surface behavior of the particle system is of importance.

$$\overline{d}_s = \left(\sum n_i d_i^2 \Big/ \sum n_i\right)^{1/2} \tag{3}$$

The diameter of average mass and surface area are quantities which involve the size raised to a power, sometimes referred to as the moment, which is descriptive of the fact that the surface area is proportional to the square of the diameter, and the mass or volume of a particle is proportional to the cube of its diameter. These averages represent means as calculated from the different powers of the diameter and mathematically converted back to units of diameter by taking the root of the moment. It is not unusual for a polydispersed particle population to exhibit a diameter of average mass as being one or two orders of magnitude larger than the arithmetic mean of the diameters. In any size distribution, the relation in equation 4 always holds.

$$\overline{d} < \overline{d}_s < \overline{d}_v \tag{4}$$

All definitions of the mean given (eqs. 1–4) are based on the number of particles being measured. As opposed to number-based means, length, surface-area, and mass-based means have also been defined. The linear, or length mean (lm) diameter, $\overline{d}_{lm}$, is defined by the following equation:

$$\overline{d}_{lm} = \frac{\sum l_i d_i}{L} = \frac{\sum n_i d_i^2}{\sum n_i d_i} \tag{5}$$

where l_i is the total length of particles within group i, as opposed to the total number in a number-based mean, and $L = \sum l_i$ the combined length of all the particles in every group.

The surface-area mean (sm) diameter, not to be confused with the number-based average surface area (eq. 3), also known as the Sauter diameter, can be calculated as follows:

$$\overline{d}_{sm} = \frac{\sum s_i d_i}{S} = \frac{\sum n_i d_i^3}{\sum n_i d_i^2} \tag{6}$$

where s_i is the total surface area, $n_i \pi d_i^2$, of the particles within group i and S is the total surface area for the entire population being measured. This diameter is related to the efficiency of liquid atomization (1). A powder sample consisting of smooth spheres may be a reasonable assumption for a wide variety of particle systems. The surface area (measured by gas adsorption techniques), the mass, and the density of a sample are sufficient to calculate the Sauter diameter, using the following relation, where ρ_p is the density of the particles and M is total mass.

$$\overline{d}_{sm} = \left(\frac{6}{\rho_p}\right)\frac{M}{S} \tag{7}$$

The mass mean (mm) diameter, also known as the De Brouckere diameter, for spherical particles of uniform density can be written as in equation 8, where m_i is the total mass of the particles in group i and M is the total mass for all groups.

$$\overline{d}_{mm} = \frac{\sum m_i d_i}{M} = \frac{\sum n_i d_i^4}{\sum n_i d_i^3} \tag{8}$$

The difference between the calculation of the various types of means is probably the most confusing aspect in averaging distributions.

The mode of distribution is simply the value of the most frequent size present. A distribution exhibiting a single maximum is referred to as a unimodal distribution. When two or more maxima are present, the distribution is called bimodal, trimodal, and so on. The mode representing a particle population may have different values depending on whether the measurement is carried out on the basis of particle length, surface area, mass, or volume, or whether the data are represented in terms of the diameter or log (diameter).

The median particle diameter is the diameter which divides half of the measured quantity (mass, surface area, number), or divides the area under a frequency curve in half. The median for any distribution takes a different value depending on the measured quantity. The median, a useful measure of central tendency, can be easily estimated, especially when the data are presented in cumulative form. In this case the median is the diameter corresponding to the fiftieth percentile of the distribution.

Another frequently used average is the geometric mean, which is particularly useful for log-normal or wider (spanning over a decade) distributions. The geometric mean diameter, d_g, is calculated using the logarithm values of the measured diameters:

$$\ln d_g = \frac{\sum n_i \ln d_i}{N} \tag{9}$$

Tabular. A typical distribution as measured by modern instrumentation can include size information on tens of thousands and even millions of individual particles. These data can be listed in a computer and then sorted into a series of successive size intervals, keeping track of the measured quantity, such as number, surface area, or mass, within each group. For narrow size distributions it may be sufficient to group the data in linear intervals, such as 0–1, 1–2, 2–3 μm, etc, and then list the intervals as a percent value of the whole.

Grouping into linear intervals has the disadvantage of not maintaining the resolution of the distribution constant across its width. For example, for an experiment where particle diameters are measured and classified in linear intervals from 0 to 200 μm, each interval two μm in width, the resolution of an interval, which can be defined as the ratio of the width of the interval to the mean interval size (2), would have a value of 0.5 for the first band (0–2 μm) and a value of 0.01 for the last band (198–200 μm). Classifying data on a geometric scale, eg, 1–2, 2–4, 4–8, 8–16, has the virtue of maintaining a consistent band resolution over the entire distribution. Typical particle size data are given in Table 1, along with the percent represented by each interval. This interval can be based on the total number of particles measured, the total sample weight, total volume, or any other basis upon which data might be acquired.

Table 1. Particle Size Distribution

Diameter range, μm	Number	Percent, %	Cumulative finer % (high size)
1–2	12	1.2	1.2
2–4	62	6.2	7.4
4–8	185	18.5	25.9
8–16	250	25.0	50.9
16–32	295	29.5	80.4
32–64	172	17.2	97.6
64–128	22	2.2	99.8
128–256	2	0.2	100.0

Cumulative frequency data, whether on a number, surface area, or mass basis, allows for easy estimates of the total number, surface area, or mass of particles less than a given size. For the data in Table 1, 12 particles out of 1000 (1.2%) had a diameter less than 2 μm, 7.4% less than 4 μm, etc.

Graphical.   A tabular presentation offers the ultimate in precision because all data are included. It is, however, inconvenient to compare tables. A graph usually offers advantages. The two common types of plots are the frequency histogram, sometimes referred to as differential plots, and the cumulative frequency plots. For example, a histogram provides a simple means for graphically presenting size distributions. Normally, the percentage of particles in a given size interval, called the frequency, is plotted against size as shown in Figure 2. It is standard practice to plot data that have been grouped in geometric progression on a logarithmic scale (Fig. 2**b**). Histograms are particularly useful for comparisons among similar distributions. Information on sizes within the individual intervals is not utilized. Differential frequency plots have the advantage of quickly conveying information on the relative amounts in the various group intervals.

The second type of graphical representation, the cumulative frequency, shown in Figure 3 for the data in Table 1, is most useful when several distributions need to be compared. Cumulative distributions by number are generated by summing the contributions of all the particles less than a certain diameter range and plotting this total contribution versus the lower boundary of the diameter range. From Table 1, because the 1.2% in the 1–2 μm interval is finer than 2 μm, the amount of finer material is 1.2%, the amount finer than 4 μm is 7.4%, and so on. Although not as frequently encountered as the cumulative less (finer than) or the cumulative greater (coarser than), a certain size can be

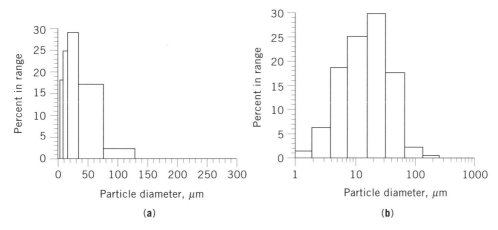

Fig. 2. Histograms of the data from Table 1, plotted on (**a**) a linear and (**b**) a logarithmic scale.

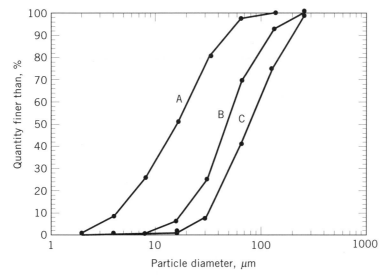

Fig. 3. Cumulative frequency distribution plotted by A, number; B, surface area; and C, volume, for the data in Table 1.

plotted by simply reversing the order of summation of the frequency histogram to obtain the amount greater than a given diameter. It should be noted that cumulative plots tend to conceal detail because of smoothing effect. This effect can be minimized by employing greater numbers of smaller increments when justified by the data. A plot of the particle diameter as a function of the cumulative surface area and mass (or volume) can also be used (see Fig. 3).

Sampling

Sampling (qv) of powders is carried out at two different levels. First there is the taking of a sample from a gross supply of powder such as a rail car or a large heap. Soot sampling is actually covered by the ASTM and British Standard Institute's protocol. Discussions of gross sampling procedures are available (3–8). Sampling from larger (ie, tons) supplies of powders can be achieved using the thief sampler (Fig. 4a) or alternatively a powder sample can be taken from a flow of powder using a device such as that shown in Figure 4b.

When a sample of ca 100 g has been obtained, a representative sample for use in size characterization equipment must then be taken. Some of the more

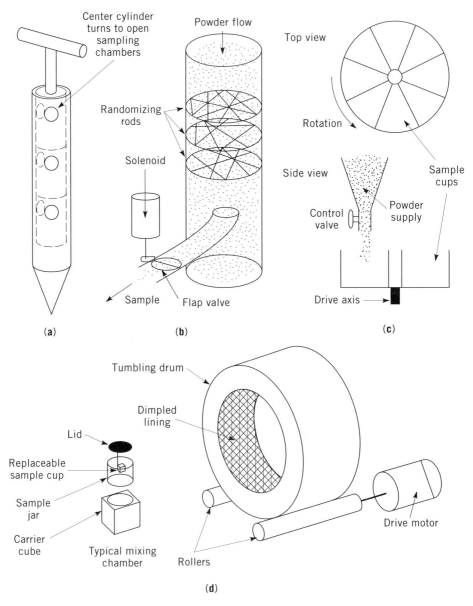

Fig. 4. Types of samplers for obtaining small (1 mg to several g) representative samples from a bulk supply of powder: (**a**) thief sampler; (**b**) passive sampler; (**c**) spinning riffler; and (**d**) free-fall tumbling mixer/sampler.

modern methods of size characterization require as little as one milligram of powder, thus obtaining a representative sample can be quite difficult. If the powder flows well and does not contain too many fines, a device known as the spinning riffler (Fig. 4c) can be used. A spinning riffler consists of a series of cups that rotate under the powder supply. The time of one rotation divided into the time of total powder flow should be as large a number as possible. Although this device has been shown to be very efficient, problems can be encountered on very small (1 mg) samples, and the powder must be processed several times. Moreover, in order to avoid cross-contamination, cleanup after each of the sampling processes can be quite difficult. Furthermore, if the powder is cohesive and does not flow well, the equipment is not easy to use. A silica flow agent can be added to the powder to enable the powder to flow (9).

A newer device, which enables a small sample to be taken efficiently for both cohesive powders and those that contain fines, is the free-fall tumbling mixer equipped with sampling scoops (Fig. 4d). The powder to be sampled is placed in a sample jar so that it fills no more than half the jar. The lid of the jar is equipped with a stirrup that holds a sampling scoop which is covered by the powder when the jar is upright. The jar of powder to be sampled is assembled and placed in a carrier cube which is then tumbled chaotically inside a rotating drum of the type shown in Figure 4d. The performance of this equipment, when sampling a cohesive calcium carbonate powder, is illustrated in Figure 5. This instrument, known as the Aerokaye, is commercially available (Amherst Process Instruments Inc., Hadley, Massachusetts). The powder sampler also works for such items as high density metal powders (10).

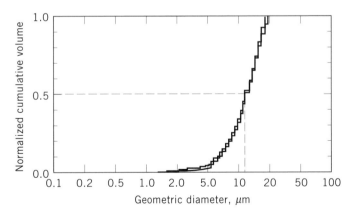

Fig. 5. Two samples of a nominally 15-μm calcium carbonate powder, tumbled in the free-fall tumbling mixer/sampler, taken nine minutes apart. The similarity between the two samples demonstrates that these are probably representative of the bulk powder.

Measurement Methods

A wide variety of particle size measurement methods have evolved to meet the almost endless variability of industrial needs. For instance, distinct technologies are required if *in situ* analysis is required, as opposed to sampling and performing the measurement at a later time and/or in a different location. In certain cases, it is necessary to perform the measurement in real time, such as in an on-line application when size information is used for process control (qv), and in other cases, analysis following the completion of the finished product is satisfactory. Some methods rapidly count and measure particles individually; other methods measure numerous particles simultaneously. Some methods have been developed or adapted to measure the size distribution of dry or airborne particles, or particles dispersed in liquids.

Every method, with the exception of imaging technologies, provides the measurement of an equivalent spherical diameter in one form or another. The spherical diameter information can be deduced indirectly from the behavior of the particles passing through restricted volumes or channels under the influence of gravity or centrifugal force fields, and from interaction with many forms of radiation.

Sieving. The oldest and still one of the most widely employed sizing methods determines particle size by the degree to which a powder is retained on a series of sieves having different opening dimensions. This technique is straightforward and requires simple equipment, but without attention to details it can lead to erroneous results (11). The sieves, particularly those of finer meshes, are often damaged by careless handling and tend to become clogged with irregularly shaped particles unless agitated, but become distorted if agitated too much (12). Furthermore, it is always a concern to determine when all the particles that might pass through the sieve have done so. Nevertheless, attempts to automate the procedure have not met with notable success (13–17).

A typical sieve is a shallow pan having a wire-mesh bottom or an electro-formed grid. Opening dimensions in any mesh or grid are generally uniform within a few percent. Sieves are available having openings from 5 μm upward in several series of size progression. Woven wire-mesh sieves have approximately square openings; electroformed sieves have round, square, or rectangular openings. Wire sieves tend to be sturdier and less expensive, and have a greater proportion of open area. They are much more frequently employed than their electroformed counterparts except in the very fine particle range where only electroformed sieves are available (18).

Dry-sieving is typically performed using a stack of sieves having openings diminishing in size from the top downward. The lowest pan has a solid bottom to retain the final undersize. Powders are segregated according to size by placing the powder on the uppermost sieve and then shaking the stack manually, using a mechanical vibrator (19,20), or with air pulses of sonic frequency (21,22) until all particles fall onto sieves through which they are unable to pass or into the bottom pan. The unit, powered by sonic energy shown in Figure 6, confines the sample using very flexible diaphragms, ensuring against loss of fines. In another device, sieves are employed one at a time within a container from which passing particles are captured by a filter. Agitation on the sieve is provided by a rotating air jet (23). The material retained by the sieve is recorded and recycled to the next coarser sieve until all the powder is exposed to the desired series of sieves or all material passes.

Wet-sieving is performed using a stack of sieves in a similar manner except that water or another liquid which does not dissolve the material is continually

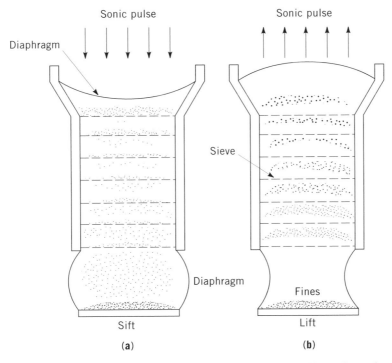

Fig. 6. Dry-siever employing sonic frequency: (**a**) particles falling through sieves on downward sonic pulse; and (**b**) particles lifted by upward sonic pulse.

applied to facilitate particle passage. A detergent is frequently added to promote particle dispersion. This enhanced dispersion is essential at the fine end because under dry conditions, electrostatic and surface forces greatly impede clean passage and isolation of sizes. To accelerate the screening process, a partial vacuum is sometimes applied (24,25). Ultrasonic energy dislodges irregular particles trapped in sieve opening, provided it is applied in moderate density; a maximum of 0.45 W/cm^2 at a frequency of 40 kHz has been recommended (12) (see ULTRASONICS).

The particle mass retained by each sieve is determined by weighing after drying when necessary, and each fraction is designated by the sieve size it passed and the size on which it was retained. The sieve diameter of a particle is therefore defined as the size of the sieve aperture through which the particle in question just passes through. Mass fractions of the particles are then presented in tabular or graphical form.

Sieve analysis is the workhorse of the mineral processing industry to assess ore crushing for mineral release (see MINERAL RECOVERY AND PROCESSING); in heavy construction work to evaluate soils, sand, and gravel for foundation stability (see SOIL STABILIZATION); in powder metallurgical operations for porosity control (see METALLURGY, POWDER METALLURGY); by the ceramic and glass (qv) industries for feedback evaluation (see CERAMICS); and in agriculture for grading seed quality and uniformity (see SIZE REDUCTION).

Computer-Automated Image Analysis. Particle characterization by image analysis consists of examining and measuring the size or shape of a relatively small number of particles that have been magnified. The pioneering studies of particle characterization by imaging technologies were carried out in the late 1960s and early 1970s (26,27). The projected area of the profiles were estimated by direct comparison with sets of reference circles, known as a reticule, engraved on the eyepiece of the microscope. The ever-increasing power of data processing capability coupled with the high performance and falling costs of television cameras and scanners has led to the development of highly sophisticated and powerful image processing and analysis systems (28). Image analyzers can extract information from negatives, photomicrographs, or directly from microscopic (both optical and electron) images by scanning or digitization techniques.

Several diameter definitions are used in particle image measurements (Fig. 7). Martin diameter, the chord length which divides the projected particle into two equal areas with respect to a fixed direction (29); Feret diameter, the projected length with respect to a reference direction (30); and the diameter of equivalent surface area, the diameter of a circle the area of which is equivalent to the projected area of the particle in question (3).

A decision-making process called thresholding establishes the particle boundary (31,32). This boundary is taken as the point where signal amplitude is midway between the optical density of the background and the optical density corresponding to the particles. A small error in boundary location is insignificant when relatively large particles are present, but can become critical for smaller particles. The magnitude of uncertainties stemming from the error in the boundary location can be considerably higher for a projected surface-area measurement as compared to a length measurement. The mathematical transformation from length or projected surface area to volume further magnifies the measurement uncertainties.

Only a limited number of particles can be acquired simultaneously in any given field of view. Thus the processing of hundreds of digitized images may be required to establish statistical significance, especially for wide size distributions, making this technique tedious and time-consuming. Results, given in terms of number, can be mathematically transformed in terms of surface area or volume. If the distribution of particle size ranges by more than a decade, multiple magnifications are usually required to cover the full range.

Optical microscopy is normally used for particles >1.0 μm in diameter that have been deposited on a microscope slide. Electron microscopy is applicable to

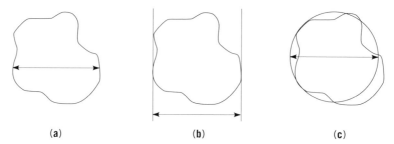

(a)   (b)   (c)

Fig. 7. Definition of diameters frequently used in image analysis: (**a**) horizontal Martin diameter, (**b**) horizontal Feret diameter, and (**c**) diameter of equal surface area.

particles having diameters from 0.002 to 15 μm. Particles are usually incorpo-rated into or deposited onto very thin membranes (33). Several sample prepa-ration methods have been used to properly disperse the particles for imaging (34–41). For accurate and repeatable particle size measurements, important sample preparation requirements such as representative sampling, uniformity of deposited particles deposited over all fields used for counting, and the mini-mum of touching particles must be considered carefully. In order to minimize the probability of touching particles, each image should not contain more than roughly 20 particles. Because the resolution limit of a digitized image corre-sponds to one pixel, the maximum resolution between two sizes is, to a large extent, governed by the magnification of the image. The precision, or repeatabil-ity of the method, is related to the total number of particles counted, the width of the distribution, and the number of pixels forming the smallest particle. The proper balance between resolution, precision, and accuracy is obtained when the amount of material dispersed in any given field of view corresponds to roughly 15–20 particles per field at the magnification where the smallest particle in the distribution covers ca 64 pixels. If the size range is such that more than a single magnification needs to be used (size range >10), the number of particles in each size class is normalized by the total area scanned.

Serious biases in the measured size distribution can be caused by particles touching the edge of the field of view. Simply ignoring the portion of any particle falling outside the field of view, or even tagging and discounting any particle touching the border, erroneously generates a finer distribution because larger particles have a much greater probability of touching the border of a finite field. Therefore, in addition to ignoring any particle touching the border of the field, it is necessary to incorporate a correction factor to adjust the count in each size class. The correction factor is dependent on the actual size of the field as compared to the particle size as follows:

$$C = \frac{Z_1 Z_2}{(Z_1 - F_1)(Z_2 - F_2)} \tag{10}$$

where C is the correction factor, Z_1 and Z_2 are the horizontal and vertical dimensions of the field of view, respectively, and F_1 and F_2 are horizontal and vertical Feret diameters, respectively.

Calibration of the image analyzer to convert pixel lengths to micrometers is necessary to ensure that all measurements are traceable back to the standard meter. The calibration is performed using certified stage micrometers, as pro-vided by such organizations as the National Institute for Testing Technologies (NIST). Distinct calibrations in both the horizontal and vertical directions may be necessary to determine the exact pixel size for every magnification used.

Microscopy (qv) is applied when particle identification and, perhaps, shape evaluation are important in addition to size. Shape characterization is used in the abrasives (qv) industries, pollution or contamination assessment, and forensic studies (see FORENSIC CHEMISTRY).

Sedimentation. Measurement of the settling rate for particles under gravitational or centrifugal acceleration in a quiescent liquid provides the basis of a variety of techniques for determining particle sizes. Gas-phase sedimentation (qv) has been investigated (42), but difficulties achieving adequate particle dispersion and the effect of electrostatic charging have restricted application. In liquid-phase sedimentation, the particles initially may be distributed uniformly throughout a liquid (homogeneous start) or concentrated in a narrow band or layer at the liquid's surface (line start). The particle movement is monitored using light and/or x-ray beams.

The particle size determined by sedimentation techniques is an equivalent spherical diameter, also known as the equivalent settling diameter, defined as the diameter of a sphere of the same density as the irregularly shaped particle that exhibits an identical free-fall velocity. Thus it is an appropriate diameter upon which to base particle behavior in other fluid-flow situations. Variations in the particle size distribution can occur for nonspherical particles (43,44). The upper size limit for sedimentation methods is established by the value of the particle Reynolds number, given by equation 11:

$$Re = \frac{dv\rho_p}{\eta} \tag{11}$$

where d is the particle diameter, ρ_p is the particle density, η is the liquid viscosity, and v is the terminal velocity of the particle, which can be determined from Stokes' law, given by equation 12:

$$d = \left(\left(\frac{18\,\eta}{(\rho_p - \rho)g}\right)\left(\frac{h}{t}\right)\right)^{1/2} \tag{12}$$

where ρ is the liquid density, g is the acceleration owing to gravity, and t is the time required to fall a distance, h. Several distinct configurations of instruments used to measure the Stokes' diameters of particles are available. These variations include gravitational vs centrifugal sedimentation (3), x-ray detection vs light detection (45), scanning vs fixed detectors (46–49), and line vs homogeneous start (3,28). All of these variations come with important consequences which are reflected in the size distribution values generated by these instruments. Therefore, close attention to the actual configuration of a system is important prior to purchase to ensure optimum performance for the particular task for which the sizer is intended. Gravitational sedimentation is intended for larger and higher density particles, which exhibit a relatively high settling rate, in order to obtain a distribution in a reasonable amount of time. The use of centrifugal force to accelerate the settling rate of particles is essential to monitor the movement of smaller, lower density particles. For particles where the diameters are close in value to the wavelength of light, the light intensity scattered by the particles is a strong function of particle diameters, and therefore optical correction factors are required for accurate measurements (50–52). Instruments using x-rays for the detection of particle concentration do not need optical correction. Scanning detectors speed up the analysis, but reduce the resolution of the measurement. Line starts yield a differential distribution characterized by a higher resolution than homogeneous starts. The latter generate a cumulative distribution in a shorter amount of time. Therefore, if resolution is preferred to throughput, a line start system having fixed detectors might be the ideal configuration.

In x-ray sedimentation, a collimated beam of x-rays permits particle concentration detection as a function of mass. The relationship between the fraction of x-rays transmitted and the mass concentration of particles of atomic weight >12 is expressed as in equation 13:

$$\ln T = -\Delta\mu L C_f \tag{13}$$

where $\ln T$ is the natural logarithm of the transmittance relative to the suspending liquid, $\Delta\mu$ is the difference in the x-ray mass-absorption coefficient of the solids and the liquid, L, is the distance through the suspension in the direction of transmission, and C_f is the particle concentration at time t. The weight percent of particles finer than size d_f, P_f, is given by equation 14:

$$P_f = 100 \frac{\ln T_f}{\ln T} \tag{14}$$

where T_f is the transmittance after time t at the distance h and T is the transmittance at the starting time.

Instead of monitoring x-ray transmission, some instruments use visible light to monitor concentration (53). This optical technique is generally referred to as turbidimetry or photoextinction. Light is usually used for low particle concentrations and detection of low molecular weight particles.

Centrifugal sedimentation permits evaluation of smaller diameters but adds mechanical complexity. Both line start and homogeneous start techniques can be used within a centrifuge tube or a disk (Fig. 8). If a centrifuge tube is used in the line start mode (54), a thin layer of suspension is established above a clear liquid filling most of the tube. The layer containing the particles must be formed of a liquid slightly less dense than the other liquid but completely miscible with it. In the homogeneous start mode, the particles are mixed with the liquid to form a homogeneous suspension. The tube is then transferred to the centrifuge and spun at a predetermined rpm, or accelerating rpm for

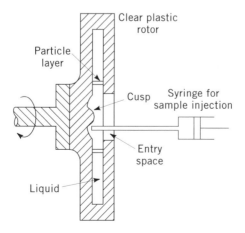

Fig. 8.   Disk profile showing injection method.

wider size distributions. A disadvantage of using a line start procedure is the possibility of streaming, in which filaments of suspension break through the interface between the suspension and clear liquid and carry down faster than they normally would fall. This problem can usually be overcome by employing a very dilute suspension and placing a second layer of intermediate density and interfacial tension beneath the suspension layer (55).

The relatively high rotation speeds of disk centrifuges (see Fig. 8) causes any liquid within the disk cavity to arrange itself to have an essentially flat air–liquid interface. Particles within the liquid migrate radially outward. The change in concentration from a homogeneous suspension at a particular radius from the center of rotation can be measured. The same is true when a thin layer of suspension is injected onto clear liquid already established in the disk. The interfacial region is somewhat perturbed by such an injection but only a small portion of clear liquid is affected. Figure 8 shows the inclusions of a cusp in the rotor face and a method of sample injection to minimize interface disturbance in a two-layer operation. In the case of the disk cavity being filled initially with homogeneous suspension, acceleration and Coriolis forces cause undesirable disturbances. Such disturbances are minor for the very small (<1 μm) particles, however, for which centrifuges are primarily used.

Sedimentation analysis is suitable for a wide variety of materials and is used for both quality control and research work, such as agglomeration studies (56), and gives well-defined, relatively high resolution results. The technique has been employed in the evaluation of soils, sediments, pigments, fillers, phosphors, clays (qv), minerals, photographic halides, and organic particles (57,58).

Field-Flow Fractionation.   Field-flow fractionation is a general name for a class of separation techniques that fractionate a particle population into groups according to size. The work in this area has been reviewed (59).

The basic principle of the technique is illustrated in Figure 9. Particulate species to be fractionated are placed in a tube or other similar device in which a liquid is flowing. The tube is subjected to either a gravitational or centrifugal force confining the particles to be separated to regions of the tube. As the liquid flows along the tube, the combined effect of the field and the flow is to separate the profiles as indicated in the diagram. A fractogram of the separation by field-flow fractionation is shown in Figure 9c.

To fractionate various species, different configurations can be used. Figure 10 shows one such configuration for centrifugal force.

Photon Correlation Spectroscopy.   Photon correlation spectroscopy (pcs), also commonly referred to as quasi-elastic light scattering (qels) or dynamic light scattering (dls), is a technique in which the size of submicrometer particles dispersed in a liquid medium is deduced from the random movement caused by Brownian diffusion motion. This technique has been used for a wide variety of materials (60–62).

The data collected in a pcs experiment essentially consist of the light scattered by the dancing colloidal particles under the influence of Brownian diffusion. The random motion of the scatterers (particles within the small measuring volume) causes a dynamic pattern of fluctuations in the scattered light arriving at the detector. Although these fluctuations are random in nature, buildup

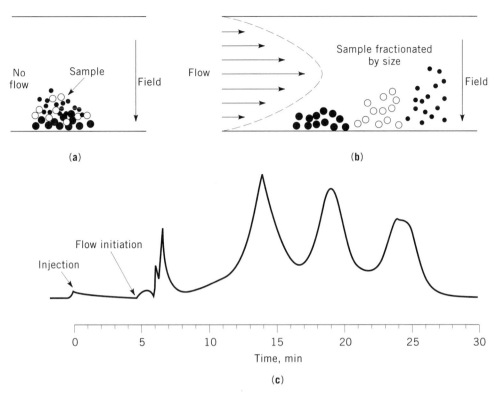

Fig. 9. Principles of field-flow fractionation: (**a**) sample equilibrium position before flow is initiated, (**b**) fractionated sample after flow initiation, and (**c**) a fractogram for silica particles subjected to a flow rate of 39.1 mL/h.

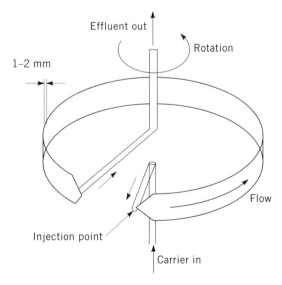

Fig. 10. Centrifugal sedimentation field-flow fractionation equipment deposits particles along the circumference of the disk by size. The fluid enters and leaves the disk axially.

271

and decay times roughly correspond to the average time required for a pair of particles to move by half a wavelength with respect to each other. The diffusion coefficient, D, of the particles under study can be obtained from the autocorrelation of the intensity fluctuations collected by the detector. The hydrodynamic radius of the particles, $d_H/2$, is then calculated from the diffusion coefficient using the Stokes-Einstein relation:

$$d_H = \frac{kT}{6\pi\eta D} \tag{15}$$

where k is Boltzmann's constant, T is the absolute temperature, η is the viscosity of the liquid medium, and d_H is the hydrodynamic diameter. Typically, this type of instrumentation is capable of measuring particles in the size range upward from ca 0.002 μm, where the scattered light intensity becomes too small to be detected, to roughly 1 μm at which size the particles tend to settle. For suspensions having broad distributions, or for multimodal distributions, the extraction of the particle size distribution from the autocorrelation function becomes extremely difficult, and not totally reliable. Pcs works well for narrow, unimodal distributions. The technique has been used for the study of microemulsions, liposomes, and many other colloidal dispersions (see COLLOIDS).

Ultrasonic Spectroscopy. Information on size distribution may be obtained from the attenuation of sound waves traveling through a particle dispersion. Two distinct approaches are being used to extract particle size data from the attenuation spectrum: an empirical approach based on the Bouguer-Lambert-Beer law (63) and a more fundamental or first-principle approach (64–66). The first-principle approach implies that no calibration is required, but certain physical constants of both phases, ie, speed of sound, density, thermal coefficient of expansion, heat capacity, thermal conductivity, attenuation of sound, viscosity for fluid phase, and shear rigidity for solid phase, are required for accurate measurements. The instrument shown in Figure 11 is capable of measuring relatively high particle concentrations over wide size ranges. A sonic wave is sent through the dispersion, which is mechanically agitated to maintain a homogeneous suspension, and its attenuation measured by the receivers. The distance traveled by the sonic wave is accurately known. The attenuation measurement is repeated for a series of frequencies ranging from 1 to 150 MHz. The attenuation of the signal results from the various extinction mechanisms: viscous, thermal, scattering, and diffraction. An iterative algorithm is used in which the theoretical attenuation spectra associated with a particular size distribution closely matches the measured spectra.

Ultrasonic spectroscopy technology, developed in the early 1990s, is proving useful in the lubricant and food industries for measurement of oil-in-water emulsions at process concentrations. This technology is anticipated to find a wide range of industrial applications.

Time-of-Flight Instrumentation

In the late 1980s and early 1990s an instrument variously known as the Galai particle size analyzer (Galai Instruments Inc., Islip, New York) or the

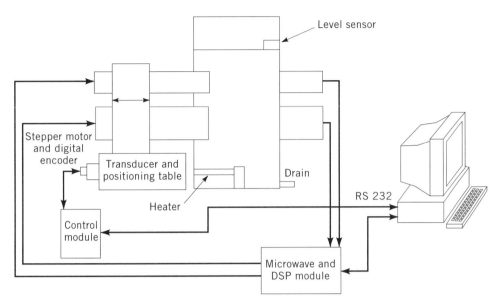

Fig. 11. Particle size analyzer based on ultrasonic spectroscopy where DSP = digital signal processor, and RS 232 is a standard serial connection that allows information to flow between two components.

Brinkmann size analyzer was used to characterize the size of particles by confining the particles to be characterized in an inspection zone scanned by a laser beam. The size of a particle is deduced from the time required for the laser beam to traverse a particle in the inspection zone. The basic configuration of the instrument is shown in Figure 12. The Galai system can also be used to provide shape information because the equipment is provided with logic modules for image analysis using a video camera to inspect the measurement zone. Another time-of-flight analyzer similar to the Galai instrument is the Lasentec Instrument (Redmond, Washington). This system is portable and has been used for on-line monitoring of particles in a slurry or a suspension.

The AeroSizer, manufactured by Amherst Process Instruments Inc. (Hadley, Massachusetts), is equipped with a special device called the AeroDisperser for ensuring efficient dispersal of the powders to be inspected. The disperser and the measurement instrument are shown schematically in Figure 13. The aerosol particles to be characterized are sucked into the inspection zone which operates at a partial vacuum. As the air leaves the nozzle at near sonic velocities, the particles in the stream are accelerated across an inspection zone where they cross two laser beams. The time of flight between the two laser beams is used to deduce the size of the particles. The instrument is calibrated with latex particles of known size. A stream of clean air confines the aerosol stream to the measurement zone. This technique is known as hydrodynamic focusing. A computer correlation establishes which peak in the second laser inspection matches the initiation of action from the first laser beam. The equipment can measure particles at a rate of 10,000/s. The output from the AeroSizer can either be displayed as a number count or a volume percentage count.

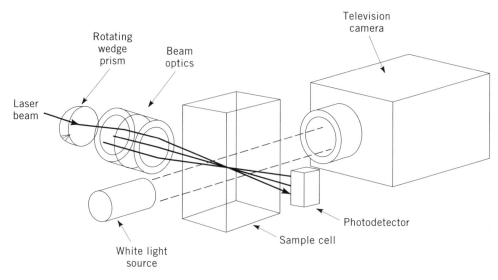

Fig. 12. Schematic representation of the Galai particle size analyzer showing both the laser-based time-of-flight sizing equipment and the television camera for direct image analysis of the suspension in the interrogation zone.

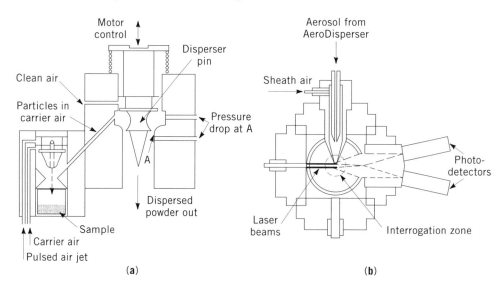

Fig. 13. (**a**) The AeroDisperser attachment for the (**b**) AeroSizer enables aerosol preparation at controlled shear rates. The aerosol is then passed to the AeroSizer and accelerated through the interrogation zone.

Another time-of-flight instrument, the Aerodynamic particle sizer (APS), is manufactured by TSI Incorporated (St. Paul, Minnesota). This system operates at subsonic flow conditions and cannot tolerate as high a flux of particles as the AeroSizer. As of 1996, the development of time-of-flight instruments is ongoing.

Resistazone Counters. The basic principles used in a resistazone counter are illustrated in Figure 14. The particles to be characterized are suspended in a conducting electrolyte and drawn through an orifice situated between two

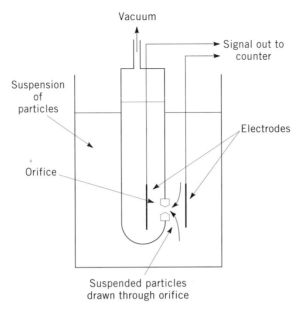

Fig. 14. Simplified schematic of an electrozone stream counter, the Coulter Counter (Coulter Counter Electronics, Hialeah, Florida).

electrodes as shown. The presence of a particle within the zone alters the electrical resistance of the electrolyte in the inspection zone. The change in the resistance of this electrolyte plus particle system can be used to measure the size of the particle. In such a system one must take into account the possibility of multiple occupation of the zone. Advanced versions of this instrument electronically edit the pulse caused by the presence of the particle to determine if it is moving directly down the middle of the inspection zone. Another resista-zone counter using the same general principles is available from Particle Data Inc. (Elmhurst, Illinois).

Laser Diffraction Equipment. A popular piece of size analysis equipment in use in the 1990s is one in which the powder to be characterized is presented as a random array to a laser beam. The diffraction pattern of this randomized array is then deconvoluted using optical theories to generate the size distribution of the powders in random array. By the mid-1990s, equipment based on the laser diffraction pattern of a random array of particles were available commercially. Manufacturers include Horiba Instruments Inc. (Irving, California), Honeywell, Inc. (Fort Washington, Pennsylvania), CILAS (Marcoussis, France), Malvern Instruments Inc. (Southboro, Massachusetts), Sympatec Inc. (Princeton, New Jersey), Shimadzu Scientific Instruments (Columbia, Maryland), and Insitec Inc. (San Raman, California). Each has their own deconvolution algorithm. As of 1996, accuracy and precision of the data generated by the equipment is difficult to generalize. Good examples of the type of information generated by such instruments may be found in References 67 and 68. Comparison of particle data from a diffractometer with that from an image analyzer (Fig. 15) indicates that there is some difference between size distributions as determined by the two methods.

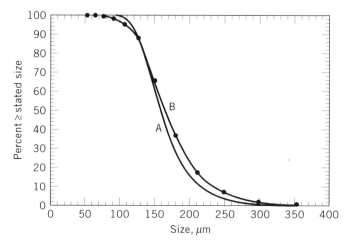

Fig. 15. Size data for a metal powder obtained by A, image analysis, and B, on a diffractometer.

Optical counters have been widely used to monitor cleanroom technology and particles in oil. Instruments manufactured by Royco Inc. (Menlo Park, California) are available for studying aerosols and particles in liquids. The HIAC counter (HIAC Instruments, Monte Claire, California) is a widely used stream counter for particles in fluid. One of the more recently developed optical counters is available from Particle Sizing Systems (Santa Barbara, California). The configuration of one of the widely used counters, the Climet counter, is shown in Figure 16. A general review of photozone counters is available (3).

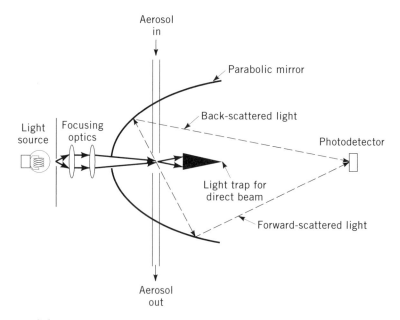

Fig. 16. Schematic representation of the internal structure of a Climet counter.

BIBLIOGRAPHY

"Size Measurement of Particles" in *ECT* 1st ed. Vol. 12, pp. 472–497, by K. T. Whitby, University of Minnesota; in *ECT* 2nd ed., Vol. 18, pp. 310–324; in *ECT* 3rd ed., Vol. 21, pp. 106–131, C. Orr, Micrometrics Instrument Corp.

1. R. A. Mugele and H. D. Evans, *Ind. Eng. Chem.* **43**, 1317 (1951).
2. *Malvern Fraunhofer Laser Diffraction User Guide*, Malvern Instruments Inc., Southboro, Mass., 1991.
3. B. H. Kaye, *Direct Characterization of Fineparticles*, John Wiley & Sons, Inc., New York, 1981.
4. Technical data, *Spinning Riffler*, The Gilson Co. Inc., Worthington, Ohio, and Quantachrome Corp., Boynton Beach, Fla.
5. C. H. Murphy, *Handbook of Particle Sampling and Analysis Methods*, VCH Publishers, Weinheim, Germany, 1984.
6. K. Sommer, *Sampling of Powders and Bulk Materials*, Springer-Verlag, New York, 1986, p. 291.
7. *British Standard Methods for the Determination of Particle Size Powders, Part I, Subdivision of Gross Sample Down to 0.2 Milliliters*, BS3406 Part I, British Standards, London, 1961.
8. T. Allen, *Particle Size Analysis*, 4th ed., Chapman and Hall, London, 1992.
9. B. H. Kaye, J. Gratton-Liimatainen, and John Lloyd, *Part. Part. Syst. Charact.* **12**(4), 194–197 (Aug. 1995).
10. B. H. Kaye and C. Turbitt, "Efficient Powder Sampling Using a Free Fall Tumbling Mixer System," *Proceedings of POWDEX, Georgia World Congress*, Atlanta, Ga., Nov. 5–7, 1996.
11. K. Leschonski, *Powder Technol.* **24**, 115 (1979).
12. J. Hidaka and S. Miwa, *Powder Technol.* **24**, 159 (1979).
13. K. Schonert, W. Schwenk, and K. Steier, *Aufbereit. Tech.* **15**, 368 (1979).
14. B. H. Kaye and N. I. Robb, *Powder Technol.* **24**, 125 (1979).
15. B. H. Kaye and M. R. Jackson, *Powder Technol.* **1**, 43 (1967).
16. R. W. Bartlet and T. H. Chin, *Trans. Am. Inst. Min. Metall. Pet. Eng.* **256**, 323 (1974).
17. C. Orr, D. K. Davis, and R. W. Camp, *Powder Technol.* **24**, 143 (1979).
18. A. Rudolph, C. Peters, and M. Schuster, *Aufbereit. Tech.* **33**, 384 (1992).
19. K. T. Whitby, *Symposium on Particle Size Measurement*, ASTM Special Technical Publication No. 234, ASTM, Philadelphia, Pa., 1959, p. 3.
20. J. E. English, *Filtr. Sep.* **11**, 195 (1974).
21. H. O. Suhm, *Powder Technol.* **2**, 356 (1968–1969).
22. C. W. Ward, *Powder Technol.* **24**, 151 (1979).
23. B. J. Wahl and P. Larouche, *Am. Ceram. Soc. Bull.* **43**, 377 (1964).
24. J. D. Zwicker, *Am. Ceram. Soc. Bull.* **45**, 716 (1966).
25. H. B. Carroll and B. Akst, *Rev. Sci. Instrum.* **37**, 620 (1966).
26. H. H. Heywood, *Proc. Second Lunar Sci. Conf.* **13**, 1989 (1971).
27. H. H. Hausner, *Symposium on Particle Size Analysis*, Loughborough, U.K., 1967.
28. T. Allen, *Particle Size Analysis*, 4th ed., Chapman and Hall, London, 1992.
29. G. Martin, C. E. Blythe, and H. Tongue, *Trans. Ceram. Soc.* **23**, 61 (1924).
30. L. R. Feret, *Assoc. Int. l'Essai Mater. Group D* (Zurich) **2** (1931).
31. J. S. Glass, *Chem. Eng. Prog.* **68**, 58 (1972).
32. J. C. Russ, *Computer Assisted Microscopy—The Measurement and Analysis of Images*, North Carolina State University, Raleigh, 1990.
33. A. M. Glauert, *Practical Methods in Electron Microscopy*, North-Holland, Amsterdam, the Netherlands, 1977.
34. N. Thaulow and E. W. White, *Powder Technol.* **5**, 177 (1972).

35. *French Standard* NF X11-661, 1984.
36. C. Orr, Jr. and J. M. Dalla Valle, *Fine Particle Measurement*, Macmillan, New York, 1959.
37. C. Orr, Jr., *Particulate Technology*, Macmillan, New York, 1966.
38. A. D. Randolph and M. A. Larson, *Theory of Particulate Processes*, Academic Press, Inc., New York, 1971.
39. F. H. Steiger, *Chemtech* **1**, 225 (1971).
40. J. K. Beddow, *Particulate Science and Technology*, Chemical Publishing Co., Inc., New York, 1980.
41. B. B. Spencer and B. E. Lewis, *Powder Technol.* **27**, 219 (1980).
42. F. S. Eadie and R. E. Payne, *Iron Age* **174**, 99 (1954).
43. D. W. Moore and C. Orr, Jr., *Powder Technol.* **8**, 13 (1973).
44. C. Bernhardt, *Part. Part. Syst. Charac.* **8**, 209 (1991).
45. B. B. Weiner, D. Fairhurst, and W. W. Tscharnuter, *ACS Symp. Ser.*, **472**(2), 184 (1991).
46. M. Weber, B. Cai, and L. Kunath, *Aufbereit. Tech.*, **31**, 351 (1990).
47. J. P. Olivier, G. K. Hickin, and C. Orr, Jr., *Powder Technol.* **4**, 257 (1970–1971).
48. P. Sennett, J. P. Olivier, and G. K. Hickin, *Tappi* **57**, 92 (1974).
49. P. K. Herrmann, *Keram. Z.* **31**, 275 (1979).
50. M. J. Devon, T. Provder, and A. Rudin, in Ref. 45, p. 134.
51. M. J. Devon, E. Meyer, T. Provder, T. Rudin, and B. B. Weiner, in Ref. 45, p. 154.
52. K. F. Hansen, in Ref. 45, p. 169.
53. C. C. McMahon, *Cer. Bull.* **49**, 794 (1970).
54. R. L. Hoffman, *J. Colloid Interface Sci.* **143**, 232 (1991).
55. B. Scarlet, M. Rippon, and P. J. Lloyd, *Proceedings of the Conference on Particle Size Analysis*, The Society for Analytical Chemistry, London, 1967, p. 242.
56. G. Staudinger and M. Hangl, *Part. Part. Syst. Charac.* **7**, 144 (1990).
57. C. J. Thomas and D. Fairhurst, in K. Sharma and F. J. Micale, eds., *Proceedings of the Fine Particle Society Symposium*, Plenum Press, New York, 1989, p. 213.
58. J. C. Thomas, A. P. J. Middelberg, J. F. Hamel, and M. A. Snoswell, *Biotechnol. Prog.* **7**, 377 (1991).
59. K. D. Caldwell, in H. G. Bart, ed., *Modern Methods of Particle Size Analysis*, Vol. 73, *Monographs on Analytical Chemistry and its Applications*, John Wiley & Sons, Inc., New York, 1984.
60. R. Pecora, ed., *Dynamic Light Scattering, Applications of Photon Correlation Spectroscopy*, Plenum Press, New York, 1985.
61. D. S. Horne, *Proc. SPIE-Int. Soc. Opt. Eng.* **1430**, 166 (1991).
62. N. Ostrosky, D. Sornette, P. Parker, and E. R. Pike, *Opt. Acta*, **28**, 1059 (1981).
63. U. Krauter and U. Riebel, *Proceedings of the First International Particle Technology Forum*, Part 1, *Particle Characterization*, American Institute of Chemical Engineers, New York, 1994, p. 30.
64. F. Alba, C. L. Dobbs, and R. Sparks, in Ref. 63, p. 36.
65. P. S. Epstein and R. R. Carhart, *J. Acoust. Soc. Am.* **25**, 533 (1953).
66. J. R. Allegra and S. A. Hawley, *J. Acoust. Soc. Am.* **S1**, 1545 (1972).

REMI TROTTIER
Aluminum Company of America

BRIAN KAYE
Laurentian University

SIZE REDUCTION

Size reduction is an extremely important unit operation, whereby materials are subjected to stress in order to reduce the size of individual pieces. The stress is applied by transmitting mechanical force to the solid.

A significant goal of size reduction processes is to improve overall efficiency. This goal of size reduction varies depending on the final application of the product but typically is covered by preparation of naturally occurring raw materials for subsequent separation processes, eg, ore preparation to allow concentration of the valuable fraction; preparation of raw material for subsequent chemically or physically reactive processes, eg, enlargement of the surface area to promote reactivity or to develop the coloring properties of pigments; production of a defined particle-size distribution necessary for a final application, eg, fillers for plastics, rubber, and paint; and preparation of waste materials for recycling, eg, shredding of old tires and waste plastic granulation.

Size reduction of solids is an extremely energy-intensive operation, with estimates of the total U.S. electricity production that is consumed by the process ranging around 2%. Minerals processing accounts for over 50% of this usage, and cement production accounts for about 25%. Only a small fraction, often below 1% of this energy, is used efficiently for size reduction, with the remainder being converted mainly into heat. This article gives an overview of the mechanics of size reduction and the equipment available. However, the science still relies heavily on experience, and it is important that requirements be discussed extensively with equipment manufacturers and large-scale test work carried out prior to decisions being made on the most suitable methods to achieve a given requirement.

Particle Breakage and Fracture Mechanics

Size reduction causes particle breakage by subjecting the material to contact forces or stresses. The applied forces cause deformation which generates internal stress in the particles, and when this stress reaches a certain level, particle breakage occurs.

It is important to differentiate between brittle and plastic deformations within materials. With brittle materials, the behavior is predominantly elastic until the yield point is reached, at which breakage occurs. When fracture occurs as a result of a time-dependent strain, the material behaves in an inelastic manner. Most materials tend to be inelastic. Figure 1 shows a typical stress–strain diagram. The section A–B is the elastic region where the material obeys Hooke's law, and the slope of the line is Young's modulus. C is the yield point, where plastic deformation begins. The difference in strain between the yield point C and the ultimate yield point D gives a measure of the brittleness of the material, ie, the less difference in strain, the more brittle the material.

The total area under the curve A–D, shown as shaded in Figure 1, is the strain energy stored in a body. This energy is not uniformly distributed throughout the material, and it is this inequality that gives rise to particle failure. Stress is concentrated around the tips of existing cracks or flaws, and crack propagation is initiated therefrom (Fig. 2) (1).

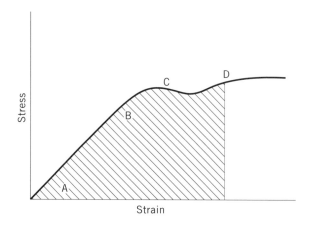

Fig. 1. Typical stress–strain diagram. See text.

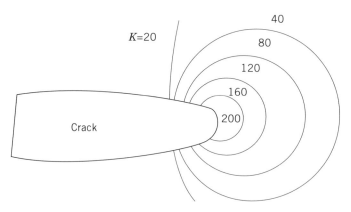

Fig. 2. Stress concentration around a crack. See text.

Stress concentration K is defined (1) as local stress/mean stress in a particle and calculated according to $K = 1 + 2(LR)^{1/2}$, where L is half the crack length and R is the radius of the crack tip.

A subsequent proposal (2) is that for a crack to propagate, the overall stress around the crack must reach a critical value. This critical value is dependent on crack length, so once stress reaches the critical value and the crack lengthens, it continues to grow. As crack propagation progresses, the strain energy released exceeds the energy associated with new surface and the excess energy concentrates around other cracks in the material, causing a multiple fracture. This is typical behavior for brittle materials.

This theory has been expanded and updated (3). With tough or plastic products the excess strain energy causes internal deformation. With decreasing particle size, materials exhibit increasing plastic behavior (4) and this explains why it is more difficult to break small particles than large particles. In small particles, the crack length is limited by the size of the particle. In practice this is seen where a limit of grindability is reached with many materials and with subjection to further grinding, no decrease in particle size can be observed.

Models Predicting Grindability and Energy Requirements

Many attempts have been made to develop models which predict the behavior of materials undergoing size reduction. One proposal is that the energy expended in size reduction is proportional to the new surface formed (5). Another theory is that the energy required to produce a given reduction ratio (feed size ÷ product size) is constant, regardless of initial feed particle size (6). Practical results show, however, that both these theories are limited in their usefulness.

A more practically useful work index based on empirical results from ball milling trials has been developed (7–9) and is expressed by equation 1:

$$E = W_i = \left(\frac{10}{\sqrt{x_p}} - \frac{10}{\sqrt{x_f}} \right) \tag{1}$$

where E is the energy required, W_i is the Bond work index, and x_f and x_p represent the particle size through which 80% of the feed and products, respectively, pass. A standard grindability test, using a specifically sized ball mill of a given design, is used to measure the work index, and many manufacturers use this to help predict energy requirements, particularly for ball mill applications.

The Hardgrove index (10,11) is more usually used when predicting energy requirements for pendulum-type roller mills. A standardized test unit is specified in ASTM 409-71 (1978) and BS 1016 Part 20 (1981). The portion of a product passing a 75-μm sieve is measured after a specified test procedure. This is related to the Hardgrove index by reference to a calibration graph. Hardgrove produced a list of indices for comparing the grindability of typical materials.

Although neither the Bond index nor the Hardgrove index can be considered to give absolute values, both are useful in predicting comparative power consumption and output when scaling up from testwork or in estimating the performance of an existing mill if new products are processed. However, in all size reduction applications, accurate prediction by calculation is extremely difficult and it is essential to carry out trials using the proposed type of equipment.

Predicting Particle-Size Distribution

The breakage process can be modeled using two basic functions. The specific breakage rate, S_j, is the probability that particles in size class j are broken. This probability can be related to either time or energy input, ie, number of mill rotations, to give S_j. The breakage distribution function, b_{ij} is the mass fraction of particles falling into size class i that are formed by the breakage of particles in size class j. The breakage distribution can also be expressed in a cumulative form B_{ij}, which is the mass fraction of particles below size class i that are produced when breaking material of size class j.

Table 1 summarizes typical data; the specific breakage rate, S_j, for particles in size class 1 is 0.6, and the breakage product mass in size classes 2–6 is as shown. The corresponding values of b_{ij} and B_{ij} are given.

Using the above concepts, models have been developed to predict size distribution from comminution devices. An assumption is that the rate of breakage of material of a particular size is proportional to the mass of that size present in

Table 1. Breakage Distribution Function by Product Size Interval Number

Size interval number	Mass	$b_{if}{}^a$	$B_{ij}{}^b$
2	1.2	0.20	1.00
3	1.5	0.25	0.80
4	1.8	0.30	0.55
5	0.9	0.15	0.25
6		0.10	0.10

[a] Basic form.
[b] Cumulative form.

the comminution zone of a machine. If the mass size distribution in the machine is m_i, where m_i is the mass of particles in size class i, then rate of breakage is given by equation 2.

$$\frac{dm_i}{dt} = \sum_{j=1}^{j=i-1} S_j m_j (b_{ij}) - S_i m_i \qquad (2)$$

The summation term is the mass broken into size interval i from all size intervals between j and i, and $S_i m_i$ is the mass broken from size internal i. Thus for a given feed material the product size distribution after a given time in a mill may be determined. In practice however, both S and b are dependent on particle size, material, and the machine utilized. It is also expected that specific rate of breakage should decrease with decreasing particle size, and this is found to be true. Such an approach has been shown to give reasonably accurate predictions when all conditions are known; however, in practical applications severe limitations are met owing to inadequate data and scale-up uncertainties. Hence it is still the usual practice to carry out tests on equipment to be sure of predictions.

Both the need to reduce experimental costs and increasing reliability of mathematical modeling have led to growing acceptance of computer-aided process analysis and simulation, although modeling should not be considered a substitute for either practical experience or reliable experimental data.

Methods for Applying Stress

Equipment for size reduction can be categorized by the method in which the necessary stress is applied to the particles. Figure 3 illustrates the different methods.

Stressing Between Two Solid Surfaces: Crushing. Either single particles (Fig. 3**a**) or a bed of particles (Fig. 3**b**) are crushed between two solid surfaces. The amount of stress that can be applied is governed by the force applied to the solid surfaces.

Stressing by Impact. Size reduction is achieved by the impact of a particle against a solid surface (Fig. 3**c**) or another particle (Fig. 3**d**). The particle can be accelerated to impact against the surface, or the surface can be accelerated to impact the particle, as in an impact mill. The momentum transferred is limited by the mass of the particle and the achievable impact velocity.

Stressing by Cutting. This method (Fig. 3**e**) is useful for materials which exhibit plastic behavior.

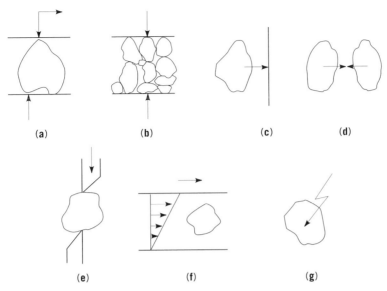

Fig. 3. Stressing mechanisms: (**a**) single particles or (**b**) a bed of particles crushed between two solid surfaces; impact of a particle against (**c**) a solid surface or (**d**) another particle; (**e**) cutting; (**f**) shearing forces or pressure waves; and (**g**) plasma reaction, an example of size reduction by nonmechanical energy.

Stressing by the Surrounding Medium. Size reduction is effected by shearing forces or pressure waves (Fig. 3**f**). The amount of energy that can be transferred is very limited and this method is used mainly to break agglomerates.

Stressing by Nonmechanical Energy. Such processes are not fully developed but examples exist of a plasma reaction (Fig. 3**g**) being used for size reduction. Such cases, however, are specialized and not in general use.

Selection Criteria for Size Reduction

Selection of the most suitable machine for a given requirement is an extremely complex process. Added to variations in the properties of the different materials, many of the machines involved have been specifically developed or adapted to perform only particular tasks. The principal factors which must be addressed are toughness/brittleness, hardness, abrasiveness, feed size, cohesity, particle shape and structure, heat sensitivity, toxicity, explodability, and specific surface.

Toughness/Brittleness. In tough materials the excess strain energy causes plastic deformation, whereas in brittle materials new cracks are propagated. Brittle materials are able to be reduced relatively easily, whereas tough materials present problems. It is sometimes possible to cool a tough material to a temperature low enough to enbrittle it.

Hardness. There are several hardness (qv) scales. In selecting size reduction equipment generally hardness is expressed according to the Mohs' scale, where a body in one range scratches a body in the immediately previous range:

Mohs' hardness	Material
1	talcum
2	gypsum
3	calcite
4	fluorite
5	apatite
6	feldspar
7	quartz
8	topaz
9	corundum
10	diamond

High speed machines such as impact mills begin to suffer high wear rates when processing materials above Mohs' hardness 3, unless very special wear-resisting measures can be taken.

Abrasiveness. This property is closely related to hardness in homogenous materials, but can be affected by particle shape, eg, the presence of sharp corners. In many cases a small proportion, as low as 0.5%, of a hard impurity is enough to cause severe wear to many high speed machines.

Feed Size. The acceptable feed size for a given machine is governed by the type of feed device and physical characteristics of the machine.

Cohesity. Many materials stick together and adhere to machine parts, depending on their condition, particle size, and temperature.

Particle Shape and Structure. Some materials exhibit particular properties owing to their particle shape or form, eg, the plate-like minerals talcum and mica or acicular wollastonite. It is often desired to maintain particle shape; in such cases, an impact-type mill is usually chosen rather than a ball mill, as the latter tends to alter the original particle shape.

Heat Sensitivity. Only around 1–2% of applied energy is effectively used for size reduction. The remainder is mainly converted to heat, which is absorbed by the grinding air, product, and equipment.

Materials containing fat can become very sticky, whereas those containing aromatics can lose flavor. In some cases it is possible to cool the process; however, there is an economic penalty. Equipment with a high air throughput is often chosen, as it provides an economical method of dissipating heat and thus limiting the temperature rise of the end product.

Toxicity. Toxicity has no influence on the actual size reduction, but equipment is often selected for ease of product containment or safe cleaning.

Explodability. Any material which is flammable in air can potentially support a dust explosion when it is finely divided and dispersed in air. Most organic materials, many metals, and other products fall into this category. Equipment has to be protected by inerting, explosion containment, explosion venting, or suppression. NFPA 68 gives guidelines for venting and VDI 3673 (Germany) and I Chem. E (U.K.) give overall guidance. The ease with which these measures can be applied can affect equipment selection.

Specific Surface. If a defined specific surface area is required, this can affect the choice of equipment. Machines that apply stress by crushing generally create more ultrafines, and hence higher surface area, than impact mills.

Equipment Survey

Crushers and Roller Mills. In this equipment group, stress is applied by either crushing single particles or a bed of particles between two solid surfaces. In general, most machines are used for coarse and medium-size reduction, with the exception of the high pressure roller mill which can achieve extremely fine particle distributions.

Jaw Crushers. Both single-toggle (Fig. 4) and double-toggle designs are still widely used. In both, the principle of operation is the same, with feed material being stressed between a stationary jaw and a reciprocating jaw, driven by an eccentric shaft. The end particle-size distribution can be varied by adjusting the width of the outlet gap. Jaw crushers are used in primary size reduction of minerals, and feed openings up to 2.5×2.5 m are available, with outputs of over 1000 t/h possible. The slow speed and absence of rubbing action minimize wear rates, enabling even very hard materials to be processed.

Gyratory and Cone Crushers. Both of these designs utilize the principle of an eccentrically driven rotor crushing material against a stationary mantle. A typical design is shown in Figure 5. Depending on the particular duty, the rotor is supported by bearings at either one or both of its ends. The size of the end product can be varied by changing the clearance between the rotor and mantle. This is usually achieved by raising or lowering the rotor or mantle. These units are used for medium-coarse size reduction, often following a jaw crusher in minerals-processing plants.

Roll Crushers. Traditional roll crushers effect size reduction by crushing single particles between two counter-rotating rolls; hence they are crushed between the two surfaces. The rolls can either be smooth or profiled to aid feeding of coarser materials. Figure 6 shows a particle of diameter x being reduced between two rolls of diameter D. For smooth rolls $D/x > 17$ is usually selected, whereas for profiled rolls this can be varied to $D/x < 10$. The feed size-to-gap width ratio is essentially limited to approximately 4.

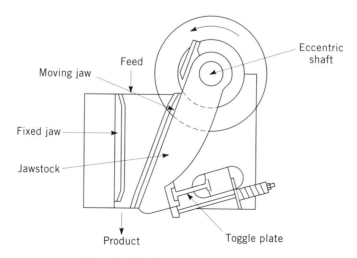

Fig. 4. Single-toggle jaw crusher.

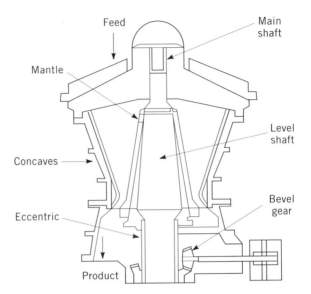

Fig. 5. Cone crusher.

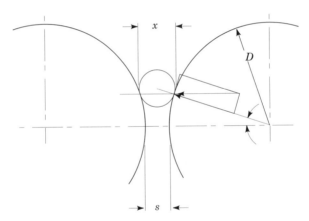

Fig. 6. Roll mill indicating the relationship between roll diameter, D, and feed size, x; s = roll gap width.

Roll mills are employed in a wide variety of applications for medium-coarse down to fine size reduction. By far the most prevalent use is in flour milling where banks of roll mills, beginning with coarse profiles and ending with smooth rolls, are used in series with sifting machines to extract the white flour from the associated bran.

In recent years high compression roll mills have become commercially important. In contrast to traditional roll mills, these machines apply stress to a bed of material rather than to single particles (Fig. 7). By applying extremely high pressures to the rolls, ranging from 50 to 500 MPa (7,250–72,500 psi), very fine particles can be produced. The end fineness is dependent on the crushing pressure, and the output is determined by the roll speed.

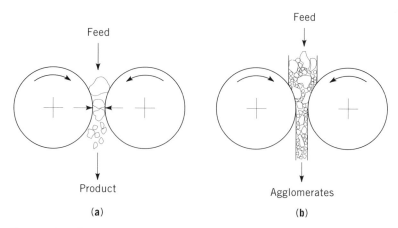

Fig. 7. Comparison between (**a**) roll crusher/mill and (**b**) high compression roller mill.

As only a small proportion of the material is in contact with the rolls and friction on the rollers is low, hard materials can be processed with little wear. The high pressure action creates a slab of ultrafine particles which usually requires a low speed impact milling system to disagglomerate. Used in closed circuit with such a disagglomerator and an air classifier, such machines can reduce the energy requirement for fine grinding many minerals.

Roller Mills: Pendulum, Table, and Bowl Type. This is a group of machines commonly applied for grinding of mineral powders down to approximately 97% below 75 μm, or even finer in some instances. The mills operate at medium speed, up to approximately 30 m/s, and can handle materials with up to Mohs' hardness 5 before wear rates become prohibitive. Many different designs are available; the two most commonly encountered variants are pendulum mills and the table roller mill.

Pendulum mills (Fig. 8) have a central, driven shaft. From this shaft several rollers are suspended on pivots. As the central shaft turns, the centrifugal force causes the pivoted rollers to press against the outer, stationary grinding ring. Material is stressed by compression between the roller and the outer ring. The grinding zone is swept by an airstream and the partially ground material is carried to the upper section of the unit, where an air classifier is located. For standard materials a stationary classifier is used, whereas to achieve 97% below 75 μm, a rotating whizzer type is employed. For finest materials possibly down to 30 μm, turbine classifiers have become common in recent years. The table roller mill employs a ring of pivoted rollers which are forced down toward a driven table by either springs or hydraulic pressure. Material is stressed between the rotating table and the rollers to achieve size reduction. Again, the unit is air-swept and the same choice of classifier units is available as for pendulum mills. Both of these roller mill types are widely employed for limestone, barytes, phosphates, dolomite, and many similar minerals.

Ball Mills and Rod Mills. Ball mills have been utilized since the late 1800s and the construction and principles remain essentially very simple. The machine consists of a cylindrical or conical tube into which loose grinding balls are filled up to a certain level (Fig. 9). Size reduction is achieved by rotating the tube so

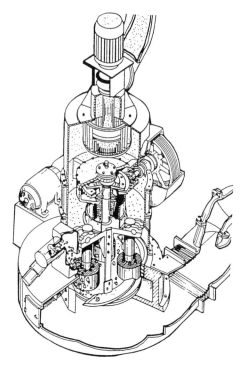

Fig. 8. Pendulum roller mill. Courtesy of Bradley Pulverizer Company.

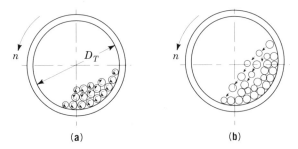

Fig. 9. Size reduction in a ball mill: (**a**) rolling balls and (**b**) tumbling balls.

that the balls either roll against each other or, if the speed is sufficient, they are lifted and fall. In general, the action is a combination of rolling and lifting and this can be influenced by the tube design, eg, using ribs to encourage lifting.

The formula used to calculate the speed limit, n_c, at which the balls begin to tumble rather than roll, is shown in equation 3, where D_T is the tube diameter and g is the acceleration due to gravity.

$$n_c = \frac{1}{2\,\eta}\left(\frac{2\,g}{D_T}\right)^{1/2} \tag{3}$$

The grinding action ensures that a very high stress can be applied to the particles so that a high portion of ultrafine product is produced in a ball mill, compared to impact grinding, for example. Generally, the most economical construction uses a steel shell and balls; however, many products demand iron-free processing, such as ceramic raw materials. In this application the balls are made from flint, aluminium oxide, or similar materials, and the mill is also lined with either one of these or a wear-resistant rubber. Variations employ cylindrical grinding media or rods or use vibration energy to excite the ball charge (Fig. 10) rather than rotate the outer cylinder.

Impact Mills. In this equipment group, stress is applied by transferring kinetic energy by either particle–particle contact or machine–particle contact. Impact mills can broadly be separated into mechanical types where high speed beaters impact the material to apply stress, and fluid energy mills, where particles are accelerated by the surrounding medium and impact against each other or a target.

Mechanical Impact Mills. The mechanical types include crushers, hammer mills, pin disk mills, turbine mills, and mills with air classifiers.

Impact Crusher. Feed material is introduced through a feed opening onto a rotor moving at between 25 and 50 m/s (Fig. 11). The initial impact by the rotor causes some size reduction, and the material is accelerated up to the speed of the rotor and flung against the impact plates, where further size reduction occurs. It is possible to wear-protect these units quite well, so that abrasive materials can be handled. The final end particle size can be varied by the inclusion of an outlet grid to vary the residence time in the machine.

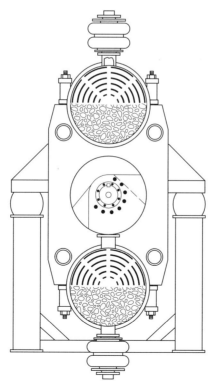

Fig. 10. Vibratory ball mill. Courtesy of KHD Humboldt Wedag AG.

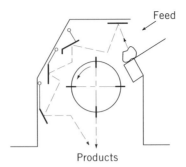

Fig. 11. Impact crusher.

Hammer Mills. One of the most versatile, economical, and widely used impact mills is the hammer mill (Fig. 12). Many variations are produced, with special types available for specialized applications, eg, quick screen change for animal feed, heavy duty for minerals, and light constructions for woodchip. The principle employed is similar to that of the impact crusher; however, the rotation speed can vary from 20 up to 100 m/s with high speed fine-grinding versions. The outlet screen is used to vary the residence time, which in turn affects final particle size. The size of the end product is an order of magnitude finer than the size of the perforations in the outlet screen.

Pin Disk Mill. Conventional pin disk mills are equipped with one rotating and one static disk. Each disk has several concentric rows of pins, and when the machine is operating the rows on the rotating disk alternate with the rows of pins on the static disk. Material is fed into the center of the unit through the static disk and is impacted by the rotating pins and the static pins. Air is swept through the machine and this action carries the ground product away to some form of collection, such as a cyclone or dust filter. Pin disk mills are particularly

Fig. 12. High speed hammer mill. Courtesy of Hosokawa Micron Powder Systems N.J.

suitable for brittle materials; however, they are very susceptible to wearing of the pins. Owing to the narrow pin diameter, mechanical strength is quickly lost as the pins wear; hence these machines are best used on material of Mohs' 3 or below. Pin speeds up to 150 m/s are typical. One special variation has pin disks that both rotate. This is an advantage in that it either increases the differential speed by rotating the disks in opposite directions or that it grinds sticky or fatty products, whereas stationary pins in the grinding zone are subject to severe buildup problems which quickly lead to the blocking of the machine.

Turbine Mills. Probably the most widely used impact mills for fine grinding down to 20 μm are turbine mills with grinding tracks or screens. Various designs of beater systems can be utilized, but all are essentially impact plates that are arranged in either a static grinding track, a perforated screen, or a combination of the two. The turbine-type rotor produces a relatively high air throughput that keeps end-product temperatures relatively cool. Impact speeds at ~120 m/s are a little lower than those of pin disk mills; however, many manufacturers produce Universal-type impact mills that can be fitted with either pin disks or a turbine and track combination, depending on the requirement.

Mechanical Mills With Air Classifiers. In order to improve the end fineness and achieve a sharper topsize cutoff point, many mechanical impact mills are fitted with integral air classifiers (Fig. 13). These can be driven separately from the mill rotor or share a common drive. The material to be ground is introduced into the mill section of the machine, where impact size reduction takes place. The airflow through the machine carries the partially ground product to the air classifier, which is usually some form of rotating turbine. The speed of rotation determines which particle size is internally recycled for further grinding and which is allowed to exit the machine with the airflow. Machines are available up to 375 kW and can achieve products with essentially all material <20 μm.

Fluid Energy and Jet Mills. Particles are accelerated rapidly in a high speed gas stream and size reduction is effected either by particle–particle attrition or by impact against a target. Although energy requirements for fluid energy mills are up to 5–10 times higher than for mechanical impact mills, the attainable fineness is much higher, with small residues of 10 μm being common. In most cases air is used as the grinding medium, but superheated steam can provide energy advantages and inert gas can be utilized where appropriate.

Spiral Jet Mill. This is the simplest form of fluid energy mill (Fig. 14) and is still widely used owing to its low cost and ease of cleaning. A flat cylindrical grinding chamber is surrounded by a nozzle ring. Material to be ground is introduced inside the nozzle ring by an injector. The jets of compressed fluid expand through the nozzles and accelerate the particles, causing size reduction by mutual impact. The expanded gas forms a free vortex spiral toward the central outlet of the machine; hence a classification effect forces the coarser particles back outward toward the jet nozzles for further grinding. Finer particles are carried through the outlet orifice with the grinding fluid. Reliance on a free vortex for classification does mean that the end fineness is affected by variations in the feed rate. The shape and size of the outlet orifice affects the final particle size, as does the pressure of the grinding fluid. Impact velocities are around 250 m/s.

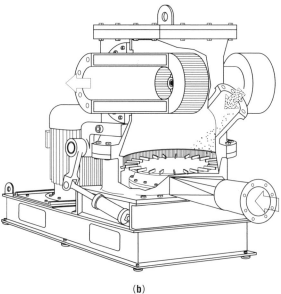

(b)

Fig. 13. (a) Mikro air classifier mill and (b) cutaway of Alpine ZPS unit. Courtesy of Hosokawa Micron Powder Systems N.J.

Fluidized-Bed Jet Mill. To achieve finer products and a better control of final particle size, many fluid energy mills are equipped with mechanical air classifiers. One popular design is the fluidized-bed jet mill (Fig. 15). The lower section of this machine is the grinding zone. The material bed is kept to a predetermined level either by load cell control, level detector, or feedback from the classifier. A ring of grinding nozzles within the material bed is focused toward a central point, and the grinding fluid accelerates the particles. Size reduction takes place within the fluidized bed of material, and this technique can

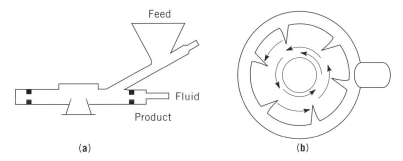

Fig. 14. Spiral jet mill elements (**a**) and grinding chamber (**b**).

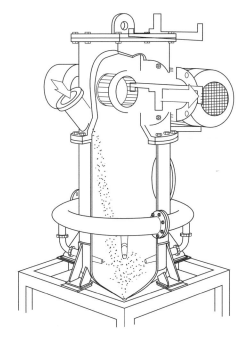

Fig. 15. Fluidized-bed jet mill. Courtesy of Hosokawa Micron Powder Systems N.J.

greatly improve energy efficiency. The partially reduced product is carried with the expanded grinding fluid upward toward the turbine air classifier. This unit is rotating at a variable speed that controls the particle size; the oversize part of the product is rejected and goes back to the fluidized bed for further grinding, and the remaining fine product can leave the machine with the expanded fluid.

Opposed Jet Mills.   These mills are, in some ways, similar to the fluidized-bed machine; however, in this case two opposed nozzles accelerate particles, causing them to collide at a central point (Fig. 16). A turbine classifier is again used to separate the product that has achieved the desired fineness from that which must be internally recycled for further grinding.

Cutting Mills.   The machines applying stress by cutting are described in Figure 3**e**. They are usually employed for size reduction of ductile materials such as plastics, vegetables, and animal products.

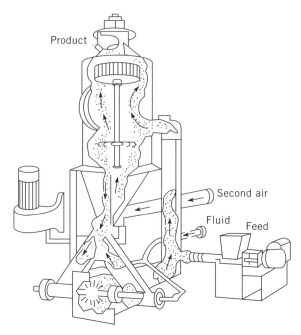

Fig. 16. Opposed jet mill.

Granulators or Knife Mills. The usual format of these machines is to have a rotor equipped with several knife blades which cut the product against stationary knife bars. A typical machine is shown in Figure 17, illustrating the size having up to 2-m knife length available. The lower section of the grinding chamber has a semicircular screen which controls the final particle size of the end product. This particular machine was delivered to cut waste electrical cable, freeing the copper conductor from the poly(vinyl chloride) (PVC) insulation for subsequent separation and re-use.

Wet Grinding. The examples given up to this point all concern size reduction in the dry state, as this is by far the predominant method. However, in certain circumstances it is advantageous to grind in a wet state: (*1*) where a suspension is required as an end product; (*2*) where the required particle size cannot be achieved in a dry state; (*3*) where toxic or flammable emissions must be avoided; and (*4*) where chemical or physical surface reactions are desired. Providing the material does not require subsequent drying after grinding, wet milling can give energy savings of ~30%; however, wear rates are usually three to five times higher.

Stirred Wet Ball Mill. The most commonly applied wet grinding device is the stirred ball mill (Fig. 18) also referred to as sand mill, pearl mill, bead mill, or agitated ball mill. The units also consist essentially of a grinding container which is partially filled with loose grinding media, usually in the size range 0.5–10 mm dia. The chamber is filled with the slurry to be ground, and some form of stirrer accelerates the grinding media. Size reduction takes place as particles are crushed between the media. These units operate either batchwise or as continuous units with a pumped flow once through or recirculated.

Fig. 17.   Granulator or knife mill. Courtesy of Hosokawa Alpine AG.

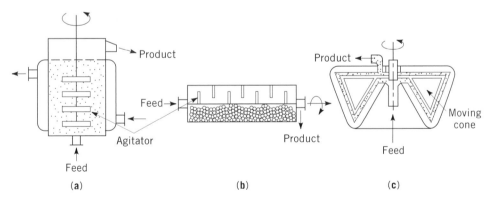

Fig. 18.   Stirred wet ball mills: (**a**) vertical and (**b**) horizontal shaft, and (**c**) annular gap.

Typical stirrer speeds are between 4 and 20 m/s. Uses range from dispersion of pigments and filler for paints, to ceramics and ultrafine minerals such as kaolin. These products usually demand a high percentage of submicrometer particles, which wet grinding can achieve. In many cases the task is to break agglomerates of primary particles that have formed owing to the ultrafine particle size, and the energy requirement for disagglomeration at this size can be considerable.

BIBLIOGRAPHY

"Size Reduction" in *ECT* 1st ed., Vol. 12, pp. 498–520, by L. T. Work, Consulting Engineer; in *ECT* 2nd ed., Vol. 18, pp. 324–365, by C. Orr, Jr., Georgia Institute of Technology; in *ECT* 3rd ed., Vol. 21, pp. 132–162, by B. P. Faulkner and H. W. Rimmer, Allis Chalmers Corp.

1. C. E. Inglis, *Inst. Naval Arch.* **55**, 220–241 (1913).
2. A. A. Griffith, *Phil. Trans. R. Soc.* **221**, 163–198 (1921).
3. H. Rumpf, *Powder Technol.* **7**, 145–159 (1973).
4. W. Hess and K. Schonert, *Inst. Chem. Eng. Symp. Ser.* (63), **D2**(2), 1–9 (1981).
5. P. R. von Rittinger, *Lehrb. Aufbereitungsk. Berlin*, (1867).
6. F. Kick, *The Laws of Proportional Resistance and Their Applications*, Arthur Felix, Leipzig, Germany, 1885.
7. F. C. Bond, *AIME Min. Eng. Trans.* **193**, 484–494 (1952).
8. F. C. Bond, *Can. Min. Metal Trans.* **62**, 286–292 (1954).
9. Ward and F. C. Bond, *Min. Eng. Trans. AIME* **187**, 37–46 (1950).
10. R. M. Hardgrove, *Trans. ASME* **54** (1932).
11. R. M. Hardgrove, *The Relation Between Pulverising Capacity, Power and Grindability*, ASME, Chicago, Ill., 1993, pp. 6–27.

General References

M. Rhodes, *Principles of Powder Technology*, John Wiley & Sons, Inc., New York, 1990.
C. L. Prasher, *Crushing and Grinding Process Handbook*, John Wiley & Sons, Inc., New York, 1987.
J. Liu and K. Schönert, *Int. J. Miner. Proc.* **44–45**, 101–115 (1996).
L. G. Austin, K. R. Weller, and I. L. Kim, "Phenomenological Modelling of the High Pressure Grinding Rolls," *XVIII International Mineral Processing Congress*, Sydney, Australia, May 1993, pp. 87–95.
A. J. Lynch, *Developments in Mineral Processing*, Vol. 1, *Crushing and Grinding Circuits: Their Simulation, Optimization, Design, and Control*, Elsevier, Amsterdam, the Netherlands, 1977, 342 pp.

MICHAEL PRIOR
Hosokawa Micron Ltd.

SLAGCERAM.   See GLASS–CERAMICS.

SLIMICIDES.   See INDUSTRIAL ANTIMICROBIAL AGENTS.

SKIN, PERCUTANEOUS ABSORPTION.   See ANTIAGING AGENTS; CONTROLLED RELEASE TECHNOLOGY; DRUG DELIVERY SYSTEMS.

SMOKES, FUMES, AND SMOG.   See AIR POLLUTION; AIR POLLUTION CONTROL METHODS.

SOAP

Soap is one of the oldest known manufactured chemical substances and was first produced thousands of years ago through the reaction of animal fats with the ashes from plants (1). Early soaps were used primarily for the cleansing of clothing, not for personal hygiene, because of cultural as well as practical reasons. For instance, these animal fat soaps possessed almost unbearable odors and probably contained excessive amounts of unreacted caustics. In the 1990s, soaps are produced using a variety of processes, including kettle boiling, continuous saponification, and hydrolysis/neutralization, as well as different fats and oils feedstocks, yielding finished materials that possess specifically desired properties for application as personal cleansing products.

Soap is one example of a broader class of materials known as surface-active agents, or surfactants (qv). Surfactant molecules contain both a hydrophilic or water-liking portion and a separate hydrophobic or water-repelling portion. The hydrophilic portion of a soap molecule is the carboxylate head group and the hydrophobic portion is the aliphatic chain. This class of materials is simultaneously soluble in both aqueous and organic phases or preferential aggregate at air–water interfaces. It is this special chemical structure that leads to the ability of surfactants to clean dirt and oil from surfaces and produce lather.

Although soaps have many physical properties in common with the broader class of surfactants, they also have several distinguishing factors. First, soaps are most often derived directly from natural sources of fats and oils (see FATS AND FATTY OILS). Fats and oils are triglycerides, that is, molecules comprised of a glycerol backbone and three ester-linked fatty oils. Other synthetic surfactants may use fats and oils or petrochemicals as initial building blocks, but generally require additional chemical manipulations such as sulfonation, esterification, sulfation, and amidation.

Second, soaps form insoluble complexes, commonly referred to as curd, in the presence of calcium and magnesium ions in solution. Calcium and magnesium ions are the principal metal ions found in water and their level is commonly referred to as the hardness of the water; hard water has high levels of both of these ions, soft water has very low levels. This curd reduces the effectiveness of soap as a surfactant and gives rise to other undesirable properties during use, eg, precipitation on surfaces. Many synthetic surfactants are considerably less susceptible to water hardness. This water hardness insensitivity has led to the replacement of soap by synthetic surfactants in a variety of applications, such as dish and laundry detergents and shampoos. Although soap is still the predominant material used in personal cleansing products, eg, facial, body, and hand cleansing, soap-based personal cleansing products are being rapidly replaced by products that contain increasing amounts of synthetic surfactants to meet changing consumer needs, such as rinsing and lather in hard water and improved mildness to the skin. The total dollar value of soaps and detergents produced in the United States for the years 1982 and 1987 are $7.9 billion and $9.5 billion, respectively (2).

Physical Properties of Surfactants

Surfactants, including soap, possess a bipolar structure, comprised of both a hydrophobic tail and a hydrophilic head group. As a result of this bifunctional structure, surfactants possess many unique physical properties. In solution, surfactants preferentially concentrate as monolayers at the interfacial region between any two phases of dissimilar dielectric constants or polarity. Examples of interfacial regions are the interfaces between oil and water, or air and water. The hydrophilic portion preferentially solubilizes in the polar or higher polarity phase whereas the hydrophobic portion preferentially solubilizes in the nonpolar or lower polarity phase. The interface between dissimilar polarity phases provides an ideal location to satisfy both preferences. The presence of surfactants at the interface provides stability to the interface by lowering the total energy associated with maintaining the boundary. Thus, surfactants facilitate stabilization of intermixed, normally immiscible phases, such as oil in water, by decreasing the energy necessary to maintain the large interfacial region associated with mixing. For example, in the absence of surfactants, an oil in water mixture commonly referred to as an emulsion, rapidly separates into two distinct layers to minimize the surface or contact area between the two phases. The ability of surfactants to lower this interfacial energy between the oil and water allows for the formation and stabilization of smaller oil droplets dispersed throughout the water. In this case, the decrease in interfacial energy offsets the increase in total surface area of the system.

Another property of surfactants is their ability to aggregate in solution to form various composite structures or phase states, such as micelles and liquid crystals, as a function of concentration and temperature. At very low surfactant levels, the surfactant exists as individual molecules in solution associating primarily with water molecules. It also concentrates or partitions at the interfacial regions as described above. However, as the concentration of surfactant in solution is increased, a point is reached where the molecules aggregate to form micelles. This concentration is defined as the critical micelle concentration (CMC). The micellar structure minimizes energy through surfactant self-association; the micelle in water is typically characterized with the hydrophobic tails pointing to the center and the head groups pointing out toward the water in spherical superstructures. As the concentration of surfactant in solution is further increased, the micelles elongate into long tubules which align with each other to form a hexagonal arrangement when viewed end-on. These structures are commonly referred to as hexagonal liquid crystals. As the surfactant concentration is further increased, the tubules expand in a second direction to form large, stacked lamellar sheets of surfactants, commonly referred to as lamellar liquid crystals. These liquid crystals are very important in soap making.

Because the core of an aqueous micelle is extremely hydrophobic, it has the ability to solubilize oil within it, as well as to stabilize a dispersion. These solubilization and suspension properties of surfactants are the basis for the cleansing ability of soaps and other surfactants. Furthermore, the ability of surfactants to stabilize interfacial regions, particularly the air–water interface, is the basis for lathering, foaming, and sudsing.

Soap Raw Materials and Their Processing

Carboxylate soaps are most commonly formed through either direct or indirect reaction of aqueous caustic soda, ie, alkali earth metal hydroxides such as NaOH, with fats and oils from natural sources, ie, triglycerides. Fats and oils are typically comprised of both saturated and unsaturated fatty acid molecules containing between eight and 20 carbons randomly linked through ester bonds to a glycerol [56-81-5] backbone. Overall, the reaction of caustic with triglyceride yields glycerol (qv) and soap in a reaction known as saponification. The reaction is shown in equation 1.

$$
\begin{array}{l}
CH_2O-\overset{\overset{\displaystyle O}{\|}}{C}(CH_2)_nCH_3 \\
\quad | \quad\;\; \overset{\displaystyle O}{\|} \\
CHO-\overset{\overset{\displaystyle O}{\|}}{C}(CH_2)_nCH_3 \; + \; 3\,NaOH \;\longrightarrow\; CHOH \; + \; 3\,NaO-\overset{\overset{\displaystyle O}{\|}}{C}(CH_2)_nCH_3 \\
\quad | \quad\;\; \overset{\displaystyle O}{\|} \\
CH_2O-\overset{\overset{\displaystyle O}{\|}}{C}(CH_2)_nCH_3
\end{array}
\qquad (1)
$$

where the glycerol product is $CH_2OH-CHOH-CH_2OH$.

Saponification can proceed directly as a one-step reaction as shown above, or it can be achieved indirectly by a two-step reaction where the intermediate step generates fatty acids through simple hydrolysis of the fats and oils and the finishing step forms soap through the neutralization of the fatty acid with caustic soda. There are practical considerations which must be addressed when performing this reaction on a commercial scale.

Compositional differences in the fats and oils give rise to their significantly different physical properties and those of the resulting fatty acids and soaps. Fats and oils are triglycerides composed of glycerol ester-linked to three fatty acids. The main compositional difference is the chain length distribution of the fatty acids associated with the fats or oils. The compositions found in some commercially important fats and oils are summarized in Table 1. High levels of unsaturated (containing double bonds) or short-chain length components produce fatty acids that are liquid and soaps that have high water solubilities at room temperature. Conversely, high levels of saturated, long-chain length components produce waxy and hard fatty acids, eg, candle wax, and soaps that are essentially insoluble at room temperature. Furthermore, unsaturated components are more susceptible to oxidative degradation, ie, the oxidation of the double bond to form a number of shorter chain components. This gives rise to undesirable odors and darker colors. A key to producing soaps with acceptable qualities is the proper blending of these fats and oils.

The quality, ie, level of impurities, of the fats and oils used in the manufacture of soap is important in the production of commercial products. Fats and oils are isolated from various animal and vegetable sources and contain different intrinsic impurities. These impurities may include hydrolysis products of the triglyceride, eg, fatty acid and mono/di-glycerides; proteinaceous materials and particulate dirt, eg, bone meal; and various vitamins, pigments, phosphatides,

Table 1. Fatty Acid Compositions of Common Fats and Oils[a]

Common name	Chemical name	Chemical formula	Symbol	Animal fats, %		Vegetable oils, %		
				Tallow	Lard	Coconut	Palm kernel	Soybean
Saturated fatty acids								
caprylic	octanoic	$C_8H_{16}O_2$	C8			7	3	
capric	decanoic	$C_{10}H_{20}O_2$	C10			6	3	
lauric	dodecanoic	$C_{12}H_{24}O_2$	C12			50	50	0.5
myristic	tetradecanoic	$C_{14}H_{28}O_2$	C14	3	1.5	18	18	0.5
palmitic	hexadecanoic	$C_{16}H_{32}O_2$	C16	24	27	8.5	8	12
margaric	heptadecanoic	$C_{17}H_{34}O_2$	C17	1.5	0.5			
stearic	octadecanoic	$C_{18}H_{36}O_2$	C18	20	13.5	3	2	4
Unsaturated fatty acids								
myristoleic	tetradecenoic	$C_{14}H_{26}O_2$	C14:1	1				
palmitoleic	hexadecenoic	$C_{16}H_{30}O_2$	C16:1	2.5	3			
oleic	octadecenoic	$C_{18}H_{34}O_2$	C18:1	43	43.5	6	14	25
linoleic	octadecadienoic	$C_{18}H_{32}O_2$	C18:2	4	10.5	1	2	52
linolenic	octadecatrienoic	$C_{18}H_{30}O_2$	C18:3	0.5	0.5	0.5		6

[a]From historical data and Procter & Gamble analyses.

and sterols, ie, cholesterol and tocopherol; as well as less descript odor and color bodies. These impurities affect the physical properties such as odor and color of the fats and oils and can cause additional degradation of the fats and oils upon storage. For commercial soaps, it is desirable to keep these impurities at the absolute minimum for both storage stability and finished product quality considerations.

There are a number of processing steps that can be utilized to improve the quality and stability of the fats and oils raw material. These include water washing, alkali refining, physical (steam) refining, deodorization, bleaching, and hydrogenation. Water washing, also called degumming when dealing with vegetable oils, is an effective means of improving the color of fats and oils through the elimination of proteinaceous solids, phosphatides, and other water-soluble impurities. Hot water, possibly containing some phosphoric acid or sodium phosphate, is mixed with the fats or oils. The water layer is allowed to separate either statically or by utilizing centrifugal force. Many of the solids and other impurities become either solubilized or suspended in the water and removed. Alkali and physical (steam) refining can be utilized to decrease the amount of fatty acid and other color bodies present in fats and oils. Alkali refining washes the fats and oils with alkaline water and converts fatty acids into soap. The resulting soap is removed with the alkaline aqueous phase through settling or centrifugation. In physical refining, volatile impurities including low boiling fatty acids are vaporized and removed from the fats and oils by steam-heating the material. Deodorization, also called steam stripping, is another steam distillation process. For deodorization, however, the distillation is performed under vacuum that allows for more efficient removal of the less volatile odor bodies. Bleaching is most commonly done using a physical adsorption process in which an activated clay, eg, bentonite, is slurried with the oil at temperatures around 100°C. The color bodies adsorb onto the clay, which is subsequently removed through a filtration process. Hydrogenation is also frequently utilized in the processing of fats and oils to improve their storage stability through reduction in the amount of polyunsaturates and unsaturates present. This is achieved by passing the fats and oils through a heated column containing a catalyst, eg, Ni or Pt, and hydrogen gas under pressure so that hydrogen adds across the double bonds.

Industry utilizes a number of analytical methods to characterize fats and oils, which include moisture, titre (solidification point), free fatty acid, unsaponifiable material, iodine value, peroxide value, and color. Moisture content of the fats and oils is an important measure for storage stability at elevated temperature because it facilitates hydrolysis which in turn impacts odor and color quality. Titre is a measure of the temperature at which the material begins to solidify, signifying the minimum temperature at which the material can be stored or pumped as a fluid. Free fatty acid is a measure of the level of hydrolysis the fats and oils have undergone. Increased fatty acid content usually negatively impacts product color stability because fatty acids are more susceptible to oxidation. Unsaponifiable material is a measure of the nontriglyceride fatty material present, which affects the soap yield of the material. The iodine value is a measure of the amount of unsaturation present in the fats and oils. Peroxide value is a measure of the amount of oxidation the fats and oils have undergone and indicates the potential for further degradation.

Fats and oils are treated as commodities in the open market and are purchased in bulk. As commodities, their prices fluctuate with supply and demand. Furthermore, fats and oils come in different grades that reflect different levels of processing and have industry-standardized specifications such as the American Fats and Oil Association. In the manufacture of soap in the United States, the source of animal fats is domestic whereas the vegetable oils are frequently obtained from Southeast Asia, primarily Malaysia and the Philippines.

Tallow [61789-97-7] is the fat obtained as a by-product of beef, and to a lesser degree sheep processing, and is the most commonly utilized animal fat in the manufacturing of soaps. The high content of longer chain length fatty acids present in tallow fat necessitates the addition of other oils, such as coconut oil, in order to produce a bar with acceptable performance.

Coconut oil [8001-31-8] is one of the primary vegetable oils utilized in the manufacture of soap products. Coconut oil is obtained from the dried fruit (copra) of the coconut palm tree. The fruit is dried either in the sun or over open fires from burning the husks of the fruit, with the oil pressed out of the dried fruit.

Palm kernel oil [8023-79-8], obtained from the nuts of the palm tree, is another frequently utilized vegetable oil and is somewhat similar in properties and composition to coconut oil (see Table 1).

Palm oil [8002-75-3] is derived from the fleshy fruit of the palm tree rather than the nut as with palm kernel oil. Palm oil has a longer chain length distribution than palm kernel oil and provides properties and compositions more similar to tallow than to other vegetable oils (see Table 1).

Other Sources.   The four oils named above are the most commonly utilized fats and oils in the soap-making industry in the United States, but other sources are also utilized throughout the world, including lard or hog fat, Babassu oil, rice brand oil, and soybean oil.

Soap Solution-Phase Properties

Commercially, soap is most commonly produced through either the direct saponification of fats and oils with caustic or the hydrolysis of fats and oils to fatty acids followed by stoichiometric (equal molar) neutralization with caustic. Both of these approaches yield workable soap in the form of concentrated soap solutions (~70% soap). This concentration of soap is the target on account of the aqueous-phase properties of soap as well as practical limitations resulting from these properties. Hence, prior to discussing the commercial manufacturing of soap, it is imperative to understand the phase properties of soap.

The Binary Soap–Water System.   Mixtures of soap in water exhibit a rich variety of phase structures, depending on temperature and concentration of the mixture (3). Phase diagrams chart the phase structures, or simply phases, as a function of temperature (on the y-axis) and concentration (on the x-axis). Figure 1 shows a typical soap–water binary phase diagram, in this case for sodium palmitate–water. Sodium palmitate is a fully saturated, sixteen-carbon chain length soap. At lower temperatures, soap crystals coexist with a dilute isotropic soap solution. Upon heating, liquid crystalline phases may form, depending on the relative concentration of soap in water. In dilute solutions, the crystals disproportionate and form simple micellar isotropic solutions (nigre

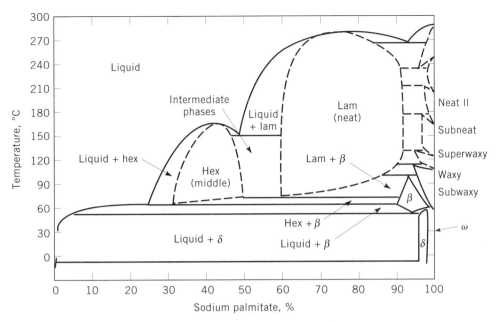

Fig. 1. Binary soap–water phase diagram for sodium palmitate (4). Courtesy of Academic Press, Ltd.

phase). As the soap concentration is increased, hexagonal (hex) liquid crystal (middle phase) is formed. At even higher soap levels, lamellar (lam) liquid crystal (neat phase) is formed. With other chain length distributions of soap, the position of the phase boundaries with respect to temperature and composition varies. However, the general features and phase progressions are similar.

At typical soap processing temperatures (80–95°C), three liquid soap phases are possible: nigre (sometimes called isotropic), middle, and neat soap (4). Nigre soap is observed in dilute soap solutions and is characterized as very fluid. However, because of the dilute nature of this phase, it is not of practical use in the manufacture of soap on account of high energy requirements for drying. Middle soap is a liquid crystalline phase that is extremely viscous and difficult to handle and work. In commercial soap-making processes, care must be taken to avoid the middle-phase region on account of the physical problems associated with it; neat phase is always approached from the more concentrated soap direction. Neat soap is considerably more fluid than middle phase and is readily pumped and mixed. This is the phase most commonly desired for soap making. Neat soap is generally found in the concentration range of 60–90% soap, with commercial processes typically targeting ~70% soap as the optimal concentration. Higher soap concentrations require increased temperatures to maintain the fully liquid crystalline properties (note the curved boundary in Fig. 1) and exhibit increased viscosities which become difficult to manage.

Ternary Soap–Water–Salt systems. A variety of components such as salt (5), fatty acid (6), and glycerol (7) can alter the general phase characteristics of the soap–water system. Ternary phase diagrams are constructed to account for the presence of a third material. These diagrams are displayed as triangles where each of the vertices defines one of the three components and each of the three sides defines the relative concentrations of the two components contained by the two vertices associated with the side. Although temperature continues to be another important variable, these ternary diagrams are often drawn for a defined temperature because of the difficulties in representing an additional dimension. Sometimes the ternary triangle is modified by increasing the angle of one of the vertices to 90°, emphasizing the most important components.

The soap–water–salt diagram is typically shown graphically with the 90° vertices (Fig. 2). At 0% salt, the phases along the axis present a slice of the binary soap–water-phase diagram at 90°C (sodium palmitate in this case). The addition of salt to the system greatly reduces the concentration ranges for the liquid crystalline phases and increases the ranges for the isotropic phases: nigre and lye (a caustic rich aqueous phase). Further increase in the salt concentration drives the system into a biphasic region in which both a concentrated soap and a nigre (or lye) phase coexist. This ability of salt to drive the system into a biphasic, neat soap–nigre/lye phase structure is the basis for the direct saponification approach to soap making. The soap can be separated at a controlled concentration from an aqueous lye or salt phase. The aqueous phase can be used to wash out the excess lye, impurities, and most importantly the glycerol, a valuable by-product of soap making.

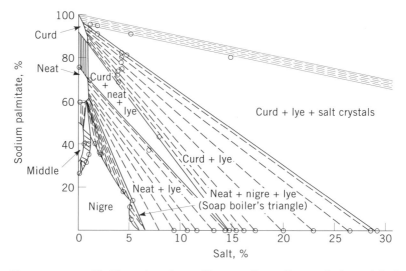

Fig. 2. Ternary soap–NaCl–water phase diagram for sodium palmitate (5). Courtesy of Academic Press, Ltd.

Soap Solid-Phase Properties and Crystallization

Soap crystallizes into bilayer structures comprised of stacks of soap layers arranged head-to-head and tail-to-tail, with water of hydration present in the head-to-head intralamellar region (8). There have been a variety of reports regarding a large number of crystalline phase structures for soap, but it is generally accepted that only four common, distinct pure sodium soap crystalline phases exist (9). These phases are designated α, β, δ, and ω and were originally identified through powder x-ray diffraction. Only ω, β, and δ are observed in conventional soaps; α-phase is found only when crystallized out of ethanol. In a strict sense, these crystalline sodium soap phases are not polymorphs, ie, different crystal arrangement of same composition; rather, they are different phase compounds, ie, compositionally different. This distinction arises not only because of the differences in crystal arrangement but also because of the different levels of hydration present in the various crystalline states (10). This fact is emphasized in Figure 1.

For sodium palmitate, δ-phase is the thermodynamically preferred, or equilibrium state, at room temperature and up to ~60°C; β-phase contains a higher level of hydration and forms at higher temperatures; and ω-phase is an anhydrous crystal that forms at temperatures comparable to β-phase. Most soap in the solid state exists in one or a combination of these three phases. The phase diagram refers to equilibrium states. In practice, the drying routes and other mechanical manipulation utilized in the formation of solid soap can result in the formation of nonequilibrium phase structure. This point is important when dealing with the manufacturing of soap bars and their performance.

X-ray diffraction (xrd) is the simplest and most reliable method for distinguishing between the different crystalline phases, including sodium soap (11). Each phase has a different crystal packing arrangement and thus produces different x-ray diffraction scattering patterns which can be used to identify and quantify the soap phase composition. The most common means of phase identification is through the measurement of the chain-to-chain packing lengths (0.25–0.4 nm), which are insensitive to the chain length of the soap being examined (9).

When processing a hot mixture of soap, no crystals are present. However, crystals form when the hot mixture is cooled to a critical temperature usually referred to as the Krafft boundary (12). The type of crystalline structure that develops initially is dependent on both the mixture composition and temperature. However, the remaining processing of the soap can further alter the crystal structure of the soap in terms of both colloidal structure, ie, distribution and size of crystals, and crystalline phase composition. The temperatures at which crystals form differ for different chain length soaps and different distributions, as well as water content and counter cation. Shorter chain length and unsaturated soaps exhibit significantly lower Krafft boundaries than longer, saturated chain length soaps. Potassium soaps, as well as ammonium soaps, possess considerably lower crystallization temperatures, whereas heavy metal and alkali earth metal soaps possess higher crystallization temperatures.

There are some general guidelines that allow the prediction of which crystal structure will form upon cooling of a hot soap mixture. The most pronounced is the fatty acid chain length distribution. Those mixtures which have more shorter or unsaturated chain lengths (the more soluble soaps) typically result in ω- or β-phase soap crystals. The cis-configuration unsaturate isomers have a greater impact than the trans-configuration isomers. Those hot soap mixtures enriched in the longer saturated chain length (palmitate and stearate) soaps typically result in the formation of δ-phase soap crystals.

Other factors also impact the type of crystals formed upon cooling of hot soap. Water activity or moisture content contribute to the final crystal state as a result of the different phases containing different levels of hydration. Any additive that changes the water activity changes the crystallization pathway. For example, the addition of salt reduces the water activity of the mixture and pushes the equilibrium state toward the lower moisture crystal structure. Additionally, the replacement of sodium with other counter cations influences the crystallization. For example, the replacement of sodium with potassium drives toward the formation of δ-phase.

Phase diagrams can be used to help understand the resulting crystal formation. Upon definition of the starting conditions (temperature and composition), the crystal phases can be estimated by following a line down in temperature at a constant composition. The first crystals to form may not be the equilibrium crystals at room temperature. Because soap phases are phase compounds and not simple polymorphs, once they form they must first disproportionate to reform into a different crystal of another phase structure. This process is inherently slow and requires energy, eg, work or additional heat, so the first crystals encountered upon cooling are likely to be frozen into the cooled, solid soap. Although in theory the crystal phase structure of soap should be predictable from knowledge of composition and temperature, in practice the use of various drying approaches, eg, chill rolls, vacuum dryers, and spray towers, often leads to the formation of a nonequilibrium, room temperature phase. It is the rapid cooling of the hot soap through these various drying techniques which, in essence, traps the soap in nonequilibrium phases. These nonequilibrium phases give rise to the phase changes that occur during processing into finished bar soap products. For the milled bar process, work and heat is put into the soap during additional processing, which results in microheating and physical mixing of the soap. Work tends to transform soap from ω-phase to either β- or δ-phase. But in addition to the work, appropriate temperature control must be maintained to allow the formation of the desired phase. For example, δ-phase is most readily achieved through the cold milling of soap; ω-phase can be maintained through the hotter milling of the soap. The common phase progression is from ω to β to δ, with the reverse progression only observed at high mill temperatures. For framed products, conditioning of the finished soap at various temperatures can result in enough energy input to facilitate the disproportionation reaction to the more thermodynamically stable crystal.

The phase of the soap can have a dramatic impact on both the in-process properties and finished product performance. Soap phase can change the ability of the soap to weld together during finishing, and induce soap stickiness, thus creating problems in milling and conveying soap to and from the various unit operations. In terms of finished product performance, phase can influence a variety of attributes. These attributes include lathering (or solubility), wet cracking, smear, and firmness. Lathering includes both the amount and type, eg, creaminess, of lather a product imparts during use. Wet cracking is a measure of the degree of cracking or fissures found in a bar after it has adsorbed water and subsequently dried out. Smear measures the total amount of partially solubilized soap which forms on the bar surface during long duration contact with water. Firmness is a measure of the physical rigidity of the bar. Simple correlation between the soap phase and these performance attributes have been noted as far back as 1929 (9). For example, β- and ω-phase soaps possess comparable firmness grades, but have significantly different lathering abilities (β much greater). However, β-phase also creates more smear or loss of soap to the water and formation of surface gel than ω-phase soap. Therefore, bar soaps are specifically designed and manufactured to create a soap phase structure that is a good balance of these properties.

In the presence of excess fatty acid, different soap crystalline phase compounds can form, commonly referred to as acid–soaps. Acid–soap crystals are composed of stoichiometric amounts of soap and fatty acid and associate in similar bilayer structures as pure soap crystals. There are a number of different documented acid–soap crystals. The existence of crystals of the composition 2 acid–1 soap, 1 acid–1 soap, and 1 acid–2 soap has been reported (13). The presence of the acid–soaps can also have a dramatic impact on the physical and performance properties of the finished soap. The presence of acid–soaps increases the plasticity of the soap during processing and decreases product firmness, potentially to the point of stickiness during processing. Furthermore, the presence of the acid–soap changes the character of the lather, decreasing the bubble size and subsequently increasing lather stability and creaminess.

It would be incomplete for any discussion of soap crystal phase properties to ignore the colloidal aspects of soap and its impact. At room temperature, the soap–water phase diagram suggests that the soap crystals should be surrounded by an isotropic liquid phase. The colloidal properties are defined by the size, geometry, and interconnectiviness of the soap crystals. Correlations between the colloid structure of the soap bar and the performance of the product are somewhat qualitative, as there is little hard data presented in the literature. However, it might be anticipated that smaller crystals would lead to a softer product. Furthermore, these smaller crystals might also be expected to dissolve more readily, leading to more lather. Translucent and transparent products rely on the formation of extremely small crystals to impart optical clarity.

Commercial Processing

Direct Saponification. Direct saponification of fats and oils is the traditional process utilized for the manufacturing of soap. Commercially this is done through either a kettle boiling batch process or a continuous process.

Kettle Boiled Batch Process. This process produces soap in large, open steel tanks known as kettles, which can hold up to 130,000 kg of material. Kettles are cylindrical tanks with conical bottoms, which contain open steam coils for heating and agitation. To make soap by this process, fats and oils, caustic soda, salt, and water are simultaneously added to the kettle. Effective mixing is important in this process because of the low miscibilities of the fats and oils and caustics. The addition of steam to the system facilities mixing and the saponification reaction. In some systems, the mixing is enhanced through the use of specially designed saponification jets, which allow for intimate mixing of the two components during the charging of the kettle. Care must be taken when blending the fats and oils with caustic soda, salt, and water to ensure a consistent reaction rate for forming the desired neat soap. It is common practice to leave some previously formed soap in the kettle before charging the kettle with the new saponification starting materials. This soap, through its surfactant properties, helps disperse the fats and oils and water phases through better emulsification, thus increasing the reaction rate. To complete the saponification process, the soap batch is boiled for a period of time using steam sparging.

Upon completion of the saponification reaction, additional salt is added to the kettle while boiling with steam to convert the mixture from a pure neat-soap phase composition into the curd soap–lye seat biphasic composition. This process is commonly called opening the grain of the soap. The lye seat is a phase having extremely high electrolytic strength (high salt and lye levels), glycerol, and small amounts of soap, which has minimal solubility in saturated salt solutions. The mixture is allowed to separate for several hours, after which the aqueous solution or lye seat is removed from the bottom of the kettle. The lye seat is usually transferred to a glycerol recovery system, where the glycerol is recovered, purified, and used for other purposes. The curd soap remaining in the kettle is typically washed a few times by adding water, converting it back into neat soap and repeating the salt addition, boiling, and separation process. This washing process provides a more complete removal of glycerol and other impurities from the soap. After the final wash, the water level in the curd soap remaining in the kettle is adjusted to achieve the proper physical properties for additional processing. This process, referred to as fitting, results in the formation of a neat soap–nigre-phase mixture, which facilitates further removal of impurities through the settling of the nigre phase. What remains in the kettle is pure neat soap at ~70% concentration with low levels of salt and glycerol. This process is time-consuming and requires several days to complete.

Continuous Saponification Systems. A relatively recent innovation in the production of soap, these systems have led to improved manufacturing efficiency and considerably shorter processing times. There are a number of commercial systems available; even though these systems are different in design aspects or specific operations, they all saponify fats and oils to finished soap using the same general process (Fig. 3).

Blended fat and oil feedstocks are continuously and accurately metered into a pressurized, heated vessel, commonly referred to as an autoclave, along with the appropriate amount of caustic, water, and salt. The concentrations of these ingredients are adjusted to yield a mixture of neat soap and a lye seat. At the temperatures (~120°C) and pressures (~200 kPa) utilized, the saponification reaction proceeds quickly (<30 min). A recirculation system ensures a residual level of soap in the autoclave to improve contact between the oil and water phases and provides additional mixing. After a relatively short resident time in the autoclave, the neat soap and lye seat reaction blend is pumped into a cooling mixer where the saponification reaction is completed and the reaction product is cooled to below 100°C. The reaction product is pumped next into a static separator, where the lye phase containing a high level of glycerol (25–30%) is separated from the neat soap through gravitational force or settling.

The neat soap is then washed with a lye and salt solution using a counter-current flow process. This is often done in a vertical column, which might be an open tube or contain mixing or baffle stages. The neat soap is introduced into the bottom of the column and the lye/salt (washing) solution is pumped into the top. The less dense neat soap rises up in the column while the lye/salt solution falls to the bottom. The washing solution removes impurities and allows for further

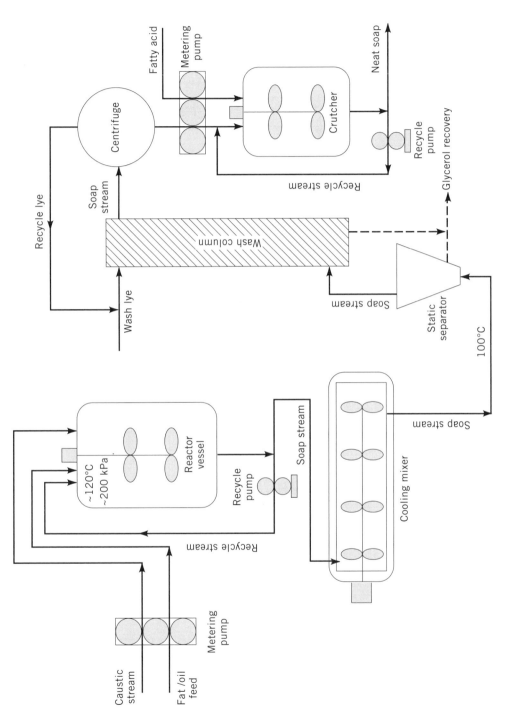

Fig. 3. Process stream diagram for a continuous saponification soap manufacturing facility. To convert kPa to psi, multiply by 0.145.

collection of the glycerol. As with the kettle process, it is important to have a proper level of electrolyte (salt and lye) for effective removal of the glycerol. Final separation of the lye seat from the neat soap is commonly achieved using centrifugation. After centrifugation, the remaining caustic or residual alkalinity in the separated neat soap is neutralized through the accurate addition of fatty acid in a steam-jacketed mixing vessel (crutcher). The soap is now ready for use in the manufacturing of soap bars.

Fatty Acid Neutralization.   Another approach to produce soap is through the neutralization of fatty acids with caustic. This approach requires a stepwise process where fatty acids are produced through the hydrolysis of fats and oils by water, followed by subsequent neutralization with appropriate caustics. This approach has a number of inherent benefits over the saponification process.

Hydrolysis Step.   The hydrolysis of fats and oils by water requires intimate mixing of these two normally immiscible phases. The reaction is carried out under conditions where water possesses appreciable solubility (10–25%) in fats and oils. In practice, this is achieved under high pressure 4–5.5 MPa (580–800 psi) and with high temperatures (~240–270°C) in stainless steel columns of around 24–31 m in height and 50–130 cm in diameter (Fig. 4). ZnO is sometimes added as a catalyst to the feedstock fats and oils to facilitate the reaction. The fat and oil feedstock is injected at the bottom and water is injected at the top of the column. The columns may be either open in design, or contain baffles to ensure better mixing through turbulent flow. High pressure steam inlets are placed at three or four different heights in the column for heating. This design establishes a countercurrent flow pattern with the water moving through the column from top to bottom and the fats and oils the opposite direction. As these materials intermix at the high temperatures and pressures employed, the ester linkages in the fats and oils hydrolyze to create fatty acids and glycerol. The newly formed fatty acids continue to rise up the column, while the resulting glycerol is carried (washed out) downward with the water phase. Because this is a reversible reaction, it is important to remove the glycerin from the mixture through the countercurrent washing process. The concentrations of glycerol and glycerides (mono, di, and tri) are lowest and the concentration of fatty acid is highest toward the top of the column.

The rate-limiting step in the reaction is the removal of glycerol from the fatty acids. This removal is reliant upon interaction with the washwater falling through the column. The Zn–soap formed by the reaction of ZnO and fatty acid acts as a phase-transfer catalyst, improving the transfer of glycerol from the oil to the water phase. The separation of the glycerol and fatty acid in the column prevents the reverse reaction from occurring. The hydrolyzer process provides around 99% efficiency for the conversion of the fats and oils to fatty acids and glycerol, and requires around 90-min residence time.

The fatty acids that emerge from the top of the column contain entrained water, partially hydrolyzed fat, and the Zn–soap catalyst. This product stream is passed into a vacuum dryer stage where the water is removed through vaporization and the fatty acid cooled as a result of this vaporization process. The dried product stream is then passed to a distillation system.

The distillation system allows for improved fatty acid quality, ie, odor and color, through the separation of the fatty acid from partially saponified fats and

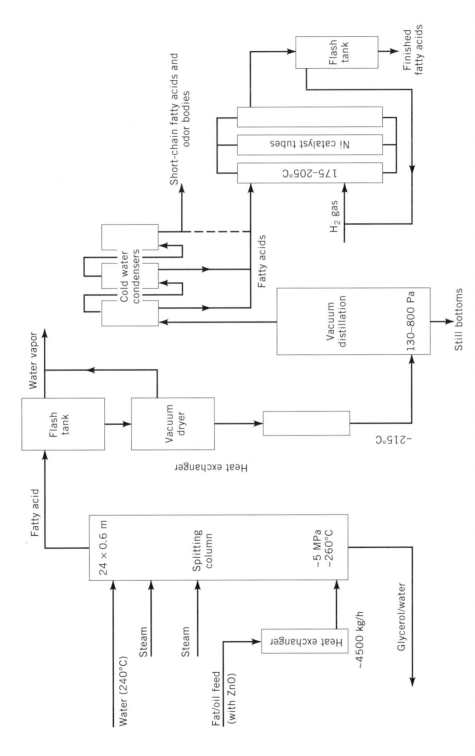

Fig. 4. Process stream diagram for the production of fatty acids through hydrolysis of fats and oils. Steam is at 5.2–6.2 MPa (750–900 psi). To convert MPa to psi, multiply by 145. To convert Pa to mm Hg, multiply by 0.0075.

oils, the Zn catalyst, and odor and color bodies. This is achieved by heating the product stream in a heat exchanger to around 205–232°C and introducing it into a vacuum chamber (flash still) at 0.13–0.8 kPa (1–6 mm Hg) absolute pressure. The fatty acids are vaporized under these conditions and removed from the undesired materials such as the partially hydrolyzed triglyceride. The vaporized fatty acids are then passed through a series of cold water condensers for fractionation and collection. Systems vary in the number of condensers but a three-condenser system is common. The fatty acids are typically separated into a heavy cut, a mid-cut, and a very light cut. Depending on the end uses for the fatty acid, these fractions may be used as separated for specialty fats such as stearic acid or reblended for desired fat ratios such as 80:20% tallow:coconut soap. The light cut is often removed from the other condensates because it contains many of the odor bodies present in the fatty acid.

The fatty acids obtained from the process can be used directly or further manipulated for improved or modified performance and stability. Hardening is an operation in which some fraction of the unsaturated bonds present in the fatty acids are eliminated through hydrogenation or the addition of H_2 across a carbon–carbon double bond. This process was initially intended to improve the odor and color stability of fatty acids through elimination of the polyunsaturated species. However, with the growth in the use of specialty fatty acids, hydrogenation is a commercially important process to modify the physical properties of the fatty acids.

Hardening is typically achieved by passing the preheated fatty acid through a series of tubes packed with catalyst in the presence of hydrogen gas. The most commonly used catalyst is Ni, but other catalysts are available. The amount of hardening is governed by the amount of hydrogen, the reaction temperature, the pressure, and the residence time. The hardened fatty acids are filtered to remove any residual catalyst and then cooled in a flash tank where the excess hydrogen gas is removed. In addition to the reduction of the unsaturate level in the fatty acid, the process may also convert some of the naturally occurring cis-configuration unsaturated fatty acids into the corresponding trans-configuration. The conversion can affect the finished product properties and is typically controlled to within desired specifications.

Neutralization Step. The formation of soap from fatty acids is achieved through the reaction of the fatty acid with the appropriate caustic. This reaction is extremely rapid for most common caustics, eg, NaOH or KOH, and requires proper stoichiometry and rigorous mixing to ensure processing effectiveness. Although this appears relatively straightforward, in practice, there are a number of processing considerations which must be addressed. First, an exact ratio of fatty acids, caustic, water, and salt must be maintained to ensure formation of the desired neat soap phase. The process is controlled to avoid the formation of middle soap, which has a high viscosity and does not dissipate rapidly, or the formation of a neat soap–nigre biphasic mixture, which may separate upon storage. Second, intimate mixing of the oil and aqueous reactants is necessary to ensure uniform neat soap phase composition. Third, because of the heat liberated by the reaction, temperature control must be maintained within certain limits to prevent overheating and boiling/foaming.

There are a variety of commercial systems for achieving neutralization. Generally, a heated fatty acid blend (~50–70°C) and caustic–salt–water (~25–30°C) streams are metered into some form of a high shear mixing system, commonly referred to as a neutralizer. The mixed stream heats to between 85 and 95°C on account of the latent heat of reaction and is pumped into a receiver tank which effectively mixes the soap through both a recirculation system and agitation. After a short residence time in the receiver tank to ensure a uniform composition, the resulting neat soap is pumped into storage tanks or to the finishing operations.

Comparison of Base Soap Manufacturing Routes. Direct saponification of fats and oils is well known, characterized, and straightforward; requires little equipment; and is relatively energy-efficient. However, it is not very effective with regard to changes in the fats and oils ratio desired for finished soap bar formulations. Furthermore, direct saponification has the drawbacks of lower glycerol yields, limited flexibility toward formation of mixed counterion soaps, and requires higher quality feedstock for good quality soaps. In contrast, the hydrolyzer/neutralizer system is more flexible with regard to formation of mixed counterions and formula changes, allows for post-hardening of fatty acids, and provides better glycerol recovery. Furthermore, the ability for both distillation and post-hardening provides greater flexibility in fats and oils feedstock selection; lesser grades can be utilized to yield comparable quality base soap. However, this process is extremely energy-intensive and requires more specialized process equipment, eg, hydrolyzer columns, stills, a hydrogenation system, and neutralizer, and necessitates the use of stainless steel on account of the corrosive nature of fatty acid.

Bar Soap Manufacturing

The conversion of wet base soap into consumer-acceptable bar soaps can be achieved using one of three commonly utilized manufacturing processes: framing, milling, and hot extrusion, all of which use a variety of processing unit operations or finishing steps. These steps include wet mixing or crutching, drying, dry mixing or compounding, and bar forming.

Framing. The framed bar process is by far the oldest and the most straightforward process utilized in the production of bar soaps. The wet base soap is pumped into a heated, agitated vessel commonly referred to as a crutcher. The minor ingredients used in soap bars such as fragrance or preservative are added to the wet soap in the crutcher or injected in-line after reduction of product stream temperature. The hot mixture is then pumped into molds and allowed to cool.

These molds can be either finished bar-shape molds or large blocks. Finished bar-shape molds can be either a mated two-piece design or a five-sided, open-top design. Upon cooling the solid bar is removed from the mold and packaged as desired. For the large blocks, the mold is pulled apart and the block of solid soap is removed. Wire cutters are employed to cut the blocks first into slabs, then into stripes, and finally into rectangular bricks representing the finished size of the bar. The rectangular brick is finished by a final stamping step which typically embosses the logo and any shape modifications into the brick. This large-block approach is only suitable for brick-like shapes, whereas the finished bar-shape molds allow for the production of much more complex shapes.

Traditionally, this process has been utilized primarily for simple soap bars because it tends to be time-consuming and thus somewhat limited for large-scale bar production. However, advances have been reported in automating this approach (14). Furthermore, the process requires fluid crutcher compositions for flow into the molds. This typically requires the formulation to contain either a high level of solvents, including water, glycerol, and alcohol, and be at elevated temperatures (>80°C) when poured into the frames. Despite these limitations, it has proven to be the preferred route to producing certain specialty products, for example, transparent bars.

Milled Bar Process.   The process utilized in the 1990s for the production of most bar soaps is the milled bar process. The milled soap process produces high quality soap bars. The process requires drying as well as milling and plodding (extrusion) of low moisture soap and is capable of high efficiency and throughput (~300–400 bars/min for a given packing line). This process, however, is also equipment-intensive on account of the number of process unit operations required.

For this process, the wet soap is pumped into a mixing vessel (crutcher) where the addition and mixing of minor ingredients may be achieved. Minor ingredients include excess fatty acids, preservatives, and potentially other synthetic surfactants. Alternatively, mixing can be achieved through the use of in-line static mixers, with the accurate addition of the minors into a flowing stream of the wet soap.

The wet soap is put into the drying operation where moldable solid soap is created by reducing the water content in the wet soap from around 30% to between 7 and 15%. This drying step can be attained through three typical approaches: atmospheric flash drying, vacuum drying, and chilled surface drying. For all three approaches the wet soap stream is first superheated in a high pressure steam heat exchanger of either a plate-and-frame or tube-and-shell design. The amount of heating the soap undergoes is dependent on the drying needs.

Atmospheric flash drying is similar to spray tower drying utilized for the formation of detergent granules. The superheated soap (~190–220°C) is sprayed at pressures of about 2.8 MPa (400 psi) (with specially designed nozzles) as small particles into the upper part of a tower (at atmospheric pressures). The high pressure spraying process causes a rapid loss of moisture from the superheated soap in the form of steam. Cooling of the hot, dry soap particles is achieved using cooling air which is blown into the bottom of the tower. The air cools the soap as it falls to the bottom of the tower. The cooling air is usually humidified to prevent overdrying of the soap. The soap is removed from the bottom of the tower.

Vacuum drying is similar to atmospheric drying but does not require as high a temperature to drive the moisture loss; it can be performed in a considerably smaller tower. Wet soap, heated to around 130–150°C in a low pressure heat exchanger, is sprayed onto the walls of an evacuated tower using a nozzle. The nozzle can either be unidirectional, mounted on a rotating shaft, or statically mounted and multidirectional. Cooling and drying is achieved in one step through the rapid release of moisture as vapor, which occurs upon introduction of superheated soap into the vacuum chamber (Joule-Thompson cooling). The dried, cooled soap is scraped off the tower wall with a scraper blade mounted on a rotating shaft. The moisture in the dried soap is dependent on the flow rate, the temperature of the soap, and the pressure in the vacuum chamber; the last also controls the final temperature of the soap. The dried soap is obtained at the bottom of the tower in the form of small pellets through an airlock created by the screw extrusion of the soap through a multiholed orifice plate. Vacuum drying has a number of advantages over atmospheric flash drying, including the lower pressure steam requirements, lower overall temperatures, and the more compact drying system/tower.

The chilled surface drying process is similar to atmospheric drying, with the cooling process being driven by a chilled surface as opposed to air flow through a tall tower. The wet soap is superheated in high pressure, nonboiling heat exchangers. Drying is achieved by the release of steam when this super-heated soap is introduced into a chamber with a slight negative pressure, which is commonly referred to as a flash chamber. The resulting hot, dry soap melt is cooled through the formation of a thin film on a chilled surfaced, commonly in the form of a roll (rotating cylinder). The hot dry soap falls into the small gap ($\sim$10–50 μm) formed at the interface between a large chilled roll and a smaller, temperature-controlled (may be heated or cooled) applicator roll that aids in uniform film formation. As the chilled roll rotates, the dry, cold soap is removed via scraping with a doctor blade and emerges in the form of flat flakes. The amount of soap drying is governed by the temperature at which the soap is introduced and the air flow in the flash chamber. This process is exceptionally good for modern synthetic surfactant containing formulations because it is amenable to more sticky in-process materials. This drying approach can also be achieved using a chilled belt in place of the chilled roll.

Additional minor ingredients, eg, pigments, fragrances, dyes, preservatives, and antibacterial actives, or some co-surfactants in modern-day bar soaps are introduced into the dried soap in either a batch or continuous fashion in a unit operation referred to as amalgamation. An amalgamator is a paddle mixer where both solids and liquids can be effectively pulverized and mixed on a macroscopic scale. In a batch process, the materials are individually weighed and placed into the amalgamator. The batch process is quite labor-intensive and time-consuming. The continuous process is more commonly utilized because it allows for higher efficiencies and throughput. In a continuous mixer, solids are typically introduced into the amalgamator through weighbelts, whereas liquid components are delivered using metering pumps. The materials are mixed and moved through the amalgamator using a helical agitator.

More intimate mixing of the soap and minor ingredients is achieved using controlled-temperature milling. Milling is an operation in which the soap is passed through a series of closely spaced, temperature-controlled steel rolls which dictates product temperature, inputs work into the soap mixture, and provides efficient micromixing. Four-roll mills are very common but three- and five-rolls are also used. Mills are designed such that successive rolls rotate in opposite directions and at slightly increased rates than the previous roll. Therefore, at the point of contact between two mill rolls, the two surfaces are moving in the same direction but with different speeds. A zone, called a bead, is created at this point of contact where material is micromixed through the high shear nature of the zone. This high shear mixing also causes heating of the product stream. The spacing between the rolls is set to generate final product of around 1.0–1.5 mm thickness. At the top roll, the soap is scraped off using a knife blade into ribbons of less than 5 cm in width. A typical four-roll mill can handle around 3600 kg/h of material, and inputs enough work and temperature to raise the dried soap temperature from 25 to 40°C.

Milling not only provides intimate mixing, but also eliminates variation in ribbon thickness and crushes lumpy materials, eg, overdried soap, which might impact finished bar texture. Milling is also utilized for the formation of the proper bar soap crystalline phase, which plays a critical role in both the performance properties of the soap bar and the handling characteristics of the in-process soap. For example, too hot a milling temperature can create sticky soap that is difficult to process further.

The formation of finished bar soaps is accomplished through the continuous extrusion of a shaped plug of soap. This extrusion is commonly referred to as plodding and is achieved using a two-stage single- or twin-worm-screw extruder. The purpose of plodding is to compact the soap noodles into a solid mass of soap which is in a manageable form and devoid of air. The first stage of the plodder pushes the soap through a multiholed orifice plate which acts as an airlock for the second stage. The second stage is under vacuum to ensure the removal of entrained air which impacts final bar appearance. The second stage also pushes the soap through a temperature-controlled barrel which terminates in a cone having a shaped orifice plate. The orifice plate yields a soap plug with proper dimensions for cutting and stamping into the desired bar shape. During this plodding step, heat may be added or removed. The worm screw and conical termination of the barrel force the soap into a plastic mass (at appropriate temperatures), which is welded together and emerges as a smoothly surfaced, continuous plug of soap.

The plug at the exit of the plodder is cut into the appropriate length and directed into the stamping and packaging operations. Product can be stamped into the desired shape on account of its intrinsic plasticity using either fixed capacity or box dies. Capacity dies are a pair of casts pressed together to form the desired shape of the bar. The dies possess a fixed capacity and excess material is pushed outside the mated die pair. The excess, on the order of 20% of the original plug mass, is recycled back to the plodder. Box dies are an arrangement of two dies that, in conjunction with a cavity referred to as a box, form the shape of the bar. The plug is placed into the cavity and the two dies push the soap to fill in the shape confined by the two dies and the box. There is very little excess because this design uses the total mass of material to fill out the shape. However, the resultant bar soap has a band around its perimeter on account of the box. Capacity dies provide greater flexibility in bar shape design, whereas box dies have the advantage of producing much lower amounts of recycle material. Dies are typically produced out of brass, highly polished to produce a high gloss, smooth bar surface, and cooled (~0–15°C) to eliminate product sticking during stamping. To further eliminate the sticking of the final bar to the die, a liquor of concentrated brine solution or glycerin is often applied to the die surface.

Stamped bars are then either wrapped or placed in cartons and bundled for sale. The entire bar finishing operation from the plodder to cases of finished product operate at rates of between 150 to 400 bar/min, depending on the stamp design and packaging equipment.

Hot Extrusion Process. The hot extrusion process was originally developed as a replacement of the traditional framed process utilized for the manufacturing of floating soap bars, ie, Ivory (15). This process utilizes a scraped-wall heat exchanger (SWHX) to provide controlled crystallization and plug forming in one step. Wet soap is first partially dried from ~30 to ~20% moisture using a high pressure heat exchanger, operating around 180–190°C, and an atmospheric flash tank to drive off the water. The high temperature and pressure drop causes the flashing. The reduced moisture soap, at ~85–95°C, is then pumped through a high shear mixer where minor ingredients such as fragrance and preservative are added. For a floating bar, air is also injected into the product stream at this point.

The product stream is then pumped into the SWHX for cooling and crystallization. The SWHX is an open tube and is jacketed for cooling using <0°C brine circulated at a high rate. Inside the open tube is a rotating shaft containing scrapers. This arrangement provides efficient cooling of the product stream as it transverses the 1–2 m of the SWHX by scraping cooled product off the wall and effectively mixing it back into the bulk product. At the end of the SWHX is a cone that causes compression of the product stream into a rectangular plug. Inside the cone are a series of rotating arms that help to ensure a homogenized product stream. At the outlet of the SWHX emerges a partially crystallized strip at ~55–70°C; this partial crystallization makes the strip firm enough to maintain its shape once extruded. The extruded strip is taken away from the outlet of the SWHX on a conveyor belt, cut to appropriate dimensions, and allowed to cool further under specific temperature and humidity conditions to ensure proper crystallization. The cut strips are then finished into packaged bars using processes similar to those described previously.

Formulation of Soaps

The formulation of bar soaps has become increasingly complex in the 1990s with changing consumer bathing habits and expectations. In the past, consumers' bathing habits were such, eg, once-a-week baths, that simple lye soaps were acceptable. However, in the 1990s, it is not uncommon to shower everyday, which puts greater demands on the performance properties of soap bar, for example, mildness to skin and formation of bathtub ring (16). Manufacturers of bar soaps have developed a variety of formulation approaches to deliver products that better meet the consumer needs of the 1990s. This is achieved through the proper balancing of soap components, inclusion of various additives, or the blending of synthetic surfactants into the formula. In addition, new forms of cleansing products have been introduced to address these changing habits and consumer needs, eg, liquid handsoaps and shower gels or body washes. For personal cleansing products, including bar soaps, performance is measured by such tests as lather, wet cracking, smear, firmness, rinsability (the amount of residue left on surfaces after rinsing with hard water), and mildness to skin (17). It is through use of these measures and exhaustive consumer research that modern soapmakers develop better products for consumer needs.

Soap Bars.   In soap bars the primary surfactant is predominantly sodium salts of fatty acids. These products typically contain between 70 and 85% soap. Occasionally, potassium soap (~5–30%) is included in the formulation to increase the solubility of the soap and, hence, the bar's lathering properties. The low Krafft temperatures for potassium soap are the basis for the lather enhancement, but also limits their content in bars.

Soap performance can be controlled through the proper blending of fats and oils to specific ratios, and the formation of the proper phase and colloidal structure. It is common to produce soap using a blend of tallow and coconut or palm kernel oils, generally in a ratio of between 85:15 to 50:50. As the amount of coconut oil is increased in the bar, the lathering profile of the product typically increases as a result of the inherent higher solubility of the soaps formed from coconut (shorter chain length soaps). However, this higher lather comes at the expense of bar smear, which increases for the same solubility reasons. Furthermore, the higher content of sodium laurate in these high coconut soaps can negatively impact the mildness of the product, because laurate soaps are intrinsically more irritating to skin than other chain lengths (18). High lathering, pure coconut oil soaps are still marketed in the United States as Castille soaps.

Additionally, soap bars typically contain between 8 and 20% water, 0.5 to 1% NaCl, and low levels of glycerol. Salt modifies the processibility of soap during plodding and milling, as well as being a carryover ingredient from the manufacture of the base soap. Glycerol is also an impurity remaining from the base soap production, but in some bars it is actually added for rinsing or skin-feel purposes.

Bar Soap Additives.   There are a variety of additives that may be formulated into soap bars to provide additional consumer benefits or modify the performance of the products.

Free Fatty Acid.   Soap bars are intrinsically alkaline in nature on account of the physical properties of soap in water and the process utilized in its manufacture, which yields base soap having a very slight excess of free caustic. Free fatty acid, commonly either coconut or palm kernel, is added into the formulation to neutralize this slight excess of caustic. Often higher fatty acid levels (1–8%) are incorporated into the formula to modify the performance of the product, referred to as superfatting. The free fatty acid associates with soap to form acid–soap crystals. The formation of acid–soap crystals changes both the texture and plasticity of the bar, as well as the lather performance. Superfatted bars are smoother and longer lasting. They also yield lathers that are more stable, creamier, and more dense than nonsuperfatted products. However, superfatting can also decrease the odor and color stability of the final product because oxidative degradation of fatty acids is faster than for the analogous soaps. Other materials can be used to achieve similar characteristics as superfatted bars, for example, waxes and triglycerides. These types of superfatting agents may also serve as emollients.

Glycerol.   This common skin care ingredient is formulated in bar soaps because of its humectant properties. Glycerol, at levels of 10%, has been shown to change significantly the consumer skin softness and smoothness perception (19). Even at low levels, glycerol can alter the skin-rinsing profile of the bar. Unfortunately, high levels of glycerol can negatively impact bar processibility through increasing product softness and stickiness.

Colorants, Dyes, and Pigments. It is quite common to modify the appearance or aesthetic properties of bar soaps through the incorporation of various colorants and opacifiers. The most commonly used material is titanium dioxide, which at low levels (<0.8%) is an effective whitener and opacifier. Most marketed bar soaps contain some level of TiO_2 as either an opacifier in conjunction with other colorants or as a whitener. A variety of dyes are also utilized in addition to TiO_2 to generate desired product colors. The dyes used are almost exclusively dyestuffs of Drug and Cosmetic or Food, Drug, and Cosmetic grades. Some producers also utilize inert, inorganic pigments for product coloration. Pigments have an advantage over dyes; they are inherently more color-stable and not water soluble. The latter attribute is important for striped or two-toned products, because water-soluble dyes can migrate in the product and eventually lessen the contrast between the two tones present.

Fragrance. A key aesthetic for consumer acceptance of personal cleansing products is the product fragrance. Fragrance is utilized by manufacturers of soaps as one of the primary means of targeting products for specific user groups and connoting different product marketing positions. A secondary purpose of fragrance is to mask the fatty base odor of the soap. Commonly, fragrance development is performed by perfume houses who focus their development on product appeal needs. For example, fragrances used for deodorant products tend to be impactful and residual to skin to provide long-lived fragrance on skin. A number of products are appearing on the market that are designed for individuals with sensitive skin. The fragrance types and levels used in these sensitive-skin products are such that they mask the base odors of the soap while providing some soft perfume notes during use, reinforcing their mildness to or compatibility with skin. Fragrance levels are typically in the range of 0.7 to 1.5%, but sensitive-skin products contain much lower levels. The level of fragrance used is a function of the target audience for the product and the odor stability of the fragrance in the product. Product odor instability results from both the loss of fragrance during storage and the propensity for oxidation of fragrance and soap components. Hence, a product may change from an acceptable to an unacceptable odor profile during its lifetime if not properly formulated.

Chelants and Antioxidants. Soaps, fatty acids, and fragrances are susceptible to oxidation during aging (20). The oxidation process is quite complex but typically results from the reaction of the unsaturated bonds in these components with oxygen in the air, resulting in the formation of shorter chain length acids, aldehydes, and ketones which are extremely odoriferous. In the case of fragrance components, oxidation can produce a change in product odor character and cause discoloration of the bar. To minimize the oxidation of the base soap and other minor ingredients in soap bars, both chelants and antioxidants are commonly used.

Chelants at concentrations of 0.1 to 0.2% improve the oxidative stability through the complexation of the trace metal ions, eg, iron, which catalyze the oxidative processes. Examples of the chelants commonly used are pentasodium diethylenetriaminepentaacetic acid (DTPA), tetrasodium ethylenediaminetetraacetic acid (EDTA), sodium etidronate (EHDP), and citric acid. Magnesium silicate, formed in wet soap through the reaction of magnesium and silicate ions, is another chelant commonly used in simple soap bars.

Antioxidants are also used in conjunction with chelants to further improve product odor and color stability. Antioxidants work by chemically trapping the free radicals formed during the oxidation process, significantly decreasing the rate of the degradation reaction. This is particularly important for fragrance components. Butylated hydroxytoluene (BHT), one of the most commonly utilized antioxidants, is usually incorporated at levels of 100–200 ppm in the formulation. BHT is frequently added directly to the fragrance to improve the storage stability of the neat material.

Mildness and Skin Additives.   The increased frequency of bathing and the changing consumer need has necessitated the development of products having skin care benefits. In addition to the two most common additives, fatty acid and glycerol, there is a wealth of other additives which are frequently used. Examples include lanolin, vitamin E, aloe vera gel, mineral oil, and baking soda.

Inert materials are sometimes used in soap bars as a means of improving the skin mildness of the product by decreasing the level of soap and surfactant in the bar. The cleansing agents at high concentrations can sometime dry and irritate skin. A variety of inert materials, both inorganic and organic, have been reported in the literature, including oatmeal, dextrin, starch, wax, and talc (21). These materials may also deposit on the skin during washing, further modifying the rinsing properties of the soap bar and impacting the consumer perception of the product and its aesthetic properties.

Newer technologies have been utilized in the manufacture of bar soaps, which truly improve the clinical mildness-to-skin of these products. One approach relies on minimizing the overall levels of the more irritating soap species such as the laurates and unsaturated species through appropriate balancing of feedstocks (22). Another approach is the incorporation of quaternary amine compounds into the formula, which effectively complexes the soap during the wash–rinse process, reducing its potential to remove oils from or interact with the skin. The amines commonly take the form of cationic polymers based on natural materials such as cellulose, guar gums, and proteins (23).

Antimicrobial Agents.   Antimicrobial agents have been used for a number of years in soap bars as a means of providing additional deodorant protection through their residual effectiveness on suppressing the growth of odor-causing bacteria. These materials actually deposit on skin during the washing process and provide a reservoir of active ingredient that is effective at suppressing bacterial growth between washings. It is widely believed that these soaps may provide additional benefits on account of their ability to control the microflora on the skin surface. One such benefit may be the reduction in the level or frequency of minor skin infections by controlling the *Staphlyococcus aureus* level on the skin surface (24). Only two active ingredients are commonly used in bar soaps: trichlorocarbanalide or TCC (Triclocarban) and trichlorohydroxydiphenyl ether or TCS (Triclosan). These compounds are typically used at concentrations of 0.25 to 1.5% in the final product and have activity against a wide range of microorganisms.

Specialty Soaps.   There are a variety of specialty soaps that require certain additives to deliver the special consumer needs for which they were developed. Scouring soaps contain an abrasive agent homogeneously distributed throughout the soap to aid in the cleaning properties of the product. The abrasives are extremely small particles of insoluble material such as pumice. Striped bar soaps are commonly produced using two soap streams with different colorant systems that are intentionally poorly mixed during extrusion through the plodder. Transparent soap bars, often called glycerin soap, are formed through the quiescent cooling of a high solubility soap system containing a high level of solvent (25). The solvent aids in the formation of a clear gel through retardation of large crystal formation. Common solvents include glycerol, triethanolamine, ethyl alcohol, and sugars. A common type of soap used in these products is triethanolamine soaps, eg, Neutrogena, which have relatively low crystallization temperatures.

Synthetic Surfactant.   Much of the development of new soap bar technologies has been focused on products containing some level of synthetic surfactants. The primary benefits of synthetic surfactants over soaps are their intrinsic lower sensitivity to water hardness, which improves their rinsing profiles, their lathering ability, and their effects on skin feel and mildness. Anionic, nonionic, and amphoteric surfactants have all been formulated into bar soaps. In most bar soaps, the synthetic surfactant serves the purpose of a secondary surfactant (at levels of 1–20%), modifying the lathering, rinsing, or skin effects profile. However, in some skin care or beauty bars, these surfactants represent the primary cleansing agent (up to 50 or 60% of formulation), with soap taking the role as the secondary surfactant, present primarily for structural purposes.

Anionic surfactants are the most commonly used class of surfactant. Anionic surfactants include sulfates such as sodium alkylsulfate and the homologous ethoxylated versions and sulfonates, eg, sodium alkylglycerol ether sulfonate and sodium cocoyl isethionate. Nonionic surfactants are commonly used at low levels (~1–2%) to reduce soap scum formation of the product, especially in hard water. These nonionic surfactants are usually ethoxylated fatty materials, such as $HOCH_2CH_2O(CH_2CH_2O)_nR$. These are commonly based on triglycerides or fatty alcohols. Amphoteric surfactants, such as cocamidopropyl betaine and cocoamphoacetate, are more recent surfactants in the bar soap area and are typically used at low levels (<2%) as secondary surfactants. These materials can have a dramatic impact on both the lathering and mildness of products (26).

These surfactants, in conjunction with soap, produce bars that may possess superior lathering and rinsing in hard water, greater lather stability, and improved skin effects. Beauty and skin care bars are becoming very complex formulations. A review of the literature clearly demonstrates the complexity of these very mild formulations, where it is not uncommon to find a mixture of synthetic surfactants, each of which is specifically added to modify various properties of the product. For example, one approach commonly reported is to blend a low level of soap (for product firmness), a mild primary surfactant (such as sodium cocoyl isethionate), a high lathering or lather-boosting cosurfactant, eg, cocamidopropyl betaine or AGS, and potentially an emollient like stearic acid (27). Such benefits come at a cost to the consumer because these materials are considerably more expensive than simple soaps.

Liquid Soaps and Body Washes.   In the late 1970s and early 1980s a new form of soap product emerged, commonly referred to as liquid handsoaps. These liquid soaps were offered as a practical replacement of soap bars for use at sinks in the bathroom and kitchen. Manufacturers have taken two basic approaches to the formulation of these products: soap-based and synthetic-based formulations. Soap-based formulas use potassium or ammonium salts to yield soaps that are highly soluble at room temperature. These soaps have typically been of either short-chain lengths, such as coconut soap, or a blend of short-chain lengths and unsaturated soaps such as oleic. More recently, these soap-based formulations have been replaced by synthetic surfactant-based formulations. Synthetic surfactant formulations have the advantages of being milder-to-skin, cleaner rinsing, higher lathering, and less sensitive to water hardness. A typical synthetic surfactant formulation is around 80% water and may contain sodium alkyl sulfate and sodium alkylethoxy sulfate as the primary surfactant, a nonionic surfactant such as lauramide DEA, and potentially a lather-building amphoteric surfactant such as cocamidopropylbetaine [61789-40-0] (28). Other commonly utilized surfactants include sodium cocoyl isethionate and sodium olefin sulfonate. The U.S. market has evolved significantly since the initial introduction of liquid handsoaps, and most products in the 1990s possess both antibacterial agents (usually TCS) and moisturizers. The latter are used to protect hands from drying and germs. The most commonly utilized moisturizing agents are cationic polymers such as those previously discussed.

Body washes are another more recent introduction into the marketplace. These products have become a mainstay in the European market and, in only a few years, have grown to be a significant fraction of the U.S. market. Body washes can be simple formulas similar to those used for liquid handsoaps or complex 2-in-1 oil-in-water emulsion, moisturizing formulations. These products contain a wide range of synthetic surfactants not typically found in bar soaps or liquid handsoaps, such as sodium monoalkyl phosphate. It is not uncommon to find over 20 different components in these formulations, with no less than six or seven different surfactants. The 2-in-1 products also contain skin conditioning agents, such as cationic polymers, and emollients or beauty oils to provide even milder-to-skin cleansing and in-use moisturization.

Economic Aspects

The personal cleansing market in the United States has changed dramatically from the mid-1980s to the mid-1990s. By the end of 1995, the total market represented over $1.8 billion in consumer sales divided among the three primary categories: bar soaps (~77%), liquid handsoaps (~13%), and body washes (10%) (2). A few large manufacturers produce the majority of these products, as indicated in Table 2. However, a number of smaller producers also manufacture bars for both general consumption and specialty uses.

Analytical Characterization of Soap

There are a variety of analytical methods commonly utilized for the characterization of neat soap and bar soaps. Many of these methods have been published as

Table 2. Primary Manufacturers of Bar Soaps, Liquid Handsoaps, and Body Washes

Producer	Product
Procter & Gamble Co.	Ivory, Oil of Olay, Zest, Coast, Safeguard, Camay
Lever Brothers Co.	Dove, Lever 2000, Caress, Shield, Lux
Colgate-Palmolive Co.	Irish Spring, Cashmere Bouquet, Softsoap, Palmolive
Dial Corp.	Dial, Pure & Natural, Tone, Spirit
Andew Jergens Co.	Jergens Mild, Naturals

official methods by the American Oil Chemists' Society (29). Additionally, many analysts choose *United States Pharmacopoeia* (USP), *British Pharmacopoeia* (BP), or *Food Chemical Codex* (FCC) methods. These methods tend to be colorimetric, potentiometric, or titrametric procedures. However, a variety of instrumental techniques are also frequently utilized, eg, gas chromatography, high performance liquid chromatography, nuclear magnetic resonance spectroscopy, infrared spectroscopy, and mass spectrometry.

Some of the traditional methods used for the characterization of bar soaps are moisture, free acid or free alkalinity, total soap, and chloride (salt). Moisture determinations are typically run by Karl Fischer titration. However, under certain circumstances, moisture can be measured gravimetrically. The measure of free fatty acid or free alkalinity is typically done via a simple pH titration using phenophthalein or potentiometric end point. Total soap is generally done using an extraction/gravimetric procedure: the soap sample is acidified in an ethanol–water mixture, the fatty acids are extracted into a hydrocarbon solvent, and the amount of extracted fatty acid is measured by gravimetric analysis after elimination of the solvent. The measurement of sodium chloride is accomplished commonly using an ion-selective electrode titration. Glycerol is analyzed using gas chromatography. The analysis of finished bar soaps include, in addition to those stated above, color and odor analysis; analysis of active ingredients, eg, antimicrobial actives; and quantitation of synthetic surfactants.

Health, Safety, and Toxicology

The manufacture of soap poses some material handling concerns because of the reaction of strong caustics with either neutral fats and oils or fatty acids at relatively high temperatures. The caustics, ie, sodium hydroxide and potassium hydroxide, represent the primary hazard. At around 50% concentrations, these caustics are extremely corrosive and may cause serious body burns and eye injuries if not removed quickly through rinsing with copious amounts of water. Appropriate protective clothing is strongly urged when handling these materials.

Soap as utilized in personal cleansing products has a long safe history of use. Modern soaps have been specifically formulated to be compatible with skin and to be used on a daily basis with minimal side effects. Excessive use of soap for skin cleansing can disrupt the natural barrier function of skin through the removal of skin oils and disruption of the lipid bilayer in skin. This can result in imperfect desquamation or a dry appearance to skin and cause an irritation response or erythema, ie, reddening of the skin. Neither of these are a permanent response and the elicitation of this type of skin reaction is quite dependent on the individual's skin type, the product formulation, and the frequency of use.

There is a considerable amount of research into the compatibility of cleansing products with skin (16,30). Modern soap manufacturers improve the skin compatibility of their products through a variety of chemical testing methods. These methods are often used to evaluate the mildness (irritation potential) of test formulations in comparison to other formulations on the basis of the dry skin and irritated (red) appearance of skin. There are many reports of comparative studies of various formulations and their mildness-to-skin; however, the results of these overly exaggerated test methods may not reflect consumer experience with products (16). Upon direct contact with other sensitive membranes such as eyes, soap may also cause irritation in the form of stinging. Again, this is a temporary response which can be rectified through rinsing with water. Ingestion of soap poses little risk at the levels of materials usually ingested. Typically, temporary minor irritation of mucous membranes and gastrointestinal disorders, eg, nausea, vomiting, and diarrhea, may be experienced.

Additional Uses of Soap

The primary use of carboxylate soaps is in the manufacture of personal cleansing products, principally bar soaps. Of the soap produced in 1987 (5.4×10^5 metric tons), approximately 80% was for use in bar soaps (2). Liquid soaps comprise a small percentage of total usage. There are also a number of other applications for both consumer use and industrial utilization. Soluble soaps such as potassium and sodium are utilized in cleansing applications where their detersive and emulsification properties can be leveraged, for example, in the textile industry where the cleansing of various in-process fibers and leather (defatting) is desired. Soaps are also utilized in emulsion polymerization. Furthermore, soaps are used in a variety of cosmetic products. Sodium stearate [822-16-2] provides the structure of many modern antiperspirant and deodorant sticks (31). Amine soaps, typically triethanolamine and ammonium stearate, are used in products where high volume, stable lathers are desired, such as cleansing and shaving creams (32).

The lower solubility metal and alkaline-earth metal soaps are also widely utilized in a variety of different applications. For example, calcium oleate is utilized in the waterproofing of cement. Metal soaps, such as calcium, magnesium, and aluminum soaps, are commonly utilized for the thickening of hydrocarbon lubricating greases, mold release, and suspending agents in paints (see DRIERS AND METALLIC SOAPS) (33). Magnesium stearate [557-04-0] is frequently used as a filler material and binder in drug tablets and as either inert binder or emulsification agent in cleansing products and cosmetics, respectively (34). Aluminum soap has been used in the manufacture of weapons of war, applying its emulsification properties to napalm. Zinc soaps have been studied and applied to specialty cleansing products because of their antibacterial and antifungal properties.

BIBLIOGRAPHY

"Soap" in *ECT* 1st ed., Vol. 12, pp. 553–598, by G. W. Busby, Lever Brothers Co.; in *ECT* 2nd ed., Vol. 18, pp. 415–432, by F. V. Ryer, Lever Brothers Co., Inc.; in *ECT* 3rd ed., Vol. 21, pp. 162–181, by F. S. Osmer, Lever Brothers Co., Inc.

1. L. Spitz, ed., *Soap Technology for the 1990's*, American Oil Chemists' Society, Champaign, Ill., 1990.
2. *1987 Census of Manufacturers*, Vol. II, U.S. Government Printing Office, Washington, D.C., Product Statistics, Table 6A; S. Colwell, *Soap Cos. Chem. Spec.* **71**, 32 (1995).
3. D. Small, ed., *Handbook of Lipid Research 4, The Physical Chemistry of Lipids from Alkanes to Phospholipids*, Plenum Press, New York, 1986, Chapts. 3 and 9.
4. R. G. Laughlin, *The Aqueous Phase Behaviour of Surfactants*, Academic Press, Ltd., London, 1994, pp. 448–451.
5. Ref. 4, pp. 377–383.
6. P. Ekwall and L. Mandell, *Kolloid Z.* **233**, 938–944 (1969).
7. P. Ekwall, L. Mandell, and K. Fontell, *J. Coll. Interfacial Sci.* **28**, 219–226 (1968).
8. D. Chapman, *The Structure of Lipids by Spectroscopic and X-ray Techniques*, John Wiley & Sons, Inc., New York, 1965.
9. R. H. Ferguson, F. B. Rosevear, and R. C. Stillman, *Ind. Eng. Chem.* **35**, 1005 (1943).
10. M. J. Buerger, L. B. Smith, A. deBretteville, and F. V. Ryer, *Proc. Nat. Acad. Sci.* **28**, 529 (1942).
11. N. Garti and K. Sato, *Crystallization and Polymorphism of Fats and Fatty Acids*, Marcel Dekker, Inc., New York, 1988.
12. Ref. 4, Chapt. 5.
13. J. W. McBain and M. C. Field, *J. Chem. Soc.*, 920 (1933); P. Z. Ekwall, *Anorg. Allem. Chem.* **210**, 337 (1933); M. L. Lynch, Y. Pan, and R. G. Laughlin, *J. Phys. Chem.* **100**, 357 (1996).
14. U.S. Pat. 4,758,370 (1988), E. Jungerman, T. Hassapis, R. A. Scott, and M. S. Wortzman (to Neutrogena Corp.).
15. U.S. Pats. 2,295,594; 2,295,595; 2,295,596 (Sept. 15, 1942), V. Mills (to Procter & Gamble Co.).
16. R. Wolf, *Dermatology* **189**, 217 (1994).
17. K. D. Ertel, B. H. Keswick, and P. B. Bryant, *J. Soc. Cos. Chem.* **46**, 67 (1995); B. H. Keswick, K. D. Ertel, and M. O. Visscher, *J. Soc. Cos. Chem.* **43**, 187 (1992).
18. C. Prottey, P. J. Hartop, and T. F. M. Ferguson, *J. Soc. Cosmet. Chem.* **24**, 473–492 (1972).
19. R. M. Dahlgren, M. F. Lukacovic, S. E. Michaels, and M. O. Visscher, in A. R. Baldwin, ed., *Proceedings of the Second World Conference on Detergents*, American Oil Chemists' Society, Champaign, Ill., 1987.
20. H. W. S. Chan, *Autoxidation of Unsaturated Lipids* Academic Press, Inc., Orlando, Fla., 1987.
21. U.S. Pat. 4,151,105 (Apr. 24, 1979), J. R. O'Roark (to Hewitt Soap Co., Inc.).
22. U.S. Pat. 5,387,362 (Feb. 7, 1995), F. R. Tollens, P. J. Kefauver, and S. W. Syfert (to Procter & Gamble Co.); U.S. Pat. 5,264,144 (Nov. 23, 1993), N. M. Moroney and co-workers (to Procter & Gamble Co.).
23. U.S. Pat. 5,296,159 (Mar. 22, 1994), D. B. Wilson and co-workers (to Procter & Gamble Co.).
24. M. B. Finkey, N. C. Corbin, L. B. Aust, R. Aly, and H. I. Maibach, *J. Soc. Cosmet. Chem.* **35**, 351 (1984).
25. U.S. Pat. 5,041,234 (Aug. 20, 1991), T. Instone and M. Bottarelli (to Lever Brothers Co.).

26. Eur. Pat. 0,472,320,A1 (Feb. 26, 1991), J. F. Ashley, A. C. Coxon, and R. S. Lee (to Unilever).

27. U.S. Pat. 4,954,282 (Sept. 4, 1990), K. J. Rys, A. P. Greene, F. S. Osmer, and J. J. Podgorsky (to Lever Brothers Co.).

28. V. World Pat. 9532705 (Dec. 7, 1995), M. Fujiwara, C. Vincent, K. Villa Anathapad-manabhan, and V. Virgilio (to Unilever PLC).

29. *Official and Tentative Methods of the American Oil Chemist's Society*, American Oil Chemist's Society, Chicago, Ill., annual publication.

30. F. R. Bettley and E. Donoghue, *Brit. J. Derm.* **72**, 67 (1960); D. D. Strube and G. Nicoll, *Cutis* **39**, 544 (1987).

31. U.S. Pat. 5,232,689 (Aug. 3, 1993), D. E. Katsoulis and J. M. Smith (to Dow Corning Corp.); U.S. Pat. 5,424,070 (June 13, 1995), P. B. Katsat and B. D. Moghe (to Mennen Co.).

32. U.S. Pat. 5,326,556 (July 5, 1994), A. G. Barnet and M. R. Mezikofsky (to Gillette Co.).

33. U.S. Pat. 5,385,682 (Jan. 31, 1994), T. Hutchings and K. M. Pilgrem (to Exxon Res. & Eng. Co.); U.S. Pat. 5,472,625 (Dec. 5, 1995), P. D. Maples; U.S. Pat. 5,096,605 (Mar. 17, 1992), J. A. Waynick (to Amoco Corp.); U.S. Pat. 5,102,565 (Apr. 7, 1992), J. A. Waynick (to Amoco Corp.); U.S. Pat. 5,279,750 (Jan. 18, 1994), T. Hanano (to Hanano Commercial Co., Ltd.).

34. U.S. Pat. 5,328,632 (July 12, 1994), B. L. Redd and co-workers (to Procter & Gamble Co.); U.S. Pat. 4,806,359 (Feb. 21, 1989), G. W. Radebaugh, R. Glinecke, and T. N. Julian (to McNeilab, Inc.).

General References

E. Woollatt, *The Manufacture of Soaps, Other Detergents and Glycerine*, John Wiley & Sons, Inc., New York, 1985.

D. Swern, ed., *Bailey's Industrial Oil and Fat Products*, 4th ed., Vols. 1 and 2, John Wiley & Sons, Inc., New York, 1979 and 1982; *ibid.*, Y. Hui, ed., 5th ed., 1996.

L. Spitz, ed., *Soap Technology for the 1990's*, American Oil Chemists' Society, Champaign, Ill., 1990.

ROBERT G. BARTOLO
MATTHEW L. LYNCH
The Procter & Gamble Company

SODA. See ALKALI AND CHLORINE PRODUCTS.

SODIUM AND SODIUM ALLOYS

SODIUM

Sodium [7440-23-5], Na, an alkali metal, is the second element of Group 1 (IA) of the Periodic Table, atomic wt 22.9898. The chemical symbol is derived from the Latin *natrium*. Commercial interest in the metal derives from its high chemical reactivity, low melting point, high boiling point, good thermal and electrical conductivity, and high value in use.

Sir Humphry Davy first isolated metallic sodium in 1807 by the electrolytic decomposition of sodium hydroxide. Later, the metal was produced experimentally by thermal reduction of the hydroxide with iron. In 1855, commercial production was started using the Deville process, in which sodium carbonate was reduced with carbon at 1100°C. In 1886 a process for the thermal reduction of sodium hydroxide with carbon was developed. Later sodium was made on a commercial scale by the electrolysis of sodium hydroxide (1,2). The process for the electrolytic decomposition of fused sodium chloride, patented in 1924 (2,3), has been the preferred process since installation of the first electrolysis cells at Niagara Falls in 1925. Sodium chloride decomposition is widely used throughout the world (see SODIUM COMPOUNDS).

Sodium was first used commercially to make aluminum by reduction of sodium aluminum chloride. The principal application as of the mid-1990s is for the manufacture of tetraethyllead (TEL), the antiknock gasoline additive. However, TEL use is declining worldwide because of the recognized toxic effects of lead (qv) released to the environment (see LEAD COMPOUNDS, INDUSTRIAL TOXICOLOGY). Sodium use is growing for manufacture of sodium borohydride and agricultural crop protection chemicals (see BORON COMPOUNDS; FUNGICIDES, AGRICULTURAL). Smaller amounts of sodium are used to produce sodium hydride, indigo dyes, tantalum metal powders, silicon, and sodium peroxide; in the preparation of many organic compounds, pharmaceuticals (qv), sodium azide, and copper; and in lead dross refining.

Sodium is not found in the free state in nature because of its high chemical reactivity. It occurs naturally as a component of many complex minerals and of such simple ones as sodium chloride, sodium carbonate, sodium sulfate, sodium borate, and sodium nitrate. Soluble sodium salts are found in seawater, mineral springs, and salt lakes. Principal U.S. commercial deposits of sodium salts are the Great Salt Lake, Searles Lake, and the rock salt beds of the Gulf Coast, Virginia, New York, and Michigan (see CHEMICALS FROM BRINE). Sodium-23 is the only naturally occurring isotope. The six artificial radioisotopes (qv) are listed in Table 1 (see SODIUM COMPOUNDS).

Physical Properties

Sodium is a soft, malleable solid readily cut with a knife or extruded as wire. It is commonly coated with a layer of white sodium monoxide, carbonate, or hydroxide, depending on the degree and kind of atmospheric exposure. In a

Table 1. Radioisotopes of Sodium

Isotope	CAS Registry Number	Half-life, s
sodium-20	[14809-59-7]	0.4
sodium-21	[15594-24-8]	23.0
sodium-22	[13966-32-0]	2.58[a]
sodium-24	[13982-04-2]	15.0[b]
sodium-25	[15760-13-1]	60.0
sodium-26	[26103-12-8]	1.0

[a]Expressed in years.
[b]Expressed in hours.

strictly anhydrous inert atmosphere, the freshly cut surface has a faintly pink, bright metallic luster. Liquid sodium in such an atmosphere looks much like mercury. Both liquid and solid oxidize in air, but traces of moisture appear to be required for the reaction to proceed. Oxidation of the liquid is accelerated by an increase in temperature, or by increased velocity of sodium through an air or oxygen environment.

Only body-centered cubic crystals, lattice constant 428.2 pm at 20°C, are reported for sodium (4). The atomic radius is 185 pm, the ionic radius 97 pm, and electronic configuration is $1s^2 2s^2 2p^6 3s^1$ (5). Physical properties of sodium are given in Table 2. Greater detail and other properties are also available (5).

Sodium is paramagnetic. The vapor is chiefly monatomic, although the dimer and tetramer have been reported (6). Thin films are opaque in the visible range but transmit in the ultraviolet at ca 210 nm. The vapor is blue, but brilliant green is frequently observed when working with sodium at high temperature, presumably because of mixing of the blue with yellow from partial burning of the vapor.

At 100–300°C sodium readily wets and spreads over many dry solids, eg, sodium chloride or aluminum oxide. In this form the metal is highly reactive (7), but it does not easily wet stainless or carbon steels. Wetting of structural metals is influenced by the cleanliness of the surface, the purity of the sodium, temperature, and the time of exposure. Wetting occurs more readily at ≥300°C and, once attained, persists at lower temperatures (5).

Sodium Dispersions. Sodium is easily dispersed in inert hydrocarbons (qv), eg, white oil or kerosene, by agitation, or using a homogenizing device. Addition of oleic acid and other long-chain fatty acids, higher alcohols and esters, and some finely divided solids, eg, carbon or bentonite, accelerate dispersion and produce finer (1–20 μm) particles. Above 98°C the sodium is present as liquid spheres. On cooling to lower temperatures, solid spheres of sodium remain dispersed in the hydrocarbon and present an extended surface for reaction. Dispersions may contain as much as 50 wt % sodium. Sodium in this form is easily handled and reacts rapidly. For some purposes the presence of the inert hydrocarbon is a disadvantage.

High Surface Sodium. Liquid sodium readily wets many solid surfaces. This property may be used to provide a highly reactive form of sodium without contamination by hydrocarbons. Powdered solids having a high surface area per unit volume, eg, completely dehydrated activated alumina powder, provide

Table 2. Physical Properties of Sodium[a]

Property	Value	Property	Value
ionization potential, V	5.12	specific heat, $kJ/(kg\cdot K)$[b]	
melting point, °C	97.82	solid	
heat of fusion, kJ/kg[b]	113	at 20°C	2.01
volume change on melting, %	2.63	mp	2.16
boiling point, °C	881.4	liquid	
heat of vaporization	3.874	at mp	1.38
at bp, MJ/kg[b]		400°C	1.28
density, g/cm^3		550°C	1.26
solid		electrical resistivity,	
at 20°C	0.968	$\mu\Omega\cdot cm$	
50°C	0.962	solid	
mp	0.951[c]	at 20°C	4.69
liquid		mp	6.60[c]
at mp	0.927	liquid	
400°C	0.856	at mp	9.64
550°C	0.820	400°C	22.14
viscosity, $mPa\cdot s(=cP)$		550°C	29.91
at 100°C	0.680	thermal conductivity,	
400°C	0.284	$W/(m\cdot K)$	
550°C	0.225	solid	
surface tension,		at 20°C	1323
$mN/m(=dyn/cm)$		mp	1193[c]
at mp	192	liquid	
400°C	161	at mp	870
550°C	146	400°C	722
		550°C	648

[a]Ref. 5.
[b]To convert J to cal, divide by 4.184.
[c]Value is estimated.

a suitable base for high surface sodium. Other powders, eg, sodium chloride, hydride, monoxide, or carbonate, can also be used.

The solid to be coated is placed in a vessel equipped with a stirrer, filled with pure, dry nitrogen or another inert gas, and heated to 110–250°C. Clean sodium is added with stirring. If enough is added, the sodium is rapidly distributed over the entire available surface. Depending on that available surface, up to 30 wt % or more sodium can be added without changing the free-flowing character of the system (7,8).

Chemical Properties

Sodium forms unstable solutions in liquid ammonia, where a slow reaction takes place to form sodamide and hydrogen, as follows:

$$Na + NH_3 \longrightarrow NaNH_2 + 0.5\ H_2$$

Iron, cobalt, and nickel catalyze this reaction. The rate depends on temperature and sodium concentration. At −33.5°C, 0.251 kg sodium is soluble in 1 kg ammonia. Concentrated solutions of sodium in ammonia separate into two liquid phases when cooled below the consolute temperature of $-41.6°C$. The composi-tions of the phases depend on the temperature. At the peak of the conjugate solutions curve, the composition is 4.15 atom % sodium. The density decreases with increasing concentration of sodium. Thus, in the two-phase region the dilute bottom phase, low in sodium concentration, has a deep-blue color; the light top phase, high in sodium concentration, has a metallic bronze appearance (9–13).

At high temperature, sodium and its fused halides are mutually soluble (14). The consolute temperatures and corresponding Na mol fractions are given in Table 3. Nitrogen is soluble in liquid sodium to a very limited extent, but sodium has been reported as a nitrogen-transfer medium in fast-breeder reactors (5) (see NUCLEAR REACTORS).

Table 3. Mutual Solubility of Sodium and Fused Sodium Halides

Compound	Consolute temperature, °C	Na concentration, mol fraction
Na–NaF	1182	0.28
Na–NaCl	1080	0.50
Na–NaBr	1025	0.52
Na–NaI	1033	0.50

The solubility–temperature relationships of sodium, sodium compounds, iron, chromium, nickel, helium, hydrogen, and some of the rare gases are impor-tant in the design of sodium heat exchangers, especially those used in liquid-metal fast-breeder reactors (LMFBR). The solubility of oxygen in sodium is par-ticularly important because of its marked effect on the corrosion of containment metals and because of problems of plugging narrow passages. This solubility S given in units of ppmwt O is

$$\log S = 6.239 - 2447/T \qquad (2)$$

for temperature, T, in Kelvin from about 400 to 825 K (5). Solubility data for many other materials in sodium are also available (5). Because metallic calcium is always present in commercial sodium, and to a lesser extent in nuclear-grade sodium, the solubilities of calcium oxide and nitride in sodium are critical to the design of heat-transfer systems. These compounds are substantially insoluble at $100-120°C$ (15) (see HEAT-EXCHANGE TECHNOLOGY).

Sodium is soluble in ethylenediamine (16,17), but solubility in other amines such as methyl- or ethylamine may require the presence of ammonia. Sodium solubility in ammonia and ethylenediamine solutions has been extensively inves-tigated (18). Sodium is insoluble in most hydrocarbons and is readily dispersed in kerosene or similar liquids toward which it is chemically inert. Such dispersions provide a reactive form of the metal.

In 1932 a class of complexes consisting of ethers, sodium, and polycyclic hydrocarbons was discovered (19). Sodium reacts with naphthalene in dimethyl ether as solvent to form a soluble, dark-green, reactive complex. The solution is electrically conductive. The reaction has been described as follows (8):

The addition product, $C_{10}H_8Na$, called naphthalenesodium or sodium naphthalene complex, may be regarded as a resonance hybrid. The ether is more than just a solvent that promotes the reaction. Stability of the complex depends on the presence of the ether, and sodium can be liberated by evaporating the ether or by dilution using an indifferent solvent, such as ethyl ether. A number of ether-type solvents are effective in complex preparation, such as methyl ethyl ether, ethylene glycol dimethyl ether, dioxane, and THF. Trimethylamine also promotes complex formation. This reaction proceeds with all alkali metals. Other aromatic compounds, eg, diphenyl, anthracene, and phenanthrene, also form sodium complexes (16,20).

In 1967, DuPont chemist Charles J. Pedersen (21) discovered a class of ligands capable of complexing alkali metal cations, a discovery which led to the Nobel Prize in Chemistry in 1987. These compounds, known as crown ethers or cryptands, allow greatly enhanced solubility of sodium and other alkali metals in amines and ethers. About 50 crown ethers having between 9–60 membered oligoether rings were described (22). Two such structures, dibenzo-18-crown-6 (**1**) and benzo-9-crown-3 (**2**), are shown.

(**1**)   (**2**)

Sodium Reactions. Sodium reacts with many elements and substances (5,16,20) and forms well-defined compounds with a number of metals. Some of these alloys are liquid below 300°C. When heated in air, sodium ignites at about 120°C and burns with a yellow flame, evolving a dense white acrid smoke. In the presence of air or oxygen a monoxide or peroxide is formed. Limited oxygen supply and temperatures below 160°C give sodium monoxide, Na_2O, as the principal reaction product. At 250–300°C in the presence of adequate oxygen, sodium peroxide, Na_2O_2, is formed along with very small amounts of superoxide, NaO_2 (see PEROXIDES AND PEROXIDE COMPOUNDS, INORGANIC PEROXIDES). Sodium superoxide is made from sodium peroxide and oxygen at high temperature and pressure. Sodium does not react with extremely dry oxygen or air, except for the possible formation of a surface film of transparent oxide (23).

The reaction of sodium and water according to the following equation

$$Na + H_2O \longrightarrow NaOH + 0.5\ H_2 + 141\ kJ/mol\ (98.95\ kcal/mol)$$

has been extensively studied as it relates to the generation of steam in sodium-cooled breeder reactors (5). Under ordinary circumstances, this reaction is very rapid. The liberated heat melts the sodium and frequently ignites the evolved hydrogen if air is present. In the absence of air and a large excess of either reactant, the reaction may be relatively nonviolent. Thus, dry steam or superheated steam may be used to clean equipment contaminated with sodium residues, but precautionary action must be taken to exclude air, avoid condensation, and design equipment to drain without leaving pockets of sodium. Any such sodium may become isolated by a layer of solid sodium hydroxide and can remain very reactive and hazardous.

Hydrogen and sodium do not react at room temperature, but at $200-350°C$ sodium hydride is formed (24,25). The reaction with bulk sodium is slow because of the limited surface available for reaction, but dispersions in hydrocarbons and high surface sodium react more rapidly (7). For the latter, reaction is further accelerated by surface-active agents such as sodium anthracene-9-carboxylate and sodium phenanthrene-9-carboxylate (26–28).

There is very little evidence of the direct formation of sodium carbide from the elements (29,30), but sodium and graphite form lamellar intercalation compounds (16,31–33). At $500-700°C$, sodium and sodium carbonate produce the carbide, Na_2C_2; above $700°C$, free carbon is also formed (34). Sodium reacts with carbon monoxide to give sodium carbide (34), and with acetylene to give sodium acetylide, $NaHC_2$, and sodium carbide (disodium acetylide), Na_2C_2 (see CARBIDES) (8).

Nitrogen and sodium do not react at any temperature under ordinary circumstances, but are reported to form the nitride or azide under the influence of an electric discharge (14,35). Sodium silicide, $NaSi$, has been synthesized from the elements (36,37). When heated together, sodium and phosphorus form sodium phosphide, but in the presence of air with ignition sodium phosphate is formed. Sulfur, selenium, and tellurium form the sulfide, selenide, and telluride, respectively. In vapor phase, sodium forms halides with all halogens (14). At room temperature, chlorine and bromine react rapidly with thin films of sodium (38), whereas fluorine and sodium ignite. Molten sodium ignites in chlorine and burns to sodium chloride (see SODIUM COMPOUNDS, SODIUM HALIDES).

At room temperature, little reaction occurs between carbon dioxide and sodium, but burning sodium reacts vigorously. Under controlled conditions, sodium formate or oxalate may be obtained (8,16). On impact, sodium is reported to react explosively with solid carbon dioxide. In addition to the carbide-forming reaction, carbon monoxide reacts with sodium at $250-340°C$ to yield sodium carbonyl, $(NaCO)_6$ (39,40). Above $1100°C$, the temperature of the Deville process, carbon monoxide and sodium do not react. Sodium reacts with nitrous oxide to form sodium oxide and burns in nitric oxide to form a mixture of nitrite and hyponitrite. At low temperature, liquid nitrogen pentoxide reacts with sodium to produce nitrogen dioxide and sodium nitrate.

Phosphorus trichloride and pentachloride form sodium chloride and sodium phosphide, respectively, in the presence of sodium. Phosphorus oxychloride, $POCl_3$, when heated with sodium, explodes. Carbon disulfide reacts violently, forming sodium sulfide. Sodium amide (sodamide), $NaNH_2$, is formed by the reaction of ammonia gas with liquid sodium. Solid sodium reacts only superficially with liquid sulfur dioxide but molten sodium and gaseous sulfur dioxide react violently. Under carefully controlled conditions, sodium and sulfur dioxide yield sodium hydrosulfite, $Na_2S_2O_4$ (41). Dry hydrogen sulfide gas reacts slowly with solid sodium, but in the presence of moisture the reaction is very rapid. The product is sodium sulfide.

Sodium reacts with dilute acids about as vigorously as it reacts with water. The reaction with concentrated sulfuric acid may be somewhat less vigorous. At 300–385°C, sodium and sodium hydroxide react according to the following equilibrium:

$$2\,Na^0 + NaOH \rightleftharpoons Na_2O + NaH$$

The reaction is displaced to the right by dissociation of sodium hydride and liberation of hydrogen. This dissociation is favored under vacuum or when the reaction zone is swept with an inert gas to remove the hydrogen (24,25). In this manner, sodium monoxide substantially free of sodium and sodium hydroxide is produced. In the more complicated reaction between sodium metal and anhydrous potassium hydroxide, potassium metal and sodium hydroxide are produced in a reversible reaction (42,43):

$$Na^0 + KOH \rightleftharpoons K^0 + NaOH$$

Superimposed on this simple equilibrium are complex reactions involving the oxides and hydrides of the respective metals. At about 400°C, the metal phase resulting from the reaction of sodium and potassium hydroxide contains an unidentified reaction product that precipitates at about 300°C (15).

Data for the free energy of formation (44,45) indicate that sodium reduces the oxides of Group 1 (IA) elements except lithium oxide. Sodium does not reduce oxides of Group 2 (IIA) elements, but does reduce the Group 12 (IIB) mercury, cadmium, and zinc oxides. Many other oxides are reduced by metallic sodium. In some cases reduction depends on the formation of exothermic complex oxides. Iron oxide is reduced by sodium below ca 1200°C. Above this temperature the reaction is reversed. Sodium reduces most fluorides except the fluorides of lithium, the alkaline earths, and some lanthanides. It reduces most metallic chlorides, although some of the Group 1 (IA) and Group 2 (IIA) chlorides give two-phase equilibrium systems consisting of fused salt and alloy layers (43). Some heavy metal sulfides and cyanides are also reduced by sodium.

Sodium reacts with many organic compounds, particularly those containing oxygen, nitrogen, sulfur, halogens, carboxyl, or hydroxyl groups. The reactions are violent in many cases, for example, organic halides. Carbon may be deposited or hydrogen liberated, and compounds containing sulfur or halogens usually form sodium sulfide or sodium halides. Alcohols give alkoxides (see ALKOXIDES, METAL). Primary alcohols react more rapidly than secondary or tertiary. The reactivity decreases with increasing number of alcohol carbon atoms, or higher concentrations.

Organosodium compounds are prepared from sodium and other organometallic compounds or active methylene compounds by reaction with organic halides, cleavage of ethers, or addition to unsaturated compounds. Some aromatic vinyl compounds and allylic compounds also give sodium derivatives.

Sodium does not react with anhydrous ethyl ether but may react with higher ethers or mixed ethers. Organic acids give the corresponding salts with evolution of hydrogen or decompose. Pure, dry, saturated hydrocarbons, eg, xylene, toluene, and mineral oil, do not react with sodium at the hydrocarbon-cracking temperature. In the presence of unsaturated hydrocarbons, sodium may add at a double bond or cause polymerization. Sodium amalgam or sodium and alcohol are employed for organic reductions. Sodium is also used as a condensing agent in acetoacetic ester and malonic ester syntheses and the Wurtz-Fittig reaction (16,21) (see MALONIC ACID AND DERIVATIVES).

Manufacture

Thermal Reduction.   Metallic sodium is produced by thermal reduction of several of its compounds. The earliest commercial processes were based on the carbon reduction of sodium carbonate (46–49) or sodium hydroxide (1,8,50):

$$2\ C^0 + Na_2CO_3 \longrightarrow 2\ Na^0 + 3\ CO$$

$$2\ C^0 + 6\ NaOH \longrightarrow 2\ Na_2CO_3 + 3\ H_2 + 2\ Na^0$$

Sodium chloride is reduced by ferrosilicon in the presence of lime:

$$4\ NaCl + 3\ CaO + (Fe)Si \longrightarrow 2\ CaCl_2 + CaO{\cdot}SiO_2 + 4\ Na^0 + Fe^0$$

This process was operated briefly in vacuum retorts by Union Carbide in 1945 (51).

The chloride is also reduced by calcium carbide at 800–1200°C under vacuum (52).

$$2\ NaCl + CaC_2 \longrightarrow 2\ C^0 + CaCl_2 + 2\ Na^0$$

A number of other thermal reductions are described in the literature (8), but it is doubtful that any have been carried out on commercial scale.

Electrolysis of Fused Sodium Hydroxide.   The first successful electrolytic production of sodium was achieved with the Castner cell (2):

Cathode	$4\ Na^+ + 4\ e^- \longrightarrow 4\ Na^0$
Anode	$4\ OH^- - 4\ e^- \longrightarrow 2\ H_2O + O_2$

The water formed at the anode diffuses to the cathode compartment where it reacts with its equivalent of sodium:

$$2\ H_2O + 2\ Na^0 \longrightarrow 2\ NaOH + H_2$$

The net change is represented as follows:

$$2\,NaOH \longrightarrow 2\,Na^0 + H_2 + O_2$$

Because the water reacts with half of the sodium produced by the electrolysis, the current yield can never be more than 50% of theoretical. Other reactions in the cell lower this yield still more.

The Castner cell was so simple in design and operation that over the years only minor changes have been made. A section of a cell used in England in the early 1950s is shown in Figure 1. The fused caustic bath is contained in the cast-iron outer pot, which rests in a brick chamber. The cylindrical copper cathode is supported on the cathode stem, which extends upward through the bottom of the cell. The cathode stem is sealed and insulated from the outer pot by a frozen

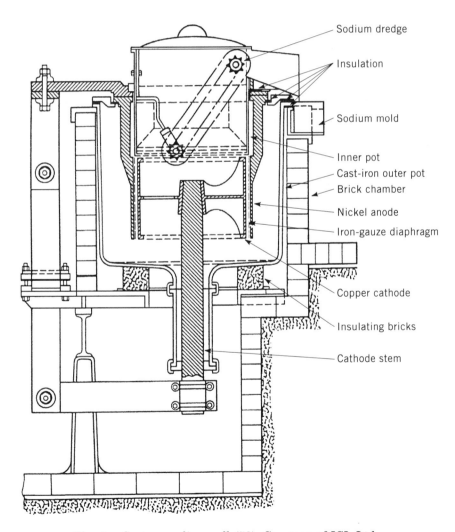

Fig. 1. Castner sodium cell (53). Courtesy of ICI, Ltd.

portion of the bath. The cylindrical nickel anode concentric with the cathode is supported from the rim of the outer pot. The cylindrical iron-gauze diaphragm, located in the 2.5-cm annular space between the electrodes, is suspended from the inner pot. Because of the difference in density, sodium rises in the hydroxide bath and collects on its surface in the inner pot; the latter is electrically insulated from the top anode ring by which it is supported. The inner pot is closed by a cover that maintains an atmosphere of hydrogen over the sodium to prevent burning. No practical way has been found to collect the hydrogen, which is vented, as is the oxygen liberated at the anode. A perforated hand ladle was used to remove sodium from early Castner cells (54). Later, the sodium was removed using mechanically driven iron-gauze buckets. The cell shown in Figure 1 can hold approximately 1 t of molten bath, consisting primarily of mercury-cell caustic soda with up to 10% each of sodium chloride and sodium carbonate. Some salt is added initially to improve the bath conductivity, but the carbonate is an unwanted impurity. After several months of operation, the chloride, carbonate, and other impurities attain concentrations that seriously impair the efficiency of the cell. The bath is then renewed. Operating characteristics are given in Table 4. Small cells are heated externally to maintain operating temperature, but large cells are heated by the electrolysis current. Of the many ingenious systems proposed to prevent the reaction of sodium with the water produced at the anode, none are known to have been applied substantially commercially.

Electrolysis of Fused Sodium Chloride. Although many cells have been developed for the electrolysis of fused sodium chloride (8,55–60), the Downs cell (3) has been most successful (see ELECTROCHEMICAL PROCESSING). In cells in general use by 1945, a single cylindrical anode constructed of several graphite blocks was inserted through the center of the cell bottom and surrounded by an iron-gauze diaphragm and a cylindrical iron cathode. In the 1940s, the single anode and cathode were replaced by a multiple electrode arrangement consisting of four anodes of smaller diameter in a square pattern, each surrounded by a cylindrical diaphragm and cathode, as shown in Figure 2. Without increasing the overall cell dimensions, this design increased the electrode area per cell, allowing increased amperage.

The cell consists of three chambers. The upper chamber is outside the chlorine dome and above the sodium-collecting ring. The other two chambers are the chlorine-collecting zone inside the dome and diaphragm, and the sodium-

Table 4. Operating Characteristics of the Castner Cell

Property	Value
bath temperature, °C	320 ± 10
cell current, kA	9 ± 0.5
cell voltage, V	4.3–5.0
cathode current density, kA/m^2	10.9
current efficiency, %	40[a]
sodium produced	
$g/(A \cdot h)$	0.4
$g/(kW \cdot h)$	90

[a]Value is approximate.

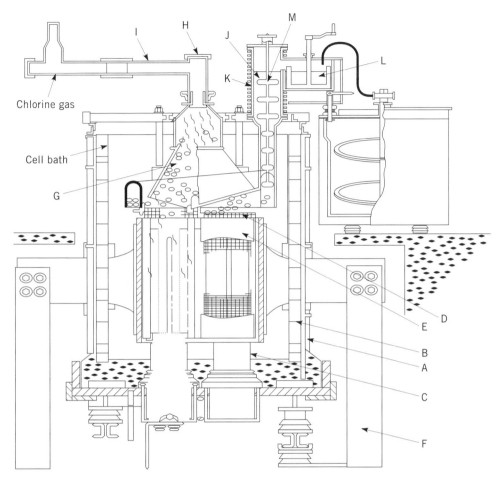

Fig. 2. Downs cell: A, the steel shell, contains the fused bath; B is the fire-brick lining; C, four cylindrical graphite anodes project upward from the base of the cell, each surrounded by D, a diaphragm of iron gauze, and E, a steel cathode. The four cathode cylinders are joined to form a single unit supported on cathode arms projecting through the cell walls and connected to F, the cathode bus bar. The diaphragms are suspended from G, the collector assembly, which is supported from steel beams spanning the cell top. For descriptions of H–M, see text.

collecting zone outside the diaphragm, and under the sodium-collecting ring of the collector unit. This arrangement prevents recombination of the sodium and chlorine. The collector is a complex assembly of inverted troughs and chambers arranged to collect the products in separate compartments as they rise through the bath.

The chlorine emerges through the nickel dome, H, and is removed through the chlorine line, I, to a header (see ALKALI AND CHLORINE PRODUCTS). Sodium, J, is channeled to a riser pipe, K, which leads to a discharge point above the cell wall. The difference in level between the overflowing sodium and the cell bath is due to the roughly 2:1 density ratio of the fused bath and liquid sodium. The upper end of the riser pipe is fitted with fins that cool the sodium and thereby precipitate dissolved calcium. The sodium, still containing some calcium, electrolyte, and oxide, overflows into a receiver, L. The calcium precipitated in the riser pipe tends to adhere to the wall from which location it is dislodged by the scraper, M, and returned to the base of the riser. The cell is fitted with a smoke-collection cover to collect particulate emissions and to protect the operators. A small area is left uncovered for visual observation, bath-level regulation, and salt-bath agitation and salt feed. Fine, dry crystalline salt is fed to the bath through a feed chute from a salt system conveyor (not shown).

The cell bath in early Downs cells (8,14) consisted of approximately 58 wt % calcium chloride and 42 wt % sodium chloride. This composition is a compromise between melting point and sodium content. Additional calcium chloride would further lower the melting point at the expense of depletion of sodium in the electrolysis zone, with the resulting complications. With the above composition, the cells operate at 580–600°C, well below the temperature of highest sodium solubility in the salt bath. Calcium chloride causes problems because of the following equilibrium reaction (56):

$$2\,Na^0 + CaCl_2 \rightleftharpoons 2\,NaCl + Ca^0$$

The alloy phase contains about 5 wt % calcium at cell conditions, an amount intolerable for most industrial uses. The bulk is removed by precipitation in the cooled riser pipe. Any precipitated calcium that adheres to the walls of the riser must be scraped off to prevent plugging. The precipitate drops to the bath–metal interface where it reacts to reform calcium chloride and sodium according to the above equilibrium. Calcium remaining in the sodium is largely removed by filtration at about 110°C. The filtered sodium contains <0.04 wt % calcium. The filtration operation produces a filter cake of calcium, sodium, chlorides, and oxides. High temperature sodium chloride reaction, and mechanical pressing of the sodium–calcium filter sludge are employed commercially in removing sodium from the calcium sludge, allowing recovery of most of the sodium content.

Characteristics of Downs cells are given in Table 5 (2).

Table 5. Characteristics of Downs Cells

Property	Country of manufacture		
	United Kingdom	United States	Germany
bath temperature, °C	580 ± 15	580–600	590 ± 5
cell current, kA	25–35	43–45	24–32
cell voltage, V	7[a]	7	5.7–6.0
cathode current density, kA/m^2	9.8	11–12	9.8
current efficiency, %	75–80	85–90	78
cell life, d	500–700	600–800	300–350
diaphragm life, d	20–100	50–90	20–30

[a]Value is approximate.

Salt that is substantially free of sulfate and other impurities is the cell feed. This grade may be purchased from commercial salt suppliers or made on site by purification of crude sea or rock salt. Dried calcium chloride or cell bath from dismantled cells is added to the bath periodically as needed to replenish calcium coproduced with the sodium. The heat required to maintain the bath in the molten condition is supplied by the electrolysis current. Other electrolyte compositions have been proposed in which part or all of the calcium chloride is replaced by other salts (61–64). Such baths offer improved current efficiencies and production of crude sodium containing relatively little calcium.

Cell life is determined by the loss of graphite from the anodes. Oxygen released at the anode by electrolysis of oxides or water in the bath reacts with the graphite to form CO and CO_2. In time, erosion of the anode increases the interelectrode spacing with corresponding increases in cell voltage and temperature. At this stage the cell is replaced.

A dimensionally stable anode consisting of an electrically conducting ceramic substrate coated with a noble metal oxide has been developed (55). Iridium oxide, for example, resists anode wear experienced in the Downs and similar electrolytic cells (see METAL ANODES).

Other commercial cells designed for the electrolysis of fused sodium chloride include the Danneel-Lonza cell and the Seward cell, both used before World War I. The former had no diaphragm and the sodium was confined to the cathode zone by salt curtains (ceramic walls); the latter utilized the contact-electrode principle, where the cathode was immersed only a few millimeters in the electrolyte. The Ciba cell was used over a longer period of time. It was an adaptation of the Castner cell to sodium chloride for fused caustic electrolysis. A mixture of sodium chloride and other chlorides, molten at 620°C, was electrolyzed in rectangular or oval cells heated only by the current. Several cells have been patented for the electrolysis of fused salt in cells with molten lead cathodes (65). However, it is difficult to separate the lead from the sodium (see ELECTROCHEMICAL PROCESSING).

Electrolysis of Amalgam.   Sodium in the form of amalgam as made by the electrolysis of sodium chloride brine in mercury cathode cells is much less expensive than any other form of the metal, but commercial use of amalgam is restricted largely to production of caustic soda (see ALKALI AND CHLORINE PRODUCTS). Many efforts have been made to develop processes for recovering sodium from amalgam (66–69). Recovery by electrolysis with the amalgam serving as anode has been the favored approach. The electrolytes were generally low melting sodium salt combinations, although liquid ammonia (70) and organic solvents (71) have been reported. The addition of lead before electrolysis using a low melting salt electrolyte has been patented (72,73).

Sodium was made from amalgam in Germany during World War II (68). The only other commercial application appears to be the Tekkosha process (74–76). In this method, preheated amalgam from a chlor–alkali cell is supplied as anode to a second cell operating at 220–240°C. This cell has an electrolyte of fused sodium hydroxide, sodium iodide, and sodium cyanide and an iron cathode. Operating conditions are given in Table 6.

The sodium produced contains 0.1–0.5 wt % mercury. This mercury is converted to calcium amalgam by treating the crude sodium with powdered

Table 6. Tekkosha Fused-Salt Electrolysis Cells[a]

Property	Value
voltage, V	3.0–3.1
current, kA	60
current density, kA/m^2	4
current efficiency, %	96–98

[a]Refs. 74–76.

anhydrous calcium chloride:

$$Hg^0 + 2\,Na^0 + CaCl_2 \longrightarrow CaHg + 2\,NaCl$$

The residual salts (CaCl$_2$, NaCl) and the calcium amalgam are removed by cooling and filtration. The Tekkosha process offers the advantages of moderate temperature, minimum corrosion, simple operation, high efficiency, low labor cost, good working conditions, and process adaptability. In the United States, these advantages would be largely offset by the environmental problems inherent in the handling of mercury and the need to produce some caustic soda to balance the in-process sodium inventory.

Electrolysis Based on Cationically Conducting Ceramics. Searching for a method for using sodium and sulfur (qv) as reactants in a secondary battery, the Ford Motor Company developed a polycrystalline β-alumina ceramic material that selectively transports sodium cations when subjected to an electric field (48,77,78) (see BATTERIES; CERAMICS AS ELECTRICAL MATERIALS). This ceramic, or any of its many variants, is useful as a diaphragm or divider in a two-compartment cell. In one compartment, the sodium is in contact with the ceramic; in the other, a suitable liquid electrolyte is in contact with the opposite side of the ceramic. Thus, the sodium is in electrochemical but not physical contact with the liquid electrolyte. Many low melting electrolytes can be used that are otherwise incompatible with sodium, eg, sodium polysulfides; sodium tetrachloroaluminate, NaAlCl$_4$; sodium hydroxide; and mixtures of sodium chloride and zinc chloride or sodium nitrite and nitrate. Because sodium is not in contact with the liquid electrolyte, the various reactions that usually lower the current efficiency of commercial cells do not occur. Cells based on this principle generally operate at close to 100% current efficiency (79). Sodium of exceptional purity is produced at satisfactory operating conditions. However, improved ceramics of predictable properties and long service life have not been commercialized as of this writing (ca 1996). Research on sodium–sulfur batteries is continuing (80–88). Operational prototypes are being tested for utility peak power load-leveling (89), and in early zero-emission electric vehicle (ZEV) trials. Solid electrolytes other than sodium beta-alumina are also reported (90,91). This technology, known as alkali metal thermal electric conversion (AMTEC), was the topic of the 1995 Intersociety Energy Conversion Engineering Conference (IECEC) (see THERMOELECTRIC ENERGY CONVERSION).

Energy Requirements. The energy requirements of several sodium manufacturing processes are compared in Table 7 (76). The data contain some ambiguities because of the allocation of energy to the coproduction of chlorine. An

Table 7. Sodium Process Energy Requirements[a]

Process	Total energy, MJ/kg Na[b]
Downs, fused NaCl	107[c]
Castner, fused NaOH	328
Tekkosha, double electrolysis	80
sodium–lead, evaporation	55

[a]Ref. 76.
[b]To convert MJ/kg to Btu/lb, multiply by 430.2.
[c]Ref. 92 gives a value of 97 MJ/kg.

independent calculation shows a somewhat lower energy consumption for the Downs process (92).

Specifications, Shipping

Sodium, generally about 99.9% Na assay, is available in two grades: regular, which contains 0.040 wt % Ca, and nuclear (low Ca), which has 0.001 wt % Ca. Both have 0.005 wt % Cl$^-$. The nuclear grade is packed in specially cleaned containers, and in some cases under special cover atmospheres. A special grade of sodium low in potassium and calcium (<10 ppm) is achievable to meet requirements for use in manufacture of the more newly developed sodium–sulfur batteries.

Sodium is commonly shipped in 36- to 70-t tank cars in the United States. Smaller amounts are shipped in 16-t tank trucks or ISO-tanks. Sodium is also available in 104- and 190-kg drums, and in bricks (0.5–5 kg). A thin layer of oxide, hydroxide, or carbonate is usually present. Sodium is also marketed in small lots as a dispersion in an inert hydrocarbon, or produced in-process via high pressure injection into a pumped stream of inert carrier fluid, such as toluene or mineral oil.

Economic Aspects

Historically, U.S. production of sodium was 70–85% of world production. As lead compounds were phased out of gasoline in North America, this situation changed (Table 8).

Table 8. North American Sodium Production and Uses[a], t × 10^3

Use	1959	1963	1967	1972	1978	1980[b]	1995[b]
gasoline additives	72	93	124	121	101	78	1
herbicides/insecticides							6
bleaching chemicals						4	8
metals reduction	5	5	12	6	14	13	8
all others	24	16	13	19	11	6	7
Totals	*101*	*114*	*149*	*146*	*126*	*101*	*30*

[a]Refs. 93 and 94.
[b]Estimates.

Analytical Methods

Sodium is identified by the intense yellow color that sodium compounds impart to a flame or spectroscopically by the characteristic sodium lines. The latter test is extremely sensitive, yet because many materials contain traces of sodium salts as impurities, it is not conclusive evidence of the presence of sodium in any considerable quantity.

The alkali metals are commonly separated from all other elements except chlorine before gravimetric determination. In the absence of other alkalies, sodium may be weighed as the chloride or converted to the sulfate and weighed. Well-known gravimetric procedures employ precipitation as the uranyl acetate of sodium–zinc or sodium–magnesium. Quantitative determination of sodium without separation is frequently possible by emission or atomic-absorption spectrometric techniques.

Metallic sodium is determined with fair accuracy by measuring the hydrogen liberated on the addition of ethyl alcohol. Sodium amalgam is analyzed by treating a sample with a measured volume of dilute standard acid. After the evolution of hydrogen stops, the excess acid is titrated with a standard base. Total alkalinity is calculated as sodium. Calcium in commercial sodium is usually determined by permanganate titration of calcium oxalate. The trace amounts of calcium present in nuclear-grade sodium, as little as 0.5 ppm, are determined by atomic absorption spectrometry. Chloride is determined as silver chloride by a turbidimetric method in which glycerol stabilizes the suspended precipitate. Sodium oxide is separated from sodium by treatment with mercury. The oxide, which is insoluble in the amalgam formed, can be separated and determined by acid titration. Methods for the determination of impurities in sodium are available (5,8,23,95).

Health and Safety Factors

The safe handling of sodium requires special consideration because of its high reactivity. Using properly designed equipment and strict safe-handling procedures, sodium is used in large- and small-volume applications without incident (96). The hazards of handling sodium are no greater than those encountered using many other industrial chemicals (5), although incidents can result in very serious injury, and fatalities have been known to occur from accidental contact with molten, burning sodium, or from contact with sodium finely dispersed in inert hydrocarbons such as mineral oil or toluene.

Direct contact of the skin with sodium can cause deep, serious burns from the action of sodium with the moisture present and the subsequent corrosive action of the caustic formed. Sodium can cause blindness on contact with the eyes. For these reasons, goggles, face shields, hard hat, hoods, long-gauntlet mittens, and multiple layers of flame-retardant protective clothing are recommended when working with molten sodium (97). All body parts should be protected and clothing needs to be designed for quick removal in case of emergencies. Contaminated clothing should first be steamed, then washed, or destroyed by burning.

Perhaps the greatest hazard presented by metallic sodium stems from its extremely vigorous reaction with water to form sodium hydroxide and hydrogen with the evolution of heat (5,14,98,99). In the presence of air this combination usually results in explosion; in a closed system where an inert atmosphere is present, the hydrogen evolved can cause a rapid increase in pressure. In the absence of air, the rate of reaction is substantially equal to the rate of mixing the reactants, and the reaction does not generally cause mechanical damage to heat-transfer equipment. In the presence of air, the results cannot be predicted. Hydrogen evolved in a closed containers should not be allowed to mix with air.

Another hazard arises from the oxidation of sodium in air. Liquid sodium can autoignite at 120°C, although under some conditions dispersed or high surface sodium may ignite at much lower temperatures (7). A small local sodium fire can be extinguished by submerging the burning mass in the remaining pool of liquid sodium using an iron blade if the bulk of the sodium has not reached the ignition point. Larger fires are more difficult to handle. The common fire extinguishers, ie, water, CO_2, CCl_4, etc, should never be used. These only aggravate the existing situation by introducing additional explosion or reaction hazards. If the vessel containing the burning sodium can be flooded with nitrogen or closed to exclude air, the fire subsides and the material can be cooled. Fires that cannot be extinguished by excluding air may be quenched by large quantities of dry salts or other dry, cold, inert powder (100). Dry light soda ash, Na_2CO_3, is excellent for this purpose but must not be used if made damp on exposure in storage, and must be carefully protected from contact with exposure to moisture in air. Process areas and equipment should be designed to confine any sodium spills and permit recovery.

Techniques for handling sodium in commercial-scale applications have improved (5,23,98,101,102). Contamination by sodium oxide is kept at a minimum by completely welded construction and inert gas-pressured transfers. Residual oxide is removed by cold traps or micrometallic filters. Special mechanical pumps or leak-free electromagnetic pumps and meters work well with clean liquid sodium. Corrosion of stainless or carbon steel equipment is minimized by keeping the oxide content low. The 8-h TWA PEL and ceiling TLV for sodium or sodium oxide or hydroxide smoke exposure is 2 mg/m^3. There is no defined ALD for pure sodium, as even the smallest quantity ingested could potentially cause fatal injury.

In the laboratory, sodium is best handled in a glove box filled with nitrogen or another inert gas, or in a water-free hood. When sodium is handled on the bench top, water and aqueous solutions must be excluded from the area. Tools for cutting or handling sodium must be clean and dry. Contact of sodium with air should be kept to a minimum because moisture in the air reacts rapidly with sodium. A metal catch pan under the equipment is essential to contain any spills or fires when breaking into pipe or equipment that previously contained liquid sodium. Provision should be made for safe removal of sodium residues from equipment and for cleaning the apparatus. Residue and sodium scrap can be destroyed by burning in a steel pan in a well-ventilated hood. Equipment may be cleaned by being opened to the air and heated until any sodium present is oxidized, or by purging thoroughly with nitrogen, then slowly admitting dry steam to the system while maintaining the nitrogen purge. The burning of sodium as part of any cleaning procedure produces an irritating and hazardous smoke of sodium oxide. This should be collected by an appropriate hood or duct and scrubbed. Dilute aqueous sodium hydroxide is a satisfactory scrubbing liquid.

Other methods for safely cleaning apparatus containing sodium residues or disposing of waste sodium are based on treatment with bismuth or lead (103), inert organic liquids (104–106), or by reaction with water vapor carried in an inert gas stream (107).

Most reactions of sodium are heterogeneous, occurring on the surface of solid or liquid sodium. Such reactions are accelerated by extending the sodium surface exposed. The sodium is generally dispersed in a suitable medium (108) or spread over a solid powder of high surface area (7,8). Dispersions in inert hydrocarbons may be briefly exposed to air and present no special hazards as long as the hydrocarbon covers the sodium. High surface sodium reacts very rapidly with air, however, and cannot be exposed without risk of fire. Dispersions of sodium spilled on cloth or other absorbent material may ignite quickly.

Uses

The largest consumption of sodium worldwide, as of the mid-1990s, is the production of tetraethyllead and tetramethyllead antiknock compounds for gasoline. This production is outside of North America. Sodium is also used for the production of other organometallic compounds such as methylcyclopentadienylmanganese tricarbonyl (MMT), another gasoline additive (see OCTANE IMPROVERS (SUPPLEMENT)).

The manufacture of refractory metals such as titanium, zirconium, and hafnium by sodium reduction of their halides is a growing application, except for titanium, which is produced principally via magnesium reduction (109–114). Typical overall halide reactions are

$$TiCl_4 + 4\,Na^0 \longrightarrow Ti^0 + 4\,NaCl$$
$$TiCl_4 + 2\,Mg^0 \longrightarrow Ti^0 + 2\,MgCl_2$$

Sodium reduction processes are also described for tantalum (115), silicon (116–118), magnesium (119), and other metals.

Metallic potassium and potassium–sodium alloys are made by the reaction of sodium with fused KCl (8,98) or KOH (8,15). Calcium metal and calcium hydride are prepared by the reduction of granular calcium chloride with sodium or sodium and hydrogen, respectively, at temperatures below the fusion point of the resulting salt mixtures (120,121).

Whereas manufacture of sodium peroxide has declined (122–126), Na_2O_2 is an excellent agent to liberate metal from complexes ores, eg, silver tetrahedrites. Sodium hydride, made from sodium and hydrogen, is employed as catalyst or reactant in numerous organic reactions and for the production of other hydrides, eg, sodium borohydride (see HYDRIDES). Sodium is used indirectly for the descaling of metals such as stainless steel and titanium (41). Sodium and hydrogen are fed to a molten bath of anhydrous caustic to generate sodium hydride, which dissolves in the melt and is the effective descaling agent (see METAL TREATMENTS), according to the following reaction (127):

$$3\ NaH + Fe_2O_3 \longrightarrow 3\ NaOH + 3\ Fe^0$$

Many sodium compounds are made from sodium. Sodium is employed as a reducing agent in numerous preparations, including the manufacture of dyes (see DYES AND DYE INTERMEDIATES), eg, indigo; herbicides (qv) (128); pharmaceuticals (qv); high molecular weight alcohols (129); perfume materials (130) (see PERFUMES); and isosebacic acid (131,132).

Sodium is a catalyst for many polymerizations; the two most familiar are the polymerization of 1,2-butadiene (the Buna process) and the copolymerization of styrene–butadiene mixtures (the modified GRS process). The alfin catalysts, made from sodium, give extremely rapid or unusual polymerizations of some dienes and of styrene (qv) (133–137) (see BUTADIENE; ELASTOMERS, SYNTHETIC; STYRENE PLASTICS).

Naphthalene sodium prepared in dimethyl ether or another appropriate solvent, or metallic sodium dissolved in liquid ammonia or dimethylsulfoxide, is used to treat polyfluorocarbon and other resins to promote adhesion (138–140). Sodium, usually in dispersed form, is used to desulfurize a variety of hydrocarbon stocks (141). The process is most useful for removal of small amounts of sulfur remaining after hydrodesulfurization.

Sodium as an active electrode component of primary and secondary batteries offers the advantages of low atomic weight and high potential (78,142,143). In addition to the secondary battery for ZEVs, a remarkable primary cell has been developed by Lockheed Aircraft Corporation in which sodium metal and water (in the form of aqueous sodium hydroxide) are the reactants (144–147). No separators or diaphragms are used, the counter-electrode is mild steel, and the interelectrode distance is small. The unexpected discovery which makes this cell possible is that, given an external circuit of reasonable resistance, hydrogen is released on the iron counter-electrode rather than on the sodium surface. Thus, sodium dissolves as NaOH in a vigorous but nonviolent manner, and the released electrons traverse the external circuit to discharge hydrogen ions at the iron electrode. Concentration cells based on amalgams of differing sodium content that are regenerated thermally have been described (148,149).

Because of the electrical conductivity, low density, low cost, and extrudability of the metal, cables were made of sodium sheathed in polyethylene (150). An earlier application used sodium-filled iron pipe as a conductor (151); a more recent patent describes a conductor of sodium contained in aluminum, copper, or steel tubing (152). A corrugated flexible thin-walled copper tube filled with sodium and particularly well-adapted for use in gas-insulated high voltage transmission lines has been developed (153). Sodium conductor distribution cables offer both economic and energy-saving advantages (154), but have never gained wide acceptance for commercial use.

Sodium is used as a heat-transfer medium in primary and secondary cooling loops of liquid-metal fast-breeder power reactors (5,155–157). Low neutron cross section, short half-life of the radioisotopes produced, low corrosiveness, low density, low viscosity, low melting point, high boiling point, high thermal conductivity, and low pressure make sodium systems attractive for this application (40).

Sodium has also been essential to new developments in heat transfer in advanced solar energy collectors (158,159) for powering systems remote from electrical distribution systems, and aerospace. A comparison of sodium with other working fluids in heat pipes is available (78,88,160,161) (see HEAT-EXCHANGE TECHNOLOGY, HEAT PIPES). Small amounts of sodium have been used widely to cool exhaust valves of heavy-duty internal combustion engines (8), affording longer valve or seat life by lower valve temperature in operation.

In metallurgical practice, sodium uses include preparation of powdered metals; removal of antimony, tin, and sulfur from lead; modification of the structure of silicon–aluminum alloys; application of diffusion alloy coatings to substrate metals (162,163); cleaning and desulfurizing alloy steels via NaH (164); nodularization of graphite in cast iron; deoxidation of molten metals; heat treatment; and the coating of steel using aluminum or zinc.

Sodium vapor lamps, in use for many years, continue to be improved, both with respect to efficiency and color of emitted light. These lamps, however, contain only a few milligrams of sodium each.

A process development known as NOXSO (DuPont) (165,166) uses sodium to purify power plant combustion flue gas for removal of nitrogen oxide, NO_x, and sulfur, SO_x compounds. This technology relies on sodium metal generated *in situ* via thermal reduction of sodium compound-coated media contained within a flue-gas purification device, and subsequent flue-gas component reactions with sodium. The process also includes downstream separation and regeneration of spent media for recoating and circulation back to the gas purification device. A full-scale commercial demonstration project was under construction in 1995.

SODIUM ALLOYS

Sodium is miscible with many metals in liquid phase and forms alloys or compounds. Important examples are listed in Table 9; phase diagrams are available (4,5,14,35).

The brittleness of metals is frequently increased by the addition of sodium to form alloys. The metals vary in their ability to dilute the natural reactivity of sodium. Most binary alloys are unstable in air and react with water. Ternary and quaternary alloys are more stable.

Sodium–potassium alloy is easily prepared by melting the clean metals in an inert atmosphere or under an inert hydrocarbon, or by the reaction of sodium with molten KCl, KOH, or solid K_2CO_3 powder.

Alloys of lead and sodium containing up to 30 wt % sodium are obtained by heating the metals together in the desired ratio, allowing a slight excess of sodium to compensate for loss by oxidation. At about 225°C, the elements react and generate enough heat to cause a rapid temperature rise. External heating is discontinued and the mixture is cooled and poured into molds. The brittle alloys can be ground to a powder and should be stored under a hydrocarbon or in air-

Table 9. Metal-Sodium Systems[a]

Metal	Alloy formation	Compound formation	Consolute temperature, °C
barium	+	+	miscible
calcium	+		ca 1200
lead	+	+	
lithium[b]	+		306
magnesium	+		>800
mercury	+	+	
potassium	+	+	miscible
rubidium	+		miscible
tin	+	+	
zinc	+	+	>800

[a]Refs. 4, 5, 14, and 37.
[b]Refs. 167 and 168.

tight containers to prevent surface oxidation. The 30 wt % sodium alloy reacts vigorously with water to liberate hydrogen, providing a convenient laboratory source of this gas. An alloy containing 10 wt % sodium may be used in controlled reactions with organic halogen compounds that react violently with pure sodium. Sodium–lead alloys that contain large amounts of sodium are used to dry organic liquids.

Sodium–lead alloys that contain other metals, eg, the alkaline-earth metals, are hard even at high temperatures, and are thus suitable as bearing metals. Tempered lead, for example, is a bearing alloy that contains 1.3 wt % sodium, 0.12 wt % antimony, 0.08 wt % tin, and the remainder lead. The German Bahnmetall, which was used in axle bearings on railroad engines and cars, contains 0.6 wt % sodium, 0.04 wt % lithium, 0.6 wt % calcium, and the remainder lead, and has a Brinell hardness of 34 (see BEARING MATERIALS).

Up to ca 0.6 wt % sodium dissolves readily in mercury to form amalgams that are liquid at room temperature (169). The solubility of sodium in mercury is ca 1 wt % at 70°C (169) and 2 wt % at 140°C (37). Alloys containing >2 wt % sodium are brittle at room temperature. Sodium-rich amalgam may be made by adding mercury dropwise to a pool of molten sodium; mercury-rich amalgam is prepared by adding small, clean pieces to sodium to clean mercury with agitation. In either case an inert atmosphere must be maintained, and the heat evolved must be removed. Solid amalgams are easily broken and powdered, but must be carefully protected against air oxidation. Amalgams are useful in many reactions in place of sodium because the reactions are easier to control (169).

Sodium amalgam is employed in the manufacture of sodium hydroxide; sodium–potassium alloy, NaK, is used in heat-transfer applications; and sodium–lead alloy is used in the manufacture of tetraethyllead and tetramethyllead, and methylcyclopentadienylmanganesetricarbonyl, a gasoline additive growing in importance for improving refining efficiency and octane contribution.

Sodium does not form alloys with aluminum but is used to modify the grain structure of aluminum–silicon alloys and aluminum–copper alloys for improved machinability. Sodium–gold alloy is photoelectrically sensitive and may be used in photoelectric cells. A sodium–zinc alloy, containing 2 wt % sodium and 98 wt % zinc, is used to deoxidize other metals.

BIBLIOGRAPHY

"Alkali Metals, Sodium" in *ECT* 1st ed., Vol. 1, pp. 435–447, by E. H. Burkey, J. A. Morrow, and M. S. Andrew, E. I. du Pont de Nemours & Co., Inc.; "Sodium" in *ECT* 2nd ed., Vol. 18, pp. 432–457 by C. H. Lemke, E. I. du Pont de Nemours & Co., Inc.; "Sodium and Sodium Alloys" in *ECT* 3rd ed., Vol. 21, pp. 181–204 by C. H. Lemke, E. I. du Pont de Nemours & Co., Inc. and University of Delaware.

1. A. Fleck, *Chem. Ind. (London)* **66**, 515 (1947).
2. D. W. F. Hardie, *Ind. Chemist*, **30**, 161–166 (1954).
3. U.S. Pat. 1,501,756 (July 15, 1924), J. C. Downs (to Roessler and Hasslacher Chemical Co.).
4. C. J. Smithells, *Metals Reference Book*, 2nd ed., Vols. I and II, Interscience Publishers, Inc., New York, 1955.
5. O. J. Foust, ed., *Sodium–NaK Engineering Handbook*, Gordon and Breach, New York, 1972.
6. J. P. Stone and co-workers, *High Temperature Properties of Sodium, NRL Report 6241*, U.S. Naval Research Laboratory, Defense Documentation Center, AD 622191, Washington, D.C., Sept. 24, 1965.
7. *High Surface Sodium*, U.S. Industrial Chemicals Company, New York, 1953 (now RMI Company, Niles, Ohio).
8. M. Sittig, *Sodium: Its Manufacture, Properties and Uses*, Reinhold Publishing Corp., New York, 1956.
9. P. B. Dransfield, *Chem. Soc. (London), Spec. Pub. No. 22,222* (1967).
10. C. A. Kraus and W. W. Lucasse, *J. Am. Chem. Soc.* **44**, 1949 (1922).
11. C. A. Kraus, *J. Chem. Educ.* **30**, 83 (1953).
12. J. F. Dewald and G. Lepoutre, *J. Am. Chem. Soc.* **76**, 3369 (1954).
13. P. D. Schettler, Jr., P. W. Doumaux, and A. Patterson, Jr., *J. Phys. Chem.* **71**(12), 3797 (1967).
14. J. W. Mellor, *The Alkali Metals*, Vol. II, Suppl. II of *Comprehensive Treatise on Inorganic and Theoretical Chemistry*, John Wiley & Sons, Inc., New York, 1961, Pt. 1.
15. Technical data, E. I. du Pont de Nemours & Co., Inc., Wilmington, Del., 1995.
16. R. E. Robinson and I. L. Mador, in N. M. Bikales, ed., *Encyclopedia of Polymer Science and Technology*, Vol. 1, Interscience Publishers, a Division of John Wiley & Sons, Inc., New York, 1970, pp. 639–658.
17. S. B. Windwer, *Solutions of Alkali Metals in Ethylenediamine*, L. C. Card No. Mic 61-707, University Microfilms, Ann Arbor, Mich., 1960, 79 pp.
18. IUPAC Suppl., *Proceedings of an International Conf. on the Nature of Metal-Ammonia Solutions, Colloque Weyl II*, pp. 1–17 (1970).
19. N. D. Scott, J. F. Walker, and V. L. Hansley, *J. Am. Chem. Soc.* **58**, 2442 (1936).
20. T. P. Whaley, in A. F. Trotman-Dickenson, ed., *Comprehensive Inorganic Chemistry*, Pergamon Press, Oxford, U.K., 1973, Chapt. 8.
21. V. L. Hansley, *Ind. Eng. Chem.* (43) 1759.
22. C. J. Pedersen, *J. Am. Chem. Soc.* **89**, 2495 (1967).
23. J. W. Mausteller, F. Tepper, and S. J. Rodgers, *Alkali Metal Handling and Systems Operating Techniques*, Gordon and Breach, New York, 1967.
24. A. C. Wittingham, *Liquid Sodium–Hydrogen System: Equilibrium and Kinetic Measurements in the 610–667 K Temperature Range*, NTIS Accession No. RD/B/N-2550, National Technical Information Service, Washington, D.C., Aug. 1974.
25. D. D. Williams, *A Study of the Sodium–Hydrogen–Oxygen System*, U.S. Naval Research Laboratory Memorandum Report No. 33, Washington, D.C., June 1952.
26. V. L. Hansley and P. J. Carlisle, *Chem. Eng. News* **23**(2), 1332 (1945).

27. P. V. H. Pascal, *Nouveau Traite de Chimie Minerale*, Vol. 11, Masson et Cie., Paris, 1966.

28. T. P. Whaley and C. C. Chappelow, Jr. in T. Moeller, ed., *Inorganic Syntheses*, Vol. 5, McGraw-Hill Book Co., Inc., New York, 1957, pp. 10–13.

29. E. W. Guernsey and M. S. Sherman, *J. Am. Chem. Soc.* **47**, 1932 (1925).

30. U.S. Pat. 2,802,723 (Aug. 13, 1957), C. H. Lemke (to E. I. du Pont de Nemours & Co., Inc.).

31. W. C. Sleppy, *Inorg. Chem.* **5**(11), 2021 (1966).

32. Ft. C. Asher and S. A. Wilson, *Nature* **181**, 409 (1958).

33. A. Herold, *Bull. Soc. Chim. France*, **999** (1955).

34. U.S. Pat. 2,642,347 (June 16, 1953), H. N. Gilbert (to E. I. du Pont de Nemours & Co., Inc.).

35. G. J. Moody and J. D. R. Thomas, *J. Chem. Educ.* **43**(4), 205 (1966).

36. E. Hohmann, *Z. Anorg. Allgem. Chem.* **251**, 113 (1948).

37. M. Hansen, *Constitution of Binary Alloys*, McGraw-Hill Book Co., Inc., New York, 1958.

38. M. J. Dignam and D. A. Huggins, *J. Electrochem. Soc.* **114**(2), 117 (1967).

39. U.S. Pat. 2,858,194 (Oct. 28, 1958), H. C. Miller (to E. I. du Pont de Nemours & Co., Inc.).

40. "The Alkali Metals," *An International Symposium, Nottingham, England, July 1966, Special Publication No. 22*, The Chemical Society, London, 1967.

41. H. N. Gilbert, *Chem. Eng. News* **26**, 2604 (1948).

42. M. I. Klyashtornyi, *Zh. Prikl. Khim.* **31**, 684 (1958).

43. E. Rinck, *Ann. Chim. (Paris)* **18**, 395 (1932).

44. A. Glassner, *The Thermochemical Properties of the Oxides, Fluorides, and Chlorides to 2500 K*, Argonne National Laboratory Report ANL-5750, U.S. Government Printing Office, Washington, D.C., 1957.

45. D. R. Stull and H. Prophet, project directors, *JANAF Thermochemical Tables*, 2nd ed., National Standards Reference Data Series, U.S. National Bureau of Standards, No. 37, June 1971, available from U.S. Government Printing Office, Superintendent of Documents, Washington, D.C.

46. G. L. Clark, ed., *Encyclopedia of Chemistry*, 2nd ed., Reinhold Publishing Corp., New York, 1966, p. 997.

47. U.S. Pat. 2,391,728 (Dec. 25, 1945), T. H. McConica and co-workers (to Dow Chemical Co.).

48. Ger. Offen. 2,243,004 (Mar. 15, 1973), V. M. Chong (to Sun Research and Development Co.).

49. Ger. Offen. 2,252,611 (May 10, 1973), E. L. Mongan, Jr. (to E. I. du Pont de Nemours & Co., Inc.).

50. U.S. Pat. 2,789,047 (Apr. 16, 1957), C. H. Lemke (to E. I. du Pont de Nemours & Co., Inc.).

51. D. J. Hansen, personal communication, Union Carbide Corp., Niagara Falls, N.Y., Apr. 19, 1968.

52. Ger. Offen. 2,044,402 (Mar. 18, 1971), C. Gentaz and G. Bienvenue (to Battelle Memorial Institute).

53. T. Wallace, *Chem. Ind. (London)*, **876** (1953).

54. H. N. Gilbert, *J. Electrochem. Soc.* **99**, 3050 (1952).

55. U.S. Pat. 4,192,794 (Mar. 11, 1980), T. Minami and S. Toda (to Chlorine Engineers Corp., Ltd., Tokyo).

56. U.S. Pat. 3,507,768 (April 21, 1970), E. I. Adaev, A. V. Blinov, G. M. Kamaryan, V. A. Novoselov, V. N. Suchkov, and L. M. Yakimenko.

57. C. L. Mantell, *Electrochemical Engineering*, 4th ed., McGraw-Hill Book Co., Inc., New York, 1960.

58. H. E. Batsford, *Chem. Metall. Eng.* **26**, 888 (1922).
59. *Ibid.*, 932 (1922).
60. W. C. Gardiner, *Office of Technical Services Report PB-44761*, U.S. Department of Commerce, Washington, D.C., 1946; Field Information Agency, Technical (FIAT) Final Report 820.
61. U.S. Pat. 2,850,442 (Sept. 2, 1958), W. S. Cathcart and co-workers (to E. I. du Pont de Nemours & Co., Inc.).
62. U.S. Pat. 3,020,221 (Feb. 6, 1962), W. H. Loftus (to E. I. du Pont de Nemours & Co., Inc.).
63. USSR Pat. 320,552 (Nov. 4, 1971), E. I. Adaev and co-workers.
64. U.S. Pat. 3,712,858 (Jan. 23, 1973), F. J. Ross (to E. I. du Pont de Nemours & Co., Inc.).
65. *Chem. Eng.* **69**(6), 90 (Mar. 1962).
66. U.S. Pat. 2,148,404 (Feb. 21, 1939), H. N. Gilbert (to E. I. du Pont de Nemours & Co., Inc.).
67. U.S. Pat. 2,234,967 (March 18, 1941), H. N. Gilbert (to E. I. du Pont de Nemours & Co., Inc.).
68. W. C. Gatdiner, *Office of Technical Services Report P8-44760*, U.S. Dept. of Commerce, Washington, D.C., 1946; Field Information Agency, Technical (FIAT) Final Report 819.
69. U.S. Pat. 3,265,490 (Aug. 9, 1966), S. Yoshizawa and co-workers (to Tekkosha Co. Ltd., Tokyo).
70. Fr. Pat. 1,457,562 (Nov. 4, 1966), (to Showa Denko KK).
71. Jpn. Pat. 71/03846 (Jan. 30, 1971), T. Ohshiba (to Showa Denko KK).
72. Jpn. Pat. 74/28322 (July 25, 1974), N. Watanabe and M. Tomatsuri (to Tekkosha Co., Ltd.).
73. N. Watanabe, M. Tomatsuri, and K. Nakanishi, *Denki Kagaku* **38**(8), 584 (1970).
74. T. Yamaguchi, *Chem. Econ. Eng. Rev.* **4**, 24 (Jan. 1972).
75. T. Nakamura and Y. Fukuchi, *J. Metall.* **24**, 25 (Aug. 1972).
76. L. E. Vaaler and co-workers, Battelle Columbus Laboratories, *Final Report on a Survey of Electrochemical Metal Winning Processes*, ANL/OEPM-79-3, Argonne National Laboratory, Mar. 1979. Available from National Technical Information Service, Washington, D.C.
77. U.S. Pat. 3,488,271 (Jan. 6, 1970), J. T. Kummer and N. Weber (to Ford Motor Co.).
78. R. M. Williams, A. Kisor, and M. A. Ryan, *J. Electrochem. Soc.* **142**(12), 4246 (1995).
79. W. E. Cowley, G. Thwaite, G. Waine, *The Selective Recovery Of Sodium From Amalgam Using β-Alumina*, Associated Octel Co. Ltd. 1978, presented at the *Second International Meeting of Solid Electrolytes*, University of St. Andrews, Scotland.
80. U.S. Pat. 4,108,743 (Aug. 22, 1978), R. W. Minck (to Ford Motor Co.).
81. Brit. Pat. 1,155,927 (June 25, 1969), A. T. Kuhn and S. F. Mellish (to Imperial Chemical Industries, Ltd.).
82. Brit. Pat. 1,200,103 (Mar. 31, 1967), A. T. Kuhn (to Imperial Chemical Industries, Ltd.).
83. U.S. Pat. 4,089,770 (May 16, 1978), C. H. Lemke (to E. I. du Pont de Nemours & Co., Inc.).
84. U.S. Pat. 4,133,728 (Jan. 9, 1979), S. A. Cope (to E. I. du Pont de Nemours & Co., Inc.).
85. U.S. Pat. 4,203,819 (May 20, 1980), S. A. Cope (to E. I. du Pont de Nemours & Co., Inc.).
86. S. Yoshizawa and Y. Ito, *Extended Abstracts of the 30th Meeting of the International Society of Electrochemistry*, Trondheim, Norway, Aug. 1979, pp. 38–40.
87. Jpn. Pat. 77/135,811 (Nov. 14, 1977), J. Koshiba and co-workers (to Toyo Soda Manufacturing Co., Ltd.).

88. M.-H. Weng and J. Newman, *J. Electrochem. Soc.* **142**(3), (1995).
89. Sadao Mori, Takashi Isozaki, *NaS Battery Development Activities By TEPCO and NGK Insulators*, Tokyo Electric Power Co., Inc., and NGK Insulators Ltd., Japan, Sept. 27, 1993. As presented at the 4th Annual Conference—Batteries For Energy Storage, Berlin, Germany.
90. U.S. Pat. 4,097,345 (June 27, 1978), R. D. Shannon (to E. I. du Pont de Nemours & Co., Inc.).
91. C. A. Levine, R. G. Heitz, and W. E. Brown, *Proceedings of the 7th Intersociety Energy Conversions and Engineering Conference*, American Chemical Society, Washington, D.C., 1972, pp. 50–53.
92. Private communication, E. I. du Pont de Nemours & Co., Inc. to Battelle Columbus Laboratories, June 1975.
93. *Sodium in Chemical Products Synopsis*, Mannsville Chemical Products, Mannsville, N.Y., Dec. 1978.
94. N. M. Levinson in *Chemical Economics Handbook, Sodium Metal*, 770.1000A-D, Stanford Research Institute, Menlo Park, Calif., Oct. 1979.
95. C. H. Lemke, N. D. Clare, and R. E. DeSantis, *Nucleonics* **19**(2), 78 (Feb. 1961).
96. *Sodium Properties, Uses, Storage, And Handling*, Bul. E-92775-1, E. I. du Pont de Nemours & Co., Inc., Wilmington, Del., 1994.
97. *Sodium Material Safety Data Sheet*, E. I. du Pont de Nemours & Co., Inc., Wilmington, Del., June 1995.
98. D. D. Adams, G. J. Barenborg, and W. W. Kendall, *Adv. Chem. Set.* **19**, 92 (1957).
99. L. F. Epstein in C. M. Nicholls, ed., *Progress in Nuclear Energy*, Ser. IV, Vol. 4, Pergamon Press, Inc., New York, 1961, pp. 461–483.
100. P. Menzenhauer, G. Ochs, and W. Peppler, *Kernforschungsz. Karlsruhe (Berlin)*, KFK 2525 (1977).
101. *Chem. Eng. News* **34**(17), 1991 (Apr. 23, 1956).
102. *Chem. Eng.* **65**(12), 63 (June 16, 1958).
103. U.S. Pat. 4,032,615 (June 28, 1977), T. R. Johnson (to U.S. Energy Research and Development Administration).
104. Jpn. Pat. 79/114,473 (Sept. 6, 1979), Y. Nishizawa (to Mitsubishi Atomic Power Industries, Inc.).
105. U.S. Pat. 3,729,548 (Apr. 24, 1973), C. H. Lemke (to E. I. du Pont de Nemours & Co., Inc.).
106. U.S. Pat. 3,459,493 (Aug. 15, 1969), F. J. Ross (to E. I. du Pont de Nemours & Co., Inc.).
107. H. P. Maffei, C. W. Funk, and J. L. Ballif, *Sodium Removal Disassembly and Examination Of the Ferrmi Secondary Sodium Pump*, Report 1974, HEDL-TC-133, available from National Technical Information Service, Washington, D.C.
108. I. Fatt and M. Tashima, *Alkali Metal Dispersions*, D. Van Nostrand Co., Inc., Princeton, N.J., 1961.
109. U.S. Pat. 2,890,111 (June 9, 1959), S. M. Shelton (to United States of America).
110. U.S. Pat. 2,828,119 (Mar. 25, 1958), G. R. Findley (to National Research Corp.).
111. U.S. Pat.. 2,890,112 (June 9, 1959), C. H. Winter, Jr. (to E. I. du Pont de Nemours & Co., Inc.).
112. Brit. Pat. 816,017 (July 8, 1959) (to National Distillers and Chemical Corp.).
113. U.S. Pat. 3,736,132 (May 29, 1973), H. H. Morse and co-workers (to United States Steel Corp.).
114. Brit. Pat. 1,355,433 (June 5, 1974), P. D. Johnston and co-workers (to Electricity Council).
115. Ger. Offen. 2,517,180 (Oct. 21, 1976), R. Haehn and D. Behrens (to Firima Hermann C. Starck, Berlin).

116. J. V. R. Heberlein, J. F. Lowry, T. N. Meyer, and D. F. Ciliberti, *Conference Proceedings of the 4th International Symposium on Plasma Chemistry*, Pt. 2, Zurich, Switzerland, 1979, pp. 716–722.

117. A. Sanjurjo, L. Nanis, K. Sancier, R. Bartlett, and V. Kapur, *J. Electrochem. Soc.* **128**, 179 (Jan. 1981).

118. D. B. Olson and W. J. Miller, Report 1978, DOE/JPL/-954777-5, AeroChern-TN-99, available from National Technical Information Service, Washington, D.C.

119. U.S. Pat. 4,014,687 (Mar. 29, 1977), N. D. Clare and C. H. Lemke (to E. I. du Pont de Nemours & Co., Inc.).

120. U.S. Pat. 2,794,732 (June 4, 1957), P. P. Alexander (to Metal Hydrides, Inc.).

121. U.S. Pat. 2,794,733 (June 4, 1957), P. P. Alexander and R. C. Wade (to Metal Hydrides, Inc.).

122. U.S. Pat. 1,796,241 (Mar. 10, 1931), H. R. Carveth (to Roessler and Hasslacher Chemical Co.).

123. U.S. Pat. 2,633,406 (Mar. 31, 1953), D. S. Nantz (to National Distillers Products Corp.).

124. U.S. Pat. 2,671,010 (Mar. 2, 1954), L. J. Governale (to Ethyl Corp.).

125. U.S. Pat. 2,685,500 (Aug. 3, 1954), R. E. Hulse and D. S. Nantz (to National Distillers Products Co.).

126. I. I. Vol'nov, *Peroxides, Superoxides, and Ozonides, of Alkali and Alkaline Earth Metals*, Plenum Publishers Corp., New York (1966).

127. *DuPont Sodium Hydride Descaling Process*, Bulletin No. SP 29-370, E. I. du Pont de Nemours & Co., Inc., Wilmington, Del., 1970.

128. *Eur. Chem. News (London)* **14**(336), 34 (1968).

129. U.S. Pat. 2,915,564 (Dec. 1, 1959), V. L. Hansley (to National Distillers and Chemical Corp.).

130. H. N. Gilbert, N. D. Scott, W. F. Zimmerli, and V. L. Hansley, *Ind. Eng. Chem.* **25**, 735 (1933).

131. M. Sittig, *Mod. Plast.* **32**(12), 150, 217 (1955).

132. U.S. Pat. 2,352,461 (June 27, 1944), J. F. Walker (to E. I. du Pont de Nemours & Co., Inc.).

133. A. A. Morton, *Ind. Eng. Chem.* **42**, 1488 (1950).

134. A. A. Morton in N. M. Bikales, ed., *Encyclopedia of Polymer Science and Technology*, Interscience Publishers, a Division of John Wiley & Sons, Inc., New York, 1964, pp. 629–638.

135. U.S. Pat. 3,966,691 (June 29, 1976), A. F. Aalasa (to Firestone Tire and Rubber Co.).

136. Jpn. Pat. 76/68,491 (June 14, 1976) T. Kitsunai, Y. Mitsuda, and S. Sato (to Denki Kagaku Kogyo KK).

137. Ger. Offen. 2,802,044 (July 20, 1978), A. Proni and A. Roggero.

138. U.S. Pat. 2,809,130 (Oct. 8, 1957), G. Rappaport (to General Motors Corp.).

139. U.S. Pat. 2,789,063 (Apr. 16, 1957), R. J. Purvis and W. R. Beck (to Minnesota Mining & Manufacturing Co.).

140. Brit. Pat. 1,078,048 (Aug. 2, 1967), R. S. Haines (to International Business Machines Corp.).

141. U.S. Pat. 3,565,792 (Feb. 23, 1971), F. B. Haskett.

142. R. D. Weaver, S. W. Smith, and N. L. Willmann, *J. Electrochem. Soc.* **109**(8), 653 (1962).

143. N. Weber and J. T. Kummer, *Proceedings of the 21st Annual Power Sources Conference*, U.S. Army, Electronics Command, Atlantic City, N.J., 1967, pp. 37–39.

144. U.S. Pat. 3,791,871 (Feb. 12, 1974), L. S. Rowley (to Lockheed Aircraft Corp.).

145. U.S. Pat. 4,053,685 (Oct. 11, 1977), L. S. Rowley and H. J. Halberstadt (to Lockheed Missiles and Space Co., Inc.).

146. H. J. Halberstadt, *Proceedings of the 8th Intersociety Energy Conversion and Engineering Conference*, American Institution of Aeronautics and Astronautics, New York, 1973, pp. 63–66.

147. R. R. Roll, *Proc. Symp. Batteries Traction Propul.*, 209 (1972).

148. I. J. Groce and R. D. Oldenkamp, *Adv. Chem. Ser.* **64**, 43 (1967).

149. L. A. Heredy, M. L. Iverson, G. D. Ulrich, and H. L. Recht, *Adv. Chem. Ser.* **64**, 30 (1967).

150. *Chem. Week*, **101**(8), 79 (Aug. 19, 1967).

151. R. H. Boundy, *Trans. Electrochem. Soc.* **62**, 151 (1932).

152. Brit. Pat. 1,188,544 (Apr. 15, 1970), (to General Cable Corp.).

153. U.S. Pat. 4,056,679 (Nov. 1, 1977), T. F. Brandt (to I. T. E. Imperial Corp.).

154. *Assessment of Sodium Conductor Distribution Cable*, DOE/ET-504-1, Westinghouse Research and Development Center, Pittsburgh, Pa., June 1979.

155. W. Peppler, *Chem. Ztg.* **103**(6), 195 (1979).

156. *Proceedings of the International Conference on Liquid Alkali Metals*, British Nuclear Energy Society, London, Apr. 4–6, 1973.

157. M. H. Cooper, ed., *Proceedings of the International Conference on Liquid Metal Technology in Energy Production*, CONF-760503-P1 and P2, Champion, Pa., May 3–6, 1976.

158. J. J. Bartel, H. J. Rack, R. W. Mar, S. L. Robinson, F. P. Gerstle, Jr., and K. B. Wischmann, *1. Molten Salt and Liquid Metal, Sandia Laboratories Materials Tast Group Review of Advanced Central Receiver Preliminary Designs*, Sept. 1979, SAND 79-8633, available from National Technical Information Service, Washington, D.C.

159. A. B. Meinel and M. P. Meinel, *Applied Solar Energy*, Addison-Wesley Publishing Co., Reading, Mass., 1976.

160. J. W. Chi, *Proc. Top. Meet. Technol. Controlled Nucl. Fusion* **2**(2), 443 (1976).

161. J. E. Kemme, J. E. Deverall, E. S. Keddy, J. R. Phillips, and W. A. Rankin, *Temperature Control with High Temperature Gravity-Assist Heat Pipes*, Los Alamos Scientific Laboratory, 1975, available from National Technical Information Service, Accession No. CONF-750812-10.

162. U.S. Pat. 3,220,876 (Nov. 30, 1965), R. D. Moeller (to North American Aviation, Inc.).

163. U.S. Pat. 3,251,719 (May 17, 1966), F. Tepper and co-workers (to M.S.A. Research Corp.).

164. U.S. Pat. 3,598,572 (Aug. 10, 1971), J. C. Robertson (to Dow Chemical Co.).

165. U.S. Pat. 4798711 (Jan. 17, 1989) (to NOXSO Corp.).

166. U.S. Pat. 4940569 (July 10, 1990) (to NOXSO Corp.).

167. M. G. Down, P. Hubberstey, and R. J. Pulham, *J. Chem. Soc. Dalton Trans.* (14), 1490 (1975).

168. R. P. Elliott, *Constitution of Binary Alloys*, 1st Suppl., McGraw-Hill Book Co., Inc., New York,

169. R. B. MacMuffin, *Chem. Eng. Prog.* **46**, 440 (1950).

General References

Reference 5 is a critical source of data on sodium properties, components, systems, handling, and safety.

References 14 and 20 cover the inorganic chemistry of sodium.

References 8 and 16 survey sodium organic chemistry.

References 37 and 167 present phase diagrams of sodium with other metals.

"Natrium," *Gmelins Handbuch der Anorganischen Chemie*, 8th ed., Vol. 2, Verlag Chemie GmbH, Weinheim, Germany, 1965, pp. 401–627, covers inorganic chemistry of sodium.

Reference 88 journal contains numerous articles on AMTEC technology in annual and monthly updates through Dec. 1995.

CHARLES H. LEMKE
E. I. du Pont de Nemours & Co., Inc.
University of Delaware

VERNON H. MARKANT
E. I. du Pont de Nemours & Co., Inc.

SODIUM CARBONATE. See ALKALI AND CHLORINE PRODUCTS.

SODIUM COMPOUNDS

SODIUM HALIDES

SODIUM CHLORIDE

Salt producers classify sodium chloride [7647-14-5], NaCl, also known as common salt, by the three methods used for its production: mechanical evaporation of solution-mined brine, ie, evaporated–granulated salt; underground mining of halite deposits, ie, rock salt; and solar evaporation of seawater, natural brine, or solution-mined brine, ie, solar salt. Salt in brine is a fourth classification, for solution-mined brine that is typically used as a feedstock for chemical production. Salt is a readily available, inexpensive bulk commodity and a basic requirement for all life. It is found throughout the world in natural underground deposits as the mineral halite and, in some locations, as mixed evaporites in saline lakes. Sodium chloride, as a compound, is the largest component of dissolved solids found in seawater, where it averages 2.6% by weight. One cubic kilometer of seawater contains nearly 458×10^6 t of sodium chloride, 66×10^6 t of magnesium chloride, and 30×10^6 t of calcium sulfate (1). Table 1 shows the composition of seawater.

Table 1. Composition of Seawater

Component	Wt %	Component	Wt %
sodium chloride	2.68	potassium chloride	0.07
magnesium chloride	0.32	sodium bromide	0.008
magnesium sulfate	0.22	water	96.582
calcium sulfate	0.12		

Salt Deposition

Salt deposits are widely distributed both in location and in geologic time. Figure 1 shows the locations of salt production sites and the principal salt deposits in the United States and Canada. Geologically, salt deposition occurred from the Precambrian through the Quaternary systems. There are 10 principal salt basins in the Western hemisphere (2): (1) the Maritime provinces of eastern Canada; (2) Appalachian (New York, Ohio, and Ontario); (3) Michigan (Michigan and Ontario); (4) Williston (North and South Dakota, Montana, and Saskatchewan); (5) Alberta (northern and eastern Alberta); (6) Mackenzie (Northwest Territories); (7) Permian (west Texas, New Mexico, Oklahoma, and Kansas); (8) Paradox (southeast Utah and southwest Colorado); (9) Supai (New Mexico and Arizona); and (10) the Gulf region (southern United States, eastern Mexico, and Cuba).

There are two basic types of salt formations: bedded deposits and salt structures or diapirs, also known as salt domes. Layered or bedded deposits can be relatively horizontal and undisturbed or they can be deformed by tectonic forces. In bedded deposits, layers of halite are separated by layers of the mineral anhydrite (calcium sulfate). Other impurities such as shale, iron pyrites, and silica may also be present. Salt structures or diapirs are formed from bedded salt deposits by isostatic salt movement (2). Typically, these structures are anticlines and salt domes or diapirs. Bedding planes in salt domes are nearly vertical as a result of the isostatic, upward movement of the salt. Over long geologic time periods, horizontal salt deposits were gradually covered by sediments and buried within the earth's crust. At great depths, the buried salt, which was less dense than the overlying rock formations, flowed upward through zones of weakness in the overlying strata. Confined salt under pressure deforms plastically and can rise as a vertically elongated plume or dome. Salt acts as a completely mobile plastic below 7600 m of overburden and at temperatures above 200°C (2). Under lesser conditions, salt domes can grow by viscous flow. Salt structures originate in horizontal salt beds at depths of 4000–6000 m or more beneath the earth's surface. The resulting salt dome or diapir is typically composed of relatively

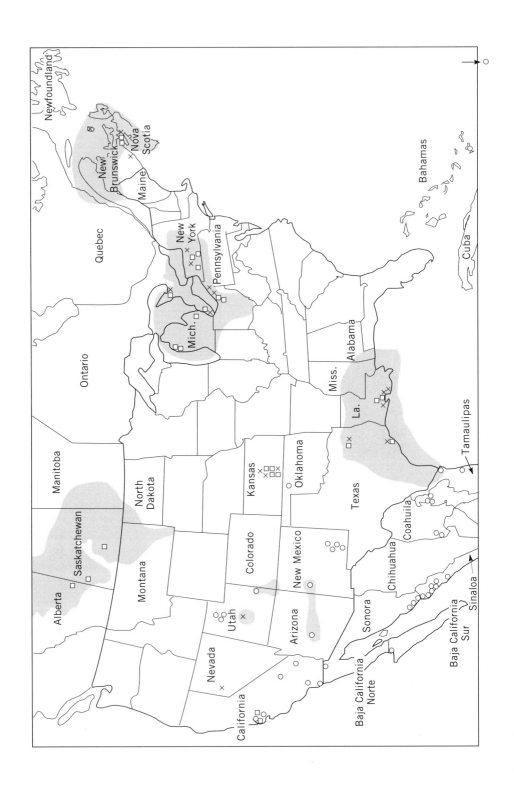

Fig. 1. Principal salt deposits and dry salt production sites in North America, where ■ represents the salt deposits and □, ×, and ○ correspond to evaporated, rock, and solar production sites, respectively. Sites in Canada, listed by company name (location), are Alberta, □, Canadian (Lindberg); Saskatchewan, □, Sifto/Namsco (Unity) and Canadian (Belle Plaine); Ontario, □ and ×, Sifto/Namsco (Goderich); □, Canadian (Windsor); ×, Canadian (Ojibway); New Brunswick, ×, Potash Corp. of Saskatchewan (Sussex); and Nova Scotia, □ and ×, Canadian (Pugwash) and Mines Seleine (Quebec); and, □, Sifto/Namsco (Amherst). Sites in the United States are Michigan, □, Morton (Manistee), and Akzo Nobel (St. Clair); New York, ×, Cargill (Lansing); □, Akzo Nobel (Watkins Glen), Cargill (Watkins Glen), and Morton (Silver Springs); Ohio: × Morton (Fairport) and Akzo Nobel (Cleveland); □, Akzo Nobel (Akron) and Morton (Rittman); California, ○ and □, Cargill (Newark); □, Morton (Newark); ○, Cargill (Redwood City), Pacific Salt & Chemical (Trona), Cargill (Amboy), Salt Products Co. (Milligan), and Western (Chula Vista); Nevada, ×, Huck (Fallon); Utah, ○, Namsco (Ogden), Akzo Nobel (Timpie), Morton (Grantsville), and Moab (Moab); ×, Redmond Clay & Salt (Redmond); Kansas: ×, Independent (Kanopolis), Lyons (Lyons), and Hutchinson (Hutchinson); □, Namsco (Lyons), Namsco (Hutchinson), Morton (Hutchinson), and Cargill (Hutchinson); Arizona, ○, Southwest/Morton (Glendale); New Mexico, ○, Zuni (Quemado) Don McKibben Trucking (Carlsbad), United (Carlsbad), New Mexico Salt & Minerals (Loving), and Sunwest Salt (Malaga); Oklahoma, ○, Cargill (Freedom); Texas, □ and ×, Morton (Grand Saline); ×, United (Hockley); □, United (Blue Ridge); and Louisiana, □, Cargill (Breaux Bridge); ×, Akzo Nobel (Avery Island) and Namsco (Cote Blanche); and □ and ×, Morton (Weeks Island), Tennessee, Namsco (New Johnsonville). Sites in Mexico are Baja California, ○, Sales de Ometepec (Mexicali); Sonora, ○, Salinas de Lobos (Ciudad Obregòn), Compañia Salinera de Yavaros (Navojoa), and Salinera del Mayo (Navojoa); Coahuila, ○, Salinera Coahuila (Torreòn) and Salinas del Rey (Torreon); Nuevo Leòn, ○, Distribuidora y Envasadora de Productos de Sodio (Monterrey); Empacadora Abelardo Martinez (Monterrey), and Industria del Alcali (Monterrey); Tamaulipas, ○, Salinera la Boladeña (Matamoros) and Salex (Tampico); Baja California Sur, ○, Exportadora de Sal (Guerrero Negro); Sinaloa, ○, Imprenta Mochis (La Mochis), Abelino Lopez Mejia y Socios (Angostura), Salinas de Elota (Culiacán), Expendio de Sal Herlinda Medina (Culiacán), Sociedad Cooperativa de Produccion Salinera Montelargo (Culiacán), Sociedad Cooperative Industrial Salinera (Culiacán), and Molino de Sal Casa Blanca (Mazatlan). Offshore sites are ○, Morton (Inagua, Bahamas), and Akzo Nobel (Bonaire, Netherlands Antilles). Courtesy of the Salt Institute.

pure sodium chloride in a vertically elongated, roughly cylindrical, or inverted teardrop-shaped mass.

History of U.S. Salt Production

Salt production in the United States can be summarized in terms of four over-lapping phases: from 1770 and continuing, salt has been obtained from solar evaporation of seawater, bay water, or natural brine; from 1788 through ca 1900, salt was obtained by open boiling, in iron kettles or iron pans, of brine obtained from salt springs and salt wells; from 1833 and continuing, evaporation of brine from salt springs and later from brine wells, first in open grainer pans and later in multiple-effect vacuum pans has been employed; and ca 1888 and continuing, conventional underground mining of rock salt deposits has occurred. U.S. salt production during the Civil War was over 225,000 t of salt.

Travelers to Onondaga, New York, in 1654 reported the presence of salt (3). The Onondaga Indians used salt springs as a source of salt, producing salt by boiling. In Kanawha, West Virginia, Native Americans made salt before 1755 by boiling brine from salt springs (4). European settlers later copied the method, making salt for their own use. Commercial salt production was underway in 1797 and the Kanawha Valley supplied the Confederacy with salt during the Civil War. Production in this area peaked during the 1860s, and by the early years of the twentieth century only one plant survived.

Beginning in the 1820s, solar salt was produced commercially using covered sheds to protect the salt from precipitation. A significant advance in evaporated salt technology occurred in 1833 when multiple-effect, open grainer pans came into use. The addition of reservoirs and settling vats clarified the brine and resulted in clean, white salt. The grainer process was further refined at Silver Springs, New York, during the nineteenth century and ultimately led to the development of steam-heated vacuum pans. Solar salt-making began on San Francisco Bay, California, in 1770 and at Salt Lake City, Utah, on the Great Salt Lake in 1847. There were 442 salt works producing salt on Cape Cod, Massachusetts, during the 1830s (5). Evaporating furnaces operated between 1790 and 1860 in Ohio, Michigan, Pennsylvania, Kentucky, Indiana, Illinois, and Missouri. In the mid-1800s salt was produced from salt springs at Saginaw and St. Clair, Michigan, where the lumber industry provided a large, inexpensive supply of waste wood products for fuel. In 1882, rock salt was found at St. Clair, Michigan, by drilling. Solution mining of rock salt deposits spread rapidly in Kansas, Louisiana, Michigan, New York, Ohio, and Texas. Conventional underground mining of rock salt began in Kansas, Louisiana, Michigan, and New York between 1880 and 1887.

Properties

Sodium chloride precipitates in cubic, crystalline form. When pure, it is colorless and consists of 60.663 wt % Cl, atomic wt 35.4527 (9) and 39.337 wt % Na, atomic wt 22.989768 (6). Sodium chloride in commercial form can appear as discrete crystals in various size ranges, fine granules or powder, and compressed pellets or blocks. If viewed with magnification, all sodium chloride is crystalline. Depending on gradation and commercial form, salt can be white, grayish, or sometimes brownish. Large crystals of apparently recrystallized halite found in some salt mines are colorless and transparent, clearly showing the mineral's characteristic cubic cleavage. Pure sodium chloride is transparent in the near- and mid-ir regions. Table 2 shows the physical properties of sodium chloride. Salt is soluble in polar solvents, insoluble in nonpolar. Solubility of pure sodium chloride in water and other solvents is given in Table 3. The aqueous solution has a pH of 7 in the absence of impurities. The phase diagram for the sodium chloride–water system is shown in Figure 2. The composition of the eutectic mixture, 23.31 wt % NaCl, may also be expressed as 37.68 wt % $NaCl \cdot 2H_2O$ and 62.32 wt % water. At low temperatures, brines more concentrated than 23.31 wt % NaCl deposit large, transparent crystals of monoclinic sodium chloride dihydrate [23724-87-0]. These crystals, although similar to ice in appearance, are birefringent and account for freezing of moist stockpiles of highway deicing salt during storage in cold weather. A good sourcebook on sodium chloride properties is available (7).

Table 2. Properties of Pure Sodium Chloride

Property	Value
molecular wt	58.44
crystalline form	cubic
color	clear to white
index of refraction, n_D^{20}	1.5442
density or specific gravity, g/cm^3	2.165
mp, °C	801
bp, °C	1413
hardness, Mohs' scale	2.5
specific heat, $J/(g \cdot K)^a$	0.853
heat of fusion, J/g^a	517.1
critical humidity at 20°C, %	75.3
heat of solution, 1 kg H_2O, 25°C, kJ/mol^a	3.757

[a]To convert J to cal, divide by 4.184.

Table 3. Solubility of Pure Sodium Chloride

Solvent	Temperature, °C	NaCl, g/100 g solvent
ethanol	25	0.065
ethylene glycol	25	7.15
formic acid	25	5.21
glycerol	25	10
hydrochloric acid		insol
liquid ammonia	−40	2.15
methanol	25	1.40
monoethanolamine	25	1.86
water	0	35.7
	100	39.12
	120	39.8

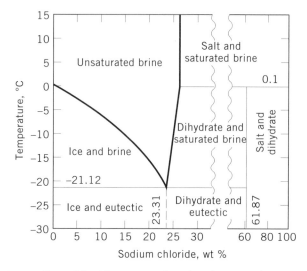

Fig. 2. The system sodium chloride–water, showing the eutectic temperature, $-21.12°C$, and composition, 23.31 wt % NaCl.

Processing

Salt production generally refers to the production of dry, crystalline sodium chloride by evaporation of brine or by conventional mining of the mineral halite. Salt in brine produced by solution mining is an intermediate product that is processed into evaporated–granulated salt or other chemical products (see CHEMICALS FROM BRINE). The characteristics of the brine are dependent on the source formation and physical and chemical treatment of the brine prior to further processing. For purposes herein, solution mining is described as preliminary to the mechanical evaporation process for producing salt. Therefore, salt production refers to the production of dry, crystalline salt from seawater, or solution-mined or natural brine, and the conventional mining of rock salt. In the strictest sense, all natural salt is oceanic in origin.

Solution Mining and Mechanical Evaporation. Bedded and domal salt deposits are solution-mined by drilling wells into halite [14762-57-7] deposits and injecting fresh and recycled water through the well casings, dissolving the salt. Salt solution mines, called Class III wells, can vary in depth from 150 to 1500 m. Solution-mined brine can be used to produce dry salt or it can be used as feedstock for the production of chemicals such as chlorine and caustic soda (see ALKALI AND CHLORINE PRODUCTS). Solution mines operate with a single well or with several wells in a brine field. Brine is produced from a single well by injecting water into the salt deposit through tubing, then extracting brine through a concentric annulus between the tubing and the well casing. Water injected into one well in a multiple-well brine field dissolves salt. The resulting brine is extracted through other wells in the same brine field. Solution-mining technology provides control over the size and shape of solution-mined caverns and minimizes the potential for surface subsidence. State-of-the-art drilling and operating techniques, well and cavern logging instruments, and other devices provide precise control over salt cavern development and use. Salt brine is extracted from the cavern and transported by pipeline to the salt refinery, usually at the same site, for processing into evaporated–granulated salt or for use as feedstock for chlor–alkali production.

Table salt is typical of the fine, evaporated–granulated salt produced in vacuum pan evaporators. Virtually all food-grade salt sold or used in the United States is produced by vacuum evaporation (qv) of brine. Prior to mechanical evaporation, the brine may be treated to remove minerals that can cause scaling in the evaporators and adversely affect salt purity. Chemical treatment of the brine, followed by settling, reduces levels of dissolved calcium, magnesium, and sulfate. Chemicals typically used are calcium hydroxide, $Ca(OH)_2$; sodium carbonate, Na_2CO_3; sodium hydroxide, NaOH; calcium chloride, $CaCl_2$; flocculating agents; and stack gas, CO_2. Sulfuric acid treatment or chlorination may be used to remove hydrogen sulfide, and hydrochloric acid neutralizes brine used in diaphragm cell production of chlorine and caustic soda. Brine purification has become increasingly important for production of high purity salt for use in chlor–alkali production, particularly in Europe where dry salt is used extensively for this purpose.

Water is evaporated from purified brine using multiple-effect or vapor recompression evaporators (Figs. 3 and 4). Multiple-effect systems typically contain three or four forced-circulation evaporating vessels (Fig. 4) connected together in series. Steam from boilers supplies the heat and is fed from one evaporator to the next to increase energy efficiency in the multiple-effect system.

Vapor-recompression forced-circulation evaporators consist of a crystallizer, a compressor, and a vapor scrubber. Feed brine enters the crystallizer vessel,

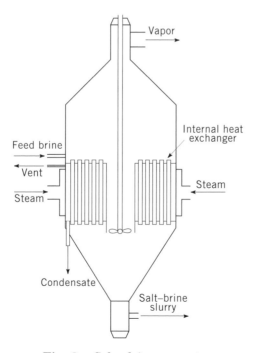

Fig. 3. Calandria evaporator.

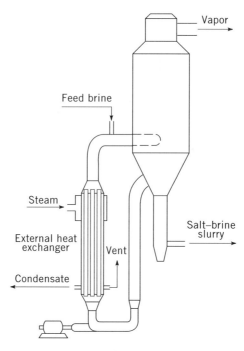

Fig. 4. Forced-circulation evaporator.

where salt is precipitated. Vapor is withdrawn, scrubbed, and compressed for reuse in the heater. Ultimately, weak brine from the process is recycled to the solution-mined cavern. Crystallized salt is removed from the elutriation leg as a slurry. Recompression evaporators are more energy-efficient than multiple-effect evaporators but require higher cost electrical power for energy input. The development of single-stage compressors has significantly reduced costs. The salt slurry from either type of evaporator is dewatered first by centrifuging or vacuum drying and then in kiln or fluidized-bed dryers, where moisture content of the final product is reduced to 0.05% or less. During the twentieth century, salt producers have made significant advances in lowering energy consumption and in reducing salting and scaling in evaporators.

Grainer Salt. Evaporated salt can also be produced with the addition of heat in open pans. The resulting grainer salt consists of flakes and is used for food applications where a coarser, flake salt product is desired. Salt crystals forming on the surface of the brine are supported by surface tension. As they become heavier, they fall to the bottom of the pan as incomplete cubes or hopper-shaped crystals. The product is collected by a mechanical rake and dried in the same way as vacuum pan salt. Many factors can adversely affect the production of grainer salt, including ambient temperature and humidity and the purity of the feed brine. The grainer process, similar to the historic process of open-pan boiling, was suitable in some areas of the United States where waste steam was available or where waste lumber products provided an inexpensive source of heat. This process is energy-inefficient and no longer in use.

Three specialty types of salt are produced by variations of the evaporating process. Alberger salt is produced with a modified grainer method and results in a combination of flakes and cubic crystals of salt. It is less dense than granulated–evaporated salt and, like grainer salt, is typically used where coarser salt is required, as in koshering. Dendritic salt is made by adding yellow prussiate of soda (YPS) (sodium ferrocyanide) to the brine; the YPS modifies the crystal structure of sodium chloride to form a branch-like, low density dendritic salt crystal. The recrystallizer process (Fig. 5) is a means of producing purified salt. Brine made by dissolving rock salt or solar salt is purified and fed to vacuum pan evaporators. The result is granulated–evaporated salt of excellent purity. Granulated–evaporated salt has many uses where purity and consistency of grain size and shape are important.

Conventional Underground Mining.   Rock salt is extracted from a conventional mine by drilling and blasting. Access is through a circular shaft, typically about 6 m dia and as deep as 700 m, depending on the depth of the salt deposit. Shafts are lined with concrete at least through the overburden and into the top of the salt deposit, and sometimes all the way to the shaft bottom. Mining methods vary somewhat depending on whether the salt deposit is a relatively horizontal bedded deposit or a vertical salt dome. The main variation is a result of the thickness and structure of the deposit. Bedded or layered deposits are mined by excavating horizontal entries or rooms (room and pillar mining), typically 3 to 8 m high and 15 m wide. Openings or cross-cuts are mined perpendicular to the length of the rooms to connect them at predetermined intervals. Rectangular

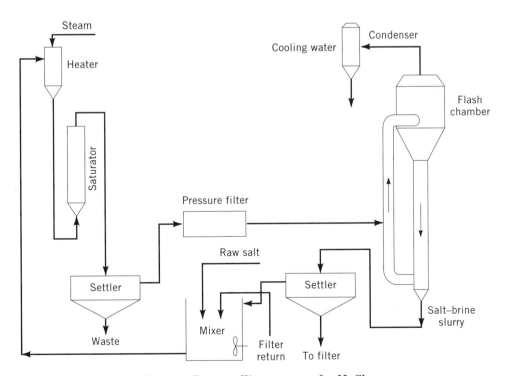

Fig. 5.   Recrystallizer process for NaCl.

salt pillars remaining in place provide structural support for the overlying formations. Mine engineers use principles of rock mechanics to size the rooms and
pillars to assure worker safety and stability of the mine openings. Confined salt
under high stress is relatively incompressible and deforms, resulting in plastic
flow or salt creep. As a result, over a long period of time, the mine floor and
roof tend to converge. Proper application of rock mechanics principles allows
convergence to occur without failure of the salt floor or roof.

Domal salt deposits, because of the potentially great thickness of the salt
in them, are mined first by using horizontal entries or rooms, followed by
excavating or benching the floor downward to a depth of 20 m or more. Excavated
rooms in domal salt mines can reach more than 30 m in height and, in some
cases, 30 m in width. Benching improves the efficiency of mining salt domes
because larger quantities of salt can be blasted in a single shot. Benching can
also be used in thicker-bedded deposits.

Salt is mined by undercutting, drilling, blasting, loading, and transporting
the broken salt for processing. Salt mining operations are highly mechanized,
with extensive use of large, diesel-powered equipment. Undercutters, drilling
machines, and explosives loading equipment use diesel power for locomotion,
but electric power is used to perform operations at the mine face. A typical
mining advance is carried out by cutting a horizontal slot 15 cm high along the
salt floor to a depth of 4 to 6 m. Salt miners then drill a series of holes to
the same horizontal depth as the undercut and load the holes with explosives
prior to blasting. Fertilizer-grade ammonium nitrate mixed with 5% No. 2 fuel
oil, known as ANFO, is commonly used for blasting. An igniter and primer,
sometimes gelex dynamite, is inserted into the drill hole and the ANFO blasting
mixture is pneumatically blown into the hole, over the primer. A single blast in
a mine face or room can bring down 350–900 t of rock salt, depending on the
mining pattern. Several rooms are blasted simultaneously to provide enough salt
to meet daily production requirements. As a safety measure, the blasted area is
inspected for loose pieces of salt in the roof or pillars. Any that are found are
removed with handheld scaling bars or with long-reach, mechanized roof-scaling
equipment.

In domal mines, after the rooms have advanced some distance, the salt
floor is benched by drilling a series of holes downward and blasting to a depth
of up to 23 m. Benching can yield several thousand tons of salt with a single
blast, resulting in high operational efficiency because the mining equipment can
remain in the same salt pile for long periods of time.

Continuous Mining.   Continuous miners or boring machines have been used
since the late 1950s for salt mine development and, in some cases, for production.
These machines have movable, rotating heads with carbide-tipped cutting bits.
Continuous miners bore into the salt and eliminate undercutting, drilling, and
blasting. Excessive fines were a problem in the past, but newer machines have
reduced fines production.

After the salt is blasted or mined by a continuous miner, it is loaded into
trucks or shuttle cars and hauled to a primary crusher. Rubber-tired end loaders having 9–14-t capacity buckets and diesel-powered trucks of 65-t capacity
are common in large, modern salt mines. In low roof mines, scoop trams (low
profile front-end loaders) of 5–10 t capacity are used. Broken salt can also be

transported with load–haul–dump (LHD) units. A LHD unit is a single piece of equipment that can collect, transport, and dump the broken salt into a primary crusher or breaker using one operator–driver. The primary crusher reduces the large pieces of blasted salt to a maximum size of about 25 cm for ease of conveyor handling. The salt is then transported by conveyor belt to second- and third-stage crushers and screening stations for further size reduction and screening into standard product grades for specific end uses. The crushing and screening stations may be either underground or on the surface, depending on the layout of the mine. The fine salt, passing 600 μm (U.S. Sieve No. 30), removed by the screening process cakes easily, is difficult to handle and has little market value. Typically it is used in the mine to construct and repair roads and to build stoppings or brattices between pillars for ventilation control. The salt, either as mine run grade or as screened grades, is then conveyed to the shaft bottom for hoisting to the surface. Large, up to 18-t capacity, rectangular skips hoist the salt to surface operations areas for further processing, packaging, and bulk shipping. Hoisting rates can reach 900 t/h at large mines.

Solar Evaporation. Solar salt is produced by natural evaporation of seawater or brine in large, diked, earthen ponds. Solar radiation and wind action concentrate the seawater or natural brine until sodium chloride crystallizes. Solar salt production requires a large area of flat, low cost land. Climatic conditions must ensure high evaporation rates and low rainfall. A deep water site for bulk loading of ocean-going vessels is desirable for high volume shipping for export. Although most solar salt production facilities use seawater as feedstock, natural brine and solution-mined brine are also used. Solar ponds are constructed to take advantage of natural ground contours to aid in brine movement. At some point in most pond systems, seawater or brine must be elevated by pumping.

The concentration of dissolved solids in seawater, although variable by location and depth, averages 3.5 wt % (3.6°Bé) and 77% of the total dissolved solids, ie, 2.7 wt % of seawater (see Table 1), is sodium chloride. Seawater must be reduced in volume by about 90% before sodium chloride begins to crystallize at a concentration of 25.8 wt % NaCl (25.4°Bé). The most favorable sodium chloride precipitation occurs between 26°Bé and 29 or 30°Bé. Above 30°Bé, high levels of magnesium reduce the evaporation rate to unacceptable levels. At 29°Bé, 72% of the NaCl has precipitated; at 30°Bé, 79% has precipitated. The high magnesium brine, called bittern because of the bitter taste, is collected for by-product removal or is discharged. Newly concentrated brine is added and the crystallization process continues. A solar salt works producing 400,000 t/yr of sodium chloride may have an area of 4,000 ha or more, depending on climatic conditions. Proper brine control during concentration and crystallization results in salt of purity >99.7% NaCl on a dry basis.

Typically, seawater enters the solar pond system and moves successively from one pond to the next. Operators control the quantity of flow with mechanical gates to maintain target brine densities and pond levels. Iron, calcium, and magnesium carbonates crystallize when the concentration of brine is 3.5–13°Bé. About 85% of the calcium sulfate crystallizes as gypsum and then anhydrite at brine concentrations from 13–25.4°Bé. Solar salt production is a form of fractional crystallization. Brine reaching the crystallizers contains in solution calcium sulfate, magnesium sulfate, magnesium chloride, and small amounts of potassium chloride, plus minuscule amounts of other elements. The saturated brine at a specific gravity of 25.4°Bé is fed onto level, rectangular crystallizing ponds to maintain a brine depth of 30 cm or less. As evaporation proceeds, sodium chloride precipitates and forms a salt layer 10–25 cm thick. In many solar salt facilities, the first crop of salt deposited remains on the crystallizer bottoms as salt floors to prevent contamination from soil and to increase the strength of the crystallizer bottoms to support harvesting equipment. When the brine reaches 25–30°Bé, the bittern is discharged. Most of the magnesium sulfate and magnesium and potassium chlorides remains in the bittern.

The salt crop is harvested using salt harvesters, elevating scrapers, or large end loaders. The harvested salt is loaded into trucks and transported to a washplant. The salt is washed with clean, nearly saturated brine to remove particulate matter and to replace magnesium-laden brine clinging to the salt crystals. Uncontaminated brine is made and recycled in a settling pond by adding seawater to dissolve fine salt collected by the wash brine. Brine made by dissolving the fine salt is free of magnesium and sulfate impurities. Weak brine or seawater is sometimes used as a final wash. However, production losses by dissolution are significant. After washing, the salt is stockpiled and allowed to drain. Limited rainfall is relied on to improve salt quality by rinsing action. Solar salt typically drains naturally to a moisture level of about 3.5%.

Similar to processing mined rock salt, solar salt may be crushed, screened, and kiln dried or fluidized-bed dried. Coarse solar salt is a premium product because of high purity and relatively large crystal size. It is in particular demand for use to regenerate the resin in cation-exchange water softeners (see ION EXCHANGE; WATER, INDUSTRIAL WATER TREATMENT).

Electrodialysis. Electrodialytic membrane process technology is used extensively in Japan to produce granulated–evaporated salt. Filtered seawater is concentrated by membrane electrodialysis and evaporated in multiple-effect evaporators. Seawater can be concentrated to a product brine concentration of 200 g/L at a power consumption of 150 kWh/t of NaCl (8). Improvements in membrane technology have reduced the power consumption and energy costs so that a high value-added product such as table salt can be produced economically by electrodialysis. However, industrial-grade salt produced in this manner cannot compete economically with the large quantities of low cost solar salt imported into Japan from Australia and Mexico.

Salt Standards and Specifications

Salt produced in the United States varies in purity from 95% NaCl for rock salt to 99.99% NaCl for mechanically evaporated salt. Mechanically evaporated salt made using purified brine generally has the highest purity; rock salt generally has the lowest. Several voluntary standards and mandatory specifications apply to salt to ensure appropriate gradation, quality, and purity for particular salt uses (Table 4).

Additives. Sodium chloride is hygroscopic at >75.3% at 20°C relative humidity (critical humidity) and individual salt crystals can adsorb enough moisture during storage to result in formation of brine on crystal surfaces. Subsequent evaporation causes recrystallization and the salt crystals bond firmly

Table 4. Specifications Applying to the Salt Industry

Type of specification	Title	Designation	Jurisdiction
analysis	*Standard Methods for Chemical Analysis of Sodium Chloride*	ASTM E534-91	American Society for Testing and Materials
food	*Sodium Chloride Monograph*	*Food Chemicals Codex*, 4th ed.	Food and Nutrition Board, National Academy of Sciences
	Standard for Food-Grade Salt	*Codex Alimentarius*	*Codex Alimentarius* Commission
highway deicing	*Standard Specification for Sodium Chloride*	ASTM D632-94	American Society for Testing and Materials
	Standard Specification for Sodium Chloride	AASHTO M143-94	American Association of State Highway Transportation Officials
medical, pharmaceutical	*Official Monograph on Sodium Chloride*	*United States Pharmacopeia 23*	United States Pharmacopeial Convention, Inc.
	Sodium Chloride	*European Pharmacopoeia*	
packaging	*Salt Packages*	ANSI/Z353.1-1983	American National Standards Institute, Inc.
reagent	*Sodium Chloride*	*Reagent Chemicals* 8th ed., 1993	American Chemical Society
soil stabilization	*Standard Test Method for Water-Soluble Chlorides Present as Admixes in Graded Aggregate Road Mixes*	ASTM D1411-92	American Society for Testing and Materials
table salt	(Draft) *Commercial Item Description* (CID), *Salt, Table, Iodized*	A-A-20041B (1995)	U.S. Department of Agriculture (USDA)
water treatment	*AWWA Standard for Sodium Chloride*	ANSI/AWWA B200-92	American National Standards Institute/American Water Works Association
	Commercial Item Description (CID), *Sodium Chloride, Technical* (Water) *Conditioning Grade*	A-A-694	General Services Administration

together. Small quantities of brine entrapped within salt crystals can also contribute to caking. Free-flow or anticaking agents are sometimes added to salt to prevent caking. Three types of free-flow or anticaking agents are typically used.

For free-flow agents, water-insoluble, finely divided adsorbents such as sodium silicoaluminate, tricalcium phosphate, calcium silicate, magnesium carbonate, and silicon dioxide are used. Concentrations between 0.5 and 2% are typically added to table salt and some industrial grades of salt for use in applications where caking may occur.

Water-soluble crystal modifiers such as yellow prussiate of soda (YPS) (sodium ferrocyanide decahydrate) or ferric ammonium citrate may also be added to some types of salt as anticaking agents. Both are approved by the U.S. Food and Drug Administration for use in food-grade salt. YPS and Prussian Blue (ferric ferrocyanide), are most commonly added to rock salt used for wintertime highway deicing. Concentrations of YPS and Prussian Blue in deicing salt vary, typically in the range of 20–100 ppm.

Humectants such as propylene glycol, glycerol, sorbitol, or calcium chloride can prevent drying and caking.

Economic Aspects

World salt production was 180,000,000 metric tons in 1994. The leading salt-producing countries are listed in Table 5. The former USSR was fourth in world production until its dissolution as a union on December 31, 1992.

Environmental pressures on chlorine production have resulted in weakening chemical salt markets, particularly in Europe, where dry salt is used extensively. Historically, deicing salt use for highway deicing increased steadily from the late 1940s through the mid-1970s but leveled off at about 9,000,000 t between the mid-1970s and 1992. Use in the United States and Canada during the winters of 1992–1993 and 1993–1994, was well above normal, however, owing to severe snow and ice conditions across the northern United States. The levels

Table 5. World Salt Production[a], t × 10^{3b}

Country	1990	1991	1992	1993	1994
United States	37,000	35,900	34,800	39,800	39,800
China	20,000	24,100	28,100	29,500	29,700
Germany	15,700	14,900	12,700	12,700	17,700
USSR[c]	14,700	14,000			
Canada	11,300	12,000	11,200	10,900	11,500
India	9,500	9,500	9,500	9,500	9,500
Australia	7,230	7,790	7,690	7,740	7,800
Mexico	7,140	7,530	7,400	7,490	7,460
United Kingdom	6,430	6,830	6,100	6,200	5,700
Brazil	5,370	4,900	5,260	5,250	5,250
all other	48,630	47,550	55,250	51,920	45,590
Total	*183,000*	*185,000*	*178,000*	*181,000*	*180,000*

[a]Ref. 9.
[b]Includes salt in brine.
[c]The USSR was dissolved Dec. 31, 1992.

of 15,500,000 t and 17,100,000 t reached in 1993 and 1994, respectively, in the United States are unlikely to be sustained.

Sales of salt for regenerating ion-exchange water softeners have grown steadily in the United States, particularly in terms of revenues. The U.S. salt industry is mature and sales are not likely to increase rapidly. The dry salt industry in the United States and Canada has undergone consolidation during the 1990s. The number of producers has continued to decline as the major salt producers have acquired smaller operations. Table 6 shows the total amount, including salt in brine, of salt sold or used in the United States for the period 1990–1994. The amount of salt sold or used in the United States in 1994 by product type is (9):

Product type	Quantity, t $\times 10^3$
vacuum and open pans	3,670
rock	14,900
solar	2,890
brine	18,000
Total	*39,500*

Statistical data on dry salt sales are available through 1994 (9). Dry salt includes salt produced as crystalline sodium chloride, but excludes salt in brine produced for production of chlor–alkali products and other chemicals. Table 7 gives United States dry salt sales for the period 1990–1994.

Table 6. Production and Consumption of Sodium Chloride^a, t $\times 10^{3b}$

Parameter	1990	1991	1992	1993	1994
	United States				
production	36,800	36,300	36,000	39,200	39,800
sold or used by producers	36,900	35,900	34,800	38,200	39,500
value, $ $\times 10^3$	827,000	802,000	803,000	904,000	956,000
exports	2,270	1,780	992	688	742
value, $ $\times 10^3$	32,900	29,900	32,200	34,800	30,200
imports for consumption	5,970	6,190	5,390	5,870	9,630
value, $ $\times 10^3$	88,400	87,400	87,800	100,000	151,000
Total consumption^c, apparent	*40,600*	*40,300*	*39,200*	*43,400*	*48,400*
	World				
production, t $\times 10^3$	183,000	185,000	178,000	181,000	180,000

^{a}Ref. 9.
^{b}Includes salt in brine.
^{c}Sold or used plus imports minus exports.

Analytical Methods

The most common impurities, depending on type of salt, are calcium sulfate, calcium chloride, magnesium chloride or magnesium sulfate, sodium sulfate,

Table 7. United States Sales of Dry Salt[a], t × 10³

Use category	1991	1992	1993	1994	1995
highway	8,830	9,490	15,540	17,130	13,088
chemical	2,860	2,320	2,390	2,480	2,766
water conditioning	2,150	2,150	2,200	2,350	2,320
agriculture	1,760	1,580	1,660	1,810	1,666
food-grade	1,030	1,160	1,160	1,200	1,228
miscellaneous	4,980	3,460	3,320	3,650	3,429
Total	*21,600*	*20,160*	*26,260*	*28,630*	*24,497*

[a]Ref. 10.

and water-insoluble material. Surface moisture is determined by drying, water-insoluble material by weighing, calcium and magnesium by EDTA titration, and sulfate gravimetrically. Salt content is expressed as the difference after converting calcium, magnesium, and sulfate to theoretical mineral combinations. Other procedures use an assay based on chloride titration, but accurate determination of salt purity by assay is difficult, especially for purified grades of salt.

Health and Safety Factors

Salt is an essential nutrient without which life could not exist (see MINERAL NUTRIENTS). Sodium is the principal cation in blood plasma and body tissue fluids. Regulation of sodium in the body maintains osmotic pressure, acid–base balance, and volume of circulating body fluids (11). Extracellular fluid volume is maintained by the total body sodium content. Brain mechanisms control the elective intake of salt, and physiological control systems regulate the conservation and excretion of sodium. The human body contains about 1840 mg (80 mEq) of sodium per kilogram of body weight, half of which is exchangeable (12). The sodium concentration of blood and other body fluids is maintained by a complex mechanism involving the kidneys and the adrenal glands. Renal excretion of sodium can be controlled so that a wide range of Na intakes can be accommodated. Sodium is conserved when intake is low, and excreted when intake is high. Chloride, too, is essential to good health. It preserves acid–base balance in the body, aids potassium absorption, supplies the essence of digestive stomach acid, and enhances the ability of the blood to carry carbon dioxide from respiring tissues to the lungs.

   Sodium Intake.   Where salt is readily available, most of the world's population chooses to consume about 6,000–11,000 mg of salt or sodium chloride a day so that average daily sodium intake from all sources is 3,450 mg (8,770 mg NaCl) (13). The U.S. FDA's GRAS review puts the amount of naturally occurring sodium in the American diet at 1000–1500 mg/d, equivalent to the amount of sodium in approximately 2500–3800 mg NaCl. Thus the average daily intake of NaCl from food-grade salt used in food processing (qv) and from salt added in cooking or at the table is from 4960–6230 mg NaCl. The requirement for salt in the diet has not been precisely established, but the safe and adequate intake for adults is reported as 1875–5625 mg (14). The National Academy of Sciences recommends that Americans consume a minimum of 500 mg/d of sodium (1250 mg/d salt) (6,15).

Sodium Restriction. The World Health Organization and the International Society of Hypertension estimate 10–15% of the adult population is hypertensive (11). There is no consensus in the scientific community about the relationship between sodium or salt intake and hypertension. Whereas there is no evidence that high levels of salt intake can cause high blood pressure in healthy, normotensive people, there is evidence that severe salt restriction can lower blood pressure in one-third to one-half of individuals with hypertension. These individuals are labeled salt-sensitive. As a result, there is disagreement over efforts by the federal government in the United States to encourage salt restriction in American diets. The principal reason for disagreement is concerns about the effectiveness or even the safety of salt restriction in the general population.

Toxicity. The U.S. Food and Drug Administration regards common salt, ie, NaCl, as GRAS for its intended use as a food additive. Oral toxicity for mammals is reported in mg/kg (16): for humans, TD_{LO}: 12,357; 23D-C (daily-continuous) for mice, LD_{50}: 4000; for rats, LD_{50}: 3000; and for rabbits, LD_{LO}: 8000. TD_{LO} and LD_{LO} are lowest level toxic and lethal dosages, respectively.

Aquatic toxicity is reported in mg/L: for *Pimephales promealas* (fathead minnow), 69-h LC_{50}: 7650 (17); for *Daphnia magna* (water flea), 48-h EC_{50}: 3310 (18); for *Myriophyllum spicatum* (water milfoil), phytotoxicity (EC_{50} for growth): 5962 (19); and for *Rana breviceps* (frog), no observed effect concentration (NOEC): 400 (20). LC_{50} and EC_{50} are lethal and effect concentrations, respectively, for 50% of the subjects tested.

Environment and Infrastructure. Both sodium and chloride are present in the environment naturally. Ecosystems supporting aquatic species and biota as well as vegetation, including roadside grasses, shrubs and trees, and food crops, can tolerate various concentrations of sodium and chloride. The U.S. Department of Agriculture classifies irrigation water according to conductivity and salinity hazard. Conductivity range can be converted to approximate mg/L dissolved solids. Salinity hazard ratings in mg/L are (21) low, 70–175; medium, 176–525; high, 526–1575; and very high, >1575.

Environmental effects of elevated salinity levels resulting from uses of salt are highly site-specific. Other site characteristics must also be considered, such as soil texture, permeability, drainage, amount of water applied, and the salt tolerance of the vegetation or crops. Deicing salt can be used to ensure traffic safety without causing permanent harm to the environment, especially where salt-tolerant trees, grasses, and shrubs are used at the roadside to minimize potential harm. Highway officials and automobile owners are less concerned about salt-related damage to infrastructure and motor vehicles. Automobiles are lasting longer without significant corrosion because of advanced designs and coatings systems; increased use of corrosion-resistant materials, eg, plastics; and improved manufacturing techniques. Highway and bridge structures are lasting longer because construction methods incorporate high quality, air-entrained concrete with epoxy-coated reinforcing steel. Programs to assure judicious salt application rates and proper salt storage methods have reduced the effects of salt on water quality and ecosystems. Other chemical deicers, widely promoted as alternatives to sodium chloride, are far more costly and less effective than salt, and are not available in quantities needed to ensure wintertime highway safety.

Uses

Salt can be classified under five principal use categories, plus a catch-all classification that includes most industrial uses, as (1) food-grade, (2) agriculture, (3) highway, (4) water conditioning, (5) chemical, and (6) miscellaneous.

Foods and Food Processing. The taste for salt is universal and an inherent characteristic of humans and other animals. Added salt enhances the flavor of foods and makes otherwise bland foods acceptable to the taste. The amount of salt available for consumption in food products, cooking, and at the table can be estimated from the quantity of food-grade salt sold in the United States. Total sales of food-grade salt in the United States during 1994 were 1,200,000 t (22), or, based on the U.S. Bureau of Census projection for 1994 resident population, just over 12.5 g/person/d. Because much salt is not incorporated into the final food product during food processing and is discharged with water during cooking and at the table, the actual intake of dietary salt is far less.

Sodium chloride performs several necessary functions in food processing and cooking. In addition to use as a flavor enhancer, salt is used in food processing as a preservative, a color developer, binder, texturizer, and as a fermentation-control agent. As a binder, salt helps extract the myofibrillar proteins in processed and formed meats (23), providing binding strength between adjacent pieces of meat. Water binding properties are increased and, as a result, cooking losses are reduced. Salt increases the solubility of muscle proteins in water. In sausage-making, stable emulsions are formed when the salt-soluble protein solutions coat the finely formed globules of fat, providing a binding gel consisting of meat, fat, and moisture.

Salt preserves foods by providing a hostile environment for certain microorganisms. Within foods, salt brine dehydrates bacterial cells, alters osmotic pressure, and inhibits bacterial growth and subsequent spoilage. Dry salt and salt brine are used in several types of curing processes. Pickles are preserved in strong brine prior to final processing.

Salt promotes the development of color in ham, bacon, hotdogs, and sauerkraut. Used with sugar and nitrate or nitrite, salt produces a color in processed meats which consumers find appealing. Salt enhances the golden color in bread crust by reducing sugar (qv) destruction in the dough and increasing carmelization.

As a texture aid, salt strengthens gluten in bread dough, providing uniform grain, texture, and dough strength. In the presence of salt, gluten holds more water and carbon dioxide, allowing the dough to expand without tearing. Salt improves the tenderness in cured meats such as ham by promoting the binding of water by protein. It also gives a smooth, firm texture to processed meats. Salt develops the characteristic rind hardness in cheese and helps produce the desirable even consistency in cheese and other foods such as sauerkraut.

In baked products, salt controls fermentation (qv) by retarding yeast activity, preventing wild fermentation, important in making a uniform product. During pickle-making, salt brine is gradually increased in concentration, reducing the fermentation rate as the process proceeds to completion. Salt is also used to control fermentation in making cheese, sauerkraut, and summer sausage. Finished products are consistent in color, flavor, and texture.

Iodized Salt. Iodized table salt has been used to provide supplemental iodine to the U.S. population since 1924, when producers, in cooperation with the Michigan State Medical Society (24), began a voluntary program of salt iodization in Michigan that ultimately led to the elimination of iodine deficiency in the United States. More than 50% of the table salt sold in the United States is iodized. Potassium iodide in table salt at levels of 0.006% to 0.01% KI is one of two sources of iodine for food-grade salt approved by the U.S. Food and Drug Administration. The other, cuprous iodide, is not used by U.S. salt producers. Iodine may be added to a food so that the daily intake does not exceed 225 μg for adults and children over four years of age. Potassium iodide is unstable under conditions of extreme moisture and temperature, particularly in an acid environment. Sodium carbonate or sodium bicarbonate is added to increase alkalinity, and sodium thiosulfate or dextrose is added to stabilize potassium iodide. Without a stabilizer, potassium iodide is oxidized to iodine and lost by volatilization from the product. Potassium iodate, far more stable than potassium iodide, is widely used in other parts of the world, but is not approved for use in the United States.

Iodine deficiency in less developed countries is still a serious problem. Whereas, iodized and iodated salt technology is readily available and relatively inexpensive, market distribution conditions, as well as a lack of understanding by consumers, prevents iodized salt from reaching much of the population in less-developed countries.

Fluoride is added to table salt in countries such as France, Mexico, and Switzerland for the prevention of dental caries.

Agriculture. Most forages provide insufficient sodium for animal feeding and may lack adequate chloride. Thus salt supplementation is a critical part of a nutritionally balanced diet for animals. In addition, because animals have a definite appetite for salt, it can be used as a delivery mechanism to ensure adequate intake of less palatable nutrients and as a feed limiter. Salt is an excellent carrier for trace minerals. Salt, either in loose form or as compressed blocks, can be mixed with feed or fed free-choice to improve animal health and productivity (see FEEDS AND FEED ADDITIVES).

Highway. Rock salt, solar salt, and in some cases in Europe, evaporated salt are used to maintain traffic safety and mobility during snow and ice conditions in snowbelt regions throughout the world. Sodium chloride melts ice at temperatures down to its eutectic point of $-21.12°C$. Most snowstorms occur when the temperature is near $0°C$, where salt is very effective. More than 40% of dry salt produced in the United States is used for highway deicing.

Water Conditioning. Well water and many public drinking water supplies contain elevated levels of calcium and magnesium. The resulting hard water reduces the sudsing action of soaps and detergents and causes a greasy, curd-like deposit when the water is used for laundering, cleaning, and bathing. Mineral scale builds up in hot water appliances and industrial boilers, reducing energy efficiency and shortening appliance and equipment life. Water is softened by removing calcium and magnesium ions from hard water in exchange for sodium ions at sites on cation-exchange resin. Water softeners typically use a gel polystyrene sulfonate cation-exchange resin regenerated with a 10% salt brine solution (25).

Chemical Uses. During 1994, 34% of the total salt distributed by producers in the United States was salt in brine produced and used by chemical manufacturers to produce chlorine and caustic soda (26). Demand for salt in chlor–alkali production has fallen steadily since 1974 from 25,900,000 t to 17,300,000 t in 1994. Chlorine and caustic soda are used by the pulp (qv) and paper (qv) industry in multistage bleaching; caustic soda is used in wood fiber processing. Reduced use of chlorine in the pulp and paper industry owing to environmental concern has contributed to the decline in chlor–alkali production. Some European countries rely more heavily on dry salt shipped to production sites near users to minimize the hazards associated with transporting chlorine. Salt is used to make sodium chlorate and metallic sodium by electrolysis and to a lesser degree is used as a reactant for sulfuric acid to produce sodium sulfate and hydrochloric acid. Downstream manufacturers produce other chemicals using chlorine as a feedstock. U.S. chemical manufacturers used salt to produce synthetic soda ash until 1986, when high production costs forced the closing of the last plant.

Industrial Uses. Salt is reported to have more than 14,000 uses (7). Salt is used by the textile and dyeing industry to fix dyes and to standardize dye batches (see DYE AND DYE INTERMEDIATES; TEXTILES); by metal processors, such as secondary aluminum refiners, to remove impurities (see ALUMINUM AND ALUMINUM ALLOYS); in rubber manufacturing to separate rubber from latex (see RUBBER, NATURAL); as a filler and grinding agent in pigment and dry-detergent processes; in ceramics (qv) manufacture for surface vitrification of heated clays; in soapmaking to separate soap (qv) from water and glycerol; in oil and gas well drilling muds to inhibit fermentation, increase density, and to stabilize rock salt strata; and in animal-hide processing and leather (qv) tanning to cure, preserve, and tan hides.

Salt Substitutes. As a result of concern about the relationship between dietary sodium and hypertension, some salt producers and food companies have developed salt substitutes or low sodium products. Mixtures of sodium chloride and potassium chloride, herbs and spices, as well as modified salt crystals of lower density are marketed in response to a limited consumer demand for reduced-sodium products. This amounts to about 2% of user salt purchases.

Encapsulated Salt. Sodium chloride coated with solid partially hydrogenated vegetable oil is used in food processing applications where controlled release of sodium chloride is desired. Encapsulation delays or prevents salt particles from dissolving in water until a suitable temperature is reached and therefore can protect against unwanted reactions. Coated salt can allow yeast activity, inhibit oxidative rancidity, and control water binding. It is used in meat products, baked goods, seasoning blends, microwaveable products, and other foods where dissolution of the salt should be delayed until the product is heated.

BIBLIOGRAPHY

"Salt" in *ECT* 1st ed., Vol. 12, pp. 67–82, by C. D. Looker, International Salt Co., Inc.; in *ECT* 2nd ed., Vol. 18, pp. 468–484, by E. J. Kuhajek and H. W. Fiedelman, Morton International, Inc.; "Sodium Halides, Sodium Chloride" under "Sodium Compounds" in

ECT 3rd ed., Vol. 21, pp. 205–223, by J. H. Heiss, Diamond Crystal Salt Co., and E. J. Kuhajek, Morton Salt.

1. F. van der Leeden, F. L. Troise, and D. K. Todd, *The Water Encyclopedia*, 2nd ed., Lewis Publishers, Chelsea, Mich., 1990, pp. 237–238.
2. M. T. Halbouty, *Salt Domes, Gulf Region, United States, Mexico*, 2nd ed., Gulf Publishing Co., Houston, Tex., 1979.
3. R. P. Multhauf, *Neptune's Gift*, The Johns Hopkins University Press, Baltimore, Md., 1978.
4. P. H. Price, C. E. Hare, J. B. McCue, and H. A. Hoskins, *Salt Brines of West Virginia*, West Virginia Geological Survey, Morgantown, W. Va., 1937.
5. W. B. Wilkinson, *Romances and Notes on Salt*, Robert Eastman, Inc., Ithaca, N.Y., 1951.
6. *Statement on the Role of Dietary Management in Hypertension Control*, National High Blood Pressure Education Program and Coordinating Committee, Bethesda, Md., 1979.
7. D. Kaufman, ed., *Sodium Chloride*, ACS Monograph Series, No. 145, The American Chemical Society, Reinhold Publishing Corp., New York, 1960.
8. M. Hamada, I. Azuma, N. Imai, and H. Ono, in *Seventh Symposium on Salt*, Vol. 2, Elsevier, Amsterdam, the Netherlands, 1993, pp. 59–64.
9. D. Kostick, *SALT*, Mineral Industry Surveys, U.S. Bureau of Mines, Washington, D.C., 1995.
10. *Statistical Report Analysis of United States Sales*, Salt Institute, Alexandria, Va., 1995. Salt Institute data represent 96–100% of U.S. salt sales.
11. D. Denton, *The Hunger For Salt*, Springer-Verlag, New York, 1982.
12. J. Elkinton and T. Danowski, *The Body Fluids*, Williams & Wilkins, Baltimore, Md., 1955.
13. Intersalt Cooperative Research Group, *Brit. Med. J.* **297**, 319–328 (1988).
14. P. Saltman, J. Furin, and I. Mothner, *The California Nutrition Book*, Little, Brown and Co., Boston, Mass., 1987.
15. W. B. Kannel and T. R. Dauber, *Hypertensive Cardiovascular Disease: The Framingham Study. Hypertension: Mechanisms and Management*, Framingham, Mass., 1973.
16. *Registry of Toxic Effects of Chemical Substances*, Vol. 5, U.S. Department of Health and Human Services, Washington, D.C., 1985–1986, p. 4433.
17. *Fisheries Res. Bd. Canada* **33**(2), 209–214 (1976).
18. *J. Water Pollution Control Fed.* **37**(g), 1308–1316 (1965).
19. *Arch. Environ. Contam. Toxicol.* **2**(4), 331–341 (1974).
20. *Indian J. Exp. Biol.* **127**(11), 1244–1245 (1979).
21. D. E. Longenecker and P. J. Lyerly, *Control of Soluble Salts in Farming and Gardening*, The Texas A&M University System Agricultural Experiment Station, College Station, Tex., 1974.
22. Technical data, Salt Institute, Alexandria, Va., 1995.
23. A. M. Pearson and F. W. Tauber, *Processed Meats*, 2nd ed., Van Nostrand Reinhold Co., Inc., New York, 1984.
24. O. P. Kimball, *Ohio State Med. J.* **35**(7), 705–708 (1939).
25. J. E. Bowie, *Water Technol.* 49, 52 (Oct. 1993).
26. Ref. 9, 1993.

General References

M. H. Alderman and B. Lamport, *Am. J. Hypertens.* **3**, 499–504 (1990).
Annual Book of ASTM Standards, The American Society for Testing and Materials, Philadelphia, Pa., 1995.

C. A. Baar, *Applied Salt-Rock Mechanics 1, The In Situ Behavior of Salt Rocks*, Elsevier Science, Inc., New York, 1977.

D. A. Benoit and C. E. Stephan, *Ambient Water Quality Criteria for Chloride—1988*, EPA 440/5-88-001., U.S. Environmental Protection Agency, Washington, D.C., 1988.

F. D'Itri, ed., *Chemical Deicers and the Environment*, Lewis Publishers, Chelsea, Mich., 1992.

I. Lerche and J. J. O'Brien, eds., *Dynamical Geology of Salt and Related Structures*, Academic Press, Inc., Orlando, Fla., 1987.

B. M. Egan and A. B. Weder, *Deleterious Effects of Short-term Dietary NaCl Restriction in Man: Relationship to Salt-Sensitivity Status*, Medical College of Wisconsin, University of Michigan, Ann Arbor, Mich., 1989.

A. Coogan and L. Hauber, eds., *Fifth Symposium on Salt*, Vols. 1 and 2, Northern Ohio Geological Society, Cleveland, Ohio, 1980.

A. Coogan, ed., *Fourth Symposium on Salt*, Vols. 1 and 2, Northern Ohio Geological Society, Inc., Cleveland, Ohio, 1974.

T. P. Guinee and P. F. Fox, *Cheese: Chemistry, Physics and Microbiology*. Chapman & Hall, London, 1987, Chapt. 7, pp. 257–302.

M. Hamada, I. Azuma, N. Imai, and H. Ono, in *Seventh Symposium on Salt*, Vol. 2, Elsevier, Amsterdam, the Netherlands, 1993, pp. 59–64.

R. M. Hanbali, *Economic Impact of Winter Road Maintenance on Road Users*, Transportation Research Record 1442, Transportation Research Board, Washington, D.C., 1994.

D. Lide, ed., *Handbook of Chemistry and Physics*, 71st ed., CRC Press, Boca Raton, Fla., 1991.

H. Hanke and C. Levin, *Influence of Winter Road Maintenance on Traffic Safety and Transport Efficiency*, Hessisches Landesamt fur Strabenbau, Technische Hochschule Darmstadt, Darmstadt, Germany, 1988.

H. Horie, Y. Tanaka, Y. Aoke, H. Shibata, and T. Kawahara, in *Seventh Symposium on Salt*, Vol. 2, Elsevier, Amsterdam, the Netherlands, 1993, pp. 71–77.

G. Hudel and J. Karoly, in *Seventh Symposium on Salt*, Vol. 2, Elsevier, Amsterdam, the Netherlands, 1993, pp. 151–157.

M. K. Jenyon, *Salt Tectonics*, Elsevier Applied Science Publishers, New York, 1986.

P. Jongema, in *Seventh Symposium on Salt*, Vol. 2, Elsevier, Amsterdam, the Netherlands, 1993, pp. 159–163.

D. S. Kostick, *Salt Annual Review 1994*, U.S. Bureau of Mines Annual Report, U.S. Bureau of Mines, Washington, D.C., Oct. 1995.

E. W. Kratz and F. E. Hoyer, in *Seventh Symposium on Salt*, Vol. 2, Elsevier, Amsterdam, the Netherlands, 1993, pp. 101–115.

D. A. Kuemmel and R. M. Hanbali, *Accident Analysis of Ice Control Operations*, presented at the Third International Symposium on Snow Removal and Ice Control Technology, Minneapolis, Minn., 1992.

S. J. Lefond, *Handbook of World Salt Resources*, Plenum Press, New York, 1969.

V. M. G. Mannar and H. L. Bradley, *Guidelines for the Establishment on Solar Salt Facilities from Seawater, Underground Brines and Salted Lakes*, United Nations Industrial Development Organization, 1983.

H. R. Hardy, Jr., and M. Langer, eds., *The Mechanical Behavior of Salt*, Gulf Publishing Co., Houston, Tex., 1984.

S. Budavari, ed., *The Merck Index*, 11th ed., Merck & Co., Rahway, N.J., 1989.

R. Berkow and A. Fletcher, eds., *The Merck Manual of Diagnosis and Therapy*, 15th ed., Merck & Co., Rahway, N.J., 1987.

K. Helrich, ed., *Official Methods of Analysis*, 15th ed., Vols. 1 and 2, Association of Official Analytical Chemists, Inc., Arlington, Va., 1990.

Official Publication, Association of American Feed Control Officials, Atlanta, Ga., 1996.

A. M. Pearson and F. W. Tauber, *Processed Meats*, 2nd ed., Van Nostrand Reinhold Co., Inc., New York, 1984.

Reagent Chemicals, 8th ed., American Chemical Society Specifications, American Chemical Society, Washington, D.C., 1993.

J. Rau, ed., *Second Symposium on Salt*, Vols. 1 and 2, Northern Ohio Geological Society, Inc., Cleveland, Ohio, 1966.

H. Kakihana, H. R. Hardy, Jr., T. Hoshi, and K. Toyokura, eds., *Seventh Symposium on Salt*, Vols. 1 and 2, Elsevier, Amsterdam, the Netherlands, 1993.

C. Schreiber and H. Harner, eds., *Sixth Symposium on Salt*, Vols. 1 and 2, Salt Institute, Alexandria, Va., 1985.

"Sodium Chloride," in *Food Chemicals Codex*, 4th ed., National Research Council Committee on Food Chemicals Codex, National Academy Press, Washington, D.C., 1996.

A. Bersticker, ed., *First Symposium on Salt*, Northern Ohio Geological Society, Inc., Cleveland, Ohio, 1963.

J. Rau and L. Delwig, eds., *Third Symposium on Salt*, Vols. 1 and 2, Northern Ohio Geological Society, Inc., Cleveland, Ohio, 1970.

U.S. Code of Federal Regulations, 21 CFR 1994, Food and Drugs, Subpart A § 182.1. Substances that are generally recognized as safe.

W. B. Wilkinson, *Romances and Notes On Salt*, Robert Eastman, Inc., Ithaca, N.Y., 1951.

BRUCE M. BERTRAM
Salt Institute

SODIUM BROMIDE

Sodium bromide [7647-15-6], NaBr, the most common and available alkali bromide, is a salt of hydrobromic acid (see BROMINE COMPOUNDS). Sodium bromide crystallizes from aqueous solution as a dihydrate [13466-08-5], NaBr·2H$_2$O, below 51°C. Above 51°C, it crystallizes as the anhydrous compound. Crystals of the dihydrate belong to the monoclinic system and have lattice parameters $a = 659$ pm, $b = 1020$ pm, and $c = 651$ pm. The anhydrous crystal belongs to the cubic system, $a = 596$ pm. Other physical properties of the anhydrous salt are listed in Table 1. The anhydrous salt is hygroscopic but not deliquescent.

Table 1. Properties of Sodium Bromide

Property	Value
molecular weight	102.89
melting point, °C	755
boiling point, °C	1390
density, g/cm^3	3.203
refractive index, n_D^{25}	1.6412
heat of fusion, J/g[a]	254.9
ΔH_f, kJ/mol[a]	−361.414
S, J/(mol·K)[a]	86.82
heat capacity, J/(mol·K)[a,b]	$47.92 + 1.331 \times 10^{-3}\ T$

[a]To convert J to cal, divide by 4.184.
[b]Temperature, T, is in Kelvin.

Sodium bromide has a very high water solubility. At 25°C a saturated solution contains 48.6% sodium bromide by weight. Values for the solubility at several temperatures are known (1). Three parameter equations for calculating the solubility in terms of mole fraction of both the anhydrous and dihydrate salts are available (2). Convenient equations for calculating the solubility in weight percent of sodium bromide in water at various temperatures, t in °C, are as follow:

$$0°C \leq t < 51°C \quad \text{soly}_{\text{NaBr}} = 44.335 + 0.1536\, t + 6.752 \times 10^{-4}\, t^2$$

$$51°C \leq t \leq 100°C \quad \text{soly}_{\text{NaBr}} = 52.207 + 3.945 \times 10^{-2}\, t - 1.307 \times 10^{-4}\, t^2$$

Both equations give the weight percent of sodium bromide in an aqueous saturated solution to within ±0.02 percentage units. One gram of sodium bromide dissolves in about 16 mL of ethanol and 6 mL of methanol (3).

Preparation and Manufacture

Small quantities of pure sodium bromide can be prepared by neutralizing solutions of either sodium hydroxide [1310-73-2] or sodium carbonate [497-19-8] using hydrobromic acid [10035-10-6] which is free of bromine, followed by evaporation and crystallization.

Commercial quantities of sodium bromide are usually prepared by adding excess bromine [7726-95-6] to a solution of sodium hydroxide, evaporating to dryness, and treating with a reducing agent such as formic acid [64-18-6] or activated carbon [7440-44-0] to reduce sodium bromate [7789-38-0] to sodium bromide. Some sodium bromide is recovered as a by-product of the bromination of organic materials. Grades of sodium bromide available are chemically pure (CP), crystal, powdered, commercial, pure, highest purity, and National Formulary (NF). Sodium bromide is available in crystalline or powdered forms and in solution (4).

Economic Aspects

U.S. sodium bromide demand accounts for 8–10% of total bromine production. In 1994 demand is estimated to have been 13,600–17,200 metric tons (5). At mid-1996, the price for technical-grade sodium bromide in truckload quantities was $1.54/kg ($0.70/lb) (6). Manufacturers of sodium bromide include Albemarle, Great Lakes Chemical, Rhône-Poulenc, and Whittaker Corporation.

Health and Safety

Sodium bromide is moderately toxic by ingestion and can affect the gastrointestinal and central nervous systems and the skin. Ingestion of large amounts in a single dose causes immediate abdominal pain. Gastrointestinal effects include nausea and vomiting, foul breath, anorexia, weight loss, dehydration, and constipation. Neurological effects include headache, apathy, slurred speech, impaired memory and intellectual capacity, and drowsiness. Skin effects include acneiform, rash, pustules, and ulcers (7). Inhalation of sodium bromide dusts may cause mucous membrane and bronchial irritation.

The LD_{50} for sodium bromide taken orally by rats is 3.5 g/kg body weight, and the TD_{LO} orally in rats is 720 mg/kg (8). RTECS lists data on reproductive effects in male and female rats. Sodium bromide is listed in the TSCA Inventory, the Canadian Domestic Substances list (DSL), the European Inventory of Existing Commercial Chemical Substances (EINECS), the Japanese Existing and New Chemical Substances (ENCS), and the Korean Existing Chemicals list (ECL). It is not regulated by the U.S. Department of Transportation.

Uses

The oil and gas drilling industry is a principal consumer of sodium bromide. Because of its high solubility in water, clear brine fluids of densities up to 1.547 g/cm^3 (12.9 lb/gal) at 25°C can be obtained. These are used directly or in blends with other clear brine fluids for well completions or workovers. Densities of sodium bromide brines containing up to 50 wt % sodium bromide and at temperatures to 100°C have been tabulated (9). The density in g/cm^3 of sodium bromide brines at temperatures in °C can be calculated from the following equation when $0°C \leq t \leq 100°C$.

$$\text{density} = (0.991 - 7.30 \times 10^{-3} \times \text{wt \% NaBr} + 4.08 \times 10^{-4}\ t)^{-1}$$

In 1988, approximately 80% of sodium bromide sales were to the drilling industry (10). This usage decreased in the early 1990s because of a decrease in drilling activity and an increase in the use of sodium bromide in the water treatment industry.

An increasingly important use of sodium bromide is as a biocide, particularly in industrial cooling water towers and in swimming pool water treatment, replacing the more hazardous chlorine [7782-50-5] (see WATER, TREATMENT OF SWIMMING POOLS, SPAS, AND HOT TUBS). Sodium bromide is usually converted to elemental bromine, bromine chloride [13863-41-7], or hypobromite through the use of activators such as hypochlorites and chlorinated isocyanurates. Some studies have suggested that bromine is three times more effective than chlorine in controlling algae in cooling towers (11). Also, the trend in cooling water treatment to operate at higher pH to minimize corrosion, leads to larger amounts of algae. Bromine is more effective under these higher pH conditions (12). Sodium bromide sales for biocides in 1995 were forecasted to be about 8600 t, up from 6800 t in 1990 reflecting about a 4% annual growth (13).

Other applications of sodium bromide include use in the photographic industry both to make light-sensitive silver bromide [7785-23-1] emulsions and to lower the solubility of silver bromides during the developing process; use as a wood (qv) preservative in conjunction with hydrogen peroxide (14); as a cocatalyst along with cobalt acetate [917-69-1] for the partial oxidation of alkyl side chains on polystyrene polymers (15); and as a sedative, hypnotic, and anticonvulsant. The FDA has, however, indicated that sodium bromide is ineffective as an over-the-counter sleeping aid for which it has been utilized (16).

BIBLIOGRAPHY

"Sodium Bromide" under "Sodium Compounds" in *ECT* 3rd ed., Vol. 21, pp. 224–225, by V. A. Stenger, The Dow Chemical Co.

1. A. Seidell and W. Linke, *Solubilities of Inorganic and Metal–Organic Compounds*, Vol. II, 4th ed., American Chemical Society, Washington, D.C., 1965, p. 831.
2. M. Broul, J. Nývlt, and O. Söhnel, *Solubility in Inorganic Two-Component Systems*, Elsevier, Amsterdam, the Netherlands, 1981, pp. 412–413.
3. S. Budavari, ed., *The Merck Index*, 12th ed., Merck & Co., Whitehouse Station, N.J., 1996, p. 1473.
4. R. J. Lewis, *Hawley's Condensed Chemical Dictionary*, 12th ed., Van Nostrand Reinhold Co., New York, 1994.
5. *Chem. Mark. Rep.* **246**(2), 37 (Jan. 14, 1994).
6. *Chem. Mark. Rep.* **249**(26), 44 (June 24, 1996).
7. R. E. Gosselin, H. C. Hodge, R. P. Smith, and M. N. Gleason, *Clinical Toxicology of Commercial Products*, 4th ed., Williams and Wilkins, Baltimore, Md., 1976, pp. 11–77.
8. "Sodium Bromide" in R. J. Lewis, ed., *SAX's Dangerous Properties of Industrial Materials*, 8th ed., Van Nostrand Reinhold Co., New York, 1994.
9. P. Novotný and O. Söhnel, *Densities of Aqueous Solutions of Inorganic Substances*, Elsevier, Amsterdam, the Netherlands, 1985, p. 245.
10. *The Economics of Bromine*, 5th ed., Roskill Information Service, Ltd., London, 1988, p. 48.
11. *Chem. Mark. Rep.* **246**, 23 (Nov. 21, 1994).
12. P. Fitzgerald, *Chem. Mark. Rep.* **248**, 12 (Oct. 16, 1995).
13. E. Brandt, *Chem. Eng.* **99**(10), 57–59 (1992).
14. Jpn. Kokai 06182715 (July 5, 1994), J. Murakami and M. Takao (to Aika Kogyo KK).
15. World Pat. 9410215 (May 11, 1994), H. D. H. Stover and co-workers (to Research Corp. Technologies).
16. *Med. World News*, 12 (Apr. 10, 1989).

ROGER N. KUST
Tetra Technologies, Inc.

SODIUM IODIDE

Sodium iodide [7681-82-5], NaI, occurs as colorless crystals or as a white crystalline solid. It has a salty and slightly bitter taste. In moist air, it gradually absorbs as much as 5% water, which causes caking or even liquefaction (deliquescence). The solid slowly becomes brown when exposed to air because some iodide is oxidized to iodine. Water solutions are neutral or slightly alkaline and gradually become brown for the same reason. Aqueous solutions are stabilized with respect to oxidation by raising the pH to 8–9.5 (see IODINE AND IODINE COMPOUNDS).

Properties

Sodium iodide crystallizes in the cubic system. Physical properties are given in Table 1 (1). Sodium iodide is soluble in methanol, ethanol, acetone, glycerol, and several other organic solvents. Solubility in water is given in Table 2.

Below 65°C, sodium iodide is present in aqueous solutions as hydrates containing varying amounts of water. When anhydrous sodium iodide is dissolved in water, heat is liberated because of hydrate formation, eg, $\Delta H = -174.4$ kJ/mol (-41.7 kcal/mol), when the dihydrate is formed. At room temperature, sodium iodide crystallizes from water as the dihydrate [13517-06-1], $NaI \cdot 2H_2O$, in the form of colorless prismatic crystals.

Table 1. Physical Properties of Sodium Iodide[a]

Property	Value
mol wt	149.895
mp, °C	651
bp, °C	1304
d_4^{25}, g/cm^3	3.667
specific heat, J/(kg·K)[b]	
at 0°C	350
50°C	360

[a]Ref. 1.
[b]To convert J to cal, divide by 4.184.

Table 2. Aqueous Solubility of Sodium Iodide

Temperature, °C	NaI g/100 g H_2O
0	158.7
20	178.7
40	205.0
60	256.8
70	294
80	296
100	302
140	321

Manufacture

Bulk production of *United States Pharmacopeia* (USP) and reagent grades is based on the reaction of sodium carbonate or hydroxide with an acidic iodide solution, typically hydriodic acid or a metal iodide. After removal of chemical impurities, the solution is filtered and concentrated. Evaporation gives the anhydrous NaI. Controlled cool-down produces either the dihydrate or the pentahydrate [81626-33-7].

Essentially no waste products are formed in the USP process if hydriodic acid and either sodium hydroxide or sodium carbonate are used as reactants. Water results from use of the former; a mole equivalent quantity of carbon dioxide is produced from the latter reagents. Higher quality grades may require some purification steps which may result in wastes from the treatment. Disposal of these impurities must then be carried out.

Economic Aspects and Uses

The market price of USP sodium iodide generally follows the price of crude iodine multiplied by a factor of 1.8–2. Higher purity grades are only marginally related to the price of crude iodine.

The principal use of sodium iodide is in scintillation crystals, which are used for gamma-ray counters (2), and in medicine as the detectors in computer-assisted tomography (CAT) scan and positron emission tomography (PET) equipment (3). A small amount is used in the wet extraction of silver, in iodized salt (see FOOD ADDITIVES), animal feeds to prevent hoofrot (see FEEDS AND FEED ADDITIVES), photographic chemicals, as an antiinfectant for body drapes in medicine, and in the manufacture of organic chemicals. It has also been used in cloud seeding and in halogen discharge lamps.

USP XXII specifies that sodium iodide contains 99–101.5% NaI, calculated on an anhydrous basis (4). It is used interchangeably with potassium iodide as a therapeutic agent, except where sodium ion is contraindicated (see POTASSIUM COMPOUNDS). Intravenous sodium iodide formulations have been used for a variety of diseases, from thyroid deficiency to neuralgia (see THYROID AND ANTITHYROID PREPARATIONS). However, these solutions are no longer listed in the *NF XVII* (4), indicating that their therapeutic value has not been satisfactorily demonstrated.

Veterinary uses of sodium iodide include the treatment of horses, cattle, sheep, swine, and dogs for various afflictions (see VETERINARY DRUGS).

BIBLIOGRAPHY

"Sodium Halides" under "Sodium Compounds" in *ECT* 1st ed., Vol. 12, pp. 603–604, by J. A. Brink, Jr., Purdue University; in *ECT* 2nd ed., Vol. 18, pp. 485–486, by F. N. Anderson, Mallinckrodt Chemical Works; in 3rd ed., Vol. 21, pp. 226–227, by P. H. Merrell, E. M. Peters, Mallinckrodt, Inc.

1. J. C. Bailar, Jr., H. J. Emelius, R. Nyholm, and A. F. Trotman-Dickenson, eds., *Comprehensive Inorganic Chemistry*, Pergamon Press, Inc., Elmsford, N.Y., 1973, Vol. 1, p. 402; Vol. 2, p. 1107.
2. N. Gehrels and R. M. Candey, *AIP Conf. Proc.*, 211, 213–223, (1990) for information on gamma-ray spectroscopy.
3. M. J. Geagan, B. B. Chase, and G. Muehllehner, *Nucl. Instrum. Methods Phys. Res., Sect. A*, 353(1–3), 379–383 (1994) for information on CAT scan and PET.
4. *The United States Pharmacopeia XXII(USP XXII-NF XVII)*, The United States Pharmacopeial Convention, Inc., Rockville, Md., 1990, p. 1261.

PHILIP H. MERRELL
ELIZABETH M. PETERS
Mallinckrodt, Inc.

SODIUM NITRATE

Sodium nitrate [7631-99-4], $NaNO_3$, is found in naturally occurring deposits associated with sodium chloride, sodium sulfate, potassium chloride, potassium nitrate, magnesium chloride, and other salts. Accumulations of sodium nitrate have been reported in several countries, but the only ones being commercially exploited are the unique nitrate-rich deposits in Chile, South America. Natural sodium nitrate is also referred to as Chilean saltpeter or Chilean nitrate.

The annual world production of sodium nitrate was steady throughout the early 1990s. About 85% is supplied by the natural product. The maximum world production of sodium nitrate occurred around 1930, at 3,000,000 t/yr, but the highest production levels attained by the Chilean nitrate industry (ca 2,900,000 t/yr) occurred in the late 1920s. Synthetic sodium nitrate production peaked in the mid-1930s at 730,000 t/yr. During that period, the Chilean industry production decreased to 1,360,000 t/yr.

Sodium nitrate is used as a fertilizer and in a number of industrial processes. In the period from 1880–1910 it accounted for 60% of the world fertilizer nitrogen production. In the 1990s sodium nitrate accounts for 0.1% of the world fertilizer nitrogen production, and is used for some specific crops and soil conditions. This decline has resulted from an enormous growth in fertilizer manufacture and an increased use of less expensive nitrogen fertilizers (qv) produced from synthetic ammonia (qv), such as urea (qv), ammonium nitrate, ammonium phosphates, ammonium sulfate, and ammonia itself (see AMMONIUM COMPOUNDS). The commercial production of synthetic ammonia began in 1921, soon after the end of World War I. The main industrial market for sodium nitrate was at first the manufacture of nitric acid (qv) and explosives (see EXPLOSIVES AND PROPELLANTS). As of the mid-1990s sodium nitrate was used in the production of some explosives and in a number of industrial areas.

Deposits

The Chilean nitrate deposits are located in the north of Chile, in a plateau between the coastal range and the Andes mountains, in the Atacama desert. These deposits are scattered across an area extending some 700 km in length, and ranging in width from a few kilometers to about 50 km. Most deposits are in areas of low relief, about 1200 m above sea level. The nitrate ore, caliche, is a conglomerate of insoluble and barren material such as breccia, sands, and clays (qv), firmly cemented by soluble oxidized salts that are predominantly sulfates, nitrates, and chlorides of sodium, potassium, and magnesium. Caliche also contains significant quantities of borates, chromates, chlorates, perchlorates, and iodates.

The nitrate deposits are made up of several layers (Fig. 1). The ore bodies are very heterogeneous and variable in size, thickness, composition, and hardness. The overburden may include *chuca*, a layer of unconsolidated sand, silt, and clay, and *panqueque*, a layer of semiconsolidated and porous material poorly cemented by salts over poorly cemented gravel. The ore composition has degraded considerably since the early days of the industry, when it was reported that ores

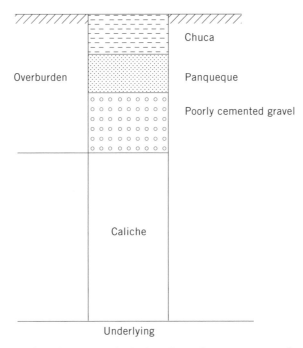

Fig. 1. Schematic of a nitrate ore bed, detailing the various overburden layers. The overburden thickness can vary from 0 to 2.5 m, where chuca = 0.1–0.5 m, panqueque = 0.1–0.4 m, and poorly cemented gravel = 1.5–2.5 m. Caliche ranges from 0.8 to 8 m.

of up to 50% sodium nitrate were mined. There are still reserves that can be commercially mined well into the twenty-first century (1).

Numerous theories exist as to how the Chilean deposits formed and survived. It has been postulated that the unique nitrate-rich caliche deposits of northern Chile owe their existence to an environment favorable to accumulation and preservation of the deposits, rather than to any unusual source of the saline materials (2). The essential conditions are an extremely arid climate similar to that of the Atacama desert in the 1990s, slow accumulation during the late Tertiary and Quaternary periods, and a paucity of nitrate-utilizing plants and soil microorganisms.

Properties

Selected physical and chemical properties of sodium nitrate are listed in Table 1. At room temperature, sodium nitrate is an odorless and colorless solid, moderately hygroscopic, saline in taste, and very soluble in water, ammonia, and glycerol. Detailed physical and chemical properties are also available (3,4).

Manufacture and Processing

Natural Sodium Nitrate. The manufacture of natural sodium nitrate is carried out by its extraction from the ore by leaching with a brine, followed by fractional crystallization. Historically the Shanks process was utilized, but

Table 1. Selected Properties of Sodium Nitrate

Property	Value
mol wt	84.99
crystal system	trigonal, rhombohedral
mp, °C	308
refractive index, n_D	
trigonal	1.587
rhombohedral	1.336
density, solid, g/cm^3	2.257
solubility in H_2O, molality ($\pm 2\%$)	
at -17.5°C	7.4
0°C	8.62
40°C	12.39
80°C	17.42
120°C	24.80
specific conductivity, at 300°C, S/cm	0.95
viscosity[a], η, mPa·s(=cP)	
at 590 K	2.85
730 K	1.53
heat of fusion, J/g^b	189.5
heat capacity, J/g^b	
solid	
at 0°C	1.035 ± 0.005
100°C	1.23 ± 0.006
liquid at 350°C	1.80 ± 0.02
aqueous solution[c]	4.138 ± 3.045

[a]Measurement method: capillary; $\eta = 25.0987 - 6.0544 \times 10^{-2}\,T + 3.8709 \times 10^{-5}\,T^2$. Precision = ca 0.6%; uncertainty = ca 3%.
[b]To convert J to cal, divide by 4.184.
[c]$C_p = 4.175 - 37.42 \times 10^{-3}\,P + 216.6 \times 10^{-6}\,P^2$, where P = wt % solute can range from 1–39%.

the last plant closed in 1977. In the Shanks process, the ore was crushed and leached in large steel vats using a solution consisting of water and a mother liquor brine having ca 450 g/L sodium nitrate. During leaching, carried out at a temperature of 70°C, the brine was concentrated to 700–750 g/L sodium nitrate and then pumped to a preliminary cooling pan where temperature was allowed to drop to ca 25°C. Sodium chloride then crystallized and the slimes were allowed to settle after the addition of a coagulant, eg, wheat flour. The clear liquor, added to $NaNO_3$-crystallizing pans, was allowed to cool overnight to ambient temperature. The brine or mother liquor was then pumped and sent to the last stages of the leaching cycle. The Shanks process made possible the recovery of ca 60% of the sodium nitrate in the ore. Fuel consumption was ca 0.154 metric tons of fuel per metric ton of $NaNO_3$.

The Guggenheim process was introduced in the late 1920s, after been developed by Guggenheim Bros., a firm engaged in developing a Chile copper company in northern Chile. The process was developed based on leaching of caliche at 40°C, which yields a fairly good extraction. Also, the sodium nitrate concentration can be as high as 450 g/L, which is more than 50% of the total dissolved solids, and it can be easily removed through crystallization. If the leaching solution contains a certain level of protective salts, eg, $MgSO_4$ and $CaSO_4$, the sparingly soluble double salt darapskite, $NaNO_3 \cdot Na_2SO_4 \cdot H_2O$, present in the caliche is broken up by magnesium action, thereby increasing $NaNO_3$ extraction. In the Guggenheim process, astrakanite, $Na_2SO_4 \cdot MgSO_4 \cdot 4H_2O$, precipitates instead of darapskite.

The Guggenheim process was developed to permit the treatment of low grade caliche ores, making it possible to mine by mechanical methods instead of by hand. Furthermore, this lower grade ore could be leached with only slightly warm solutions, as opposed to the boiling solutions required for high grade ore, and the nitrate was precipitated by refrigeration, replacing the process of evaporation and cooling in open tanks. In order to provide low cost heat to warm the leach solutions and make subsequent mechanical refrigeration economical, a carefully balanced system of power-plant heat recovery was put into operation (5). The utilization of the Guggenheim process has made it possible commercially to exploit reserves having as low as 7% $NaNO_3$ ore. The process used in the mid-1990s is basically the same as the original one. Only a few modifications had been made, eg, those that permit obtaining anhydrous sodium sulfate and iodine as by-products.

SQM Nitratos (Chile) operates two sodium nitrate plants in northern Chile: Pedro de Valdivia and María Elena, about 30 km distant from one another. The caliche is mined in open-pit areas. A solar evaporation plant, Coya Sur, lies in between. A flow sheet of the processing operations for sodium nitrate production is shown in Figure 2.

Because the ore quality is variable, large open-pit mining areas are first identified by general exploration; specific mining strips are later identified by further exploration and testing. Surface mining methods are used. The overburden is drilled, blasted, and removed, and the waste from a given strip is dumped into a previously worked-out strip. After removal of the overburden, the exposed caliche is drilled, blasted, and loaded into 80-metric ton trucks that deliver the ore to a transfer rail station for transportation to the plants.

At the plant, the ore cars pass through a rotary-car dumper, and the ore is dumped into crushing units. The mineral brought from the mine varies in size from fine particles to chunks of 3–5 metric tons. Crushing is carried out in three stages by means of jaw and cone-type crushers. Before crushing, a selecting screen rejects material smaller than 0.42 mm. About 80% of the crushed material, ie, that having diameter >0.42 mm, is sent to the leaching plant. The remaining 20% is sent to the fines treatment ponds. These two fractions have to be leached separately because clay present in the fine fraction swells when in contact with rich brines, occurrence of which would retard the rate of leaching of the coarser fraction.

Once the coarser fraction is transferred to the leaching plant (see Fig. 2), leaching takes place in a series of 10,000-m^3 leaching vats built of reinforced concrete. The process consists of countercurrent leaching, with one cycle involving 10 vats, one of which is being loaded with crushed ore while another one is being unloaded. Leaching time is 20 h per vat; one total cycle takes 120 h. The leaching is carried out at a temperature of 40°C, using a mother liquor entering the process at a concentration of 320 g/L of sodium nitrate and ending at a concentration of 440–450 g/L. The leaching process terminates with a

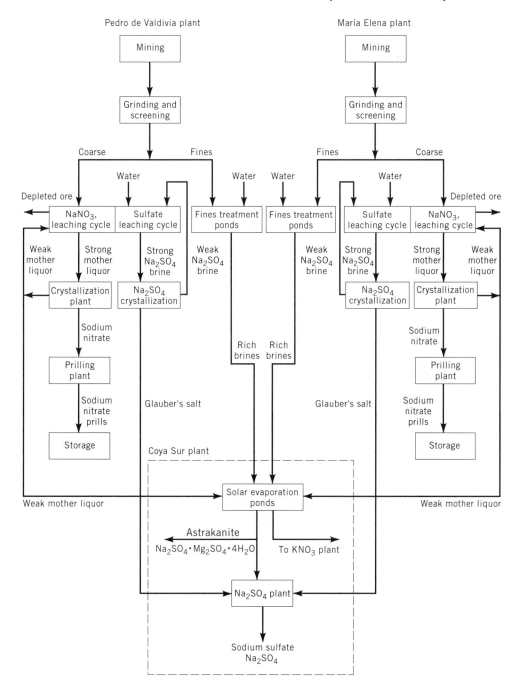

Fig. 2. Processing operations flow sheet of the María Elena and Pedro de Valdivia plants.

final washing with water, where Glauber's salt, $Na_2SO_4 \cdot 10H_2O$, is obtained by crystallization (qv). The depleted ores, containing ca 1% sodium nitrate, are removed by means of electromechanical dredgers into trucks that haul them to tailing disposal areas.

The strong solution from the leaching process is cooled to crystallize sodium nitrate in a series of shell-and-tube heat exchangers. Cooling is produced first by using the already cold spent solution. Further cooling is obtained by ammonia refrigeration. The final temperature is 8–12°C. The slurry containing sodium nitrate crystals is sent to thickening tanks and the resulting sludge is pumped to continuous centrifuges. The centrifuged product is rinsed to displace entrained brines, and crystallized sodium nitrate having 4–6% moisture is obtained. The crystallized sodium nitrate can be dried and prilled to produce fertilizer-grade sodium nitrate, or recrystallized to remove impurities and obtain technical grades of sodium nitrate for industrial uses. Otherwise, crystallized sodium nitrate can be sent to the potassium nitrate plant, where potassium nitrate is produced through a direct reaction of sodium nitrate and potassium chloride (see POTASSIUM COMPOUNDS). Crystallized sodium nitrate can also be combined with existing potassium nitrate to obtain a potassium sodium nitrate.

The sodium nitrate crystals are carried by conveyor belt to granulation plants where they are melted at 315–325°C. The melting system consists of reverberatory furnaces and indirect warming boilers. The melted salt is pumped to granulation (prilling) towers 30 or 70 m high and sprayed. The droplets solidify while falling against an upward air current, forming spherical granules (prills). The granules are collected and screened, then cooled to 35°C by means of shell-and-tube exchangers. The final product is transported by railway cars and stored in silos in the SQM Nitratos facilities at the seaport of Tocopilla.

Synthetic Sodium Nitrate. Sodium nitrate can be synthetically obtained by absorption of nitrous gases or by neutralization of nitric acid (qv). Whereas low NO_x content in waste gases from nitric acid plants makes these gases useless for producing nitric acid, one way to avoid emission of nitrous gases to the atmosphere consists of using an alkaline solution of NaOH or Na_2CO_3 to absorb them. Products are mainly sodium nitrite or sodium nitrate. The general reactions involved are as follows:

$$NO_2 + NO + 2\,NaOH \longrightarrow 2\,NaNO_2 + H_2O$$

$$2\,NO_2 + 2\,NaOH \longrightarrow NaNO_2 + NaNO_3 + H_2O$$

$$NO_2 + NO + Na_2CO_3 \longrightarrow 2\,NaNO_2 + CO_2$$

$$2\,NO_2 + Na_2CO_3 \longrightarrow NaNO_2 + NaNO_3 + CO_2$$

The kinetics of these reactions markedly favor sodium nitrite production. If some sodium nitrate is formed, it can be easily separated by differential crystallization. Otherwise, sodium nitrate can be formed by treating sodium nitrite with nitric acid:

$$3\,NaNO_2 + 2\,HNO_3 \longrightarrow 3\,NaNO_3 + 2\,NO + H_2O$$

The resulting nitrous oxide can be recirculated to the nitric acid plant or be used for other purposes. Free acid remaining in the impregnation water of sodium nitrate crystals is neutralized by adding some NaOH to the washing water. Whereas several nitric acid plants utilize absorption of nitrous gases to treat tail gases, almost all of these plants produce small volumes of sodium nitrate.

Sodium nitrate can also be produced by neutralizing nitric acid with sodium hydroxide or sodium carbonate:

$$2\ HNO_3 + 2\ NaOH \longrightarrow 2\ NaNO_3 + 2\ H_2O$$
$$2\ HNO_3 + Na_2CO_3 \longrightarrow 2\ NaNO_3 + CO_2$$

In the United States, Olin Corporation had a plant using this method that closed in 1988.

Economic Aspects

As of 1996 world production of sodium nitrate was about 520,000 metric tons annually. Of this quantity, some 450,000 t (86%) are produced in Chile from natural deposits by SQM Nitratos and distributed worldwide by several affiliates, eg, Chilean Nitrate Corporation in the United States and Nitrate Sales International in Belgium. The remainder, ca 70,000 t, is manufactured mainly in Europe, Japan, and Russia, generally as a by-product of nitric acid production. Additionally, China is known to manufacture some unknown but significant volumes of sodium nitrate for domestic use.

The product is offered in two main grades: technical and agricultural, the former being of somewhat higher purity. Consumption is about 240,000 t/yr of technical grades and some 280,000 t/yr of the agricultural grade. The latter is produced only by the Chilean industry. World consumption of technical grades has been steady since the late 1980s.

Technical Grades. Chile's SQM Nitratos is the largest producer of technical-grade $NaNO_3$, making ca 170,000 t/yr. About 70,000 t/yr are manufactured in countries such as Germany (BASF), Japan (Mitsubishi), and others. Sodium nitrate has not been manufactured in the United States since Olin Corporation stopped production in 1988.

SQM Nitratos produces three technical grades, industrial, technical, and refined, in increasing order of purity. The refined grade has the trademark Niterox in the United States. Chilean Nitrate Corporation distributes these grades in the United States. Prices (fob warehouse) during 1994 were ca $225/t for industrial grade, $258/t for technical grade, and $340/t for refined grade. The main markets are the manufacture of glass, explosives, and charcoal briquettes. The largest consumption regions are North America, Europe, and Asia.

Agricultural Grade. Some 280,000 t of sodium nitrate are used annually as fertilizer, although production for fertilizer use has declined from the usage that prevailed prior to the mid-1980s, mainly because since 1987 SQM Nitratos has increasingly been using part of its sodium nitrate production as a raw material for manufacturing growing volumes of potassium nitrate. The company also uses part of its sodium nitrate to manufacture a potassium sodium nitrate fertilizer. Sodium nitrate is used for specific crops and soils, mostly sugar beet, vegetables, and tobacco, and mainly acid soils. Most important markets are in regions of Chile, the southeastern United States, Holland, Belgium, and Japan. The price of agricultural sodium nitrate in the United States in 1994 was about $175/t (fob warehouse).

Standards and Specifications

Typical analyses of some commercial grades of sodium nitrate are given in Table 2. Chilean sodium nitrate is a prilled product having a bulk density of 1.22 g/cm^3, a mp of 301°C, and a water solubility of 91 g/100 mL of water at 25°C. The specifications for reagent-grade sodium nitrate are given in Table 3, *Food Chemicals Codex* specifications are given in Table 4; U.S. Federal chemical specifications are listed in Table 5. U.S. Federal specifications for particle size refer to U.S. standard sieves and are as follows: minimum 98.0% to pass through sieve No. 60 for class 1 sodium nitrate; minimum 98.0% to pass through sieve No. 100 for class 2 sodium nitrate, and minimum 80.0% retained sieve No. 20 for class 3 sodium nitrate (7).

Table 2. Analyses of Commercial Grades of Sodium Nitrate^a, wt %^{b}

Parameter	Agricultural	Industrial	Technical	Refined (Niterox)
	Components			
sodium nitrate	96.8–97.9	98.0^c	99.2^c	99.6^c
nitrogen	16^c	na	na	na
sodium	26	na	na	na
potassium, as K$_2$O	0.4–1.2	na	na	na
magnesium, as MgO	0.10	na	na	na
sulfur	0.10	na	na	na
sodium sulfate	na	0.35	0.10	0.10
chlorine	1	na	na	na
sodium chloride	na	0.8	0.30	0.13
sodium nitrite	na	0.05	0.005	0.005
boron	0.35	0.025	na	na
sodium tetraborate	na	na	0.03	0.03
insolubles	0.15	0.15	0.02	0.02
moisture	0.15	0.15	0.15	0.15
	Screen analysis			
screen size, mm (Tyler mesh)				
on 2.38 (8)	10	8	5	5
on 1.68 (10)	70	65	45	45
on 1.19 (14)	96	96	90	90
through 0.841 (20)	1.5	2	8	8

^{a}Data correspond to Chilean sodium nitrate. Courtesy of SQM Nitratos (Chile).
^{b}Value given is maximum unless otherwise noted.
^{c}Value given is minimum.

Health and Safety Factors

The acceptable daily intake by human adults for nitrates suggested by the World Health Organization (WHO) is 5 mg/kg body wt per day (expressed as sodium nitrate). Some studies have suggested that this figure might be increased to 25 mg/kg. High doses of nitrates are lethal. Accidental ingestion of ca 8–15 g or more of sodium or potassium nitrate causes severe abdominal pain, bloody stools and urine, weakness, and collapse. Victims of sodium nitrate or potassium nitrate poisoning contract severe gastroenteritis. Outbreaks of poisoning from the ingestion of meats containing sodium nitrate and sodium nitrite have occurred from

Table 3. ACS Specifications for Reagent-Grade Sodium Nitrate

Specification	Value
insoluble matter, wt %	0.005
pH of 5 wt % solution at 25°C	5.5–8.3
total chlorine, wt %	0.001
iodate, IO_3^-, wt %	ca 0.0005
nitrite, NO_2^-, wt %	ca 0.001
phosphate, PO_4^{3-}, wt %	0.0005
sulfate, SO_4^{2-}, wt %	0.003
calcium, magnesium, and R_2O_3, wt %	0.005
precipitate, wt %	
heavy metals, as Pb	0.0005
iron, Fe	0.0003

Table 4. *Food Chemicals Codex* Specifications for Sodium Nitrate[a]

Specification	Value
$NaNO_3$ assay after drying, wt %	99.0
limits of impurities, wt %	
arsenic, as As	0.0003
heavy metals, as Pb	0.001
total chlorine	ca 0.2[b]

[a]Ref. 6.
[b]Passes the test.

Table 5. U.S. Federal Government Chemical Specifications for Technical-Grade Sodium Nitrate[a,b]

Parameter	Specification, wt %		
	Grade A	Grade B	Grade C
nitrates, as $NaNO_3$	97.0[c]	99.5[c]	97.0[c]
alkalinity, as Na_2O	0.05	none	0.06
chlorates, as $KClO_3$	0.06	none	0.06
calcium, as CaO	0.3	0.1	0.3
magnesium, as MgO	0.15	0.06	0.15
sulfates, as Na_2SO_4	0.5	0.2	0.45
chlorides, as NaCl	na	0.15	0.15

[a]Ref. 7.
[b]Values are maximum unless otherwise noted.
[c]Value is minimum.

the accidental incorporation of excessive amounts of nitrate–nitrite mixtures, ie, 0.5 wt % nitrite as compared to maximum ingredient specifications of 0.05 and 0.02 wt % of nitrates and nitrites, respectively. The health hazards associated with nitrates result mainly from the bacterial conversion of ingested nitrates to nitrites. Infants in the first three months of life are particularly susceptible to nitrite-induced methemoglobinemia. In most cases of illness and death, well water has been used to add to powdered milk. WHO has recommended a maximum permissible concentration of nitrates of 45 mg/L in drinking water. The U.S. Public Health Service standard for nitrates is 10 mg/L. Nitrate toxicity in ruminants is considered to depend on the ability of nitrate to reduce to nitrite, because nitrites are considerably more toxic. Maximum safe levels of nitrate in water for livestock consumption are <100 ppm, although levels in excess of 500 ppm nitrate are required to produce acute poisoning. Levels in feeds should not exceed 5,000 ppm nitrate, and death may result from 15,000 ppm nitrate in the total diet. For further information, see Reference 8.

Uses

Sodium nitrate is used as a fertilizer as well as in a number of industrial applications. As a fertilizer, it provides nitrogen, needed in large quantities by plants and commonly in shortage in soils. Sodium nitrate is a special fertilizer because all the nitrogen is in the most highly oxidized form and thus different from fertilizers providing ammonium nitrogen. Nitrate nitrogen, in contrast to the ammonium form, acts more quickly, has a neutralizing effect on soil and subsoil acidity, does not volatilize to the atmosphere in the form of ammonia, and does not interfere with absorption of potassium, magnesium, and calcium by plants. Nitrate nitrogen acts quickly even in cold weather, under low rainfall conditions, and in acid soils promoting better-quality vegetable and fruit crops and tobacco. Nitrate nitrogen is well suited to supply nitrogen to growing crops at the right time, and to replenish nitrogen promptly when it is lost by leaching from sandy soils.

Sodium is an indispensable element for some crops (notably sugar beet), can partially substitute for potassium in several crops, contributes to neutralizing soil and subsoil acidity, and has a positive effect on soil phosphorus solubility. Sodium is an essential nutrient for cattle, and sodium application to soil increases its content in pastures. Sodium nitrate is particularly effective as a nitrogen source for sugar beet, vegetable crops, tobacco, and cotton (qv), and for any crop in acid soils.

Sodium nitrate is used in a number of industrial processes, in most of them acting primarily as an oxidizing agent. A primary use is in the manufacture of medium and high quality glass (qv), such as optical and artistic glass, television and computer screens, and fiber glass. Sodium nitrate has a role in the elimination of bubbles, in the oxidation of organic matter, and in the oxidation of ferrous oxide and of arsenic or antimony trioxide. Additionally, it is a source of sodium oxide and a fluidizing agent. The amount of sodium nitrate added in the glass manufacture process is up to 2 wt % of the total raw material.

In the manufacture of explosives, sodium nitrate is used mainly in blasting agents. In slurries and emulsions, sodium nitrate improves stability and sensitivity. It also improves the energy balance because sodium nitrate replaces water, so that more fuel can be added to the formulation. Sodium nitrate reduces crystal size of slurries, which in turn increases detonating speed. In dynamites sodium nitrate is used as an energy modifier. Typical content of sodium nitrate is 20–50 wt % in dynamites, 5–30 wt % in slurries, and 5–15 wt % in emulsions. Sodium nitrate is used also in permissible dynamites, a special type of dynamite for coal (qv) mining.

Another large application is as an ingredient in the production of charcoal briquettes. The amount of sodium nitrate used in charcoal briquette manufacture depends on the type and amount of wood and coal used. Typically charcoal briquettes contain up to almost 3% sodium nitrate. Sodium nitrate is also used in the manufacture of enamels and porcelain as an oxidizing and fluxing agent. In porcelain–enamel frits used for metal coating, the amount of sodium nitrate in a batch varies with the various metal bases to be coated, typically from about 3.8 to 7.8 wt %.

Sodium nitrate is also used in formulations of heat-transfer salts for heat-treatment baths for alloys and metals, rubber vulcanization, and petrochemical industries. A mixture of sodium nitrate and potassium nitrate is used to capture solar energy (qv) to transform it into electrical energy. The potential of sodium nitrate in the field of solar salts depends on the commercial development of this process. Other uses of sodium nitrate include water treatment (qv), ice melting, adhesives (qv), cleaning compounds, pyrotechnics, curing bacons and meats (see FOOD ADDITIVES), organics nitration, certain types of pharmaceutical production, refining of some alloys, recovery of lead, and production of uranium.

BIBLIOGRAPHY

"Sodium Nitrate" under "Sodium Compounds" in *ECT* 1st ed., Vol. 12, pp. 605–606, by J. A. Brink, Jr., Purdue University; in *ECT* 2nd ed., Vol. 18, pp. 486–498, by L. C. Pan, Chemical Construction Corp.; in *ECT* 3rd ed., pp. 228–239, by S. Maya, Beecham Products, and M. Laborde, CODELCO.

1. Grossling and G. E. Ericksen, *Computer Studies of the Composition of Chilean Nitrates Ores*, U.S. Geological Survey, Washington, D.C., Dec. 1970.
2. G. E. Ericksen, *Geology and Origin of the Chilean Nitrate Deposits*, U.S. Geological Survey, Professional Paper 1188, U.S. Government Printing Office, Washington, D.C., 1981, 37 pp.
3. *International Critical Tables*, McGraw-Hill Book Co., Inc., New York, 1928.
4. G. J. Janz, C. B. Allen, N. P. Bansal, R. M. Murphy, and R. P. T. Tomkins, *Physical Properties Data Compilations Relevant to Energy Storage. II. Molten Salts: Data on Single and Multi-Component Salt Systems*, U.S. Department of Commerce, National Bureau of Standards, Washington, D.C., Apr. 1979, pp. 142–154.
5. V. Sauchelli, ed., *Fertilizer Nitrogen, Its Chemistry and Technology*, Reinhold Publishing Corp., New York, 1964.
6. *Food Chemicals Codex*, 3rd ed., National Academy Press, Washington, D.C., 1981, p. 292.
7. *Sodium Nitrate*, U.S. Military Technical Specifications, MIL-S-322C, U.S. Government Printing Office, Washington, D.C., Feb. 5, 1968.
8. J. N. Hathcock, ed., *Nutritional Toxicology*, Vol. 1, Academic Press, Inc., New York, 1982, pp. 312–315, 327–381.

LUDWIK POKORNY
IGNACIO MATURANA
SQM Nitratos SA

SODIUM NITRITE

Sodium nitrite [7362-00-0], $NaNO_2$, a stable, odorless, pale yellow or straw-colored compound of molecular weight 69.00, is the sodium salt of nitrous acid [7782-77-6], HNO_2. Sodium nitrite has been produced commercially in the United States since the 1920s, and is available in dry granular or flake forms, as well as in water solutions. Most of the common package types are offered, from bags to drums to bulk. Sodium nitrite is used in dye, rubber chemicals and pharmaceuticals manufacture, as a corrosion inhibitor, in heat treating and heat-transfer salts, in meat curing, and several other applications. The U.S. market is served primarily by domestic producers, with some imports. Production is by absorption of oxides of nitrogen into sodium carbonate or sodium hydroxide solutions. Sodium nitrite is an oxidizer and is toxic; as such, it requires care in its handling, storage, and use.

Properties

Pure anhydrous crystalline sodium nitrite has a specific gravity of 2.168 at 0°C/0°C (1). The crystal structure is body-centered orthorhombic, having the unit cell dimensions $a = 0.355$ nm, $b = 0.556$ nm, and $c = 0.557$ nm (2). Sodium nitrite melts at ~284°C and decomposition begins above 320°C, yielding N_2, O_2, NO, and Na_2O. The heat of formation is -362.3 kJ/mol (-86.6 kcal/mol) at 25°C (3). Sodium nitrite has a transition point at 158–165°C and displays significant changes in physical properties within this temperature range. The specific heat increases gradually from ~980 J/(kg·K) (0.234 cal/(g·°C)) at 60°C to ~1160 J/(kg·K) (0.277 cal/(g·°C)) at 200°C, but exhibits a peak value of ~2290 J/(kg·K) (0.547 cal/(g·°C)) at 161°C corresponding to this transition temperature (1).

Sodium nitrite is hygroscopic and very soluble in water. Dissolution of sodium nitrite in water results in the absorption of heat in the amount of 15.1 kJ/mol (3.6 kcal/mol) at 18°C (3). Water solubility characteristics are displayed in Figure 1. Sodium nitrite has limited solubility in most organic solvents (Table 1). The pH of a 1% solution of sodium nitrite is ~9. A hemihydrate, $NaNO_2·1/2H_2O$ [82010-95-5], reported at temperatures below -5.1°C, is of no known commercial significance. The eutectic composition is 28.1% $NaNO_2$, although some supercooling of solutions up to ~38% may occur.

Sodium nitrite is stable in alkaline solutions. Acidification liberates nitrous acid which is unstable. The decomposition of nitrous acid yields nitric acid [7697-37-2], HNO_3, according to the following reaction:

$$3\ HNO_2\ (aq) \longrightarrow HNO_3\ (aq) + 2\ NO\ (g) + H_2O$$

Colorless nitric oxide [10102-43-9], NO, spontaneously oxidizes, in the presence of atmospheric oxygen, to brown-colored nitrogen dioxide [10102-44-0], NO_2. The resulting mixture of NO and NO_2, commonly referred to as NO_x gases, is corrosive and toxic and its generation should be avoided. Nitrous acid is not an article of commerce owing to its inherent instability. Sodium nitrite

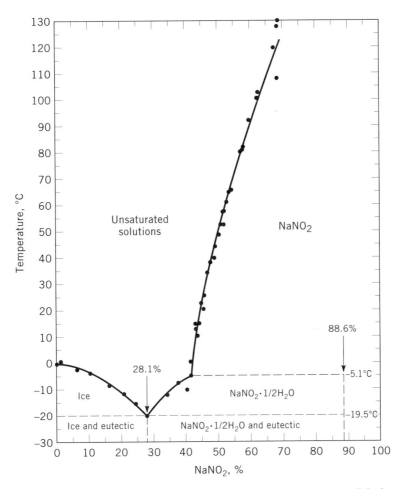

Fig. 1. Sodium nitrite solubility in water where (———) represents solid-phase boundaries (1,2,4,5).

Table 1. Solubility of Sodium Nitrite in Nonaqueous Solvents[a]

Solvent	g NaNO₂/100 g solvent	Temperature, °C
acetone	insoluble	
ammonia, anhydrous	very soluble	−77 to 172
ethanol		
absolute	0.31	19.5
94.9%	1.424	25
ethylenediamine	12.60	30
ethylene glycol	16.78	25
methanol, absolute	4.43	19.5
methyl ethyl ketone	insoluble	
monoethanolamine	8.74	30
propylene glycol	8.47	25
pryridine	0.34	25

[a]Refs. 4 and 6.

serves as the primary industrial source for nitrous acid in organic syntheses, for instance in the diatozation and nitrosation of aromatic amines. Under controlled conditions of acidification, the nitrous acid generated can react before excessive decomposition occurs.

As an oxidizer, sodium nitrite can convert ammonium ion to nitrogen, urea to carbon dioxide and nitrogen, and sulfamate to sulfate and nitrogen. The oxidizing properties of sodium nitrite contribute to its application as a corrosion inhibitor (see CORROSION AND CORROSION INHIBITORS), in detinning of scrap tinplate, and phosphating of metal surfaces. Because it is a strong oxidizer, sodium nitrite is capable of supplying oxygen and thus accelerating the combustion of organic matter. It can undergo vigorous, perhaps violent reactions with certain inorganic compounds such as ammonium salts, acidic materials, thiocyanates, and thiosulfates. It functions as a reducing agent to more powerful oxidizers such as dichromate, permanganate, chlorate, and chlorine. At ambient temperatures, sodium nitrite is stable; it slowly oxidizes to sodium nitrate at elevated temperatures (see SODIUM COMPOUNDS, SODIUM NITRATE). References 1 and 2 provide comprehensive data on the physical and chemical properties of sodium nitrite.

Manufacturing

Sodium nitrite has been synthesized by a number of chemical reactions involving the reduction of sodium nitrate [7631-99-4], $NaNO_3$. These include exposure to heat, light, and ionizing radiation (2), addition of lead metal to fused sodium nitrate at 400–450°C (2), reaction of the nitrate in the presence of sodium ferrate and nitric oxide at ~400°C (2), contacting molten sodium nitrate with hydrogen (7), and electrolytic reduction of sodium nitrate in a cell having a cation-exchange membrane, rhodium-plated titanium anode, and lead cathode (8).

Industrial production of sodium nitrite is by absorption of nitrogen oxides (NO_x) into aqueous sodium carbonate or sodium hydroxide. NO_x gases originate from catalytic air oxidation of anhydrous ammonia, a practice common to nitric acid plants:

$$4\ NH_3 + 5\ O_2 \longrightarrow 4\ NO + 6\ H_2O$$

Gas contact is typically carried out in absorption towers over which the alkaline solutions are recirculated. Strict control over the conditions of absorption are required to efficiently capture the NO_x and convert it predominantly to sodium nitrite according to the following reaction, thereby minimizing the formation of by-product sodium nitrate. Excessive amounts of nitrate can impede the separation of pure sodium nitrite from the process.

$$2\ NaOH + NO + NO_2 \longrightarrow 2\ NaNO_2 + H_2O$$

Solutions of sodium nitrite thus produced are concentrated and a slurry of crystals obtained in conventional evaporation (qv) and crystallization (qv) equipment. Much of this equipment can be of mild steel construction because sodium nitrite functions as a corrosion inhibitor toward most ferrous metals. The crystals are typically separated from the mother liquor by centrifugation and subsequently dried. Because of its tendency to lump and cake rapidly in storage, dry sodium nitrite products are frequently treated with an anticaking agent to keep them free-flowing. Alternatively, larger flakes or pellets are prepared from the granular material through a compaction process. The limited surface contact between these larger particles allows them to remain uncaked for extended periods. Technical solutions for commerce can be obtained directly from the process; higher purity solution products are prepared by dissolving crystals.

Shipment and Storage

Dry products of sodium nitrite are most commonly packaged into 22.7 kg (50 lb) or 45.4 kg (100 lb) multi-ply paper bags which contain a polyethylene moisture barrier. Fiber drums and semibulk sacks are also utilized. Bulk shipments are limited to flake material in specially designed sparger cars which allow the material to be unloaded as a solution. Dry sodium nitrite is regulated by the U.S. Department of Transportation (DOT) and classified as an Oxidizer, Hazard Class 5.1, UN1500, UN Packaging Group III (9). Containers must bear the Class 5.1 Oxidizer label, and bulk shipments must be placarded appropriately.

Liquid sodium nitrite products are typically 40–42% $NaNO_2$ and can be shipped in tank cars or tank trucks when volume and freight considerations allow. Sodium nitrite solutions are also regulated by the DOT and classified as Nitrites, Inorganic, Aqueous Solution, NOS, Hazard Class 5.1 Oxidizer, UN3219, UN Packaging Group III (9). Liquid product must also carry the Oxidizer label. Lesser quantities of liquid product may also be available in drums from local chemical distributors. Solution products are often preferred because of more convenient, efficient, and cost-effective handling versus bagged material and compatibility with inexpensive mild steel equipment.

Care must be exercised in using sodium nitrite near other chemicals. It is incompatible with ammonium salts, thiocyanates, thiosulfates, and strong reducing agents. In acid solutions, sodium nitrite evolves toxic NO_x; in the presence of secondary amines it can form nitrosamines which are suspected carcinogens.

Sodium nitrite exhibits good shelf-life characteristics if stored in secure containers in a cool, dry place, segregated from combustible and incompatible materials. Sodium nitrite does not burn, but its decomposition in fire promotes burning by furnishing additional oxygen. In the case of fire, water flooding should be used and the runoff kept away from streams and sewers to the extent possible. If spilled, care should be exercised to avoid contact with any acidic materials, as toxic NO_x could evolve. Under current regulations, spills in excess of 45.4 kg (100 lb) of dry sodium nitrite equivalent are reportable to the U.S. EPA (10). Additional state and/or local regulations may also apply.

Health and Safety Factors

Sodium nitrite is poisonous and prolonged contact with dry sodium nitrite or its solutions can cause irritation to the skin, eyes, and mucus membranes. The

LD_{50} (oral, rat) is 85 mg per kg body weight (11). Inhalation or ingestion of significant quantities of dust or mist may result in acute toxic effects such as nausea, cyanosis, and low blood pressure, which can lead to possible collapse, coma, and even death.

Persons responsible for the procurement, use, or disposal of sodium nitrite products should become familiar with safety information contained in the manufacturer's Product Safety Data Sheet (PSDS) (12). For handling dry products, a hard hat, safety glasses, impervious gloves, and long sleeves should be worn as a minimum. Where dusty or misty conditions prevail or when handling solutions, a NIOSH-approved respirator, chemical goggles, and full impervious clothing may be required. Contact lenses should not be worn. Persons should wash thoroughly after handling sodium nitrite. Eating or smoking in areas where sodium nitrite is being handled should be prohibited.

In case of skin contact, the area should be washed thoroughly with water and examined by a physician if irritation persists. If exposed, the eyes should be flushed with water for at least 15 minutes. Remove inhalation victims to fresh air and administer artificial respiration if the victim is not breathing. If ingested, vomiting should be induced. All incidents should be followed by prompt medical attention.

Specifications, Analysis, and Quality

Dry sodium nitrite is offered in several grades: technical, drug (*U.S. Pharmacopeia* (USP)), food (*Food Chemicals Codex* (FCC)), and reagent (American Chemical Society (ACS)). Granular product has a tendency to lump and cake into an unmanageable mass during storage. Flake or treated granular types overcome this caking tendency and exhibit excellent shelf life. The most commonly used anticake is sodium mono- and dimethyl naphthalene sulfonates at a typical dosage of ≤0.1%. This is the only such agent allowed in the food-grade product in the United States (13). The USP and reagent ACS grades contain no anticake, and in granular form harden within weeks. Dry products typically contain >99% sodium nitrite and <1% sodium nitrate; a small amount of residual alkalinity is also present. The specifications for technical, USP, food, and ACS reagent grades are given in Table 2. Typically liquid products are of a technical or purified grade containing between 35 and 45% sodium nitrite and varying levels of nitrate.

Sodium nitrite products can be analyzed using methods that accompany specifications for the particular grade of product used. Assay methods are typically based on oxidation of the sample by a known excess of standard potassium permanganate solution, which is in turn reduced with a known excess of standard oxalic acid or ferrous ammonium sulfate solution. The excess is then titrated to a pink color end point and calculation gives the percent sodium nitrite in the sample. Standard laboratory analytical equipment such as hot plates, glassware, reagents, and analytical balances are required to run the tests. Careful attention to the standardization of reagents is important in obtaining reliable results. In contaminated or process samples, other substances present, which may be oxidized by potassium permanganate, give positive interferences. The analysis should be carried out in a proper fume hood to avoid the introduction of toxic

Table 2. Specifications for Dry Grades of Sodium Nitrite

Parameter[a]	Technical[b]	USP[c]	Food[d]	Reagent[e]
assay, as $NaNO_2$, wt %[f]	97.0	97.0–101.0	97.0	97.0
loss on drying, wt %[f]		0.25	0.25	
heavy metals, as Pb, %		0.002	0.002	0.001
arsenic, as As, ppm			3	
lead, as Pb, ppm			10	
sodium sulfate, as Na_2SO_4, wt %	0.2			
sodium chloride, as NaCl, wt %	0.2			
insolubles, wt %	0.5			0.01
pH	8 ± 1			
chloride, as Cl, wt %				0.005
sulfate, as SO_4, wt %				0.01
calcium, as Ca, wt %				0.01
iron, as Fe, wt %				0.001
potassium, as K, wt %				0.005

[a]Value is maximum unless noted. [b]Ref. 14. [c]Ref. 15. [d]Ref. 16. [e]Ref. 17. [f]Value is minimum.

gases into the workplace. The same safety precautions apply to handling sodium nitrite in the laboratory as previously described for the production area. Complete details on assay procedures and impurities testing are available (4,14–17).

Nitrite can be estimated in the field by using one of the many available test kits offered by a number of companies specializing in this area including Chemetrics, Inc. (Calverton, Virginia) and the Hach Company (Loveland, Colorado). These kits are designed for specific concentration ranges, involve simple procedures, and provide accuracy reasonable for field work by using color comparators or by counting drops of titrant to a color-change end point. Facilities producing the drug and food grades must follow a rigid set of guidelines for cleanliness and product reliability known as Good Manufacturing Practices (GMPs) and avail their plants to inspections by the U.S. FDA. Accurate and thorough recordkeeping is also required in the production of these grades. General Chemical Corporation's sodium nitrite facility has had its quality system registered (October 1993) as complying with the internationally recognized quality standard ISO 9002.

Uses

The many industrial uses for sodium nitrite primarily are based on its oxidizing properties or its liberation of nitrous acid in acidic solutions.

Dyes.   Sodium nitrite is a convenient source of nitrous acid in the nitrosation and diatozation of aromatic amines. When primary aromatic amines react with nitrous acid, the intermediate diamine salts are produced which, on coupling to amines, phenols, naphthols, and other compounds, form the important azo dyes (qv). The color center of the dye or pigment is the $-N{=}N-$ group and attached groups modify the color. Many dyes and pigments (qv) have been manufactured with shades of the entire color spectrum.

Rubber Chemicals. Sodium nitrite is an important raw material in the manufacture of rubber processing chemicals. Accelerators, retarders, antioxidants (qv), and antiozonants (qv) are the types of compounds made using sodium nitrite. Accelerators, eg, thiuram [137-26-8], greatly increase the rate of vulcanization and lead to marked improvement in rubber quality. Retarders, on the other hand (eg, N-nitrosodiphenylamine [156-10-5]), delay the onset of vulcanization but do not inhibit the subsequent process rate. Antioxidants and antiozonants, sometimes referred to as antidegradants, serve to slow the rate of oxidation by acting as chain stoppers, transfer agents, and peroxide decomposers. A commonly used antioxidant is N,N'-disubstituted p-phenylenediamine which can employ sodium nitrite in its manufacture (see RUBBER CHEMICALS).

Heat Treatment and Heat-Transfer Salts. Mixtures of sodium nitrite, sodium nitrate, and potassium nitrate are used to prepare molten salt baths and heat-transfer media. One of the most widely used eutectic mixtures uses 40% $NaNO_2$, 7% $NaNO_3$, and 53% KNO_3 [7757-79-1] to give a melting point of 143°C. Its advantages are low melting point, high heat-transfer rate, thermal stability to 538°C, and a noncorrosive effect on steel (qv) at high temperature. The salts can be used for indirect heating or cooling or as quenching baths in the annealing of iron and steel.

Corrosion Inhibition. Sodium nitrite acts as an anodic inhibitor toward ferrous metals by forming a tightly adhering oxide film over the steel, preventing the dissolution of metal at anodic areas. When used in mixed metal systems that may include, for instance, copper, brass, or aluminum (as in automobile cooling systems), synergistic additives may be required for complete system protection. Some renewed interest in furthering nitrite use has been spawned from reduction in widespread use of carcinogenic hexavalent chromium-based inhibitors. Loss of protection owing to biological consumption of nitrite has been addressed by the use of higher initial concentrations (18). Sodium nitrite is used in boiler water treatment, as a dip or spray for protection of metals in process and storage, and in concrete. The EPA has ruled that sodium nitrite should not be used as a corrosion inhibitor in amine-based metalworking fluids because of the formation of potentially carcinogenic nitrosamines (19).

Metal Finishing. In phosphating solutions, sodium nitrite performs as an accelerator and oxidizer, serving to reduce processing times and control buildup of ferrous ions in solution, respectively. Phosphate coatings are applied to steel as a base coating prior to painting. In gold–sulfite-plating baths, sodium nitrite functions in the formation a gold–sulfite–nitrite complex. $Na_4Au(SO_3)_2NO_2$ [51846-25-4], from which the gold can be electrolytically deposited (20) (see GOLD AND GOLD COMPOUNDS). This bath is considered to be safer than the poisonous cyanide baths traditionally used for gold plating. Sodium nitrite is also used in the recovery of tin from scrap tinplate (see TIN AND TIN ALLOYS). It functions as an oxidizer in converting tin to sodium stannate with caustic soda; high purity tin can then be electroplated directly from the stannate solution.

Meat Curing. Sodium nitrite is used extensively in curing meat and meat products (qv), particularly pork products such as ham, bacon, frankfurters, etc. As an ingredient in curing brines, sodium nitrite acts as a color fixative and inhibits bacteria growth, including *Clostridium botulinum*, the source of the botulism toxin. Certain fish and poultry products are also cured with brines containing sodium nitrite. All food uses of sodium nitrite are strictly regulated by the FDA (21) and USDA.

Other Uses. Other applications for sodium nitrite include the syntheses of saccharin [81-07-2] (see SWEETENERS), synthetic caffeine [58-08-2] (22), fluoroaromatics (23), and other pharmaceuticals (qv), pesticides (qv), and organic substances; as an inhibitor of polymerization (24); in the production of foam blowing agents (25); in removing H_2S from natural gas (26); in textile dyeing (see TEXTILES); as an analytical reagent; and as an antidote for cyanide poisoning (see CYANIDES).

Sodium nitrite has played a key role in the invention of the following: a freezing point depressant in a large steel thermal storage tank for a district cooling system (27), antifungal agent for treatment of skin diseases (28); vapor-phase corrosion inhibitor–desiccant material (29); method for estimating methyl anthranilate [134-20-3] (30); process for preparing optically active benzoic acid (qv) derivatives (31); process for dewaxing oil-producing formations (32); method for treating chelated metal wastewaters (33); method of sewage sludge treatment (34); and preparation of iron nitrosyl carbonyl catalysts (35).

Economic Aspects

Sodium nitrite is manufactured in the United States by General Chemical Corporation at their plant in Solvay, New York, and by E. I. du Pont Company's Gibbstown, New Jersey plant. Recent U.S. demand is estimated to be between 50,000 and 60,000 metric tons per year, the vast majority of which is produced domestically. Imports primarily have been from Germany and Poland; lesser amounts originate from the United Kingdom, the Netherlands, and China. The quantity of product imported into the United States has varied in the 1990s from ~1000 to 4000 t/yr. Currency exchange rates have had a decided influence on import volumes. The list price of sodium nitrite in 45.4 kg (100 lb) bags, fob works in August 1996, was reported to be $0.907/kg ($0.412/lb) versus $0.66/kg ($0.30/lb) in 1980, on the same basis. Liquid products sell at discounted equivalent-basis prices.

BIBLIOGRAPHY

"Sodium Nitrite" under "Sodium Compounds" in *ECT* 1st ed., Vol. 12, p. 606, by J. A. Brink, Jr., Purdue University; in *ECT* 2nd ed., Vol. 18, pp. 498–502, by L. C. Pan, Chemical Construction Corp.; in *ECT* 3rd ed., Vol. 21, pp. 240–245, by J. Kraljic, Allied Corp.

1. "Natrium," in *Gmelins Handbuch der Anorganischen Chemie*, System 21, Vol. 3, Verlag Chemie, Weinheim, Germany, 1966.
2. J. W. Mellor, *A Comprehensive Treatise on Inorganic and Theoretical Chemistry*, Vol. 8, Longmans, Green & Co., London, 1928; J. W. Mellor, *Supplement to Mellor's Treatise on Inorganic and Theoretical Chemistry*, Vol. VIII, Suppl. II, Part II, John Wiley & Sons, Inc., New York, 1967.
3. R. H. Perry, ed., *Chemical Engineer's Handbook*, 5th ed., McGraw-Hill Book Co., Inc., New York, 1973.

4. A. Seidel, *Solubilities of Inorganic and Metal Organic Compounds*, 4th ed., Vol. 2, American Chemical Society, Washington, D.C., 1965.

5. *International Critical Tables*, Vol. 4, McGraw-Hill Book Co., Inc., New York, 1928.

6. *Sodium Nitrite*, Product brochure, GC7767, General Chemical Corp., Parsippany, N.J., 1989.

7. U.S. Pat. 2,294,374 (Sept. 1, 1945), J. R. Bates (to Houdry Process Corp.).

8. Ger. Offen. 2.940,186 (Apr. 24, 1980), M. Yoshida (to Asahi Chemical Industry Co., Inc.).

9. *Code of Federal Regulations*, Title 49, Part 172, Section 101; Part 173, Sections 201,213, 241 (49 *CFR* 172.101, 173.201, 173.213, 173.241), U.S. Government Printing Office, Washington, D.C., 1989.

10. Ref. 9, Title 40, Parts 116–117.

11. "Nitrous Acid, Sodium Salt," in *NIOSH Registry of Toxic Effects of Chemical Substances* (*RTECS*), Accession No. RA-1225000, Dept. of Health and Human Services National Institute for Occupational Safety & Health (NIOSH), Cincinnati, Ohio, 1979.

12. *Sodium Nitrite*, Product Safety Data Sheet, GC3061, General Chemical Corp., Parsippany, N.J., 1994.

13. Ref. 9, Title 21, Part 172, Section 824.

14. *Sodium Nitrite*, U.S. Military Specification MIL-S-24521, Washington, D.C., Sept. 2, 1975.

15. *USP 23/NF 18*, United States Pharmacopoeial Convention, Inc., Rockville, Md., Jan. 1, 1995.

16. *Food Chemicals Codex*, 3rd. ed., National Academy of Sciences, National Academy Press, Washington, D.C., 1981.

17. *Reagent Chemicals*, 8th ed., American Chemical Society, Washington, D.C., 1993.

18. U.S. Pat. 5,558,772 (Sept. 24, 1996), S. Bean and W. Bortle (to General Chemical Corp.).

19. Ref. 10, Part 721, Section 4740.

20. S. Afr. Pat. 73 7671 (July 16, 1974), C. Bradford and H. Middleton (to Johnson Matthey and Co., Ltd.).

21. Ref. 13, Sections 175–177.

22. G. Austin, *Shreeve's Chemical Process Industries*, 5th ed., McGraw-Hill Book Co., Inc., New York, 1984, pp. 802–803.

23. W. Sheppard and C. Sharts, *Organic Fluorine Chemistry*, W. A. Benjamin, Inc., New York, 1969, pp. 92, 168.

24. U.S. Pat. 3,714,008 (Jan. 30, 1973), T. Masaaki and co-workers (to Japan Atomic Energy Research Institute).

25. Jpn. Kokai 74 02,868 (Jan. 11, 1974), S. Murakami and co-workers (to Eiwa Chemical Industrial Co., Ltd).

26. U.S. Pat. 4,515,759 (May 7, 1985), E. Burnes and K. Bhatia (to NL Industries, Inc.); reissued (Nov. 2, 1992) (to Exxon Chemical Patents, Inc.).

27. U.S. Pat. 5,465,585 (Nov. 14, 1995), G. Mornhed, J. Young, and H. Thompson (to Trigen Energy Corp.).

28. U.S. Pat. 5,427,801 (June 27, 1995), U. Kazutoyo (to Japan Lotion Co.).

29. U.S. Pat. 5,344,589 (Sept. 6, 1994), J. M. Foley, B. A. Miksic, and T-Z. Tzou (to Cortec Corp.).

30. U.S. Pat. 5.250,441 (Oct. 5, 1993), B. Becraft and P. F. Vogt.

31. U.S. Pat. 5,229,032 (July 20, 1993), C. Inoue and co-workers (to Showa Denko KK).

32. U.S. Pat. 5,183,581 (Feb. 2, 1993), C. N. Khalil, A. Rabinovitz, and R. K. Romeu (to Petroleo Brasileiro SA).

33. U.S. Pat. 5,160,631 (Nov. 3, 1992), J. G. Frost and K. J. Snyder (to Halliburton Co.).

34. U.S. Pat. 5,147,563 (Sept. 15, 1992), R. D. Blythe and co-workers (to Long Enterprises, Inc.).

35. U.S. Pat. 5,096,870 (Mar. 3, 1992), D. E. Heaton (to The Dow Chemical Co.).

WALTER H. BORTLE
General Chemical Corporation

SODIUM SULFATES

Sulfates of sodium are industrially important materials commonly sold in three forms (Table 1). In the period from 1970 to 1981, over one million metric tons were consumed annually in the United States. Since then, demand has declined. In 1988 consumption dropped to 890,000 t, and in 1994 to 610,000 t (1,2). Sodium sulfate is used principally (40%) in the soap (qv) and detergent industries. Pulp and paper manufacturers consume 25%, textiles 19%, glass 5%, and miscellaneous industries consume 11% (3). About half of all sodium sulfate produced is a synthetic by-product of rayon, dichromate, phenol (qv), or potash (see CHROMIUM COMPOUNDS; FIBERS, REGENERATED CELLULOSICS; POTASSIUM COMPOUNDS). Sodium sulfate made as a by-product is referred to as synthetic. Sodium sulfate made from mirabilite, thenardite, or naturally occurring brine is called natural sodium sulfate. In 1994, about 300,000 t of sodium sulfate were produced as a by-product; another 300,000 t were produced from natural sodium sulfate deposits (4).

Common names have been given to sodium sulfate as a result of manufacturing methods. In rayon production, by-product sodium sulfate is separated from a slurry by filtration where a 7–10-cm cake forms over the filter media. Thus rayon cake was the term coined for this cake. Similarly, salt cake, chrome cake, phenol cake, and other sodium sulfate cakes were named. Historically, sulfate cakes were low purity, but demand for higher purity and controlled particle size has forced manufacturers either to produce higher quality or go out of business. Sodium sulfate is mined commercially from three types of mineral evaporites: thenardite, mirabilite, and high sulfate brine deposits (see CHEMICALS FROM BRINE).

Minerals of sodium sulfate occur naturally throughout the world. The deposits result from evaporation of inland seas and terminal lakes. Colder climates, such as those found in Canada and the former Soviet Union, favor formation

Table 1. Sulfates of Sodium

Chemical name	CAS Registry Number	Mineral name	Common name	Formula
sodium sulfate	[7757-82-6]	thenardite	salt cake[a]	Na_2SO_4
sodium sulfate decahydrate	[7727-73-3]	mirabilite	Glauber's salt	$Na_2SO_4 \cdot 10H_2O$
sodium bisulfate	[7681-38-1]		niter cake	$NaHSO_4$

[a] Whereas at one time the name salt cake implied a low grade material, increasingly it is used as another name for both high and low grade Na_2SO_4.

of mirabilite. Warmer climates, such as those found in South America, India, Mexico, and the western United States, favor formation of thenardite. In areas where other anions and cations are present, double salts can be found of the kinds shown in Table 2, which lists nearly all naturally occurring minerals containing sodium sulfate. Except for mirabilite, thenardite, and astrakanite, these mineral deposits play a minor role in sodium sulfate production.

Table 2. Minerals Containing Sodium Sulfate

Mineral name	Composition
aphthitalite (glaserite)	$Na_2SO_4 \cdot 3K_2SO_4$
astrakanite (bloedite)	$Na_2SO_4 \cdot MgSO_4 \cdot 4H_2O$
burkeite	$2Na_2SO_4 \cdot Na_2CO_3$
d'ansite	$9Na_2SO_4 \cdot MgSO_4 \cdot 3NaCl$
ferrinatrite	$3Na_2SO_4 \cdot Fe_2(SO_4)_3 \cdot 6H_2O$
glauberite	$Na_2SO_4 \cdot CaSO_4$
Glauber's salt (mirabilite)	$Na_2SO_4 \cdot 10H_2O$
hanksite	$9Na_2SO_4 \cdot 2Na_2CO_3 \cdot KCl$
hydro-glauberite	$5Na_2SO_4 \cdot 3CaSO_4 \cdot 6H_2O$
loweite	$Na_2SO_4 \cdot MgSO_4 2.5H_2O$
thenardite	Na_2SO_4
tychite	$Na_2SO_4 \cdot 2Na_2CO_3 \cdot 2MgCO_3$
vanthoffite	$3Na_2SO_4 \cdot MgSO_4$

Physical and Chemical Properties

Physical and chemical properties of the three most important forms of sodium sulfate are summarized in Table 3. The solubility of sodium sulfate in water from 0 to 360°C is shown in Figure 1 (5). The solubility of the $NaCl \cdot Na_2SO_4 \cdot H_2O$-saturated system is also shown. The aqueous solubility of sodium sulfate changes rapidly from 0 to 40°C, and addition of NaCl to a saturated solution of Na_2SO_4 dramatically suppresses this solubility. These two effects are exploited by all manufacturers of sodium sulfate.

Table 3. Properties of Sodium Sulfates

Property	Sodium sulfate		Sodium hydrogen sulfate
	Anhydrous	Decahydrate	
mol wt	142.05	322.21	120.06
mp, °C	884	32.4	315
specific gravity	2.664	1.464	2.435
specific heat, $J/(g \cdot K)^a$	0.845		
heat of formation, kJ/mol[a]	−1385	−4322	−1125
heat of solution, kJ/mol[a]	1.17	−78.41	7.28
heat of crystallization, kJ/mol[a]	−8.8	78.2	
refractive index	1.464	1.394	1.459
crystalline form	rhombic, monoclinic, and hexagonal	monoclinic	triclinic

[a]To convert J to cal, divide by 4.184.

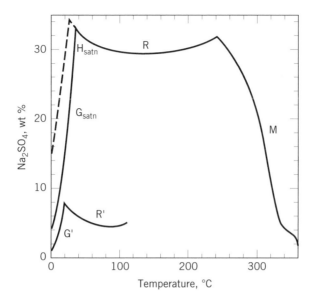

Fig. 1. Solubility system of (——) $Na_2SO_4 \cdot H_2O$ where R and M refer to rhombic and monoclinic Na_2SO_4, respectively, in H_2O; G_{satn} and H_{satn} represent Glauber's salt and sodium sulfate hemihydrate, $Na_2SO_4 \cdot 7H_2O$, respectively, at saturation in H_2O; and (—) $Na_2SO_4 \cdot NaCl \cdot H_2O$ where R' and G' represent the rhombic form and Glauber's salt, both saturated with NaCl. The dashed line represents a metastable form.

The reactivity of Na_2SO_4 is relatively low at room temperature with the exception of sulfuric acid, with which various other acid compounds are formed at temperatures below 100°C. At higher temperatures Na_2SO_4 is very reactive (6).

Sodium sulfate decahydrate melts incongruently at 32.4°C to a sulfate liquid phase and an anhydrous sulfate solid phase. The presence of other salts, such as NaCl, can depress the melting point to 17.9°C.

Sodium sulfate crystallized from solution has an attraction for iron and iron compounds and for various organics. Glauber's salt does not show this attraction and in fact rejects most impurities. Thus higher quality Na_2SO_4 is made from Glauber's salt.

Manufacture and Processing

Nearly all manufacturers of sodium sulfate use Glauber's salt in an intermediate process step. Glauber's salt is then converted to anhydrous sodium sulfate. In 1990, there were only three significant producers of natural sodium sulfate: Ozark-Mahoning (Texas), North American Chemical (California), and Great Salt Lake Minerals (Utah).

In Texas, subterranean sulfate brines are pumped to the surface where the brines are first saturated with NaCl before they are cooled by mechanical refrigeration to form Glauber's salt (7,8). This salt is then separated from its mother liquor, melted, and dehydrated with mechanical vapor recompression evaporators (9).

Processing at Searles Lake, California, by North American Chemical is similar to that of Texas brines. Brine is cooled to 16°C to remove borax crystals, then cooled to 4°C which precipitates Glauber's salt. This salt is then separated from its mother liquor, melted in multi-effect vacuum crystallizers to form anhydrous sodium sulfate, and dried. Both processes produce crystals that are 99.3–99.7% pure (9).

At Great Salt Lake Minerals Corporation (Utah), solar-evaporated brines are winter-chilled to −3°C in solar ponds. At this low temperature, a relatively pure Glauber's salt precipitates. Ponds are drained and the salt is loaded into trucks and hauled to a processing plant. At the plant, Glauber's salt is dissolved in hot water. The resulting liquor is filtered to remove insolubles. The filtrate is then combined with solid-phase sodium chloride, which precipitates anhydrous sodium sulfate of 99.5–99.7% purity. Great Salt Lake Minerals Corporation discontinued sodium sulfate production in 1993 when it transferred production and sales to North American Chemical Corporation (Trona, California).

Figure 2 shows a general process flow diagram for almost all production of natural sodium sulfate. Glauber's salt can be converted to anhydrous sodium sulfate by simply drying it in rotary kilns. Direct drying forms a fine, undesirable powder, and any impurities in the Glauber's salt become part of the final product. This process is not used in the United States but is used in other countries.

The Mannheim process produces sodium sulfate by reaction of sodium chloride and sulfuric acid.

$$NaCl + H_2SO_4 \longrightarrow Na_2SO_4 + HCl$$

This reaction takes place in a fluidized-bed reactor or a specially made furnace called a Mannheim furnace. This method was last used in the United States in the 1980s. In another process, SO_2, O_2, and H_2O react with NaCl.

$$4\,NaCl + 2\,SO_2 + O_2 + 2\,H_2O \longrightarrow 2\,Na_2SO_4 + 4\,HCl$$

This is called the Hargreaves process. Only a minor amount of sodium sulfate is made in the United States using the Hargreaves process, but both the Hargreaves and the Mannheim processes are used widely in the rest of the world. Table 4 lists U.S. producers and capacity information for natural and synthetic sulfate in 1992.

Economic Aspects

Half of all sodium sulfate made is actually a waste by-product and may even present a disposal problem. For this reason, manufacturers sell it at a low price to ensure quick sale. This in turn sets the price for all naturally produced sodium sulfate and tends to keep values low. Figure 3 shows price fluctuations from 1970 to 1995.

In 1980, over one million tons of sodium sulfate were consumed in the United States, but this had declined to <600,000 t by the end of 1994. The decline is partly a result of higher energy prices and more efficient use of

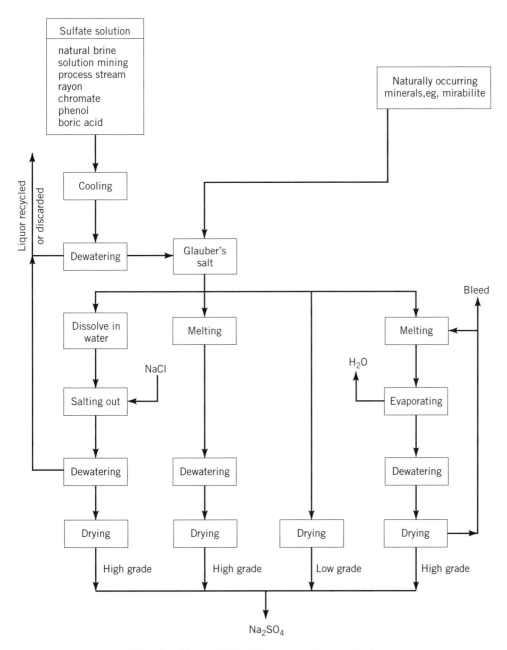

Fig. 2. Natural Na_2SO_4 processing methods.

Na_2SO_4 by the paper industry. At one time the kraft paper industry consumed two-thirds of sodium sulfate production. Pressures on paper producers to clean effluent streams and reduce energy forced improvements in internal processes and recycling of sodium sulfate (11,12).

Table 4. U.S. Producers and Production Capacity of Sodium Sulfate

Company	Capacity, $t \times 10^3$	Location	Source
	Natural		
Great Salt Lake Minerals Corp.	45	Ogden, Utah	Great Salt Lake brine
North American Chemical Co.	218	Trona, Calif.	Searles Lake dry bed[a]
Ozark-Mahoning Co.	141	Seagraves, Tex.	dry lake bed[a]
Total natural	*404*		
	Synthetic		
Lenzing AG	34	Lowland, Tenn.	rayon manufacture
Courtaulds North America, Inc.	45	La Moyne, Ala.	rayon manufacture
Flour Corp., Doe Run Co.	9	Boss, Mo.	battery recycling
W.R. Grace & Co.	8	Nashua, N.H.	chelating agents
J.M. Huber Corp.	32	Etowah, Tenn.	silica pigments
	14	Harve de Grace, Md.	silica pigments
INDSPEC Chemical Corp.	35	Petrolia, Pa.	resorcinol manufacture
North American Rayon Corp.	14	Elizabethton, Tenn.	rayon manufacture
Occidental Chemical Corp.	109	Castle Hayne, N.C.	sodium dichromate
Public Service Co. of New Mexico	6	Waterflow, N.Mex.	flue gas desulfurization
Teepak, Inc.	6	Danville, Ill.	cellulose manufacture
Star Enterprise	3	Delaware City, Del.	flue gas desulfurization
Total synthetic	*315*		

[a]Underground brine.

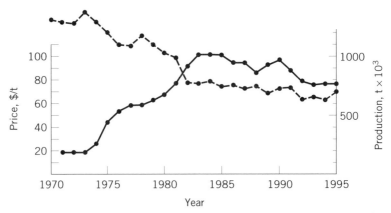

Fig. 3. Price (– – –) and production (——) of Na_2SO_4. See Refs. 2 and 10.

Sodium sulfate is also used as a filler in powdered soap and detergents. Introduction of liquid detergents, in which Na_2SO_4 is not used, has also resulted in lower consumption of the sulfate.

The demand for sodium sulfate leveled off in the first half of the 1990s and is expected to begin to rise again with increased U.S. population and as the demand for kraft paper products and powdered detergents rises.

Specifications, Standards, and Quality Control

Specifications vary with use. The paper and detergent industries are concerned with whiteness and specify various methods to describe color and black or dark specks. It is also important in the detergent industry that sodium sulfate has a particle size and density compatible with other components in the blend to eliminate segregation when it is handled. A typical specification for detergent-grade sodium sulfate is given in Table 5.

Moisture content must be held at low levels or bulk material hardens or cakes. Material that cakes causes severe handling problems anywhere it is stored and reclaimed.

Components of sodium sulfates are easy to analyze using standard procedures. Except for normal analytical care in handling samples, no special precautions or procedures are required.

Table 5. Sodium Sulfate Specifications[a]

Parameter	Value
Chemical properties	
component, %	
Na_2SO_4	99.4
Ca	0.02
Cl	0.52
K	0.01
moisture	0.01
water insolubles	0.03
pH, 5% solution	9.0[b]
Physical properties	
bulk density, kg/m³	
loose	120
tapped	144
solution	clear
color	white
angle of repose, deg	34

Tyler sieve sizes[c]	Retained on sieve, %
28 (640)	2
80 (177)	61
200 (74)	93
270 (53)	97

[a]Ref. 13. [b]Value is typical. [c]Mesh size (approximate μm sizes).

Health and Safety

Environmental concerns have influenced Na_2SO_4 manufacturing. For example, whereas waste Na_2SO_4 was historically discharged into waterways, regulations prohibit discharges of liquids containing sulfate into the environment. This affects paper producers as well as by-product producers. In general, Na_2SO_4 is not considered an environmentally dangerous material, but the Mannheim and Hargreaves processes are practically nonexistent in the United States because it is difficult to keep emissions of particulate Na_2SO_4 low and keep HCl from escaping to the atmosphere.

Sodium sulfate in moderation is used as a diuretic and cathartic for humans and animals (14) (see GASTROINTESTINAL AGENTS). It is also used in consumer products such as laxatives, antacids, and as a natural filler it is used extensively in powdered laundry detergents (see DETERGENCY).

Sodium bisulfate, $NaHSO_4$, is mildly acidic. Appropriate precautions should be taken when using it.

Uses

The principle uses of Na_2SO_4 are in the manufacture of paper, soaps, and detergents. These accounted for 65% of U.S. consumption from 1990 to 1995, representing a significant shift from 1980 when paper production alone consumed 67%. Pulp (qv) and paper consume only 25% (2). The kraft paper process uses a mixture of sodium sulfide and sodium hydroxide to digest wood chips. Both the sulfide and hydroxide are generated, starting with sodium sulfate as the raw material.

At low temperatures, Na_2SO_4 is nonreactive; because of this and given its relatively low cost, it is used as a filler in household soaps and detergents. Detergents average about 20% Na_2SO_4, but some grades have much higher content. Besides adding bulk to the detergent, particle size and whiteness of the sulfate improve appearance, handling characteristics, and assist the detergent in its ability to clean.

Properties of sodium sulfate help speed up the melting process in glassmaking. Its use reduces the tendency for alkaline gas bubbles to form in the glass and provides a less expensive form of Na_2O than soda ash (6). Sodium sulfate improves the working properties of high silica glasses.

Both Na_2SO_4 and $NaHSO_4$ are used to adjust pH and dilute dyes. Sodium sulfate is used in cattle feed (see FEEDS AND FEED ADDITIVES), in cellulose-sponge, as a cement and plaster hardener, and as an aid in metallurgy refining. Glauber's salt, $Na_2SO_4 \cdot 10H_2O$, has a high heat of crystallization. This fact together with its low cost have made it a candidate as a means of storing solar energy (qv). When heated, it melts, absorbing large quantities of heat. When cooled, it recrystallizes, releasing its heat (15,16). Because of its incongruent property, the energy-storing capability of the salt has not as of this writing (ca 1996) met with commercial application. Researchers are working to develop ways to use it as an effective energy-storing salt.

Sodium bisulfate, $NaHSO_4$, is a convenient mild acid and is safe for uses as a household toilet-bowl cleaner, automobile-radiator cleaner, and for swimming pool pH adjustment. It is used for metal pickling, as a dye-reducing agent, for soil disinfecting, and as a promoter in hardening certain types of cement.

BIBLIOGRAPHY

"Sodium Sulfates" under "Sodium Compounds" in *ECT* 1st ed., Vol. 12, pp. 607–609, by J. A. Brink, Jr., Purdue University; in *ECT* 2nd ed., Vol. 18, pp. 502–510, by J. J. Jacobs, Jacobs Engineering Co.; in *ECT* 3rd ed., pp. 245–256, by T. F. Canning, Kerr-McGee Chemical Corp.

1. D. S. Kostick, *Annual Report 1992*, U.S. Dept. of the Interior, Bureau of Mines, Washington, D.C., p. 7.
2. U.S. Bureau of Mines FaxBack, *Sodium Sulfate*, 1966, see *General Reference*.
3. *Mineral Commodity Summaries*, U.S. Bureau of Mines, Washington, D.C., Jan. 1995, p. 158.
4. *Soda Ash and Sodium Sulfate*, U.S. Department of the Interior, Bureau of Mines, Washington, D.C., Jan. 1995.
5. W. F. Linky and A. Seidell, *Solubilities of Inorganic and Metal Organic Compounds*, 4th ed., Vol. II, American Chemical Society, Washington, D.C., 1965, pp. 982–983.
6. *Mellor's Comprehensive Treatise on Inorganic and Theoretical Chemistry*, Vol. II, Suppl. II, John Wiley & Sons, Inc., New York, 1961.
7. W. I. Weisman and R. C. Anderson, *Min. Eng.* **5**, 711 (1953).
8. W. I. Weisman, *Chem. Eng. Prog.* **60**(11), 47 (1964).
9. Ref. 1, p. 2.
10. Ref. 1, p. 9.
11. *Ind. Minerals Glass Surv.* **77**(125), 81 (Feb. 1978).
12. *Chem. Week* **120**, 37 (Mar. 2, 1977).
13. *Specification for Anhydrous Sodium Sulfate*, Great Salt Lake Minerals Corp., Ogden, Utah, Jan. 1996.
14. M. Windholtz, ed., *The Merck Index*, 9th ed., Merck & Co., Inc., Rahway, N.J., 1976.
15. Brit. Pat. 900,970 (July 11, 1962), (to E. I. du Pont de Nemours & Co., Inc.).
16. U.S. Pat. 3,714,008 (Jan. 30, 1973), T. Massaaki and co-workers (to Japan Atomic Energy Research Institute).

General References

U.S. Bureau of Mines FaxBack, Document on Demand, on-line 24 h/d, 7 d/wk (703) 548 4999.
Chemical Engineer's Handbook, 6th ed., McGraw-Hill Book Co., Inc., New York, 1984, Sect. 3.

DAVID BUTTS
Great Salt Lake Minerals Corporation

SODIUM SULFIDES

The sodium sulfides have many diverse uses and are expected to experience modest demand growth into the twenty-first century. Sodium sulfide [*1313-82-2*],

Na$_2$S, mol wt 78.05; sodium hydrosulfide [16721-80-5] (sodium sulfhydrate, sodium bisulfide, sodium hydrogen sulfide), NaHS, mol wt 56.06; and sodium tetrasulfide [12034-39-8], Na$_2$S$_4$, mol wt 174.24, are somewhat interchangeable in many applications. These compounds are used in the pulp (qv) and paper (qv) industries, in mining and leather (qv) tanning applications, as chemical intermediates, and in dye production (see DYES AND DYE INTERMEDIATES; MINERALS RECOVERY AND PROCESSING). Environmental applications of these sulfides, including heavy metal precipitation from wastewater and the removal of nitrogen oxides from emissions, are of particular interest to many industrial chemical consumers.

Sodium Hydrosulfide

Properties. Pure sodium hydrosulfide is a white, crystalline solid, mp 350°C, sp gr 1.79. The commercial product, available in flake form at approximately 73% strength, is yellow in color and highly deliquescent. The color of the commercial product varies under the influence of minor changes in manufacturing and is not an indicator of purity beyond the range of a few parts per million. The melting point of the commercially available flake is approximately 52°C; the bulk density is approximately 0.64 g/cm^3. The average water of hydration may be expressed as NaHS·0.81H$_2$O. The flake is highly soluble in water, alcohol, or ether (1).

The heat of formation of NaHS is -237.6 kJ/mol (-56.79 kcal/mol); the heat of solution is 15.9 kJ/mol (10.7 kcal/mol) (2). In aqueous solution NaHS has an alkaline pH. Boiling points, densities, and freezing points for solutions of varying strengths are shown in Figure 1 (1). When exposed to air, sodium hydrosulfide undergoes autoxidation and gradually forms polysulfur, thiosulfate, and sulfate. It also absorbs carbon dioxide, forming sodium carbonate.

Manufacture. Sodium hydrosulfide was first produced in Germany and then in the United States in 1938. Production is closely related to the supply of hydrogen sulfide, which reacts with sodium hydroxide to form NaHS. Hydrogen sulfide can also be obtained by the reaction of hydrogen and sulfur, as a by-product of the carbon disulfide process, barium and strontium chemical manufacturing, or from desulfurizing petroleum fractions (see SULFUR REMOVAL AND RECOVERY). High purity NaHS production is very dependent on a supply of high purity H$_2$S and NaOH. The reaction produces high yields based on both reactants.

$$H_2S + NaOH \longrightarrow NaHS + H_2O \qquad (1)$$

NaHS, marketed as 71.5–74.5 wt % flakes and 43–60 wt % liquor in the high purity grades, is also available as 10–40 wt % liquor from the oil refining desulfurization process. NaHS is sold commercially in 22.7-kg bags, 181.4-kg drums, in tank trucks, and in rail cars.

Economic Aspects and Uses. Production and sales values for high purity sodium hydrosulfide are listed in Table 1. These figures exclude the low purity

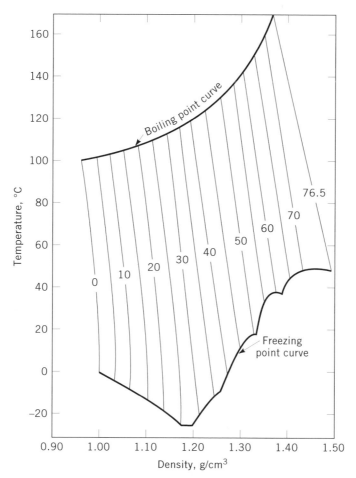

Fig. 1. Density of sodium hydrosulfide solutions in water, where the numbers on the vertical lines represent concentrations of NaSH in wt % (1).

material produced by oil refiners, believed to be sold primarily to pulp mills. Estimates of uses of NaHS in the United States for 1994 are as shown (3,4):

Use	Percentage
pulp processing (kraft)	40
ore flotation	28
chemicals and dyes	15
leather processing	12
miscellaneous, includes rayon and cellophane desulfurization	5

Use of NaHS in ore flotation (qv) has decreased over the years owing to the substitution of more environmentally sound methods, but this usage may fluctuate

Table 1. U.S. Production and Prices of Sodium Hydrosulfide

Year	Production, t $\times 10^3$	Price, \$/t
1965	39.9	118
1970	26.3	133
1975	21.8	236
1980	41.7	277
1985	56.2	454
1991	45.4	530

along with copper (qv) production. Use in dye production has remained constant, whereas the use of NaHS in pulp processing is increasing as the demand for paper products increases. Modest growth in the leather (qv) tanning sector has resulted from strengthening of demand in the automotive industry and increases in leather exports (3). The engineering plastic poly(phenylene sulfide) uses NaHS as a raw material and has the potential of becoming a significant growth area in this market (5) (see ENGINEERING PLASTICS; POLYMERS CONTAINING SULFUR).

In many applications sodium hydrosulfide and sodium sulfide are interchangeable. Where either chemical may be used, 28% less sodium hydrosulfide is required by weight to achieve a given level of sulfidity and is therefore the more economical choice. If desired, the sodium hydrosulfide can be converted to sodium sulfide by the addition of sodium hydroxide:

$$\text{NaHS} + \text{NaOH} \longrightarrow \text{Na}_2\text{S} + \text{H}_2\text{O} \qquad (2)$$

Sodium sulfides are very effective as heavy metal precipitants owing to the extremely low solubilities exhibited by metal sulfides. Sulfides, effective for precipitating most of the priority pollutant metals, which include Ag, As, Be, Cd, Cu, Cr, Hg, Ni, Pb, Se, Tl, and Zn, have been adopted by the EPA as a best available technology (BAT) application (6). Sulfide application offers several advantages over the conventional hydroxide processes. The lower solubility of sulfide metals provides improved metal removal efficiency (Fig. 2). Sulfide precipitation of metals remains effective over a wide pH range without sacrifice to minimum solubilities. It is effective in treating complexed metals, and the final disposal of the precipitated metal sulfides is easier and safer than the hydroxide process owing to lower acid (pH 5) leaching characteristics (9).

Sodium Sulfide

Properties. Pure sodium sulfide is a white, crystalline solid, mp 1180°C, sp gr 1.856. The commercial product is available in flake form at approximately 60% strength, is a light tan-to-yellow color, and is deliquescent. Figure 3 shows the boiling points, densities, and freezing points for various solution strengths (1). The heat of formation for the crystalline state is -373 kJ/mol (-89.1 kcal/mol), and the heat of solution is -63.5 kJ/mol (-15.2 kcal/mol) (10). In solution, Na_2S is strongly alkaline. The bulk density of the commercially available flake is approximately 0.64 g/cm^3. The flake exists as a concentrated mixture of several hydrated forms and contains 59–62 wt % sodium sulfide. The

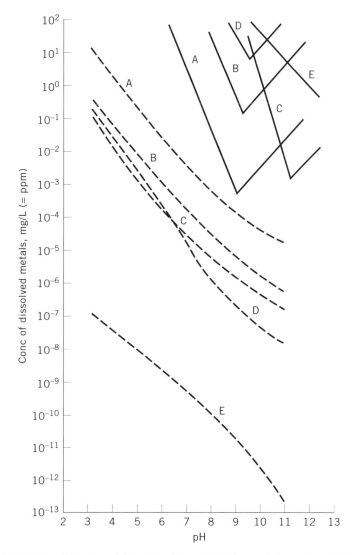

Fig. 2. Solubilities of (——) metal hydroxides, $M(OH)_2$, and (– – –) metal sulfides, MS, as a function of pH, where A–D correspond to M = Cu, Zn, Cd, and Pb, respectively (7). For E, the formulas are (——) AgOH and (– – –) Ag_2S. Curves for metal sulfides are based on experimental data listed in Reference 8.

average water of hydration may be expressed as $Na_2S \cdot 2.71H_2O$. Sodium sulfide crystallizes from aqueous solutions as the nonahydrate [1313-84-4], $Na_2S \cdot 9H_2O$. The flake is readily soluble in water, slightly soluble in alcohol, and insoluble in ether. When exposed to air, sodium sulfide undergoes autoxidation to form polysulfur, thiosulfate, and sulfate. It also absorbs carbon dioxide to form sodium carbonate. Reactions with strong oxidizing agents form elemental sulfur.

Manufacture.   The oldest method for producing Na_2S is by the reduction of sodium sulfate with carbon in a refractory oven at 900–1000°C. Whereas

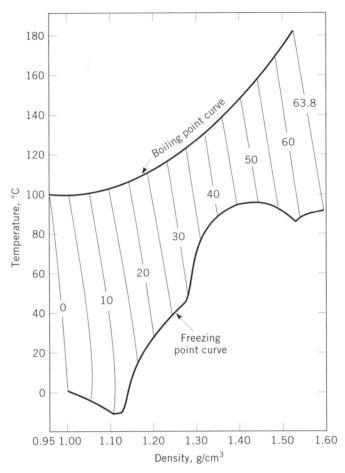

Fig. 3.   Density of sodium sulfide solutions in water, where the numbers on the vertical lines represent concentrations of Na_2S in wt % (1).

this method is no longer used commercially in the United States, a variation is used to produce sodium sulfide captively during kraft pulp processing to replace lost sodium and sulfur values that were initiated into the system by merchant-supplied sodium sulfide. In this method, sodium sulfate is added to the system in the recovery furnace, where it is reduced by carbon from the wood pulp to produce sodium sulfide.

$$Na_2SO_4 + 4\,C \longrightarrow Na_2S + 4\,CO \tag{3}$$

$$Na_2SO_4 + 4\,CO \longrightarrow Na_2S + 4\,CO_2 \tag{4}$$

One commercial process for producing sodium sulfide is as a by-product of barium carbonate production (see BARIUM COMPOUNDS). Barite ore, $BaSO_4$, is reduced with carbon at 800°C to produce crude barium sulfide (black ash), which is then leached to dissolve the barium sulfide in solution. The solution is then reduced using sodium carbonate to produce barium carbonate, leaving a weak sodium sulfide solution as the by-product. The sodium sulfide solution may then be concentrated and flaked or crystallized.

$$BaS + Na_2CO_3 \longrightarrow BaCO_3 + Na_2S \tag{5}$$

Another process involves two steps. Sodium hydrosulfide from equation 1 reacts with sodium hydroxide to yield sodium sulfide (eq. 2). Concentration by evaporation to 60 wt % is practiced unless concentrated sodium hydroxide is used.

Economic Aspects and Uses. Production and pricing information for Na_2S through 1991 are listed in Table 2. U.S. production of sodium sulfide increased rapidly from 1965 through 1972 and then began to decrease. The last year that the U.S. Bureau of the Census released official production figures was in 1974 because at that time there were only three producers of sodium sulfide. Estimates indicate that 1991 production fell to the levels of the late 1950s. List prices have increased since 1974 as sulfur and sodium hydroxide prices have increased.

About 65% of 1990 U.S. sodium sulfide usage was in the leather industry for dehairing hides before tanning. This application is similar to that of sodium hydrosulfide. The production of miscellaneous chemicals, which took up to 25% in 1990, included the production of polysulfide elastomers and plastics as well as a variety of organic chemicals. In the dye industry, Na_2S is used as a solvent for water-soluble dyes and as a reducing agent. In ore flotation, the mining industry uses Na_2S to form insoluble metal sulfides of copper, lead, molybdenum, nickel, and cobalt. Other uses of Na_2S include wood (qv) digestion, the preparation of lubrication oils, and in the preparation of rayon and cellophane by removing sulfur.

Technology has been developed for the absorption of nitrogen oxides from gas streams via sodium sulfide scrubber systems. The nitrogen oxide streams from various processes can be converted to elemental nitrogen, whereas the sulfide is oxidized to the sulfate ion (12).

Sodium Tetrasulfide. Sodium tetrasulfide is prepared by the reaction of sodium sulfide with sulfur. The 34 wt % solution is normally dark red, solidifies at -15 to $-9°C$, boils at $113°C$, and has a specific gravity at $15.5°C$ of 1.268. The chemical formula is written Na_2S_4, but the product is better regarded not as a compound but as a mixture of sodium sulfide with free, elemental sulfur, ie, $Na_2S \cdot S_3$. Sodium tetrasulfide is available in 249.5-kg drums of 34 wt % solution.

Table 2. U.S. Production and Prices of Sodium Sulfide

Year	Production, t $\times$ 10^3	Flake price, $/t
1965	39	113
1970	68	124
1972	80	127
1974	64	132
1991	25	476[a]

[a]Ref. 11.

It is used in leather processing, dye manufacturing, wastewater treatment, in metals finishing, ore manufacturing, and in lubricant manufacturing. No commercial polysulfide of significance is produced other than the tetrasulfide.

Analysis

A double end point, acid–base titration can be used to determine both sodium hydrosulfide and sodium sulfide content. Standardized hydrochloric acid is the titrant; thymolphthalein and bromophenol blue are the indicators. Other bases having ionization constants in the ranges of the indicators used interfere with the analysis. Sodium thiosulfate and sodium thiocarbonate interfere quantitatively with the accuracy of the results. Detailed procedures to analyze sodium sulfide, sodium hydrosulfide, and sodium tetrasulfide are available (1).

Health and Safety

The combination of organic matter and the sodium sulfides can cause combustion to occur. Zinc, aluminum, and copper should not be used where these can come into direct contact with any of the sodium sulfides. Personnel handling sodium sulfides must be equipped with goggles, a full face shield, and rubber or plastic protective clothing. Chemical cartridge escape respirators should be available at all storage and use locations, owing to the potential for hydrogen sulfide formation. The sodium sulfides are similar to sodium hydroxide and other alkalies as corrosive substances on animal tissues. Contact with skin can be very irritating and is especially harmful to soft tissues.

Inhalation of sodium hydrosulfide mist causes irritation of the respiratory tract and possible systemic poisoning. Hydrogen sulfide gas, which may be given off when acid is present, causes headache, dizziness, nausea, and vomiting. Continued exposure can lead to loss of consciousness, respiratory failure, and death. Extreme care should be taken to avoid mixing any of the sodium sulfides and acids. Storage areas and sewers should be segregated so as to avoid any possibility of these chemicals being mixed. It is recommended that a continuous-monitoring hydrogen sulfide gas detection and alarm system be installed in any area where hazardous levels of hydrogen sulfide gas may occur. Concentrations of 4.3 to 46% hydrogen sulfide by volume in air are explosive and self-ignite above 260°C. Burning hydrogen sulfide gas produces toxic sulfur dioxide. Firefighters must wear self-contained breathing apparatus to prevent breathing sulfur dioxide gas (13).

BIBLIOGRAPHY

"Sodium Sulfides" in *ECT* 1st ed., Vol. 12, pp. 609–611, by J. A. Brink, Jr., Purdue University; "Sodium Sulfides" under "Sodium Compounds," in *ECT* 2nd ed., Vol. 18, pp. 510–515, by J. S. Sconce, Hooker Chemical Corp.; in *ECT* 3rd ed., Vol. 21, pp. 256–262, by C. Drum, PPG Industries, Inc.

1. *Sulfur Chemicals*, PPG Industries, Inc., Pittsburgh, Pa., 1992.
2. D. D. Wagman, *Selected Values of Chemical Thermodynamic Properties*, Circular 500, National Bureau of Standards, Washington, D.C., 1952, pp. 456–460.

3. *Chem. Mark. Rep.* (Nov. 14, 1994).

4. *Miscellaneous Sulfur Chemicals—United States, Chemical Economics Handbook*, 780.4000M-X, Stanford Research Institute International, Menlo Park, Calif., May 1992.

5. U.S. Pat. 3,354,129 (Nov. 21, 1967), J. T. Edmonds and co-workers (to Phillips Petroleum).

6. *Development Document for Effluent Limitations Guidelines and Standards for the Inorganic Chemicals Manufacturing, Point Source Category* (*proposed*), EPA Publication 440/1-79/007, Washington, D.C., June 1990.

7. *Ind. Water Eng.*, 31 (Jan.–Feb. 1973).

8. A. Seidell, *Solubilities of Inorganic and Metal Organic Compounds*, 3rd ed., Vol. 1, D. Van Nostrand Co., Inc., New York, 1940.

9. *Sodium Sulfide and Sodium Hydrosulfide as Heavy Metal Precipitants*, Technical Service Bulletin, PPG Industries, New Martinsville, W. Va., 1995.

10. *J. Phys. Chem. Ref. Data* **11**(S2), (June 1990).

11. Ref. 4, 780.4001F-M.

12. *Sodium Sulfide: Wet Scrubber for Oxides of Nitrogen (NO$_x$) Absorption*, Technical Service Bulletin, PPG Industries, New Martinsville, W. Va., 1995.

13. *Chemtox Database*, on DIALOG File 337, Dialog Information Services, Palo Alto, Calif., last updated June 6, 1993.

DAVID R. BUSH
PPG Industries, Inc.

SODIUM HYDROXIDE.   See ALKALI AND CHLORINE PRODUCTS.

SODIUM SILICATE.   See SILICON COMPOUNDS.

SODIUM SULFITES.   See SULFUR COMPOUNDS.

SODIUM TRIPOLYPHOSPHATE.   See PHOSPHORIC ACIDS AND PHOSPHATES.

SOIL CHEMISTRY OF PESTICIDES

Pesticide residues in foods have been a matter of public interest since the publication of Rachel Carson's *Silent Spring* in 1962 (1). The detection of trace amounts of organic pesticides (qv) in surface and groundwater has been a significant environmental issue since the early 1980s. The simultaneous detection of the nematicide 1,2-dibromo-3-chloropropane (DBCP) in groundwater in California and the insecticide aldicarb in well-water on Long Island, New York, in 1979 triggered the controversy over the safety of the U.S. water supply. From a national perspective, particular concern was focused on the rural drinking water supplies for which groundwater is the principal source (see GROUNDWATER MONITORING). Public debate about the safety of agricultural chemicals in drinking water has also involved nitrates from fertilizers and other sources. The scope of the pesticide issue in water is so large and complex that this article can only address the more important classes of organic pest control chemicals. Soils play a significant role in modifying the amounts and kinds of pesticides ultimately detected in water. Intensive research on the dynamic interactions between pesticides, soils, and water has led to an increase in understanding of the physical, chemical, and biochemical processes that impinge on all three systems.

Pesticide Usage

There is a strong relationship between the amount of pesticide applied and the amount detected in soil and water. Some background information on pesticide usage and terminology is useful in understanding their impact on the environment.

Pesticide is a generic name for compounds used in pest control (see PESTICIDES). The three principal groups of pesticides, and the pests they control, are insecticides for insects, herbicides (qv) for weeds, and fungicides (qv) for plant diseases (see INSECT CONTROL TECHNOLOGY). There is also a smaller group of conventional pesticide chemicals, including rodenticides, nematicides, fumigants, molluscicides, and plant growth regulators. This latter group of pesticides is of relatively low volume use compared to the three principal groups, but some have been involved in important episodes of water contamination. There is also a group of nonconventional pesticides which include important industrial compounds that have pesticidal properties. This last group includes the wood preservatives, disinfectants (excluding chlorine), and sulfur.

Pesticides are further subdivided into classes of compounds. Historically, insecticides included the organochlorine, methyl carbamate, and organophosphate classes of pesticides. Herbicides comprise about 10–12 principal classes of compounds. Within each class of pesticide there may be several hundred active ingredients.

Agriculture is the largest user of pesticides on a weight basis (77%), but significant amounts are also used by the industrial, commercial, and government sectors (16%) and for home and garden use (6%) (2). The last two categories are significant because each consumed 93 and 35 million kg of pesticides, respectively, in 1995.

There has been a dramatic shift in the types of pesticides used in American agriculture since the 1950s. In the late 1950s and early 1960s, the organic insecticides dominated the market. One of the largest classes of insecticides in use at that time were the organochlorines. The environmental era that started with the publication of *Silent Spring* and the following regulatory legislation led to the ultimate demise of these hard pesticides. Most uses of the organochlorine insecticides, including aldrin [309-00-2] and dieldrin [60-57-1] (1974), BHC [58-89-9] (1976), chlordane [59-74-9] and heptachlor [76-44-8] (1980), DDT [50-29-3] (1982), lindane [58-89-9] (1984), strobane [8001-50-1] (1976), and toxaphene [8001-35-2] (1982), have been canceled in the United States (see CHLOROCARBONS AND CHLOROHYDROCARBONS, TOXIC AROMATICS). These persistent, nonpolar materials are extremely lipophilic and tend to accumulate in the fatty tissues of many wildlife species. As of 1996, the chlorinated hydrocarbons are used in certain countries. Despite being banned in the 1980s, the chlorinated hydrocarbon insecticides were still being detected in the 1990s, albeit at low levels, in air, sediment, and water samples. Since the mid-1970s, organic herbicides have been the leading class of pesticides used in the United States from both a sales and tonnage basis.

The total pesticide usage in the United States almost doubled between 1964 and 1977 and has been quite stable since that time, at about 500,000 t of active ingredient (3). Most of the increase in usage has been for agriculture, increasing from 145,000 t in 1964 to 439,000 t in 1995.

Approximately 21,000 formulated pesticide products are registered by the U.S. Environmental Protection Agency (EPA) for marketing and use in the United States (2). This large number of products occurs because the same active ingredient, for example, 2,4-dichlorophenoxyacetic acid (2,4-D), may be formulated, packaged, and sold under a number of different brand names. There are about 860 active ingredients registered under the Federal Insecticide, Fungicide, and Rodenticide Act (FIFRA), which was first enacted in 1947. FIFRA was amended in 1964 to add a cancellation process for those pesticides deemed to pose an unacceptable risk, in 1972 to establish the modern registration process by the newly established EPA, and in 1988 for the reregistration process.

Total sales of pesticides in 1995 were estimated at $10.4 billion. The distribution of sales among various classes of pesticides is shown in Table 1. The herbicides continue to dominate both the amount and total cash value of pesti-

Table 1. 1995 U.S. and World User Level Pesticide Sales[a]

Group	U.S. market Quantity	U.S. market %	World market Quantity	World market %	U.S. share of world market, %
User expenditures, 10^6 $					
herbicides	5,927	57	13,400	47	44
insecticides	3,091	30	8,350	29	37
fungicides	768	7	5,600	20	14
other	635	6	1,350	5	36
Total	*10,421*	*100*	*28,700*	*101*	*36*
Volume of active ingredient, 10^6 kg					
herbicides	301	53	1,002	47	30
insecticides	153	27	767	36	20
fungicides	74	13	256	12	30
other	40	7	107	5	36
Total	*568*	*100*	*2,132*	*100*	*27*

[a]Ref. 2.

cides sold in the United States. The leading pesticides used (by weight) in the United States are shown in Table 2.

One reason for the extensive use of herbicides in the 1990s was the significant change in farming practices. No-till or conservation tillage is being used on larger and larger acreages of U.S. croplands. Instead of plowing and harrowing fields prior to planting, seeds are drilled directly into the soil containing plant residues from the previous crop. Prior to drilling the seed, all weedy vegetation is killed using a contact herbicide such as paraquat, and full-season weed control is achieved with a soil-applied herbicide such as atrazine. No-till generally requires more herbicide usage than conventional tillage, but reduces soil erosion, permits greater water infiltration, and is more economical from a labor standpoint.

In addition to conventional pesticides such as insecticides, herbicides, and fungicides, there are other chemicals classified as pesticides and regulated under FIFRA. These chemicals include wood preservatives, disinfectants (excluding chlorine), and sulfur. In the United States these chemicals have annual usage of about 500,000 t, which is equal to conventional pesticides.

Table 2. Quantities of Pesticides Most Commonly Used in U.S. Agricultural Crop Production in 1995[a]

Pesticide	CAS Registry Number	Type[b]	Rank	Usage, 10^6 kg ai[c]
atrazine	[1912-24-9]	H	1	31–33
sulfur	[7704-34-9]	F	2	27–29.5
metolachlor	[51218-45-2]	H	3	27–29
methyl bromide	[74-83-9]	N	4	25.5–28
petroleum oil		I, H	5	23–25
metam sodium	[137-42-8]	SF	6	22–24.5
dichloropropene	[542-75-6]	N	7	17–19.5
2,4-D	[94-75-7]	H	8	14–16.4
glyphosate	[1071-83-6]	H	9	11.4–13.6
cyanazine	[21725-46-2]	H	10	11–13
pendimethalin	[40487-42-1]	H	11	10.5–12.7
trifluralin	[1582-09-8]	H	12	10.5–12.7
acetochlor	[34256-82-1]	H	13	10–12.3
alachlor	[15972-60-8]	H	14	8.6–11
EPTC	[759-94-4]	H	15	4.1–5.9
chlorpyrifos	[2921-88-2]	I	16	4.1–5.9
chlorothalonil	[1897-45-6]	F	17	3.6–5.5
copper hydroxide	[20427-59-2]	F	18	3.2–5
propanil	[709-98-8]	H	19	2.7–4.5
dicamba	[1918-00-9]	H	20	2.7–4.5
terbufos	[13071-79-9]	I	21	2.7–4.1

[a]Ref. 2.
[b]Pesticide type: H = herbicide; I = insecticide; SF = soil fumigate; F = fungicide; and N = nematicide.
[c]ai = active ingredient.

Pesticide Properties and Detection

One of the first problems encountered by scientists attempting to get a national perspective on the potential magnitude of the groundwater pollution problem was the large number of soil types and pesticides involved. It is estimated that there are about 10,000 soil types in the United States and about 860 active ingredients registered under FIFRA. The use of models to predict the potential movement of pesticides in soils under a variety of conditions began in earnest about 1980. An integral component of these models deals with chemical and physical properties of the pesticides.

An extensive pesticide properties database was compiled, which includes six physical properties, ie, solubility, half-life, soil sorption, vapor pressure, acid pK_a, and base pK_b, for about 240 compounds (4). Because not all of the properties have been measured for all pesticides, some values had to be estimated. By early 1995, the Agricultural Research Service (ARS) had developed a computerized pesticide property database containing 17 physical properties for 330 pesticide compounds. The primary user of this data has been the USDA's Natural Resources Conservation Service (formerly the Soil Conservation Service) for leaching models to advise farmers on any combination of soil and pesticide properties that could potentially lead to substantial groundwater contamination.

Limits of Detection. One reason for the concern about pesticides in groundwater has been the ability to detect trace amounts of these compounds by more sophisticated analytical methodology. Based on the past usage rates and levels of production, pesticides must have occurred in groundwater prior to the 1980s, when significant efforts were made to detect, quantify, and rectify the drinking water problem. Limits of residue detection have increased progressively from parts per million (ppm), parts per billion (ppb), to parts per trillion (ppt). For an excellent review on pesticide analysis, see Reference 5.

Monitoring Studies. The highly effective nematicide, 1,2-dibromo-3-chloropropane [76-12-8] (DBCP), has aided in the past growers of citrus, peaches, grapes, cotton, and numerous other fruit and vegetable crops with no apparent environmental or toxicological consequences. In 1977, however, DBCP was discovered to cause temporary sterility among male production plant workers and, at about the same time, the chemical was identified as a potential carcinogen. Use of DBCP in California was suspended in 1977. A monitoring study conducted in May of 1979 revealed that 59 of 119 wells tested in the San Joaquin Valley contained DBCP residues at levels of 0.1–39 ppb and averaged 5 ppb (6). DBCP had been used on these sandy soils from ca 1960 to 1977. Although residues were highest (0.3 ppb) in shallower wells, DBCP was reported in two wells at 180-m deep. DBCP use was subsequently suspended throughout the United States following these findings.

Residues of the insecticide/nematicide aldicarb were detected in a domestic well located close to irrigated potato fields in Suffolk County on Long Island, New York in August 1979 (7). This discovery was followed by extensive survey of other wells in the vicinity and regulatory actions that canceled the use of aldicarb on Long Island. A number of agronomic and geological conditions on Long Island led to the penetration of aldicarb into local groundwater aquifers. First, application rates of aldicarb [116-06-3] were high, 5.6–7.9 kg/ha (5–7 lb/acre), to ensure adequate control of two pests, the Colorado potato beetle and the Golden nematode. Second, potatoes were grown on irrigated sandy soils with high water tables on Long Island. Finally, the contaminated Long Island aquifer is largely a shallow confined aquifer and the pH and alkalinity of the water are low.

The DBCP and aldicarb episodes sparked intensive monitoring activity on a national level. Aldicarb field studies were conducted in 16 states over a period of six years involving approximately 20,000 soil and water samples. National surveys showed evidence that other pesticides were being detected in wellwater samples. A 1984 review of leaching and monitoring data found 12 different pesticides in groundwater in 18 states as a result of agricultural activities (8); two years later a similar survey found at least 17 different pesticides in 23 states (9). A chronology of selected monitoring studies in various states has been summarized in Table 3.

The various surveys reported between 1979 and 1988 gave some valuable clues about the magnitude and extent of groundwater contamination on a national basis; taken together, they presented a challenge to policy makers on developing regulations to reduce pesticide residues in groundwater. There were variations in sampling techniques, statistical design, and analytical methodology among studies. Problems also arose in defining the source of pesticides, ie, nonpoint (normal agricultural use) vs point sources (spills), and the integrity of the wells sampled.

The most comprehensive national survey on pesticide in public and private wells has been conducted by the U.S. Environmental Protection Agency beginning in 1985 (11). The purpose of the National Pesticide Survey was both to determine the frequency and concentration of pesticides in drinking water wells nationwide, and to improve understanding of the association of the patterns of pesticide use and the vulnerability of groundwater to contamination. Extensive planning went into the statistical design for the selection of sampling sites and analytical methods for this national survey. Samples were taken from 540 community water wells in all 50 states and from 752 rural domestic wells in 38 states. One hundred and twenty six pesticides and degradation/metabolic products were analyzed in this survey.

The most frequently detected analyte was tetrachloroterephthalate, a degradation product from the herbicide dacthal [1861-32-1] or dimethyl tetrachloroterephthalate [1861-32-1] (DCPA). This product was detected in 6.4% of the community wells and in 2.5% of the rural wells at concentrations well below the health advisory levels of 4000 mg/L. Health advisory levels (HAs) are defined as contaminate concentrations in drinking water that would have no adverse health effects over specified exposure periods. Dacthal has been used as a herbicide on

Table 3. Pesticides in Groundwater from Normal Agricultural Use[a]

Year	Number of pesticides found	Number of states where pesticides found
1979	aldicarb	New York
1979	DBCP	California
1984	12	18
1985	17	23
1985	56	California[b]
1988	67	33
1988	46	26

[a]Ref. 10.
[b]Only California was studied.

lawns, turf, and golf courses, but finds greatest use in fruit and vegetable production. The second most widely detected pesticide was the herbicide atrazine, used widely in corn and sorghum production. Atrazine was detected in 1.7 and 0.7% of the 1292 community and rural wells, respectively. Other pesticides detected included simazine [122-34-9], prometon [1610-18-0], hexachlorobenzene [118-74-1], DBCP, dinoseb [88-85-7], ethylene dibromide [106-93-4], lindane, bentazon [25057-89-0], ethylene thiourea [96-45-7] (a product of the ethylenebisdithiocarbamate (EBDC) fungicides), alachlor [15972-60-8], chlordane [12789-03-6], and 4-nitrophenol [100-02-7] (a degradation product of parathion).

A large database has been compiled from groundwater samples collected by industry (Ciba, Monsanto), EPA, and three Midwestern states (Minnesota, Iowa, and Wisconsin) (12). Atrazine was the product of significant interest in the database on account of its extensive use (see Table 2). The database includes wells in general areas, which were randomly picked, independent of herbicide use, and wells from sensitive areas of high atrazine use or where groundwater was particularly vulnerable to pesticide transport. Eight years of collective monitoring have shown relatively few atrazine detections above the maximum contaminant level (MCL) of 3.0 ppb, which is a Federal Safe Drinking Water Act calculation that sets the annual average level of a chemical allowed in water (Table 4).

Table 4. Groundwater Database, Atrazine[a]

Wells	General areas	Sensitive areas	Total wells
number	10,200	5,300	15,500
above maximum contaminant level	78	103	181
above maximum contaminant level, %	0.76	1.94	1.17

[a]Ref. 12.

Pesticide Metabolism and Chemical Degradation

Pesticides are susceptible to a variety of transformations in the environment, including both chemical degradation and microbial metabolism. Microbial transformations are catalyzed exclusively by enzymes, whereas chemical transformations are mediated by a variety of organic and inorganic compounds. Many pesticide transformations can occur either chemically or biologically. Consequently, most pesticide dissipation studies include sterile treatments to distinguish between chemical degradation vs microbial metabolism. Common sterilization treatments include autoclaving; fumigation, eg, with ethylene oxide; addition of microbial inhibitors, eg, azide, mercuric chloride, and antibiotics; and gamma irradiation.

Microbial Metabolism. Studies indicate that, for many pesticides, metabolism by microorganisms is the most important environmental fate. Pesticide-degrading microorganisms are found in soils, aquatic environments, and wastewater treatment plants, although the greatest number and variety of microorganisms are probably in agricultural soils. A wide variety of pesticide-degrading microorganisms have been identified, including over 100 genera of bacteria and fungi (13). This is indicative of the extraordinary metabolic diversity of microorganisms as well as the extreme variety in pesticide structural chemistry.

The rate and extent of pesticide metabolism can vary dramatically, depending on chemical structure, the number of specific pesticide-degrading microorganisms present and their affinity for the pesticide, and environmental parameters. The extent of metabolism can vary from relatively minor transformations which do not significantly alter the chemical or toxicological properties of the pesticide, to mineralization, ie, degradation to CO_2, H_2O, NH_4^+, Cl^-, etc. The rate of metabolism can vary from extremely slow (half-life of years) to rapid (half-life of days).

The majority of pesticides used, although generally susceptible to enzymatic transformations, are not utilized as growth substrates by microorganisms, ie, as sources of carbon, nitrogen, and/or energy; this phenomenon is termed cometabolism (14). Consequently, population densities of most pesticide degraders are stable, or fluctuate in response to variables other than pesticide applications. In some instances, however, microorganisms are able to utilize pesticides as growth substrates. In the case of foliar-applied pesticides this may be desirable; however, in the case of soil-applied pesticides this typically leads to enhanced or accelerated rates of biodegradation, resulting in losses of efficacy (15). It should be noted that only a portion of the pesticide molecule needs to be mineralized in order to observe enhanced rates of biodegradation. In addition, pesticides may also be utilized as growth substrates by consortia (two or more distinct strains) of microorganisms.

Transformations/Metabolic Pathways. The initial enzymatic transformation of most pesticides can be generically characterized as oxidative, reductive, or hydrolytic. In general, oxidative and hydrolytic reactions are typical of both fungi and bacteria, whereas reductive reactions are most typical of bacteria. Oxidative reactions occur only under aerobic conditions, ie, in the presence of oxygen; reductive reactions typically occur under anaerobic conditions, ie, in the absence of oxygen; hydrolytic reactions occur under both. The extent and/or pathway of pesticide metabolism can be highly variable, depending on the mix of pesticide-degrading microorganisms present at a particular site. Many, if not most, pesticides are susceptible to several kinds of transformations and some are susceptible to complete mineralization. Consequently, it is difficult to predict the fate of any given pesticide at any given site.

Oxidative Reactions. The majority of pesticides, or pesticide products, are susceptible to some form of attack by oxidative enzymes. For more persistent pesticides, oxidation is frequently the primary mode of metabolism, although there are important exceptions, eg, DDT. For less persistent pesticides, oxidation may play a relatively minor role, or be the first reaction in a metabolic pathway. Oxidation generally results in degradation of the parent molecule. However, attack by certain oxidative enzymes (phenol oxidases) can result in the condensation or polymerization of the parent molecules; this phenomenon is referred to as oxidative coupling (16). Examples of some important oxidative reactions are ether cleavage, alkyl-hydroxylation, aryl-hydroxylation, N-dealkylation, and sulfoxidation.

Ether Cleavage. This is commonly observed as the initial step in the metabolism of the phenoxy herbicides 2,4-D (**1**), (2,4,5-trichlorophenoxy)acetic acid (2,4,5-T), and mecoprop (17). A wide variety of bacteria have been isolated which are able to catalyze this reaction (eq. 1), including *Alcaligenes*, *Azotobacter*, *Pseudomonas*, *Acinetobacter*, *Xanthobacter*, *Flavobacterium*, and *Arthrobacter*.

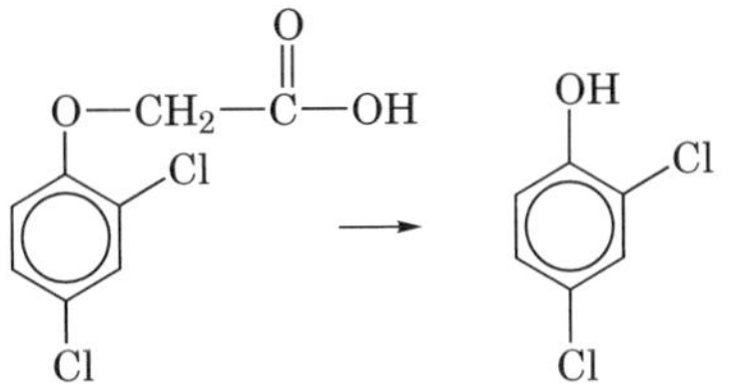

(1)

(**1**)

Alkyl-Hydroxylation. This is commonly observed as the initial transformation of alkyl-substituted aromatic pesticides such as alachlor [*15972-60-8*] and metolachlor [*51218-45-2*] (eq. 2) (**2**) (16). These reactions are typically catalyzed by relatively nonspecific oxidases found in fungi and actinomycetes.

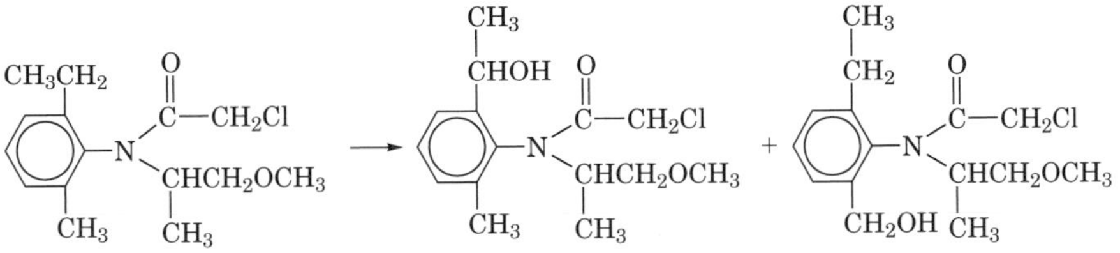

(2)

(**2**)

Aryl-Hydroxylation. This is occasionally observed as the initial transformation of aromatic pesticides. The vast majority of aromatic pesticide degradation products are susceptible to aryl-hydroxylation, representing either cometabolism or the initial step in mineralization (17). Numerous genera of bacteria and fungi possess the monooxygenases and dioxygenases responsible for hydroxylation of aromatic products. Examples of aromatic products susceptible to aryl-hydroxylation include 2,4-dichlorophenol [*120-83-2*] (from 2,4-D) (eq. 3), 4-nitrophenol (from parathion) (eq. 4), 3,4-dichloroaniline [*95-76-1*] (from propanil), and 3,6-dichlorosalicylic acid [*3401-80-7*] (from dicamba).

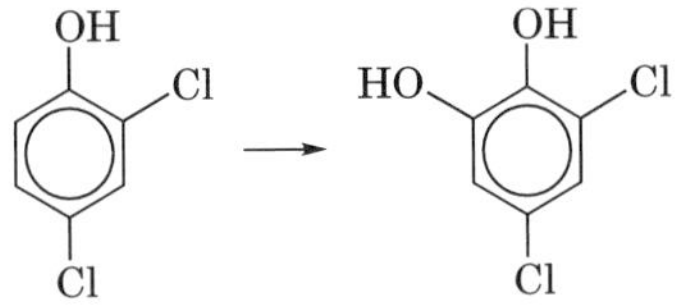

(3)

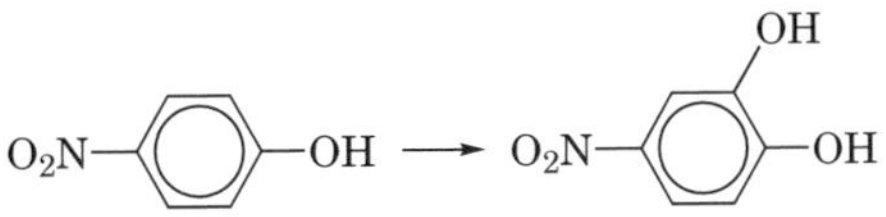

(4)

N-Dealkylation. This is commonly observed as a primary transformation of pesticides with *N*-alkyl substituents, such as atrazine [*1912-24-9*] (**3**) (eq. 5), trifluralin [*1582-09-8*] (**4**) (eq. 6) (16), and *S*-ethyl dipropylthiocarbamate [*759-94-4*] (EPTC) (**5**) (eq. 7) (18). These reactions are catalyzed by a variety of bacterial strains, including *Nocardia*, *Pseudomonas*, *Rhodococcus*, and *Streptomyces*.

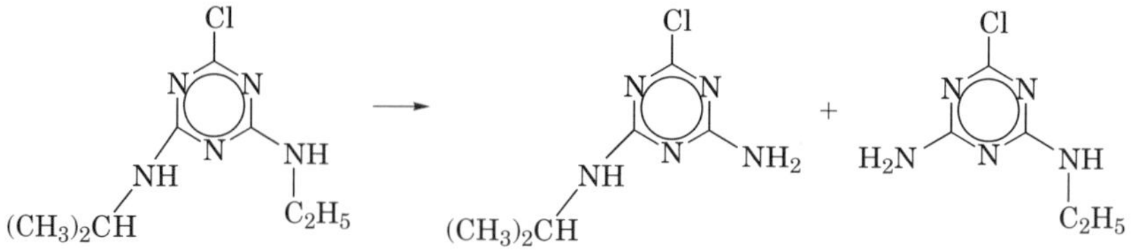

(5)

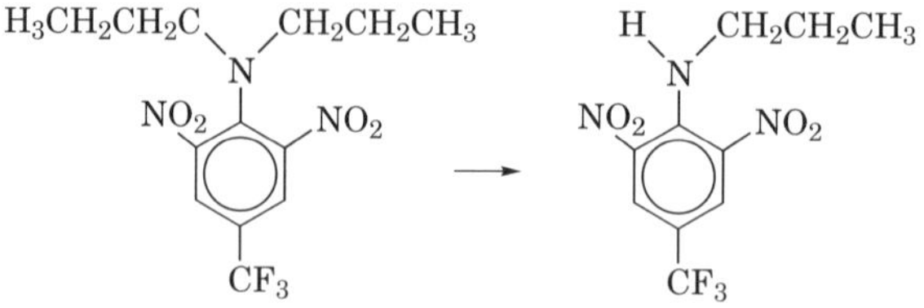

(3)

(6)

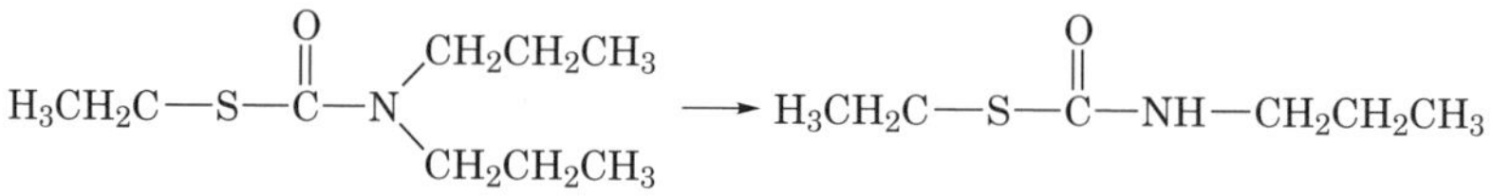

(4)

(7)

(5)

Sulfoxidation. This is a fairly common transformation of sulfur-containing pesticides such as aldicarb (**6**) (eq. 8) and EPTC (19).

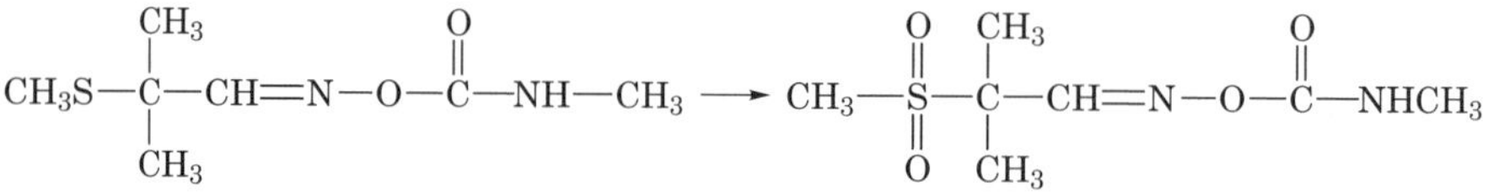

(8)

(6)

Reductive Reactions. A number of pesticides are susceptible to reductive reactions under anaerobic conditions, depending on the substituents present on the molecule. Reductive reactions can be either chemically or enzymatically mediated. Because biologically generated reductants, eg, cysteine and porphyrins, are frequently the electron donors for both chemical and enzymatic reactions, results from sterile controls are not necessarily conclusive in distinguishing between the two mechanisms. The only definitive means of distinguishing between chemical vs biological (enzymatic) reactions is to determine whether the reaction rate is consistent with enzyme kinetics. The most common reductive reactions are the reduction of nitro substituents and reductive dechlorination.

Reduction of Nitro Substituents. These reactions are very common in anaerobic environments and result in amine-substituted pesticides; anaerobic bacteria capable of reducing nitrate to ammonia appear to be primarily responsible.

All nitro-substituted pesticides appear to be susceptible to this transformation, eg, methyl parathion (**7**) (eq. 9), trifluralin, and pendimethalin.

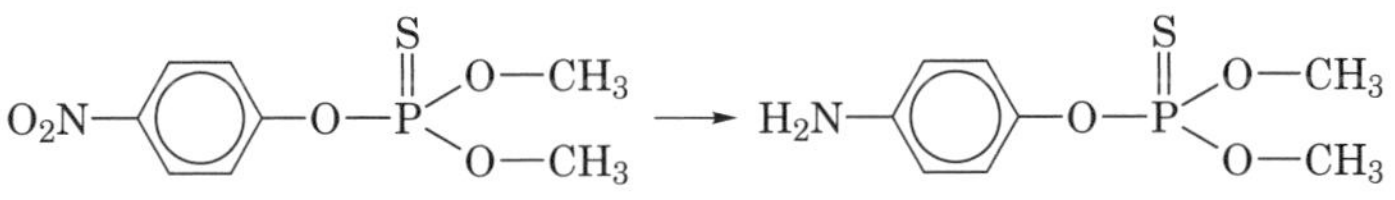

(9)

(**7**)

Reductive Dechlorination. Such reduction of chlorinated aliphatic hydrocarbons, eg, lindane, has been known since the 1960s. More recently, the dechlorination of aromatic pesticides, eg, 2,4,5-T, or pesticide products, eg, chlorophenols, has also been documented (eq. 10) (20). These reactions are of particular interest because chlorinated compounds are generally persistent under aerobic conditions.

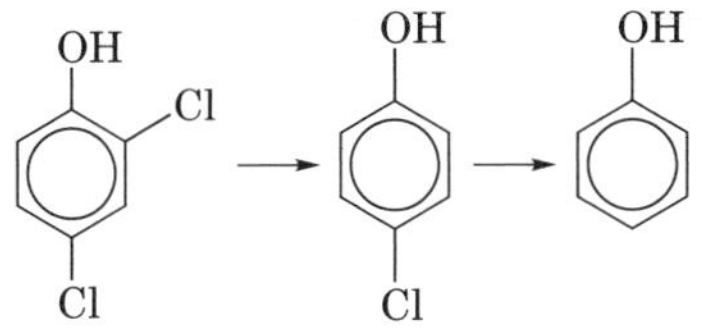

(10)

Hydrolytic Reactions.   Many pesticides possess bonds that are susceptible to hydrolytic attack. These reactions are most easily characterized according to the type of bond hydrolyzed: carboxylic acid ester, carbamate, organophosphate, urea, or chlorine (hydrodechlorination). In many instances the specific hydrolytic enzymes have been purified and characterized and the genes encoding for the enzymes isolated and cloned. It is commonly observed that there are multiple forms of the enzymes catalyzing a particular hydrolytic reaction, which suggests that these catalytic functions have evolved independently in different bacteria (19).

Carboxylic acid ester hydrolysis is frequently observed as the initial reaction for pesticides with ester bonds, such as 2,4-D esters, pyrethroids, and DCPA (dacthal) (**8**) (eq. 11) (16).

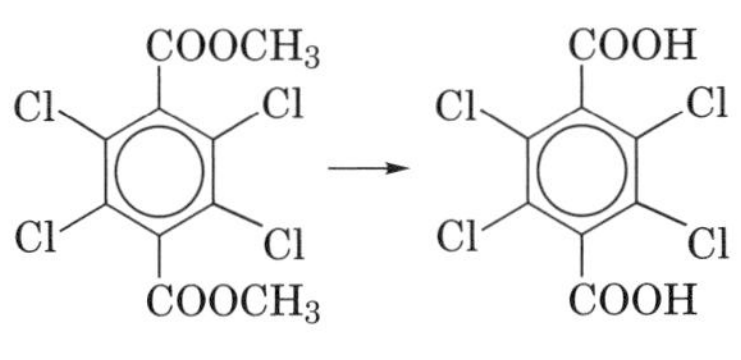

(11)

(**8**)

Carbamate hydrolysis is frequently observed as the initial reaction for pesticides having carbamate bonds, such as aldicarb, carbofuran, carbaryl, and benomyl (eq. 12) (19). Numerous genera of carbamate-hydrolyzing bacteria have been identified, including *Pseudomonas, Arthrobacter, Bacillus, Nocardia, Achromobacter, Flavobacterium, Streptomyces, Alcaligenes, Azospirillum, Micrococcus,* and *Rhodococcus.*

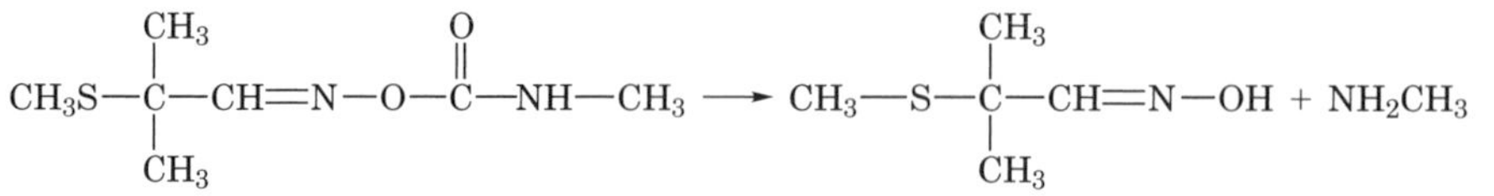

$$
\mathrm{CH_3S{-}\underset{\underset{CH_3}{|}}{\overset{\overset{CH_3}{|}}{C}}{-}CH{=}N{-}O{-}\overset{\overset{O}{\|}}{C}{-}NH{-}CH_3} \longrightarrow \mathrm{CH_3{-}S{-}\underset{\underset{CH_3}{|}}{\overset{\overset{CH_3}{|}}{C}}{-}CH{=}N{-}OH} + \mathrm{NH_2CH_3} \tag{12}
$$

Organophosphate hydrolysis is frequently observed as the initial reaction for pesticides having organophosphate bonds, such as methyl parathion, chlorpyrifos (**9**) (eq. 13), diazinon, and coumaphos (19). Several genera of organophosphate-hydrolyzing bacteria have been identified, including *Pseudomonas, Arthrobacter, Bacillus,* and *Flavobacterium.*

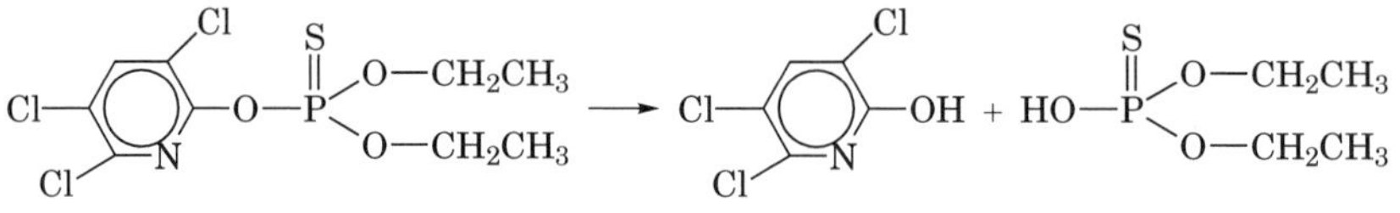

$$\tag{13}$$

(**9**)

Urea hydrolysis is frequently observed as the initial reaction for pesticides having urea bonds, such as linuron, diuron, and chlorsulfuron (**10**) (eq. 14) (16).

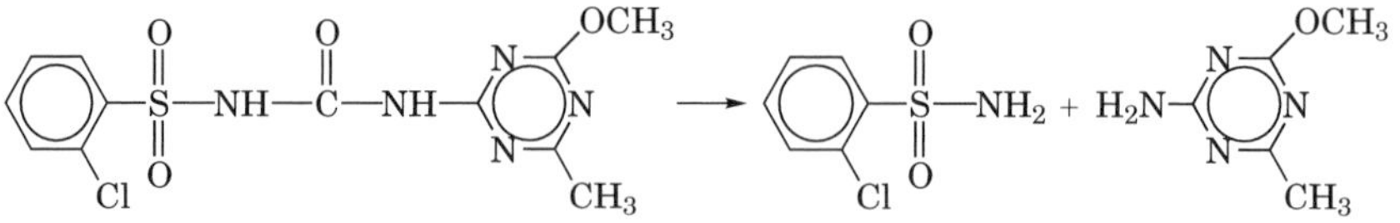

$$\tag{14}$$

(**10**)

Hydrodechlorination has long been recognized as an important chemical transformation. However, the enzymatic hydrodechlorination of atrazine (**3**) by soil microorganisms has also been demonstrated (eq. 15) (21).

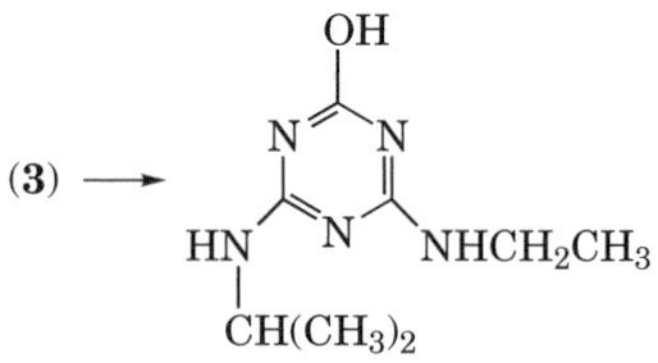

$$(\mathbf{3}) \longrightarrow \tag{15}$$

Metabolic Pathways.   Some pesticides are susceptible to complete degradation, ie, mineralization. This typically requires a sequence of enzymatic transformations, ie, metabolic pathway in which the product(s) are utilized as growth substrates by microorganisms or consortia of microorganisms. The mineralization of pesticides by the white rot fungi, eg, *Phanerochaete chrysosporium*, is apparently an exception to this scenario; these fungi mineralize pesticides via extracellular peroxidases without necessarily utilizing the products as growth substrates. Most pesticides are susceptible to mineralization only under aerobic conditions, although a few, eg, dinoseb, can also be mineralized under anaerobic conditions. One of the first pesticides demonstrated to be mineralized by soil microorganisms was 2,4-D (22). The metabolic pathway of 2,4-D biodegradation has been elucidated and shown to consist of the steps shown in Figure 1 (23).

Other representative pesticides that have also been shown to be mineralized include glyphosate, parathion, carbaryl, EPTC, isofenphos, and propachlor. Pesticides that are susceptible to mineralization are not typically found in, or considered to be a threat to, groundwater supplies because of their rapid degra-

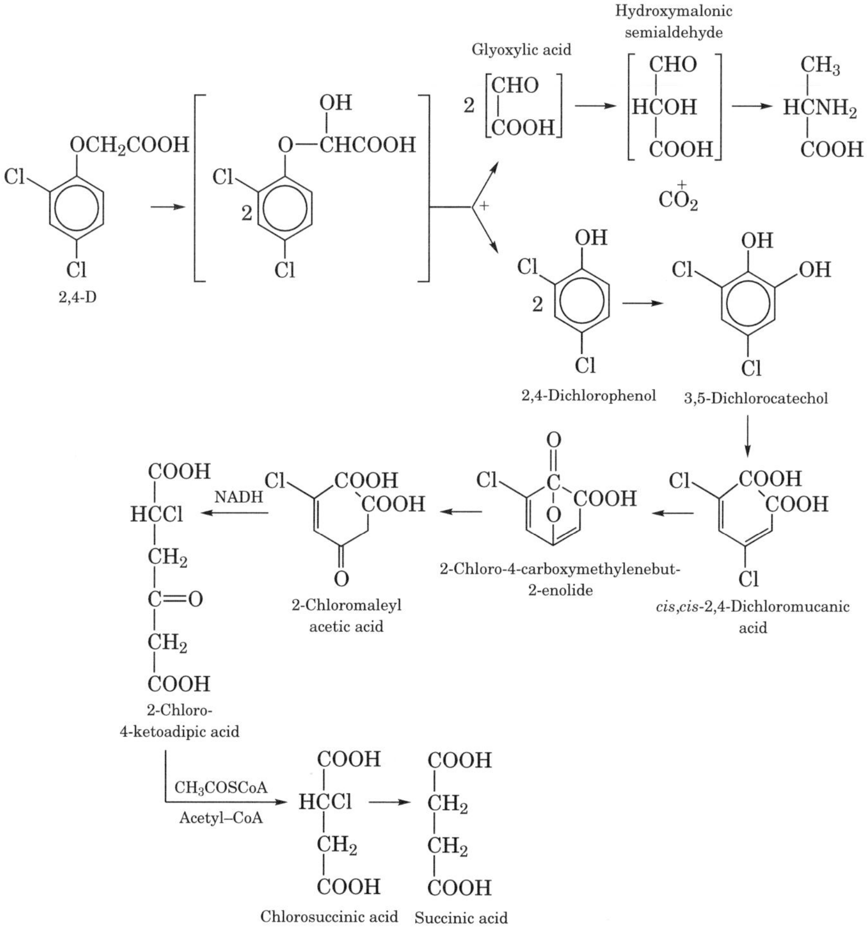

Fig. 1.   Metabolic pathway of 2,4-D biodegradation (23).

dation, ie, nonpersistence. Microorganisms can evolve, that is, develop metabolic pathways for the mineralization of previously persistent compounds. For example, there have been several reports documenting the existence of atrazine-mineralizing microorganisms (21).

Kinetics of Pesticide Biodegradation. Rates of pesticide biodegradation are important because they dictate the potential for carryover between growing seasons, contamination of surface and groundwaters, bioaccumulation in macrobiota, and losses of efficacy. Pesticides are typically considered to be biodegraded via first-order kinetics, where the rate is proportional to the concentration. Figure 2 shows a typical first-order dissipation curve.

For those pesticides that are cometabolized, ie, not utilized as a growth substrate, the assumption of first-order kinetics is appropriate. The more accurate kinetic expression is actually pseudo-first-order kinetics, where the rate is dependent on both the pesticide concentration and the numbers of pesticide-degrading microorganisms. However, because of the difficulties in enumerating pesticide-transforming microorganisms, first-order rate constants, or half-lives, are typically reported. Based on kinetic constants, it is possible to rank the relative persistence of pesticides. Pesticides with half-lives of <10 days are considered to be relatively nonpersistent; pesticides with half-lives of >100 days are considered to be relatively persistent.

For those pesticides which are utilized as microbial growth substrates, sigmoidal rates of biodegradation are frequently observed (see Fig. 2). Sigmoidal data are more difficult to summarize than exponential (first-order) data because of their inherent nonlinearity. Sigmoidal rates of pesticide metabolism can be described using microbial growth kinetics (Monod); however, four kinetics constants are required. Consequently, it is more difficult to predict the persistence of these pesticides in the environment.

Variability (spatial and temporal) in the rate of biodegradation of specific pesticides is frequently observed. Rates of biodegradation tend to be site-specific because of the differences in the numbers of specific pesticide degraders, pesticide bioavailability, and soil parameters such as temperature, moisture, and pH. Rates of metabolism are directly proportional to the population densities of pes-

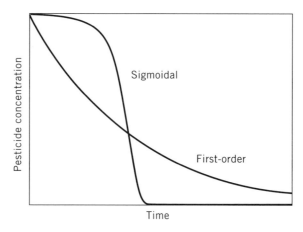

Fig. 2. Kinetics of pesticide biodegradation.

ticide degraders as well as the bioavailable, ie, soluble, concentrations. Studies indicate that pesticides sorbed to soil/sediment particles are not readily available for biodegradation; they must desorb into the solution phase before being metabolized. Within certain boundary conditions, there is a positive correlation between rates of metabolism and soil temperature, moisture, and pH, although there are exceptions, eg, oxidative reactions are less likely to occur in water-logged soils as a result of the slow rates of oxygen diffusion, whereas transformations catalyzed by fungi are more likely to occur at lower pH and/or soil moistures.

Chemical Degradation

Chemical, or abiotic, transformations are an important fate of many pesticides. Such transformations are ubiquitous, occurring in either aqueous solution or sorbed to surfaces. Rates can vary dramatically depending on the reaction mechanism, chemical structure, and relative concentrations of such catalysts as protons, hydroxyl ions, transition metals, and clay particles. Chemical transformations can be generically classified as hydrolytic, photolytic, or redox reactions (transfer of electrons).

Hydrolytic and Substitution Reactions. A variety of functional groups common to many pesticides are susceptible to hydrolysis. Hydrolysis reactions are catalyzed by acids (low pH), bases (high pH), and/or transition metals (Cu^{2+}, Fe^{3+}, Mn^{2+}). Consequently, environmental parameters such as pH, mineral composition and concentration, and clay content can have dramatic effects on rates of hydrolysis. In addition, the reaction mechanism in conjunction with chemical structure is of critical importance in dictating the rate of reaction. For instance, in the case of aromatic pesticides, if the reaction mechanism involves attack by a nucleophile (OH^-), then the presence of electron-withdrawing substituents such as NO_2^- and Cl^- causes the bond to be more electron-poor (more positive), resulting in faster rates of hydrolysis, whereas the presence of electron-donating substituents such as NH_2^- and CH_3^- causes the bond to be more electron-rich (more negative), resulting in slower rates of hydrolysis. If the reaction mechanism involves attack by an electrophile ($OH\cdot$), then electron-withdrawing substituents cause the rate of hydrolysis to be slower, whereas electron-donating substituents cause the rate of hydrolysis to be faster. Pesticides possessing bonds that are susceptible to chemical hydrolysis include carboxylic acid esters, carbamates, organophosphates, and ureas. Chlorinated compounds are susceptible to substitution (hydrodechlorination).

Carboxylic acid ester, carbamate, organophosphate, and urea hydrolysis are important acid/base-catalyzed reactions. Typically, pesticides that are susceptible to chemical hydrolysis are also susceptible to biological hydrolysis; the products of chemical vs biological hydrolysis are generally identical (see eqs. 8, 11, 13, and 14). Consequently, the two types of reactions can only be distinguished based on sterile controls or kinetic studies. As a general rule, carboxylic acid esters, carbamates, and organophosphates are more susceptible to alkaline hydrolysis (24), whereas sulfonylureas are more susceptible to acid hydrolysis (25).

Hydrodechlorination is a common reaction of chlorinated pesticides such as atrazine (eq. 15), alachlor, and metolachlor (**2**) (eq. 16). These reactions are catalyzed primarily by transition metals or by soil surfaces (clays or humic substances).

$$(2) \longrightarrow \qquad (16)$$

The kinetics of hydrolysis reactions may be first-order or second-order, depending on the reaction mechanism. However, second-order reactions may appear to be first-order, ie, pseudo-first-order, if one of the reactants is not consumed in the reaction, eg, OH^-, or if the concentration of active catalyst, eg, reduced transition metal, is a small fraction of the total catalyst concentration.

Photolytic Reactions. Much of the early research on photolysis of pesticides was conducted in organic solvents at high concentrations using powerful light sources. Both high and low pressure mercury vapor arcs, which emit uv light in sharp spectral lines, were frequently employed in these studies. These earlier studies yielded useful data on the mechanisms and products of pesticide photodegradation. More recently, there has been considerable interest in photolysis in natural systems; an excellent review of this research has appeared (26). Extensive pesticide photodegradation in soil is problematic for many compounds because light penetration into soils is extremely limited, often to depths of only 0.5 mm or less. The most likely candidate pesticides for soil photolysis are those that are water-soluble, weakly sorbed to soil surfaces, and have low vapor pressure; such compounds are most likely to rise with capillary water to the soil–atmospheric interface where photodegradation can occur. Napropamide and imazaquin are two pesticides that have been demonstrated to exhibit this behavior (27).

Studies have appeared where photolysis in natural bodies of water under normal sunlight conditions has been examined. For example, metolachlor was slowly photodegraded by sunlight in lake water, with a half-life of 22 days in summer and 205 days in winter (28). Addition of a 5% solution of dissolved organic matter to the water extended the half-lives two to three times longer, depending on the season (see PHOTOCHEMICAL TECHNOLOGY, PHOTOCATALYSIS).

Redox Reactions. Oxidative reactions typically occur as a consequence of the light-mediated production of singlet oxygen or hydroxyl radical, which are both potent oxidants. This process, termed indirect photolysis, involves the initial absorption of light energy by organic molecules, eg, humic substances, which either is directly transferred to oxygen (sensitization) or results in a chain reaction leading to the formation of oxidants. In contrast, soil organic matter has also been shown to quench photolysis of certain sorbed molecules. Chemical oxidative reactions in soil are generally of less environmental importance than biological oxidative reactions because observed reaction rates are slower on account of competition for oxidants by organic matter. Although these may appear to be pseudo-first-order, the kinetics of redox reactions are typically second-order because either an oxidizing or a reducing species is required.

S-oxidation of sulfur-containing pesticides such as aldicarb, parathion, and malathion can be of importance in the absence of microbial activity (29). The products of chemical vs biological oxidation are generally identical (eq. 8).

Reductive reactions typically occur in anaerobic environments where there is an abundant supply of electron donors. Electron donors are typically of microbial origin, eg, porphyrins or cysteine, which sometimes leads to confusion regarding the nature, ie, chemical vs enzymatic, of the reductive reaction. By definition, all reductive reactions which are not enzymatically catalyzed are chemical. The most significant chemical reductive reaction is reductive dechlorination.

Reductive dechlorination of chlorinated aliphatic hydrocarbons, eg, lindane (**11**) (eq. 17) is extremely facile and occurs almost exclusively via chemical mechanisms, although microorganisms are typically the source of electron donors (30).

$$\text{(11)} \qquad\qquad \tag{17}$$

(11)

The reductive dechlorination of chlorinated aromatics is more complicated in that the initial dechlorination of more highly chlorinated compounds may be either chemical or enzymatic, eg, PCP, whereas the dechlorination of less chlorinated compounds or dechlorinated products is typically enzymatic. For example, the first dechlorination of 2,4-dichlorophenol (ortho position) can occur either chemically or enzymatically; the second dechlorination (para position) is enzymatic (eq. 10).

Physical Processes Affecting Pesticides in Soil and Water

Persistence of pesticides in the environment is controlled by retention, degradation, and transport processes and their interaction. Retention refers to the ability of the soil to bind a pesticide, preventing its movement either within or outside of the soil matrix. Retention primarily refers to the sorption process, but also includes absorption into the soil matrix and soil organisms, both plants and microorganisms. In contrast to degradation that decreases the absolute amount of the pesticide in the environment, sorption processes do not affect the total amount of pesticide present in the soil but can decrease the amount available for transformation or transport.

Transport processes describe movement of the pesticide from one location to another or from one phase to another. Transport processes include both downward leaching, surface runoff, volatilization from the soil to the atmosphere, as well as upward movement by capillary water to the soil surface. Transport processes do not affect the total amount of pesticide in the environment; however, they can move the pesticide to sites that have different potentials for degradation. Transport processes also redistribute the pesticide in the environment, possibly contaminating sites away from the site of application such as surface and groundwater and the atmosphere. Transport of pesticides is a function of both retention and transport processes.

Many factors affect the mechanisms and kinetics of sorption and transport processes. For instance, differences in the chemical structure and properties, ie, ionizability, solubility in water, vapor pressure, and polarity, between pesticides affect their behavior in the environment through effects on sorption and transport processes. Differences in soil properties, ie, pH and percentage of organic carbon and clay contents, and soil conditions, ie, moisture content and landscape position; climatic conditions, ie, temperature, precipitation, and radiation; and cultural practices, ie, crop and tillage, can all modify the behavior of the pesticide in soils. Persistence of a pesticide in soil is a consequence of a complex interaction of processes. Because the persistence of a pesticide can govern its availability and efficacy for pest control, as well as its potential for adverse environmental impacts, knowledge of the basic processes is necessary if the benefits of the pesticide are to be maximized.

Sorption and Desorption Processes. Sorption is a generalized term that refers to surface-induced removal of the pesticide from solution; it is the attraction and accumulation of pesticide at the soil–water or soil–air interface, resulting in molecular layers on the surface of soil particles. Experimentally, sorption is characterized by the loss of pesticide from the soil solution, making it almost impossible to distinguish between sorption in which molecular layers form on soil particle surfaces, precipitation in which either a separate solid phase forms on solid surfaces, covalent bonding with the soil particle surface, or absorption into soil particles or organisms. Sorption is generally considered a reversible equilibrium process.

Desorption is the reverse of the sorption process. If the pesticide is removed from solution that is in equilibrium with the sorbed pesticide, pesticide desorbs from the soil surface to reestablish the initial equilibrium. Desorption replenishes pesticide in the soil solution as it dissipates by degradation or transport processes. Sorption/desorption therefore is the process that controls the overall fate of a pesticide in the environment. It accomplishes this by controlling the amount of pesticide in solution at any one time that is available for plant uptake, degradation or decomposition, volatilization, and leaching. A number of reviews are available that describe in detail the sorption process (31–33); desorption, however, has been much less studied.

Pesticides are sorbed on both inorganic and organic soil constituents. The sorptive reactivity of soil organic and inorganic surfaces to pesticides is dependent on the number and type of functional groups at accessible surfaces. When a pesticide reacts with the surface functional groups, either an inner- or an outer-sphere surface complex is formed. Although functional groups account for much of the reactivity of soil to pesticide retention, accessibility of the functional groups to the pesticide is also an important factor. For instance, steric hinderance caused by a large neighboring substituent or chemical may preclude the pesticide from interacting with the functional group. The intimate association among different soil minerals and between soil minerals and organic matter (Fig. 3) makes many functional groups inaccessible to pesticide molecules, although some functional groups are accessible to molecules that move through tiny soil pores, clay interlayers, or polymeric soil organic matrix.

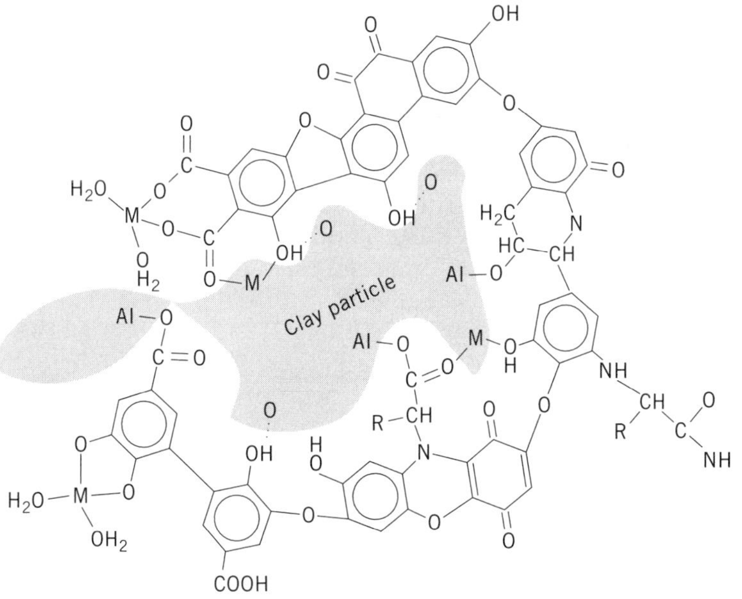

Fig. 3. Association of clay particles and the functional groups of organic matter (32).

Inorganic solids are composed of crystalline and noncrystalline amorphous minerals. The key features of clay minerals in relation to clays as sorbents for pesticides have been described (33). The principal functional groups on inorganic surfaces contributing to the sorptive capacity are siloxane ditrigonal cavities in phyllosilicate clays and inorganic hydroxyl groups generally associated with metal (hydrous) oxides.

Organic components of the solid phase include polymeric organic solids, decomposing plant residues, and soil organisms. The exact structure of humic materials in soil is largely unknown, but it is suggested that humic materials may contain a variety of functional groups, including carboxyl, carbonyl, phenylhydroxyl, amino, imidazole, sulfhydryl, and sulfonic groups. The variety of functional groups in soil organic matter and the steric interactions between functional groups lead to a continuous range of reactivities in soil organic matter.

The relative importance of organic vs inorganic constituents on pesticide sorption depends on the amount, distribution, and properties of these constituents, and the chemical properties of the pesticide. Soil organic matter is the principal sorbent for many organic compounds (31,32) such as the unionized weak acid pesticides 2,4-D, chlorsulfuron, and picloram; the nonionizable pesticides linuron and trifluralin; and the unionized weak base pesticide metribuzin. It has been suggested that the retention mechanism of nonionic organic chemicals in soil is a partitioning of the chemical between the aqueous phase and the hydrophobic organic matter (34). However, the mechanism may not be that simple (35). For example, some clays have hydrophobic sites and many nonionic organic chemicals sorb extensively on the clay mineral fraction of soil (36).

Of the various inorganic soil constituents, smectites (montmorillonite clays) have the greatest potential for sorption of pesticides on account of their large surface area and abundance in soils. Weak base pesticides, both protonated and neutral species, have been shown to be sorbed as interlayer complexes. Sorption of atrazine on smectites ranges from 0 to 100% of added atrazine, depending on the surface charge density of the smectite (36).

The intramolecular forces that can attract molecules to the interface and retain them on the surface have been classified according to the mechanism involved (31–33,37). Organic compounds can be sorbed with varying degrees of strengths of interactions by physical/chemical bonding such as van der Waals forces, hydrogen bonding, dipole–dipole interactions, ion exchange, and covalent bonding (Fig. 4).

For any given compound, there is likely a continuum of mechanisms with differing energy relationships that is responsible for sorption onto soil. For example, an organic molecule may be sorbed initially by sites that provide the strongest mechanism, followed by progressively weaker sites as the stronger sorption sites become filled.

London and van der Waals forces are short-range interactions resulting from a correlation in electron movement between two molecules to produce a small net electrostatic attraction. These interactions are particularly important for neutral high molecular weight compounds. Hydrogen bonds are dipole–dipole interactions involving an electrostatic attraction between an electropositive hydrogen nucleus on functional groups such as –OH and –NH and exposed electron pairs on electronegative atoms such as –O and –N. Hydrogen bonding is probably most prevalent in the bonding of pesticides to organic surfaces in the soil. For instance, hydrogen bonding has been proposed to be a significant soil binding mechanism for chlorsulfuron, fluazifop, and triazines such as atrazine.

Cation and water bridging involve complex formation between an exchangeable cation and an anionic or polar functional group on the pesticide. Cation and water bridging have been proposed as sorption mechanisms for fluazifop-butyl, picloram, glyphosate, and chlorthiamid. Protonation of a pesticide, or formation of charge-transfer complexes, at a mineral surface occurs when an organic functional group forms a complex with a surface proton. This retention mechanism is

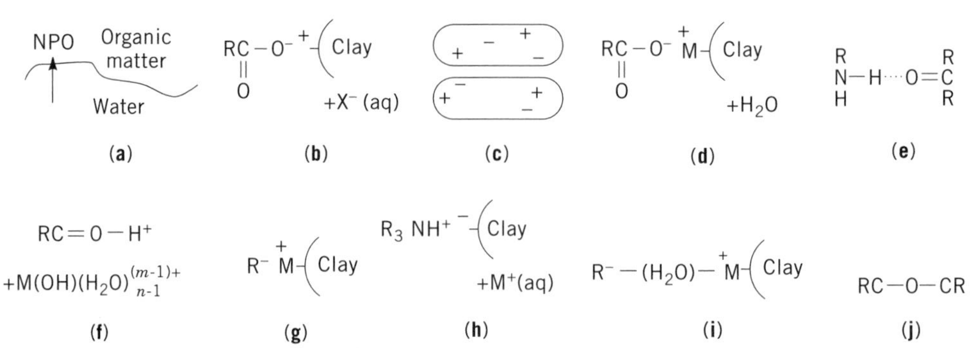

Fig. 4. Sorption mechanisms for pesticides on soil, where R = H or side-chain, M = exchangeable cation, X = exchangeable inorganic anion, and NPO = nonpolar organic compound: (**a**) hydrophobic bonding; (**b**) anion exchange; (**c**) London–van der Waals; (**d**) ligand exchange; (**e**) hydrogen bonding; (**f**) protonation; (**g**) cation bridging; (**h**) cation exchange; (**i**) water bridging; and (**j**) covalent bonding (37).

particularly important for basic functional groups at acidic mineral surfaces at low pH and low water content, particularly in the presence of aluminum or other metal cations. Protonation may be a mechanism for sorption of some s-triazines, chlorthiamid, fluazifop and fluazifop-butyl, and chlorsulfuron on various substrates.

Anion-exchange mechanisms involve a nonspecific electrostatic attraction of an anion to a positively charged site on the soil surface, involving the exchange on one anion for another at the binding site. Ligand exchange is a sorption mechanism that involves displacement of an inorganic hydroxyl or water molecule from a metal ion at a hydrous oxide surface by a carboxylate or hydroxyl on an organic molecule. For instance, this has been proposed as a mechanism for chlorsulfuron sorption on iron oxides. Cation exchange is an electrostatic attraction that involves the exchange of a cation for a cation sorbed at a negatively charged site on the soil surface. Herbicides can be permanently cationic, such as paraquat and diquat; however, weakly basic herbicides that have functional groups such as amines and heterocyclic nitrogen compounds may also protonate to form the cationic form. Cation exchange has been observed with paraquat and diquat, fluridone, and s-triazines.

Hydrophobic interactions and trapping of molecules in a molecular sieve formed by humic materials have been hypothesized as retention mechanisms for prometryn. It has been shown that fluridone, fluazifop, and bipyridylium herbicides penetrate into interlamellar spaces of smectites and can become trapped.

A variety of mechanisms or forces can attract organic chemicals to a soil surface and retain them there. For a given chemical, or family of chemicals, several of these mechanisms may operate in the bonding of the chemical to the soil. For any given chemical, an increase in polarity, number of functional groups, and ionic nature of the chemical can increase the number of potential sorption mechanisms for the chemical.

Ionizable compounds such as basic compounds (triazines and pyridinones) and acidic compounds (carboxylic acids and phenols) can sorb by ionic mechanisms when they are ionized. Weakly basic compounds may sorb by cation exchange; weakly acidic compounds may sorb by anion exchange. For these chemicals ion exchange is not the sole sorption mechanism. For instance, sorption of bipyridylium cations, ie, diquat and paraquat, is primarily the result of cation exchange. Other physicochemical forces, such as charge-transfer interactions, hydrogen bonding, and van der Waals forces, can also be involved in the sorption process.

Triazines are weakly basic chemicals that can be easily protonated at low soil pH levels. The pK_a values for triazines range from about 1.7 for atrazine to 4.3 for prometon. There is abundant evidence for cation exchange as the bonding mechanism for triazines to soil. On the other hand, at soil pH values greater than two pH units above the pK_a, triazines are not protonated to a great extent and other mechanisms become more important, such as hydrogen bonding and hydrophobic attractions. Pyridinones, such as fluridone, are also weakly basic compounds. With a pK_a of 1.7, fluridone sorption can involve cation exchange only in low pH soils. Sorption on soil at pH 5 to 6 is suggested to be by the same mechanisms for sorption on both soil organic matter and montmorillonite, ie, charge-transfer interactions, hydrogen bonding, and van der Waals forces.

Depending on the pH of the system, weakly acidic organic chemicals (carboxylic acids and phenols) exist either as the undissociated molecule or the corresponding anion. Numerous studies have shown that the anion of such herbicides as 2,4-D is readily sorbed by anion-exchange resins, but sorption of organic anions by soils via anion exchange is not likely because clays and organic matter are generally either noncharged or negatively charged. Sorption of weakly acidic organics probably involves physical adsorption of the undissociated molecule and is not site-specific. Other sorption mechanisms for weakly acidic organics are also possible. Charge-transfer and hydrogen bonding were postulated as the sorption mechanisms for the weak acid chlorsulfuron.

Sorption of nonionic, nonpolar hydrophobic compounds occurs by weak attractive interactions such as van der Waals forces. Net attraction is the result of dispersion forces; the strength of these weak forces is about 4 to 8 kJ/mol (~1–2 kcal/mol). Electrostatic interactions can also be important, especially when a molecule is polar in nature. Attraction potential can develop between polar molecules and the heterogeneous soil surface that has ionic and polar sites, resulting in stronger sorption.

Although most nonionic organic chemicals are subject to low energy bonding mechanisms, sorption of phenyl- and other substituted-urea pesticides such as diuron to soil or soil components has been attributed to a variety of mechanisms, depending on the sorbent. The mechanisms include hydrophobic interactions, cation bridging, van der Waals forces, and charge-transfer complexes.

Sorption in the soil is generally controlled by the rate of molecular diffusion into soil aggregates and the rate of reaction (rate of sorption) at the soil–water interface. Diffusion has been found to be the rate-limiting step (38,39). Solute moves from mobile pore water to the sorbent surface surrounded by immobile pore water, limiting the initial rate of sorption as sorption slows down (38). The actual retention reactions tend to be relatively rapid, particularly the exchange-type reactions; however, it has been proposed that two types of sorption sites may be involved that are controlled by the kinetics of the sorption process (40). In one report (39), sorption and desorption of atrazine and linuron on sediments reached 75% of the equilibrium value within 3 to 60 min; labile sites filled before restricted sites (Fig. 5). A pesticide may be retained on the soil surface sorption site initially by a rapid low energy binding mechanism and over time may bind to more stable high energy sites (41).

Sorption Modeling.   Pesticide sorption is characterized by describing sorption isotherms using the Freundlich equation, $S = K_f C^N$, where S is the pesticide sorbed concentration, C is the pesticide solution concentration after equilibration, and K_f and N are constants. Although other equations have been used, the Freundlich has satisfactorily described experimental sorption results for a wide range of pesticides in a variety of soils. The value of N is usually < 1 and between 0.75 and 0.95, which indicates that pesticides are proportionally more sorbed at low solution concentration than at high solution concentration.

For many modeling purposes, N has been assumed to be 1 (42), resulting in a simplified equation, $S = K_d C$, where K_d is the linear distribution coefficient. This assumption usually works for hydrophobic polycyclic aromatic compounds sorbed on sediments, if the equilibrium solution concentration is $<10^{-5}$ M (43).

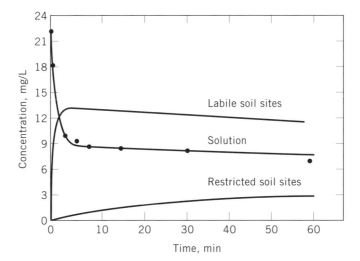

Fig. 5. Sorption and desorption of atrazine and linuron on soil sediments (39).

For many pesticides, the error introduced by the assumption of linearity depends on the deviation from linearity.

Because many studies have shown a direct relationship between pesticide sorption and organic carbon content of soil, attempts have been made to develop a universal sorption coefficient based on sorption of the pesticide to soil organic carbon (44). Sorption based on soil organic carbon is expressed as $S_{oc} = K_{oc}C$, where S_{oc} is pesticide sorbed per unit mass soil organic carbon, and C is pesticide solution concentration after equilibration. If f_{oc} is the fraction of organic carbon, K_{oc} can be obtained from K_d in the equation $K_{oc} = K_d f_{oc}$. Assumptions in the use of this approach include sorption–desorption equilibrium, linearity of sorption isotherm, reversible sorption, sorption limited to the organic component of soil, and soil organic carbon having the same sorption capacity for different soils (44).

Because none of the assumptions is valid in the strict sense, the magnitude of error introduced in using this approach depends on how severely the assumptions are violated. For instance, imazethapyr is a weak acid herbicide and has both carboxylic acid and basic quinoline and pyridine functional groups. Imazethapyr is weakly sorbed. For four soils with different pH, % organic carbon, and % clay, K_f ranged from 0.53 to 1.4; however, the range in K_{oc} was from 14 to 79, thus indicating that sorption is not limited to the organic component (45). Once sorbed, imazethapyr was only partially desorbable from all soils (N-desorption $\ll N$-sorption). The hysteresis observed in desorption (Fig. 6) may be responsible for the difference between mobility estimations made from laboratory sorption studies and the limited mobility observed in the field.

Sorption coefficients are used in a variety of applications, ranging from sophisticated pesticide transport models to simplistic mobility screening models. The degree of rigor required in the characterization of sorption depends on the accuracy required of the intended use. In general, the greater the sorption coefficient, the greater the retardation of the pesticide while leaching through

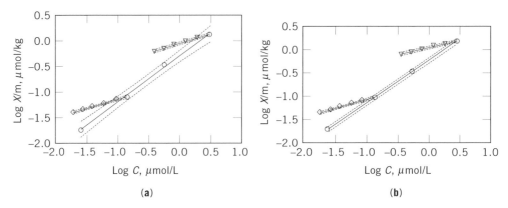

Fig. 6. Sorption and desorption of imazethapyr on (**a**) Sassfras sandy loam and (**b**) Webster clay loam (45): ○ = sorption isotherm, ◇ = desorption isotherm at 0.1728 μmol/L, and ▽ = desorption isotherm at 3.4561 μmol/L.

soil. For instance, in mobility screening models, a $K_{oc} < 500$ indicates that the pesticide is a leacher; a $K_{oc} > 500$ indicates a nonleacher.

Indirect methods of estimating sorption have been used when actual measurement of sorption isotherm is impossible (44). For instance, sorption coefficients have been estimated from soil organic carbon and a specific surface of soil, and from semiempirical equations using pesticide properties.

Pesticide Transport Mechanisms

Pesticides can be transported away from the site of application either in the atmosphere or in water. The process of volatilization that transfers the pesticide from the site of application to the atmosphere has been discussed in detail (46). The off-site transport and deposition can be at scales ranging from local to global. Once the pesticide is in the atmosphere, it is subject to chemical and photochemical processes, wet deposition in rain or fog, and dry deposition.

Water leaves the field either as surface runoff, carrying pesticides dissolved in the water or sorbed to soil particles suspended in water, or as water draining through the soil profile, carrying dissolved pesticides to deeper depths. The distribution of water between drainage and runoff is dependent on the amount of water applied to the field, the physical and chemical properties of the soil, and the cultural practices imposed on the field. These factors also impact the retention and transformation processes affecting the pesticide.

The documented occurrence of pesticides in surface water is indicative that runoff is an important pathway for transport of pesticide away from the site of application. An estimated 160 t of atrazine, 71 t of simazine, 56 t of metolachlor, and 18 t of alachlor enter the Gulf of Mexico from the Mississippi River annually as the result of runoff (47). Field application of pesticides inevitably leads to pesticide contamination of surface runoff water unless runoff does not occur while pesticide residues remain on the surface of the soil. The amount of pesticides transported in a field in runoff varies from site to site. It is controlled by the timing of runoff events, pesticide formulation, physical–chemical properties of the pesticide, and properties of the soil surface (48). Under worst-case conditions, 10% or more of the applied pesticide can leave the edge of the field where it was applied.

It appears that pesticides with solubilities greater than 10 mg/L are mainly transported in the aqueous phase (48) as a result of the interaction of solution/sediment ratio in the runoff and the pesticide sorption coefficient. For instance, on a silt loam soil with a steep slope (>12%), >80% of atrazine transport occurs in the aqueous phase (49). In contrast, it has been found that total metolachlor losses in runoff from plots with medium ground slopes (2–9%) were <1% of applied chemical (50). Of the metolachlor in the runoff, sediment carried 20 to 46% of the total transported pesticide over the monitoring period.

There are three basic strategies to control the amount of pesticide in runoff: reduce the amount of runoff of soil and water; lower the amount of pesticide in the runoff; and retard the field-to-stream delivery of pesticide (49). Cultural practices can reduce the pesticide in runoff. For instance, pesticide incorporation and contour plowing can lead to significant reductions in dissolved and sorbed pesticide concentrations, and in total metolachlor loss in runoff, relative to application as a pre-emergence spray with cross-contour plowing (50). Pesticide runoff is also affected by timing of application. For example, twice as much atrazine-applied pre-emergent was in the runoff water and sediment compared to that of applied post-emergent (51).

Tillage practices have also been shown to have a dramatic impact on soil erosion and water runoff. Conservation tillage systems and contouring have been shown in numerous studies to reduce sediment loss in runoff, but water runoff is not necessarily affected. The percentage of applied alachlor lost from up-and-down slope moldboard plow, strip-till, and no-till were 6, <1, and 2%, respectively (52). Total loss from contoured moldboard plow, strip-till, and no-till were 2, <0.1, and 1%, respectively. The greater amounts of alachlor transported in no-till compared to strip-till were the result of pesticide washoff from plant foliage and residue. It was also found that alachlor and carbofuran were transported from plots largely in moving water, but terbufos and metabolites were recovered mainly in eroded sediment.

Conservation tillage increased atrazine and metolachlor surface runoff by 42% and decreased tile discharge by 15% compared with conventional tillage, but total field runoff was the same from all treatments (53). Runoff events shortly after herbicide application produced the greatest herbicide concentrations and losses in both surface runoff and subsurface drainage.

The influence of conventional and soil-specific management on leaching and runoff losses of soil-applied alachlor was studied across a soil catena (landscape) with varied slope and drainage characteristics. Averaged across soils and events, the concentrations of alachlor in runoff (water, sediments, and water) were less for soil-specific application rates than for the uniform application rate (54). There is little information on the removal of pesticide from field runoff. However, water containing pesticide moving through a vegetative filter strip resulted in reductions of up to 70% for 2,4-D and up to 96% for trifluralin (55).

Pesticide Runoff Modeling.   Obtaining the field data necessary to understand the potential runoff of pesticides under a variety of conditions and soils would be an expensive and time-consuming process. As a result, a variety of simulation models that vary in their conceptual approach and degree of complexity have been developed. Models are influenced by their intended purpose, the biases of the developer, and the scale at which they are used.

One of the first complete, continuous simulation models was the pesticide runoff transport model (PRT) (56). Improvements in the PRT model led to the hydrologic simulation program—Fortran model (57). A number of other models have been developed (58,59). These models represent a compromise between the available data and the ability to encompass a wide range in soils, climates, and pesticides. These models have had mixed success when extended beyond the data with which they were calibrated. No model has yet been developed that can be proven to give accurate predictions of pesticide runoff on an absolute basis, ie, predicting pesticide concentrations in runoff consistently to within a few percent (60).

Leaching. Numerous studies of pesticides in groundwater have indicated that pesticides are present as the result of agricultural practices and may be the product of both point source and nonpoint source pollution. The movement of pesticides in soil water depends on rainfall or irrigation water, the macroscopic and microscopic structure of the soil, and the sorption–desorption characteristics of the pesticide on the soil. Water moves through the soil under both saturated and unsaturated conditions. When the soil is saturated with water, the pores are filled with water and transport occurs at the maximal rate. Movement of water and pesticide occurs at much slower rates under unsaturated conditions because only the smaller pores are filled and water moves in response to water potential gradients. Generally, coarse-textured soils have greater rates of water movement than fine-textured soils when saturated. However, under unsaturated conditions, fine-textured soils may have greater transport rates.

In water, herbicides are transported by mass flow. Highly soluble pesticides have a greater initial potential for movement than insoluble pesticides, assuming that sorption of both chemicals to soil is similar. Simultaneously, the chemical process of diffusion affects the distribution of herbicide in the water. Dispersion resulting from differential flow rates within pores and sorption to soil retards the movement of pesticides relative to that of water, or a noninteracting tracer such as bromide. Soil pores have a wide range of sizes and lengths, and are highly interconnected. A portion of the pore space usually is not part of the continuous flow path or flows at much slower than average rates. Pesticides can enter and exit these spaces through diffusion. Pesticide diffusion coefficients are inversely related to sorption, except for pesticides of high vapor pressure, such as trifluralin, which have a high degree of movement as a volatile compound through air-filled pore space (61).

Because many pesticides are applied to the soil surface, the transport of pesticide during water infiltration is important. Water infiltration is characterized by high initial infiltration rates which decrease rapidly to a nearly constant rate. Dry soils have greater rates of infiltration than wet soils during the initial application of water. Thus, perfluridone movement after application of 3.8 cm of water was considerably greater in soil at a water content of <1% of field capacity than at 50% of field capacity (62). Fluometuron moved deeper into the soil in response to greater rainfall intensity or after rainfall onto a dry rather than a moist soil (63).

Sorbed pesticides are not available for transport, but if water having lower pesticide concentration moves through the soil layer, pesticide is desorbed from the soil surface until a new equilibrium is reached. Thus, the kinetics of sorption and desorption relative to the water conductivity rates determine the actual rate of pesticide transport. At high rates of water flow, chances are greater that sorption and desorption reactions may not reach equilibrium (64). Nonequilibrium models may describe sorption and desorption better under these circumstances. The prediction of herbicide concentration in the soil solution is further complicated by hysteresis in the sorption–desorption isotherms. Both sorption and dispersion contribute to the substantial retention of herbicide found behind the initial front in typical breakthrough curves and to the depth distribution of residues.

Pesticide Leaching Modeling. Modeling of pesticide movement through soil has received considerable attention beginning about the time when groundwater contamination in numerous locations was found. Simulation models can provide assessment of potential contamination of groundwater and alternative cultural practices; they can be used in lieu of expensive monitoring studies. Numerous pesticide transport simulation models have been developed, including PRZM (59), LEACHM (65), CMLS (66), GLEAMS (67), and Opus (68).

These models have met with varying degrees of success as a result of several factors. Spatial variability in soil properties and climatic conditions cause problems in characterizing pesticide leaching as well as limitations in the model's characterization of the soil/crop system. For instance, most models assume homogeneous unsaturated fields. However, many field studies have shown the existence of macropores resulting from soil aggregated structure, earthworm burrows, wetting and drying cracks, and decaying plant roots. Models are being developed in the 1990s to include preferential flow through the macropores (69).

For pesticides to leach to groundwater, it may be necessary for preferential flow through macropores to dominate the sorption processes that control pesticide leaching to groundwater. Several studies have demonstrated that large continuous macropores exist in soil and provide pathways for rapid movement of water solutes. Increased permeability, percolation, and solute transport can result from increased porosity, especially in no-tillage systems where pore structure is still intact at the soil surface (70). Plant roots are important in creation and stabilization of soil macropores (71).

Preferential flow through root-mediated soil pores has been demonstrated for chloride, nitrate, and other ions that are not sorbed onto soil organic matter and clays. However, pesticide sorption onto soil affects both mobility of the pesticide as well as its residual life in the soil. Pesticide sorption onto root organic matter or organic linings of worm burrows may also slow transport of pesticides relative to water (72), thus countering the effects of increased permeability caused by roots.

Sorption and transport processes are directly or indirectly affected by soil properties such as soil moisture, temperature, pH, and organic carbon content. Tillage systems affect these same soil properties (73,74). Conventional tillage systems can decrease moisture compared to conservation tillage systems, resulting in decreased degradation, volatilization, and leaching of the pesticide. Different tillage systems have different effects on soil temperature. In general, increased tillage increases soil temperature in the spring and summer. Rates of chemical and microbiological reactions and volatilization are temperature-dependent.

Continuous application of ammonium fertilizer in conservation tillage systems decreases pH substantially. Tillage dramatically affects organic carbon content of soil. Once a soil is plowed for the first time, the organic carbon content begins to decrease. Leaving a residue on the surface, as in conservation tillage, increases the organic carbon content of the soil. Soil microbial populations can be substantially greater in conservation tillage systems.

The interaction of all these factors makes it difficult to predict an overall effect of conservation tillage on the potential leaching of a pesticide compared to that in a conventionally tilled field. However, it was found that a prolonged rain immediately after application resulted in short-term levels of pesticide in groundwater to be greater under no-till than under conventional till plots, which suggested that preferential transport in no-till had occurred (75). In contrast, it was suggested that there can be greater leaching losses of surface-applied pesticides to groundwater under plow-tillage than under no-till (76).

Future Trends

The year-to-year variation in pesticide usage depends on weather, area planted, farm income, interest rates, previous year's inventory, extended labeling of existing products, deleted registrations, crop prices, strength of the U.S. dollar (exports), patent expirations, lawsuits, levels of pest infestation, and federal legislation and enforcement. In addition, there are several important trends that influence the production and sales of pesticides and in turn have a significant impact on the detection of pesticides in the U.S. surface and groundwater supply.

New Herbicides.   There are also several significant developments that will have longer-term impact on pesticide usage and residues in water. There has been a steady decrease in the amount of herbicide needed to control weeds since the 1940s (Fig. 7).

Extensive use of two more recently developed classes of herbicides will further dramatically reduce the amount of applied to control weeds. The sulfonyl-

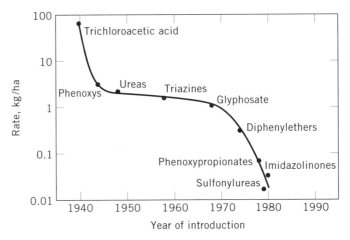

Fig. 7.   Historical trends in recommended herbicide application rates, by chemical class (77). Courtesy of the American Chemical Society.

urea herbicides are extremely active compounds first discovered in the mid-1970s at DuPont; they have been discussed extensively (78). Sulfonylurea herbicides have experienced a rapid and widespread success since their commercial introduction in 1982 with chlorsulfuron (Table 5). The sulfonylureas are applied at rates of 2–75 g/ha. The chemistry of the sulfonylurea molecule permits the synthesis of a very large number of useful analogues, consequently many new herbicides are anticipated for crop production. As of this writing (1996), over 350 patents have been issued to about 27 agricultural companies covering tens of millions of structures known or expected to be herbicidally active.

Another class of herbicides, the imidazolines, was discovered at American Cyanamid in the early 1980s. Extensive research has led to the development of four commercial compounds: imazapyr, imazamethabenz-methyl, imazethapyr, and imazaquin (see Table 5). Like the sulfonylureas, the imidazolines are extremely active at low rates.

Biotechnology. A second factor that will impact the future of the pesticide market are the advances made in biotechnology. Although biotechnology has played a significant role in medical technology, it has yet to make an important impact in pest control programs. One of the most widely researched genetic engineering approaches involves inserting an insecticidal gene from *Bacillus thuringiensis* (Bt) into plants. Bt is a common gram-positive soil microorganism that produces insecticidal protein crystals, called delta-endotoxins, that control a number of lepidopteran (caterpillar), dipteran (fly and mosquitos), and coleopteran (beetle) pests. The crystals are highly condensed, high molecular weight proteins (typically 75–150 Kd in mass). The genes that encode the toxic protein are found on plasmids in the *Bacillus*, ie, small, extrachromosomal, self-replicating, circular pieces of DNA. By the use of plant vector systems, Bt genes have been inserted into tomato, potato, and tobacco plants (79). These transgenic plants express (produce) enough of the endotoxin to be protected from feeding damage by insect larva. Insect resistance is stably inherited in subsequent plant generations. The Bt genetic approach is particularly attractive because the toxin is extremely safe to humans, fish, animals, and other nontarget organisms. Genetic engineering (qv) research is also underway on soybeans to make them tolerant to the herbicide glyphosate, a broad-spectrum weed control agent generally considered extremely safe from an environmental standpoint.

Sustainable Agriculture. The third factor that will influence the future of pesticide sales is the emphasis on sustainable agriculture systems that rely on more natural pest control methods and reduced pesticide usage. These are integrated systems that require nutrients and crop protection chemicals from on-farm natural sources and cultural methods. Many current sustainable farms are site-specific systems that may depend on the soils in a particular region and the availability of large volume, cheap nutrient sources, ie, cover crops or manure. Composting is under renewed interest in the 1990s as a source of nutrients and natural pesticides. In part the success of sustainable agriculture will depend on how well useful genes can be manipulated in economic crops via biotechnology.

New Legislation. A fourth factor that should reduce pesticide usage on a global basis are the regulations being passed in the 1990s in many countries, which are to limit the use of agricultural chemicals by specific deadlines. This trend is most apparent among the European Community (EC) of nations in

Table 5. Application Rates of Sulfonylurea and Imidazoline Herbicides

Name	CAS Registry Number	Structure	Application rate, g/ha
chlorsulfuron; 2-chloro-N-(4-methoxy-1,3,5-triazin-2-yl)-aminocarbonyl)benzenesulfonamide	[64902-72-3]		4–26
chlorimuron ethyl; ethyl 2-(((4-chloro-6-methoxy-pyrimidin-2–yl)aminocarbobnyl)aminosulfonyl)-benzoate	[90982-32-4]		8–13
imazapyr; 2-(4,5-dihydro-4-methyl-4-(1-methylethyl)-5-oxo-1H-imidazol-2-yl)-3-pyridinecarboxylic acid	[81335-77-5]		150–200
imazaquin; 2-(4,5-dihydro-4-methyl-4-(1-methylethyl)-5-oxo-1H-imidazol-2-yl)-3-quinolinecarboxylic acid	[81335-37-7]		70–140

448

Western Europe (80). Pesticide usage is very high in the EC countries; Western Europe is the largest agrochemical sales market in the world. In 1991, for example, the EC held a 31% share of global agrochemical sales. Per hectare annually, Dutch farmers use about 20 kg of pesticides, considerably more than the Belgian producer, at 12.4 kg. French and Swiss farmers use 6 kg; German farmers, 4 kg; and U.S. growers, 2.2 kg (2 lb/acre). Holland has used pesticides intensively because of climatic conditions that are conducive to fungal and bacterial diseases, limited diversity in their crop rotations that would otherwise disrupt the life cycle of pests, and large exports of propagated materials, eg, bulbs, that require high plant sanitary standards. In 1991, the Dutch government instituted a plan to decrease pollution pressures from agrochemicals. One goal of the Dutch program is to reduce pesticide use by 50% by the year 2000.

Another objective is to reduce the input of soil sterilization products to control insects, nematodes, and fungal pathogens by making purchase of the chemicals on a prescription basis with the stipulation that they can only be applied every four years to a particular field. Denmark has passed similar regulations that would reduce pesticide use by 50% by 1995. The interim goal of 25% reduction by 1990 was reached in 1988. The French government is beginning to enact legislation that would protect water. In April 1991, the French Environmental Ministry introduced a water plan to better manage water resources and decrease both point and nonpoint sources. Germany is working with individual growers to reduce agricultural production by reducing pesticide usage. Local German authorities are prohibiting pesticide usage on nonagricultural fields.

Biotechnology, use of highly active chemicals at lower rates, wide adoption of sustainable agriculture, and enactment of more restrictive use regulations will lower the pesticide burden in the environment and improve water quality.

BIBLIOGRAPHY

"Soil Chemistry" in *ECT* 1st ed., Vol. 12, pp. 614–633, by E. R. Graham, University of Missouri; "Soil Chemistry of Pesticides" in *ECT* 2nd ed., Vol. 18, pp. 515–540, by P. C. Kearney, J. R. Plimmer, and C. S. Helling, U.S. Dept. of Agriculture; in *ECT* 3rd ed., Vol. 21, pp. 263–294, by P. C. Kearney, C. S. Helling, and J. R. Plimmer, U.S. Dept. of Agriculture.

1. R. L. Carson, *Silent Spring*, Houghton-Mifflin Co., Boston, Mass., 1962, p. 368.
2. A. Aspelin, *Pesticides Industry Sales and Usage: 1994 and 1995 Market Estimates*, U.S. EPA No. 733-K-96-001, Washington, D.C., 1996.
3. A. Aspelin, A. H. Grube, and A. H. Toria, *Pesticide Industry Sales and Usage: 1990 and 1991 Market Estimates*, U.S. EPA No. 733-K-92-001, Washington, D.C., 1992.
4. R. D. Wauchope and co-workers, *Rev. Envir. Contam. Tox.* **123**, 1 (1992).
5. T. Cairns and J. Sherma, *Emerging Strategies For Pesticide Analysis*, CRC Press, Inc., Boca Raton, Fla., 1992.
6. S. A. Peoples and co-workers, *Bull. Environ. Contam. Toxicol.* **24**, 611 (1980).
7. M. H. Zaki, D. Moran, and D. Harris, *Am. J. Public Health*, **72**, 1395 (1982).
8. S. Z. Cohen and co-workers, *Am. Chem. Soc. Symp. Ser.* **259**, 297 (1984).
9. S. Z. Cohen, C. Eiden, and M. N. Lober, *Am. Chem. Soc. Symp. Ser.* **465**, 170 (1986).
10. R. J. Nash and co-workers, *Am. Chem. Soc. Symp. Ser.* **465**, 1 (1991).
11. J. S. Briskin, *Mechanisms of Pesticide Movement into Ground Water*, Lewis Publishers, Boca Raton, Fla., 1994, pp. 143.

12. CIBA, *Atrazine Concerns Poses No Unreasonable Risk*, OGA-560-00180-A, 1994.

13. I. C. MacRae, *Rev. Environ. Contam. Toxicol.* **109**, 1 (1989).

14. R. S. Horvath, *Bacteriol. Rev.* **36**, 146 (1972).

15. K. Racke and J. R. Coats, *Am. Chem. Soc. Ser.* **426** (1990).

16. J.-M. Bollag and S.-Y. Liu, in H. H. Cheng, ed., *Pesticides in the Soil Environment: Processes, Impacts, and Modeling*, Soil Science Society of America, Madison, Wis., 1990, pp. 168–175.

17. M. M. Haggblom, *FEMS Microbiol. Rev.* **103**, 29 (1992).

18. W. A. Dick and co-workers, *Am. Chem. Soc. Symp. Ser.* **426**, 98 (1990).

19. S. Chapalamadugu and G. R. Chaudry, *Critical Rev. Biotechnol.* **12**, 357 (1992).

20. W. W. Mohn and J. M. Tiedje, *Microbiol. Rev.* **56**, 482 (1992).

21. R. T. Mandelbaum, L. P. Wackett, and D. L. Allan, *Environ. Sci. Technol.* **27**, 1943 (1993).

22. L. J. Audus, *Plant Soil*, **3**, 170 (1951).

23. G. S. Sayler, *Microbial Decomposition of Chlorinated Aromatic Compounds*, USEPA 600/2-86/090, Washington, D.C., 1986.

24. J. R. Coats, *Am. Chem. Soc. Symp. Ser.* **459**, 10 (1991).

25. A. M. Blair and T. D. Martin, *Pesticide Sci*, **22**, 195 (1988).

26. J. R. Helz, R. C. Zepp, and D. G. Crosby, *Aquatic and Surface Photochemistry*, Lewis Publishers, Boca Raton, Fla., 1994.

27. G. C. Miller and S. G. Donaldson, in Ref. 26.

28. J. Kochany and R. J. Maquire, *J. Agric. Food Chem.* **42**, 406 (1994).

29. C. J. Miles, *Am. Chem. Soc. Symp. Ser.* **459**, 61 (1991).

30. I. Sceunert, *Terrestrial Behavior of Pesticides*, Springer-Verlag, Berlin, Germany, 1992, pp. 25–75.

31. J. W. Hamaker and J. M. Thompson, in C. A. I. Goring and J. W. Hamaker, eds., *Organic Chemicals in the Soil Environment*, Marcel Dekker, Inc., New York, 1972, pp. 49–143.

32. W. C. Koskinen and S. S. Harper, in Ref. 16, pp. 51–77.

33. G. Sposito, The surface chemistry of Soils, Oxford University Press, New York, 1984.

34. C. T. Chiou, L. J. Peters, and V. H. Freed, *Science*, **206**, 831 (1979).

35. U. Mingelgrin and Z. Gerstl, *J. Environ. Qual.* **12**, 1 (1983).

36. D. A. Laird and co-workers, *Soil Sci. Soc. Am. J.* **56**, 62 (1992).

37. S. S. Harper, *Rev. Weed Sci.* **6**, 207 (1994).

38. M. T. van Genuchten and P. J. Wierenga, *Soil Sci. Soc. Am. Proc.* **40**, 473 (1976).

39. R. D. Wauchope and R. S. Myers, *J. Environ. Qual.* **14**, 132 (1985).

40. P. S. C. Rao and co-workers, *Soil Sci. Soc. Am. J.* **43**, 22 (1979).

41. W. C. Koskinen and co-workers, *Intern. J. Environ. Anal. Chem.* **58**, 379 (1995).

42. P. S. C. Rao and J. M. Davidson, in M. R. Overcash and J. M. Davison, eds., *Environmental Impact of Nonpoint Source Pollution*, Ann Arbor Science Publisher, Ann Arbor, Mich., 1980, pp. 23–67.

43. S. W. Karickoff, *Chemosphere*, **10**, 833 (1981).

44. R. E. Green and S. W. Karickoff, in Ref. 16, pp. 79–101.

45. J. Gan and co-workers, *Weed Sci.* **42**, 92 (1994).

46. A. W. Taylor and W. F. Spencer, in Ref. 16, pp. 213–269.

47. W. E. Pereira and F. D. Hostettler, *Environ. Sci. Technol.* **27**, 1542 (1993).

48. R. D. Wauchope, *J. Environ. Qual.* **7**, 459 (1978).

49. J. L. Baker and H. P. Johnson, *Trans. ASAE*, **22**, 554 (1979).

50. J. M. Buttle, *J. Environ. Qual.* **19**, 531 (1990).

51. D. J. Pantone and co-workers, *J. Environ. Qual.* **21**, 567 (1992).

52. A. S. Felsot, J. K. Mitchell, and A. L. Kenimer, *J. Environ. Qual.* **19**, 539 (1990).

53. J. D. Gaynor, D. C. MacTavish, and W. I. Findlay, *J. Environ. Qual.* **24**, 246 (1995).

54. B. R. Khakural, P. C. Robert, and W. C. Koskinen, *Soil Use Manage.* **10**, 158 (1994).
55. W. A. Rhode and co-workers, *J. Environ. Qual.* **9**, 37 (1980).
56. N. H. Crawford and A. S. Donigian, *Pesticide Transport and Runoff Model for Agricultural Lands*, U.S. EPA No. EPA-660/2-74-013, Washington, D.C., 1973.
57. A. S. Donigan Jr., J. C. Imhoff, and B. R. Bichnell, in F. W. Schaller and G. W. Bailey, eds., *Agricultural Management and Water Quality*, Iowa State University Press, Ames, Iowa, 1983, pp. 200–249.
58. W. G. Knisel, G. R. Foster, and R. A. Leonard, in F. W. Schaller and G. W. Bailey, eds., *Agricultural Management and Water Quality*, Iowa State University Press, Ames, Iowa, 1983, pp. 178–199.
59. R. F. Carsel and co-workers, Users' Manual for the *Pesticide Root Zone Model (PRZM)*, Release I, USEPA 600/3-84-109, Washington, D.C., 1984.
60. R. A. Leonard, in Ref. 16, pp. 303–349.
61. H. D. Scott and R. E. Phillips, *Soil Sci. Soc. Am. Proc.* **36**, 714 (1972).
62. M. L. Ketchersid and M. G. Merkle, *Weed Sci.* **23**, 344 (1975).
63. F. L. Baldwin, P. W. Santelmann, and J. M. Davidson, *J. Environ. Qual.* **4**, 191 (1975).
64. A. G. Hornsby and J. M. Davidson, *Soil Sci. Soc. Am. Proc.* **37**, 823 (1973).
65. R. J. Wagenet and J. L. Hutson, *J. Environ. Qual.* **15**, 315 (1986).
66. D. L. Nofziger and A. G. Hornsby, *CMLS: Interactive Simulation of Chemical Movement in Layered Soils*, Institute of Food and Agriculture Sciences, Circular 786, University of Florida, Gainesville, Fl., 1986.
67. R. A. Leonard, W. G. Knisel, and D. A. Still, *Trans. ASAE*, **30**, 1403 (1987).
68. R. E. Smith, *Opus: An Integrated Simulation Model for Transport of Nonpoint-Source Pollutants at the Field Scale*, Vol. I, Documentation, USDA ARS-98, U.S. Dept. of Agriculture, Washington, D.C., 1992.
69. H. H. Gerke and M. T. van Genuchten, *Water Resour. Res.* **29**, 304 (1993).
70. V. K. Quisenberry and R. E. Phillips, *Soil Sci. Soc. Am. J.* **40**, 484 (1976).
71. R. Tippkotter, *Geoderma*, **29**, 355 (1983).
72. A. B. Zins, D. L. Wyse, and W. C. Koskinen, *Weed Sci.* **39**, 262 (1991).
73. R. E. Phillips and S. H. Phillips, eds., *No-tillage Agriculture: Principles and Practices*, Van Nostrand Reinhold Co., Inc., New York, 1984.
74. W. C. Koskinen and C. G. McWhorter, *J. Soil Water Conserv.* **41**, 365 (1986).
75. A. R. Isensee, R. G. Nash, and C. S. Helling, *J. Environ. Qual.* **19**, 434 (1990).
76. D. Levanon and co-workers, *J. Environ. Qual.* **22**, 155 (1993).
77. R. G. Nash and A. R. Leslie, eds., *Groundwater Residue Sampling Design*, American Chemical Society, Washington, D.C., 1991.
78. E. M. Beyer and co-workers, *Herbicides: Chemistry, Degradation and Mode of Action*, Vol. 3, Marcel Dekker, Inc., New York, 1988, pp. 117–189.
79. M. Vaeck and co-workers, *Am. Chem. Soc. Symp. Ser.* **379**, 280 (1988).
80. P. Szmedra, *Agri. Outlook*, **198**, 28 (July 1993).

PHILIP C. KEARNEY
University of Maryland

DANIEL R. SHELTON
WILLIAM C. KOSKINEN
USDA-Agricultural Research Service

SOIL STABILIZATION

Chemical grouting is the practice of injecting liquid solutions of cement or organic materials into soil, rock, or concrete in order to form solid inorganic or organic masses that impart desirable permanent physical characteristics in the soil, rock, or concrete. The solutions that are injected, ie, chemical grouts, undergo either polymerization of monomers or cross-linking of soluble polymers to form insoluble polymer masses. Chemical grouting is often used in construction or repair of buildings, dams, shafts, etc, when it is desired to restrict or reroute the flow of water through a formation, or to strengthen the formation. Repair of sewer systems is an important application of chemical grouting. Chemical grouts, other cement-like materials, as well as the so-called geosynthetics, ie, nonwovens, membranes, grids, and honeycombs, are commonly used individually and in combinations to reinforce soil as foundations for roads, buildings, etc (see GEOTEXTILES). This article focuses on chemical grouts.

Soil conditioners are materials that measurably improve the physical characteristics of the soil as a plant growth medium. Typical uses include erosion control, prevention of surface sealing, and improvement of water infiltration and drainage. Many natural materials such as peat and gypsum are used alone or in combination with synthetics for soil conditioning. This article is concerned with synthetic soil conditioners, many of which are introduced as polymeric systems similar to the gels and foams formed *in situ* by chemical grouts.

Chemical Grouting

The early history of chemical grouting has been reviewed (1,2). The first true chemical grouting technique involved injection of sodium silicate solution and calcium chloride solution through two adjacent bore holes to form a precipitate where the two liquids came into contact. Until the 1950s, all chemical grout materials were variations on the sodium silicate–calcium chloride system and these materials became synonymous with chemical grouting. Portland cement (qv) was used on a large scale in construction of several dams in the United States. Practices, specifications, and theories that became grouting standards were developed empirically, based on procedures used by various government agencies in large construction projects in the first third of the twentieth century.

The ideal chemical grout is a low viscosity solution capable of penetrating finely divided profiles as easily and to the same extent as water. The grout solution viscosity remains low for a predetermined time to allow the desired penetration of the profile. Rapid gelation then occurs to form a water-impermeable barrier filling all the voids in the formation, thereby waterproofing it. The barrier's durability should be adjustable. Long-lived barriers are needed in applications such as repair of sewers or tunnels but short-lived grouts are useful for temporary stabilization of excavations such as in construction. Grouts that are too viscous or that contain particulates are usually not useful in grouting of finely divided formations. Adjustable gel times are important because very short gel times are useful to make seals in the presence of running water and longer gel times may be needed to penetrate profiles to the desired depth. Grout systems must be chemically compatible with the profiles to be grouted and capable of being applied with available equipment. Safe and environmentally sound mixing, use, and disposal of organic grouts are other factors to be considered. Chemical grouts based on several different inorganic and organic polymer systems are commercially available.

Silicate Grouts. Sodium silicate [1344-09-8] has been most commonly used in the United States. Its properties include specific gravity, 1.40; viscosity, 206 mPa·s(= cP) at 20°C; $SiO_2:Na_2O = 3.22$. Reaction of sodium silicate solutions with acids, polyvalent cations, such organic compounds as formamide, or their mixtures, can lead to gel formation at rates which depend on the quantity of acid or other reagent(s) used.

In the Joosten or two-shot method (3), successive injections are made of a concentrated solution of sodium silicate and a calcium chloride [10043-52-4] solution into a single pipe. Unconfined compressive tests indicate the stabilized strengths of injected sands to be 2.1–4.1 MPa (300–600 psi). Reaction is almost instantaneous; there is no ground heave and only minimal travel of the grout from the injection point. The relatively high viscosity of undiluted sodium silicate, 50 to 200 mPa·s(= cP), and the fast reaction with $CaCl_2$ require closely spaced grout injection points, resulting in high installation costs.

In the Siroc or one-shot method (4), formamide is used to coagulate sodium silicate. The silicate solution used in the Joosten method can be diluted with water to lower its viscosity. Concentrations of sodium silicate between 10–70% are used (viscosities of 2.5–50 mPa·s). Concentrations of formamide are between 2 and 30%. Other reactants such as $CaCl_2$ and sodium aluminate are used in concentrations between 2.4–12 g/L of silicate solution.

For waterproofing, sodium silicate concentrations below 30% are adequate; concentrations between 35 and 70% are used for strength improvement. Grouts having 35 vol % or higher silicate resist deterioration on freeze–thaw or wet–dry cycles. Water permeability of sands can be reduced from 10^{-2} to 10^{-8} cm/s. Unconfined compressive strengths of stabilized sand can vary from 103 to 4130 kPa (15–600 psi); the normal range is between 690 and 1380 kPa.

Soils containing up to 20% silt or clay can be treated by injection. Low injection rates are used with finer soils to avoid fracturing the profile and forming lenses of chemical grout. The optimum injection rate of a soil depends on such factors as soil texture, density, void ratio, permeability, depth of overburden, as well as the viscosity of the grouting solution. The pH of the grouted material should be between 5 and 11; more acid or basic soils may have to be neutralized before grouting. Sodium aluminate [1302-42-7] reactant is used when acid soil is to be grouted.

Various additives can impart desired handling and performance properties. Sodium bicarbonate used with sodium silicate grouts produces low strength semipermanent grouts to stabilize formations during construction projects. Portland cement is used as a reactant with sodium silicate grouts to obtain short gel times, useful to stop flowing water and seal grouting cavities. Silicate-based grouts containing added Portland cement cannot be injected into medium sands or finer soils. Addition of particulates to grouts generally reduces their ability to penetrate finely divided formations. Several patents describe the use of organic grouting systems in combination with silicate-based grouts.

The Siroc grouting system is considered nonhazardous and nonpolluting. Sodium silicate is essentially nontoxic. Formamide is toxic and corrosive, but does not present a serious hazard if normal safety precautions are followed. Siroc chemical grout materials are two to five times more expensive than Portland cement, depending on the sodium silicate to formamide concentration ratios. Installed costs are generally more similar to those for cement grouts.

Syneresis of sodium silicate gels may occur under some conditions, eg, in pure gels or coarse formations. Cement grouting should then precede chemical grouting. Leaching that results from dissolution under water-saturated conditions may be eliminated by use of proper reagent proportions.

Organic Polymer Grouts. There are several types of organic grouting systems (5).

Acrylamide. Aqueous acrylamide solution grouts were introduced in the United States in 1955 and rapidly became popular because of lower cost, better flexibility, and superior performance compared to other grouts then commercially available. Acrylamide grouts came closest to matching performance requirements for an ideal grout and set the technical standard for performance of other grouting systems. Specialized equipment was designed for injection of acrylamide grouts. Acrylamide has been the most-often selected grout for use in sewer applications. An acrylamide-based grout was used in construction of the final grout curtain of the Rocky Reach Dam (Wenatchee, Washington); almost 950 m^3 was required. In 1989, about 295,500 kg of acrylamide grout was consumed, approximately 43% of the total grout used.

Polymer gels resulting from acrylamide polymerization are generally regarded as nonhazardous. Acrylamide (qv) monomer, however, has been reported to have potential neurotoxic effects in higher animals (6). This led to a search, promoted by the U.S. Environmental Protection Agency (EPA) (7), for less toxic organic grouting systems. Progress had been relatively slow on account of the difficulties involved in matching the performance of acrylamide grouts with other systems, especially because specialized equipment designed for acrylamide application was not easy to adapt to the requirements of other systems. In the 1990s, systems based on polyurethanes and polyacrylamides offer performance similar to that achievable with acrylamide-based grouts. Citing the reported neurotoxicity of acrylamide [*79-06-1*] and the availability of other grouting systems, the EPA proposed in 1991 (8) to ban the use of grouts based on acrylamide and N-methylolacrylamide [*924-42-5*] (NMA). Despite opposition from the National Association of Sewer Service Companies (NASSCO) and others, this ban was in the final-rule stage as of September 1996 (9).

Acrylamide grouts generally comprise a 19:1 mixture of acrylamide and methylenebisacrylamide [*110-26-9*] or some other cross-linking agent. Catalysts include persulfates, peracids, or peroxides; ammonium persulfate is commonly used. Accelerators such as triethanolamine are used; dimethylaminopropionitrile is better from a control viewpoint but is a suspect carcinogen. Sometimes inhibitors, such as potassium ferricyanide, are also added. Acrylamide grouts are available in either solid or liquid form. Liquid forms were developed to facilitate handling and reduce exposure of workers to acrylamide. Packaging (qv) of measured quantities of acrylamide in water-soluble bags has also been described as a method to minimize user contact (10). The monomer and cross-linker are maintained in one solution and catalyst in another. The two solutions are mixed at a Y-junction and injected into the path of leakage that is to be stopped. Chemical reaction then begins.

Advantages of acrylamide grouts include low cost, quick and controllable gel times from a few seconds to several hours, low viscosity of $1-2$ mPa·s($=$ cP), and a long history of reliable performance. Gel times can be accurately controlled by varying the concentration of one or both parts of the catalyst system, or by adding inhibitors such as potassium ferricyanide. Dissolved salts normally present in groundwater have little effect on gel time, especially if groundwater is used to dissolve the reagents. Gel times of a few seconds have been used to shut off flowing water in construction operations; short gel times down to $5-10$ seconds are obtained by omitting potassium ferricyanide and increasing the persulfate and amine components. A low viscosity grout is advantageous for penetration of finely divided formations. Acrylamide grouts are among the lowest viscosity grouts available.

Unconfined compressive strengths of soils stabilized with acrylamide-based grouts are typically in the range of $345-1380$ kPa ($50-200$ psi). Strength increases with increasing density and decreasing grain size of the formation. The stiffness of the grout obtained can be varied by altering the acrylamide:methylenebisacrylamide ratio. A 97:3 ratio, a transparent, sticky, elastic, low strength gel is obtained; at 9:1 ratio, a harder, stiffer, opaque gel is obtained. Gels typically contain approximately 90% water, trapped mechanically in the polymer matrix. Gels are permanent when located below the water table or in very humid environments. Mechanical deterioration of the gels occurs on exposure to freeze–thaw or wet–dry cycles where dry periods predominate. Long exposure to dry conditions causes gels to shrink. When rehydrated, gels return to their former volume; cracks close but do not necessarily heal. Acrylamide gels have a water permeability of approximately 10^{-10} cm/s. The permeability of unstabilized soils fall in the range of $10^{-1}-10^{-5}$ cm/s. When polymer fills the soil voids completely, the permeabilities of treated soil and gel are similar.

N-Methylolacrylamide. NMA-based grouts accounted for approximately 3% (27,300 kg) of the total grout market in 1989. The equipment and processes necessary for sewer rehabilitation and manhole sealing, which are the main uses of NMA-based grouts, are the same as those for acrylamide-based grouts, although a different persulfate catalyst is typically used. In the United States, the ban proposed by the EPA on the use of NMA grouts is in the final-rule stage (8,9).

Acrylates. Acrylate grouts (11–13) such as Geo/Chem AC-400 Chemical Grout (Geochemical Corporation) were developed specifically as acrylamide grout replacements intended for sewer rehabilitation and similar applications. The objective was to provide low viscosity, low toxicity grouting systems that could be used in the same equipment and with the same catalysts as acrylamide grouts to provide controllable gel times and strong, durable, and water-impermeable grouts. Acrylate grouts are formulated using calcium acrylate [6292-01-9] and/or magnesium acrylate [5698-98-6] with lesser amounts of lithium acrylate [13270-28-5], 2-hydroxyethyl acrylate [818-61-1], or polyethylene glycol diacrylate. The catalyst system comprises ammonium persulfate [7727-54-0] and triethanolamine [102-71-6]; the cross-linker is a few percent of methylenebisacrylamide. Use of acrylate grouts in combination with silicic acid [1343-98-2] has been reported (14). The viscosity of a 10% acrylate grout formulation is approximately the same as for acrylamide grouts. The gel is formed by polymerization of the monomers on mixing of two equal volume solutions, one containing monomers and triethanolamine in water, the other a solution of ammonium persulfate in water. Compressive strengths of acrylate gels are somewhat lower than for acrylamide gels. Grouts formed are reported to be roughly comparable to those obtained with acrylamide grouts. Field experience has given mixed results, with some reports of corrosivity, swelling in water, poor durability, etc.

Polyacrylamides. Polyacrylamide gels were designed to obtain the advantages of the performance of acrylamide-based gels while substantially avoiding exposure to acrylamide monomer (15,16). Low molecular weight polyacrylamides, which form low viscosity solutions in water, are cross-linked on demand to form insoluble gels (see ACRYLAMIDE POLYMERS). An early product developed by Dow, but withdrawn in the mid-1980s, comprised a mixture of 20% polyacrylamide and 40% glyoxal (5,17). Cyanagel 2000 Chemical Grout (Cytec Industries) is an anionic polyacrylamide grout in which gel formation is based on complex formation between anionic (carboxylate) sites on the polymer and ferric ions, generated by oxidation of ferrous ions with a mixture of sodium chlorate and sodium bromate. Gel rates increase with increasing bromate:chlorate ratio. As with acrylamide grouts, ethylene glycol can be added to protect against extremes of temperature and repeated wet–dry cycles. This system forms gels similar to those obtained with acrylamide grout and can thus be operated with equipment designed to handle the latter.

Urethanes. Urethane grouts described in the patent literature (18–25) comprise low molecular weight prepolymers of polyethylene or polypropylene glycol, end-terminated with toluene diisocyanate [*1321-38-6*] (TDI) or methylenebis(phenyl isocyanate) [*101-68-8*] (MDI). Gel-strengthening agents, such as aqueous polymeric latex (26), may be added to reduce shrinkage (see URETHANE POLYMERS). Fillers such as Celite, silicates, gypsum, or related substances may also be added. Cure rate is controlled through design of the polyol to make it more hydrophilic for faster rates and also through addition of catalysts such as trimethylolpropane, 1-isopropylimidazole, diazabicyclooctane, or organometallic compounds. Biocides, root control preparations, and other similar chemicals may also be incorporated. Polymerized forms of these grouts are considered nonhazardous. Isocyanate monomers are toxic but are converted to nonhazardous ureas on contact with the environment (27) (see ISOCYANATES, ORGANIC).

Foams and gels are both available. For example, Scotch Seal Chemical Grout 5600 (foam) and 5610 (gel) (trademarks of 3M Company) products are available through Avanti International, and are often used in combination for sewer repair. Large cracks are sealed by placing oakum, saturated in a water-activated foam product, into the cracks where it foams and expands to seal the crack as it cures. A urethane gel is then used to seal smaller cracks with an impermeable, flexible mass; to fill any existing voids behind leaking structures; and to permeate soil to create a grouted soil mass outside the leaking joint or crack. The urethane gels are used diluted six to ten times with water, together with any desired catalysts, strengtheners, fillers, etc. Gel grouts can be applied with automated equipment or pumped through and around structures to be grouted using hand-held injection equipment or high pressure multiratio pump systems. Urethane grouts have shooting viscosities 20 or more times higher than acrylamide or acrylate grouts.

Published comparisons (28) indicate that properly used urethane grouts offer cost–performance equivalent or superior to that of acrylamide grouts in many applications. Urethane grouts have good tensile strength, elongation, recovery properties, shrink resistance, and resistance to wet–dry cycles. Relatively low viscosities in diluted formulations are compatible with grouting of finely divided formations. There are some practical difficulties involved, however. For instance, it is difficult to deliver 1:5 or higher diluted mixtures of urethane grout in equipment designed to deliver 1:1 acrylamide grout:water mixtures. A urethane grout designed to be used in 1:1 ratios with water was introduced by 3M in 1991 but is no longer commercially available. Many urethane grouts contain some free organic isocyanates and either contain solvents such as acetone or ethylene glycol, or require them for cleanup.

Epoxies. Grouts based on epoxy resins (qv) are commercially available both as coatings and as gels. Two-part systems are mixed on-site and applied promptly. Application to wet surfaces is possible and sewer repair has been demonstrated. The Part A component can comprise epoxy resin, with inorganic additives such as titanium dioxide, kaolin, silicates, or ceramics. Part B can comprise cycloaliphatic amines and aliphatic polyamines. Injections of epoxy have been used to strengthen large mechanical structures such as the undersides of bridges. Compressive, tensile, and shear strengths greater than concrete make epoxies desirable for such applications.

Other Grouting Materials. Several other organic polymer systems have been used as grouts. These include urea–formaldehyde resins (aminoplasts) (29), phenol–formaldehyde resins (phenoplasts) (2), and lignosulfonates (2) (see AMINO RESINS; PHENOLIC RESINS). When strengthening of soil to support the weight of roads, railroads, or building foundations is desired, combinations may be used of grouts, admixtures with other materials such as fluidized-bed ash (30), concrete (31), and geosynthetics, usually polyolefin- or polyester-based nonwovens, membranes, honeycombs, and grids (32). Such soil reinforcements to support foundations are widely used but long-term durability is a concern and has prompted detailed study of degradation mechanisms and ways to detect resulting degradation (33).

Organic grouts dominate many grouting applications but growth has been slowed by concerns over worker and environmental exposure to toxic monomers and solvents, as well as resulting regulations on use and disposal of the grouts. Improved products are needed which provide the performance of acrylamide grouts in nonhazardous material systems. Combinations of organic grouts and inorganics, including cement and concrete, are promising, especially for containing hazardous materials. For this application organic grouts alone are considered to be neither sufficiently effective nor sufficiently durable.

Soil Conditioners

Agricultural Applications. The emphasis in soil conditioning for agriculture is on formation of soil aggregates that support seed germination, seedling emergence, efficient use of irrigation water, and erosion prevention; and stabilization of these aggregates against the impact of wind, rain, and irrigation water. Chemical treatments can be a useful supplement to other methods used to prepare fields for agriculture. However, chemical treatment alone cannot be used to recover or prepare fields that are too wet or otherwise unsuitable for agriculture (34,35).

Slaking of weak soil aggregates leads to formation of finely divided material that is deposited on the surface of the soil, forming seals and blocking soil pores. Surface seals impede water infiltration, promote ponding, runoff, and erosion, and reduce water use efficiency, soil aeration, and root respiration. When surface seals dry, they form hard crusts that mechanically impede the emergence of seedlings, reduce stands, lower yields, and require expensive overplanting and thinning or even replanting of crops. Poor water infiltration and internal drainage are common problems on arable soils of the arid southwestern United States (36). Surface crusting and plugging of soil pores caused by fine clay particles (37) and swelling in clay heavy soils (38) result in poor infiltration and drainage. This impairs management of salty or sodic soils, which require adequate leaching and drainage to prevent accumulation of salt and sodium. Inadequate infiltration and leaching result in lower crop yields (39).

Synthetic organic polymers were introduced as agricultural soil conditioners in the early 1950s. A bibliography from 1950 to ca 1995 is available (40). Many different polymer systems have been used, including copolymers of maleic anhydride with either vinyl acetate or isobutene, neutral and anionic polyacrylamides, poly(ethylene glycol), poly(vinyl alcohol), and poly(urea–formaldehyde), as well as polyurethanes and cellulose xanthate. Natural polymers such as polysaccharides and copolymers have also been used (32). Some dramatic improvements in soil properties were possible using polymer systems available in the 1950s, but only at application rates that were not cost-effective. By the mid-1960s, initial enthusiasm had disappeared and there was little activity in polymer soil conditioning. Since that time, improvements in the understanding of soil structure and in organic polymer science have led to better polymers and more efficient ways to apply them for soil conditioning. Cost-effective commercial soil conditioners are emerging; some of the new products produce the same results as the 1950s products did, but at 1% or less of the dose.

Erosion Control.   The serious levels of erosion associated with irrigation and especially with furrow irrigation have been recognized (41). For example, in the northwestern United States, approximately 1.5 million hectares are surface-irrigated. Soils are derived from ash and loess, are low in organics and clay, have weak structure, and contain few durable aggregates. From 5 to 50 t/yr of soil per hectare can be lost from irrigated fields (42). Known erosion-control practices coupled with conservation tillage and selected crop sequences can substantially eliminate erosion. Furrow erosion can be reduced using settling ponds (43), minibasins and buried pipe to control runoff (44), straw placed in furrows (45), and sodded furrows (46). Unfortunately, farmers have been reluctant to use methods that may reduce usable acreage and be cumbersome or expensive to employ.

Overland water flow applies shear forces to soil surfaces. When shear forces exceed the stress required to overcome cohesive forces between soil particles, the particles are detached and suspended in the flow. Suspended particles are carried into surface soil with infiltrating water where they block pores and initiate seal formation (47). Thus, erosion results in reduced water infiltration as well as loss of soil from the field and consequent downstream water pollution. If erosion is controlled, good water infiltration is maintained.

Both synthetic organic polymers and polysaccharides have been shown in laboratory tests to maintain structure and permeability of soils under artificial rainfall conditions (48). Several field studies have demonstrated that 1–10-ppm levels of polyacrylamides dissolved in irrigation water, approximately 1 lb/acre, can eliminate most erosion during furrow irrigation (49). Soluble high molecular weight anionic polyacrylamides have been more effective in erosion control applications than lower molecular weight anionic, neutral, or cationic polyacrylamides (50). The preferred polymers have molecular weights in excess of 10 million and contain approximately 20% carboxylate groups. The USDA group in Kimberly, Idaho recommended an application rate of ≤10 ppm in irrigation water in the early stages of irrigation. Treatments using this protocol minimize loss of both silt and minerals from the field, while minimizing polyacrylamide in the irrigation effluent. The fate of polyacrylamides in the environment and possible environmental impacts have been reviewed (51,52). Both dry polyacrylamides and liquid forms have been used to prepare the necessary solutions in irrigation water. Synergies of the polyacrylamides with other chemical agents such as gypsum (53,54) and certain polysaccharides (55) have been reported. Use of anionic polyacrylamides in erosion control is synergistic with nonchemical erosion control strategies and is a recommended erosion control practice of the Natural Resources Conservation Service (NRCS) in the United States (56).

Different soils, terrains, and irrigation practices may require different application strategies for polyacrylamides, but their effective use to eliminate most silt and mineral, eg, nitrate and phosphate, losses from irrigated fields has been demonstrated at many test sites in the northwestern United States and in Arizona, California (57), and Colorado. Reports from 1996 on the benefits of polyacrylamides are similar to those reported in many studies published since the 1950s (40) except that generally the application rates are much lower. Field productivity is maintained in areas such as the U.S. Northwest where topsoil is thin. Downstream river pollution is reduced. There are other reported benefits, for instance, water infiltration is improved. This effect can be seen clearly in better crop yields on problem fields where high or variable slopes make it difficult to distribute water uniformly by the furrow irrigation method. Better infiltration may also make it possible to use more efficient irrigation strategies. Preventing the formation of soil crusts is another possible benefit. Two commercially available anionic polyacrylamides suitable for erosion control applications are Soiltex G1 soil conditioner (Allied Colloids) and Superfloc A836 Soil Erosion Polymer (Cytec Industries). If polyacrylamide is used on each irrigation throughout the season (not always necessary), the total cost of treatments each season in the mid-1990s would be in the $49–$74/ha range.

Prevention of Soil Crusting. Acid-based fertilizers such as Unocal's N/Furic (a mixture of urea with sulfuric acid), acidic polymers such as FMC's Spersal (a poly(maleic acid) derivative originally developed to treat boiler scale) (58), the anionic polyacrylamides described previously, as well as lower molecular weight analogues such as Cytec's Aerotil L Soil Conditioner, have all been used successfully in at least some circumstances to prevent the formation of soil crusts. It is difficult to prove benefits in the laboratory, and field tests may give variable results depending on local weather conditions. The results of 86 trials of crust prevention agents in Europe and Africa under conditions where rainfall supplied water for agriculture have been summarized (34). In this circumstance, crusts do not form or impede seedling emergence unless it rains between the time of planning and seedling emergence. Individual results were therefore variable but on average the crop yields were increased 16%.

Improvement of Water Retention.   Sodium acrylate–acrylamide copolymers cross-linked with methylenebisacrylamide, the so-called superabsorbent polymers, have been used to improve soil properties, specifically water distribution, availability, and drainage characteristics in situations where soil texture is coarse (sandy) or rainfall is marginal for agriculture. The cross-linked gels absorb large volumes of water, swelling and preventing gravity-induced downward flow in the soil. The absorbed water can later be lost to evaporation or extracted by plant roots. When intimately mixed with soils at application rates of 49–74 kg/ha (more may be needed in sandier soils), the gels may make it easier to cultivate crops under marginal conditions of soil and rainfall. Proper placement of the polymers by spraying (59) or other means (60) can disrupt undesirable soil capillary action while providing a water reservoir (61) in the right location to promote desirable root system growth.

Typical products are supplied as white powders having particle diameters ranging in size from a few micrometers to a few millimeters, depending on the grade. Gel powders can be mixed dry with the soil or partially hydrated and then sprayed onto the soil. Some manufacturers also sell liquid forms. The gels resist aerobic biodegradation and persist in soil for several years even under wet conditions. Decomposition occurs in several months under exposure to sunlight. The particles hydrate when in contact with water, absorbing from 40 (seawater) to 500 (distilled water) times their weight of water and swelling proportionately. Full absorption requires a few hours. The pH of absorbed water is neutral. Hydration and swelling are reversible. Plant root systems create a sufficient pressure gradient to extract more than 90% of the water held by the polymer. The polymers, which may be used in combinations with mulches (62) or other treatments designed to slow evaporation, improve water retention and location in porous soils and help to protect plants where rainfall is marginal. The gels are considered nonhazardous and are compatible with many fertilizers, although polymer dose rates may have to be increased.

Commercial products that are available include Alcosorb 400 Water-Retaining Polymer (Allied Colloids), Aquasorb PR 3005 Superabsorbent Polymer (S. N. Floerger), and Aquastore Absorbant Polymer (Cytec Industries). These products are also sold by distributors, often combined with synergists, under various trade names. A saponified starch-graft polyacrylonitrile copolymer originally developed by the USDA is also available as a biodegradable super slurper (63). Application rates and hence application costs for these products are too high to allow cost-effective use in most large-scale agriculture. Urban uses such as golf courses in arid climates are common. Additional uses include hydroseeding, hydroponics, transplanting of annual plants, and mulching for trees and shrubs.

Other Applications. Construction of highways creates many steep slopes which must be stabilized against erosion by water or wind. Excavations as part of construction projects and natural disasters such as the Oakland, California fire storm of 1991 also create severe erosion problems. Many types of inert structures are available for slope stabilization and erosion control, including retaining structures of various types, revetment systems, and ground covers such as artificial mulches (cellulose fibers, fiber glass); chemical systems such as tackifiers and emulsions; blankets, mats, and nettings to cover slopes (64); and cellular confinement systems. Many of these systems offer very high strength in surprisingly light structures, such as polyethylene-based honeycombs formed into blocks. Although designed for long life, these inert systems, whether based on steel, concrete, or synthetic polymers (65), slowly degrade (66) with time. Hence, reestablishment of vegetation is highly desirable and is possible with porous retaining structures, revetments, or ground covers (67).

Hydroseeding is widely used in slope stabilization. A mixture comprising grass seeds, fertilizer, synthetic polymer, and water is sprayed onto banks. Polymers that include poly(vinyl alcohol), poly(vinyl acetate), polyacrylamide, methacrylates, acrylate–acetate copolymers are used by different manufacturers in formulations for hydroseeding. Many products include a cohesive binding agent (tackifier), such as a latex copolymer emulsion (68), mixed at 2–4% levels in water and sprayed onto the soil, where it forms a thin water-resistant crust that reduces dust formation and controls erosion and then gradually decomposes as vegetation becomes reestablished. Applications of polymer mixed with seed may range from 0.5 to over 5.0 m^3/ha, depending on circumstances (69).

Other chemical systems used in stabilization of highways and embankments have been described (70). Commercial products include $CaCl_2$ and $MgCl_2$ for dust control, soil agglomerating systems that may include active ingredients such as ammonium laureth sulfate, and lignosulfonates used for dust control and road stabilization which act by agglomerating clay particles in soil. Cementitious products, such as the lime-based product produced by Chemical Lime Corporation, are used at a rate of about 5 t/ha, applied with about 4000 liters of water per ton of dry product and sprayed on the soil. The product is applied after the hydromix to hold soil in place, gradually breaking down as vegetation is reestablished. U.S. Gypsum produces a soil stabilizer containing calcium sulfate hydrate which forms a crust after it sets. This material requires 4–6 hours to set; it gradually dissolves, supplying calcium and sulfur to the soil.

Control of surface-irrigation-induced erosion using dissolved anionic polyacrylamides at ppm levels is generally regarded as effective, safe, and affordable in large-scale agriculture. Possible ancillary benefits such as improved crop yields, prevention of soil crusting, and improved water infiltration into soil are under active study. Chemical soil and water conditioning used in combination with other established methods for soil preparation and erosion control is likely to play a key role in the development of truly sustainable irrigated agriculture. Many of the polymer systems used in soil conditioning in the 1990s were originally developed for other purposes. Development of next-generation systems specifically designed for soil conditioning is expected. Use of geosynthetics to support, direct, and control soil profiles is also on the rise.

BIBLIOGRAPHY

"Chemical Grouts" in *ECT* 3rd ed., Vol. 5, pp. 368–374, by R. H. Karol, Rutgers University, and J. Welsh, Haywood Baker Co.

1. R. H. Borden, R. D. Holtz, and I. Juran, eds., *Proceedings of the 1992 ASCE Specialty Conference on Grouting, Soil Improvement, and Geosynthetics, New Orleans, La., Feb. 25–28, Geotechnical Special Publication* Vol. 1, no. 30, ASCE, New York, 1992.
2. R. H. Karol, in *Proceedings of the Conference on Grouting in Geochemical Engineering*, American Society of Civil Engineers, New York, 1982, pp. 359–377.
3. U.S. Pat. 1,827,238 (Oct. 13, 1924) and Brit. Pat. 322,182 (May 24, 1928), H. Joosten (to Tiefban and Kalteinindustrie AG).
4. U.S. Pat. 2,968,572 (Jan. 17, 1961), C. E. Peeler, Jr. (to Diamond Alkali Co.).
5. W. B. Jacques, *Civ. Eng. (New York)* **51**, 59–62 (1981).
6. S. H. Stevens, *J. Am. Coll. Toxicol.* **10**, 193–202 (1991); summarizes toxicity data for polyacrylamides.
7. R. H. Sullivan and W. B. Thompson, *Assessment of Sewer Sealants*, Report EPA-600/8-81-012, Municipal Environmental Research Laboratory, U.S. Environmental Protection Agency, Cincinnati, Ohio, July 1982.
8. *Fed. Reg.* **56**(191), 49863 (Oct. 2, 1991).
9. *Fed. Reg.* **60**(88), 23958 (May 8, 1995).
10. U.S. Pat. 4,429,804 (Jan. 31, 1984), D. M. Piccirilli.
11. W. J. Clarke, in W. H. Baker, ed., *Proceedings of the Conference on Grouting in Geomechanical Engineering, New Orleans, La., Feb. 10–12, 1982*, American Society of Civil Engineers (ASCE), New York, 1982, pp. 418–432.
12. U.S. Pat. 4,295,762 (Oct. 20, 1981), M. Slovinsky (to Nalco Chemical Co.).
13. Brit. Appl. 2,081,281 (Feb. 17, 1982), W. J. Clarke (to Hayward Baker Co.).
14. Fr. DE 2,580,659 (Oct. 24, 1986), A. Bertalan and co-workers (to Magyar Tudomanyos Akademia Termeszettudomanyi Kutato Laboratoriumai, Hungary).
15. Eur. Pat. EP 586911A1 (Mar. 16, 1994), E. E. Miller (to American Cyanamid Co.).
16. World Pat. WO 9402567 (Feb. 3, 1994), K. G. Goodhue, Jr. and M. M. Holmes (to KB Technologies, Ltd.).
17. R. M. Berry, "Inejectite-80 Polyacrylamide Grout" in Ref. 11; U.S. Pat. 4,199,625 (Apr. 22, 1980), R. J. Pilney and co-workers (to The Dow Chemical Co.).
18. U.S. Pat. 3,985,688 (Oct. 12, 1976), S. R. Speech (to 3M Co.).
19. U.S. Pat. 3,719,050 (Mar. 6, 1973), H. Asao and co-workers (to Toho Chemical Industry Co., Ltd.).
20. U.S. Pat. 5,037,879 (Aug. 6, 1991), G. P. Roberts (to 3M Co.).
21. Jpn. Kokai Tokkyo Koho JP 93-147109 (May 26, 1993) (to Nippon Polyurethane Kogyo KK).
22. Jpn. Kokai Tokkyo Koho JP 05 306321 (Nov. 19, 1991) (to Bridgestone Corp.).
23. Jpn. Kokai Tokkyo Koho JP 05 078655 (Mar. 30, 1993) (to Tokai Rubber Ind. Ltd.).
24. U.S. Pat. 4,329,436 (May 11, 1982) and 4,439,552 (May 11, 1982), R. R. Dedolph; U.S. Pat. 4,495,310 (Jan. 22, 1985), R. R. Dedolph (to Gravi Mechanics).
25. U.S. Pat. 4,454,252 (June 12, 1984), F. Meyer (to Bergwerksverband GmbH).
26. U.S. Pat. 4,315,703 (Feb. 16, 1982), A. J. Gasper (to 3M Co.).
27. D. S. Gilbert, *J. Cell Plast* **24**(2), 178–192 (1988).
28. *Water Control Quarterly*, Avanti International, Webster, Tex., Apr. 1991, p. 4; Oct. 1991, pp. 4–5; July 1992, p. 3.
29. A. N. Pukhovitskaya and co-workers, *Plast. Massy*, 60–61 (1987).
30. K. K. Pandey and co-workers, *Proceedings of the Specialty Conference on Geotechnical Practice in Waste Disposal*, Part 2, Geotechnical Special Publication no. 46/2, 1995, ASCE, New York, pp. 1422–1436.

31. M. Morris, *Concrete (London)*, **21**(2), 25–26 (1987).

32. *Innovator's Digest*, Innovator's Digest, Inc., Plantation, Fla., Aug. 14, 1990.

33. *ASTM Special Technical Publication*, no. 1190, ASTM, Philadelphia, Pa., 1993, state-of-the-art technologies and new developments in geosynthetic soil reinforcement and testing.

34. M. F. De Boodt, *Soil Colloids and Their Association in Aggregates*, NATO ASI Ser. B 215, 1990, pp. 517–556, applications of polymeric substances as physical soil conditioners.

35. M. F. De Boodt, *Pedologie* **XLIII-1**, 157–195 (1993).

36. C. Sanchez, private communication, 1994–1995.

37. W. H. Frenkel, J. O. Goertzen, and J. D. Rhoades, *Soil Sci.. Soc. Am. J.* **42**, 32–39 (1978); D. S. McIntyre, *Soil Sci.* **85**, 185–189 (1958).

38. B. L. McNeal, W. A. Novell, and N. T. Coleman, *Soil Sci. Soc. Am. Proc.* **30**, 313–317 (1966).

39. L. Bernstein, *Saline Tolerance of Plants*, Bulletin No. 283, U.S. Dept. of Agriculture, Washington, D.C., 1964.

40. A. Wallace and G. A. Wallace, in *Soil Conditioner and Amendment Technologies*, Vol. 1, Wallace Laboratories, El Segundo, Calif., 1995.

41. D. L. Carter, in B. A. Stewart and D. R. Nielsen, eds., *Irrigation of Agricultural Crops*, Agron. Monogr. 30, ASA, CSSA, and SSSA, Madison, Wis., 1990, pp. 1143–1171; B. F. Hajek and co-workers, in W. E. Larson and co-workers, eds., *Proceedings of the Soil Erosion Productivity Workshop Bloomington, Minn., Mar. 13–15, 1989*, University of Minnesota, St. Paul, 1990.

42. R. D. Berg and D. L. Carter, *J. Soil Water Conserv.* **35**, 267–270 (1980); W. D. Kemper and co-workers, *Trans. ASAE* **28**, 1564–1572 (1985).

43. M. J. Brown, *J. Soil Water Conserv.* **40**, 389–391 (1985).

44. D. L. Carter, *Soil Eros. Conserv.*, 355–364 (1985).

45. M. J. Brown, *J. Soil Water Conserv.* **40**, 389–391 (1985).

46. J. W. Cary, *Soil Sci. Soc. Am. J.* **50**, 1299–1302 (1986).

47. A. G. Segeren and T. J. Trout, *Soil Sci. Soc. Am. J.* **55**, 640–646 (1991).

48. A. M. Helalia and J. Letey, *Soil Sci. Soc. Am. J.* **52**, 247–250 (1980); I. Shainberg, D. N. Warrington, and P. Rengaswamy, *Soil Sci.. Soc. Am. J.* **149**, 301–307 (1990); M. Ben Hur, J. Letey, and I. Shainberg, *Soil Sci. Soc. Am. J.* **54**, 1092–1095 (1990); A. R. Mitchell, *Soil Sci.* **141**, 353–358 (1986); I. Shainberg and G. J. Levy, *Soil Sci.* **158**, 267–273 (1994).

49. R. D. Lentz, I. Shainberg, R. E. Sojka, and D. L. Carter, *Soil Sci. Soc. Am. J.* **56**, 1926–1931 (1992); H. McCutchan, P. Osterli, and J. Letey, *Calif. Agric.* **47**(5), 10–11 (1993); R. D. Lentz and R. E. Sojka, in D. G. Watson, F. S. Zazueta, and T. V. Harrison, eds., *Computers in Agriculture, 1994, 5th International Conference on Computers in Agriculture*, Orlando Fla., Feb. 6–9, 1994, AWAE, St. Joseph, Mich., 1994; R. D. Lentz and R. E. Sojka, *Soil Sci.* **158**, 274–282 (1994); R. D. Lentz, J. A. Forester, and R. E. Sojka, *Agron. Abs.*, 358 (1994); A. M. Nadler, M. Magaritz, and L. Leib., *Soil Sci.* **158**, 249–254 (1994).

50. R. D. Lentz, R. E. Sojka, and D. L. Carter, *Proceedings of the 24th International Erosion Control Association Conference, Indianapolis, Ind., Feb. 23–26, 1993*, International Erosion Control Association, Steamboat Springs, Colo., 1993, pp. 161–168.

51. C. A. Seybold, *Commun. Soil Sci. Plant Anal.* **25**(11–12), 2171–2185 (1994).

52. F. Barvenik, *Soil Sci.* **158**(4), 235–243 (1994); discusses polyacrylamide characteristics related to soil applications.

53. A. Wallace, *Commun. Soil Sci. Plant Anal.* **25**(1–2), 109–116 (1994).

54. M. F. Zahow and C. Amhrein, *Soil Sci. Soc. Am. J.* **56**(4), 1257–1260 (1992).

55. U.S. Pat. 4,797,145 (Jan. 10, 1989), G. A. Wallace and A. Wallace.

56. *Irrigation Erosion Control (Polyacrylamide)*, National Resources Conservation Service, Interim Conservation Practice Standard WNTC 201-4, 1995.

57. H. McCutchan, P. Osterli, and J. Letey, *Calif. Agric.* **47**(5), 10–11 (1993).

58. T. V. Sylling and S. L. Allen, U.S. Pat. 4,687,505 (Aug. 18, 1987); U.S. Pat. 4,923,500 (May 8, 1990); U.S. Pat. 5,106,406 (Apr. 21, 1992) (to Sotac Corp.); WO 92/21452 (Dec. 10, 1992) (to Sotac Corp.).

59. U.S. Pat. 5,185,024 (Feb. 9, 1993), S. R. Siemer and co-workers (to Aqua Source, Inc.).

60. U.S. Pat. 5,303,663 (Apr. 19, 1994), R. D. Salestrom (to Soil Injection Layering Systems, Inc.).

61. K. S. Kazanskii and S. A. Dubrovskii, *Advan. Polymer Sci.* **104**, 98–133 (1992).

62. D. J. Wofford, Jr. and M. D. Orzolek, *Am. Veg. Grower*, 20 (Nov. 1993).

63. L. Cook, *Agric. Res.*, 16 (Jan. 1994).

64. R. H. Manz, *TAPPI Proceedings: 1985 Nonwovens Symposium*, TAPPI Press, Atlanta, Ga., 1985, pp. 187–190.

65. N. E. Wrigley, *Mater. Sci. Technol.* **3**(3), 161–170 (1987); J. Templeman, *Plast. Rubber Int.* **10**(4), 20–24 (1985); M. Morris, *Concrete (London)* **21**(2), 25–26 (1987).

66. D. G. Bright, *ASTM Special Technical Publication*, No. 1190, ASTM, Philadelphia, Pa., pp. 218–227; Y. G. Hsuan, R. Koerner, and A. E. Lord, *ibid.*, pp. 228–243; A. N. Netravali, R. Krstic, J. L. Crouse, and L. E. Richmond, *ibid.*, pp. 207–217.

67. D. H. Gray, *Erosion Control*, 33–39 (Sept.–Oct. 1994); J. Jesitus, *ibid.*, 28–33 (Nov.–Dec. 1995).

68. *Better Roads* **54**(2), 28–29 (1994).

69. B. Karg, *Erosion Control*, 52–61 (May–June 1995).

70. D. Snow, *Erosion Control*, 25–31 (Sept.–Oct. 1994).

General References

"Scotch Seal Chemical Sealants," product literature 98-0701, 3M Co., St. Paul, Minn., 1995.

"Alcosorb 400 Water-Retaining Polymer," undated product literature, Allied Colloids; describes agricultural uses of cross-linked polyacrylamide gels.

R. M. Berry and F. D. Magill, "Chemical Grouting: What's New and What's Being Done This Decade," unpublished manuscript, 1992; contains tables of grouting products, formulation viscosities, as well as pricing information; available from Avanti International, Webster, Tex.

S. N. Floerger, "Polymer Soil Conditioners," undated product literature, Saint Etienne, France; describes functions and uses of cross-linked anionic polyacrylamides in agriculture.

"Geo/Chem AC-400 Chemical Grout," undated product bulletin, Geochemical Corp., Ridgewood, N.J.

DONALD VALENTINE, JR.
Cytec Industries Inc.

SOLAR ENERGY

Solar energy represents a potentially limitless source of energy. Roughly 10,000 times as much solar energy falls on the surface of the Earth each year as is consumed in the form of fossil and nuclear fuel. The United States consumed approximately 88 EJ (8315×10^{15} Btu) of primary energy in 1993, about a quarter of the world total. Petroleum (qv) accounted for 40% of U.S. energy needs, coal (qv) and natural gas (see GAS, NATURAL) 25% each, and nuclear (see NUCLEAR REACTORS) 8% in 1993. In its various forms, solar energy provided for approximately 4% of the annual energy requirements of the United States in the mid-1990s (1). The actual contribution of solar energy to U.S. and world energy needs over the coming decades is expected to be strongly influenced by the economic and societal attractiveness of the technologies developed to convert the solar resource to useful energy forms. The rate of commercialization of solar technologies is dependent on the adoption of attractive, albeit novel, technology systems.

Solar energy is most certainly expected to make increasing contributions to the world energy supply in the twenty-first century. However, progress in research and development is needed to assure that solar technology achieves its potential. Greater reliance on solar energy provides secure indigenous supplies of energy that can be utilized with minimal environmental impact. Solar energy production usually involves the creation of local jobs, and most of the technologies can be implemented using relatively lower cost workers and materials.

Whereas much of the research and development effort has focused on techniques for generating electricity, solar energy is also capable of producing process or space heat, chemical energy, or even high value fuels and chemicals. In 1970 essentially all utilization of solar energy came from hydroelectric power, used to generate electricity, and wood (qv), used for cooking, space heating, industrial processes, and power generation (qv) at rather low thermodynamic efficiency. As of the mid-1990s, many of the advanced solar energy technologies such as wind power plants, solar thermal power plants, municipal waste energy plants, and alcohol transportation fuels are being added to the traditional forms (see ALCOHOL FUELS; FUELS FROM BIOMASS; FUELS FROM WASTE; RENEWABLE ENERGY RESOURCES).

Part of the desire to utilize solar energy results from the technologies being thought of as environmentally benign. These technologies use sunlight, rainwater, or wind as the energy resource and thus generally do not produce gaseous emissions or waste materials having adverse environmental impact (see AIR POLLUTION). Many of the most promising solar technologies, however, are still in the early stages of development. In order to reduce the environmental impact of growing world energy use, additional technological advances and cost reductions must be accomplished.

The intensity of incident sunlight is diffuse, having a peak power density of only 1 kW/m^2 at the Earth's surface at noon in the tropics. The efficiency of conversion of solar energy varies from a few percent for photosynthetic production of biomass to as much as 15–20% for production of electricity for some photovoltaic modules (see PHOTOCHEMICAL TECHNOLOGY; PHOTOVOLTAIC CELLS). For this reason, progress in reduction of the intensity of energy demand is important and makes widespread use of solar energy more practical. The energy intensity of the U.S. economy declined by fully 30% between 1970 and 1990. During the same period the electricity sector share of the energy mix increased from 25 to 36%, and the U.S. Energy Information Administration projects a further increase to 36% by 2010. The trend toward electrification worldwide militates in favor of a shift to solar or renewable energy because many of the most promising solar conversion technologies naturally produce electricity.

Great progress in both cost and reliability of a number of emerging solar technologies was made during the 1980s and early 1990s. In addition, a much better understanding of the remaining technical hurdles to be surmounted in order to bring solar energy systems to market was provided. A number of innovative means of efficient conversion of resources into usable energy have been identified and are being explored. These include thin-film photovoltaics, high performance wind turbines, fast pyrolysis of biomass, anaerobic digestion of biomass, simultaneous saccharification and fermentation of cellulosic materials, and genetically engineered biomass production (see CHEMURGY; FUELS FROM BIOMASS; GENETIC ENGINEERING; THIN FILMS).

The technology and cost progress of the emerging solar technologies as of the mid-1990s is discussed herein. A significant level of private sector interest has led to advances in wind, solar thermal, and photovoltaic electricity systems, as well as various possibilities for advanced biomass utilization including gasification/electricity generation and the production of transportation fuels and chemicals from biomass feedstocks.

Wind Energy Technology

The use of wind as a renewable energy source involves the conversion of power contained in moving air masses to rotating shaft power. These air masses represent the complex circulation of winds near the surface of Earth caused by Earth's rotation and by convective heating from the sun. The actual conversion process utilizes basic aerodynamic forces, ie, lift or drag, to produce a net positive torque on a rotating shaft, resulting in the production of mechanical power, which can then be used directly or converted to electrical power.

The scope of the wind resource is widespread and less dependent upon latitude than other solar technologies. The accessible resource in the United States has been conservatively estimated to be capable of providing more than 10 times the electricity consumed therein. The intermittency of the wind resource, however, makes it impractical to base more than 10–20% of electricity generation on this resource until a suitable storage technology is developed. Wind is a very complex resource, existing in three dimensions, rather than the two associated with other solar resources. It is intermittent and strongly influenced by terrain effects. Moreover, there is a nonlinear (cubic) relationship between wind speed and power or energy available. This last factor is best illustrated by comparing good, excellent, and outstanding wind sites having average wind velocities of 5.5, 7.0, and 8.5 m/s, respectively. This 1.5-m/s difference results in the excellent site having 106% more available energy per unit than the good site for conversion to electricity; the outstanding site has 269% more available energy than the good site.

Wind machines can be classified as either horizontal-axis or vertical-axis designs and typically utilize either two or three airfoils, as shown in Figure 1. Vertical-axis wind machines include both the Darrieus and Savonius designs. The Darrieus machine requires an auxiliary starting mechanism in order to produce useful energy. Commercial interest has centered on horizontal-axis machines more recently, partly because of the need to elevate most of the structure of the vertical-axis machines for maximum effectiveness. During the 1970s, large government-supported demonstration projects in the United States focused on the design and testing of very large machines of capacity from 2–5 MW. More recent commercial designs are evolving toward machines having capacities between 200–500 kW each. These smaller machines are usually grouped into wind farms of total capacity of 20 MW or more. Wind turbines of much smaller (10 kW) capacity are finding increased application for rural electrification, particularly in developing countries.

Wind energy, which has proved to be the most cost-competitive and utilized solar technology for the bulk power market, is one of the fastest-growing electricity generating sources worldwide. In 1995 alone, wind generating capacity grew by 32% to a total of 4900 MW (2). Although the United States has led in

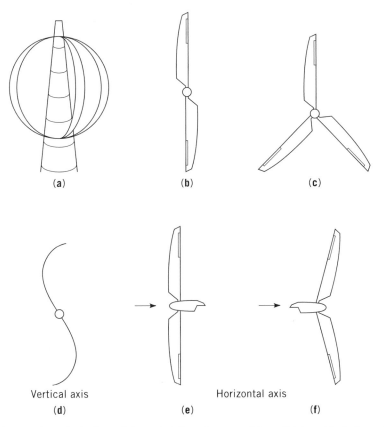

Fig. 1. Categories of wind machines: (**a**) Darrieus machine; (**b**) two-bladed machine; (**c**) three-bladed machine; and (**d**) Savonius machine. Parts (**e**) and (**f**) represent upwind and downwind placement, respectively.

installed wind capacity, having 1650 MW at the end of 1995, most recent growth, ca 1996, has come from projects in Europe and Asia. As a result of early operating experience in California, the industry has improved turbine technology and operating strategies such that installations reliably produce energy at costs that as of 1996 rivaled those of most conventional resources.

Wind technology provides economical energy to remote areas and for specialized applications. The modularity and wide size range of wind turbines available enable wind energy to serve many applications. Examples are small radio transmitters, offshore oil rigs, navigational aids, and remote communities. By combining the intermittent wind energy with backup power sources such as diesel generators or storage devices, most loads can be reliably and competitively served.

An issue for traditional electricity generation and distribution grids centers on how wind and other intermittent solar technologies should be considered relative to capacity credits. Historically, little if any capacity credit was given because wind cannot be considered dispatchable nor can the output be relied on to coincide with utility loads. Improved weather forecasting techniques as well as greater geographic dispersal of wind farms are expected to help mitigate this concern, but consistent methodologies are needed. In the longer term, the cost of generating wind energy could fall to the point where it is less expensive than the cost of the fuel for a fossil-fired power plant connected to the same utility grid. In that eventuality, wind energy would be used to extend fossil fuel reserves; conventional power plants would then be used to firm up the intermittent wind resource.

Wind energy has little or no impact on flora, fauna, climate, materials, or in terms of human health hazards. It does, however, have a potential negative impact on land use. On the negative side, three siting considerations require mention: the visual impact of large, rotating structures; the nearby acoustic disturbance associated primarily with the generation of aerodynamic forces on the rotating airfoils (see NOISE POLLUTION AND ABATEMENT (SUPPLEMENT)); and concerns about the possibility of bird kills from the rotating blades. In general, experience has shown the first two factors to be minimal, as long as the turbines are not located in proximity to populated areas. Considerable study has gone into the issue of bird kills (3). Mitigation of this factor may simply depend on a redesign of the wind turbine support tower so as to minimize available perches for avian raptors.

On the positive side, the three-dimensional nature of the resource provides it with a distinct advantage compared to other solar technologies. Specifically, because siting usually involves placing the individual turbines as high as possible, typically spacing turbines about two to three blade diameters apart crosswind and 10 diameters apart downwind, only a small fraction of a wind farm area is actually occupied. The rest of the land remains available for other applications, such as crop production or livestock grazing.

Performance of wind turbines, as well as other sources of energy, must be judged by the cost of energy (COE), ie, the levelized cost per kilowatt hour of electricity produced. For wind turbines, this cost can be determined from only a few parameters: the capital cost (in $/m^2, including all balance of system costs), the annual energy capture (in kWh/m^2), and the operation and maintenance/replacement costs (annualized to $/kWh). Costs in California in 1995 were $0.07 to $0.10/kWh, derived from capital costs of about $450/m^2, annual energy capture of 600–800 kWh/m^2, and operation and maintenance/replacement costs of $0.012 to $0.014/kWh. These cost and energy figures represent a significant improvement over the values for machines installed in the early 1980s, particularly with respect to capital costs. To tap a significant portion of the accessible resource mentioned, however, performance must be improved.

Potential incremental technological advancements that in aggregate would represent a dramatic improvement in turbine performance are shown in Figure 2. Whereas these improvement areas are broadly defined because of the uncertainty associated with long-term research, near-term improvement possibilities are fairly well defined. Initial improvements to existing designs are expected to occur through the use of advanced design tools, including turbulence codes, aerodynamic and structural codes, and fatigue life models (see

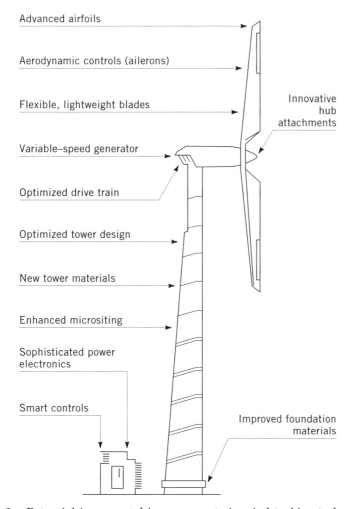

Fig. 2.   Potential incremental improvements in wind turbine technology.

COMPUTER-AIDED DESIGN AND MANUFACTURING (CAD/CAM)). Also included is the use of advanced airfoils designed specifically for wind turbines to increase both energy output and rotor fatigue life. Site tailoring refers to the optimization of system designs for site-specific characteristics. Examples might include tall towers for locations having a strong vertical wind shear and control strategies optimized for different turbulence levels that would maximize power output while minimizing operation in damaging wind conditions. Operating strategies include the possible use of power electronics that allow the speed of the rotor to vary with wind speed while maintaining constant frequency power output thus allowing the turbine to operate at optimum efficiency over a wide range of wind speeds. Possible array spacing strategies are also being investigated to maximize energy capture over large arrays of wind turbines. The optimum spacing of turbines, both within and between rows, is dependent on the terrain as well as predominant atmospheric conditions. Other strategies include varying the heights of adjacent turbines to promote mixing in the boundary layer. This would reduce wake energy deficits and turbulence effects for downstream rows of turbines.

For the longer term, configurations and advanced designs that achieve dramatically improved reliability and manufacturability are sought. Rotors designed to withstand the fatigue loads that are only beginning to be understood would provide the reliability needed. Advanced designs might include new, highly flexible, lightweight rotors that are relatively insensitive to high wind turbulence levels. Greater strength or flexibility at reduced weights and costs are desired, as is optimum manufacturability of the advanced turbines. The goal of the improvements is to reduce the COE from wind energy. A reduction to 30–40% of 1995 levels should be possible.

Achievement of development goals is expected to lead to cost reductions such that cost-effective machines at a good site should produce electricity at ~$0.05/kWh.

The issues of facilitating options such as energy storage and transmission may prove to be important to the success of wind energy technology. Cost-effective storage coupled to wind systems would yield capacity credit benefits. In addition, because sites are often isolated, the value of wind energy would benefit from transmission/distribution access.

Solar Thermal Electric Technology

Use of concentrated sunlight to generate electricity by thermodynamic processes is well documented (4). Reflective surfaces concentrate incident sunlight onto a receiver, where it is absorbed into a working fluid that powers a thermal conversion–generator device. Solar thermal systems, operating either with storage or in a hybrid mode with an auxiliary fuel, offer significant potential as capacity to meet utility peaking or intermediate electric power-generation needs.

Three main types of concentrating collectors have evolved for use in solar thermal systems: low concentration parabolic troughs, high concentration parabolic dishes, and central receivers (Fig. 3). Higher concentration produces higher temperatures in a working fluid and makes electrical generation more efficient.

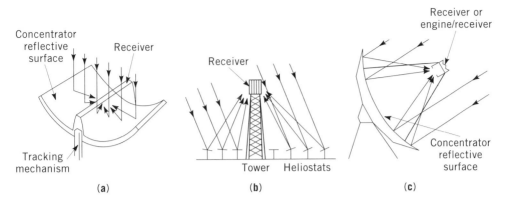

Fig. 3. Solar thermal designs: (**a**) parabolic trough; (**b**) central receiver; and (**c**) parabolic dish.

Parabolic trough systems use surface reflectors to concentrate sunlight onto a fluid-filled receiver tube that is positioned along the line of focus. Concentration ratios of more than 100 times are typically used to generate temperatures of 400–500°C. Troughs are modular and many can be grouped together to produce large amounts of heated fluid. The fluid is then transported to a nearby facility to generate electricity.

The modular parabolic troughs and dishes are classified as distributed systems, whereas central receiver systems, in which heliostats are deployed in a central receiver configuration by placing large numbers of them around a tower-mounted receiver, are more centralized. All concentrating systems have their best annual output in regions where direct insolation is highest. Examples are the southwestern United States and other semiarid regions of the world. These systems also can be utilized, at slightly higher cost, in other geographical areas having somewhat lower levels of direct insolation.

All solar thermal electric system types have been demonstrated in industrial-like settings. A 10-MWe experimental central receiver power plant, Solar One, was deployed by a joint government–industry team and operated successfully by the Southern California Edison Utility on its grid for six years. Whereas system efficiency (7.4%) for this plant was somewhat below the initial predictions, extensive operational experience was gained and the plant delivered more than 37,000 MWh net energy to the grid, achieving 99% heliostat availability and 96% overall availability for the entire plant. The next generation plant, Solar Two, has been designed to incorporate modifications to the central receiver and heliostat design. The central receiver is to use molten salt as the working fluid and should mitigate most of the problems attributable to inter-mittency of sunlight encountered using Solar One. Annual system efficiencies of 14–15% and costs of $0.08–0.12/kWh have been projected for this plant.

Prototype parabolic dish electric systems totaling about 5 MWe have also been operated in utility settings in Georgia and in southern California. More recent development of a dish-mounted engine–generator has led to significant increases in system performance as compared to the earlier designs which collect the heat as thermal energy and transport it to a central location for electric generation. Indeed, a dish/Stirling engine–generator model has achieved a 29% overall system conversion of sunlight to electricity (5).

As of this writing (ca 1996), 354 MWe of privately funded, parabolic-trough electric generating capacity was operating in California. These trough systems operate in a hybrid mode, using natural gas. Collectively they accounted for more than 90% of worldwide solar electric capacity. The cost of these systems fell steadily from $0.24/kWh for the first 14-MW system to an estimated $0.08/kWh for the 80-MW plant installed in 1989 (5).

Impressive technologies have reduced technical and financial risks in solar thermal electric technology. Although first-generation solar thermal systems have proved successful, the 1991 bankruptcy of the primary solar trough development company, Luz International, Ltd., offered insight into the challenges faced in the introduction of a new electricity generation technology. This bankruptcy was the result of a complex interplay between the unanticipated elimination of federal and state tax benefits and a sharp drop in the levelized cost of electricity beginning in the mid-1980s as advanced gas turbines using natural gas became available (6). Following this bankruptcy, solar thermal trough plants have been successfully operated at Kramer Junction, Daggett, and Harper Lake, California by separate operating companies.

International markets also provide an opportunity for solar thermal technology. Small systems, such as a prototype 7-kW parabolic dish/Stirling engine system under development by the Cummins Power Generation Company, have the potential to be competitive in either grid-connected or stand-alone applications in many Third World countries. Science Applications International has developed a similar design using a 25-kW dish system which is to be tested as part of a government–industry joint venture in the United States.

The cost of energy from solar thermal electric systems, which was $0.24/kWh in 1984, was reduced to $0.08–0.12/kWh by the mid-1990s. Components that provide further improvement have been developed and are being evaluated. Dish electric systems utilizing a stretched-membrane dish integrated with a reflux receiver and a reliable Stirling engine, when developed and mass-produced, are projected to achieve $1200/kWe. Cost estimates for energy from such a dish electric system are projected to reach $0.05/kWh, low enough to be competitive with fossil fuel power generation.

Photovoltaics

Photovoltaic devices typically consist of a series of thin semiconductor layers that are designed to convert sunlight to direct-current electricity (see SEMICONDUCTORS). As long as the device is exposed to sunlight, a photovoltaic (PV) cell produces an electric current proportional to the amount of light it receives. The photovoltaic effect, first observed in 1839, did not see commercial application until the 1950s when photovoltaic modules were used to power early space satellites. Many good descriptions of the photovoltaic phenomenon are available (7).

Photovoltaic devices produce electricity from incident direct or diffuse sunlight. Figure 4 shows a schematic of the operation of a photovoltaic device. These devices have no moving parts and thus are extremely reliable. Moreover, their

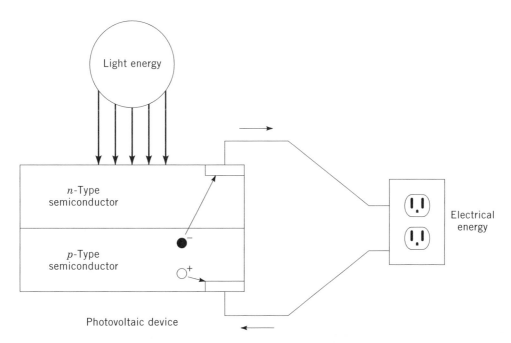

Fig. 4. Schematic illustrating operation of a photovoltaic cell.

operation does not release any effluent to the atmosphere. Costs, however, were relatively high compared to the operation of bulk electricity generation technologies as of the mid-1990s and the output of photovoltaics were intermittent because of variations in sunlight. Despite these limitations, the growth in demand for photovoltaic systems averaged about 30% per year throughout the 1980s and early 1990s.

The smallest unit of a PV system is called the PV cell. Cells are manufactured using crystalline and amorphous forms of silicon, copper indium diselenide [12018-95-0] (CIS), cadmium telluride [1306-25-8], and gallium arsenide [1303-00-0] as well as even more exotic materials. Photovoltaic systems generally consist of a flat layer of semiconductor material encapsuled by a glass or plastic cover, or of individual high efficiency PV cells incorporated in an optical arrangement to concentrate the sunlight. This latter arrangement often requires a solar tracking system, whereas the former, flat plate arrangement is normally installed at a fixed angle determined by the latitude of the site. Both types of PV devices are progressing about equally toward reduced cost.

Modules, the building blocks of large PV systems, are aggregates of PV cells large enough to provide convenient levels of electrical power. These modules can be 0.1 m^2 to more than 2 m^2 and can be expected to produce from 0.5 to 2 W/m^2 of power during a clear midday, depending on the conversion efficiency of the cell material. The efficiency is defined as the ratio of electricity produced to the amount of sunlight incident on the PV device and is a critical figure-of-merit characterizing all PV cells, modules, and systems. The output power of a module at noon on a clear day is called its peak-watt power because it represents a maximum typical output. A module characterized as 100 Wp produces 100 W of power during a clear midday.

The efficiency of the flat-plate modules ranged from about 5% to almost 15% in 1995. Concentrator modules had correspondingly higher efficiencies and costs. The improvement in efficiency for a number of promising PV materials is shown in Figure 5. Two other important parameters are lifetime and module cost. Lifetimes range from a few years, for some of the newer technologies, to 20 years or more for crystalline silicon modules. Module costs were in the range of $200–$500/m^2 as of 1995 and showed a continual decline that was expected to be ongoing. Indeed, photovoltaic modules were increasingly being mass-produced and therefore likely to benefit from the economies of mass production. Figure 6 demonstrates the kind of manufacturing learning curve which would be expected for such a technology into the year 2000.

Unlike solar thermal systems or PV concentrator systems, the PV flat plate systems work well in cloudy locations because these latter convert diffuse as well as direct sunlight to electricity. On an annualized basis, the energy produced by a photovoltaic array varies by only about ±25% from an average value for the contiguous 48 states of the United States. As a result, it is practical to use photovoltaic systems in normally cloudy locations such as Seattle or northern Maine.

The terrestrial PV market has three principal segments: consumer products, remote power, and utility generation. The consumer product market was one of the first economic applications of the technology and is characterized by

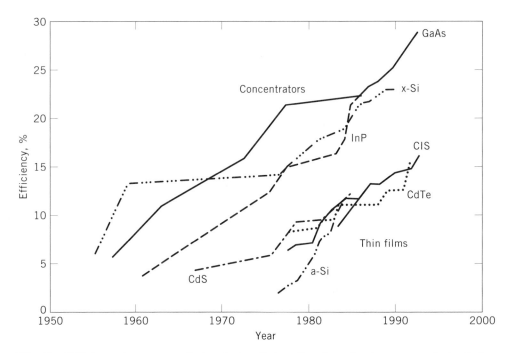

Fig. 5. Efficiency improvements in photovoltaic cells where (——) corresponds to GaAs; (- - -) InP; (– · – ·) CdS; (· · · ·) CdTe; (· · – · ·) amorphous silicon; and (– · · · –) crystalline silicon. Courtesy of the National Renewable Energy Laboratory.

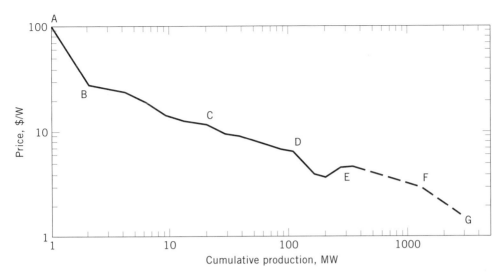

Fig. 6. Photovoltaic module experience where the dashed line corresponds to projected values. Point A represents the 1970 price of $100/W; B, the 1975 price of $30/W; C, the 1980 price of $12/W; D, the 1985 price of $6.75/W; E, the 1992 price of $5/W; F, the 1995 price at $3/W; and G, the year 2000, at <$2/W. Courtesy of the National Renewable Energy Laboratory.

millions of small, milliwatt-sized cells powering calculators and watches. This market, which had been more than 5 MW/yr in the early to mid-1990s, is expanding to larger systems, eg, for battery charging and walkway lighting, reaching power levels near those used for the remote power application market (see BATTERIES).

The largest use of PV is the remote power market. The self-contained and modular nature of PV systems has led to their adoption to meet power loads remote from the electric utility. These applications have come to be referred to as stand-alone because all of the energy needed by the load must come from on-site sources. Typical stand-alone uses are power for telecommunications, lighting, security systems, water supply, battery charging, cathodic protection, vaccine refrigeration, remote monitoring, rural housing, and small villages. The systems are economical because there is no reasonable alternative, such as for a microwave repeater on an inaccessible mountaintop, or because the alternative (often diesel generators) is too costly to install, operate, and refuel. The remote power market accounts for the vast majority of the sales by U.S. industry and is split about equally between international and domestic applications. The domestic customers are remote homeowners, companies purchasing for telecommunications, cathodic protection, and literally hundreds of other uses, as well as governmental agencies like the U.S. Coast Guard (navigation aids), the Department of Defense (battery chargers), and state highway departments (emergency call boxes). The international customers are governments or donor agencies involved in rural electrification and development. The mid-1990s rapid growth rate in PV sales was almost exclusively a result of the increase in sales for remote power. Photovoltaics have gained acceptance and recognition as a reliable and economical remote power source.

Biomass and Biofuels

Biomass is the term used to describe all plant-derived materials, whether wood (qv) or wood wastes, residue of wood-processing industries, food industry waste products, sewage or municipal solid waste (MSW), herbaceous or other biological materials cultivated as energy crops, or other biological materials. Biomass is both a principal and a prospective source of energy. Green plants use the sun's energy to convert CO_2 from the atmosphere to sugars during photosynthesis. Hence biomass is considered a form of solar or renewable energy. Unlike direct solar or wind, the solar energy in biomass is stored for later use. The conversion efficiency of photosynthesis is very low, however. The key feature of the biomass technology is the rapid (≤ 20 yrs) recycling of the carbon fixed in the biological process. Unlike the burning of fossil fuels, Earth's reserves of which were ultimately derived from solar energy because these materials consist of degraded residues of plants and animals, combustion of biomass merely recycles the carbon fixed by photosynthesis in the growth phase and typically has no net impact on global carbon dioxide levels.

Biomass has been used as a source of energy throughout history, representing as of 1995 up to 35% of the primary energy used for cooking and heating in developing countries. The use of biomass or biofuels as a source of energy for space heating, process heat, electricity production, transportation fuels, or as an intermediate gaseous fuel is attractive not only for economic reasons wherever the fuel is readily available at low cost, but also for economic development and environmental reasons. The systems that convert biomass into usable energy can be modular and efficient on a relatively small scale. Both thermal supply and electric generation systems provide 24-hour, base-load (dispatchable) output. Biomass is a renewable and indigenous resource that requires no foreign exchange. The agricultural and forestry industries that supply feedstocks also provide substantial economic development opportunities in rural areas. The pollutant emissions from combustion of biomass are usually lower than those from fossil fuels. Furthermore, commercial use of biomass may avoid or reduce problems of waste disposal in other industries, such as forestry and wood products, food processing (qv), and particularly MSW in urban centers. In addition, recycling (qv) of paper, glass, plastics, and metal products, at times performed in conjunction with municipal waste collection and combustion, can conserve energy resources required for the primary manufacture of these energy-intensive materials.

There are four principal ways in which biomass is used as a renewable energy resource. The first, and most common, is as a fuel used directly for space and process heat and for cooking. The second is as a fuel for electric power generation. The third is by gasification into a fuel used on the site. The fourth is by conversion into a liquid fuel that provides the portability needed for transportation and other mobile applications of energy. Figure 7 shows the varied pathways which can be followed to convert biomass feedstocks to useful fuels or electricity.

Thermal Combustion of Biomass. Direct combustion in air is the principal mechanism used to convert biomass into useful energy. The heat or steam (qv) produced is used to generate electricity or thermal requirements for indus-

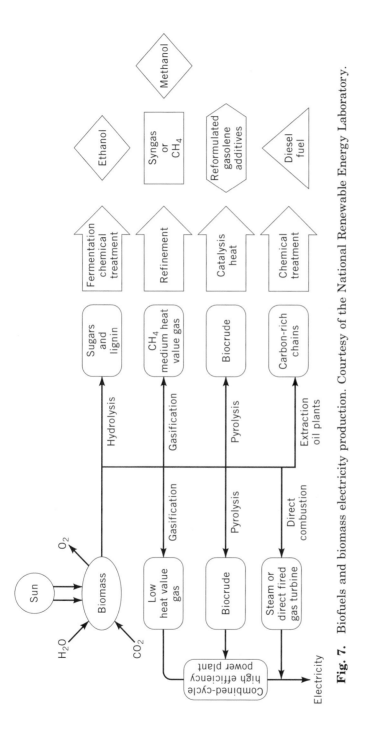

Fig. 7. Biofuels and biomass electricity production. Courtesy of the National Renewable Energy Laboratory.

trial processes, building heating, cooking, or district heating in municipalities. The thermal combustion of biomass for cooking, space heating, or the production of process heat, either directly or in the form of steam, may be attractive where biofuels are available at economic prices or, particularly in rural areas, where the fuel may be available for gathering by the consumer. Small-scale use, such as for home cooking and fireplace use, has traditionally been inefficient.

Larger furnaces and boilers have been designed and are available for burning various types of biomass such as wood, wood wastes, chips, black liquor from pulping operations, food industry wastes, and MSW. The larger units can be very efficient, nearly matching the performance of fossil fuel furnaces. The greater moisture content of most biomass, as well as the wide range of particle size and composition, make it difficult to achieve comparable efficiencies at reasonable costs. The economic advantages, however, make installation of cogeneration facilities attractive for most industrial consumers having available biomass feedstock.

The residential sector uses biomass for direct applications such as cooking and space heating. Considerable progress has been made in the design of household cooking and heating appliances that use biomass, primarily wood, to improve efficiency and reduce CO and particulate emissions. Energy-efficient cooking appliances as well as heating equipment, such as stoves and fireplaces, have been developed, although costs and appearance of fireplace equipment, for example, have deterred use, particularly in potential retrofits. Restrictions on allowable emissions from the burning of wood or increased conventional energy prices could stimulate additional conversions of existing wood-fired installations and fossil-fueled or electric equipment.

The industrial sector uses biomass for both process and space heating, as well as power generation, often jointly in cogeneration projects. The technology available to these larger consumers, however, is equivalent to that used in burning conventional fossil fuels and has been widely implemented. Somewhat lower efficiency is obtained compared to fossil fuels as the result of the high moisture content of biomass. Moreover, derating of the furnace or boiler used might occur for a plant not designed with the flexibility to handle biomass. However, the lower efficiency of the combustion process is often more than offset by the low cost of the fuel, particularly if it is waste or by-product material that otherwise would be marked for disposal. The small amount of ash produced is usually suitable as a soil supplement.

Generation of Electric Power Using Biomass.   As of this writing (ca 1996), electric utilities were making only limited use of biomass as a fuel for power generation, although the utilities often bought power from cogenerators who used biomass as fuel. Most power generation from biomass, whether generated by industry or utilities, was via steam turbines. Research is continuing to develop gas clean-up technologies that would permit use of gasified biomass as a fuel for gas turbines. Wood and wood wastes and by-products are the principal fuels.

Electric power generation using biomass as a fuel is economic in situations where the cost of the fuel is competitive with that of fossil fuels. The cost of a commercially available biomass steam–electric power plant is about $1500/kW for a wood-fired facility. If wood can be obtained at a cost of $2.00/GJ ($2.10 $\times$ 10^{-6}/Btu), the total cost of power for base-load operation would be about

$0.05/kWh. If wood or agricultural wastes are available at lower costs, the cost of electricity would be significantly lower. Similarly, if the low pressure steam from the turbine exhaust can be used (cogeneration), the overall efficiency would be higher and the costs would be lower.

Greater use of biomass resources (exclusive of MSW) in electricity generation is constrained by delivered resource costs. Wood or other biomass resources must generally be procured within no more than a 80-km radius of the power plant to be economical. Transportation costs for biomass are high. All generation capacity (other than MSW plants) has been sited where readily accessible waste resources are available at low cost. Biomass-fueled plants which are to be competitive with coal, oil, and natural gas, the fossil fuel feedstocks, must be available for $2.00/GJ or less, although a higher fuel cost may be competitive in isolated locations where low cost coal is not a viable alternative as a fuel source. At high levels of utilization, competition among energy and nonenergy uses may tend to bid up biomass resource prices. To stabilize price and availability, a potential long-term solution is the development of dedicated high productivity herbaceous or short-rotation woody crops for feedstock production.

Gasification of Biomass. The third energy conversion mechanism is the production of biogas, a mixture of methane and carbon dioxide, CO_2, which can be produced from either thermal conversion or the biological anaerobic digestion of biomass materials. The methane can be subsequently separated from the CO_2 using conventional technology and the resultant gas supplied to a natural gas system or other consumer. Processes have been developed and tested; some have been applied in commercial operations where biomass feedstocks were available at low cost. MSW may be processed anaerobically to produce methane from the digestible components. The volume of gasification residue, which includes materials such as burnable plastics, is greater than the residues from combustion processes. Combustible plastics and similar materials could be separated if required before feeding the remaining material to the digester. Landfills are also a source of methane produced from the decomposition of MSW although the economics of recovery of the naturally occurring methane are not universally favorable. A lower heat-content gas, syngas, consisting primarily of carbon monoxide and hydrogen (CO and H_2), can also be produced for use as a fuel or as an intermediate feedstock.

Interest in the development of a high efficiency biogas-fueled power plant has increased to the point that the Global Environmental Facility (an arm of the World Bank) has decided to fund a demonstration project in Brazil. The Biomass Integrated Gasification–Gas Turbine (BIG–GT) project is proposed to use trees from a eucalyptus plantation in the state of Bahia to demonstrate high efficiency biomass electricity technology with a 25–30-MW plant. The project team, composed of representatives of the Brazilian Ministry of Science and Technology, Electrobras and several of its regional utilities, and Shell Brazil, has decided to use a low pressure gasification technology developed by the Swedish company TPS. The BIG–GT technology could eventually achieve efficiencies on the order of 50% and promises to revolutionize the field of biomass electricity production.

Biofuels. Biofuels are liquid fuels, primarily used in transportation (qv), produced from biomass feedstocks. Identified liquid fuels and blending components include ethanol (qv), methanol (qv), and the ethers ethyl *t*-butyl ether (ETBE) and methyl *t*-butyl ether (MTBE), as well as synthetic gasoline, diesel, and jet fuels.

Ethanol.   Ethanol can be produced from sugar (qv), starch (qv), or cellulosic feedstocks, ie, from wood, energy crops, and municipal and other wastes. In the United States, the primary pathway for conversion of biomass to alcohol fuels is the fermentation of corn to ethanol. In the biochemical conversion process, the biomass feedstock is first separated into its three main components, cellulose (qv), hemicellulose (qv), and lignin (qv). The cellulose is hydrolyzed to sugars, primarily glucose, which are then fermented easily to produce ethanol. The hemicellulose portion is more readily converted to sugars, primarily xylose; however, xylose is more difficult to ferment to ethanol. Finally, the lignin, although it cannot be fermented, can be converted to a high octane liquid fuel or, as is more common, burned to provide process energy.

The cost of ethanol produced from corn in the United States is about $0.34/L. The corn feedstock represents roughly half of this cost. Revenues from animal feed coproducts include about half the total costs. At this cost, ethanol production is not competitive with gasoline in the absence of federal and state tax credits. Laboratory research, utilizing biotechnology and genetic engineering, has reduced the estimated cost of cellulose-derived ethanol to about $0.36/L. Research plans, based on the use of enzymatic hydrolysis technology, suggest that a goal of $0.16/L may be achievable as early as 1998 for ethanol from cellulosic and hemicellulosic feedstocks. This cost would be competitive with the projected prices of gasoline without tax credits.

Methanol.   Methanol is made from biomass by first gasifying the feedstock to form a syngas, a mixture of CO, H_2, CO_2, higher hydrocarbons, and tar. A gas shift reaction is employed to adjust the chemical structure of the components of the gas mixture to the requisite H_2-to-CO ratio. The syngas is then cleaned and conditioned before being converted, in the presence of standard commercial catalysts, to form methanol. Research, development, and demonstration have produced several gasifiers that make syngas. Biomass gasifiers are specifically designed to take advantage of the superior characteristics of biomass feedstocks as compared to coal, ie, very little sulfur, high volatility, greater hydrogen content, and low ash content, for the production of syngas.

Although biomass-to-methanol technology has yet to be commercialized, laboratory technology suggests that commercial production would be feasible at a cost of about $0.20/L. Assuming that expected improvements in syngas cleanup and a reduction in feedstock costs are realized, the costs may be reduced to the target of $0.15/L as early as 1998.

Synthetic Hydrocarbon Fuels.   The basic approach used in converting biomass to traditional hydrocarbon fuels is to first pyrolyze the biomass feedstock to form an intermediate biocrude liquid product. The second step is to catalytically convert the biocrude to gasoline (see FUELS, SYNTHETIC). The technology uses a fast pyrolysis step that obtains higher yields of desired liquid components than those achieved in longer residence time processes. The fast pyrolysis process has been demonstrated using three different reactor designs. There are two potential routes for the second step: hydrogenation at high pressures and zeolite cracking at low pressures. As of the mid-1990s, attention is focusing on the potentially less costly, lower pressure process.

An alternative method of producing hydrocarbon fuels from biomass uses oils that are produced in certain plant seeds, such as rape seed, sunflowers, or oil palms, or from aquatic plants (see SOYBEANS AND OTHER OILSEEDS). Certain aquatic plants produce oils that can be extracted and upgraded to produce diesel fuel. The primary processing requirement is to isolate the hydrocarbon portion of the carbon chain that closely matches diesel fuel and modify its combustion characteristics by chemical processing.

Biomass-to-hydrocarbon fuel processes have yet to be commercialized. Based on research results, the cost estimate for the pyrolysis process is $0.42/L for gasoline. The cost target of $0.22/L by 2005 is based on expected achievement of improvements in the second-stage catalytic conversion process as well as the availability of feedstock at a cost of $2.00/GJ. At the mid-1990s stage of development of the technology, diesel fuel from algal oil was estimated to cost about $1.85/L. Process improvements and research accomplishments are needed to achieve the projected cost goal of $0.26/L by 2010. Substantial improvements in feedstock costs and in extraction technologies would be required in the same period to achieve similar cost reductions in recovering oil from algae or plant seeds.

BIBLIOGRAPHY

"Solar Energy" in *ECT* 3rd ed., Vol. 21, pp. 294–342, by D. Halacy, Solar Energy Research Institute.

1. *Annual Energy Outlook 1994*, U.S. Department of Energy, Energy Information Administration, Washington, D.C., Jan. 1994.
2. C. Flavin, *Worldwatch*, Worldwatch Institute, Washington, D.C., Sept./Oct. 1996.
3. *A Pilot Golden Eagle Study in the Altamont Pass Wind Resource Area California*, National Renewable Energy Laboratory, Golden, Colo., NREL/TP-441-7821, May 1995.
4. C.-J. Winter, R. L. Sizmann, and L. L. Vant-Hull, *Solar Power Plants*, Springer-Verlag, Berlin, 1991.
5. R. B. Diver, in *Progress in Solar Energy Technologies and Applications: An Authoritative Review*, American Solar Energy Society, Boulder, Colo., Jan. 1994.
6. N. D. Becker, *Solar Today*, 24–26 (Jan./Feb. 1992).
7. G. Cook, L. Billman, and R. Adcock, *Photovoltaic Fundamentals*, DOE/CH10093-117-Rev 1, U.S. Department of Energy, Washington, D.C., Feb. 1995.

General References

K. W. Boer, *Advances in Solar Energy: An Annual Review of Research and Development*, Vol. 7, American Solar Energy Society, Boulder, Colo., 1992.
Renewing Our Energy Future, Office of Technology Assessment, U.S. Congress, Washington, D.C., Sept. 1995.
K. Ahmed, *Renewable Energy Technologies: A Review of the Status and Costs of Selected Technologies*, World Bank Technical Page Number 240, Washington, D.C., Jan. 1994.
T. B. Johansson and co-workers, eds., *Renewable Energy—Sources for Fuels and Electricity*, Island Press, Washington, D.C., and Covelo, Calif., 1993.

ROBERT A. STOKES
Stokes Associates

SOLDERS AND BRAZING FILLER METALS

Metals joining by brazing and soldering is an ancient practice that has become a sophisticated modern technology owing to the contribution of modern materials science. The ultimate goal of these process is to join parts into an assembly through metallurgical bonding (1–4). A relatively low melting temperature alloy, a filler metal (FM), is placed in the clearance (gap) between the pieces of base materials (BM) to be joined and the assembly is subsequently heated until the FM has melted and spread throughout the gap. The molten metal fills the gap and reacts with parts to be brazed, forming after solidification an integral solid whole. Assembly heating can be carried out by various means. These include electromagnetic induction, Joule heating, or use of an oven, flame, etc. Joining temperatures above 723 K are arbitrarily associated with brazing rather than with soldering. However, these processes are essentially similar. In order to distinguish joining materials, those used at temperatures below 723 K are called solders; those above, brazing filler metals.

There are three principal stages in any brazing or soldering process (Fig. 1). The first occurs during the heating of an assembled workpiece. At this stage, the FM melts and flows, filling completely the gap normally existing between the parts. This gap is usually rather small, ie, on the order of 10 to a maximum of a few hundred micrometers, and sometimes even close to zero. Thus, the degree of penetration of a molten alloy into the gap is determined largely by capillary effects and by wetting. Because physical parameters such as part dimensions and surface tension forces play the primary role, this stage may be conditionally called a physical one.

The second stage, which normally sets in at a given joining temperature, is characterized by an intensive solid–liquid interaction accompanied by a substantial mass transfer through the interface at strongly uneven rates. Indeed, BM immediately adjoining the liquid filler metal dissolves in this stage. At the same time, a small amount of material from the liquid phase penetrates into the solid BM. Such mass-transfer unbalance results from dramatically different diffusion rates in the solid and liquid phases. This redistribution of components in the joint area leads to changes of phase composition and sometimes to the onset of crystallization, or, in other words, to joint solidification. All the processes of this stage may be properly called a metallurgical one. In a majority of cases in soldering, all the processes proceed but to a much lesser degree because of a substantial difference in the melting temperatures of a solder and a BM. However, even in soldering, sometimes the BM surface must be protected from erosion by molten solder.

The final stage of the brazing (soldering process) overlaps with the second and is characterized by the formation of the final joint microstructure, progressing vigorously during assembly cooling while the liquid phase is still present. Here, the crystallization process takes place. Subsequent cooling results in only minor changes such as annealing of the joint and partial relief of thermal stresses induced on cooling. The final stage has very similar features in both brazing and soldering processes.

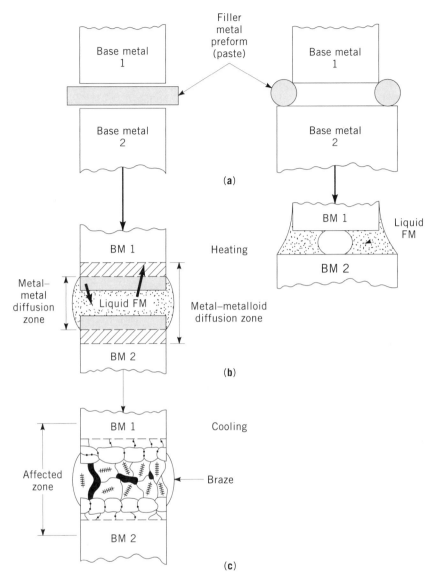

Fig. 1. The essence of the brazing/soldering processes: (**a**) before brazing/soldering; (**b**) heating to the brazing/soldering temperature, $T_{br/sold}$; and (**c**) cooling to the use of service temperature after brazing/soldering, T_{serv}. The two thick arrows in (**b**) denote mass transfer.

   In any brazing/soldering process, a molten alloy comes in contact with a surface of solid, which may be an alloy, a ceramic, or a composite material (see CERAMICS; COMPOSITE MATERIALS). For a molten alloy to advance over the solid surface a special relationship has to exist between surface energies of the liquid–gas, solid–gas, and liquid–solid interfaces. The same relationships should, in principle, hold in joining processes where a molten alloy has to fill the gaps existing between surfaces of the parts to be joined. In general, the molten alloy should have a lower surface tension than that of the base material.

All transition metals, particularly iron (qv), which is the base element of steel (qv), and the majority of alloys, have surface tension higher than 1500 mN/m, whereas nontransition metals such as silver, tin, and lead, and particularly metalloids such as silicon, have surface tension substantially lower than those of the principal groups of steels and other BM alloys. The addition of phosphorus (qv), silicon, boron (qv), etc, to aluminum, nickel, and copper (qv) results in a precipitous drop in the surface tension of the resultant material. Therefore, for low temperature joining, tin- and lead-based alloys have been used as solders, whereas aluminum-, copper-, and nickel-based alloys having additions of P, Si, and B have been used as brazing filler metals. Both groups of alloys in the molten state have surface tension values favorable for strong wetting (see ALUMINUM AND ALUMINUM ALLOYS; COPPER ALLOYS; LEAD ALLOYS; NICKEL AND NICKEL ALLOYS).

Because the surface tension of oxides such as Fe_xO_y, Al_2O_3, Cr_2O_3, etc, is much lower than that of the majority of metals and alloys, the presence of an oxide film on a BM surface impedes wetting. It is of paramount importance to keep the BM surface clean from the so-called virgin oxides and oxides that may have formed during the brazing operation if the heat-treating atmosphere is oxidizing. This is particularly important in brazing stainless steels and high temperature alloys (qv) containing active metals, such as chromium, titanium, and aluminum, that form corresponding protective oxide films. The wettability of base metals is strongly affected by cleanliness not only with respect to oxides but also by the presence of traces of grease, oil, etc, on part surfaces. Therefore, cleaning of the BM pieces to be joined and their subsequent protection from oxidation during heating are essential steps in joining operations.

Chemically active substances (fluxes) are used to accomplish cleaning and to provide protective shielding when vacuum or protective atmospheres are not available (see METAL SURFACE TREATMENTS). However, the use of nonmetallic agents in fluxes is not always desirable owing to the increased propensity for their entrapment and the subsequent formation of voids in the brazed joint. On the other hand, if the brazing environment is not oxidizing, then boron, silicon, and phosphorus, which are used as inherent constituents of FM alloys, can play the role of a flux, ie, reducing original oxide films and making the FM self-fluxing.

Basic Forms of Filler Metal

Filler metal forms include solid preforms and powders, used mostly in the compound form of paste, plastic-bonded tape, and, in the case of soldering, rosin core wire. In special soldering applications, solders may be used as a liquid medium of the soldering baths in which electronic boards are immersed for a short time to solder multiple joints. Conventional paste forms of joining alloys, eg, FM powder plus fluxing agent plus binder/solvent, are applied at externally accessible locations of the clearance between BM pieces. Such practice requires substantial FM fluid flow during joining to achieve fusion of constituent FM phases and binder/solvent extraction. In addition, organic binders decompose when compound FM forms are used in the high vacuum brazing of parts intended for critical high temperature service. Such decomposition of binders can result in the formation of soot in the joint, which can act to degrade the performance of expensive vacuum equipment.

Joint Requirements

A basic requirement for any joint is that its strength and ductility be equal or at least close to that of the BM. In general, the strength of a material increases with decreasing grain size, whereas ductility is affected by the presence of brittle phases. Therefore, in addition to limiting porosity in joints, it is important to limit grain size and the amounts of brittle phases. Ideally, the maximum size that the grains in a joint may achieve is equal to half the clearance between the BM pieces. Thus, the use of a smaller clearance during BM joining necessarily limits maximum grain size, promotes higher cooling rates of the FM alloy, and thereby results in a refined joint microstructure. A smaller clearance in brazing also promotes improved retention of BM properties because of curtailed BM erosion by the use of a smaller volume of FM. For these reasons, a preplaced self-fluxing thin FM foil used as a preform is superior to the use of powder-containing paste. The latter requires larger clearances for filling joint cross sections, thus resulting in deleterious effects on properties owing to a coarser joint grain size, more fully developed intermetallic compounds, and the presence of substantial amounts of contaminants.

The grain size and brittle intermetallic size and extent in the parent FM have a direct effect on the strength of the formed joint. This consideration is particularly important under transient heating/cooling conditions such as torch or belt oven brazing and automatic solder die bonding. The rapid dissolution of constituent FM alloy phases allows the use of higher joining throughput speeds and/or lower joining temperatures. The newer rapidly solidified microcrystalline and amorphous joining materials are thus the epitome of FM microstructural possibility. These materials possess ultimate uniformity of their elemental components, which is beneficial for formation of a fine microcrystalline joint microstructure in both soldering and brazing. In the case of brazing, amorphous foil has the ability of self-fluxing because practically all brazing amorphous foils contain a large amount of phosphorus, boron, and silicon (5).

Solder and Brazing Filler Metals

Solders.   In spite of the wide use and development of solders for millennia, as of the mid-1990s most principal solders are lead- or tin-based alloys to which a small amount of silver, zinc, antimony, bismuth, and indium or a combination thereof are added. The principal criterion for choosing a certain solder is its melting characteristics, ie, solidus and liquidus temperatures and the temperature spread or pasty range between them. Other criteria are mechanical properties such as strength and creep resistance, physical properties such as electrical and thermal conductivity, and corrosion resistance.

According to International Organization for Standardization (ISO)/DIS 9543 specifications (Table 1), the majority of solders can be divided into three

Table 1. ISO/DIS 9453 Specification for Tin–Lead and Other Lead–Alloy Solders[a]

Alloy number	Alloy designation	Melting or solidus/liquidus temperature[b], °C	Chemical composition[c,d], %										Sum of all impurities except Sb, Bi, and Cu, %
			Sn	Sb	Cd	Zn	Al	Bi	As	Fe	Cu	Ag	
							Tin–lead alloys						
1	S–Sn63Pb37	183	62.5–63.5	0.12	0.002	0.001	0.001	0.10	0.03	0.02	0.05		0.08
1a	E–Sn63Pb37	183	62.5–63.5	0.05	0.002	0.001	0.001	0.05	0.03	0.02	0.05		0.08
2	S–Sn60Pb40	183–190	59.5–60.5	0.12	0.002	0.001	0.001	0.10	0.03	0.01	0.05		0.08
2a	E–Sn60Pb40	183–190	59.5–60.5	0.05	0.002	0.001	0.001	0.05	0.03	0.02	0.05		0.08
3	S–Pb50Sn50	183–215	49.5–50.5	0.12	0.002	0.001	0.001	0.10	0.03	0.02	0.05		0.08
3a	E–Pb50Sn50	183–215	49.5–50.5	0.05	0.002	0.001	0.001	0.05	0.03	0.02	0.05		0.08
4	S–Pb55Sn45	183–226	44.5–45.5	0.50	0.005	0.001	0.001	0.25	0.03	0.02	0.08		0.08
5	S–Pb60Sn40	183–235	39.5–40.5	0.50	0.005	0.001	0.001	0.25	0.03	0.02	0.08		0.08
6	S–Pb65Sn35	183–245	34.5–35.5	0.50	0.005	0.001	0.001	0.25	0.03	0.02	0.08		0.08
7	S–Pb70Sn30	183–255	29.5–30.5	0.50	0.005	0.001	0.001	0.25	0.03	0.03	0.08		0.08
8	S–Pb90Sn10	268–302	9.5–10.5	0.50	0.005	0.001	0.001	0.25	0.03	0.02	0.08		0.08
9	S–Pb92Sn8	280–305	7.5–8.5	0.50	0.005	0.001	0.001	0.25	0.03	0.02	0.08		0.08
10	S–Pb98Sn2	320–325	1.5–2.5	0.12	0.002	0.001	0.001	0.10	0.03	0.02	0.05		0.08
							Tin–lead–antimony alloys						
11	S–Sn63Pb37Sb	183	62.5–63.5	0.12–0.50	0.002	0.001	0.001	0.10	0.03	0.02	0.05		0.08
12	S–Sn60Pb40Sb	183–190	59.5–60.5	0.12–0.50	0.002	0.001	0.001	0.10	0.03	0.02	0.05		0.08
13	S–Pb50Sn50Sb	183–216	49.5–50.5	0.12–0.50	0.002	0.001	0.001	0.10	0.03	0.02	0.05		0.08
14	S–Pb58Sn40Sb2	185–231	39.5–40.5	2.0–2.4	0.005	0.001	0.001	0.25	0.03	0.02	0.08		0.08
15	S–Pb69Sn30Sb1	185–250	29.5–30.5	0.5–1.8	0.005	0.001	0.001	0.25	0.03	0.02	0.08		0.08
16	S–Pb74Sn25Sb1	185–263	24.5–25.5	0.5–2.0	0.005	0.001	0.001	0.25	0.03	0.02	0.08		0.08
17	S–Pb78Sn20Sb2	185–270	19.5–20.5	0.5–3.0	0.005	0.001	0.001	0.25	0.03	0.02	0.08		0.08
							Lead–silver and lead–tin–silver alloys						
32	S–Pb98Ag2	304–305	0.25	0.10	0.002	0.001	0.001	0.10	0.03	0.02	0.05	2.0–3.0	0.2
33	S–Pb95Ag5	304–365	0.25	0.10	0.002	0.001	0.001	0.10	0.03	0.02	0.05	4.5–6.0	0.2
34	S–Pb93Sn5Ag2	296–301	4.8–5.2	0.10	0.002	0.001	0.001	0.10	0.03	0.02	0.05	1.2–1.8	0.2

[a]Ref. 1. [b]Temperatures given are for information purposes and are not specified requirements for the alloys. [c]The balance of material is lead. [d]Single values given are maximums.

categories: tin–lead alloys, tin–lead alloys with antimony, and lead–tin–silver alloys. The tin–lead eutectic and near-eutectic alloys are the most commonly used solders that find application in electronics, particularly for soldering of circuit boards, and electrical connections and general purposes such as joining pipelines (qv) for a wide range of gases and fluids. Small (ca 0.25 wt %) additions of antimony are used to suppress the formation of the low temperature allotrope of tin, whereas 2–4 wt % additions are used to improve monotonic and creep strength of the solder. Because of relatively high melting temperature, lead–silver solders are used in applications where joint strength at moderately high temperatures is needed. The addition of tin improves wetting and flow and reduces corrosion in a humid atmosphere. These solders are used in cryogenic apparatuses and for soldering of fine copper wires because of their lower tendency to dissolve copper.

The presence of the so-called heavy metals, eg, lead, cadmium, and antimony, in traditional solders has become an important environmental issue owing to concerns for health and safety. As a result, solders containing no lead and antimony such as tin–silver are finding a growing number of applications (Table 2).

Brazing Filler Metals. The choice of both specific FM alloy composition and optimal brazing conditions is mostly determined by joint performance under specific service conditions. Ideally, the composition of an FM alloy must be such that the following four functions are achieved: (1) the FM melting temperature must be lower than that of the corresponding BM; (2) the FM surface tension for both solid and liquid states must be lower than that of the BM to provide a driving force for wetting; (3) the FM must be compatible with the BM in order to form good metallic bonding, ie, the FM and BM structure, composition, and properties should be similar; and (4) the FM-containing elements should be able to bring about chemical reduction/decomposition or physical removal of BM oxide film.

Five families of brazing FM alloy compositions have emerged with respect to metallurgical nature or type. Of these, the first four families of conventional brazing FM have been classified by the American Welding Society (AWS) into eight well-defined classes as indicated in Table 3. The preferred base metals with which each specific family is most compatible as well as the principal areas of FM applications are also given. Compositions of specific alloys are given in Tables 4–8.

The first and largest family of FM alloys, Family I, contains eutectic-type alloys having aluminum, nickel, cobalt, and copper as a base to which silicon–boron (aluminum- and nickel-base alloys) and phosphorus (copper- and nickel-base alloys) are added. Family I encompasses AWS Classes 1, 4, 6, and 7 as well as the nonclassified nickel- and palladium-based alloys. The presence of one or more of the metalloid elements in alloys tends to impart the required characteristics to the FM. Silicon combined with a small amount of magnesium is successfully used in aluminum-based FM alloys, the surface tension and melting temperature of which are depressed and good bonding is promoted. The magnesium addition serves as a fluxing agent. Boron and silicon are used in high temperature brazing alloys in which the presence of another potential melting temperature depressant and surfactant element, ie, phosphorus, can cause unacceptable joint brittleness. Of the alloy additions that promote self-fluxing of the FM during

Table 2. ISO/DIS 9453 Specification for Tin–Silver Solders[a]

Alloy number	Alloy designation	Melting or solidus/liquidus temperature[b], °C	Chemical composition[c–e], %												
			Pb	Sb	Bi	Cd	Cu	In	Ag	Al	As	Fe	Zn
28	S–Sn96Ag4	221	0.10	0.10	0.10	0.002	0.05	0.05	3.5–4.0	0.001	0.03	0.02	0.001
29	S–Sn97Ag3	221–230	0.10	0.10	0.10	0.005	0.10	0.05	3.0–3.5	0.001	0.03	0.02	0.001

[a]Ref. 1.
[b]Temperatures given are for information purposes and are not specified requirements for the alloys.
[c]The balance of material is tin.
[d]Single values given are maximum.
[e]The sum of all impurities is 0.2%.

488

Table 3. AWS Classes of Brazing Filler Metals

Class	Alloy type (and family)	AWS designation	Forms[a]	Base materials joined	Applications
1	Al–Si, eutectic (I)	BAlSi	preforms, wire, rods, foil, powder, RS foil[b]	aluminum and aluminum alloys, steel to aluminum and aluminum to beryllium	car radiators, heat exchangers, honeycomb aircraft structures, structural parts
2	Cu–X, solid solution (II)	BCu	preforms, wire, rods, foil, powder	copper and copper alloys, copper to mild steel, copper to stainless steel	heat exchangers, structural parts, automotive parts
3	Cu–Zn, peritectic (III) Cu–Sn, peritectic (III)	RBCuZn	same as Class 2 same as Class 2 and RS foil	same as Class 2 same as Class 2	same as Class 2 same as Class 2
4	Cu–P, eutectic (I)	BCuP	preforms, wire, rods, foil, powder, RS foil	copper to copper, copper to silver/oxide-powdered metal composites	electrical contacts, bus bars, heat exchangers
5	Cu–Ag, eutectic (IV)	BAg	preforms, foil, powder	most ferrous and nonferrous metals, except aluminum and magnesium	most widely used utility filler metals
6	TM–Si–B[c], eutectic (I) (Ni/Fe + Cr)–Si–B (I)	BNi	powder, tape[d], RS foil	AISI 300 and 400 series steels and nickel- and cobalt-base superalloys, carbon steels, low alloy steels, and copper	aircraft turbine components, automotive parts, heat exchangers, honeycomb structure

489

Table 3. (Continued)

Class	Alloy type (and family)	AWS designation	Forms[a]	Base materials joined	Applications
7	(Co, Cr)–Si–B	BCo	powder, tape, RS foil	cobalt-base heat-resistant alloys, steels	aircraft engines, honeycomb marine structures
	(Ni, Pd)–Si–B		powder, tape, RS foil	AISI 300 series stainless steels, cemented carbide, superalloys	honeycomb structures, cemented carbide/polycrystalline diamond tools, orthodontics, catalytic converters
8	Au–Ni, solid solution (II)	BAu	preforms, wire, rods, foil, tape	nickel-base heat-resistant alloys, steels	honeycomb structures, structural turbine parts
	Cu–(Ti, Zr)–Ni eutectic and peritectic (V)		clad strip, RS foil	titanium/zirconium-base alloys	titanium tubing, aircraft engines, honeycomb aircraft structures, aircraft structural parts, chemical reactors

[a]RS = rapid solidification.

[b]May be produced as rapidly solidified, ductile, amorphous/microcrystalline foil.

[c]This family includes alloys based on transition metals, such as nickel, iron, cobalt, and palladium.

[d]Brazing filler metal is carried on a plastic-bonded tape.

490

Table 4. Chemical Composition Requirements for Aluminum and Magnesium Filler Metals[a]

AWS classification	UNS number	Composition[b], wt %						
		Si	Cu	Mg	Fe	Zn	Mn	Other[c]
Aluminum-based alloys[d]								
BAlSi-2	A94343	6.8–8.2	0.25		0.8	0.2	0.1	
BAlSi-3	A94145	9.3–10.7	3.3–4.7	0.15	0.8	0.2	0.15	0.15 Cr
BAlSi-4	A94047	11.0–13.0	0.3	0.1	0.8	0.2	0.15	
BAlSi-5	A94045	9.0–11.0	0.3	0.05	0.8	0.2	0.05	0.2 Ti
BAlSi-7	A94004	9.0–10.5	0.25	1.0–2.0	0.8	0.2	0.1	
BAlSi-9	A94147	11.0–13.0	0.25	0.1–0.5	0.8	0.2	0.1	
BAlSi-11	A94104	9.0–10.5	0.25	1.0–2.0	0.8	0.2	0.1	0.02–0.2 Bi
Magnesium-based alloys[e]								
BMg-1	M19001	0.05	0.05	f	0.005	1.7–2.3	0.15–1.5	0.005 Ni; 0.0002–0.0008 Be;[g]

[a]Ref. 2.
[b]Single values are maximum.
[c]The filler metal is analyzed for those specific elements for which values are shown. If the presence of other elements is indicated in the analysis, the amount of those elements is determined to ensure that the maximum for each is ≤0.05 wt % and the maximum total of other elements is ≤0.15 wt %.
[d]Remainder of material is Al.
[e]Remainder of material is Mg.
[f]Contains 8.3–9.7 Al.
[g]The maximum allowed for other elements is 0.3 wt %.

Table 5. Chemical Composition Requirements for Copper-Based Filler Metals[a]

Alloys		Composition[b], wt %		
AWS classification	UNS number	Cu	P	Other[c]
Copper-based alloys				
BCu-1	C14180	99.90 min	0.075	0.02 Pb; 0.01 Al*
BCu-1a[d]		99.90 min		
BCu-2[e]		86.50 min		
Copper–zinc-based alloys[f]				
RBCuZn-A	C47000	57.0–61.0		0.25–1.00 Sn; 0.05 Pb*; 0.01 Al*
RBCuZn-C	C68100	56.0–60.0		0.80–1.10 Sn; 0.25–1.20 Fe; 0.01–0.50 Mn; 0.05 Pb*; 0.01 Al*; 0.04–0.15 Si
RBCuZn-D	C77300	46.0–50.0	0.25	9.0–11.0 Ni; 0.05 Pb*; 0.01 Al*; 0.04–0.25 Si
Copper–phosphorus-based alloys[g]				
BCuP-1	C55180		4.8–5.2	
BCuP-2	C55181		7.0–7.5	
BCuP-3	C55281		5.8–6.2	4.8–5.2 Ag
BCuP-4	C55283		7.0–7.5	5.8–6.2 Ag
BCuP-5	C55284		4.8–5.2	14.5–15.5 Ag
BCuP-6	C55280		6.8–7.2	1.8–2.2 Ag
BCuP-7	C55282		6.5–7.0	4.8–5.2 Ag

[a]Ref. 2.
[b]Single values are maximum.
[c]The filler metal is analyzed for those specific elements for values where asterisks (*) are shown. If the presence of other elements is indicated in the analysis, the amount of those elements is determined to ensure that the maximum total is ≤0.10 for BCu-1; ≤0.30 for BCu-1a; ≤0.50 for BCu-2, RBCuZn-A, RBCuZn-C, and RBCuZn-D; and ≤0.15 for all others.
[d]The balance is oxygen, present as cuprous oxide.
[e]These chemical composition requirements pertain only to the cuprous oxide powder and do not include requirements for the organic vehicle in which the cuprous oxide is suspended, when applied in paste form.
[f]Remainder of material is Zn.
[g]Remainder of material is Cu.

brazing, boron has the greatest penetrating power. On the other hand, phosphorus is a beneficial fluxing element for use in copper (low temperature) brazing, whereas another element, eg, silicon, would cause unacceptable copper joint brittleness. Most of these alloys are brittle because various intermetallic phases precipitate when processed by conventional technology. This brittle character has limited the available forms of conventionally produced brazing alloys to powder. The presence, however, of silicon, phosphorus, and boron in many conventional FM alloys having near-eutectic compositions facilitates the conversion of such alloys into ductile, thin amorphous alloy foil form when rapid solidification (RS) technology is used (5).

The second family of brazing FM, Family II, consists of solid solution alloys based on copper (Class 2) and gold–nickel (Class 8). These alloys are used mainly in vacuum-brazing applications and therefore require no alloying

Table 6. Chemical Composition Requirements for Silver Filler Metals[a]

Alloy AWS classification	UNS number	Composition, wt %			
		Ag	Cu	Zn	Other[b]
BAg-1	P07450	44.0–46.0	14.0–16.0	14.0–18.0	23.0–25.0 Cd
BAg-1a	P07500	49.0–51.0	14.5–16.5	14.5–18.5	17.0–19.0 Cd
BAg-2	P07350	34.0–36.0	25.0–27.0	19.0–23.0	17.0–19.0 Cd
BAg-2a	P07300	29.0–31.0	26.0–28.0	21.0–25.0	19.0–21.0 Cd
BAg-3	P07501	49.0–51.0	14.5–16.5	13.5–17.5	15.0–17.0 Cd; 2.5–3.5 Ni
BAg-4	P07400	39.0–41.0	29.0–31.0	26.0–30.0	1.5–2.5 Ni
BAg-5	P07453	44.0–46.0	29.0–31.0	23.0–27.0	
BAg-6	P07503	49.0–51.0	33.0–35.0	14.0–18.0	
BAg-7	P07563	55.0–57.0	21.0–23.0	15.0–19.0	4.5–5.5 Sn
BAg-8[c]	P07720	71.0–73.0			
BAg-8a[c]	P07723	71.0–73.0			0.25–0.50 Li
BAg-9	P07650	64.0–66.0	19.0–21.0	13.0–17.0	
BAg-10	P07700	69.0–71.0	19.0–21.0	8.0–12.0	0.15[d] Mn
BAg-13[c]	P07540	53.0–55.0		4.0–6.0	0.5–1.5 Ni
BAg-13a[c]	P07560	55.0–57.0			1.5–2.5 Ni
BAg-18[c]	P07600	59.0–61.0			9.5–10.5 Sn
BAg-19[c]	P07925	92.0–93.0			0.15–0.30 Li
BAg-20	P07301	29.0–31.0	37.0–39.0	30.0–34.0	
BAg-21	P07630	62.0–64.0	27.5–29.5		2.0–3.0 Ni; 5.0–7.0 Sn
BAg-22	P07490	48.0–50.0	15.0–17.0	21.0–25.0	4.0–5.0 Ni; 7.0–8.0 Mn
BAg-23[e]	P07850	84.0–86.0			
BAg-24	P07505	49.0–51.0	19.0–21.0	26.0–30.0	1.5–2.5 Ni
BAg-26	P07250	24.0–26.0	37.0–39.0	31.0–35.0	1.5–2.5 Ni; 1.5–2.5 Mn
BAg-27	P07251	24.0–26.0	34.0–36.0	24.5–28.5	12.5–14.5 Cd
BAg-28	P07401	39.0–41.0	29.0–31.0	26.0–30.0	1.5–2.5 Sn
BAg-33	P07252	24.0–26.0	29.0–31.0	26.5–28.5	16.5–18.5 Cd
BAg-34	P07380	37.0–39.0	31.0–33.0	26.0–30.0	1.5–2.5 Sn

[a]Ref. 2.
[b]The brazing filler metal is analyzed for those specific elements for which values are shown. If the presence of other elements is indicated in the analysis, the amount of those elements is determined to ensure that the maximum total of each is ≤0.15 wt %.
[c]Remainder of material is Cu.
[d]Value represents maximum.
[e]Remainder of material is Mn.

elements playing the role of fluxing agents (see VACUUM TECHNOLOGY). Family III consists of alloys having a phase diagram where a peritectic reaction exists, such as copper–zinc alloys (Class 3) and nonclassified Cu–Sn alloys. Alloys of Family IV (Class 5) are probably the most widely used. Family IV is based on the copper–silver binary eutectic system modified by substantial additions of zinc and cadmium, both of which provide fluxing activity, and minor additions of tin and nickel. The fifth family of brazing FM alloys, Family V, although so far

Table 7. Chemical Composition Requirements for Nickel and Cobalt Filler Metals[a]

Alloy AWS classification	UNS number	Cr	B	Si	Fe	C	P	S	Al	Ti	Zr	Co	Se	Other[c]
						Composition[b], wt %								
					Nickel-based alloys[d]									
BNi-1	N99600	13.0–15.0	2.75–3.50	4.0–5.0	4.0–5.0	0.60–0.90	0.02	0.02	0.05	0.05	0.05	0.10	0.005	
BNi-1a	N99610	13.0–15.0	2.75–3.50	4.0–5.0	4.0–5.0	0.06	0.02	0.02	0.05	0.05	0.05	0.10	0.005	
BNi-2	N99620	6.0–8.0	2.75–3.50	4.0–5.0	2.5–3.5	0.06	0.02	0.02	0.05	0.05	0.05	0.10	0.005	
BNi-3	N99630		2.75–3.50	4.0–5.0	0.5	0.06	0.02	0.02	0.05	0.05	0.05	0.10	0.005	
BNi-4	N99640		1.50–2.20	3.0–4.0	1.5	0.06	0.02	0.02	0.05	0.05	0.05	0.10	0.005	
BNi-5	N99650	18.5–19.5	0.03	9.75–10.50		0.06	0.02	0.02	0.05	0.05	0.05	0.10	0.005	
BNi-6	N99700					0.06	10.0–12.0	0.02	0.05	0.05	0.05	0.10	0.005	
BNi-7	N99710	13.0–15.0	0.01	0.10	0.2	0.06	9.7–10.5	0.02	0.05	0.05	0.05	0.10	0.005	0.04 Mn
BNi-8	N99800			6.0–8.0		0.06	0.02	0.02	0.05	0.05	0.05	0.10	0.005	21.5–24.5 Mn; 4.0–5.0 Cu
					Cobalt-based alloys[e]									
BNi-9	N99612	13.5–16.5	3.25–4.0		1.5	0.06	0.02	0.02	0.05	0.05	0.05	0.10	0.005	
BNi-10	N99622	10.0–13.0	2.0–3.0	3.0–4.0	2.5–4.5	0.40–0.55	0.02	0.02	0.05	0.05	0.05	0.10	0.005	15.0–17.0 W
BNi-11	N99624	9.0–11.75	2.2–3.1	3.35–4.25	2.5–4.0	0.30–0.50	0.02	0.02	0.05	0.05	0.05	0.10	0.005	11.5–12.75 W
BCo-1	R39001	18.0–20.0	0.70–0.90	7.5–8.5	1.0	0.35–0.45	0.02	0.02	0.05	0.05	0.05	*f*	0.005	3.5–4.5 W

[a]Ref. 2.

[b]Single values are maximum.

[c]The filler metals are analyzed for those specific elements for which values are shown. If the presence of other elements is indicated in the analysis, the amount of those elements is determined to ensure that the maximum total for each is ≤0.50 wt %, except for BNi-1, which is ≤0.05 wt %.

[d]Remainder of material is Ni.

[e]Remainder of material is Co.

[f]Contains 16.0–18.0 Ni.

Table 8. Chemical Composition Requirements for Gold Filler Metals[a]

| Alloy | | Composition[b], wt % | | |
AWS classification	UNS number	Au	Pd	Ni
BAu-1[c]	P00375	37.0–38.0		
BAu-2[c]	P00800	79.5–80.5		
BAu-3[c]	P00350	34.5–35.5		2.5–3.5
BAu-4[d]	P00820	81.5–82.5		
BAu-5	P00300	29.5–30.5	33.5–34.5	35.5–36.5
BAu-6	P00700	69.5–70.5	7.5–8.5	21.5–22.5

[a]Ref. 2.
[b]The brazing filler metal is analyzed for those specific elements for which values are shown. If the presence of other elements is indicated in the analysis, the amount of those elements is determined to ensure that the maximum total is ≤ 0.15 wt %.
[c]Remainder of material is copper.
[d]Remainder of material is nickel.

unclassified, consists of purely metallic eutectic/peritectic titanium–zirconium-base alloys to which copper and/or nickel are added.

The majority of all these classes, even noneutectic alloys, have been processed successfully by rapid solidification technology. This technology provides a beneficial alternative in the form of a flexible ductile foil when materials that are inherently brittle are used. Examples are the nickel–boron–silicon alloys and many others, when produced using conventional technology (5).

Joining Process Technology

Joint design must ensure a variety of service criteria such as joint mechanical strength; resistance to service environment; electrical conductivity, which is of prime importance in soldering of electrical circuitries; ease of manufacturing; and economics. Compatibility of the parts to be joined (BM) with the forming braze (FM) is always considered from minimization of mechanical stresses, which may appear after brazing/soldering owing to the difference in coefficient of thermal expansion that always exists between the BM and the FM. These stresses may be very high. Specifics of joint design are available for brazing (2) and for soldering of electronic components (1).

Preparation and Protection of the Parts.   One, if not the most, important technological step to guarantee success of a joining operation is the preparation of the parts to be joined. Cleaning part surfaces of oxide films, oil, grease, and dirt includes mechanical and/or chemical means, solvent usage, and, finally, rinsing and drying. If not properly cleaned, the joined assemblies may leak or lack the necessary strength owing to incomplete joining (see METAL SURFACE TREATMENTS).

Parts must be protected from oxidation during heat treating in joining. Only then can complete wetting of parts by the molten filler metal be ensured and a quality joint result. Protection from oxidation may be accomplished by using self-fluxing brazing filler metals such as the Cu–P-based alloys, protective fluxes, protective atmospheres, and simply by using vacuum furnaces that have no trace of oxygen.

Fluxes. Fluxes often play multiple roles. Not only do fluxes protect parts from oxidation during heating but they also clean up virgin surfaces from existing tarnishing oxide films by reducing or scaling the oxides. In addition, fluxes also decrease the surface tension of molten metals and improve capillary flow. Rosin-based fluxes, for example, address these phenomena in soldering where these fluxes are often used for joining electrical circuitries. Organic-acid or water-soluble fluxes are also frequently preferred in soldering operations in which a final cleaning of soldered boards can be achieved using noncorrosive water solutions.

Heat Treatments. Heating methods in joining can be divided into two principal categories: the local one, where heat is supplied predominantly to the joint area, and the overall one, where the brazed assembly is heated to a certain temperature. Gas torches, radiation heaters, contact heaters such as soldering irons, and induction sources are used for local heating; electrical and gas furnaces are used for complete heating. Sophisticated and highly productive methods such as wave soldering and electrical resistance soldering have been applied successfully in mass production soldering (6,7).

Health and Safety Factors

Brazing safety is subject to the requirements of the American National Standard (8). The specifics of brazing, which should be considered in addition to the conventional safety requirements applied to manufacturing environments, mostly relate to the metal fumes and metal and oxide particulars evolved during processing. Many brazing filler metals contain a wide variety of harmful metallic components. Among these, cadmium, lead, and zinc are the most insidious, which also evaporate easily upon melting the alloys that contain these materials. The exposure limits to these substances and to the various active fluxes and the solvents used are primary safety parameters regulating the workplace. Limits for these materials are available (1,9).

The principal goal of safety provision in any brazing and soldering shop is the cleanliness of the atmosphere. Sufficient ventilation must be provided and proper working practices are regulated by OSHA as well as various state and federal agencies (9). Additional concerns include the high temperatures and gases under high pressure.

Stringent OSHA composition limits exist for applications of brazing filler metals and solders. For example, only no-lead solders are permitted for joining parts that may come in contact with potable water.

BIBLIOGRAPHY

"Solders and Brazing Alloys" in *ECT* 1st ed., Vol. 12, pp. 634–640, by C. H. Chatfield, Handy & Harman; in *ECT* 2nd ed., Vol. 18, pp. 541–549, by C. H. Chatfield, Handy & Harman; in *ECT* 3rd ed., Vol. 21, pp. 342–355, by G. Sistare, Consultant, and F. Disque, Alpha Metals, Inc.

1. *Welding, Brazing and Soldering*, American Society for Metals, Metals Park, Ohio, 1993, pp. 328–372, 617–640, 903–1002.
2. *Brazing Handbook*, American Welding Society, Miami, Fla., 1991.
3. M. M. Schwartz, *Brazing*, American Welding Society, Miami, Fla., 1987.
4. G. Hupston and D. Jacobson, *Principles of Soldering and Brazing*, American Society for Metals, Metals Park, Ohio, 1993.
5. A. Rabinkin and H. H. Liebermann, in H. H. Liebermann, ed., *Rapidly Solidified Alloys*, Marcel Dekker, Inc., New York, 1993, pp. 691–736.
6. F. G. Yost, F. M. Hosking, and D. R. Frear, *The Mechanics of Solder Alloy Wetting and Spreading*, Van Nostrand Reinhold Co., Inc., New York, 1993.
7. J. H. Lau, *Handbook of Fine Pitch Surface Mount Technology*, Van Nostrand Reinhold Co., Inc., New York, 1994.
8. American National Standard Z49.1, *Safety in Welding and Cutting*, AWS Publications, American Welding Society, Miami, Fla., 1994.
9. *Code of Federal Regulations*, 1910 Title 29, U.S. Printing Office, Washington, D.C., 1996.

ANATOL RABINKIN
AlliedSignal Inc.

SOL–GEL TECHNOLOGY

Traditional ceramic and glass processing use high temperatures to transform inorganic powders into dense objects by melting or sintering. The high temperatures and agglomeration of powders often limit control of the microstructure, properties, shape, and surface features obtained. The goal of sol–gel technology is to use low temperature chemical processes to produce net-shape, net-surface objects, films, fibers (qv), particulates, or composites that can be used commercially after a minimum of additional processing steps (see COMPOSITE MATERIALS; THIN FILMS). Traditional ceramic processing produces materials having microstructures typically in the range of 1–100-micrometer diameter. Sol–gel processing can provide control of microstructures in the nanometer size range, ie, 1–100 nm (0.001–0.1 μm), which approaches the molecular level. These materials often have unique physical and chemical characteristics (see NANOTECHNOLOGY (SUPPLEMENT)).

Although the origins of chemical-based ceramic processes may be dated to as early as 4000 BC, the concept of control of shape and molecular structure of ceramics (qv) and glasses (see GLASS) by use of sol–gel chemistry probably dates from Bergman's studies on water glasses in 1779, Ebelman's and Graham's studies on silica (qv) gels in 1847 and 1864, respectively, and a large body of work on the science of colloids (qv) in the mid-1800s (1).

These early studies led to the following definitions:

Sols are dispersions of colloidal particles in a liquid. Sol particles are typically small enough to remain suspended in a liquid by Brownian motion.

Colloids are nanoscaled entities dispersed in a fluid.

Gels are viscoelastic bodies that have interconnected pores of submicrometric dimensions. A gel typically consists of at least two phases, a solid network that entraps a liquid phase. The term gel embraces numerous combinations of substances, which can be classified into the following categories (2): (*1*) well-ordered lamellar structures; (*2*) covalent polymeric networks that are completely disordered; (*3*) polymer networks formed through physical aggregation that are predominantly disordered; and (*4*) particular disordered structures.

Sol–gel technology is the preparation of ceramic, glass, or composite materials by the preparation of a sol, gelation of the sol, and removal of the solvent (1). Potential advantages of the sol–gel process (3,4) include (*1*) high homogeneity owing to the use of molecularly tailored, usually liquid, compounds; (*2*) high purity owing to the use of chemically prepared precursor; (*3*) lower temperatures of processing that lead to savings in energy, reduction of risk of crystallization and phase separation, and combinations between ceramics and low temperature materials such as organic compounds and polymers; (*4*) preparation of new crystalline phases and new noncrystalline solids; (*5*) special products such as fibers, films, and aerogels (qv); (*6*) ultrastructural control of materials by manipulating network formation from early stages of sol formation; (*7*) production of net-shape optics such that complex geometries and surface replication can be achieved, and lightweight optics formed involving reduced grinding and polishing operations; (*8*) preparation of materials having improved physical properties (fully dense silica) such as lower coefficients of thermal expansion and lower uv cutoff and higher optical transmission; and (*9*) preparation of transparent porous materials (porous silica) that allow impregnation using organic compounds and polymers at controlled chemical doping and oxidation states of dopants, as well as production of graded refractive index.

One of the earliest examples of sol–gel processing to make a new material at low temperatures was the hydrolysis of the precursor tetraethyl orthosilicate [78-10-4] (TEOS), $Si(OC_2H_5)_4$, under acidic conditions, which yields silica, SiO_2, in the form of a glass-like material. Fibers could be drawn from the viscous gel and even from the monolithic optical lenses or composites formed (3). However, extremely long drying times of one year or more were necessary to avoid fracture of the silica gels into a fine powder. Consequently, there was little technological interest in the process for many years.

From the late 1800s through the 1920s, many noted chemists have investigated the periodic precipitation phenomenon that leads to the formation of Liesegang rings and the growth of crystals from gels (5). Much descriptive literature resulted from these studies but little was understood of the physical chemistry principles. In the early 1900s, sol–gel methods were used to produce highly porous silica gels that could be used for desiccants (qv), adsorbents, and catalysts (1). Silica-supported catalysts were prepared by impregnation of partially dried silica gels using Pt or Fe compounds, followed by stabilization of the gels by drying and firing. Sol–gel processes for a wide range of oxides were developed, including Fe, Cr, Mn, Cu, Bi, Pb, Th, Ni, V, and mixed gels of silica with Fe, Cu, Ni, Sn, W, or Al oxides (1).

Some of the earliest (1930s) commercial sol–gel technology was the use of partially hydrolyzed alkoxysilane solutions for impregnation of porous stone, concrete, and brick; in dental cements; and for coating silica on the interior of light bulbs and vacuum tubes (1). In the 1940s, commercial applications of sol–gel-derived antireflective coatings on glass emerge, along with a growing industrial use of sol–gel-derived catalysts. Emphasis in catalyst development was on the use of metal alkoxides as precursors to help control the homogeneity of multiphase systems (see ALKOXIDES, METAL). Work from the 1950s onward led to commercial processes for antireflection thin-film coatings of $TiO_2/SiO_2-TiO_2-SiO_2$ on automobile rear-view mirrors and sunshielding windows; by the mid-1980s, as many as 50 different optical products were being made with sol–gel coatings (1). Work on sol–gel coatings has been reviewed (6).

In the mid-1950s (7), the potential for achieving very high levels of chemical homogeneity in colloidal gels was recognized and the sol–gel method was used to synthesize many novel ceramic oxide compositions, involving Al, Si, Ti, Zr, etc. These could not be made using traditional ceramic powder methods. During the same period, pioneering work in silica chemistry (8) led to the commercial development of colloidal silica powders, ie, the Ludox spheres. Using ammonia (qv) as a catalyst for the TEOS hydrolysis reaction was shown to control both the morphology and size of the powders, yielding the so-called Stober spherical silica powder (9). The final size of the spherical silica powder is a function of the initial concentration of water and ammonia; the type of silicon alkoxide, eg, methyl, ethyl, pentyl, and esters; the alcohol, ie, methyl, ethyl, butyl, and pentyl, mixture used; and the reactant temperature.

During the 1980s, stimulated by several discoveries, interest in the sol–gel process increased greatly. Based on early work (10), very low density silica monoliths, called aerogels, have been made by supercritical drying (11). Supercritical drying of silica gels has been shown to yield large, fully dense silica glass monoliths (12). Another important development was the preparation of large, ie, pieces of several mm, monolithic pieces of optically transparent transition alumina by sol–gel methods (13). These demonstrations of potentially practical routes for production of new net-shape materials having unique properties coincided with the growing recognition that powder processing of materials had inherent limitations in homogeneity owing to difficulty in controlling agglomeration.

Compositions

Nucleation of particles in a very short time followed by growth without supersaturation yields monodispersed colloidal oxide particles that resist agglomeration (9,10). A large range of colloidal powders having controlled size and morphologies have been produced using these concepts (3,14). Materials include oxides, eg, TiO_2, α-Fe_2O_3, Fe_3O_4, $BaTiO_3$, and CeO_2; hydroxides, eg, $AlOOH$, $FeOOH$, and $Cr(OH)_3$; carbonates, eg, $Cd(OH)CO_3$, $Ce_2O(CO_3)_2$, and $Ce(III)/YHCO_3$; sulfides, eg, CdS and ZnS; metals such as Fe(III), Ni, and Co; as well as various mixed phases or composites, eg, Ni, Co, and Sr ferrites; mixed sulfides such as Zn–CdS and Pb–CdS; and coated particles such as Fe_3O_4 with $Al(OH)_3$ or $Cr(OH)_3$. Controlled hydrolysis of alkoxides has also been used to produce submicrometer TiO_2, doped TiO_2, ZrO_2, doped ZrO_2, doped SiO_2, $SrTiO_3$, and even cordierite powders (1,3). Emulsions have been employed to produce spherical powders of mixed cation oxides, such as yttrium aluminum garnets (YAG), and many other systems (15). Sol–gel powder processes have also been applied to fissile elements (16). Spray-formed sols of UO_2 and UO_2–PuO_2 were formed as rigid gel spheres during passage through a column of heated liquid. Abrasive grains based on sol–gel-derived mixed alumina are important commercial products (1). Powders for superconductors, eg, the YBaCuO system, and magnetic ceramics were also developed using the sol–gel technology (see MAGNETIC MATERIALS; SUPERCONDUCTING MATERIALS).

Glass and polycrystalline ceramic fibers have been prepared using the sol–gel method. Compositions include (1) TiO_2–SiO_2 and ZrO_2–SiO_2 glass fibers, high purity SiO_2 waveguide fibers, and Al_2O_3, ZrO_2, ThO_2, MgO, TiO_2, $ZrSiO_4$, and $3Al_2O_3 \cdot 2SiO_2$ fibers. Table 1 summarizes the extensive compositions of ceramics prepared by alkoxide-derived gel methods.

Sol–Gel Process Steps

Overview. Three approaches are used to make most sol–gel products: method 1 involves gelation of a dispersion of colloidal particles; method 2 employs hydrolysis and polycondensation of alkoxide or metal salts precursors followed by supercritical drying of gels; and method 3 involves hydrolysis and polycondensation of alkoxide precursors followed by aging and drying under ambient atmospheres.

Production of net-shape silica (qv) components serves as an example of sol–gel processing methods. A silica gel may be formed by network growth from an array of discrete colloidal particles (method 1) or by formation of an interconnected three-dimensional network by the simultaneous hydrolysis and polycondensation of a chemical precursor (methods 2 and 3). When the pore liquid is removed as a gas phase from the interconnected solid gel network under supercritical conditions (critical-point drying, method 2), the solid network does not collapse and a low density aerogel is produced. Aerogels can have pore volumes as large as 98% and densities as low as 80 kg/m^3 (12,19).

When the pore liquid is removed at or near ambient pressure by thermal evaporation, ie, by drying (methods 1 and 3), shrinkage occurs and the monolith is termed as xerogel. If the pore liquid is primarily alcohol-based, the monolith is often termed an alcogel. The generic term gel is usually applied to either xerogels or alcogels, whereas aerogels are usually specified as such. A gel is defined as dried when the physically adsorbed solvent is completely evacuated. This occurs between 100 and 180°C.

The surface area of dried gels made by method 3 is very large (>400 m^2/g) and the average pore radius is very small (<10 nm). Larger pore radii can be produced by thermal treatment (3), by chemical washing during aging, or by additions of HF to the sol (3,9). The small pore radii can lead to large capillary pressures (eq. 1) during drying or when the dried gel is exposed to liquids as described by Laplace's equation (20,21) for the drying of gels:

$$p = \frac{-2\gamma \cos \theta}{r} \tag{1}$$

Table 1. Sol–Gel Multicomponent Materials Prepared by Method 2[a]

System[b]	Precursors[c]
Binary silicates	
SiO$_2$, B$_2$O$_3$	TEOS, TMOS, B(OCH$_3$)$_3$, (NH$_4$)$_2$B$_4$O$_7$·4H$_2$O
SiO$_2$–Al$_2$O$_3$	*sec*-Al(OC$_4$H$_9$)$_3$
SiO$_2$, (20–35)[d] GeO$_2$	TMOS, Ge(OC$_2$O$_5$)$_4$
SiO$_2$, TiO$_2$	TEOS, TMOS, Ti(OC$_2$O$_5$)$_4$, Ti(OC$_3$H$_7$)$_4$
SiO$_2$, (7–48) ZrO$_2$	TEOS, Zr(OC$_3$H$_7$)$_4$
SiO$_2$, (1, 5, and 10) SrO	TEOS, Sr(NO$_3$)$_2$
SiO$_2$, (5–40) Fe$_2$O$_3$	TEOS, Fe(OC$_2$H$_5$)$_3$
Ternary silicates	
SiO$_2$, (5–30) La$_2$O$_3$, (80–85) SiO$_2$, 10 La$_2$O$_3$, 5 to 10 Al$_2$O$_3$	silica hydrosol Ludox AS, La(NO$_3$)$_3$, Al(NO$_3$)$_3$·9H$_2$O, ZrOCl$_2$·8H$_2$O
(80–94) SiO$_2$, 5 to 10 La$_2$O$_3$, 1 to 10 ZrO$_2$	TEOS, TMOS, La(NO$_3$)$_3$, Al(OC$_3$H$_7$)$_3$, Al(OC$_4$H$_9$)$_3$, ZrOCl$_2$·8H$_2$O
(60–76) SiO$_2$, 17 to 33 ZrO$_2$, 7 Na$_2$O	TEOS, Zr(OC$_3$H$_7$)$_4$, NaOCH$_3$
SiO$_2$, Al$_2$O$_3$/MgO 0.5[e], 0.71, 1.3; TiO$_2$ (up to 10) and (5.5) LiO$_2$	Al(OC$_4$H$_9$)$_3$, (CH$_3$COO)$_2$Mg·4H$_2$O, (CH$_3$COO)Li·2H$_2$O, Ti(OC$_2$H$_5$)$_4$
SiO$_2$, TiO$_2$, ZrO$_2$	TEOS, Ti(OC$_4$H$_9$)$_4$, Zr(NO$_3$)$_4$·5H$_2$O
SiO$_2$, Al$_2$O$_3$, B$_2$O$_3$	TEOS, AlCl$_3$, H$_3$BO$_3$
SiO$_2$, CaO, P$_2$O$_5$	TEOS, (C$_2$H$_5$O)$_3$P(O), Ca(NO$_3$)$_2$·4H$_2$O or Ca(O$_2$C$_3$H$_7$)$_2$
Multicomponent silicates	
66 SiO$_2$, 18 B$_2$O$_3$, 7 Al$_2$O$_3$, 3 BaO, 6 Na$_2$O	TEOS, B(OCH$_3$)$_3$, Al(OC$_4$H$_9$)$_3$, Ba(OC$_2$H$_5$)$_2$
86.9 SiO$_2$, 5.9 B$_2$O$_3$, 2.5 Al$_2$O$_3$, 3.9 Na$_2$O, 0.7 K$_2$O	TMOS, H$_3$BO$_3$, *sec*-Al(OC$_4$H$_9$)$_3$, NaOCH$_3$, KOC$_2$H$_5$
Nonsilicate oxide materials	
ZrO$_2$, Y$_2$O$_3$	Zr(OC$_4$H$_9$)$_4$, Y(NO$_3$)$_3$
ZrO$_2$, CaO	Zr(OC$_3$H$_7$)$_4$, Ca(CH$_3$COO)$_2$
ZrO$_2$, CeO$_2$	Zr(OC$_3$H$_7$)$_4$, Ce(NO$_3$)$_3$
BaTiO$_3$	Ti(OC$_3$H$_7$)$_4$, Ba(OC$_2$H$_5$)$_3$
LiNbO$_3$	Li(OC$_2$H$_5$), Nb(OC$_2$H$_5$)$_5$
YBa$_2$Cu$_3$O$_{(7-x)}$	Y(OC$_4$H$_9$)$_3$, Ba(OC$_4$H$_9$)$_2$, Cu (OC$_4$H$_9$)$_2$
P$_2$O$_5$, Na$_2$O	TEOS, P(OC$_2$H$_5$)$_3$, NaOCH$_3$
B$_2$O$_3$, (14.2 and 17.9) Li$_2$O	*n*-B(OC$_4$H$_9$)$_3$, LiOH$_3$

[a]Refs. 1, 5, 17, and 18.
[b]Values given are in wt % unless otherwise indicated.
[c]TEOS = Si(OC$_2$H$_5$)$_4$; TMOS = Si(OCH$_3$)$_4$.
[d]Values are in vol %.
[e]Value is a mol ratio.

where p = the pressure difference in the capillaries, γ = specific surface energy of the vapor–liquid interface, θ = contact angle, and r = pore radius.

A dried gel still contains a very large concentration of chemisorbed hydroxyls on the surface of the pores. Thermal treatment in the range of 500–800°C desorbs the hydroxyls and thereby decreases the contact angle and the sensitivity of the gel to rehydration stresses, resulting in a stabilized gel. Heat treatment of a gel at elevated temperatures substantially reduces the number of pores and their connectivity owing to viscous phase sintering. This is termed densification. The density of the material increases and the volume fraction of porosity decreases during sintering. The porous gel is transformed to a dense glass when all pores are eliminated. Densification is complete at 1250–1500°C for silica gels made by method 1 and as low as 1000°C for gels made by method 3. The densification temperature decreases as the pore radius decreases and surface area of the gels increases, as illustrated in Figure 1. Silica glass made by densification

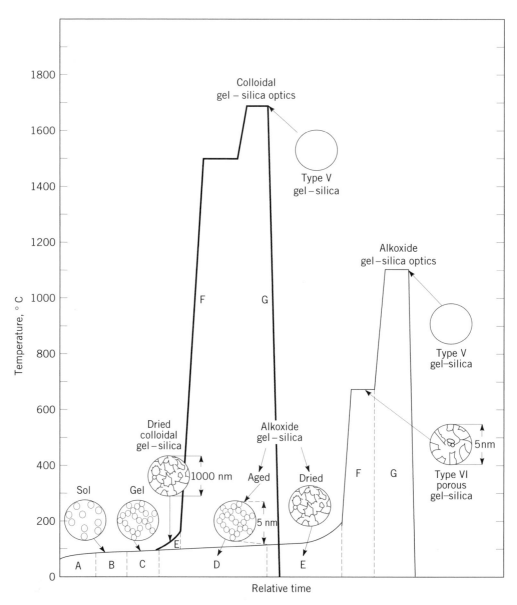

Fig. 1. Gel–silica glass process sequence showing the process steps: A, mixing; B, casting; C, gelation; D, aging; E, drying; F, dehydration or stabilization; and G, densification.

of porous silica gel is amorphous and nearly equivalent in structure and density to vitreous silica made by fusing quartz crystals or sintering of SiO_2 powders made by chemical vapor deposition (CVD) of $SiCl_4$ (3).

The seven processing steps shown schematically in Figure 1 are involved to various degrees in making sol–gel-derived silica monoliths by methods 1, 2, and 3. The emphasis herein is primarily on net-shape sol–gel-derived silica monoliths made by the alkoxide process (method 3) prepared under ambient pressures.

Mixing. In method 1, a suspension of colloidal powders, or sol, is formed by mechanical mixing of colloidal particles in water at a pH that prevents precipitation (step A in Fig. 1) (8). In method 2 or 3, a liquid alkoxide precursor such as $Si(OR)_4$, where R is CH_3 (TMOS), C_2H_5 (TEOS), or C_3H_7, is hydrolyzed by mixing with water (eq. 2).

Hydrolysis

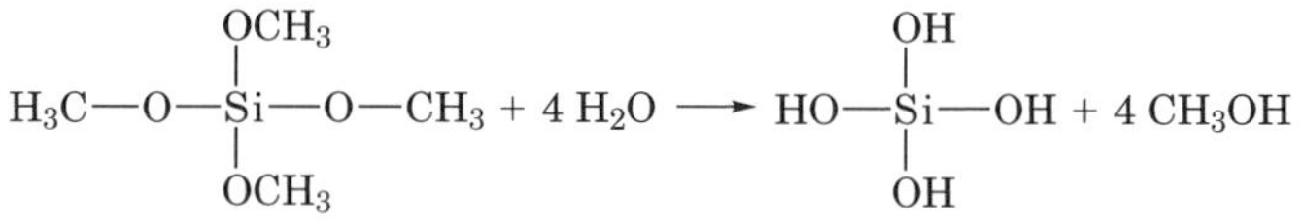

$$H_3C-O-\underset{\underset{\displaystyle OCH_3}{|}}{\overset{\overset{\displaystyle OCH_3}{|}}{Si}}-O-CH_3 + 4\,H_2O \longrightarrow HO-\underset{\underset{\displaystyle OH}{|}}{\overset{\overset{\displaystyle OH}{|}}{Si}}-OH + 4\,CH_3OH \qquad (2)$$

As soon as any hydrolyzed specie is present, condensation proceeds. The hydrated silica tetrahedra interact in a condensation reaction (eq. 3), forming $\equiv Si{-}O{-}Si\equiv$ bonds.

Condensation

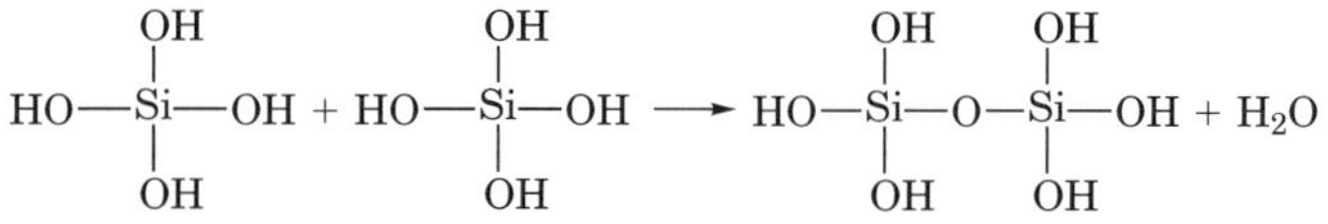

$$HO-\underset{\underset{\displaystyle OH}{|}}{\overset{\overset{\displaystyle OH}{|}}{Si}}-OH + HO-\underset{\underset{\displaystyle OH}{|}}{\overset{\overset{\displaystyle OH}{|}}{Si}}-OH \longrightarrow HO-\underset{\underset{\displaystyle OH}{|}}{\overset{\overset{\displaystyle OH}{|}}{Si}}-O-\underset{\underset{\displaystyle OH}{|}}{\overset{\overset{\displaystyle OH}{|}}{Si}}-OH + H_2O \qquad (3)$$

Linkage of additional $\equiv Si{-}OH$ tetrahedra occurs as a polycondensation reaction (eq. 4) and eventually results in a SiO_2 network. The H_2O and alcohol expelled from the reaction remain in the pores of the network.

Polycondensation

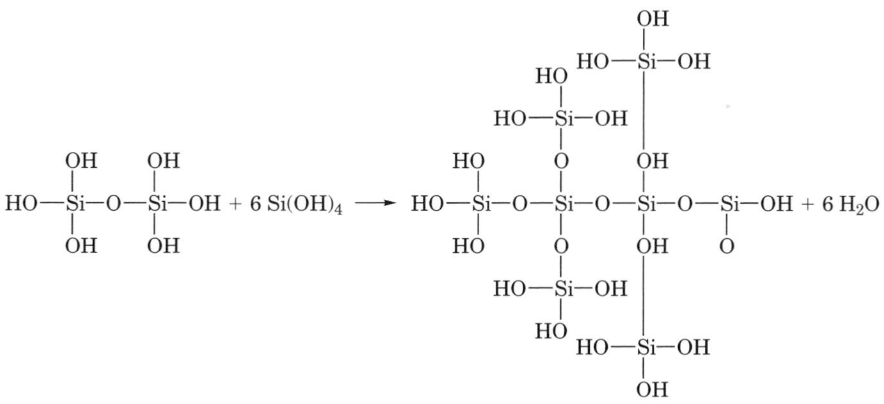

$$HO-\underset{\underset{\displaystyle OH}{|}}{\overset{\overset{\displaystyle OH}{|}}{Si}}-O-\underset{\underset{\displaystyle OH}{|}}{\overset{\overset{\displaystyle OH}{|}}{Si}}-OH + 6\,Si(OH)_4 \longrightarrow \cdots + 6\,H_2O \qquad (4)$$

The hydrolysis and polycondensation reactions initiate at numerous sites within the TMOS/H$_2$O solution as mixing occurs. When sufficient interconnected Si–O–Si bonds are formed in a region, the material responds cooperatively as colloidal (submicrometer) particles or a sol. The size of the sol particles and the cross-linking within the particles, ie, the density, depends on the pH and R ratio, where R = [H$_2$O]/[Si(OR)$_4$].

Casting. Because the sol is a low viscosity liquid, it can be cast into a mold (step B, Fig. 1). The mold must be selected to avoid adhesion of the gel. The sol can be applied as a coating on a substrate or drawn into fibers or emulsified.

Gelation. After some time the colloidal particles and condensed silica species link together to become a three-dimensional network (step C, Fig. 1). The physical characteristics of the gel network depend greatly on the size of particles and extent of cross-linking prior to gelation. At gelation, the viscosity increases sharply and a solid object results in the shape of the mold. Using appropriate control of the time-dependent change of viscosity of the sol, fibers can be pulled or spun. Particles having controlled morphology and structure can be created during gelation.

Aging. The process that involves a continuous change in structure and properties of a completely immersed gel in liquid after the gel point is called aging (step D, Fig. 1). The shrinkage of the gel and the resulting expulsion of liquid from the pores during aging is called syneresis. During aging, polycondensation continues along with localized solution and reprecipitation of the gel network, which increases the thickness of interparticle necks and decreases the porosity. The strength of the gel thereby increases with aging. An aged gel must develop sufficient strength to resist cracking during drying.

Drying. During drying (step E, Fig. 1), the liquid is removed from the interconnected pore network. Large capillary stresses can develop during drying when the pores are small (<20 nm). These stresses can cause gels to crack catastrophically unless the drying process is controlled by either decreasing the liquid surface energy by addition of surfactants, elimination of very small pores (method 1), supercritical evaporation which avoids the vapor–liquid interface (method 2), or producing a homogenous structure free of defects by controlling the rates of hydrolysis and condensation (method 3). After supercritical drying (method 1), an aerogel of very low density is a very good thermal insulator when sandwiched between glass plates and evacuated (12,22).

Dehydration or Chemical Stabilization. The removal of surface silanol (Si–OH) bonds from the pore network results in a chemically stable ultraporous solid (step F, Fig. 1). Porous gel–silica made in this manner by method 3 is optically transparent, having both interconnected porosity and sufficient strength to be used as unique optical components when impregnated with optically active polymers, such as fluors, wavelength shifters, dyes, or nonlinear polymers (3,23).

Densification. Heating the porous gel at high temperatures causes densification (step G, Fig. 1) to occur. The pores are eliminated and the density ultimately becomes equivalent to quartz or fused silica. The densification temperature depends considerably on the dimensions of the pore network, the connectivity of the pores, and surface area, as illustrated in Figure 1 (3,21). Alkoxide gels (method 3) have been densified as low as 1000°C (24), whereas gels made by the commercial colloidal process (5) require heating to 1500–1720°C.

Colloidal silica gels that have carefully controlled dense packing can also be densified at temperatures as low as 1000°C (3,25). The purity and homogeneity of dense gel–silica made by method 3 is superior to other silica glass processing methods. The ability to produce optics having nearly theoretical limits of optical transmission, lower coefficients of thermal expansion, and greater homogeneity, along with net-shape casting, are significant advances resulting from sol–gel processing of monoliths (3,5).

Hydrolysis and Polycondensation. As shown in Figure 1, at gel time (step C), events related to the growth of polymeric chains and interaction between colloids slow down considerably and the structure of the material is frozen. Post-gelation treatments, ie, steps D–G (aging, drying, stabilization, and densification), alter the structure of the original gel but the resultant structures all depend on the initial structure. Relative rates, k, of hydrolysis, k_H, (eq. 2) and condensation, k_C, (eq. 3) (k_H/k_C) determine the structure of the gel. Many factors influence the kinetics of hydrolysis and condensation because both processes often occur simultaneously. The most important variables are temperature, nature and concentration of electrolyte, nature of solvents, and type of alkoxide precursor. Pressure also influences the gelation process. The k_H increases with pressure but the latter is usually not a processing variable.

Hydrolysis of TEOS in various solvents is such that for a particular system k_H increases directly with the concentration of H^+ or H_3O^+ in acidic media and with the concentration of OH^- in basic media. The dominant factor in controlling the hydrolysis rate is pH (21). However, the nature of the acid plays an important role, so that a small addition of HCl induces a 1500-fold increase in k_H, whereas acetic acid has little effect. Hydrolysis is also temperature-dependent. The reaction rate increases 10-fold when the temperature is varied from 20 to 45°C. Nmr experiments show that k_H varies in different solvents as follows: acetonitrile > methanol > dimethylformamide > dioxane > formamide, where the k_H in acetonitrile is about 20 times larger than the k_H in formamide. The nature of the alkoxy groups on the silicon atom also influences the rate constant. The longer and the bulkier the alkoxide group, the lower the k_H (3).

The ^{29}Si-nmr is one of the most useful techniques to follow the hydrolysis and first-stage polymerization of silicon alkoxides. This technique allows determination of the concentration of the different $Si(OR)_x(OH)_y$ and $(OH)_u(OR)_v Si-O-Si(OR)_x(OH)_y$ species. Each of the monomer and dimer species has a specific chemical shift with respect to the metal alkoxide. A typical sequence of condensation products is dimer, linear trimer, cyclic trimer, cyclic tetramer, and a higher order generation of discrete colloidal particles which are commonly observed in aqueous systems. A k_C value of 3.3×10^{-6} L/(mol·s) has been reported (26) for the dimerization of monosilicic acid [1343-98-2]. Application of 500 MPa (5 kbar) pressure increased the polycondensation rate constant by 10-fold (26). This sequence of condensation requires both depolymerization (ring opening) and the availability of monomers, ie, species which may be produced by depolymerization. However, in alcoholic solutions, especially at low pH, the depolymerization rate is very low. Condensation may result in a spectrum of structures ranging from molecular networks to colloidal particles. Under acidic conditions, more linear structures are formed prior to gelation. Under basic conditions, the distribution of polysilicate species is much broader and characteristic of branched polymers having a high degree of cross-linking, whereas for acidic conditions there is a lower degree of cross-linking. Thus, the shape and size of polymeric structural units are determined by the relative values of the rate constants for hydrolysis and polycondensation reactions (k_H and k_C, respectively). Fast hydrolysis and slow condensation favor formation of linear polymers, whereas slow hydrolysis and fast condensation results in larger, bulkier, and more ramified polymers (3,21).

Chemical additions to a sol–gel silica system, such as formamide, decrease the hydrolysis rate and slightly increase the condensation rate. This is attributed to the ability of $HCONH_2$ to form hydrogen bonds and to its high dielectric constant ($\epsilon = 110$). The presence of formamide also decreases the time of gelation (t_g) and increases the pH. The change in hydrolysis and condensation rates greatly alters the texture of the gel and its fractal character, as determined from Raman spectroscopy and small-angle x-ray scattering, and modifies the physical properties of the resulting gel greatly (3,5).

Gelation. A sol becomes a gel when it can support a stress elastically, defined as the gelation point or gelation time, t_g. A sharp increase in viscosity accompanies gelation. A sol freezes in a particular polymer structure at the gel point (27). This frozen-in structure may change appreciably with time, depending on the temperature, solvent, and pH conditions or on removal of solvent.

Gelation Time. The time of gelation changes significantly with sol–gel chemistry (9,21,23). One method of measuring t_g determines the viscoelastic response of the gel as a function of shear rate (27). Figure 2 shows the large change in the loss tangent at the gelation time along with the changes in storage modulus, G', and loss modulus, G'' (27). The rapid increase in G' near t_g is consistent with the concept that the interconnection of the particles becomes sufficient to support a load elastically. Gelation may occur at different extents of reaction completion. For example, in the polymerization of TMOS, more silicon alkoxide must be hydrolyzed to reach t_g, when the experimental conditions favor a ramified polymer, rather than a linear one.

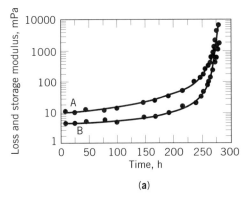
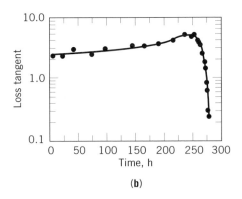

(a) (b)

Fig. 2. Loss tangent as a measure of gelation time for a silica sol (27): (**a**) loss (A) and storage (B) modulus as a function of aging time for H_2O/TEOS mol ratio of 20 and HNO_3/TEOS mol ratio of 0.01; and (**b**) loss tangent as a function of aging time. To convert mPa to mm Hg, multiply by 7.50×10^{-6}.

The curve that represents the dependence of gelation time on solution pH has a maximum around the isoelectric point of silica (pH ≈ 2) and a minimum near pH 5–6 (8). The longer and the larger the solvent molecule, the longer the gelation time. Also, the longer and the larger the alkoxy group, the longer the t_g.

The amount of water employed for hydrolysis also has a dramatic influence on gelation time. For an R, ie, mole ratio of water to silicon alkoxide, of 2, t_g is about seven hours when the gelation process is held at 70°C and HF is present as catalyst; for $R = 8$, t_g decreases to 10 minutes (3). For low water contents, an increase in the amount of hydrolysis water decreases the gelation time, although there is a dilution effect. For higher water content, the gelation time increases with the quantity of water.

Gel Structure. The system evolves from a sol, where individual particles interact more or less weakly with each other, to a gel, which is a continuous network occupying the entire volume. The techniques available to follow structural evolution at the nanometer scale of sol–gel networks include small-angle x-ray scattering (saxs), neutron scattering, light scattering, and transmission electron microscopy. Small-angle x-ray scattering allows the determination of a characteristic length of the particle (Guinier's radius of gyration, or electronic radius of gyration) and a fractal dimension, which gives some information on the structure of the polymer (branched vs linear) and on the growth mechanism. Small-angle neutron scattering (sans) has also been applied to the study of silica sols giving similar results to those of saxs (3). Scattering studies show that acid-catalyzed sols develop a more linear structure with less branching, whereas base-catalyzed systems have highly ramified structures (21).

Theories of Gelation. The classical or mean field theory of polymerization (4) is useful for visualizing the conditions for gelation. This model yields a degree of reaction, p_c, of one-third at the time of gelation for chemical species having functionality equal to four. Two-thirds of the possible connections are still available and therefore may play a role in subsequent processing. This value is lower than the experimental evidence but represents the minimum degree of reaction before gelation can occur. Experimental results indicate that $0.6 < p_c \leq 0.84$ for silica sol–gel systems (20). Percolation theory is also used to represent gelation. Its relationship to gelation has been reviewed (28,29).

Diffusion-limited aggregation of particles results in a fractal object. Growth processes that are apparently disordered also form fractal objects (30). Sol–gel particle growth has also been modeled using fractal concepts (3,20). The nature of fractals requires that they be invariant with scale, ie, the fractal must look similar regardless of the level of detail chosen. The second requirement for mass fractals is that their density decreases with size. Thus, the fractal model overcomes the problem of increasing density of the classical models of gelation, yet retains many of its desirable features. The mass of a fractal, M, is related to the fractal dimension and its size or radius, R, by equation 5:

$$M \propto R^{d_f} \tag{5}$$

Fractal objects are quantified by their fractal dimension, d_f. For linear-like structures, $1 < d_f < 2$. Fractally rough structures have a mass fractal dimension, $2 < d_f < 3$. Uniform nonfractal objects have a fractal dimension, $d_f = 3$.

A fractal gelation model was developed (31) based on sol particles growing from partially hydrolyzed TEOS, $Si(OC_2H_5)_4$. Using saxs, the fractal dimension of gels were shown to increase as the ratio between H_2O/TEOS used in the gel preparation increased. It was assumed that the nonhydrolyzed ethyl groups on TEOS molecules would dictate the fractal nature of growing particles by yielding more linear polymers. When fully hydrolyzed TEOS was used in the model, a fully dense particle having no fractal nature was formed. A range of fractal dimensions from 1.6 to 2.4 depends on the degree of hydrolysis. Partially hydrolyzed TEOS yields more linear structures that result in lower values for d_f. More linear structures can also be obtained by using acids as catalysts instead of bases. The fractal dimension also increases when the ratio between rate of hydrolysis and rate of condensation is decreased, as shown experimentally (3) for sols catalyzed with and without formamide.

A primary sol particle in an acid-catalyzed sol has radius between 1 and 2 nm (3). The secondary fractal particle has a radius, R_g, of 5 to 20 nm as seen from saxs (3). For the TMOS-based sols investigated by saxs, d_f increases with time, as does the Guinier radius, R_g. The structure reaches a fractal dimension around 2.3 at the gelation point.

Theoretical Studies.   Theoretical models for the $Si(OR)_4$ hydrolysis, poly-condensation, and dehydration reactions involved in sol–gel processes have been developed using semiempirical molecular orbital models. These have been re-viewed (3,5).

Aging.   When a gel is maintained in its pore liquid, the structure and prop-erties continue to change long after the gel point. This process is called aging. Four aging mechanisms can occur, singly or simultaneously: polycondensation, syneresis, coarsening, and phase transformation (9,21).

Polycondensation reactions (eqs. 3 and 4), continue to occur within the gel network as long as neighboring silanols are close enough to react. This increases the connectivity of the network and its fractal dimension. Syneresis is the spontaneous shrinkage of the gel and resulting expulsion of liquid from the pores. Coarsening is the irreversible decrease in surface area through dissolution and reprecipitation processes.

Polycondensation.   The number of bridging bonds in a silica gel increases long after gelation. The condensation reaction continues to occur because of the large concentration of silanol groups in a newly formed gel. As the hydroxyls are lost during aging, new bonds are formed, thus creating more highly cross-linked structures. The ^{29}Si-nmr show substantial amounts of Q^2 species at the gel point (32); the proportions of Q^3 and Q^4 species increase with time long after gelation. The Q^n terminology is used to represent an Si atom bonded through a bridging oxygen to n other Si atoms. Monomers may also be present at the gel time, mostly for gels prepared in conditions where depolymerization reactions are enhanced (base catalysis). Because the chemical reaction is faster at higher temperature, aging can be accelerated by hydrothermal treatment, which increases the rate of the condensation reaction (8).

Syneresis.   The shrinkage of a gel and the resulting expulsion of liquid from the pores is called syneresis (9,21). Syneresis in alcoholic gel systems is attributed to formation of new bonds through condensation reactions, which increase the number of bridging bonds and cause contraction of the gel network. In aqueous gel systems or colloidal gels, the structure is controlled by the balance between electrostatic repulsion and attractive van der Waals forces. The extent of shrinkage is controlled by additions of electrolyte. The rate of contraction of silica gel during syneresis has a minimum at the isoelectric point (IEP). For silica, this point is at a pH of 2, at which the silicate species are uncharged (8). Shrinkage is driven by the condensation reaction in equation 3. The syneresis contraction rate increases with concentration of silica in the sol and with temperature. When organic solvents are present, these may form hydrogen bonds with the silanol groups, which inhibit condensation and slow syneresis. The rate of syneresis decreases with time, probably owing to increasing stiffness of the network as more bridging bonds are formed. Total syneresis strain is greater at lower temperatures, although the contraction rate is slower.

Coarsening. Coarsening, or Ostwald ripening, results from convex surfaces being more soluble than concave surfaces. If a gel is immersed in a liquid in which it is soluble, dissolved material tends to precipitate into regions of negative curvature (8). Necks between particles grow and small pores are filled, increasing the average pore size of the gel and decreasing the specific surface area.

The solubility of silica increases at high pH, so does the rate of coarsening of silicate gels. The pore size distribution in a silica xerogel increases when aged in a basic solution. At a given normality, the effect of solution on aging decreases in the following order: $HCl > H_2SO_4 > H_3PO_4$. However, the rate of aging is independent of the type of acid if the activity of the proton is the same in each solution. Silica gels even coarsen in concentrated mineral acids ($1 < pH < 2$). Reduction in surface area is produced by dissolution and reprecipitation. A high pore volume results because the stiffer gel produced by aging does not shrink as much under the influence of capillary pressure. The texture of a gel can be affected in every stage of its preparation, including gelation, after-treatment of the hydrogel by aging and washing with various liquids, and drying (5). Time, temperature, and pH are the primary variables that alter the aging process. Washing the pore liquor out of a gel is also an aging step and the pH of the washwater is critical in the case of gels made from acid-catalyzed precursors. The final properties of such gels depend on both pH at which the gel was formed and the pH in which it was washed before drying.

Soaking a silica gel in dilute ammonium hydroxide solution at 50–85°C can result in significant coarsening of the gel texture (5). Aging and thermal treatments result in a one-way process, ie, loss of specific surface area and in increase in pore size. The pore size can also be enlarged by dissolution of some of the silica. Treating a silica gel with 0.5-N KOH or dilute HF can enlarge the pores from 0.7 to 3.7 nm (3).

Properties. During aging, there are changes in most textural and physical properties of the gel. The change in mechanical properties and pore size during aging is most important. Inorganic gels are viscoelastic materials responding to a load with an instantaneous elastic strain and a continuous viscous deformation. Because the condensation reaction creates additional bridging bonds, the stiffness of the gel network increases, as does the elastic modulus, the viscosity, and the modulus of rupture. The modulus of rupture of a silica gel aged at 105°C in its mother liquor reaches 40 Pa (0.30 mm Hg) after 40 days and increases logarithmically between 1 and 32 days. The strength increases exponentially with aging temperature between 25 and 105°C. The strongest gels had moduli of rupture ≈400 Pa (3 mm Hg). Shear modulus and strength of silica gels increases for up to nine months of aging.

Drying.   For porous systems, there are three stages of drying. During the first stage of drying the decrease in volume of the gel is equal to the volume of liquid lost by evaporation. The compliant gel network is deformed by the large capillary forces that cause shrinkage of the object. In classical large-pore systems, this first stage of drying is called the constant rate period because the evaporation rate per unit area of the drying surface is independent of time (21,33,34). This behavior is applicable to gels made by colloidal precipitation (method 1) or base-catalyzed alkoxide gels (method 3) that have average pore diameters of >20 nm. The drying kinetics of acid-catalyzed alkoxide gels having pores <20 nm is such that the rate of drying in stage 1 is not constant, but decreases owing to changes in liquid composition within the pores. Pore liquid is usually composed of a mixture of solvents having different levels of volatility. Pore size has also been shown to influence the decrease in drying rate during the first period (35). The drop on the rate of drying for isolated pore liquid is much steeper and ends shorter than for liquid within very small (<5 nm) pores. Thus, changes in pore size during drying as well as a shift in composition of pore liquid can affect the rate of drying in stage 1. For large- or small-pore gels the greatest changes in volume, weight, density, and structure occur during stage-1 drying. Stage 1 ends when shrinkage ceases.

The second stage begins when the critical point is reached. Classical drying theory calls this the leatherhard point (36). The critical point occurs when the strength of the network has increased owing to the greater packing density of the solid phase, which is sufficient to resist further shrinkage. As the network resistance increases, the radius of the meniscus is reduced. Eventually, at the critical point, the meniscus equals the radius of the pore. This condition creates the highest capillary pressure and, unable to compress the gel any further, the pores begin to empty, which is the start of stage 2. In stage 2, liquid transport occurs by flow through the surface films that cover partially empty pores. The liquid flows to the surface where evaporation takes place. The flow is driven by the gradient in capillary stress (21). Because the rate of evaporation decreases in stage 2, this is termed the first falling rate period (37).

The third stage of drying is reached when the pores have substantially emptied and surface films along the pores cannot be sustained. The remaining liquid can escape only by evaporation from within the pores and diffusion of vapor to the surface. During this stage, called the second falling rate period, there are no further dimensional changes, only a slow progressive loss of weight until equilibrium is reached, which is determined by the ambient temperature and partial pressure of water. The factors in this stage that are relevant to the drying of gels have been reviewed (21,38).

A comparison of results on the drying of an alumina gel (33), pores ⩾20 nm, with those for an acid-catalyzed alkoxide silica gel (35), pores <20 nm, illustrates the importance of pore size in gel drying behavior. The rate of water loss was constant from the large-pore gels during stage 1 and similar to the rate of evaporation from a free layer of distilled water. The gel lost pore liquid at a constant rate until the critical point was reached at about 35% water remaining in the gel, or onset of stage 2. The constant rate period was independent of gel thickness and controlled by surface vaporization fed by capillary flow of pore liquid to the surface (33). The drying rate of the large-pore gels decreases considerably from the critical point onward in stage 2, the first falling rate period, until the onset of stage 3. 87% of the initial liquid present in the gel was removed in stage 1; 10% in stage 2; and only 3% in stage 3. Liquid transport in stage 3 was by diffusion in the gas phase. Cracking occurred at some point in stage 2, where the water–air interface moved inward, and depended on thickness of the gel. At dried thicknesses less than 80 μm, the gel shrinkage was primarily perpendicular to the surface and cracks seldom occurred. At 40-μm thickness, no cracks occurred. At thicknesses >80 μm, shrinkage was both radial and perpendicular and cracking was prevalent. The thickness sensitivity of cracking results from the pressure gradient built inside the body.

It has been shown quantitatively (21) that the difference in shrinkage rates between the inside and outside of a drying body indeed results in a tensile drying stress, σ_x. This tensile stress is a function of thickness, L:

$$\sigma_x(L) \approx L\eta_L V_E/3D \tag{6}$$

where η_L is viscosity of the liquid, V_E is the evaporation rate, and D is the permeability of the network. As the thickness of a gel or the evaporation rate increases, the probability of cracking increases. For a typical gel having strength after casting of ca 0.1 MPa (14.5 psi), $\eta_L = 1$ mPa·s(=cP), $L = 1$ cm, and $D = 10^{-14}$ cm^2, the strength is exceeded for any drying rate, $V_E > 0.03$ μm/s. This corresponds to a minimum drying time of $L/V_E \approx 4$ days, which is consistent with the observation that a gel spontaneously fractures when exposed to the atmosphere after casting. It is also consistent with the fact that traditional porous ceramic bodies, having pores ⩾20 nm, are much more crack-resistant during drying. Because the permeability of a 50% porous body having 1-μm pores has a D-value of ca 10^{-10} cm^2, the stress is 10^4 smaller than in a gel drying at the same rate.

When the pores in a gel are <5 nm, ie, characteristic of acid-catalyzed alkoxide gels, the analysis of drying stresses is complicated by the fact that stage 1 is not a constant rate period of drying. As shown in Figure 3, during stage 1, the evaporation rate from the silica gel having pores <5 nm dropped from a maximum of 0.034 to 0.013 g/(h·cm^2) at the critical point, although the meniscus remained at the surface throughout this stage, as required by the definition of stage 1. The relationship between relative weight (W/W_o) and relative volume (V/V_o) of the small pore gel during the various stages of drying is shown in Figure 4. The change between stage 1 and stage 2 is very clear.

The evaporation rate from the gel is dictated by the difference between the vapor pressure at the evaporating surface, P_s, and the vapor pressure of the ambient atmosphere, P_a. Evaporation continues as long as $P_s > P_a$ at a rate, V_e, of

$$V_e = K_e(P_s - P_a) \tag{7}$$

where K_e is a constant.

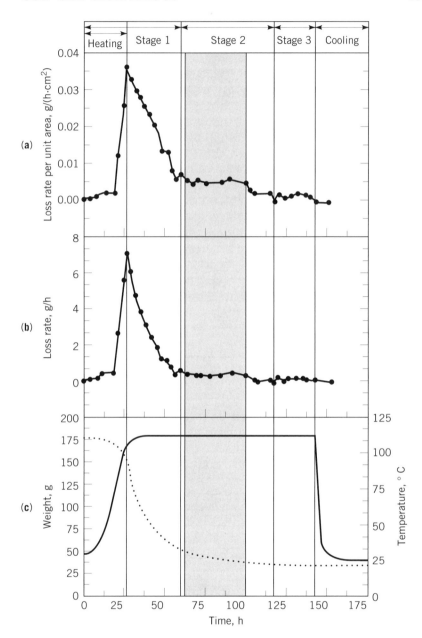

Fig. 3. Time dependence per unit of area for a silica–gel, where the shaded area represents an opaque ministage: (**a**) loss rate; (**b**) absolute loss rate; and (**c**) (——) temperature and (···) weight.

The vapor pressure of a surface composed of a large number of very small pores, P_v, is influenced by the radius of the pores, as described by the Gibbs-Kelvin equation:

$$\ln(P_s/P_o) = (B\gamma 2)/RTr_m \qquad (8)$$

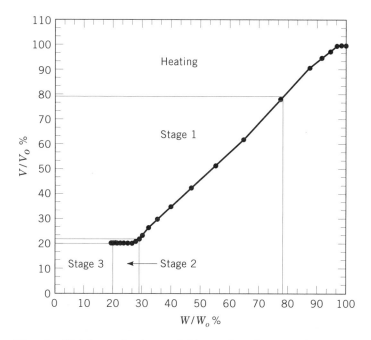

Fig. 4. Weight and volume % for a gel undergoing drying (35).

where P_s is the vapor pressure over the meniscus of the pore, P_o is the vapor pressure over a flat surface at 101.3 kPa (760 Torr), B is the molar volume of the liquid (0.18 g/cm³), R is the universal gas constant (8.314 × 10³ kPa·cm³/(mol·K)), T is temperature in K, r_m is the radius of curvature of the meniscus in m, and γ is the difference in the solid–vapor and liquid–vapor interfacial energies, ie, $\gamma = 0.072$ J/m² (3.42 × 10⁻⁵ ft·lbf/in.²). The depressed evaporation rate from the small pore gel results from a bound layer of water on the pore surfaces, which has a more ordered structure than free water and possesses properties, such as melting point, that are markedly different than bulk water (3,5). Bound water thickness depends on pore radius and temperature.

A schematic of a small pore silica gel surface at the beginning of stage-1 drying is shown in Figure 5a. At this point, the radius of the meniscus is larger than the pore radius. The surface area is initially largely free water, separated by a relatively small areal fraction of gel network with a transition zone of bound water. As the free water evaporates, the solid network is drawn together by the capillary stresses which decrease the areal fraction of free water and increase the areal fraction of solid. At the critical point, end of stage 1, the structural schematic is as depicted in Figure 5b. The bound water is not removed until higher temperatures result in the stabilization regime.

Shortly after entering the second stage, the gel turns opaque, starting at the edges and progressing linearly toward the center. This phenomenon is caused by light scattering from isolated groups of pores in the process of emptying (39). These are of such a dimension that they are able to scatter light. After transparency is regained, Figure 3 shows that the loss rate of the acid-catalyzed alkoxide silica gel monolith gradually falls to a value of ca 0.001 g/(h·cm²) until

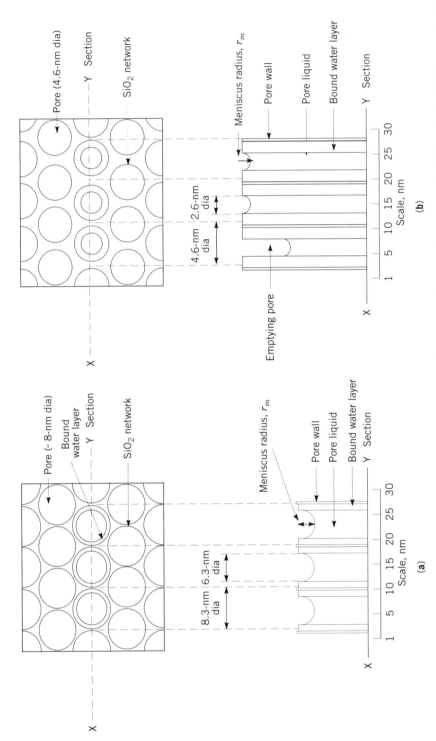

Fig. 5. Schematic representation of gel surface: (**a**) at the beginning of stage 1, where total pore area is 91% and total SiO₂ area is 9%; and (**b**) at the end of stage 1 (critical point), where total pore area is 53% and total SiO₂ area is 47% (35).

no further weight changes occur. The transition to stage 3 is the hardest to identify and its start is probably best defined as the end of the opaque stage. Stage 3 has been described (21) as the second falling rate period, where the temperature of the body is not as strongly suppressed as when evaporation rates were higher. The remaining liquid evaporates within the pores and is removed by diffusion of its vapor to the surface. It is unaffected by local changes in temperature, ambient vapor pressure, or flow rate (35). By the start of stage 3, the gel is dry enough that it can be removed from the drying chamber and dehydrated under much more severe conditions, eg, 180°C at 13.3 Pa (0.1 torr), without risk of cracking. Stress birefringence measurements indicate a gradual reduction in residual stress during stage 3, and the end of stage 2.

When gels crack, they do so at distinct points within the drying sequence. Cracking during stage 1 is rare but can occur when the gel has had insufficient aging and strength (21), and therefore does not possess the dimensional stability to withstand the increasing compressive stress. If the loss rate is increased by lowering the vapor pressure of the ambient atmosphere or increasing the draft rate, there then comes a point when the loss rate exceeds the maximum rate of shrinkage. If this occurs, localized pore emptying results and surface cracks develop. Most failures occur during the early part of stage 2, the point at which the gel stops shrinking. At the start of stage 2 the modulus of the gel is very high and the compressive stress is in the order of 100 MPa (145,000 psi). The possibility of cracking at this point is great owing to the high stresses and low strain tolerance of the material. Cracking during stage 3 seldom occurs (3,35). Defects introduced during gel processing, such as gas bubbles, dirt, or precipitates, create stress concentrations and cause crack initiation. Successful drying of films, fibers, and gel monoliths requires control of the drying rate through the opaque stage and elimination of processing defects during mixing, casting, and gelation.

Stabilization. A critical step in preparing sol–gel products and especially Type VI silica optical components is stabilization of the porous structure as indicated in Figure 1. Both thermal and chemical stabilization is required in order for the material to be used in an ambient environment. The reason for the stabilization treatment is the large concentration of hydroxyls on the surface of the pores of these high (>400 m^2/g) surface area materials.

Chemical stabilization involves removing the concentration of surface hydroxyls and surface defects, such as metastable three-membered rings, below a critical level so that the surface is not stressed by rehydroxylation in use. Thermal stabilization involves reducing the surface area sufficiently to enable the material to be used at a given temperature without reversible structural changes. The mechanisms of thermal and chemical stabilization are interrelated because of the extreme effects that surface hydroxyls and chemisorbed water have on structural changes. Full densification of gels, such as the transformation of gel–silica to a glass, is nearly impossible without dehydration of the surface prior to pore closure (23,40).

Many chlorine compounds, including methyl chlorosilanes, such as ClSi(CH$_3$)$_3$, Cl$_2$Si(CH$_3$)$_2$, Cl$_3$Si(CH$_3$); tetrachlorosilane [10026-04-7], SiCl$_4$; chlorine, Cl$_2$; and carbon tetrachloride, CCl$_4$, can completely react with molecular surface hydroxyl groups to form hydrochloric acid (40), which then desorbs from the gel body in a temperature range of 400–800°C, where the pores are still interconnected. Carbon tetrachloride can yield complete dehydration of ultrapure gel–silica optical components (3,23).

Dehydration of a gel requires removal of two forms of water: free water within the ultraporous gel structure, ie, physisorbed water, and hydroxyl groups associated with the gel surface, ie, chemisorbed water. The amount of physisorbed water adsorbed to the silica particles is directly related to the number of hydroxyl groups existing on the surface. A summary of the hydration/dehydration characteristics of the silica gel/water system follows (41). (1) Physisorbed water can be eliminated and surface silanol, Si–O–H, groups condensed, starting at about 170°C. (2) Dehydration is completely reversible, up to about 400°C. Irreversible decomposition of organic residuals occurs up to 400°C. (3) Above 400°C, dehydration is irreversible as a result of shrinkage and sintering to close pores. The amount of hydroxyl groups on the gel surface is an inverse function of the temperature of densification. Above 800°C, only single hydroxyls remain on the surface, which prevents rehydroxylation. (4) Viscous flow occurs above 850°C; the exact temperature depends on the pore size and composition of the gel. Isolated hydroxyl groups on the gel surface react with each other and bring particles together, thereby eliminating voids within the gel. If surface water is not desorbed prior to pore closure, it is trapped inside the densified gel and can cause bloating. (5) Above 850°C, single hydroxyl groups depart from the gel surface until the gel is densified. When a silica gel has been completely dehydrated, there are no surface hydroxyl groups to adsorb the free water, and the surface is hydrophobic (41). It is the realization of this critical point that is the focus for making stable gel products.

The vibrational overtones and combinations of hydroxyl groups and their associated molecular water occurring in the spectra of various gel silica materials are summarized in Table 2 and discussed in References 3, 5, and 22. These peaks and bands found in the preparation of alkoxide-derived silica gel monoliths are identical to those described for silica gel powders (41).

One of the complications associated with the use of chlorine compounds in dehydration of gels is the incorporation of chlorine ions in the densified gel–glass structure. A dechlorination treatment using an oxygen atmosphere at 1000–1100°C after chlorination at 800°C to remove the hydroxyl ions has been described (40). Reducing the surface area by a presintering process is useful for reducing both the hydroxyl and chlorine content in the dehydrated, densified silica glass.

Using Raman spectroscopy, the structure of alkoxide-derived silica gels has been examined from the dry gel through fully dense amorphous SiO_2 (3,21,42). There are interrelated changes in intensity of the SiOH peaks at 980 cm^{-1} and 3750 cm^{-1}, with the cyclotrisiloxane D_2 and cyclotetrasiloxane D_1 defect peaks at 495 cm^{-1} and 605 cm^{-1}, respectively, and the main SiO_2 structural vibrations at 440 cm^{-1}, 800 cm^{-1}, 1060 cm^{-1}, and 1195 cm^{-1}. Large concentrations of cyclotrisiloxane D_2 rings are formed on the internal pore surface as the hydroxyl concentration and the internal pore surface area decrease with increasing temperature. The three-membered D_2 rings are strained in comparison to the four-membered D_1 rings and, consequently, can only form above 250°C on the surface of the gels via the condensation of adjacent isolated surface silanols.

Table 2. Absorption Peaks of Gel–Silica Monolith Pore Water and Surface Hydroxyl Groups[a]

Wavelength, nm	Band		Observations
	Stretch[b]	Bend[c]	
2919.70	v_4		broad peak on broad band
2816.88	v_3		tiny peak on broad band
2732.24	v_2		joint of two small peaks at 2768.90 and 2698.90 nm
2668.80	v_1		very sharp symmetric peak
2262.48	v_3	v_{OH}	broad band; no peak
2207.51	v_2	v_{OH}	high broad asymmetric peak
1890.35	v_3	$2v_{OH}$	high broad asymmetric peak
1459.85	$2v_4$		tiny peak on broad band
1808.44	$2v_3$		small peak on broad band
1366.12	$2v_2$		very sharp symmetric peak
1237.85	d		small peak
1131.21	$2v_3$	$2v_{OH}$	tiny peak
938.95	$3v_3$	$2v_{OH}$	small peak
843.88	$3v_3$	v_{OH}	no peak observed
704.22	$4v_3$		tiny peak

[a] Ref. 3.
[b] v_1 and v_2 correspond to stretches of isolated and adjacent Si–O–H bonds, respectively; v_3, to an Si–O–H bond which is hydrogen-bonded to water; and v_4, to adsorbed water.
[c] v_{OH} corresponds to an out-of-plane bending vibration of an Si–O–H bond.
[d] $[(2v_3 + v_{OH}) + (2v_2 + v_{OH})]/2$.

In contrast, the four-membered D_1 rings form initially in the sol stage and are retained until the gel is dense (21). The strained three-membered silica rings are easily hydrolyzed to form three-membered chains with an accompanying dilation. Thermal stabilization treatments at 800–1000°C must eliminate the D_2 rings in order for the porous matrix to be used without subsequent cracking from moisture adsorption.

The change in structural density of alkoxide-derived silica gels during thermal processing is apparently caused by at least four interrelated mechanisms: elimination of metastable three-membered silica rings, loss of hydroxyls, loss of organic groups, and relaxation of the SiO_2 structure. Figure 6 shows the structural density, ρ_s, of a series of silica gels held for 12 hours at each temperature. Average hydraulic pore radii, R, were 1.21, 4.18, and 8.98 nm. Structural density starts out below that for amorphous silica, ie, $<\rho_s = 2.20$ g/mL, and goes through a maximum of about 2.30 g/mL before equilibrating at full densification at about 2.22 g/mL.

Expansion and contraction of silica gel monoliths show hysteresis upon heating and cooling. As long as the sample was cycled below approximately 500°C, there was no hysteresis and the monolith was thermally stable (6,23).

Densification. Densification, the final treatment process of gels, occurs between 1000–1700°C, depending on the radii of the pores and the surface area (see Fig. 1). Controlling the gel–glass or gel–ceramic transition to retain the initial shape of the starting material is difficult. It is essential to eliminate

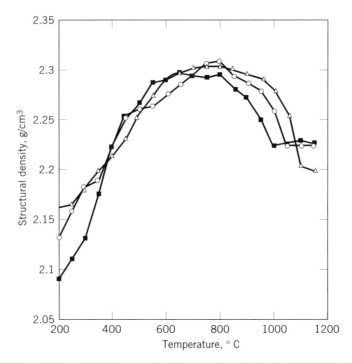

Fig. 6. Dependence of structural density of alkoxide-derived porous silica gels as a function of sintering temperature held for 12 h for three different average hydraulic pore radii, R, of (■) 1.21; (○) 4.18; and (△) 8.98 nm (3).

volatile species prior to pore closure and density gradients owing to nonuniform thermal or atmosphere gradients. Initially, gel-derived glasses were made by melting the chemically derived gel (1,13). Molecular scale homogeneity of the gels yielded glasses that ordinarily devitrify at low temperatures. Hot pressing densifies gels at lower temperatures and produced glasses which otherwise would have crystallized. Using successful stabilization treatments, it is possible to produce monolithic, dense, gel-derived glasses without pressure or heating to temperatures above the melting point (3,23,24).

The amount of water in the gel influences sintering behavior. The viscosity is strongly affected by the concentration of water, which in turn determines the temperature of the beginning of densification. A silica gel prepared under acidic conditions has a higher surface area and water content than a gel prepared in basic conditions, and starts to densify about 200°C lower than a base-catalyzed gel (43). Simultaneously with the removal of water, the structure and texture of the gel evolves. Gels have higher free energy than glasses or ceramics because of their high specific surface area. During sintering, the driving force is a reduction in surface area. Small pores close first for gels because these have a higher capillary pressure owing to their small radii of curvature. After a gel has been densified and heated above the glass-transition temperature, its structure and properties become indistinguishable from those of a melt-derived glass (3,21).

There are at least four mechanisms responsible for the shrinkage and densification of gels (21): capillary contraction, condensation, structural relaxation, and viscous sintering. It is likely that several mechanisms operate at the same time, eg, condensation and viscous sintering. Using three different models, the sintering behavior of a gel can be described. Frenkel's theory (3,21), which is derived for spheres, is valid for the early stages of sintering. Scherer (21) developed a model for describing the early stage and the intermediate stage of sintering, assuming that the microstructure consists of cylinders intersecting in a cubic array. To reduce surface area, the cylinders become shorter and thicker. The last stage of densification is represented by the Mackenzie-Shuttleworth model, which is only applicable to systems having closed porosity (44).

Structural evolution of the gel–glass transition has been described using topological concepts that characterize the interconnected pore structures (5). This is suitable for diffusion, doping, catalysis, and impregnation procedures. Knowledge about the volume or surface area of pores is not enough to characterize the structure; for transport, the extent of interconnection of the structure must be known. In a sol–gel-processed material, the interconnected structure is developed during gelation, aging, and drying. In topological terms, these processes correspond to the first stage of densification. During the second stage of densification, the genus (connectivity) decreases but the number of nodes (pores interceptions) remains constant and the number of branches decreases. During the third stage of densification, as the numbers of nodes and branches decrease, the coordination number, ie, interconnections, of the pores actually goes to zero (5).

Physical Properties.   Processing optimization has been achieved for commercial production of gel-derived silica optical components termed Type V (fully dense) gel–silica and Type VI (optically transparent porous gel–silica) (45). The properties of Types V and VI have been compared with commercial fused quartz optics (Types I and II) and synthetic fused silica optics (Types III and IV). Type V gel–silica has excellent transmission from 160 to 4200 nm and no OH absorption peaks. As shown in Figure 7, the uv cutoff is shifted to lower wavelengths by removal of –OH from the gel-derived glass, as predicted by quantum calculations (5). Other physical properties and structural characteristics of Type V gel–silica are similar to high grade fused silica but offer the advantages of near net-shape casting, including internal cavities, and a lower coefficient of thermal expansion, 0.2×10^{-6} cm/cm compared with 0.55×10^{-6} cm/cm. Optically transparent porous gel silica (Type VI) has a uv cutoff ranging from 250–300 nm. Type VI gel–silica optics have a density as low as 60% of Types I–V silica and can be impregnated with up to 30–40% by volume of a second phase of an optically active organic or inorganic compounds. Mechanical properties of the Type VI porous gel–silica monoliths have been determined (5) and related to topological and metric features.

Alumina Derived from Sol–Gel

Aluminum oxide [1344-28-1] (alumina), Al_2O_3, has high technological value. Sol–gel processing of alumina has created novel applications and improved some of its properties. Products such as catalyst carriers, abrasives, fibers, films for electronic applications, aerogels, and membranes for molecular filtration have been developed on a laboratory scale based on sol–gel processing. Aluminum

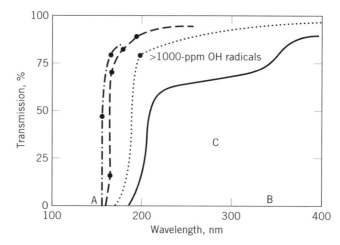

Fig. 7.   Improvements in uv transmission of alkoxide gel–silicas: (——) DYNASIL; (······) 7940 Corning; (– – –) GELSIL 1987; and (–•–) GELSIL 1988, compared to the quantum mechanical predictions of uv cutoff wavelength for a theoretical ring of five silica tetrahedra. Point A is at 144 nm and a 6 mol % OH in silica; point B, at 345 nm. Region C represents an area of increasing OH• concentration (3).

alkoxide hydrolysis has been studied for several decades to understand the diverse structures of aluminum hydroxides (46). In 1975, monolithic alumina samples derived from aluminum alkoxide were produced using four steps (13): hydrolysis of aluminum alkoxide, peptization of the hydrolyzed mixture, gel formation, and drying and pyrolysis (see ALUMINUM COMPOUNDS, ALUMINA).

Hydrolysis and Condensation Reactions of Aluminum Alkoxides. Aluminum alkoxides may be coordinatively unsaturated with three possible coordination numbers. Aluminum is less electronegative than silicon, causing it to be more electrophilic and thus less stable toward hydrolysis (21). These features are responsible for the greater rates of hydrolysis and condensation of aluminum alkoxides when compared to the rates of silicon alkoxide reactions. Hydrolysis and condensation reactions probably occur by nucleophilic addition, followed by proton transfer and elimination of either alcohol or water under neutral conditions. Both reactions are catalyzed by the addition of acid or base. When acids are added, they protonate organic or hydroxyl groups, creating reactive species and eliminating the requirement for proton transfer as an intermediate step. Bases deprotonate water or OH groups, leading to the formation of strong nucleophiles.

Condensation reactions tend to proceed along the reaction pathway that maximizes the hydrogen bonds of the aluminum hydroxides produced during hydrolysis (21). Because the pathway for condensation reactions is not random, formation of crystalline species is allowed. Aluminum alkoxides react rapidly with water, producing aluminum hydroxides having structure and shapes determined by the conditions in which hydrolysis was performed. The influence of hydrolysis temperature on the structure of the aluminum hydroxide formed has been shown (13). When high (80°C) hydrolysis temperatures are used, boehmite [1318-23-6] is formed as fibers and plates. When hydrolysis of the aluminum alkoxide is performed at low (room temperature) temperatures, an amorphous aluminum hydroxide is formed that has a higher residual content of organic radicals. The aging of this amorphous hydroxide at room temperature and in the mother liquor allows the beginning of a dissolution process followed by bayerite precipitation.

Peptization.   Aggregation of small inorganic polymeric chains and clusters produced from hydrolysis and condensation reactions of aluminum alkoxides forms macroparticles and nonuniform aggregates. These aggregates that are broken during the peptization step effect the formation of a clear sol having a narrow distribution of particle size. Another mechanism associated with peptization is the production of surface charges on the colloids (47) that eventually leads to either gelation or dispersion. Variables influencing peptization of aluminum alkoxide sols are type and concentration of inorganic acids, temperature of hydrolysis, and the water/alkoxide ratio. Peptization of sols prepared at 90°C occurs when inorganic acids are added having acid/alkoxide molar ratios greater than 0.03. This leads to the formation of inorganic polymers having greater sizes and ramified chains. Otherwise, peptization only occurs in low temperature sols when additions of larger quantities of acid than those required to peptize mixtures hydrolyzed with hot water are provided. The resulting polymers have lower molecular weight (48).

Gelation.   Mechanisms of gelation of alumina sols derived from alkoxides differ from gelation of silicon alkoxide sols. Whereas the sol–gel transition for silica sols is basically a consequence of interactions between long inorganic chains, the gel transition in alumina sols results from colloidal growth by dissolution and reprecipitation processes (Ostwald ripening), followed by formation of linkages between particles. These linkages are initialized by physical–chemical interactions between surface-charged colloids that eventually produce a three-dimensional network, formed by interconnected colloids. The initial contact points between colloids are responsible for neck formation by a coarsening process, followed by particle reshaping and densification. Gelation can be induced either by eliminating excess of water added during hydrolysis or by adding electrolytes to the peptized sol prepared in temperatures above 60°C that leads to sol flocculation.

The effects of both pH and temperature of aluminum alkoxide hydrolysis on gelation is shown in Figure 8. Addition of acid into the mixture hydrolyzed at 90°C, and by consequence reduction of pH, reduces the gelation time of the samples, whereas in mixtures hydrolyzed at room temperature, acidic addition increases gelation time.

X-ray diffraction of alumina xerogels prepared at temperatures of hydrolysis higher than 50°C shows a pseudoboehmite phase. Xerogels prepared at room temperature have an amorphous structure (49). The presence of a large amount of residual organic groups, large content of acid, and large concentration of water stabilizes the amorphous structure of aluminum hydroxide. Structural analysis of gels can be used to explain the gelation behavior of sols prepared at different temperatures. Owing to the presence of a cleavage plane in pseudoboehmite colloids, edges of the colloidal plates have positive surface charges in acidic conditions, whereas faces that do not have primary broken bonds have negatively

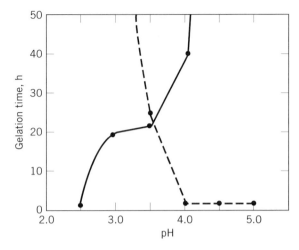

Fig. 8. Gelation behavior of alumina sols prepared from hydrolysis of aluminum iso-propoxide at (——) 90°C and (– – –) room temperature (49).

or less positively charged surfaces. Thus, face–edge attractions among colloids are increased with a reduction of pH, because pH reduction adds positive surface charges to the edges of plates. Mixtures hydrolyzed and peptized at room temperature produce amorphous colloids that do not have cleavage planes and are spherical in shape. Thus, surface charges are uniformly distributed. Addition of acid increases the positive surface charge density of colloids (increase in zeta potential), provoking a greater repulsion among the particles and extending the gelation time.

Drying of Gels.   Drying of alumina gels prepared using high temperatures of hydrolysis consists of two steps: sol concentration and pore liquor removal. The rate of sol concentration can dictate some gel properties. High rates of sol concentration lead to less efficiency in packing of colloids.

Transparent monoliths can be prepared by drying high density gels (49,50). Control of initial conditions of sol preparation is essential to achieve success during the drying step.

Phase Transformations and Properties of Gels from Aluminum Alkoxide. Pure alumina xerogels, when submitted to heat treatments at 500 and 1100°C, give rise respectively to γ-alumina and α-alumina (47). This sequence of phase transformation is traditionally obtained for pure aluminum hydroxides. The effects of gel preparation conditions on the volumetric density and surface area are also compatible with the mechanisms of sol formation and gelation (49) described. Gels prepared at room temperature show a volumetric density much greater than gels prepared at 90°C as a result of both different magnitudes of capillary pressure and efficiency of packing. Moreover, the pH of mixtures hydrolyzed at high (>50°C) temperatures affects the gel structure, and a reduction in pH produces a reduction of the surface area of the gels prepared at 90°C. Short gelation times at progressively lower pHs (higher magnitude of face–edge attractions) allow the stabilization of a significant disorganization in the arrangement of the pseudoboehmite particles.

Economic Aspects

The sol–gel process can be utilized to yield products within a wide range of applications. Some of these applications include production of nanocomposites, films, fibers, porous and dense monoliths, and biomaterials.

The commercialization of sol–gel-derived products grew in the early to mid-1990s and is expected to grow faster in the twenty-first century (50).

Table 3 summarizes the value of markets for sol–gel products worldwide. Applications commanding the market as of 1995 were those related to abrasives (qv), coatings (qv), catalysts, and optical components, among others. A significant growth in the market from optical applications as well as from chemical sensors (qv) was expected for the late 1990s. Other products expected to emerge and have high importance are those related to biomedical applications and organic–inorganic hybrid composites.

Table 3. Markets for Sol–Gel Products[a], $ \times 10^6$

Markets	Year			Average annual growth rate, %
	1996	2001	2006	
United States	158.76	504	1600	26
rest of world	529	1064	2140	15

[a]Ref. 50.

Applications

Sol–Gel Processing of Thin Films. The sol–gel method enables the production of ceramic films having thickness from 10–1000 nm (see THIN FILMS). The rheological characteristics of the sol allow the deposition of a film by several procedures: dip coating, spin coating, electrophoresis, thermophoresis, and settling (see COATING PROCESSES). Dip and spin coating are the most frequently used procedures. Dip coating can be divided into five stages (21): immersion, startup, deposition, drainage, and evaporation. The fluid mechanical boundary layer, which is pulled with the substrate, splits into two. The inner layer moves upward with the substrate, while the outer layer is returned to the bath. The thickness where the split occurs is responsible for the thickness of the film. Five events govern the split phenomenon (21) and consequently the thickness of the film: viscosity of the solution, force of gravity, resultant force of surface tension in the concavely curved meniscus, velocity of pulling, and reactivity between the sol particles and the substrate. A good correlation among viscosity, η; substrate velocity, V; and film thickness, h, can be given by the following equation:

$$h = c_1(\eta V/rg)^{1/2} \tag{9}$$

where c_1 is a constant, r is density, and g is gravitational acceleration. Thus low viscosities and low substrate withdrawal speed are necessary to produce thin films.

The spin coating process can be divided into four stages (51): deposition, spinup, spinoff, and evaporation. In the first stage, an excess of liquid is distributed along the surface that is to undergo deposition. The spinup stage is related to flow of liquid along all the surface, driven by centrifugal force. In the spinoff stage, the excess of liquid flows to the perimeter and is eliminated as droplets. The solvent is eliminated in the fourth stage by evaporation, which leads to the thinning of the film.

Sol–Gel Fibers. During sol-to-gel evolution, changes in the rheology of the sol can be used to allow fiber pulling. Formation of elongated polymers in a solution is a requirement for spinnability, ie, the ability to form fibers (18,21,27). Reduced viscosity for solutions of chain-like or spherical polymers is independent of concentration, whereas linear polymers give a direct relation between reduced viscosity and concentration (18). Acidic pHs and low values for the molar ratio between water and alkoxide result in the production of linear polymers that exhibit spinnability. High molar ratios of water/alkoxide and basic media lead to the production of spherical and ramified polymers that form a three-dimensional network. Low molar ratios of water/alkoxide allow the production of a functionality of two in the inorganic polymers, whereas an acidic medium reduces the immiscibility gap in the alcohol–alkoxide–water system and provides a catalytic effect important in the development of linear polymers.

The procedure for pure silica sol–gel fiber processing (18) includes mixing TEOS, water, ethanol, and hydrochloric acid in the following molar ratios: H_2O/TEOS of 1.5–2, HCl/TEOS of 0.01, and ethanol/TEOS of 3. The mixture is kept at 80°C for three hours in an open container for solvent evaporation and a consequent increase in viscosity. Spinnability can be tested by dipping a small rod in the solution and withdrawing a fiber. Continuous fibers can be obtained by extruding the solution through a spinneret and rolling it. After fiber pulling, a sintering step is necessary to allow decomposition of organics and densification. Spinnability is obtained when a sol shows a pseudoplastic but not thixotropic behavior (27).

Multicomponent sol–gel fibers have been successfully developed (1,52). The early stages of sol formation and gelation are crucial for controlling the fiber microstructure. Aluminosilicates, zirconates, and aluminates (1,18,52) can be prepared by sol–gel methods. Mullite [55964-99-3], $3Al_2O_3 \cdot 2SiO_2$, fibers are difficult to produce by conventional processes, because mullite has a very high melting point and great tendency to crystallize. The sol–gel process enhances making this type of fiber. Production of mullite fibers from the reaction of aluminum and silicon alkoxides involves the use of a chelating agent prior to the hydrolysis of the aluminum alkoxide. The chelating agent serves to make the rates of hydrolysis of the silicon and aluminum alkoxides compatible. In nonsilicate systems, control of the polymerization of the inorganic chains cannot be easily obtained and the use of agents to control viscosity is usually required.

Organic–Inorganic Hybrids. Ceramics and polymers have been combined into high performance composites. The association between high modulus and high strength ceramic fibers, such as glass, carbon, and boron fibers, having the inherent ductility and toughness of some polymers, enables the fabrication of materials having special properties. The integration of different types of materials is restricted by the high temperature processing conditions usually employed in ceramic fabrication. The sol–gel method enables preparation of ceramic materials in a temperature range compatible with organic polymer stability, and involves mechanisms of network formation such as hydrolysis and polycondensation reactions, that are similar to the polymerization reactions of polymers. Thus, sol–gel can be used to produce new types of composites involving ceramics and polymers. These are called organic–inorganic hybrids or ceramers. These materials have been reviewed (53), which emphasizes the potential uses of this technology for development of coatings, sealing materials, and materials having rare optical properties, such as nonlinear behavior. The interface of the nanometric phases must be tailored to building the properties of an inorganic–organic hybrid composite.

A combination of silica derived from silicon alkoxide and different types of polymers leads to transparent composites having microphase (in the range of 10–100 nm) separation, a higher elastic modulus, and a greater strength than the polymer. Sol–gel silica–PDMS (polydimethyl siloxane) hybrids are prepared by *in situ* precipitation of sol–gel-derived silica in a swollen PDMS network. The process includes the preparation of a cross-linked PDMS, swelling of the network in tetraethyl orthosilicate, and precipitation of the silica by introducing the PDMS filled with TEOS in an aqueous solution containing ethylamine. Mechanical performance of these composites is better than PDMS filled with fumed silica as in conventional PDMSs. Another approach to the preparation of sol–gel silica–PDMS hybrids uses the interaction between silanol-terminated PDMS and silanol groups formed during hydrolysis of TEOS (54). This gives a more homogeneous structure, generated by covalent bonds between the silica network and PDMS chains. The material is optically transparent, showing the absence of macrophase separation, and exhibits a high degree of flexibility.

Organic–inorganic hybrids using poly(vinyl acetate) (PVAc) and silica derived from silicon alkoxide involved dissolution of commercial PVAc in tetrahydrofuran (THF) (20 wt %), addition of TEOS, and subsequent introduction of an acidic solution having the stoichiometric amount of water required for the hydrolysis of TEOS. Other procedures include use of a vinyl acetate, vinyl triethoxisilane copolymer (55). The composites are optically clear, formed by nanometric phases rich in silica and polymer, as shown by the transmission electron microscopy (tem). Dielectric and dynamic mechanical spectroscopy (dms) studies performed on the composites showed no substantial increase in the onset of glass transition or activation energy for this transition with the introduction of silica. Extensive hydrogen bonding between silanol and carbonyl groups in the PVAc was observed. Organic–inorganic hybrids prepared from polystyrene and silica-derived TEOS used a copolymer prepared by free-radical polymerization of styrene and 3-(trimethoxysilyl)propyl methacrylate. This copolymer then reacted with TEOS in the presence of an acidic water solution and in a THF environment to produce a covalently bonded composite through the interaction between silanol groups on the polymer chains and on the silica network (56). For composites fabricated having more than 22 wt % of (trimethoxysilyl)propyl methacrylate, optically transparent composites were obtained. For these composites, no visible phase separation could be noted with scanning electron microscopy (sem) and no glass transition was detected by differential scanning calorimetry (dsc), which suggests the presence of polymer chains uniformly distributed and covalently bonding with the silica network. Some properties of the system as a function of TEOS content, such as refractive index, density, and hardness, showed values intermediate to the pure components that could be tailored by composite composition.

Several types of organic–inorganic hybrids have been prepared by using different polymers coupled with TEOS (57). The basic procedure involves dissolution of the polymer in THF (20 wt %) followed by addition of TEOS with an acidic water solution. Some of the polymers, ie, poly(methyl methacrylate), poly(vinyl acetate), poly(vinyl pyrrolidone), and poly(N,N-dimethylamide), yielded transparent films, this demonstrating the absence of macrophase separation. Polycarbonate, poly(acrylic acid), and Nylon trogamid lead to the production of opaque films. Therefore, in order to restrict macrophase separation, the formation of hydrogen bonding between the inorganic and organic chains is required, because for those polymers capable of producing hydrogen bonding, no macrophase separation was observed. Comparable rates of gelation and solvent removal can also be useful in reducing phase separation.

One of the problems of combining organic polymers with TEOS is the immiscibility of a large number of polymers in aqueous solutions. Polymers having better water miscibility, such as poly(tetramethylene oxide) (PTMO), alleviate this problem. To increase compatibility between the silica network and PTMO, the polymeric chains can be end-capped using triethoxysilane and mixed with TEOS in an isopropanol/THF solution. After that, an acidic aqueous solution can be introduced into the mixture. Optically clear materials were prepared by this technique. These, however, exhibited microphase domains identified by saxs. PTMO has also been combined with ceramic materials other than silica to produce hybrids. A PTMO–alumina, derived from aluminum butoxide, and PTMO–ZrO_2–TiO_2 were prepared by a procedure similar to that used for silica hybrids. To avoid premature hydrolysis of the aluminum butoxide, which has a higher rate of hydrolysis than silicon alkoxides, a chelating agent (ethyl acetoacetate) was introduced to the Al alkoxide solution prior to hydrolysis.

Extensive work has been done in terms of combining nanometric clay particles using either Nylon or polyimide. Montmorillonite [1318-93-0] has been modified by cation exchange using aminolauric acid, and the new groups attached on the surface of the clay bonded to the polymer by initiation of polymerization. Analysis of the composite structure showed silicate layers of the clay dispersed among polymer chains. The composite could be injection-molded after the synthesis, and the resulting material showed a large degree of orientation of the clay particles in relation to the direction of flow during injection. A 50% increase in elastic modulus and strength, both at 25°C and high temperature, were obtained by adding 4–10 wt % of clay to the polymer.

Other organic–inorganic hybrids include poly(ethyloxazoline)–silica, poly(vinyl alcohol)–silica, poly(arylene ether) ketone–silica, polyimide–silica, polyozoline–silica, poly(ethylene oxide)–silica, and polymers–modified alkoxysilane.

Sol–Gel Bioactive Glasses. Bioactive glasses and ceramics bond to both soft and hard tissue (58). The chemical bond formed between the implant and tissue can provide the desired adhesion required in many medical and dental applications. Two types of sol–gel processing yield bioactive materials in the

SiO_2–CaO–P_2O_5 system having high bioactivity index (58) (see DENTAL MA-TERIALS; PROSTHETIC AND MEDICAL DEVICES). The first procedure (59) involves hydrolysis of TEOS in acidic medium, introduction to the hydrolyzed TEOS of triethyl phosphate, addition of hydrated calcium nitrate to the previous mixture, gelation and aging at 60°C, and drying of the aged gels at 180°C and partial densification at 700°C.

In the second procedure, calcium nitrate was replaced by calcium alkoxide (60). Calcium and silicon alkoxides have very different rates of hydrolysis. To avoid the production of inhomogeneities, a slow and controlled hydrolysis of a mixture of silicon, calcium, and phosphorous alkoxide was performed. The resulting materials were highly homogenous, and monolithic pieces could be produced. The bioactivity of the gel-derived materials is equivalent or greater than melt-derived glasses.

BIBLIOGRAPHY

1. T. E. Wood and H. Dislich, *Ceramic Transactions*, **55**, 1 (1994).
2. P. J. Flory, *Principles of Polymer Chemistry*, Cornell University Press, New York, 1953, Chapt. 9.
3. L. L. Hench and J. K. West, *Chem. Rev.* **90**, 33–72 (1990).
4. J. D. Mackenzie, in L. L. Hench and D. R. Ulrich, eds., *Ultrastructure Processing of Ceramics, Glasses and Composites*, John Wiley & Sons, Inc., New York, 1984.
5. L. L. Hench and W. L. Vasconcelos, *Ann. Rev. Mater. Sci.*, **20**, 269 (1990).
6. J. Wenzel, in A. F. Wright and A. F. Dupuy, eds., *Glass.Current Issues*, Martinus Nijhoff, Dordrecht, Netherlands, 1985, p. 224.
7. G. J. McCarthy, R. Roy, and J. M. McKay, *J. Am. Ceram. Soc.* **54**, 637 (1971).
8. R. K. Iler, *The Chemistry of Silica*, John Wiley & Sons, Inc., New York, 1979.
9. W. Stober, A. Fink, and E. Bohn, *J. Colloid Interface Sci.* **26**, 62 (1969).
10. S. S. Kistler, *Nature*, **127**, 742 (1981).
11. J. Fricke, ed., *Aerogels*, Vol. 6, Springer-Verlag, Heidelberg, Germany, 1986.
12. J. Phalippou, M. Prassas, and J. Zarzycki, *J. Non-Cryst. Solids*, **48**, 17 (1982).
13. B. E. Yoldas, *Bull. Am. Ceram. Soc.* **54**, 286 (1975).
14. E. Matijevic, in J. D. Mackenzie and D. R. Ulrich, eds., *Ultrastructure Processing of Advanced Ceramics*, John Wiley & Sons, Inc., New York, 1988, p. 429.
15. A. Hardy and co-workers, in Ref. 4, p. 407.
16. M. E. A. Hermans, *Sci. Ceram.* **5**, 523 (1970).
17. E. M. Rabinovich, in L. C. Klein, eds., *Sol–Gel Optics: Processing and Applications*, Kluwer Academic Publishers, Boston, Mass., 1994, p. 1.
18. S. Sakka and T. Yoko, *J. Non-Cryst. Sol.* **147–148**, 394–403 (1992).
19. L. W. Hrubesh and co-workers, *Chemical Processing of Advanced Materials*, John Wiley & Sons, Inc., New York, 1992, p. 19.
20. J. Zarzycki, in L. L. Hench and D. R. Ulrich, eds., *Science of Ceramic Chemical Processing*, John Wiley & Sons, Inc., New York, 1986, p. 21.
21. C. J. Brinker and G. W. Scherer, *Sol–Gel Science*, Academic Press, New York, 1990.
22. J. Fricke and co-workers, in Ref. 19, p. 3.
23. L. L. Hench and S. H. Wang, *Phase Transitions*, **24–26**, 785–834, 1990.
24. L. C. Klein and G. J. Garvey, in Ref. 4, p. 88.
25. M. D. Sacks and T. Y. Tseng, *J. Am. Ceram. Soc.* **67**, 532 (1984).
26. I. Artaki and co-workers, *J. Non-Cryst. Solids*, **72**, 391 (1985).
27. M. D. Sacks and R. S. Sheu, in Ref. 20, p. 102.

28. R. Zallen, *The Physics of Amorphous Solids*, John Wiley & Sons, Inc., New York, 1983, Chapt. 4.
29. D. Stauffer, A. Coniglio, and M. Adam, *Adv. Polym. Sci.* **44**, 103 (1982).
30. T. A. Witten and M. E. Cates, *Science*, **232**, 1607 (1986).
31. K. D. Keefer, in Ref. 20, p. 131.
32. L. W. Kelts, N. J. Effinger, and S. M. Melpolder, *J. Non-Cryst. Solids*, **83**, 353 (1986).
33. R. K. Dwivedi, *J. Mater. Sci. Lett.* **5**, 373 (1986).
34. A. R. Cooper, in G. Y. Onoda, Jr., and L. L. Hench, eds., *Ceramics Processing Before Firing*, John Wiley & Sons, Inc., New York, 1978, p. 261.
35. L. L. Hench and M. J. R. Wilson, *J. Non-Cryst. Solids*, **121**, 234 (1990).
36. R. W. Ford, *Ceramics Drying*, Pergamon, New York, 1986.
37. M. Fortes and M. R. Okos, in A. S. Mujumdar, ed., *Advances in Drying*, Vol. 1, Hemisphere, New York, 1980, p. 119.
38. S. Whitaker, in Ref. 37, p. 23.
39. T. M. Shaw, in C. J. Brinker, D. E. Clark, and D. R. Ulrich, eds., *Better Ceramics Through Chemistry II*, Vol. 73, Materials Research Society, Pittsburgh, Pa., 1986, p. 215.
40. K. Susa and co-workers, *J. Non-Cryst. Solids*, **79**, 165 (1986).
41. M. L. Hair, *Infrared Spectroscopy in Surface Chemistry*, Marcel Dekker, Inc., New York, 1967.
42. J. L. Rousset and co-workers, *J. Non-Cryst. Solids*, **107**, 27 (1988).
43. M. Nogami and Y. Moriya, *J. Non-Cryst. Solids*, **37**, 191 (1980).
44. J. D. Mackenzie and R. Shuttleworth, *Proc. Phys. Soc. London* **62**, 833 (1949).
45. L. L. Hench, S. H. Wang, and J. L. Nogues, in R. J. Gunshor, ed., *Multifunctional Materials*, V. 878, SPIE, Bellingham, Wash., 1988, p. 76.
46. B. C. Lippens and J. J. Steggerda, *Physical and Chemical Aspects of Adsorbents and Catalysts: Active Alumina*, Academic Press, Inc., New York, 1970, p. 171.
47. D. Clark and J. J. Lannutti, in Ref. 4, p. 126.
48. A. L. Pierre and D. R. Uhlmann, *J. Am. Ceram. Soc.* **70**, 28 (1987).
49. R. L. Oréfice and W. L. Vasconcelos, in J. D. Mackenzie, ed., *SPIE Proceedings: Sol–Gel Optics III*, SPIE, San Diego, 1994, p. 733.
50. L. Sheppard and T. Abraham, *Mater. Tech.* **10**, 256 (1995).
51. C. Brinker and co-workers, in L. L. Hench and J. K. West, eds., *Chemical Processing of Advanced Materials*, John Wiley & Sons, Inc., New York, 1992, pp. 395–413.
52. H. G. Sowman, *Ceramic Bulletin*, **67**, 1911, 1969.
53. H. Schmidt, *J. Sol–Gel Sci. Tech.* **1**, 217–231 (1994).
54. H.-H. Huang, B. Orler, and G. L. Wilkes, *Macromolecules*, **20**, 1322–1330 (1987).
55. S. Yano and co-workers, *J. Applied Pol. Sci.* **54**, 163–176 (1994).
56. Y. Wei and co-workers, *J. Mater. Res.* **8**, 1143–1152 (1993).
57. J. T. Landry and co-workers, *Polymer*, **33**, 1496–1506 (1992).
58. L. L. Hench, *J. Am. Ceram. Soc.* **74**(7) 1487–1510 (1991).
59. R. Li and co-workers, in Ref. 51, p. 627.
60. M. M. Pereira, A. E. Clark, and L. L. Hench, *J. Biomed. Mater. Res.* **28**, 693 (1994).

LARRY L. HENCH
RODRIGO OREFICE
University of Florida

SOLVENTS, INDUSTRIAL

The term industrial solvents is generally applied to liquid organic compounds used on a large scale to perform numerous functions in industry. Compared to the large number of potential organic compounds that could be used as solvents, relatively few find use in large amounts. Solvents that are used in large quantities have been selected because these can be produced economically and have attractive safety and use characteristics in manufacturing and application environments. Useful solvents are found in various chemical classes, the most widely used of which are aromatic and aliphatic hydrocarbons. Other classes include alcohols, ketones, esters, ethers, glycols, glycol ethers, chlorinated hydrocarbons, amines, and aldehydes.

Industrial solvent applications are broad, varied, and complex, and each has its own set of characteristics and requirements. Proper solvent selection and blend development have a large impact on the success of the operation in which the solvent is used, from the perspectives of economic effects, technical adequacy, safety issues, and environmental impacts.

In most applications, solvents serve a transitory function by facilitating a process or performing a task, then exiting the process. Even though the solvent may have a temporary function, the success of the industrial process and the quality of the product from the process depends on solvent selection. For example, an automobile coating applied using the proper solvent combination has the gloss, hardness, adhesion, and durability that enhance the appearance of the automobile. Conversely, inadequate cleaning of a machined part because of poor cleaning solvent selection can interfere with subsequent processing steps, resulting in a product that does not have the quality expected by the purchaser. Poor solvent performance can also manifest itself as a process economic penalty. For instance, in an oilseed extraction process, the wrong hexane blend can increase solvent losses and energy consumption as well as lower oil extraction efficiency. In addition to the process economic problems, poorer product quality can also occur.

Thus, innumerable production processes and products depend on the proper solvent. Solvents perform numerous functions, and the characteristics of the specific application determine which function or functions are most important (Table 1).

Regulations

Regulations have a significant impact on the selection and use of industrial solvents. Since 1966, new and more restrictive regulations have been adopted at both the federal and state level, which influence the kind of solvents used, the amount used, and the way they are used. Regulations have given impetus to the development of new technologies and products which decrease solvent emissions to minimize environmental impact.

National Ambient Air Quality Standards (NAAQS) have been established for ozone (qv), nitrogen oxides, lead (qv), carbon monoxide (qv), sulfur dioxide (qv), and particulates. The standards have been set to safeguard human health and the environment. Many areas of the country violate one or more of the NAAQS.

Table 1. Functions of Industrial Solvents in End Uses

Solvent function	Purpose
dissolving	prepare solutions of polymers, resins, and other substances
softening	used as tackifiers, improve adhesion to substrate for better bonding
suspension/dispersion	pigments and other particulates
extraction	separate one material from another by selective dissolution
viscosity reduction	thin coatings to application viscosity
chemical intermediate	react with other compounds to form new substances
manufacturing/processing	improved workability during processing
heat-transfer fluid	remove heat of reaction in chemical manufacturing processes
reaction medium	an inert medium in which other compounds react

To achieve healthful air for all citizens, the states establish an implementation plan for each area that violates the NAAQS for any of the six criteria pollutants. The plan is a strategy designed to achieve sufficient emission reductions to meet the NAAQS within the deadline. Based on the implementation plan, specific regulations are written which govern the operations emitting the pollutant.

Of the NAAQS standards, ozone nonattainment has the greatest impact on solvent operations. Although solvent operations do not emit ozone directly, solvents, as volatile organic compounds (VOCs), react with nitrogen oxides in the atmosphere under the influence of sunlight to produce photochemical smog, of which ozone is a significant component. Implementation plans reduce ozone levels by reducing VOC emissions from all sources. As VOC emitters, solvent-using operations are also regulated to attain ozone compliance. Lower emissions are usually attained by either lowering solvent usage in the product or operation, capturing and/or destroying the solvent by various means, or adopting a nonsolvent or reduced solvent technology.

As different regulations have been adopted, the approach to controlling pollution has evolved. Initial regulations controlled solvent composition; later regulations were concerned primarily with overall VOC reduction. More recent regulations have combined VOC reduction with composition constraints.

Rule 66 Approach. This regulation was adopted by the Los Angeles Air Pollution Control District in 1966 and took effect in 1967. It has since been referred to as Rule 102 of the South Coast Air Quality Management District. Smog chamber testing established that some solvent types are more reactive than others and thus contribute more to ozone pollution. The Rule 66 approach restricts the concentration of the more reactive solvent types used in solvent operations according to a three-class designation. Class I, olefinic compounds, is restricted to 5% by volume of the solvent blend. Class II, consisting of C8+ aromatics excepting ethylbenzene, is limited to 8% by volume. Class III, comprised of branched ketones, toluene, ethylbenzene, and trichloroethylene, is limited to 20% by volume. In addition to the individual class limits, the sum of all three classes is limited to 20% by volume.

Because branched ketones and aromatic hydrocarbons are used for their cost/performance benefits, they became the solvents of choice for many applications. Numerous solvent systems had to be reformulated to comply with

Rule 66. This usually meant an increase in cost, sometimes accompanied by performance degradation. Rule 66-type regulations were adopted in many other states and cities as well. Federal specifications for coatings and many other solvent-containing materials also incorporate Rule 66 requirements.

VOC Emissions Reduction Approach. The Rule 66-type approach focuses on solvent composition; further developments have led to regulatory approaches that emphasize overall VOC emission reduction. Even though the more reactive solvents react near their emission point, all VOC compounds eventually react to form ozone pollution. This may occur some distance downwind, increasing ozone levels in areas which have low artificial emissions.

A series of Control Technique Guideline documents (CTGs) generated by the EPA have established Reasonably Available Control Technology (RACT) standards for many solvent uses, including numerous industrial solvent applications. This represents emission levels achievable with widely used technology and formed the basis for allowed emission levels in state implementation plans. Because different technology is used in different industries, the CTGs established different emission levels between industries. Even within an industry, different emission levels were established for different parts of the process, reflecting the difference in technology development. Not only have the RACT guidelines reduced solvent emissions, further technology developments have resulted in polymers that require less solvent for successful use. Improved capture and destruction technology, as well as better application and handling methods, have also reduced solvent emissions.

Clean Air Act as Amended in 1990. The Clean Air Act Amendments of 1990 represent a more recent effort by the U.S. Congress to address clean air concerns. The first Clean Air Act, passed in 1967, provided authority to establish air quality standards. Further legislation passed in 1970, 1974, and 1977 extended and modified the original act. The seven titles of the 1990 Act not only extended previous measures, but also broke new conceptual ground.

VOC Emissions Reduction/Ozone Attainment. Title I of the 1990 Amendments continues the process of diminishing VOC emissions from all sources to reduce ozone concentrations. A compliance timetable by category has been established, which depends on the level of current ozone concentration. The definition of a major source also depends on the ozone nonattainment category:

Classification	Deadline to attain	Major VOC source, t/yr
marginal	Nov. 15, 1993	100
moderate	Nov. 15, 1996	100
serious	Nov. 15, 1999	50
severe	Nov. 15, 2005 to Nov. 15, 2007	25
extreme	Nov. 15, 2010	10

In addition to achieving the ozone standard by the deadline, moderate and higher areas must demonstrate a total net VOC emissions reduction below the base year in accordance with an aggressive schedule of percentage reductions.

These reductions are 15% in the first six years (through November 15, 1996), and 3%/yr thereafter.

The persistence of ozone pollution may result in additional emission restrictions for industry. Because of aggressive emissions reduction programs in the past, industrial emissions represent a minority share of total emissions, typically in the range of 15–20%. Despite this success in reducing emissions, it is likely that large industrial emission sources, including industrial solvent emissions, will have additional controls imposed because those plants still represent the easiest targets to obtain further reductions.

Hazardous Air Pollutants. Title III of the Amendments extends the air toxics component of the Clean Air Act. It attacks the issue of air toxics by establishing a long list of toxic pollutants to be regulated. The EPA is commanded to impose tight controls in two stages. The first stage is based on technology standards which require companies to install maximum achievable control technology (MACT). The second stage may subject some facilities to further regulation once MACT standards have been met, setting tighter emissions requirements based on an assessment of residual risk.

The MACT standard for each HAP compound will be established for each source category. The MACT standards imposed require controls equivalent to the emissions levels achieved at the average of the most tightly controlled 12% of existing facilities in each source category. All standards must be promulgated by November 15, 2000, with attainment three years later. Although VOC controls vary somewhat according to the severity of ozone pollution in an area, the MACT regulations are uniform across the country. Industrial solvents, including some used in very specialized applications, which are listed as HAP compounds, are as follows (1): acetonitrile, benzene, carbon tetrachloride, chlorobenzene, chloroform, mixed cresols, cumene, 1,2-dichloroethane, dimethylformamide, 1,4-dioxane, ethylbenzene, ethylene glycol, 1,1-dichloroethane, ethylene glycol (EG) ethers, EG ether esters, *N*-hexane, isophorone, methyl alcohol, methyl ethyl ketone, methyl isobutyl ketone, methyl *t*-butyl ether, nitrobenzene, 2-nitropropane, perchloroethylene, 1,1,2,2-tetrachloroethane, toluene, 1,1,1-trichloroethane, 1,1,2-trichloroethane, trichloroethylene, 2,2,4-trimethylpentane, mixed xylenes, *o*-xylene, *p*-xylene, and *m*-xylene. The hydrocarbon compounds cumene, ethylbenzene, *n*-hexane, toluene, 2,2,4-trimethylpentane, and the xylene isomers may be found as components in mixed hydrocarbon solvents.

Stratospheric Ozone Protection. Title VI of the 1990 Amendments deals with stratospheric ozone protection. Certain chlorinated fluorocarbon (CFC) compounds, eg, 1,1,1-trichloroethane and carbon tetrachloride, will be phased out over a scheduled time period. Knowingly venting CFCs from household appliances, commercial refrigerators, and air conditioners is prohibited. Since 1992, repair shops must use certified recycling equipment when servicing motor vehicle air conditioning units. Nonessential CFC-containing consumer products are to be banned, including plastic party streamers, noise horns, and cleaning fluids for electronic and photographic equipment. Among industrial solvent users, the largest impact will be felt in the area of vapor-phase degreasing.

Proposition 65. The formal designation of Proposition 65 is the Safe Drinking Water and Toxic Enforcement Act of 1986. This act was passed overwhelmingly by California's voters on November 4, 1986. As of this writing (1996), California is the only state that has adopted this type of regulation. The principal objectives of Proposition 65 are to protect the State's drinking water sources from toxic contamination and to warn the public of possible carcinogenic and reproductive hazards associated with certain identified chemicals.

The Act requires the Governor of California to identify and publish a list of those chemicals known to the state of California to cause cancer or reproductive harm. Additions are made to the list periodically, typically on a quarterly basis. Those who knowingly and intentionally discharge, release, or otherwise expose the public to the listed chemicals are required to take two actions. First, a "clear and reasonable" warning must be given to the person exposed. The warning requirement takes effect 12 months from the date the chemical is listed. Second, any "knowing" discharge or release of a listed chemical is prohibited where the chemical may pass into sources of drinking water. The discharge requirement takes effect 20 months after the chemical is listed.

The warning requirement may be met by different techniques which alone or in combination meet the requirements of Proposition 65. This can include a hazard warning label consisting of required labeling language for listed chemicals. The MSDS for a product may also convey the appropriate warning for all products used in workplaces covered by the OSHA Hazard Communications Standard. Proposition 65 allows companies to advertise the required warnings for listed chemicals on media such as newspaper, television, and radio as long as it is "clear and reasonable." Signs and posters in the workplace or at the point of sale, or billboards in the vicinity of the plant may also be employed.

Some industrial solvents are found on the Proposition 65 list. As of January 1, 1993 (2), benzene, carbon tetrachloride, chloroform, methylene chloride, 1,4-dioxane, 2-nitropropane, and perchloroethylene have been listed as substances known to the state of California to cause cancer. Ethyl alcohol in alcoholic beverages, ethylene glycol monomethyl ether, ethylene glycol monoethyl ether, ethylene glycol monomethyl ether acetate, ethylene glycol monoethyl ether acetate, and toluene are listed as substances known to the state of California to cause reproductive toxicity. Because the list is being expanded periodically, it is necessary for a user of industrial solvents to maintain awareness of the Proposition 65 list.

Solvent users and formulators may choose to avoid notification requirements for their workplaces and products by selecting solvents that do not contain Proposition 65 substances. As a result, Proposition 65 considerations can also influence the composition of solvents used in industrial processes and solvent-containing products.

Reg Neg Process.   Over the years, more than a dozen regulatory negotiations (reg negs) have been held to develop consensus for regulations for diverse, complex industries. In nearly all cases, the negotiations achieved consensus and led to more realistic, efficient regulations. Under charter from the EPA, a reg neg committee conducts negotiations to establish the basis for regulations for a particular industry. Parties to the negotiations include the EPA, state and local environmental authorities, public interest groups, and industry. The primary objective is to develop a regulation agreeable to all parties which recognizes the realities of the industry and provides an improved environment. The reg neg process may then help develop a meaningful, effective regulation more quickly. Because all affected parties participate in the development of the basis for the regulation and their concerns are heard, this process also minimizes the possibility of legal action challenging the regulation.

Two more recent regulatory negotiations involve industries that use a lot of industrial solvents. The Wood Furniture reg neg reached consensus and suggested VOC and HAP emission levels for the different segments of the wood furniture industry, as well as for each application within the individual segment. The Architectural and Industrial Maintenance (AIM) reg neg did not achieve consensus, but the negotiations established a basis which the EPA can use to write a realistic, significant regulation. Both industries have many segments that employ different practices and different kinds of solvent. In both of these reg negs, detailed information was developed concerning industry practices and this has led to more realistic regulations.

Solvent Selection and Formulation

Practical Solubility Concepts. Solution theory can provide a convenient, effective framework for solvent selection and blend formulation (3). When a solute dissolves in a solvent, a change in free energy occurs as a result of solvent–solute interactions. The change in free energy of mixing must be negative for dissolution to occur. In equation 1,

$$\Delta G = \Delta H - T\Delta S \qquad (1)$$

where ΔG is free energy; ΔH, enthalpy; and ΔS, entropy. One way to evaluate the enthalpy of mixing, ΔH, is through the concept of cohesive energy density (CED), the potential energy per cm^3 of solvent in equation 2, where ΔE is the energy of vaporization and V is the molar volume.

$$\delta^2 = \text{CED} = -\Delta E/V \qquad (2)$$

The Hildebrand solubility parameter, δ, is the square root of the CED. It is a measure of all interactions that occur between molecules of the solvent.

The solubility parameter concept can provide some useful guidelines, but it does not by itself differentiate polar effects and hydrogen bonding forces. For many solutes, polar interactions and hydrogen bonding are important, and a solvent blend designed on the basis of just the solubility parameter may not be adequate. Three parameter approaches, in which the root mean square of dispersion, and polar and hydrogen bonding components of the solubility parameter equals the solubility parameter, have been devised (4,5). Polar interactions can be described based on dipole moment (6) or on fractional polarity (7). A hydrogen bonding index based on infrared measurements, such as the Shell net hydrogen bonding index (8), can describe the hydrogen bonding capabilities of a solvent.

Computerized optimization using the three-parameter description of solvent interaction can facilitate the solvent blend formulation process because numerous possibilities can be examined quickly and easily and other properties can also be considered. This approach is based on the premise that solvent blends with the same solvency and other properties have the same performance characteristics. For many solutes, the lowest cost-effective solvent blends have solvency that is at the border between adequate and inadequate solvency. In practice, this usually means that a solvent blend should contain the maximum amount of hydrocarbon the solute can tolerate while still remaining soluble.

Tables 2, 3, 4, 5, and 6 contain information concerning the physical properties, solvent characteristics (9–11), economic and production data (12,13), and health and safety characteristics of common industrial solvents (14).

Solvent Characteristics. Solvents are conveniently classified as hydrocarbons or nonhydrocarbons. The latter are generally oxygenated compounds.

Hydrocarbon Solvents. Most hydrocarbon solvents are mixtures. Few commercial hydrocarbon solvents are single compounds. Toluene is an exception. Hydrocarbon solvents are usually purchased and supplied on specification. The most important specification properties are distillation range, solvency as expressed by aniline cloud point and Kauri-Butanol (KB) value, specific gravity, and flash point. Composition requirements such as aromatic content and benzene concentration are also important in many applications.

Hydrocarbon solvents marketed by each manufacturer differ in composition from those of other manufacturers, even if the specification properties are similar. This means that hydrocarbon solvents are not specified on the basis of molecular content. The composition of a hydrocarbon solvent depends on the crude feed to the process as well as the specific process steps the solvent undergoes during manufacture. Because each manufacturer uses a different feed and a somewhat different manufacturing scheme, hydrocarbon solvents differ somewhat in their properties, even in situations where the solvent performs the same.

The key to selecting a hydrocarbon solvent is understanding the requirements of the application. The solubility parameter and evaporation time are useful in designing blends containing both oxygenated solvents and hydrocarbons, but the preferred properties for hydrocarbon solvent blends are aniline cloud point, KB value, and distillation range. In many cases, the solvent manufacturer produces only a limited number of base hydrocarbon solvent building blocks, and prepares many more formulas by blending the building blocks. In this manner, the specific requirements of a customer can be met without undue manufacturing and logistical complexity.

Nonhydrocarbon and Oxygenated Solvents. Most industrial solvents that are not hydrocarbons are pure chemical compounds. As such, they have sharp boiling points and well-defined properties. Specifications for these solvents focus mostly on impurities such as water and other contaminants. This also means that a solvent from one manufacturer should perform the same as the same solvent from another manufacturer; any differences are probably the result of impurities, stabilizer content, etc, rather than the properties of the overall solvent.

Solvent Selection. A thorough knowledge of the requirements of each solvent application is necessary to formulate a solvent system successfully and meet all needs at the lowest possible cost. The most important properties are solvency, evaporation rate, flash point, and solvent balance. In nearly every

Table 2. Physical Properties of Common Industrial Solvents

Solvent name	CAS Registry Number	Common name (trade name)	Empirical formula	KB^a value	Solubility parameter δ, $MPa^{1/2b}$
Alcohols					
methanol	[67-56-1]	methyl alcohol, wood alcohol	CH_4O		29.7
ethanol	[64-17-5]	ethyl alcohol, (Tecsol)	C_2H_6O		26.0
1-propanol	[71-23-8]	*n*-propyl alcohol	C_3H_8O		24.3
2-propanol	[67-63-0]	isopropanol, 2-propanol	C_3H_8O		23.5
1-butanol	[71-36-3]	*n*-butyl alcohol, butanol	$C_4H_{10}O$		23.3
2-butanol	[78-92-2]	*sec*-butanol	$C_4H_{10}O$		22.1
2-methyl-1-propanol	[78-83-1]	isobutanol	$C_4H_{10}O$		21.9
2-methyl-2-propanol	[75-65-0]	*tert*-butanol	$C_4H_{10}O$		20.9
furfuryl alcohol	[98-00-0]	furyl alcohol	$C_5H_7O_2$		25.6
tetrahydrofurfuryl alcohol	[97-99-4]	tetrahydro-2-furylmethanol	$C_5H_{10}O_2$		
1-pentanol	[71-41-0]	*n*-amyl alcohol	$C_5H_{12}O$		22.3
3-methyl-1-butanol	[123-51-3]	isopentanol	$C_5H_{12}O$		
allyl alcohol	[107-18-6]	3-hydroxy-propene	C_3H_6O		24.1
cyclohexanol	[108-93-0]	cyclohexyl alcohol	$C_6H_{12}O$		23.3
1-hexanol	[111-27-3]	*n*-hexanol	$C_6H_{14}O$		21.9
4-methyl-2-amyl alcohol	[108-11-2]	methyl isobutyl carbinol, methyl amyl alcohol	$C_6H_{14}O$		20.5
2-ethylbutyl alcohol	[97-95-0]	*sec*-hexyl alcohol	$C_6H_{14}O$		21.5
hexyl alcohol (mixture)	[111-27-3]		$C_6H_{14}O$		
benzyl alcohol	[100-51-6]	phenylcarbinol	C_7H_8O		24.7
2-octanol	[123-96-6]		$C_8H_{18}O$		20.2
2-ethylhexanol	[104-76-7]	2-ethylhexyl alcohol	$C_8H_{18}O$		19.4
2-ethyl-4-methyl-1-pentanol	[106-67-2]	2-ethylisohexanol	$C_8H_{18}O$		
diisobutyl carbinol	[108-82-7]	2,6-dimethyl-4-heptanol	$C_9H_{20}O$		19.2
Aliphatic hydrocarbons					
pentane	[109-66-0]	*n*-pentane	C_5H_{12}	33.8	14.3
cyclohexane	[110-82-7]		C_6H_{12}	54.3	16.9
hexane	[110-54-3]	*n*-hexane	C_6H_{14}	26.5	14.9
heptane	[142-82-5]	*n*-heptane	C_7H_{16}	25.4	15.1
n-octane	[111-65-9]		C_8H_{18}	24.5	15.6
α-pinene	[80-56-8]	2-pinene	$C_{10}H_{16}$		
β-pinene	[127-91-3]	pseudopinene	$C_{10}H_{16}$		

Water solubility at 25°C[c], wt %		Water azeotrope, wt %/°C	Boiling range[d], °C	Vapor pressure at 25°C, kPa[e]	Specific gravity at 20°C[c]	Refractive index at 20°C[c]	Freezing point, °C
In water	Water in						

Alcohols

In water	Water in	Water azeotrope	Boiling range	Vapor pressure	Specific gravity	Refractive index	Freezing point
∞	∞		64.5	16.9	0.79	1.3284	−97.8
∞	∞	95.6/78.2	78.32	7.9	0.794	1.36143	−114.1
∞	∞	72/87	97.15	2.6	0.8036	1.385	−127.0
∞	∞	88/79.5	82.33	5.8	0.7864	1.3772	−88.43
7.7	20.1	55.5/93	117.7	0.8	0.8109	1.3993	−89
8.7	15.0	73.2/87	107.8	2.2	0.8034	1.3859	−108
15.4	65.1	67/90	99.5	1.7	0.8079	1.3969	−114.7
∞	∞	88/80	82.6	5.9	0.7793_{26}	1.3841	25.66
∞	∞	9/99	170	0.1	1.1285	1.4869	−14.63
			178	0.1	1.0485	1.4599	<−80
2.2	7.5	50/95	137.9	0.3	0.8160	1.4079	−78.2
3.6_{30}			130.5	0.4	0.8071_{25}	1.4052	−117.2
∞	∞	72/89	96.9	3.7	0.854	1.4110	−190 glass
0.13	11.8	31/98	160.65	0.2	0.9493	1.4656	−25.15
0.31_{20}	5.4_{20}	33/98	157.1	0.1	0.8212	1.4181	−44.6
1.64	6.35	55.6/94	131.8	0.8	0.8078	1.4110	−90 glass
0.43	4.6	41/97	147		0.8328	1.4025	−114.4
4.7	8.4_{20}	9/100	204.7	0.01	1.043	1.5396	15.3
0.096			173−182.5	0.03	0.835_{15}	1.4256	−38
0.07	2.6	20/99.1	184.8	0.02	0.8338	1.4328	−76
				0.03			
0.06	0.99	29.6/98.5	178.1	0.06	0.8121	1.4229	−65

Aliphatic hydrocarbons

In water	Water in	Water azeotrope	Boiling range	Vapor pressure	Specific gravity	Refractive index	Freezing point
		98/35	36	68.3	0.626	1.357	−130
		91/69	81	13	0.7786	1.4262	7
		94.4/62	69	20.1	0.6594	1.375	−95
		87/79	98	6.1	0.6838	1.3876	−91
		74.5/90	126	1.9	0.7025	1.3974	−57
			156.9	0.6	0.8539_{25}	1.4632	−64
			166	0.4	0.8667_{25}	1.4768	−61.5

Table 2. (*Continued*)

Solvent name	CAS Registry Number	Common name (trade name)	Empirical formula	KB[a] value	Solubility parameter δ, MPa$^{1/2b}$
		Aliphatic hydrocarbons			
rubber solvent				34	15.1
VM&P naphtha				35	15.5
high flash VM&P naphtha					
short-range mineral spirits				32	15.1
Rule-66 mineral spirits				35	15.3
regular mineral spirits				37	15.5
140 solvent				30	15.1
odorless mineral spirits				26	15.1
low odor 140 solvent					
mineral seal oil					
		Alkyl halides			
methylene chloride	[75-09-2]	dichloromethane, (Aerothene MM)	CH_2Cl_2	136	19.8
chloroform	[67-66-3]	trichloromethane	$CHCl_3$		19.0
carbon tetrachloride	[56-23-5]	tetrachloromethane	CCl_4		17.6
perchloroethylene	[127-18-4]	1,1,2,2-tetra-chloroethylene	C_2Cl_4	90	19.0
1,1,2-trichloro-1,2,2-trifluoroethane (TTE)	[76-13-1]	(Freon 113TR-T)	$C_2Cl_3F_3$	32	14.9
trichloroethylene	[79-01-6]	ethylene trichloride	C_2HCl_3	129	19.0
1,1,2,2-tetrachloro-ethane	[79-34-5]	tetrachloroethane	$C_2H_2Cl_4$		21.3
1,1,1-trichloroethane	[71-55-6]	methyl chloroform, (Chlorothene NU)	$C_2H_3Cl_3$	124	17.8
1,1,2-trichloroethane	[79-00-5]	ethane trichloride	$C_2H_3Cl_3$		19.6
1,2-dichloroethane	[107-06-2]	ethylene dichloride	$C_2H_4Cl_2$		20.0
1,2-dibromoethane	[106-93-4]	ethylene dibromide	$C_2H_4Br_2$		21.3
ethyl chloride	[75-00-3]	chloroethane	C_2H_5Cl		18.8
ethyl bromide	[74-96-4]	bromoethane	C_2H_5Br		19.6
propylene dichloride	[78-87-5]	1,2-dichloro-propane	$C_3H_6Cl_2$		
1,2,4-trichloro-benzene	[120-82-1]		$C_6H_3Cl_3$		
o-dichlorobenzene	[95-50-1]	1,2-dichloro-benzene (Dowtherm E)	$C_6H_4Cl_2$		20.5
chlorobenzene	[108-90-7]		C_6H_5Cl		19.4
p-chlorotoluene	[106-43-4]	p-tolyl chloride	C_7H_7Cl		18.0

Water solubility at 25°C[c], wt %		Water azeotrope, wt %/°C	Boiling range[d], °C	Vapor pressure at 25°C, kPa[e]	Specific gravity at 20°C[c]	Refractive index at 20°C[c]	Freezing point, °C
In water	Water in						
Aliphatic hydrocarbons							
			64–114	20.7	0.684		
			119–139	1.7	0.749		
				0.4			
			159–176	0.3	0.769		
			162–201	0.2	0.779		
			164–202	0.2	0.792		
			190–207	0.1	0.787		
			179–204	0.1	0.755		
				0.1			
Alkyl halides							
2.0	0.2	98.5/38.1	40.7	57.3	1.336	1.4237	−96.7
0.82	0.061	97.8/56	61.2	26.0	1.489	1.4422	−63.5
0.08	0.008	95.9/66	76.7	15.0	1.595	1.4631	−22.6
0.015	0.0105	84.2/88.7	121	2.5	1.623	1.5045	−19 to 22.35
0.028_{21}	0.009_{21}	99.0/44.5	47.6	44.5	1.574_{21}		
0.10	0.035	94/73.6	87	9.5	1.462	1.4782	−86
0.32	0.03	66/94.3	146.1	0.9	1.600	1.4940	−43
	0.05	95.7/65	74.1	16.6	1.332_{25}	1.4377_{21}	−30.6
0.48	0.032	83.6/86	113.7	2.8	1.443	1.4711	−36.7
0.9	0.15	91.8/70.5	83.6	10.5	1.2550	1.4443	−35
0.431_{30}			131.4	1.9	2.1805	1.5357_{25}	9.3
0.6_{20}	0.754_{20}		12.3		1.001_{25}	1.3790	−138.3
0.91_{20}		98.7 (vol)/37	38.2		1.440	1.4276	−118.6
0.26	0.07	88/78	95.9		1.160	1.4418	−70
			213		1.451_{25}	1.5732_{19}	17
<0.1		33/98	179.6	0.2	1.3048	1.5518_{22}	4.5
0.049_{30}		71.6/90	131.6	1.6	1.117	1.5275	−45
			162.3	0.5	1.069_{25}	1.5184_{22}	6.8

Table 2. (*Continued*)

Solvent name	CAS Registry Number	Common name (trade name)	Empirical formula	KB^a value	Solubility parameter δ, MPa$^{1/2b}$
		Amines			
n-propylamine	[107-10-8]	1-aminopropane	C_3H_9N		19.7
isopropylamine	[75-31-0]	2-aminopropane	C_3H_9N		
butylamine (mixture)	[109-73-9]		$C_4H_{11}N$		
tert-butylamine	[75-64-9]	2-aminoiso-butane	$C_4H_{11}N$		
diethylamine	[109-89-7]	2-aminopentane	$C_4H_{11}N$		16.3
diethylenetriamine	[111-40-0]	bis(2-amino-ethyl)amine	$C_4H_{13}N_3$		25.8
cyclohexylamine	[108-91-8]	aminocyclo-hexane, cyclo-hexanamine	$C_6H_{13}N$		18.9
triethylamine	[121-44-8]	(diethylamino)-ethane	$C_6H_{15}N$		
diisopropylamine	[108-18-9]		$C_6H_{15}N$		
toluidines (mixture)	[26915-12-8]		C_7H_9N		
dibutylamine	[111-92-2]	di-*n*-butylamine	$C_8H_{19}N$		
ethanolamine	[141-43-5]	2-aminoethanol	C_2H_7ON		31.5
morpholine	[110-91-8]	tetrahydro-1,4-oxazine	C_4H_9ON		21.5
ethylaminoethanol	[110-73-6]	2-*N*-monoethyl-aminoethanol	$C_4H_{11}ON$		
dimethylethanol-amine	[108-01-0]	*N*-dimethyl-aminoethanol	$C_4H_{11}ON$		
diethanolamine	[111-42-2]	bis(2-hydroxy-ethyl) amine (DEA)	$C_4H_{11}O_2N$		
diisopropanolamine	[110-97-4]	bis(2-hydroxy-propyl)amine	$C_6H_{15}O_2N$		
triethanolamine	[102-71-6]	trihydroxy-triethylamine	$C_6H_{15}O_3N$		
		Aromatic hydrocarbons			
benzene	[71-43-2]		C_6H_6		18.8
toluene	[108-88-3]	methylbenzene	C_7H_8	105	18.2
ethylbenzene	[100-41-4]	phenylethane	C_8H_{10}		18.0
xylene (mixture)	[1330-20-7]	dimethylbenzene	C_8H_{10}	95	18.1
o-xylene	[95-47-6]	*o*-dimethyl-benzene	C_8H_{10}		18.4
m-xylene	[108-38-3]	*m*-dimethyl-benzene	C_8H_{10}		18.0
p-xylene	[106-42-3]	*p*-dimethyl-benzene	C_8H_{10}		17.9
aromatic 100 solvent				92	18.0
aromatic 150 solvent				89	18.0
		Esters			
methyl acetate	[79-20-9]	acetic acid methyl ester	$C_3H_6O_2$		19.6

Water solubility at 25°C[c], wt %		Water azeotrope, wt %/°C	Boiling range[c], °C	Vapor pressure at 25°C, kPa[e]	Specific gravity at 20°C[c]	Refractive index at 20°C[b]	Freezing point, °C
In water	Water in						
			Amines				
∞	∞		48.7	42.1	0.718	1.3910	−83
∞	∞		32.4	78	0.686	1.3711	−101.2
∞	∞		44.4	49	0.6908_{25}	1.3761	−72.7
∞	∞		55.0	30.1	0.711	$1.3873_{17.6}$	−50
∞	∞		207.1		0.9542	1.4810	−39
∞	∞	44.2/96.4	134.5	1.2	0.8647_{25}	1.4565	−17.7
5.5_{20}	4.6_{20}	90/75	89.5	7.7	0.730	1.4003	−114.8
		91/74.1	83.9	10.7	0.7153	1.3924	−96.3
∞	∞	49.5/97	160		0.768	1.419	<−50
∞	∞		172.2	0.04	1.0180	1.4539	10.5
∞	∞		128.9	1.3	0.9998	1.4544	−4.9
			169.5		0.9140	1.4440	−9
		7.4/99	133.5		0.8866	1.4300	<−70
∞	∞		268.0		1.0985	1.4747	28
∞	∞		360		1.1242	1.4852	21.2
			Aromatic hydrocarbons				
		91.2/69	86	12.7	0.8790	1.50112	5.5
0.047_{16}	0.05_{16}	79.8/85	110−111	3.8	0.862	1.49693	−94.9
0.014_{15}		67/92	136	1.3	0.8670	1.49588	−94.9
			139−142	1.1	0.863		
			144	0.9	0.8802	1.50545	−25.2
		64.2/92	139	1.1	0.8642	1.49722	−47.8
			138	1.2	0.8611	1.49582	13.3
			160−176	0.3	0.870		
			173−208	0.1	0.888		
			Esters				
24.5	8.2		57.1	28.7	0.9353	1.3594	−98.1

Table 2. (*Continued*)

Solvent name	CAS Registry Number	Common name (trade name)	Empirical formula	KB^a value	Solubility parameter δ, $MPa^{1/2b}$
		Esters			
ethyl acetate	[141-78-6]	ethyl ethanoate	$C_4H_8O_2$		18.6
n-propyl acetate	[109-60-4]	1-propyl acetate	$C_5H_{10}O_2$		17.9
isopropyl acetate	[108-21-4]	2-propyl acetate	$C_5H_{10}O_2$		17.6
methyl acetoacetate	[105-45-3]	methyl-3-oxobutyrate	$C_5H_8O_3$		
ethyl acetoacetate	[141-97-9]	ethyl-3-oxobutanoate	$C_6H_{10}O_3$		
n-butyl acetate	[123-86-4]	1-butyl acetate	$C_6H_{12}O_2$		17.6
sec-butyl acetate	[105-46-4]	2-butanol acetate	$C_6H_{12}O_2$		16.8
isobutyl acetate	[110-19-0]	2-methyl-1-propyl acetate	$C_6H_{12}O_2$		17.0
amyl acetate	[628-63-7]	primary amyl acetate	$C_7H_{14}O_2$		18.2
isoamyl acetate	[123-92-2]	3-methyl-1-butyl acetate	$C_7H_{14}O_2$		16.0
4-methyl-2-pentanol, acetate	[108-84-9]	methylamyl acetate	$C_8H_{16}O_2$		16.8
isobutyl isobutyrate	[97-85-8]	2-methylpropyl isobutyrate	$C_8H_{16}O_2$		15.7
benzyl acetate	[140-11-4]		$C_9H_{10}O_2$		
methyl propionate	[554-12-1]	methyl propanoate	$C_4H_8O_2$		18.6
ethyl propionate	[105-37-3]	ethyl propanoate	$C_5H_{10}O_2$		17.8
isopropyl propionate			$C_6H_{12}O_2$		16.6
n-propyl propionate	[106-36-5]		$C_6H_{12}O_2$		17.6
n-butyl propionate	[590-01-2]		$C_7H_{14}O_2$		17.4
sec-butyl propionate			$C_7H_{14}O_2$		16.6
pentyl propionate	[624-54-4]		$C_8H_{16}O_2$		17.2
ethylene glycol monomethyl ether acetate	[110-49-6]	2-methoxyethyl acetate (Methyl Cellosolve Acetate)	$C_5H_{10}O_3$		18.8
ethylene glycol monoethyl ether acetate	[111-15-9]	2-ethyoxyethanol acetate (Cellosolve Acetate)	$C_6H_{12}O_3$		17.8
ethylene glycol monobutyl ether acetate	[112-07-2]	2-butoxyethanol acetate (Butyl Cellosolve Acetate)	$C_8H_{16}O_3$		16.8
ethylene glycol monopropyl ether acetate	[20706-25-6]	2-propoxyethyl acetate	$C_7H_{14}O_3$		17.2
diethylene glycol monomethyl ether acetate	[629-38-9]	(Methyl Carbitol Acetate)	$C_7H_{14}O_4$		
diethylene glycol monoethyl ether acetate	[112-15-2]	(Carbitol Acetate)	$C_8H_{16}O_4$		19.4

Water solubility at 25°C[c], wt %		Water azeotrope, wt %/°C	Boiling range[d], °C	Vapor pressure at 25°C, kPa[e]	Specific gravity at 20°C[c]	Refractive index at 20°C[c]	Freezing point, °C
In water	Water in						
Esters							
2.9	3.0	92/70.4	77.1	12.6	0.897	1.3725	−83.6
2.3	2.6	86/82.4	101.6	4.4	0.885	1.3844	−92.5
2.9	1.9	89/75.4	88.7	8.0	0.866	1.3770	−73.1
			171.7		1.0747	1.4186	−80
4.9_{20}	13_{17}		145.5	0.1	1.025	1.4198	−89.5
0.78	2.88	73/90.7	126.1	1.5	0.881	1.3951	−73.5
2.6	1.3	77.5/87	112.2	3.2	0.960	1.3915	−98.9
0.67	1.65	78/88.2	117.2	2.6	0.865	1.3880	−99.85
		59/95.2	146		0.870	1.4013	−100 glass
0.16	1.8	63.7/93.8	142.5	0.8	0.8699_{25}	1.400_{21}	−78.5
0.18	0.08	63.3/94.8	140−147	0.7	0.857	1.4008	
1.3	0.8	61/95.5	147.3	0.6	0.859	1.3999	−81.0
		12.5/99.6	215.5	0.01	1.0550	1.5232	−51.5
6.5_{20}		91.8/71	78.7	11.5	0.9151	1.3779	−87.5
		90/81.2	99.1	5.0	0.8899	1.3839	−73.9
		80.1/85		3.3			
		77/88.9	122.5	1.9	0.833	1.6015	−76
0.2	<0.02	59/94.8	145.5	0.5	0.8818	1.3982	−89.5
				1.0			
<0.02	<0.03		168.7	0.2	0.871	1.4096	−73.1
∞	∞	48.5/97	143	0.5	1.0067	1.4025	−65
23.8	6.5	44.4/97.5	156.3	0.2	0.970	1.4030	−61.7
1.1	1.6		191.6	0.05	0.938	1.4200	−64.6
			209.1		1.0396		
∞	∞		217.7	0.02	1.011	1.4230	−25

Table 2. (*Continued*)

Solvent name	CAS Registry Number	Common name (trade name)	Empirical formula	KB[a] value	Solubility parameter δ, MPa$^{1/2b}$
		Esters			
diethylene glycol monobutyl ether acetate	[124-17-4]	(Butyl Carbitol Acetate)	$C_{10}H_{20}O_4$		18.0
propylene glycol monomethyl ether acetate	[108-65-6]	(Arcosolv PM Acetate)	$C_6H_{12}O_3$		18.8
dipropylene glycol monomethyl ether acetate	[88917-22-0]	(Arcosolv DPM Acetate)	$C_8H_{16}O_4$		16.8
ethyl 3-ethoxy propionate	[763-69-9]	(Ektasolve EEP)	$C_7H_{14}O_3$		17.8
propylene carbonate	[108-32-7]	propylene glycol cyclic carbonate	$C_4H_6O_2$		27.2
dibasic ester (mixture)			mixture		19.8
		Glycol ethers			
ethylene glycol monomethyl ether	[109-86-4]	(Methyl Cellosolve)	$C_3H_8O_2$		22.1
ethylene glycol monoethyl ether	[110-80-5]	(Cellosolve Solvent)	$C_4H_{10}O_2$		20.2
diethylene glycol monomethyl ether	[111-77-3]	(Methyl Carbitol)	$C_5H_{12}O_3$		20.9
ethylene glycol dimethyl ether	[110-71-4]	monoglyme, 1,2-dimethoxy-ethane	$C_4H_{10}O_2$		17.6
ethylene glycol monobutyl ether	[111-76-2]	(Butyl Cellosolve, Butyl OXITOL)	$C_6H_{14}O_2$		18.2
diethylene glycol monoethyl ether	[111-90-0]	(Carbitol Solvent)	$C_6H_{14}O_3$		19.8
ethylene glycol monopropyl ether	[2807-30-9]	(Propyl Cellosolve)	$C_5H_{12}O_2$		19.2
ethylene glycol monophenyl ether	[122-99-6]	(Phenyl Cellosolve)	$C_8H_{10}O_2$		22.3
diethylene glycol monobutyl ether	[112-34-5]	(Butyl Carbitol, Butyl DIOXITOL)	$C_8H_{18}O_3$		18.2
propylene glycol monomethyl ether	[1589-49-7]	(Arcosolv PM)	$C_4H_{10}O_2$		20.9
propylene glycol monoethyl ether	[1569-02-4]	1-ethoxy-2-propanol	$C_5H_{12}O_2$		18.4
propylene glycol monobutyl ether	[5131-66-8]	(Dowanol PnB)	$C_7H_{16}O_2$		17.6
propylene glycol monotertiary butyl ether	[57018-52-7]	(Arcosolv PTB)	$C_7H_{16}O_2$		16.6
dipropylene glycol monomethyl ether	[34590-94-8]	(Arcosolv DPM)	$C_7H_{16}O_3$		19.6

Water solubility at 25°C[c], wt %		Water azeotrope, wt %/°C	Boiling range[d], °C	Vapor pressure at 25°C, kPa[e]	Specific gravity at 20°C[c]	Refractive index at 20°C[c]	Freezing point, °C
In water	Water in						
Esters							
6.5	3.7	8/99.5	246.7	0.0006	0.980	1.4262	−32.2
20	5.9		145.7	0.5	0.97	1.40	
				0.02			
2.9	2.9	37/97	165–172	0.2	0.946	1.4050	< −50
				0.004	1.2069		
5.3	3.1		196–225	0.03		1.4213	−20
Glycol ethers							
∞	∞	15/99.9	124.5	1.3	0.963	1.4021	−85.1
∞	∞	15.3/99.9	135.1	0.8	0.928	1.4076	−100
∞	∞	28.8/99.4	194.2	0.04	1.021	1.4263	−85
		89.5/76	83.5	10.4	0.8629	1.3796	−58
∞	∞	20.8/99.8	171.2	0.15	0.901	1.4193	−75
∞	∞		201.9	0.04	0.988	1.4273	−76
∞	∞		150.1	0.4	0.916	1.4136	−90
10.8_{20}	2.7_{20}		245.2	0.0009	1.1020_{22}	1.5386	14
∞	∞		230.4	0.002	0.956	1.4316	−68.1
∞	∞	51.5/98.3	120.1	1.6	0.919	1.4011	−95
				1.3			
5.6	15.5	28/98.6	170.1	0.16	0.884	1.4173	−90
14.5	20.1	22/95	151	0.17	0.872	1.4116	−56
∞	∞		188.3	0.07	0.956	1.4198	−80

Table 2. (*Continued*)

Solvent name	CAS Registry Number	Common name (trade name)	Empirical formula	KB[a] value	Solubility parameter δ, MPa$^{1/2b}$
		Glycol ethers			
propylene glycol monophenyl ether	[770-35-4]		$C_9H_{12}O_2$		
		Ketones			
acetone	[67-64-1]	dimethyl ketone	C_3H_6O		20.5
methyl ethyl ketone	[79-93-3]	2-butanone	C_4H_8O		19.0
methyl-*n*-propyl ketone	[107-87-9]	2-pentanone	$C_5H_{10}O$		17.8
mesityl oxide	[141-79-7]	2-methyl-2-penten-4-one	$C_6H_{10}O$		18.4
cyclohexanone	[108-94-1]		$C_6H_{10}O$		20.2
methyl *n*-butyl ketone	[591-78-6]		$C_6H_{12}O$		17.4
methyl isobutyl ketone	[108-10-1]		$C_6H_{12}O$		17.2
diacetone alcohol	[123-42-2]	4-hydroxy-4-methylpen-tanone-2	$C_6H_{12}O_2$		18.8
methyl amyl ketone	[110-43-0]	2-heptanone	$C_7H_{14}O$		17.4
methyl isoamyl ketone	[110-12-3]	5-methyl-2-hexanone	$C_7H_{14}O$		17.0
diisobutyl ketone	[106-83-8]	2,6-dimethyl-4-heptanone	$C_9H_{18}O$		16.0
isophorone	[78-59-1]	1,1,3-trimethyl-3-cyclohexene-5-one	$C_9H_{14}O$		18.6
		Others			
ethylene glycol	[107-21-1]	1,2-ethanediol	$C_2H_6O_2$		30.1
propylene glycol	[57-55-6]	1,2-propanediol, 1,3-propanediol	$C_3H_8O_2$		25.8
diethylene glycol	[111-46-6]	2,2″-oxydiethanol	$C_4H_{10}O_3$		24.7
dipropylene glycol	[110-98-5]	1,1′-oxydi-2-propanol	$C_6H_{14}O_3$		20.9
triethylene glycol	[112-27-6]		$C_6H_{14}O_4$		22.5
tripropylene glycol	[24800-44-0]		$C_9H_{20}O_4$		18.4
hexylene glycol	[107-41-5]	2-methyl-2,4-pentanediol	$C_6H_{14}O_2$		19.8
acetonitrile	[75-05-8]	methyl cyanide	C_2H_3N		24.3
n-butyronitrile	[109-74-0]	propyl cyanide	C_4H_7N		21.5
N-methyl-2-pyrrolidone	[872-50-4]	(M-Pyrol)	C_5H_9ON		23.1
nitromethane	[75-52-5]		CH_3O_2N		25.8
1-nitropropane	[105-03-2]	1-NP	$C_3H_7O_2N$		21.1
2-nitropropane	[79-46-9]	2-NP	$C_3H_7O_2N$		20.2
nitrobenzene	[98-95-3]		$C_6H_5O_2N$		
m-nitrotoluene	[99-08-1]	3-nitrotoluene	$C_7H_7O_2N$		

Water solubility at 25°C[c], wt %		Water azeotrope, wt %/°C	Boiling range[d], °C	Vapor pressure at 25°C, kPa[e]	Specific gravity at 20°C[c]	Refractive index at 20°C[c]	Freezing point, °C
In water	Water in						
Glycol ethers							
			242.7	0.003			13
Ketones							
∞	∞		56.1	30.8	0.79	1.3590	−94.9
27.1	12.5	88.7/73	79.64	12.1	0.805−0.807	1.3788	−86.69
4.3	3.3	80.5/83.3	102.3	4.7	0.81	1.3904	−77.5
3.0_{20}	3.4_{20}	65.2/91.8	130	1.2	0.853	1.4456	−47
2.3	8.0	45/96	156.7	0.6	0.944−0.950	1.4507	−47.0
3.5	3.7	74 (vol)/90.5	127.5	1.7	0.818	1.4024	−56.9
2.04	2.41	75.7/88	116.2	2.6	0.80	1.3957	−83.5
∞	∞	13/99.6	169.2	0.18	0.9382	1.4234	−44.0
0.43	1.5	52/95	150.5	0.5	0.817	1.4110	−35.0
0.55	1.4	56/94.7	145.4	0.7	0.814	1.4069	−74.21
<0.05	0.75		169.3	0.3	0.807−0.814	1.4230	−41.5
1.2	4.3	16.1/99.5	215.2	0.06	0.923	1.4775	−12
Others							
∞	∞		197.3	0.01	1.1155	1.4318	−12.7
∞	∞		187.3	0.03	1.0381	1.431_{26}	−60.0
∞	∞		245.0	0.0003	1.118	1.4472	−7.8
∞	∞		231.8	0.004	1.025	1.439_{25}	−40 pour
∞	∞		287.4	0.0002	1.125	1.4559	−7.2
∞	∞		267	0.0002	1.023_{25}		
∞	∞		198.27	0.01	0.921	1.4276	−50 glass
∞	∞	84/76	81.6	12.2	0.7766_{25}	1.3416	−43.8
0.43	0.87	67.5/89	117.9	2.7	0.7865_{25}	1.3860_{25}	−111.9
∞	∞		202.0	0.06	1.027_{25}	1.469_{25}	−24.4
9−10	2.2	76.4/84	101.2	4.8	1.312_{25}	1.3796	−28.5
1.4	1.7	65/96.6	132	1.3	1.002	1.3996	−108
1.7	0.6	71/88.6	120.3	2.3	0.987	1.3941	−93.0
0.10_{20}		−/98.6	210−211	0.03	1.205_{18}	1.5546	6
0.05_{20}			230.5		1.157	1.5475	15.1

Table 2. (*Continued*)

Solvent name	CAS Registry Number	Common name (trade name)	Empirical formula	KB^a value	Solubility parameter δ, MPa$^{1/2b}$
		Others			
N,N-dimethyl- formamide (DMF)	[68-12-2]	dimethyl formamide	C_3H_7ON		24.7
ethyl ether	[60-29-7]	diethyl ether	$C_4H_{10}O$		15.1
isopropyl ether	[108-20-3]	diisopropyl ether	$C_6H_{14}O$		14.3
n-butyl ether	[142-96-1]	di-*n*-butyl ether	$C_8H_{18}O$		16.0
diphenyl oxide	[101-84-8]	diphenyl ether	$C_{12}H_{10}O$		20.7
propylene oxide	[75-56-9]	epoxypropane	C_3H_6O		18.8
tetrahydrofuran	[109-99-9]	tetramethylene oxide	C_4H_8O		20.2
1,4-dioxane	[123-91-1]	*p*-dioxane	$C_4H_8O_2$		20.5
γ-butyrolactone	[96-48-0]	4-butanolide	$C_4H_6O_2$		25.8
furfural	[98-01-1]	furfuraldehyde	$C_5H_4O_2$		22.9
pine oil	[8002-09-3]	oil of pine			17.6
turpentine	[8006-64-2]			61	16.6
cresylic acid (mixture of *o*-, *m*-, and *p*-cresol)	[1319-77-3]	hydroxytoluol	C_7H_8O		
m-cresol	[108-39-1]	*m*-toluol	C_7H_8O		20.9

^{a}Kauri-Butanol.
^{b}To convert MPa to bar, multiply by 0.1.
^{c}Temperature as indicated except where noted by subscript in °C.
^{d}At 101.3 kPa (=1 atm).
^{e}To convert kPa to psi, multiply by 0.145.

application, these properties are important even though the specific requirements differ greatly from one application to another. Each potential solvent has a particular set of properties, and the solvent chosen and the amount of each depend on the specific application requirements.

In terms of general solvency, solvents may be described as active solvents, latent solvents, or diluents. This differentiation is particularly popular in coatings applications, but the designations are useful for almost any solvent application. Active solvents are strong solvents for the particular solute in the application, and are most commonly ketones or esters. Latent solvents function as active solvents in the presence of a strong active solvent. Alcohols exhibit this effect in nitrocellulose and acrylic resin solutions. Diluents, most often hydrocarbons, are nonsolvents for the solute in the application.

The specific solvents that make up the three solvency categories depend on the solute in question. For example, an aliphatic hydrocarbon may have adequate solvency for a long oil alkyd, but would be a diluent for an acrylic or vinyl resin, which require stronger solvents such as ketones or esters. The formulator must understand the solvency requirements of the solute to know which category a particular solvent would occupy.

Water solubility at 25°C^c, wt %		Water azeotrope, wt %/°C	Boiling range^d, °C	Vapor pressure at 25°C, kPa^e	Specific gravity at 20°C^c	Refractive index at 20°C^c	Freezing point, °C
In water	Water in						
Others							
∞	∞		153.0	0.5	0.944_{25}	1.4269_{25}	−61.0
7.5_{20}	1.3_{20}	34.1/98.7	34.6	71.2	0.7135	1.3526	116.2
0.90_{20}	0.57_{20}	95.5/62.2	68.5	20.0	0.720	1.3682	−85.5
0.05_{20}	0.19_{20}	67/92.9	142	0.8	0.769	1.3968	−95
		3.3/99.3	257		1.0728	1.5763	28
			33.9	71.7	0.8287	1.3660	−111.9
		93/63	66.0	21.6	0.884	1.4073	−108.5
		82/87.8	106.1	5.0	1.0353	1.4203	12
			206		1.124_{25}	1.4348	−44
8.3_{20}	4.8_{20}	34.5/98	161.7	0.2	1.1598	1.5261	−36.5
			200−220				
			154−170		0.857		
			191−203		$1.030-1.038_{25}$		10.9−35.5
2.35_{20}			202.8	0.02	1.034	1.5396	10.9

Solvent Blend Design. Similar guidelines apply regardless of the specific application. The base of the solvent system should be a strong active solvent for the solute in question. If that is a hydrocarbon solvent, then specific hydrocarbon solvent components can be chosen on the basis of aniline cloud point and KB value. If the solute requires a stronger solvent, the most commonly used nonhydrocarbon strong solvents are ketones and esters, which function as true solvents for many solutes. Depending on the solute, a latent solvent, typically an alcohol, may be included, which contributes solvency while lowering blend cost. The remainder of the blend would be composed of diluent solvents, usually hydrocarbons. The maximum amount of hydrocarbon should be formulated into the blend commensurate with adequate solvency and other performance characteristics. Solvency costs money because high solvency solvents are more expensive than low solvency solvents, and excess solvency results in blends that cost more than needed. Given the tight profit margins and strong competition of many industries using solvents, excessive solvent cost can have a strong negative impact.

The evaporation profile of the solvent blend is also important, as well as the related property, flash point. The evaporation rate or boiling range selected should be appropriate for the needs of the application method. In coatings formulations, this would mean a fast evaporating blend for a coating applied by spray, and a slow evaporating blend for brush or roller application. As a safety-related property, flash point must be a consideration. Although it may be tempting to put a large amount of a fast-evaporating active solvent such as acetone in the blend because acetone is cheaper than slower-evaporating active

Table 3. Solvent Characteristics of Common Industrial Solvents

Solvent name	Viscosity, neat, mPa·s(=cP)$_{°C}$	Surface tension, mN/m(=dyn/cm)$_{°C}$	Coeff. of expansion at 20°C, $\Delta V/(V \cdot \Delta T)$, °C^{-1}
Alcohols			
methanol	0.56$_{25}$	22.6$_{20}$	0.00119
ethanol	1.1$_{25}$	22.27$_{20}$	0.0011
1-propanol	2.0$_{25}$	23.8$_{20}$	0.00096
2-propanol	2.4$_{25}$	21.35$_{20}$	0.00104
1-butanol	2.6$_{25}$	24.6$_{20}$	0.00090
2-butanol	1.8$_{25}$	22.8$_{20}$	0.00096
2-methyl-1-propanol	2.5$_{25}$	23.0$_{20}$	0.00101
2-methyl-2-propanol	3.35$_{30}$	20.7$_{20}$	0.00133
furfuryl alcohol	5$_{25}$	38.2$_{25}$	0.000852
tetrahydrofurfuryl alcohol	6.24$_{20}$	37$_{25}$	0.00052
allyl alcohol	1.072$_{30}$	25.68$_{20}$	0.00101
cyclohexanol	52.7$_{25}$	35.1$_{20}$	0.00077
1-hexanol	5.4$_{20}$	23.6$_{20}$	
4-methyl-2-amyl alcohol	3.8$_{25}$	22.8$_{20}$	0.00103
2-ethylbutyl alcohol	5.63$_{20}$	28.05$_{28}$	0.000842
benzyl alcohol	5.58$_{20}$		
2-octanol	8.2$_{25}$		
2-ethylhexanol	7.7$_{25}$	97.71$_{20}$	0.00088
diisobutyl carbinol	15.45$_{20}$	26$_{20}$	
Aliphatic hydrocarbons			
pentane	0.28$_{25}$		
cyclohexane	0.28$_{25}$		
hexane	0.28$_{25}$		
rubber solvent	0.56$_{25}$		
VM&P naphtha	0.56$_{25}$		
short-range mineral spirits	1.00$_{25}$		
Rule-66 mineral spirits	1.02$_{25}$		
regular mineral spirits	1.10$_{25}$		
140 solvent	1.52$_{25}$		
odorless mineral spirits	1.50$_{25}$		
Alkyl halides			
methylene chloride	0.425$_{20}$	28.2	
chloroform	0.563$_{20}$	27.14$_{20}$	0.001399
carbon tetrachloride	0.96$_{20}$	26.8$_{20}$	0.00127
perchloroethylene	0.88$_{20}$	32.32$_{20}$	0.001079
1,1,2-trichloro-1,2,2-trifluoro-ethane (TTE)	0.694$_{20}$	18.8$_{20}$	
trichloroethylene	0.550$_{25}$	32.0$_{25}$	0.00115
1,1,2,2-tetrachloroethane	1.77$_{20}$		0.000998
1,1,1-trichloroethane	0.79$_{25}$	25.6$_{25}$	
1,2-dichloroethane	0.78$_{25}$	37.5$_{25}$	0.00116
1,2-dibromoethane		38.71$_{20}$	
ethyl chloride	0.279$_{10}$		
ethyl bromide		24.5$_{20}$	
propylene dichloride	0.865$_{20}$	31.4$_{25}$	0.001108
o-dichlorobenzene			0.00083

Table 3. (Continued)

Solvent name	Viscosity, neat, mPa·s(=cP)$_{°C}$	Surface tension, mN/m(=dyn/cm)$_{°C}$	Coeff. of expansion at 20°C, $\Delta V/(V \cdot \Delta T)$, °C^{-1}
Alkyl halides			
chlorobenzene	0.844_{15}	33.08_{25}	0.00092
p-chlorotoluene		34.6_{25}	
Amines			
isopropylamine	0.36_{25}		
diethylamine	0.36_{20}		
triethylamine	0.36_{20}	20.7_{20}	
diisopropylamine	0.39_{20}	19.7_{20}	
dibutylamine	0.89_{20}	24.7_{20}	0.00170
ethanolamine	3.40_{20}	51_{20}	0.000770
morpholine	2.23_{20}	37.5_{20}	
dimethylethanolamine	3.4_{25}		0.00120
diethanolamine		51.5_{20}	
triethanolamine	10_{20}	53_{25}	0.00048
Aromatic hydrocarbons			
benzene	0.654_{20}	31.6_{20}	
toluene	0.62_{25}		0.00107
ethylbenzene	0.6687_{20}		0.00101
xylene (mixture)	0.67_{25}		
o-xylene			0.000972
m-xylene			0.000990
p-xylene			0.001008
aromatic 100 solvent	0.88_{25}		
aromatic 150 solvent	1.08_{25}		
Esters			
methyl acetate	0.381_{20}	24.6_{20}	0.00139
ethyl acetate	0.45_{25}	23.9_{20}	0.00133
n-propyl acetate	0.58_{25}	23.9_{20}	0.00126
isopropyl acetate	0.525_{20}	24.5_{20}	0.00131
ethyl acetoacetate		32.5_{20}	0.00106
n-butyl acetate	0.693_{25}	27.6_{27}	0.00121
sec-butyl acetate	0.65_{25}	22.8_{20}	0.00118
isobutyl acetate	0.70_{25}	23.3_{20}	0.00119
amyl acetate	0.9_{25}	28.5_{20}	0.00115
isoamyl acetate		24.7_{20}	0.00199
4-methyl-2-pentanol, acetate	0.87_{25}		0.00109
isobutyl isobutyrate	0.83_{25}		
methyl propionate	0.44_{25}		
ethyl propionate	0.51_{25}	24.2_{20}	0.00125
isopropyl propionate	0.58_{25}		
n-propyl propionate	0.66_{25}		
n-butyl propionate	0.9_{20}	25.3_{20}	0.001
sec-butyl propionate	0.68_{25}		
pentyl propionate	1.0_{20}		0.00108
ethylene glycol monomethyl ether acetate	1.1_{25}		0.00110

Table 3. (Continued)

Solvent name	Viscosity, neat, $mPa \cdot s (=cP)_{°C}$	Surface tension, $mN/m (=dyn/cm)_{°C}$	Coeff. of expansion at 20°C, $\Delta V/(V \cdot \Delta T)$, $°C^{-1}$
Esters			
ethylene glycol monoethyl ether acetate	1.3_{25}	31.8_{25}	0.0012
ethylene glycol monobutyl ether acetate	1.8_{20}	27.4_{25}	0.00104
ethylene glycol monopropyl ether acetate	1.45_{25}		
diethylene glycol monoethyl ether acetate	2.8_{25}		0.00102
diethylene glycol monobutyl ether acetate	3.6_{20}	28.5_{25}	0.00101
propylene glycol monomethyl ether acetate	1.17_{20}	28.2_{20}	0.00096
dipropylene glycol mono-methyl ether acetate	2.10_{25}		
ethyl 3-ethoxy propionate	1.2_{25}	24.2_{20}	0.001176
propylene carbonate	2.44_{25}		
dibasic ester	2.39_{25}		
Glycol ethers			
ethylene glycol monomethyl ether	1.20_{25}	30.6_{20}	0.00095
ethylene glycol monoethyl ether	1.35_{25}	27.9_{20}	0.00097
diethylene glycol monomethyl ether	3.9_{20}	34.8_{20}	0.0086
ethylene glycol monobutyl ether	2.00_{25}	27.3_{20}	0.00092
diethylene glycol monoethyl ether	1.75_{25}	35.5_{20}	0.00090
diethylene glycol monobutyl ether	2.15_{25}	30.0_{20}	0.00085
ethylene glycol monophenyl ether	30.5_{20}		
ethylene glycol monopropyl ether	2.76_{20}	29.2_{20}	0.00095
propylene glycol monomethyl ether	1.71_{25}	27.7_{20}	0.00099
propylene glycol monoethyl ether	1.30_{25}		
propylene glycol monobutyl ether	3.47_{20}	27.4_{20}	0.00087
propylene glycol monotertiary butyl ether	3.30_{25}	24.2_{25}	
dipropylene glycol mono-methyl ether	3.42_{25}	28.8_{20}	0.00094
Ketones			
acetone	0.31_{25}	22.32_{20}	0.00143

552

Table 3. (*Continued*)

Solvent name	Viscosity, neat, mPa·s(=cP)°C	Surface tension, mN/m(=dyn/cm)°C	Coeff. of expansion at 20°C, $\Delta V/(V\cdot\Delta T)$ °C^{-1}
Ketones			
methyl ethyl ketone	0.41_{25}	24.6_{20}	0.00131
methyl propyl ketone	0.68_{25}	26.6_{20}	0.0012
mesityl oxide	0.60_{20}		0.00108
cyclohexanone	2.0_{25}	27.7_{20}	0.00094
methyl *n*-butyl ketone	0.584_{25}	25.2_{20}	0.00099
methyl isobutyl ketone	0.55_{25}	23.64_{20}	0.00115
diacetone alcohol	2.9_{25}	28.9_{20}	0.00094
methyl amyl ketone	0.77_{25}		0.00104
methyl isoamyl ketone	0.73_{25}	28.5_{20}	0.00107
diisobutyl ketone	0.95_{25}	22.5_{20}	0.00102
isophorone	2.3_{25}	32.3_{20}	0.00085
Others			
ethylene glycol	17.4_{20}	48.4_{20}	0.000566
propylene glycol	43.0_{20}	36.0_{20}	0.000695
diethylene glycol	17.4_{20}	48.4_{20}	0.000566
dipropylene glycol	48_{25}	32_{25}	
triethylene glycol	38.2_{25}	45.2_{20}	0.00171
hexylene glycol	29.8_{25}	33.1_{20}	0.00072
acetonitrile	3.6_{25}	29.3_{20}	0.00137
n-butyronitrile		25.2_{25}	0.00107
N-methyl-2-pyrollidone	1.7_{25}	40.7_{25}	
nitromethane		37_{25}	
2-nitropropane		30_{25}	
N,N-dimethylformamide (DMF)	0.82_{25}	35.2_{20}	
ethyl ether	0.2332_{20}	17.0_{20}	0.00164
isopropyl ether	0.379_{20}	32_{20}	0.00149
n-butyl ether		22.9_{20}	
tetrahydrofuran	0.50_{25}	26.4_{25}	
1,4-dioxane	1.19_{25}		
furfural	1.49_{25}	43.5_{20}	0.00087
turpentine			0.00089

solvents, such practice will decrease the flash point significantly and could impair worker safety.

Composition changes during evaporation must also be considered. This characteristic, referred to as solvent balance, is important because adequate solvency must be maintained throughout the entire processing cycle. If a diluent solvent is the last solvent component remaining, it is likely that the solute will precipitate from solution, resulting in inadequate performance. In most solvent blends, the highest-boiling, slowest-evaporating solvent should be a strong solvent for the solute.

Formulator's Dilemma. The regulatory discussion included a listing of solvents designated as HAP compounds. Emissions of these solvents are to be

Table 4. Blush Resistance and Dilution Ratio of Common Industrial Solvents

Solvent name	Blush resistance[a] at 27°C, % rh	Dilution ratio[b]	
		Toluene	Aliphatic naphtha
Alcohols			
methanol		2.9	0.3
tetrahydrofurfuryl alcohol		2.6 (xylene)	
Esters			
methyl acetate		2.9	0.9
ethyl acetate	37	3.0	1.0
n-propyl acetate	76	3.2	1.5
isopropyl acetate	69	2.7	0.92
n-butyl acetate		3.05	1.4
sec-butyl acetate	76	2.6	1.3
isobutyl acetate	80	2.7	1.1
amyl acetate	91	2.3	1.3
isoamyl acetate		2.5	1.4
4-methyl-2-pentanol, acetate		1.7	1.0
isobutyl isobutyrate		1.5	0.8
ethyl propionate		2.1	0.8
n-butyl propionate	90	2.1	1.2
pentyl propionate	95	1.4	0.7
ethylene glycol monomethyl ether acetate		2.3	0.6
ethylene glycol monoethyl ether acetate		2.5	0.9
ethylene glycol monobutyl ether acetate		1.8	1.2
diethylene glycol monoethyl ether acetate		2.2	0.6
diethylene glycol monobutyl ether acetate	96+	1.8	0.9
propylene glycol monomethyl ether acetate	60	2.6	0.8
Glycol ethers			
ethylene glycol monomethyl ether	50	3.4	0.2
ethylene glycol monoethyl ether	67	4.9	1.1
diethylene glycol mono- methyl ether		2.3	
ethylene glycol monobutyl ether	96	3.3	1.8
diethylene glycol monoethyl ether	<50	4.8	immiscible[c]
diethylene glycol monobutyl ether	85	3.9	1.9
ethylene glycol monopropyl ether		4.0	2.0
propylene glycol monomethyl ether		5.2	0.9

Table 4. (*Continued*)

Solvent name	Blush resistance[a] at 27°C, % rh	Dilution ratio[b]	
		Toluene	Aliphatic naphtha
Glycol ethers			
propylene glycol monobutyl ether		1.9	0.9
propylene glycol mono-tertiary butyl ether		2.3	1.2
dipropylene glycol mono-methyl ether		4.4	0.8
Ketones			
acetone	<20	4.4	0.8
methyl ethyl ketone	36	4.3	0.9
methyl propyl ketone	70	3.9	1.0
cyclohexanone	92	5.8	1.3
methyl *n*-butyl ketone	80	4.0	1.1
methyl isobutyl ketone	78	3.6	1.0
diacetone alcohol		2.3	0.3
methyl amyl ketone	93	3.9	1.2
methyl isoamyl ketone	89	3.8	1.1
diisobutyl ketone		1.5	0.8
isophorone	97	6.2	
Others			
acetonitrile		2.0	
n-butyronitrile		3.1	0.8
N,N-dimethylformamide (DMF)		7.7	0.2
tetrahydrofuran		2.9	1.1

[a]Relative humidity at which water condenses on the applied film as a result of evaporative cooling effects.
[b]Volume ratio (ASTM D1720) of diluent to solvent that just fails to dissolve completely 8 g of nitrocellulose in 100 mL of solvent. The higher the dilution ratio, the more diluent the solution tolerates.
[c]Aliphatic naphtha and this solvent are immiscible in each other.

significantly reduced. For many applications this means that less is to be allowed. In a situation where the allowed VOC emission levels are also being reduced, the formulator would like to use the most effective solvents available. In the past, MEK and MIBK were frequently used as active solvents and aromatic hydrocarbons as diluents. These solvents have been popular because they are cost-effective.

Reformulating to reduce HAP solvents frequently means that solvent blend costs increase. The newer blends are generally not be as effective. For example, many coatings were usually formulated using ketones as the active solvents with aromatic hydrocarbons as diluents. This combination produced the most cost-effective formulations. However, when MEK, MIBK, toluene, and xylene became HAP compounds, less-effective solvents had to be used for reformulation. Esters are the most common ketone replacements, and aliphatic diluents would

Table 5. Cost and Production Data for Industrial Solvents

Solvent name	1970 Production, t × 10³	1970 Cost, $/kg	1979 Production, t × 10³	1979 Cost, $/kg	1991 Production, t × 10³	1991 Cost, $/kg	1995 Cost, $/kg
Alcohols							
methanol	2237	0.01	3342	0.03	3948	0.15	0.51
ethanol	887	0.03	639	0.09	125	0.52	0.93
1-propanol	27	0.05	85	0.12	78.7	0.81	1.14
2-propanol	870	0.03	862	0.08	609	0.54	0.64
1-butanol			347	0.10	599	0.55	0.92
2-butanol							1.09
2-methyl-1-propanol			65	0.07	61	0.53	0.92
2-methyl-2-propanol							1.69
furfuryl alcohol							1.76
tetrahydrofurfuryl alcohol							2.33
1-pentanol							1.18
3-methyl-1-butanol							2.18
allyl alcohol							2.20
cyclohexanol	5	0.06					1.83
1-hexanol	7	0.05	20	0.15			1.94
4-methyl-2-amyl alcohol							1.56
benzyl alcohol		0.14	3	0.40			1.85
2-octanol							1.94
2-ethylhexanol	207	0.04	144	0.10	298	0.75	1.10
diisobutyl carbinol							3.12
Aliphatic hydrocarbons							
pentane	518	0.01	447	0.05			
cyclohexane	835	0.01	1100	0.10			0.43
hexane	147	0.02	176	0.06	176	0.29	0.34
heptane					52.4		
α-pinene							1.32
β-pinene							2.40

rubber solvent							0.43
VM&P naphtha							0.35
short-range mineral spirits							0.41
Rule-66 mineral spirits							0.35
regular mineral spirits							0.35
140 solvent							0.47
odorless mineral spirits							0.61
Alkyl halides							
methylene chloride	182	0.04	287	0.09	177	0.35	0.77
chloroform	108	0.03	161	0.09	229	0.41	1.01
carbon tetrachloride	459	0.02	324	0.05	143	0.15	0.79
perchloroethylene	320	0.03	351	0.05	109	0.22	0.64
1,1,2-trichloro-1,2,2-trifluoroethane (TTE)							5.61
trichloroethylene	277	0.03	145	0.05			0.99
1,1,2,2-tetrachloroethane							0.68
1,1,1-trichloroethane	166	0.05	325	0.07	292	0.50	1.65
1,1,2-trichloroethane							0.92
1,2-dichloroethane	3383	0.01	5350	0.05	6220	0.10	0.37
1,2-dibromoethane	134	0.08	264	0.08			1.25
ethyl chloride							0.72
ethyl bromide							2.79
propylene dichloride	47		31				
1,2,4-trichlorobenzene							2.75
o-dichlorobenzene	30	0.05	26	0.14			1.47
chlorobenzene	220	0.03	147	0.12			
p-chlorotoluene							2.20

Table 5. (Continued)

Solvent name	1970 Production, t × 10³	1970 Cost, $/kg	1979 Production, t × 10³	1979 Cost, $/kg	1991 Production, t × 10³	1991 Cost, $/kg	1995 Cost, $/kg
Amines							
n-propylamine							2.44
isopropylamine							1.98
butylamine (mixture)	8	0.14					
tert-butylamine							3.45
diethylamine	4	0.10			7.4	1.93	3.67
diethylenetriamine					33	2.35	3.08
cyclohexylamine							
triethylamine					9.7	1.89	
diisopropylamine	1	0.10					
toluidines (mixture)							2.09
dibutylamine					3.7	1.81	1.12
ethanolamine	39	0.05			122	0.87	2.31
morpholine					24	1.69	2.92
dimethylethanolamine							
diethanolamine	42	0.05			90	0.80	1.14
triethanolamine	38	0.06			86	0.95	1.17
Aromatic hydrocarbons							
benzene	3765	0.01	5579	0.08	5209	0.37	0.30
toluene	2730	0.01	3323	0.07	2857	0.28	0.40
ethylbenzene	2190	0.02	3832	0.07			0.40
xylene (mixture)	1761	0.01	3187	0.07	2866	0.26	0.40
o-xylene							0.53
m-xylene							0.99
p-xylene							0.66
aromatic 100 solvent							0.45
aromatic 150 solvent							0.51

Esters							
ethyl acetate	94	0.08	143	0.15	118	0.76	1.25
n-propyl acetate	10	0.05	23	0.13	36	1.02	1.43
isopropyl acetate					24	0.92	1.34
methyl acetoacetate							2.31
ethyl acetoacetate							2.31
n-butyl acetate	34	0.05	63	0.13	168	0.80	1.32
isobutyl acetate					26	0.73	1.28
amyl acetate							1.61
isobutyl isobutyrate	1	0.19	1	0.49			1.06
benzyl acetate							3.30
n-butyl propionate							1.65
pentyl propionate							1.74
ethylene glycol monomethyl ether acetate							1.14
ethylene glycol monoethyl ether acetate							1.67
ethylene glycol monobutyl ether acetate							1.89
diethylene glycol monoethyl ether acetate							2.15
diethylene glycol monobutyl ether acetate							2.00
propylene glycol monomethyl ether acetate							1.72
ethyl 3-ethoxy propionate							1.65
propylene carbonate							1.41
dibasic ester							1.10
Glycol ethers							
ethylene glycol monomethyl ether	44	0.05	46	0.13	32	0.67	1.96
ethylene glycol monoethyl ether	62	0.06	112	0.14	15	0.94	1.58
diethylene glycol monomethyl ether	6	0.05	9	0.15	156	0.67	1.74
ethylene glycol monobutyl ether	48	0.07	99	0.15	13	1.31	0.84
diethylene glycol monoethyl ether	16	0.06	15	0.13	109	1.15	1.91
diethylene glycol monobutyl ether	7	0.08	19	0.16			1.78

Table 5. (Continued)

Solvent name	1970 Production, t × 10³	1970 Cost, $/kg	1979 Production, t × 10³	1979 Cost, $/kg	1991 Production, t × 10³	1991 Cost, $/kg	1995 Cost, $/kg
Glycol ethers							
ethylene glycol monopropyl ether							1.64
propylene glycol monomethyl ether							1.50
propylene glycol monoethyl ether							1.65
propylene glycol monobutyl ether							1.28
propylene glycol mono-*tert*-butyl ether							1.17
dipropylene glycol monomethyl ether							1.52
Ketones							
acetone	733	0.02	782	0.07	1065	0.51	0.66
methyl ethyl ketone	218	0.04	298	0.10	233	0.75	0.88
mesityl oxide			10	0.13			
cyclohexanone	324	0.05	3967	0.17			1.61
methyl isobutyl ketone	90	0.05	86	0.13	82	1.05	1.14
diacetone alcohol		0.06		0.15	8.8	1.23	1.43
methyl amyl ketone							1.57
methyl isoamyl ketone							1.57
diisobutyl ketone							1.50
isophorone							1.94
Others							
ethylene glycol	1378	0.03	396	0.17			1.61
propylene glycol	194	0.09	277	0.12	302	0.92	1.23
diethylene glycol	155	0.03	178	0.10	221	0.52	0.57
dipropylene glycol							1.25
triethylene glycol					53	0.96	1.08
tripropylene glycol							1.67
hexylene glycol							1.43
acetonitrile					10		2.20

n-butyronitrile					4.70
N-methyl-2-pyrollidone					3.83
nitromethane					3.34
1-nitropropane					4.40
2-nitropropane					1.21
nitrobenzene	248		432	0.11	0.73
m-nitrotoluene					2.53
N,N-dimethylformamide (DMF)					1.36
ethyl ether					1.31
isopropyl ether					1.01
diphenyl oxide					2.99
propylene oxide	5.35	0.04	1020		1.32
tetrahydrofuran			54	0.36	2.93
1,4-dioxane					2.97
γ-butyrolactone					3.48
furfural					1.74
pine oil					1.76
cresylic acid (mixture of o-, m-, and p-cresol)	1	0.01	1	0.04	1.80
m-cresol					2.53

Table 6. Health and Safety Data for Industrial Solvents[a]

Solvent name	PEL, ppm	Flash point, °C	LEL, vol %	UEL, vol %	Evaporation rate[b]	Autogenous ignition temperature, °C	Vapor density[c]
Alcohols							
methanol	200	11	6.0	36.5	0.56	470	1.1
ethanol	1000	13	3.3	19.0	1.60	423	1.59
1-propanol	200	15	2.1	13.5	0.86	440	2.07
2-propanol	400	12	2.5	12	1.44	456	2.07
1-butanol	100	37	1.4	11.2	0.43	365	2.55
2-butanol		22	1.7	9.8	0.81	406	2.55
2-methyl-1-propanol	100	28	1.2	10.9	0.62	427	2.55
2-methyl-2-propanol	100	10	2.4	8.0	1.05	480	2.55
furfuryl alcohol	10	83 PMCC	1.8	16.3		391	3.37
tetrahydrofurfuryl alcohol		84	1.5	9.7	0.07	282	3.5
1-pentanol		33	1.2	10	0.20		3.04
3-methyl-1-butanol	100	43	1.2	9.0	0.19	350	3.04
allyl alcohol	2	21	2.5	18		378	2.00
cyclohexanol	50 (skin)	68			0.05	300	3.45
1-hexanol		63					3.52
4-methyl-2-amyl alcohol	25	41					3.52
2-ethylbutyl alcohol		57 COC					3.4
hexyl alcohol (mixture)							3.52
benzyl alcohol		101				436	3.72
2-octanol							4.49
2-ethylhexanol		81			0.02		4.49
Alcohols							
2-ethyl-4-methyl-1-pentanol							4.49
diisobutyl carbinol		65			0.03		4.97

Aliphatic hydrocarbons

pentane	600	<−40	1.5	7.8	12.4	309	2.48
cyclohexane	300	−17	1.3	8.4	4.4	245	2.90
hexane	50	−23	1.2	7.5	7.1	225	2.97
heptane	400	−4	1.05	8.7	4.8	223	3.45
n-octane		13				220	
α-pinene		33				255	4.7
β-pinene		31					
rubber solvent					7.6		
VM&P naphtha					1.5		
short-range mineral spirits					0.27		
Rule-66 mineral spirits					0.14		
regular mineral spirits					0.10		
140 solvent					0.05		
odorless mineral spirits					0.09		

Alkyl halides

methylene chloride	500	none	12	19	9.2	662	2.93
chloroform	2	none			7.6		4.12

Alkyl halides

carbon tetrachloride	2	none			4.5		5.30
perchloroethylene	25	none			1.5		5.83
1,1,2-trichloro-1,2,2-trifluoroethane (TTE)	1000				14.5	680	6.2
trichloroethylene	50	32	8	10.5	3.0	420	4.53
1,1,2,2-tetrachloroethane	1 (skin)	none			0.57		5.79
1,1,1-trichloroethane	350	none	7	16	4.6	537	4.60
1,1,2-trichloroethane		none					4.60
1,2-dichloroethane	1	13	6.2	15.9		413	3.35
1,2-dibromoethane	20	none			2.8		6.48
ethyl chloride	1000	−50	3.8	15.4		519	2.22
ethyl bromide	200	<−20	6.7	11.3		511	3.76
propylene dichloride	75	16	3.4	14.5		557	3.9
1,2,4-trichlorobenzene	5	110					6.26
o-dichlorobenzene	50	66	2.2	9.2		648	5.05
chlorobenzene	75	29	1.3	7.1		638	3.88
p-chlorotoluene		140 COC					4.37

Table 6. (*Continued*)

Solvent name	PEL, ppm	Flash point, °C	LEL, vol %	UEL, vol %	Evaporation rate[b]	Autogenous ignition temperature, °C	Vapor density[c]
Amines							
n-propylamine		−37	2.0	10.4		318	2.04
isopropylamine		−18			36.3	402	2.04
butylamine (mixture)		−12 OC				312	2.53
tert-butylamine	10	−18	1.8	10.1	16.6	312	2.53
diethylamine		−18				312	2.52
diethylenetriamine	1	102 OC				399	3.48
cyclohexylamine	10	21				293	3.42
triethylamine	10	8	1.2	8.0	5.6		3.48
diisopropylamine	5 (skin)	13			6.4		3.5
toluidines (mixture)	2 (skin)	85–89				482	3.9
dibutylamine		58 OC					4.46
ethanolamine	3	93 OC					2.11
morpholine	20 (skin)	38 OC				310	3.00
ethylaminoethanol		71 OC					3.06
dimethylethanolamine		33					3.03
diethanolamine	3	152 OC				662	3.65
diisopropanolamine		127 OC					
triethanolamine		179					5.14
Aromatic hydrocarbons							
benzene	1	−11	1.4	8.0	5.1	562	2.77
toluene	100	4	1.27	7	2.0	536	3.14
ethylbenzene	100	15	1.2	6.8	0.8	432	3.66
xylene (mixture)	100	26			0.73	516	3.66
o-xylene	100	25	1.1	7.0	0.58	530	3.66
m-xylene	100	17	1.0	6.0	0.71		3.66
p-xylene	100	25	1.1	7.0	0.76	530	3.66
aromatic 100 solvent		44	1	7	0.21	468	
aromatic 150 solvent		61	1	6	0.09	454	

			Esters				
methyl acetate	200	−10	8	16	5.9	502	2.55
ethyl acetate	400	−4	2.2	11	3.9	427	3.04
n-propyl acetate	200	14	2	8	2.1	450	3.52
isopropyl acetate	250	4	1.8	7.8	3.4	460	3.52
methyl acetoacetate		77				280	4.00
ethyl acetoacetate		85				295	4.48
n-butyl acetate	150	22	1.4	7.5	1.00	425	4.0
sec-butyl acetate	200	−8	1.3	7.5	1.8		4.0
isobutyl acetate	150	18	2.4	10.5	1.5	423	4.0
amyl acetate	100	25			0.38	379	4.5
isoamyl acetate	100	25	1	7.5		360	4.49
4-methyl-2-pentanol acetate	50	45			0.46		4.97
isobutyl isobutyrate		38	1	7.6	0.47	432	4.97
benzyl acetate		102				461	5.1
methyl propionate		−2	2.5	13	4.0	469	3.03
ethyl propionate		12	1.9	11	2.3	440	3.52
isopropyl propionate		13			2.0		4.0
n-propyl propionate		23			1.2		4.0
n-butyl propionate		38			0.43		4.49
sec-butyl propionate		27			0.85		4.49
pentyl propionate		57			0.18		4.97
ethylene glycol monomethyl ether acetate	25 (skin)	49	1.7	8.2	0.31	392	4.07
ethylene glycol monoethyl ether acetate	100 (skin)	52	1.7	19.4	0.18	379	4.72
ethylene glycol monobutyl ether acetate		74			0.03		5.52
ethylene glycol monopropyl ether acetate		61			0.09		5.03
diethylene glycol monomethyl ether acetate		82 OC					5.59
diethylene glycol monoethyl ether acetate		91			0.01		6.07
diethylene glycol monobutyl ether acetate		104			<0.01	299	7.04
propylene glycol monomethyl ether acetate		47	1.3	13.1	0.32	354	4.55
			Esters				
dipropylene glycol monomethyl ether acetate		86			0.01		6.56
ethyl 3-ethoxy propionate		58			0.12		5.03
propylene carbonate		122			<0.01		3.52
dibasic ester		100			0.01		

Table 6. (Continued)

Solvent name	PEL, ppm	Flash point, °C	LEL, vol %	UEL, vol %	Evaporation rate[b]	Autogenous ignition temperature, °C	Vapor density[c]
Glycol ethers							
ethylene glycol monomethyl ether	25 (skin)	39	2.5	14	0.53	285	2.62
ethylene glycol monoethyl ether	200 (skin)	94	1.8	14	0.39	235	3.11
diethylene glycol monomethyl ether		93			0.02		4.14
ethylene glycol dimethyl ether		4					
ethylene glycol monobutyl ether	25 (skin)	59			0.07	243	4.07
diethylene glycol monoethyl ether		89			0.02	216	4.62
diethylene glycol monobutyl ether		97			<0.01	228	5.58
ethylene glycol monophenyl ether		121			<0.01		4.77
ethylene glycol monopropyl ether		50			0.18	235	3.59
propylene glycol monomethyl ether		32			0.78		3.11
propylene glycol monoethyl ether		42			0.49		3.59
propylene glycol monobutyl ether		59	1.1	14.4	0.09		4.56
propylene glycol mono-*tert*-butyl ether		45			0.26		4.56
dipropylene glycol monomethyl ether		75			0.02		5.11
propylene glycol monophenyl ether					<0.01		5.24
Ketones							
acetone	750	−18	2.6	12.8	5.7	465	2.00
methyl ethyl ketone	200	−6	1.8	11.5	3.9	516	2.42
methyl *n*-propyl ketone	200	7	1.56	8.7	2.3	449	2.97
mesityl oxide	15	31			0.87	344	3.38
cyclohexanone	25	44			0.30	420	3.4
methyl *n*-butyl ketone		30	1.3	8	1.1	424	3.45
methyl isobutyl ketone	50	17	1.4	7.5	1.6	459	3.45
diacetone alcohol	50	58	1.8	6.9	0.12	603	4.00
methyl amyl ketone	100	39	1.11	7.9	0.34	533	3.94
methyl isoamyl ketone	50	37	1.05	8.2	0.46	425	3.94
diisobutyl ketone	25	60	0.8	6.2	0.19	396	4.9
isophorone	4	84	0.8	3.8	0.02	462	4.77

Compound		*Others*					
ethylene glycol	ceiling[b] 50 (vapor)	111	3.2	12.6	<0.01	400	2.14
propylene glycol		99	2.6		<0.01	371	2.62
diethylene glycol		149			<0.01	229	3.66
dipropylene glycol		138			<0.01		4.63
triethylene glycol		152	0.9	9.2	<0.01	371	5.17
tripropylene glycol		142			<0.01		6.63
hexylene glycol		100			<0.01		4.08
acetonitrile	40	6 COC	4.4	16		524	1.42
n-butyronitrile		26 OC					2.38
N-methyl-2-pyrrollidone		96			0.03		3.42
nitromethane	100	35	7.3		1.3	287	2.11
1-nitropropane	25	34	2.2			418	3.06
2-nitropropane	10	28	2.6		1.1	421	3.06
nitrobenzene	1 (skin)	88	1.8			428	4.25
m-nitrotoluene	2 (skin)	112				482	4.72
N,N-dimethylformamide (DMF)	10 (skin)	58	2.2	15.2	0.21	445	2.51
ethyl ether	400	−45	1.85	36	11.1	160	2.56
isopropyl ether	500	−28	1.4	7.9	8.2	443	3.52
n-butyl ether		25	1.5	7.6		194	4.48
diphenyl oxide	1	115	0.8	1.5		620	5.86
propylene oxide	20	−37 TOC	2.8	37			2.0
tetrahydrofuran	200	−17	1.8	11.8	4.8	321	2.5
1,4-dioxane	25 (skin)	12	2.0	22.2	6.7	180	3.03
γ-butyrolactone		98 OC					3.0
furfural	2 (skin)	60	2.1	19.3		316	3.31
pine oil		104					
turpentine	100	35	0.8		<0.01	253	4.84
cresylic acid (mixture of o-, m-, and p-cresol)	5 (skin)	81			<0.01		3.72
m-cresol	5 (skin)	94	1.1		<0.01	559	3.72

[a] Closed cup unless otherwise specified. PMCC = Pensky-Martens closed cup; COC = Cleveland open cup; OC = open cup; and TOC = Tag open cup; PEL = personal exposure limit; LEL = lower exposure limit; UEL = upper exposure limit.
[b] Ceiling limit is the concentration that should not be exceeded during any part of the working exposure; relative vs n-butyl acetate = 1.
[c] Relative to air = 1.

replace the aromatic hydrocarbons. In this situation, more strong solvent is required compared to the ketone/aromatic formulation and costs increase. The combination of reduced VOC emissions and composition constraints in the form of HAP restrictions have complicated the formulator's task.

Solvent Uses

Although regulatory pressures are gradually reducing the amount of solvents used overall, solvents will continue to be used in significant amounts. A gradual shift away from HAP solvents, combined with a general decrease in solvent usage resulting from VOC emission reduction requirements is expected. The degree to which each occurs depends on the solvents and the specifics of the solvent application.

Coatings. Paints (qv) and coatings (qv) comprise the largest single category of solvent consumption, accounting for nearly half the solvent used. Lower solvent emission technologies such as high solids solvent-based coatings, water-based coatings, and nonsolvent technologies such as powder coatings and uv/electron beam coatings will continue to increase in usage at the expense of the traditional low solids, higher solvent content coating. No single lower solvent technology will dominate, but each will find a place. In situations where emission controls are cost-effective, coating formulations will be optimized for performance rather than reduced emissions.

Solvent-Based. Solvents serve multiple functions in solvent-based paints and coatings, including solubilization, wetting, viscosity reduction, adhesion promotion, and gloss enhancement. Solvent dissolves the resin or polymer forming the continuous coating phase. Dyes and pigments, which provide hiding and coloration, are then dispersed in the solution. Solvent promotes pigment wetting and helps the resin to coat the pigment particles, stabilizing the dispersion. Prior to application, it is common practice to add solvent thinner to attain the desired viscosity for the particular application technique being used. Solvent begins to evaporate as soon as coatings material atomizes; evaporation is completed after the coating impacts the surface. As the solvent evaporates, film formation occurs and a continuous, compacted film develops. Solvent can also increase adhesion by softening the primer coat. In many coatings applications, the solvent system includes a slow-evaporating active solvent to promote gloss and smoothness. Single solvents are rarely used in coatings formulations, most are blends of several solvents, each serving a particular purpose.

Water-Based. Even though the dominant volatile is water, most water-based coatings still contain some solvent. Latex water-based coatings dominate the architectural market, interior as well as exterior, flat, semigloss, and gloss coatings. More water-based systems are being used in industrial maintenance coatings which are based mostly on water-soluble or water-dispersible resin systems.

Latex and water-soluble polymers dry by different mechanisms and use different solvents. Latexes are composed of dispersed polymer particles in water (see LATEX TECHNOLOGY); film formation occurs when the particles deform to establish a continuous film. The particles must be soft enough to flow and adhere to each other at application temperature. Two different kinds of solvents are used in latex coatings. Coalescents, also known as filming aids, act as temporary plasticizers during the film formation process, promoting deformation and flow, then evaporating after film formation has occurred. The most common coalescents are slow-evaporating glycol ethers and glycol ether esters. Glycols such as ethylene glycol or propylene glycol are commonly added for storage stability and resistance to freezing. Propylene glycol is used in much larger amounts in glossy trim enamels, where the glycol promotes film formation, gloss development, and flowout of brush marks following application.

Water-dispersible resins contain carboxylic groups which are neutralized using base or amine compounds. This solubilizes the resin in solution and also promotes pigment wetting. Film formation occurs by the evaporation of volatiles followed by cross-linking through ambient cure oxidative reactions or elevated temperature reactions. Solvents, most commonly glycol ethers, are used to promote film formation and improve film quality.

Relative humidity can have a significant impact on drying behavior and film quality. Water-based formulations that perform well when applied under dry conditions may be deficient under high humidity application conditions. The rate of water evaporation is much slower at high humidity, but solvent evaporation continues. This results in solvent depletion during the critical phases of film formation and consequent poor film development.

Cleaning. *Dry Cleaning.* Perchloroethylene remains the dominant dry-cleaning solvent, accounting for more than 80% of solvent usage. This solvent exhibits good solvency, easy recovery, and a lack of flammability that makes it very suitable for dry cleaning. Emissions have decreased significantly with the adoption of improved recovery techniques, modified operating procedures designed to reduce emissions, and improved equipment. These changes have been implemented as a result of a consent decree. Even though there are flammability concerns, hydrocarbon solvents are used in some dry-cleaning plants. In these plants, there is a shift underway to a 140-type solvent with a higher flash point than Stoddard solvent.

Cold Cleaning. Cold solvent cleaning is used to degrease metal parts and other objects in many operations, including automotive repair facilities. Mineral spirits have been popular in cold cleaning, but are being supplanted by higher flash point hydrocarbon solvents on account of emissions and flammability concerns. Hydrocarbon solvents used in cold cleaning should have a narrow distillation range, which results in the highest flash point commensurate with fast drying (see METAL TREATMENTS).

Vapor Degreasing. In vapor degreasing, the solvent is vaporized and the cold part is suspended in the vapor stream. The solvent condenses on the part, and the liquid dissolves and flushes dirt, grease, and other contaminants off the surface. The part remains in the vapor until it is heated to the vapor temperature. Drying is almost immediate when the part is removed and solvent residues are not a problem. The most common solvent used in vapor degreasing operations has been 1,1,1-trichloroethane. As production and use of 1,1,1-trichloroethane is phased out, vapor degreasing with that solvent can no longer be used. This has stimulated interest in returning to trichloroethylene, which can be an excellent solvent choice. Hydrocarbon solvents can also be used, but flammability concerns inhibit adoption of vapor-phase hydrocarbon solvent degreasing. Some operations adopt semiaqueous cleaning techniques as the cleaning procedure.

Printing Inks. Printing ink preparation is similar to many coating systems. The resin is dissolved in the solvent, followed by pigment dispersion to produce the ink. In most printing operations, the solvent must evaporate fast for best production speed. Alcohol–hydrocarbon solvent combinations are used with polyamide resins for some printing processes (see INKS).

Agricultural Products. Pesticides are frequently applied as emulsifiable concentrates. The active insecticide or herbicide is dissolved in a hydrocarbon solvent which also contains an emulsifier. Hydrocarbon solvent selection is critical for this application. It can seriously impact the efficacy of the formulation. The solvent should have adequate solvency for the pesticide, promote good dispersion when diluted with water, and have a flash point high enough to minimize flammability hazards. When used in herbicide formulas, low solvent phytotoxicity is important to avoid crop damage. Hydrocarbon solvents used in post-harvest application require special testing to ensure that polycyclic aromatics are absent.

Reaction and Heat-Transfer Solvents. Many industrial production processes use solvents as reaction media. Ethylene and propylene are polymerized in hydrocarbon solvents, which dissolves the gaseous reactant and also removes the heat of reaction. Because the polymer is not soluble in the hydrocarbon solvent, polymer recovery is a simple physical operation. Ethylene oxide production is exothermic and the catalyst-filled reaction tubes are surrounded by hydrocarbon heat-transfer fluid.

Process Raw Material. Industrial solvents are raw materials in some production processes. For example, only a small proportion of acetone is used as a solvent, most is used in producing methyl methacrylate and bisphenol A. Alcohols are used in the manufacture of esters and glycol ethers. Diethylenetriamine is also used in the manufacture of curing agents for epoxy resins. Traditionally, chlorinated hydrocarbon solvents have been the starting materials for fluorinated hydrocarbon production.

Solvent Extraction. Extraction processes, used for separating one substance from another, are commonly employed in the pharmaceutical and food processing industries. Oilseed extraction is the most widely used extraction process on the basis of tons processed. Extraction-grade hexane is the solvent used to extract soybeans, cottonseed, corn, peanuts, and other oilseeds to produce edible oils and meal used for animal feed supplements. Tight specifications require a narrow distillation range to minimize solvent losses as well as an extremely low benzene content. The specification also has a composition requirement, which is very unusual for a hydrocarbon, where the different components of the solvent must be present within certain ranges (see EXTRACTION).

BIBLIOGRAPHY

"Solvents, Industrial" in *ECT* 1st ed., Vol. 12, pp. 654–686, by A. K. Doolittle, Carbide and Carbon Chemicals Co., A Division of Union Carbide and Carbon Corp.; in *ECT* 2nd ed.,

Vol. 18, pp. 564–588, by J. W. Wyart, Celanese Chemical Co., and M. F. Dante, Shell Chemical Co.; in *ECT* 3rd ed., Vol. 21, pp. 377–401, by C. F. Parrish, Indiana State University.

1. *Fed. Reg.* **59**, 19465 (Apr. 22, 1994).
2. *California Regulat. Not. Reg.* **93**(1–Z), 18–26 (Jan. 1, 1993).
3. A. F. M. Barton, *CRC Handbook of Solubility Parameters and Other Cohesion Parameters*, CRC Press, Inc., Boca Raton, Fla., 1983.
4. K. L. Hoy, *J. Paint Technol.* **42**, 76 (1970).
5. C. M. Hansen, *Ind. Eng. Chem. Prod. Res. Dev.* **8**, 2 (1969).
6. E. P. Lieberman, *Offic. Dig.* **34**, 30 (1962).
7. J. L. Gardon, *J. Paint Technol.* **38**, 43 (1966).
8. R. C. Nelson, R. W. Hemwall, and G. D. Edwards, *J. Paint Technol.* **42**, 636 (1970).
9. E. W. Flick, ed., *Industrial Solvents Handbook*, 3rd ed., Noyes Data Corp., Park Ridge, N.J., 1985.
10. J. Gmehling and U. Onken, *Chemistry Data Series, Vapor–Liquid Equilibrium Data Collection*, Dechema, Frankfurt, Germany, 1977ff.
11. S. Ohe, *Computer Aided Data Book of Vapor Pressure*, Data Book Publishing Co., Tokyo, Japan, 1976.
12. *Synthetic Organic Chemicals, United States Production and Sales, 1991*, U.S. International Trade Commission, Washington, D.C., 1993.
13. *Chem. Mark. Rep.* (May 8, 1995).
14. R. J. Lewis, Sr., *Sax's Dangerous Properties of Industrial Materials*, 8th ed., Van Nostrand Reinhold Co., New York, 1992.

DON A. SULLIVAN
Shell Chemical Company

SOMATOTROPINS. See GENETIC ENGINEERING, ANIMALS; GROWTH REGULATORS; HORMONES, HUMAN GROWTH HORMONE.

SONOCHEMISTRY. See SUPPLEMENT.

SORBIC ACID

Sorbic acid [*110-44-1*] is a white crystalline solid first isolated in 1859 by hydrolysis of the oil distilled from unripened mountain-ash berries (1). The name is derived from the scientific term for the rowan tree, *Sorbus aucuparia* Linne, which is the parent plant of the mountain ash. Sorbic acid was first synthesized in 1900 (2). Interest in this compound was minimal until independent researchers, E. Mueller of Germany and C. M. Gooding of the United States, discovered its antimicrobial effect in 1939 and 1940, respectively. Early interest in manufacturing sorbic acid centered around its use as a tung oil replacement when tung oil supplies were curtailed in the United States during World War II. High manufacturing costs prohibited expanded use until its approval as a food preservative in 1953. Sorbic acid is widely used in foods having a pH of 6.5 or below, where control of bacteria, molds, and yeasts is essential for obtaining safe and economical storage life.

Physical Properties

The sorbic acid crystal has a well-ordered morphology as a result of its hydrogen bonding and trans,trans structure (1).

(1)

Physical properties are given in Table 1, along with those of the commercially most used salt, E,E-potassium sorbate [24634-61-5]. Table 2 shows the solubility

Table 1. Physical Properties of Sorbic Acid

Properties	Sorbic acid
mol wt	112.13
melting point, °C[a]	134.5
boiling point at 101.3 kPa	228
(=1 atm), °C	
density, g/cm^3, at 19°C[b]	1.204
flash point, °C	126–130
dissociation constant at 25°C,	1.73×10^{-5}
mol/L	
pK$_a$^{25}	
H$_2$O	4.76
50 wt % ethanol	4.62
pK$_a$, 0.1 M NaCl	4.51
dissociation constant of	1.96×10^{-4}
dimer, K^{24} (CCl$_4$), mol/L	
specific heat, J/(g·K)[c]	1.84
latent heat of fusion, kJ/mol[c]	13.6
heat of combustion, kJ/mol[c]	3107
heat of neutralization, kJ/mol[c]	6.07
vapor pressure, kPa[d]	
130°C	1.3
150°C	3.7
170°C	9.3

[a] Potassium sorbate, mol wt 150.22, decomposes at 270°C.
[b] Density for potassium sorbate at 20°C = 1.363 g/cm^3.
[c] To convert J to cal, divide by 4.184.
[d] To convert kPa to mm Hg, multiply by 7.5.

Table 2. Solubility of Sorbic Acid and Potassium Sorbate

Solvent	Temperature, °C	Solubility, g/100 g solvent		Reference
		Sorbic acid	Potassium sorbate	
water	0	0.14		3,4
	20	0.15	58.2	
	40	0.34		
	60	0.72		
	80	1.6		
	100	3.9		
pH[a]				
4.25	20	0.33		5
6.25	20	3.1		
7.25	20	12.0		
acetic acid (glacial)	23	11.5		4
acetone	20	9.2	0.1	4
butyl alcohol	25	11.3		3
carbon tetrachloride	20	1.3	<0.01	
cyclohexane	20	0.28		
ethyl alcohol				
anhydride	20	12.9	2.0	3,4
60 wt %	20	6.4		4
ethyl ether	20	5.0	0.1	
glycerol	20	0.31	0.2	
isopropyl alcohol	20	12.9		3
methyl alcohol				
anhydride	20	12.9	16	4
50 wt %	20	1.6		4,5
corn oil	20	0.7	0.01	5
soybean oil	20	0.52		5
propylene glycol	20	5.5	20	4
sodium chloride, 15 wt %	20	0.038	15	3,4,5

[a]Controlled pH.

in various solvents. More extensive data on solubility are given in References 3 and 5. Sorbic acid dust, as well as any organic dust, can accumulate a static charge and become an explosion hazard, particularly when mixed with highly flammable solvents or oxidizing agents. Minimum explosive limits are 0.02 g/L of air for sorbic acid (6).

Chemical Properties

The chemical reactivity of sorbic acid is determined by the conjugated double bonds and the carboxyl group.

Conjugated Double Bonds. Sorbic acid is brominated faster than other olefinic acids (7). Reaction with hydrogen chloride gives predominately 5-chloro-3-hexenoic acid (8).

Reactions with amines at high temperatures under pressure lead to mixtures of dehydro-2-piperidinones (9):

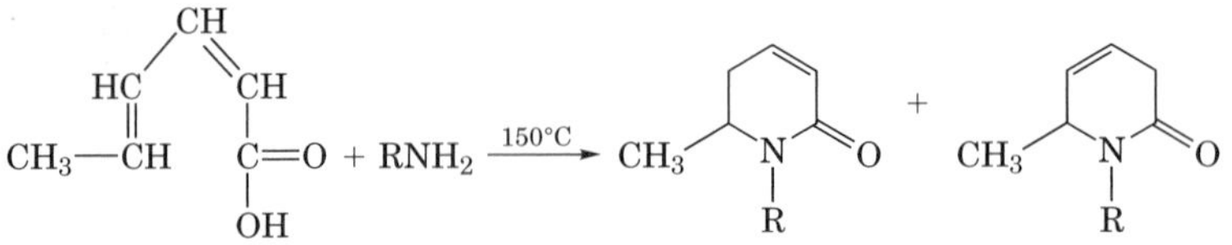

A yellow crystalline complex (**2**) melting at 198°C is formed from sorbic acid and iron tricarbonyl (10):

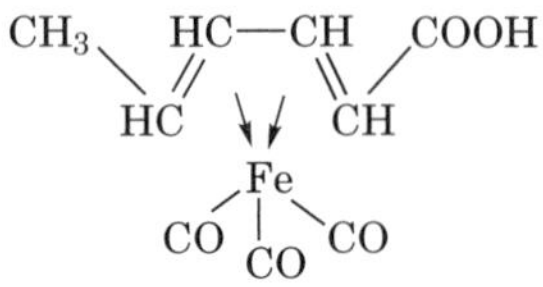

(**2**)

Similar coordination occurs also in the presence of other di- and trivalent metals. Reduction of the double bonds can produce various hexenoic acid mixtures.

Sorbic acid is oxidized rapidly in the presence of molecular oxygen or peroxide compounds. The decomposition products indicate that the double bond farthest from the carboxyl group is oxidized (11). More complete oxidation leads to acetaldehyde, acetic acid, fumaraldehyde, fumaric acid, and polymeric products. Sorbic acid undergoes Diels-Alder reactions with many dienophiles and undergoes self-dimerization, which leads to eight possible isomeric Diels-Alder structures (12).

Polymerization catalyzed by free radicals occurs with sorbic acid. The polymers (**3**) formed have high molecular weights with linear structures; the trans form of the residual double bond is preserved (13).

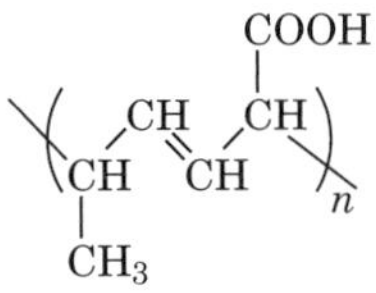

(**3**)

Copolymers with acrylonitrile, butadiene, isoprene, acrylates, piperylene, styrene, and polyethylene have been studied. The high cost of sorbic acid as a monomer has prevented large-scale uses. The ability of sorbic acid to polymerize, particularly on metallic surfaces, has been used to explain its corrosion inhibition for steel, iron, and nickel (14).

Carboxylic Acid Group.   Sorbic acid undergoes the normal acid reactions forming salts, esters, amides, and acid chlorides. Industrially, the most important compound is the potassium salt because of stability and high water solubility. Sodium sorbate [7757-81-5] (*E,E* form [42788-83-0]) is less stable and not commercially available. The calcium salt [7492-55-9], which has limited solubility, has use in packaging (qv) materials.

Sorbic acid anhydride [13390-06-2] can be prepared by heating the polyester of 3-hydroxy-4-hexenoic acid with sorboyl chloride [2614-88-2] or by reaction of sorbic acid with oxalyl chloride (15,16). Preparation of the esters of sorbic acid must be controlled to prevent oxidation and polymerization. The lower sorbic acid esters have a pleasant odor.

Synthesis and Manufacture

The first synthesis of sorbic acid was from crotonaldehyde [4170-30-3] and malonic acid [141-82-2] in pyridine in 32% yield (2,17,18). The yield can be improved with the use of malonic acid salts (19). One of the first commercial methods involved the reaction of ketene and crotonaldehyde in the presence of boron trifluoride in ether at 0°C (20,21). A β-lactone (4) forms and then reacts with acid, giving a 70% yield.

(4)   (5)

Most commercial sorbic acid is produced by a modification of this route. Catalysts composed of metals (zinc, cadmium, nickel, copper, manganese, and cobalt), metal oxides, or carboxylate salts of bivalent transition metals (zinc isovalerate) produce a condensation adduct with ketene and crotonaldehyde (22–24), which has been identified as (5).

An excess of crotonaldehyde or aliphatic, alicyclic, and aromatic hydrocarbons and their derivatives is used as a solvent to produce compounds of molecular weights of 1000–5000 (25–28). After removal of unreacted components and solvent, the adduct referred to as polyester is decomposed in acidic media or by pyrolysis (29–36). Proper operation of acidic decomposition can give high yields of pure *trans,trans*-2,4-hexadienoic acid, whereas the pyrolysis gives a mixture of isomers that must be converted to the pure trans,trans form. The thermal decomposition is carried out in the presence of alkali or amine catalysts. A simultaneous codistillation of the sorbic acid as it forms and the component used as the solvent can simplify the process scheme. The catalyst remains in the reaction batch. Suitable solvents and entraining agents include most inert liquids that boil at 200–300°C, eg, aliphatic hydrocarbons. When the polyester is split thermally at 170–180°C and the sorbic acid is distilled directly with the solvent, production and purification can be combined in a single step. The solvent can be reused after removal of the sorbic acid (34). The isomeric mixture can be converted to the thermodynamically more stable trans,trans form in the presence of iodine, alkali, or sulfuric or hydrochloric acid (37,38).

Food-grade specifications require further purification in the form of carbon treatments and recrystallization from aqueous or other solvent systems. The

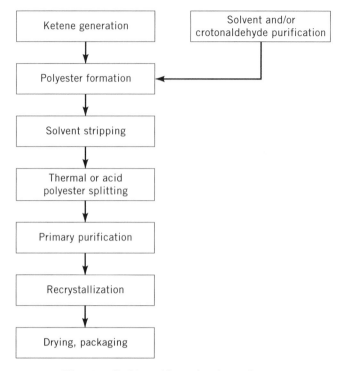

Fig. 1. Sorbic acid production scheme.

illustrated flow scheme for sorbic acid production in Figure 1 has been greatly simplified.

The ketene–crotonaldehyde route through polyester with various modifications and improvements is reportedly practiced by Hoechst Celanese, Cheminova, Daicel, Ueno, Chisso, Nippon Gohsei, and Eastman Chemical Company. Differences in their processes consist mostly in the methods of polyester splitting and first-stage purification. Production of the potassium salt can be from finished sorbic acid or from a stream in the sorbic acid production route before the final drying step. Several patents on the process for producing sorbic acid and potassium sorbate from this route are given in the literature.

Union Carbide abandoned the ketene–crotonaldehyde route in 1953 in favor of the oxidation of 2,4-hexadienal made by acetaldehyde condensation. A silver compound used as the catalyst prevented peroxidation of the ethylenic bonds (39,40). Their plant operated until 1970.

Preparing sorbic acid by reaction of crotonaldehyde and acetone followed by oxidation of the crotonylidenacetone is of interest in the former Soviet Union (41,42):

$$CH_3CH{=}CHCHO + CH_3\overset{\overset{\displaystyle O}{\|}}{C}CH_3 \longrightarrow CH_3CH{=}CHCH{=}CH\overset{\overset{\displaystyle O}{\|}}{C}CH_3 \xrightarrow{\text{NaOCl}}$$

$$CH_3CH{=}CHCH{=}CH\overset{\overset{\displaystyle O}{\|}}{C}CCl_3 \xrightarrow{\text{NaOH}} CH_3CH{=}CHCH{=}CH\overset{\overset{\displaystyle O}{\|}}{C}ONa \xrightarrow{\text{HCl}} \textbf{(1)}$$

Other methods include ring opening of parasorbic acid [*108-54-3*] (δ-lactone of 5-hydroxy-2-hexenoic acid) in hydrochloric acid or in alkaline solutions (43,44), the ring opening of γ-vinyl-γ-butyrolactone in various catalysts (45,46), or isomerization of 2,5-hexadienoic acid esters (47,48). Other methods are described in the literature (6,49,50).

Economic Aspects

Sorbic acid is produced and marketed in the United States in the dust-free powder form. The 1995 truckload (TL) list price was U.S. $8.60–$8.80/kg for food grade and $12.41/kg for National Formulary (NF) grade. Water-soluble potassium sorbate is marketed as a powder or as granules. The 1995 TL list price was U.S. $7.90–$8.15/kg for food grade and $11.75/kg for NF grade. In addition to Eastman, the only U.S. producer, there are four Japanese producers and two producers in the European Union. Worldwide consumption in 1994 was approximately 2700–3000 metric tons and the nameplate productive capacity was approximately 2900–3100 metric tons.

Purification Specifications

Sorbic acid and its salts are highly refined to obtain the necessary purity for use in foods. The quality requirements are defined by the *Food Chemicals Codex* (Table 3). Codistillation or recrystallization from water, alcoholic solutions, or acetone is used to obtain sorbic acid and potassium sorbate of a purity that passes not only the *Codex* requirements but is sufficient for long-term storage. Measurement of the peroxide content and heat stability can further determine the presence of low amounts of impurities. The presence of isomers, other than the trans,trans form, causes instability and affects the melting point.

Analytical Techniques. Sorbic acid and potassium sorbate are assayed titrimetrically (51). The quantitative analysis of sorbic acid in food or beverages, which may require solvent extraction or steam distillation (52,53), employs various techniques. The two classical methods are both spectrophotometric (54–56).

Table 3. Specifications for Sorbic Acid and Potassium Sorbate[a]

Specification[b]	Sorbic acid	Potassium sorbate
assay (dry basis), wt %	99–101	98–101
arsenic, ppm	3	3
heavy metals (as lead), ppm	10	10
melting range, °C	132–135	
residue on ignition, wt %	0.2	3
water, wt %	0.5	
acidity, as sorbic acid		passes
alkalinity, as K_2CO_3		passes
loss on drying, wt %		1
lead, mg/kg		5

[a]Ref. 51.
[b]Values are maximum unless noted otherwise.

In the ultraviolet method, the prepared sample is acidified and the sorbic acid is measured at ~250–260 nm. In the colorimetric method, the sorbic acid in the prepared sample is oxidized and then reacts with thiobarbituric acid; the complex is measured at ~530 nm. Chromatographic techniques are also used for the analysis of sorbic acid. High pressure liquid chromatography with ultraviolet detection is used to separate and quantify sorbic acid from other ultraviolet-absorbing species (57–59). Sorbic acid in food extracts is determined by gas chromatography with flame ionization detection (60–62).

Uses

Sorbic acid and its potassium salt, collectively called sorbates, are used primarily in a wide range of food and feed products (63) and to a lesser extent in certain cosmetics (64), pharmaceuticals, and tobacco products. There are limited applications of the calcium and sodium salts, but the acid and its potassium salt are used almost exclusively.

Since the first demonstration of antimicrobial activity in the 1940s, sorbates have been shown to inhibit a wide spectrum of yeasts, molds, and bacteria (Table 4) (65), including most foodborne pathogens (66–70). As bacterial inhibitors, sorbates are least effective against lactic acid bacteria. Although this can be a problem for foods that suffer from lactic spoilage, it has proven to be a positive point in cases of yeast and mold suppression during lactic fermentations (71–74). The effectiveness of sorbates can be influenced by a number of factors, including pH, microbial load, water activity, temperature, and atmosphere (75). The antimicrobial activity of sorbates has been reviewed (75,76).

The inhibitory activity of sorbates is attributed to the undissociated acid molecule. The activity, therefore, depends on the pH of the substrate. The upper limit for activity is approximately pH 6.5 in moist applications; the degree of activity increases as the pH decreases. The upper pH limit can be increased in low water activity systems. The following indicates the effect of pH on the dissociation of sorbic acid, ie, percentage of undissociated sorbic acid at various pH levels (76,77).

pH	Wt %
3.0	98
3.5	95
4.0	86
4.5	65
4.76 (pK_a)	50
5.0	37
5.5	15
6.0	6
6.5	2
7.0	<1

The activity of the sorbates at a higher pH is one distinct advantage over the two other most commonly used food preservatives, benzoic and propionic acids,

Table 4. Genus Names of Molds, Yeasts, and Bacteria Inhibited by Sorbates[a]

Molds

Alternaria citri	M. sp.
A. tenuis	Papularia arundinis
A. spp.	Penicillium atromentosum
A. cucumis	P. chermesinum
A. sp.	P. chrysogenum
Aspergillus clavatus	P. citrinum
A. elegans	P. digitatum
A. flavus	P. duclauxi
A. fumigatus	P. expansum
A. glaucus	P. frequentans
A. niger	P. funiculosum
A. ocraceus	P. gladioli
A. parasiticus	P. herquei
A. sydowi	P. implicatum
A. terreus	P. italicum
A. unguis	P. janthinellum
A. versicolor	P. notatum
Botrytis cinerea	P. oxalicum
Cephalosporium sp.	P. patulum
Cercospora sp.	P. piscarium
Chaetomium globosum	P. purpurogenum
Cladosporium cladosporiodes	P. restrictum
Colletotrichum lagenarium	P. roquefortii
Cunninghamella echinulata	P. rugulosum
Curvularia trifolii	P. sublateritium
Fusarium episphaeria	P. thomii
F. moniliforme	P. urticae
F. oxysporum	P. variabile
F. roseum	P. spp.[b]
F. rubrum	Pestolotiopsis macrotricha sp.
F. solani	Phoma sp.
F. tricinctum	Pullularia pullulans
Geotrichum candidum	Rhizoctonia solani
G. sp.[b]	Rhizopus arrhizus
Gliocladium roseum	R. nigricans
Helminthosporium sp.[b]	Rosellinia sp.
Heterosporium terrestre	Sporotrichum pruinosum
Humicola fusco-atra.	Stagonospora sp.
Mucor silvaticus	Stysanus sp.
M. spp.[c]	Thielavia basicola
Myrothecium roridum	Trichoderma viride
M. verrucaria	Truncatella sp.

Yeasts

Brettanomyces clausenii	Rhodotorula flava
B. versatilis	R. glutinis
Candida albicans	R. rubra
C. krusei	R. spp.
C. tropicalis	Saccharomyces cerevisiae
C. mycoderma	S. cerevisiae var. ellipsoideus

579

Table 4. (*Continued*)

Cryptococcus terreus	*S. carlsbergensis*
C. neoformans	*S. fragilis*
C. sp.	*S. rouxii*
Debaryomyces membranaefaciens	*S. delbrueckii*
D. membranaefaciens var. *hollandicus*	*S. lactis*
D. spp.	*Schizosaccharomyces octosporus*
Endomycopsis ohmeri	*Sporobolomyces* sp.
Hansenula anomala	*Torulaspora rosei*
H. saturnus	*Torulopsis candida*
H. subpelliculosa	*T. caroliniana*
Oospora sp.	*T. minor*
Pichia alcoholophila	*T. polcherrima*
P. membranaefaciens	*T. versitalis lipofera*
P. polymorpha	*Zygosaccharomyces globiformis*
P. silvestris	*Z. halomembranis*
P. sp.	

Bacteria

Acetobacter aceti	*Micrococcus* sp.
A. xylinum	*Propionibacterium zeae*
Achromobacter sp.	*P. freundenreichii*
Alcaligenes faecalis	*Proteus vulgaris*
Azotobacter agilis	*Pseudomonas fragi*
Bacillus coagulans	*P. fluorescens*
B. cereus	*P.* sp.
B. polymyxa	*Salmonella heidelberg*
B. stearothermophilus	*S. montevideo*
B. subtilis	*S. typhimurium*
Clostridium perfringens	*S. enteritidis*
C. sporogenes	*Sarcina lutea*
Enterobacter aerogenes	*Serratia marcescens*
Escherichia coli	*Staphylococcus aureus*
E. freundii	*Vibrio parahaemolyticus*
Lactobacillus brevis	

[a]Ref. 65. [b]Two strains tested. [c]Five strains tested.

because the upper pH limits for activity of these compounds are approximately pH 4.5 and 5.5, respectively. Although the effect of sorbates can be microbiocidal under certain conditions, activity is most often manifested as a microbial growth retardant.

The exact mechanism of inhibition by sorbic acid has not been thoroughly elucidated, even though it has been the subject of extensive research and numerous hypotheses. A number of enzyme systems in fungi and bacteria have been designated as sites of sorbate inhibition (78–84). Sorbic acid has been shown to inhibit the transport of carbohydrates into yeast cells, inhibit oxidative and fermentative assimilation, and uncouple oxidative phosphorylation in a variety of bacteria in studies conducted for various systems, including whole cells, cell-free extracts, and isolated enzyme systems (83,85–87). Although all events may occur under specific conditions, no single proposed mechanism seems to account wholly for the inhibitory activity of sorbates. More likely, microbial inhibition by sorbates is the result of a combination of events that may differ from one organism to another and from one set of conditions to another.

As of this writing (1996), there has been no evidence to indicate that microorganisms can develop resistance to sorbates, as occurs with antibiotics and certain other antimicrobial chemicals. However, there is variation in the sorbate sensitivity of microorganisms from one genus to another, between different species in the same genus, and even between different strains of the same species. *Saccharomyces bailii* is resistant to sorbates, benzoates, and other short-chain monocarboxylic acids because of an inducible enzyme system that transports these compounds out of the cell (88). Some molds, when present in extremely high numbers, can metabolize sorbates. This has been attributed to typical β-oxidation as occurs with other fatty acids (78). Species of *Penicillium*, particularly *P. roquefortii*, can decarboxylate sorbates to 1,3-pentadiene, resulting in a hydrocarbon odor (89,90). *Desulfoarculus baarsii* was found to oxidize sorbic acid completely to carbon dioxide, whereas at higher concentrations the bacteria produced pentanone-2 and isopentanone-2 (91).

Food applications of sorbates expanded rapidly after issuance of the original patents in 1945 (92). The first uses were based on their excellent fungistatic properties and thus involved foods with low pH and/or low water activity in which yeasts and molds are the primary spoilage agents. More recent application research has been directed toward utilizing the bacteriostatic properties of sorbates.

Sorbates are classified as generally recognized as safe (GRAS) in the United States, with no upper limit set for foods that are not covered by Standards of Identity. They are also allowed in more than 70 food products having Standards of Identity. Examples of products that often contain sorbates are natural and processed cheeses, other cheese products, salad dressings, bakery products, prepared salads, fermented vegetable products, dried fruits, fruit juices, margarine, wine, fish products, jams, and jellies (65). Use levels in food products are 0.01–0.5 wt % (Table 5). Compared with other antimicrobial preservatives, sorbates can be used in higher concentrations without affecting the flavor of foods.

Table 5. Sorbate Concentration in Food Products[a], Wt %

Product	Typical concentration, %
cheese and cheese products[b]	0.2–0.3
fruit drinks	0.025–0.075
beverage syrups	0.1
cider	0.05–0.1
wine	0.02–0.04
cakes and icings	0.05–0.1
pie fillings	0.05–0.1
margarine (unsalted)[b]	0.1
prepared vegetable salads	0.05–0.1
dried fruits	0.02–0.05
semimoist pet food	0.1–0.3
salad dressings (pour-type)[b]	0.05–0.1

[a]Ref. 65.
[b]Maximum use level allowed by Standard of Identity.

The level of sorbates necessary for preservation of a specific product depends on numerous factors, including product composition (pH, moisture, presence of other inhibitors, fat content), initial contamination level, packaging, and storage temperature. Maximum shelf life extension with sorbates is achieved when products have low initial levels of microbial contamination and are properly handled and stored. Therefore, the preservative cannot be used to mask poor quality product or poor handling practices.

Sorbates can be applied to food by any of several methods, including direct addition, dipping in or spraying with an aqueous sorbate solution, dusting with sorbate powder, or addition to food packaging materials. The potassium salt is used in applications where high water solubility is desired.

Margarine. Improvements in the technology of making margarine have greatly reduced the associated spoilage problems. Sorbates and benzoates are both used in margarine; sorbates is the more effective because of the high product pH and the comparative oil:water distribution coefficients for the two compounds, ie, sorbic acid 3:1, benzoic acid 6:1 (93,94). A greater proportion of the sorbate thus remains in the water phase, where microbial spoilage occurs. Sorbates are generally used at 0.1 wt % and are most often used in product forms with higher spoilage potential, eg, unsalted or reduced-, low-, or no-fat margarine, or product packaged in plastic tubs leaving headspace.

Wine. Sorbic acid is used in table wines to prevent secondary fermentation of residual sugar. It is used at 0.01–0.025 wt % in addition to sulfur dioxide. Adding sorbic acid affords protection against recontamination by yeasts for wines that have been heated or filter-sterilized, but at those low levels it does not provide adequate protection against undesirable malolactic or acetic acid bacteria (95,96). It has been found that bacteria in red wine containing residual sugar can metabolize potassium sorbate, resulting in off-odors and -flavors even in trace amounts (97). The most offensive by-product was 2-ethoxy-3,5-hexadiene, which produces a geranium-like odor. Other odors identified were 1-ethoxyhexa-2,4-diene (mint/garlic) and ethyl sorbate (honey/apple).

Dairy Products. The dairy industry is the largest commercial user of sorbates, with the largest portion used in processed cheeses. Data on the antimicrobial efficacy of sorbates in cheese have been published (98,99). The most common application methods include dipping or spraying with potassium sorbate solutions for natural cheeses and direct addition of sorbic acid to processed cheeses and cold-pack cheeses. Sorbate-impregnated wrapping material can be used for packaged cheese slices and pieces. For cottage cheese, sorbic acid is added to the cream dressing prior to pasteurization to a level of 0.075 wt % in the finished product (100,101). Common cheese processing and storage conditions, particularly heat treatment, do not affect the stability or efficacy of sorbic acid (102). Most cheese products are covered by Standards of Identity and, except for cottage cheese, a maximum use level is set (see Table 5).

Seafood. Sorbates are used to extend the shelf life of many seafood products, both fresh and processed (103,104). For smoked or dried fish, an instantaneous dip in 5 wt % potassium sorbate or a 10-minute dip in 1.0 wt % potassium sorbate prior to drying or smoking inhibits the development of yeast and mold (105,106). For fresh fish, sorbates can be incorporated at approximately 0.5 wt % into the ice, refrigerated seawater, or ice-water slush in which fish are packed, or applied as a 2.5–5.0 wt % potassium sorbate dip for fillets (107,108). Sorbates inhibit the growth of psychrotrophic spoilage bacteria, but the treatment must be applied while the product is fresh (109,110). Sorbates can extend shelf life and delay *Clostridium botulinum* type-E toxigenesis in fish that has been packaged in modified atmospheres (111).

Fruit and Vegetable Products.   Sorbates are applied at 0.05–0.1 wt % as a fungistat for prunes, pickles, relishes, maraschino cherries, olives, and figs (64,112). The same levels extend shelf life of prepared salads such as potato salad, cole slaw, and tuna salad (99). In fermented vegetables, sorbates protect the finished product by retarding yeasts during fermentation or in the cover brine (65,72–74,94).

Sorbates reduce post-harvest losses of fresh citrus fruit, particularly when the spoilage fungi are resistant to chemical treatments (113,114). Post-harvest treatment of apples and apple juice with potassium sorbate decreases spoilage and may prevent mycotoxin production (115). A combination of potassium sorbate and sodium benzoate greatly reduces mold caused by mold during drying and gives the additional benefit of reducing the drying time required (116–118).

Sorbate combined with mild heat has a synergistic effect with regard to microbial destruction; thus, in the presence of 0.025–0.06 wt % sorbate, products such as apple juice, peach and banana slices, fruit salads, and strawberries can be treated with less severe heat treatments to extend shelf life (119,120). Sorbates increase the heat sensitivity of various spoilage fungi under varying conditions of pH and water activity (121–124). A similar synergistic effect has been reported for the combination of sorbate with irradiation (125).

Bakery Products.   Sorbates are used in and/or on yeast-raised and chemically leavened bakery products. The internal use of sorbates in yeast-raised products at one-fourth the amount of calcium–sodium propionate that is normally added provides a shelf life equal to that of propionate without adversely affecting the yeast fermentation. Sorbates added at one-tenth the propionate level reduce the mix time by 30% (126). This internal treatment combined with an external spray of potassium sorbate can provide the same or an increased shelf life of pan breads, hamburger and hot-dog buns, English muffins, brown-and-serve rolls, and tortillas. The total sorbate useful in or on these baked goods ranges from 0.03 wt % for pan breads to 0.5 wt % for tortillas; 0.2–0.3 wt % sorbic acid protects chemically leavened yellow and chocolate cakes (127). Fruit-pie fillings and icings can be protected with 0.03–0.1 wt % sorbates.

Meat and Poultry.   The only sorbate treatment of meat permitted by the United States Department of Agriculture (USDA) is a potassium sorbate dip for dry sausages to prevent mold growth. Numerous research studies support increased sorbate use in meat and poultry, but most of these applications have not been approved for use in the United States. A combination of 0.26 wt % potassium sorbate with 0.004 wt % sodium nitrite in curing bacon has been shown to reduce nitrosamine formation during frying and provide a safe antibotulinal shelf life (128,129). In cooked cured sausages (beef, pork, and chicken frankfurter emulsions), 0.004 wt % nitrite/0.20 wt % sorbic acid delayed germination and outgrowth of *Clostridium botulinum* (130,131). In sausage or tryptose broth, potassium sorbate alone or combined with sodium nitrate caused an initial minimal reduction of *L. monocytogenes* (132).

For fresh poultry, a potassium sorbate dip significantly reduces total viable bacteria and doubles the refrigerated shelf life of ice-packed broilers (133). In cooked, uncured, vacuum-packaged turkey and poultry stored at 4°C, 0.2–0.25 wt % potassium sorbate suppresses microbial growth for up to 10 days (134). Sorbic acid at 0.5% in a marinade mixture for chicken drummettes extends refrigerator shelf life (135). Country-cured hams sprayed with a 10 wt % potassium sorbate solution showed no mold growth for up to 30 days (136). A review of sorbate use in meat and fish products has been published (137).

Pet Foods and Commercial Animal Feeds.   For many years, it has been known that stable, long-shelf-life, intermediate-moisture pet foods can be prepared through the use of 0.1–0.3 wt % sorbates. In these products, the antimicrobial effectiveness of sorbates is enhanced by a combination of moderate heat treatment, pH adjustment, and reduced water activity via humectants such as propylene glycol, or by adjusting sugar and salt content. These techniques have been reviewed extensively (138,139).

As energy costs have escalated in recent years, the use of high moisture food by-products in commercial animal feeds has also escalated, particularly in beef cattle and dairy rations, as a means of reducing production costs. Because of the broad activity spectrum, sorbates are extremely effective in the preservation of wet by-products, eg, brewers' and distillers' grains, beet pulp, citrus pulp, and condensed whey (139).

Treating alfalfa hay with potassium sorbate and potassium carbonate at cutting decreases drying time and improves preservation (140). Potassium sorbate applied to tobacco at a pH below 5.0 inhibits a number of spoilage fungi (141). A longer, safe storage period results in high moisture corn from the treatment of potassium sorbate alone or combined with propylene glycol (142).

Sorbic acid is not only suitable for preservation of feedstuffs but also improves the feed utilization and weight gain of chickens. This has proven to be of economic value under practical conditions when sorbic acid is added to the feed at 0.02–0.04 wt % (143–145). Similar effects have been observed for the use of sorbic acid in swine feeds (146).

Regulatory Status

United States.   Sorbic acid and potassium sorbate are generally recognized as safe (GRAS) for use in food under U.S. food ingredient regulations in the *Code of Federal Regulations* (147,148) when used in accordance with current good manufacturing practice and where permitted by applicable U.S. FDA Food Standards of Identity. U.S. regulations, including the Food Standards of Identity, cover about 70 foods or food product categories. Among them are cheese (limit 0.3%) (149), baked goods (limit 0.23%) (150), and margarine (limit 0.1% individually or 0.2% in combination with other preservatives) (151,152). Other permitted food categories include salad dressings (153) and condiments, where the limit of addition is generally 0.2%. Thus users must ascertain whether their particular food products are subject to a standard and, if so, what limitations apply.

Potassium sorbate, but not sorbic acid, is generally recognized as safe for use in animal feeds (154) and in paper and paperboard products for food packaging (155).

Potassium sorbate may also be used in meat and poultry food products under USDA regulations (156,157). The USDA regulations allow its use only to retard mold growth in sausage, including beef jerky, when applied from a 10% water solution to casings before or after stuffing.

Both sorbic acid and potassium sorbate may be used under U.S. Department of Treasury, Bureau of Alcohol, Tobacco, and Firearms regulations as sterilizing and preservative agents for treatment of wine and juice at up to 300 ppm/L (158).

Other U.S. regulatory clearances for sorbic acid allow its use as an antimicotic in the manufacture of food packaging materials (159), as a pesticide adjuvant when applied to growing crops (160), and with methylcellulose and dimethylpolysiloxane as an antifoaming agent (161).

Canada. Sorbic acid and potassium sorbate are cleared in Canada as Class II and Class III preservatives (Table XI, Parts II and III, Food and Drug Regulations) (162). They are cleared for use in the same food types. As in the United States, their lawful use is predicated upon conformity with published food standards. Otherwise they may be used in bread and unstandardized foods, except meat (Divisions 14 and 21 of the regulations), fish, and poultry, at levels up to 1000 ppm, in cider and wine at 500 ppm, and in cheeses at 3000 ppm in accordance with the food standards for cheese (Section B of the regulations).

Japan. *The Japanese Standards for Food Additives* allow sorbic acid and potassium sorbate to be used as antimicrobial agents in a variety of foods; specific limitations depend on the food (163).

European Union. As of this writing (1996), sorbic acid (E 200) and potassium sorbate (E 202) may be used without restriction in foods under European Union legislation (164). This legislation provides food use limitations similar to those under United States, Canadian, and Japanese food regulations. Specific food use limitations and restrictions are given in the legislation in Annex III, Part A, and allow use of the sorbates in food at levels ranging from 200 to 2000 ppm, depending on the food type.

Health and Safety Factors

The health effects of sorbic acid and sorbates have been reviewed (165–167). The extremely low toxicity of sorbic acid enhances its desirability as a food preservative. The oral LD_{50} for sorbic acid in rats is 7–10 g/kg body weight compared to 5 g/kg for sodium chloride (165–169). In subacute and chronic toxicity tests in rats, 5% sorbic acid in the diet results in no abnormal effects after 90 days or lifetime feeding studies. A level of 10% in rat diets results in a slight enlargement of the liver, kidneys, and thyroid gland (170). This same dietary level fed to mice also resulted in an increase in liver and kidney weight (171). These increased organ weights were not associated with any histopathological changes and are attributed to the energy utilization of sorbic acid (165,166). Studies of the long-term toxicity of sorbic acid in mice and in rats indicate no carcinogenic effects at dietary levels up to 10% (170,171).

Literature reports indicate that sodium sorbate causes weak genotoxic effects such as chromosomal aberrations and mutations in mammalian cells (172,173). This effect is thought to be caused by oxidative products of sodium sorbate in stored solutions (173–175). The main oxidation product of sodium sorbate, 4,5-oxohexenoate, is mutagenic in a *Salmonella*/mammalian-microsome test (176). Sorbic acid and potassium sorbate were not genotoxic under the same test procedures (167,172,174–177).

Sorbic acid is metabolized to carbon dioxide and water in the same way as other fatty acids, releasing 27.6 kJ/g sorbate (6.6 kcal/g) (165). As a result of the favorable toxicological and physiological aspects, the World Health Organization (WHO) has allowed sorbic acid at the highest acceptable daily intake of all food preservatives, 25 mg/kg body weight (178).

BIBLIOGRAPHY

"Sorbic Acid" in *ECT* 1st ed. Vol. 6, pp. 272–274, by J. A. Field, Union Carbide and Carbon Corp.; Suppl. 1, pp. 840–849, by A. E. Montagna, Union Carbide Chemicals Co.; in *ECT* 2nd ed., Vol. 18, pp. 589–599, by S. W. Moline, C. E. Colwell, and J. E. Simeral, Union Carbide Corp.; in *ECT* 3rd ed., Vol. 21, pp. 402–416, by C. L. Keller, S. M. Balaban, C. S. Hickey, and V. G. DiFate, Monsanto Co.

1. A. W. Hofmann, *Lieb. Ann. Chem. Pharm.* **110**, 129 (1859).
2. O. Doebner, *Ber. Dtsch. Chem. Gas.* **33**, 2140 (1900).
3. N. G. Polyanskii and co-workers, *Zh. Prikl. Khim. (Leningrad)*, **39**, 2005 (1966).
4. E. Lück, *Sorbinsäure*, Vol. 1, B. Bher's Verlag GmbH, Hamburg, Germany, 1969.
5. Trolle-Lassen Co., *Arch. Pharm. Org. Chem.* **66**(23), 1235 (1959).
6. *U.S. Bureau of Mines*, Report 7132, May 1968.
7. J. J. Sudborough and J. Thomas, *J. Chem. Soc.* **97**, 2450 (1910).
8. C. K. Ingold, G. J. Pritchard, and H. G. Smith, *J. Chem. Soc.* 79 (1934).
9. R. Fittig, *Lieb. Ann. Chem. Pharm.* **161**, 307 (1872).
10. U.S. Pat. 3,126,401 (Mar. 24, 1964), G. Ecke (to Ethyl Corp.).
11. P. Heinänen, *Ann. Acad. Sci. Fenn. Ser.* **A49**(4), 112 (1938).
12. Technical data, J. J. Bloomfield, Monsanto Co., St. Louis, Mo., 1981.
13. K. Fugjiwana and co-workers, *Nippon Hoshasen Kobunshi Kenkyu Kyokai Nempo*, **4**, 183 (1962).
14. I. N. Putilova, *Tr. Mezhd. Kongr. Korrozii Metallov (Moscow)*, **2**, 32 (1966).
15. Ger. Pat. 1,283,832 (Jan. 27, 1965), H. Fernholz and H. J. Schmidt (to Hoechst).
16. R. Adams and L. Ulich, *J. Am. Chem. Soc.* **42**, 599 (1920).
17. O. Doebner, *Ber. Dtsch. Chem. Gas.* **23**, 2372 (1890).
18. *Ibid.*, **35**, 1136 (1902).
19. Pol. Pat. 47,632 (Oct. 14, 1963), I. Nagrodzka and co-workers.
20. U.S. Pat. 2,484,067 (Oct. 11, 1949), A. B. Boese, Jr. (to Union Carbide Corp.).
21. H. J. Hagemeyer, *Ind. Eng. Chem.* **41**, 765 (1949).
22. Jpn. Pat. 39-13,849 (Aug. 5, 1967), H. Nakamura (to Nippon Gohsei).
23. Jpn. Pat. 45-9,368 (Apr. 14, 1970), O. Nakamura (to Nippon Gohsei).
24. Ger. Pat. 1,042,573 (Nov. 6, 1958), H. Fernholz (to Hoechst).
25. U.S. Pat. 3,022,342 (Feb. 20, 1962), H. Fernholz, K. Ruths, and K. Heinmann-Trosien (to Hoechst).
26. U.S. Pat. 3,021,365 (Feb. 13, 1962), H. Fernholz and E. Munolos (to Hoechst).
27. Ger. Pat. 1,150,672 (June 27, 1963), O. Probst (to Hoechst).
28. U.S. Pat. 3,499,029 (Mar. 3, 1970), H. Fernholz and H. Neu (to Hoechst).
29. Jpn. Pat. 44-26,646 (Nov. 7, 1969), I. Nakajima (to Nippon Gohsei).
30. U.S. Pat. 3,759,988 (Sept. 18, 1973), G. Kunstle (to Wacker).
31. Can. Pat. 8,982,73 (Apr. 18, 1972), R. Smith and E. Jeans (to Chemcell).
32. Ger. Pat. 2,203,712 (Aug. 16, 1973), H. Fernholz, H. J. Schmidt, and F. Wunder (to Hoechst).

33. Ger. Pat. 1,153,742 (Sept. 5, 1963), K. Ruths and O. Probst (to Hoechst).
34. Ger. Pat. 1,059,899 (June 25, 1959), H. Fernholz (to Hoechst).
35. U.S. Pat. 3,461,158 (Aug. 12, 1969), L. Hörnig and H. Neu (to Hoechst).
36. N. G. Polyanskii and co-workers, *Tr. Tombovsk Inta Khim. Mashinostr.* **2**, 94 (1968).
37. U.S. Pat. 3,642,885 (Feb. 15, 1972), L. Hörnig and co-workers (to Hoechst).
38. Ger. Pat. 1,281,439 (Oct. 31, 1968), L. Hörnig and O. Probst (to Hoechst).
39. U.S. Pat. 2,887,496 (May 19, 1959), E. Lashley (to Union Carbide).
40. G. F. Woods and co-workers, *J. Am. chem. Soc.* **77**, 1800 (1955).
41. V. S. Markevich and S. M. Markevich, *Khim. Promst. (Moscow)*, **12**, 898 (1973).
42. Rus. Pat. 169,520 (Mar. 17, 1965), S. M. and V. S. Markevich.
43. R. Joley and C. Amiaro, *Bull. Soc. Chem. Fr.*, 139 (1947).
44. U. Eisner, J. Elvidoe, and R. Lindstead, *J. Chem. Soc.* 1372 (1953).
45. U.S. Pat. 4,022,822 (May 10, 1977), Y. Tsu Jino (to Nippon Gohsei).
46. U.S. Pat. 4,158,741 (June 19, 1979), M. Goi (to Nippon Gohsei).
47. Ital. Pat. 719,380 (Nov. 2, 1966), G. P. Chiusoli, S. Merzoni, and G. Cometti (to Montecatini).
48. G. P. Chiusoli, *Angew. Chem.* **72**, 74 (1960).
49. B. N. Utkin, *Khimiya Sorbinovoi Kislofy*, M. NIITE Khim. (1970).
50. N. G. Polyanskii, *Khim. Promst. (Moscow)*, **1**, 20 (1963).
51. *Food Chemicals Codex*, 3rd ed. and 1st, 2nd, 3rd, and 4th Supplements to the 3rd ed., National Academy of Science, National Academy Press, Washington, D.C., 1981.
52. *Official Methods of Analysis*, 15th ed., Sections 971.15, 974.10, 975.10, and 983.16, Association of Official Analytical Chemists, Arlington, Va., 1990.
53. *Pearson's Composition and Analysis of Foods*, 9th ed., Longman Scientific and Technical, London, U.K., 1991.
54. A. Caputi and P. A. Stafford, *J. Assoc. Off. Anal. Chem.* **60**, 1044–1047 (1977).
55. G. Wilamowski, *J. Assoc. Off. Anal. Chem.* **57**, 675–677 (1974).
56. G. Wilamowski, *J. Assoc. Off. Anal. Chem.* **54**, 663–665 (1971).
57. M. C. Bennett and D. R. Petrus, *J. Food Sci.* **42**, 1220–1221 (1977).
58. M. S. Ali, *J. Assoc. Off. Anal. Chem.* **68**, 488–492 (1985).
59. H. Terada and Y. Sakabe, *J. Chroma.* **346**, 333–340 (1985).
60. B. K. Larsson, *J. Assoc. Off. Anal. Chem.* **66**, 775–780 (1983).
61. A. Graveland, *J. Assoc. Off. Anal. Chem.* **55**, 1024–1026 (1972).
62. D. E. LaCroix and N. P. Wong, *J. Assoc. Off. Anal. Chem.* **54**, 361–363 (1971).
63. E. Lück, *Food Add. Contam.* **7**(5), 711–715, 1990.
64. R. Woodford and E. Adams, *Am. Perfum. Cosmet.* **85**, 25 (1970).
65. *Sorbic Acid and Potassium Sorbate for Preserving Food Freshness*, Eastman Chemical Co. Publication ZS-1C, Kingsport, Tenn., 1995.
66. E. S. Beneke and F. W. Fabian, *Food Technol.* **9**, 486 (1955).
67. T. A. Bell, J. L. Etchells, and A. F. Borg, *J. Bacteriol.* **77**, 573 (1959).
68. R. H. Vaughn and L. O. Emard, *Bacteriol. Proc.* **5**, 38 (1951).
69. L. O. Emard and R. H. Vaughn, *J. Bacteriol.* **63**, 487 (1952).
70. G. K. York, Ph.D. dissertation, *Studies on the Inhibition of Microbes by Sorbic Acid*, University of California, Davis, 1960.
71. G. F. Phillips and J. O. Mundt, *Food Technol.* **4**, 291 (1950).
72. R. N. Costilow, W. E. Ferguson, and S. Ray, *Appl. Microbiol.* **3**, 341 (1955).
73. R. N. Costilow and co-workers, *Food Res.* **21**, 27 (1956).
74. R. N. Costilow and co-workers, *Appl. Microbiol.* **5**, 373 (1957).
75. M. B. Liewen and E. H. Marth, *J. Food Prot.* **4**(48), 364–375 (1985).
76. J. N. Sofos and F. F. Busta, *J. Food Prot.* **4**, 614 (1981).
77. F. Sauer, *Food Technol.* **31**, 66 (1977).
78. D. Melnick, F. H. Luckmann, and C. M. Gooding, *Food Res.* **19**, 44 (1954).

79. J. J. Azukas, R. N. Costilow, and H. L. Sadoff, *J. Bacteriol.* **81**, 189 (1961).
80. G. K. York and R. H. Vaughn, *Bacteriol. Proc.* **55**, 20 (1955).
81. W. Martoadiprawito and J. R. Whitaker, *Biochem. Biophys. Acta,* **77**, 526 (1963).
82. J. R. Whitaker, *Food Res.* **24**, 37 (1959).
83. G. K. York and R. H. Vaughn, *J. Bacteriol.* **88**, 411 (1964).
84. J. A. Troller, *Can. J. Microbiol.* **11**, 611 (1965).
85. T. Deak and E. K. Novak, *Yeasts: The Proceedings of the Second Symposium on Yeasts*, Slovac Academy of Sciences, Bratislava, Czech Republic, 1966, pp. 533–536.
86. A. G. Man, in *Bacteria: A Treatise on Structure and Function*, Vol. 1, Academic Press, New York, 1960.
87. G. K. York and R. H. Vaughn, *Bacteriol. Proc.* **60**, 47 (1960).
88. A. D. Warth, *J. Appl. Bacteriol.* **43**, 215 (1970).
89. E. H. Marth and co-workers, *J. Dairy Sci.* **49**, 1197 (1966).
90. J. L. Kinderlerer and P. V. Hatton, *Food Add. Contam.* **7**(5), 657–669 (1990).
91. S. Schnell and co-workers, *Biodeg.* **2**, 33–41 (1991).
92. U.S. Pat. 2,379,294 (June 26, 1945), C. M. Gooding (to Best Foods, Inc.).
93. C. J. Doherty, *Technical Service Report*, Union Carbide Corp., South Charleston, W. Va., 1965.
94. E. Beker and I. Roeder, *Fette Seifen Austrichm*, **49**, 321 (1957).
95. G. Wurdig, *Brew. Distill. Int.* **6**, 42 (1976).
96. R. C. Auerbach, *Wines Vines*, **40**, 26 (1959).
97. M. G. Chisholm and J. M. Samuels, *J. Agric. Food Chem.* **40**(4), 630–633 (1992).
98. M. D. Bonner and L. G. Harmon, *J. Dairy Sci.* **40**, 1599 (1957).
99. E. Lüeck and K. H. Remmert, *Indus. Lech.* **60**, 10–14,16 (1980).
100. *Sorbic Acid and Potassium Sorbate for Use in Cottage Cheese*, Publication No. IC/FI-20, Monsanto, Co., St. Louis, Mo., 1977.
101. E. B. Collins and H. H. Moustafa, *J. Dairy Sci.* **52**, 439 (1969).
102. J. A. Torres, J. O. Bouzas, and M. Karel, *J. Food Process. Preserv.* **13**(6), 409–415 (1989).
103. B. R. Thakur and T. R. Patel, *Food Rev. Inter.* **10**(1), 93–107 (1994).
104. A. Pedrosa-Menabrito and J. M. Regenstein, *J. Food Qual.* **13**, 129–146 (1990).
105. J. J. Geminder, *Food Technol.* **13**, 459 (1959).
106. J. W. Boyd and H. L. A. Tarr, *Food Technol.* **9**, 411 (1953).
107. J. P. H. Wessels and co-workers, *J. Food Technol.* **7**, 303 (1972).
108. N. Tomlinson and co-workers, *Technical Report No. 783*, Canadian Fisheries and Marine Service, Vancouver, B.C., Canada, 1978.
109. J. M. Debevere and J. P. Voets, *J. Appl. Bacteriol.* **35**, 351 (1972).
110. M. S. Fey, Ph.D. dissertation, *Extending the Shelf Life of Fresh Fish by Potassium Sorbate and Modified Atmospheres at 0–1° Celsius*, Cornell University, Ithaca, N.Y., 1980.
111. R. C. Lindsay, *Appl. Environ. Micro.* **44**(5), 1212–1221 (1982).
112. *CRC Handbook of Food Additives*, 2nd ed., CRC Press, Boca Raton, Fla., 1972.
113. J. J. Smoot and A. A. McCormack, *Proc. Fla. State Hortic. Soc.* **91**, 119 (1978).
114. *Use of Potassium Sorbate in Protecting Citrus Fruit*, Publication No. IC/NC-602, Monsanto Co., St. Louis, Mo., 1981.
115. D. Ryu and D. L. Holt, *J. Food Prot.* **56**, 862–867 (1993).
116. J. N. Bizri and I. A. Wahem, *J. Food Sci.* **59**, 130–134 (1994).
117. C. B. Hall, *Proc. Fla. State Hortic. Soc.* **72**, 280 (1959).
118. *Use of MP-11 for Mold Protection of Field-Drying Raisins*, Publication No. IC/NC-603, Monsanto, Co., St. Louis, Mo., 1981.
119. J. R. Robinson and C. H. Hills, *Food Technol.* **16**, 77 (1962).
120. U.S. Pat. 2,992,114 (July 11, 1961), E. A. Weaver (to U.S. Government).

121. L. R. Beuchat, *J. Food Sci.* **46**, 771 (1981).

122. L. R. Beuchat, *J. Food Prot.* **44**, 450 (1981).

123. L. R. Beuchat, *Appl. Environ. Microbiol.* **41**, 472 (1981).

124. L. R. Beuchat, *J. Food Prot.* **44**, 765 (1981).

125. C. F. Niven, Jr. and W. R. Chesbro, *Antibiot. Annu.* 855 (1956–1957).

126. *Potassium Sorbate Surface Treatment for Yeast-Raised Bakery Products*, Publication No. IC/FI-21, Monsanto Co., 1977.

127. D. Melnick, H. W. Vahlteich, and A. Hackett, *Food Res.* **21**, 133 (1956).

128. *Shelf Life Sensory, Cooking, and Physical Characteristics of Bacon Cured with Varying Levels of Sodium Nitrate and Potassium Sorbate*, U.S. Dept. of Agriculture, Washington, D.C., July 1979.

129. F. J. Ivey and co-workers, *J. Food Prot.* **41**, 621 (1978).

130. M. C. Robach, *Appl. Environ. Microbiol.* **38**, 840 (1978).

131. J. N. Sofos, F. F. Busta, and C. E. Allen, *Appl. Environ. Microbiol.* **37**, 1103 (1979).

132. Y. A. Hefnawy, S. I. Moustafa, and E. H. Marth, *Lebensm. Wiss. U. Technol.* **26**, 167–170 (1993).

133. E. C. To and M. C. Robach, *J. Food Technol.* **15**, 543 (1980).

134. M. C. Robach and co-workers, *J. Food Sci.* **45**, 638 (1980).

135. J. T. Chuang and co-workers, *Korean J. Food Sci. Technol.* **23**(6), 717–722 (1991).

136. J. D. Baldock and co-workers, *J. Food Prot.* **42**, 780 (1979).

137. M. C. Robach and J. N. Sofos, *J. Food Prot.* **45**, 374 (1982).

138. R. Davis, G. G. Birch, and J. J. Parker, eds., *Intermediate Moisture Foods*, Applied Science Publishers, Ltd., London, U.K., 1976.

139. N. W. Desrosier, *The Technology of Food Preservation*, AVI Publishing Co., Westport, Conn., 1970, pp. 365–383.

140. E. H. Jaster and K. J. Moore, *Anim. Feed Sci. Technol.* **38**, 175–186 (1992).

141. E. S. Mutasa, N. Magan, and K. J. Seal, *Mycol. Res.* **94**(7), 971–978 (1990).

142. M. Yasin, M. A. Hanna, and L. B. Bullerman, *Am. Soc. Agric. Eng.* **35**(4), 1229–1233 (1992).

143. G. Dust, in E. Lück, eds., *Sorbinsäure*, Vol. 3, B. Berhr's Verlag GmbH, Hamburg, Germany, 1970.

144. B. C. Dilworth and co-workers, *Poult. Sci.* **6**, 1445 (1979).

145. M. Suwathep and co-workers, *Poult. Sci.* **7**, 1741 (1981).

146. Technical data, T. Veum, University of Missouri, Columbia, Mo., 1981.

147. 21 CFR 182.3089, Apr. 1994.

148. 21 CFR 182.3640, Apr. 1994.

149. 21 CFR Part 133, Apr. 1994.

150. 21 CFR Part 136, Apr. 1994.

151. 21 CFR Part 166, Apr. 1994.

152. 9 CFR 318.7, *Miscellaneous*, 219, Jan. 1995.

153. 21 CFR Part 169, Apr. 1994.

154. 21 CFR 582.3640, Apr. 1994.

155. 21 CFR 182.90, Apr. 1994.

156. 9 CFR 318.7, Jan. 1995.

157. 9 CFR 381.147, Jan. 1995.

158. 27 CFR 24.246, Apr. 1988.

159. 21 CFR 181.23, Apr. 1994.

160. 21 CFR 182.99, Apr. 1994.

161. 21 CFR 177.2260, Apr. 1994.

162. *The Food and Drugs Act and Regulations*, Departmental Consolidation of the Food and Drugs Act and the Food and Drug Regulations, with amendments to Dec. 15, 1994, Minister of Supply and Services, Canada, 1994.

163. *The Japanese Standards for Food Additives*, 6th ed., Japan Food Additives Association, Tokyo, 1994.
164. European Parliament and Council Directive 95/2/EC, Feb. 1995.
165. J. N. Sofos and F. F. Busta, in P. M. Davidson and A. L. Branen, eds., *Antimicrobials in Foods*, 2nd ed., Marcel Dekker, New York, 1993, pp. 49–94.
166. J. N. Sofos, *Sorbate Food Preservatives*, CRC Press, Inc., Boca Raton, Fla., 1989, pp. 205–224.
167. R. Walker, *Food Add. Contam.* **7**, 671–676 (1990).
168. H. J. Deuel and co-workers, *Food Res.* **19**, 1–12 (1954).
169. H. F. Smyth and C. P. Carpenter, *J. Indus. Hygiene Toxic.* **30**, 63–68 (1948).
170. I. F. Gaunt, K. R. Butterworth, and S. D. Gangolii, *Food Cosm. Toxic.* **13**, 31–45 (1975).
171. R. J. Hendy and co-workers, *Food Cosm. Toxic.* **14**, 381–386 (1976).
172. M. M. Hasegawa and co-workers, *Food Chem. Toxic.* **22**, 501–507 (1984).
173. R. Munzner, C. Guigas, and H. W. Renner, *Food Chem. Toxic.* **28**, 397–401 (1990).
174. D. Schiffmann and J. Schlatter, *Food Chem. Toxic.* **30**, 669–672 (1992).
175. J. Schlatter and co-workers, *Food Chem. Toxic.* **30**, 843–851 (1992).
176. R. Jung and co-workers, *Food Chem. Toxic.* **30**, 1–7 (1992).
177. F. E. Wurgler, J. Schlatter, and P. Maier, *Mut. Res.* **283**, 107–111 (1992).
178. Joint FAO/WHO Expert Committee on Food Additives, *WHO Food Add. Ser.* **5**, 121–127 (1974).

CATHERINE L. DORKO
GEORGE T. FORD, JR.
MADELYN S. BAGGETT
ALISON R. BEHLING
HAROLD E. CARMAN
Eastman Chemical Company

SORBITE. See STEEL.

SORBITOL. See SUGAR.

SORGHUMS. See WHEAT AND OTHER CEREAL GRAINS.

SOYBEANS AND OTHER OILSEEDS

Soybeans, cottonseed, peanuts, and sunflowers are the four principal oilseed crops grown in the United States. Except for cottonseed, these are consumed directly as foods to varying extents, but all serve as sources of edible oils commonly referred to as vegetable oils (qv). Although not normally considered to be an oilseed, corn (see STARCH) provides the second largest quantity of vegetable oil used domestically in foods. Selected data on corn oil are therefore included for comparison with oilseed oils. After removal of the oils, the resulting oilseed meals are rich in proteins (qv) and find widespread use as animal feeds (see FEEDS AND FEED ADDITIVES). Food uses of oilseed proteins derived from meals are relatively small and are limited to soybeans and peanuts. Table 1 summarizes botanical names, geographic distribution, and the main uses of these four oilseeds.

Soybeans, the principal oilseed crop in the United States, are believed to have been domesticated in the eastern half of northern China around the eleventh century BC or earlier. They were later introduced and established in Japan and other parts of Asia, brought to Europe, and introduced to North

Table 1. Botanical Classification, Area of Production, and Uses of Oilseeds

Botanical classification		Principal production areas	Uses
Family	Genus and species		
Soybean			
Leguminosae (legume)	*Glycine max Merrill*	United States, Brazil, People's Republic of China, Argentina	edible oil, animal feed, food, edible proteins, industrial oils and proteins
Cottonseed			
Malvaceae (mallow)	*Gossypium arboreum*, Sri Lanka cotton; *G. herbaceum*, Levant cotton; *G. barbadense*, Sea Island cotton; *G. hirsutum*, Upland cotton	People's Republic of China, United States, India, FSU[a]	edible oil, animal feed
Peanut			
Leguminosae (legume)	*Arachis hypogaea*	People's Republic of China, India, United States	food, edible oil, animal feed, edible protein
Sunflower			
Compositae (composite)	*Helianthus annuus*	Argentina, FSU[a], EU-15[b], United States	edible oil, animal feed, food

[a] Former Soviet Union.
[b] European Union.

America in 1765 (1). Soybeans became an established oilseed crop in the late 1920s, attaining commercial importance during World War II. Cotton (qv) has a long history that can be traced back as far as 3000 BC through spun cotton yarn found in the Indus valley. It is indigenous to many parts of the world, and its establishment as an oilseed in the United States is associated with the invention of the cotton gin by Eli Whitney in 1794. Peanuts or groundnuts likely originated in South America and were later introduced to Africa and Asia. Subsequent cultivation of peanuts in North America was started with plants imported from Africa (see NUTS). Sunflowers are native to North America and probably originated in the southwestern United States. They were introduced to Spain by the early Spanish explorers and then spread to Russia, where they became established as an oilseed crop. Sunflowers became a significant U.S. oilseed crop as late as 1967, upon the development of varieties having high oil content and improved agronomic characteristics such as increased resistance to diseases and pests.

Physical Characteristics

Plants and seeds of the four oilseeds vary in growth habit, size, shape, and other features. A common feature of the four oilseeds is storage of the bulk of the protein and oil in distinct membrane-bound, subcellular organelles called protein bodies and lipid bodies, respectively, as illustrated for soybeans in Figure 1. Although not shown, the protein bodies contain inclusions referred to as globoids (≤ 0.1 μm dia) that are storage sites for phytate and cations such as potassium, magnesium, and calcium (3). During germination, the contents of the storage organelles are mobilized and utilized by the growing seedling.

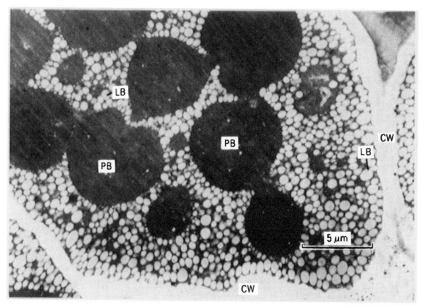

Fig. 1. Transmission electron micrograph of a section of a mature, hydrated soybean cotyledon. Protein bodies (PB), lipid bodies (LB), and cell wall (CW) are identified (2).

Soybean. *Plant.* Soybeans grow on erect, bushy annual plants, 75–125 cm high, having hairy stems and trifoliate leaves. The flowers are white or purple or combinations thereof. Growing season varies with latitude and is 120–130 d in central Illinois.

Seeds. Seeds are produced in pods, usually containing three almost spherical-to-oval seeds weighing 0.1–0.2 g. Commercial varieties have a yellow seed coat plus two cotyledons, plumule, and hypocotyl-radicle axis. The cotyledons contain primarily protein and lipid bodies (see Fig. 1).

Cottonseed. *Plant.* Cotton (qv) grows as an annual or perennial herb or shrub, sometimes tree-like, 0.25–3 m or more in height depending on species, with three-, five-, or seven-lobed leaves with white to yellow or purple-red flowers. The herbage is irregularly dotted with pigment glands. The growing season for Upland cotton grown in the United States is 120 d.

Seeds. The seeds are produced in leathery capsules (bolls), covered with lint and fuzz fibers. The seeds are ovoid, 8–12 mm long having brown to nearly black seed coats. The interior of the seed consists of a radicle, hypocotyl, epicotyl, and two cotyledons. Cotyledon tissue contains protein and lipid bodies. Most varieties contain pigment glands (storage sites of gossypurpurin and gossypol) 100–400 μm long (4). Glandless varieties of cottonseed, which lack the toxic gossypol, are available but are not grown commercially despite having undergone extensive research and development (5).

Peanut. *Plant.* Peanuts grow on annual herbaceous plants, bushy upright (erect) or spreading (runner) types, 25–50 cm high with bijugate leaves, hairy stems, and bright yellow flowers having peduncles that bend after fertilization and push the pods underground, where they develop and ripen. The growing season is 120–140 d.

Seeds. The seeds are produced in pods containing two or three seeds. The kernels are almost spherical to roughly cylindrical (0.4–1.1 g each) and consist of a thin coat (testa) containing two cotyledons and the embryo. Cotyledons contain protein bodies, lipid bodies, and starch granules.

Sunflower. Two types of sunflowers are grown in the United States. Varieties grown for oilseed production, ca 85% of crop, are generally black-seeded, having thin seed coats that adhere to the kernels. These contain 40–50% oil and ca 20% protein. Nonoilseed varieties, ca 15% of crop, sometimes referred to as confectionery, striped, or large-seeded sunflowers, have striped, relatively thick hulls that do not adhere to the kernels. These contain 20–30% oil and are usually larger than seeds of oilseed varieties.

Plant. The sunflower is an erect annual, 1–4 m high, having alternating leaves. The heads are ≤30 cm wide having orange-yellow rays and dark center disks. The growing season is ca 120 d.

Seeds. The sunflower seed (achene) is four-sided and flattened, ca 9 mm long × 4–8 mm wide, having a black or striped gray and black seed coat (pericarp) enclosing a kernel. The kernel contains protein and lipid bodies.

Chemical Composition

Compositions of the four oilseeds are given in Table 2. All except soybeans have a high content of seed coat or hull. Because of the high hull content, the crude fiber

Table 2. Composition^a of Oilseeds, Wt %^{b}

Oilseed	Hulls	Oil	Protein^c	Ash	Protein in dehulled, defatted meal^d
soybean	8	20	43	5.0	52
cottonseed					
acid delinted	36	21.6	21.5	4.2	
kernels		36.4	32.5	4.7	63
peanut	20–30				
kernels	2–3.5^e	50.0	30.3	3.0	57
sunflower					
arrowhead variety, low oil type^f	47	29.8	18.1		~68^g
armavirec variety					~60^g
high oil type^f	30	48.0	16.9		
kernels^f		64.7	21.2		

^{a}Approximate; moisture-free basis. ^{b}Ref. 6, except as otherwise noted. ^{c}As nitrogen, N, × 6.25. ^{d}Data vary with efficiency of dehulling and oil extraction, variety of seed, and climatic conditions during growth. ^{e}Red skins or testa. ^{f}Ref. 7. ^{g}Calculated from kernels on oil-free basis.

content of the other oilseeds is also high. Confectionery varieties of sunflower seed may contain up to 28% crude fiber on a dry basis (8). Soybeans differ from the other oilseeds in their high protein and low oil content. All these oilseeds, however, yield high protein meals when dehulled and defatted.

Proteins. The proteins found in the four oilseeds are complex mixtures consisting of four characteristic fractions having molecular weights of ca 9,000–700,000, as illustrated by the ultracentrifuge pattern for soybean proteins (Fig. 2). The 7S and 11S fractions usually predominate, but in sunflower seeds, the 2S and 11S (helianthinin) proteins are the main fractions (11). The

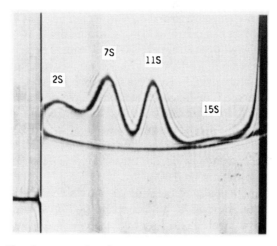

Fig. 2. Ultracentrifugal pattern for the water-extractable proteins of defatted soybean meal in pH 7.6, 0.5 ionic strength buffer. Numbers above peaks are approximate sedimentation coefficients in Svedberg units, S. Molecular weight ranges for the fractions are 2S, 8,000–50,000; 7S, 100,000–180,000; 11S, 300,000–350,000; and 15S, 600,000–700,000 (9). The 15S fraction is a dimer of the 11S protein (10).

7S and 11S fractions are considered to be storage proteins and are located in the protein bodies (12). Of the four oilseeds, the proteins of soybeans are the best characterized. The principal portion of soybean 7S fraction, β-conglycinin, consists of at least seven isomers resulting from various combinations of three subunits, ie, α, α', and β. These are $\alpha\beta_2$, $\alpha'\beta_2$, $\alpha'\alpha\beta$, $\alpha_2\beta$, $\alpha\alpha_2'$, α_3, and β_3 (13,14). Based on the sizes of the subunits, the β-conglycinin isomers have molecular weights between 126,000 and 171,000 (see PROTEINS).

The 11S molecule in soybeans, also called glycinin, is more complex than β-conglycinin. It consists of six subunits (ca 60,000 mol wt) each of which contains an acidic polypeptide (37,000–44,000 mol wt) linked to a basic polypeptide (17,000–22,000 mol wt) by a single disulfide bond. The subunits are synthesized as single polypeptides, proglycinin precursors, which undergo proteolytic cleavage to form the two polypeptide chains after the precursors enter the protein bodies (15). Five glycinin genes have been cloned and sequenced (16). The relative sizes of the primary polypeptides constituting the principal soybean proteins are shown in the gel electrophoresis diagram of Figure 3.

Arachin, the counterpart of glycinin in peanuts, consists of subunits of 60,000–70,000 mol wt which on reduction with 2-mercaptoethanol yield polypeptides of 41,000–48,000 and 21,000 mol wt (17) analogous to the behavior of glycinin. In addition to the storage proteins, oilseeds contain a variety of minor

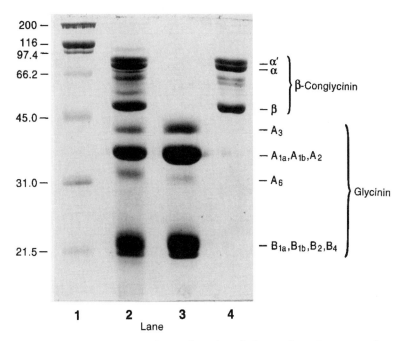

Fig. 3. Sodium dodecyl sulfate–polyacrylamide gel electrophoretic pattern for molecular weight standards (lane 1); water-extractable proteins of defatted soybean meal (lane 2); purified 11S (glycinin) (lane 3); and purified 7S (β-conglycinin) (lane 4) where the numbers represent mol wt $\times 10^3$. The gel was run in the presence of 2-mercaptoethanol, resulting in the cleavage of the disulfide bond linking the acidic (A bands) and basic (B bands) polypeptides of the 11S molecule.

proteins, including trypsin inhibitors, hemagglutinins, and enzymes. Examples of the last are urease and lipoxygenase in soybeans.

Amino acid compositions of the four oilseeds are given in Table 3 along with the amino acid (see AMINO ACIDS) requirements for humans suggested by a Joint FAO/WHO/UNU Expert consultation.

Lipids. Representative fatty acid compositions of the unprocessed triglyceride oils found in the four oilseeds are given in Table 4 (see FATS AND FATTY OILS). Cottonseed, peanut, and sunflower oils are classified as oleic–linoleic acid oils because of the high (>50%) content of these fatty acids. Although the oleic and linoleic acid content of soybean oils is high, it is distinguished from the others by a content of 4–10% of linolenic acid, and hence is called a linolenic acid oil.

In addition to the triglycerides, the four oilseeds also contain phosphatides. For example, soybean oil containing 1.47% phosphatides consists of 48.9% phosphatidylcholine, 27.0% phosphatidylethanolamine, 21.9% phosphatidylinositol and 2.2% phosphatidic acid (24). Total phosphatides of cottonseed and peanut kernels are estimated to be 1.5–1.9 and 0.8%, respectively (25).

Sterols are present in concentrations of 0.2–0.4% in the oils. Compositions are given in Table 5. The sterols exist in the seeds in four forms: free, esterified, nonacylated glucosides, and acylated glucosides. Soybeans contain a total of 0.16% of these sterol forms in the ratio of ca 3:1:2:2 (27) (see STEROIDS).

Table 3. Amino Acid Composition of Defatted Oilseed Meals and Amino Acid Requirements, mg / g Crude Protein

| | | | | | Requirements[e] | | |
| | | | | | Child, age | | |
Amino acid	Soybean[a]	Cottonseed[b]	Peanut[c]	Sunflower[d]	2–5	10–12	Adult
lysine	64	44	30	38	58	44	16
histidine	26	27	23	25	19	19	16
arginine	73	116	113	89			
aspartic acid	118	92	141	87			
threonine	39	30	25	32	34	28	9
serine	55	42	49	39			
glutamic acid	186	217	199	210			
proline	55	36	44	50			
glycine	43	41	56	51			
alanine	43	39	42	41			
valine	46	45	45	48	35	25	13
cystine	14	26	13	18 ⎫	25	22	17
methionine	11	15	9	19 ⎬			
isoleucine	46	31	41	40	28	28	13
leucine	78	58	67	61	66	44	19
tryosine	38	31	41	27 ⎫	63	22	19
phenylalanine	50	54	52	47 ⎬			
tryptophan	14	12	10	11	11	9	5

[a]Means based on 32 hydrolysates except for proline, cystine, and tryptophan (18). [b]Means for eight-glanded seed varieties (19). [c]Ref. 20. [d]Means for seven varieties (7). [e]Essential amino acid requirements for humans recommended by FAO/WHO/UNU (21). An essential amino acid is one that cannot be synthesized by humans at a sufficiently rapid rate to meet metabolic needs.

Table 4. Fatty Acid Composition of Unprocessed Oilseed Oils, Wt %

Carboxylic acid[a]	Soybean[b]	Cottonseed[b]	Peanut[c]	Sunflower[b]
		Saturated fatty acids		
10:0		0.48		
12:0	0.10	0.38		
14:0	0.16	0.79		0.1
16:0	10.7	22.0	10.5	5.81
18:0	3.87	2.24	3.2	4.11
20:0	0.22	0.19	1.4	0.29
22:0			2.1	
24:0			0.7	
		Unsaturated fatty acids		
16:1	0.29	0.78		0.10
18:1	22.8	18.1	50.3	20.7
18:2	50.8	50.3	30.6	63.5
18:3	6.76	0.40		0.32
20:1			1.0	0.10

[a]Carboxylic acid nomenclature designates number of carbons:number of double bonds. For example, 10:0 indicates a fully saturated C_{10} acid.
[b]Ref. 22.
[c]Mean values for 1968 crop of 82 peanut genotypes (23).

Table 5. Composition of Oilseed Sterols[a], %

Oil CAS Registry Number	Campesterol[b] [474-62-4]	Stigmasterol[c] [83-48-7]	β-Sitosterol[d] [83-46-5]	Δ⁵-Avenasterol[e]	Δ⁷-Stigmasterol [83-45-4]
soybean	20	20	53	3	3
cottonseed	4	1	93	2	trace
peanut	15	9	64	8	3
sunflower	8	8	60	4	15

[a]Ref. 26.
[b]Also known as ergost-5-en-3-ol (**1**).
[c]Stigmasterol is stigmasta-5,22-dien-3β-ol.
[d]Also known as stigmasta-5-en-3-ol (**2**).
[e]Δ⁵-avenasterol is stigmasta-5,24(28)-dien-3β-ol.

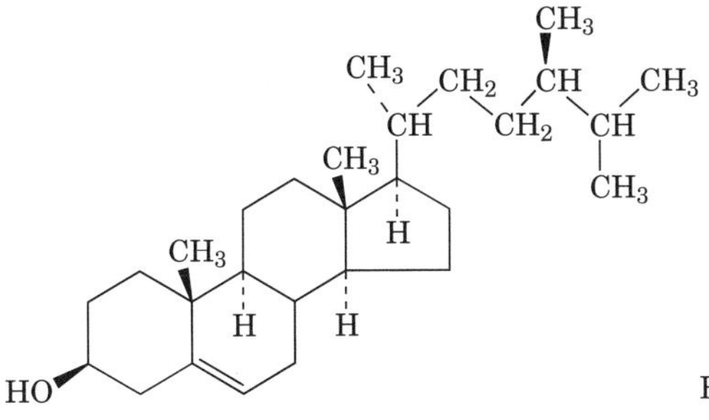

(**1**)

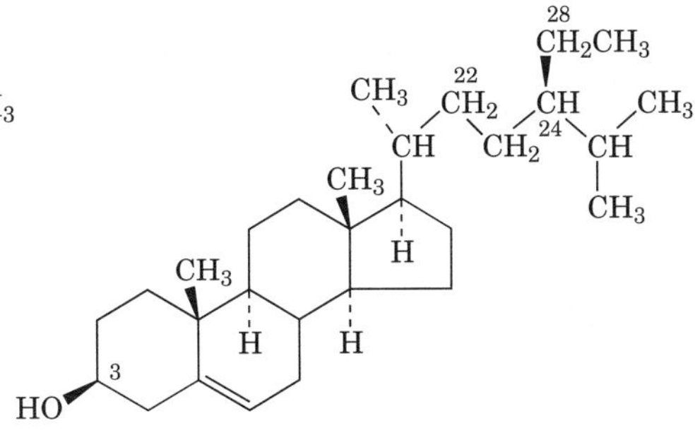

(**2**)

Carbohydrates. Oilseeds contain two types of carbohydrates (qv): soluble mono- and oligosaccharides and largely insoluble polysaccharides. Contents of oligosaccharides of defatted meals are given in Table 6. Sucrose, raffinose, and stachyose are the principal sugars present. The polysaccharide content, not including crude fiber, is roughly equal to the total carbohydrates minus the total oligosaccharides, and these values range from about 12 to 22% of the flours.

Minor Constituents. All four oilseeds contain minor constituents that affect the use of the defatted seeds, especially in feeds and foods. Percentages of phytic acid [83-86-3] (**3**), for example, are soybean, 1.0–1.5 (30); cottonseed kernels, 2.2–3.8 (25); peanut kernels, 0.8 (25); and sunflower, 1.6–1.7 (31).

$$X = O\overset{\displaystyle O}{\overset{\displaystyle \|}{P}}(OH)_2$$

(**3**)

Glanded cottonseed kernels contain 1.1–1.3% gossypol (19) plus related pigments that affect nutritional properties and color of the oil and meal. Cottonseed also contains the cyclopropenoid acids, malvalic and sterculic acids, which exist as glycerides and are concentrated in the seed axis (32).

Soybeans contain only 1–25 ppm of phenolic acids (33) whereas defatted sunflower meal contains 2.7% chlorogenic acid, 0.38% quinic acid, and 0.2% caffeic acid (34). Chlorogenic acid turns from yellow to green and finally to brown as the pH is raised from 7 to 11. Interaction of chlorogenic acid with sunflower proteins results in the formation of green-to-brown isolates, thereby limiting use in foods (35) (see ALKALOIDS; COFFEE).

The isoflavones genistein [446-72-0] (**4**), daidzein [486-66-8] (**5**), and glycetein (**6**) occur in soybeans in the form of malonyl glucosides, acetylated glucosides, glucosides, and aglycones; total concentrations range from 1200–4200 μg/g for different varieties (36).

Table 6. Oligosaccharide Contents of Defatted Oilseed Meals[a], %

Constituent	Soybean	Cottonseed	Peanut	Sunflower
sucrose	6.42	1.64	8.1	6.5
raffinose	1.26	6.91	0.33	3.09
stachyose	4.34	2.36	0.99	0.14
verbascose	trace	trace	trace	
Total	*12.02*	*10.91*	*9.42*	*9.73*
Total carbohydrate[b]	*34.0*	*22.5*	*22.4*	*24.2*

[a]Ref. 28.
[b]Ref. 29. Estimated by difference: 100 − (protein + oil + ash + crude fiber) = nitrogen-free extract.

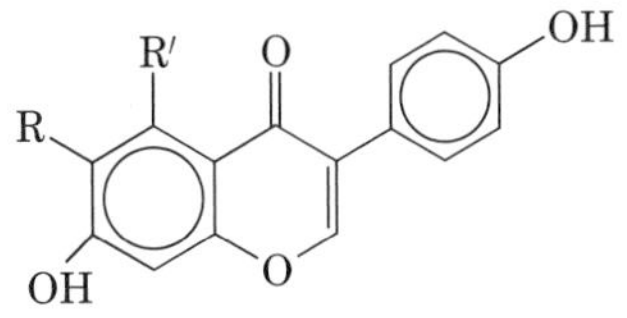

(**4**) R = H, R′ = OH

(**5**) R = R′ = H

(**6**) R = OCH$_3$, R′ = H

Soybeans and peanuts also contain saponins, which are glucoside derivatives of triterpenoid alcohols (37). Saponins range from 0.09 to 0.32% in 457 soybean varieties (38).

Harvesting and Storage

Soybeans. The U.S. soybean crop is grown from Minnesota to Florida. Primary production occurs in the Midwest, ie, in Illinois, Iowa, Minnesota, Indiana, Ohio, and Missouri. Harvest begins in September or October. Ideal moisture for harvesting is 13%, and the crop can be successfully stored at this moisture content until the following summer. Soybeans at ≤12% moisture can be stored for two years or more with no significant deterioration although the entire crop is usually processed in little over a year after harvest. Beans having moisture above safe storage limits are dried or placed in aerated bins for gradual moisture reduction.

Soybeans are trucked from the farm to country elevators and are then moved by truck or rail to processing plants, subterminal elevators, or terminal elevators. From the terminal elevators the beans are shipped by rail or barge to export elevators or processors.

Soybeans are stored in concrete silos 6–12 m dia with heights of ≥46 m. The silos are often arranged in multiple rows, and the resulting interstitial silo areas are likewise used for storage. In bulk storage, seasonal temperature changes cause variations in temperature between the different portions of the grain mass; for example, in the winter, soybeans next to the outer walls are colder than those in the center of the silo. Such temperature differences initiate air currents that transfer moisture from warm to cold portions of the seed mass. Thus, bulk soybeans originally at safe moisture concentrations may, after storage, have localized regions of higher moisture that cause growth of microorganisms, which in turn can lead to heating. If these conditions persist, the beans turn black and may eventually ignite. Such seasonal moisture transfer also occurs with other oilseeds. In commercial practice the temperature is carefully monitored and, when it rises, the beans are either remixed or processed. Aflatoxin contamination is not a problem as it is with cottonseed and peanuts. Although fungi invade soybeans stored at high moisture and temperatures, *Aspergillus flavus* does not grow well on soybeans and aflatoxin levels are negligible. Other mycotoxins, eg, zearalenone, zearalenol, diacetoxyscirpenol, deoxynivalenol, and T-2 toxin, can be found in soybeans damaged by molds in the field when abnormally warm and humid weather prevails and delays harvesting (39).

Cottonseed.   In the United States, cotton is grown in the southern and western states, mainly in Texas, California, Mississippi, Arizona, and Louisiana. Harvesting begins in late July when the ripened boll bursts and the cotton dries and fluffs. The harvest is usually completed by the end of December. After picking, the cotton is processed in gins to separate the lint from the seed. Moisture, temperature, initial quality, previous history, and length of storage determine how long cottonseed can be held before processing. Moisture is the most important factor. For safe storage, moisture should be <10%. Cottonseed is usually stored in Muskogee-type warehouses, ie, low, flat metal buildings having roofs that slope close to the angle of repose of cottonseed, and that are equipped with aeration and temperature-monitoring systems. High temperatures cause rapid deterioration of the seed through formation of free fatty acids; hence the need for ventilation fans and aeration ducts when the temperature rises. At high moisture and temperatures of 28–37°C, cottonseed is also prone to invasion by the fungus, *Aspergillus flavus*, that produces aflatoxins (40). Aflatoxins are highly toxic and produce liver cancer in animals (see FOOD TOXICANTS, NATURALLY OCCURRING).

Peanuts.   When the kernels are fully developed and taking on a mature color, the plants are dug mechanically, shaken to remove the soil, and inverted into windrows to dry (cure) and mature completely. Ideally, the peanuts are left to cure for several days until the moisture content drops to ca 10%. They are harvested mechanically. Green harvesting is practiced under adverse weather conditions, yielding peanuts with 18–25% moisture; artificial drying reduces the moisture to ca 10%. After the moisture is equilibrated between the kernels and hulls, the former contain 7–8% moisture, which is safe for storage.

The cured peanuts are stored or shelled and <10% of the crop is retailed in the shell. Shelled peanuts should have a moisture content of ca 7% (6.5% is optimum) for safe storage and are best stored under refrigeration. Moisture control is critical to maintain quality. At high moisture levels molds grow, including *Aspergillus flavus*, that produce aflatoxins. The Agricultural Marketing Service and the FDA limit aflatoxin to 25 ppb in raw peanuts and 20 ppb in processed peanut food products.

Sunflowers.   Grown primarily in North Dakota, which is the principal producer, South Dakota, Minnesota, Kansas, Colorado, Texas, and Nebraska, sunflower seeds are harvested after a killing frost in late September and October, ca 120 d after planting. In rainy fall conditions, the seed may contain >20% moisture. Such seed must be dried rapidly to ≤10% moisture. The crop dries easily because air readily passes through beds of the large seed. A moisture content of 9.5% is safe for short-term storage, but ≤7% is recommended for long-term storage without aeration (41). Marketing channels lead from the farm to country elevators to processors and export terminals.

Processing

Soybeans.   Virtually all soybeans processed in the United States are solvent-extracted with hexane to recover the oil. This traditional process is outlined in Figure 4. Beans arriving at the plant are cleaned and dried, if necessary, before storage. When the beans move from storage to processing, they are cleaned further and may be dried and allowed to equilibrate at 10–11% moisture to facilitate loosening of the seed coat or hull. They are then cracked, dehulled by screening and aspiration, and conditioned by treatment with steam (qv) to facilitate flaking. The conditioned meats are flaked and extracted with hexane to remove the oil. Hexane and the oil in the miscella are separated by evaporation and the hexane is recovered. Residual hexane in the flakes is removed by steam treatment in the desolventizer-toaster. The moist heat treatment also inactivates antinutritional factors, such as trypsin inhibitors and lectins, in the raw flakes and increases protein digestibility (42–44). Innovations in soybean processing include more efficient methods for dehulling (45) and the use of expanders (46,47). Expanders are extruders used to convert the flakes into collets, ie, porous, sponge-like extrudates. Collets are larger, denser, and less fragile than flakes and thus extract more rapidly and drain more completely, reducing the amount of hexane that needs to be recovered in the desolventizer (see SOLVENT RECOVERY). The result is to increase plant capacity. Fifty percent or more of the soybean plants use expanders, and adoption is expected to continue.

A metric ton of soybeans yields 181 kg of crude oil, 794 kg of 44% protein meal (hulls present), or 722 kg of 48% protein meal (hulls removed) plus 72 kg of hulls. Shrinkage is 25 kg.

Cottonseed. In the United States, cottonseed is processed into oil and meal by screw-pressing or solvent extraction. In screw-pressing the seed is

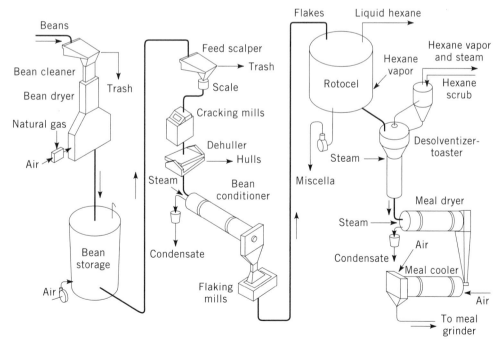

Fig. 4. Schematic outline for processing soybeans into oil and meal by hexane extraction. Courtesy of Dravo Corporation.

cleaned, delinted, dehulled, flaked, and cooked prior to pressing. Screw-pressing yields a cake containing 2.5–4.0% residual oil. The cake is ground into a meal, and ground cottonseed hulls are blended back to adjust protein content to trading standards. In the solvent extraction procedure (Fig. 5) the flakes are often processed through an expander to rapidly cook the flakes and to form collets, which are then extracted with hexane (46,47). Meal emerging from the solvent extractors is freed of hexane by heating. The resulting meal contains about 1% residual oil (see EXTRACTION, LIQUID–SOLID).

Screw-pressed oil is allowed to stand to settle out suspended solids, filtered through plate filter presses, and then pumped to storage. The oil-rich solvent (miscella) from the solvent-extraction process is filtered or clarified, and most of the solvent is removed in a long tube evaporator. Finally, the concentrated oil passes through a stripping column where sparging steam is injected to remove the residual solvent. A metric ton of cottonseed yields ca 91 kg linters, 247 kg hulls, 162 kg oil, and 455 kg meal.

Peanuts. Only 15–20% of the U.S. peanut crop is converted into oil and meal. Processing is carried out by screw-pressing or prepressing, followed by solvent extraction (48). In screw-pressing, the peanuts are shelled, cooked, and

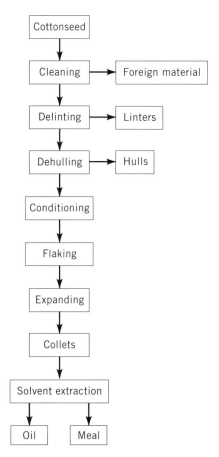

Fig. 5. Schematic outline for processing of cottonseed into oil and meal.

pressed to yield a crude oil plus a cake containing ca 5% residual oil. The cake is ground, and the ground peanut hulls are blended back to adjust protein content. In prepressing–solvent extraction, the cooked meats are screw-pressed at low pressure to remove a portion of the oil and then extracted with hexane to reduce the residual oil to ca 1%. Residual hexane in the meal is recovered by applying jacket or live steam in a desolventizer. Hexane in the miscella is recovered by evaporation, as in the processing of cottonseed. A metric ton of peanuts yields ca 317 kg oil and 418 kg meal. The remainder is shells, foreign matter, and shrinkage.

Sunflowers.   Processing of sunflowers consists of screw-pressing, direct extraction with hexane, or prepress–solvent extraction. The latter is most commonly used in the United States. The first step is cleaning, followed by dehulling. The dehulled seed is conditioned by heating and then goes to screw presses or is flaked as in the case of direct solvent extraction with hexane. The screw-pressed cake is ground for use in feeds or granulated and extracted by hexane to recover the remaining oil. In some plants the hulls are separated and burned to generate part of the power used in processing (49). On processing without dehulling, a metric ton of sunflower seed yields ca 400 kg of oil and 550 kg of high fiber meal.

Economic Aspects

Soybeans are the predominant oilseed crop in the world, providing one-half of the total oilseed production (Table 7). Cottonseed is second. Worldwide, rapeseed is in third place, but this is still a minor oilseed in the United States. Peanuts and sunflower are fourth and fifth, respectively. Since becoming established as a crop in the United States, soybeans have shown phenomenal growth and contribute significantly to the agricultural economy (Table 8).

Average prices of the principal U.S. oilseeds and derived products for 1990–1994 are summarized in Table 9. The United States is the principal producer of soybeans; China is the largest grower of cottonseed and peanuts; Argentina leads in the production of sunflower seed (Table 10).

Soybeans are the most important oilseed in international trade. The United States, Brazil, and Argentina are the main suppliers to the export markets. In 1994–1995, United States exports of 22.0×10^6 metric tons represented 32%

Table 7. World Production of Oilseeds[a], t $\times$ 10⁶

Oilseed	Crop year			
	1991–1992	1992–1993	1993–1994	1994–1995
soybean	107.38	117.23	117.32	137.83
cottonseed	36.62	31.65	29.77	32.99
peanut	22.18	23.08	23.81	26.45
sunflower	21.82	21.31	20.88	23.05
rapeseed	28.27	25.31	26.67	30.22
others[b]	8.14	8.92	9.01	9.55
Total	*224.41*	*227.50*	*227.46*	*260.09*

[a]Ref. 50.
[b]Includes copra and palm kernel.

Table 8. U.S. Soybean Production and Crop Value[a]

Year	Production, t $\times 10^3$	Crop value, $ $\times 10^6$
1930	379	19
1940	2,124	70
1950	8,145	738
1960	15,109	1,185
1970	30,678	3,205
1980	48,926	13,560
1990	52,421	11,042
1995	60,992	13,221

[a]Refs. 50 and 51.

Table 9. Average U.S. Prices of Oilseeds and Derived Products, $/t[a]

Product	1990–1991	1991–1992	1992–1993	1993–1994	1994–1995[b]
			Soybeans		
seed	211	205	204	235	200
meal	200	209	214	213	174
oil	463	421	472	597	601
			Cottonseed		
seed	133	78	108	125	111
meal	144	155	178	181	124
oil	492	443	551	612	573
			Peanut		
seed	765	624	661	670	639
			Sunflower		
seed	240	192	215	284	233
meal	97	85	98	104	66
oil	520	476	558	683	601

[a]Ref. 50.
[b]Values are estimated.

of the domestic production, 16% of the world crop, and 69% of the international trade in soybeans. The European Union, Japan, Mexico, and Taiwan were the largest importers of soybeans (51). Japan, which produces <5% of its soybean needs, is the largest single soybean customer of the United States. Japan imported 3.4 $\times 10^6$ t in 1994 (51). The United States also exports processed soybean products, eg, oil and meal. For 1990–1993, an average of 47% of the crop was exported as whole beans and processed products.

U.S. exports of the other oilseeds are smaller and follow different patterns (50). Exports accounted for only 2–3% of the cottonseed crop and 16–31% of the processed oil production for 1991–1994. Of the peanut crop ca one-half is consumed domestically as whole nut products and 15–20% is exported. Only about 10% of the sunflower seed crop is exported. From 1991 to 1994 an average of 70% of the oil was exported, but only about 11% of the meal was shipped overseas.

Table 10. World Oilseed, Oil, and Meal Production, 1994–1995[a], t × 10³

Country	Soybean			Cottonseed			Peanut			Sunflower		
	Seed	Oil	Meal	Seed	Oil	Meal	Seed	Oil	Meal	Seed	Oil	Meal
Argentina	12,500	1,590	7,469	602	92	269	280	35	44	5,800	1,838	1,976
Brazil	25,900	4,023	16,820	960	114	430	150	14	20	6	2	2
China	16,000	1,220	6,553	7,704	897	3,013	9,682	1,354	2,165	1,500	233	615
EU-15[b]	1,031	2,645	11,566	698	94	270	2	8	14	4,009	2,167	2,815
FSU[c]	556	80	452	3,466	488	1,419	0	0	0	4,443	1,067	976
Indonesia	1,600	17	80	6	0	0	880	16	18	0	0	0
India	3,300	495	2,200	4,615	508	1,627	8,561	1,975	2,880	1,270	415	530
Pakistan	14	6	29	2,830	245	1,120	85	0	0	106	30	37
Paraguay	2,200	160	668	239	35	120	40	8	8	4	1	2
Senegal	0	0	0	20	3	9	720	88	105	0	0	0
United States	68,493	7,082	30,178	6,898	595	1,660	1,927	143	188	2,194	528	653
others	5,688	2,787	12,535	4,915	651	1,859	4,257	450	542	4,202	1,653	1,805
World total	*137,282*	*20,105*	*88,550*	*32,953*	*3,722*	*11,796*	*26,584*	*4,091*	*5,984*	*23,534*	*7,934*	*9,411*

[a]Ref. 52.
[b]European Union.
[c]Former Soviet Union.

About 62 mills having an annual capacity of 47×10^6 t process soybeans in the United States. These operate at an average of ca 80% of capacity (53).

Although soybeans contribute about one-half of the world production of oilseeds, they supply less than one-third of the total edible vegetable fats and oils (Table 11) because of their relatively low oil content. Nonetheless, production of soybean oil exceeds the combined production of cottonseed, peanut, and sunflower seed oils.

The bulk of the oil obtained from the four oilseeds is consumed in food products. For many uses, the various oils can be substituted one for the other. Consequently, the proportions used in a product may depend on small price differences. Estimates for 1993–1994 of the largest domestic uses of vegetable oils, in thousand metric tons, are shown in Table 12 (50). For 1992–1993 totals for peanut and sunflower oils were 75×10^3 and 59×10^3 t, respectively.

Industrial, ie, nonfood, utilization of vegetable oils is much smaller than food usage. Soybean oil usage in the United States for industrial purposes was ca 2% of total production in 1993. Quantities are given in Table 13 (55).

Free substitution of protein meals in feeds is much more restricted than interchange of oils in foods. Because of a good balance of essential amino acids, soybean meal is an indispensable ingredient for efficient feeding of nonruminants, eg, poultry and swine. Soybeans provide ca 60% of the world's protein meals, including fish meal (Table 14). Of the 30.0×10^6 t of soybean meal produced in the United States in 1994–1995, 24.2×10^6 t was used domestically, primarily in feeds, and 5.7×10^6 t was exported (50). In the United States, poultry consume the largest share of soybean meal, followed by swine. Lesser

Table 11. World Production of Edible Vegetable Oils[a], t$\times 10^3$

Oil	1991–1992	1992–1993	1993–1994	1994–1995[b]
soybean	17,472	18,226	19,973	19,695
cottonseed	3,644	3,391	3,717	3,814
peanut	3,597	3,618	4,107	3,820
sunflower	7,328	6,933	7,938	8,925
rapeseed	8,393	9,066	10,051	11,161
others[c]	19,619	20,086	21,492	21,813
Total	*60,053*	*61,320*	*67,278*	*69,228*

[a]Ref. 54.
[b]Values are estimated.
[c]Includes olive, coconut, palm, and palm kernel oils.

Table 12. Food Use of Vegetable Oils, 1993–1994, t $\times 10^3$

Product	Soybean	Cottonseed	Corn
baking and frying fats	1165	98	39
margarine	835	[a]	[a]
salad and cooking oils[b]	2268	131	187
other	100	[a]	
Total	*5367*	*253*	*294*

[a]Usage not disclosed.
[b]Peanut oil also used, but usage not disclosed.

Table 13. Nonfood U.S. Use of Soybean Oil, 1993[a]

Product	Quantity, $t \times 10^3$
resins and plastics	45
paint and varnish	18
feed	11
lubricants	3
others[b]	57
Total	*134*

[a]Ref. 55.
[b]Includes soaps, fatty acids, SoyDiesel fuel and ink.

Table 14. World Production of Oilseed Protein Meals[a], $t \times 10^6$

Protein meal	1990–1991	1991–1992	1992–1993	1993–1994	1994–1995[b]
soybean	69.5	73.1	75.9	80.8	87.1
cottonseed	12.2	13.3	11.5	10.7	11.8
peanut	4.8	4.8	5.1	5.2	6.0
sunflower	8.9	8.6	8.3	8.1	9.4
others[c]	23.8	25.5	23.5	25.2	28.1
Total	*119.2*	*125.3*	*124.3*	*130.0*	*142.4*

[a]Ref. 51.
[b]Preliminary.
[c]Includes rapeseed, fish, copra, and palm kernel meals.

amounts are fed to beef and dairy cattle. Soybean meal is a principal ingredient in many pet foods (see FEEDS AND FEED ADDITIVES).

The edible oilseed protein industry is comparatively small and is restricted to peanut and soybean proteins. One company manufactures partially defatted peanut flours made by hydraulic pressing. The products contain 40–42% protein. Production estimates for edible soybean proteins in the United States in 1993–1994 (56) and wholesale prices as of November 1995 are given in Table 15.

Table 15. U.S. Edible Soybean Protein Production, 1993–1994[a]

Product	Quantity, $t \times 10^{3b}$	Price[c], $/kg
Soybean protein products		
flours and grits	316	0.40–0.81
protein concentrates	127	1.43–1.69
protein isolates	164	2.90–3.72
fiber	5	1.10–5.50
Traditional soy foods[d]		
tofu (regular and fried)	9[e]	
soymilk	4[e]	
others	37[e]	

[a]Ref. 56.
[b]Unless otherwise noted.
[c]Wholesale, as of Nov. 1995.
[d]Ref. 57.
[e]Metric tons, 1994.

Nutritional Properties and Antinutritional Factors

Oil. Because of their high linoleic acid [60-33-3] contents, unhydrogenated and partially hydrogenated soybean, cottonseed, peanut, and sunflower oils are good sources of this essential fatty acid. Soybean oil is the principal vegetable oil consumed in the United States (see Table 12), and approximately three-fourths is partially hydrogenated to improve flavor stability, increase resistance to oxidation, and increase the melting point. The last is important in margarines, bakery and confectionery fats, and shortenings. Linoleic and linolenic acid [463-40-1] contents of soybean oil are reduced by hydrogenation, but the process is also accompanied by migration of double bonds up and down the fatty acid chain and the conversion of cis to trans isomers, ie, positional and geometrical isomerization. Although epidemiological evidence has suggested a relationship between trans fatty acid consumption and coronary heart disease risk, a more recent review of the data concludes that such a conclusion is equivocal (58). Further study is needed to clarify the possible role of trans fatty acids in risk of heart disease. Interesterification of fats of different compositions is being used in Europe and Canada as an alternative to hydrogenation to modify the physical properties of oils. Interest in this process is increasing in the United States (59).

Heating of fats during the frying of foods results in hydrolysis and oxidation reactions that generate various compounds, including free fatty acids, alcohols, aldehydes, ketones, dimer acids, and polymeric fatty acids. Under extreme conditions of heating for prolonged periods of time, products toxic to laboratory animals can be formed. Foods produced in such situations are, however, unpalatable. Thus consumption of such foods becomes self-limiting (60).

Cyclopropenoid fatty acids found in cottonseed oil are biologically active in several animal species (61). Upon ingestion by laying hens, these fatty acids are deposited in the egg yolks. On storage of the eggs, the yolks become rubbery and the whites turn pink (62). Cyclopropenoid acids act as synergists with aflatoxins and as liver carcinogens when fed to trout (63). The long history of cottonseed oil use in the human diet, however, has not revealed any adverse effects. Hence, there is presumptive evidence that humans are not affected at past and present levels of ingestion (64). Crude cottonseed oils contain 0.6–1.0% cyclopropenoid acids (65) but on refining, particularly during deodorization, the levels drop to 0.04–0.4% (66).

Proteins and Meals. Nutritional properties of the oilseed protein meals and their derived products are determined by the amino acid compositions, content of biologically active proteins, and various nonprotein constituents found in the defatted meals. Phytic acid (**3**), present as salts in all four meals, is believed to interfere with dietary absorption of minerals such as zinc, calcium, and iron (67) (see FOOD TOXICANTS, NATURALLY OCCURRING; MINERAL NUTRIENTS).

Soybeans. Numerous studies have demonstrated that methionine is the first limiting amino acid in soybean proteins. That is, methionine is in greatest deficit for meeting the nutritional requirements of a given species. Although it is common practice to add synthetic methionine to broiler feeds to compensate for this deficiency, methionine supplementation is not necessary for humans except for infants (68). Table 3 shows that soybean proteins meet or exceed the essential amino acid requirements of FAO/WHO/UNU for children from age two to adults.

The presence of trypsin inhibitors in soybeans is well-documented, and when ingested by laboratory animals, these inhibit growth and affect organs such as the pancreas (42,43). The inhibitors are largely inactivated by moist heat, and there are no documented cases where ingestion of soybeans by humans has affected the pancreas.

Cottonseed.   When compared with FAO/WHO/UNU essential amino acid requirements (see Table 3), cottonseed proteins are low in lysine, threonine, and leucine for two to five-year-old children, yet meet all requirements for adults.

Raw defatted cottonseed flours contain 1.2–2.0% gossypol [*303-45-7*] (**7**) (19). When cottonseed is treated with moist heat, the ϵ-amino group of lysine and gossypol forms a derivative that is biologically unavailable thereby inactivating gossypol but further lowering the effective content of lysine.

(**7**)

Gossypol has other adverse effects. It is toxic to monogastric animals and, when present in rations of laying hens, causes yolk discoloration (4,62). Ruminants are generally considered to be capable of detoxifying gossypol, but when large quantities of gossypol are ingested for several weeks, toxicosis can occur (69). Cottonseed flour containing gossypol has been fed to humans in a number of studies, but no instances of toxicity have been reported (4). The U.S. Food and Drug Administration (FDA) limits the content of free gossypol, the portion extractable with acetone:water, 70:30 (vol/vol), in edible cottonseed flour to 450 ppm. Gossypol content of cottonseed flour can be lowered by liquid or air classification to remove the pigment glands but neither method is used commercially (5). Glandless gossypol-free cottonseed was formerly produced on a limited scale.

The cyclopropenoid fatty acids, malvalic acid and sterculic acid, exist in hexane-defatted meal at levels of 21–76 ppm (70). In rainbow trout, the cyclopropenoid acids cause cancer of the liver either alone or by acting synergistically with aflatoxin B_1. However, similar effects in mammals or humans have not been demonstrated (63).

Contamination of cottonseed by aflatoxins is a perennial concern. The FDA limits the amount of aflatoxin in cottonseed meal intended for beef cattle, swine, and poultry to 300 ppb and for dairy cattle the limit is 20 ppb.

Peanuts.   The proteins of peanuts are low in lysine, threonine, cystine plus methionine, and tryptophan when compared to the amino acid requirements for children but meet the requirements for adults (see Table 3). Peanut flour can be used to increase the nutritive value of cereals such as cornmeal but further improvement is noted by the addition of lysine (71). The trypsin inhibitor content of raw peanuts is about one-fifth that of raw soybeans, but this concentration is sufficient to cause hypertrophy (enlargement) of the pancreas in rats. The inhibitors of peanuts are largely inactivated by moist heat treatment (48). As for cottonseed, peanuts are prone to contamination by aflatoxin. FDA regulations limit aflatoxin levels of peanuts and meals to 100 ppb for breeding beef cattle, breeding swine, or poultry; 200 ppb for finishing swine; 300 ppb for finishing beef cattle; 20 ppb for immature animals and dairy animals; and 20 ppb for humans.

Sunflower Seed. Compared to the FAO/WHO/UNU recommendations for essential amino acids, sunflower proteins are low in lysine, leucine, and threonine for two to five-year-olds but meet all the requirements for adults (see Table 3). There are no principal antinutritional factors known to exist in raw sunflower seed (35). However, moist heat treatment increases the growth rate of rats, thereby suggesting the presence of heat-sensitive material responsible for growth inhibitions in raw meal (72). Oxidation of chlorogenic acid may involve reaction with the ϵ-amino group of lysine, thus further reducing the amount of available lysine.

Oilseed Products and Uses

Oil. Most crude oil obtained from oilseeds is processed further and converted into edible products. Only a small fraction of the total oil from soybeans, cottonseed, peanuts, and sunflower seed is used for industrial (nonedible) purposes.

Edible Oil. For edible uses, oilseed oils are processed into salad and cooking oils, shortenings, margarines, and confectionery fats such as for candy, toppings, icings, and coatings (73). These products are prepared by a series of steps, as outlined for soybean oil in Figure 6.

Degumming removes the phosphatides and gums, which are refined into commercial lecithin (qv) or returned to the defatted flakes just before the desolventizing-toasting step (see Fig. 4, hexane extraction). Then, free fatty acids, color bodies, and metallic prooxidants are removed using aqueous alkali. Some processors omit the water-degumming step and remove the phosphatides and free fatty acids with alkali in a single operation. High vacuum steam distillation in the deodorization step removes undesirable flavors to yield a product suited for salad oil. Partial hydrogenation, under conditions where linolenate is selectively hydrogenated, results in greater stability to oxidation and flavor deterioration. After winterization, ie, cooling and removal of solids that crystallize in the cold, the product is suited for use as salad and cooking oils. Alternatively, soybean oil can be hydrogenated under selective or nonselective conditions to increase its melting point and produce hardened fats. Such a partially hydrogenated soybean oil, by itself or in a blend with other vegetable oils or animal fats, is used for shortening and margarine. Blends of soybean or other oils of varying melting point ranges are utilized to obtain desired physical characteristics, eg, mouth feel and plastic melting ranges, and the least expensive formulation.

Polyunsaturated fatty acids in vegetable oils, particularly linolenic esters in soybean oil, are especially sensitive to oxidation. Even a slight degree of oxi-

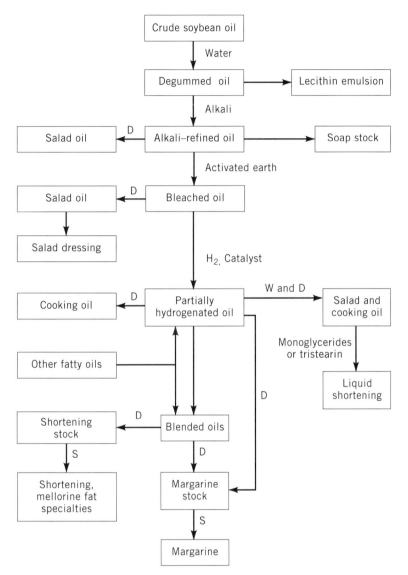

Fig. 6. Schematic outline for manufacture of edible soybean oil products, where D = deodorization, W = winterization, and S = solidification (73). Courtesy of the American Soybean Association and the American Oil Chemists' Society.

dation, commonly referred to as flavor reversion, results in undesirable flavors, eg, beany, grassy, painty, or fishy. Oxidation is controlled by the exclusion of metal contaminants, eg, iron and copper; addition of metal inactivators such as citric acid; minimum exposure to air, protection from light, and selective hydrogenation to decrease the linolenate content to ca 3% (74). Careful quality control is essential for the production of acceptable edible soybean oil products (75).

Nonfood Uses. Vegetable oils are utilized in a variety of nonedible applications, but only a few percent of the U.S. soybean oil production is used for such products (see Table 13). Soybean oil is converted into alkyd resins (qv) for protective coatings, plasticizers, dimer acids, surfactants (qv), printing inks, SoyDiesel fuel (methyl esters used to replace petroleum-based diesel fuel) and other products (76).

Protein Products. The bulk of the meal obtained in processing of oilseeds is used as protein supplements in animal feeds. Since the 1960s appreciable amounts have been also converted into products for human consumption, the majority of which have been derived from defatted soybean flakes.

Feeds. The high protein content of oilseed meals has made them essential ingredients of poultry and livestock feeds. Approximate compositions are shown in Table 16. Soybean meal, especially the dehulled product, is low in crude fiber and high in lysine. Although limiting in methionine, soybean meal is a key ingredient for blending with corn in formulating feeds for nonruminants, eg, poultry and swine. The two proteins complement each other; soy supplies the lysine and corn provides some of the methionine. Poultry rations are routinely supplemented with synthetic methionine to provide a balanced ration at relatively low cost. Because of its gossypol content, high fiber, and low lysine content, cottonseed meal is used primarily for beef and dairy feeds. Less than one-third of the U.S. cottonseed meal is used for poultry feeds. Peanut meal also is high in fiber but limiting in lysine and methionine, and low in tryptophan; its main outlet is therefore for dairy and beef cattle feeds. Sunflower meal is high in fiber and low in lysine and threonine, and therefore also fed mainly to ruminants, ie, sheep, beef, and dairy cattle. It is sometimes used to partially replace soybean meal in poultry and swine rations (see FEEDS AND FEED ADDITIVES).

Edible Protein Ingredients. As of the mid-1990s only peanuts and soybeans are converted into protein ingredients for use in food products. Peanuts are hydraulically pressed to remove about 55% of the oil and the pressed peanuts are then ground into flours and sold raw or roasted for use in baked products, snacks, and confections.

Starting materials for soybean protein ingredients are defatted flakes prepared essentially as outlined in Figure 4, except that greater attention is paid to sanitation than in processing for feed use, and the hexane is removed by vapor desolventizing–deodorizing or flash desolventizing, thereby yielding flakes

Table 16. Compositions of Oilseed Protein Meals[a,b], wt %

Meal	Dry matter	Crude Protein	Fat	Fiber	Ash
Solvent process					
soybean					
with hulls	89.6	44.0	0.5	7.0	6.0
dehulled	89.3	47.5	0.5	3.0	6.0
peanut with hulls	92.3	47.0	1.0	13.0	4.8
Prepress-solvent process					
cottonseed	89.9	41.0	0.8	12.7	6.4
sunflower[c]	89.4	31.0	2.5	21.8	6.0

[a]Ref. 77, unless otherwise noted.
[b]Analytical values, on an as-is basis, are approximate.
[c]Ref. 78.

ranging from raw to fully cooked (79). Degree of cooking is determined by measuring the amount of water-soluble protein remaining after moist heat treatment with the protein dispersibility index (PDI) or nitrogen solubility index (NSI) test (80). A raw flake has a PDI or NSI of ca 90, whereas a fully cooked flake has a value of 5–15.

Defatted soybean flakes are processed into three classes of products differing in minimum protein content (expressed on a dry basis): flours and grits (50% protein); protein concentrates (65% protein); and protein isolates (90% protein). Table 17 shows typical analyses. Flours and grits are made by grinding and sieving flakes. Concentrates are prepared by extracting and removing the soluble sugars from defatted flakes by leaching with dilute acid at pH 4.5 or by leaching with aqueous alcohol (82). Protein isolates are obtained by extracting the soluble proteins from defatted flakes with water at pH 8–9, precipitating with acid at pH 4.5, centrifuging the resulting protein curd, washing, redispersing in water (with or without adjusting the pH to ca 7), and finally spray-drying (83,84) (Fig. 7). Flours, concentrates, and isolates are also processed, commonly by extrusion, to texturize them for use as meat extenders and substitutes (85).

Oilseed proteins are used as food ingredients at concentrations of 1–2% to nearly 100%. At low concentrations, the proteins are added primarily for their functional properties, eg, emulsification, fat absorption, water absorption, texture, dough formation, adhesion, cohesion, elasticity, film formation, and aeration (86) (see FOOD PROCESSING). Because of high protein contents, textured flours and concentrates are used as the principal ingredients of some meat substitutes.

Table 17. Compositions of Soybean Protein Products and Their Uses[a], wt %

Constituent	Defatted flours and grits[b]	Protein	
		Concentrates[c]	Isolates[d]
protein[e]	56.0	72.0	96.0
fat	1.0	1.0	0.1
fiber	3.5	4.5	0.1
ash	6.0	5.0	3.5
carbohydrates			
soluble	14.0	2.5	0
insoluble	19.5	15.0	0.3

[a]Analytical values on a moisture-free basis (81).

[b]Used in baked goods (breads, cakes, cookies, crackers, doughnuts), pasta products, emulsified and coarsely ground meat products, meat analogues, breakfast cereals, dietary foods, infant foods, confections, milk replacers, and pet foods.

[c]Used in baked goods (breads, cakes, cookies, snack items), pasta products, infant formulas, milk replacers, emulsified and coarsely ground meat items, meat analogues, dietary foods, and soup mixes and gravies.

[d]Used in baked goods (breads, cakes and cake mixes, cookies, crackers, snacks), pasta products, dairy-type products (beverage powders, coffee whiteners, whipped toppings), infant formulas, milk replacers for young animals, emulsified and coarsely ground meat items, meat analogues, hams, poultry breasts, dietary food items, and soup mixes and gravies.

[e]As nitrogen, N, $\times$ 6.25.

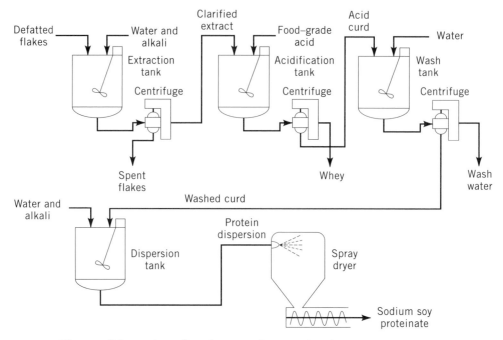

Fig. 7.   Schematic outline for manufacture of soybean protein isolates.

Use of some oilseed proteins in foods is limited by flavor, color, and flatus effects. Raw soybeans, for example, taste grassy, beany, and bitter. Even after processing, residues of these flavors may limit the amounts of soybean proteins that can be added to a given food (87). The use of cottonseed and sunflower seed flours is restricted by the color imparted by gossypol and phenolic acids, respectively. Flatus production by defatted soy flours has been attributed to raffinose and stachyose, which are removed by processing the flours into concentrates and isolates (88).

Industrial Products.   Among the oilseeds, only soybean protein isolates are used for industrial applications in the United States. These are available commercially in unmodified, partially hydrolyzed, and chemically modified forms. The hydrolyzed types are made by suspending the acid-precipitated curd (see Fig. 7) in alkali and heating to dissociate the proteins into subunits and partially hydrolyze the polypeptide chains. The reaction is terminated by acidifying to pH of ca 4.5 to precipitate the modified protein, which is then recovered by centrifuging and drying. The chemically modified isolates are made by proprietary processes. Industrial isolates are used primarily as adhesives for binding pigments to paper and paperboard to make the surfaces suitable for printing (76,89).

Food Products.   Soybeans, peanuts, and sunflower seeds are consumed as such or are processed into edible products.

Soybeans.   Soybeans are not eaten raw because they are too hard and have an unpalatable grassy–beany flavor. Small amounts are roasted and salted for snacks. Nut substitutes for baked products and confections are also manufactured from soybeans. Larger amounts are used in Oriental foods, some of which are increasingly popular in the United States.

Soymilk. In the traditional process, soybeans are soaked in water, ground into a slurry, cooked, and filtered to remove the insoluble cell wall and hull fractions. A number of modifications have been made in the process since the 1960s, including heat treatment before or during grinding to inactivate the enzyme lipoxygenase and thus prevent formation of grassy and beany flavors. The soymilks are available in plain and flavored, eg, vanilla and chocolate, forms (90,91).

Tofu. Tofu is prepared by adding a coagulant such as calcium sulfate to soymilk to precipitate the protein and oil into a gelatinous curd. The curd is then separated from the soluble portion (whey), pressed, and washed to yield a market-ready product. Tofu, a traditional food in Japan (90), was popularized in the United States in the late 1970s and is available in many U.S. supermarkets.

Miso. Miso is a paste-like food having the consistency of peanut butter. It is made by fermenting cooked soybeans and salt with or without a cereal such as rice or barley (92). It is used as a base for soups in Japan, and as a seasoning in southern and eastern Asia. It is produced on a small scale in the United States.

Tempeh. Dehulled cooked soybeans are inoculated with the mold, *Rhizopus oligosporus*, packed in perforated plastic bags and allowed to ferment for 18 h. The mold mycelium overgrows the soybean cotyledons and forms a compact cake. When sliced and deep-fried in oil, a crisp and golden brown product is obtained. Although native to Indonesia, tempeh has become popular with vegetarians in the United States and other Western countries (93).

Soy Sauce. Soy sauce is a well-known condiment made by fermentation or acid hydrolysis. In the fermentation process defatted soybean meal is cooked and then mixed with roasted, coarsely ground wheat and mixed with a culture of *Aspergillus oryzae* or *Aspergillus sojae*. After the mold grows for 2–3 d to form koji, brine is added, and the mixture is allowed to ferment for 6–8 m. The product is then filtered and pasteurized (94). Popularization of fermented soy sauce in the U.S. began in the late 1940s with imports from Japan, followed by construction of a plant in Wisconsin in 1973. Soy sauce is widely available in U.S. supermarkets and restaurants. In the acid hydrolysis process, defatted soybean flour is refluxed with hydrochloric acid to hydrolyze the proteins. The hydrolysate is then filtered, neutralized, and bottled.

Peanuts. About 65% of U.S. peanuts are consumed directly in the form of peanut butter, roasted peanuts, and confections. Peanut butter is made and consumed primarily in the United States. It is prepared by shelling the nuts, dry roasting, removing the skins, and grinding finely. The ground material is then blended with salt and other ingredients that may include stabilizers, eg, vegetable oil, mono- and diglycerides; sweeteners, eg, glucose, corn syrup solids, or honey; lecithin; and antioxidants (95) (see NUTS).

Sunflower Seed. Confectionery-type sunflower seed, ca 15% of the U.S. crop, is cleaned and sized by screening. The large seed is dry-roasted, salted, and sold in the shell. The medium sized seeds are dehulled, roasted dry or in oil and used in cookies, salad toppings, ice cream toppings, trail and snack mixes, and breads and rolls. The small-sized, off-sized, and broken seed is sold as bird and pet feeds (96).

BIBLIOGRAPHY

"Soybeans" in *ECT* 1st ed., Vol. 12, pp. 689–701 by J. C. Cowan, Northern Regional Research Center, U.S. Dept. of Agriculture; in *ECT* 2nd ed., Vol. 18, pp. 599–614, by J. C. Cowan, U.S. Dept. of Agriculture; "Soybeans and Other Oilseeds" in *ECT* 3rd ed., Vol. 21, pp. 417–442, by W. J. Wolf, U.S. Dept. of Agriculture.

1. T. Hymowitz and J. R. Harlan, *Econ. Bot.* **37**, 371 (1983).
2. K. Saio and T. Watanabe, *Nippon Shokuhin Kogyo Gakkai-Shi* **15**, 290 (1968).
3. J. N. A. Lott, in N. E. Tolbert, ed., *The Biochemistry of Plants, A Comprehensive Treatise*, Vol. 1, *The Plant Cell*, Academic Press, Inc., New York, 1980, pp. 589–623.
4. L. C. Berardi and L. A. Goldblatt, in I. E. Liener, ed., *Toxic Constituents of Plant Foodstuffs*, 2nd ed., Academic Press, Inc., New York, 1980, pp. 183–237.
5. E. W. Lusas and G. M. Jividen, *J. Am. Oil Chem. Soc.* **64**, 839 (1987).
6. A. K. Smith, in A. M. Altschul, ed., *Processed Plant Protein Foodstuffs*, Academic Press, Inc., New York, 1958, pp. 249–276.
7. F. R. Earle, C. H. Van Etten, T. F. Clark, and I. A. Wolff, *J. Am. Oil Chem. Soc.* **45**, 876 (1968).
8. P. J. Wan, G. W. Baker, S. P. Clark, and S. W. Matlock, *Cereal Chem.* **56**, 352 (1979).
9. W. J. Wolf, *J. Agric. Food Chem.* **18**, 969 (1970).
10. W. J. Wolf, *J. Agric. Food Chem.* **44**, 785 (1996).
11. E. H. Rahma and M. S. Narasinga Rao, *J. Food Sci.* **44**, 579 (1979).
12. E. Derbyshire, D. J. Wright, and D. Boulter, *Phytochemistry* **15**, 3 (1976).
13. G. E. Sykes and K. R. Gayler, *Arch. Biochem. Biophys.* **210**, 525 (1981).
14. V. H. Thanh and K. Shibasaki, *Biochem. Biophys. Acta* **490**, 370 (1977).
15. N. E. Tumer, V. H. Thanh, and N. C. Nielsen, *J. Biol. Chem.* **256**, 8756 (1981).
16. N. C. Nielson and co-workers, *Plant Cell* **1**, 313 (1989).
17. T. G. Krishna and R. Mitra, *Phytochemistry*, **26**, 897 (1987).
18. J. F. Cavins, W. F. Kwolek, G. E. Inglett, and J. C. Cowan, *J. Assoc. Off. Anal. Chem.* **55**, 686 (1972).
19. J. T. Lawhon, C. M. Cater, and K. F. Mattil, *J. Am. Oil Chem. Soc.* **54**, 75 (1977).
20. E. W. Lusas, *J. Am. Oil Chem. Soc.* **56**, 425 (1979).
21. *Energy and Protein Requirements*, FAO/WHO/UNU Expert Consultation report, Tech. Rep. Ser. No. 724, World Health Organization, Geneva, Switzerland, 1985.
22. C. A. Brignoli, J. E. Kinsella, and J. L. Weihrauch, *J. Am. Diet. Assoc.* **68**, 224 (1976).
23. R. E. Worthington, R. O. Hammons, and J. R. Allison, *J. Agric. Food Chem.* **20**, 727 (1972).
24. S. L. Abidi, T. L. Mounts, and K. A. Rennick, *J. Liq. Chromatogr.* **17**, 3705 (1994).
25. W. A. Pons, Jr., M. F. Stansbury, and C. L. Hoffpauir, *J. Assoc. Off. Agric. Chem.* **36**, 492 (1953).
26. T. Itoh, T. Tamura, and T. Matsumoto, *J. Am. Oil Chem. Soc.* **50**, 122 (1973).
27. T. Hirota, S. Goto, M. Katayama, and S. Funahashi, *Agric. Biol. Chem.* **38**, 1539 (1974).
28. T. M. Kuo, J. F. Van Middlesworth, and W. J. Wolf, *J. Agric. Food Chem.* **36**, 32 (1988).
29. G. F. Cegla and K. R. Bell, *J. Am. Oil Chem. Soc.* **54**, 150 (1977).
30. G. M. Lolas, N. Palamidis, and P. Markakis, *Cereal Chem.* **53**, 867 (1976).
31. M. Saeed and M. Cheryan, *J. Food Sci.* **53**, 1127 (1988).
32. G. S. Fisher and J. P. Cherry, *Lipids* **18**, 589 (1983).
33. M. B. V. Ramakrishna, B. K. Mital, K. C. Gupta, and N. K. Sand, *J. Food Sci. Technol.* **26**, 154 (1989).
34. C. M. Cater, S. Gheyasuddin, and K. F. Mattil, *Cereal Chem.* **49**, 508 (1972).
35. E. W. Lusas, in A. M. Altschul and H. L. Wilcke, eds., *New Protein Foods*, Vol. 5, Academic Press, Inc., New York, 1985, pp. 393–433.

36. H. Wang and P. A. Murphy, *J. Agric. Food Chem.* **42**, 1674 (1994).

37. K. R. Price, I. T. Johnson, and G. R. Fenwick, *Crit. Rev. Food Sci. Nutr.* **26**, 27 (1987).

38. M. Shiraiwa, K. Harada, and K. Okubo, *Agric. Biol. Chem.* **55**, 323 (1991).

39. B. J. Jacobsen and co-workers, *Plant Dis.* **79**, 86 (1995).

40. T. A. P. Hamsa and J. C. Ayres, *J. Am. Oil Chem. Soc.* **54**, 219 (1977).

41. E. H. Gustafson, *J. Am. Oil Chem. Soc.* **55**, 751 (1978).

42. G. Grant, *Prog. Food Nutr. Sci.* **13**, 317 (1989).

43. I. E. Liener, *Crit. Rev. Food Sci. Nutr.* **34**, 31 (1994).

44. M. L. Kakade, D. E. Hoffa, and I. E. Liener, *J. Nutr.* **103**, 1772 (1973).

45. H. Schumacher, in T. H. Applewhite, ed., *Proceedings of the World Congress on Vegetable Protein Utilization in Human Foods and Animal Feedstuffs*, American Oil Chemists' Society, Champaign, Ill., 1989, pp. 37–40.

46. E. W. Lusas and L. R. Watkins, *J. Am. Oil Chem. Soc.* **65**, 1109 (1988).

47. M. A. Williams, *Inform* **6**, 289 (1995).

48. K. C. Rhee, in Ref. 35, pp. 359–391.

49. D. Lilleboe, ed., *Sunflower Handbook*, National Sunflower Association, Bismarck, N.D., 1991, 35 pp.

50. *Oil Crops Yearbook*, OCS-1995, U.S. Dept. of Agriculture, Economic Research Service, Washington, D.C., July 1995.

51. *Soya Bluebook Plus*, Soyatech, Inc., Bar Harbor, Maine, 1996.

52. *Commodity and Marketing Programs*, U.S. Dept. of Agriculture, Foreign Agricultural Service, Oilseeds and Products Division, Washington, D.C., Mar. 1996.

53. Technical data, National Oilseed Processors' Association, Nov. 1995.

54. *Agricultural Statistics*, U.S. Dept. of Agriculture, Washington, D.C., 1995.

55. *Industrial Uses of Agricultural Materials Situation and Outlook*, IUS-5, U.S. Dept. of Agriculture, Economic Research Service, Washington, D.C., Sept. 1995.

56. Technical data, United Soybean Board, Chesterfield, Mo., Mar. 1995.

57. W. Shurtleff, "Breeding and Marketing Soybeans for Food Uses: A Blueprint for Changing Our Seed Company's Basic Mission," presented at *Incoming Soybean Technical Mission*, sponsored by Ontario Soybean Growers' Marketing Board, Harrow Research Station, Harrow, Ontario, Canada, Oct. 18, 1994.

58. P. M. Kris-Etherton, ed., *Am. J. Clin. Nutr.* **62**(3S), 655S (1995).

59. B. F. Haumann, *Inform* **5**, 668 (1994).

60. W. L. Clark and G. W. Serbia, *Food Technol.* **45**(2), 84 (1991).

61. A. A. Andrianaivo-Rafehivola, E. M. Gaydou, and L. H. Rakotovao, *Oleagineux* **49**, 177 (1994).

62. R. A. Phelps, F. S. Shenstone, A. R. Kemmerer, and R. J. Evans, *Poultry Sci.* **44**, 358 (1965).

63. J. D. Hendricks, R. O. Sinnhuber, P. M. Loveland, N. E. Pawlowski, and J. E. Nixon, *Science* **208**, 309 (1980).

64. F. H. Mattson, in *Toxicants Occurring Naturally in Foods*, 2nd ed., National Academy of Sciences, Washington, D.C., 1973, pp. 189–209.

65. A. V. Bailey, J. A. Harris, E. L. Skau, and T. Kerr, *J. Am. Oil Chem. Soc.* **43**, 107 (1966).

66. J. A. Harris, F. C. Magne, and E. L. Skau, *J. Am. Oil Chem. Soc.* **41**, 309 (1964).

67. N. R. Reddy, S. K. Sathe, and D. K. Salunkhe, *Adv. Food Res.* **28**, 1 (1982).

68. H. R. Churella, M. W. Borschel, M. R. Thomas, M. Breen, and J. Jacobs, *J. Am. Coll. Nutr.* **13**, 262 (1994).

69. L. A. Kerr, *Am. Assoc. Bovine Practitioners Comp. Contin. Edu. Prac. Vet.* **11**, 1139 (1989).

70. R. S. Levi, H. G. Reilich, H. J. O'Neill, A. F. Cucullu, and E. L. Skau, *J. Am. Oil Chem. Soc.* **44**, 249 (1967).

71. G. N. Bookwalter, K. Warner, R. A. Anderson, and E. B. Bagley, *J. Food Sci.* **44**, 820 (1979).

72. H. E. Amos, D. Burdick, and R. W. Seerley, *J. Anim. Sci.* **40**, 90 (1975).

73. D. R. Erickson, ed., *Practical Handbook of Soybean Processing and Utilization*, American Oil Chemists' Society Press, Champaign, Ill. and United Soybean Board, St. Louis, Mo., 1995, Chapts. 9–15, pp. 161–276 and Chapts. 18–20, pp. 314–379.

74. E. N. Frankel, in D. R. Erickson, E. H. Pryde, O. L. Brekke, T. L. Mounts, and R. A. Falb, eds., *Handbook of Soy Oil Processing and Utilization*, American Soybean Association, St. Louis, Mo., and American Oil Chemists' Society, Champaign, Ill., 1980, pp. 229–244.

75. Ref. 73, Chapt. 24, pp. 483–503.

76. Ref. 73, Chapt. 21, pp. 380–427.

77. R. D. Allen, *Feedstuffs* **53**(30), 25 (1981).

78. D. H. Kinard, *Feed Manage.* **32**(6), 16 (1981).

79. K. W. Becker, *J. Am. Oil Chem. Soc.* **60**, 216 (1983).

80. D. Firestone, ed., *Official Methods and Recommended Practices of the American Oil Chemists' Society*, 4th ed., American Oil Chemists' Society, Champaign, Ill., 1989, Methods Ba 10-65 and Ba 11-65.

81. F. E. Horan, *J. Am. Oil Chem. Soc.* **51**, 67A (1974).

82. M. F. Campbell, C. W. Kraut, W. C. Yackel, and H. S. Yang, in Ref. 35, pp. 301–337.

83. D. W. Johnson and S. Kikuchi, in Ref. 45, pp. 66–77.

84. D. H. Waggle, F. H. Steinke, and J. L. Shen, in R. H. Matthews, ed., *Legumes— Chemistry, Technology, and Human Nutrition*, Marcel Dekker, Inc., New York, 1989, pp. 99–138.

85. F. E. Horan, in A. M. Altschul, ed., *New Protein Foods*, Vol. 1A, *Technology*, Academic Press, Inc., New York, 1974, pp. 366–413.

86. J. E. Kinsella, S. Damodaran, and B. German, in Ref. 35, pp. 107–179.

87. G. MacLeod and J. Ames, *Crit. Rev. Food Sci. Nutr.* **27**, 219 (1988).

88. J. J. Rackis, *J. Am. Oil Chem. Soc.* **58**, 503 (1981).

89. R. A. Olson and P. T. Hoelderle, in R. Strauss, ed., *Protein Binders in Paper and Paperboard Coating*, Tappi Monograph Series No. 36, Tappi Press, Atlanta, Ga., 1975, pp. 75–96.

90. W. Shurtleff and A. Aoyagi, *The Book of Tofu*, Vol. 2, *Tofu and Soymilk Production*, New-Age Foods Study Center, Lafayette, Calif., 1979, 336 pp.

91. S. Chen, in E. W. Lusas, D. R. Erickson, and W.-K. Nip, eds., *Food Uses of Whole Oil and Protein Seeds*, American Oil Chemists' Society, Champaign, Ill., 1989, pp. 40–86.

92. Ref. 91, Chapt. 9, pp. 131–147.

93. Ref. 91, Chapt. 7, pp. 102–117.

94. Ref. 91, Chapt. 8, pp. 118–130.

95. Ref. 91, Chapt. 12, pp. 171–190.

96. Ref. 91, Chapt. 14, pp. 205–217.

General References

A. M. Altschul and H. L. Wilcke, eds., *New Protein Foods*, Vol. 5, *Seed Storage Proteins*, Academic Press, Inc., New York, 1985, 471 pp.

J. F. Carter, ed., *Sunflower Science and Technology*, American Society of Agronomy, Madison, Wis., 1978, 505 pp.

D. R. Erickson, ed., *Edible Fats and Oil Processing: Basic Principles and Modern Practices*, American Oil Chemists' Society, Champaign, Ill., 1990, 442 pp.

D. R. Erickson, ed., *Practical Handbook of Soybean Processing and Utilization*, American Oil Chemists' Society Press, Champaign, Ill. and United Soybean Board, St. Louis, Mo., 1995, 584 pp.

D. R. Erickson, E. H. Pryde, O. L. Brekke, T. L. Mounts, and R. A. Falb, eds., *Handbook of Soy Oil Processing and Utilization*, American Soybean Association, St. Louis, Mo. and American Oil Chemists' Society, Champaign, Ill., 1980, 598 pp.

Y. H. Hui, ed., *Bailey's Industrial Oil and Fat Products*, 5th ed., John Wiley & Sons, Inc., New York, 1996, 3000 pp.

R. J. Kohel and C. F. Lewis, eds., *Cotton*, American Society of Agronomy, Madison, Wis., 1984, 605 pp.

E. W. Lusas, D. R. Erickson, and W.-K. Nip, eds., *Food Uses of Whole Oil and Protein Seeds*, American Oil Chemists' Society, Champaign, Ill., 1989, 401 pp.

H. B. W. Patterson, *Handling and Storage of Oilseeds, Oils, Fats and Meal*, Elsevier Applied Science, New York, 1989, 394 pp.

G. Röbbelen, R. K. Downey, and A. Ashri, eds., *Oil Crops of the World, Their Breeding and Utilization*, McGraw-Hill Book Co., Inc., New York, 1989, 553 pp.

D. K. Salunkhe, J. K. Chavan, R. N. Adsule, and S. S. Kadam, *World Oilseeds— Chemistry, Technology, and Utilization*, Van Nostrand Reinhold, Co., Inc., New York, 1992, 554 pp.

W. O. Scott and S. R. Aldrich, *Modern Soybean Production*, 2nd ed., S & A Publications, Inc., Champaign, Ill., 1983, 230 pp.

H. E. Snyder and T. W. Kwon, *Soybean Utilization*, Van Nostrand Reinhold Co., New York, 1987, 346 pp.

D. P. S. Verma and R. C. Shoemaker, eds., *Soybean: Genetics, Molecular Biology and Biotechnology*, CAB International, Wallingford, Oxon, U.K., 1996, 288 pp.

R. Wilcox, ed., *Soybeans: Improvement, Production, and Uses*, 2nd ed., American Society of Agronomy, Inc., Madison, Wis., 1987, 888 pp.

R. F. Wilson, ed., *Designing Value-Added Soybeans for Markets of the Future*, American Oil Chemists' Society, Champaign, Ill., 1991, 135 pp.

J. G. Woodroof, ed., *Peanuts: Production, Processing, Products*, 3rd ed., AVI Publishing Co., Inc., Westport, Conn., 1983, 414 pp.

Copra (coconut), palm fruit, and palm kernel

J. Am. Oil Chem. Soc. **62**(2) (1985).

Flaxseed

J. F. Carter, *Cereal Foods World* **38**, 753 (1993).

C. L. Lay and C. D. Dybing in G. Röbbelen, R. K. Downey, and A. Ashri, eds., *Oil Crops of the World, Their Breeding and Utilization*, McGraw-Hill Book Co., Inc., New York, 1989, pp. 416–430.

B. D. Oomah and G. Mazza, *Food Chem.* **48**, 109 (1993).

Rapeseed

J. K. Daun, *Cereal Foods World* **29**, 291 (1984).

R. Ohlson, in A. M. Altschul and H. L. Wilcke, eds., *New Protein Foods*, Vol. 5, John Wiley & Sons, Inc., New York, 1985, pp. 339–358.

R. Ohlson, *J. Am. Oil Chem. Soc.* **69**, 195 (1992).

WALTER J. WOLF
U.S. Department of Agriculture

SPACE CHEMISTRY. See EXTRATERRESTRIAL MATERIALS; SPACE PROCESSING.

SPACE PROCESSING

The idea of processing materials in space originated at the NASA Marshall Space Flight Center (MSFC) during the Apollo program as the confluence of several different and essentially unrelated disciplines. Engineers responsible for propulsion systems needed to understand the behavior of liquids in tanks of orbiting stages. This led to the study of fluid mechanics in low gravity, where liquid behavior is dominated by interfacial forces. The need for welding and brazing in space was recognized for the eventual building of large space structures. Thermal designers considering the use of phase-change materials for thermal control of spacecraft were also interested in understanding how materials solidify in the virtual absence of gravity. A group of solid-state physicists, searching for new superconducting mechanisms, became interested in producing fine *in situ* dispersions of semiconductors (qv) in metals and performed the first low gravity monotectic alloy solidification experiments using the MSFC drop tower (1).

Because the residual accelerations resulting from atmospheric drag and gravity gradient effects in low Earth orbit are typically on the order of 10^{-6} times Earth-gravity, the term microgravity generally is used to describe this acceleration environment. Conferences were held at MSFC in 1968 and 1969 to develop concepts for using the space environment for the actual manufacturing of unique materials. The prospects for using microgravity to prevent convective mixing, reduce sedimentation, and eliminate contamination by containerless processing were recognized, as was the possibility of using space to create a nearly perfect vacuum. The disciplines that could benefit from space processing were identified as alloy solidification, crystal growth, and separation of biological materials.

Some simple preliminary microgravity experiments were carried out during return trips from the moon on Apollo 14 and 17, but the Skylab and Apollo–Soyuz missions in the mid-1970s offered the first opportunities to conduct materials processing experiments in dedicated facilities. The results of these experiments, though interesting, were not spectacular, and it became apparent that a much greater understanding of the subtle effects of fluid behavior in the virtual absence of gravity was needed in order to properly utilize this new environment. Further, it was realized that some of the early hopes of actually producing unique materials in space on a commercial basis would remain impractical ones for the foreseeable future.

On the other hand, it was recognized, at least by some, that the microgravity environment of low Earth orbit provided a unique laboratory in which various processes could be studied under greatly simplified transport conditions. It was believed that such experiments, if properly done, could lead to a better fundamental understanding of the role of transport in various processes that could suggest improved control strategies for use on Earth. Also, the prospect of being able to produce benchmark materials to serve as paradigms for Earth-based technologies may be useful for evaluating potential improvements in performance that could be expected from better control of heat and mass transport during the process. Thus, the goal of the original space processing program was redirected from one of space manufacturing toward a more scientific materials science and applications program for the U.S. Space Shuttle era.

Because by the mid-1990s the space shuttle has been operational for more than a decade, a large number of microgravity experiments have been conducted on various flights, as opportunities warrant. Several microgravity-emphasis missions, in which the shuttle is flown in an attitude that minimizes acceleration disturbances, have also been flown to accommodate experiments that are exceptionally sensitive to accelerations. The more academically oriented experiments designed to address fundamental issues of materials processing are sponsored by the NASA Office of Microgravity Science and Applications; experiments designed to address issues of more direct interest to industrial research are sponsored by the NASA Office of Space Access and Technology (formerly the Office of Commercial Programs).

Materials Experiments in Space

Protein Crystal Growth. As of the mid-1990s, the protein crystal growth experiments produced the most spectacular results of all the space processing experiments. The importance of x-ray crystallography as a mechanism for determining three-dimensional structure of complex macromolecules has placed new demands on the ability to grow large (ca 0.5 mm on a side), highly ordered crystals of a vast variety of biological macromolecules in order to obtain high resolution x-ray diffraction data. There are numerous difficulties encountered in attempts to grow macromolecular crystals of biological interest, and the ability to grow such crystals of sufficient size and quality has become an important step for advancement in this field.

The difficulties encountered prompted some investigators to consider growing protein crystals in reduced gravity. The motivation for these experiments was first the ability to suspend the growing crystals in the growth solution to provide a more uniform growth environment, and secondly to reduce the convective mass transport so that growth could take place in a diffusion-controlled environment. The effect of convection on the growth of crystals is not well understood, but it is generally accepted that nonuniform growth conditions that can result from convective flows are not conducive to good internal order.

The first protein growth experiment in reduced gravity was carried out in 1983 (2) on Spacelab 1; crystals of lysozyme and β-galactosidase grew substantially larger in space than in ground control experiments. These results prompted the large number of other space protein crystal growth experiments conducted following that flight.

The first U.S. protein crystal growth experiment under reasonably well-controlled conditions was carried out in September 1988 on a space transportation system (STS-26), the first shuttle flight after the Challenger accident (3). The γ-interferon sample and the porcine elastase sample grew much larger than the ground control samples. The isocitrate lyase sample grew as discrete prisms, whereas the ground control crystals always grew dendritically. Crystals of all three of these proteins exhibited significantly higher x-ray diffraction resolution than any produced on Earth (4). This was true even when some of the smaller space-grown crystals were compared with larger Earth-grown crystals.

Following STS-26, there have been many other attempts to grow a variety of protein crystals in space. Considered as a whole, these experiments have produced a mixed set of results. In some cases the space experiments yielded no crystals or produced crystals that were inferior to those grown on Earth. However, there have been a number of cases in which the space-grown crystals were larger and better ordered than the best ever grown on Earth. In fact, the improvement in internal order obtained in several important proteins grown in reduced gravity has been so dramatic that structures have been solved or refined to higher resolution than had been possible using the diffraction data from the best available Earth-grown crystals. For example, a single crystal of human serum albumin (HSA) was grown on the first International Microgravity Lab (IML-1) that provided a data set with 15% more observations than the combined and averaged data sets from the best Earth-grown crystals, including those grown in gels (5). Crystals of satellite tobacco mosaic virus (STMV) grew 10-fold larger (by volume) than the largest ever grown on Earth and extended the resolution from 0.23 to 0.18 nm (6). Crystals of lysozyme, grown on Spacelab-1 and on IML-2, exhibited a reduction in mosaicity by a factor of 3, as determined both from rocking curve widths and Laue spot size (7).

The ability of a payload specialist on the first United States Microgravity Lab (USML-1) to mix the protein solutions actively and monitor the nucleation and early growth paid large dividends in the ability to grow crystals of several systems that had not been successful in earlier flight experiments (5). Malic enzyme that diffracted to 0.26 nm was grown in space, whereas the best crystals grown on Earth diffracted to only 0.32 nm. Similar results were obtained for Factor D, human α-thrombin, and HIV-1 reverse transcriptase. Bovine brain prolyl-isomerase crystals grown on Earth often form clusters and twins. The space-grown crystals were substantially larger and showed no clustering, twinning, or variations in diffraction quality.

However, there is still the question of why only some of the protein growth experiments in space were able to produce superior results, while many others did not. It should be remembered that not all experiments on the ground are successful; some are not reported. One possible explanation is that the growth process is developed and optimized on the ground before committing the experiment to flight. However, the conditions that are optimum under normal gravity may not take advantage of the microgravity environment. Therefore, it may be necessary to actually develop the optimal growth processes in space in order to improve the yield of protein crystal growth experiments in microgravity.

Solution Growth of Small-Molecule Crystals.   At least some of the advantages obtained from growth of macromolecular crystals in microgravity appear to carry over to the growth of small-molecule crystals. Triglycine sulfate (TGS) crystals were grown from solution on Spacelab-3 and again on IML-1 using a novel cooled sting method (8). Supersaturation was maintained by extracting heat through the seed, which was mounted on a small heat pipe, and in turn was attached to a thermoelectric device. By growing under diffusion-controlled transport conditions, it was hoped it would be possible to avoid liquid–vapor inclusions. These inclusions are the most common types of defect in solution-grown crystals and are believed to be caused by unsteady growth conditions resulting from convective flows.

Typically, TGS crystals are grown on $\langle 001 \rangle$ oriented seeds, because growth on the $\langle 010 \rangle$ face tends to be nonuniform and multifaceted. However, in the absence of convection, growth on the $\langle 001 \rangle$ seeds on Spacelab-3 was mostly around the periphery of the seed. Therefore, seeds with a natural $\langle 010 \rangle$ face were cut from a polyhedral TGS crystal for the experiments on IML-1. The crystal was grown in space with a 4°C undercooling, which produced a growth rate of 1.6 mm/d. Even though this is somewhat larger than typical growth rates (because of the limited time available) the quality of the space-grown crystal was extremely good. Growth was very uniform and the usual growth defects in the vicinity of the seed that form during the transition from dissolution to growth, known as ghost of the seed, were notably absent. High resolution x-ray topographs taken at Brookhaven National Laboratory (Brookhaven, New York) using the National Synchrotron Light Source indicated a crystal of exceptional quality. The only inclusions were from the incorporation of the polystyrene marker particles which had been added to the solution for flow visualization. As a pyroelectric detector for the far infrared, the detectivity, D^*, of the space-grown crystal was significantly higher than that of the seed crystal, and the loss tangent was reduced from 0.12–0.18 for the seed to 0.007 for the space-grown material.

Bridgman Growth of Electronic and Photonic Materials. The early Skylab experiments demonstrated that growth striations in doped semiconductors (qv), believed to be caused by unsteady convective flows, could be eliminated in microgravity and that diffusion-controlled growth conditions could be established (9). This prompted a number of attempts to grow bulk multicomponent alloy-type systems with the objective of obtaining better compositional homogeneity necessary to achieve uniform electronic and optical properties. This goal, as of ca 1996, has not yet been realized because of the small (micro-g) quasisteady residual accelerations from atmospheric drag and gravity gradient effects. Many systems of interest have Schmidt numbers (ratio of viscosity to chemical diffusivity) that are on the order of 10^2–10^3. In order to achieve good homogeneity, it is necessary to keep the product of Grashof and Schmidt number on the order of unity (10). This requires an acceleration level below the typical micro-g available in low Earth orbit. For this reason, NASA is considering the development of furnaces with imposed magnetic fields for the space station, to further suppress the small residual flows.

In many of the Bridgman growth experiments in reduced gravity, however, the solidified ingot was found to be smaller than the growth ampul, and the melt appears to have pulled away from the ampuls during the solidification process. A satisfactory explanation for this effect has been offered (11). According to this model, the newly grown crystal initially pulls away from the ampul wall because of differential thermal expansion. In the absence of hydrostatic pressure, a meniscus forms between the solid and the wall and the subsequent growth front tracks this meniscus, causing the diameter of the growing crystal to become progressively smaller. This continues until the meniscus angle reaches the value required for constant diameter growth. This effect certainly alters the thermal boundary conditions and opens the possibility for surface tension-driven (Marangoni) convection to disturb the diffusion-controlled transport conditions.

However, crystals grown in this manner frequently exhibit fewer dislocations, twins, and other growth defects (12). In fact, it has been suggested that this low gravity effect may provide more important benefits related to Bridgman growth than the establishment of diffusion-limited growth conditions (13).

Vapor Crystal Growth. For materials that lend themselves to physical or chemical vapor transport, growth from the vapor offers some attractive alternatives to growth from the melt. Growth can take place at temperatures considerably lower than the melting point, thus avoiding some of the higher temperature problems associated with melt growth. Gravity-driven convection definitely influences the growth process, perhaps in ways that are not yet completely understood or appreciated. For example, it has been shown that compositional gradients arising from the interaction of multicomponent systems with any vertical wall always results in horizontal density gradients which produce buoyancy-driven convective flows (14). However, since the Schmidt numbers characteristic of the vapor growth process are ca 1, diffusion-limited growth conditions can be obtained under far less stringent acceleration conditions than those required for melt growth (15).

Several crystal growth experiments on the shuttle have produced provocative results that are not at all understood, eg, growth of unseeded GeSe crystals by physical vapor transport using an inert noble gas as a buffer in a closed tube on STS-7 (16). In the ground control experiment, many small crystallites formed a crust inside the growth ampul at the cold end. The flight experiment produced dramatically different results; the crystals apparently nucleated away from the walls and grew as thin platelets that eventually became entwined with one another, forming a web that was loosely contained by the ampul. Even more striking was the appearance of the surfaces of the space-grown crystals. These were mirror-like and almost featureless, exhibiting only a few widely spaced growth terraces. By contrast, the crystallites in the ground control experiments conducted under identical thermal conditions had many pits and irregular closely spaced growth terraces.

Another vapor growth experiment was carried out on USML-1 in which $Hg_{0.4}Cd_{0.6}Te$ was grown by closed-tube chemical vapor deposition on HgCdTe substrates using HgI_2 as the transport agent. Again, considerable improvements in the flight samples were observed in terms of surface morphology, chemical microhomogeneity, and crystalline perfection (17). These improvements were attributed to the sensitivity of the $Hg_{1-x}Cd_xTe–HgI_2$ vapor transport system to minute fluid dynamic disturbances that are unavoidable in normal gravity.

When thin films of copper phthalocyanine, CuPc, were grown on Cu substrates by physical vapor deposition in a similar experiment on an earlier shuttle flight, a dramatic difference in appearance of the space-grown film as compared with the ground control experiment was also found (18). The central portion of the space-grown samples had a distinct opalescent quality (18). Detailed analysis of the films using a variety of optical, x-ray diffraction, and sem techniques revealed not only a significant difference in the growth morphology, but also a new polymorph of CuPc that had never been seen. The film on the ground control samples, when viewed by sem, had the appearance of a shag carpet in which the individual columns of crystalline CuPc were relatively sparse and were quite irregular. By contrast, the central portion of the space-grown films had the appearance of a thick pile carpet. The columns of CuPc were closely spaced and vertically aligned.

Mercuric iodide crystals grown by physical vapor transport on Spacelab-3 exhibited sharp, well-formed facets indicating good internal order (19). This was confirmed by γ-ray rocking curves which were approximately one-third the width of the ground control sample. Both electron and hole mobility were significantly enhanced in the flight crystal. The experiment was repeated on IML-1 with similar results (20).

Mercuric iodide forms a layered structure, similar to graphite, in which the A–B planes are bonded by van der Waals forces. Consequently, the crystalline structure is very weak, especially at the growth temperature, and it was believed that the performance of the materials as a room-temperature nuclear spectrometer was limited by defects caused by self-deformation during the growth process. This was the original motivation for the flight experiments. It is not clear as of ca 1996 whether the improved quality of the flight crystals was the result of the elimination of the weight of the crystal during its growth, or of the diffusion-controlled transport conditions that produced a more uniform growth environment.

Test of Dendritic Growth Models.   The microgravity environment provides an excellent opportunity to carry out critical tests of fundamental theories of solidification without the complicating effects introduced by buoyancy-driven flows. This advantage was used to carry out a series of experiments to elucidate dendrite growth kinetics under well-characterized diffusion-controlled conditions in pure succinonitrile (SCN) (21). This constituted a rigorous test of various nonlinear dynamical pattern formation theories that provide the basis for the prediction of the microstructure and physical properties achieved in a solidification process.

Dendrite tip velocities were measured as a function of undercooling over a range from 0.05–1.5 K. Comparing these measurements with ground-based measurements, it was possible to show that effects of convection are more significant at the smaller undercoolings and are still important up to undercoolings as large as 1.3 K. Even in microgravity, a slight departure in the data was noted at the smallest undercooling, which was attributed to the residual acceleration of the spacecraft. These data also allow the determination of the scaling constant important in the selection of the dynamic operating state, which the present theories have been unable to provide.

Electrodeposition.   Electrodeposition experiments in reduced gravity have produced some intriguing results. An early experiment on the German TEXUS rocket, using higher current densities than can normally be used on Earth, reported the deposition of amorphous Ni on Au substrates (22). In a series of experiments on the Consort Rocket, it was possible to repeat this result and a mechanism by which this occurs was proposed (23). Careful analysis of the x-ray-transparent Ni films produced in low gravity revealed that a significant amount of hydrogen had been incorporated into the deposit to form NiH. This distorted the lattice, destroying long-range order to the extent that x-ray diffraction peaks could no longer be observed. It was conjectured that the difference in the morphology of the hydrogen bubbles collecting on the cathode surface owing to buoyancy effects must have been responsible for the increased hydrogen incorporation into the film in the low gravity experiments. Ground-based experiments using a frit between the electrodes were partially successful in duplicating some of the effects observed in low gravity (24).

This group also carried out a series of codeposition experiments in which diamond dust was incorporated in Ni coatings and chromium carbide was incorporated in Co coatings. These experiments led directly to a new ground-based technique for electrodepositing a bone-like hydroxyapatite coating on prosthetic implants, which provides significantly better adhesion, and unlike the presently used plasma spray coatings, does not damage the hydroxyapatite (25–27) (see PROSTHETIC AND BIOMEDICAL DEVICES). This is highly significant in developing implants that do not loosen with time. Also, during this investigation, a new plating process was also developed using Cr(III) which poses significantly few environmental problems than the more common Cr(VI) process (28). Exciting spin-offs such as these are likely to be the most important return from the microgravity program.

BIBLIOGRAPHY

1. L. L. Lacy and G. H. Otto, *AIAA J.* **13**, 219 (1975).
2. W. Littke and C. Johns, *Science* **225**, 203 (1984).
3. L. J. DeLucas and co-workers, *J. Cryst. Growth* **110**, 302 (1991).
4. L. J. DeLucas and co-workers, *Science* **246**, 651 (1989).
5. L. J. DeLucas and co-workers, *J. Cryst. Growth* **135**, 172 (1994).
6. J. Day and A. McPherson, *Protein Sci.* **1**(10), 1254 (1992).
7. J. R. Helliwell, E. H. Snell, S. Weisgerber, F. Weckert, K. Holzer, and R. Schroer, *Acta Crystallog. Sect.* **D51**, 1099 (1995).
8. R. B. Lal, A. K. Batra, J. D. Trolinger, W. R. Wilcox, and B. Steiner, *Ferroelectrics* **158**, 81 (1994).
9. A. F. Witt, H. C. Gatos, M. Lichtensteiger, M. C. Lavine, and C. J. Herman, *J. Electrochem. Soc.* **125**, 276 (1975).
10. R. J. Naumann, *J. Cryst. Growth* **142**, 253 (1994).
11. W. R. Wilcox and L. L. Regel, *Microgravity Sci. Technol.* **VIII/I**, 56 (1995).
12. D. J. Larson, "Growth of CdZnTe Compound Semiconductors on Earth and in Space," in Ref. 7.
13. D. T. J. Hurle, G. Mueller, and R. Nitsche, in H. U. Walter, ed., *Fluid Sciences and Materials Science in Space*, Springer-Verlag, Berlin, 1987, Chapt. X, p. 340.
14. F. Rosenberger, in J. Zierep and H. Oertel, eds., *Convective Transport and Instability Phenomena*, Verlag G. Braun, Karlsruhe, Germany, 1982, p. 469.
15. A. Nadarajah, F. Rosenberger, and J. I. D. Alexander, *J. Cryst. Growth* **118**, 49 (1992).
16. H. Wiedemeier, S. B. Trivedi, X. R. Zhong, and R. C. Whiteside, *J. Electrochem. Soc.* **133**, 1015 (1986).
17. H. Wiedemeier, *Vapor Transport of HgCdTe in Microgravity*, NASA Conference Publication 3272, Marshall Space Flight Center, Alabama, May 1994, p. 263.
18. M. K. Debe and co-workers, *Thin Solid Films* **186**, 257 (1990).
19. L. van den Berg and W. F. Schnepple, *Nucl. Instr. Meth. Phys. Res.* **A283**, 335 (1989).
20. L. van den Berg, *Proc. Mater. Res. Soc.* **303**, 73 (1993).
21. M. E. Glicksman, M. B. Koss, and E. A. Winsa, *Phys. Rev. Lett.* **73**, 573 (July 25, 1994).
22. J. Ehrhardt, *Galvanotecnik* **72**, 13 (1981).
23. C. Riley, H. Abi-Akar, B. Benson, and G. Maybee, *J. Spacecraft Rockets* **29**, 386 (1990).

24. J. H. Lee, *Electrodeposition of Ni and Ni–Co Alloys in Low Gravity*, Ph.D. dissertation, University of Alabama in Huntsville, 1994.

25. H. Dasarathy, *Development of Metal–Hydroxyapatite Composite Surface Coatings*, Ph.D. dissertation, University of Alabama in Huntsville, 1994.

26. U.S. Pat. 5,330,826 (July 19, 1994), T. E. Taylor, C. Riley, W. R. Lacefield Jr., H. D. Coble, and G. W. Maybee (to McDonnell Douglas).

27. U.S. Pat. 5,338,433 (Aug. 16, 1994), T. E. Taylor, C. Riley, W. R. Lacefield Jr., H. D. Coble, G. W. Maybee, E. Field, and H. Dasarathy (to McDonnell Douglas).

28. H. Dasarathy, C. Riley, and H. D. Coble, *J. Electrochem. Soc.* **141**, 1773 (1994).

ROBERT J. NAUMANN
University of Alabama in Huntsville

SPANDEX AND OTHER ELASTOMERIC FIBERS. See FIBERS, ELASTOMERIC.

SPECTROSCOPY, OPTICAL

Spectroscopy, the study of electromagnetic radiation and its interaction with matter as a function of frequency or wavelength, is a versatile and powerful tool for investigating atomic and molecular structure, as well as for qualitative and quantitative analysis. Optical spectroscopy conventionally implies the ultraviolet, visible, and infrared spectral regions. Herein coverage is extended to include shorter and longer wavelengths that interact with matter by the same basic mechanism of coupling with the electric vector of the electromagnetic field; mass, acoustic, particle energy, and magnetic resonance spectroscopies are thus excluded (see MASS SPECTROMETRY; MAGNETIC SPIN RESONANCE). The analytical applications of spectroscopy, which range from bench analyses of chemical samples in the laboratory, to process monitoring in chemical plants, to the detection and monitoring of pollutants in the atmosphere, are discussed herein.

The objective in any analytical procedure is to determine the composition of the sample (speciation) and the amounts of different species present (quantification). Spectroscopic techniques can both identify and quantify in a single measurement. A wide range of compounds can be detected with high specificity, even in multicomponent mixtures. Many spectroscopic methods are noninvasive, involving no sample collection, pretreatment, or contamination (see NONDESTRUCTIVE EVALUATION). Because only optical access to the sample is needed, instruments can be remotely situated for environmental and process monitoring (see ANALYTICAL METHODS; PROCESS CONTROL). Spectroscopy provides rapid real-time results, and is easily adaptable to continuous long-term monitoring. Spectra also carry information on sample conditions such as temperature and pressure.

In spectroscopic analysis, species are identified by the frequencies and structures of absorption, emission, or scattering features, and quantified by the intensities of these features. The many applications of optical methods to chemical analysis rely on just a few basic mechanisms of light–matter interaction.

Absorption spectroscopy records depletion by the sample of radiant energy from a continuous or frequency-tunable source, at resonance frequencies that are characteristic of various energy levels in atoms or molecules. The basic law of absorption, credited to Bouguer-Lambert-Beer, states that in terms of the incident, I_0, and transmitted, I_t, light intensities, the absorbance, A (or transmittance, T), is given by equation 1:

$$A(\nu) = -\log T(\nu) = \log(I_0/I_t) = a(\nu)cl \tag{1}$$

where c is the concentration of the absorbing species, l is the path length through the sample, and $a(\nu)$ is an intensive property that specifies the absorption strength of the analyte at frequency ν. Terminology for $a(\nu)$ varies depending on the units of c, on whether A is a decadic or Napierian logarithm, and on conventions employed in different spectral regions. The terms absorption coefficient, absorptivity, extinction coefficient, molar absorptivity, and molar extinction coefficient may each be encountered in different contexts for $a(\nu)$. In multicomponent systems, $A(\nu)$ is simply the sum of the contributions from the individual absorbers. The linear relationship between A and c in equation 1 usually holds over a wide range of c, making this law the basis for quantitative spectroscopy. In practice, deviations from Beer's law are not infrequent, arising from chemical interactions in the sample, multiple scattering in dense or inhomogeneous media, and instrumental effects such as insufficient resolving power or stray light. These deviations simply require that calibration curves be established if quantitative information is desired.

Emission spectroscopy is the analysis, usually for elemental composition, of the spectrum emitted by a sample at high temperature, or that has been excited by an electric spark or laser. The direct detection and spectroscopic analysis of ambient thermal emission, usually in the infrared or microwave regions, without active excitation, is often termed radiometry. In emission methods the signal intensity is directly proportional to the amount of analyte present.

Scattering techniques record the change of a usually monochromatic probe signal scattered by a sample. It can involve elastic (energy-conserving) interactions, such as Rayleigh scattering, where photons undergo only a change in momentum, or the inelastic (energy-changing) Raman effect, in which scattering is accompanied by discrete changes in frequency. Rayleigh scattering occurs for all species having dimensions much smaller than the wavelength of the probe light, Mie scattering occurs from larger dielectric particles, and Tyndall scattering from discontinuities such as interfaces. These elastic processes provide little chemical information, but in atmospheric applications can furnish a return signal for laser infrared radar (lidar) sounding. The Raman effect is weaker by factors of $\sim 10^3$, but spectroscopic analysis of scattered Raman light reveals spectral shifts characteristic of different chemical species (see INFRARED TECH-NOLOGY AND RAMAN SPECTROSCOPY).

Fluorescence and phosphorescence are types of luminescence, ie, emission attributed to selective excitation by previously absorbed radiation, chemical reaction, etc, rather than to the temperature of the emitter. Laser-induced and x-ray fluorescence are important analytical techniques (see LUMINESCENT MATERIALS, CHEMILUMINESCENCE).

Many schemes for exploiting processes of absorption, emission, and scattering have been developed around the experimental details of available light sources, detectors, and spectral analyzers. Spectroscopic analysis has been strongly impacted by the development of lasers (qv) (1,2), used both as powerful monochromatic excitation sources and as broadly tunable spectroscopic probes (3–7). Lasers are routinely used both in laboratory analysis and for active remote sensing. Their extremely high spectral intensity (photons per unit bandwidth) and spatial coherence (low divergence, allowing tight focusing) make even weak scattering processes such as the Raman effect useful. Lasers have been exploited for many of the newer nonlinear responses (8–11) and for ultrasensitive procedures that allow the detection of single atoms and molecules (12) (see TRACE AND RESIDUE ANALYSIS).

Herein optical spectroscopy for laboratory analysis, giving some attention to remote sensing using either active laser-based systems (13–16) or passive (radiometric) techniques (17–20), is emphasized.

Background

The Electromagnetic Spectrum. Electromagnetic radiation is characterized by its wavelength, λ, frequency, ν, or wavenumber, $\tilde{\nu}$, which are related by equation 2:

$$\nu = c/\lambda \qquad \tilde{\nu} = 1/\lambda = \nu/c \qquad (2)$$

where c is the speed of light. Units for wavelength are commonly nm or μm (1 nm = 10 Å = 10^{-3} μm = 10^{-9} m); for frequency, some multiple of cycles per second (hertz); and for wavenumber, cm^{-1} (1 cm^{-1} ≈ 30 GHz). The photon energy is $E = h\nu = hc\tilde{\nu}$, where h is Planck's constant, and so is proportional to frequency and wavenumber (1 eV ≈ 8066 cm^{-1}).

The electromagnetic spectrum is conventionally divided into several energy regions characterized by the different experimental techniques employed and the various nuclear, atomic, and molecular processes that can be studied; these are summarized in Table 1.

Atomic and Molecular Energy Levels. Absorption and emission of electromagnetic radiation can occur by any of several mechanisms. Those important in spectroscopy are resonant interactions in which the photon energy matches the energy difference between discrete stationary energy states (eigenstates) of an atomic or molecular system: $\Delta E_{system} = E_{photon} = h\nu$. This is known as the Bohr frequency condition. Transitions between different types of eigenstates have characteristic energies (see Table 1), and so occur in different spectral regions. All of these regions have at least some applications to chemical analysis, but

Table 1. Regions of the Electromagnetic Spectrum

Region	Wavelength limits[a]	Frequency or photon energy[a]	Transitions observed or excited
radio waves	>30 cm	<1,000 MHz	hyperfine structure from nuclear spins and isotopic shifts
microwaves	1 mm–30 cm	300–1 GHz	rotation and inversion of molecules; cyclotron resonance of electrons in solids
far-infrared (fir) (sub-mm waves)	50–1000 μm	200–10 cm^{-1}	molecular rotations and certain low frequency bending, torsional, and skeletal vibrations; lattice modes in solids
mid-infrared	2.5–50 μm	4,000–200 cm^{-1}	fundamental molecular vibrations (rovibrational spectra)
near-infrared (nir)	0.8–2.5 μm	12,500–4,000 cm^{-1}	vibrational overtones and combinations
visible (vis)	400–800 nm (0.4–0.8 μm)		valence electrons
near-ultraviolet (uv)	200–400 nm		valence electrons (atomic and rovibronic molecular spectra)
vacuum ultraviolet (vuv)	10–200 nm	125–6 eV	inner-shell electrons; ionization
x-rays	0.01–10 nm (0.1–100 Å)	125–0.125 keV	inner-shell electrons; nuclear
gamma-rays	<0.01 nm	>0.125 MeV	nuclear

[a]Values are approximate.

most useful are rotational, vibrational, and electronic transitions. Molecules and molecular ions exhibit all three types of spectra; atoms and atomic ions undergo only electronic transitions.

Rotational transitions in gaseous molecules occur in the far-infrared and microwave regions, generally $\lambda > 100$ μm. Very light species absorb at shorter wavelengths, and in fact the strong, dense rotational spectrum of water vapor for $\lambda > 15$ μm makes operation in the fir difficult. Microwave spectroscopy is an important discipline oriented more toward molecular structure research than chemical analysis (see MICROWAVE TECHNOLOGY). The radar region, which includes longer microwaves and shorter radio waves ($\lambda \approx 0.54-133$ cm), is used for the detection and ranging of extended objects, from raindrops to aircraft to large weather systems. Microwave and radio wave spectroscopy is useful for detecting molecules in astronomical sources (radio astronomy).

Molecules vibrate at fundamental frequencies that are usually in the mid-infrared. Some overtone and combination transitions occur at shorter wavelengths. Because infrared photons have enough energy to excite rotational motions also, the ir spectrum of a gas consists of rovibrational bands in which each vibrational transition is accompanied by numerous simultaneous rotational transitions. In condensed phases the rotational structure is suppressed, but the vibrational frequencies remain highly specific, and information on the molecular environment can often be deduced from linewidths, frequency shifts, and additional spectral structure owing to phonon (thermal acoustic mode) and lattice effects.

Shorter-wavelength radiation promotes transitions between electronic orbitals in atoms and molecules. Valence electrons are excited in the near-uv or visible. At higher energies, in the vacuum uv (vuv), inner-shell transitions begin to occur. Both regions are important to laboratory spectroscopy, but strong absorption by O_2 and O_3 make the vuv unsuitable for atmospheric monitoring. Electronic transitions in molecules are accompanied by structure from vibrational and, in gases, rotational transitions (vibronic and rovibronic bands). Deep inner-shell electronic transitions can be induced by x-ray excitation, useful for elemental analysis (see X-RAY TECHNOLOGY). Electronic transitions typically have larger absorption cross sections than vibrational transitions, and hence greater analytical sensitivity, but spectral overlap and interferences are more likely to be problems.

Transition Widths and Strengths. The widths and strengths of spectroscopic transitions determine the information that can be extracted from a spectrum, and are functions of the molecular parameters summarized in Table 2. Detectivity is determined by spectral resolution and transition strength. Resolution, the ability to distinguish transitions of nearly equal wavelength, depends on both the widths of the spectral features and characteristics of the instrumentation. Unperturbed transitions have natural, $\Delta\nu_N$, widths owing to the

Table 2. Line-Shape Parameters

	Transitions		
Parameter	Rotational	Vibrational	Valence electronic
line frequency, $\tilde{\nu}$, cm^{-1}	1–100	100–4,000	<50,000 (>200 nm)
natural linewidth, $\Delta\tilde{\nu}_N$, cm^{-1}	$<10^{-11}$	$<10^{-7}$	$<3 \times 10^{-3}$
Doppler width at 300 K, $\Delta\tilde{\nu}_D$, cm^{-1}	<0.0005	<0.01	0.01–0.2 (~0.0005 nm)
natural radiative lifetime, τ_N, s	$>10^{-1}$	$>2 \times 10^{-4}$	$>2 \times 10^{-9}$
peak Doppler-broadened absorption cross section, σ_A, cm^2	$<10^{-20}$	$\leq 10^{-18}$	$10^{-11} - 10^{-16\,a}$ $<10^{-17\,b}$
peak differential scattering cross section, $d\sigma/d\Omega$, cm^2/sr			
Rayleigh	negligible	negligible	$<2 \times 10^{-13\,a,c}$ $<10^{-26\,b,d}$
Raman	$<10^{-27}$	$<10^{-28}$	$\sim 10^{-24\,a}$
fluorescence	negligible	negligible	$<5 \times 10^{-16\,a,e}$ $10^{-20} - 10^{-25\,b}$

^{a}Values are for atoms.
^{b}Values are for molecules.
^{c}Values are for resonant scattering by atomic vapors.
^{d}Values are for nonresonant scattering of visible and near-uv radiation by atmospheric gases; resonant Raman scattering approaches 10^{-24} cm^2/sr.
^{e}Values are for STP atmospheric conditions.

intrinsic lifetimes of the states involved. The full width at half-maximum (fwhm), is defined as $\Delta\nu_N = (2\pi\tau)^{-1}$, where τ is the natural lifetime for a spontaneous transition in the absence of external perturbations. Natural linewidths can be resolved by modern laser spectroscopies under laboratory conditions.

Natural linewidths are broadened by several mechanisms. Those effective in the gas phase include collisional and Doppler broadening. Collisional broadening results when an optically active system experiences perturbations by other species. Collisions effectively reduce the natural lifetime, so the broadening depends on a characteristic impact time, τ_C, that is typically 1 ps at atmospheric pressure:

$$\Delta\nu_C = \frac{1}{2\pi}\left(\frac{1}{\tau} + \frac{1}{\tau_C}\right) \tag{3}$$

Doppler broadening arises from the random thermal agitation of the active systems, each of which, in its own rest frame, sees the applied light field at a different frequency. When averaged over a Maxwellian velocity distribution, ie, assuming noninteracting species in thermal equilibrium, this yields a linewidth (fwhm) in cm^{-1}:

$$\Delta\tilde{\nu} = 7.16 \times 10^{-7}\,\tilde{\nu}_0(T/M)^{-1/2} \tag{4}$$

where T is the sample temperature in Kelvin, M the molecular weight of the species in amu, and $\tilde{\nu}_0$ is the transition energy in cm^{-1}. The Doppler broadening of a transition represents the fundamental lower limit for resolution unless special nonlinear spectroscopies are exploited.

Natural and collisional broadening are homogeneous processes because all radiators experience the same local effects. These produce, for both gases and liquids, a Lorentzian line shape, with $\Delta\nu$ (fwhm) $= (2\pi\tau')^{-1}$, where τ' is the effective lifetime of a radiator's uninterrupted oscillation period. In gases, this shape is distorted by Doppler broadening, an inhomogeneous effect having a Gaussian distribution. The contour of a gas-phase transition is thus the mathematical combination (convolution) of the Lorentzian and Gaussian functions, the Voigt profile.

For condensed species, additional broadening mechanisms from local field inhomogeneities come into play. Short-range intermolecular interactions, including solute-solvent effects in solutions, and matrix, lattice, and phonon effects in solids, can broaden molecular transitions significantly.

Finally, instrumental broadening results from resolution limitations of the equipment. Resolution is often expressed as resolving power, $\nu/\Delta\nu$, where $\Delta\nu$ is the probe linewidth or instrumental bandpass at frequency ν. Unless $\Delta\nu$ is significantly smaller than the spectral width of the transition, the observed line is broadened, and its shape is the convolution of the instrumental line shape (apparatus function) and the true transition profile.

With inadequate resolution much of the information in a complex spectrum can be lost leading to consequent degradation of specificity and loss of quantitative accuracy. In condensed phases, demands on instrumental resolution are modest. In gases at atmospheric pressure, collisional broadening is the dominant mechanism. There is usually a linear relationship between pressure and linewidth such that pressure-broadening coefficients are typically $0.0001-0.005$ $(\text{cm} \cdot \text{kPa})^{-1}$ $(0.01-0.5$ $(\text{cm} \cdot \text{atm})^{-1})$ for infrared rovibrational transitions. Thus ir spectrometers used in remote atmospheric sensing might need 0.1 cm^{-1} resolution for scanning the troposphere (lower atmosphere), but analysis in the stratosphere (>25 km) may require <0.01 cm^{-1}. Somewhat lower resolution may suffice for the uv–vis because of larger collisional and Doppler broadening. For laboratory gas samples, low pressures may be used to obtain the narrowest linewidths. Under these conditions, instrumental resolution becomes an important consideration. The most advanced techniques of high resolution interferometry and tunable-laser spectroscopy are required.

The strength of a photon–molecule interaction is determined by the frequency-dependent cross section $\sigma(\nu)$, expressed in cm^2 for absorption and related to $a(\nu)$ in equation 1; or by the differential cross section $d\sigma(\nu)/d\Omega$ in units of cm^2/sr for scattering (14). The latter specifies the likelihood that active species scatter some portion of the incident laser fluence (photons/cm^2) into a viewing solid angle, $\Delta\Omega$, measured in steradians (Fig. 1). The cross sections can be expressed as in equation 5:

$$\sigma(\nu) = \sigma_0 \cdot L(\nu - \nu_0, \Delta\nu) \tag{5}$$

where σ_0 (or $d\sigma_0/d\Omega$) is the peak value for absorption (scattering) and $L(\nu - \nu_0, \Delta\nu)$ is a symmetrical line-shape function parameterized by a center frequency, ν_0, and linewidth $\Delta\nu$.

An important semiclassical measure of the frequency-integrated absorption cross section, known as Ladenburg's formula, is equation 6:

$$\int \sigma(\nu)\, d\nu = \pi r_c c f \equiv S \quad (\text{cm}^2 \cdot \text{frequency}) \tag{6}$$

where r_c is the classical electron radius (2.8×10^{-13} cm) and f is the oscillator strength for the transition (21,22). Similar expressions can be written for the differential scattering or fluorescence cross sections (14). The f-values or line strengths, S, can in principle be calculated from quantum mechanics, but are generally obtained empirically. S is related to a matrix element, d_{ij}, connecting the initial and final states of the transition, ie, $S \propto (d_{ij})^2$, where d_{ij} has dimensions typical of an atom ($\sim 10^{-8}$ cm). The maximum value of the line shape function is inversely proportional to the linewidth, so the peak absorption (or angle-integrated scattering) cross sections, σ_0, can be approximated as in equation 7:

$$\sigma_0 \approx S/\Delta\nu \tag{7}$$

The peak absorption (scattering) cross sections are thus useful comparative measures of detectivity because the latter is a product of the line strength and the practical line resolution.

General Instrumental Considerations. A spectroscope disperses light for visual observation, using a slit to define the source, a collimating lens, a disper-

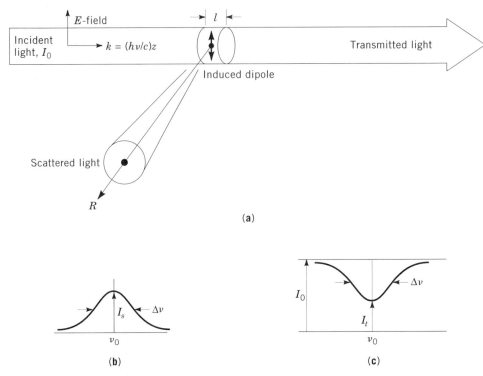

Fig. 1. An incident electromagnetic field of intensity, I_0, having an associated electric field, E, induces dipole oscillation in the absorbers. The transmitted intensity, I_t, is reduced by absorption; of the radiated (scattered) light, a portion, I_s, is shown here detected in a conventional right-angle configuration with collection solid angle $\Delta\Omega_c$. (**a**) Schematic of the absorption/scattering where $I_0 = (1/2)ce_0|E|^2$ in W/m²; (**b**) plot of scattered light, I_s, vs frequency where $I_s = I_0 Nl(d\sigma/d\Omega)\Delta\Omega_c$ and $\Delta\Omega_c$ is proportional to (detector area)/R^2; and (**c**) plot of transmitted light, I_t, vs frequency where $I_t = I_0 \exp(-N\sigma l) \approx I_0(1 - N\sigma l)$ for $N\sigma l \ll 1$.

sive prism or grating, and an objective lens or telescope. Spectroscopes fitted with wavelength scales and cameras are termed spectrometers and spectrographs, respectively. A monochromator is a spectrometer having an exit slit in the focal plane to isolate a narrow wavelength region. If the output is focused on a detector for quantitative intensity measurements it becomes a spectrophotometer. The use of such instrumentation constitutes the field of spectroscopy (sometimes called spectrometry), although spectroscopes themselves are, as of the 1990s, of little importance, for the electromagnetic spectrum extends from ~10 to ~10^{24} Hz, compared to the visible region of only $(3.8 - 7.5) \times 10^{14}$ Hz.

Selecting suitable materials and components for a very broad frequency range presents special problems. Front-surface reflecting optics are usually required because of the lack of suitable achromatic optical materials. Strong absorption by the atmosphere in many regions requires instrumentation that operates in a vacuum or an optical path that can be purged using a transparent gas. Table 3 indicates components and materials typically used in the important near-uv, visible, and ir regions.

Table 3. Components and Materials Used in Uv–Vis–Ir Spectroscopy[a]

Instrumental parameter	Far-infrared[b]	Mid-infrared[c]	Near-uv, visible, and near-ir[d]
broad-band thermal sources	plasma emission from high pressure mercury arc lamp	blackbodies: Nernst glower (zirconia), globar (silicon carbide), nichrome wire coil	deuterium lamp (160–370 nm), Xe arc lamp (300 nm–1.3 μm), quartz-envelope tungsten–halogen lamp (350 nm–2.5 μm)
continuously tunable laser sources	CO_2 laser frequency difference generation (65–1000 μm), microwave sideband mixing (>70 μm)	OPO (<16 μm), color-center lasers (1.5–4 μm), nonlinear optical mixing (2–9 μm), semiconductor diodes (3–28 μm), spin-flip Raman (5–6 μm), waveguide CO_2 (9–11 μm)	organic dye lasers (320 nm–1.3 μm), OPO (>410 nm), Ti:sapphire laser (700–1000 nm), frequency conversion of other lasers (>190 nm), diode lasers (>670 nm)
detectors	pyroelectric (DTGS, LT), doped-germanium and InSb bolometers, Golay pneumatic cell	photoconductors (<40 μm) (InSb; doped germanium; MCT), thermal (thermocouples, bolometers), pyroelectric (DTGS, LT)	photomultipliers (120 nm–1.1 μm), photodiodes (InAs, InGaAs, Si, Ge), photoconductors (Si, Ge)
array detectors	none	PtSi (<5 μm), InSb (<5 μm), MCT (<12 μm), doped Si (10–30 μm)	silicon arrays (175 nm–1.1 μm), photographic emulsions (<1.2 μm), Ge arrays (700 nm–1.5 μm), InSb, PbSe, PtSi
general optical materials	polyethylene, Mylar, quartz, diamond (>6 μm)	alkali halide crystals, fluorite, CaF_2 (<9 μm), ZnSe (800 nm–20 μm), AgCl (420 nm–25 μm), KRS-5 (550 nm–40 μm)	alkali halide crystals, glass (300 nm–2.2 μm), quartz (170 nm–3.5 μm), diamond (220 nm–4 μm), sapphire (150 nm–6 μm), fluorite (120 nm–9 μm)
fiber optics	none	fluoride glasses (900 nm–5 μm), chalcogenide glasses (2.2–12 μm), hollow metal fibers (~10 μm)	silica glasses, quartz (250 nm–1.3 μm)

[a]DTGS = deuterated triglycine sulfate; KRS-5 = mixed thallium bromide–iodide; LT = lithium tantalate; MCT = mercury cadmium telluride; and OPO = optical parametric oscillator. [b]50–1000 μm. [c]2.5–50 μm. [d]200 nm–2.5 μm.

Radio Wave and Microwave Spectroscopy

The longest wavelengths of the electromagnetic spectrum are sensitive probes of molecular rotation and hyperfine structure. An important application is radio astronomy (23–26), which uses both radio and microwaves for chemical analysis on galactic and extragalactic scales. Herein the terrestrial uses of microwave spectroscopy are emphasized (27–29).

Instrumentation. Microwaves have dimensions comparable to those of the experimental apparatus, so neither geometrical optics nor electrical circuit theory applies. Analysis by electromagnetic wave theory is required. Microwave sources are electronic rather than thermal. Klystrons, used most frequently, induce radio-frequency (r-f) fields in a resonant cavity using a modulated electron beam, providing coherent output tunable over a 10% frequency range. Magnetrons, traveling-wave tubes, and backward-wave oscillators are also useful sources. Klystrons and magnetrons cover ~5–50 GHz but can be tuned into the submillimeter region using crystal multipliers. No dispersion is required, because these sources are essentially monochromatic. Microwaves are transmitted through metal waveguides that also serve as gas sample cells. Thin mica (qv) windows are used when necessary. Detectors are crystal rectifiers and sometimes bolometers.

Applications. Molecules couple to an electromagnetic field through their electric dipoles, so only those having a permanent dipole moment exhibit significant rotational spectra. For such species, microwave spectroscopy yields highly precise moments of inertia and details of centrifugal distortion. Applied electric or magnetic fields induce Stark or Zeeman spectra, respectively; the former provide the magnitude of the dipole moment; the latter yield its sign and also the quadrupole moment, a measure of the asymmetry of the molecular charge cloud. Degeneracies of electron and nuclear spin states can be removed using applied magnetic fields, revealing energy differences that typically correspond to microwave frequencies. This is the basis of such important techniques as nuclear magnetic resonance (nmr) and electron paramagnetic resonance (epr) (also called electron spin resonance, esr), which fall outside the definition herein of optical spectroscopy (see MAGNETIC SPIN RESONANCE).

Microwave spectroscopy is used for studying free radicals and in gas analysis (30). Much laboratory work has been devoted to molecules of astrophysical interest (31). The technique is highly sensitive: 10^{-12} mole may suffice for a spectrum. At microwave resolution, frequencies are so specific that a single line can unambiguously identify a component of a gas mixture. Tabulations of microwave transitions are available (32,33). Remote atmospheric sensing (34) is illustrated by the analysis of trace ClO, O_3, HO_2, HCN, and N_2O at the part per trillion level in the stratosphere, using a ground-based millimeter-wave superheterodyne receiver at 260–280 GHz (35).

Infrared Spectroscopy

Infrared spectroscopy has broad applications for sensitive molecular speciation. Infrared frequencies depend on the masses of the atoms involved in the various vibrational motions, and on the force constants and geometry of the bonds connecting them; band shapes are determined by the rotational structure and hence by the molecular symmetry and moments of inertia. The rovibrational spectrum of a gas thus provides direct molecular structural information, resulting in very high specificity. The vibrational spectrum of any molecule is unique, except for those of optical isomers. Every molecule, except homonuclear diatomics such as O_2, N_2, and the halogens, has at least one vibrational absorption in the infrared. Several texts treat infrared instrumentation and techniques (22,36–38) and their applications (39–42).

Instrumentation. The ir region was developed using dispersive techniques adapted as appropriate from uv–vis spectroscopy. Unfortunately, ir sources and detectors tend to be inefficient compared to those for other spectral regions.

In early ir spectrometers (43), mechanically chopped light from a broad-band source passed through a sample cell, was dispersed by a monochromator by either refraction in a prism or diffraction from a reflection grating, and converted into an a-c electrical signal by a detector. This signal was amplified with a lock-in amplifier, thus distinguishing the modulated source beam from thermal ir radiation naturally emitted by the sample. Rotation of the dispersing element scanned the frequencies across the detector, and proper mechanical and electrical manipulation yielded a plot of signal intensity vs wavelength or wavenumber. Highly sophisticated grating spectrophotometers became available commercially in the 1950s, covering the mid-infrared and having resolutions of ~ 1 cm^{-1}. Many of these instruments employed a double-beam configuration, in which a set of rotating mirrors switched the source beam alternately through the sample and through an equivalent reference path many times per second. These signals were compared at the detector, and their ratio provided a spectrum free from artifacts owing to atmospheric absorption, variations in source output, and absorption and scattering by cell windows.

Greatly improved performance was achieved using the Fourier-transform spectrometer (fts) (44–46), essentially a Michelson interferometer in which a collimated light beam divided by a partially reflecting beam splitter is recombined after the optical delay (retardation) of one arm is changed by a scanning mirror. The resulting signal strength as a function of mirror travel is an interferogram, from which the desired spectrum (intensity vs wavenumber) can be obtained by performing a Fourier transform. The discovery of the Cooley-Tukey fast Fourier transform (fft) algorithm in 1962 and the availability of powerful and inexpensive computers, led to Fourier spectroscopy becoming a practical technique.

There are several reasons for the high performance of Fourier instruments. Whereas a scanning spectrometer records a spectrum sequentially, one spectral resolution element at a time, an interferometer processes information from all frequencies simultaneously giving the multiplex or Fellgett advantage. Interferometers also have a throughput (Jacquinot) advantage, accepting a large solid angle of radiation and hence passing a much greater light flux than can slit-limited monochromators. These two advantages can be converted into orders-of-magnitude improvements in resolution, scan time, and/or signal-to-noise (S/N) ratio, the three related instrumental parameters that are of the greatest practical interest to the analyst. The resolution of an fts is approximately the reciprocal of the maximum retardation, or twice the mirror travel. A resolution of 0.1 cm^{-1} thus requires a mirror travel of only 5 cm. Many commercial interferometers offer at least this capability, and research instruments are marketed that can resolve better than 0.002 cm^{-1}. For kinetics studies and monitoring unstable species, scan rates of 50 Hz are available for short mirror travel (low resolution). Step-scan instruments can achieve nanosecond time resolution (47).

Besides bench and research-grade Fourier-transform infrared (ftir) instruments, which have largely replaced dispersive spectrometers, compact and robust ftir analyzers are available specifically for on-line process monitoring. Permanently aligned, industrially hardened instruments are vibration- and shock-resistant and sealed against dust and moisture. Units are available that can monitor many dozens of gases at < ppm sensitivities.

The development of nearly monochromatic lasers that can be continuously frequency-tuned throughout much of the ir (see Table 3) has revolutionized vibrational spectroscopy (48–50). By far the most used are lead salt semiconductor tunable diode lasers (TDLs), which can be tuned over $50–100$ cm^{-1} intervals in continuous scans of >1 cm^{-1}. Commercial TDL systems are modular, consisting of a cryogenic or liquid nitrogen-cooled laser source assembly, collimating optics, a simple mode-selecting monochromator, and a photodetector with lock-in amplifier. Because of the limited tuning ranges, TDLs are not suited for general survey spectroscopy, but do offer the highest resolution (<0.0003 cm^{-1}). In addition to continuous tuning, these can be operated at discrete frequencies, tuning on and off a resonance many times per second for instant quantification. Industrial analytical applications of TDLs include stack-gas monitors for various pollutants. Other tunable sources such as color-center lasers (51), optical parametric oscillators (OPOs) (52), and far-ir lasers (53,54), are also useful in high resolution spectroscopy.

Sampling.   Almost any sample can be prepared for ir analysis (55,56). Cells for gases and liquids, typically 10 cm and 0.1 mm paths, respectively, are available in many configurations. For trace gases, compact small-volume folded-path cells offer adjustable optical paths up to 200 m (57). Gaseous species can be isolated in crystal lattices by dilution in an inert gas that is condensed at cryogenic temperatures (matrix isolation) (58–60). This eliminates Doppler and collisional broadening and suppresses rotational structure, concentrating all the intensity of a band in one sharp absorption line for increased sensitivity. Solutions and gels can be coated directly on ir-transmitting salt plates or microporous plastic films (61). Immersion probes are used for *in situ* analysis in chemical reaction vessels and process streams.

Solids can be observed directly in thin sections, as mulls in mineral or halocarbon oils, as finely ground dispersions in pressed disks of an ir-transparent salt such as KBr, as films cast from solutions, or as solutions in solvents having few infrared absorptions, such as CCl_4 or CS_2. Specular reflection from a solid surface samples the outer ~ 10 μm of materials such as coatings (qv) and films. Nonspecularly reflected light is collected by mirrors in diffuse reflectance spectroscopy, useful for powdered samples and rough surface solids. Examples of such techniques are the characterization of monolayers by specular reflectance (62) and of catalysts by diffuse reflectance (63). Infrared light can be polarized by wire-grid or Brewster's angle pile-of-plates polarizers for the study of oriented and crystalline solids (64).

Specialized sampling techniques include ir microscopes that focus down to 10 μm (approximately the ir diffraction limit) for spectra of less than nanogram samples. This is useful in fiber analysis and forensics (65–68). In attenuated total reflection (ATR), an ir beam is transmitted inside a thin, highly refractive crystal (ZnSe or Ge) by multiple internal reflection; the evanescent wave that penetrates some fraction of a wavelength into the surrounding lower index medium is absorbed by a sample in optical contact with the crystal, providing a useful sampling technique for gels, slurries, strongly absorbing liquids, and coatings (69,70). External reflection methods are also used for solids and films (71). Spectra of small samples under pressures of up to 100 GPa (10^6 atm) can be obtained using diamond anvil cells (72). Many devices have been designed for holding samples at cryogenic or elevated temperatures (73). Time-resolved techniques are available for the study of transient species (74–76).

For process monitoring and analysis in chemical plants, transmission of the ir beam through fiber optics allows safe access to a small sample region in harsh environments far removed from the spectrometer (77–81). Mid-ir transmitting fibers are available (see Table 3), although usefulness is limited by moisture sensitivity, brittleness, and impurity absorptions. Systems employing low power diodes and silica fibers have been demonstrated for monitoring explosive gases and NH_3, using near- and mid-ir absorption.

For open-path atmospheric monitoring, the two basic optical arrangements are point-to-point (bistatic), where the source and detector are separated and the volume to be sampled lies between them, and single-ended (monostatic), in which a probe beam is returned from a retroreflecting target (a mirror or topographic feature) and traverses the sampled volume twice. The optical path length through the sample may be ill-defined, yielding a column density averaged over the total path, so results are usually given in units of concentration $\times$ length. Sea-level atmospheric sounding is limited to regions free of H_2O and CO_2 absorption. Significant portions of the ir ($\sim$1.4, $\sim$1.9, 2.5–2.9, 4.2–4.4, 5.5–7.5, and >15 μm) are unusable (82). High altitude observatories and balloon-borne spectrometers, however, can conduct useful upper atmospheric and astronomical observations even in the far-ir (53).

Null-Background Techniques. In conventional absorption spectroscopy the difference between two large quantities, the incident and transmitted intensities, is measured, thus limiting the minimum detectable absorbance, A_{min}, to $\sim 10^{-3}$. This can be greatly improved using null-background techniques, where the detected signal is (within limits) proportional to the source intensity. An example is harmonic or derivative spectroscopy (48), in which a narrow-band light source is frequency-modulated and synchronously amplified at the modulation frequency (or at a nth order harmonic), yielding the first (nth order) derivative of the absorption profile. This eliminates the background, reduces effects of low frequency drifts in the source, and discriminates against spurious signals lacking a sharp wavelength dependence at the modulated frequency. For TDLs the drive current is modulated using standard radio-frequency (r-f) sources; balloon-borne TDL spectrometers have reached $A_{min} \approx 10^{-5}$ (83), and laboratory values of 10^{-8} have been reported.

Another such technique is direct calorimetric measurement of the radiant energy absorbed as the latter is converted into kinetic motion (heat) (4). In photoacoustic spectroscopy (pas) (84–87) pressure modulation caused by absorption of a modulated source is synchronously detected by a sensitive microphone transducer (spectrophone). A closed gas cell is required (for maximum sensitivity, an acoustically resonant one), and the microphones used are very small, so pas is suited for the analysis of small samples. Liquids or solids can be placed in direct contact with piezoelectric or pyroelectric transducers that convert pressure waves or temperature changes into an electrical signal. Typical applications of pas are for opaque solids, highly scattering media such as biological specimens, and low concentration gases, including pollutants (88). Pas detection has been demonstrated for volatile organics at <ppb levels using line-tunable CO and CO_2 lasers. A_{min} can theoretically reach 10^{-10}. Achievable sensitivity is illustrated by the analysis of ethylene (qv), C_2H_4, at the 20 parts per trillion level in air at 10 μm (89).

A closely related technique useful for localized gas concentrations and leaks is photoacoustic detection and ranging (padar) (90). A laser pulse tuned to an absorption line generates an acoustic signal that is detected by a parabolic microphone. A range resolution of 1 cm out to 100 m is feasible.

Applications. Infrared spectroscopy is broadly applicable to analytical problems of molecular speciation and quantification (39,91,92), for most molecules have strong fundamental vibrational transitions in the mid-infrared. In the region 3–8 μm, many chemical functional moieties exhibit characteristic group frequencies that are relatively independent of the molecular environment, providing information on the chemical nature of the absorber (93,94). An example is the strong C=O stretching mode in saturated aliphatic ketones at $1705-1725$ cm^{-1} (see KETONES). At longer wavelengths the frequencies are influenced more by the skeletal vibrations of the molecule. This is known as the fingerprint region where even similar species may have sufficiently different spectra to be readily distinguished.

Single-component unknowns can be identified by simply comparing their spectra with reference spectra, of which many catalogues are available (95). Reference spectra are also available in digitized versions, and searches of databases (qv) can be made rapidly by computer. Even if no reference spectrum of the unknown is available, the group frequencies may provide enough information for an experienced spectroscopist to make a full identification. The rovibrational structure in gaseous samples further increases specificity, and also furnishes an estimate of the sample temperature that may be useful in remote sensing.

Mixtures can be identified with the help of computer software that subtracts the spectra of pure compounds from that of the sample. For complex mixtures, fractionation may be needed as part of the analysis. Commercial instruments are available that combine ftir, as a detector, with a separation technique such as gas chromatography (gc), high performance liquid chromatography (hplc), or supercritical fluid chromatography (96,97). Instruments such as gc/ftir are often termed hyphenated instruments (98). Pyrolyzer (99) and thermogravimetric analysis (tga) instrumentation can also be combined with ftir for monitoring pyrolysis and oxidation processes (100) (see ANALYTICAL METHODS, HYPHENATED INSTRUMENTS; THERMAL AND GRAVIMETRIC ANALYSIS).

Quantitative analysis based on Beer's law (eq. 1) is performed by measuring the absorption at a peak of known strength. In n-component mixtures, n such features are measured and the resulting set of simultaneous linear equations is solved for the n concentrations. Detection limits depend on the intrinsic band strengths, the optical configuration employed (especially the path length), and the spectral resolution. For solutions these are typically in the range of 0.1–1%, with precisions of a few percent of the quantity measured. For narrow absorptions, sensitivity improves with increasing resolution until the latter becomes somewhat less than the absorption width. Tunable lasers, because of their narrow linewidths and easy adaptability to derivative detection, provide excellent sensitivity. Low power TDLs monitoring sharp rovibrational lines in gaseous samples can detect molecules at the ppb level at atmospheric pressures over a single-pass 10-cm path length.

The ir spectral region is particularly useful for organic compounds, which have sufficiently distinctive spectra to be easily identified and quantified. Areas of application include drug analysis (101), biological systems (102–104), surface analysis (105), isotopic analysis (106), polymers (107,108), and electrode processes in solutions (spectroelectrochemistry) (109). Ir remote sensing can be as varied as monitoring volatile organics in indoor air by ftir (110), ethylene emissions from a petrochemical plant at the ppb level using a monostatic CO_2 laser system (111), general atmospheric monitoring with TDLs (112), and upper atmospheric studies using airborne ftir (113,114) (see AIR POLLUTION; INFRARED TECHNOLOGY AND RAMAN SPECTROSCOPY, INFRARED TECHNOLOGY).

Near-Infrared Spectroscopy. Many vibrational overtones and combinations, especially of hydrogen-containing functional groups, appear in the near-infrared (nir). Although these bands are broad and weak, transmission and reflection nir spectroscopies have emerged as important probes for industrial and process analysis (115–119), such as monitoring moisture, saturation of oils, and protein and fat content in the food and agricultural industries. Nir radiation can probe, with minimal sample preparation, long-path, concentrated, and aqueous samples that would totally absorb longer wavelengths. The potential loss of sensitivity is offset by instrumental advantages such as dependable low cost tunable laser sources, sensitive detectors, and fiber optics suitable for industrial environments. Usually grating or ftir spectrometers are used, but low resolution bandpass filters may suffice. Filters can be either fixed-frequency interference or tunable acousto-optic (120,121). Because nir bands often overlap, chemometric techniques (122–124) are useful in data reduction (see CHEMOMETRICS).

Solid-state multi-element detector arrays in the focal planes of simple grating monochromators can simultaneously monitor several absorption features. These devices were first used for uv–vis spectroscopy. Infrared coverage is limited (see Table 3), but research continues to extend the response to longer wavelengths. Less expensive nir array detectors have been applied to on-line process instrumentation (125) (see PHOTODETECTORS).

Examples of nir analysis are polymer identification (126,127), pharmaceutical manufacturing (128), gasoline analysis (129,130), and on-line refinery process chemistry (131). Nir fiber optics have been used as immersion probes for monitoring pollutants in drainage waters by attenuated total internal reflectance (132). The usefulness of nir for aqueous systems has led to important biological and medical applications (133).

Radiometry. Radiometry is the measurement of radiant electromagnetic energy (17,18,134), considered herein to be the direct detection and spectroscopic analysis of ambient thermal emission, as distinguished from techniques in which the sample is actively probed. At any temperature above absolute zero, some molecules are in thermally populated excited levels, and transitions from these to the ground state radiate energy at characteristic frequencies. From Wien's displacement law, $\lambda_{\max}T = 2898 \ \mu\text{m·K}$, the emission maximum at 300 K is near 10 μm in the mid-ir. This radiation occurs at just the energies of molecular rovibrational transitions, so thermal emission carries much the same information as an ir absorption spectrum. Detection of the emissions of remote thermal sources is the ultimate passive and noninvasive technique, requiring not even an optical probe of the sampled volume.

When the spectral characteristics of the source itself are of primary interest, dispersive or ftir spectrometers are readily adapted to emission spectroscopy. Commercial instruments usually have a port that can accept an input beam without disturbing the usual source optics. Infrared emission spectroscopy at ambient or only moderately elevated temperatures has the advantage that no sample preparation is necessary. It is particularly applicable to opaque and highly scattering samples, anodized and painted surfaces, polymer films, and atmospheric species (135). The Voyager interferometric spectrometer (IRIS) spectra from the outer planets demonstrated the analytical capabilities of ftir emission spectroscopy. As an example of industrial monitoring, smokestack effluents have been analyzed by ftir at a range of 74 m, using H_2O/CO_2 concentration ratios to distinguish gas and oil combustion (136). Field-deployable commercial instruments achieve sensitivities of the order of ppm·m out to 1 km using retroreflectors.

Direct analysis of weak ambient thermal radiation in the presence of an intense solar background is difficult. A sensitive technique for detecting these signals is laser heterodyne radiometry, in which one measures not the emitted frequency itself, but the beat frequency between this and another, accurately known, frequency. The incident radiation is combined with the output of a coherent local oscillator (LO), usually a fixed-frequency laser, in a high speed photomixer, thus generating a difference frequency called the intermediate frequency (IF), which is synchronously detected and amplified. The IF preserves the spectral characteristics of the source, but shifts this information into the radio-frequency region where sensitive radio detection techniques can be used. The tuning range is limited by the IF bandwidth of the mixer. The HgCdTe photodiodes used in the infrared provide a spectrum covering $\pm 0.08 \ \text{cm}^{-1}$ around the LO frequency. The requirement of finding a molecular transition this close to a gas laser frequency is highly restrictive, so the technique is not suitable for speciation, but rather for monitoring one or a few specific molecular features. Tunable diode lasers have sufficient power to serve as local oscillators, permitting continuous tunability, but any single TDL is limited to a tuning range of some 100 cm^{-1}. Heterodyne radiometry provides excellent sensitivity, and has been used with appropriate receiving telescopes for detection of constituents of planetary and stellar atmospheres.

Molecular Uv–Vis Absorption Spectroscopy

Spectroscopy in the uv–vis detects electronic transitions, and so is applicable to both atoms and molecules. This is a mature technique having important qualitative and quantitative applications (137–141).

Instrumentation and Sampling.   Quartz spectrographs and photographic plates long served for the near-uv–vis–nir region (142), but modern commercial recording instruments employ holographic gratings and photomultiplier tubes or (in the nir) avalanche photodiode detectors to cover ca 190 nm to 3 μm (see HOLOGRAPHY; PHOTODETECTORS). A typical benchtop spectrophotometer has 2 nm resolution in the uv–vis, but dual-grating research instruments can resolve 0.05 nm. In the vacuum ultraviolet (vuv), special techniques are required (143–145). Molecular oxygen absorbs below ~190 nm, and below the transmission limit of lithium fluoride, LiF, (105 nm) no bulk optical materials are suitable, even as window materials. This specialized region is not used for routine chemical analysis.

Simple uv–vis monochromators are widely used with solid-state imaging arrays (146,147), called optoelectronic imaging devices (OIDs) or optical multichannel analyzers (OMAs), placed in the focal plane to record a spectrum nearly instantaneously. These are the modern equivalent of the photographic plate, and have the advantage over an emulsion of rapid response and real-time results. An array detector consists of a set of photodiodes together with an integral electronic readout scheme. Arrays having time-gated windows as short as 5 ns are useful for spectroscopy and kinetics of short-lived and unstable species, for which arrays have obvious advantages over mechanically scanned spectrophotometers, and are often superior to Fourier-transform spectroscopy. Two-dimensional arrays designed for image recording can be used for time-resolved spectroscopy by rastering a temporally changing spectrum across the second dimension of the array.

Interferometry is difficult in the uv because of much greater demands on optical alignment and mechanical stability imposed by the shorter wavelength of the radiation (148). In principle any fts interferometer can be operated in the uv when the proper choice of source, beam splitter, and detector is made, but in practice good performance at wavelengths much shorter than the visible has proved difficult to obtain. Some manufacturers have claimed operating limits of 185 nm, and Fourier transform laboratory instruments have reached 140 nm (145).

Tunable uv–vis lasers are well developed (3). Optically pumped organic dye lasers provide especially useful continuously tunable high power sources, widely used in laboratory research on spectroscopy and photochemistry (see PHOTOCHEMICAL TECHNOLOGY). A single output frequency is selected from the broad-band fluorescence by a dispersive optical cavity. Pulsed dye lasers, pumped with fixed-frequency Nd:YAG, excimer, or Cu-vapor lasers, or by flashlamps, can provide high peak powers at repetition rates to 100 Hz and linewidths of 0.1–1 cm^{-1}, which can be improved to <0.01 cm^{-1} with an intracavity etalon (a pair of parallel plates acting as an interferometer). A given dye–solvent combination can typically be tuned continuously over a 40–80 nm range. Using Ar$^+$ or Kr$^+$ pump lasers, continuous-wave (cw) operation is possible having output powers of 0.1–1 W. Multimode cw cavities can achieve linewidths of a few GHz, and commercial ring-laser geometries improve the resolution to <0.5 MHz over a 1-cm^{-1} continuous scan. Dye lasers have certain drawbacks. These are complex devices requiring a separate pump laser (or at least a flashlamp) and having demanding optical and alignment requirements, especially the synchronization of the tuning elements. Moreover, dye lasers usually require a flowing dye system (for cw operation, a liquid dye jet) to dissipate heat generated by the pump. Other useful laser sources are available (see Table 3), including conversion of longer-wavelength tunable lasers by Raman shifting or by frequency doubling in nonlinear media.

Most of the specialized ir sampling methods described have uv equivalents with appropriate modifications. Sample cells (cuvettes) for solutions, typically of 1-cm path lengths, may be of glass or quartz, and are available for sample volumes as small as a few μL. There is a wide choice of suitable solvents, including (for $\lambda > 200$ nm) water and most saturated organic compounds. Difference spectroscopy with double-beam instruments is used in the uv–vis not only to eliminate background effects, but to assess the effects of changes in pH, temperature, or solvent on one of two otherwise identical samples. Specialized hardware allows the direct scanning of electrophoresis gels and films. Derivative spectroscopy (149,150) by either numerical differentiation or beam modulation methods is important in determining absorbances of weak features obscured by stronger peaks. Photoacoustic methods are employed for strongly absorbing samples. Glass and polymer fiber optics are available for process monitoring.

Applications. Uv–vis absorption results from transitions between outer electron shells. The specific moiety or structure responsible is termed the chromophore. Compounds having only single σ-bonds generally absorb only in the vuv. Saturated organic compounds containing heteroatoms exhibit uv transitions, and the π-electrons of an unsaturated bond or aromatic nucleus are strong chromophores, absorbing more strongly and at longer wavelengths (to the visible) in conjugated and fused-ring systems. Inorganic species having incomplete electron shells, notably transition-metal cations, also absorb in the uv–vis.

Uv–vis spectra do not offer the unique group frequencies and fingerprinting ability of the ir, but different chromophores exhibit absorptions at specific wavelengths, λ, and have characteristic intensities. These are tabulated in handbooks as λ_{max} and ϵ_{max}, where ϵ_{max} is a molar decadic absorption coefficient equivalent to the $a(\nu)$ of equation 1, but in units typically of L/(mol·cm). Thus the ketone C$=$O has a strong absorption at 195 nm and a much weaker one at 270–285 nm. Spectral atlases and catalogues are available (151–153), as are specialized treatments of laboratory analysis in such fields as pharmaceuticals (qv) (154,155) and biomedical diagnostics (156). Scanning uv–vis diode-array absorption detectors are used in high performance liquid chromatography (hplc), and can cover 190–600 nm in 0.1 s using a 10-μL sample.

Photon energies sufficient to promote electronic transitions can also excite vibrational and rotational transitions, so the electronic spectra of gaseous molecules consist of highly structured rovibronic bands, having very high specificity. In condensed phases, rotational structure is suppressed, but the molecules still vibrate, resulting in vibronic bands with progressions of characteristic vibrational frequencies that accompany each electronic transition. The uv offers high sensitivity and excellent quantitative accuracy. The greater uv absorption cross sections (see Table 2) and more efficient uv sources and quantum detectors result in detection limits several orders of magnitude better than in the ir. On the other hand, spectroscopic congestion may be a problem. Many strong transitions occur in a relatively narrow wavelength region. Spectral overlap, and the greater susceptibility of shorter-wavelength radiation to Rayleigh and particulate scattering and turbulence, limits the use of uv spectroscopy for identifying complex mixtures in process streams or in the atmosphere.

Atmospheric and remote-sensing applications of uv–vis spectroscopy have been reviewed (157,158). Many volatile organics absorb in the near-uv, and the vapors exhibit rovibronic structure suitable for identification and quantification if interferences can be avoided. Especially suitable for uv monitoring are the strong Huggins (300–370 nm) and Hartley (210–300 nm) bands of O_3; other inorganics such as NO_2 and SO_2 have been successfully analyzed at kilometer distances (see OZONE).

Atomic Uv/Vis Spectroscopy

Narrow-line uv–vis spectra of free atoms, corresponding to transitions in the outer electron shells, have long been employed for elemental analysis using both atomic absorption (AAS) and emission (AES) spectroscopy (159,160). Atomic spectroscopy is sensitive but destructive, requiring vaporization and decomposition of the sample into its constituent elements. Some of these techniques are compared, together with mass spectrometry, in Table 4 (161,162).

Atomic Absorption Spectroscopy. AAS (163,164) developed into an important technique in the 1960s (165). Samples can be prepared by either flame or electrothermal atomization. In the former (166), the sample is dissolved if necessary, nebulized into an aerosol by a high velocity gas jet, and sprayed into a burner where dissociation produces neutral free atoms. The flame fuel and oxidant are chosen to optimize the number of neutral atoms and minimize interferences from emission and ionization. Flame temperatures are 2050°C (hydrogen/air) to 3150°C (oxyacetylene). In electrothermal atomization the sample is placed usually in a graphite tube furnace (167) that follows a programmed heating cycle to dry, ash, and finally atomize it at ~2400°C. Free-atom production is most efficient in a furnace, resulting in greater sensitivity.

Table 4. Comparison of Atomic Spectroscopy Techniques[a,b]

Technique[c]	Detection limits, ppb	Precision, %	Sample size, mL	Economy	Multielement	Dynamic range	Matrix interference	Spectral interference	Refractories
AAS									
flame	1–100	0.5	10	+	−	−	−	+	−
furnace	0.01–1	5	0.010	+	−	−		+	−
AES									
flame	100–10,000	0.5	5	+	−		−		−
ICP	0.1–100	1	5	−	+	+	+	−	+
ICP-MS	0.001–0.1	2	1	−	+	+		+	+

[a]Typical values are shown.
[b]Advantages and drawbacks are indicated by + and −, respectively. Thus, + implies inexpensive, multielement capability, wide dynamic range, relative freedom from interferences, and the ability to analyze refractory elements.
[c]ICP = inductively coupled plasma.

Furnaces require smaller samples, can accept solids without pretreatment, and have lower background noise. Flames, however, are cheaper, more convenient for routine use, and offer better precision.

For quantitative analysis, the resolution of the spectral analyzer must be significantly narrower than the absorption lines, which are ~ 0.002 nm at 400 nm for $M = 50$ amu at 2500°C (eq. 4). This is unachievable with most spectrophotometers. Instead, narrow-line sources specific for each element are employed. These are usually hollow-cathode lamps, in which a cylindrical cathode composed of (or lined with) the element of interest is bombarded with inert gas cations produced in a discharge. Atoms sputtered from the cathode are excited by collisions in the lamp atmosphere and then decay, emitting very narrow characteristic lines. More recently semiconductor diode arrays have been used for AAS (168) (see SEMICONDUCTORS).

The most sensitive analytical frequency is the first resonance line, representing absorption from the ground to the first excited electronic state. For all metals and most metalloids these occur at >200 nm. Most nonmetals fall below 185 nm, and are less suitable for AAS. A typical AAS spectrometer uses a grating monochromator to isolate the appropriate resonance line, and covers 190–900 nm. Each element analyzed requires its own source, but some two- or three-element hollow cathodes are available, as are dual-channel spectrophotometers that can analyze for two elements simultaneously. Turrets holding up to 16 lamps are made for rapid sequential multielement analysis (169). Some instruments incorporate a broad-band source such as a deuterium lamp, allowing corrections to be made for background absorption. Especially in flames, there may be matrix effects in which a sample component affects the rate of analyte vaporization. These can be minimized using hotter flames, or by pretreatment with releasing agents that preferentially bind possible interferents. Examples of AAS analysis include the determination of heavy metals in blood and urine (170) and in aerosols (171).

Atomic Emission Spectroscopy. AES (172) is an important analytical technique that has enjoyed vigorous commercial instrument development (173). Dispersive spectrographs used for AES are conceptually no different from those discussed. Traditionally, large and expensive concave- or plane-grating, high resolution instruments have been needed to resolve narrow and possibly interfering atomic lines. Echelle gratings, designed to be used in high orders with a prism cross-dispersing element, allow for more compact short focal length spectrographs. These produce a two-dimensional output of wavelength vs grating order in the focal plane. Photographic plates, read using microdensitometers, were still used as detectors as of 1996, for these plates provide simultaneous recording of the whole spectrum and allow for long integration times for detection of weak lines. Photomultiplier tubes are quicker and more sensitive, but can record only a single frequency at a time. These tubes are used in either a scanning mode, recording different lines sequentially, or a direct-reading configuration (quantometer) in which as many as 60 tubes are precisely positioned in the focal plane for simultaneous recording of as many lines. The obvious trade-off is analysis speed vs cost and complexity. Photodiode arrays are also used, but the resolution limits imposed by their pixel size usually require larger and more expensive spectrometers.

Many techniques exist for volatilizing and exciting samples. Flame emission spectroscopy uses much the same source apparatus as AAS, but records the emission lines that in the latter method must be minimized. Line interference is more serious than for absorption, and chemiluminescence in the flame may be a problem. AES is most appropriate for alkali metals, for which it offers the best detectability, alkaline earths, rare earths, and trace metals. Electrical discharge methods include high temperature (4000–6000°C) a-c or d-c arc and a-c spark discharges in air between high purity graphite electrodes, one of which holds the sample. These were long the principal AES techniques, but suffer from problems with reproducibility and interferences. The most notable interference is the cyanogen molecular emission spectrum (360–420 nm). Arc and spark emission were still used for rapid semiquantitative analysis of ferrous metals and other industrial samples as of this writing.

Noncombustion plasma sources offer improved accuracy, less background interference, better dynamic range, and easier sample handling (especially for solutions), and have replaced flame and arc/spark excitation in many applications. In the inductively coupled plasma (ICP) (174–177), a r-f field applied through an induction coil excites an argon gas flow containing the sample aerosol. The resulting eddy current of Ar^+ and electrons seeded from a Tesla coil generates a plasma at 6,000–10,000 K. Complete dissociation reduces matrix effects, and the background is limited to argon lines and some weak diatomic band emission from OH, NO, NH, and CN. Ionized species contribute significantly, which increases the complexity of the spectra and resolution requirements on the spectrometer, but offers flexibility in choosing an appropriate analytical line. Many commercial ICP spectrometers are marketed, and methods of evaluating their performance have been developed (178). The direct-current argon plasma (DCP) (179) provides somewhat better detection limits and greater stability than ICP, but is less suited to automated operation and not as widely marketed. Lower temperature microwave-induced plasmas are being developed with significantly better detection limits but greater matrix interferences than either ICP or DCP. Proper sample introduction into plasma is critical for accurate quantitative results. Pneumatic nebulizers are most used, but depending on sample characteristics, electrothermal vaporizers (180) or flow injection may be appropriate (see PLASMA TECHNOLOGY).

Laser-generated plasmas are also useful sources for AES (181–185). In laser-induced breakdown spectroscopy (libs) the plasma emission is dispersed and analyzed directly (186), but laser ablation can be used simply as a sampling process followed by ICP analysis (187), cross-excitation with a spark discharge, or mass spectrometry. It provides rapid qualitative or semiquantitative results. Laser microprobes using focused coherent radiation can sample areas as small as 50 μm in diameter for surface profiling. Finally, samples can be placed in low pressure hollow-cathode lamps to generate emission spectra, providing narrow lines but requiring more sample handling. Various low pressure glow (cold cathode) discharges have somewhat similar characteristics (188–190).

The variety of AES techniques requires careful evaluation for selecting the proper approach to an analytical problem. Table 4 only suggests the various characteristics. More detailed treatment of detection limits must include consideration of spectral interferences (191). AES is the primary technique for metals analysis in ferrous and other alloys; geological, environmental, and biological samples; water analysis; and process streams (192).

Scattering Techniques

Spectroscopic examination of light scattered from a monochromatic probe beam reveals the expected Rayleigh, Mie, and/or Tyndall elastic scattering at unchanged frequency, and other weak frequencies arising from the Raman effect. Both types of scattering have applications to analysis.

Elastic Scattering.   Elastic scattering is not, in its simplest form, a spectroscopic technique, and conveys no chemical information about the sample, though it can provide physical properties useful in analytical applications (193–195). Rayleigh scattering and polarization measurements on Mie-scattered light furnish particle size and distribution, absolute number density, and mean molecular weight. Maps of velocity fields in flowing process streams and temperature distributions in flames can be recovered from Doppler shifts of the scattered light.

Elastic scattering is also the basis for lidar, in which a laser pulse is propagated into a telescope's field of view, and the return signal is collected for detection and in some cases spectral analysis (14,196). The azimuth and elevation of the scatterers (from the orientation of the telescope), their column density (from the intensity), range (from the temporal delay), and velocity (from Doppler shifts) can be determined. Such accurate, rapid three-dimensional spatial information about target species is useful in monitoring air mass movements and plume transport, and for tracking aerosols and pollutants (197).

An important modification of lidar is two-frequency differential absorption lidar (dial) (14,198,199), which combines lidar and absorption spectroscopy, using two alternating frequencies from a tunable laser, one on and one off an absorption resonance. This generates a differential absorption signal from the Raleigh or Mie scattering returns, which are relatively insensitive to small changes in frequency, and yields range-resolved concentrations of specific molecules. The back-scatter can be from natural aerosol and dust, or from a retroreflecting target, thereby sacrificing range information for increased sensitivity. Detectivities of 1–10 ppb·km can be achieved; ie, a few parts per billion over integrated ranges of several km.

Spontaneous Raman Spectroscopy.   The Raman effect is the inelastic scattering of an incident photon $\tilde{\nu}_i$ to a new frequency $\tilde{\nu}_R = \tilde{\nu}_i \pm \tilde{\nu}_{mol}$, where $\tilde{\nu}_{mol}$ is the energy acquired (−) or lost (+) by the molecule during a vibrational, rotational, or (less commonly) electronic transition. It is an important technique for the investigation of molecular structure, providing information similar, and often complementary, to ir spectroscopy, and for chemical analysis, both in the laboratory (40–42,200–203) and for remote atmospheric monitoring (204). Raman selection rules differ from those of single-photon absorption, depending on change in polarizability rather than change in dipole moment. There is at least one Raman-active vibrational fundamental for every molecule. Relative intensities can vary greatly between absorption and scattering. Vibrations of highly polar groups are often strong in the ir and weak in the Raman. The reverse is true of C — H and C = C stretches and aromatic ring breathing modes.

When the exciting frequency $\tilde{\nu}_i$ is nonresonant (distant from any electronic transition), the differential scattering cross section at wavelength λ is as in equation 8:

$$\frac{d\sigma}{d\Omega} \cong \left(\frac{2\,\pi}{\lambda}\right)^4 A_{if}^2 \sin^2\theta \tag{8}$$

where θ is the scattering angle and A_{if}, the polarizability matrix element connecting the initial and final states, has values of $<10^{-24}$ cm^3, comparable to the volume of the molecule.

The spectrum of the scattered light contains a strong Rayleigh line at the exciting frequency, and much weaker (by factors of 10^3 or more) red-shifted Raman lines (Stokes lines) from interactions in which the molecules have gained rovibrational energy from the photons, and corresponding but even weaker blue-shifted anti-Stokes lines, where the molecules have lost energy to the photons. The resulting rovibrational spectra permit species identification and quantification. The relative strengths of the Stokes and anti-Stokes lines yield the sample temperature. Each Raman feature is characterized by a depolarization ratio ρ, which depends on the nature of the molecular vibration involved, and is useful in making vibrational assignments (see INFRARED TECHNOLOGY AND RAMAN SPECTROSCOPY, RAMAN SPECTROSCOPY).

Raman Instrumentation and Sampling. Raman instrumentation must provide an intense monochromatic source, means for resolving the desired frequency-shifted signal from the strong elastically scattered background, and provision for recording the Stokes Raman intensity as a function of $\tilde{\nu}_{mol}$. For much of its history (205), Raman spectroscopy employed mercury-arc sources. The scattered radiation was dispersed by prisms and detected using photographic plates or photomultipliers. The inherent weakness of the effect, inadequate sources and detectors, and problems with colored samples, fluorescence, and stray Rayleigh-scattered light, combined to make it a difficult technique, used typically to extract structural information in connection with ir spectroscopy rather than for analytical purposes. The development of lasers revitalized Raman spectroscopy. In fact, lasers found perhaps their first practical use as Raman sources.

As of the 1990s, both high throughput grating spectrometers, often having double monochromators, and Fourier-transform interferometers are used. The latter (206–208) can be stand-alone instruments or accessories for ftir spectrometers. The relative advantages of the dispersive and ft approaches have been discussed (209). Source selection must balance the λ^{-4} cross-section dependence in equation 8, which favors uv excitation, against increased interference from fluorescence at shorter wavelengths. Common sources are He:Ne, Ar$^+$, Kr$^+$, and Nd:YAG lasers; the last, at 1.064 μm, is especially useful with fluorescent samples. Detectors are InGaAs photodiodes or multichannel charge-coupled (CCD) array detectors (210,211). Raman spectra can routinely be acquired in a few seconds using a dispersive spectrograph and CCD detector; ft instrumentation typically takes some minutes. Spectra have been recorded having 10 ps temporal resolution using a pulsed source at 588 nm and a streak camera (212).

Sample preparation is straightforward for a scattering process such as Raman spectroscopy. Sample containers can be of glass or quartz, which are weak

Raman scatterers, and aqueous solutions pose no problems. Raman microprobes have a spatial resolution of ~ 1 μm, much better than the diffraction limit imposed on ir microscopes (213). Fiber-optic probes can be used in process monitoring (214).

Special Raman Spectroscopies. The weakness of Raman scattering results typically in the conversion of no more than 10^{-8} of the incident laser photons into a usable signal, limiting the sensitivity of conventional spontaneous Raman spectroscopy. This situation can be improved using alternative approaches (8,215,216).

In resonance Raman (RR) scattering (204,217,218), the excitation frequency is tuned to within a few linewidths of an allowed electronic absorption. This increases the intensities of both fluorescence and Raman scattering by several orders of magnitude, allowing the highly selective enhancement of a particular component of a mixture, with sensitivity adequate for monitoring trace constituents. It requires an analyte with an accessible electronic transition. Applications include pigment analysis, biochemical systems, and atmospheric O_2. Surface-enhanced Raman scattering (sers) (219–221) exploits certain imperfectly understood mechanisms that can enhance by more than 10^6 the Raman signal from molecules at colloidal metal surfaces. Detection limits in solutions can reach the ppm level.

Under intense laser excitation, the dipole moment and polarization induced in a molecule may vary nonlinearly with the electric field, giving rise to new nonlinear spectroscopic phenomena (3,8,11). High conversion efficiencies are achieved, and the coherent and directional scattered radiation can be efficiently collected and separated from incoherent emissions such as fluorescence and Rayleigh scattering. Such stimulated scattering requires very intense laser irradiance (>1 MW/cm^2), even near resonance. Important nonlinear and coherent Raman processes include coherent anti-Stokes Raman spectroscopy (cars) (222); Raman-induced Kerr-effect spectroscopy (rikes); stimulated Raman scattering (srs), including stimulated Raman gain spectroscopy (srgs) (223) and inverse Raman scattering; the hyper-Raman effect (224); and four-wave mixing processes. These are useful in molecular research, and some important analytical applications have been developed (225). Cars employs two excitation frequencies, ν_1 and ν_2, which generate intense coherent anti-Stokes emission at $\nu_3 = 2\nu_1 - \nu_2$ as ν_2 is tuned to bring the frequency difference $\nu_1 - \nu_2$ into resonance with a Raman-active vibrational frequency. It is especially useful in mapping species distributions and temperatures in combustion and flame diagnostics with a spatial resolution of better than 1 mm^3.

A coherent Raman process of great promise for sensitive analysis is degenerate four-wave mixing (DFWM) (226), which uses a strong nonresonant pump and weaker probe beam from the same laser to induce a large back-scattered coherent signal. DFWM is simpler to align than cars, requires 10^3-fold less irradiance, and produces an intense and highly directional return even in inhomogeneous or turbulent media. The process can employ a Doppler-free pump–probe scheme, making it especially useful for studying high temperature combustion and plasma processes. DFWM has been used for high fidelity spatial mapping of analytes in flames, and has shown excellent detection sensitivities for molecular species; for example, OH at concentrations of 10^{10} cm^{-3}. It is even more sensitive to strong atomic transitions. Detection of excited-state Na atoms has been reported with a sensitivity of $\sim 10^6$ atoms in a 10-mm^3 interaction volume (227).

Applications. The broad range of Raman analytical applications is covered elsewhere (200,202) (see INFRARED TECHNOLOGY AND RAMAN SPECTROSCOPY, RAMAN SPECTROSCOPY). Its suitability for aqueous solutions has led to important biological applications (228,229). Low frequency vibrational modes important in inorganic and organometallic chemistry can often be studied more easily by Raman than by far-ir spectroscopy (see ORGANOMETALLICS). Using modern instrumentation Raman spectra can be recorded to within a few cm^{-1} of the exciting line. Sampling ease and the ability to probe through packaging materials makes ft-Raman useful in industries such as pharmaceuticals (230).

Scattering techniques have important applications to atmospheric and environmental monitoring. Lidar requires an intense, monochromatic laser source, which can also excite dial, Raman, and laser-induced fluorescence (lif), depending on how the source is tuned and whether the return signal is spectrally analyzed. Much of the instrumentation can be shared by the various techniques, and they are often combined into one instrument. Many field-deployable systems have been described (231–233). There are advantages in using probe wavelengths of 200–370 nm, where sunlight is absorbed by stratospheric ozone (qv) but oxygen absorption is insignificant. Such solar-blind operation eliminates background radiation, allowing reliable daytime operation. Typically the returned signal is detected as a function of time, and its intensity yields concentration versus range. Concentrations can be calibrated by ratioing the Raman signal of interest against that from N_2.

Optical Activity

Although the usual absorption and scattering spectroscopies cannot distinguish enantiomers, certain techniques are sensitive to optical activity in chiral molecules. These include optical rotatory dispersion (ORD), the rotation by the sample of the plane of linearly polarized light, used in simple polarimeters; and circular dichroism (CD), the differential absorption of circularly polarized light.

Circular dichroism employs standard dispersive or interferometric instrumentation, but uses a thermal source that is rapidly modulated between circular polarization states using a photoelastic or electro-optic modulator. Using phase-sensitive detection, a difference signal proportional to the absorption difference between left- and right-polarized light, $\Delta A = A_L - A_R$, is recorded as a function of wavenumber. Relative differential absorptions (dissymmetry factors) $\Delta A/A$ at absorption maxima are typically 0.1–0.01 for uv–vis electronic transitions and 10^{-4}–10^{-5} for vibrational modes in the ir.

As the stronger effect, electronic circular dichroism (234–236) is well established in stereochemistry, but is applicable only to easily accessible electronic transitions. Although weaker, vibrational optical activity (41,237) can be observed on many bands, and helps in elucidating short-range structural features. In the ir it is designated vibrational circular dichroism (VCD) and is applied, for example, to solution studies of conformations in pharmaceuticals (238) and biomolecules (239). Raman optical activity (ROA), the differential scattering of circularly polarized light, is an analogous but more complex process with several experimental variations (237).

Luminescence Spectroscopies

Luminescence is spontaneous light emission during a transition from a nonthermally excited level to a lower state (240,241). Such processes are classified as short-lived (on a ns time scale) emission or fluorescence (242–246) that continues only as long as the sample is being excited; and longer-lived phosphorescence (247,248), persisting after excitation ceases. For the important case of electronic excitation, fluorescence typically represents temperature-independent spin-allowed transitions from singlet excited states; phosphorescence is longer-wavelength spin-forbidden emission from lower lying triplet states. The distinction between fluorescence and scattering under near-resonant conditions is somewhat arbitrary, especially at high pump intensities. Fluorescence can be distinguished as noncoherent emitted radiation that is broadened by the natural lifetime and collisions of the excited state, whereas resonant Raman and Rayleigh scattering are not so broadened.

The strength of a fluorescence signal is directly related to the absorption cross section σ_A:

$$\frac{d\sigma}{d\Omega} = \frac{1}{4\pi}\sigma_A F \tag{9}$$

where the geometric factor, $(4\pi)^{-1}$, accounts for the isotropic re-emission in all directions, and F is a fluorescence yield factor. Generally $F \ll 1$, owing to nonradiative dissipative quenching of the excited states and fluorescence branching to several final states. At atmospheric pressure, most quenching occurs by collisional transfer, so fluorescence intensities strongly depend on pressure, temperature, and composition, and the resulting signal may be difficult to interpret quantitatively. Fluorescence is typically more intense in the uv than in the ir, owing to greater uv absorption cross sections and longer ir radiative lifetimes. The latter favor collisional de-excitation. Typically, $F < 10^{-3}$ for atomic and molecular transitions in the uv and is progressively smaller for longer wavelength transitions.

Fluorometry and Phosphorimetry. Modern spectrofluorometers can record both fluorescence and excitation spectra. Excitation is furnished by a broad-band xenon arc lamp followed by a grating monochromator. The selected excitation frequency, λ_{ex}, is focused on the sample; the emission is collected at usually 90° from the probe beam and passed through a second monochromator to a photomultiplier detector. Scan control of both monochromators yields either the fluorescence spectrum, ie, emission intensity as a function of wavelength λ_{em} for a fixed λ_{ex}, or the excitation spectrum, ie, emission intensity at a fixed λ_{em} as a function of λ_{ex}. Fluorescence and phosphorescence can be distinguished from the temporal decay of the emission.

Molecular fluorescence extends over a broad wavelength region, limiting its usefulness for analyzing multicomponent systems. By scanning both the excitation and fluorescence wavelengths synchronously with an empirically chosen fixed wavelength separation, $\lambda_{ex} - \lambda_{em}$, the fluorescence can be reduced to a narrow signal at the region of overlap between the excitation and fluorescence spectra. This allows the components in complex mixtures to be distinguished. Independent scanning of the excitation/fluorescence spectra can produce a highly specific contour plot of fluorescence as function of both λ_{ex} and λ_{em}, ie, multidimensional fluorescence or total luminescence spectroscopy (249).

Samples are usually in solution, but solids (often frozen solutions) yield narrow-line spectra that are useful in distinguishing components of mixtures. In phosphorimetry solid sampling may be necessary to minimize quenching processes.

As null-background techniques, fluorometry and phosphorimetry are orders of magnitude more sensitive than absorption spectroscopy, having achievable detectivities of ng/L. Fluorescence is strongest in species with pronounced uv spectra, specifically aromatic and conjugated molecules, which have readily excited delocalized π-electrons. Substituents that favor delocalization, eg, $-NH_2$, $-OH$, and $-OCH_3$, enhance fluorescence and electrophilic groups, eg, $-Cl$ and $-NO_2$, tend to quench it. Compounds especially suitable for fluorescence analysis include drugs, certain pesticides, and pollutant carcinogens and other toxic species. Several monographs treat applications for specific types of samples (250,251).

Atomic fluorescence spectroscopy (afs) using either a laminar-flow flame similar to AAS or an inductively coupled plasma (ICP) source provides multielement analysis having relative freedom from background and interferences, and detection limits similar to AAS (see Table 4). It is appropriate for any element having strong absorption and a dominant nonbranching transition to the ground state, and is especially useful in the analysis of trace metals. A commercial afs–icp instrument may employ up to a dozen hollow-cathode lamps, each focused on that portion of the plasma that provides the optimum signal, and each with its own photomultiplier tube to detect the resonance fluorescence from that element. Afs is the basis for some environmental sensors (252).

A useful specialized type of analytical instrumentation is the fiber-optic sensor or optrode (253,254), in which an optical transducer monitors some chemically selective change. These are often based on the fluorescence of a reagent immobilized at the distal end of the fiber (255).

Laser-Induced Fluorescence. Laser-induced fluorescence (lif) provides, much as does ir spectroscopy, fingerprints of different organic molecules, which can be quantified by measuring fluorescence intensities. Selectivity is excellent, as both pump and fluorescence frequencies can be individually chosen for optimum performance, and it can be improved with measurements of fluorescence lifetimes and polarization behavior. The enhanced null-background sensitivity can achieve single-atom or single-molecule detection (256–258). Lif has important applications in gas analysis (259) and combustion and plasma diagnostics (260).

Lif has some disadvantages that limit use in atmospheric monitoring, including collisional quenching, solar background interference, and the lack of sensitive photomultiplier detectors for operation in the infrared beyond 1 μm. Despite this, lif may be appropriate for certain species that fluoresce strongly even at atmospheric pressure, including both atoms (Na, K, Li, Ca, Ca^+) and radicals (OH, CN, NH). Particularly important are polycyclic aromatic hydrocarbons such as anthracene, pyrene, chrysene, and benzo[a]pyrene, potent carcinogens produced by engine and incinerator combustion processes. Many of these compounds are near-ideal fluorophores, having large fluorescence quantum yields and long fluorescence lifetimes. Detection limits of a few parts per trillion in water solution have been reported using lif (261). In the stratosphere, lif detection is easier because of reduced collisional redistribution at pressures <4 kPa (30 torr), and trace metals have been detected using ground-based lif lidar at the level of a few atoms/cm^3 (14).

Chemiluminescence. Chemiluminescence (262–265) is the emission of light during an exothermic chemical reaction, generally as fluorescence. It often occurs in oxidation processes, and enzyme-mediated bioluminescence has important analytical applications (241,262). Chemiluminescence analysis is highly specific and can reach ppb detection limits with relatively simple instrumentation. Nitric oxide has been so analyzed from reaction with ozone (266–268), and ozone can be detected by the emission at 585 nm from reaction with ethylene.

X-Ray Spectroscopy

X-rays provide an important suite of methods for nondestructive quantitative spectrochemical analysis for elements of atomic number $Z > 12$. Spectroscopy involving x-ray absorption and emission (269–273) is discussed herein. X-ray diffraction and electron spectroscopies such as Auger and electron spectroscopy for chemical analysis (esca) or x-ray photoelectron spectroscopy are discussed elsewhere (see X-RAY TECHNOLOGY).

Instrumentation. An x-ray tube accelerates electrons from a heated cathode through a high voltage field to a target anode, where kinetic energy is converted to a continuum of x-rays having wavelengths $\lambda > hc/eV = 1240/V$ for λ in nm and exciting voltage V in volts. The electrons can also ionize an atom by ejecting an inner-shell electron. The vacancy is immediately filled by another bound electron, accompanied by emission of characteristic lines, of which the most prominent represent transitions from the L and M electronic levels to the lowest K level (the $K\alpha$ and $K\beta$ lines, respectively). A tube operating at 50 kV using a Cu or Mo target provides x-rays of $\lambda > 0.025$ nm, sufficient to generate K spectra of elements up to the rare earths and L spectra to the transuranium elements.

X-ray spectroscopy by wavelength dispersion employs a scanning crystal spectrometer conceptually not unlike visible and ir spectrometers. The beam is collimated using parallel slits or a bundle of fine-diameter tubes, and dispersed by Bragg's law diffraction from a goniometer-mounted crystal where the lattice spacing is chosen to act as a grating for the wavelengths of interest. Focusing optics can be used instead, with a curved analyzing crystal that makes collimation unnecessary. Typically lithium fluoride [7789-24-4], LiF, is used for 0.03–0.38 nm and pentaerythritol and potassium acid phthalate for longer wavelengths. Detectors are gas proportional counters for $\lambda > 0.2$ nm; or, for $\lambda < 0.3$ nm, scintillation detectors such as thallium-doped NaI, which convert ionization energy into visible light for amplification by photomultipliers. In industrial applications requiring rapid, routine analyses, 20 or more crystals may be aligned in an automated spectrometer for simultaneous determination of many elements.

An alternative spectroscopic technique is energy dispersion, which exploits the energy resolution of solid-state detectors. The beam is not dispersed, but strikes a high purity (intrinsic) germanium or silicon semiconductor, where generation of electron–hole pairs allows the energy of each photon to be recorded by a multichannel pulse-height analyzer. This has both multiplex and throughput advantages (similar to those offered by interferometers in the ir), but in this case at a significant sacrifice in resolution. Relative merits of the two dispersion methods for process analysis have been discussed (274).

X-Ray Absorption Spectroscopy. As the excitation energy incident on a sample is increased, sharp rises in absorption occur at the $K, L, \ldots$, absorption edges where the energy just matches that required for ionization by ejection of an electron from the $K, L, \ldots$, shells. These energies are characteristic for each element. The absorption follows Beer's law (eq. 1), which in this region is usually written as $A(\lambda) = \mu_m(\lambda)\rho x$, where ρ is the density, x the path length, and $\mu_m(\lambda)$ the mass absorption coefficient at wavelength λ. Between absorption edges, $\mu_m(\lambda)$ is proportional to $Z^4\lambda^3$ and is nearly independent of physical or chemical state. An absorption measurement on each side of an absorption edge is required for each element analyzed. X-ray absorption is especially useful in determining heavy elements in mixed materials of lower Z, such as lead in gasoline and uranium in aqueous solution.

X-Ray Emission and Fluorescence. X-ray analysis by direct emission following electron excitation is of limited usefulness because of inconveniences in making the sample the anode of an x-ray tube. An important exception is the x-ray microphobe (275), in which an electron beam focused to ~1 μm diameter excites characteristic x-rays from a small sample area. Surface corrosion, grain boundaries, and inclusions in alloys can be studied with detectability limits of ~10^{-14} g (see SURFACE AND INTERFACE ANALYSIS).

X-ray fluorescence (275–277) employs sample excitation using radiation from a standard x-ray tube, providing weaker (by factors of ~10^3) signals than electron excitation, but having greater flexibility. The characteristic fluorescence emission is analyzed using a crystal spectrometer. Samples can be liquid or solid, including thin films and particulates. Powders are pressed into wafers or fused with borax. Typical detectivity is ~1 ppm for middle-Z elements, poorer for low Z elements, and precision is ~1% for principal constituents. For trace analysis, grazing (278) or variable-angle (279) geometries may offer improved sensitivity. Detection limits of $<10^{-12}$ g have been reported using the former. Where only a few elements are present, energy-dispersive analysis without a crystal may provide sufficient resolution with increased sensitivity, called energy-dispersive x-ray fluorescence (EDXRF). Examples of x-ray fluorescence applications are in specialty chemicals (280), hazardous waste analysis (281,282), thin films (283), and geochemistry (162).

X-radiation can also be induced by high energy (several MeV) proton beams from ion accelerators. Such particle-induced x-ray emission (PIXE) (284) is useful for thin samples and particulates, having detection limits of ~10^{-12} g. Intense synchrotron x-ray sources have found applications in chemical investigations (285), using toroidal holographic gratings for dispersion.

Although x-rays probe inner rather than valence electrons, in light elements the chemical state of the emitting atom may affect inner-shell energies enough to be detected at high resolution. Thus the $K\alpha$ lines of sulfur at 0.537 nm shift by 0.3 pm between the oxidation states S^{6+} and S^{2-}.

Gamma-Ray Spectrometry

A nuclear spectrum represents the energy distribution of particles and radiation emitted during nuclear processes such as natural radioactive decay, nuclear reaction, and fission. These emissions may include alpha-particles (helium nuclei), beta-particles (electrons or positrons), neutrinos, and gamma-rays (high energy photons), of which the last is of interest as electromagnetic radiation subject to spectrometric analysis (286). Gamma-ray energies correspond to transitions between energy levels in atomic nuclei, and provide a unique signature of nuclear processes.

A gamma-ray spectrum is produced nondispersively by pulse-height (multichannel) analysis using scintillation or semiconductor detectors. Resolving power, typically ~100 at 100 keV and ~700 at 2 MeV, is quite modest compared with that achievable in other spectral regions, but is sufficient to identify nuclides unambiguously.

Gamma-ray spectrometry is a probe of nuclear rather than chemical processes, but its high specificity and sensitivity have applications in analysis of materials (286). It is especially suited for activation analysis. Unstable nuclides produced by nuclear bombardment can be identified by their characteristic gamma-ray decay emissions. An important example is slow neutron capture by nitrogen with subsequent decay of $^{15}N^{*}$ detected from its 1.7–10.8 MeV gamma lines, a signature useful for remote, nondestructive detection of possible hidden explosives (see EXPLOSIVES AND PROPELLANTS). Gamma-ray astronomy (287), using balloon and satellite instrumentation, has detected line emissions from nuclear reactions on the solar surface, from nucleosynthesis elsewhere in the galaxy, and from supernovas (288).

Mössbauer Spectroscopy. The low resolution of gamma-ray spectrometry can be dramatically improved, greatly increasing its usefulness, by exploiting the Mössbauer effect (289,290), also known as recoilless gamma-ray resonance absorption. In certain solids the decaying nucleus can be fixed in the crystalline lattice to eliminate energy loss by recoil, narrowing the Doppler emission linewidth sufficiently to achieve resolving powers of ~10^{13}. Differences in the chemical environment of emitting and absorbing nuclei of the same isotope result in small differences in gamma-ray energy, which can be detected at this resolution by Doppler shifts as the source is moved relative to the stationary absorber by a transducer at varying velocities of up to a few cm/s.

Chemical applications of Mössbauer spectroscopy are broad (291–293): determination of electron configurations and assignment of oxidation states in structural chemistry; polymer properties; studies of surface chemistry, corrosion, and catalysis; and metal-atom bonding in biochemical systems. There are also important applications to materials science and metallurgy (294,295) (see SURFACE AND INTERFACE ANALYSIS).

Ultrasensitive Techniques

Some very high resolution or ultrasensitive spectroscopies emerging as of ca 1996 were beam spectroscopy, multiphoton absorption and ionization, and frequency-modulation spectroscopy (8). Most of these were used primarily for laboratory research as of the mid-1990s, but eventual application to analytical or sensor applications is expected.

Beam Spectroscopy.   Both specificity and sensitivity can be greatly enhanced by suppressing collisional and Doppler broadening. This is accomplished in supersonic atomic and molecular beams (296) by probing the beam transversely to its direction of flow in a near-collisionless regime. When a gas adiabatically expands through a nozzle into a vacuum, its thermal energy is converted into kinetic energy of mass flow, cooling the gas to effective temperatures of <10 K, and collapsing any rovibrational structure into a few transitions originating from the lowest lying quantum states. Beam work has emphasized basic atomic and molecular structure and aggregation, but the methods may eventually be applied to analysis. Besides I_t/I_0 absorption measurements, absorption can be followed in beams by monitoring the beam energy using sensitive pyrometers or other calorimeters. Laboratory detection limits for strongly absorbing species can reach the ppb level.

The greatest sensitivities are achieved in uv–vis fluorescence using atomic beam sources, having demonstrated limits of <1 atom/mm^3 and excellent isotopic discrimination (261). Lif/beam techniques are difficult for quantification because of complications of collisional and radiative redistribution. These complications can be reduced by careful experimental design. Lif has provided useful information about collisional transfer. Applications to kinetic modeling of atmospheric chemistry and gaseous discharges are important (see ATMOSPHERIC MODELING; KINETIC MEASUREMENTS).

Multiphoton Absorption and Ionization.   High laser powers can induce the simultaneous absorption of two or more photons that together provide the energy necessary to excite a transition; this transition may be one that is forbidden as a single-photon process (8,297). Such absorption can be made Doppler-free by propagating two laser beams of frequency ν in opposite directions, so the Doppler shifts cancel and a two-photon transition occurs at 2ν for any absorber velocity. The signal is strong because all absorbers contribute, and peak amplitudes are enhanced by $\sim\Delta\nu_D/\nu_N$, which may represent a 1000-fold improvement in selectivity and sensitivity.

Photoionization of atoms or molecules permits sensitive detection, for the ionic signal is a clear null-background signature (3,4,8,298–300). Tunable dye lasers have made this a practical technique. Multiphoton ionization (mpi) has been applied to liquids and interfaces, but is most useful in probing low pressure gases and molecular beams, where the charged photofragments can be extracted with high efficiency and analyzed by electron or mass spectrometers. Mpi and its many variants can be considered as null-background absorption techniques, but are often treated as a distinct category of nonlinear optical spectroscopies, requiring initiation by intense laser irradiance (typically >100 MW/cm^2).

Resonance photoionization techniques (301) such as resonantly enhanced multiphoton ionization (rempi) are important spectroscopic probes of high lying and normally forbidden levels. Rempi is especially sensitive for detecting trace absorbers, because charged photofragments, unlike photons, can easily be collected with near unity efficiency. Single atoms of an analyte like Cs, even in the presence of an atmosphere of noble gas, can be detected (299). Rempi became a practical analytical technique as electron−ion multipliers and mass spectrometers were used to detect the charged photofragments. Resonant ionization mass spectrometry (rims) is a mature analytical methodology, and usually the most efficient method to induce ionization of trace analytes. Under laboratory conditions, cascading rempi with high resolution mass spectrometry has demonstrated detection limits of 10^{-17} g, representing the detection of less than 50,000 trace atoms (300).

Frequency-Modulation Spectroscopy. Frequency-modulation spectroscopy (fms) is a high sensitivity null-background infrared technique for measuring absorbances down to 10^{-8} with fast acquisition speeds. Fms involves frequency-modulating a laser source at ω_0 to produce a carrier frequency having sidebands at $\omega_0 \pm n\omega_m$, where $n\omega_m$ is an integral multiple of the modulation frequency. Dye lasers and many other single-line sources can be modulated using an external electro-optic modulator (EOM); various EOM materials are available for 0.42−5.2 μm (LiTaO$_3$, LiNbO$_3$), and in the deeper ir (GaAs, 2.0−11 μm; CdTe, 1.0−25 μm). Sensitivity scales as $(I_0)^{1/2}$ and can be improved by increasing the probe intensity. Shot noise-limited detection has been demonstrated for visible cw dye lasers (302) and lead salt laser diodes in the mid-ir (303). Fms has also been demonstrated using single-line sources (most notably with a CO$_2$ laser) and pulse lasers, and after second harmonic conversion to the uv.

The frequency modulated (FM) laser light is passed through the sample, where near-resonant absorption (or dispersion) leads to differential absorption (scattering) of the sidebands. This produces an amplitude modulated (AM) beat frequency that is coherently detected by standard phase-sensitive mixing. If the sideband modulation ω_m is small compared to the spectral linewidth, the observed signature is a derivative of the line shape. Greater signal-to-noise ratio (S/N) is achieved when the sideband spacing is much larger than the spectral linewidth. This requires high speed detectors having bandwidths exceeding 3 GHz if Doppler- or collision-broadened lines are to be detected at the highest sensitivity. This drawback can be avoided by using two-tone FM, in which a pair of closely spaced sidebands having large average modulation frequency are demodulated at their smaller relative separation, yielding second-derivative line shapes. The demodulation frequency is still high (50−100 MHz), but is within the limits imposed by standard detector technology. The advantage of using these high radio frequencies is that the signal is recovered in a higher frequency band where the noise of the laser amplitude fluctuations is usually negligible, so that the detection sensitivity is limited by amplifier noise and/or quantum fluctuations. Balloon-borne *in situ* laser sensors using FM absorption have demonstrated ~0.5 ppb sensitivities over 500-m path lengths for soundings made in the stratosphere above 20 km.

Other Techniques. Under proper conditions, a strong laser probe can interact with only a particular velocity subset of gaseous absorbers, allowing homogeneous line shapes to be resolved in static cells (8). This phenomenon, saturated absorption, offers excellent selectivity at higher sample densities than in molecular beams. Saturated absorption is used mainly as a sensitive null-background spectroscopy to measure precise line positions. It has the advantage of requiring only modest cw laser powers (<100 mW), thus minimizing light field-induced level shifts and broadening. The technique is easiest in the uv–vis because of the availability of suitable lasers and stronger transitions.

Sensitivity can be improved by factors of 10^6 using intracavity absorption, placing an absorber inside a laser resonator cavity and detecting dips in the laser emission spectrum. The enhancement results from both the increased effective path length, and selective quenching of laser modes that suffer losses by being in resonance with an absorption feature.

Photothermal spectroscopy (85,304) includes two related high sensitivity approaches, thermal lensing and photothermal deflection. The former (305,306) monitors thermal focusing or defocusing: a modulated cw laser induces thermal pulses in a sample, and the transmission of a coaxial probe laser is synchronously measured using a pinhole-detector combination. It can detect absorbances of <10^{-7}, but is limited to the study of flat, homogeneous, and reasonably transparent samples and substrates. Photothermal deflection is more adaptable and is especially suited to analyzing thin films (qv) and solid–liquid interfaces. Quantification is difficult because sharp thermal gradients at the edge of the pump beam are being observed.

A promising technique is cavity ringdown laser absorption spectroscopy (307), in which the rate of decay of laser pulses injected into an optical cavity containing the sample is measured. Absorption sensitivities of 5×10^{-7} have been measured on a μs time scale. Applications from the uv to the ir have been made in studies of reaction kinetics and *in situ* combustion chemistry, and in trace gas analysis.

Future Trends.   Methods of laser cooling and trapping are emerging as of the mid-1990s that have potential new analytical uses. Many of the analytical laser spectroscopies discussed herein were first employed for precise physical measurements in basic research. Applications to analytical chemistry occurred as secondary developments from 10 to 15 years later.

Two important consequences of Ladenburg's formula (eq. 6) are that the peak absorption cross section, σ_0, for any radiative process varies inversely with the resolved linewidth of the transition, and when the limit of the natural linewidth is achieved, it scales simply as $\sigma_0 \approx \lambda_0^2$, where λ_0 is the peak transition wavelength. In principle, then, any transition can absorb strongly if both the laser linewidth and environmental perturbations of the absorber are reduced below the natural linewidth. Transitions having natural linewidths of ≥ 1 MHz can be resolved as of the mid-1990s using commercial lasers by Doppler-free beam spectroscopy. Much weaker (electric dipole-forbidden) transitions could be resolved if the atoms/molecules could be trapped and cooled to reduce residual Doppler and collisional broadening to ~100 Hz. Ions and neutral atoms can be cooled as of this writing to ~1 μK. Even this temperature represents a Doppler linewidth of >40 kHz in the visible. Tunable cw dye lasers and laser diodes can be stabilized to linewidths ~1 Hz and ≤ 10 kHz, respectively, as of the mid-1990s. Ultrastable laser sources, used with atom/molecule laser cooling/trapping concepts, are anticipated to become the absorption spectrometers for chemical and elemental analysis of the twenty-first century.

BIBLIOGRAPHY

1. P. W. Milonni and J. H. Eberly, *Lasers*, John Wiley & Sons, Inc., New York, 1988.
2. J. Hecht, *The Laser Guidebook*, 2nd ed., TAB Books, Blue Ridge Summit, Pa., 1992.
3. W. Demtröder, *Laser Spectroscopy: Basic Concepts and Instrumentation*, 2nd ed., Springer-Verlag, Berlin, 1996.
4. D. S. Kliger, ed., *Ultrasensitive Laser Spectroscopy*, Academic Press, Inc., New York, 1983.
5. J. R. Murray, in L. J. Radziemski, R. W. Solarz, and J. A. Paisner, eds., *Laser Spectroscopy and Applications*, Marcel Dekker, Inc., New York, 1987, pp. 91–174.
6. D. L. Andrews, ed., *Applied Laser Spectroscopy: Techniques, Instrumentation, and Applications*, VCH Publishers, New York, 1992.
7. E. R. Menzel, *Laser Spectroscopy: Techniques and Applications*, Marcel Dekker, Inc., New York, 1995.
8. M. D. Levenson and S. S. Kano, *Introduction to Nonlinear Laser Spectroscopy*, rev. ed., Academic Press, Inc., Boston, Mass., 1988.
9. S. Mukamel, *Principles of Nonlinear Optical Spectroscopy*, Oxford University Press, New York, 1995.
10. N. Bloembergen, *Nonlinear Optics*, 4th ed., World Scientific Publishing Co., River Edge, N.J., 1996.
11. R. L. Sutherland, *Handbook of Nonlinear Optics*, Marcel Dekker, Inc., New York, 1996.
12. B. L. Fearey, ed., *Advanced Optical Methods for Ultrasensitive Detection, SPIE Proc.* **2385** (1995).
13. E. D. Hinkley, ed., *Laser Monitoring of the Atmosphere*, Springer-Verlag, Berlin, 1976.
14. R. M. Measures, *Laser Remote Sensing*, John Wiley & Sons, Inc., New York, 1984.
15. R. M. Measures, ed., *Laser Remote Chemical Analysis*, John Wiley & Sons, Inc., New York, 1988.
16. W. B. Grant, in Ref. 5, pp. 565–621.
17. J. T. Houghton, F. W. Taylor, and C. D. Rodgers, *Remote Sounding of Atmospheres*, Cambridge University Press, Cambridge, U.K., 1984.
18. R. Beer, *Remote Sensing by Fourier Transform Spectrometry*, John Wiley & Sons, Inc., New York, 1992.
19. K. Narahari Rao and A. Weber, eds., *Spectroscopy of the Earth's Atmosphere and Interstellar Medium*, Academic Press, Inc., Boston, Mass., 1992.
20. J. Ballard, *Advances Spectrosc.* **24**, 49–84 (1995).
21. A. C. G. Mitchell and M. W. Zemansky, *Resonance and Radiation and Excited Atoms*, Cambridge University Press, Cambridge, U.K., 1971.
22. J. T. Houghton and S. D. Smith, *Infra-Red Physics*, Oxford University Press, Oxford, U.K., 1966.
23. J. D. Kraus, *Radio Astronomy*, 2nd ed., Cygnus-Quasar Books, Powell, Ohio, 1986.
24. G. L. Verschuur and K. I. Kellermann, eds., *Galactic and Extragalactic Radio Astronomy*, 2nd ed., Springer-Verlag, New York, 1988.
25. K. Rohlfs and T. L. Wilson, *Tools of Radio Astronomy*, 2nd ed., Springer-Verlag, New York, 1996.
26. B. F. Burke and F. Graham-Smith, *An Introduction to Radio Astronomy*, Cambridge University Press, New York, 1996.
27. C. H. Townes and A. L. Schawlow, *Microwave Spectroscopy*, McGraw-Hill Book Co., Inc., New York, 1955; corrected reprint, Dover, New York, 1975.
28. G. W. Chantry, ed., *Modern Aspects of Microwave Spectroscopy*, Academic Press, London, 1979.

29. W. Gordy and R. L. Cook, *Microwave Molecular Spectra*, 3rd ed., John Wiley & Sons, Inc., New York, 1984.

30. R. Varma and L. W. Hrubesh, *Chemical Analysis by Microwave Rotational Spectroscopy*, John Wiley & Sons, Inc., New York, 1979.

31. S. Saito, *Appl. Spectrosc. Rev.* **25**, 261–296 (1989–1990).

32. P. F. Wacker and P. Kisliuk, eds., *Microwave Spectral Tables*, 5 vol., U.S. National Bureau of Standards, Washington, D.C., 1964–1969.

33. F. J. Lovas, *Frequencies for Interstellar Molecular Microwave Transitions*, Physics Laboratory, National Institute of Standards and Technology, Gaithersburg, Md., 1996; on Internet at http://physics.nist.gov.

34. M. A. Janssen, ed., *Atmospheric Remote Sensing by Microwave Radiometry*, John Wiley & Sons, Inc., New York, 1993.

35. A. Parrish, R. L. deZafra, P. M. Solomon, and J. W. Barrett, *Radio Sci.* **23**, 106–118 (1988).

36. H. A. Willis, J. H. van der Maas, and R. G. J. Miller, eds., *Laboratory Methods in Vibrational Spectroscopy*, 3rd ed., John Wiley & Sons, Inc., New York, 1987.

37. I. J. Spiro and M. Schlessinger, *Infrared Technology Fundamentals*, 2nd ed., Marcel Dekker, Inc., New York, 1995.

38. W. L. Wolfe, *Introduction to Infrared System Design*, SPIE Press, Bellingham, Wash., 1996.

39. A. L. Smith, *Applied Infrared Spectroscopy: Fundamentals, Techniques, and Analytical Problem-Solving*, John Wiley & Sons, Inc., New York, 1979.

40. N. B. Colthup, L. H. Daly, and S. E. Wiberly, *Introduction to Infrared and Raman Spectroscopy*, 3rd ed., Academic Press, Inc., Boston, Mass., 1990.

41. M. Diem, *Introduction to Modern Vibrational Spectroscopy*, John Wiley & Sons, Inc., New York, 1993.

42. B. Schrader, *Infrared and Raman Spectroscopy: Methods and Applications*, VCH Publishers, New York, 1994.

43. F. A. Miller, *Anal. Chem.* **64**, 824A–831 (1992); P. A. Wilks, Jr., *ibid.*, pp. 833A–838A; P. R. Griffiths, *ibid.*, pp. 868A–875A; N. Sheppard, *ibid.*, pp. 877A–883A.

44. J. R. Ferraro and L. J. Basile, eds., *Fourier Transform Infrared Spectroscopy*, Vols. 1–4, Academic Press, Inc., New York, 1978–1985.

45. P. R. Griffiths and J. A. de Haseth, *Fourier Transform Infrared Spectroscopy*, John Wiley & Sons, Inc., New York, 1986.

46. J. R. Ferraro and K. Krishnan, eds., *Practical Fourier Transform Infrared Spectroscopy: Industrial and Laboratory Chemical Analysis*, Academic Press, Inc., San Diego, Calif., 1990.

47. R. A. Palmer, *Spectroscopy* **8**(2), 26–36 (Feb. 1993).

48. C. Webster, R. Menzies, and E. D. Hinkley, in Ref. 15, pp. 163–272.

49. R. Grisar and co-workers, eds., *Monitoring of Gaseous Pollutants by Tunable Diode Lasers*, 3 Vols., Kluwer Academic Publishers, Boston, Mass., 1987–1991.

50. R. S. McDowell, *Vibrational Spectra Struct.* **10**, 1–151 (1981).

51. L. F. Mollenauer, in L. F. Mollenauer and J. C. White, eds., *Tunable Lasers*, Springer-Verlag, Berlin, 1987.

52. U. Simon and F. K. Tittel, *Laser Focus World* **30**(5), 99–110 (May 1994).

53. G. A. Blake, K. B. Laughlin, R. C. Cohen, K. L. Busarow, D.-H. Gwo, C. A. Schmuttenmaer, D. W. Steyert, and R. J. Saykally, *Rev. Sci. Instrum.* **62**, 1693–1700 (1991).

54. K. M. Evenson, in G. W. F. Drake, ed., *Atomic, Molecular, & Optical Physics Handbook*, American Institute of Physics, Woodbury, N.Y., 1996, pp. 473–478.

55. P. B. Coleman, ed., *Practical Sampling Techniques for Infrared Analysis*, CRC Press, Boca Raton, Fla., 1993.

56. T. J. Porro and S. C. Pattacini, *Spectroscopy* **8**(7), 40–47 (Sept. 1993); **8**(8), 39–44 (Oct. 1993).

57. J. Altmann, R. Baumgart, and C. Weitkamp, *Appl. Optics* **20**, 995–999 (1981).

58. E. L. Wehry and G. Mamantov, *Prog. Anal. Spectrosc.* **10**, 507–527 (1987).

59. M. J. Almond and A. J. Downs, *Spectroscopy of Matrix Isolated Species*, John Wiley & Sons, Inc., New York, 1989; *Adv. Spectrosc.* **17** (1989).

60. N. K. Wilson and J. W. Childers, *Appl. Spectrosc. Rev.* **25**, 1–61 (1989).

61. T. A. Dirksen and J. E. Gagnon, *Spectroscopy* **11**(2), 58–62 (Feb. 1996).

62. R. M. Crooks, C. Xu, L. Sun, S. L. Hill, and A. J. Ricco, *Spectroscopy* **8**(7), 28–39 (Sept. 1993).

63. J. P. Blitz and S. M. Augustine, *Spectroscopy* **9**(8), 28–34 (Oct. 1994).

64. E. W. Thulstrup and J. Michl, *Elementary Polarization Spectroscopy*, VCH Publishers, New York, 1989.

65. R. G. Messerschmidt and M. A. Harthcock, eds., *Infrared Microspectroscopy: Theory and Applications*, Marcel Dekker, Inc., New York, 1988.

66. J. E. Katon, A. J. Sommer, and P. L. Lang, *Appl. Spectrosc. Rev.* **25**, 173–211 (1989–1990).

67. J. E. Katon and A. J. Sommer, *Anal. Chem.* **64**, 931A–940A (1992).

68. H. J. Humecki, ed., *Practical Guide to Infrared Microspectroscopy*, Marcel Dekker, Inc., New York, 1995.

69. F. M. Mirabella, Jr., ed., *Internal Reflection Spectroscopy: Theory and Applications*, Marcel Dekker, Inc., New York, 1993.

70. M. W. Urban, *Attenuated Total Reflectance Spectroscopy of Polymers: Theory and Practice*, American Chemical Society, Washington, D.C., 1996.

71. R. Mendelsohn, J. W. Brauner, and A. Gericke, *Ann. Rev. Phys. Chem.* **46**, 305–334 (1995).

72. J. R. Ferraro, *Vibrational Spectroscopy at High External Pressures: The Diamond Anvil Cell*, Academic Press, Inc., New York, 1984.

73. R. J. H. Clark and R. E. Hester, eds., *Molecular Cryospectroscopy*, John Wiley & Sons, Inc., New York, 1995; *Adv. Spectrosc.* **23** (1995).

74. R. J. H. Clark and R. E. Hester, eds., *Time Resolved Spectroscopy*, John Wiley & Sons, Inc., New York, 1989; *Adv. Spectrosc.* **18** (1989).

75. M. J. Wirth, *Anal. Chem.* **62**, 270A–277A (1990).

76. P. F. Bernath, *Ann. Rev. Phys. Chem.* **41**, 91–122 (1990).

77. U. Krull and R. S. Brown, in Ref. 15, pp. 505–532.

78. P. Klocek and G. H. Sigel, Jr., *Infrared Fiber Optics*, SPIE Optical Engineering Press, Bellingham, Wash., 1989.

79. P. K. Cheo, *Fiber Optics and Optoelectronics*, 2nd ed., Prentice-Hall, Englewood Cliffs, N.J., 1990.

80. R. D. Driver, G. L. Dewey, D. A. Greenberg, and J. D. Stark, *Spectroscopy* **9**(4), 36–40 (May 1994).

81. A. Ganz and J. P. Coates, *Spectroscopy* **11**(1), 32–38 (Jan. 1996).

82. R. Beer, in Ref. 15, pp. 85–162.

83. C. R. Webster and R. D. May, *J. Geophys. Res.* **92**, 11931–11950 (1987).

84. A. Rosencwaig, *Photoacoustics and Photoacoustic Spectroscopy*, Wiley-Interscience, New York, 1980.

85. P. Hess and J. Pelzl, eds., *Photoacoustic and Photothermal Phenomena*, Springer-Verlag, Berlin, 1987.

86. V. P. Zharov and V. S. Letokhov, *Laser Optoacoustic Spectroscopy*, Springer-Verlag, Berlin, 1986.

87. J. R. Small and E. Kurian, *Spectroscopy* **10**(9), 27–33 (Nov.–Dec. 1995).

88. P. L. Meyer and M. W. Sigrist, *Rev. Sci. Instrum.* **61**, 1779–1807 (1990).

89. F. J. M. Harren, J. Reuss, E. J. Woltering, and D. D. Bicanic, *Appl. Spectrosc.* **44**, 1360–1368 (1990).

90. D. J. Brassington, *J. Phys.* **D15**, 219–228 (1982).
91. J. R. Durig, ed., *Chemical, Biological and Industrial Applications of Infrared Spectroscopy*, John Wiley & Sons, Inc., New York, 1985.
92. J. L. Koenig, *Anal. Chem.* **66**, 515A–521A (1994).
93. D. Lin-Vien, N. B. Colthup, W. G. Fateley, and J. G. Grasselli, *The Handbook of Infrared and Raman Characteristic Frequencies of Organic Molecules*, Academic Press, Inc., San Diego, Calif., 1991.
94. G. Socrates, *Infrared Characteristic Group Frequencies*, 2nd ed., John Wiley & Sons, Inc., New York, 1994.
95. J. P. Coates, *Spectroscopy* **10**(7), 14–17 (Sept. 1995).
96. R. White, *Chromatography/Fourier Transform Infrared Spectroscopy and Its Applications*, Marcel Dekker, Inc., New York, 1990.
97. C. Fujimoto and K. Jinno, *Anal. Chem.* **64**, 476A–481A (1992).
98. B. W. Cook, *Spectroscopy* **6**(6), 22–26 (July–Aug. 1991).
99. J. W. Washall and T. P. Wampler, *Spectroscopy* **6**(4), 38–42 (May 1991).
100. T. Provder, M. W. Urban, and H. G. Barth, eds., *Hyphenated Techniques in Polymer Characterization: Thermal-Spectroscopic and Other Methods*, American Chemical Society, Washington, D.C., 1994.
101. K. S. Kalasinsky, B. Levine, M. L. Smith, and G. E. Platoff, Jr., *Crit. Rev. Anal. Chem.* **23**, 441–457 (1993).
102. I. W. Levin and E. N. Lewis, *Anal. Chem.* **62**, 1101A–1111A (1990).
103. R. J. H. Clark and R. E. Hester, eds., *Biomolecular Spectroscopy*, 2 Vols., John Wiley & Sons, Inc., New York, 1993; *Adv. Spectrosc.* **20–21** (1993).
104. H. H. Mantsch and D. Chapman, eds., *Infrared Spectroscopy of Biomolecules*, John Wiley & Sons, Inc., New York, 1996.
105. W. Suëtaka, *Surface Infrared and Raman Spectroscopy: Methods and Applications*, Plenum Press, New York, 1995.
106. T. B. Sauke, J. F. Becker, M. Loewenstein, T. D. Gutierrez, and C. G. Bratton, *Spectroscopy* **9**(5), 34–40 (June 1994).
107. J. L. Koenig, *Spectroscopy of Polymers*, American Chemical Society, Washington, D.C., 1992.
108. W. W. Urban and T. Provder, eds., *Multidimensional Spectroscopy of Polymers: Vibrational, NMR, and Fluorescence Techniques*, American Chemical Society, Washington, D.C., 1995.
109. K. Ashley, *Talanta* **38**, 1209–1218 (1991).
110. K. A. B. Lee, A. L. Hood, A. L. Clobes, J. A. Schroeder, G. P. Ananth, and L. H. Hawkins, *Spectroscopy* **8**(5), 24–29 (June 1993).
111. H. Ahlberg, S. Lundqvist, and B. Olsson, *Appl. Opt.* **24**, 3924–3928 (1985).
112. D. J. Brassington, *Adv. Spectrosc.* **24**, 85–148 (1995).
113. M. T. Coffey and W. G. Mankin, *Spectroscopy* **8**(6), 22–27 (July–Aug. 1993).
114. M. J. Persky, *Rev. Sci. Instrum.* **66**, 4763–4797 (1995).
115. P. Williams and K. Norris, eds., *Near-Infrared Technology in the Agricultural and Food Industries*, American Association of Cereal Chemists, St. Paul, Minn., 1987.
116. J. K. Drennan, E. G. Kraemer, and R. A. Lodder, *Crit. Rev. Anal. Chem.* **22**, 443–475 (1991).
117. D. A. Burns and E. W. Ciurczak, eds., *Handbook of Near-Infrared Analysis*, Marcel Dekker, Inc., New York, 1992.
118. W. F. McClure, *Anal. Chem.* **66**, 43A–53A (1994).
119. G. Downey, *Analyst* **119**, 2367–2375 (1994).
120. C. D. Tran, *Anal. Chem.* **64**, 971A–981A (1992).
121. J. Soos, *Laser Focus World* **30**(8), 87–92 (Aug. 1994).
122. M. A. Sharaf, D. L. Illman, and B. R. Kowalski, *Chemometrics*, John Wiley & Sons, Inc., New York, 1986.

123. R. G. Brereton, *Chemometrics: Applications of Mathematics and Statistics to Laboratory Systems*, E. Horwood, New York, 1990.
124. S. J. Haswell, ed., *Practical Guide to Chemometrics*, Marcel Dekker, Inc., New York, 1992.
125. J. Coates, T. Davidson, and L. McDermott, *Spectroscopy* **7**(9), 40–49 (Nov.–Dec. 1992).
126. C. E. Miller, *Appl. Spectrosc. Rev.* **26**, 277–339 (1991).
127. P. Castro and H. Mark, *Spectroscopy* **9**(1), 27–32 (Jan. 1994).
128. E. W. Ciurczak and J. K. Drennen, *Spectroscopy* **7**(6), 12–14 (July–Aug. 1992).
129. J. J. Kelly, C. H. Barlow, T. M. Jinguji, and J. B. Callis, *Anal. Chem.* **61**, 313–320 (1989).
130. S. J. Swarin and C. A. Drumm, *Spectroscopy* **7**(7), 42–49 (Sept. 1992).
131. J. P. Coates, *Spectroscopy* **9**(9), 36–40 (Nov.–Dec. 1994).
132. J.-P. Conzen, J. Bürck, and H.-J. Ache, *Appl. Spectrosc.* **47**, 753–763 (1993).
133. R. J. Dempsey, D. G. Davis, R. G. Buice, Jr., and R. A. Lodder, *Appl. Spectrosc.* **50**(2), 18A–34A (Feb. 1996).
134. H. S. Chen, *Space Remote Sensing Systems: An Introduction*, Academic Press, Inc., Orlando, Fla., 1985.
135. S. V. Compton, D. A. C. Compton, and R. G. Messerschmidt, *Spectroscopy* **6**(6), 35–39 (July–Aug. 1991).
136. R. C. Carlson, A. F. Hayden, and W. B. Telfair, *Appl. Opt.* **27**, 4952–4959 (1988).
137. A. Knowles and C. Burgess, eds., *Practical Absorption Spectrometry*, Chapman & Hall, London, 1984.
138. R. C. Denney and R. Sinclair, *Visible and Ultraviolet Spectroscopy*, John Wiley & Sons, Inc., New York, 1988.
139. R. Lobinski and Z. Marczenko, *Crit. Rev. Anal. Chem.* **23**, 55–111 (1992).
140. H.-H. Perkampus, *UV–VIS Spectroscopy and Its Applications*, Springer-Verlag, New York, 1992.
141. B. J. Clark, T. Frost, and M. A. Russell, eds., *UV Spectroscopy: Techniques, Instrumentation, Data Handling*, Chapman & Hall, London, 1993.
142. I. R. Altemose, *J. Chem. Educ.* **63**, A216–A223, A262–A266 (1986).
143. J. A. R. Sampson, *Techniques of Vacuum Ultraviolet Spectroscopy*, John Wiley & Sons, Inc., New York, 1967.
144. A. N. Zaidel' and E. Ya. Shreider, *Vacuum Ultraviolet Spectroscopy*, Humphrey Science Publishers, Ann Arbor, Mich., 1970.
145. G. Stark and P. L. Smith, in Ref. 54, pp. 487–498.
146. J. V. Sweedler, R. D. Jalkian, and M. B. Denton, *Appl. Spectrosc.* **43**, 953–962 (1989).
147. R. E. Fields, M. E. Baker, D. A. Radspinner, R. S. Pomeroy, and M. B. Denton, *Spectroscopy* **7**(9), 28–35 (Nov.–Dec. 1992).
148. R. Williams, *Appl. Spectrosc. Rev.* **25**, 63–79 (1989).
149. G. Talsky, *Derivative Spectrophotometry*, VCH Publishers, New York, 1994.
150. C. B. Ojeda, F. S. Rojas, and J. M. C. Pavon, *Talanta* **42**, 1195–1214 (1995).
151. *The Sadtler Handbook of Ultraviolet Spectra*, Sadtler Research Laboratories, Philadelphia, Pa., 1979, and updates.
152. R. M. Silverstein, G. C. Bassler, and T. C. Morrill, *Spectrometric Identification of Organic Compounds*, 5th ed., John Wiley & Sons, New York, 1991.
153. H.-H. Perkampus, *UV–VIS Atlas of Organic Compounds*, 2nd ed., VCH Publishers, New York, 1992.
154. I. Sunshine, *CRC Handbook of Spectrophotometric Data of Drugs*, CRC Press, Boca Raton, Fla., 1981.
155. S. Görög, *Ultraviolet–Visible Spectrophotometry in Pharmaceutical Analysis*, CRC Press, Boca Raton, Fla., 1995.

156. E. Sevick-Muraca and D. Benaron, eds., *Biomedical Optical Spectroscopy and Diagnostics*, Optical Society of America, Washington, D.C., 1996.

157. R. E. Huffman, *Atmospheric Ultraviolet Remote Sensing*, Academic Press, Inc., Boston, Mass., 1992.

158. J. M. C. Plane and N. Smith, *Adv. Spectrosc.* **24**, 223–262 (1995).

159. J. W. Robinson, *Atomic Spectroscopy*, Marcel Dekker, Inc., New York, 1990.

160. L. H. J. Lajunen, *Spectrochemical Analysis by Atomic Absorption and Emission*, Royal Society of Chemistry, Cambridge, U.K., 1992.

161. W. Slavin, *Spectroscopy* **6**(8), 16–21 (Oct. 1991).

162. C. Grégoire, *Spectroscopy* **9**(6), 12–17 (July–Aug. 1994).

163. W. J. Price, *Spectrochemical Analysis by Atomic Absorption*, Heyden & Son, Ltd., London, 1979.

164. Z. Fang, *Flow Injection Atomic Absorption Spectrometry*, John Wiley & Sons, Inc., New York, 1995.

165. B. V. L'vov, *Anal. Chem.* **63**, 924A–931A (1991); A. Walsh, *ibid.*, pp. 933A–941A.

166. M. S. Cresser, *Flame Spectrometry in Environmental Chemical Analysis: A Practical Guide*, Royal Society of Chemistry, Cambridge, U.K., 1994.

167. N. J. Miller-Ihli, *Anal. Chem.* **64**, 964A–968A (1992).

168. K. Niemax, A. Zybin, C. Schnürer-Patschau, and H. Groll, *Anal. Chem.* **68**, 351A–356A (1996).

169. R. K. Gillette, *Spectroscopy* **9**(4), 42–44 (May 1994).

170. L. C. Schrier and S. E. Manahan, *Spectroscopy* **9**(2), 24–29 (Feb. 1994).

171. J. Sneddon, M. V. Smith, S. Indurthy, and Y.-I. Lee, *Spectroscopy* **10**(1), 26–29 (Jan. 1995).

172. K. Slickers, *Automatic Atomic-Emission Spectroscopy*, 2nd ed., Brühl-Universitäts Druckerei, Giessen, Germany, 1993.

173. A. T. Zander, *Spectroscopy* **9**(3), 16–32 (Mar.–Apr. 1994).

174. P. W. J. M. Boumans, ed., *Inductively Coupled Plasma Emission Spectroscopy*, 2 Vols. (*Methodology, Instrumentation, and Performance*; *Applications and Fundamentals*), John Wiley & Sons, Inc., New York, 1987.

175. M. Thompson and J. N. Walsh, eds., *Handbook of Inductively Coupled Plasma Spectrometry*, 2nd ed., Chapman & Hall, New York, 1989.

176. D. Beauchemin, *Spectroscopy* **7**(7), 12–18 (Sept. 1992).

177. A. Montaser and D. W. Golightly, eds., *Inductively Coupled Plasmas in Analytical Atomic Spectrometry*, 2nd ed., VCH Publishers, New York, 1992.

178. J.-M. Mermet and E. Poussel, *Appl. Spectrosc.* **49**(10), 12A–18A (Oct. 1995).

179. A. T. Zander, *Anal. Chem.* **58**, 1139A–1149A (1986).

180. J. M. Carey and J. A. Caruso, *Crit. Rev. Anal. Chem.* **23**, 397–439 (1992).

181. D. A. Cremers and L. J. Radziemski, in Ref. 5, pp. 351–415.

182. L. J. Radziemski and D. A. Cremers, eds., *Laser-Induced Plasmas and Applications*, Marcel Dekker, Inc., New York, 1989.

183. V. Majidi and M. R. Joseph, *Crit. Rev. Anal. Chem.* **23**, 143–162 (1992).

184. L. J. Radziemski, *Microchem. J.* **50**, 218–234 (1994).

185. E. Fogarassy, D. Geohegan, and M. Stuke, eds., *Laser Ablation*, Elsevier Science BV, Amsterdam, the Netherlands, 1996; *Appl. Surf. Sci.* **96–98** (1996).

186. J. Sneddon, *Spectroscopy* **9**(6), 34–38 (July–Aug. 1994).

187. R. E. Russo, *Appl. Spectrosc.* **49**(9), 14A–28A (Sept. 1995).

188. R. K. Marcus, ed., *Glow Discharge Spectroscopies*, Plenum Press, New York, 1993.

189. R. K. Marcus, T. R. Harville, Y. Mei, and C. R. Shick, Jr., *Anal. Chem.* **66**, 902A–911A (1994).

190. J. A. C. Broekaert, *Appl. Spectrosc.* **49**(7), 12A–19A (July 1995).

191. P. W. J. M. Boumans, *Anal. Chem.* **66**, 459A–467A (1994).
192. G. A. Meyer, *Spectroscopy* **8**(9), 28–34 (Nov.–Dec. 1993).
193. B. Chu, *Laser Light Scattering: Basic Principles and Practice*, 2nd ed., Academic Press, Inc., Boston, Mass., 1991.
194. G. D. J. Phillies, *Anal. Chem.* **62**, 1049A–1057A (1990).
195. D. Langevin, *Light Scattering by Liquid Surfaces and Complementary Techniques*, Marcel Dekker, Inc., New York, 1992.
196. L. Thomas, *Adv. Spectrosc.* **24**, 1–47 (1995).
197. D. K. Killinger and N. Menyuk, *Science* **235**, 37–45 (1987).
198. K. Fredriksson, in Ref. 15, pp. 273–332.
199. K. A. Fredriksson, *Appl. Opt.* **24**, 3297–3304 (1985).
200. D. P. Strommen and K. Nakamoto, *Laboratory Raman Spectroscopy*, John Wiley & Sons, Inc., New York, 1984.
201. D. J. Gardiner and P. R. Graves, eds., *Practical Raman Spectroscopy*, Springer-Verlag, New York, 1989.
202. J. G. Grasselli and B. J. Bulkin, eds., *Analytical Raman Spectroscopy*, John Wiley & Sons, Inc., New York, 1991.
203. J. R. Ferraro and K. Nakamoto, *Introductory Raman Spectroscopy*, Academic Press, Inc., Boston, Mass., 1994.
204. H. Inaba, in Ref. 13, pp. 153–236.
205. J. R. Ferraro, *Spectroscopy* **11**(3), 18–25 (Mar.–Apr. 1996).
206. P. Hendra, C. Jones, and G. Warnes, *Fourier Transform Raman Spectroscopy: Instrumentation and Chemical Applications*, Ellis Horwood, New York, 1991.
207. D. B. Chase and J. F. Rabolt, eds., *Fourier Transform Raman Spectroscopy: From Concept to Experiment*, Academic Press, Inc., San Diego, Calif., 1994.
208. P. J. Hendra and co-workers, *Analyst* **120**, 985–991 (1995).
209. S. F. Parker, N. Conroy, and V. Patel, *Spectrochim. Acta* **49A**, 657–666 (1993).
210. J. V. Sweedler, K. L. Ratzlaff, and M. B. Denton, eds., *Charge-Transfer Devices in Spectroscopy*, VCH Publishers, New York, 1994.
211. C. Adjouri, A. Elliasmine, and Y. Le Duff, *Spectroscopy* **11**(6), 44–49 (July–Aug. 1996).
212. T. Tahara and H. Hamaguchi, *Appl. Spectrosc.* **47**, 391–398 (1993).
213. K. P. J. Williams, I. C. Wilcock, I. P. Hayward, and A. Whitley, *Spectroscopy* **11**(3), 45–50 (Mar.–Apr. 1996).
214. C. L. Schoen, *Laser Focus World* **30**(5), 113–120 (May 1994).
215. R. J. H. Clark and R. E. Hester, eds., *Advances in Non-Linear Spectroscopy*, John Wiley & Sons, Inc., New York, 1988; *Adv. Spectrosc.* **15** (1988).
216. J. J. Valentini, in Ref. 5, pp. 507–564.
217. T. J. Vickers, C. K. Mann, J. Zhu, and C. K. Chang, *Appl. Spectrosc. Rev.* **26**, 341–375 (1991).
218. S. A. Asher, *Anal. Chem.* **65**, 59A–66A, 201A–210A (1993).
219. R. L. Garrell, *Anal. Chem.* **61**, 401A–411A (1989).
220. J. J. Laserna, *Anal. Chim. Acta* **283**, 607–622 (1993).
221. J. M. E. Storey, T. E. Barber, R. D. Shelton, E. A. Wachter, K. T. Carron, and Y. Jiang, *Spectroscopy* **10**(3), 20–25 (Mar.–Apr. 1995).
222. J. J. Valentini, in G. A. Vanasse, ed., *Spectrometric Techniques*, Vol. 4, Academic Press, Inc., Orlando, Fla., 1985, pp. 1–62.
223. E. P. C. Lai and J. M. Harris, *Appl. Spectrosc.* **45**, 1590–1597 (1991).
224. S. Nie, L. A. Lipscomb, and N.-T. Yu, *Appl. Spectrosc. Rev.* **26**, 203–276 (1991).
225. A. B. Harvey, ed., *Chemical Applications of Nonlinear Raman Spectroscopy*, Academic Press, Inc., New York, 1981.

226. R. L. Farrow and D. J. Rakestraw, *Science* **257**, 1894–1900 (1992).
227. P. Ewart and S. V. O'Leary, *J. Phys.* **B15**, 3669–3677 (1982); **B17**, 4609–4616 (1984).
228. T. G. Spiro, ed., *Biological Applications of Raman Spectroscopy*, Vols. 1–2, John Wiley & Sons, Inc., New York, 1987.
229. A. J. Aller, *Appl. Spectrosc. Rev.* **26**, 59–71 (1991).
230. C. J. Petty, D. E. Bugay, W. P. Findlay, and C. Rodriguez, *Spectroscopy* **11**(5), 41–45 (June 1996).
231. J. D. Houston, S. Sizgoric, A. Ulitsky, and J. Banic, *Appl. Opt.* **25**, 2115–2121 (1986).
232. H. Edner, K. Fredriksson, A. Sunesson, S. Svanberg, L. Unéus, and W. Wendt, *Appl. Opt.* **26**, 4330–4338 (1987).
233. F. J. Barnes, R. R. Karl, K. E. Kunkel, and G. L. Stone, *Remote Sens. Environ.* **32**, 81–90 (1990).
234. N. Purdie and H. G. Brittain, eds., *Analytical Applications of Circular Dichroism*, Elsevier, Amsterdam, the Netherlands, 1994.
235. K. Nakanishi, N. Berova, and R. W. Woody, eds., *Circular Dichroism: Principles and Applications*, VCH Publishers, New York, 1994.
236. G. D. Fasman, ed., *Circular Dichroism and the Conformational Analysis of Biomolecules*, Plenum Press, New York, 1996.
237. L. A. Nafie, *Appl. Spectrosc.* **50**(5), 14A–26A (May 1996).
238. P. L. Polavarapu, *Spectroscopy* **9**(9), 48–55 (Nov.–Dec. 1994).
239. M. Diem, *Spectroscopy* **10**(4), 38–43 (May 1995).
240. S. G. Schulman, ed., *Molecular Luminescence Spectroscopy: Methods and Applications*, Vols. 1–3, John Wiley & Sons, Inc., New York, 1985–1993.
241. W. R. G. Baeyens, D. de Keukeleire, and K. Korkidis, eds., *Luminescence Techniques in Chemical and Biochemical Analysis*, Marcel Dekker, Inc., New York, 1991.
242. J. R. Lakowicz, *Principles of Fluorescence Spectroscopy*, Plenum Press, New York, 1983.
243. G. G. Guilbault, ed., *Practical Fluorescence*, 2nd ed., Marcel Dekker, Inc., New York, 1990.
244. J. R. Lakowicz, ed., *Topics in Fluorescence Spectroscopy*, Vols. 1–4 (*Techniques*; *Principles*; *Biochemical Applications*; and *Probe Design and Chemical Sensing*), Plenum Press, New York, 1991–1994.
245. O. S. Wolfbeis, ed., *Fluorescence Spectroscopy: New Methods and Applications*, Springer-Verlag, New York, 1993.
246. F. V. Bright, *Appl. Spectrosc.* **49**(1), 14A–19A (Jan. 1995).
247. T. Vo-Dinh, *Room Temperature Phosphorimetry for Chemical Analysis*, John Wiley & Sons, Inc., New York, 1984.
248. R. J. Hurtubise, *Phosphorimetry: Theory, Instrumentation, and Applications*, VCH Publishers, New York, 1990.
249. K. A. Destrampe and G. M. Hieftje, *Appl. Spectrosc.* **47**, 1548–1554 (1993).
250. E. A. Permyakov, *Luminescent Spectroscopy of Proteins*, CRC Press, Boca Raton, Fla., 1993.
251. L. Munck, ed., *Fluorescence Analysis in Foods*, John Wiley & Sons, Inc., New York, 1989.
252. P. B. Stockwell and W. T. Corns, *Analyst* **119**, 1641–1645 (1994).
253. S. M. Angel, *Spectroscopy* **2**(4), 38–48 (Apr. 1987).
254. M. A. Arnold, *Anal. Chem.* **64**, 1015A–1025A (1992).
255. A. W. Czarnik, ed., *Fluorescent Chemosensors for Ion and Molecule Recognition*, American Chemical Society, Washington, D.C., 1993.
256. C. L. Stevenson and J. D. Windfordner, *Appl. Spectrosc.* **46**, 407–419, 715–724 (1992).

257. M. D. Barnes, W. B. Whitten, and J. M. Ramsey, *Anal. Chem.* **67**, 418A–423A (1995).

258. R. A. Keller, W. P. Ambrose, P. M. Goodwin, J. H. Jett, J. C. Martin, and M. Wu, *Appl. Spectrosc.* **50**(7), 12A–32A (July 1996).

259. J. Pfab, *Adv. Spectrosc.* **24**, 149–222 (1995).

260. R. P. Lucht, in Ref. 5, pp. 623–676.

261. J. H. Richardson, *Mod. Fluores. Spectrosc.* **4**, 1–24 (1981).

262. M. A. DeLuca and W. D. McElroy, eds., *Bioluminescence and Chemiluminescence: Basic Chemistry and Analytical Applications*, Academic Press, Inc., New York, 1981.

263. K. Van Dyke, ed., *Bioluminescence and Chemiluminescence: Instruments and Applications*, CRC Press, Boca Raton, Fla., 1985.

264. K.-D. Gundermann and F. McCapra, *Chemiluminescence in Organic Chemistry*, Springer-Verlag, New York, 1987.

265. A. K. Campbell, *Chemiluminescence: Principles and Applications in Biology and Medicine*, VCH Publishers, New York, 1988.

266. B. A. Ridley and L. C. Howlett, *Rev. Sci. Instrum.* **45**, 742–746 (1974).

267. H. I. Schiff, D. Pepper, and B. A. Ridley, *J. Geophys. Res.* **84**, 7895–7897 (1979).

268. A. Torres, *J. Geophys. Res.* **90**, 12875–12880 (1985).

269. J. G. Ferreira and M. T. Ramos, eds., *X-Ray Spectroscopy in Atomic and Solid State Physics*, Plenum Press, New York, 1988.

270. A. Meisel, G. Leonhardt, and R. Szargan, *X-Ray Spectra and Chemical Binding*, Springer-Verlag, New York, 1989.

271. B. K. Agarwal, *X-Ray Spectroscopy: An Introduction*, 2nd ed., Springer-Verlag, New York, 1991.

272. R. E. van Grieken and A. A. Markowicz, eds., *Handbook of X-Ray Spectrometry: Methods and Techniques*, Marcel Dekker, Inc., New York, 1993.

273. R. Jenkins, R. W. Gould, and D. Gedcke, *Quantitative X-Ray Spectrometry*, 2nd ed., Marcel Dekker, Inc., New York, 1995.

274. B. J. Price, N. S. Robson, K. M. Field, and B. Boyer, *Spectroscopy* **10**(9), 34–38 (Nov.–Dec. 1995).

275. K. L. Williams, *An Introduction to X-Ray Spectrometry: X-Ray Fluorescence and Electron Microprobe Analysis*, Allen & Unwin, Boston, Mass., 1987.

276. R. Jenkins, *X-Ray Fluorescence Spectrometry*, John Wiley & Sons, Inc., New York, 1988.

277. G. R. Lachance and F. Claisse, *Quantitative X-Ray Fluorescence Analysis: Theory and Application*, John Wiley & Sons, Inc., New York, 1995.

278. R. Klockenkämper, J. Knoth, A. Prange, and H. Schwenke, *Anal. Chem.* **64**, 1115A–1123A (1992).

279. T. Scimeca, *Crit. Rev. Anal. Chem.* **21**, 225–235 (1989).

280. G. P. Darsey, *Spectroscopy* **8**(6), 28–31 (July–Aug. 1993).

281. B. D. Wheeler, *Spectroscopy* **8**(5), 34–39 (June 1993).

282. R. P. Swift, *Spectroscopy* **10**(6), 31–35 (July–Aug. 1995).

283. T. A. Anderson, K. L. Evans, F. Carney, G. Carney, H. Berg, and R. Gregory, *Spectroscopy* **6**(6), 28–34 (July–Aug. 1991).

284. S. A. E. Johansson, J. L. Campbell, and K. G. Malmqvist, eds., *Particle-Induced X-Ray Emission Spectrometry (PIXE)*, John Wiley & Sons, Inc., New York, 1995.

285. J. V. Smith, *Analyst* **120**, 1231–1245 (1995).

286. G. Gilmore and J. D. Hemingway, *Practical Gamma-Ray Spectrometry*, John Wiley & Sons, Inc., New York, 1995.

287. D. L. Bertsch, C. E. Fichtel, and J. I. Trombka, *Space Sci. Rev.* **48**, 113–168 (1988).

288. H. Bloemen, *Ann. Rev. Astron. Astrophys.* **27**, 469–516 (1989).

289. T. E. Cranshaw, *Mössbauer Spectroscopy and its Applications*, Cambridge University Press, New York, 1985.

290. D. P. E. Dickson and F. J. Berry, eds., *Mössbauer Spectroscopy*, Cambridge University Press, New York, 1986.

291. R. H. Herber, ed., *Chemical Mössbauer Spectroscopy*, Plenum Press, New York, 1984.

292. G. J. Long, ed., *The Application of Mössbauer Spectroscopy in Inorganic Chemistry*, Vols. 1–3, Plenum Press, New York, 1984–1989.

293. S. Mitra, ed., *Applied Mössbauer Spectroscopy: Theory and Practice for Geochemists and Archeologists*, Elsevier Science Inc., Tarrytown, N.Y., 1993.

294. G. J. Long and J. G. Stevens, eds., *Industrial Applications of the Mössbauer Effect*, Plenum Press, New York, 1986.

295. G. J. Long and F. Grandjean, eds., *Mössbauer Spectroscopy Applied to Magnetism and Materials Science*, Plenum Press, New York, 1993.

296. G. Scoles, ed., *Atomic and Molecular Beam Methods*, Vols. 1–2, Oxford University Press, New York, 1988–1992.

297. S. H. Lin, Y. Fujimura, H. J. Neusser, and E. W. Schlag, *Multiphoton Spectroscopy of Molecules*, Academic Press, Inc., Orlando, Fla., 1984.

298. V. S. Letokhov, *Laser Photoionization Spectroscopy*, Academic Press, Inc., Orlando, Fla., 1987.

299. G. S. Hurst, M. G. Payne, S. D. Kramer, and J. P. Young, *Rev. Mod. Phys.* **51**, 767–819 (1979).

300. B. A. Bushaw, *Prog. Anal. Spectrosc.* **12**, 247–276 (1989).

301. J. A. Paisner and R. W. Solarz, in Ref. 5, pp. 175–260.

302. M. Gehrtz, G. C. Bjorklund, and E. A. Whittaker, *J. Opt. Soc. Am. B* **2**, 1510–1526 (1985).

303. C. B. Carlisle, D. E. Cooper, and H. Preier, *Appl. Optics* **28**, 2567–2576 (1990).

304. S. Bialkowski, *Photothermal Spectroscopy Methods for Chemical Analysis*, John Wiley & Sons, Inc., New York, 1996.

305. J. Georges, *Talanta* **41**, 2015–2023 (1994).

306. R. D. Snook and R. D. Lowe, *Analyst* **120**, 2051–2068 (1995).

307. J. J. Scherer, J. B. Paul, C. P. Collier, A. O'Keefe, D. J. Rakestraw, and R. J. Saykally, *Spectroscopy* **11**(5), 46–50 (June 1996).

General References

A. Corney, *Atomic and Laser Spectroscopy*, Oxford University Press, Oxford, U.K., 1977.

J. I. Steinfeld, *Molecules and Radiation*, 2nd ed., MIT Press, Cambridge, Mass., 1985.

G. W. Ewing, *Instrumental Methods of Chemical Analysis*, 5th ed., McGraw-Hill, New York, 1985.

A. P. Thorne, *Spectrophysics*, 2nd ed., Chapman and Hall, London, 1988.

H. H. Willard, L. J. Merritt, Jr., J. A. Dean, and F. A. Settle, Jr., *Instrumental Methods of Analysis*, 7th ed., Wadsworth Publishing Co., Belmont, Calif., 1988.

J. D. Ingle, Jr., and S. R. Crouch, *Spectrochemical Analysis*, Prentice-Hall, Englewood Cliffs, N.J., 1988.

W. S. Struve, *Fundamentals of Molecular Spectroscopy*, John Wiley & Sons, Inc., New York, 1989.

J. M. Hollas, *Modern Spectroscopy*, 3rd ed., John Wiley & Sons, Inc., New York, 1996.

Chemical Analysis, John Wiley & Sons, Inc., New York, monograph series on optical applications to analytical chemistry.

Springer Series on Optical Sciences, periodical of reviews on new developments in lasers and their applications.

Spectroscopy, annual reviews of new analytical instrumentation from the Pittsburgh Conference on Analytical Chemistry and Applied Spectroscopy.

Analytical Chemistry, "Fundamental Reviews" (June 1994, June 1996), analytical applications of infrared, ultraviolet, atomic absorption, emission, Raman, fluorescence, phosphorescence, chemiluminescence, and x-ray spectroscopy.

Atomic Spectroscopy and *Journal of Analytical Atomic Spectrometry*, regular and occasional topical bibliographies.

ROBIN S. McDOWELL
JAMES F. KELLY
Pacific Northwest National Laboratory

SPRAYS

A spray comprises a cloud of liquid droplets randomly dispersed in a gas phase. Depending on the application, sprays may be produced in many different ways. The purposes of most sprays are (*1*) creation of a spectrum of droplet sizes to increase the liquid surface-to-volume ratio, (*2*) metering or control of the liquid throughput, (*3*) dispersion of the liquid in a certain pattern, or (*4*) generation of droplet velocity and momentum.

The mechanical devices designed to generate sprays are commonly called atomizers. Liquid atomizers have a profound impact on modern industry. They can be found in many industrial, agricultural, and propulsion systems. Processes that require atomizing systems include spray drying, cooling, fuel combustion, spray painting and coating, application of herbicides and pesticides, food processing, molten-metal solidification, medical nebulizers, and aerosol sprays for consumer products.

Although sprays are useful in numerous commercial applications, they may create some serious environmental problems because of inefficient atomization or through misuse. For example, there is a growing concern over pollutant emissions from aircraft and automotive engines that utilize atomizers. Pollutants from engines include carbon monoxide, unburned hydrocarbons, oxides of nitrogen, and smoke. These pollutants can cause photochemical smog, depletion of the ozone layer, acid rain, and other conditions harmful to human life. Atmospheric pollution may also occur in spray painting and coating processes. Devices used to atomize solvents and coating formulations must be designed to meet air pollution standards. Because of the potential problems associated with sprays, it has become increasingly important to understand the process of atomization. Liquid atomizers must be properly designed and selected to minimize unnecessary hazards.

In the past, the design of atomizers and spray processes was based on traditional fluid dynamic principles and empirical methods. Optimum performance was generally sought by trial-and-error methods, and performance was evaluated by relatively crude experimental techniques. In the 1990s, however, the technology of spraying has advanced rapidly through computer-aided design procedures, mathematical modeling, and sophisticated instrumentation. The mechanics of liquid dispersion is better understood, and prediction of important spray parameters with improved accuracy is possible.

Because high quality, low cost, and optimum performance are required for spray equipment, improved analytical and experimental tools are indispensable for increasing productivity in many competitive industries. In most instances, it is no longer adequate to characterize a spray solely on the basis of flow rate and spray pattern. Information on droplet size, velocity, volume flux, and number density is often needed and can be determined using advanced laser diagnostic techniques. These improvements have benefited a wide spectrum of consumer and specialized industrial products.

This article covers basic information in the areas of atomizer design, the physics of atomization, spray characteristics, and instrumentation. It also includes some advanced topics to assist research and design engineers in their studies. Atomization technology will continue to expand because more development will be needed to address specific concerns in a wide variety of applications. Industry standards must be established in spray terminology, testing, and design procedures to avoid confusion and misunderstanding. Innovative concepts will also be required to meet more stringent environmental standards. The development of microprocessor or electronically controlled atomizers responsive to changes in operating conditions also appears to have a growing demand from industry.

Liquid Atomizers

Because of the diverse applications involving liquid atomizers, a large vocabulary of terms has evolved in the spray community. The American Society for Testing and Materials, ASTM Subcommittee E29.04 on Liquid Particle Characterization, has attempted to standardize the terminology relating to atomizing devices (1). The definitions adopted by ASTM are used herein.

The transformation of bulk liquid to sprays can be achieved in many different ways. Basic techniques include applying hydraulic pressure, electrical, acoustic, or mechanical energy to overcome the cohesive forces within the liquid.

Atomizers can be classified according to the energy source used to achieve liquid breakup. For example, when liquid pressure is used as the primary energy source, the device can be called a pressure atomizer. When an electric charge is the primary source of energy, it is called an electrostatic atomizer. If high frequency acoustic energy is imparted to the liquid for breakup, it is called an ultrasonic atomizer. A pneumatic atomizer utilizes the momentum of gas as the primary source of energy. Table 1 summarizes liquid atomizers based on this source of energy.

Liquid atomizers may also be classified according to their distinct design features and spray characteristics. More detailed information on various atomizers is available (1–3).

Physics of Liquid Atomization

Liquid atomization involves a series of complicated physical processes. These processes can generally be divided into three different flow regimes: internal

Table 1. Summary of Atomizers Based on Source of Energy

Atomizer	Description	Design and spray features
air-assisted	pneumatic atomizer in which pressurized air is utilized to enhance atomization produced by pressurized liquid	requires external source of pressurized air; device tends to be energy inefficient, but can produce very fine droplets; commonly used in industrial furnaces or gas turbines
airblast	pneumatic atomizer that utilizes a relatively large volume of low pressure air	liquid is spread into a thin conical sheet exposed to high velocity air on both sides of sheet; widely used in aircraft gas turbine engines
centrifugal	rotating solid surface is the primary source of energy utilized to produce spray	liquid is fed into center of spinning disk, cup, or wheel, and spreads out toward rim; produces a 360° spray pattern and relatively uniform drop size; used in spray drying and cooling applications
electrostatic	electric charge is the primary source of energy utilized to produce spray	some devices use capillary tubes or conical disks directly charged at high voltage; others charge liquid film or jet by electrodes outside atomizer; produces very fine droplets, but cannot handle high flow rates; used in printing or painting processes
piloted airblast	airblast atomizer combined with a lower capacity pressure atomizer	typically a small pressure atomizer tip surrounded by an annular orifice; atomizing air flows between and outside the two concentric sprays; used in gas turbine engines
pneumatic (twin-fluid)	movement of gas/vapor is primary souce of energy utilized to produce spray	gas stream directed through various configurations to impinge or lift a liquid stream; tangential slots and a chamber often used to enhance mixing and fluid interaction; large variety of commercial applications
pressure	pressurized liquid is primary source of energy utilized to produce spray	circular orifice outlet preceded by swirl chamber having ≥1 tangential inlets; hollow conical pattern with spray angles between 30 and 120°; flow rate varies approximately with square root of operating

Table 1. (*Continued*)

Atomizer	Description	Design and spray features
		pressure; widely used in oil heating equipment
sonic	pneumatic or vibratory atomizer in which energy is imparted (frequencies <20 KHz) to liquid	accelerated gas directed to impinge on plate or resonant cavity to produce high frequency sound waves; produces droplets <50 μm, but acoustic noise may be problem
ultrasonic	pneumatic or vibratory atomizer in which energy is imparted at high frequency to liquid	liquid is fed through or over transducer and horn, which vibrates at ultrasonic frequencies; widely used in medical inhalation therapy
vibratory (piezoelectric)	oscillating solid surface is primary source of energy	consists of hypodermic needle vibrating at controlled frequency; can produce uniform droplet sizes to 30 μm, determined by liquid flow rate, resonant frequency, and needle orifice size

flow, breakup, and droplet dispersion. The internal flow regime extends from the atomizer inlets to the discharge orifice where liquid emerges. The liquid breakup regime starts at the atomizer exit plane and ends at a certain distance downstream where primary atomization is complete. The final process of atomization is the dispersion regime where spherical droplets gradually evolve into a particular spray pattern.

Internal Flow. Depending on the atomizer type and operating conditions, the internal fluid flow can involve complicated phenomena such as flow separation, boundary layer growth, cavitation, turbulence, vortex formation, and two-phase flow. The internal flow regime is often considered one of the most important stages of liquid atomization because it determines the initial liquid disturbances and conditions that affect the subsequent liquid breakup and droplet dispersion.

The flow characteristics inside liquid atomizers have been studied by numerous investigators (4–8). Of special interest to designers is the work reported on swirl atomizers (4), fan spray atomizers (6,7), and plain jets (8). The following discussion focuses on the flow characteristics of a swirl atomizer.

Swirl Atomizer. Many atomizers utilize tangential slots or passages to produce liquid swirl to facilitate atomization. Figure 1 is a schematic diagram of a swirl atomizer. Under high pressure, liquid flow is forced into the swirl chamber through several tangential slots on the distributor. As liquid enters into the swirl chamber, it spins in the manner of a whirlpool. The liquid swirling effect creates a central low pressure region that draws external air into the chamber to form an air vortex. This air vortex extends from the rear of the chamber through the center of the exit orifice. Because of this air core, the exit

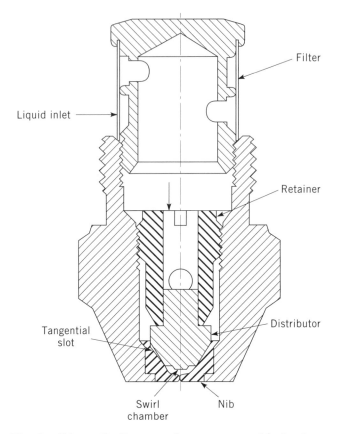

Fig. 1. Schematic diagram of a pressure swirl atomizer.

orifice is not completely filled with liquid. Figure 2 shows the air core structure inside a typical swirl chamber.

A thorough description of the internal flow structure inside a swirl atomizer requires information on velocity and pressure distributions. Unfortunately, this information is still not completely available as of this writing (1996). Useful insights on the boundary layer flow through the swirl chamber are available (9–11). Because of the existence of an air core, the flow structure inside a swirl atomizer is difficult to analyze because it involves the solution of a free-surface problem. If the location and surface pressure of the liquid boundary are known, however, the equations of motion of the liquid phase can be applied to reveal the detailed distributions of the pressure and velocity.

One proposed simplified theory (4) provides reasonably accurate predictions of the internal flow characteristics. In this analysis, conservation of mass as well as angular and total momentum of the liquid is assumed. To determine the exit film velocity, size of the air core, and discharge coefficient, it is also necessary to assume that a maximum flow through the orifice is attained.

Numerous studies for the discharge coefficient have been published to account for the effect of liquid properties (12), operating conditions (13), atomizer geometry (14), vortex flow pattern (15), and conservation of axial momen-

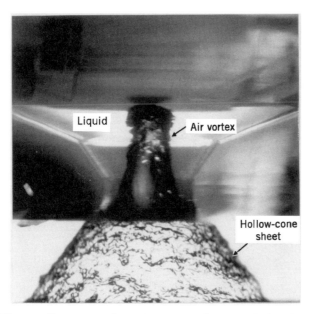

Fig. 2. Structure of an air core inside a swirl chamber.

tum (16). From one analysis (17), the following empirical equation appears to correlate well with the actual data obtained for swirl atomizers over a wide range of parameters, where the discharge coefficient is defined as $C_d = Q/(A_2(2g\Delta P/\rho_L)^{1/2})$; typical values of C_d range between 0.3 and 0.5.

$$C_d = 0.45\left(\frac{d_0\rho_L U}{\mu_L}\right)^{-0.02}\left(\frac{l_0}{d_0}\right)^{-0.03}\left(\frac{L_s}{D_s}\right)^{0.05}\left(\frac{A_p}{D_s d_0}\right)^{0.52}\left(\frac{D_s}{d_0}\right)^{0.23} \qquad (1)$$

Liquid Breakup. In the breakup regime, high magnification photography reveals that liquid atomization is associated with the phenomena of wave formation and propagation, rupture of ligaments, ligament collision and coalescence, and continuous disintegration caused by shear, rotation, impingement, and pulsation. Depending on the atomizer types and operating conditions, liquid breakup can be governed by different mechanisms.

The fundamental principle of liquid disintegration lies in the balance between disruptive and cohesive forces. The common disruptive forces in atomizer systems include kinetic energy, turbulent fluctuation, pressure fluctuation, interface shearing, friction, and gravity. The cohesive forces within the liquid are molecular bonding, viscosity, and surface tension.

Fan Sprays. It was demonstrated around the 1950s that instability theory can be used to analyze the wave growth on a thin liquid sheet (18). This analysis predicted the existence of an optimum wavelength at which a wave would grow rapidly. This optimum wavelength, λ_{opt}, corresponds to a condition that leads to liquid sheet disintegration. It can be expressed as in equation 2:

$$\lambda_{\mathrm{opt}} = \frac{4\pi\sigma}{\rho_A U_R^2} \qquad (2)$$

The theory has been extended to evaluate sheet breakup (19). This model (19) assumes that the fastest growing wave detaches at the leading edge in the form of a ribbon with a width of a half-wavelength. The ribbon immediately contracts into multiple ligaments, which subsequently reshape themselves into spherical droplets. The characteristic dimension of the ligament, D_L, is as follows, where t is the sheet thickness at the breakup location.

$$D_L = \left(\frac{2}{\pi} \lambda_{\mathrm{opt}} t \right)^{0.5} \tag{3}$$

In accordance with Rayleigh's analysis (20), the characteristic dimension, D_L, of a ligament is related to the droplet mean diameter in the form of $\overline{D} = 1.89 D_L$. Using this analysis, the predicted mean diameter, $\overline{D}$, is roughly 30% higher than the measured values.

Hollow-Cone Sprays.   In swirl atomizers, the liquid emerges from the exit orifice in the form of a conical sheet. As the liquid sheet spreads radially outward, aerodynamic instability immediately takes place and leads to the formation of waves which subsequently disintegrate into ligaments and droplets. Figure 3 illustrates the breakup process in an annular liquid sheet.

In the analysis of hollow-cone sprays, two basic physical models are required to describe the sheet breakup process. First, a film dynamics model is needed to simulate the motion of the liquid film as it gradually develops into a conical shape. Second, an instability model is necessary to describe the growth and propagation of waves on the film surface. If the amplitude of the waves gradually increases from a small initial disturbance as they progress downstream, liquid disintegration can occur at a point where the wave amplitude exceeds one-half of the wavelength or the local film thickness. A characteristic equation (21) for studying the growth rate of unstable waves on a film surface can be expressed

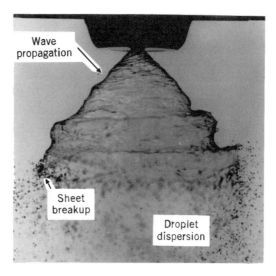

Fig. 3.   Wave formation and breakup of an annular sheet.

as follows, where $G_1 = \rho_L h$, $G_2 = \mu_L h n^2$, and $G_3 = \rho_A U_R^2 n - \sigma n^2$.

$$G_1 \alpha^2 + G_2 \alpha = G_3 \tag{4}$$

Equation 4 establishes the relationship between the growth rate, α, and local film parameters. This model provides a reasonable foundation for further analysis of droplet formation.

Jet Spray. The mechanism that controls the breakup of a liquid jet has been analyzed by many researchers (22,23). These studies indicate that liquid jet atomization can be attributed to various effects such as liquid–gas aerodynamic interaction, gas- and liquid-phase turbulence, capillary pinching, gas pressure fluctuation, and disturbances initiated inside the atomizer. In spite of different theories and experimental observations, there is agreement that capillary pinching is the dominant mechanism for low velocity jets. As jet velocity increases, there is some uncertainty as to which effect is most important in causing breakup.

It has been postulated that jet breakup is the result of aerodynamic interaction between the liquid and the ambient gas. Such theory considers a column of liquid emerging from a circular orifice into a surrounding gas. The instability on the liquid surface is examined by using first-order linear theory. A small perturbation is imposed on the initially steady liquid motion to simulate the growth of waves. The displacement of the surface waves can be obtained by the real component of a Fourier expression:

$$\xi = R(\xi_0 e^{inx + \alpha t}) \tag{5}$$

where R = the real part of a complex expression, x denotes the axial distance along the jet surface, n is the wave number, ξ represents the displacement of the surface in the radial direction from its undisturbed position, and ξ_0 is the initial amplitude of the disturbance. In addition to the initial disturbance, there are disturbances associated with pressure and velocity fluctuations in both the liquid and gas phases. These fluctuations can be described by the continuity and momentum equations that relate the growth of an initial perturbation to its wavelength and the other physical and dynamical parameters of the liquid and surrounding gas.

Droplet Dispersion. The primary feature of the dispersed flow regime is that the spray contains generally spherical droplets. In most practical sprays, the volume fraction of the liquid droplets in the dispersed region is relatively small compared with the continuous gas phase. Depending on the gas-phase conditions, liquid droplets can encounter acceleration, deceleration, collision, coalescence, evaporation, and secondary breakup during their evolution. Through droplet and gas-phase interaction, turbulence plays a significant role in the redistribution of droplets and spray characteristics.

After breakup, droplets continue to interact with the surrounding environment before reaching their final destination. In theory (24), each droplet group produced during primary breakup can be traced by using a Lagrangian calculation procedure. Droplet size and velocity can be determined as a function of spatial locations.

Spray Characteristics

Spray characteristics are those fluid dynamic parameters that can be observed or measured during liquid breakup and dispersal. They are used to identify and quantify the features of sprays for the purpose of evaluating atomizer and system performance, for establishing practical correlations, and for verifying computer model predictions. Spray characteristics provide information that is of value in understanding the fundamental physical laws that govern liquid atomization.

In the breakup regime, spray characteristics include film angle, film velocity and thickness, breakup length, breakup rate, surface wave frequency, wavelength, growth rate, and penetration distance. These quantities, however, are extremely difficult to measure on account of the very small size and rapidly changing features of disintegrating liquid jets or films.

In the dispersed regime, the physical and instrumental limitations are not as stringent as those in the breakup regime because primary atomization has been completed and the droplets have been dispersed into a much larger volume of space. Therefore, the dynamic quantities can be measured by most instruments. Parameters that are useful for describing the dispersed regime include droplet size and velocity, number density, volume flux, turbulence, gas dynamics quantities, spray pattern and angle, skewness, droplet arrival statistics, droplet trajectories and angles of flight, and vapor concentration.

During the formation of a spray, its properties vary with time and location. Depending on the atomizing system and operating conditions, variations can result from droplet dispersion, acceleration, deceleration, collision, coalescence, secondary breakup, evaporation, entrainment, oxidation, and solidification. Therefore, it may be extremely difficult to identify the dominant physical processes that control the spray dynamics and configuration.

Spray Parameters.   The more common spray parameters are as follows.

Droplet Size Distribution.   Most sprays comprise a wide range of droplet sizes. Some knowledge of the size distribution is usually required, particularly when evaluating the overall atomizer performance. The size distribution may be expressed in various ways. Several empirical functions, including the Rosin-Rammler (25) and Nukiyama-Tanasawa (26) equations, have been commonly used.

Most distribution functions contain an average size and a variance parameter typically based on the cumulative droplet number or volume distributions. For example, the Rosin-Rammler function uses the cumulative liquid volume as a means of expressing the distribution. It can be expressed as follows, where V is the fraction of the total volume contained in droplets of diameter less than D_i, n is a measure of the spread in the reported diameters, and $\overline{D}$ is an average diameter.

$$V = 1 - \exp\left[-\left(\frac{D_i}{\overline{D}}\right)^n\right] \tag{6}$$

The larger the value of n, the more uniform is the size distribution. Other types of distribution functions can be found in Reference 1. Distribution functions based on two parameters sometimes do not accurately match the actual distributions.

In these cases a high order polynomial fit, using multiple parameters, must be considered to obtain a better representation of the raw data.

Mean Diameters. Several mean diameters are frequently used to represent the statistical properties of droplets produced by liquid atomizers. These mean diameters may be expressed according to the following proposed notation (27):

$$\overline{D}_{pq} = \left[\frac{\sum\limits_{i=1}^{k} (N_i D_i^p)}{\sum\limits_{i=1}^{k} (N_i D_i^q)} \right]^{\frac{1}{(p-q)}} \qquad (7)$$

The p and q denote the integral exponents of $\overline{D}$ in the respective summations, and thereby explicitly define the diameter that is being used. N_i and D_i are the number and representative diameter of sampled drops in each size class i. For example, the arithmetic mean diameter, $\overline{D}_{10}$, is a simple average based on the diameters of all the individual droplets in the spray sample. The volume mean diameter, $\overline{D}_{30}$, is the diameter of a droplet whose volume, if multiplied by the total number of droplets, equals the total volume of the sample. The Sauter mean diameter, $\overline{D}_{32}$, is the diameter of a droplet whose ratio of volume-to-surface area is equal to that of the entire sample. This diameter is frequently used because it permits quick estimation of the total liquid surface area available for a particular industrial process or combustion system. Typical values of $\overline{D}_{32}$ for pressure swirl atomizers range from 50 to 100 μm.

Median Diameter. The median droplet diameter is the diameter that divides the spray into two equal portions by number, length, surface area, or volume. Median diameters may be easily determined from cumulative distribution curves.

Number Density and Volume Flux. The determination of number density and volume flux requires accurate information on the sample volume cross-sectional area, droplet size and velocity, as well as the number of droplets passing through the sample volume at any given instant of time. Depending on the instrumentation, the sample volume may vary with the optical components and droplet sizes. The number density represents the number of droplets contained in a specified volume of space at a given instant. It can be expressed as follows, where $\overline{u}$ is the mean droplet velocity, t the sample time, and A the representative cross-sectional area at the sampling location.

$$N = \frac{\sum\limits_{i=1}^{k} N_i}{\overline{u} t A} \qquad (8)$$

Volume flux is the volume contained by the droplets passing through a unit cross-sectional area per unit interval of time. It can be calculated as follows, where $\overline{D}_{30}$ is the volume mean diameter and n is the total number of droplets.

$$F = \frac{\frac{\pi}{6} n \overline{D}_{30}^3}{t A} \qquad (9)$$

Measurements of local volume flux distributions may be used to establish the degree of symmetry of a spray. Flux values must be integrated across the measurement planes and verified against the liquid flow rate of the atomizer.

Cone Angle.   The spray cone angle is one of the most important parameters in the specification of atomizers. Unfortunately, it is very difficult to define and measure because typical sprays have curved and fuzzy boundaries. A common method of defining the spray cone angle is to draw two tangent lines originating at the orifice and extending to the outermost spray edges at a specified axial distance. Several devices are commonly used to make quick estimates of spray angle. These devices include goniometers, needle probes equipped with linear displacement transducers, and projectors for back-lighting spray images.

Patternation.   The spray pattern provides important information for many spray applications. It is directly related to the atomizer performance. For example, in spray drying, an asymmetric spray pattern may cause inadequate liquid–gas mixing, thereby resulting in poor efficiency and product quality. Instruments that provide quantitative information on spray patterns are therefore essential for many processes. The pattern information must be able to reveal characteristics such as skewness, degree of pattern hollowness, and the uniformity of liquid flux over the entire cross-sectional area.

Patternators may comprise an array of tubes or concentric circular vessels to collect liquid droplets at specified axial and radial distances. Depending on the patternator, various uniformity indexes can be defined using the accumulated relative values between the normalized flow rate over a certain sector or circular region and a reference value that represents a perfectly uniform distribution. For example, using an eight-sector pie-shaped collector, the reference value for a perfectly uniform spray would be 12.5%. The uniformity index (28) could then be expressed as follows, where w_i is the normalized volume or mass flow rate percentage in each 45-degree sector.

$$\text{uniformity index} = \sum_{i=1}^{8} |w_i - 12.5| \tag{10}$$

Spray Dynamic Structure.   Detailed measurements of spray dynamic parameters are necessary to understand the process of droplet dispersion. Improvements in phase Doppler particle analyzers (PDPA) (29) permit *in situ* measurements of droplet size, velocity, number density, and liquid flux, as well as detailed turbulence characteristics for very small regions within the spray. Such measurements allow designers to evaluate differences in atomizers, changes in droplet size distributions, radial and circumferential symmetry, size–velocity correlations, interactions between droplets and the surrounding gas, droplet time-dependent behavior, and droplet drag and trajectories. In addition, the information can be extremely valuable for verifying physical models and establishing general correlations for atomizer performance.

Spray dynamic structures vary significantly depending on the operating conditions and atomizer types. One of the most common patterns, hollow spray, is discussed here in detail.

Hollow Sprays.   Most atomizers that impart swirl to the liquid tend to produce a cone-shaped hollow spray. Although swirl atomizers can produce varying degrees of hollowness in the spray pattern, they all seem to exhibit similar spray dynamic features. For example, detailed measurements made with simplex, duplex, dual-orifice, and pure airblast atomizers show similar dynamic structures in radial distributions of mean droplet diameter, velocity, and liquid volume flux. Extensive studies have been made (30,31) on the spray dynamics associated with pressure swirl atomizers. Based on these studies, some common features were observed. Test results obtained from a pressure swirl atomizer spray could be used to illustrate typical dynamic structures in hollow sprays. The measurements were made using a phase Doppler spray analyzer.

Figure 4 shows a three-dimensional distribution of the Sauter mean diameter, $\overline{D}_{32}$, measured 38.1-mm downstream from the nozzle using a Delavan 1 GPH-80°A pressure atomizer. The operating pressure was 690 kPa (100 psi). Typically, the mean diameters gradually increase with an increase in radial distance. This indicates that, in hollow sprays, the small droplets are mainly distributed in the center region, whereas the large droplets are found near the outer edge of the pattern.

Figure 5 shows the variation of the droplet mean axial velocity at the same axial location. The primary feature of this velocity profile is that the maximum velocity peaks at the centerline. The velocity magnitude and direction in the center region tend to be related to the liquid swirl strength and axial distance. A reverse (recirculation) flow with negative velocity is possible if the swirl is intense. Under such conditions, the maximum velocity tends to shift away from the centerline.

Spray Correlations. One of the most important aspects of spray characterization is the development of meaningful correlations between spray parameters and atomizer performance. The parameters can be presented as mathematical expressions that involve liquid properties, physical dimensions

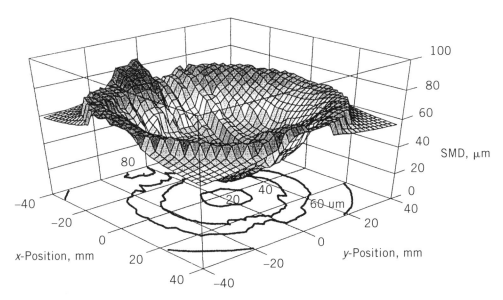

Fig. 4. Three-dimensional distribution of Sauter mean diameter (SMD) in a typical hollow-cone spray.

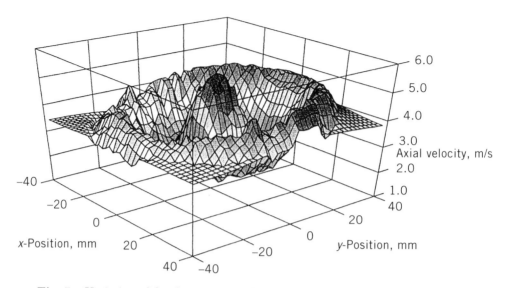

Fig. 5. Variation of droplet mean axial velocity in a typical hollow-cone spray.

of the atomizer, as well as operating and ambient conditions that are likely to affect the nature of the dispersion. Empirical correlations provide useful information for designing and assessing the performance of atomizers. Dimensional analysis has been widely used to determine nondimensional parameters that are useful in describing sprays. The most common variables affecting spray characteristics include a characteristic dimension of atomizer, d; liquid density, ρ_L; liquid dynamic viscosity, μ_L; surface tension, σ; pressure, ΔP; liquid velocity, v_L; gas density, ρ_G; and gas velocity, v_G.

Based on such analyses, the Reynolds and Weber numbers are considered the most important dimensionless groups describing the spray characteristics. The Reynolds number, Re, represents the ratio of inertial forces to viscous drag forces.

$$Re = \frac{\rho_L v_L d}{\mu_L}$$

The Reynolds number is sufficient as a parameter for describing the internal flow characteristics, such as discharge coefficient, air core ratio, and spray angle at the atomizer exit.

The Weber number, We, is defined as follows and represents the ratio of the disruptive aerodynamic forces to the restoring surface tension forces.

$$We = \frac{\rho_L d(v_L - v_G)^2}{\sigma}$$

The Weber number becomes important at conditions of high relative velocity between the injected liquid and surrounding gas. Other dimensionless parameters, such as the Ohnesorge ($(We)^{1/2}/Re$), Euler ($\Delta P/\rho_L v_L^2$), and Taylor (Re/We) numbers, have also been used to correlate spray characteristics. These parameters, however, are not used as often as the Reynolds and Weber numbers.

Many empirical correlations have been published in the literature for various types of liquid atomizers, eg, one book (2) provides an extensive collection of empirical equations. Unfortunately, most of the correlations share some common problems. For example, they are only valid for a specific type of atomizer, thereby imposing strict limitations on their use. They do not represent any specific physical processes and seldom relate to the design of the atomizer. More important, they do not reveal the effect of interactions among key variables. This indicates the difficulty of finding a universal expression that can cover a wide range of operating conditions and atomizer designs.

Droplet Size Correlations. The majority of correlations found in the literature deal with mean droplet diameters. A useful equation for Sauter mean diameters produced by pressure swirl atomizers (eq. 11) has been proposed (32). It consists of two separate terms, one dominated by liquid viscosity and pressure, the other by film thickness. To estimate the Sauter mean diameter, it is necessary to calculate first the film thickness, t_0, at the discharge orifice of the atomizer. Equation 12 may be used, where t_0 is the initial film thickness in meters, d the orifice diameter in meters, m_L the mass flow rate in kg/s, μ_L the dynamic viscosity in kg/m/s, ρ_L the liquid density in kg/m^3, and ΔP the pressure drop in Pascal. In equation 11, $\overline{D}_{32}$ is in meter, σ is the surface tension in N/m, and ρ_G is the gas density in kg/m^3.

$$\overline{D}_{32} = 2.29\left(\frac{\sigma\mu_L^2}{\rho_G}\right)^{0.25}\Delta P^{-0.5}t_0^{0.25} + 0.89\left(\frac{\sigma\rho_L}{\rho_G}\right)^{0.25}\Delta P^{-0.25}t_0^{0.75} \qquad (11)$$

$$t_0 = 3.66\left(\frac{dm_L\mu_L}{\rho_L\Delta P}\right)^{0.25} \qquad (12)$$

Using equations 11 and 12, the estimated Sauter mean diameters agree quite well with experimental data obtained for a wide range of atomizer designs. Note that the two constants in equation 11 differ from those shown in Lefebvre's equation (32). These constants have been changed to fit a wide range of experimental data.

For airblast-type atomizers, it has been speculated (33) that the Sauter mean diameter is governed by two factors, one controlled by air velocity and density, the other by liquid viscosity. Equation 13 has been proposed for the estimation of $\overline{D}_{32}$. In equation 13, A and B are constants whose values depend on atomizer design; ALR is the air–liquid mass ratio.

$$\overline{D}_{32} = L\left[1 + \frac{1}{\text{ALR}}\right]\left[A\left(\frac{\sigma}{\rho_G U_R^2 L}\right)^{0.5} + B\left(\frac{\mu_L^2}{\rho_L\sigma L}\right)^{0.5}\right] \qquad (13)$$

The first term is essentially the reciprocal of the Weber number and the second term is a function of the Ohnesorge number. Equation 13 may be invalid for airblast atomizers operating at high pressures, >1 MPa (>10 atm), or with high viscosity liquids.

Effect of Variables on Mean Droplet Size. Some of the principal variables affecting the mean droplet diameters for pressure swirl atomizers may be expressed by equation 14.

$$\overline{D}_{32} \propto \sigma^a \mu_L^b m_L^c \Delta P^{-d} \tag{14}$$

Because of the wide range of applications and complexity of the physical phenomena, the values of the exponents reported in the literature vary significantly. Depending on the range of Reynolds and Weber numbers, constant a ranges between 0.25 and 0.6, constant b between 0.16 and 0.25, constant c between 0.2 and 0.35, and constant d from 0.35 to 1.36.

Equation 14 indicates that liquid pressure has a dominant effect in controlling the mean droplet sizes for pressure atomizers. The higher the liquid pressure, the finer the droplets are. An increase in liquid viscosity generally results in a coarser spray. The effect of liquid surface tension usually diminishes with an increase in liquid pressure. At a given liquid pressure, the mean droplet size typically increases with an increase in flow capacity. High capacity atomizers require larger orifices and therefore produce larger droplets.

The principal parameters affecting the size of droplets produced by twin-fluid atomizers have also been discussed (34). These parameters include liquid viscosity, surface tension, initial jet diameter (or film thickness), air density, relative velocity, and air–liquid ratio. However, these parameters may have an insignificant effect on droplet size if atomization occurs very rapidly near the atomizer exit.

Most studies indicate that air velocity has a profound influence on mean droplet size in twin-fluid atomizers. Generally, the droplet size is inversely proportional to the atomizing air velocity. However, the relative velocity between the liquid and air stream is more important than the absolute air velocity.

Liquid viscosity generally produces adverse effects on drop size. It increases the initial film thickness and hinders the growth of unstable waves. Both effects can produce coarser atomization. However, the influence of liquid viscosity on atomization appears to diminish for high Reynolds or Weber numbers. Liquid surface tension appears to be the only parameter independent of the mode of atomization. Mean droplet size increases with increasing surface tension in twin-fluid atomizers (34). $\overline{D}_{32}$ is proportional to σ^n, where the exponent n varies between 0.25 and 0.5. At high values of Weber number, however, drop size is nearly proportional to surface tension.

The practice of establishing empirical equations has provided useful information, but also exhibits some deficiencies. For example, a single spray parameter, such as $\overline{D}_{32}$, may not be the only parameter that characterizes the performance of a spray system. The effect of cross-correlations or interactions between variables has received scant attention. Using the approach of varying one parameter at a time to develop correlations cannot completely reveal the true physics of complicated spray phenomena. Hence, methods employing the statistical design of experiments must be utilized to investigate multiple factors simultaneously.

Spray Instrumentation

An ideal droplet measurement instrument should (*1*) not interfere with the spray pattern or breakup process, (*2*) provide for large representative samples, (*3*) permit rapid sampling or counting, (*4*) have adequate resolution and accuracy over a wide range of droplet sizes, and (*5*) accommodate variations in the liquid and ambient gas properties. Significant advances have been made in the development of laser diagnostic techniques for measuring sprays. Prior to selecting such an instrument, users should have a thorough understanding of its capabilities and limitations.

Existing droplet measurement techniques may be classified into three broad categories: (*1*) optical nonimaging techniques; (*2*) imaging techniques; and (*3*) nonoptical methods. A comprehensive review of these techniques are available (35–39).

Droplet Measurements by Optical Nonimaging Techniques. A number of sizing methods are based on the principle of passing a single or multiple laser beams through the spray and analyzing the optical patterns produced when the light is scattered by the liquid particles. Instruments utilizing this principle are denoted as optical nonimaging systems because they do not generate images of droplets. Two types of such instruments employ laser diffraction or Doppler techniques.

Laser Diffraction. This family of instruments utilizes Fraunhofer diffraction to determine droplet size. A collimated laser beam is directed through the spray, and is focused on a spot in the focal plane of the receiving lens located in the forward direction. Diffraction patterns produced by droplets passing through the beam are detected by multiple-element photodetectors. Because the intensity of the scattered light at various distances from the optical axis is a function only of droplet size and the scattering angle, measurement of the scattered light energy at various radial locations in the near-forward direction allows the determination of the entire droplet size distribution.

Diffraction techniques can be used only if three conditions are satisfied. The droplet diameters must be larger than the optical wavelength, the refractive index of the liquid particles must be different from that of the surrounding medium, and the scattering angles must be relatively narrow. Diffraction techniques are usually effective and reliable when obtaining global measurement of droplet ensembles. The diffraction patterns are not affected by the location or refractive index of the particles. Also, measurements are possible over a wide range of droplet velocities.

In practical applications, diffraction instruments may exhibit certain problems. For example, there may be poor resolution for the larger droplets. Also, it is not possible to obtain an absolute measure of droplet number density or concentration. Furthermore, the Fraunhofer diffraction theory cannot be applied when the droplet number density or optical path length is too large. Errors may also be introduced by vignetting, presence of nonspherical droplets, large gradients of refractive index, and multiple scattering.

Phase Doppler Interferometry. In the phase Doppler technique, interference fringe patterns are produced by the reflected and refracted components of scattered light as droplets pass through the intersection of two laser beams. The light rays emerging from the droplet will have different optical path lengths, depending on the scattering angle and droplet diameter. Pairs of detectors are positioned in the receiver plane at off-axis angles so as to detect the Doppler signals and phase shift resulting from the different path lengths. From optics theory, it may be shown that the change in phase is directly proportional to the droplet diameter.

The phase Doppler method utilizes the wavelength of light as the basis of measurement. Hence, performance is not vulnerable to fluctuations in light intensity. The technique has been successfully applied to dense sprays, highly turbulent flows, and combustion systems. It is capable of making simultaneous measurements of droplet size, velocity, number density, and volume flux.

Phase Doppler particle analyzers are essentially single-particle counters because they measure one particle at a time within a small sampling volume. This volume must be kept small to minimize the probability of having more than one droplet in the volume at any given instant. This probability increases as the concentration of droplets becomes greater, and there is more risk of measurement errors.

Numerous technical papers (38,39) describing the accuracy and limitations of phase Doppler instrumentation have been published. Reports indicate that droplet trajectories through the sample volume can have a significant effect on the relative intensity between the reflected and refracted light components. Detection of the wrong component can result in erroneous measurements. This problem, however, can be minimized by changing the intersection angle of the laser beams. The phase Doppler method should not be used to measure irregular-shaped particles or droplets exhibiting severe deformation. Nonspherical particles tend to produce irregular fringe patterns and can cause false readings.

Droplet Measurements by Imaging Techniques. Imaging techniques require various optical and computer equipment for image formation, storage, and analysis in sprays so as to provide droplet size and velocity information. These methods include photography, cinematography, video, holography, and image-scanning systems. With such methods, moving droplets may be observed directly, and the measurements are not affected by the shape or optical properties of the liquid particles, or by their collision or coalescence.

Like optical nonimaging systems, imaging methods have potential limitations. One of the more serious problems is optical resolution, which typically is no better than 5 μm. Because manual droplet sizing is extremely time-consuming, automated image processing is indispensable in most procedures. Other concerns are associated with the depth of focus, size of the viewing volume, magnification, working range, light source, recording media, data processing speed, and memory.

Droplet Measurements by Nonoptical Techniques. Nonoptical techniques involve the use of mechanical or electrical devices to determine droplet size. The mechanical methods involve the collection of droplet samples on a solid surface or in cells containing a special immersion fluid whose density is slightly lower than that of the sprayed liquid. The captured droplets are then photographed at high magnification to provide images that can be sized manually or by an automatic scanning machine. Most of the electrical methods are based on the detection and analysis of electronic pulses generated by the droplets as they contact an electrically charged or heated wire. Many of the nonoptical techniques have gradually become obsolete because of the broader capabilities of the laser optical instruments.

Industrial Applications

Although atomizers are usually small components in many industrial spray applications, they play an important role in determining the performance and efficiency of the entire process. It has long been recognized that atomizers must be properly selected to achieve optimum performance. More recently it has become necessary to comply with stringent environmental regulations to reduce waste and pollution. These requirements have prompted research and design engineers in many industries to improve the design, manufacture, quality control, and testing of atomizers.

Some concerns directly related to atomizer operation include inadequate mixing of liquid and gas, incomplete droplet evaporation, hydrodynamic instability, formation of nonuniform sprays, uneven deposition of liquid particles on solid surfaces, and drifting of small droplets. Other possible problems include difficulty in achieving ignition, poor combustion efficiency, and incorrect rates of evaporation, chemical reaction, solidification, or deposition. Atomizers must also provide the desired spray angle and pattern, penetration, concentration, and particle size distribution. In certain applications, they must handle high viscosity or non-Newtonian fluids, or provide extremely fine sprays for rapid cooling.

Because of the complexity of designs and performance characteristics, it is difficult to select the optimum atomizer for a given application. The best approach is to consult and work with atomizer manufacturers. Their technical staffs are familiar with diverse applications and can provide valuable assistance. However, they will usually require the following information: properties of the liquid to be atomized, eg, density, viscosity, and surface tension; operating conditions, such as flow rate, pressure, and temperature range; required mean droplet size and size distribution; desired spray pattern; spray angle requirement; ambient environment; flow field velocity requirements; dimensional restrictions; flow rate tolerance; material to be used for atomizer construction; cost; and safety considerations.

Though spray requirements differ from one application to another, the spray pattern or shape appears to be a sensible criterion for selecting liquid atomizers for certain processes. Table 2 lists a variety of applications that are based on the pattern of the spray.

In practical applications, the final test of an atomizer spray lies in the results of process performance and quality of the end product. In most cases, it is necessary to fine-tune the sprays through trial and error to achieve the goals of low cost and high performance. An understanding of the physics of atomization as well as the principles of operation in atomizers will be extremely beneficial in optimizing spray performance. A few examples are given to illustrate the technical challenges that engineers are faced with in various applications.

In the oil heating industry, standards have been established for manufacturing tolerances, spray quality, atomizer performances, as well as combustion emission. The 1990s standards of burner performance include a combustion

Table 2. Summary of Atomizer Sprays for Specific Applications

Atomizer spray	Special application
cone spray, hollow or solid	aerating water, brine sprays, chemical processing, coil defrosting, dust control, evaporative condensers, evaporative coolers, industrial washers, roof cooling, spray ponds, spray coating, spray drying, gas scrubbing and washing, humidification, gas cooling, cooling towers, coal washing, degreasing, gravel washing, dish washing, foam control, suspensions and slurries for food and chemical products, pollution control, and oil heating
flat spray	asphalt or tar laying, bottle washing, coal and gravel washing, foam control, degreasing, metal cleaning–rinsing, spray coating, vehicle washing and water misting, descaling, roll cooling, quenching, and agricultural spraying
plain jet spray	rocket engines, diesel engines, agitation, mixing of liquids, cataphoresis plants, and cutting
air atomizng spray	chemical processing, continuous casting, cooling casting and molds, curing concrete products, evaporative coolers, foam control, incineration, quenching, spray coating, spray painting, spray drying, flue gas desulfurization, pollution control, gas turbine engines, and medical spray

efficiency of more than 90%, a smoke number of less than 0.5, and an emission level of NO_x and CO in the range of 35 ppm. To meet these stringent requirements, the atomizers need to produce fine droplets, be insensitive to changes in fuel properties, be free of plugging and carboning, provide proper fuel distribution throughout the burner, and permit easy ignition. In addition, they have to meet the goals of long service life, low cost, tight tolerance, and compatibility. A typical hollow-cone pressure swirl atomizer used in the oil burner provides a flow capacity of 3.785 L/h at 700-kPa (7-bar) pressure. This type of atomizer will generate sprays with a Sauter mean diameter of about 70 μm when using No. 2 fuel oil. In this case, design improvement might be necessary to meet industry standards.

In spray drying, the selection of atomizers is largely determined by dryer designs, feed materials, the number and relative location of nozzles, and the spatial volume to be filled by the spray droplets. Advancements have been made in the design of spray drying nozzles to provide required bulk density, moisture content, and solubility. An ideal spray drying nozzle must offer the following benefits: tight tolerances, ability to atomize difficult feed materials, fewer nozzles, longer wear life, reduced pump pressure, low maintenance cost, less plugging, uniform droplet size, and lower percentage of fine droplets. The centrifugal pressure atomizer is commonly used in the spray drying industry. This type of atomizer may have a spiral swirl chamber containing a single inlet. It is available in flow capacities from 50 to 3000 L/h based on water rating at 7-MPa (70-bar) pressure. This type of atomizer is extremely versatile, allowing the users to switch swirl chamber and orifice plates for numerous flow rate and spray angle combinations. Sauter mean diameters in spray drying typically range from 50 to 250 μm.

In the rubber and printing industries, carbon black is used to enhance the mechanical properties of rubber and to provide pigments for printing inks. The production of carbon black involves the partial combustion of feedstock oil in a high temperature, low pressure environment. Fuel oil is injected into the reactor using air- or steam-assisted atomizers. The key issues relating to the processes of carbon black production include the protection of refractory lining, spray impingement by the quenching nozzles, and carbon buildup. In more recent years, manufacturers continue to add new grades of carbon blacks for product improvement and competition. The control of spray droplet size, size distribution, and spray angle becomes extremely important to obtain the desired quality. Typical carbon black furnaces utilize either pressure- or air-assisted (internal or external) atomizers. The choice is largely determined by the type of black being produced, flow rate, available fuel and air pressure, fuel properties, and fuel preheating capability. A typical plant handles fuel flow capacity between 500 to 3000 L/h and fuel pressure from 0.3 to 1.5 MPa (3–15 bar). The atomizing air can range from 18.9 to 37.8 L/s (40–80 SCFM) with air pressure about 700 kPa (7 bar). If the atomizing medium is steam, the flow capacity can run between 180 to 350 kg/h using a line pressure up to 1.5 MPa. A typical mean droplet size for this type of operation is around 100 μm.

In agricultural spraying, one of the biggest concerns is the drifting of small droplets. Drifting sprays not only lead to waste and environment problems, but also could endanger other nearby crops. Droplets smaller than 150 μm can be easily blown away from the intended target area by a cross wind. A typical herbicide atomizer produces a spray with 15–20% of the liquid volume contained in droplets less than 150 μm. Atomizer improvements must be made so that the spray contains a narrow droplet size distribution with liquid volume less than 5% contributed by the smaller droplets. Correlations between droplet size and surrounding field conditions are also important.

NOMENCLATURE

Symbol	Definition	Units
ALR	air–liquid ratio by mass	
A_p	cross-sectional area of inlet ports	m^2
A_2	cross-sectional area of orifice	m^2
C_d	discharge coefficient	
d_0	orifice diameter	m
D_s	swirl chamber diameter	m
$\overline{D}$	mean droplet diameter	μm
F	liquid volume flux	cm^3/(cm^2·s)
g	gravitational constant	m/s^2
h	one-half of the film thickness	m
l_0	orifice length	m
L	characteristic dimension of atomizer	
L_s	swirl chamber length	m

m	mass flow rate of gas or liquid	kg/s
n	wave number, $n = 2\pi/\lambda$.	
N_i	number of droplets in the size class i	
ΔP	pressure drop across atomizer	Pa
Q	volume flow rate	m^3/s
t	film thickness, or time scale	m or s
U	mean axial velocity at exit	m/s
U_R	relative velocity	m/s
w_i	normalized volume or mass flow rate percentage in each sector	
x	axial distance	m
α	wave growth rate	1/s
λ	wavelength	m
μ_L	liquid dynamic viscosity	kg/(m·s)
ξ	coordinate along the film trajectory, or wave amplitude	m
ρ_A	air density	kg/m^3
ρ_L	liquid density	kg/m^3
σ	liquid surface tension	kg/s^2

BIBLIOGRAPHY

"Sprays" in *ECT* 1st ed., Vol. 12, pp. 703–721, by W. E. Meyer and W. E. Ranz, Pennsylvania State University; in *ECT* 2nd ed., Vol. 18, pp. 634–654, by R. W. Tate, Delavan Manufacturing Co.; in *ECT* 3rd ed., Vol. 21, pp. 466–483, by J. Fair, University of Texas.

1. ASTM E1620-96, Terminology Relating to Liquid Particles and Atomization, ASTM, Philadelphia, Pa., 1996.
2. A. H. Lefebvre, *Atomization and Sprays*, Hemisphere Publishing Corp., New York, 1989.
3. *Proceedings of International Conferences on Liquid Atomization and Spray Systems*, ICLASS-1978, ICLASS-1982, ICLASS-1985, ICLASS-1988, ICLASS-1991, and ICLASS-1994, Begell House, Inc., New York.
4. M. Doumas and R. Laster, *Chem. Eng. Prog.* **49**(10), 519–526 (Oct. 1953).
5. N. K. Rizk and A. H. Lefebvre, *J. Propulsion.* **1**(3), 193–199 (1985).
6. N. Dombrowski, D. Hasson, and D. E. Ward, *Chem. Eng. Sci.* **12**, 35–50 (1960).
7. H. Zhu and co-workers, *Atom. Sprays*, **5**(3), 343–356 (1995).
8. V. I. Asihmin, Z. I. Geller, and Y. A. Skobel'cyn, *Oil Ind. (Moscow)*, **9** (1961).
9. J. C. Cooke, "Numerical Solution of Taylor's Swirl Atomizer Problem," Technical Report No. 66128, Royal Aircraft Establishment, U.D.C. No. 532.527, U.K. Government, 1966.
10. C. Dumouchel and co-workers, *Atom. Sprays*, (2), 225–237 (1992).
11. G. I. Taylor, *Proc. 7th Int. Cong. Appl. Mechanics*, **2**, 280–285 (1948).
12. A. Radcliffe, *Proc. Inst. Mech. Eng.* **169**, 93–106 (1955).
13. M. Suyari and A. H. Lefebvre, *J. Propulsion*, **2**(6), 528–533 (1986).
14. N. Dombrowski and D. Hasson, *AIChE J.* **15**, 604 (1969).
15. K. R. Babu, M. V. Narasimhan, and K. Narayanaswamy, *Proceedings of the 2nd International Conference on Liquid Atomization and Spray Systems*, Madison, Wis., (1982), pp. 91–97.

16. A. J. Yule and J. J. Chinn, *International Conferences on Liquid Atomization and Spray Systems*, ICLASS-94, Rouen, France, July 1994.
17. A. R. Jones, in Ref. 15, pp. 181–185.
18. H. B. Squire, *Brit. J. Appl. Phys.* **4**, 167–169 (1953).
19. R. P. Fraser and co-workers, *AIChE J.* **8**(5), 672–680 (1962).
20. Lord Rayleigh, *Proc. London Math. Soc.* **10**, 4–13 (1878).
21. N. Dombrowski and W. R. Johns, *Chem. Eng. Sci.* **18**, 203–214 (1963).
22. M. J. McCarthy and N. A. Malloy, *Chem. Eng. J.* **7**, 1–20 (1974).
23. S. P. Lin and Z. W. Lian, *AIAA J. Propulsion* **28**(1), 120–126 (1990).
24. C.-P. Mao, S. Chuech, and A. J. Przekwas, *Atom. Sprays*, **1**(2), 215–235 (1991).
25. P. Rosin and E. Rammler, *J. Inst. Fuel*, **7**(31), 29–36 (1933).
26. S. Nukiyama and Y. Tanasawa, *Trans. Soc. Mech. Eng. Japan*, **5**(18), 62–67 (1939).
27. R. A. Mugele and H. D. Evans, *Ind. Eng. Chem.* **43**(6), 1317–1324 (1951).
28. R. W. Tate, *Ind. Eng. Chem.* **52**(10), 49–52 (1960).
29. W. D. Bachalo and M. J. Houser, *Opt. Eng.* **23**(5), 583 (1984).
30. J. Ortman and A. H. Lefebvre, *AIAA J. Propulsion*, **1**(1), 11–15 (1985).
31. C.-P. Mao, G. Wang, and N. Chigier, *Atom. Spray Tech.* **2**(2), 151–169 (1986).
32. A. H. Lefebvre, *Atom. Spray Tech.* **3**(1), 37–51 (1987).
33. A. H. Lefebvre, *Prog. Energy Combust. Sci.* **6**, pp. 233–261 (1980).
34. A. H. Lefebvre, *Proceedings of 5th International Conference on Liquid Atomization and Spray Systems*, Gaithersburg, Md., 1991, pp. 49–64.
35. N. Chigier, *Prog. Energy Combust. Sci.* **9**, 155–177 (1983).
36. J. M. Tishkoff, R. D. Ingebo, and J. B. Kennedy, eds., *Liquid Particle Size Measurement Techniques*, ASTM STP 848, Publication Code Number 04-848000-41, ASTM, Philidelphia, Pa., 1984.
37. R. A. Jones, *Prog. Energy Combust. Sci.* **3**, 225–234 (1977).
38. B. J. Lazaro, United Technology Research Center, Technical Report No. UTRC91-11, East Hartford, Conn., 1991.
39. V. G. McDonell and G. S. Samuelsen, *Liquid Particle Size Measurement Techniques*, Vol. 2, ASTM STP1083, American Society for Testing and Materials, Philadelphia, Pa., 1990, pp. 170–189.

CHIEN-PEI MAO
ROGER TATE
Delavan Inc

SPUTTERING. See THIN FILMS, FILM FORMATION TECHNIQUES; METALLIC COATINGS.

STAINS, INDUSTRIAL

With few exceptions, stains used in wood finishing are formulated to improve the appearance of the substrate. Unlike paints, sealers, and topcoats, stains are utilized either to accentuate the natural beauty of the wood or to hide inherent defects found in most species of wood.

The application of stains to substrates includes spray, flow, dip, roll-coat, or hand application, ie, brush or rag methods. The type of application method used depends on the function of the stain itself as well the type of product being manufactured.

There are various types of wood stains and criteria used to choose the right type for a job. Although there are several distinct groups or types of stains, it is safe to categorize wood stains into two groups: dye stains and pigmented stains.

Dye Stain or Soluble Colors

A defining characteristic of dyes is the ability to dissolve in a given medium. Dissolution leaves no particles to refract or scatter light and thus a dye solution is transparent. A distinct advantage of a soluble-type stain is this transparency and brightness afforded by use of various dye types. Solubility is increased by agitation or heat, or a combination of the two.

Solvent Dyes. As the name implies, these dyes are popular because of their solubility in a variety of industrial solvents. Color strength, brilliance, lightfastness, and the ability to formulate with solvents other than alcohol and water have helped to increase their use and popularity. Because these dyes are soluble in solvents other than alcohol or water, the finisher may lessen the effect to the substrate itself. The moisture content of alcohols and the tendency of alcohol to draw moisture from the air during the drying process result in grain raising of the wood. Using solvents other than alcohols lessens this effect to some degree. These "smoothcoat" stains are used on certain species of wood to lessen the amount of grain rupture associated with moisture. Based on cell structure or quality, certain species of wood are prime candidates for solvent stains.

Acid Dyes. These dyes are sodium salts of color acids and are the most widely used dyes for nongrain-raising wood stains. Generally speaking, acid dyes have excellent brightness, strength, and lightfastness. They are partially soluble in alcohol, but perform best when initially incorporated in ethylene glycol or propylene glycol and later reduced in methanol. They exhibit excellent transparency.

Theory and Practice of Wood Staining

Color Mixing. The various types of dye powders used to make dye stains are blended to achieve the desired color. Most finishers purchase wood stains premixed to specified colors. In the wood-finishing industry, various shades of brown are the most common. These colors are usually blended from primary colors. Color-matching skills can be acquired only by practice, but the basic theory of color matching is relatively simple and easily understood. The basic theory of color matching can be demonstrated by using the color circle shown in Figure 1 (see COLOR).

In Figure 1 the three primary colors consisting of red, yellow, and blue are spaced equally, 120° apart. Between them are the secondary colors: orange, green, and purple. The secondary colors are made by combining the adjacent primary colors. Thus, a mixture of red and yellow produces orange, a mixture of yellow and blue produces green, and blue and red gives purple. Colors that are directly opposite one another on this circle are complementary colors that when mixed become varying shades of brown, the most commonly used color in wood stains. By slightly altering the levels in which each of these colors is present, literally thousands of variations of brown can be achieved. In most wood stains, black or a blue-black is used in place of a pure blue.

Types of Wood Stains. The term wood stain in this article applies only to those stains that are applied directly to the wood or over another wood stain. The types of stains used are dependent on the application, finished product, and regional terminology. Multiple wood stains are frequently used in the residential furniture industry to achieve a desired look. They include equalizers and nongrain-raising (NGR) stains.

Equalizers. In the majority of wood-finishing systems, there is a problem of matching up varying shades of colors in the natural wood. The lack of uniformity in the substrate may result from sapwood areas or mineral streaks. In the case of finishing assembled furniture, it may also be the result of incorporating different species of wood into the same piece. It is not unusual, for example, to see casegoods with cherry tops, maple drawer fronts, and poplar or gum rails. These differences in substrate color must be addressed during the early stages of finishing to ensure uniform appearance of the finished case.

Equalizers can be either pigmented or dye-type stains used to tone down or lighten dark areas of wood prior to finishing. Although it is not as effective, equalizing is sometimes done in place of bleaching. Because there are no white dyes, white pigment or pearl essence is usually incorporated with the dyes to achieve the desired look.

Another form of equalizing stains is through the use of sap stain. Color differences between the sapwood and late-growth area of the lumber can be made uniform by using this type of stain. Sap stains are usually alcohol-based dye

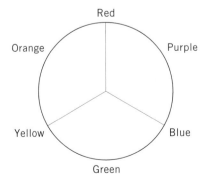

Fig. 1. Color circle, showing the relationship of colors.

stains that tie lighter areas of the wood into darker areas. Transparency of sap stain and equalizers is important to ensure a natural, nonpainted appearance.

NGR Stains. Nongrain-raising stains are usually sprayed overall and contribute the greatest to the overall undertone color of a finished piece. These stains are also sometimes referred to as body stains or overall stains. They are usually dye-type and can be formulated, depending on the type of dye powder and solvent system, to minimize the effect of the stain on the substrate.

The color and effect produced by NGR stains and any stain mixture depend on several factors other than the colors or type of dyes used. Those factors include strength of the mixture, the amount applied, the type of substrate, and the solvent system used for the stain. The role of the wood stain is not to provide protection; rather, the primary function of the stain is to impart color effects by accentuating grain patterns. The transparency and brightness needed to enhance the natural beauty of the wood are optimized by using dye-type stains for wood.

Pigmented Stains

Pigment colors are finely divided color particles which do not dissolve in any available solvent; they can only be dispersed by grinding them in a liquid. The wood stains discussed thus far, with the exception of equalizers that incorporate white or pearl essence, generally are more transparent and brighter than pigmented stains. There are, however, certain wood stains which can be formulated with pigments and still give the finish a desirable look. These pigmented wood stains are used in a variety of applications, and in many instances such use shortens the number of steps needed to achieve a desired appearance. Pigmented wood stains are used commonly in the finishing of kitchen cabinets, where shorter finishing systems are desirable. Also, pigmented wood stains are popular with "do-it-yourselfers."

There are a large number of pigments to choose from when formulating a stain. Three basic groups of pigments were used in the industry, ca 1996: (*1*) natural or earth colors consisting of umbers, siennas, yellow oxide, and red oxide. Vandyke brown is also a natural pigment, but it is of organic rather than mineral origin.

(*2*) Organic pigments are carbon compounds which are usually derived from coal-tar bases or petroleum. Organic pigments include carbon and lampblack, which are derived from natural gas by incomplete combustion. Other examples of organic pigments include lithols, toluidines, and phthalocyanines (see PIGMENTS, ORGANIC).

(*3*) Chemical pigments or synthetics may be metal compounds. A good example is white titanium dioxide. Other chemical pigments include cadmium sulfide colors, iron blue, and several synthetic versions of iron oxides.

Penetrating Stains. Penetrating or no-wipe stains are used in a variety of applications. The most common use of these direct-to-wood stains is on small-pore species of wood. Maple, cherry, and coniferous species such as pine are good candidates for penetrating stains.

As the name implies, these stains are sprayed on and require little if any wiping. The solvent itself penetrates into the pore and allows the pigment and a small amount of binder to remain on the surface. These stains usually are composed of an oil-type vehicle and a combination of earth pigments reduced in a combination of aliphatic and aromatic hydrocarbons such as naphtha and toluene. The solvent system itself plays a big role in the appearance of the stain owing to the varying degrees to which solvents penetrate. Restrictions on the use of certain aromatic hydrocarbons have affected the manner in which these stains work.

The transparency and brightness of dyes cannot usually be duplicated through the use of pigments. However, optimum clarity and look can be achieved when using pigmented stains by consideration of such factors as the amount of pigment contained in the stain and the type. Application also plays a significant role in the strength and transparency of a stain. Usually the finisher wants to avoid a painted look when using penetrating stains.

Advantages of penetrating stains are numerous. The stains are commonly applied by either spraying or dipping, the latter being the most economical method of staining. Additionally, pigments used in penetrating stains are generally less expensive than dyes. Because of the uniform appearance provided by penetrating stains, many finishing applications utilize penetrating stains as the only color step.

Toners and Tinted Sealers. These materials are usually pigments dispersed in nonpenetrating lacquers. In some instances, solvent or spirit dyes can be used to improve clarity. The solids of the vehicle can be adjusted to control the depth of penetration into the substrate.

Toners can be used in a variety of applications, depending on the substrate and the effect desired. They can be sprayed either directly onto the wood or over other wood stains. The solids of toners are usually low (5–15 wt %) and generally require the use of a washcoat or sealer prior to topcoating.

Although the difference between toners and tinted sealers may not be clearly defined, it is usually the role of the tinted sealer to provide both color and sealing properties. Therefore the tinted sealer usually is higher in solids and provides the majority of color to the finish. There has been a resurgence of popularity of tinted sealers, owing to the appeal of blonde or natural finishes. The fact that tinted sealers are becoming more popular may be the result in part of their ability to fill the roles of both stain and film builder within a finishing system.

Both toners and tinted sealers provide uniform pigment distribution without penetration. The use of more transparent pigments can also provide acceptable clarity, but with limited depth and contrast.

Overtone Stains

Washcoats. Although washcoats are not classified as stains, the role they play in the staining procedures makes it important to examine the properties they bring and the contribution they make to the finishing process. Washcoats for wood finishing can be defined as thin coats of sealer applied to control the amount of penetration and subsequent staining from overtone stains and fillers. Naturally, the solids content of a washcoat determines the amount of penetration of an overtone stain. Washcoats are usually sanded prior to application of a glaze, wiping stain, or filler. Therefore the extent of the sanding also plays a role in the penetration and staining action.

Washcoats perform the following roles in the finishing process: (1) they form a thin film to protect the wood stain(s) from the effects of handling, or the solvent from subsequent finishing steps; (2) they raise and stiffen wood fibers to enable the surface to be sanded prior to application of wiping stains and fillers, and this sanded, smooth, hard surface facilitates the wiping and clean-up of fillers and other wiping steps; (3) they seal the surface of the wood and wood pores to prevent pinholes and blistering when fillers are used on large-pore substrates; and (4) by controlling the solids of a washcoat, the amount of penetration of overtone stains can be controlled.

The solids content of a washcoat is usually 5–12 wt %. It is important that washcoats leave a very thin film of material to allow the proper amount of sealing and to prevent bridging the pore, which would result in blistering of subsequent build coats.

Fillers.   Wood fillers are applied directly over the washcoat in a multistep operation. First the filler is reduced according to the manufacturer's specifications and applied, usually by spray. The filler is then worked into the pores of the wood in a circular motion using a rag or pad, either by hand or by machine. Following this vigorous step to ensure that each pore is thoroughly filled, the filler is then wiped in the direction of the grain to clean up residue or wiping marks. At this point, it is imperative that the filler be allowed to dry. This is accomplished by either allowing it to dry overnight or by force-drying it in an oven. Fillers that are not properly dried will eventually shrink, and full fill is not achieved.

Fillers (qv) perform two significant functions: they fill pores and give color to the pores.

Glazes and Wiping Stains.   Some applications such as kitchen cabinet finishing utilize wiping stains direct-to-the-wood. In most fine furniture applications, wiping stains and glazes are applied over the washcoat or sealer step.

These overtone stains are normally composed of pigments, oils, solvents, and driers. The important quality of glazes and wiping stains is the ability to apply a color coat which can be wiped on and then highlighted to add depth and contrast to the overall appearance of the finish.

These wiping coats are usually sprayed on and then wiped with rags to varying degrees, depending on how much color and effect the finisher desires from the wiping coat. In many instances, the material is then brush-blended into corners and recesses to give uniform coverage and appearance. After the glaze has been wiped and brush-blended, the finisher usually highlights or strikes through areas of glaze, using steel wool or some other abrasive material to give contrast or accentuate certain grain patterns or characteristics such as cathedrals or knots.

The wiping and blending properties of a wiping stain are important considerations when formulating these types of products. A good glaze must not be sticky or bite into the sealer and must stay open long enough to be wiped, brushed, and worked to the desired level by the finisher. Long working times are important, yet these stains must be recoatable within a reasonable period of time. Unless the stains are formulated correctly, the glaze or wiping stain may cause poor intercoat adhesion or discoloration when lacquer topcoats are applied.

Shade Stains. These stains are usually applied after the sealer or first topcoat and are typically sprayed on specific areas to compensate for uneven color distribution during the initial finishing process. For instance, perhaps the glaze was wiped too clean on an edge. Rather than going back to restain or glaze that small area, the finisher can spray a small amount of this shade stain on the desired area and achieve the same result in a fraction of the time.

Another common use of the shade stain is to enhance the contrast or depth of the finish. This is usually the role of the glaze, but when additional contrast is needed, the finisher may elect to shade certain areas. Care must be taken not to apply too much shade, which can result in a painted or artificial appearance.

Pad Stains. More progressive or higher end furniture finishers add color or pad stains to enhance grain patterns, produce shadows, and create hues found only in exceptionally fine veneers and woods. These pads are applied at varying levels to create the illusion of the third dimension.

Pad stains are divided into two groups: spray pads and hand pads. The difference between them, not surprisingly, can only be defined by the means of application. Like the shade stains, they are usually applied over the sealer or first topcoat and subsequently are topcoated themselves.

The primary purpose of the pad stains is to accent figures. An example of this may be following a V-shaped grain pattern or cathedral or adding artistic value to an otherwise uninteresting piece of wood. Grain patterns of wood sometimes may be too geometric, as is often the case with oak or ash. In such instances, a pad stain may be used to break up the monotony of the grain. Spray pads are frequently used when attempting to add some relief to an otherwise monotonous grain configuration. Crossfire or mottled patterns on bland substrates can simulate Crossfire Maple or English Oak.

Distressing Stains. Interest and charm can be added to furniture by the deliberate infliction of imperfections. These imperfections may be caused by physical distress such as hammers, files, nails, chains, or rocks or caused by finish distress such as "fly specking" or staining to simulate past abuse, such as waterstains caused by the careless placement of a drinking glass.

Physical distressing is usually a part of the initial finishing procedure and is done in the white wood stage. Finish distressing is normally part of the latter phases of the finishing operation. Like the pad and shade stains, distressing stains are usually applied over the sealer or between coats of topcoat.

Environmental Considerations

Most of the stains discussed are solvent-borne and therefore possess inherent properties such as flammability, toxicity, and reactivity. Those properties make it necessary to observe safe handling practices when using or storing wood stains. Adequate ventilation and avoidance of any source of possible ignition are key in the use and storage of these materials. Personal protection when using stains should include safety glasses or goggles, respirators, aprons, and rubber gloves. Overexposure to certain solvent vapors may cause respiratory damage or impaired judgment, whereas prolonged skin contact may result in skin irritations or even neurological effects.

Provisions of the Clean Air Act have resulted in the regulation of certain wood finishing applications. Residential and Institutional Furniture and Kitchen Cabinet Industries are under regulations which specify the amount and types of solvent emissions allowed. The size and location of a finishing facility determine the extent of the effect stemming from the regulations.

These regulations are based in part on the amount of solvents in relation to the amount of solids. Most wood stains are low solids materials which rely on their transparency and their ability to penetrate and dry fast. Those characteristics themselves put great emphasis on the type of solvents that are used to formulate stains. The low solids content of wood stains limits the scope of solvent substitution or reformulation.

A good example of the effect of regulations on wood stains is the issue surrounding methanol (qv). Methanol is the most widely used solvent for wood stains because of its fast-drying properties, low cost, and the solubility of dyes in methanol. Because methanol is listed by the U.S. EPA as a hazardous air pollutant (HAP), and because of the extremely low solids of wood stains, it is most likely that wood stains such as NGR, body stains, and sap stains will need to be reformulated before the end of the twentieth century.

BIBLIOGRAPHY

"Stains, Industrial" in *ECT* 1st ed., Vol. 12, pp. 722–740, by W. H. Peacock, American Cyanamid Co.; in *ECT* 2nd ed., Vol. 18, pp. 654–671, by W. H. Peacock, American Cyanamid Co.; in *ECT* 3rd ed., Vol. 21, pp. 484–491, by R. S. Bailey, Lilly Industrial Coatings, Inc.

General References

Assorted publications on related subject matter, including writings of the late J. A. Hager, Guardsman Products Inc., High Point, N.C.
Technical data, Sandoz Chemical Corp., Charlotte, N.C., 1996.
Technical data, Crompton and Knowles Corp., Charlotte, N.C., 1996.

RON W. TUCKER
Lilly Industries, Inc.

STARCH

Starch [9005-25-8], $(C_6H_{10}O_5)_n$, the main reserve food of plants, constitutes two-thirds of the carbohydrate caloric intake of most humans but only 47% of the carbohydrate caloric intake by Americans, who also get about 52% of their carbohydrate calories from sugar. Commercial starches are obtained from seeds, particularly corn, waxy corn, high amylose corn, wheat, and rice, and from tubers or roots particularly potato, sweet potato, and tapioca (cassava). Their principal use is in foods; the major nonfood uses are in sizing of paper and textiles, and as adhesives (qv).

Egyptian papyrus bonded with a starchy adhesive has been dated to 3500–4000 BC. Pliny the Elder (23–74 AD) described Egyptian use of wheat starch modified by boiling in vinegar to produce a smooth surface for papyrus documents.

Enzyme technology, especially immobilized-enzyme technology, has allowed production of sweeteners (qv) from corn starch, ie, corn syrup or high fructose corn syrup (HFCS), and often also provides sources of chemical feedstocks to replace petroleum. Enzymes, notably α-amylase, glucoamylase, and glucose isomerase, catalyze specific polysaccharide-degrading reactions to produce various glucose syrups. Immobilized enzyme technology has significant advantages because of the recoverability of the enzymes and their re-use in a continuous system, rather than in batch systems. Starch conversion technology, especially conversion to glucose and fructose-containing syrups, has been reviewed (1) (see ENZYME APPLICATIONS; SYRUPS).

The quasi-crystalline structure of natural starch granules causes them to be insoluble in water at normal room temperature and gives them relative resistance to carbohydrases other than α-amylase and glucoamylase unless the granules become swollen. Three-dimensional arrangements of crystalline and amorphous zones in starch granules have been suggested (2).

Physical Properties

Starch granules in plants vary in diameter from 1–150 μm (3). Among commercial starches, rice starch (3–9 μm) has among the smallest granules and potato starch (15–100 μm), the largest (4). Corn starch granules are 5–26 μm with an average diameter of 15 μm. Amaranth starch granules are 1–3 μm in diameter. Microscopic examination of corn starch granules reveals a distinct growth point known as the hilum (3), ie, the nucleus from which granule growth begins. Polarized light microscopy reveals a birefringence which, along with x-ray diffraction, is evidence of granule semicrystallinity. Between crossed Nicol prisms of the microscope, a black maltese cross (cross of isoclines) is observed centered on the hilum. Cereal starches give an A-type x-ray pattern, tuber starches a B-type pattern, and a few starches give an intermediate, C-type diffraction pattern (5). From x-ray fiber diffraction patterns it has been determined that the starch molecule(s) fiber axis is perpendicular to the visible growth rings in the granule (6). The fiber period in starch is generally around 1.06 nm and a model has been proposed (7,8) for both A and B x-ray patterns in which the starch molecules are left-handed parallel double-helices in hexagonal packing. Diffraction patterns may also be used to provide estimates of the relative amount of crystalline and amorphous phases, ie, x-ray crystallinity. Other methods of instrumental analysis have been applied to the solid-state structure of starch including optical and electron microscopy, small-angle light scattering, thermal analysis (differential scanning calorimetry, dsc), and ^{13}C-nmr spectroscopy (9). Cross-polarization/magic angle spinning experiments allow distinction between double-helical arrangements in starch and amorphous single chains. Such information is crucial in understanding the ultrastructure of raw starch granules.

Undamaged starch granules are insoluble in cold water but imbibe water reversibly accompanied by a slight swelling. With continued uptake of water at ambient temperature granule diameter increases 9.1% for corn and 22.7% for waxy corn (10). In hot water a larger irreversible swelling occurs producing gelatinization, which takes place over a discrete temperature range that depends on starch type.

Starch	Range, °C
potato	59–68
tapioca	58.5–70
corn	62–72
waxy corn	63–72
wheat	58–64

At a specific temperature during heating (lower limit of gelatinization temperature), the kinetic energy of the molecules is sufficient to overcome intermolecular hydrogen bonding in the interior of the starch granule. The amorphous regions of the granule are initially solvated and the granule swells rapidly, eventually to many times its original size. During swelling, some of the linear amylose molecules leach out of the granule into the enveloping solution. When a cooked starch paste containing a mixture of linear amylose molecules, swollen granules, and granule fragments is cooled, the dispersion thickens, and if sufficiently concentrated may form a gel. This property of forming thick pastes or gels is the basis of most starch uses. The effect of thermal treatment on starch depends strongly on whether it occurs in excess water, limited water, under pressure, or in extrusion cooking. In excess water it appears that starch swelling is a two-stage process consisting of initial granule swelling followed by granule dissolution; both steps are irreversible (11). In limited water, thermal responses have been interpreted as being due to starch crystallite melting (12,13). In extrusion cooking, granules are physically torn apart, allowing more rapid penetration of water into the granule (14). Starch fragmentation (dextrinization) appears to be the predominant reaction during extrusion, in contrast to normal gelatinization (15). It has been reported that during extrusion the average molecular weight of amylose decreases by a factor of 1.5 and the average molecular weight of amylopectin decreases by a factor of 15 (16). Effects of starch structure on starch rheological properties during gelatinization or extrusion have been well reviewed (17).

The starch gelatinization temperature range begins with the onset of granule swelling and ends at the point where nearly 100% of the granules are gelatinized. The designated range over which gelatinization occurs depends on the method of measurement. The most sensitive microscopic method follows the loss of birefringence of a starch–water mixture on a Kofler hot stage (18). Chemical additives affect the gelatinization temperature range in a predictable way that may be important in specific industrial applications. Chemicals such as sodium sulfate, sucrose, and D-glucose inhibit gelatinization and increase the gelatinization temperature range, probably by competing for available water. Other chemicals, such as sodium nitrate, sodium hydroxide, and urea, lower the gelatinization temperature range, possibly by disrupting the granular inter-molecular hydrogen bonds.

Physical properties of starch can be altered by mechanical treatments. If granular integrity is disrupted, as by grinding dry starch, the starch gelatinizes more easily, perhaps even in cold water. Damaged granules are quite reactive to chemicals and enzymes. Underivatized gelatinized granules are fragile. Even agitation of a cooked paste ruptures a majority of the swollen granules. As a result, a cooled paste loses viscosity and gelling ability. Physical properties of starches may also be altered by chemical modification of the starch. Acid treatment, oxidation, cross-linking, and esterification are some of the chemical means used to modify peak viscosity, hot cooked paste stability, gelation, and freeze–thaw stability (19).

The concept and use of food polymer science in describing the behavior of starch during and after thermal treatment has been developed (20,21). In this theory a fringed micelle model is used in describing starch as being composed of microcrystalline regions covalently cross-linked by flexible chain segments. In such a quasi-crystalline system the important thermal transitions are a glass-transition, T_g, for the amorphous component, and a phase-transition (crystalline melting temperature), T_m, for the micellar component. Using this model and theory, the antiplasticizing effect of sugars on starch gelatinization as well as gelation and retrogradation (staling) mechanisms have been explained (21). This theory and supporting experimental data (largely dsc) have been applied in investigating structure–function relationships of cookie and cracker ingredients (22). Use of dsc as a diagnostic tool in the investigation of cookie dough as well as in the analysis of staling in such low moisture baked goods has been described (22).

Specific optical rotation values, $[\alpha]_D$, for starch pastes range from 180 to 220° (5), but for pure amylose and amylopectin fractions $[\alpha]_D$ is 200°. The structure of amylose has been established by use of x-ray diffraction and infrared spectroscopy (23). The latter analysis shows that the proposed structure (23) is consistent with the proposed ground-state conformation of the monomer D-glucopyranosyl units. Intramolecular bonding in amylose has also been investigated with nuclear magnetic resonance (nmr) spectroscopy (24).

Chemical Properties

Most normal starches contain two distinct types of D-glucopyranose polymers. Amylose is an essentially linear polymer of α-D-glucopyranosyl units linked $(1\rightarrow4)$ as shown in Figure 1. Although amylose molecules are generally thought of as being linear chains of α-D-glucopyranosyl units, most amylose preparations contain amylose molecules with two to five branches. These long branches allow

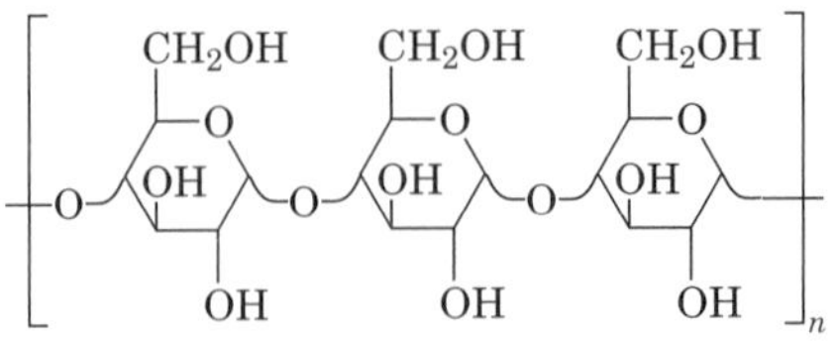

Fig. 1.   Structure of amylose.

the molecules to possess nearly the same properties as truly linear molecules. Starch gives a characteristic blue color when stained with iodine, due to insertion of iodine into an amylose helical structure to form a complex with iodine on the inside of the helix. Amylopectin [9037-22-3] also forms a complex, but its color is purple to reddish brown, depending on the source of the amylopectin (25). In the presence of amylose this color reaction is usually obscured by the amylose–iodine blue. The characteristic blue color of the iodine–amylose complex has been employed both as a qualitative and quantitative test for starch. Amylose [9005-82-7] may be isolated by complete aqueous gelatinization and dispersion of starch and mixing the hot starch solution with an organic complexing agent such as 1-butanol (26) in water. On cooling, the amylose–butanol complex crystallizes and is removed by centrifugation. Recrystallization of the amylose–1-butanol complex and subsequent removal of the 1-butanol produces a pure amylose. Fractionation of amylose and amylopectin has been reported employing gel filtration chromatography and elution with aqueous sodium chloride (27). However, some starches contain only highly branched molecules. These are termed waxy starches because of the vitreous sheen of waxy corn grains when cut. Alternatively, some mutant seed varieties produce starches having up to 85% linear molecules (high amylose starch), although most starches have about 25% linear and 75% branched molecules. Starches high in amylose are subject to significant intermolecular association, leading to what is known as resistant starch (28–30). Resistant starch is that fraction not extracted from dietary fiber unless the sample is treated with dimethyl sulfoxide or alkali; the starch is resistant owing to retrogradation of the amylose. However, other types of resistant starch occur, such as the physically segregated starch in some beans, the ungelatinized B-type starch granules from potatoes or green bananas, and chemically modified starch (30). Resistant, retrograded, high amylose starches find some use as fat replacers in a variety of food types.

Amylopectin is a highly branched polymer of α-D-glucopyranosyl units containing $1 \rightarrow 4$ links with $1 \rightarrow 6$ branch points (Fig. 2). When starch is fully converted to the methyl ether and its fractions are hydrolyzed (31), 4.67% tetra-O-methyl-D-glucopyranose is obtained from the amylopectin and 0.32% from the amylose. These particular methyl ethers can only come from nonreducing end groups, and show that the average branch consists of ~25 D-glucopyranosyl units in amylopectin and that amylose consists of ~350 D-glucopyranosyl monomers per chain.

Knowledge of the fine structure of starch molecules has benefitted from the use of starch-degrading enzymes such as α-amylase, isopullulanase, neopullulanase, β-amylase, glucoamylase, cyclomaltodextrin glucanotransferase, and

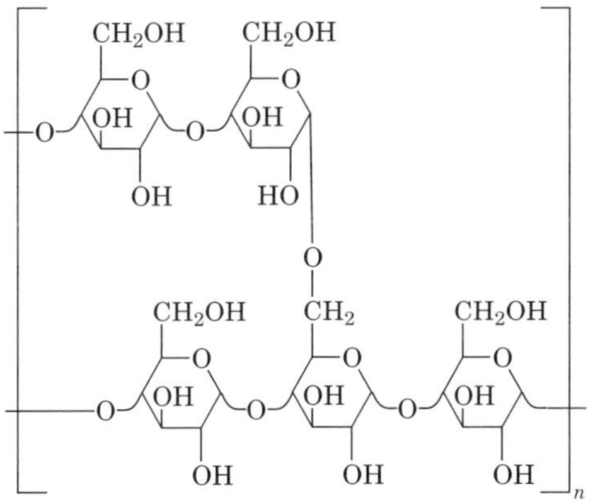

Fig. 2. Branch point structure in amylopectin.

phosphorylase. Debranching enzymes that catalyze hydrolysis of $(1 \rightarrow 6)$-α-D-glucosidic linkages, such as isoamylase, pullulanase, (R)-enzyme (limit dextrinase), and amylo-1,6-glucosidase 4-α-D-glucanotransferase, have also found much application in starch structural studies. Separation of amylose and amylopectin, and products from enzymatic hydrolysis, by size exclusion chromatography (sec) is critical for determination of molecular weight distribution. High performance anion-exchange chromatography has also been employed in the examination of homoglucan oligomers from starch hydrolysis. The specific analytical structural features of amylose, amylopectin, and the starch granule itself have been reviewed (32,33).

When dispersed in solution, amylose behaves as a random coil or double helix. On cooling a dispersion, molecules associate via hydrogen bonding to form an insoluble precipitate. This mechanism of reassociation, called retrogradation, is significant in many food systems. In baked goods such as bread it is believed that most amylose is retrograded by the time the bread has cooled to room temperature and any subsequent changes in texture, typically referred to a staling, are as a result of reassociation of the longer outer chains of the wheat amylopectin (34). Because of its retrogradation, corn starch is not used in starch-thickened frozen cream pies or gravies because of its changed physical properties on cooling and freezing. Such retrograded products from regular corn starch have an unacceptable spongy texture and appearance. Amylopectin and its simple modifications do not as easily retrograde and are usable in food systems requiring freeze–thaw stability. Waxy varieties of corn, barley, and rice starch which contain no amylose are the thickeners of choice for starch-thickened frozen foods.

Starch not only varies in polymer structure but also in molecular weight distribution. Several techniques have been used to determine the molecular weight of starch molecules, including end-group analysis for amylose, osmotic pressure determination, and light-scattering methods. Low angle laser light scattering (aqueous size exclusion chromatography) is useful in the investigation of molecular weight distribution and the degree of branching in starch (35). Osmotic pressure measurements provide a weight range of 10,000–60,000 for amylose

(36). The degree of polymerization (DP) of amylose usually falls in the range of 200–20,000 DP, although there are some exceptions (33). Using anaerobic techniques to prevent oxidative degradation, measurements suggest that amylose has a molecular weight range of $(1.6–7.0) \times 10^5$ (37). Amylose extracted from starch with dimethyl sulfoxide has a molecular weight of 1.9×10^6 daltons, measured by light-scattering techniques. Amylopectin, a much larger molecule, has a molecular weight range of $4–5 \times 10^8$ (38). Measurements based on light scattering indicate that large amylopectin molecules (250×10^6 daltons) are present in Easter lily starch (39), and in potato starch have a molecular weight of 36×10^6 (40). Molecular weight distributions in starch and methods used to measure them have been reviewed (33).

Starch hydrolysis is accomplished in industry by using acid, enzymes, or both sequentially. Commercially, starch usually is hydrolyzed with thermoduric α-amylase or glucoamylase at elevated temperatures to produce commercial D-glucose. Partial hydrolysis with acids produces molecular fragments called dextrins and low molecular weight sugar-like fragments (41). On acid hydrolysis, D-glucose is produced in amounts that increase throughout the reaction (42). Acid treatment of starch causes random cleavage of α-$1\rightarrow4$ and α-$1\rightarrow6$ linkages. Other products are oligosaccharides and acid breakdown products of D-glucose such as 5-hydroxymethylfurfural and levulinic acid. Oligosaccharides are produced by incomplete hydrolysis of starch and by reversion, that is, by acid-catalyzed recombination of two or more D-glucose monomers to produce numerous oligosaccharides.

α-Amylase hydrolyzes starch mainly to a mixture of D-glucose and maltose, and limits dextrins derived from the amylopectin. α-Amylase randomly cleaves α-$(1\rightarrow4)$ glucosidic bonds. The α-$(1\rightarrow6)$ bonds of amylopectin are resistant to α-amylase so when a branch point is encountered, hydrolysis ceases and the molecular remnant is an α-amylase limit dextrin. α-Amylase is a common enzyme present in human and animal digestive tracts and in plants and microorganisms. Plant and animal enzymes are involved in starch conversion *in vitro*, but the principal enzymes in commercial starch digestion are microbial in origin. Such enzymes are employed in distilling, brewing, baking, and textile and paper manufacture.

β-Amylase occurs in many plants, such as barley, wheat, rye, soy beans, and potatoes, where it is generally accompanied by some α-amylase. β-Amylase initiates hydrolysis at the nonreducing end of an amylose or amylopectin chain, and removes maltose units successively until the reducing end of the molecule is encountered in amylose or a branch is met in amylopectin. β-Amylase is used commercially in the preparation of maltose syrups. After β-amylase hydrolysis of amylopectin there remains a β-amylase limit dextrin. β-Amylase has been used as a probe of the fine structure of amylopectin (43–46).

Cyclomaltodextrin glucanotransferase (CGTase) produces commercially important cyclomaltodextrins from starch by a cyclization reaction. α-Cyclodextrins (six D-glucopyranosyl units in a ring) are produced by *Bacillus macerans, Klebsiella pneumoniae,* and *B. stearothermophilus*; β-cyclodextrins (seven-membered rings) by *B. megaterium, B. circulans, B. ohbensis,* and alkalophilic *Bacillus*; and γ-cyclodextrins (eight-membered rings) by *Bacillus* sp. AL6 and *B. subtilis* No. 313 (47). CGTase also catalyzes transfer of D-glucosyl residues, using starch as the donor, to the C-4 position of D-glucose, D-xylose, and various other deoxy and methylated glucoses to produce oligosaccharides.

Glucoamylase degrades both amylose and amylopectin, yielding D-glucose as the only product. Thus, the enzyme splits both $1 \rightarrow 4$-α-D- and $1 \rightarrow 6$-α-D-glucosidic bonds, although the $1 \rightarrow 4$ bonds are hydrolyzed more rapidly. Glucoamylase removes single D-glucose units from the nonreducing end(s) of the molecule. Glucoamylases are produced by some species of fungi in the *Aspergillus* and *Rhizopus* genus, and also by specific yeasts and bacteria.

Depolymerization of starch in alkaline solution proceeds more slowly than in acid and produces isosaccharinic acid derivatives rather than D-glucose as a major product. The mechanism involves a β-elimination-type reaction (48).

Oxidation of starch hydroxyl groups by hypochlorite gives aldehydes, ketones, or carboxylic acids. Periodic acid treatment results in opening the sugar ring with formation of aldehyde groups at C2 and C3 positions. Starch is oxidized with other reagents including nitrogen dioxide, chlorine, permanganate, dichromate, and ozone to produce various aldehydes, ketones, and carboxylic acids and derivatives.

Etherification and esterification of hydroxyl groups produce derivatives, some of which are produced commercially. Derivatives may also be obtained by graft polymerization wherein free radicals, initiated on the starch backbone by ceric ion or irradiation, react with monomers such as vinyl or acrylyl derivatives. A number of such copolymers have been prepared and evaluated in extrusion processing (49). A starch–acrylonitrile graft copolymer has been patented (50) which rapidly absorbs many hundred times its weight in water and has potential applications in disposable diapers and medical supplies.

Starch derivatives are also used to encapsulate pesticides (qv) in a cross-linked starch–xanthate to improve safety in handling and reduce water leaching losses (51,52). Little or no pesticide is lost during drying. The encapsulated formulations have excellent shelf life when dry, but when placed in water or soil the pesticide is readily released.

Manufacture

Wet-Milling of Corn Starch. Milling of corn, *Zea mays*, provides a quality starch, used in food and nonfood applications. Other grains are milled for starch, but corn is the principal source because of its steady price. Corn wet-milling processes are fully automated to provide well-separated grain components (53). At least eight new plants have been constructed in the United States since the 1970s. Corn may be dry-milled using screening and air classified for separation of particle size, but this process incompletely separates oil, protein, starch, and hull (54).

To understand the milling process, it is necessary to examine the structure of the corn kernel (Fig. 3) (55). Principal parts of the kernel are the tip cap (0.8%), pericarp or hull (5%), germ or embryo (11%), and the endosperm (82%). The tip cap and pericarp are separated in the fiber fraction in wet-milling, or in the bran fraction in dry-milling. The germ is comprised mainly of protein and lipids, whereas the endosperm consists of starch granules embedded in a proteinaceous cellular matrix. The principal U.S. corn crop, dent corn, has

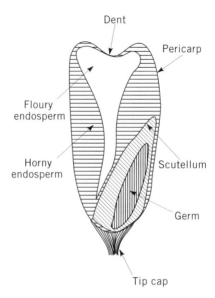

Fig. 3.   Schematic cross-sectional view of a corn kernel.

two distinct regions of endosperm, floury and horny. Floury endosperm at the grain center has loosely packed starch in less dense proteinaceous cells, whereas horny endosperm at the grain periphery contains densely packed starch granules in a region of high protein content. Starch granules in the more dense horny endosperm are polygonal as opposed to the more round granules in the floury endosperm. Most dent corn has a ratio of floury to horny endosperm of about 1:2. Horny endosperm requires thorough steeping to soften the protein matrix and ensure maximum starch recovery. The germ next to the endosperm is the scutellum, a repository for enzymes required for hydrolyzing the endosperm during embryonic development during germination to produce a new corn plant. Because the scutellar epithelium is strongly bound to the endosperm, long steeping times are required for separation. The average composition of corn grain on a dry basis is 71.3% starch, 9.91% protein, and 4.45% fat (56,57). Normal water content is 10–15%.

A flow chart for a typical corn-milling operation is shown in Figure 4. Corn is first cleaned by screening to remove cob, sand, and other foreign material, followed by aspiration to remove lighter dust and chaff. The grain is then transferred with water containing ~0.1% sulfur dioxide at pH 3–4 to large vats (steeps) which softens the kernels for milling. The steeping process requires careful control of countercurrent water flow at 48–52°C. Corn is introduced into the steeps at a moisture content of 15% and attains a final moisture content of 45% at the end of 30–40 h. Water absorption is accelerated by sulfur dioxide (58) in the steep water and results in a 55–65% increase in kernel volume (59–61).

Sulfur dioxide was initially added to corn steep water to prevent the growth of degradative microorganisms, but it is now known to be indispensable in maximizing starch recovery. It acts on the nitrogen-containing components of corn which include 10% albumin and nonprotein nitrogen, 10% globulin, 38% zein, and 42% glutelin (62). Sulfur dioxide softens and then disperses the glutelin matrix

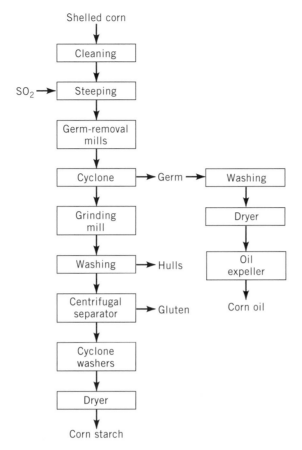

Fig. 4. Typical corn-milling operation.

(61), allowing maximum starch release and recovery, especially from the horny endosperm. Although sulfur dioxide inhibits the growth of microorganisms, after several hours its concentration decreases and lactic acid bacteria resembling *Lactobacillus bulgaricus* begins to grow. Because of toxicity problems with sulfur dioxide it has largely been replaced with sodium bisulfite.

Steeped corn is coarsely ground in an attrition mill to free the germ. The mill gap during this step must be adjusted to maximize the amount of germ freed but with a minimum of breakage of the germ, which would cause oil loss and present problems in later purification steps. Germ is removed from the aqueous slurry in a hydroclone (cyclone separator) (Fig. 5), where suspended components are separated by density (63,64), the endosperm and fiber exiting in the hydroclone underflow and the germ from the center. The germ fraction is then screened, washed to remove residual starch, dewatered to 50–55% water content, and processed to produce corn oil.

The cyclone underflow is re-milled to complete the release of starch granules. Some starch factories use a Bauer mill, a combination attrition–impact mill (65); others favor the Entoleter mill, an impact mill only (66). After the second milling, the suspension contains starch, gluten, and fiber. Fiber is removed

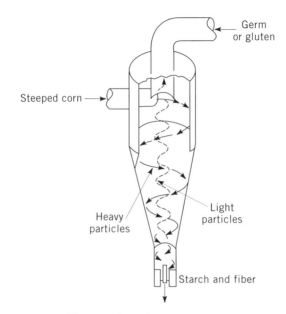

Fig. 5. A cyclone separator.

by flowing the slurry over fixed concave washing screens. It is retained on the screen while starch and gluten pass through. Collected fiber is slurried and re-screened to remove any residual starch and protein. This fiber is later combined with the grain's gluten to a content of 21% for animal feed use.

The resultant starch–gluten suspension, known as mill starch, is concentrated by centrifugation. The low density of gluten, compared to starch, permits easy separation. As a result, protein content is reduced to 1–2%. The starch suspension from the centrifugal separator is diluted and subjected to 8–14 stages of hydroclone washing (67,68). Concentrated starch underflow from these processes is again diluted and passed through a final battery of hydroclones to wash the starch and remove most of the remaining protein.

This starch suspension may be dried and marketed as unmodified corn starch, modified by any of a number of chemical or physical means, gelatinized and dried, or hydrolyzed to produce corn syrup. During processing of corn, wet-milling uses about 0.2 m^3 H$_2$O/100 kg corn (20 gal/100 lb) with the water removed before marketing. Starch is usually dewatered by centrifugation, followed by injection into a column of hot air (200–260°C). Starch granules dry rapidly and are collected in cyclones (69,70). Large amounts of energy used in drying starch makes the wet-milling industry the second most energy-intensive food industry in the United States (71). The main product of milling is unmodified corn starch, a white powder with a pale yellow tint. Unless the corn used is a white variety, absolute whiteness requires chemical bleaching. The final product usually has a moisture content of about 11% and may contain 1% protein, ash, lipids, and fiber (18,72–76).

Chemical Modification.   Production and use of acid-modified starch have been reviewed (77). Treating starch with acid under nongelatinizing conditions was explored in the late nineteenth century (78) and many acid modifications

were patented (79). Acidic treatment below the gelatinization temperatures initially attacks amorphous regions of the granule but leaves the crystalline regions
relatively unaffected (5,80,81). In corn starch modification, amylopectin is more
extensively depolymerized than amylose. Properties of acid-treated starches,
as compared to the unmodified starches, include decreased hot paste viscosity (82–84), decreased intrinsic viscosity (85,86), lower gel strength (84), and
higher gelatinization temperature range (83,87). However, mild acid treatment
of starch produces what is known as thin-boiling starch which has excellent
cooled gel strength. In fact, this product is used in starch gum candy (eg, orange
slices) manufacture. Similar acid-modified starches are also employed in textiles,
gypsum board, and paper and paperboard manufacture (88).

In industrial production of acid-modified starches, a 40% slurry of normal
corn starch or waxy maize starch is acidified with hydrochloric or sulfuric acid at
25–55°C. Reaction time is controlled by measuring loss of viscosity and may vary
from 6 to 24 hours. For product reproducibility, it is necessary to strictly control
the type of starch, its concentration, the type of acid and its concentration, the
temperature, and time of reaction. Viscosity is plotted versus time, and when the
desired amount of thinning is attained the mixture is neutralized with soda ash
or dilute sodium hydroxide. The acid-modified starch is then filtered and dried.
If the starch is washed with a nonaqueous solvent (89), gelling time is reduced,
but such drying is seldom used. Acid treatment may be used in conjunction with
preparation of starch ethers (90), cationic starches, or cross-linked starches. Acid
treatment of 34 different rice starches has been reported (91), as well as acidic
hydrolysis of wheat and corn starches followed by hydroxypropylation for the
purpose of preparing thin-boiling and nongelling adhesives (92).

Starch oxidation was investigated as early as 1829 by Liebig. The objective,
as with other modifications, was to obtain a modified granular starch. The
oxidant commonly employed is sodium hypochlorite, prepared from chlorine and
aqueous sodium hydroxide. This reaction is exothermic and external cooling must
be provided during preparation of the oxidant.

$$2\,NaOH + Cl_2 \longrightarrow NaOCl + H_2O + NaCl$$

To produce oxidized starch, a slurry of starch granules is treated with alkaline hypochlorite for an appropriate period, neutralized, washed to remove
the inorganic salts, and finally dried to a moisture content of 10–12%. Temperature, an important process variable, is usually in the range of 21–38°C.
Structure of the unmodified starch influences the type and extent of oxidation
reactions (93,94). As with acid hydrolysis, most oxidation occurs primarily in
the loosely organized, amorphous region of the starch granule (93–95). Chemical
changes (93,94) include formation of carboxyl and carbonyl groups and breaking of some D-glucosyl linkages. The changes result in a decrease in polymer
molecular weight. Oxidized starches are bleached white by the hypochlorite. As
they are anionic, they can strongly absorb cationic pigments such as methylene
blue. Oxidation results in lower gelatinization temperatures, decreased hot-paste
viscosity, and lower paste setback (96).

Corn and rice starches have been oxidized and subsequently cyanoethylated
(97). As molecular size decreases due to degradation during oxidation, the degree

of cyanoethylation increases. The derivatized starch shows pseudoplastic flow in water dispersion; at higher levels of cyanoethylation the flow is thixotropic. Corn and rice starches have been oxidized and subsequently carboxymethylated (98). Such derivatives are superior in the production of textile sizes. Potato starch has been oxidized with neutral aqueous bromine and fully chemically (99) and physically (100) characterized. Amylose is more sensitive to bromine oxidation than amylopectin and oxidation causes a decrease in both gelatinization temperature range and gelatinization enthalpy.

Sodium chlorite oxidation of corn and rice starches is recommended for the production of textile sizes (101) and oxidized starch is recommended as a hardening agent in the immobilization of microbial cells within gelatin (102).

Changes in starch due to heat treatment were investigated in the 1930s (103,104). Later work on dextrinization showed that the products of dry heat treatment, the pyrodextrins, were soluble in cold water and did not retrograde (105). In dextrinization, starch is thermally degraded with production of low molecular weight dextrins, some of which recombine to form a more highly branched structure. Incipient pyrolysis eliminates water from a few D-glucopyranosyl groups, producing C–C double bonds and internal carbohydrate anhydrides such as anhydroglucose.

In manufacture of pyrodextrins, dry starch is sprayed with dilute inorganic acid, usually hydrochloric, nitric, or acid salts, and re-dried to 1–5% water content. To produce a white dextrin, well-mixed acidified starch is hydrolyzed and heated to a final temperature of 110–150°C; at 135–160°C, a canary yellow dextrin is produced. When made with small amounts of acid, and a longer heating time to a temperature of 150–180°C, the product is known as a British gum. The rate of temperature increase and residual moisture content achieved during heating are crucial to proper conversion. The product must be cooled rapidly to prevent over-conversion. Acid may be neutralized at this point if desired. Chemical changes produced by acid–heat treatment lead to starches with lower water holding ability, lower dispersion viscosity, and greater hot water solubility compared to the original starch. Such preparations and their applications as adhesives have been thoroughly reviewed (106).

Hydroxylalkyl Starch Ethers.   Starch hydroxyethyl ethers with a degree of substitution (DS) of 0.05–0.10 are produced in various ways, but usually their preparation begins at the end of the wet-milling process, utilizing a high solids–starch suspension. The ether modification of ungelatinized starch is filterable and can be produced economically in a pure form.

During corn wet-milling, a 40–50% solids–starch suspension is treated with a metal hydroxide and ethylene oxide at approximately 50°C to produce DS of 0.1 and the product is purified by filtration and washing.

Cationic Starches.   The two general categories of commercial cationic starches are tertiary and quaternary aminoalkyl ethers. Tertiary aminoalkyl ethers are prepared by treating an alkaline starch dispersion with a tertiary amine containing a β-halogenated alkyl, 3-chloro-2-hydroxypropyl radical, or a 2,3-epoxypropyl group. Under these reaction conditions, starch ethers are formed that contain tertiary amine free bases. Treatment with acid easily produces the cationic form. Amines used in this reaction include 2-dimethylaminoethyl chloride, 2-diethylaminoethyl chloride, and *N*-(2,3-epoxypropyl) diethylamine.

Commercial preparation of low DS derivatives employ reaction times of 6–12 h at 40–45°C for complete reaction. The final product is filtered, washed, and dried.

Quaternary ammonium alkyl ethers are prepared similarly: an alkaline starch is reacted with a quaternary ammonium salt containing a 3-chloro-2-hydroxypropyl or 2,3-epoxypropyl radical. Alternatively, such derivatives can be prepared by simple quaternization of tertiary aminoalkyl ethers by reaction with methyl iodide. Sulfonium (107) and phosphonium (108) starch salts have also been prepared and investigated. Further work has explained the synthesis of diethylaminoethyl starch (109) as well as the production of cationic starches from the reaction of alkaline starch with 3-chloro-2-hydroxypropylamine, 1,1,1-N-tris(3-chloro-2-hydroxypropyl)amine, and glycidyl trimethylammonium acetate (110). A dry cationization process for producing these materials, as opposed to the well known slurry process, has also been described (111).

Starch Phosphates. Starch phosphate monoesters may be prepared by heating a dry mixture of starch and acid salts of ortho-, pyro-, or tripolyphosphoric acid at 50–60°C for one hour. DS is generally low (<0.15), but higher DS derivatives can be prepared by increasing temperature, phosphate salt concentration, and reaction time. Phosphorylation of corn starch with sodium tripolyphosphate (STP) can be done in a Brabender single-screw extruder (112). Highest degree of substitution was obtained at an extruder temperature of 200°C, STP concentration of 1.4 g/100 mL of water, and a pH of 8.5. Highly phosphorylated, cross-linked potato starch is produced by heating starch in benzene or pyridine with phosphorus pentoxide (113). The DS of this material is approximately 1.0.

Starch in aqueous suspension may react to form diesters with phosphorus oxychloride, phosphorus pentachloride, and thiophosphoryl chloride (114). Cross-bonded starches can also be manufactured by reaction with trimetaphosphates (115), but these require more vigorous conditions than phosphorus oxychloride. Typically, a starch slurry and 2% trimetaphosphate salt react at pH 10–11 and 50°C for one hour.

Low DS starch acetates are manufactured by treatment of native starch with acetic acid or acetic anhydride, either alone or in pyridine or aqueous alkaline solution. Dimethyl sulfoxide may be used as a cosolvent with acetic anhydride to make low DS starch acetates; ketene or vinyl acetate have also been employed. Commercially, acetic anhydride–aqueous alkali is employed at pH 7–11 and room temperature to give a DS of 0.5. High DS starch acetates are prepared by the methods previously detailed for low DS acetates, but with longer reaction time.

Of particular importance for modifications of starch are the enzyme degradation products such as glucose syrups, cyclodextrins, maltodextrins, and high fructose corn syrups (HFCS). Production of such hydrolysis products requires use of selected starch-degrading enzymes such as α-amylase, β-amylase, and debranching enzymes. Conversion of D-glucose to D-fructose is mediated by glucose isomerase, mostly in its immobilized form in columns. Enzymic degradation of starch to syrups has been well reviewed (116–118), and enzymic isomerization, especially by immobilized glucose isomerase, has been fully described (119) (see SYRUPS).

New Starches

A factory for the first production of banana starch is under development by a United States corporation in Costa Rica. This large-granule starch has many properties of potato starch but also has other interesting and unique properties that are said to give it extensive usage in the food industry. In addition, it is expected to be low in cost because it is made from cull bananas, those that are cut from bunches because they are too small for marketing or those that have been injured during harvest. The total amount of cull bananas averages 25% of the crop.

Another likely commercial starch is that from amaranth seed, an expanding crop for food use, particularly its flour. Amaranth starch granules (1–3 micrometers dia) have potential for numerous food applications, one of which is as a fat replacer because of their small size and especially after minor surface hydrolysis with α-amylase or glucoamylase to produce a fluffy surface (see FAT REPLACERS).

Economic Aspects

Commercial starch is mainly corn starch, but smaller amounts of sorghum, wheat, and potato starch are also produced. In 1992, 1303 million bushels (45.8 $\times$ 10^6 m^3) of corn were ground for starch and other products (120); 1 m^3 corn weighs ~721 kg and yields 438 kg starch, 26 kg oil, and 142 kg combined gluten and hulls. In the United States in 1994–1995, 462 million bushels were used to produce high fructose corn syrup, 231 million bushels went to produce D-glucose, 533 million bushels were used for alcohol production, and 247 million bushels were converted to starch (121).

In 1991 U.S. starch production was ca 5 $\times$ 10^6 t. Of this figure 1.5 $\times$ 10^4 t were exported to Canada, 1.0 $\times$ 10^4 t went to the European Community (EC), and 7.5 $\times$ 10^3 t were sold to Pacific Rim countries. Additionally, Canada purchased another 4.8 $\times$ 10^4 t of variously modified starches (122). Total starch consumption in the EC was about 5.0 $\times$ 10^6 t in 1989 of which 42% (2.1 $\times$ 10^6 t) were used by the chemical industry. The continuing development of high fructose corn syrup (HFCS) further stimulated the growth of the corn-refining industry. In 1979 HFCS was consumed at 6.8 kg per capita, accounting for about 12% of the total corn-sweetener market (123). By 1994 the amount consumed per capita had grown to 25.4 kg (124) and accounted for 54% of the corn sweetener market. The growth rate of corn-derived sweeteners has been phenomenal and is expected to continue to grow in the future. Growth potential for the corn wet-milling industry continues to be excellent because of predicted increases in HFCS production and demand for corn-derived alcohol. Ethanol production has utilized more than 5% of total U.S. corn production from 1990 to 1995.

Uses

Nonfood Uses. Native corn starch is principally used in nonfood applications in mining, adhesives, and paper industries. Pregelatinized starch is chemically unmodified, but it is physically modified. Pregelatinized starches are used to decrease water losses in oil-well drilling muds, in cold water-dispersable wallpaper pastes, and in papermaking as an internal fiber adhesive.

Modified starches may be acid-modified, oxidized, or heat-treated. Acid-modified (thin-boiling) starches are used mainly in textiles as warp sizes and fabric finishes. Here they increase yarn strength and abrasion resistance and improve weaving efficiency. Thin-boiling starches also have selected applications in paper and laundry starch preparations.

Oxidized starches, usually those prepared by hypochlorite oxidation, are used in paper coatings and adhesives (qv) to improved surface characteristics for printing or writing. Oxidized starches may also be employed as textile warp sizes and finishes, in manufacture of insulation and wallboard, and in laundry spray starch.

Starch pyrodextrins and British gums have the ability, in aqueous dispersion, to form films capable of bonding like or unlike materials. Thus, they have uses as adhesives for envelopes, postage stamps, and other products. These dextrins are used in glass-fiber sizing to protect the extruded fiber from abrasion, and as binders for metal core castings, water color paints, briquettes, and many other composite materials (qv).

Various organic chemicals, eg, ethanol, isopropyl alcohol, n-butanol, acetone, 2,3-butylene glycol, glycerol, fumaric acid, citric acid, and lactic acid, are derived from starch by fermentation (125). Other compounds derived from starch include D-glucose, sorbitol, methyl α-D-glucopyranoside, and glycerol or glycol D-glucopyranosides.

Food Uses. Unmodified starch is used in foods that require thickening or gelling. Such foods include puddings, salad dressings, pie fillings, and candies. Pregelatinized starch is used in products where thickening is required but cooking must be avoided, such as instant pudding, pie fillings, and cake frostings.

Acid-modified starches are used in the manufacture of gum candies because they form hot concentrated pastes that form strong gels on cooling. Thermalized starches are used in foods to bind and carry flavors and colors. Sweetening agents (corn syrup, HFCS) are made from starch by enzymatic or acid treatment as previously noted.

Derivatives

Starches, as organic polyhydroxy compounds, undergo many reactions characteristic of alcohols, such as esterification and etherification. Because D-glucopyranosyl monomers contain, on average, three free hydroxyl groups, the degree of substitution (DS) may be a maximum of three. Commercial starch derivatives are generally very lightly derivatized (DS < 0.1). Such modification produces distinct changes in colloidal properties and generally produces polymers with properties useful under a variety of conditions.

Hydroxyethyl group introduction at low DS results in distinct modification of physical properties. Among these are decreased gelatinization temperature range (126), increased granule swelling rate (127), and decreased ability of starch pastes to gel and retrograde.

Low DS hydroxyethyl starches are used as paper coatings and sizes to improve sheet strength and stiffness. They are also employed as paper-coating color adhesives, and to increase fiber bonding in paper products. Hydroxyethylstarches are also used as textile warp sizes. Hydroxyethylstarch [9005-27-0] has been investigated by hydrodynamic and magnetic spectroscopic methods (128) because of its increasing use as a plasma volume expander.

Hydroxypropyl and other hydroxylalkyl starches find uses as additives in salad dressings, pie fillings, and in other food thickening applications. Examination of hydroxypropylstarch [9049-76-7] by light microscopy/staining techniques (129) shows that the central granular region has a large proportion of the hydroxypropyl groups, possibly because of its low density. Fourier transform infrared spectroscopy (ftir) can be used to quantitate the degree of starch hydroxypropylation (130).

Cationic starches show decreased gelatinization temperature range and increased hot paste viscosity. Pastes remain clear and fluid even at room temperatures and show no tendency to retrograde. This stability is due to Coulombic repulsion between positively charged starch molecules in dispersion.

Quaternary ammonium starches, like tertiary ammonium derivatives, show lower gelatinization temperature ranges, increased paste clarity and viscosity, and reduced retrogradation. Quaternary ammonium starch salts exhibit cold water swelling at a DS as low as 0.07. Cationic starches are used in paper manufacture principally for fiber and pigment retention. They also improve bursting strength and fold endurance, and have been employed as emulsifiers for water-repellent paper sizes. Because of their relatively high cost, cationic starches are not widely used in textile sizes, although they have been employed in ore refining as flocculating agents (qv).

Compared to native starches, monophosphate esters have a decreased gelatinization temperature range and swell in cold water at a DS of 0.07. Starch phosphates have increased paste viscosity and clarity and decreased retrogradation. Their properties are in many ways similar to those of potato starch, which naturally contains phosphate groups.

Starch monophosphates are quite useful in foods because of their superior freeze–thaw stability. As thickeners in frozen gravy and frozen cream pie preparations, they are preferred to other starches. A pregelatinized starch phosphate has been developed (131) which is dispersible in cold water, for use in instant dessert powders and icings and nonfood uses such as core binders for metal molds, in papermaking to improve fold strength and surface characteristics, as a textile size, in aluminum refining, and as a detergent builder.

In contrast to monophosphates, starch phosphate diesters contain crosslinks between two or more starch chains. This covalent linkage in the granule produces a starch product which swells less but is more resistant to heat, agitation, and acid than natural starch.

Starch phosphate diesters show a significant increase in stability of the swollen granule. Depending on the degree of derivatization, the hot paste viscosity may be more or less than that of the parent starch. In contrast to starch phosphate monoesters, diester pastes do not produce increased clarity when in aqueous dispersion. Starches with high DS have exceptional stability to high temperatures, low pH, and mechanical agitation. If cross-linking is sufficient, swelling and gelatinization can be completely inhibited, even in boiling water.

Cross-linked starches are used as thickeners and stabilizers in baby foods, salad dressings, fruit pie filling, and cream-style corn. They are superior to native starches in their ability to keep food particles in suspension after cooking, greater resistance to gelling and retrogradation, freeze–thaw stability, and lack of syneresis. They are also used to produce high wet-strength paper (114), as ion exchangers, and as metal sequesterants to prevent oxidative rancidity in food-grade oils.

Starch acetates may have low or high DS. The industrial importance of low DS acetates results from their ability to stabilize aqueous polymer sols. Low DS acetates inhibit association of amylose polymers and reduce the association of the longer outer chains of amylopectin. These properties are important in food applications. Highly derivatized starches (DS 2–3) are useful because of their solubility in organic solvents and ability to form films and fibers.

Low DS starch acetates have reduced gelatinization temperature ranges and reduced tendency to retrograde after pasting and cooling. Gelling may be completely inhibited if the DS is sufficiently high. Low DS starch acetate polymers also form films which are useful in textile and paper manufacture.

Lightly derivatized starch acetates are employed in food because of the clarity of their gels and their stability. Applications include frozen fruit pies and gravies, baked goods, instant puddings, and pie fillings. Starch acetates are used in textiles as warp sizes and in paper to improve printability, surface strength, and solvent resistance.

In general, high DS starch acetates and amylopectin acetates give weak and brittle films and fibers. Amylose triacetate can be spun into strong fibers and cast into strong, clear films, although these have not yet been commercialized. It is soluble in organic solvents including acetic acid, pyridine, and chloroform. Films of such a high DS acetate, cast from chloroform solution, are pliable, lustrous, transparent, and colorless. These properties are useful in packaging materials (qv).

BIBLIOGRAPHY

"Starch" in *ECT* 1st ed., Vol. 12, pp. 764–778, by R. W. Kerr, Corn Products Refining Co.; in *ECT* 2nd ed., Vol. 18, pp. 672–691, S. M. Parmerter, Corn Products Co.; in *ECT* 3rd ed., Vol. 21, pp. 492–507, R. L. Whistler and J. R. Daniel, Purdue University.

1. G. M. A. Van Beynum and J. A. Roels, eds., *Starch Conversion Technology*, Marcel Dekker, Inc., New York, 1985.

2. A. Imberty, A. Buleon, V. Tran, and S. Perez, *Starch* **43**, 375 (1991).

3. R. W. Kerr, in R. W. Kerr, ed., *Chemistry and Industry of Starch*, Academic Press, Inc., New York, 1950, pp. 3–25.

4. O. B. Wurzburg, in T. E. Furia, ed., *Handbook of Food Additives*, Vol. 1, The Chemical Rubber Co., Boca Raton, Fla., 1973.

5. D. French, in Ref. 3, pp. 157–178.

6. D. French, in R. L. Whistler, J. N. BeMiller, and E. F. Paschall, eds., *Starch: Chemistry and Technology*, 2nd ed., Academic Press, Inc., New York, 1984, pp. 183–247.

7. A. Imberty and Z. Perez, *Biopolymers* **27**, 1205 (1988).

8. A. Imberty, H. Chanzy, S. Perez, A. Buleon, and V. Tran, *J. Mol. Biol.* **201**, 365 (1988).

9. M. J. Gidley and S. M. Bociek, *J. Am. Chem. Soc.* **107**, 7040 (1985).

10. N. N. Hellman, T. F. Boesch, and E. H. Melvin, *J. Am. Chem. Soc.* **74**, 348 (1952).

11. G. Baumann, D. Hartenthaler, and K. Breslauer, *Center for Advanced Food Technology—Physical Forces Research Accomplishments*, January Report, Rutgers University, New Brunswick, N.J., 1988.
12. J. W. Donovan, *J. Food Agric.* **28**, 571 (1979).
13. J. W. Donovan, *Biopolymers*, **18**, 263 (1979).
14. B. C. Burros, L. A. Young, and P. A. Carroad, *J. Food Sci.* **52**, 1372 (1987).
15. M. H. Gomez and J. M. Aguilera, *J. Food Sci.* **49**, 40 (1984).
16. P. Colonna, J. L. Doublier, J. D. Melcion, F. deMonredon, and C. Mercier, *Cereal Chem.* **61**, 538 (1984).
17. J. L. Kokini, L.-S. Lai, and L. L. Chedid, *Food Technol.* **46**(6), 124 (1992).
18. T. J. Schoch and E. C. Maywald, in R. L. Whistler and E. F. Paschall, eds., *Starch: Chemistry and Technology*, Vol. II, Academic Press, Inc., New York, 1967, pp. 637–647.
19. M. W. Rutenberg and D. Solarek, in Ref. 6, pp. 311–388.
20. L. Slade and H. Levine, in A. M. Peason, T. R. Dutson, and A. Bailey, eds., *Advances in Meat Research*, Vol. 4, AVI Publishing Co., Inc., New York, 1987, p. 251.
21. L. Slade and H. Levine, in R. P. Millane, J. N. BeMiller, and R. Chandrasekaran, eds., *Frontiers in Carbohydrate Research—1: Food Applications*, Elsevier Applied Science, New York, 1989, p. 215.
22. L. Slade and H. Levine, in H. Faridi, ed., *The Science of Cookie and Cracker Production*, Chapman and Hall, New York, 1994, p. 23.
23. B. Casu and M. Reggiana, *Stärke* **18**, 218 (1966).
24. M. St. Jacques, P. R. Sundararajan, J. K. Taylor, and R. H. Marchassault, *J. Am. Chem. Soc.* **98**, 4386 (1976).
25. D. Grebel, *Mikrokosomos* **57**, 111 (1968).
26. T. J. Schoch, *Cereal Chem.* **18**, 121 (1941).
27. J. F. Kennedy, Z. S. Rivera, L. L. Lloyd, and F. P. Warner, *Starch* **44**, 53 (1992).
28. H. N. Englyst, H. S. Wiggins, and J. H. Cummings, *Analyst*, **107**, 307 (1982).
29. H. N. Englyst, S. M. Kingman, and J. H. Cummings, *Eur. J. Clin. Nutr.* **46**, S33 (1992).
30. N.-G. Asp, *Am. J. Clin. Nutr.* **61** (suppl), 930S (1995).
31. W. Z. Hassid and R. M. McCready, *J. Am. Chem. Soc.* **65**, 1157 (1943).
32. W. R. Morrison and J. Karkalas, in P. E. Dey, ed., *Starch, Methods in Plant Biochemistry*, Vol. 2, *Carbohydrates*, Academic Press, Inc., New York, 1990, p. 324.
33. S. Hizukuri, in A.-C. Eliasson, ed., *Carbohydrates in Food*, Marcel Dekker, Inc., New York, 1996, p. 347.
34. K. Kulp and J. G. Ponte, *Crit. Rev. Food Sci. Nutr.* **15**, 1 (1981).
35. L.-P. Yu and J. E. Rollings, *J. Appl. Polym. Sci.* **33**, 1909 (1987).
36. K. H. Meyer, in H. F. Mark and A. V. Tobolsky, eds., *Physical Chemistry of High Polymeric Systems*, 2nd ed., Interscience Publishers, Inc., New York, 1950, p. 456.
37. C. T. Greenwood, *Stärke* **12**, 169 (1960).
38. W. Banks and C. T. Greenwood, *Starch and Its Components*, Edinburgh University Press, Scotland, 1975, p. 45.
39. B. H. Zimm and C. D. Thurmond, *J. Am. Chem. Soc.* **74**, 1111 (1952).
40. L. P. Witnauer, F. R. Senti, and M. D. Stern, *J. Polym. Sci.* **16**, 1 (1955).
41. G. S. C. Kirchoff, *Acad. Imp. Sci., St. Petersbourg, Mem.* **4**, 27 (1811).
42. T. deSaussere, *Bull. Pharm.* **6**, 499 (1814).
43. G. N. Bathgate and D. J. Manners, *Biochem. J.* **101**, 3C (1966).
44. E. Y. C. Lee, C. Mercier, and W. J. Whelan, *Arch. Biochem. Biophys.* **125**, 1028 (1968).
45. C. Mercier, *Stärke* **25**, 78 (1973).
46. J. J. Marshall and W. J. Whelan, *Arch. Biochem. Biophys.* **161**, 234 (1974).

47. S. Kitahata, *Handbook of Amylase and Related Enzymes: Their Sources, Isolation, Methods, Properties, and Applications*, Pergamon Press, Oxford, U.K., 1988, p. 154.

48. J. N. BeMiller, in R. L. Whistler and E. F. Paschall, eds., *Starch: Chemistry and Technology*, Vol. I, Academic Press, Inc., New York, 1965, p. 521.

49. E. B. Bagley, G. F. Fanta, R. C. Burr, W. M. Doane, and C. R. Russell, *Polym. Eng. Sci.* **17**, 311 (1977).

50. U.S. Pat. 3,935,099 (Apr. 3, 1974); U.S. Pats. 3,981,100, 3,985,616, and 3,997,484 (Sept. 8, 1975), M. O. Weaver, E. B. Bagley, G. F. Fanta, and W. M. Doane (to United States of America as represented by Secretary of Agriculture).

51. B. S. Shasha, W. M. Doane, and C. R. Russell, *J. Polym. Sci. Polym. Lett. Ed.* **14**, 417 (1976).

52. W. M. Doane, B. S. Shasha, and C. R. Russell, *Am. Chem. Soc. Symp. Ser.* **53**, 74 (1977).

53. S. A. Watson, in Ref. 18, pp. 1–51.

54. O. L. Brekke, in G. E. Inglett, ed., *Corn: Culture, Processing, Products*, AVI Publishing Co., Westport, Conn., 1970, pp. 262–291.

55. M. J. Wolf, C. L. Buzan, M. M. MacMasters, and C. E. Rist, *Cereal Chem.* **29**, 321 (1952).

56. G. Bianchi, P. Avato, and G. Mariani, *Cereal Chem.* **56**, 491 (1979).

57. G. C. Shove, in Ref. 52, pp. 60–72.

58. L. T. Tan, H. C. Chen, J. A. Shellenberger, and D. S. Chung, *Cereal Chem.* **42**, 385 (1965).

59. L. T. Tan, P. S. Chu, and J. A. Shellenberger, *Biotechnol. Bioeng.* **4**, 311 (1962).

60. R. A. Anderson, *Cereal Chem.* **39**, 406 (1962).

61. M. J. Cox, M. M. MacMasters, and G. E. Hilbert, *Cereal Chem.* **21**, 447 (1964).

62. J. S. Wall, in Y. Pomeranz, ed., *Advances in Cereal Science and Technology*, Vol. II, American Association of Cereal Chemists, St. Paul, Minn., 1978, pp. 135–219.

63. Brit. Pat. 701,613 (1953) (to Stamicarbon BV, Heerland, the Netherlands).

64. U.S. Pat. 2,913,122 (Nov. 17, 1959), F. P. L. Stavenger and D. E. Wuth (to Dorr-Oliver, Inc.).

65. U.S. Pat. 3,040,996 (June 26, 1962), M. E. Ginaven (to The Bauer Brothers Co.).

66. U.S. Pat. 3,029,169 (Apr. 10, 1962), D. W. Dowie and D. Martin (to Corn Products Co.).

67. U.S. Pat. 2,689,810 (Sept. 21, 1954), H. J. Vegter (to Stamicarbon NV).

68. U.S. Pat. 2,778,752 (Jan. 22, 1957), H. J. Vegter (to Stamicarbon NV).

69. F. Baunack, *Stärke* **15**, 299 (1963).

70. U.S. Pat. 4,021,927 (May 10, 1977), L. R. Idaszak (to CPC International, Inc.).

71. *Industrial Energy Study of Selected Food Industries for the Federal Energy Office*, Final Report, U.S. Dept. of Commerce, Washington, D.C., 1974.

72. R. L. Sims, *Sugar Azucar*, 50 (Mar. 1978).

73. *Corn Starch*, 3rd ed., Corn Industries Research Foundation, Washington, D.C., 1964.

74. W. R. Morrison, in Ref. 59, p. 221.

75. T. J. Schoch and E. C. Maywald, *Anal. Chem.* **28**, 382 (1956).

76. W. Bergthaller and G. Tegge, *Stärke* **24**, 348 (1972).

77. P. Shildneck and C. E. Smith, in Ref. 18, p. 217.

78. C. J. Litner, *J. Prakt. Chem.* **34**, 378 (1886).

79. U.S. Pat. 675,822 (Jan. 12, 1899), C. B. Duryea; U.S. Pat. 696,949 (May 24, 1901), C. B. Duryea.

80. K. H. Meyer and P. Bernfeld, *Helv. Chim. Acta* **23**, 890 (1940).

81. W. C. Mussulman and J. A. Wagoner, *Cereal Chem.* **45**, 162 (1968).

82. G. V. Casar and E. E. Moore, *Ind. Eng. Chem.* **27**, (1935).

83. W. Gallay and A. C. Bell, *Can. J. Res. Sect. B* **14**, 381 (1936).

84. W. G. Bechtel, *J. Colloid Sci.* **5**, 260 (1950).
85. R. W. Kerr, in Ref. 3, p. 682.
86. S. Lansky, M. Kooi, and T. J. Schoch, *J. Am. Chem. Soc.* **71**, 4066 (1949).
87. H. W. Leach and T. J. Schoch, *Cereal Chem.* **39**, 318 (1962).
88. R. G. Rohner and R. E. Klem, in Ref. 6, pp. 529–541.
89. U.S. Pat. 3,446,628 (May 27, 1969), T. J. Schoch, D. F. Stella, and H. J. Wolfmeyer (to Corn Products Co.).
90. E. T. Hjermstad, in Ref. 18, p. 425.
91. B. O. Juliano and C. M. Perez, *Starch* **42**, 49 (1990).
92. K. M. Chung and P. A. Seib, *Starch* **43**, 441 (1991).
93. J. Schmorak, D. Mejzler, and M. Lewin, *Stärke* **14**, 278 (1962).
94. J. Schmorak and M. Lewi, *J. Polymer Sci.* **A1**, 2601 (1963).
95. F. F. Farley and R. M. Hixon, *Ind. Eng. Chem.* **34**, 677 (1942).
96. T. J. Schoch, *Tappi* **35**, 4 (1952).
97. I. A. El-Thalouth, A. Ragheb, R. Refai, and A. Hebeish, *Starch* **42**, 18 (1990).
98. A. Hebeish, M. I. Khalil, and A. Hashem, *Starch* **42**, 185 (1990).
99. L. J. Torneport, B. A.-C. Salomonsson, and O. Theander, *Starch* **42**, 413 (1990).
100. P. Muhrbeck, A.-C. Eliasson, and A.-C. Salomonsson, *Starch* **42**, 418 (1990).
101. A. Hebeish, F. El-Sisy, S. A. Abdel-Hafiz, A. A. Abdel-Rahman, and M. H. El-Rafie, *Starch* **44**, 388 (1992).
102. E. de Alteriis, P. Parascandola, and V. Scardi, *Starch* **42**, 57 (1990).
103. J. R. Katz, *Rec. Trav. Chim.* **53**, 555 (1934).
104. J. R. Katz and A. Weidinger, *Z. Phys. Chem. Abt. A* **184**, 100 (1939).
105. B. Brimhall, *Ind. Eng. Chem.* **36**, 72 (1944).
106. H. M. Kennedy and A. C. Fisher, Jr., in Ref. 6, p. 593.
107. U.S. Pat. 2,989,520 (Apr. 22, 1959), M. W. Rutenberg and J. L. Volpe (to National Starch and Chemical Corp.).
108. U.S. Pat. 3,077,469 (June 28, 1961), A. Aszalos (to National Starch and Chemical Corp.).
109. E. A. El-Alfy, S. H. Samaha, and F. M. Tera, *Starch* **43**, 235 (1991).
110. M. I. Khalil, S. Farag, and A. Hashem, *Starch* **45**, 226 (1993).
111. G. Hellwig, D. Bischoff, and A. Rubo, *Starch* **44**, 69 (1992).
112. E. Salay and C. F. Ciacco, *Starch* **42**, 15 (1990).
113. K. Marusza and P. Tomasik, *Starch* **43**, 66 (1991).
114. U.S. Pat. 2,328,537 (Aug. 9, 1940), G. F. Felton and H. H. Schopmeyer (to American Maize Products Co.).
115. U.S. Pat. 2,938,901 (Nov. 23, 1956), R. W. Kerr and F. C. Cleveland, Jr. (to Corn Products Co.).
116. N. E. Lloyd and W. J. Nelson, in Ref. 6, p. 611.
117. P. D. Fullbrook, in S. Z. Dziedzic and M. W. Kearsley, *Glucose Syrups: Science and Technology*, Elsevier Applied Science Publishers, London, 1984, p. 65.
118. P. J. Reilly, in G. M. A. Van Beynum and J. A. Roels, eds., *Starch Conversion Technology*, Marcel Dekker, Inc., New York, 1985, p. 101.
119. R. van Tilburg, in Ref. 118, p. 175.
120. *Census of Manufactures*, U.S. Dept. of Commerce, Bureau of the Census, Washington, D.C., 1992.
121. *Feed Situation and Outlook Yearbook*, FDS-1995, U.S. Dept. of Agriculture, Washington, D.C., Nov. 1995.
122. J. R. Daniel, R. L. Whistler, A. C. J. Voragen, and W. Pilnik, in B. Elvers, S. Hawkins, and W. Russey, eds., *Ullmann's Encyclopedia of Industrial Chemistry*, 5th ed., Vol. A25, VCH, Weinheim, Germany, 1994, p. 21.
123. *Sugar and Sweetener Report*, SSR Vol. 4, No. 12, U.S. Dept. of Agriculture, Washington, D.C., 1979.

124. *Sugar and Sweetener Situation and Outlook Yearbook*, SSSV20N4, U.S. Dept. of Agriculture, Economic Research Service, Washington, D.C., Dec. 1995.
125. G. E. Tong, *Chem. Eng. Prog.* **74**, 70 (1978).
126. T. J. Schoch and E. C. Maywald, *Anal. Chem.* **28**, 385 (1956).
127. A. Harsveldt, *Tappi* **45**, 85 (1962).
128. W.-M. Kulicke, D. Roessner, and W. Kull, *Starch* **45**, 445 (1993).
129. H. R. Kim, A.-M. Hermansson, and C. E. Eriksson, *Starch* **44**, 111 (1992).
130. B. Forrest, *Starch* **44**, 179 (1992).
131. U.S. Pat. 2,884,346 (Dec. 24, 1957), J. A. Korth (to Corn Products Co.).

General References

Refs. 3, 18, 48, and 122 are also general references.
W. Banks and C. T. Greenwood, *Starch and Its Components*, Edinburgh University Press, Edinburgh, Scotland, 1975.
C. T. Greenwood, *Adv. Carbohydr. Chem.* **22**, 483 (1967).
C. T. Greenwood and E. A. Milne, *Adv. Carbohydr. Chem.* **23**, 282 (1968).
R. V. MacAllister, *Adv. Carbohydr. Chem.* **36**, 15 (1979).
D. J. Manners, *Adv. Carbohydr. Chem.* **17**, 371 (1962).
J. J. Marshall, *Adv. Carbohydr. Chem.* **30**, 257 (1974).
J. A. Radley, ed., *Examination and Analysis of Starch and Starch Products*, Applied Science Publishers, Ltd., London, 1976.
J. A. Radley, ed., *Industrial Uses of Starch and Its Derivatives*, Applied Science Publishers, Ltd., London, 1976.
J. A. Radley, ed., *Starch Production Technology*, Applied Science Publishers, Ltd., London, 1976.
G. M. A. Von Beynum and J. A. Roels, eds., *Starch Conversion Technology*, Marcel Dekker, Inc., New York, 1985.
R. L. Whistler, J. N. BeMiller, and E. F. Paschall, eds., *Starch: Chemistry and Technology*, 2nd ed., Academic Press, Inc., New York, 1984.
R. L. Whistler, ed., *Methods in Carbohydrate Chemistry*, Vol. 4, Academic Press, Inc., New York, 1964.
O. B. Wurzburg, ed., *Modified Starches: Properties & Uses*, CRC Press, Boca Raton, Fla., 1986.

ROY L. WHISTLER
JAMES R. DANIEL
Purdue University

STEAM

Steam [*7732-18-5*], gaseous H_2O, is the most important industrially used vapor and, after water (qv), the most common and important fluid used in chemical technology. Steam can be generated by evaporation (qv) of water at subcritical pressures, by heating water above the critical pressure, and by sublimation of ice. Steam is used in electric power generation (see POWER GENERATION); for driving mechanical devices; for distribution of heat; as a reaction medium; as a solvent; as a cleaning, blanketing, or smothering agent; and as a distillation (qv) aid. It is so widely used because water is generally available and steam is easy to generate and distribute. It has high latent heat, moderate density, and, except for thermal characteristics, nonpolluting properties. Steam provides easy control of temperature in processes and heating applications because the temperature is a function of pressure, which is easy to control.

Steam is generated from water by boiling, flash evaporation, and throttling from high to low pressure. The phase change occurs along the saturation line such that the specific volume of steam is larger than that of the boiling water. Thermal energy, ie, the heat of evaporation, is absorbed during the process. At the critical and supercritical pressures, the water–steam distinction disappears, and the fluid can go from water-like properties to steam-like properties without an abrupt change in density or enthalpy. The heat of evaporation under these conditions becomes zero.

Properties of steam can be divided into thermodynamic, transport, physical, and chemical properties. In addition, the molecular structure and chemical composition of steam are of interest. It was at the start of industrialization, ca 1763, that thermodynamic relationships were first measured by Watt. A century later, in 1859, Rankine published his *Manual of the Steam Engine*, which gave a practical thermodynamic basis for the design and performance of steam engines.

The thermodynamic and physical properties of pure steam are well established over the range of pressures and temperatures used. The chemical properties of steam and of substances in steam, their molecular structures, and interactions with the solid surfaces of containments need to be more fully explored.

Physical Properties

Official Properties. The physical properties of steam have long had considerable commercial importance. The expected efficiency of steam turbines depends on them. The first steam tables for practical use were based on Regnault's data (1) and began to appear toward the end of the nineteenth century. A thermodynamically consistent set of equations for fitting data was devised in 1900 by H. L. Callendar and was adopted by Mollier and others. The library of the United States National Institute of Standards and Technology (NIST) contains six different steam tables published between 1897 and 1915. The necessity of international property formulations was recognized as early as 1929, when the First International Steam Table Conference was held in London. As of this writing (1996), 12 international conferences on the properties of steam have been held. In 1972, the International Association for the Properties of Steam (IAPS) was formed. At the 12th International Conference on Properties of Water and Steam (ICPWS), IAPS changed its name to the International Association for Properties of Water and Steam (IAPWS), an association of national committees that maintains the official standard properties of steam and water for power cycle use. In the United States, the national committee is sponsored by the American Society of Mechanical Engineers (ASME).

IAPWS maintains two formulations of the properties of water and steam. The first is an industrial formulation, the official properties for the calculation of steam power plant cycles (2). This formulation is appropriate from 0.001 to 100 MPa (0.12–1550 psia) and from 0 to 800°C (32–1472°F). Because there is considerable industrial cost in changing from one set of properties to another, this formulation is revised every 25 to 30 years. This formulation is used in the design of steam turbines and power cycles. Millions of calls to the steam properties routines based on this formulation may be made in a single design calculation, thus the computational speed of the industrial formulation remains a significant issue. IAPWS maintains a second formulation of the properties of water and steam for scientific and general use from 0.01 MPa (extrapolating to ideal gas) at 0°C (1.45 psia at 32°F) to the highest temperatures and pressures for which reliable information is available. This latter formulation is revised as new data become available. Because the normal use of this formulation is for modest numbers of calculations, computational speed is not an issue. This second formulation agrees with the industrial formulation within the tolerance of the industrial formulation. The general and scientific formulation is made available in the United States by NIST. IAPWS conducts a conference every five years, and the official properties formulations and releases as well as guidelines on other properties of steam and water are published in the conference proceedings (3).

Thermodynamic Properties. Ordinary water contains three isotopes of hydrogen [1333-74-0] (qv), ie, ^{1}H, ^{2}H, and ^{3}H, and three of oxygen [7782-44-7] (qv), ie, ^{16}O, ^{17}O, and ^{18}O. The bulk of water is composed of ^{1}H and ^{16}O. Tritium [15086-10-9], ^{3}H, and ^{17}O are present only in extremely minute concentrations, but there is about 200-ppm deuterium [16873-17-9], ^{2}H, and 1000-ppm ^{18}O in water and steam (see DEUTERIUM AND TRITIUM). The thermodynamic properties of heavy water are subtly different from those of ordinary water. IAPWS has special formulations for heavy water. The properties given herein are for ordinary water having the usual mix of isotopes.

Vapor Pressure. Vapor pressure is one of the most fundamental properties of steam. Figure 1 shows the vapor pressure as a function of temperature for temperatures between the melting point of water and the critical point. This line is called the saturation line. Liquid at the saturation line is called saturated

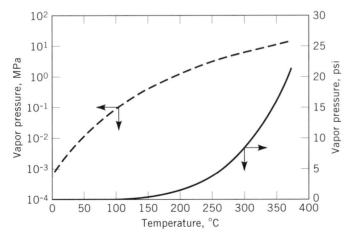

Fig. 1. Vapor pressure of ordinary water, where (——) represents linear and (– – –) logarithmic scale. To convert MPa to psi, multiply by 145.

liquid; liquid below the saturation line is called subcooled. Similarly, steam at the saturation line is saturated steam; steam at higher temperature is superheated. Properties of the liquid and vapor converge at the critical point, such that at temperatures above the critical point, there is only one fluid. Along the saturation line, the fraction of the fluid that is vapor is defined by its quality, which ranges from 0 to 100% steam.

Density.   The density of saturated water and steam is shown in Figure 2 as a function of temperature on the saturation line. As the temperature approaches the critical point, the densities of the liquid and vapor phase approach each other. This fact is crucial to boiler construction and steam purity because the efficiency of separation of water from steam depends on the density difference.

Internal Energy, Enthalpy, and Entropy.   The enthalpies and internal energies of steam and water on the saturation line are shown in Figure 3. Like the density, these converge at the critical point. The entropy along the saturation line is shown in Figure 4. Many plots of enthalpy, internal energy, and entropy are possible, but the Mollier chart (Fig. 4) combines these in a useful way. Before the advent of computer programs to generate values of properties, Mollier charts printed in very large sizes allowed accurate reading of values for equipment design. The heat capacity at constant pressure, C_p, defined as the derivative

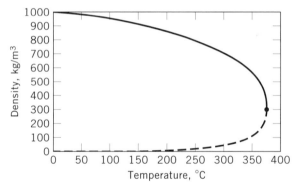

Fig. 2.   Density on saturation line of (——) water and (– – –) steam, where (●) represents the critical point.

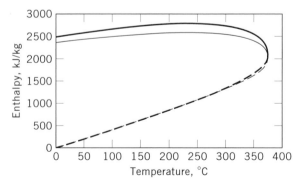

Fig. 3.   Enthalpy (bolded line) and internal energy of (– – –) water and (——) steam. To convert kJ to kcal, divide by 4.184.

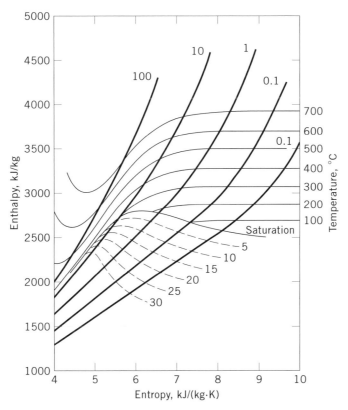

Fig. 4. Mollier chart, where the numbers on the solid lines represent pressure in MPa, and the values associated with the dashed lines represent percentages of moisture. To convert MPa to psi, multiply by 145. To convert kJ to kcal, divide by 4.184.

of enthalpy with respect to temperature, is shown in Figure 5. The value of C_p becomes very large in the vicinity of the critical point. The variation is much smaller for the heat capacity at constant volume, C_v, also given in Figure 5.

Transport Properties.   Viscosity, thermal conductivity, the speed of sound, and various combinations of these with other properties are called steam transport properties, which are important in engineering calculations. The speed of sound (Fig. 6) is important to choking phenomena, where the flow of steam is no longer simply related to the difference in pressure. Thermal conductivity (Fig. 7) is important to the design of heat-transfer apparatus (see HEAT-EXCHANGE TECHNOLOGY). The viscosity, ie, the resistance to flow under pressure, is shown in Figure 8. The sharp declines evident in each of these properties occur at the transition from liquid to gas phase, ie, from water to steam. The surface tension between water and steam is shown in Figure 9.

Miscellaneous Properties.   *Dielectric Constant.*   The dielectric constant, a physical property having great importance to the chemical properties of hot water and steam, is shown as a function of temperature in Figure 10. Along the saturation line, the steam and water values converge at the critical point. The ability of water to dissolve salts results from the high dielectric constant. The precipitous drop in water dielectric constant in the region of the critical point is

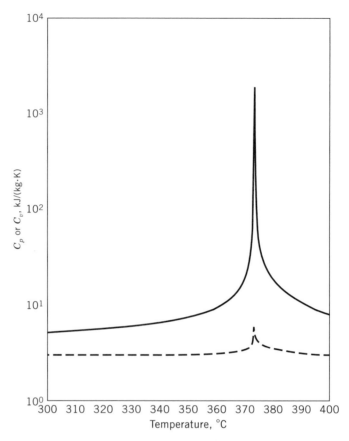

Fig. 5. Heat capacities at (——) constant pressure, C_p, and at (– – –) constant volume, C_v, on the 220-MPa (31,900-psi) isobar. To convert kJ to kcal, divide by 4.184.

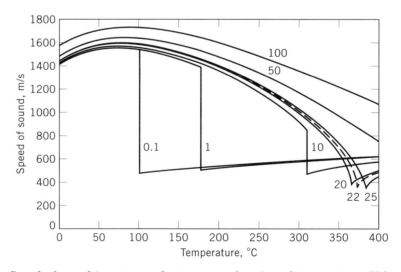

Fig. 6. Speed of sound in water and steam as a function of temperature. Values given correspond to pressures in MPa. To convert MPa to psi, multiply by 145.

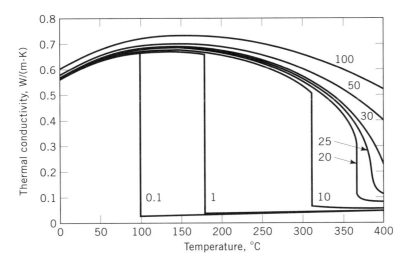

Fig. 7. Thermal conductivity of water and steam as a function of temperature. Values given correspond to pressures in MPa. To convert MPa to psi, multiply by 145.

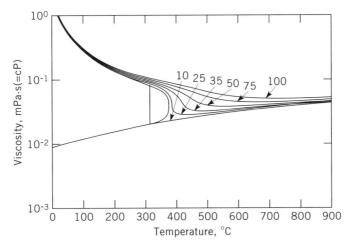

Fig. 8. Viscosity of water and steam as a function of temperature. Values given correspond to pressures in MPa. To convert MPa to psi, multiply by 145.

very important to the solubility of salts in water near the critical temperature. Many salts exhibit declining solubilities as the critical temperature is approached and then exceeded. The drop in dielectric constant is largely a result of the decline in density (see Fig. 2). The dielectric constant on the 1-kg/dm^3 isochore, ie, the density at 25°C, declines with the increase in temperature but not as much as does the dielectric constant on the saturation line. The isochore does not show the sharp drop at the critical temperature.

    Ion Product. The ion product of water is the product of the molality of the hydrogen and hydroxide ions, $K_w = m_{H^+} m_{OH^-}$. Its temperature variation is shown in Figure 11. The ion product increases with temperature to 250°C and then declines. The initial increase is the temperature effect, and the later decline

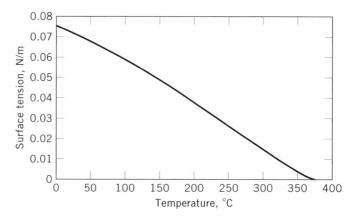

Fig. 9. Surface tension between liquid water and steam along the saturation line.

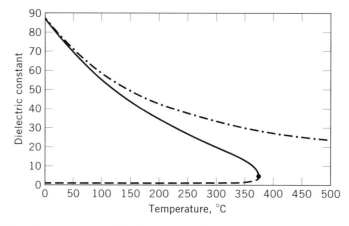

Fig. 10. Dielectric constants of (——) water and (— — —) steam on the saturation line, where (●) corresponds to the critical point, and (— · —) along the 1-kg/dm³ isochore.

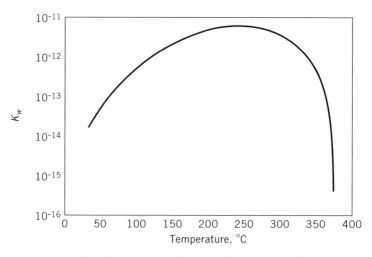

Fig. 11. Ion product of water at saturation line.

is on account of the decline in the dielectric constant of water. This variation means that neutral pH, which is the square root of the ion product, varies with temperature.

Chemical Properties

Molecular Nature of Steam. The molecular structure of steam is not as well known as that of ice or water. During the water–steam phase change, rotation of molecules and vibration of atoms within the water molecules do not change considerably, but translation movement increases, accounting for the volume increase when water is evaporated at subcritical pressures. There are indications that even in the steam phase some H_2O molecules are associated in small clusters of two or more molecules (4). Values for the dimerization enthalpy and entropy of water have been determined from measurements of the pressure dependence of the thermal conductivity of water vapor at 358–386 K (85–112°C) and 13.3–133.3 kPa (100–1000 torr). These measurements yield the estimated upper limits of equilibrium constants, K_n, for cluster formation in steam, where n is the number of molecules in a cluster.

n	K_n, $kPa^{-(n-1)}$
3	1.9×10^{-8}
4	5.8×10^{-11}
5	3.8×10^{-13}
6	2.8×10^{-15}

Solvent. The solvent properties of water and steam are a consequence of the dielectric constant. At 25°C, the dielectric constant of water is 78.4, which enables ready dissolution of salts. As the temperature increases, the dielectric constant decreases. At the critical point, the dielectric constant is only 2, which is similar to the dielectric constants of many organic compounds at 25°C. The solubility of many salts declines at high temperatures. As a consequence, steam is a poor solvent for salts. However, at the critical point and above, water is a good solvent for organic molecules.

Solubility of Salts in Steam. Although the solubility of salts in steam is small, it has great significance to corrosion of steam system components, particularly steam turbines. Much of this information, also of interest to geochemists and cosmochemists, is published in geochemistry journals such as *Geochimica et Cosmochimica Acta*. There has been considerable investigation of the equilibrium distribution of salts across the vapor–liquid phase boundary. Historically, this work was entirely for single solute systems. For many solutes, the distribution ratio in dilute solutions can be described by equation 1:

$$\log D = m \log(\rho_v/\rho_l) \tag{1}$$

where $D = c_v/c_l$, the ratio of the concentration of solute in the vapor phase, c_v, to the concentration of solute in the liquid phase, c_l; ρ_v is the density of the vapor; ρ_l, the density of the liquid; and m, a coefficient, which has been related to a coordination number (5). In careful work, the change in the critical point pressure and temperature with concentration of dissolved species must be considered. The Ray diagram (Fig. 12) shows the distribution of solutes as a function of density ratio for various metal oxides, salts, and silica (5,6). The slopes in Figure 12 are the m values for the distribution equation (eq. 1). At the critical point, the distribution ratio is 1. As the density ratio declines, the distribution ratio declines. The data were collected at relatively high concentrations and for pure materials only. Dissociation and common ion effects are not included in this treatment. When the system contains more than one solute, the distribution cannot be represented so simply, as common ion effects occur. Only a few of such systems have been studied.

Metal Oxides.   The metal oxides tend to have very low solubilities at the critical point, evidenced by the small slope in Figure 12. For the most part, these

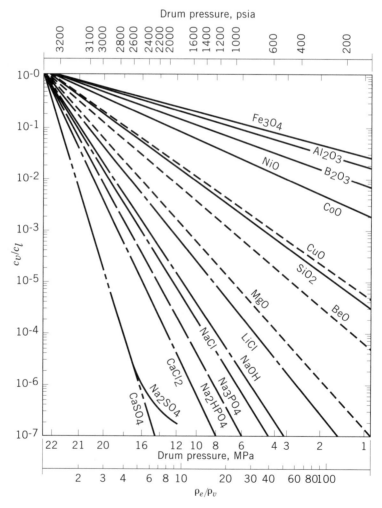

Fig. 12.   Ray diagram of carryover coefficients of salts and metal oxide contaminants in steam (6). To convert MPa to psi, multiply by 145.

oxides may be considered as solid suspensions. Copper oxides and alumina under some conditions are exceptions to this rule. Solubilities of the metal oxides are available (7).

Ammonium Chloride.   Work on the distribution of ammonium chloride [12125-02-9] between the vapor and liquid phases (8) suggests that the Ray diagram is sometimes an oversimplification. In most steam systems, there is much more ammonia than any other impurity. In particular, there is more ammonia than hydrogen chloride. The volatility of ammonium chloride is therefore expressed by the following chemical equation:

$$\mathrm{NH_4^+\ (aq) + Cl^-\ (aq) \longrightarrow NH_4Cl\ (g)} \qquad K = a(\mathrm{NH_4Cl})/a(\mathrm{NH_4^+})a(\mathrm{Cl^-})$$

where a is the activity of the species in the parentheses. The distribution coefficient for ammonium chloride is shown in Figure 13 for equation 2:

$$\log K = -20.24 + 18207/T - 10.48 \log(\rho_l/\rho_v) \tag{2}$$

where T is in K and the densities are of pure water or steam. Investigations of sodium salts at low concentrations suggest that similar treatments may be required for most species (8).

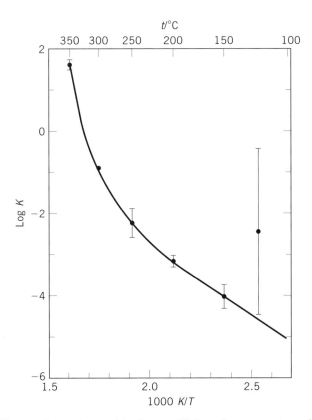

Fig. 13.   Thermodynamic partitioning coefficient for ammonium chloride (8).

Sodium Chloride.   Sodium chloride, a corrodent of many materials, is the archetype contaminant and has been studied more than other salts. The solubility of sodium chloride in superheated steam is shown at the conditions of a typical steam turbine expansion in Figure 14. The solubilities were measured in the region of higher solubility (9). As the steam expands, sodium chloride becomes considerably less soluble. The solubility, S, in parts per billion (ppb) can be represented by equation 3:

$$\log S = 3.36 \log V + 1760/T + 4.45 \tag{3}$$

where V is the molar volume of steam in L/mol and T is the temperature in K. There is a combined effect of decreasing density and decreasing temperature (9).

Silica.   Silica is not actually a corrodent of turbines. However, it can deposit on and cause blocking of turbine passages, thus reducing turbine capacity and efficiency. As little as 76 μm (3 mils) of deposit can cause measurable loss in turbine efficiency. Severe deposition can also cause imbalance of the turbine and vibration. The solubility in steam and water is shown in Figure 15, as is a typical steam turbine expansion. Silica is not a problem except in low pressure turbines unless the concentrations are extraordinarily high.

Moderately Volatile Materials.   For moderately volatile materials, such as the amines commonly used in feedwater and boiler water chemical treatment, the distribution ratios vary from 0.1 to 30; for gases, the ratios are much higher. The distribution ratios of amines and organic acids are generally temperature-dependent. The distribution ratios for ammonia [7664-41-7], morpholine [110-91-8], and acetic acid [64-19-7] are shown in Figure 16 as examples.

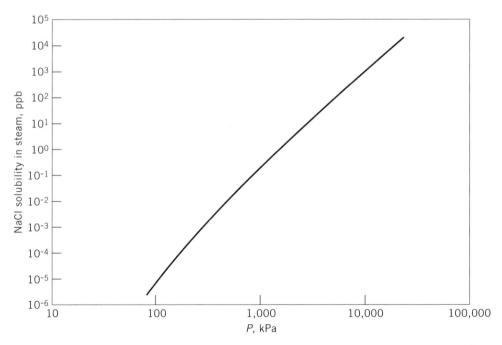

Fig. 14.   Solubility of NaCl as a function of pressure on typical turbine expansion line (9). To convert kPa to psi, multiply by 0.145.

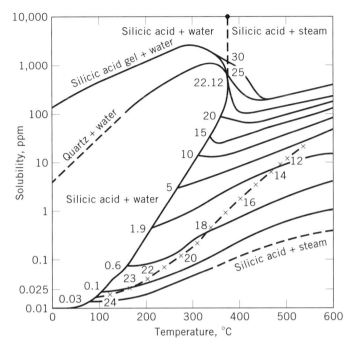

Fig. 15. Silica solubility diagram, where (–×–) is the turbine expansion line and (●) is the critical point (10). Numbers represent pressure in MPa. To convert MPa to psi, multiply by 145.

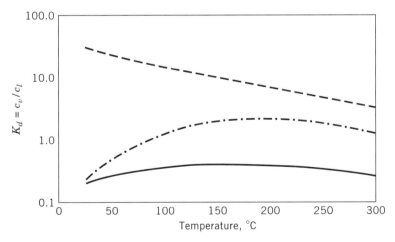

Fig. 16. Distribution of (–––) ammonia, (–·–) morpholine, and (——) acetic acid between water and steam (7).

Gases. At low temperatures and pressures, most gases are relatively insoluble in water and tend to appear in the steam phase. Only those gases that ionize to some extent violate this rule. However, as the pressure approaches the critical point, the solubility of gases increases.

Reactant. Steam can behave as an oxidant. The partial pressure of oxygen generated by the dissociation of steam into hydrogen and oxygen is shown in Figure 17.

Steam reacts with salts so that the salts dissociate into the respective hydroxide and acid. For sodium salts, the sodium hydroxide is largely in a liquid solution and the acid is volatile. Figure 18 shows the concentration of hydrochloric acid [7647-01-0], HCl, in steam owing to hydrolysis of sodium chloride. Although the amount is not large, it can be measured (9).

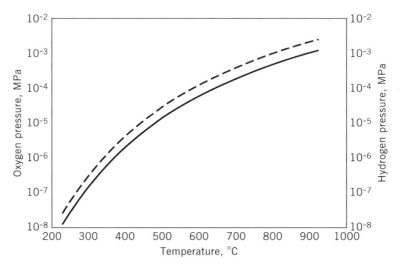

Fig. 17. Oxygen (——) and hydrogen (– – –) partial pressure in steam at 3 MPa (425 psi) (10). To convert MPa to psi, multiply by 145.

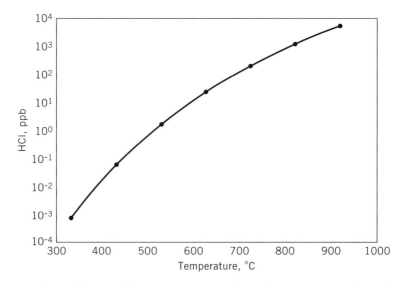

Fig. 18. Hydrochloric acid in steam owing to hydrolysis of sodium chloride (10).

Generation

Equipment. *Simplified Cycle.* A simplified fossil steam cycle appears in Figure 19. The water accumulates in the bottom of the condenser, called the hotwell. It goes through a feed pump to pressurize it. The pressurized water passes through one or more feedwater heaters, which raise the temperature. The water then enters the boiler where heat from the fuel converts it to steam. The steam expands through the engine, usually a turbine, which extracts work. In the middle of the turbine some of the steam is extracted to supply heat to the feedwater heater. The remainder expands through the turbine and is condensed. The rejected heat is carried away by the condenser coolant, which is usually water, but sometimes air. The condensed steam then returns to the hotwell to repeat the cycle.

Steam Generators or Boilers. Steam is produced in a boiler or steam generator. The term steam generator is used when the heat source is nuclear power (see NUCLEAR REACTORS) and is often used for fossil-fired boilers, particularly supercritical once-through units, where the fluid changes gradually from liquid-like to vapor-like properties without really boiling. Boilers using hot gas as a heat source are generally called heat recovery steam generators (HRSGs). The design (12,13) of steam generators and boilers is complicated and is treated herein in an overview.

In fossil fuel-fired boilers there are two regions defined by the mode of heat transfer. Fuel is burned in the furnace or radiant section of the boiler. The walls of this section of the boiler are constructed of vertical, or near vertical, tubes in which water is boiled. Heat is transferred radiatively from the fire to the waterwall of the boiler. When the hot gas leaves the radiant section of the boiler, it goes to the convective section. In the convective section, heat is transferred to tubes in the gas path. Superheating and reheating are in the convective section of the boiler. The economizer, which can be considered as a gas-heated feedwater heater, is the last element in the convective zone of the boiler.

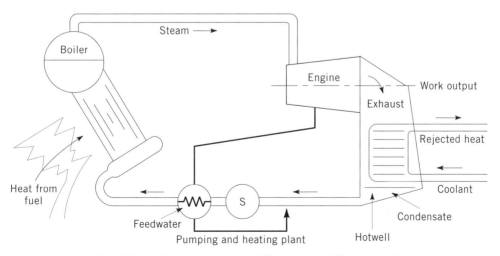

Fig. 19. Schematic of a simplified steam (S) cycle (12).

Recirculating or Drum Boilers.   There are two basic types of boilers: recirculating and once-through. The difference between the boiler types is the manner in which heat is absorbed from the fuel to generate steam. In a recirculating boiler, shown schematically in Figure 20, water enters the boiler at

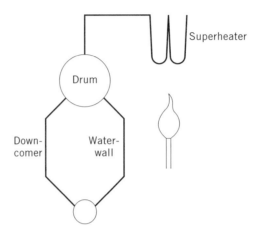

Fig. 20.   Schematic of a simplified fossil-fuel-fired recirculating steam generator.

the bottom of the waterwall. Approximately 25 to 33% of the water boils by the time it reaches the top of the waterwall. In the steam drum, the steam is separated from the remaining water. The mixture of steam and water passes through steam-separating devices, eg, baffles, cyclones, or corrugated plates, where the direction of flow changes rapidly. Steam follows the change in direction but water hits the metal and is drained away. The efficiency of this process depends on the density difference between the steam and the water. The density difference decreases with increasing pressure, as does the efficiency of the separation. The steam goes to the superheater and then to the turbine or process. The water separated from the steam is recirculated to the bottom of the boiler. The water flow in a recirculating boiler may be either convective or forced. Convective flow relies on the difference in the average densities of the water in the downcomer and the water and steam in the waterwall. In forced circulation, the natural circulation is aided by a pump. Because the steam is separated from the water, nonvolatile impurities, usually solids, concentrate in the water. A small portion of the water, usually 2% or less, is removed (blown down) to reduce these impurities. Blowdown is a dilution process because the concentration of the impurities in the makeup water is lower than in the boiler water.

Once-Through Steam Generators.   In a once-through system the feedwater enters the steam generator at the bottom of the waterwall and passes out of the boiler through the superheater. Most once-through steam generators are supercritical (>23 MPa (3300 psia)). Once-through steam generators require extremely pure feedwater, because no purification of the water occurs in the boiler. Subcritical once-through boilers require the same high purity feedwater, but do not provide as high efficiency as the supercritical unit.

Nuclear Steam Generators.   There are three basic types of nuclear heat sources. Most common are pressurized water reactors having steam generators (Fig. 21). In this system, the nuclear reactor heats water in a high pressure loop,

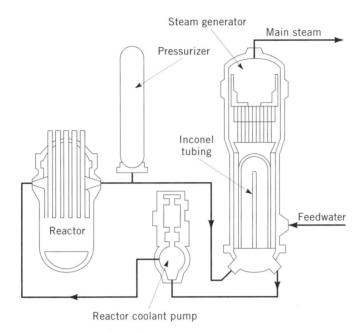

Fig. 21. Schematic of a pressurized-water-loop reactor coolant system.

typically 14.1 to 17.2 MPa (2050–2500 psia). The water is circulated through tubes in a steam generator. On the outside of the tubes, water is boiled to steam which goes to the steam turbine. In recirculating steam generators, the steam is saturated. In once-through steam generators, the steam has about 28°C (50°F) superheat. In boiling-water reactors, the second type, the nuclear heat is used to boil the feedwater directly. In gas-cooled reactors, the third type, the gas takes the same role as the pressurized water in a pressurized-water reactor (PWR) and transfers the heat to the steam generator. Typical nuclear turbine inlet conditions range from 5.0 to 7.3 MPa (725 to 1059 psia) at saturation temperatures of 264–289°C (507–552°F) and approximately 0.25% moisture.

Heat Recovery Steam Generators. Heat recovery steam generators, a special class of boilers where essentially all heat transfer takes place convectively, are used to extract energy from hot gas streams. One of the principal uses is extraction of heat from the exhaust gases of a combustion turbine. Turbine exhaust gas is typically 540°C (1004°F). Careful design of the HRSG allows the gas leaving the HRSG to approach the feedwater temperature within 28°C (50°F). High heat-transfer rates in modern HRSGs are assisted by finned tubing.

Feedwater Heaters. Feedwater heaters use steam from the turbine to preheat the feedwater before it reaches the boiler. Feedwater heaters increase efficiency of steam cycles because heat comes from a source having a lower temperature than the fire in the boiler. Although feedwater heaters may be either shell and tube or direct-contact, most are of shell-and-tube construction. Water flows through the inside of the tubes. Steam is admitted to the shell where it condenses, warming the water inside the tubes. The steam entering the feedwater heater first passes over the tubes containing the water that is about to leave the heater, then to tubes containing the cooler water entering the heater. The condensed steam is collected in a drain, which is either cascaded to a lower temperature feedwater heater or collected and pumped forward into the main feedwater stream. In a direct-contact or deaerating heater, the water is sprayed through a steam space and flows down trays in thin layers. Steam is admitted at the bottom and flows upward countercurrent to the water. Dissolved gases are removed in direct-contact heaters and must be vented.

Condenser. Water-cooled condensers are constructed of tubes between tubesheets (Fig. 22). Cooling water is pumped from the source into the inlet waterbox, through the tubes, and through the outlet waterbox. Tight joints between the tubes and tubesheet are necessary to prevent ingress of the cooling water. Total cooling water leakage as small as 4 L/d of fresh water can produce unacceptable impurity concentrations in feedwater for nuclear steam generators. Careful selection of condenser materials is required for long-term reliability. Double tubesheets having condensate between the two sheets are a method of reducing the likelihood of cooling water leaks. The cooling water may be drawn from lakes or rivers or may be recirculated from cooling towers. Cooling water chemistry is complex because of precipitation and microbiological considerations. It is treated in handbooks by water treatment chemical vendors (22–24) (see WATER, INDUSTRIAL WATER TREATMENT).

Air-cooled condensers are used more frequently in the 1990s as cooling water sources are being exhausted. Air-cooled condensers consist of large numbers of finned tubes. Air is forced past the outside of the tubes and the steam condenses on the inside. Cooling water leakage is not a problem when air-cooled condensers are used, but air in-leakage is a bigger problem than for water-cooled condensers.

Water Chemistry in Steam-Generating Systems. *Specifications.* *Steam Purity.* The usual function of steam purity limits is to protect the turbine from deposition and subsequent corrosion. In systems where the steam is used

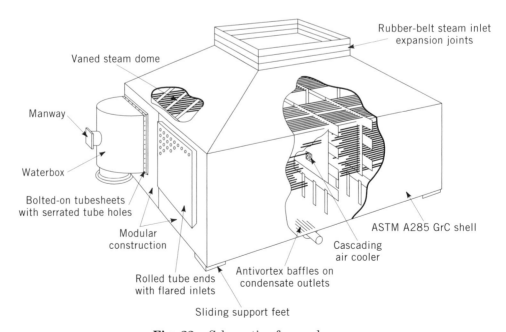

Fig. 22. Schematic of a condenser.

for chemical processes, the specific process may create additional requirements for steam purity. For instance, in food processing (qv), regulations may limit or prohibit hydrazine in the steam, so ascorbic acid or sodium sulfite must be used as an oxygen scavenger. When steam is sent to processes and returned as condensate, it may be necessary to add pH control agents to the steam to control corrosion at various points in the process and condensate return system. Table 1 gives typical steam purity recommendations for steam turbine protection. The recommendations vary, depending on whether the feedwater treatment generates an oxidizing or a reducing environment. In addition to the normal limits, most vendors have time-limited excursion ranges, which provide tolerance for upset conditions and the higher impurity concentrations commonly found on the startup of a boiler system. The recommendations are derived from a blend of theoretical considerations of solubility of salts in steam, practical limits on water purity, and experience with what has proved successful. The development of steam purity limits is available (14,15).

Boiler Water. The steam purity limits define boiler-water limits because the steam cannot be purified once it leaves the boiler. For a once-through boiler, the boiler water must have the same specifications as the steam. A recirculating boiler is a still, and there can be considerable purification of the steam as it boils and is separated from the water in the steam drum. The process of separation is not perfect, however, and some water is entrained in the steam. This water, called mechanical carryover, contains impurities in the same proportions as the boiler water, and its contribution to steam impurity is in those proportions. Typical mechanical carryover is less than 0.25% and often less than 0.1%, but operating conditions in the boiler can affect the mechanical carryover. In addition to mechanical carryover, chemicals can be carried into the steam because of solubility. This is called vaporous carryover. Total carryover is the sum of mechanical and vaporous carryover. The boiler-water specification must be such that the total carryover conforms to the steam purity requirements. For salts, such as sodium phosphate and sodium chloride, vaporous carryover is not a significant problem below approximately 15 MPa (2175 psia). As boiler pressures approach the critical point, vaporous carryover increases rapidly. Above 15 MPa (150 bar), boiler solids concentrations must be carefully controlled to minimize vaporous carryover. Most boilers operating over 18 MPa (180 bar) use all volatile

Table 1. Steam Purity Recommendations

Parameter	Environments	
	Reducing	Oxidizing
conductivity, μS/cm		
hydrogen cation-exchanged	0.1–0.3	<0.2
degassed cation	0.15–0.35	
sodium, ppb	3–10	
silica, ppb	10–20	
iron, ppb	20	
copper, ppb	2	
chloride, ppb	5–10	
sulfate, ppb	3–6	
oxygen, ppb		50–150

treatment to prevent deposition of salts in turbines. Boiler-water limits for utility boiler are listed in Table 2. Recommendations from American Boiler Manufacturers' Association (ABMA) for boiler-water limits for drum-type boilers and associated steam purity for watertube boilers are listed in Table 3.

In addition to the requirement to conform to steam purity needs, there are concerns that the boiler water not corrode the boiler tubes nor produce deposits, known as scale, on these tubes. Three important components of boiler tube scale are iron oxides, copper oxides, and calcium salts, particularly calcium carbonate [471-34-1]. Calcium carbonate in the feedwater tends to produce a hard, tenacious deposit. Sodium phosphate is often added to the water of recirculating boilers to change the precipitate from calcium carbonate to calcium phosphate (see also WATER, INDUSTRIAL WATER TREATMENT).

Feedwater. The feedwater for a steam cycle must be purified. The degree of purity depends on the pressure of the boiler. Higher pressure boilers require higher feedwater purity. There is some trade-off between feedwater purity and boiler blowdown rate. However, increasing blowdown rate to compensate for lower feedwater purity is expensive, because blowdown water has been heated to the saturation temperature. Typical feedwater specifications for utility boilers are given in Table 4. To some extent turbine steam purity requirements determine the feedwater purity requirements. The boiler-water silica required to maintain adequate steam purity for higher pressure steam turbines is considerably less than the boiler could tolerate if deposition in the boiler were the only issue.

Makeup. Makeup water is the water supplied to replenish the steam system for any losses. In most systems it is introduced into the condenser or the feed pump suction. In steam systems where the makeup is a small fraction of the total feedwater, its purity may be somewhat lower than the feedwater requirement because it is diluted by condensate. In systems where there is little condensate return, such as heating steam supplies, the makeup purity must be essentially the same as the feedwater.

Water Treatment. Water and steam chemistry must be rigorously controlled to prevent deposition of impurities and corrosion of the steam cycle. Deposition on boiler tubing walls reduces heat transfer and can lead to overheating, creep, and eventual failure. Additionally, corrosion can develop under the deposits and lead to failure. If steam is used for chemical processes or as a heat-transfer medium for food and pharmaceutical preparation there are limitations on the additives that may be used. Steam purity requirements set the allowable impurity concentrations for the rest of most cycles. Once contaminants enter the steam, there is no practical way to remove them. Thus all purification must be

Table 2. Boiler-Water Limits for Utility Boiler[a,b]

Elements, ppm	All-volatile treatment	Phosphate
sodium	0.7	1.7
chloride	0.28	0.9
sulfate	0.28	1.6
silica	13	13
phosphate		2.2

[a]At 17.2 MPa (2500 psia).   [b]Ref. 14.

Table 3. ABMA Recommended Boiler-Water Limits[a]

Drum pressure, MPa[b]	Boiler-water limits, ppm			Steam limits, ppm
	Total dissolved solids[c]	Total alkalinity	Suspended solids[c]	Expected total dissolved solids[c,d]
	Drum boilers			
0.1–2.2	700–3500	140–170	15	0.2–1.0
2.2–3.2	600–3000	120–600	10	0.2–1.0
3.2–4.2	500–2500	100–500	8	0.2–1.0
4.2–5.3	400–2000	80–400	6	0.2–1.0
5.3–6.3	300–1500	60–300	4	0.2–1.0
6.3–7.0	250–1250	50–250	2	0.2–1.0
7.0–12.5	100	e	1	0.1
12.5–16.3	50		f	0.1
16.3–18	25		f	0.05
18.0–20.1	15		f	0.05
	Once-through boilers			
≥9.8	0.05	f	f	0.05

[a]Ref. 16.
[b]To convert MPa to psi, multiply by 145.
[c]Values are maximum.
[d]Values are exclusive of silica.
[e]Dictated by boiler-water treatment.
[f]Not applicable.

739

Table 4. Utility Feedwater Specifications for Normal Operation

Parameter	Reducing chemistry		Oxidizing chemistry
	Recirculating boiler	Once-through boiler	Once-through boiler
pH			
all-ferrous metallurgy	9.0–9.6	9.0–9.6	8.0–8.5
mixed Fe–Cu metallurgy	8.8–9.3	8.8–9.3	
cation-exchanged conductivity, μS/cm	≤0.2	≤0.2	≤0.15
iron, ppb	≤10	≤10	≤5
copper, ppb	≤2	≤2	
oxygen, ppb	≤5	≤5	50–150

[a]Refs. 14 and 17.

carried out in the boiler or preboiler part of the cycle. The principal exception is in the case of nuclear steam generators, which require very pure water. These tend to provide steam that is considerably lower in most impurities than the turbine requires. A variety of water treatments are summarized in Table 5. Although the subtleties of water treatment in steam systems are beyond the scope of this article, uses of various additives may be summarized as follows:

Water treatment additive	Effect
sodium orthophosphate	pH control; hardness precipitation
sodium hydroxide, lithium hydroxide	pH control (acid neutralization)
neutralizing volatile amines (ammonia, morpholine, cyclohexylamine, diethanolamine, ethanolamine, etc)	once-through and pressurized-water reactor cycles, high heat flux steam generators; control of preboiler corrosion-product generation and transport; no control of scale formation by feedwater contaminants
hydrazine	oxygen scavenging
sodium sulfite	oxygen scavenging
sludge conditioners	dispersion of sludge for easy removal by blowdown in drum boilers, inhibiting of scale formation
synthetic sulfonated polymer, synthetic carboxylated polymer, polyacrylic acid	<6.9 MPa (1000 psi)
carboxymethyl cellulose, organophosphonate	<4.1 MPa (600 psi)
lignin	<2.1 MPa (300 psi)
oxygen or hydrogen peroxide	improve surface passivation in all-ferrous high purity water systems
filming amines (octadecylamine and some of its salts)	surface protection against condensate corrosion
antifoams (polyglycols, polyamides)	reduce foaming and carryover in boilers

Table 5. Water Treatment Schemes

Program	Characteristics	
	Favorable	Unfavorable
phosphate		
conventional, Na_3PO_4(+NaOH)	hardness salts converted to a form readily removed by bottom blowdown; relatively high levels of suspended solids successfully controlled; acids neutralized; surface passivation by PO_4^{3-}	high pressure boilers cannot tolerate intentional formation of boiler sludge; required alkalinity levels are too high for operation >10.4 MPa (1500 psig); oil or organic contamination produces highly adherent deposit; possible under-deposit corrosion by NaOH
coordinated, [Na]:[PO_4] < 3	caustic corrosion may be eliminated; deposit form makes for easy removal; produces low solids levels and high steam purity; acids neutralized; surface passivation by PO_4^{3-}	in boilers containing deposits, chemical interaction of iron and phosphate can lead to caustic corrosion; at very low [Na]:[PO_4] molar ratios (<2.1), corrosion by phosphoric acid possible
congruent, 2.6 < [Na]:[PO_4] < 2.8	caustic and acid corrosion eliminated; deposit form makes for easy removal; produces low solids levels and high steam purity; acids neutralized; surface passivation by PO_4^{3-}	control of the [Na]:[PO_4] molar ratio may be difficult; continuous feed and blowdown may be required
NaOH or LiOH	acid neutralization	can cause rapid corrosion when concentrated in high quality regions or under deposits; vaporous carryover in steam at higher pressures; difficult to analyze
chelates, ethylenediaminetetraacetic acid (EDTA), nitrilotriacetic acid (NTA)	optimum heat-transfer and boiler efficiency obtained under good feedwater-quality conditions; elimination of boiler sludge prevents formation of adherent deposits involving oil/organics	inability to analyze free chelant residual accurately can lead to overfeed and subsequent corrosion[a]; does not complex iron or copper under normal boiler-water pH conditions; presence of oxygen in boiler water can cause dechelation and excessive corrosion; limited to lower pressure boilers (see text)

Table 5. (*Continued*)

Program	Characteristics	
	Favorable	Unfavorable
all-volatile[b]	near-zero solids in boiler water and high purity steam realized under ideal feedwater conditions; having some corrosion protection; no carryover of solids; no surface concentration; condenser leakage detection by sodium measurement; boiler deposition of corrosion products easy to remove by chemical cleaning	feedwater contamination may exceed inhibiting ability of volatile fed, leading to boiler corrosion; introduction of contaminants into feed-water produces deposits that may be hard to remove; marginal acid neutralization, particularly when using ammonia in wet steam regions; interference with condensate polishing; corrosion of copper alloys by ammonia and oxygen; organic acid decom-position products of organic amines
oxygenated treatment	no interference of additives with condensate polishing; low corrosion rates of ferritic steels	requires extremely low concentrations of impurities in feed-water; no corrosion protection in case of an upset; corrosion of copper alloys; precise control required

[a]Condition is heightened when treatment is applied to deposit-bearing boiler.
[b]Such as ammonia, morpholine, cyclohexylamine, and ethanolamine (ETA).

Reducing Chemistry. Most steam systems are maintained in a reducing state to avoid oxidation of the steel piping and other components. Oxidation is further suppressed by raising the pH, which is commonly controlled by using ammonia, although organic amines, such as morpholine, cyclohexylamine, ethanolamine, and dimethylamine, are also used. Organic amines are not generally employed for systems having superheat temperatures $\geq538°C$ (1000°F) because these tend to decompose to organic acids, which can be corrosive. Most feedwater is treated with hydrazine to reduce the oxygen concentration to the 1–5-ppb range. Carbohydrazide, hydroquinone, and methylethylketoxime are also used as oxygen scavengers. In plants where the steam may be in contact with food, sodium sulfite, ascorbic acid, or erythorbic acid is commonly used as an oxygen scavenger.

Feedwater treatment is designed to protect the feedwater system and, to some extent, the boiler. Most systems contain carbon steel piping. Carbon steel corrosion (Fig. 23**a**) is considerably slower at a pH between 9.0 and 11.0. In

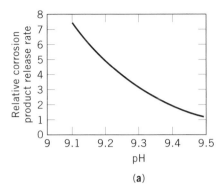 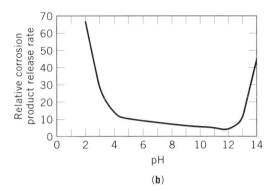

Fig. 23. Corrosion rates of carbon steel as a function of pH of (**a**) the feedwater (18) and (**b**) boiler conditions (19).

all-ferrous feedwater systems, the preferred pH range is therefore 9.2 to 9.6, although some systems are operated at a pH as high as 10. In systems where copper alloys are present, high concentrations of ammonia accelerate corrosion of the copper alloys. In those systems the preferred pH is 8.8–9.2.

For once-through boilers, the treatment must be without solid residues, so all-volatile treatment (AVT) is used. AVT, which is also used in some drum boiler systems, relies on the feedwater chemical additives, ammonia and hydrazine, to provide water appropriate to the boiler. Because the additives are volatile, they do not accumulate in the boiler and provide only minimal protection during contaminant ingress. Most plants using AVT have some form of condensate polisher to remove impurities from the condensate.

Boiler tubes are susceptible to corrosion owing to contaminant ingress. Adding solid buffers to the boiler water can reduce corrosion. Figure 23**b** shows the rate of carbon steel corrosion as a function of pH at boiler conditions. If feedwater impurities or faulty chemical additions cause the boiler pH to move outside the acceptable range, corrosion can be rapid and boiler tube failure may occur in a matter of hours. If the boiler does not fail immediately, it can also be damaged by hydrogen embrittlement, which makes it subject to failures during thermal transients, such as startup.

In drum boilers sodium hydroxide (caustic), sodium phosphate, or both may be added for pH and scale control. Sodium hydroxide is used more in Europe than in the United States, where sodium phosphate treatment is usually preferred. In boilers operating above 4 MPa (580 psia), caustic concentrations must be carefully controlled to prevent highly corrosive deposits from forming. In the lowest pressure boilers, phosphate treatment may be used to compensate for lower purity feedwater. As the boiler pressure increases, the allowable phosphate concentration decreases, and at 16.5 MPa (2400 psia) or above, equilibrium phosphate treatment may be used. In this treatment, caustic is added to a low phosphate concentration in the boiler to maintain the proper pH (20).

In lower pressure boilers a variety of additional treatments may be appropriate, particularly if the steam is used in chemical process or other nonturbine application. Chelants and sludge conditioners are employed to condition scale and enable the use of less pure feedwater. When the drum pressure is less than

7 MPa (1015 psia), sodium sulfite may be added directly to the boiler water as an oxygen scavenger. It has minimal effect on the oxygen concentration in the system before the boiler.

The selection of boiler-water treatment is also dependent on the type of cooling water. When cooling water reaches the boiler, various compounds precipitate before others. For instance, seawater contains considerable magnesium chloride. When the magnesium precipitates as the hydroxide, hydrochloric acid remains. In some lake waters, calcium carbonate is a significant impurity. When it reaches the boiler, carbon dioxide is driven off in the steam and calcium hydroxide is formed. If the cooling water tends to form acid in the boiler, either caustic or phosphate may be added to counteract the effect. When the cooling water tends to form base, only phosphate treatment is appropriate. Many drum boilers that are operated on AVT have provision for phosphate or caustic treatment during condenser leaks because the amines used in AVT have neither significant buffering capacity nor precipitate conditioning properties in the boiler.

Oxidizing Chemistry.   In high pressure boilers systems having the ability to maintain hydrogen cation-exchanged conductivity near or below 0.1 μS/cm, the feedwater may be treated with oxygen. Oxygen is added either as gaseous oxygen or as hydrogen peroxide. The pH may be neutral or elevated with ammonia. The goal of this treatment is to maintain all iron alloy surfaces in a passivated state. The rate of carbon steel corrosion as a function of oxygen is shown in Figure 24. At the low (0.1 μS/cm) cation conductivity, the corrosion rate drops over an order of magnitude as the oxygen concentration is increased from 0.02 to 0.2 ppm. When higher anion concentrations inhibit passivation, the increased oxygen causes faster corrosion. Oxygenated water treatment has been used in Germany since the 1970s and is becoming widespread throughout the world. This treatment regime requires systems having no copper alloys. This practice is most advantageous for once-through boiler systems. In recirculating boiler systems, it is very difficult to maintain the hydrogen cation-exchanged conductivity low enough in the boiler water. Thus the oxygen must be flashed in the drum to prevent boiler corrosion. Nonetheless, some recirculating systems have been successfully operated on oxygenated water treatment.

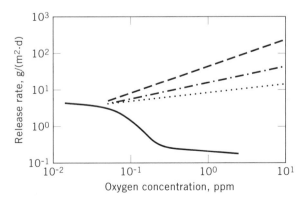

Fig. 24. Corrosion rates of carbon steel as a function of oxygen concentration at cation conductivity values of (——) 0.1, (⋯⋯) 7, (– ⋅ –) 87, and (– – –) 850 μS/cm (21).

Other Water Treatment Issues. *Attemperation.* Maintaining steam temperatures at correct values may require a process called attemperation, where water is sprayed into the inlet of the superheater or reheater to lower the temperature of the entering and, consequently, exiting steam. Attemperation bypasses the drum of a recirculating boiler. Any impurities in the feedwater are sent directly to the turbine. Copper fouling of turbines is commonly associated with high rates of attemperation. Attemperation can also allow salts from a condenser leak to bypass the boiler and deposit on the turbine.

Condensate Polishing. In order to maintain the feedwater purity required for once-through boilers, but also as an aid to maintaining feedwater purity for drum boilers and nuclear steam generators, condensate polishers are used. Condensate polishers are large ion-exchange (qv) systems designed to pass high flows. These reduce contamination, but do not protect a system against large contaminant ingress. When there is massive contaminant ingress, condensate polishers provide time for an orderly shutdown.

Makeup. Makeup treatment depends extensively on the source water. Some steam systems use municipal water as a source. These systems may require dechlorination followed by reverse osmosis (qv) and ion exchange. Other systems use wellwater. In hard water areas, these systems include softening before further purification. Surface waters may require removal of suspended solids by sedimentation (qv), coagulation, flocculation, and filtration. Calcium may be reduced by precipitation softening or lime softening. Organic contaminants can be removed by absorption on activated carbon. Details of makeup water treatment may be found in many handbooks (22–24) as well as in technical literature from water treatment chemical suppliers.

Water and Steam Purity Monitoring. To maintain appropriate steam and water purity requires analysis at the concentrations of interest. Details of monitoring systems, including the need for special nozzles to sample steam can be found in a supplement to the ASME Power Test Code (25). Feedwater and steam purity for most boiler systems approaches the detection limit of on-line monitoring instruments. One of the most reliable monitoring devices is electrical conductivity. Pure water has a resistivity of 18 MΩ·cm or a conductivity of 0.055 μS/cm (1 siemen = 1/Ω). Ammonia or an amine added to raise the pH also raises the conductivity. The agreement between the pH predicted from measured conductivity and the measured values of pH is one measure of water purity. By passing the sample stream through a cation resin in the hydrogen form, the ammonia or amine is removed and all the other cations are converted to hydrogen. The conductivity of the resulting stream (hydrogen-cation-exchanged conductivity, commonly called cation conductivity) is a measure of the total anions in the solution. Carbon dioxide may contribute a large fraction of this conductivity. Degassing the stream after the cation exchange removes the carbon dioxide and other volatile species. The resulting conductivity is essentially a measure of the dissolved solids. Conductivities less than 1 μS/cm should be measured on continuous streams with on-line instrumentation.

Sodium and chloride may be measured using ion-selective electrodes (see ELECTROANALYTICAL TECHNIQUES). On-line monitors exist for these ions. Silica and phosphate may be monitored colorimetrically. Iron is usually monitored by analysis of filters that have had a measured amount of water flow through them. Chloride, sulfate, phosphate, and other anions may be monitored by ion chromatography using chemical suppression. On-line ion chromatography is used at many nuclear power plants.

Uses

Power Production. Steam cycles for generation of electric power use various types of boilers, steam generators, and nuclear reactors; operate at subcritical or supercritical pressures; and use makeup and often also condensate water purification systems as well as chemical additives for feedwater and boiler-water treatment. These cycles are designed to maximize cycle efficiency and reliability. The fuel distribution of sources installed in the United States from 1990–1995 are as follow: coal, 45%; combined cycle, 27%; miscellaneous, 14%; nuclear, 11%; solar, oil, and geothermal, 1% each; and natural gas, 0.3%. The 1995 summer peak generation in the United States was 620 GW (26). The combined cycle plants are predominantly fired by natural gas. The miscellaneous sources include bagasse, black liquor from paper mills, landfill gas, and refuse (see FUELS FROM BIOMASS; FUELS FROM WASTE).

Modern power cycles have turbine throttle pressures of 16.3 MPa (2400 psig) or 24.1 MPa (3500 psig). The lower pressure is used for subcritical recirculating boilers, the higher pressure for supercritical once-through boilers. A cycle schematic for a typical 16.3 MPa (2400 psig) plant is shown in Figure 25. The number of feedwater heaters, which may vary between five and seven and is occasionally eight, is a compromise between efficiency and capital cost. The cycle schematic shows P, T, and flows at multiple locations. Fossil turbines generally operate at 3600 rpm (60 Hz) or 3000 rpm (50 Hz) using two-pole generators. Nuclear turbines, and some older large fossil low pressure turbines, generally operate at 1800 (60 Hz) or 1500 (50 Hz) using four-pole generators, which produce double the frequency of two-pole generators for a given speed.

Turbines. The structure of steam turbines varies with the size. Very large (>500 MW) utility turbines have several individual turbines, usually classified as high pressure (HP), from throttle pressure to typically 4.8 MPa (696 psig); intermediate pressure (IP), typically from 4.8 MPa (696 psig) to 1.0 MPa (145 psig); and low pressure (LP), typically from 1.0 MPa (145 psig) to condenser pressure of 5.0 kPa (38 mm Hg). On smaller turbines, typically in the range of 125 to 420 MW, a combined HP–IP turbine is joined with an LP turbine. Still smaller turbines, typically below 125 MW, have a total combination of HP–IP–LP in a single cylinder or case.

Turbine Construction. Large steam turbines are generally designed as a rotor inside one or two cylinders or casings. The low pressure nuclear turbine having a partial integral rotor (Fig. 26) illustrates many of the features of both older and newer turbines. The rotor carries multiple rows of blades (or buckets); the cylinder acts as a pressure vessel and carries the same number of diaphragms of stationary blades. The stationary blades may also be called vanes or nozzles. A row of stationary blades followed by a row of rotating blades is called a stage. A typical turbine has 20 to 25 stages varying in height from approximately 2 cm at the inlet to more than 130 cm at the exhaust. The low pressure stages may be multiplied in a double, quadruple, or sextuple flow. Each

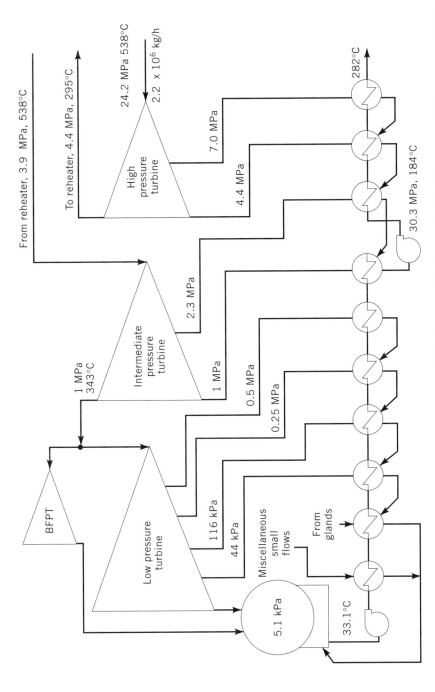

Fig. 25. Schematic of large fossil steam turbine system where BFPT = boiler feed pump turbine. To convert MPa to psi, multiply by 145.

747

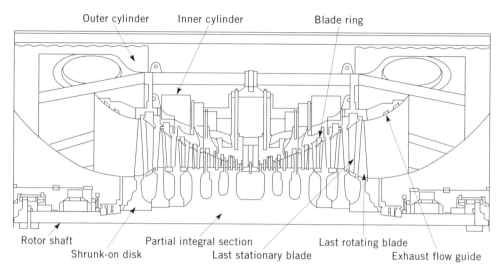

Fig. 26. Schematic of a nuclear low pressure turbine having a partial integral rotor.

blade row, stationary and rotating, contains from 80 to 150 blades. This amounts to thousands of blades, each of which is a precision airfoil. The stationary blades direct the expanding steam in the proper directions so that the rotating blades efficiently convert the energy into torque. The difference between the rotating and stationary blades is significant. The rotating blades experience centrifugal stress, whereas the stationary blades do not. Because the principal source of operating stress in the rotating parts is usually the rotation, stress corrosion cracking thus occurs more in rotating parts than in stationary ones.

Modern utility turbines are generally rather massive devices. The longest turbine systems may exceed 30 m. Turbine cylinders are pressure vessels (see TANKS AND PRESSURE VESSELS). The high pressure cylinder of a fossil unit must withstand 16.5 to 24.1 MPa (2400–3500 psig) at 538°C (1000°F), at which temperatures and pressures creep becomes significant during the design life, even for the thick metal walls. To reduce the pressure and temperature differentials, either the exhaust or the extraction steam from the HP turbine is used to pressurize and cool the cavity between the inner and outer cylinders. In spite of the various cooling and pressure distribution features, the high pressure inner and outer cylinders are still massive castings and do not tolerate rapid temperature changes without experiencing low cycle fatigue and cracking. Because the cylinders cannot be operated above rated temperature for extended times without significantly reducing the creep life of the parts, careful control of steam temperature is necessary. High and intermediate pressure cylinders are usually made from chromium–molybdenum, CrMo, steel. To allow the opening of the turbine for maintenance, cylinders are cast in upper and lower halves and bolted together for operation. The bolts are high strength, high temperature materials that are susceptible to stress corrosion cracking if a liquid phase is present. The intermediate pressure inner and outer cylinders are also cast. Low pressure turbines have an inner cylinder that contains pressure and an outer shell that contains a vacuum. Low pressure inner and outer shells are usually fabricated structures. The rotary seals between the turbine shaft and cylinders are steam glands. In a gland, steam is leaked through a labyrinth to a low (subatmospheric) pressure point (28). Air also goes to the low pressure point, which is the gland condenser.

Rotor bodies are generally made from low alloy steels. A nuclear low pressure rotor weighs over 90 metric tons. High pressure and intermediate pressure rotor bodies, made from 0.9–1.5% Cr–0.7–1.5% Mo–0.2–0.35% V steels in single forgings (28), are machined to accept blade attachments of various types. Low pressure rotor bodies are generally made from 3.25–4.0% Ni–1.25–2.0% Cr–0.25–0.6% Mo–0.05–0.15% V steel (28). When first designed, low pressure rotor bodies (everything except the blades) have often been larger than the largest available forgings. The early bodies of a given design are built up, whereas the later bodies of a design are single forgings. One method of buildup is to weld disks together; the other is to shrink disks on a shaft. The first method has all the problems associated with welding (qv); the second creates crevices in high stress regions. The partial integral rotor (Fig. 26) shows the intermediate stage where the center part is a single forging but the last several disks are shrunk-on. As of the 1990s, forging technology permits monoblock rotors in all applications. To reduce weight, low pressure rotor bodies are forged with a disk on shaft shape. The outer rims of the disks are machined to accept the blades.

Low pressure stationary blades are made from 12% Cr steel and their supporting structures are usually made from carbon steel. Stationary blades are commonly welded into diaphragms. The diaphragms are caulked either into blade rings which mount in the turbine cylinder, or directly into the cylinder. Rotating blades are commonly made of 12 or 13% Cr alloy. One manufacturer uses 17–4 PH (17% Cr and 4% Ni, precipitation hardened with Cu) for long low pressure blades. Occasionally, titanium is used for corrosion resistance, stress reduction owing to its light weight, or both. Blades in high pressure, intermediate pressure, and the first stages of low pressure turbine are generally machined. The airfoils of long (28 cm or longer) blades in the last several stages of the LP turbine are generally precision-forged and the attachment area machined to complement the attachment of the rotor body. The attachment area of rotating blades is a crevice and prone to the problems associated with crevices. Details of turbine design and the aerodynamics of turbines are available (28).

Rankine Cycle Thermodynamics. Carnot cycles provide the highest theoretical efficiency possible, but these are entirely gas phase. A drawback to a Carnot cycle is the need for gas compression. Producing efficient, large-volume compressors has been such a problem that combustion turbines and jet engines were not practical until the late 1940s.

The Rankine cycle overcomes the problem of an efficient gas compressor by compressing the liquid. Efficient pumps (qv) are much easier to construct than efficient compressors. Figure 27**a** shows the Rankine cycle on a pressure–volume (P–V) diagram; Fig. 27**b**, the cycle on a temperature–entropy (T–S) diagram for a cycle between 5 kPa (0.74 psia) and 10,000 kPa (1450 psia; 1435 psig). This is near the highest pressure for which a simple Rankine cycle is practical. First the pressure is increased adiabatically from 5 kPa to 10 MPa. This compression of the liquid water raises the temperature and enthalpy slightly (0.7°C or 2.8 kJ/kg), but the amounts are so small that the difference cannot be easily dis-

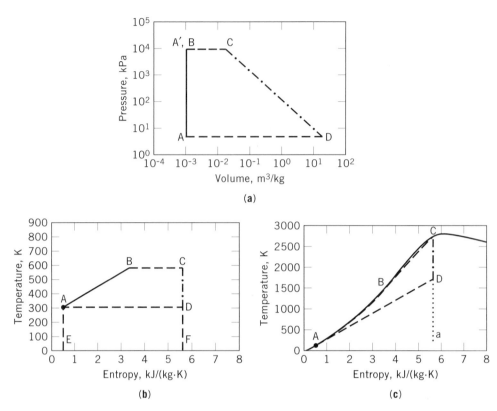

Fig. 27. Rankine cycle in terms of (**a**) pressure and volume; (**b**) temperature and entropy; and (**c**) Mollier (enthalpy vs entropy) chart, where adiabatic (isentropic) compression is represented by line AA′ in (**a**) and by (●) in (**b**) and (**c**). (– – –) represents isobaric heating or heat rejection, including vaporization; (– · –) represents isentropic expansion; and (—) in (**c**) is the saturation line. Heat rejection is from point D to point A. To convert kPa to psi, multiply by 0.145. To convert kJ to kcal, divide by 4.184.

tinguished on most temperature–entropy or Mollier (enthalpy–entropy) charts (Fig. 27c) that represent the rest of the cycle. This process is represented as points A and B in Figure 27. The compressed liquid is heated along the isobar to the boiling point 310°C (492°F) at point B. The fluid is then boiled isothermally to point C. The steam is expanded reversibly (isentropically) to D, where the pressure is 5 kPa and the temperature is 32.9°C. The steam is then isothermally condensed back to point A.

The efficiency of the Rankine cycle, $\eta = W/Q$, can be derived (29,30) as follows:

$$\eta = (h_C - h_D)/(h_C - h_A) \tag{4}$$

where h is the enthalpy at the points A, C, and D. Graphically, in Figure 27**b**, the efficiency is the ratio of the area of the region ABCDA to region EABCFE. The efficiency equation (eq. 4) emphasizes the importance of Mollier charts, where the ideal work is simply the distance along the vertical axis. On the Mollier chart, Figure 27**c**, the reversible work in the cycle is described as the length of the vertical line CD. The efficiency is the length of CD divided by the length of Ca, where point a is simply the projection of A to the same entropy as the turbine expansion.

Superheat and Reheat. One way to increase the area of the work regions on the T–S diagrams is to superheat the steam (Fig. 28). The pressure for the cycle defined by ABCC′D′A is again 10 MPa. The primed letters correspond to points in Figure 27**b**, but the points are not identical. The saturated steam at point C (311°C (592°F)) is heated to 536°C (998°F) at point C′. The superheated steam is expanded to point D′. The increase in work is area CC′D′D. The increase in rejected heat is area DD′F′F. The ratio of these two areas is greater than the ratio of the areas in the simple Rankine cycle, so the efficiency of the total cycle is improved. Similar improvement can be achieved by reheating the steam. After the steam has expanded to about 25% of its original pressure in the high pressure turbine, it can be returned to the boiler and reheated to approximately the same temperature as the inlet steam. This high temperature steam can then be expanded through the intermediate and low pressure turbines to the condenser.

Regenerative Rankine Cycle. A primary inefficiency in the Rankine cycle is caused by the heat added to the cycle to heat the water from the condensate temperature to the boiling point. Regenerative heating addresses this loss. In regenerative heating, some of the steam is removed from the expansion in the turbine and used to heat the water (30). In essence, regeneration uses some of the latent heat of the steam vaporization to heat feedwater, rather than rejecting this heat to the condenser. Figure 29 shows the effect of this process in the theoretical limit of an infinite number of regenerative heating steps. The area ABD is removed from the cycle, the heat rejected as area EADF is eliminated, and the increase in efficiency is substantial. In practice, a maximum of eight feedwater heaters is economical, and line DB is no longer straight. When a cycle includes regeneration, the amount of work generated by each kilogram of steam at the turbine inlet is reduced because some of the potentially available work is used to heat the feedwater. When power capacity is temporarily more important

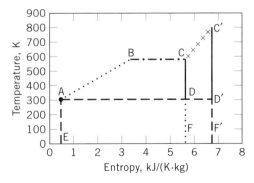

Fig. 28. Rankine cycle for superheat, where (●) represents adiabatic (isentropic) compression; (····), isobaric heating; (─·─), vaporization; (××××), superheating; (──), turbine expansion; and (───), heat rejection. To convert kPa to psi, multiply by 0.145. To convert kJ to kcal, divide by 4.184.

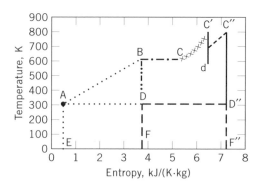

Fig. 29. Theoretical regenerative cycle at 16 MPa (160 bar), showing reheat and efficiency improvement resulting from regeneration, where (⋯⋯) is the region removed from cycle; (— · —), vaporization; (×××), superheating; (—) turbine expansion; (———), heat rejection; (- - -), reheating; and (— ·· —), regeneration. To convert kPa to psi, multiply by 0.145. To convert kJ to kcal, divide by 4.184.

than efficiency, cycles may be operated with one or more heaters out of service, provided that total steam flow does not exceed the turbine capacity at any stage.

Real Fossil Fuel Cycles. Modern fossil fuel cycles start at 16.6 MPa (2415 psia; 2400 psig) or supercritical 23.9 MPa (3515 psia; 3500 psig). Usually throttle pressures are given in gauge pressures. In a typical subcritical fossil fuel cycle (Fig. 25), the boiler feed pump suction comes from the deaerating heater. The more feedwater heaters, the more efficient is the cycle, but eight is usually the economically practical limit. Some cycles have as few as three feedwater heaters. The boiler feed pump turbine for large units is usually driven by the steam from the low pressure crossover and is thus a low pressure turbine. However, at low loads, it is driven by high pressure steam. Some cycles have a motor-driven boiler feed pump, so the boiler feed pump turbine is omitted from the diagram and the steam normally used to drive it goes into the main turbine. There is a transition from dry to wet steam. In Figure 25, the last (lowest pressure) extraction is essentially at the saturation line, which is the typical situation. In some units there is a small amount of superheat at the last extraction; in others there may be a small amount of moisture.

Nuclear Steam Cycles. A cycle diagram for a nuclear power plant is shown in Figure 30. Although most nuclear power plant cycles begin with saturated steam, this diagram is for a unit with a once-through nuclear steam generator. The throttle steam has 17.2°C (31°F) of superheat. This steam is expanded through the high pressure turbine. In the middle of this expansion, steam is extracted for the highest pressure feedwater heater. After heating the feedwater, the water in the shell side of the heater is drained to the shell side of the second highest pressure heater. Because the inlet steam has such low enthalpy, the steam becomes moist in the high pressure turbine, typically 7–13% moisture at the HP exhaust. Moisture reduces the efficiency of the turbine because the droplets do not follow the steam flow. They impinge on the turbine and must be accelerated. Typically, 10% efficiency is lost in wet steam. In addition, the droplets erode the turbine. After the high pressure turbine, the steam is dried

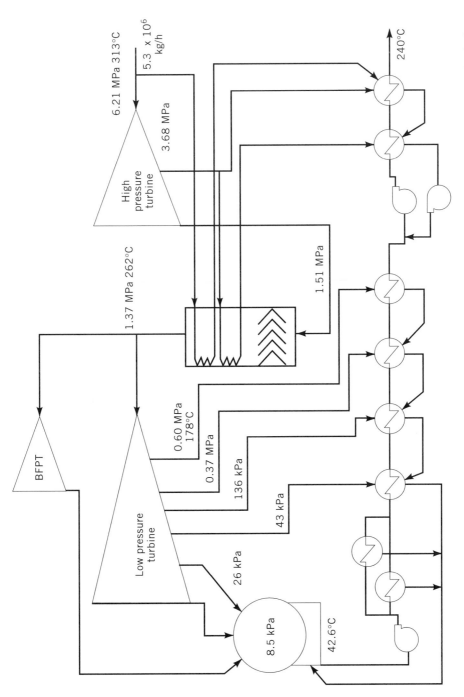

Fig. 30. Schematic of nuclear steam cycle where BFPT = boiler feed pump turbine. To convert MPa to psi, multiply by 145.

753

and superheated in a moisture separator reheater (MSR). The moisture separation is thermodynamically equivalent to isothermal reheating. The two-phase fluid with a typical enthalpy of 2680 kJ/kg (1152 Btu/lb) is separated into steam at 2790 kJ/kg (1199 Btu/lb) and water at 846 kJ/kg (361.6 Btu/lb). The water goes to the second highest pressure heater, where it is merged with steam from the outlet of the first-stage reheater. The water from the second heater shell is then pumped into the feedwater. This arrangement creates a high pressure recirculating loop. The heat source for reheating is the main steam or a combination of the main steam and steam extracted from the high pressure turbine. The superheat is typically 28°C (50°F), which usually keeps three to five stages of the low pressure turbine dry.

Nuclear Cycle Water Treatment Issues. Relatively volatile components, such as amines and organic acids, can appear at rather high concentrations in the high pressure loop of nuclear power steam cycles. Because of the variation in distribution between water and steam phases with temperature and the generation of significant moisture, the chemical relationships can be quite complex. A full model of the equilibriums occurring in the turbine is required to understand the chemical effects (31). Feedwater pH is not a good indicator of the pH of the liquid phase in the turbine. The pH of the turbine liquid depends on feedwater pH, the amines present, and the acids present. The equilibriums can be such that an increase in the boric acid concentration in the feedwater raises the pH of the liquid phase at the end of the high pressure turbine. This apparent chemical inconsistency occurs because increased boric acid requires additional amine to maintain the feedwater pH.

Combined Cycle. Combined cycles use a steam turbine system as a bottoming cycle for a combustion turbine. Exhaust gases leave the combustion turbine at 510–593°C (950–1100°F). This heat is wasted in the simple combustion turbine cycle. However, it can be used to boil water in a heat recovery steam generator (HRSG). This process lowers the temperature of much of the rejected heat to that leaving the steam turbine. However, to prevent HRSG corrosion, the stack gas temperature is nominally 150°C (302°F), causing additional heat rejection to the environment. Because the combustion turbine does not need preheated air, there is no air heater in combined cycles. The economizer of the HRSG can extract as much heat as possible from the combustion gas. Thus the feedwater heating of conventional utility cycles is not needed. There is usually only a single direct-contact heater to remove oxygen from the feedwater. Combined cycle power plants have achieved 58% and newer designs are expected to approach 60% thermal efficiency (32).

As of the mid-1990s, many older conventional steam plants have been converted to combined cycle. The old boiler is removed and replaced by a combustion turbine and heat recovery steam generator. Although the cycle efficiency is not as high as completely new plants, substantial capital cost is avoided by the modification and reuse of existing steam turbine and auxiliary equipment. In many combined cycle power plants, steam is injected into the combustors of the combustion turbine to lower peak flame temperatures and consequently lower NO_x.

Cogeneration. Cogeneration is another modification of the Rankine cycle used when steam is required to heat a process. It is common in pulp (qv) and paper (qv) mills, chemical plants, and municipal or district heating systems. The steam is produced at higher pressure and temperature than would be required for the process. The steam is expanded through a turbine and steam at the desired pressure is extracted. The turbine may be used to drive a generator or other machinery. Figure 31, a schematic diagram of the steam system of a pulp and paper mill, shows drive turbines, turbine-generators, and multiple boilers. One of the boilers is a heat recovery boiler, another is fueled by bark. There are pressure manifolds at 6.2 MPa (885 psig) and 538°C (1000°F); 2.8 MPa (385 psig) and 229°C (445°F); and 0.41 MPa (45 psig) and 145°C (292°F). Turbogenerator 2 exhausts to condensate, but turbogenerator 1 exhausts to the 0.41-MPa header, where the entire exhaust is used for heating. This steam would be generated in a low pressure boiler if it did not come from the turbine. For turbogenerator 1, the thermal efficiency of the electrical generation process approaches the mechanical efficiency of the turbogenerator (90+%), because there is no heat rejected by this cycle.

Mechanical Drives. Steam turbines are very efficient at high load ratings and, depending on the steam balance in a plant, are normally considered as drive units if more than 37 kW (50 hp) is required. If many small loads are to be handled separately in a plant, it may be preferable to generate electric power by passing the steam through a back-pressure (no condenser) turbine connected to an electric generator. The generated electricity in turn can be fed into the motors throughout the plant. Steam turbines operate very effectively at high speeds (3,000–10,000 rpm) and thus lend themselves well to large power output, high speed drives for gas compressors, multistage high pressure pumps such as boiler feed pumps, and other high speed rotating equipment. In such installations, the steam can be fed to the turbine and extracted at various stages of its expansion corresponding to the desired process operating pressures. Steam turbine drives having rpm reduction are used in large commercial ships and navy vessels for propulsion.

Steam Heating. Wet and saturated steam has a definite pressure for each fixed boiling or condensing temperature. Therefore, the control of the desired temperature for any process heating requirement may be fixed by choosing the steam pressure. It is customary for steam to be generated under slightly superheated conditions if the steam generator is to be located at a considerable distance from the various users.

When higher temperatures are needed, the higher corresponding pressure becomes important in the design of equipment. Steam pressure rises quite rapidly with temperature and is of significant concern for heating conditions at temperatures above 180°C. The use of steam for heating is normally limited to pressures less than 2.2 MPa (300 psig) because the cost of the heat-transfer equipment becomes too high for sizable heating units as higher pressures are required. When temperatures in excess of 180–205°C are required, other heating media are usually preferred (see HEAT-EXCHANGE TECHNOLOGY).

Evaporators. Steam heating systems have often been installed in a cascade system, as in multiple-effect evaporators (see EVAPORATION). This arrangement makes possible the recovery of heat at several successive levels merely by reducing the pressure at each of the stages. Condensing all the steam vaporized in one stage by heating and vaporizing the material present in the succeeding stage at lower operating pressure produces good economy of heat. This system

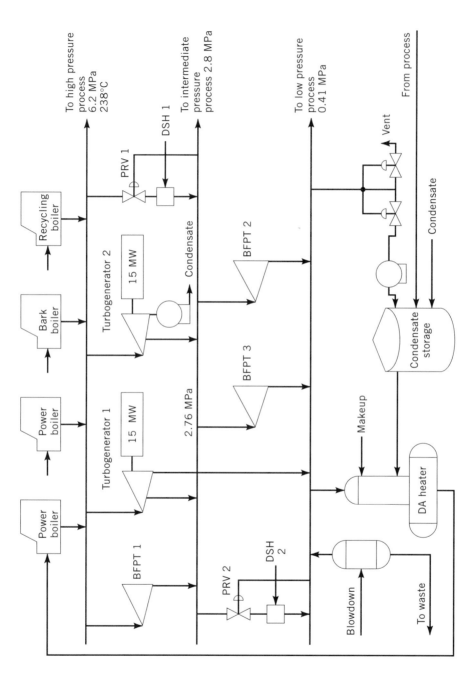

Fig. 31. Steam system of a pulp and paper mill where PRV = pressure reducing valve, DSH = desuperheater, and DA = deaerating. To convert MPa to psi, multiply by 145 (33).

of heating operates best in the low pressure range because at higher pressures the equilibrium temperature changes more slowly with the pressure.

Control. When close temperature control is required in order to prevent overheating of material being processed or to ensure a high heating density, steam is the medium normally used. A pressure regulator controlling the steam pressure of a heating unit maintains temperatures usually within degrees of the design conditions on the process side.

Industrial Processes. *Steam Reforming Processes.* In the steam reforming process, light hydrocarbon feedstocks (qv), such as natural gas, liquefied petroleum gas, and naphtha, or in some cases heavier distillate oils are purified of sulfur compounds (see SULFUR REMOVAL AND RECOVERY). These then react with steam in the presence of a nickel-containing catalyst to produce a mixture of hydrogen, methane, and carbon oxides. Essentially total decomposition of compounds containing more than one carbon atom per molecule is obtained (see AMMONIA; HYDROGEN; PETROLEUM).

Steam reforming in industrial practice falls into two main classes according to catalyst type and reactor equipment use: hydrogen (qv) production by high (generally $>700°C$) temperature reforming, and methane manufacture by low ($<550°C$) temperature reforming. The amount of steam consumed by reaction in the overall process depends on the choice of product gas. A general hydrocarbon, C_xH_{2x+2}, reacting with steam gives in the low temperature case the following:

$$4\,C_xH_{2x+2} + 2x\,H_2O \longrightarrow 3x\,CH_4 + x\,CO_2 + 4\,H_2$$

and in the high temperature case:

$$C_xH_{2x+2} + 2x\,H_2O \longrightarrow (3x + 1)\,H_2 + x\,CO_2$$

In practice, two competing reactions combine to limit the process.

$$CH_4 + H_2O \rightleftharpoons CO + 3\,H_2$$
$$CO + H_2O \rightleftharpoons CO_2 + H_2$$

The equilibrium composition of the product gas can be altered by choice of suitable temperature, pressure, and steam to feedstock ratio to produce a gas mixture consisting largely of methane or largely of hydrogen having varying proportions of carbon monoxide (qv). In each case, some carbon dioxide (qv) is produced, which can be removed. It is often convenient to use more than one reaction stage to modify the gas mixture produced from the primary gasification step toward the desired composition.

Synthetic Ammonia. Steam-methane reforming is used to produce hydrogen for the production of ammonia (qv), which is synthesized by using nitrogen from the air at high pressure and temperature over a suitable catalyst. In this process, approximately 0.75 kg of steam is required to produce 1 kg of ammonia. A very large synthetic ammonia plant uses a single-train process employing centrifugal compressors driven by steam turbines. The exhaust steam is used for the gas reforming reaction. The high pressure steam thus generated from waste heat is the heart of the energy system for the process, which in turn is the key to the process economics (34) (see PROCESS ENERGY CONSERVATION).

Coal Gasification.   Coal gasification processes involve the reaction of coal at high temperature with steam and air or oxygen to produce a mixture of gases, typically carbon monoxide, carbon dioxide, hydrogen, and methane. Producer gas, a mixture of carbon monoxide and hydrogen, has been made by this method since early in the twentieth century. Sulfur (qv) in the coal (qv) reacts to form hydrogen sulfide. The gases from the gasifier can be further upgraded in several steps. By reaction with steam, the carbon monoxide in the gas is converted to hydrogen and carbon dioxide. Hydrogen sulfide and carbon dioxide can be removed in a purification system and the hydrogen converted to methane by reaction with carbon monoxide. There are numerous variations on the basic process. Because the gas produced in a coal gasification process can serve as a fuel for combustion turbines, coal gasification is becoming increasingly important. The combustion turbines used in the high efficiency combined-cycle power generation systems require gas or liquid fuel (see COAL CONVERSION PROCESS, GASIFICATION; FUELS, SYNTHETIC–GASEOUS FUELS).

Coal Liquefaction.   Steam is used to produce hydrogen for the liquefaction of coal. In the liquefaction process, coal is crushed, dried, pulverized, and then added to a solvent to produce a slurry. The slurry is heated, usually in the presence of hydrogen to dissolve the coal. The extract is cooled to remove hydrogen, hydrocarbon gases, and hydrogen sulfide. The liquid is then flashed at low pressure to separate condensable vapors from the extract. Mineral matter and organic solids are separated and used to produce hydrogen for the process. The extract may be desulfurized. The solvent is separated from the products. There are at least six different liquefaction processes (see COAL CONVERSION PROCESS, LIQUEFACTION; FUELS, SYNTHETIC–LIQUID FUELS).

Petroleum Recovery.   Steam is injected into oil wells for tertiary petroleum recovery. Steam pumped into the partly depleted oil reservoirs through input wells decreases the viscosity of crude oil trapped in the porous rock of a reservoir, displaces the crude, and maintains the pressure needed to push the oil toward the production well (see PETROLEUM, ENHANCED RECOVERY). Steam is also used in hot-water extraction of oil from tar sands (qv) in the caustic conditioning before the separation in a flotation tank (35).

Evaporation and Distillation.   Steam is used to supply heat to most evaporation (qv) and distillation (qv) processes, such as in sugar-juice processing and alcohol distillation. In evaporation, pure solvent is removed and a low volatility solute is concentrated. Distillation transfers lower boiling components from the liquid to the vapor phase. The vapors are then condensed to recover the desired components. In steam distillation, the steam is admitted into direct contact with the solution to be evaporated and the flow of steam to the condenser is used to transport distillates of low volatility. In evaporation of concentrated solutions, there may be substantial boiling point elevation. For example, the boiling point of an 80% NaOH solution at atmospheric pressure is 226°C.

Desalination.   A special case of distillation is water desalination. In places where energy is abundant but fresh water is not, eg, the Arabian Peninsula, water may be produced from seawater in flash evaporators. Low pressure turbine steam is extracted to provide heat for the evaporators. Condensed steam is returned to the cycle. Such units may be particularly prone to corrosion by salts. Sizes vary, but a plant scheduled for completion in 1996 had six units and a total capacity of 345,600 m³/d. Power generation was expected to be 17,500 MW (36).

Steam Cleaning. High pressure steam can be used to produce a high velocity jet with some superheating by expansion through a suitable nozzle to atmospheric pressure. The high velocity is effective in removing dirt and loose scale from solid surfaces. The high temperature encourages the melting or vaporization of oil and grease deposits, thus releasing the solid deposits for mechanical blast removal. Some condensation of steam on the initially cold surfaces also takes place, which may help to dissolve and release dirt and scale.

Hydrothermal Treatment of Wastes. Hydrothermal processing (qv) of materials appears to be a promising method of disposal for many noxious materials and of conversion of some wastes to valuable by-products. This method consists of mixing reactants and water; pressurizing, heating to reaction temperatures, and cooling the products; and then carrying out secondary processing of the products. For efficient oxidation, oxygen is added to the water. Typical pressures and temperatures are 22 MPa (3190 psi) and 550°C (1022°F) for destruction of wastes. Somewhat lower pressures and temperatures are used for conversions. Contact times are on the order of a few minutes. Organic chemicals are oxidized to carbon dioxide and water. Because the processes are below 1000°C, nitrogen is usually reduced to the elemental state. Other common heteroatoms are converted to corresponding acids, such as HCl and H_2SO_4. Reaction at high temperatures and pressures is an active field of research. There is consensus that the hydrothermal oxidation processes proceed by free-radical mechanisms.

Along the saturation line and the critical isobar (22.1 MPa (3205 psi)), the dielectric constant of water declines with temperature (see Fig. 10). In the last 24°C below the critical point, the dielectric constant drops precipitously from 14.49 to 4.77; in the next 5°C, it further declines to 2.53; and by 400°C it has declined to 1.86. In the region of the critical point, the dielectric constant of water becomes similar to the dielectric constants of typical organic solvents (Table 6). The solubility of organic materials increases markedly in the region near the critical point, and the solubility of salts tends to decline as the temperature increases toward the critical temperature.

Table 6. Dielectric Constants of Solvents[a]

Substance	Dielectric constant at 20–25°C
n-pentane	1.844
cyclohexane	2.22
benzene	2.284
chloroform	4.8
acetone	20.7
ethanol	24.3
methanol	32.63
water	78.45

[a] Ref. 37.

At temperatures near the critical temperature, many organic degradation reactions are rapid. Halogenated hydrocarbons loose the halogen in minutes at 375°C (38). At temperatures typical of nuclear steam generators (271°C (520°F)), the decomposition of amines to alcohols and acids is well known (39). The pressure limits for the treatment of boiler waters using organic polymers reflect the rate of decomposition.

An important advantage of hydrothermal processing over incineration is that the former system is fully contained. If a process must be shut down, there is no discharge of material to the environment. Furthermore, the products may be analyzed or further processed before being discharged. The possibility of further processing makes hydrothermal processing an important candidate for reduction of wastes containing both organic molecules and radionuclides. Radionuclides can be recovered (see WASTE REDUCTION). Another advantage is that the input waste stream need not be highly characterized. The oxygen demand and the approximate amount of acid produced are the only principal environmental parameters to be considered.

The formation of acids from heteroatoms creates a corrosion problem. At the working temperatures, stainless steels are easily corroded by the acids. Even platinum and gold are not immune to corrosion. One solution is to add sodium hydroxide to the reactant mixture to neutralize the acids as they form. However, because the dielectric constant of water is low at the temperatures and pressure in use, the salts formed have low solubility at the supercritical temperatures and tend to precipitate and plug reaction tubes. Most hydrothermal processing is oxidation, and has been called supercritical water oxidation.

There are two general goals in hydrothermal processing of wastes. First is the total destruction of organic material. Pilot plants for destruction of chemical warfare (qv) agents and explosives have been constructed and are operating (see EXPLOSIVES AND PROPELLANTS). In benchtop experiments, destruction and removal of these agents may exceed 99.9999%. As of this writing (1996), a small commercial system to dispose of 0.019 m^3/d (5 gal/d) of wastes from pilot plants is in use at one chemical company (40).

The second goal of hydrothermal processing is the conversion of waste materials to useful chemicals. This goal is usually pursued at lower temperatures and pressures. Glucose and cellulose (qv) can be pyrolysed in supercritical steam, yielding carbon dioxide and hydrogen. At room temperature and atmospheric pressure, the process produces char, hydrogen, and hydrocarbons. At intermediate temperatures, supercritical and hot water dissolve hemicellulose (qv) and lignin (qv) from wood. Ether linkages can also be cleaved.

Alcohols undergo dehydration in supercritical and hot water (41). Tertiary alcohols require no catalyst, but secondary and primary alcohols require an acid catalyst. With 0.01 M H_2SO_4 as a catalyst, ethanol eliminates water at 385°C and 34.5 MPa to form ethene. Reaction occurs in tens of seconds. Only a small amount of diethyl ether forms as a side reaction.

The use of supercritical and hot water as a solvent is still largely experimental. Because supercritical technology is well known in the power industry, this use of water is likely to increase in the future. Corrosion control may be an important limiting consideration. General process economics are the second potential limit (see SUPERCRITICAL FLUIDS).

Corrosion in Steam

Use of metals in hot steam is limited by oxidation rate, mechanical strength, and creep resistance (see CORROSION AND CORROSION INHIBITORS). Temperature and stress limits and corrosion allowances are specified in national standards and pressure vessel codes (42,43). General corrosion rates in pure steam (Table 7) are about the same as in high purity deoxygenated water, except for gray iron, nickel, lead, and zirconium, which corrode faster in steam. Iron-base alloys, including the austenitic and ferritic stainless steels, are used extensively in contact with steam. These oxidize to form a protective film of the spinel oxide, Fe_3O_4 (magnetite), or, in the case of stainless steels, M_3O_4, where M is iron, chromium, or nickel. Gamma-Fe_2O_3 has also been found on ferrous alloys in degassed high temperature water and steam. Its physical properties are very similar to those of Fe_3O_4. It is magnetic and has an almost identical crystal structure.

Table 7. Corrosion rates of metals in steam[a]

Material	Temperature, °C	Corrosion rate, mm/yr	Comments
cast irons			
gray	25–112	<0.5	
nickel	25–350	<0.05	
silicon	25–300	<0.05	
mild steel	25–510	<0.05	
austenitic stainless steel alloys			
302, 304, 321, 347	25–790	<0.05	intergranular crack
316, 317	25–350	<0.05	
copper	25–250	<0.05	
copper alloys			
brass			
70–80 Cu + Zn, Sn, or Pb	25–212	<0.05	
57–93 Cu + Al, Zn, or As	25–212	<0.05	
cupro-nickel 66–88:11–13	25–300	<0.05	no Zn
nickel 99	25–425	0.5–1.3	
Ni–Cu 66–32	25–350	<0.05	
Ni–Cr–Fe T6–16–7	25–815	0.5–1.3	stress cracks
Ni–Mo 62–28 + Fe, V	25–350	<0.05	
Ni–Cr–Mo 54–15–16 + Fe, W	25–350	<0.05	
aluminum	50–250	0.5–1.3	
aluminum alloys			
gold	25–350	<0.05	
lead	25–150	0.5–1.3	
	300	>1.3	
platinum	25–350	<0.05	
silver	25–350	<0.05	
tantalum	25–250	<0.05	
titanium	25–350	0.0	aerated steam
zirconium	25–350	>1.3	

[a]Extracted from Ref. 44.

Only Fe_3O_4 and γ-Fe_2O_3 are considered to be protective films. Both are adherent and good electronic conductors. Alpha-Fe_2O_3, which forms in water and steam containing oxygen, is not adherent, is less protective, and is an insulator. FeO, which does not form at temperatures below 570°C, is nonprotective.

A good summary of the behavior of steels in high temperature steam is available (45). Calculated scale thickness for 10 years of exposure of ferritic steels in 593°C and 13.8 MPa (2000 psi) superheated steam is about 0.64 mm for 5 Cr–0.5 Mo steels, and 1 mm for 2.25 Cr–1 Mo steels. Steam pressure does not seem to have much influence. The steels form duplex layer scales of a uniform thickness. Scales on austenitic steels in the same test also form two layers but were irregular. Generally, the higher the alloy content, the thinner the oxide scale. Excessively thick oxide scale can exfoliate and be prone to under-the-scale concentration of corrodents and corrosion. Exfoliated scale can cause solid particle erosion of the downstream equipment and clogging. Thick scale on boiler tubes impairs heat transfer and causes an increase in metal temperature.

Where corrosive impurities from steam or water concentrate on metal surfaces, corrosion can be severe (6,46–48). Concentration occurs in boilers in the waterwalls, where boiling occurs. It occurs in turbines near the saturation line, where the ability of salts to elevate the boiling point of water creates a salt solution zone (49). In the salt solution zone, a 35% NaCl solution is produced from steam containing a few parts per billion NaCl. General and pitting corrosion, stress–corrosion cracking, corrosion fatigue, corrosion–erosion, caustic gouging, and hydrogen embrittlement and cracking have been observed. Combined corrosion effects and tensile stresses can produce sudden failures of materials. Corrosion is not usually caused by the steam, but by the concentrated steam impurities, such as chlorides, caustic, inorganic and organic acids, carbonates, sulfates, hydrogen sulfide, and their mixtures. Oxygen and oxides of copper, lead, and nickel can aggravate the corrosion.

Economic Aspects

Increasing fuel costs and sizes of industrial and utility installations have forced the emphasis in economical considerations to shift to high thermal efficiency, reliability, and availability. The investment, operating, maintenance, transmission, insurance, and other costs as well as depreciation must also be considered, but these are often less important.

Thermal efficiency and heat rate directly influence the fuel cost. Increasing cycle pressure and temperature and using superheat, reheat, and condensing cycles result in a significant increase of efficiency and lower heat rate. In the process industry, this can be best achieved by cogeneration; in the electric utility industry, by combined gas turbine–steam turbine cycles and further increase of cycle parameters. Using advanced power cycles (50), a 1000-MWe (1.34×106 hp) unit can save coal at a rate of 150,000 t/yr, or 10^7 at $65/t, if the cycle parameters can be increased from 24.1 MPa (3500 psi)/540°C inlet/540°C reheat to 34.5 MPa (5000 psi)/650°C inlet/593°C double reheat. This change would result in an average gross plant efficiency improvement from 39.83 to 42.59% and in a heat rate improvement from 7.728 to 7.222 kJ (1.847 to 1.726 kcal)/kWh.

BIBLIOGRAPHY

"Steam" in *ECT* 1st ed., Vol. 12, pp. 778–793, by H. N. La Croix, Foster Wheeler Corp.; in *ECT* 2nd ed., Vol. 18, pp. 692–715, by J. K. Rice, Cyrus Wm. Rice and Co.; in *ECT* 3rd ed., Vol. 21, pp. 507–551, by O. Jonas, Westinghouse Electric Corp.

1. Regnault, *Mém. l'Inst. France*, **21**, 465 (1847).
2. C. A. Meyer and co-workers, *Steam Tables*, 6th ed., ASME, New York, 1993.
3. H. J. White and co-workers, *Proceedings of the 12th International Conference on the Properties of Water and Steam, Orlando, Fla., 1994*, Begell House, New York, 1995.
4. L. A. Curtis, D. J. Frurip, and M. Blander, in J. Straub and K. Schaffler, *Proceedings of the 9th International Conference on the Properties of Steam*, Pergamon Press, Oxford, U.K., 1980, p. 521.
5. O. I. Martynova, *Russian J. Phys. Chem.* **38**, 587 (1964).
6. O. Jonas, *Combustion*, **50**, 11 (1978).
7. J. W. Cobble and S. W. Lin, in P. Cohen, *ASME Handbook on Water Technology for Thermal Power Systems*, ASME, New York, 1989, Chapt. 8.
8. J. M. Simonson and D. A. Palmer, "An Experimental Study of the Volatility of Ammonium Chloride from Aqueous Solutions to High Temperatures," *Proceedings of the 13th International Water Conference*, Pittsburgh, Pa., Oct. 21–23, 1991, p. 253.
9. J. F. Galobardes, D. R. Van Hare, and L. B. Rogers, *J. Chem. Eng. Data*, **26**, 363 (1981).
10. M. W. Chase and co-workers, *JANAF Thermochemical Tables*, 3rd ed., American Chemical Society, Washington, D.C., 1986.
11. P. M. Goodall, ed., *The Efficient Use of Steam*, IPC Science and Technology Press, Guilford, Surrey, U.K., 1980.
12. *Steam: Its Generation and Use*, 39th ed., Babcock and Wilcox Co., New York, 1978.
13. J. G. Singer, ed., *Combustion: Fossil Power Systems*, Combustion Engineering, Inc., Windsor, Conn., 1981.
14. A. F. Aschoff and co-workers, *Interim Consensus Guidelines on Fossil Plant Cycle Chemistry*, Electric Power Research Institute, CS-4629, 1986.
15. J. C. Bellows, *Proceedings of the 56th International Water Conference*, Pittsburgh, Pa., 1995, p. 155.
16. *Boiler-Water Limits and Steam Purity Recommendations for Watertube Boilers*, American Boiler Manufacturers Association, Arlington, Va., 1981.
17. B. Dooley and co-workers, *Proceedings of the 53rd International Water Conference*, Pittsburgh, Pa., 1992, p. 154.
18. F. J. Pocock, J. A. Lux, and R. V. Seibel, *Proc. Am. Power Conf.* **24**, 758 (1966).
19. E. P. Partridge and R. E. Hall, *Trans. ASME*, **61**, 597 (1939).
20. J. Stodola, *Proceedings of the 47th International Water Conference*, Pittsburgh, Pa., 1986, p. 234.
21. R. K. Freier, in Hamburg Power Works, *Chemistry and Physics*, HEW, Hamburg, Germany, 1972.
22. F. N. Kemmer, ed., *NALCO Water Handbook*, McGraw-Hill Book Co., Inc., New York, 1979.
23. *Betz Handbook of Industrial Water Conditions*, Betz, Trevose, Pa., 1962.
24. A. R. Cantafio, ed., *Principles of Industrial Water Treatment*, Drew Industrial Division, Ashland Chemical Co., Boonton, N.J., 1994.
25. ASME Power Test Code, PTC 19.11, ASME, New York, 1974 (revision in press).
26. Technical data, North American Electric Reliability Council, Princeton, N.J., 1995.
27. Technical data, Westinghouse Electric, Orlando, Fla., 1996.
28. P. Schofield, in Ref. 7, Chapt. 3.
29. G. J. Silvestri, Jr., *Steam Cycle Performance*, Power Division, ASME, New York.

30. G. J. Silvestri, Jr., in Ref. 7, Chapt. 1.
31. J. C. Bellows, *Proceedings of the 54th International Water Conference*, Pittsburgh, Pa., 1993, p. 127.
32. M. S. Briesch and co-workers, *Trans. ASME* **117**, 734 (1995).
33. C. H. Cho, *Efficient Allocation of Steam*, MIT Press, Cambridge, Mass., 1982, p. 3.
34. *Chem. Eng.* **74**, 112 (1967).
35. R. Loftness, *Energy Handbook*, Van Nostrand Reinhold Co., Inc., New York, 1978.
36. C. Sommariva, *Desalination and Water Reuse*, **6**(1) 30 (1996).
37. D. R. Lide, ed., *Handbook of Chemistry and Physics*, 75th ed., CRC Press, Ann Arbor, Mich., 1994.
38. V. A. Korovin and co-workers, *Energetik* **1980**(5), 24.
39. R. Gilbert, C. Lamarre, and Y. Dundar, *Proceedings of the 10th Annual Canadian Nuclear Society Conference*, 1989.
40. R. N. McBrayer and J. W. Griffith, *Proceedings of the 56th International Water Conference*, Pittsburgh, Pa., 1995, p. 499.
41. M. J. Antal, in Ref. 3, p. 24.
42. *Brit. Standards*, BS749, pp. 1113, 2790.
43. *ASME Poiler and Pressure Vessel Code*, ASME, New York, 1977.
44. N. E. Hammer, *Metals Section Corrosion Data Survey*, 5th ed., National Association of Corrosion Engineers, Houston, Tex., 1974, pp. 174–175.
45. G. E. Lien, ed., *Behavior of Superheater Alloys in High Temperature, High Pressure Steam*, ASME, New York, 1971.
46. O. Jonas, *Identification and Behavior of Turbine Steam Impurities, Corrosion 77*, National Association of Corrosion Engineers, San Francisco, Calif., 1977.
47. O. Jonas, W. T. Lindsay, Jr., and N. A. Evans, in J. Straub and K. Schaffler, *Proceedings of the 9th International Conference on the Properties of Steam*, Pergamon Press, Oxford, U.K., 1980, p. 595.
48. R. J. Lindinger and R. M. Curran, *Experience with Stress Corrosion Cracking in Large Steam Turbines, Corrosion 81*, National Association of Corrosion Engineers, Toronto, Ontario, Canada, 1981.
49. W. T. Lindsay, Jr., *Proceedings of the Westinghouse Steam Turbine-Generator Symposium*, Charlotte, N.C., 1978.
50. T. Suzuki, *The Development of Coal Firing Power Unit with Ultra High Performance in Japan*, EPRI Fossil Heat Rate Improvement Workshop, Charlotte, N.C., 1981.

JAMES BELLOWS
Westinghouse Electric Corporation

STEARIC ACID.   See CARBOXYLIC ACIDS.

STEATITE.   See TALC.

STEEL

Steel, the generic name for a group of ferrous metals composed principally of iron (qv), is the most useful metallic material known on account of its abundance, durability, versatility, and low cost. In the form of bars, plates, sheets, structural shapes, wire pipe and tubing, forgings, and castings, steel is used in the construction of buildings, bridges, railroads, aircraft, ships, automobiles, tools, cutlery, machinery, furniture, household appliances, and many other articles on which the convenience, comfort, and safety of society depends. Steel is also an essential material for spacecraft and supporting facilities as well as in practically every kind of material needed for national defense (see BUILDING MATERIALS; TOOL MATERIALS).

The durability and versatility of steel are shown by its wide range of mechanical and physical properties. By the proper choice of carbon content and alloying elements, and by suitable heat treatment, steel can be made so soft and ductile that it can be cold-drawn into complex shapes such as automobile bodies. Conversely, steel can be made extremely hard for wear resistance, or tough enough to withstand enormous loads and shock without deforming or breaking. In addition, some steels are made to resist heat and corrosion by the atmosphere and by a wide variety of chemicals.

About 800 million metric tons of raw steel is produced annually throughout the world. Its usefulness is enhanced by the fact that it is inexpensive. The price as of the mid-1990s ranged from ca 44¢/kg for the common grades to several dollars per kg for special steels such as certain tool steels. Prices have remained relatively constant since the early 1980s.

The uses of steel are too diverse to be listed completely or to serve as a basis of classification. Inasmuch as grades of steel are produced by more than one process, classification by method of manufacture is not advantageous. The most useful classification is by chemical composition into the large groups of carbon steels, alloy steels, and stainless steels. Within these groups are many subdivisions based on chemical composition, physical or mechanical properties, or uses.

It has been known for many centuries that iron ore, embedded in burning charcoal, can be reduced to metallic iron (1,2). Iron was made by this method as early as 1200 BC. Consisting almost entirely of pure iron, the first iron metal closely resembled modern wrought iron, which is relatively soft, malleable, ductile, and readily hammer-welded when heated to a sufficiently high temperature. This metal was used for many purposes, including agricultural implements and various tools.

It is not known when the first metal resembling modern carbon steel was made intentionally. It has been speculated that steel was made in India around 100 BC. Remains of swords with steel blades have been found in Luristan, a western province of Iran (formerly Persia), at sites dated at ca 800 BC.

Most modern steels contain >98% iron. However, steel also contains carbon, which, if present in sufficient amounts (up to ca 2%), gives a property unmatched by any of the metals available to the ancients. This property is the ability to become extremely hard if cooled very quickly (quenched) from a high enough temperature, as by immersing it in water or some other liquid. Its hardness and the ability to take and hold a sharp cutting edge make steel an extremely valuable metal for weapons, tools, cutlery, surgical instruments, razors, and other special forms.

It was not until the eighteenth century that carbon was recognized as a chemical element, and it is quite certain that no early metallurgist was aware of the basis of the unique properties of steel as compared to those of wrought iron. Carbon can be alloyed with iron in a number of ways to make steel, and all methods described herein have been used at various times in many localities for perhaps 3000 or more years.

For example, low carbon wrought-iron bars were packed in air-tight containers together with charcoal or other carbonaceous material. By heating the containers to a red heat and holding them at that temperature for several days, the wrought iron absorbed carbon from the charcoal; this method became known as the cementation process. If the iron is made in primitive furnaces, carbon can be absorbed from the charcoal fuel during and after its reduction. The Romans, for instance, built and operated furnaces that produced a steel-like metal instead of wrought iron. In neither of the two foregoing cases did the iron or steel become molten.

In ancient India, a steel called wootz was made by placing very pure iron ore and wood or other carbonaceous material in a tightly sealed pot or crucible heated to high temperature for a considerable time. Some of the carbon in the crucible reduced the iron ore to metallic iron, which absorbed any excess carbon. The resulting iron–carbon alloy was an excellent grade of steel. In a similar way, pieces of low carbon wrought iron were placed in a pot along with a form of carbon and melted to make a fine steel. A variation of this method, in which bars that had been carburized by the cementation process were melted in a sealed pot to make steel of the best quality, became known as the crucible process.

Before the invention of the Bessemer process for steelmaking in 1856, only the cementation and crucible processes were of any industrial importance. Although both of the latter processes had been known in the ancient world, their practice seems to have been abandoned in Europe before the Middle Ages. The cementation process was revived in Belgium around 1600, whereas the crucible process was rediscovered in the British Isles in 1740. Both processes were practiced in secret for some time after their revival, and little is known of their early history. The cementation process flourished in the United Kingdom during the eighteenth and nineteenth centuries and continued to be used to a limited extent into the early part of the twentieth century. At red heat, a low carbon ferrous metal, in contact with carbonaceous material such as charcoal, absorbed carbon that, up to the saturation point of about 1.70%, varied in amount according to the time the metal was in contact with the carbon and the temperature at which the process was conducted. A type of muffle furnace or pot furnace was used and the iron and charcoal were packed in alternate layers.

In the softer grades (average carbon content ca 0.50%), the composition of the bar was uniform throughout. In the harder grades (average carbon content as high as 1.50%), the outside of a bar might have a carbon content of 1.50–2.00%, whereas the center contained 0.85–1.10%. Steels made by this method were called cement steels.

The crucible process gave steels that were not only homogeneous throughout but were free from occluded slag originating in the wrought iron used to make cement steel. Crucible steel was so superior to cement steel for many purposes that the crucible process quickly became the leader for the production of the finest steels. Its drawback was, however, that each crucible held only ca 50 kg steel.

The various steelmaking processes were all eventually supplanted (3,4). The first of the newer techniques was the historic pneumatic or Bessemer process, introduced in 1856. Shortly thereafter, the regenerative-type furnace, known in the 1900s as the open-hearth furnace, was developed in the United Kingdom. Adapted to steelmaking, the open-hearth process was the principal method for producing steel throughout the world until 1970. As of this writing (1996), it is still used in China and the CIS.

Since 1970, the most widely used processes for making liquid steel have been the oxygen steelmaking processes, in which commercially pure oxygen (99.5% pure) is used to refine molten pig iron. In the top-blown basic oxygen process, oxygen is blown down onto the surface of the molten pig iron and such coolants as scrap steel or iron ore. In the bottom-blown basic oxygen process, the oxygen is blown upward through the molten pig iron. The direction in which the oxygen is blown has important effects on the final steel composition and on the amount of iron lost in the slag. Some plants use a combination of top and bottom blowing.

Liquid steel is also made by melting steel scrap in an electric-arc furnace. This process is a significant contributor to large-scale production (see RECYCLING, METALS).

Pig iron and iron and steel scrap are the sources of iron for steelmaking in basic-oxygen furnaces. Electric furnaces have relied on iron and steel scrap, although newer iron sources such as direct-reduced iron (DRI), iron carbide, and even pig iron are becoming both desirable and available (see IRON BY DIRECT REDUCTION). In basic-oxygen furnaces, the pig iron is used in the molten state as obtained from the blast furnace; in this form, pig iron is referred to as hot metal.

Pig iron consists of iron combined with numerous other elements. Depending on the composition of the raw materials used in the blast furnace, principally iron ore (beneficiated or otherwise), coke, and limestone, and the manner in which the furnace is operated, pig iron may contain 3.0–4.5% carbon, 0.15–2.5% or more manganese, as much as 0.2% sulfur, and 0.025–2.5% phosphorus; silicon can be as low as 0.15% with modern techniques and is almost always less than 0.8%. Sulfur, phosphorus, and silicon can be reduced significantly by treating the hot metal between the blast furnace and the steelmaking vessel. During the steelmaking process, many but not all solutes are reduced, often drastically.

Steelmaking processes were historically either acid or basic, but acid processes have virtually disappeared. Basic slags, rich in lime, are able to absorb much of the unwanted sulfur and phosphorus, and do not react vigorously with the furnace linings, which are predominantly magnesite [546-93-0] or dolomite [17069-72-6]. In electric melting, when the power loading is high, it may be necessary to water-cool these linings.

Oxidation is employed to convert a molten bath of pig iron and scrap, or scrap alone, into steel. Each steelmaking process has been devised primarily to provide some means by which controlled amounts of oxygen can be supplied to the molten metal undergoing refining. The increase in oxygen potential in the iron bath may be thought of most simply as burning out impurities either as a gas, eg, CO, or an oxide, eg, MnO, P_2O_5, or SiO_2, which dissolved in the slag. More exactly, it is an ionic liquid–metallic liquid equilibration where the distribution coefficient depends on compositions of slag and metal and also on temperature. The impurities, including the majority of the sulfur, exist largely as ions in the slag. As the purification of the pig iron proceeds, owing to the removal of carbon, the melting point of the bath is raised, and sufficient heat must be supplied to keep the bath molten.

In general, steel having similar chemical compositions have similar mechanical and physical properties, no matter by which process they are made, unless the patterns of inclusions (oxides, silicates, and sulfides) are very different.

Methods exist to make impure iron directly from ore, ie, to make DRI without first reducing the ore in the blast furnace to make pig iron which has to be purified in a second step. These processes, generally referred to as direct-reduction processes, are employed where natural gas is readily available for the reduction (see also IRON BY DIRECT REDUCTION). Carbonization of iron ore to make iron carbide as an alternative source of iron units is in its infancy as of the mid-1990s but may grow.

Electric Furnace Processes

The principal electric furnace steelmaking processes are the electric-arc furnace, induction furnace, consumable-electrode melting, and electroslag remelting. The main raw material for all these processes is solid steel. As implied by the term steelmaking, in the first two processes steel is made that is different in composition and shape from the starting material, which is usually steel scrap and/or DRI. The starting material used for the last two processes closely resembles the desired steel ingot subsequently rolled or forged, and yields very high quality steel for applications with extremely strict requirements.

Electric-Arc Furnace. The electric-arc furnace is by far the most popular electric steelmaking furnace. The carbon arc was discovered by Sir Humphry Davy in 1800, but it had no practical application in steelmaking until Sir William Siemens of open-hearth fame constructed, operated, and patented furnaces operating on both direct- and indirect-arc principles in 1878. At that early date, the availability of electric power was limited and very expensive. Furthermore, carbon electrodes of the quality to carry sufficient current for steel melting had not been developed (see FURNACES, ELECTRIC).

The first successful direct-arc electric furnace, patented by Heroult in France, was placed in operation in 1899. The patent covered single- or multiphase furnaces with the arcs placed in series through the metal bath. This type of furnace, utilizing three-phase a-c power, was historically the most common for steel production.

The first direct-arc furnace in the United States was a single-phase two-electrode rectangular furnace of 4-t capacity at the Halcomb Steel Company (Syracuse, New York), which made its first heat in 1906. A similar but smaller furnace was installed two years later at the Firth-Sterling Steel Company in McKeesport, Pennsylvania. In 1909, a 15-t three-phase furnace was installed in the South Works of the Illinois Steel Company, in Chicago, Illinois, which was, at that time, the largest electric steelmaking furnace in the world. It was the first round instead of rectangular furnace and operated on 25-cycle power at 2200 V.

Electric-arc furnaces offer the advantages of low construction costs, flexibility in the use of raw materials, and the ability to produce steels over a wide range of compositions (carbon, alloy, and stainless) and to operate below full capacity.

The biggest change in steelmaking in the last quarter of the twentieth century is the fraction of steel made by remelting scrap or, increasingly, other iron units in an electric-arc furnace (EAF). This change was originally to serve a relatively nondemanding local market, but has increased in both quality and quantity of products to compete with mills using the blast furnace/oxygen steelmaking route (5,6). As of the mid-1990s, market share is approaching 40%. Because the cost of entry into the market is attractive for electric melting relative to integrated plants, and because the market has been strong, a plethora of new plants has been built or is being planned as of the mid-1990s. From 10 to 18 million metric tons of new capacity having advanced features are estimated. Competition is expected to be fierce.

Early processes used three-phase a-c, but increasingly the movement is to a single d-c electrode with a conducting hearth. The high power densities and intense temperatures necessitate both water cooling and improved basic refractory linings. Scrap is charged into the furnace, which usually contains some of the last heat as a liquid heel to improve efficiency. Oxygen is blown to speed the reactions, which do not differ in any significant way from basic-oxygen furnace (BOF) steelmaking. Oxy-fuel burners are also used to accelerate scrap heating and reduce electricity consumption. Various techniques are applied to ensure adequate separation of melt and slag to permit effective ladle treatment. A fairly common practice is to use eccentric bottom tapping (Fig. 1), which minimizes vortexing.

Historically, much of the refining was done in the furnace, including a second slag made after the first meltdown and refining by oxygen blowing. As of the 1990s, the furnace is usually strictly a melting unit of 50–200 t. Final treatment takes place in a ladle furnace, which allows refining, temperature control, and alloying additions to be made without delaying the next heat. The materials are continuously cast with various degrees of sophistication, including slabs down to 50 mm in thickness. Direct casting to sheet seems likely for some products.

The degree to which electric melting can replace more conventional methods is of great interest and depends in large part on the availability of sufficiently pure scrap or other iron units at an attractive price. Some improvements in surface quality are also needed to make the highest value products. Advances in electric furnace technology are being aggressively countered by developments and cost control in traditional steelmaking.

The energy needed to melt steel is much less than that required to reduce iron oxide to a molten product. The latter can be well over 2000 kWh/t for the chemical reaction alone. To melt steel from room temperature takes about 390 kWh/t. By using some preheat from waste gases, actual electrical usages in

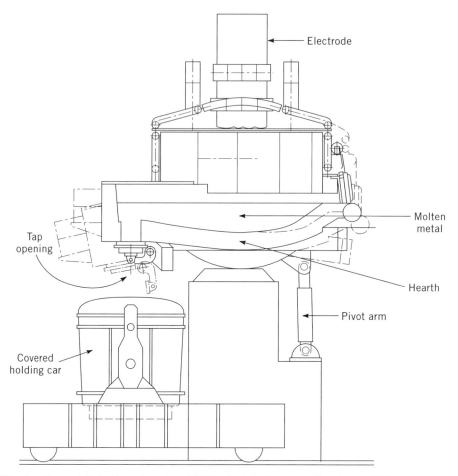

Fig. 1. Eccentric bottom tapping (EBT) electric furnace. Courtesy of Mannesmann.

best practice can be <390 kWh/t, an advance from 450–500 kWh/t needed in the 1980s and still characteristic of many furnaces.

The best labor practices in integrated mills struggle to approach 3.5 h/t for sheet, whereas electric furnaces are easily <1 h/t. The best plants are around 0.5 h/t. At labor costs of $18–$28/h, this represents a real pecuniary advantage. Thus electric melting has taken the segments of the market wherever the product is good enough and by continual improvements it is nibbling away at other segments. As of this writing (1996), deep drawing sheet, tinplate, electro-galvanized steel, and some high quality bars are completely in the integrated sector.

Induction Furnace. The high frequency coreless induction furnace is used in the production of complex, high quality alloys such as tool steels. It is used also for remelting scrap from fine steels produced in arc furnaces, for melting chrome–nickel alloys and high manganese scrap, and more recently for vacuum steelmaking processes.

The induction furnace was first patented in Italy in 1877 as a low frequency furnace. It was first commercially applied, installed, and operated in Sweden. The first installation in the United States was made in 1914 by the

American Iron and Steel Company in Lebanon, Pennsylvania; however, it was not successful. Other low frequency furnaces have been operated successfully, especially for stainless steel.

Most commonly, the melting procedure is essentially a dead-melt process, that is, the solid constituents of the charge melt smoothly and mix with each other. Little if any refining is attempted in ordinary induction melting, and no chemical reactions occur where gaseous products agitate the molten bath. The charge is selected to produce the composition desired in the finished steel with a minimum of further additions, except possibly small amounts of ferroalloys as final deoxidizers. For ordinary melting, the high frequency induction furnace is not used extensively.

Vacuum and Atmosphere Melting. A coreless high frequency induction furnace is enclosed in a container or tank which can be either evacuated or filled with a gaseous atmosphere of any desired composition or pressure. Provision is made for additions to the melt, and tilting the furnace to pour its contents into an ingot mold also enclosed in the tank or container without disturbing the vacuum or atmosphere in the tank (Fig. 2).

Although vacuum melting has often been employed as a remelting operation for very pure materials or for making electrodes for the vacuum consumable-electrode furnace, it is generally more useful in applications that include refining. Oxygen, nitrogen, and hydrogen are removed from the molten metal in vacuum melting, as well as carbon when alloys of very low carbon content are being produced (7,8).

Consumable-Electrode Melting. This refining process produces special-quality alloy and stainless steels by casting or forging the steel into an electrode that is remelted and cast into an ingot in a vacuum (Fig. 3). These special steels include bearing steels, heat-resistant alloys, ultrahigh strength missile and aircraft steels, and rotor steels.

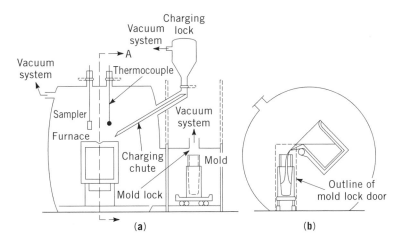

Fig. 2. Schematic arrangement of a furnace in a vacuum chamber equipped with charging and mold locks for vacuum induction melting (1): (**a**) front cross section; and (**b**) section A during pouring.

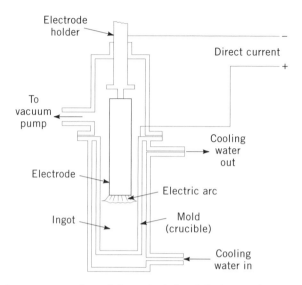

Fig. 3. Schematic representation of the principle of design and operation of a consumble-electrode furnace for melting steels in a vacuum (1).

A consumable-electrode furnace consists of a tank that encloses the electrode and a water-cooled copper mold. After the furnace has been evacuated, power is turned on and an arc is struck between the electrode and a starting block that is placed in the mold before operation begins. Heat from the arc progressively melts the end of the electrode. Melted metal is deposited in a shallow pool of molten metal on the top surface of the ingot being built up in the mold. Rate of descent of the electrode is automatically controlled to maintain the arc. The remelting operation removes gases (hydrogen, oxygen, and nitrogen) from the steel, improves its cleanliness, produces an ingot that exhibits practically no center porosity or segregation, and improves hot workability and the mechanical properties of the steel at both room and high temperatures. A second remelting can further improve quality.

Electroslag Remelting.   Remelting of electroslag has the same general purpose as consumable-electrode melting, and a conventional air-melted ingot serves as a consumable electrode. No vacuum is employed. Melting takes place under a layer of slag that removes unwanted impurities. Grain structure and orientation are governed by controlled cooling during solidification.

Powder Techniques.   Highly alloyed materials made by the processes described are particularly susceptible to segregation of alloying elements during solidification both on a macro- and a microscale. Much plastic working was necessary to minimize this susceptibility before service applications.

An ingenious method to avoid or reduce segregation of alloying elements involves preparing small spheres of material by the atomization of a liquid stream through a nozzle to produce a powder. This powder can be compacted, often hot and triaxially by gas pressure, to form a material where, on further heating, the residual pores close by diffusion to approach 100% density.

Oxygen Steelmaking Processes

In oxygen steelmaking, 99.5% pure oxygen gas is mixed with hot metal, causing the oxidation of the excess carbon and silicon in the hot metal and thereby producing steel. In the United States, this process is called the basic-oxygen process (BOP) (4,9,10). The first U.S. commercial installation began operation in 1955.

Blowing with oxygen was investigated in Germany and Switzerland some years before the first commercial steelmaking plants to use this method began operation in Austria in the early 1950s. This operation was designed to employ pig iron produced from local ores that were high in manganese and low in phosphorus. The process spread rapidly throughout the world, largely displacing open hearths.

Top-Blown Basic Oxygen Process. The top-blown basic oxygen process is conducted in a cylindrical furnace somewhat similar to a Bessemer converter. This furnace has a dished bottom without holes and a truncated cone-shaped top section in which the mouth of the vessel is located. The furnace shell is made of steel plates ca 50-mm thick; it is lined with refractory 600–1200-mm thick (11).

A jet of gaseous oxygen is blown at high velocity onto the surface of a bath of molten pig iron and scrap at the bottom of the furnace by a vertical water-cooled retractable lance inserted through the mouth of the vessel (Fig. 4). The furnace is mounted in a trunnion ring and can be tilted backward or forward.

When the furnace is tilted toward the charging floor, which is on a platform above ground level, solid scrap is dumped by an overhead crane into the mouth of the furnace. Scrap can form up to 30% of the charge unless it is preheated, when up to 45% may be used. The crane then moves away from the furnace and another crane carries a transfer ladle of molten pig iron to the furnace and pours the molten pig iron on top of the scrap.

The manganese residue of the blown metal before ladle additions is generally higher than in an open-hearth process and is closely related to the amount of manganese in the furnace charge. High slag fluidity and good slag–metal contact promote transfer of phosphorus from the metal into the slag, even before the carbon reduction is complete. Efficiency of sulfur elimination is as good

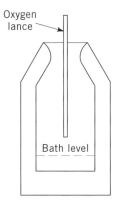

Fig. 4. Sketch of a basic-oxygen steelmaking furnace, where an oxygen lance is inserted through the mouth of the furnace (1).

as or better than that in the basic open-hearth process because the bath action is more vigorous, the operating temperatures are higher, and the fuel does not contain sulfur.

Because the basic-oxygen process uses a refining agent containing practically no nitrogen, the product has a low nitrogen content. Oxygen residues depend on the carbon content.

Residual alloying elements such as copper, nickel, or tin are usually considered undesirable. Their main source is purchased scrap. Because of the generally high consumption of hot metal in the basic-oxygen process, the residual alloy content is usually sufficiently low, depending on the quality of the purchased scrap.

No external heat source is required. In all types of steelmaking that employ pig iron, which melts at temperatures well below low carbon steel, the heat balance between exothermic oxidation of elements, such as C, Si, and Mn, and the cooling provided by scrap or sometimes other endothermic coolants, such as iron ore, are critical issues. The numerical factors are well understood and are routinely contained in computer programs used by operators. If the balance is such that the temperature after blowing is too high, refractory consumption is increased significantly.

After charging, the furnace is immediately returned to the upright position, the lance is lowered into it to the desired height above the bath, and the flow of oxygen is started. Striking the surface of the liquid bath, the oxygen immediately forms iron oxide, part of which disperses rapidly through the bath. Carbon monoxide generated by the reaction of iron oxide and carbon is evolved, giving rise to a violent circulation that accelerates refining. Some of the sensible heat of combustion of this carbon monoxide can be captured enabling more scrap to be melted for a given bath composition.

Slag-forming fluxes, chiefly burnt lime, fluorspar, and mill scale (iron oxides), are added in controlled amounts from an overhead storage system shortly before or after the oxygen jet is started. These materials, which produce a slag of the proper basicity and fluidity, are added through a chute built into the side of a water-cooled hood positioned over the mouth of the furnace. This hood collects the gases and the dense reddish brown fume emitted by the furnace during blowing, and conducts them to a cleaning system where the solids are removed from the effluent gas before it is discharged to the atmosphere. Without access to sintering machines, which are largely closed by environmental constraints but used to be valuable facilities to recycle dusts of various types, easy disposal is a problem (12).

The oxidizing reactions take place so rapidly that a 300-metric-ton heat, for example, can be processed in ca 30 min. The intimate mixing of oxygen with the molten pig iron permits this rapid refining. Mixing oxygen with the molten pig iron gives a foamy emulsion of oxygen, carbon monoxide, slag, and molten metal which occupies a volume ca five times that of the slag and molten metal. The furnace must be big enough to accommodate this foamy mixture.

The mechanism of carbon elimination is similar to those of the earlier open-hearth processes, ie, oxidation of carbon to carbon monoxide and carbon dioxide. The chemical reactions and results are the same in both cases. The progress of the reaction is plotted in Figure 5.

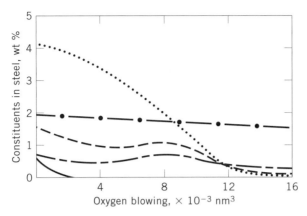

Fig. 5. Progress of refining in the basic-oxygen steelmaking process, where (——) corresponds to Si; (— — —) to Mn; (·····) to C; (— — — —) to P; and (——•——) to S. The values for phosphorus are one-tenth those indicated on the ordinate; the values for sulfur are one-hundredth those indicated. The time elapsed is ~20 minutes in most cases.

The general reaction, whereby silicon is oxidized to silica and transferred to the slag, applies to the basic-oxygen process. Oxidation of silicon is important mainly because of its thermal effects. Only a trace of silicon remains in the steel at the end of the refining period. The amount of silicon in pig iron varies from around 0.8% to as low as 0.15%. Lower silicon values lead to lower slag volumes but take more energy in the blast furnace, and produce less heat during steelmaking.

Refractory lining life was at one time an important maintenance cost, but developments using a nitrogen lance to splash slag over the walls have increased lining life by a factor of ~10 (2000 to 20,000 heats).

After the oxygen blowing is completed, the lance is withdrawn and the furnace is tilted to a horizontal position. Historically, temperature and composition were measured at this point, and if the reaction was not perfect, a further, short reblow or scrap addition was carried out. As of the mid-1990s sufficient sensors (qv) and models are available so that the number of reblows is virtually zero, thereby contributing to productivity. In general, rather than aiming to make a variety of compositions during blowing, a generic low carbon steel is made and modified in the ladle. The furnace is tilted toward its taphole side and the steel is tapped into a waiting ladle. After the steel has been tapped, the furnace is inverted by tilting toward its opposite side and the remaining slag is dumped into a slag pot. The furnace is then turned to charging position and is ready for the next heat.

Bottom-Blown Basic-Oxygen Process. The bottom-blown basic-oxygen process, called oxygen bottom metallurgy (OBM) in Europe (Maxhülte) and Quelle basic-oxygen process (Q-BOP) (*quelle* is "fountain" or "gusher" in German) in the United States and Japan, is also conducted in a furnace similar to a Bessemer converter. The furnace comprises two parts, the bottom and the barrel. Both the bottom and barrel consist of an outer steel shell that is about 50-mm thick and lined with refractory that is 600–1200-mm thick. The bottom contains

6–24 tuyeres or double pipes (Fig. 6). Oxygen is blown into the furnace through the center pipes and natural gas or some other hydrocarbon is blown into the furnace through the annular space between the two pipes (13–15).

Considerable heat is generated when the oxygen gas oxidizes the carbon, silicon, and iron in the molten pig iron. If special precautions are not taken, the temperature of the refractory immediately adjacent to the oxygen inlet increases when the oxygen is blown upward through the furnace bottom, as in a Bessemer converter. No commercial refractory has been found that withstands such temperatures without cooling. Surrounding the oxygen jet with a cylindrical stream of a hydrocarbon such as methane imparts the required cooling by endothermic decomposition of the hydrocarbon (Fig. 7). The bottom-blown oxygen process is critically dependent on this technique.

Both the OBM and the Q-BOP processes are operated in about the same way as the top-blown process. The furnace is charged with steel scrap and hot metal, and oxygen is blown into this mixture to produce steel of the desired composition. However, the method of adding lime to the furnace is different. In the bottom-blown processes, powdered lime is added to the oxygen before it is blown into the furnace. Thus, the oxygen serves as a carrier gas to transport the powdered lime pneumatically into the furnace. The powdered lime dissolves rapidly in the slag, permitting increased oxygen-blowing flow rate and thus a decrease in blowing time. Furthermore, slopping is avoided, ie, the sudden ejection of appreciable

Fig. 6. Liquid blast-furnace iron (hot metal) being charged into a 200-metric ton Q-BOP furnace in preparation for making a heat of steel.

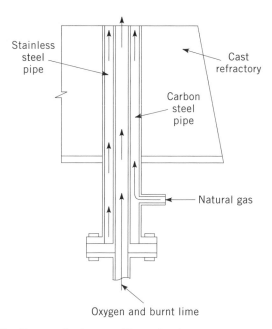

Fig. 7. Tuyere for bottom-blown basic-oxygen process (16).

amounts of slag and metal from the furnace. In the top-blown process, pebble-sized lime is used with pieces 2.5–7.5 cm in diameter.

The chemistry of the bottom-blown processes is similar to that of the top-blown process. However, the slags in the bottom-blown processes contain significantly less iron oxide, causing a ~2% increase in yield, that is, in the amount of liquid steel produced from a given charge of hot metal and scrap. Because bottom blowing is generally closer to thermodynamic equilibrium, the final carbon can be as low as 0.01% without excessive oxidation of iron into the slag, a source of serious yield loss. In top-blown vessels, the carbon is usually 0.03%.

Decrease in the iron oxide content of the slag also causes an increase in the amount of manganese remaining in the liquid steel at the end of the oxygenation stage (oxygen blow). Manganese is desirable in the finished steel and is almost always added after the steel is tapped from the furnace into a ladle. Thus, a smaller amount of manganese can be added to bottom-blown steel during tapping, which results in decreased costs. However, the low iron oxide content of the slag increases the slag viscosity, and liquid steel is entrapped in the slag (also a yield loss), unless sufficient fluorspar or other slag fluidizer is added to the charge.

Top- and Bottom-Blown Basic-Oxygen Processes. During the 1970s, several combinations of top and bottom blowing were developed. In the lance bubbling equilibrium (LBE) process, nitrogen or argon is injected through a number of porous refractory plugs installed in the furnace bottom, while oxygen is top-blown into the furnace through a lance. The bottom injection of nitrogen or argon causes more intimate mixing of slag and metal, and hence most of the advantages of the Q-BOP bottom-blowing process are obtained in furnaces designed for top blowing. The Linz-Donawitz-Kawasaki gas (LD-KG) process is similar to the LBE process but the nitrogen or argon is injected through tuyeres in the furnace bottom rather than through porous refractory plugs (15).

Determination of Chemical Composition

The chemical composition of a given steel is generally specified by the customer within rather narrow limits for each element (other than residual elements), so the chemical composition of the steel during refining is followed closely. However, it is no longer necessary to obtain nearly instantaneous analysis during a blow. By blowing fully and then using the ladle to add alloys and control temperature, accurate composition control has become much simpler. Robots are used to carry out analysis in some shops (see ROBOTICS, MANUFACTURING (SUPPLEMENT)).

Scrap as Raw Material

Scrap consists of the by-products of steel fabrication and worn-out, broken, or discarded articles containing iron or steel (17,18). It is a principal source of the iron for steelmaking; the other source is iron from blast furnaces, either molten as it comes from the furnace (hot metal) or in solid pig form. Scrap is of great practical value. Every ton of scrap consumed in steelmaking is estimated not only to displace and conserve for future use 3.5–4 t of natural resources, including iron ore, coal, and limestone, but also to contribute positively to the goal of total recycling. The amount of home scrap, ie, the various trimmings that accumulate during production, has been greatly reduced. The introduction of almost-complete continuous casting has limited the discards from ingots that made up a large part of home scrap.

Purchased scrap comes in two basic categories. Industries that have a steady supply of discards from such operations as stamping, drawing, machining, and forging count on resale of what is often a premium product (if identity has been maintained) as part of their routine costs. Less economically desirable material comes from salvaging the values in products such as ships, automobiles, railroad equipment, and buildings. A whole industry has grown up to recycle the many millions of scrapped automobiles, a mixture of steels, polymers, copper, etc. Following removal of some discrete components such as batteries, the hulk passes through a shredder, the products are given further separation, and a usable scrap becomes available. The shredded polymers give rise to a product known as fluff, which, as of this writing, is not recyclable and must be landfilled (see RECYCLING, METALS).

Scrap can have many compositions and sizes, which translate to charge density, and its economic value is based on these. During the latter 1980s and early 1990s, the price spiked as high as $190/t for the best-quality scrap. At such prices, electric melting becomes marginal, and the smaller amount used in BOF melting helps to cushion the price to some degree. Another issue arises in scrap quality. For the most demanding brands of sheet steel, certain impurities affect the product and thus are made in integrated shops. As a result, there is increasing demand for sources of iron units other than scrap that is free of impurities. Examples are iron carbide, which is in the early demonstration stage, and iron ore treated by reducing gases to give direct-reduced iron. Several million metric tons per year of the latter are on the world market as of the mid-1990s.

The choice of metallic charge is a complex issue, involving the markets to be served and the prices expected. Nevertheless, scrap of all types is expected to remain important because the energy of melting is so much lower than that for reduction from ore. Preheating of scrap using the sensible heat of the previous melt is being practiced increasingly and can be significant in the thermal balance. For instance, the amount of scrap in a BOF can approach 45% rather than the 30% if cold scrap is used.

Scrap is a worldwide commodity. The United States alone exports some 10,000,000 t/yr. The total circulation of scrap in the United States is in the range of $70-90 \times 10^6$ t/yr, including that for foundry operations, which is ca 10×10^6 t/yr, and modest quantities charged to blast furnaces when this is economical.

The ability to tap an array of iron units to provide flexibility in all parts of the business cycle is a topic of great interest in keeping costs of raw steel down in the international arena. This is being combined with efforts to reduce the time for a given process by even a few seconds.

Chemical Composition. A standard problem in purchased scrap is the accuracy of analysis of composition. For home scrap, this may not be much of an problem, but sampling (qv) difficulties and analytical costs can be troublesome in other cases. The difficulty of estimating the alloy content, whether this is to be used to reduce the amount of additional additives necessary in the ladle or to avoid possible troubles from copper or tin, is a continuing worry. Elements such as lead in free machining steels, which can harm furnace linings, are becoming of less concern as alternative elements replace lead. Volatile elements such as lead, zinc, and antimony vaporize and may require adequate capture mechanisms. Elements such as aluminum can add to the heat of oxidation in the furnace and allow higher amounts of scrap. Finally, there is a long-standing interest in purifying scrap to allow higher economic value but these efforts have only been effective sporadically.

Addition Agents

In steelmaking, various elements are added to the molten metal to effect deoxidation, control of grain size, improvement of the mechanical, physical, thermal, and corrosion properties, and other specific results. Originally, the chemical element to be incorporated into the steel was added to the bath in the form of an alloy that consisted mainly of iron but was rich in the desired element. Such alloys, because of their high iron content, became known as ferroalloys and were mostly produced in iron blast furnaces. Later, the production of alloys for steelmaking was carried out in electric-reduction and other types of furnaces, and a number of these alloys contain very little iron. For this reason, the term addition agent is preferred when describing the materials added to molten steel for altering its composition or properties; ferroalloys are a special class of addition agents.

Included in the ferroalloy class are alloys of iron with aluminum, boron, calcium, chromium, niobium, manganese, molybdenum, nitrogen, phosphorus, selenium, silicon, tantalum, titanium, tungsten, vanadium, and zirconium.

Some of these chemical elements are available in addition agents that are not ferroalloys, as well as in almost pure form. These include relatively pure metals such as aluminum, calcium, cobalt, copper, manganese, and nickel; oxides of molybdenum, nickel, and tungsten; carbon, nitrogen, and sulfur in various forms; and alloys consisting principally of combinations of two or more of the foregoing elements. Some rare-earth alloys are also used for special purposes, but to a minor extent. Addition agents are predominantly introduced in the ladle.

The economical manufacture of alloy steels requires consideration of the relative affinity of the alloying elements for oxygen as compared with the affinity of iron for oxygen. For example, copper, molybdenum, or nickel may be added with the charge or during the working of the heat and are fully recovered. The advent of ladle metallurgy has led to the easily oxidized elements, such as Al, Cr, Mn, B, Ti, V, and Zr, being added after the oxygen content of the bath has been reduced. A marked increase in yield has resulted.

Ladle Metallurgy

The finished steel from any furnace, whether basic-oxygen or electric, is tapped into ladles. Most ladles hold all the steel produced in one furnace heat. Some slag is allowed to float on the surface of the steel in the ladle to form a protective blanket. Excess slag is prevented from entering the ladle by controlling its exit from the furnace taphole.

A ladle consists of a steel shell, lined with refractory brick, having an off-center opening in its bottom equipped with a nozzle (Fig. 8). A valve makes it possible to enlarge or close the opening to control the flow of steel through the nozzle. Simple stopper-rod valves have been replaced by more elaborate slide gate systems involving spring-loaded refractory disks, each containing a central hole and sliding relative to the other. Better flow control and positive shutoff are achieved.

Ladle metallurgy, the treatment of liquid steel in the ladle, is a field in which several new processes, or new combinations of old processes, continue to be developed (19,20). The objectives often include one or more of the following

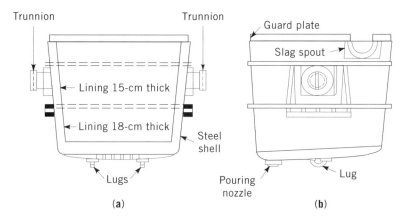

Fig. 8.   Steel ladle: (**a**) vertical section through the trunnions, and (**b**) side view.

on a given heat: more efficient methods for alloy additions and control of final chemistry; improved temperature and composition homogenization; inclusion flotation; desulfurization and dephosphorization; sulfide and oxide shape control; and vacuum degassing, especially for hydrogen and carbon monoxide to make interstitial-free (IF) steels. Electric arcs are normally used to raise the temperature of the liquid metal (ladle arc furnace).

Argon Treatment. In early ladle furnaces, argon was used to provide better mixing of additives, such as ferroalloys, to even out temperature gradients and to float out inclusions such as oxides and sulfides. Argon could be added through a porous plug in the bottom of the ladle or through a lance inserted in the bath. Expensive ferroalloys can be added through the area where the slag is displaced by rising bubbles of argon without significant oxidation loss. These are the composition adjustment by sealed argon bubbling (CAS) and sealed argon bubbling (SAB) processes. At first, there was no cover on the ladle, but as of the 1990s almost all ladles have covers to prevent atmospheric oxidation.

In the argon–oxygen decarburization (AOD) process, argon or argon–oxygen mixtures are blown in through bottom tuyeres to lower the carbon content of the bath by making escape of carbon monoxide in the argon bubbles easier. Although this process works on carbon steel, better alternatives are available. The predominant use of AOD is in making stainless steels more weldable by lowering the carbon content to 0.03% max.

When the ladle is fitted with a tight cap, other processes become attractive, especially if the steelmaking slag, which contains considerable iron oxide and hinders sulfur removal, can be substantially separated from the melt before further treatment. Sulfur in solution can be removed as sulfide by injecting CaSi or Mg (the Thyssen Niederrhein (TN) process) or 90% CaO–10% CaF_2 (Kimitsu injection (KIP) process). Thermodynamically, calcium is more effective than magnesium in reducing S; CaF_2 increases sulfide capacity of slags, as does increasing the basicity, ie, increasing Al_2O_3 and reducing SiO_2. CaF_2 also increases fluidity. When the proper oxygen–sulfur balance is present in liquid steel, calcium additions can give hard, spherical inclusions which do not deform during rolling, as does, eg, MnS. Thus Ca additions can lead to better properties in the direction normal to the rolling direction. With a capped ladle, addition of ferroalloys into the plume of argon bubbling up through the steel through a synthetic slag (with little or no FeO) is also effective (capped argon bubbling (CAB) process).

Vacuum Processes. More complete control over ladle treatment is achieved by the ability to seal a vessel well enough so that a good vacuum can exist over the steel. Although the expense can be justified for steels with the most difficult property requirements, for many purposes less elaborate treatments are adequate. Many possible configurations exist (21).

Some early processes involved stream degassing, where on entering the vacuum chamber the molten steel fragmented into large numbers of droplets, thus offering a large surface for loss of gases such as hydrogen. This was not very effective in inclusion or carbon removal. Another process was the Dortmund-Hoerder (DH) process, where an evacuated vessel can be lowered into the ladle for 10 to 20 cycles and thus the steel is exposed to an adequately low vacuum (0.13 kPa (1 mm Hg)). Ferroalloy additions can be made *in situ*.

A more complicated system is found in the Ruhrstahl-Heraeus (RH) process (Fig. 9), which has two hollow legs as compared to only one in the DH. By bubbling argon into one leg, a pressure difference is created circulating steel though the vacuum chamber and back into the vessel. Oxygen and carbon dissolved in the steel react to form CO, which escapes. Reduction of carbon from 0.08 to 0.03% and oxygen from 400 to 50 ppm occurs, at which point the steel can be economically killed with aluminum. The higher carbon contents at entry improve the yield of iron in the BOF vessel. In the Ruhrstahl-Heraeus oxygen blowing (RH-OB) process, even higher carbon contents (0.10%) can be treated, with oxygen blown into the degasser, thus further improving yield. For interstitial-free steels, the start carbon, which may be 0.03%, is reduced to ≤20 ppm. Oxygen and nitrogen are also held at levels near 20 ppm. A very high quality sheet steel is produced with exceptional deep-drawing properties.

The simpler vacuum oxygen decarburization (VOD) process used originally for stainless steels can be used for carbon steel by placing the ladle in an evacuated tank, introducing argon into the bottom of the ladle, and blowing oxygen through a lance. The carbon can be reduced to 0.01%.

In other systems, extra energy to maintain temperature can be supplied by an arc, chemical heating by injecting Al and O_2, or by induction stirring with or without vacuum. Sulfur and oxygen are reduced even without a vacuum; hydrogen is reduced if a vacuum is present.

A principal purpose of ladle metallurgy is to produce a well-stirred and homogeneous bath. Thus considerable effort has been spent on the fluid dynamics necessary to ensure this. Water has been a valuable tool as a modeling agent.

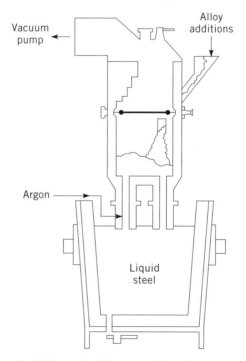

Fig. 9.   Schematic diagram of the RH degasser.

All special processes involve extra capital and operating costs, time, and a realistic analysis of markets and procedures. Because of increasing quality demands, some form of ladle treatment has become essential in steelmaking of the 1990s.

Casting

Continuous Casting. Formerly, all steel for hot working was cast into ingots. As of the mid-1990s, however, very little steel is cast as ingots in the United States, except for limited quantities where the typical 250-mm cast slab thickness is inadequate for the final product, eg, heavy plate. For certain special property requirements, bottom pouring of the ingot through a refractory riser is common.

The possibility of casting molten steel continuously into useful shapes equivalent to conventional semifinished shapes, such as slabs and billets, and thus eliminating the ingot and primary-mill stages of rolled-steel production, led to continuous attempts to solve the problem by many investigators using a variety of machine designs. Because of the high melting point, high specific heat, and low thermal conductivity of steel, success did not come easily in the case of ferrous metals. For nonferrous metals, on the other hand, continuous-casting quickly proved practical.

The obvious rewards available to successful casting provided a strong driving force. Early machines built in the 1930s and 1940s were not practical, but progress was made in the 1960s, which included such novel ideas as following the caster with a rolling mill to change the otherwise fixed cross section, a constraint that limits yield. Around 1970, a great surge in capacity was nearly complete in Japan, and casters had been an integral concept. The decision was made by Nippon Steel to forbid any ingot casting at their then-new sheet works in Oita. Thus the casters had to work without fail and new levels of operating confidence were achieved. Consortia of steel companies and equipment builders arose to provide reliable casters, and to permit second generation improvements such as molds having widths that could be varied during operation and sensors that could measure necessary variables of interest to long strings of casts. Slowly it became possible to reach truly continuous operation with runs of a week or more, making width and composition chances along the way. Improvements in the steel supplied were critical (22–24).

For various reasons, including financial ones, the United States was slower than Japan and Europe to install continuous slab casters for the production of sheet. Electric melters cast billets continuously from about 1975 onward. Casting was done crudely at first but the sophistication increased rapidly, culminating in the operation of a thin-slab caster at one of Nucor's plants. This opened up another avenue of attack on sheet markets, once the province of integrated mills.

Although the continuous casting of steel appears deceptively simple in principle, many difficulties are inherent to the process. When molten steel comes into contact with a water-cooled mold, a thin solid skin forms on the wall (Fig. 10). However, because of the physical characteristics of steel, and because

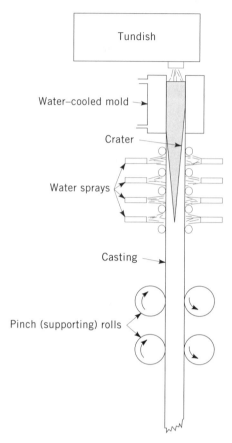

Fig. 10.   Schematic representation of the process of cooling of steel (■) during continuous casting. The tundish is a buffer between the ladle and the mold.

thermal contraction causes the skin to separate from the mold wall shortly after solidification, the rate of heat abstraction from the casting is low enough that molten steel persists within the interior of the section for some distance below the bottom of the mold. The thickness of the skin increases because the action of the water sprays as the casting moves downward and, eventually, the whole section solidifies.

The mass of the solidifying section is supported as it descends by driven pinch rolls that control the speed of descent by a roller apron and is withdrawn from the mold. Any tendency for the casting to adhere to the mold wall may cause the skin to rupture. However, molds that move up and down for a predetermined distance at controlled rates during casting have practically eliminated the sticking problem (Fig. 11). The continuous-cast section issuing from the withdrawal or supporting rolls may be disposed of in several ways, some of which are illustrated in Figure 12.

A number of other designs of continuous-casting machines are being used in the production of semifinished sections. For example, bars and light structural

Fig. 11. Continuous casting of square steel blooms. The liquid steel is poured into molds (not shown) at top of photograph. The solidifying steel blooms move downward and then toward the camera. In the foreground, the blooms are cut to the proper length.

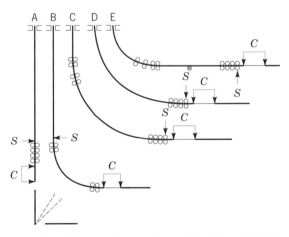

Fig. 12. Development of casting machines where S is the end of the supported length and C the cutoff zone; circles represent rolls. A is a vertical mold; B a vertical mold with bending; C a vertical mold with progressive bending; D a curved mold with straightening; and E a curved mold with progressive straightening (1).

785

sections are made from billets, ie, rounds or round-cornered squares having sizes in the range of 100 to 250 mm, and often having several (typically up to six) strands in parallel. At one time, rounds were cast in a rotating mold but the hoped-for better quality can be obtained more easily using a stationary mold and more care with the liquid feed.

Many other casting schemes have been tried. There has been some measure of success but limited general application. Examples are horizontal casting, pressure casting (Fig. 13**a**) (25), and wheel-belt casting. The most important development in the 1980s was the thin-slab (~50 mm) caster, which has the potential of reducing the number of hot rolling stands from eight or nine down to perhaps five or less. However, the full potential has not yet been realized because of shape problems (Fig. 13**b,c**). A variant on the thin-slab caster would use 85- to 100-mm thick slabs. There are some technical issues to be solved in balancing the ease of feeding the mold and the greater area of surface produced, which can lead to problems in high quality material such as deep-drawing sheet. However, all electric shops making flat rolled products are expected to use thin-slab casters until direct-sheet casting arrives.

This logical extension of thin slabs to cast directly to sheet in the 1-mm range is a possibility long recognized but difficult to realize. Progress in both stainless and carbon steel has been rapid and early prototypes are expected by the year 2000. Several schemes have been tried in laboratories all over the world.

Continuous casting is almost universal. It gives a higher yield than ingot casting and avoids the cost of rolling ingots into slabs because the slabs are produced directly from liquid steel. Excessive segregation during solidification can create problems of low ductility in regions of high carbon and alloy content, and cause cracking during processing. In general, segregation increases with increasing thickness, ie, with slower cooling rates. Continuous-cast slabs are only ca 20–25-cm thick and exhibit much less segregation than slabs rolled from ingots that are 3–5 times as thick. Consequently, there is usually less variation in steel composition from one heat to the next in continuous-cast steel than there is from the top to the bottom in a single ingot.

Continuous-cast aluminum-killed steel is similar in composition and properties to ingot-cast aluminum-killed steel. It contains 0.025–0.060% aluminum and has excellent deep-drawing characteristics. Whereas earlier developments merely involved care in shielding the liquid stream from the atmosphere which could produce undesirable inclusions, great care is taken in later efforts not only to prevent chemical contamination but to understand and control the fluid flow in the mold and the tundish, the latter a buffer between the ladle and the mold, and holding up to 70 t of liquid steel. The aim is to avoid mixing of metal and slag and also to give time and opportunity for any inclusions that do form in the mold to float to the top and be absorbed in a special slag.

A vast array of sensors provides data to the operators so that accidents such as breakouts from thin spots in the solidifying shell very rarely occur and the internal and surface quality are extremely good. In fact, the surface is good enough so that cooling to room temperature for defect removal is often not necessary if certain criteria are met, and slabs can be charged while still hot for rolling, thereby saving much of their sensible heat.

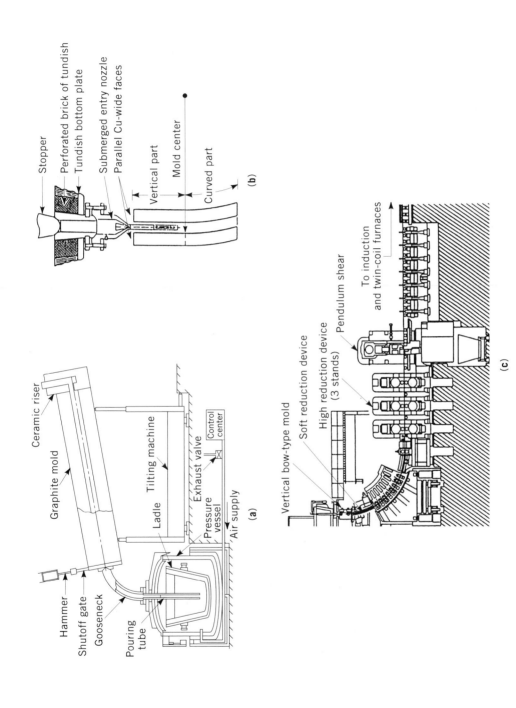

Fig. 13. (**a**) The bottom-pressure casting method as applied to slabs; (**b**) curve mold and submerged entry nozzle; and (**c**) portion of Arvedi caster for in-line strip production (ISP). Courtesy of the Iron & Steel Society.

Plastic Working of Steel

Plastic working of a metal such as steel is the permanent deformation accomplished by applying mechanical forces to a metal surface. The primary objective is usually the production of a specific shape or size (mechanical shaping), although increasingly it also involves the improvement of certain physical and mechanical properties of the metal (mechanical treatment). These two objectives can be readily attained simultaneously.

Plastic deformation of steel can be accomplished by hot working or cold working. Prior to hot working, the steel is heated to 1090–1310°C, depending on the grade and the work to be accomplished. The force required to deform the metal is very sensitive to the rate of application and the temperature of working; however, after deformation, the basic strength of the steel is essentially unchanged. In cold working, on the other hand, steel is not heated before working, and the force required to cause deformation is relatively insensitive to the application rate and temperature variations, but the yield strength of the steel is increased.

The principal hot-working techniques are hammering, pressing, extrusion, and rolling (Fig. 14), the first two of which are called forging. Other methods include rotary swaging, hot spinning, hot deep-drawing, roll forging, and die forging. Although at high temperatures the material is not very strong when compared to its room temperature strength, the shape changes are made in several stages for reasons of mechanics. For example, the maximum reduction that can be obtained in a single roll pass is 10–40% because the ability of the rolls to drag the piece through depends on geometric factors and the coefficient of friction. Thus, to reduce a 250-mm slab to a 2-mm hot band takes about 10 separate reductions. During this process the austenite normally present may recrystallize several times and the recrystallization can have a significant effect on final properties (26–31).

Cold working is generally applied to bars, wire, strip, sheet, and tubes. It reduces the cross-sectional area of the piece being worked on by cold rolling,

Fig. 14.   Hot rolling of steel slabs to plate. In the foreground is a steel plate mill; behind the plate is the next slab to be rolled.

cold drawing, or cold extrusion. Similar limitations on reduction per pass apply, although recrystallization does not occur. Trying to get reductions too large per pass can lead to loads high enough to break rolls. Cold working imparts improved mechanical properties, better machinability, good dimensional control, bright surface, and production of thinner material than can be accomplished economically by hot working (see METAL TREATMENTS). The thickness of available hot-rolled material is decreasing steadily and hot-rolled material is becoming a less-expensive competitor for some material traditionally cold-rolled and annealed.

Metallography and Heat Treatment

The great advantage of steel as an engineering material is its versatility. Properties can be controlled and changed by heat treatment. Thus, if steel is to be formed into some intricate shape, it can be made very soft and ductile by heat treatment; on the other hand, alternative heat treatments can also impart high strength.

The physical and mechanical properties of steel depend on its microstructure, that is, the nature, distribution, and amounts of its metallographic constituents as distinct from its chemical composition. The amount and distribution of iron and iron carbide determine most of the properties, although most plain carbon steels also contain manganese, silicon, phosphorus, sulfur, oxygen, and traces of nitrogen, hydrogen, and other chemical elements such as aluminum and copper. These elements may modify, to a certain extent, the main effects of iron and iron carbide, but the influence of iron carbide always predominates. This is true even of medium alloy steels, which may contain considerable amounts of nickel, chromium, and molybdenum.

There are two allotropic forms of iron: ferrite and austenite. Ferrite in both its low or α (up to 910°C) and high or δ (1390°C to mp 1536°C) is body-centered cubic (bcc). The spaces between atoms (interstices) are larger in the face-centered cubic (fcc) austenite than in ferrite and can accommodate more small interstitial atoms such as carbon and nitrogen. Thus the solubility in austenite is 2 wt % for carbon and 2.8 wt % for nitrogen, as opposed to a few hundredths of a percent for ferrite. This effect is critical in the heat treatment of steels where ferrite and austenite may also contain alloying elements such as manganese, silicon, or nickel. The atomic arrangement in the two allotropic forms of iron is shown in Figure 15.

Cementite, the term for iron carbide in steel, is the form in which carbon appears in steels. It has the formula Fe_3C, and thus consists of 6.67 wt % carbon and the balance iron. Cementite is very hard and brittle. As the hardest constituent of plain carbon steel, it scratches glass and feldspar, but not quartz. It exhibits about two-thirds the induction of pure iron in a strong magnetic field, but has a much lower Curie temperature.

Most commercial steels contain at most a few tenths of a percent of carbon. In fact the great bulk of sheet steel which accounts for about 60% of commercial production has less than 0.1%. Thus when these materials are heated to temperatures where austenite is stable, the carbon readily dissolves. Iron–nitrogen alloys are essentially not commercial. The limiting temperature for austenite

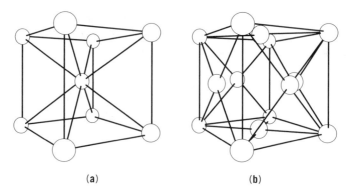

(a)   (b)

Fig. 15. Crystalline structure of allotropic forms of iron. Each white sphere represents an atom of (**a**) α- and δ-iron in bcc form; and (**b**) γ-iron in fcc form (1).

are given by phase diagram (Fig. 16). By controlling the cooling rate from the austenite range, the carbon comes out of solution in a range of microstructures that have very specific properties (32–35).

   A nonalloyed carbon steel having 0.76% carbon, the eutectoid composition, consists of austenite above its lowest stable temperature, 727°C (the eutectoid temperature). On reasonably slow cooling from above 727°C, transformation of the austenite occurs above about 550°C to a series of parallel plates of α plus cementite known as pearlite. The spacing of these plates depends on the temperature of transformation, from 1000 to 2000 nm at about 700°C and below 100 nm at 550°C. The corresponding Brinell hardnesses (BHN), which correspond approximately to tensile strengths, are about BHN 170 and BHN 400, respectively.

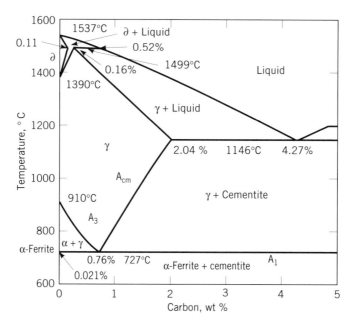

Fig. 16. Iron–iron carbide phase diagram (1). See text.

Eutectoid steels are not common, although small volumes are used as rails, high strength wire, and in some inexpensive tool steels. Above (hyper-) or below (hypo-) eutectoid, a primary precipitation of cementite or ferrite occurs until the austenite reaches the eutectoid composition when it transforms as before. Normally, these proeutectoid components form at grain boundaries. At lower temperatures of precipitation, however, they may also form on certain well-defined planes in the austenite crystals. The amount of proeutectoid constituents offers a rough guide to the composition of the steel. This guide is more accurate at slow-cooling rates.

If small specimens are prepared in which the austenite can be cooled to 250–500°C sufficiently rapidly to avoid the above microconstituents, and transformed at temperatures in this range, the formation of a completely different phase, a bcc α-phase supersaturated with carbon and containing small cementite particles (bainite), which is both strong and tough, occurs. Bainite is rarely found in plain carbon steels, but it can be obtained in commercial practice by judicious alloying and is increasing in importance.

Finally, if all of the above sets of phases can be suppressed by sufficiently rapid cooling, the austenite transforms to a phase known as martensite. This phase contains all of the carbon in the austenite. The result is a body-centered structure supersaturated in carbon. The crystal structure is body-centered tetragonal, ie, a cube extended in one direction, where the degree of tetragonality, and thus lattice strain, increases linearly with carbon. Martensite begins to form at a well-defined temperature known as M_s and continues to form as the temperature continues to drop. For most steels, the transformation is complete (defined as M_f) before reaching room temperature, but in high carbon and certain alloy steels, some austenite may remain at room temperature and requires special consideration. Martensite forms by a lattice shear reaction without diffusion of the carbon atoms, and usually has a plate or needle-like morphology. Because of its lattice strain, it is normally too brittle for service until some of the carbon is precipitated by a second heating between 200 and 700°C, ie, tempering. Material given this treatment can have unsurpassed combinations of strength and toughness.

Iron–Iron Carbide Phase Diagram

The iron–iron carbide phase diagram (see Fig. 16) shows the ranges of compositions and temperatures in which the various stable or metastable phases, such as austenite, ferrite, and cementite, are present in slow-cooled steels. This diagram covers the temperature range from 600°C to the melting point of iron, and carbon contents from 0–5%. In steels and cast irons, carbon can be present either as iron carbide (cementite) or as graphite. Under equilibrium conditions, only graphite is present because iron carbide is metastable with respect to iron and graphite. However, in commercial steels, iron carbide is present instead of graphite. When a steel containing carbon solidifies, the carbon in the steel usually solidifies as iron carbide. Although the iron carbide in a steel can change to graphite and iron when the steel is held at ca 900°C for several days or weeks, iron carbide in steel under normal conditions is quite stable for many years, eg, in power stations.

The portion of the iron–iron carbide diagram of interest herein is that part extending from 0–2% carbon. The range from 3 to 4.5% carbon covers most of the cast irons, a fascinating set of materials used as-cast, sometimes heat-treated, but not deformed either hot or cold. Application of the phase diagram to heat treatment can be illustrated by considering the changes occurring on heating and cooling steels of selected carbon contents.

Iron occurs in two allotropic forms, α or δ and γ (see Fig. 15). The temperatures at which these phase changes occur are known as the critical temperatures. For pure iron, these temperatures are 910°C for the $\alpha-\gamma$ phase change and 1390°C for the $\gamma-\delta$ phase change. The boundaries in Figure 16 show how these temperatures are affected by composition.

Changes on Heating and Cooling Pure Iron. The only changes occurring on heating or cooling pure iron are the reversible changes at ca 910°C from bcc α-iron to fcc γ-iron and from the fcc γ-iron to bcc δ-iron at ca 1390°C.

Changes on Heating and Cooling Eutectoid Steel. Eutectoid steels are those that contain 0.76% carbon. The diagram shows that at and below 727°C the constituents are α-ferrite and cementite. At 600°C, the α-ferrite may dissolve up to 0.007% carbon. Up to 727°C, the solubility of carbon in the ferrite increases until, at this temperature, the ferrite contains about 0.02% carbon. The phase change on heating a eutectoid carbon steel occurs at 727°C, which is designated as A_1, the eutectoid or lower critical temperature. On heating such a steel just above this temperature, all ferrite and cementite transform to austenite, albeit slowly, and on slow cooling the reverse change occurs.

When a eutectoid steel is slowly cooled from the austenite range, the ferrite and cementite form in alternate layers of microscopic thickness. Under the microscope at low magnification, the diffraction effects from this mixture of ferrite and cementite give an appearance similar to that of a pearl, hence the material is called pearlite.

Changes on Heating and Cooling Hypoeutectoid Steel. Hypoeutectoid steels are those that contain less carbon than the eutectoid steels. If the steel contains more than 0.02% carbon, the constituents present at and below 727°C are usually ferrite and pearlite. The relative amounts depend on the carbon content. As the carbon content increases, the amount of ferrite decreases and the amount of pearlite increases.

The first phase change on heating, if the steel contains more than 0.02% carbon, occurs at 727°C. On heating just above this temperature, the pearlite slowly changes to austenite. The excess ferrite, called proeutectoid ferrite, remains unchanged. As the temperature rises further above A_1, the austenite dissolves more and more of the surrounding proeutectoid ferrite, becoming lower and lower in carbon content until all the proeutectoid ferrite is dissolved in the austenite at the upper critical temperature, A_3, which now has the same average carbon content as the steel.

On slow cooling the reverse changes occur. Ferrite precipitates, generally at the grain boundaries of the austenite, which becomes progressively richer in carbon. Just above A_1, the austenite is substantially of eutectoid composition, 0.76% carbon.

Changes on Heating and Cooling Hypereutectoid Steel. The behavior on heating and cooling hypereutectoid steels (steels containing >0.76% carbon) is similar to that of hypoeutectoid steels, except that the excess constituent is cementite rather than ferrite. Thus, on heating above A_1, the austenite gradually dissolves the excess cementite until at the A_{cm} temperature, ie, the highest temperature at which austenite and cementite can co-exist (see Fig. 16), the proeutectoid cementite has been completely dissolved and austenite of the same carbon content as the steel is formed. Similarly, on cooling below A_{cm}, cementite precipitates and the carbon content of the austenite approaches the eutectoid composition. On cooling below A_1, this eutectoid austenite changes to pearlite and the room temperature composition is therefore pearlite and proeutectoid cementite.

Some confusion occurred in early iron–carbon equilibrium diagrams, which indicated a critical temperature at ca 768°C. Although there are changes in specific heat in this vicinity, leading to changes in slope of cooling curves, this is the Curie temperature, Θ_c, where ferromagnetic α becomes paramagnetic. This is a second-order phase transformation as contrasted to the first-order transformation, which is of interest and importance in heat treatment. In older literature the Curie temperature was called A_2, a terminology no longer used.

Effect of Alloys on the Equilibrium Diagram. The iron–carbon diagram may be significantly altered by alloying elements, and its quantitative application should be limited to plain carbon and low alloy steels. The most important effects of the alloying elements are that the number of phases which may be in equilibrium is no longer limited to two as in the iron–carbon diagram; that the temperature and composition range, with respect to carbon, over which austenite is stable may be increased or reduced; and that the eutectoid temperature and composition may change.

Alloying elements either enlarge the austenite field or reduce it. The former include manganese, nickel, cobalt, copper, carbon, and nitrogen and are referred to as austenite stabilizers.

The elements that decrease the extent of the austenite field include chromium, silicon, molybdenum, tungsten, vanadium, tin, niobium, phosphorus, aluminum, and titanium. These are known as ferrite stabilizers.

Manganese and nickel lower the eutectoid temperature, whereas chromium, tungsten, silicon, molybdenum, and titanium generally raise it. All these elements seem to lower the eutectoid carbon content.

Grain Size. The crystal structures shown in Figure 15 exist as a regular array over a distance of 10,000 or more unit cells in a given direction to give a crystal commonly called a grain. Eventually, however, these run into a region where the orientation of the cells differs by rotation in three directions. The region where the grains abut is a surface where no atoms are at their equilibrium spacing and is known as a grain boundary. Grain boundaries can usually be observed readily in an optical microscope, especially if they are etched in a mild reagent for accentuation. Typical grain sizes, ie, the average distance between boundaries, are in the range of 500–5000 nm.

Austenite. A significant aspect of the behavior of steels on heating is the grain growth that occurs when the austenite, formed on heating above A_3 or A_{cm} (see Fig. 16), is heated even higher. The austenite, like any metal, consists of polyhedral grains. As formed at a temperature just above A_3 or A_{cm}, the size of the individual grains is small, but as the temperature is increased above the critical temperature, the grain size increases. The final austenite grain size depends primarily on the maximum temperature to which the steel is heated.

In practice, either deliberately or otherwise, phases other than iron carbide may be present. These are typically carbides (qv) or nitrides (qv) of aluminum, vanadium, niobium, and/or titanium, which do not dissolve easily in austenite. They tend to be found at grain boundaries and serve to prevent the boundaries from moving as easily as such boundaries would in the absence of these phases. As long as these particles exist, growth in size of austenite grains is severely constrained. When the temperature is increased so that the particles go into solution, the grains coarsen rapidly. The grain size of the austenite has a marked influence on transformation behavior during subsequent cooling and on the size of the constituents of the final microstructure.

During much conventional processing, especially to sheet and strip, an austenitic slab is being reduced in cross section typically by deformation through a series of rolls each of which imposes 10 to 40% decrease in thickness. At high temperatures, the deformed structure recrystallizes, ie, new small grains form and may coarsen even before reaching the next roll. As the temperature drops upon passage through the mill, recrystallization and grain growth occur with increasing difficulty. By controlling the rolling schedule, it is possible to obtain very fine grain sizes in austenite prior to its subsequent transformation. The general effects of austenite grain size on the properties of heat-treated steel are summarized in Table 1.

Microscopic Grain Size Determination. The microscopic grain size of steel is customarily determined from a polished plane section prepared in such a way as to delineate the austenite grain boundaries. Grain size can be estimated by several methods. Results can be expressed as diameter of average grain in millimeters (reciprocal of the square root of the number of grains per mm^2), number of grains per unit area, number of grains per unit volume, or a grain size number obtained by comparing the microstructure of the sample at a fixed magnification to a series of standard charts.

Phase Transformations. *Austenite.* Close to equilibrium, that is with very slow cooling, austenite transforms to pearlite when cooled below the A_1 (see Fig. 16) temperature. When austenite is cooled more rapidly, this transformation occurs at a lower temperature. The faster the cooling rate, the lower

Table 1. Trends in Heat-Treated Products

Property	Coarse-grained austenite	Fine-grained austenite
	Quenched and tempered products	
hardenability	increasing	decreasing
toughness	decreasing	increasing
distortion	more	less
quench cracking	more	less
internal stress	higher	lower
	Annealed or normalized products	
machinability		
rough finish	better	inferior
fine finish	inferior	better

the temperature at which transformation occurs. Furthermore, the structure of the ferrite–carbide aggregate formed when the austenite transforms varies markedly with the transformation temperature, and the properties are found to vary correspondingly. Thus, heat treatment involves a controlled supercooling of austenite. In order to take full advantage of the wide range of structures and properties that this treatment permits, a more detailed knowledge of the transformation behavior of austenite and the properties of the resulting aggregates is essential. For most specimens, heat flow from the interior of the piece to the surface controls the cooling rate of internal points. Thus understanding the effects of this issue becomes essential.

Isothermal Transformation Diagram. To separate the effects of transformation temperature from those of heat flow, it is essential to understand the nature of the transformation of austenite at a given, preselected temperature below the A_1. Information needed includes the starting time, the amount transformed as a function of time, and the time for complete transformation. A convenient way to accomplish this is to form austenite in specimens so thin (usually about 1-mm thick) that heat flow is not an issue, rapidly transfer the specimens to a liquid bath at the desired temperature, and follow the transformation with time. The experiment is repeated at several other transformation temperatures. On the same specimens, the microstructure and properties of the transformation products can be assessed. These data can be summarized on a single graph of transformation temperature versus time known as an isothermal transformation (IT) diagram or, more usually, a time–temperature–transformation (TTT) diagram. A log scale is used for convenience. This concept, put forward in 1930, revolutionized the understanding of heat treatment.

Figure 17 shows a typical curve for a eutectoid steel, along with representative microstructures and hardness as a surrogate for mechanical properties. For noneutectoid steels, ferrite or cementite begins to form at austenite grain boundaries at temperatures below the A_3 or A_{cm}, respectively, but these reactions do not go to completion. Rather, the precipitation is interrupted by formation of pearlite giving a two-phase structure. Schematically, that would give, for example, a ferrite start time to the left of the pearlite start time, merging at around 550°C, ie, the knee of the TTT curve.

There are several important points relating to these curves. (*1*) The start of transformation occurs at increasingly short times down to about 550°C, during which interval pearlite forms. The pearlite has a finer spacing as formation temperature decreases, becoming thus harder and stronger. During the period from beginning to end of transformation, colonies of pearlite form and grow slowly until all the austenite is consumed. (*2*) Below 550°C to just over 200°C, the start of transformation takes longer and longer and the product of transformation is bainite. The hardness increases as the transformation temperature decreases. (*3*) At the M_s temperature, martensite begins to form more or less instantaneously. The fraction that forms depends on the undercooling below M_s, as shown schematically by the indications of 50 and 90% transformation of austenite. For this steel, M_f is below room temperature so some untransformed austenite is present. Martensite, usually acicular, is harder and stronger than any of the other constituents but is much too brittle in this steel to put in service without further treatment (tempering). (*4*) This curve is only valid for eutectoid steel

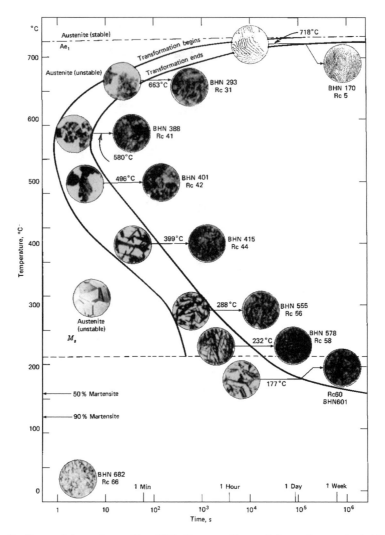

Fig. 17. Isothermal transformation (IT) diagram for a plain carbon eutectoid steel (1). Ae_1 is A_1 temperature at equilibrium; BHN, Brinell hardness number; Rc, Rockwell hardness scale C. C is 0.89%; Mn, 0.29%; austenitized at 885°C; grain size is 4–6. Photomicrographs originally ×2500.

having a specific austenite grain size. Other compositions and grain sizes have different TTT curves. Curves are collected in various reference books (32–34). (5) The effect of austenite grain size is real but relatively small. The effects of alloying elements can be very large (Fig. 18) and provide a practical tool to enable desired structures to be produced in large sections where heat flow is a factor.

Because almost no steel undergoes isothermal transformation, the effect of heat transfer is critical. Heat is lost from the surface, either to air or to a liquid such as water or oil. Although this leads to a complex set of equations, these are well known and characterized. Basically, all transformations are shifted to lower temperatures as a result of heat transfer. A primary concern in some cases is that for certain steels and cooling rates, it may be impossible to avoid the formation

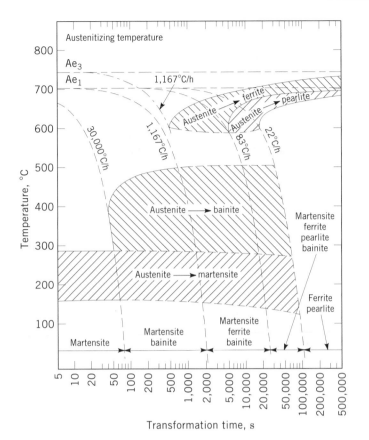

Fig. 18. Continuous-cooling transformation diagram for a Type 4340 alloy steel, with superimposed cooling curves illustrating the manner in which transformation behavior during continuous cooling governs final microstructure (1). Ae₃ is critical temperature at equilibrium. Ae₁ is lower critical temperature at equilibrium.

of ferrite and pearlite in parts of the specimen. This may be undesirable from a property standpoint. Increasing the surface cooling rate, superficially attractive as a way of avoiding ferrite and pearlite formation, has drawbacks because of possible distortion or even cracking during the quench. The solution is to change the position of the TTT curve to longer times along the x-axis, usually by alloying because increasing austenite grain size has more negatives than positives.

Constituent Properties. *Pearlite.* Pearlites, softer than bainites or martensites, are less ductile than the lower temperature bainites and, for a given hardness, far less ductile than tempered martensite. As the transformation temperature decreases within the pearlite range, the interlamellar spacing decreases, and these fine pearlites, formed near the nose of the isothermal diagram, are both harder and more ductile than the coarse pearlites formed at higher temperatures. Thus, although as a class pearlite tends to be soft and not very ductile, its hardness and toughness both increase markedly with decreasing transformation temperatures.

Bainite. In a given steel, bainite microstructures are generally found to be both harder and tougher than pearlite, although less hard than martensite. Bainite properties generally improve as the transformation temperature decreases. Lower bainite compares favorably with tempered martensite at the same hardness and can exceed it in toughness. Upper bainite, on the other hand, may be somewhat deficient in toughness as compared to fine pearlite of the same hardness (33).

Martensite. Martensite is the hardest and most brittle microstructure obtainable in a given steel. The hardness of martensite increases with increasing carbon content up to the eutectoid composition. The hardness of martensite at a given carbon content varies only very slightly with the cooling rate.

Although for some applications, particularly those involving wear resistance, the hardness of martensite is desirable in spite of the accompanying brittleness, this microstructure is mainly important as starting material for tempered martensite structures, which have definitely superior properties for most demanding applications.

Tempered Martensite. Martensite is tempered by heating to a temperature ranging from 170–700°C for 30 min to several hours. This treatment causes the martensite to transform to ferrite interspersed with small particles of cementite. Higher temperatures and longer tempering periods cause the cementite particles to increase in size and the steel to become more ductile and lose strength. Tempered martensitic structures are, as a class, characterized by very desirable toughness at almost any strength. Figure 19 describes, within $\pm 10\%$, the mechanical properties of tempered martensite, regardless of composition. For example, a steel consisting of tempered martensite having an ultimate strength of 1035 MPa (150,000 psi) might be expected to exhibit elongation of 16–20%, reduction of area of between 54 and 64%, yield point of 860–980 MPa (125,000–142,000 psi), and Brinell hardness of about 295–320. Because of its high ductility at a given hardness, this is the structure that is generally preferred.

Transformation Rates. The main factors affecting transformation rates of austenite are composition, grain size, and homogeneity. In general, increasing carbon and alloy content as well as increasing grain size tend to lower transformation rates. These effects are reflected in the isothermal transformation curve for a given steel. In practice, it is generally desirable to use as low a carbon content as possible for achieving the desired mechanical properties. Toughness, internal stress, distortion, and weldability are thus improved.

Continuous Cooling. The basic information depicted by an isothermal transformation diagram illustrates the structure formed if the cooling is interrupted and the reaction completed at a given temperature. The information is also useful for interpreting behavior when the cooling proceeds directly without interruption, as in the case of annealing, normalizing, and quenching. In these processes, the residence time at a single temperature is generally insufficient for the reaction to go to completion. Instead, the final structure consists of an association of microstructures that were formed individually at successively lower temperatures as the piece cooled. However, the tendency to form the various structures is still capable of being represented usefully on a modified TTT diagram.

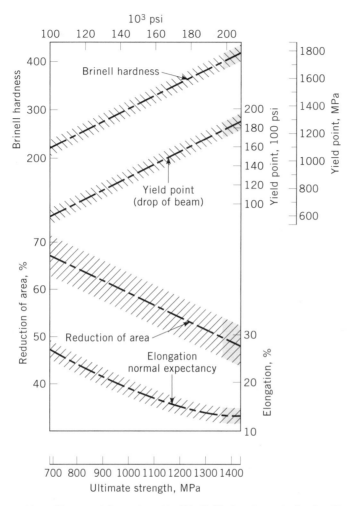

Fig. 19. Properties of tempered martensite (1). Fully heat-treated miscellaneous analyses, low alloy steels; 0.30–0.50% carbon. To convert MPa to psi, multiply by 145.

The final microstructure after continuous cooling depends on the time spent at the various transformation temperature ranges through which a piece is cooled. The transformation behavior on continuous cooling thus represents an integration of these times by constructing a continuous-cooling transformation diagram at constant rates similar to the isothermal transformation diagram (see Fig. 18). This diagram lies below and to the right of the corresponding isothermal transformation diagram if plotted on the same coordinates. That is, transformation on continuous cooling starts at a lower temperature and after a longer time than the intersection of the cooling curve and the isothermal diagram would predict. This displacement is a function of and increases in direct proportion with the cooling rate.

Figure 18 contains several superimposed cooling-rate curves. The changes occurring during these cooling cycles illustrate the manner in which diagrams of this type can be correlated with heat-treating processes and used to predict the resulting microstructure in a commonly used Cr–Ni–Mo alloy steel (Type 4340).

Considering first a relatively low cooling rate (<22°C/h), the steel is cooled through the regions in which transformations to ferrite and pearlite occur for the constitution of the final microstructure. This cooling rate corresponds to a slow cooling in the furnace, such as might be used in annealing.

At a higher cooling rate (22–83°C/h), such as might be obtained on normalizing a large forging, the ferrite, pearlite, bainite, and martensite fields are traversed and the final microstructure contains all these constituents. At cooling rates of 1,167–30,000°C/h, the microstructure is free of proeutectoid ferrite and consists largely of bainite and a small amount of martensite. A cooling rate of at least 30,000°C/h is necessary to obtain the fully martensitic structure desired as a starting point for tempered martensite. This rate, 30,000°C/h, corresponds to the cooling rate at the center of a 60-mm bar quenched in agitated oil. Thus, the final microstructure, and therefore the properties of the steel, depends on the transformation behavior of the austenite and the cooling conditions, and can be predicted if these factors are known.

Hardenability

Hardenability refers to the depth of hardening or to the size of a piece that can be hardened under given cooling conditions, and not to the maximum hardness that can be obtained in a given steel (36). The maximum hardness depends almost entirely on the carbon content, whereas the hardenability (depth of hardening) is far more dependent on the alloy content and, to a lesser degree, on the grain size of the austenite. Steels in which IT diagrams indicate a long time interval before the start of transformation to pearlite are useful when large sections are to be hardened, because if steel is to transform to bainite or martensite, it must escape any transformation to pearlite. Therefore, the steel must be cooled through the high temperature transformation ranges at a rate rapid enough for transformation not to occur even at the nose of the IT diagram. This rate, which just permits transformation to martensite without earlier transformation at a higher temperature, is known as the critical cooling rate for martensite. It furnishes one method for expressing hardenability. For example, in the steel of Figure 19, the critical cooling rate for martensite is 30,000°C/h or 8.3°C/s.

Although the critical cooling rate can be used to express hardenability, cooling rates ordinarily are not constant but vary during the cooling cycle. Especially when quenching in liquids, the cooling rate of the steel always decreases as the steel temperature approaches that of the cooling medium. In fact, quoted cooling rates are properly those that occur at a specified temperature, normally 700°C. It is therefore customary to express hardenability in terms of depth of hardening in a standardized quench. The quenching condition used in this method of expression is a hypothetical one in which the surface of the piece is assumed to come instantly to the temperature of the quenching medium. This condition implies no barrier to heat transfer at the steel–bath interface, ie, an infinite heat-transfer coefficient at this surface. This is known as an ideal quench. The diameter of a round steel bar, which is quenched to the desired microstructure, usually 50% martensite and 50% softer products, or a corresponding hardness value, at the center in an ideal quench, is known as the ideal diameter, D_I. The relationships between the cooling rates of the ideal quench and those of other cooling conditions are known. Thus, the hardenability values in terms of ideal diameter can be used to predict the size of round or other shape which have the same cooling rate when cooled in actual quenches where cooling severities are known. The cooling severities (usually referred to as severity of quench) which form the basis for these relationships are called H values. The H value for the ideal quench is infinity. Those for some commonly used cooling conditions are given in Table 2.

Hardenability is most conveniently measured by a test in which a steel sample is subjected to a continuous range of cooling rates. In the end-quench or Jominy test, a round bar, 25-mm dia and 100-mm long, is heated to the desired austenitizing temperature to control the austenite grain size and quenched in a fixture by a stream of water impinging on only one end. Hardness measurements are made on flats that are ground along the length of the bar after quenching. The results are expressed as a plot of hardness versus distance from the quenched end of the bar. The relationships between the distance from the quenched end and cooling rates in terms of D_I are known, and the hardenability can be evaluated in terms of D_I by noting the distance from the quenched end at which the hardness corresponding to the desired microstructure occurs and using this relationship to establish the corresponding cooling rate or D_I value. Published heat flow tables or charts relate the ideal diameter value to cooling rates in quenches or cooling conditions where H values are known. Thus, the ideal diameter value can be used to establish the size of a piece in which the desired microstructure can be obtained under the quenching conditions of the heat treatment to be used. The hardenability of steel is such an important property that it has become common practice to purchase steels to specified hardenability limits. Composition thus becomes a secondary specification. Such steels are called H steels.

Table 2. *H* **Values Designating Severity of Quench for Commonly Used Cooling Conditions**[a]

Degree of medium agitation	Quenching medium		
	Oil	Water	Brine
none	0.25–0.30	0.9–1.0	2
mild	0.30–0.35	1.0–1.1	2.0–2.2
moderate	0.35–0.40	1.2–1.3	
good	0.40–0.50	1.4–1.5	
strong	0.50–0.80	1.6–2.0	
violent	0.80–1.1	4.0	5.0

[a]H values are proportional to the heat-extracting capacity of the medium.

Heat-Treating Processes

In almost all heat-treating processes, steel is heated above the A_3 point and then cooled at a rate that results in the microstructure that gives the desired properties (34,37). Process annealing and stress-relieving are exceptions.

Austenitization. The steel is first heated above the temperature at which austenite becomes stable. The actual austenitizing temperature should be high enough to dissolve the carbides completely and take advantage of the hardening effects of any alloying elements present. In some cases, such as tool steels or high carbon steels, undissolved carbides may be retained for wear resistance. The temperature should not be high enough to produce pronounced grain growth. It is important to heat long enough for complete solution. For low alloy steels in a normally loaded furnace, 1.8 min/mm of diameter or thickness usually suffices.

Excessive heating rates may create high stresses, resulting in distortion or cracking. Certain types of continuous furnaces, salt baths, and radiant-heating furnaces provide very rapid heating, but preheating of the steel may be necessary to avoid distortion or cracking, and sufficient time must be allowed for uniform heating throughout. Unless special precautions are taken, heating causes scaling or oxidation, and may result in loss of carbon near the surface (decarburization). Controlled-atmosphere furnaces or salt baths can minimize these effects.

Quenching. The primary purpose of quenching is to cool rapidly enough to suppress at least some, and perhaps all, transformation at temperatures above the M_s temperature. For material to be used in bending or torsion, where the maximum stress is at the surface, it is often extravagant in alloy use to produce martensite throughout the piece. There can be other advantages. By careful selection and processing, which control the timing of volume changes during austenite transformation, compressive stresses can be retained in the surface and thereby contribute to improved fatigue life. For material where the stresses are applied in tension or compression, transformation to martensite throughout the piece is generally advisable.

The cooling rate required depends on the size of the piece and the hardenability of the steel. The preferred quenching media historically are water, oils, and brine. The temperature gradients set up by quenching create high thermal and transformational stresses that may lead to cracking and distortion. A quenching rate no faster than necessary should be employed to minimize these stresses. Water and brine are often too stringent for satisfactory use, whereas oil can be a fire hazard. Oils also deteriorate upon use and can be a difficult disposal problem. Thus polymer–water mixtures have found application. These mixtures have a range of heat abstraction rates and some of the problems with oil can be avoided. Agitation of the cooling medium accelerates cooling and improves uniformity. Cooling should be long enough to permit complete transformation to martensite. Then, in order to minimize cracking from quenching stresses, the article should be transferred immediately to the tempering furnace (Fig. 20).

Tempering. Quenching forms hard, brittle martensite having high residual stresses. Tempering relieves these stresses and precipitates excess carbon as carbides; it improves ductility, although at some expense of strength and hardness. The operation consists of heating at temperatures below the lower critical temperature, A_1.

Measurements of stress relaxation on tempering indicate that, in a plain carbon steel, residual stresses are significantly lowered by heating to temperatures as low as 150°C, but that temperatures of 480°C and above are required to reduce these stresses to adequately low values. The times and temperatures required for stress relief depend on the high temperature yield strength of the steel, because stress relief results from the localized plastic flow that occurs when the steel is heated to a temperature where its yield strength is less than

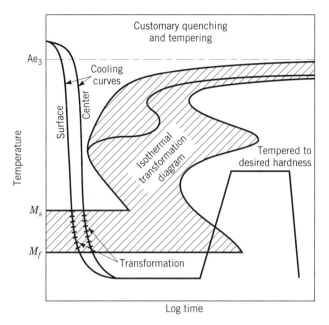

Fig. 20. Transformation diagram for quenching and tempering martensite. The product is tempered martensite.

the internal stress. This phenomenon may be affected markedly by composition, and particularly by alloy additions.

The toughness of quenched steel, as measured by the notch impact test, first increases on tempering up to 200°C, then decreases on tempering between 200 and 310°C, and finally increases rapidly on tempering at 425°C and above. This behavior is characteristic and, in general, temperatures of 230–310°C should be avoided. Where this range is unavoidable because of strength requirements, eg, for aircraft landing gear, additions of 1 to 2% silicon change the rate of carbide formation and can move the brittle range to higher tempering temperatures to provide acceptable toughness.

In some more highly alloyed steels, there is the possibility of precipitation of alloy carbides from the ferrite around 500°C and an actual increase in hardness. For steels needing high temperature strength, eg, tools and pressure vessels in refineries, these are often an attractive choice. In order to minimize cracking, tempering should follow quenching immediately. Any appreciable delay may promote cracking.

The tempering of martensite results in a contraction, and if the heating is not uniform, stresses result. Similarly, heating too rapidly may be dangerous because of the sharp temperature gradient set up between the surface and the interior. Recirculating-air furnaces can be used to obtain uniform heating. Oil baths are commonly used for low temperature tempering, salt baths can be used over a bigger range at higher temperatures. Some steels lose toughness on slow cooling from ca 540°C and above, a phenomenon known as temper brittleness. Rapid cooling after tempering is desirable in these cases.

Martempering. A modified quenching procedure known as martempering minimizes the high stresses created by the transformation to martensite during the rapid cooling characteristic of ordinary quenching (Fig. 21). In practice, it is ordinarily carried out by quenching in a molten salt bath just above the M_s temperature. Transformation to martensite does not begin until after the piece reaches the temperature of the salt bath and is then allowed to cool relatively slowly in air. Because the large temperature gradient characteristic of conventional quenching is absent, the stresses produced by the transformation are much lower and a greater freedom from distortion and cracking is obtained. After martempering, the piece may be tempered to the desired strength.

Austempering. Lower bainite is generally as strong as and somewhat more ductile than tempered martensite. Austempering, which is an isothermal heat treatment that results in lower bainite, offers an alternative heat treatment for obtaining optimum strength and ductility if the specimens are sufficiently small.

In austempering the article is quenched to the desired temperature in the lower bainite region, usually in molten salt, and kept at this temperature until transformation is complete (Fig. 22). Usually, the piece is held twice as long as the period indicated by the isothermal transformation diagram. The article may then be quenched or air-cooled to room temperature after transformation is complete, and may be tempered to lower hardness if desired.

Normalizing. In this operation, steel is heated above its upper critical temperature (A_3) and cooled in air. The purpose of this treatment is to refine the hot-rolled structure (often quite inhomogeneous), depending on the finishing temperature, and to obtain a carbide size and distribution that is more favorable

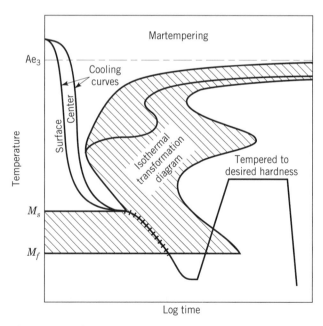

Fig. 21. Transformation diagram for martempering. The product is tempered martensite.

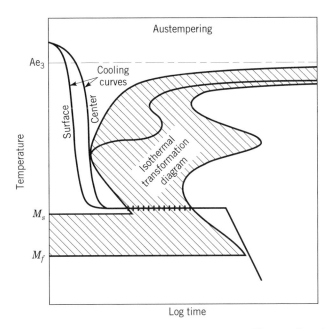

Fig. 22. Transformation diagram for austempering. The product is bainite.

for carbide solution on subsequent heat treatment than the earlier as-rolled structure.

In alloy steels, particularly if these have been slowly cooled after rolling, the carbides in the as-rolled condition tend to be massive and are difficult to dissolve on subsequent austenitization. The carbide size is subject to wide variations, depending on the rolling and slow cooling. Here, again, normalizing tends to establish a more uniform and finer carbide particle size that facilitates subsequent heat treatment. Although an expense, this process provides more uniform quality in the finished product.

The usual practice is to normalize at 50–80°C above the upper critical temperature. For some alloy steels, however, considerably higher temperatures may be used. Heating may be carried out in any type of furnace that permits uniform heating and good temperature control.

Annealing. Annealing has two different purposes: to relieve stresses induced by hot- or cold-working, and to soften the steel to improve its machinability or formability. It may involve only a subcritical heating to relieve stresses, recrystallize cold-worked material, or spheroidize carbides; alternatively, it may involve heating above the upper critical temperature (A_3) with subsequent transformation to pearlite or, less commonly, directly to a spheroidized structure on cooling.

The most favorable microstructure for machinability in the low or medium carbon steels is coarse pearlite. The customary heat treatment to develop this microstructure is a full annealing, illustrated in Figure 23. It consists of austenitizing at a relatively high temperature to obtain full carbide solution, followed by slow cooling to give transformation exclusively in the high temperature end of the pearlite range. This simple heat treatment is reliable for most steels. It is,

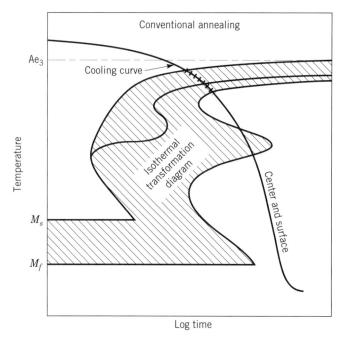

Fig. 23. Transformation diagram for full annealing. The product is ferrite and pearlite.

however, rather time-consuming because it involves slow cooling over the entire temperature range from the austenitizing temperature to a temperature well below that at which transformation is complete.

Isothermal Annealing. Annealing to coarse pearlite can be carried out isothermally by cooling to the proper temperature for transformation to coarse pearlite and holding until transformation is complete. This method, called isothermal annealing and illustrated in Figure 24, may save considerable time over the full-annealing process described previously. Neither the time from the austenitizing temperature to the transformation temperature, nor the one from the transformation temperature to room temperature, is critical. Both may be shortened as desired. If extreme softness of the coarsest pearlite is not necessary, the transformation may be carried out near the nose of the IT curve, where the transformation is completed rapidly and the operation further expedited. The pearlite in this case is much finer and harder.

Isothermal annealing can be conveniently adapted to continuous annealing, usually in specially designed furnaces. This is commonly referred to as cycle annealing.

Spheroidization Annealing. Coarse pearlite microstructures are too hard for optimum machinability in the higher carbon steels. Such steels are customarily annealed to develop spheroidized microstructures by holding the as-rolled, slowly cooled, or normalized materials just below the lower critical temperature range. Such an operation is known as subcritical annealing. Full spheroidization may require long holding times at the subcritical temperature. The method may be slow, but it is simple and may be more convenient than annealing above the critical temperature.

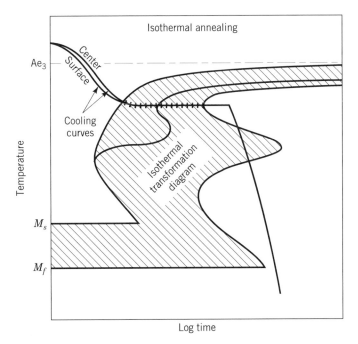

Fig. 24. Transformation diagram for isothermal annealing. The product is ferrite and pearlite.

Using some modifications, the annealing procedures described to produce pearlite can give spheroidized microstructures. If free carbide remains after austenitizing, transformation in the temperature range where coarse pearlite ordinarily would form proceeds to spheroidized rather than pearlitic microstructures. Thus, heat treatment to form spheroidized microstructures can be carried out like heat treatment for pearlite, except for the lower austenitizing temperatures. Spheroidization annealing may thus involve a slow cooling similar to the full-annealing treatment used for pearlite, or it may be a treatment similar to isothermal annealing. An austenitizing temperature not more than 55°C above the lower critical temperature is customarily used for this type of annealing.

Process Annealing. Process annealing is the term used for subcritical annealing of cold-worked materials. It customarily involves heating at a temperature high enough to cause recrystallization of the cold-worked material and to soften the steel. The most important example of process annealing is the box annealing of cold-rolled low carbon sheet steel. The sheets are enclosed in a large box which can be sealed to permit the use of a controlled atmosphere to prevent oxidation. Annealing is usually carried out between 590 and 715°C. The operation usually takes ca 24 h, after which the charge is cooled slowly within the box. The entire process takes several days. Developments using a hydrogen atmosphere have largely restored the competitiveness of box annealing against continuous annealing processes.

An alternative way to anneal the low carbon sheet and certain higher strength steels used in large tonnages for automobiles, appliances, and containers is to pass the sheet, after cold rolling, continuously through a long furnace having a carefully designed temperature profile. Usually it is desirable to have a certain crystallographic texture in the sheet after annealing, and to exercise control over the amounts of carbon and nitrogen remaining in solution in the ferrite. The control requirements are thus severe but the savings in time over box annealing, as well as the very small scatter in properties, makes this an attractive option when capital is available.

Carburizing. In carburizing, low carbon steel acquires a high carbon surface layer by heating in contact with carbonaceous materials. On quenching after carburizing, the high carbon skin forms martensite, whereas the low carbon core remains comparatively soft. The result is a highly wear-resistant exterior over a very tough interior. This material is particularly suitable for gears, camshafts, etc. Carburizing is most commonly carried out by packing the steel in boxes with carbonaceous solids, sealing to exclude the atmosphere, and heating to about 925°C for a period of time depending on the depth desired. This method is called pack carburizing. Alternatively, the steel may be heated in contact with carburizing gases, in which case the process is called gas carburizing; or, least commonly, the steel may be heated in liquid baths of carburizing salts, in which case it is known as liquid carburizing. The M_f of the high carbon case may be below room temperature, which means that some soft retained austenite may be present. Further heat treatments may be necessary to eliminate this or minimize its effects.

Nitriding. The nitrogen case-hardening process, termed nitriding, consists of subjecting machined and, preferably, heat-treated Cr–Mo steel parts at about 500°C to the action of a nitrogenous medium, commonly ammonia gas, under conditions whereby surface hardness is imparted without requiring any further treatment. Wear resistance, retention of hardness at high temperatures, and resistance to certain types of corrosion are also imparted by nitriding. The hardness of nitrided steels arises primarily from the presence of 0.9 to 1.4% aluminum, which leads to large numbers of small precipitates of aluminum nitride [24304-00-5].

An intermediate treatment that adds both carbon and nitrogen to steel surfaces can be obtained by exposing the parts to a bath of molten cyanide at just above the critical temperature of the core for about one hour followed by direct quenching. The hardened area is about 0.25-mm deep.

Carbon Steels

Plain carbon steels, by far the largest volume of steel produced, have the most diverse applications of any metallic engineering materials. These include castings, forging, tubular products, plates, sheet and strip, wire and wire products, structural shapes, bars, and railway materials such as rails, wheels, and axles. Carbon steels are made by all modern steelmaking processes and, depending on their carbon content and intended purpose, may be rimmed, semikilled, or fully killed (33). Rimmed steels having good surfaces had been desirable when ingots were common, but are difficult to produce by modern continuous casting process.

The American Iron and Steel Institute (AISI) has published standard composition ranges for plain carbon steel. Each composition range is assigned an identifying number according to an accepted method of classification (Table 3).

Table 3. Standard Numerical AISI–SAE[a] Designations of Plain Carbon and Constructional Alloy Steels

Series designation[b]	Types	Series designation[b]	Types
10xx	nonresulfurized carbon steel grades	47xx	1.05% Ni–0.45% Cr–0.20% Mo
11xx	resulfurized carbon steel grades	48xx	3.50% Ni–0.25% Mo
12xx	rephosphorized and resulfurized carbon steel grades	50xx	0.50% Cr
13xx	1.75% Mn	51xx	1.05% Cr
23xx	3.50% Ni	5xxxx	1.00% C–1.45% Cr
25xx	5.00% Ni	61xx	0.95% Cr–0.15% V
31xx	1.25% Ni–0.65% Cr	86xx	0.55% Ni–0.65% Cr–0.20% Mo
33xx	3.50% Ni–1.55% Cr	87xx	0.55% Ni–0.50% Cr–0.25% Mo
40xx	0.25% Mo	92xx	0.85% Mn–2.00% Si
41xx	0.95% Cr–0.20% Mo	93xx	3.25% Ni–1.20% Cr–0.12% Mo
43xx	1.80% Ni–0.80% Cr–0.25% Mo	98xx	1.00% Ni–0.80% Cr–0.25% Mo
46xx	1.80% Ni–0.25% Mo		

[a]The AISI (American Iron and Steel Institute) and the SAE (Society of Automotive Engineers) specifications are essentially the same. The list is simplified to show typical compositions. For ranges, see original tables.

[b]The first figure indicates the class to which the steel belongs; 1xxx indicates a carbon steel, 2xxx a nickel steel, and 3xxx a nickel–chromium steel. In the case of alloy steels, the second figure generally indicates the approximate percentage of the principal alloying element. Usually, the last two or three figures (represented in the table by x) indicate the average carbon content in points or hundredths of 1 wt %. Thus, a nickel steel containing ca 3.5% nickel and 0.30% carbon would be designated as 2330.

In this system, carbon steels are assigned to one of three series: 10xx (non-resulfurized), 11xx (resulfurized), and 12xx (rephosphorized and resulfurized). Modern low carbon steels are often of such low carbon that the lowest Society of Automotive Engineers (SAE) designation, 1005, is not helpful. Material is ordered to property and performance specifications. Normal practice makes an upper limit of 0.015% for sulfur and phosphorus easily attainable, and by ladle metallurgy this unit can readily be reduced to 0.003%. Sulfur in amounts as high as 0.33% max may be deliberately added to the 11xx and as high as 0.35% max to the 12xx steels to improve machinability. In addition, phosphorus up to 0.12% max may be added to the 12xx steels to increase stiffness.

In identifying a particular steel, the letters x are replaced by two digits representing average carbon content. For example, an AISI 1040 steel would have an average carbon content of 0.40%, with a tolerance of ±0.03%, giving a range of 0.37–0.44% carbon.

Properties. The properties of plain carbon steels are governed principally by carbon content and microstructure. These properties can be controlled by heat treatment as discussed. About half the plain carbon steels are used in the hot-rolled form, although increasingly the property combinations are enhanced by controlled cooling following the last stand of the hot mill for structural shapes, sheet, and strip. The other half are cold-rolled to thin sheet or strip and used directly or with an annealing treatment such as described.

Certain properties of plain carbon steels may be modified by residual elements other than the carbon, manganese, silicon, phosphorus, and sulfur that are always present, as well as gases, especially oxygen, nitrogen, and hydrogen, and their reaction products. These incidental elements are usually acquired from scrap, deoxidizers, or the furnace atmosphere. The gas content depends mostly on melting, deoxidizing, ladle treatments, and casting procedures. Consequently, the properties of plain carbon steels depend heavily on the manufacturing techniques.

The average mechanical properties of as-rolled 2.5-cm bars of carbon steels as a function of carbon content are shown in Figure 25. This diagram is an illustration of the effect of carbon content when microstructure and grain size are held approximately constant. Although the diagram is representative, the

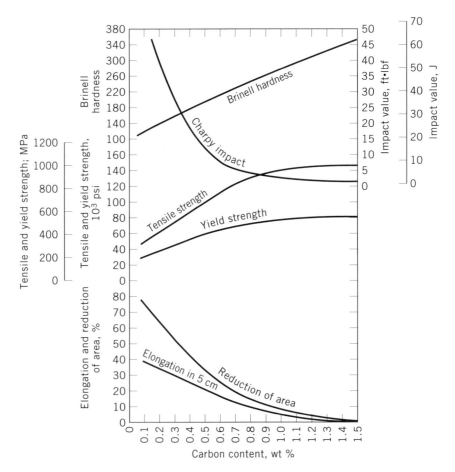

Fig. 25. Variations in average mechanical properties of as-rolled 2.5-cm bars of plain carbon steels, as a function of carbon content (1).

great bulk of tonnage used is at low carbon levels. These levels are often below 0.2% C; much is below 0.1% C.

Microstructure and Grain Size. The carbon steels having relatively low hardenability do not contain martensite or bainite in the cast, rolled, or forged state. The constituents of the hypoeutectoid steels are therefore ferrite and pearlite, and of the hypereutectoid steels, cementite and pearlite.

As of the 1990s, most of the steels do not rely primarily on pearlite for strength. Although the pearlite does contribute somewhat, the resulting reduction in ductility and toughness make it much more attractive to improve strength, ductility, weldability, and toughness simultaneously by concentrating on reducing the ferrite grain size in a lower carbon material (38). The yield strength, σ_y, has been shown (39) to increase with the inverse square root of the ferrite grain size, d.

$$\sigma_y = A + B{\cdot}d^{-1/2}$$

Thus where 25 to 50 μm was historically considered quite appropriate for a ferrite grain size, 5 μm has become not at all unusual. These smaller grain sizes can be produced directly from hot rolling by controlled recrystallization of the austenite to a fine grain size, which in turn leads to many more independently nucleated ferrite grains. The addition of small amounts of carbide and nitride formers, such as Nb, V, and Ti, helps to keep the grain sizes small, and by precipitation in the ferrite at lower temperatures, assisted by water cooling of the material out of the last rolling stand, gives extra strengthening. These materials, a relatively recent and important development, are commonly known as microalloyed or high strength low alloy (HSLA) steels (38,40,41).

Microstructure of Cast Steels. Cast steel is generally coarse-grained because austenite forms at high temperature and the pearlite is usually coarse, in as much as cooling through the critical range is slow, particularly if the casting is cooled in the mold. In hypoeutectoid steels, ferrite ordinarily precipitates at the original austenite grain boundaries during cooling, although ferrite often precipitates as plates within individual austenite grains, especially if these are large. In hypereutectoid steels, cementite is similarly precipitated. Such large-grained mixtures of ferrite or cementite and coarse pearlite have poor strength, toughness, and ductility properties, and heat treatment is usually necessary to reduce the grain size and thereby obtain suitable microstructures and properties in cast steels.

Hot Working. Many carbon steels are used in the form of as-rolled finished sections. The microstructure and properties of these sections are determined largely by composition, rolling procedures, and cooling conditions. The rolling or hot working of these sections is ordinarily carried out in the temperature range in which the steel is austenitic, with four principal effects: considerable homogenization occurs during the heating for rolling, tending to eliminate some, but never all, of the dendritic segregation present in the ingot; the dendritic structure is broken up during rolling; recrystallization takes place during rolling, and the final austenitic grain size is determined by the temperature at which the last passes are made (the finishing temperature); and dendrites and inclusions are reoriented, with markedly improved ductility, in the rolling direction.

The inhomogeneities in composition inherent in the solidification process are in practice never totally removed. Rolling serves to reduce the length scales over which these inhomogeneities occur to a few tens of micrometers normal to the rolling direction. Although this can lead to some anisotropy of properties, such as toughness and ductility, in the directions parallel and normal to the rolling direction, it rarely is seriously deleterious to engineering applications.

The austenite grain size at the end of hot rolling, during which the steel recrystallizes several times, is sensitive to undissolved particles. These may be products of the steelmaking process during deoxidation, or occur as deliberate additions of carbide and/or nitride formers as in HSLA. The ultimate microstructure, and thus properties and performance, depends on how and at what temperature this austenite transforms during the subsequent cooling. Section size has important ramifications. A large I-beam cools slowly, but hot-rolled sheet is normally coiled from the mill directly into a large torus weighing several tons and thus also cools slowly.

Cold Working.   The manufacture of wire, sheet, strip, bar, and tubular products often includes cold working, with effects that may be eliminated by annealing. However, some products, particularly wire, are used in the cold-worked condition. The most pronounced effects of cold working are increased strength and hardness and decreased ductility. The effect of cold working on the tensile strength of plain carbon steel is shown in Figure 26. The yield strength also increases, generally more rapidly than tensile strength, so cold working is a cost-effective method of strengthening if ductility is not critical.

Upon reheating cold-worked steel to the recrystallization temperature (~450°C) or above, depending on composition, extent of cold working, and other variables, the original microstructure and properties may be substantially restored.

Heat Treatment.   Although many wrought (rolled or forged) carbon steels are used without a final heat treatment, this may be employed to improve the microstructure and properties for specific applications if the cost can be justified.

The option of annealing a hot- or cold-rolled structure depends on the type of further processing to be applied. For example, the forces required to cold draw a section prior to final machining may be too high if the specimen is not given a full anneal above the A_3, followed by slow cooling to produce a fairly coarse ferrite–pearlite structure. The moderate cold work produced by the drawing to a final size is helpful to obtaining good chip formation during machining.

At the other extreme, the exceptional ductility requirements of a material to be shaped by drawing, flanging, or other large plastic deformation modes necessitate careful combinations of advanced steelmaking, hot rolling, cold rolling, and annealing. The demands of consumers for aesthetic and functional requirements are passed through such product manufacturers as autos, appliances, and the steelmaker. The increase in quality and reproducibility in these materials during the 1980s and 1990s has been impressive.

The tendency toward lower carbon steels has minimized the utilization of pearlite as a structural constituent. The growing use of HSLA steels has greatly reduced the quenching and tempering of carbon steels to provide slightly stronger materials without the use of expensive alloys. Normalizing is still used to reduce variability in hot-rolled material if the economics can be justified.

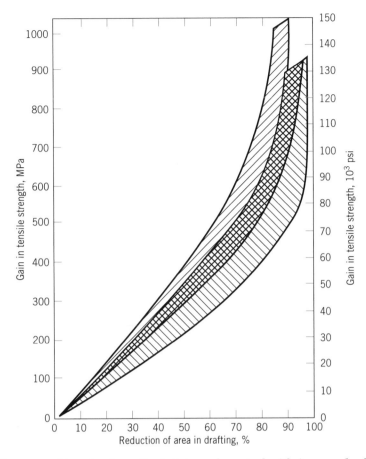

Fig. 26. Increase of tensile strength of plain carbon steel with increased cold working, where (▨) represents 0.05–0.30 wt % carbon; (▨) 0.30–0.60 wt %; and (▨) 0.60–1.00 wt %.

Residual Elements. In addition to carbon, manganese, phosphorus, sulfur, and silicon which are always present, carbon steels may contain small amounts of hydrogen, oxygen, or nitrogen, introduced during the steelmaking process; nickel, copper, molybdenum, chromium, and tin, which may be present in the scrap; and aluminum, titanium, vanadium, or zirconium, which may have been introduced during deoxidation.

Nitrogen can be a relatively harmful element unless special precautions are taken. A smaller atom than carbon, nitrogen is more soluble and has a higher diffusivity. Thus nitrogen can be particularly deleterious to the properties of sheet steel, where, unless care is taken, deformation can be quite inhomogeneous. This can lead to ugly surface markings totally unsuitable in automobile or appliance applications. The effect can be masked for a time by a light skin pass prior to leaving the rolling mill, but as nitrogen diffuses over a period of several days at room temperature during shipment or storage, the effect reappears. To avoid this, the steel was normally killed with aluminum, but at an extra cost. Nitrogen is not easy to remove by vacuum treatment.

During the 1990s, the ability to produce interstitial-free steels (42) low in carbon and nitrogen has been achieved, whereby the carbon and oxygen left in solution after steelmaking react to form carbon monoxide under a vacuum in the ladle degasser. The bubbles leaving also serve to carry off some of the nitrogen. Steels made in the electric furnace tend to have higher nitrogen and are thus not able to make the most difficult types of deep-drawing sheet as of this writing (1996).

An embrittling effect, the mechanism of which is still debated, is possible with a hydrogen content of more than ca 3 ppm. The content of hydrogen and other gases can, however, be reduced by vacuum dagassing.

Alloying elements such as nickel, chromium, molybdenum, and copper, which may be introduced with scrap, can increase the hardenability, although only slightly, because the concentrations are ordinarily low. However, the heat-treating characteristics may change, and for applications in which ductility is important, as in low carbon steels for deep drawing, the increased hardness and lower ductility imparted by these elements may be harmful.

Tin, even in low amounts, is harmful in steels for deep drawing. For most applications, however, the effect of tin in the quantities ordinarily present is negligible.

Aluminum is generally desirable as a grain refiner and tends to reduce the susceptibility of carbon steel to aging associated with strain. Unfortunately, aluminum tends to promote graphitization and is therefore undesirable in steels used at high temperatures. The other elements that may be introduced as deoxidizers, such as titanium, vanadium, or zirconium, are ordinarily present in such small amounts as to be ineffective, although in slightly larger amounts these can lead to HSLA steels. Copper and tin can create a rough surface condition during hot rolling through a process called hot shortness. Here the copper and tin form a liquid alloy that can penetrate grain boundaries and cause surface cracks.

Alloy Steels

For slightly less than 10% of products, alloying elements are introduced to produce properties not available for carbon steels where the functional elements are usually considered to be carbon, silicon (to 0.6%), and manganese (to 1.65%). Copper, which may be present up to 0.6 wt %, is relatively rare compared to the ubiquitous silicon and manganese.

The principal classes of alloy steels in decreasing order of volume are HSLA (usually hot- or cold-rolled), AISI alloy steels (usually quenched and tempered), stainless steels (cast or wrought), electrical steels (largely iron–silicon), alloy tool steels, and nonstainless heat-resistant steels. One or more alloying elements may be critical to the resulting properties. Some very specialized steels have been developed to meet exceptionally harsh service requirements. Even though the tonnage of these specialized steels may be small, the importance to modern technology may be huge if the task is only feasible with the specific alloy (see HIGH TEMPERATURE ALLOYS; TOOL MATERIALS).

Functions of Alloying Elements. Alloy steels may contain up to ca 50% of alloying elements which directly enhance properties, although 30% is a more common upper limit. Examples are the increased corrosion resistance of high chromium steels, the enhanced electric properties of silicon steels, the improved strength of the high strength low alloy steels, and the improved hardenability and tempering characteristics of the AISI alloy steels.

High Strength Low Alloy Steels. Until the 1970s, increasing the strength of steel involved producing dispersions of carbides by quenching and tempering (accompanied by cost and joining problems), increasing the fraction of pearlite (resulting in ductility and toughness problems), or cold working (ductility problems). As the understanding of the origin of properties developed, other mechanisms began to be attractive to produce inexpensive steels having good combinations of the properties essential for service needs. The most important advance was to refine the ferrite grain size, and be able to disperse small amounts of very fine alloy carbides or carbonitrides. In this way, toughness and ductility were improved simultaneously. Processing improvements on reliable composition control and tight control of the rolling schedules lead to very attractive and economical combinations of properties in the hot-rolled condition (38,40–43). Because the carbon content is lower, the weldability is also improved, and in a few cases the corrosion resistance in ambient air has also been better.

Several types of processing that contribute to these desirable properties are as follows. (1) Controlled rolling of steels having small amounts of carbon and nitrogen containing V, Ti, Nb, and/or Zr in the austenite produces fully recrystallized fine austenite grains; rolling in the austenite–ferrite range produces deformed pancake grains. Subsequent cooling leads to very fine ferrite grains. The pancake grain structure can lead to anisotropy of properties or in extreme cases to delamination. (2) Accelerated cooling from the hot mill after controlled rolling in the range produces equiaxed ferrite while avoiding acicular ferrite, not so slowly, however, that growth of carbonitride precipitates occurs during coiling, thus leading to a strength loss. (3) Sufficiently rapid cooling of low (<0.08%) carbon in steels having enough hardenability produces low carbon bainite. Adequate toughness in these steels requires controlled rolling (38). (4) Simple normalizing for some steels, eg, those containing vanadium, can give sufficient ferrite grain refinement for these steels to be attractive. (5) Intercritical annealing, ie, in the $\alpha + \gamma$ range, of low carbon steels containing 1.5% manganese gives martensite islands in a ferrite matrix after rapid cooling. This is a product with a low initial yield strength but rapid work hardening and increased ductility. These dual-phase steels are useful in parts needing good yield strength which are to be deformed before service, eg, auto wheels.

There are several hundred types of steels that use these principles separately or in combination. Some of these are generic, many are proprietary. Among them are various microalloyed ferrite–pearlite steels having limited pearlite amounts and the low carbon bainites, as well as the dual-phase steels.

Some other steels that fall into related categories are the various long-standing proprietary weathering steels, eg, USS's Cor-Ten A, which uses small amounts of copper and phosphorus to improve atmospheric corrosion resistance. These elements also produce significant solid solution hardening. Phosphorus is also used for its strengthening effects alone, although the toughness is degraded.

Other alloy effects in these steels are in the control of the shape of sulfide inclusions. Manganese(II) sulfide [18820-29-6], MnS, is a common form in which sulfur is present; this is ductile at rolling temperatures and is thus in the form of stringers. This leads to poor through thickness ductility; ie, the ductility perpendicular to the rolling direction. Other sulfides, eg, calcium sulfide [20548-54-3], CaS, and titanium(II) sulfide [12039-07-5], TiS, are sufficiently stronger so that during hot rolling these remain essentially spherical and give isotropic mechanical properties. There is a clear market for these inclusion-shape-controlled steels.

Finally, hydrogen in relatively small amounts can lead to cracking, especially as the strength increases. If the presence of hydrogen cannot be adequately lowered by the ladle treatment, combinations of low carbon and silicon, inclusion shape control, and more than 0.25% copper can be helpful in minimizing loss of toughness.

Interstitial-Free Steels.   In some ways, interstitial-free (IF) steels are primarily carbon steels having deliberately low yield strength. Many of the same principles apply as for other alloy steels (42). It has long been known that the maximum ductility in steel sheet, which is so important in formability, depends on the carbon and nitrogen in solution in the ferrite. If these can be reduced below 50 ppm, not only does the ductility approach 50% and the yield strength decrease, but the harmful effects of strain aging and quench aging essentially disappear. By vacuum treatment in the ladle, carbon and nitrogen in solution are reduced; by adding small quantities of carbide and nitride stabilizers, any residual interstitials are tied up as precipitates. These IF steels are becoming increasingly important both on their own and as a base for the hot dip galvanizing commonly used for increased corrosion resistance.

AISI Alloy Steels.   The American Iron and Steel Institute defines alloy steels as follows: "steel is considered to be alloy steel when the maximum of the range given for the content of alloying elements exceeds one or more of the following limits: manganese, 1.65%; silicon, 0.60%; copper, 0.60%; or in which a definite range or a definite minimum quantity of any of the following elements is specified or required within the limits of the recognized field of constructional alloy steels: aluminum, boron, chromium up to 3.99%, cobalt, columbium (niobium), molybdenum, nickel, titanium, tungsten, vanadium, zirconium, or any other alloying element added to obtain a desired alloying effect" (44). Steels that contain 4% or more of chromium are included by convention among the special types of alloy steels known as stainless steels (45,46).

Steels that fall within the AISI definition have been standardized and classified jointly by AISI and SAE (see Table 3). These represent by far the largest alloy steels production and are generally known as AISI alloy steels. They are also referred to as constructional alloy steels.

The effect of the alloying elements on AISI steels is indirect because alloying elements control microstructure through their effect on hardenability. These elements permit the attainment of desirable microstructures and properties over a much wider range of sizes and sections than is possible with carbon steels.

Quenched and Tempered Low Carbon Constructional Alloy Steels.   A class of quenched and tempered low carbon constructional alloy steels has been very extensively used in a wide variety of applications such as pressure vessels, mining and earth-moving equipment, and in large steel structures (see TANKS AND PRESSURE VESSELS).

As a general class, these steels are referred to as low carbon martensites to differentiate them from constructional alloy steels of higher carbon content, such as the AISI alloy steels, that develop high carbon martensite upon quenching. The low carbon martensites are characterized by a relatively high strength, minimum yield strengths of 690 MPa (100,000 psi), good toughness down to $-45°C$, and weldability with joints showing full joint efficiency when welded with low hydrogen electrodes. They are most commonly used in the form of plates, but also as sheet products, bars, structural shapes, forgings, and semifinished products. Several steel-producing companies manufacture such steels under various trade names. Compositions are proprietary.

Alloy Tool Steels. Alloy tool steels are classified roughly into three groups. The first consists of alloy tool steels, to which alloying elements have been added to impart hardenability higher than that of plain carbon tool steels. Accordingly, these steels may be hardened in heavier sections or using less drastic quenches to minimize distortion. The second group is that of intermediate alloy tool steels, which usually contain elements such as tungsten, molybdenum, or vanadium. These form hard, wear-resistant carbides, often by secondary hardening during tempering. The last are high speed tool steels that contain large amounts of carbide-forming elements, which serve not only to furnish wear-resisting carbides, but also to increase secondary hardening thereby allowing operation at red heat.

Stainless Steels. Stainless steels are more resistant to rusting and staining than plain carbon and low alloy steels (47–50). This superior corrosion resistance results from the presence of chromium. Although other elements, such as copper, aluminum, silicon, nickel, and molybdenum, also increase corrosion resistance; these are limited in their usefulness.

Germany, the United Kingdom, and the United States shared alike in the early development of stainless steels. In the United Kingdom in 1912, during the search for steel that would resist fouling in gun barrels, a corrosion-resistant composition containing 12.8% chromium and 0.24% carbon was reported. It was suggested that this composition be used for cutlery. In fact, the composition of AISI Type 420 steel (12–14% chromium, 0.15% carbon) is similar to that of the first corrosion-resistant steel.

The higher chromium–iron alloys were developed in the United States from the early twentieth century on, when the effect of chromium on oxidation resistance at 1090°C was first noticed. Oxidation resistance increased markedly as the chromium content was raised above 20%. For steels containing appreciable quantities of nickel, 20% chromium seems to be the minimum amount necessary for oxidation resistance at 1090°C.

The austenitic iron–chromium–nickel alloys were developed in Germany around 1910 in a search for materials for use in pyrometer tubes. Further work led to the widely used versatile 18% chromium–8% nickel steels, the so-called 18–8.

The many compositions would appear to make the role of individual elements difficult to understand. There is, however, an understandable pattern. The corrosion resistance is primarily dependent on the chromium content, which is in four general ranges, reflecting the nature of the protective oxide film formed. These ranges fall around 5%, 10–13%, 18%, and 25%, and the grades listed contain these amounts of chromium. Increasing the chromium content is valuable for resisting general (uniform) corrosion. However, under some circumstances very localized and potentially damaging corrosion can occur. Examples are (*1*) pitting corrosion, which can be deep enough to penetrate sheet and which can be mitigated by the addition of 2–3% molybdenum, hence Grades 316 and 317; (*2*) grain boundary attack where carbides have formed at grain boundaries, which can be prevented by low carbon contents, hence the L grades, by avoiding heat treatments, which permit carbide precipitation or by adding carbide formers such as Ti or Nb, hence Grades 321 and 347; and (*3*) stress corrosion cracking (SCC), whereby cracks, usually transgranular, form under the simultaneous action of stress, from an applied load or more insidiously from residual stresses, and specific environments, especially chlorides in contact with austenitic steels. Composition changes do not give effective control cheaply and great care is necessary to avoid SCC in process units operating at a few hundred degrees.

The next step to comprehending the many grades of stainless steel lies in understanding the crystal structure of the iron-rich matrix. The austenite field in iron exists over an increasingly small temperature range as chromium is added and disappears at about 12% chromium. Thus, to make the martensitic grades, it is important to be able to form 100% austenite first. Fortunately, carbon extends this range so it is possible to have all austenite prior to quenching in a 12% chromium carbon steel, or, if the carbon is high enough, even in a 17% chromium steel.

For the ferrite grades, it is necessary to have at least 12% chromium and only very small amounts of elements that stabilize austenite. For these materials, the structure is bcc from room temperature to the melting point. Some elements, such as Mo, Nb, Ti, and Al, which encourage the bcc structure, may also be in these steels. Because there are no phase transformations to refine the structure, brittleness from large grains is a drawback in these steels. They find considerable use in structures at high temperatures where the loads are small.

Austenitic steels are extremely valuable structural components in a wide variety of applications. Adding several percent of nickel to an iron–chromium alloy can allow austenite to exist metastably or stably down to ambient temperature. Depending on composition, the M_s can be near or even above room temperature. A bcc phase, or sometimes a hexagonal close-packed phase, can then form martensitically, especially if the material is plastically deformed, eg, Type 301, and can give very high strengths. Using more nickel, the austenite is stable and formable and increasingly finds both industrial and domestic uses. The several types in the 300 series have other specific alloys largely to control the various forms of localized corrosion or to improve the creep strength.

The standard AISI types with the Unified Numbering System (UNS) designation are identified in Table 4. A number of proprietary types have not been listed. These are for specific uses, which justify their substantial cost.

Martensitic Stainless Steels.   Martensitic stainless steels include Types 403, 410, 414, 416, 420, 431, 440A, 440B, 440C, 501, and 502 (see Table 4). The most widely used is Type 410, which contains 11.50–13.50% chromium and <0.15% carbon. In the annealed condition, this grade may be drawn or formed. It is an air-hardening steel, offering a wide range of properties by heat treatment. In sheet or strip form, Type 410 is used extensively in the petroleum industry for

Table 4. Compositions of Standard Stainless Steels[a]

| Type | UNS desig-nation | Composition, wt %[b] | | | | | | | |
		C	Mn	Si	Cr	Ni	P	S	Other
					Austenitic				
201	S20100	0.15	5.5–7.5	1.00	16.0–18.0	3.5–5.5	0.06	0.03	0.25 N
202	S20200	0.15	7.5–10.0	1.00	17.0–19.0	4.0–6.0	0.06	0.03	0.25 N
205	S20500	0.12–0.25	14.0–15.5	1.00	16.5–18.0	1.0–1.75	0.06	0.03	0.32–0.40 N
301	S30100	0.15	2.00	1.00	16.0–18.0	6.0–8.0	0.045	0.03	
302	S30200	0.15	2.00	1.00	17.0–19.0	8.0–10.00	0.045	0.03	
302B	S30215	0.15	2.00	2.0–3.0	17.0–19.0	8.0–10.00	0.045	0.03	
303	S30300	0.15	2.00	1.00	17.0–19.0	8.0–10.00	0.20	0.15[c]	0.6 Mo[d]
303Se	S30323	0.15	2.00	1.00	17.0–19.0	8.0–10.0	0.20	0.06	0.15[c] Se
304	S30400	0.08	2.00	1.00	18.0–20.0	8.0–10.5	0.045	0.03	
304H	S30409	0.04–0.10	2.00	1.00	18.0–20.0	8.0–10.5	0.045	0.03	
304L	S30403	0.03	2.00	1.00	18.0–20.0	8.0–12.0	0.045	0.03	
304LN	S30453	0.03	2.00	1.00	18.0–20.0	8.0–12.0	0.045	0.03	0.10–0.16 N
302Cu	S30430	0.08	2.00	1.00	17.0–19.0	8.0–10.0	0.045	0.03	3.0–4.0 Cu
304N	S30451	0.08	2.00	1.00	18.0–20.0	8.0–10.5	0.045	0.03	0.10–0.16 N
305	S30500	0.12	2.00	1.00	17.0–19.0	10.5–13.0	0.045	0.03	
308	S30800	0.08	2.00	1.00	19.0–21.0	10.0–12.0	0.045	0.03	
309	S30900	0.20	2.00	1.00	22.0–24.0	12.0–15.0	0.045	0.03	
309S	S30908	0.08	2.00	1.00	22.0–24.0	12.0–15.0	0.045	0.03	
310	S31000	0.25	2.00	1.50	24.0–26.0	19.0–22.0	0.045	0.03	
310S	S31008	0.08	2.00	1.50	24.0–26.0	19.0–22.0	0.045	0.03	
314	S31400	0.25	2.00	1.5–3.0	23.0–26.0	19.0–22.0	0.045	0.03	
316	S31600	0.08	2.00	1.00	16.0–18.0	10.0–14.0	0.045	0.03	2.0–3.0 Mo
316F	S31620	0.08	2.00	1.00	16.0–18.0	10.0–14.0	0.20	0.10[c]	1.75–2.5 Mo
316H	S31609	0.04–0.10	2.00	1.00	16.0–18.0	10.0–14.0	0.045	0.03	2.0–3.0 Mo
316L	S31603	0.03	2.00	1.00	16.0–18.0	10.0–14.0	0.045	0.03	2.0–3.0 Mo

Table 4. (Continued)

Type	UNS designation	Composition, wt %[b]							
		C	Mn	Si	Cr	Ni	P	S	Other
316LN	S31653	0.03	2.00	1.00	16.0–18.0	10.0–14.0	0.045	0.03	2.0–3.0 Mo; 0.10–0.16 N
316N	S31651	0.08	2.00	1.00	16.0–18.0	10.0–14.0	0.045	0.03	2.0–3.0 Mo; 0.10–0.16 N
317	S31700	0.08	2.00	1.00	18.0–20.0	11.0–15.0	0.045	0.03	3.0–4.0 Mo
317L	S31703	0.03	2.00	1.00	18.0–20.0	11.0–15.0	0.045	0.03	3.0–4.0 Mo
321	S32100	0.08	2.00	1.00	17.0–19.0	9.0–12.0	0.045	0.03	5 × % C Ti[c]
321H	S32109	0.04–0.10	2.00	1.00	17.0–19.0	9.0–12.0	0.045	0.03	5 × % C Ti[c]
330	N08330	0.08	2.00	0.75–1.5	17.0–20.0	34.0–37.0	0.04	0.03	
347	S34700	0.08	2.00	1.00	17.0–19.0	9.0–13.0	0.045	0.03	10 × % C Nb[c]
347H	S34709	0.04–0.10	2.00	1.00	17.0–19.0	9.0–13.0	0.045	0.03	8 × % C[c]–1.0 Nb
348	S34800	0.08	2.00	1.00	17.0–19.0	9.0–13.0	0.045	0.03	0.2 Co; 10 × % C[c] Nb; 0.10 Ta
348H	S34809	0.04–0.10	2.00	1.00	17.0–19.0	9.0–13.0	0.045	0.03	0.2 Co; 8 × % C[c]–1.0 Nb; 0.10 Ta
384	S38400	0.08	2.00	1.00	15.0–17.0	17.0–19.0	0.045	0.03	
Ferritic									
405	S40500	0.08	1.00	1.00	11.5–14.5		0.04	0.03	0.10–0.30 Al
409	S40900	0.08	1.00	1.00	10.5–11.75	0.50	0.045	0.045	6 × % C[c]–0.75 Ti
429	S42900	0.12	1.00	1.00	14.0–16.0		0.04	0.03	
430	S43000	0.12	1.00	1.00	16.0–18.0		0.04	0.03	
430F	S43020	0.12	1.25	1.00	16.0–18.0		0.06	0.15[c]	0.6 Mo[d]
430FSe	S43023	0.12	1.25	1.00	16.0–18.0		0.06	0.06	0.15 Se[c]
434	S43400	0.12	1.00	1.00	16.0–18.0		0.04	0.03	0.75–1.25 Mo
436	S43600	0.12	1.00	1.00	16.0–18.0		0.04	0.03	0.75–1.25 Mo; 5 × % C[c]–0.70 Nb
439	S43035	0.07	1.00	1.00	17.0–19.0	0.50	0.04	0.03	0.15 Al; 12 × % C[c]–1.10 Ti
442	S44200	0.20	1.00	1.00	18.0–23.0		0.04	0.03	

Type	UNS	C	Mn	Si	Cr	Ni	P	S	Composition, % other
444	S44400	0.025	1.00	1.00	17.5–19.5	1.00	0.04	0.03	1.75–2.50 Mo; 0.025 N; 0.2 + 4 (% C + % N)[c] –0.8 (Ti + Nb)
446	S44600	0.20	1.50	1.00	23.0–27.0		0.04	0.03	0.25 N
Duplex (ferritic–austenitic)									
329	S32900	0.20	1.00	0.75	23.0–28.0	2.50–5.00	0.04	0.03	1.00–2.00 Mo
Martensitic									
403	S40300	0.15	1.00	0.50	11.5–13.0		0.04	0.03	
410	S41000	0.15	1.00	1.00	11.5–13.5		0.04	0.03	
414	S41400	0.15	1.00	1.00	11.5–13.5	1.25–2.50	0.04	0.03	
416	S41600	0.15	1.25	1.00	12.0–14.0		0.06	0.15[c]	0.6 Mo[d]
416Se	S41623	0.15	1.25	1.00	12.0–14.0		0.06	0.06	0.15 Se[c]
420	S42000	0.15 min	1.00	1.00	12.0–14.0		0.04	0.03	
420F	S42020	0.15 min	1.25	1.00	12.0–14.0		0.06	0.15[c]	0.6 Mo[d]
422	S42200	0.20–0.25	1.00	0.75	11.5–13.5	0.5–1.0	0.04	0.03	0.75–1.25 Mo; 0.75–1.25 W; 0.15–0.3 V
431	S43100	0.20	1.00	1.00	15.0–17.0	1.25–2.50	0.04	0.03	
440A	S44002	0.60–0.75	1.00	1.00	16.0–18.0		0.04	0.03	0.75 Mo
440B	S44003	0.75–0.95	1.00	1.00	16.0–18.0		0.04	0.03	0.75 Mo
440C	S44004	0.95–1.20	1.00	1.00	16.0–18.0		0.04	0.03	0.75 Mo
Precipitation-hardening									
PH 13-8 Mo	S13800	0.05	0.20	0.10	12.25–13.25	7.5–8.5	0.01	0.008	2.0–2.5 Mo; 0.90–1.35 Al; 0.01 N
15-5 PH	S15500	0.07	1.00	1.00	14.0–15.5	3.5–5.5	0.04	0.03	2.5–4.5 Cu; 0.15–0.45 Nb
17-4 PH	S17400	0.07	1.00	1.00	15.5–17.5	3.0–5.0	0.04	0.03	3.0–5.0 Cu; 0.15–0.45 Nb
17-7 PH	S17700	0.09	1.00	1.00	16.0–18.0	6.5–7.75	0.04	0.04	0.75–1.4 Al

[a] Ref. 51. Courtesy of ASM International.
[b] Single values are maximum values unless otherwise indicated.
[c] Value is minimum.
[d] Optional.

ballast trays and liners. It is also used for parts of furnaces operating below 605°C, and for blades and buckets in steam turbines.

Type 420, with ca 0.35% carbon and a resultant increased hardness, is used for cutlery. In bar form, it is used for valves, valve stems, valve seats, and shafting where corrosion and wear resistance are needed. Type 440 may be employed for surgical instruments, especially those requiring a durable cutting edge. The necessary hardness for different applications can be obtained by selecting grade A, B, or C, with increasing carbon content in that order.

Other martensitic grades are Types 501 and 502. The former has >0.10% and the latter <0.10% carbon. Both contain 4–6% chromium. These grades are also air-hardened, but do not have the corrosion resistance of the 12% chromium grades. Types 501 and 502 have wide application in the petroleum industry for hot lines, bubble towers, valves, and plates.

Ferritic Stainless Steels. These steels are iron–chromium alloys not hardenable by heat treatment. In alloys having 17% chromium or more, an insidious embrittlement occurs in extended service around 475°C. This can be mitigated to some degree but not eliminated. They commonly include Types 405, 409, 430, 430F, and 446 (see Table 4); newer grades are 434, 436, 439, and 442.

The most common ferritic grade is Type 430, containing 0.12% carbon or less and 14–18% chromium. Because of its high chromium content, the corrosion resistance of Type 430 is superior to that of the martensitic grades. Furthermore, Type 430 may be drawn, formed, and, using proper techniques, welded. At one time widely used for automotive and architectural trim, it is also employed in equipment for the manufacture and handling of nitric acid (qv), to which it is resistant. Type 430 does not have high creep strength but is suitable for some types of service up to 815°C, and thus has application in combustion chamber for domestic heating furnaces.

Type 409, developed as a less expensive replacement for Type 430 in automotive applications such as trim and catalytic converters, has become the principal alloy in this area.

The high (23–27%) chromium content of Type 446 imparts excellent heat resistance, although its high temperature strength is only slightly better than that of carbon steel. Type 446 is used in sheet or strip form up to 1150°C. Some variants of Type 446 are available for severe applications. This grade does not have the good drawing characteristics of Type 430, but may be formed with care. Accordingly, Type 446 is widely used for furnace parts such as muffles, burner sleeves, and annealing baskets. Its resistance to nitric and other oxidizing acids makes it suitable for chemical-processing equipment.

Austenitic Stainless Steels. These steels, based on iron–chromium–nickel alloys, are not hardenable by heat treatment and are predominantly austenitic. They include Types 301, 302, 302B, 303, 304, 304L, 305, 308, 309, 310, 314, 316, 316L, 317, 321, and 347. The L refers to 0.03% carbon max, which is readily available. In some austenitic stainless steels, all or part of the nickel is replaced by manganese and nitrogen in proper amounts, as in one proprietary steel and Types 201 and 202 (see Table 4).

The most widely used austenitic stainless steel is Type 304, known as 18–8. It has excellent corrosion resistance and, because of its austenitic structure, excellent ductility. It may be deep-drawn or stretch formed. It can be readily welded, but carbide precipitation must be avoided in and near the weld by cooling rapidly enough after welding. Where carbide precipitation presents problems, Types 321, 347, or 304L may be used. The applications of Types 304 are wide and varied, including kitchen equipment and utensils, dairy installations, transportation equipment, and oil-, chemical-, paper- (qv), and food-processing (qv) machinery.

The low nickel content of Type 301 causes it to harden faster than Type 304 on account of reduced austenite stability. Accordingly, although Type 301 can be drawn successfully, its drawing properties are not as good as those of Type 302 or 304. Type 301 can be cold-rolled to very high strength. Type 301, because of its lower carbon content, is not as prone as Type 304 to give carbide precipitation problems in welding, and can be used to withstand severe corrosive conditions in the paper, chemical, and other industries. The austenitic stainless steels are widely used for high temperature services.

Types 321 and 347 have additions of titanium and niobium, respectively, and are used in welding applications and high temperature service under corrosive conditions. Type 304L may be used as an alternative for Types 321 and 347 in welding (qv) and stress-relieving applications below 426°C.

The addition of 2–4% molybdenum to the basic 18–8 composition produces Types 316 and 317, which have improved corrosion resistance. These grades are employed in the textile, paper, and chemical industries where strong sulfates, chlorides, and phosphates, and reducing acids such as sulfuric, sulfurous, acetic, and hydrochloric acids, are used in such concentrations that the use of corrosion-resistant alloys is mandatory. These are also used in some surgical implants (see PROSTHETICS AND BIOMEDICAL DEVICES). Types 316 and 317 have the highest rupture strengths of any commercial stainless steels, although Types 347 and 310 may show lower creep rates in certain ranges.

The austenitic stainless steels most resistant to oxidation are Types 309, 310, and 314 where the higher Si contributes significantly. Because of their high chromium and nickel contents, these steels resist scaling at temperatures up to 1090 and 1150°C and consequently are used for furnace parts and heat exchangers. They are somewhat harder and not as ductile as the 18–8 types, but may still be drawn and formed. Types 309 and 310 can be welded readily and have increasing use in the manufacture of industrial furnace equipment (see HIGH TEMPERATURE ALLOYS). For applications requiring better machinability, Type 303 containing sulfur or selenium may be used.

High Temperature Service, Heat-Resisting Steels. The term high temperature service covers many types of operations in many industries. Conventional high temperature equipment includes steam (qv) boilers and turbines, gas turbines, cracking stills, tar stills, hydrogenation vessels, heat-treating furnaces, and fittings for diesel and other internal-combustion engines. Numerous steels are available. Where unusual conditions occur, modification of the chemical composition may adapt an existing steel grade to service conditions. In some cases, however, entirely new alloy combinations must be developed to meet service requirements. For example, the aircraft and missile industries have encountered design problems of increased complexity, requiring metals of great high temperature strength for both power units and structures. Steels are constantly under development to meet these requirements (52).

The temperatures needed for high performance turbines are so high that iron base alloys are not contenders. The melting point of iron (1536°C) is high but not high enough. For several years, nickel-based alloys strengthened by compounds such as Ni_3Al have been used in single crystal form. The next generation seems likely to be based on intermetallic compounds, such as titanium aluminide, TiAl, if ductility problems can be controlled.

A number of steels suitable for high temperature service are given in Table 5.

The design of load-bearing structures for service at room temperature is generally based on the yield strength or for some applications on the tensile strength. The metal is expected to behave essentially in an elastic manner, that is, the structure undergoes an elastic deformation immediately upon load application and no further deformation occurs with time. When the load is removed, the structure returns to its original dimensions.

At high temperature, the behavior is different. A structure designed according to the principles employed for room temperature service continues to deform with time after load application, even though the design data may have been based on tension tests at the temperature of interest. This deformation with time is called creep because the design stresses at which it was first recognized occurred at a relatively low rate.

In spite of the fact that plain carbon steel has lower resistance to creep than high temperature alloy steels, its price allows it to be widely used in such applications up to 540°C, where rapid oxidation commences and a chromium-bearing steel must be employed. Low alloy steels containing small amounts of chromium and molybdenum have higher creep strengths than carbon steel and are employed where materials of higher strength are needed. Above ca 540°C, the amount of chromium required to impart oxidation resistance increases rapidly. The 2% chromium steels containing molybdenum are useful up to ca 620°C, whereas 10–14% chromium steels may be employed up to ca 700–760°C. Above this temperature, the austenitic 18–8 steels are commonly used. Their oxidation resistance is considered adequate up to ca 815°C. For service between 815 and 1090°C, steels containing 25% chromium and 20% nickel, or 27% chromium,

Table 5. Alloy Composition of High Temperature Steels

Ferritic steels	Stainless steels	AISI type
0.5% Mo	18% Cr–8% Ni[a]	304
0.5% Cr–0.5% Mo	18% Cr–8% Ni with Mo[a]	316
1% Cr–0.5% Mo	18% Cr–8% Ni with Ti[a]	321
2% Cr–0.5% Mo	18% Cr–8% Ni with Nb[a]	347
2.25% Cr–1% Mo	25% Cr–12% Ni[a]	309
3% Cr–1% Mo	25% Cr–20% Ni[a]	310
3% Cr–0.5% Mo–1.5% Si	12% Cr[b]	410
5% Cr–0.5% Mo–1.5% Si	17% Cr[b]	430
5% Cr–0.5% Mo, with Nb added	27% Cr[b]	446
5% Cr–0.5% Mo, with Ti added		
9% Cr–1% Mo		

[a]Austenitic.
[b]Ferritic.

are used. The behavior of steels at high temperature is quite complex, and only the simplest design considerations have been mentioned herein (see HIGH TEMPERATURE ALLOYS).

Miscellaneous High Strength Steels. Strengths above 1400–1700 MPa (203,000–246,500 psi) are not often used or required in steels because of difficulties in joining, ductility, and/or toughness. For specialized uses, however, it is possible to achieve such strength (see HIGH PRESSURE TECHNOLOGY) (53).

One of the earliest materials for ultrahigh strength was piano wire, heavily cold-drawn fine pearlite where the strength can be over 3500 MPa (507,500 psi). For many steels, any form of carbon as a contributor to strength needs great care. It is possible to quench and temper a 0.40% carbon alloy steel to give a usable yield in the range of 1800–2000 MPa (261,000–290,000 psi) by melting and adding enough silicon so that the tempering temperature can be raised to give adequate ductility. This process has been used, for instance, for aircraft landing gear, where weight is critical and high costs are acceptable. This is far from common, however.

A different way of reaching high strengths is to use alternative precipitates to cementite and alloy carbides. Early efforts in this area came from making stainless steels of very low carbon, by using an unstable austenite (M_s above room temperature) so a body-centered phase can exist in large amounts after cooling. After dissolving the precipitable elements, eg, Mo, Al, Nb, and Cu, in austenite at around 1000°C, the material can be cooled without quenching to ambient temperature and subsequently aged around 500°C to form the hardening compound in the bcc phase. Yield strengths up to 1590 MPa (230,550 psi) can be obtained, although this can be reduced to optimize other properties in these precipitation hardening (PH) steels.

An important item in this array of materials is the class known as maraging steels. This group of high nickel martensitic steels contain so little carbon that they are often referred to as carbon-free iron–nickel martensites (54). Carbon-free iron–nickel martensite with certain alloying elements is relatively soft and ductile and becomes hard, strong, and tough when subjected to an aging treatment at around 480°C.

The first iron–nickel martensitic alloys contained ca 0.01% carbon, 20 or 25% nickel, and 1.5–2.5% aluminum and titanium. Later an 18% nickel steel containing cobalt, molybdenum, and titanium was developed, and still more recently a series of 12% nickel steels containing chromium and molybdenum came on the market.

By adjusting the content of cobalt, molybdenum, and titanium, the 18% nickel steel can attain yield strengths of 1380–2070 MPa (200,000–300,000 psi) after the aging treatment. Similarly, yield strengths of 12% nickel steel in the range of 1035–1380 MPa (150,000–200,000 psi) can be developed by adjusting its composition.

Silicon Steel Electrical Sheets. The silicon steels are characterized by relatively high permeability, high electrical resistance, and low hysteresis loss when used in magnetic circuits (see MAGNETIC MATERIALS). First patented in the United Kingdom around 1900, the silicon steels permitted the development of more powerful electrical equipment and have furthered the rapid growth of the electrical power industry. Steels containing 0.5–5% silicon are produced in sheet form for the laminated magnetic cores of electrical equipment and are referred to as electrical sheets (54).

The grain-oriented steels, containing ca 3.25% silicon, are used in the highest efficiency distribution and power transformers and in large turbine generators. They are processed in a proprietary way and have directional properties related to orientation of the large crystals in a preferred direction.

The nonoriented steels are subdivided into low, intermediate, and high silicon steels. The first contain about 0.5–1.5% silicon, used mainly in rotors and stators of motors and generators. Steels containing ca 1% silicon are used for reactors, relays, and small intermittent-duty transformers.

Intermediate silicon steels (2.5–3.5% Si) are used in motors and generators of average to high efficiency and in small- to medium-size intermittent-duty transformers, reactors, and motors.

High silicon steels (historically 3.75–5.00% but more recently 6% Si) are used in power transformers and high efficiency motors, generators, and transformers, and in communications equipment.

Economic Aspects

The production of steel is of great importance in most countries because modern civilization depends heavily on steel, the raw material for many industries. As a result, most countries have an active steel industry, which at one time was heavily subsidized but as of this writing is increasingly privatized. The world trade in steel was frequently a source of hard currency where the United States was the main contributor. Trade is much more at market prices that reflect the real cost of production. Under these conditions, the United States in its own market is very often the low cost producer following massive cost reduction in the 1980s. The United States can export a few million tons per year at a profit.

Production. World production of raw steel, largely continuously cast, is shown in Table 6. U.S. production has approached 9.0×10^7 t/yr, over 90% of which is continuously cast. The amount melted in electric furnaces approached 40% and in 1995 was still increasing.

The leading producers in 1979 were the CIS, United States, Japan, Germany, and China. By the mid-1990s, China and South Korea had grown enormously in capacity. United States, Germany, and Japan have dropped somewhat and the CIS significantly. Countries such as India and those in the Pacific rim and Latin America have become increasingly important steel producers as they build infrastructure. World growth in the latter part of the twentieth century was small but the redistribution of producers was significant.

Prices. Through the 1970s the world price of steel continued to increase with inflation. Around 1980, however, overcapacity was around 25% and prices became flat at best. On a basis depreciated for inflation they mostly dropped and a huge restructuring of the industry took place. Plants closed worldwide and capacity slowly came into better alignment with demand by about 1990. Quality and customer service improved dramatically and, by 1995, modest price increases began to stick. The challenge from competitive materials intensified, but on the whole the steel industry has not lost a principal market since beverage

Table 6. World Raw-Steel Production, 1000 Metric Tons

Country	1969	1979	1995
Argentina	1,690	3,200	3,700
Australia	7,032	8,200	8,500
Austria	3,926	4,900	5,000
Belgium	12,832	13,100	11,600
Brazil	4,925	13,900	25,100
Bulgaria	1,515	2,500	
Canada	9,351	16,000	14,400
China	16,000	32,000	93,000
CIS and the former USSR	110,315	153,000	73,300[a]
Czechoslovakia	10,802	15,500	10,900[b]
France	22,510	23,600	18,100
Germany	50,456[c]	53,800[c]	41,800
Hungary	3,032	3,800	1,870
India	6,557	10,100	20,300
Italy	16,428	24,000	27,700
Japan	82,166	112,000	101,600
North Korea	2,000	5,100	
South Korea	373	7,600	36,700
Luxembourg	5,521	5,000	2,600
Mexico	3,467	7,000	12,100
Netherlands	4,712	5,800	6,400
Poland	11,251	19,500	11,900
Romania	5,540	12,000	6,500
South Africa	4,625	8,900	8,500
Spain	5,982	12,100	13,900
Sweden	5,322	4,700	4,900
United Kingdom	26,896	22,000	17,700
United States	128,151	124,000	83,100
Yugoslavia	2,220	3,500	
others	9,038	21,500	59,100
Total	*574,635*	*748,300*	*730,355*

[a] Russia and Ukraine.
[b] Includes Czech Republic and Slovak Republic.
[c] Both FRG and GDR.

cans were lost to aluminum in the 1970s. Much tonnage of carbon steel was available in the mid-1990s at $0.15–$0.30/#, not very different from the 1979 price and of much better quality.

Products. The production amounts of various products in the United States are shown in Table 7. Sheet, both hot- and cold-rolled, dominates at about 60% of production. Coated material, mostly galvanized, is playing an increasingly important role as demands for longer life become an issue (Fig. 27) (56). Dominant customers remain the automotive and construction industries.

At one point in the mid-1980s, imports reached 26% of consumption. As of 1995, imports were much less than 20% and included some semifinished slabs that were finished in the United States because of insufficient domestic capacity for cast slabs to make sheet.

Table 7. Net Shipments of U.S. Steel Products, 1000 Metric Tons[a]

Steel products	1960	1970	1980	1994
ingots and steel castings	307	833	404	192
blooms, slabs, billets, sheet bars	1,286	4,328	1,982	2,104
skelp	118	1	22	
wire rods	833	1,525	2,438	4,319
structural shapes (heavy)	4,387	5,049	4,410	4,996
steel piling	384	449	314	396
plates	5,562	7,316	7,330	7,764
rails				
standard (over 30 kg/m)	606	779	984	449
all other	42	38	48	4
joint bars	24	23	15	120[b]
tie plates	113	160	186	
track spikes	41	69	77	
wheels (rolled and forged)	221	226	144	
axles	101	147	175	
bars				
hot-rolled (includes light shapes)	6,274	7,355	6,270	7,966
reinforcing	2,009	4,437	4,249	4,473
cold finishing	1,257	1,352	1,438	1,621
tool steel	79	80	72	61
pipe and tubing				
standard	1,949	2,317	1,608	1,036
oil country goods	1,086	1,186	3,277	1,211
line	2,440	1,900	1,082	1,031
mechanical	692	955	1,051	939
pressure	246	215	135	34
structural		455	516	176
stainless		29	37	23
wire				
drawn	2,213	2,143	1,263	715
nails and staples	291	272	157	
barbed and twisted	43	89	64	
woven wire fence	94	109	111	
bale ties and bailing wire	57	108	83	
black plate	523	553	457	288
tin plate	4,958	5,176	3,779	2,518
tin-free steel		777	887	857
all other tin mill products		65	55	91
sheets				
hot-rolled	7,249	11,175	10,991	14,205
cold-rolled	13,124	12,927	12,077	11,811
sheets and strip				
galvanized	2,773	4,359	4,757	13,275
all other metallic coated	237	500	655	1,554
electrical	515	663	545	398
strip				
hot-rolled	1,209	1,173	607	646
cold-rolled	1,203	1,045	844	887
Total steel products	*64,544*	*82,358*	*75,786*	*86,283*

[a]All grades, including carbon, alloy, and stainless steels.
[b]Including joint bars, tie plates, track spikes, wheels (rolled and forged), and axles.

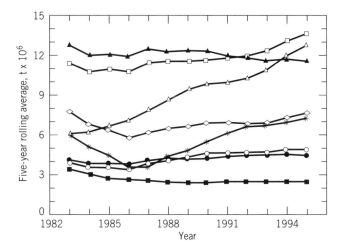

Fig. 27. U.S. shipments of steel mill products, where (—•—) represents bars-reinforcing, (—◇—) hot-rolled and cold-finished bars, (—■—) tinplate, (—▲—) sheet-cold-rolled, (—○—) structural shapes, (—*—) plate, (—□—) sheet-hot-rolled, and (—△—) sheet-coated products.

Health and Safety Factors

The hazards and environmental problems associated with steelmaking are many and varied, but have been sharply reduced by industrywide efforts. Since the early 1980s, increasingly stringent environmental regulations have been imposed by local, state, and federal government in the United States (57). There have been many arguments as to the cost-effective nature of some of these. Some other nations have largely equivalent regulations, but many do not, giving them a competitive advantage.

    Accidents. In an industry that has massive equipment, high temperature operations, and many moving objects, the potential for accidents is always present. Safety is taken seriously and is the direct responsibility of the plant superintendent. Accidents do occur, although in the United States the rate in the steel industry compares favorably with heavy industry as a whole.

    Health Hazards. There are many sources of industrial hygiene (qv) and occupational health hazards. Some are direct, eg, radiant heat, noise, and exposure to local concentrations of noxious gases such as carbon monoxide near blast furnaces. Others are long-term, eg, the acid fumes from pickling baths or the possible leakage of organic vapors from coke ovens. The use of protective clothing, masks, eye shields, ear protection, etc, is routine and rigorously enforced. Even though constant vigilance and training are essential, the hazards are contained effectively when sensors monitoring potential dangers are employed along with good work practices and engineering controls such as radiation shields, ventilation, air conditioning of control stations, mufflers, and soundproof enclosures. Networking between companies in the safety area, which makes a further contribution to overall safety, is excellent in the United States.

BIBLIOGRAPHY

"Steel" in *ECT* 1st ed., Vol. 12, pp. 793–843, by W. Carroll, Republic Steel Corp.; in *ECT* 2nd ed., Vol. 18, pp. 715–805, by H. E. McGannon, United States Steel Corp.; in *ECT* 3rd ed., Vol. 21, pp. 552–625, by R. J. King, United States Steel Corp.

1. U.S. Steel Corp., *Making, Shaping and Treating of Steel*, 10th ed., Association of Iron and Steel Engineers, Pittsburgh, Pa., 1985.
2. R. F. Tylecote, *A History of Metallurgy*, Metals Society, London, U.K., 1976.
3. Ref. 1, Chapt. 1.
4. Technical data from the proceedings of various *Iron & Steel Society Annual Steelmaking Conferences*, to 1996.
5. Technical data from the proceedings of various *Iron & Steel Society Electric Furnace Conferences*, to 1996.
6. *Enhancement of EAF Performance by Injection Technology, Conference Proceedings*, McMaster University, Hamilton, Ont., Canada, 1994.
7. "Ultra-Refining for Specialty Steels," *Proceedings from the 11th Iron and Steel Society Process Technology Conference*, Iron & Steel Society, Warrendale, Pa., 1993.
8. *Proceedings from the Iron and Steel Society Vacuum Metallurgy Conference*, Iron & Steel Society, Warrendale, Pa., 1989; 1992.
9. R. D. Pehlke and co-workers, eds., *BOF Steelmaking*, Iron and Steel Society, Warrendale, Pa., 1975.
10. B. Deo and R. Boom, *Fundamentals of Steelmaking Metallurgy*, Prentice Hall, N.J., 1993.
11. Ref. 1, pp. 414–427.
12. *Pretreatment and Reclamation of Dusts, Sludges, and Scales in Steel Plants, Conference Proceedings*, McMaster University, Hamilton, Ont., Canada, 1993.
13. Fr. Pat. 1,450,718 (July 18, 1966), G. Savard and R. Lee (to Air Liquide).
14. K. Brotzmann, *Technik Forschung*, **47**, 718 (1968).
15. N. L. Kotraba, *Pneumatic Steelmaking*, Vol. 1, Iron and Steel Society, Warrendale, Pa., 1988.
16. G. Derge, ed., *Basic Open-Hearth Steelmaking*, 3rd ed., AIME, Warrendale, Pa., 1964.
17. R. J. Fruehan, *Iron and Steelmaker* **12**(5), 36 (1985).
18. *Handbook, Including Specifications for Steel Scrap*, Institute for Iron and Steel Scrap, Washington, D.C., 1979.
19. *Development in Ladle Steelmaking and Continuous Casting*, Canadian Institute of Mining, Metallurgy and Petroleum, Montreal, Quebec, Canada, 1990.
20. R. J. Fruehan, *Ladle Metallurgy Principles and Practices*, Iron and Steel Society, Warrendale, Pa., 1985.
21. R. J. Fruehan, *Vacuum Degassing*, Iron and Steel Society, Warrendale, Pa., 1990.
22. *ISS Continuous Casting Series*, Vols. 1–7, Iron and Steel Society, Warrendale, Pa., 1983–1995.
23. *Proceedings from the 13th ISS Process Technology Conference*, Iron & Steel Society, Warrendale, Pa., 1995.
24. *Near Net Shape Casting in Minimills*, Canadian Institute of Mining, Metallurgy and Petroleum, Montreal, Quebec, Canada, 1995.
25. W. J. Link, *Iron Steel Eng.* **55**, 29 (July 1978).
26. *Flat Rolling*, Metals Society, London, U.K., 1979.
27. C. M. Sellars and G. J. Davies, eds., *Hot Working and Forming Processes*, Metals Society, London, U.K., 1980.
28. F. A. A. Crane, *Mechanical Working of Metals*, Macmillan, New York, 1964.
29. W. L. Roberts, *Hot Rolling of Metals*, Marcel Dekker, New York, 1983.

30. R. A. Grange, *Fundamentals of Deformation Processing*, Syracuse University Press, Syracuse, New York, 1962, p. 299.
31. Technical data from proceedings of various ISS Mechanical Working and Steel Processing Annual Conferences.
32. W. C. Leslie, *The Physical Metallurgy of Steels*, McGraw-Hill Book Co., Inc., New York, 1981.
33. E. C. Bain and H. W. Paxton, *Alloying Elements in Steel*, ASM International, Materials Park, Ohio, 1961.
34. G. Krauss, *Principles of the Heat Treatment of Steel*, ASM International, Metals Park, Ohio, 1980.
35. *Proceedings from the ISS Gilbert R. Speich Symposium*, Iron & Steel Society, Warrendale, Pa., 1992.
36. D. V. Doane and J. S. Kirkaldy, *Hardenability Concepts with Applications to Steel*, TMS-AIME, Warrendale, Pa., 1978.
37. M. Atkins, *Atlas of Continuous Cooling Transformations for Engineering Steels*, British Steel Corp., Sheffield, U.K., 1978.
38. *Microalloying '75*, Union Carbide, New York, 1977, p. 9.
39. N. J. Petch, *Prog. Metal Physics*, **5**, 1 (1954).
40. *HSLA Steels, Technology and Applications*, ASM International, Materials Park, Ohio, 1983.
41. R. A. Kot and B. L. Bramfitt, *Fundamentals of Dual-Phase Steels*, TMS-AIME, Warrendale, Pa., 1981.
42. *IF Steel Sheet, Processing, Fabrication and Properties*, Canadian Institute of Mining, Metallurgy and Petroleum, Montreal, Quebec, Canada, 1991.
43. *International Symposium on Microstructure and Properties of Microalloyed and Other Modern HSLA Steels*, Iron and Steel Society, Warrendale, Pa., 1992.
44. *Steel Products Manual: Strip Steel*, American Iron and Steel Institute, Washington, D.C., 1978.
45. *Metals Handbook*, Vol. 1, 10th ed., ASM International Materials Park, Ohio, 1990, pp. 140–239.
46. *Annual Book of ASTM Standards*, Volumes 1–6, ASTM, Philadelphia, Pa.
47. Ref. 45, pp. 841–949.
48. F. B. Pickering, *The Metallurgical Evolution of Stainless Steels*, ASM International, Materials Park, Ohio, 1979.
49. D. Peckner and I. M. Bernstein, *Handbook of Stainless Steels*, McGraw-Hill Book Co., Inc., New York, 1977.
50. *Alloying*, ASM International, Materials Park, Ohio, 1988, p. 225.
51. Ref. 45, p. 843.
52. Ref. 45, pp. 950–1006.
53. Ref. 45, pp. 391–398.
54. Ref. 45, pp. 793–800.
55. B. D. Cullity, *Introduction to Magnetic Materials*, Addison Wesley, Reading, Mass., 1972.
56. *Proceedings from International Galvatech Conference*, Iron & Steel Society, Warrendale, Pa., 1995.
57. *Proceedings from 12th ISS Process Technology Conference*, Iron & Steel Society, Warrendale, Pa., 1993.

General References

The Iron and Steel Society (ISS-AIME) based in Warrendale, Pa., publishes probably the most complete current technical materials on the steel industry. A listing can be obtained directly from their publications office or at http://www.issource.org. All references from ISS-AIME and Canadian Institute of Mining, Metallurgy, and Petroleum are listed at this source. Much is available on microfilm through ISS or directly from the University of Michigan. The *ISS Transactions* are published annually. For a general discussion of the scientific basis of steel making, two useful volumes published by ISS are: (*1*) *The John Elliott Symposium*, 1990; and (*2*) *The E. T. Turkdogan Symposium*, 1994.

HARRY PAXTON
Carnegie Mellon University

STERILIZATION TECHNIQUES

Sterilization, the act of sterilizing or the condition of being sterilized, is defined as rendering a substance incapable of reproduction. Whereas this is often taken to mean total absence of living organisms, a more accurate representation is that the substance is free from living microorganisms with a probability previously agreed to be acceptable.

Sterilization technology is of primary importance in industries as diverse as food processing (qv) and space exploration. Generally, however, it is more readily associated with the health-care profession and industry or with the electronics industry (see CONTAMINATION CONTROL TECHNOLOGY (SUPPLEMENT)). Many of the former industries employing sterilization technology are regulated by some federal agency. In the United States, the introduction of foods, pharmaceuticals (qv), and medical devices into interstate commerce is regulated by the FDA. The registration of chemical sterilants is regulated by the EPA in the United States.

The U.S. market in hospital sterilizing equipment is ca \$70 $\times$ 10^6. Manufacturers of various types of equipment are given in Table 1.

A distinction must be made between sterilization and certain other processes often called sterilization as a result of popular misconception. Methods and procedures less rigorous than sterilization, such as disinfection, sanitization, and the use of antiseptics and bacteriostats (see DISINFECTANTS AND ANTISEPTICS; FOOD PROCESSING; INDUSTRIAL ANTIMICROBIAL AGENTS), are often applied to render the object safe for certain applications. In most instances, a judgment on the suitability of a sterilization or a substitute process can only be made by a microbiologist. Well-known exceptions are the common practice of boiling infants' feeding bottles under ambient atmospheric pressure for 5–15 min, being the aim to sterilize them, or the practice of soaking objects in 70% alcohol (1). Boiling at atmospheric pressure does not result in sterilization, nor does soaking in alcohol. Neither procedure kills bacterial spores.

Table 1. Manufacturers of Steam, Ethylene Oxide Sterilizers, and Dry Heat Sterilizers

Manufacturer	Location
Amsco/American Sterilizer Co.[a]	Erie, Pa.
H. W. Anderson Products, Inc.	Oyster Bay, N.Y.
Bard International CR	Durham, U.K.
Baumer Equipment Medico Hospitalar S/A	Sao Paulo, Brazil
Be Venue Laboratories, Inc.	Bedford, Ohio
British Sterilizer Co., Ltd.	Essex, U.K.
Britains Hospital Supplies, Ltd.	Cheddleton, U.K.
MDT-Castle Co.	Rochester, N.Y.
Consolidated Stills & Sterilizers	Boston, Mass.
Danspital, Ltd., Turn-Key Hospitals	Roedovre, Denmark
Dent & Hellyer, Ltd.	Andover, U.K.
Downs Surgical, Ltd.	Surrey, U.K.
Drayton Castle, Ltd.	Middlesex, U.K.
Electrolux Wascator	Alingas, Sweden
Environmental Tectonics Corp.	Southampton, Pa.
Getinge AG	Getingen, Sweden
Harsanyi Labor Mim	Budapest, Hungary
Intermed	Stafford, U.K.
Labotal, Ltd.	Jerusalem, Israel
MMM Munchener Medizin Mechanik	Munich, Germany
O.C.R.A.S. Zambelli S.A.S.	Torino, Italy
Nishimoto Sangyo Co., Ltd.	Osaka, Japan
Royal Adlinden	Wyndyecht, the Netherlands
Sakura Finetechnical Co., Ltd.	Tokyo, Japan
Scardi Construzioni Sanitario	Milano, Italy
Surgical Equipment Supplies, Ltd.	London, U.K.
3M (Minnesota Mining and Manufacturing Co.)	St. Paul, Minn.
Vacudyne Altair	Chicago Heights, Ill.
Webecke & Co.	Bad Schwartau, Germany
Atomic Energy of Canada, Ltd.[b]	Ottawa, Canada

[a]As of 1996, purchased by Steris Corp. (Ohio).
[b]Radiation sterilizer.

It is necessary to establish a criterion for microbial death when considering a sterilization process. With respect to the individual cell, the irreversible cessation of all vital functions such as growth, reproduction, and in the case of viruses, inability to attach and infect, is a most suitable criterion. On a practical level, it is necessary to establish test criteria that permit a conclusion without having to observe individual microbial cells. The failure to reproduce in a suitable medium after incubation at optimum conditions for some acceptable time period is traditionally accepted as satisfactory proof of microbial death and, consequently, sterility. The application of such a testing method is, for practical purposes, however, not considered possible. The cultured article cannot be retrieved for subsequent use and the size of many items totally precludes practical culturing techniques. In order to design acceptable test procedures, the kinetics and thermodynamics of the sterilization process must be understood.

Kinetics

An overwhelming body of evidence, starting with the earliest investigations (2), supports the contention that the rate of destruction of microorganisms is logarithmic, ie, first order with respect to the concentration of microorganisms. The process can be described by the following expression:

$$\frac{N_o}{N_t} = e^{-kt} \tag{1}$$

in which N_t = the number of organisms alive at time t, N_o = the initial number of organisms, and k = the kinetic rate constant. It can be seen that N_t approaches zero as t approaches infinity. Absolute sterility, accordingly, is impossible to attain.

First-order kinetics yield linear plots on semilogarithmic graphs when plotted against time. It has been found convenient to express the rate of microbial kill in terms of a decimal reduction rate or D-value. The D-value represents the time of exposure (at given conditions for a given microorganism) required for a 10-fold decrease in the viable population. This principle has found great utility in the health-care industry as well as in food processing (3). The practical significance of the D-value is that it simplifies the design of sterilization cycles. A sufficiently large D for any process results in a negative log N_t which, in a practical sense, represents the probability of survival of the last remaining microorganism. In the health sciences, a 10^{-6} residual concentration of microorganisms is generally regarded as an acceptable criterium for sterility. Exceptions in the direction of higher or lower values do exist, however, depending on the type of sterilization process used, or the purpose of the sterile product.

Thermodynamics

The Arrhenius rate theory, an empirical derivation, holds for the sterilization process:

$$k = Ae^{-\Delta E/RT} \tag{2}$$

where k = kinetic constant, R = gas-law constant, T = absolute temperature, ΔE = the activation energy, and A = the Arrhenius rate constant.

The Eyring equation for the theory of absolute reaction rates is more accurate:

$$k = \frac{k_B T}{h} e^{(T\Delta S - \Delta H/RT)} \tag{3}$$

where k_B = Boltzman's constant, h = Plank's constant, and ΔS and ΔH are the standard entropy and enthalpy changes, respectively. Determinations of ΔE or ΔH and ΔS usually yield values of 167–335 J/mol (40–80 cal/mol). Such values are often characteristic of protein denaturation. Microbial death may involve irreversible denaturation of some or even all of the cell's proteins.

The relationship between the D-value and k can be derived by considering the meaning of D:

$$D = \frac{t_2 - t_1}{\log_{10} N_1 - \log_{10} N_2} \tag{4}$$

where $N_1 = 10\ N_2$.

Substituting into equation 1:

$$D = 2.3/k \tag{5}$$

Of great practical value is the derivation for the effect of temperature, the z-value, defined as the temperature change necessary to effect a 10-fold reduction ($1D$):

$$z = \frac{T_2 - T_1}{\log_{10} D_1 - \log_{10} D_2} \tag{6}$$

A convenient multiple of D is the thermal death time F_o. It is the exposure time required for less than 1×10^{-6} probability of survival (4,5). The relationship between F_o and z becomes equation 7:

$$z = \frac{T_2 - T_1}{\log_{10} F_o^1 - \log_{10} F_o^2} \tag{7}$$

The comparison of the effectiveness of sterilization cycles at different temperatures becomes possible. For example, for steam sterilization,

$$F_o^{121} = t \Big/ \text{antilog}\, \frac{121 - t}{z} \tag{8}$$

F_o^{121}, the thermal death time at 121°C, is accepted to be 18 min (6) and the z-value used is 10°C. Substituting into equation 8, an equivalent thermal death time t of 1.35 min at 132°C or 0.75 min at 135°C is obtained.

Testing and Monitoring

Direct testing for sterility by culturing is a destructive test method, ie, the product is rendered useless for food or medical purposes. Indirect testing methods usually rely on a statistically valid sampling (qv) pattern for a product. In the case of sterilization, where the desired outcome has to demonstrate a $<10^{-6}$ probability of failure, even a sample size of 500,000 cultures could only provide a 50% chance of detecting a failure in the process. Accordingly, product monitoring for sterilization using a test sample is a limited value. Product monitoring is only utilized if no other information is available for the particular process cycle for a given lot of products, eg, a screening test. Because sterilization is a highly reproducible and well-understood process, it has been found that process monitoring is far more suitable for purposes of sterility assurance. Process monitoring can be accomplished by measurement of individual parameters, each known to be critical for success, or by methods that are capable of integrating all critical conditions into a single display which can be observed or measured.

Biological Monitoring. Biological indicators are preparations of specific microorganisms particularly resistant to the sterilization process they are intended to monitor. The organisms are inoculated onto filter paper specified in the *U.S. Pharmacopeia* for specific purposes. Organisms considered to be particularly suitable are specified in appropriate sections of the ISO standards and the *U.S. Pharmacopeia*, *European Pharmacopeia*, or pharmacopeias of other nations. It is possible to prepare indicators using organisms other than those, or to inoculate product or packaging samples selected to resemble as closely as possible the actual product being sterilized.

When designing industrial sterilization cycles, the bioburden or bioload is determined first. The bioload, the average number of organisms present on or in an article that is to be sterilized, has been found to be highly reproducible from lot to lot for mass-produced items. An appropriate number of biological indicators is added to the load, reflecting the bioload and the desired degree of safety. If the bioload were small, eg, 10^2, the number of organisms on the biological indicator would have to be 10^6. All of the organisms must be killed during the process to indicate that the process produced the 10^{-6} concentration considered necessary.

The carriers containing the biological indicators are retrieved following exposure, transferred aseptically into sterile culture media, and incubated for the required length of time, usually two to seven days. Some unexposed indicators are also incubated to prove that the spores were viable. If no growth is observed while the viability control displays the required growth, the conclusion is made that the sterilization cycle was successful. In order for the biological indicator concept to work, it is necessary to place the indicators into that portion of the sterilizer load considered the most difficult for the sterilant to reach. The lot of products is quarantined to prevent dissemination in case of incomplete sterilization.

Hospital and health-care institutions face a different problem. The sterilizer loads are diverse and generally prepared manually. Therefore, the bioburden varies and it is impossible to determine it for each load. Some prior assumptions are made about the bioburden when designing hospital cycles, and design includes a sufficient degree of additional safety factors. A problem is the inability of hospitals to quarantine sterilized supplies, both for lack of an adequate number of instruments and storage space. For hospital sterilization, biological indicators are considered inadequate and the measurement of physical parameters is essential.

Spores of *Bacillus stearothermophilus* are frequently used for testing steam sterilization because of their high resistance to this type of sterilization. *Bacillus subtilis* is used for dry-heat, ethylene oxide, or other types of vapor/gas sterilization. For radiation, *Bacillus pumilus* is used. The viability of any spore preparation is known to change with time, and such preparations have limited shelf lives. It is always advisable to test the viability of a given lot of spore preparations by culturing an unexposed sample from the same lot alongside the exposed samples. The labeling of commercially prepared products usually includes information on the number of organisms present, the *D*-values, performance characteristics, instructions for culturing, and the expiration date.

Typical performance characteristics for some of the most widely used biological indicators are given in Table 2.

According to the D-values shown in Table 2, a 6-D reduction using the 1.5-min requirement for steam at 121°C results in a 9-min cycle. Yet a minimum 12-min exposure is recommended, and with the safety factor, an 18-min steam contact is required (Table 3) for hospital cycles at 121°C.

Culturing is time-consuming and quarantining of packages sterilized in the hospital is not feasible. Under such circumstances, it is not possible to obtain information about a specific sterilizer load by means of biological indicators before the contents of that load are used. Nevertheless, authoritative sources such as the Joint Commission on Accreditation of Hospitals (JCAH) (8), the Sterilization

Table 2. Performance Characteristics of Some Biological Indicators[a]

Culture spores	Sterilization process	Approximate D-value
Bacillus subtilis	ethylene oxide at 50% rh and 54°C	
	600 mg/L	3 min
	1200 mg/L	1.7 min
Bacillus stearothermophilus	saturated steam at 121°C	1.5 min
Bacillus pumilus	gamma-radiation	
	wet preparations	2×10^{-6}Gy[b]
	dry preparations	1.5×10^{-6}Gy[b]
Bacillus subtilis	dry heat at 121–170°C	60–1 min
Clostridium sporogenes	saturated steam at 112°C	3.5–0.7 min

[a]Ref. 6.
[b]To convert Gy to rad, multiply by 100.

Table 3. Minimum Exposure Periods for Sterilization of Hospital Supplies[a]

Load	Temperature, °C	Holding time, min	Air removal method	Heat-up time, min	Safety factor	Exposure time, min
hard goods						
unwrapped	121–123	12	gravity	1	2	15
	133–135	2	gravity[b]	<1	0.5	3
	133–135	2	pulsing	<1	1	4
	133–135	2	prevacuum	1	1	4
wrapped	121–123	12	gravity	5	3	20
	133–135	2	gravity	7	1	10
	133–135	2	pulsing	<1	1	4
	133–135	2	prevacuum	1	1	4
fabrics, packs	121–123	12	gravity	12	6	30
	133–135	2	pulsing	<1	1	1
	133–135	2	prevacuum	1	1	4
	141–142	0.5	pulsing	<1	0.5	2
liquids	121–123	12	gravity	c	c	c

[a]Ref. 7. Courtesy of Charles C. Thomas, Publisher.
[b]High speed.
[c]Depending on liquid volume and container.

Standards Committee of the Association for the Advancement of Medical Instrumentation (AAMI) (9), and the Association of Operating Room Nurses (AORN) (10) recommend that biological testing of all sterilizers be conducted at least once a week, but preferably every day for steam sterilizers, every cycle for ethylene oxide sterilizers, and in every cycle when implantable devices are sterilized. If possible, the implantables should be quarantined until the results of the biological test become available. These tests are carried out by placing the biological indicators in test packs of specific constructions, and placing the test packs in otherwise normally loaded sterilizers.

Hospital sterilizer loads vary in composition, thus the challenge presented to the test organism can vary considerably, depending on the type and contents of packages in which they are placed. The benefits of a standardized test-pack construction and test protocol are obvious, and such recommendation is made by AAMI for steam and ethylene oxide sterilizers (11). More recently in European (CEN) and International (ISO) standards, biological indicators are considered as additional information supplemental to the measurement of physical parameters. Indeed, for sterilization using moist heat (steam), the biological indicator information is not considered to be relevant.

Monitoring by Electromechanical Instrumentation. According to basic engineering principles, no process can be conducted safely and effectively unless instantaneous information is available about its conditions. All sterilizers are equipped with gauges, sensors (qv), and timers for the measurement of the various critical process parameters. More and more sterilizers are equipped with computerized control to eliminate the possibility of human error. However, electromechanical instrumentation is subject to random breakdowns or drifts from calibrated settings and requires regular preventive maintenance procedures.

An inherent problem is the location of the sensors. It is not possible to locate the sensors inside the packages which are to be sterilized. Electromechanical instrumentation is, therefore, capable of providing information only on the conditions to which the packages are exposed; but cannot detect failures as the result of inadequate sterilization conditions inside the packages. Such instrumentation is considered a necessary, and for dry and moist heat sterilization, a sufficient, means of monitoring the sterilization process.

Chemical Monitoring. Chemical indicators are devices employing chemical reactions or physical processes designed in such a way as to permit observation of changes in a physical condition, such as color or shape, and to monitor one or more process parameters. Chemical indicators can be located inside the packages, and the results are observable immediately when the package is opened for use. Most available chemical indicator types, however, are not capable of fully integrating all the critical process parameters. They are, therefore, not accepted as a guarantee of sterility, although some can indicate if conditions were adequate for sterilization.

A particular advantage of chemical indicators is the manufacturer's ability to formulate a relatively large, homogeneous batch of reagent mixture that can be deposited on inexpensive substrates, such as paper, by high speed printing techniques, resulting in relatively low unit costs. Because hospital sterilization cycles are standardized, hospitals can benefit considerably from the use of chemical indicators. Moreover, the use of chemical indicators makes it easy to distinguish those materials which have gone through a sterilization procedure from those that have not.

Industrial sterilization cycles tend to vary considerably, not only from manufacturer to manufacturer, but often from product type to product type, depending on the bioburden present on a given load. Chemical indicators have historically been used only to differentiate between sterilized and nonsterilized packages. More recent developments have resulted in the availability of chemical dosimeters of sufficient accuracy to permit their application either as total monitors or as critical detectors of specific parameters.

Dry-Heat Sterilization

Dry-heat sterilization is generally conducted at 160–170°C for ≥ 2 h. Specific exposures are dictated by the bioburden concentration and the temperature tolerance of the products under sterilization. At considerably higher temperatures, the required exposure times are much shorter. The effectiveness of any cycle type must be tested. For dry-heat sterilization, forced-air-type ovens are usually specified for better temperature distribution. Temperature-recording devices are recommended.

It is an axiom of sterilization technology that appropriate conditions must be established throughout the material to be sterilized. The time-at-temperature conditions are critical when considering any sterilization method using heat.

Chemical indicators for dry-heat sterilization are available either in the form of pellets enclosed in glass ampuls, or in the form of paper strips containing a heat-sensitive ink. The former displays its end point by melting, the latter by a color change (see CHROMOGENIC MATERIALS).

Steam Sterilization

Steam (qv) sterilization specifically means sterilization by moist heat. The process cannot be considered adequate without assurance that complete penetration of saturated steam takes place to all parts and surfaces of the load to be sterilized (Fig. 1). Steam sterilization at 100°C and atmospheric pressure is not considered effective. The process is invariably carried out under higher pressure in autoclaves using saturated steam. The temperature can be as low as 115°C, but is usually 121°C or higher.

Great care is needed in the design of autoclaves and sterilization cycles because of the requirement for the presence of moisture. The autoclave must be loaded to allow complete steam penetration to occur in all parts of the load before timing of the sterilization cycle commences. The time required for complete penetration, the so-called heat-up time, varies with different autoclave construction and different types of loads and packaging materials. The time may not exceed specific limits in order to guarantee reproducibility and, for porous loads, saturated steam. The volume of each container has a considerable effect on the heatup time whenever fluids are sterilized. Thermocouples led into the chamber through a special connector are often employed to determine heatup times and peak temperatures. The pressure is relieved at the end of each

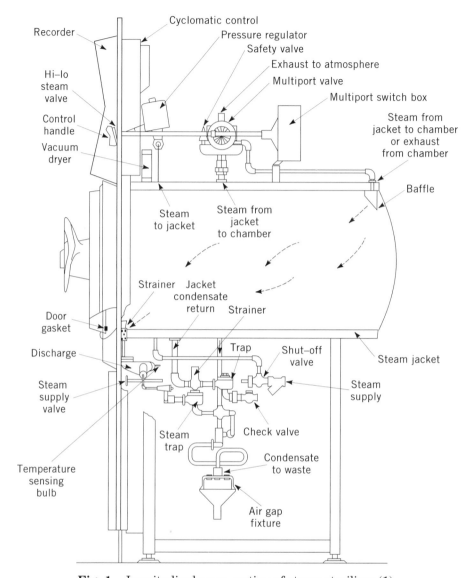

Fig. 1.   Longitudinal cross section of steam sterilizer (1).

sterilization cycle. Either vented containers must be used or specific provisions be made to allow for the safe removal of nonvented, hermetically sealed containers from the autoclave.

The elimination of air from the chamber and complete steam penetration of the load is of critical importance. This may be accomplished by gravity displacement or prevacuum techniques.

The gravity-displacement-type autoclave relies on the relative nonmiscibility of steam and air to allow the steam that enters to rise to the top of the chamber and fill it. The air is pushed out through the steam-discharge line located at the bottom of the chamber. Gravity-displacement autoclaves are utilized for the sterilization of liquids and for unwrapped nonhollow medical instruments at 134°C.

The prevacuum technique, as its name implies, eliminates air by creating a vacuum. This procedure facilitates steam penetration and permits more rapid steam penetration. Consequently this results in shorter cycle times. Prevacuum cycles employ either a vacuum pump/steam (or air) ejector combination to reduce air residuals in the chamber or rely on the pulse-vacuum technique of alternating steam injection and evacuation until the air residuals have been removed. Pulse-vacuum techniques are generally more economical; vacuum pumps or vacuum-pump–condenser combinations may be employed. The vacuum pumps used in these systems are water-seal or water-ring types, because of the problems created by mixing oil and steam. Prevacuum cycles are used for fabric loads and wrapped or unwrapped instruments (see VACUUM TECHNOLOGY).

In prevacuum autoclaves, problems are created by the removal of air and the air-insulation systems. A specific test called the Bowie & Dick test (12), was developed to evaluate the ability of prevacuum sterilizers to provide rapid and even steam penetration which includes the ability to eliminate air from the chambers, and prevent air from leaking back during the prevacuum phase. The test utilizes a pack of specific construction (or its proven equivalent) placed in the empty chamber and exposed to specific test conditions (9,12). The pack contains a chemical indicator sheet. A correctly functioning sterilizer produces a uniform color change. A nonuniform color change indicates poor steam penetration and the possible presence of air which requires the attention of a qualified mechanic. The daily testing of all porous load sterilizers is recommended (1,9,10).

The critical parameters of steam sterilization are temperature, time, air elimination, steam quality, and the absence of superheating. Temperature and time are interrelated, as shown in equation 8. The success of steam sterilization is dependent on direct steam contact which can be prevented by the presence of air in the chamber. The ability of steam to heat a surface to a given temperature is considerably reduced by the presence of air. Air elimination, therefore, is regarded as an absolute parameter. If the required amount of air has not been eliminated from the chamber and the load, no combination of time and temperature results in complete sterilization.

The term steam quality refers to the amount of dry steam present relative to liquid water in the form of droplets. The steam delivered from the boiler usually contains some water. Excessive amounts can result in air entrapment, drying problems following exposure, and unacceptable steam levels (>3% water or <97% quality steam). Excessive amounts of water deposits dissolve boiler chemicals onto the load to be sterilized. Boiler chemicals are used to prevent corrosion in the lines. Inappropriate boiler chemicals, also called boiler amines, may introduce toxicity problems (see CORROSION AND CORROSION CONTROL).

Superheated steam results when steam is heated to a temperature higher than that which would produce saturated steam. The equilibrium between liquid and vapor is destroyed, and the steam behaves as a gas. It loses its ability to condense into moisture when in contact with the cooler surface of the article to be sterilized. This process resembles dry-heat sterilization more than steam sterilization and, under ordinary time–temperature conditions for steam sterilization, does not produce sterility.

The selection of an appropriate steam-sterilization cycle must be made after a careful study of the nature of the articles to be sterilized, the type and number of organisms present, type and size of each package, and type of packaging material used. Cycle-development studies may be conducted using full autoclave loads.

Because hospital loads are not uniform, certain assumptions were made by the manufacturers of sterilizers in arriving at recommended cycles. These recommendations include a safety factor, as well as allowance for heatup time. Exposure recommendations for various types of articles are available (12).

Biological indicators for steam sterilization utilize *Bacillus stearothermophilus*. In monitoring industrial cycles, a sufficient number of preparations each having a known degree of resistance are added to the load and retrieved after exposure, and cultured.

Electromechanical monitors for steam sterilization include pressure, temperature, time-recording charts, and pressure–vacuum gauges. Most recording charts are also capable of displaying pressure–vacuum values. The temperature sensor is generally located in the chamber-drain line, considered to be the coolest area in the autoclave because air exits from the chamber via the drain line. There is no way of locating the sensors inside the packages being sterilized except under specialized test conditions.

Chemical indicators for steam sterilization can be classified into four categories. Some indicators integrate the time–temperature of exposure. Some of these operate throughout the temperature range utilized for steam sterilization (121–141°C); others function in specific time–temperature cycles. Some are capable of monitoring the safety factor in the exposure; others only minimal conditions. All are capable of indicating incomplete steam penetration. The results are indicated by a color change. Certain types change from a specific initial color through a series of intermediate shades to a specific final color. An incomplete color change is an indication of incomplete processing. Other types function by having a color column advancing along the length of the test strip, in a manner reminiscent of paper chromatography. Inadequate processing is indicated by an advance that falls short of a predesignated finish line.

Some indicators can determine whether a specific temperature has been achieved. Because the entrapment of large amounts of air can result in the lowering of steam temperatures, these indicators react to some critical defect in sterilization conditions. For each different temperature, a different indicator must be used.

Indicators can determine if uniform steam penetration has been achieved during a Bowie & Dick-type test. Produced in the form of sheets (23 × 30 cm), chemical indicators are capable of uniform color change over their entire surface when exposed to pure saturated steam under test conditions. Nonuniform color development is an indication of failure of the test. U.S. and international stands for the performance and accuracy of chemical indicators have been published (13,14).

Indicators can be utilized to distinguish packages that have been processed from those that have not been processed. These are external indicators that do not have the capability to detect critical shortcomings in cycle parameters because they are not located inside the packages. A well-known example of this type is autoclave tape, which is also used to hold together packages wrapped in muslin or other kinds of wrap-type packaging materials (qv).

Gas Sterilization

When articles that cannot withstand the temperatures and moisture of steam sterilization or exposure to radiation require sterilization, gaseous sterilants that function at relatively low temperatures offer an attractive alternative. Although many chemical compounds or elements can be considered sterilants, an obvious practical requirement is that the gas selected should allow safe handling, and that any residue should volatilize relatively quickly if absorbed by components of the article sterilized. When properly applied, ethylene oxide [75-21-8] satisfies most of these requirements, and is the most frequent choice (1,6,15–17) (Fig. 2). Because it is highly flammable, ethylene oxide (qv) must be used in a carefully controlled manner. It is either dispensed from a single-use cartridge or diluted with inert gases until no longer flammable. The most frequently used diluents are hydrofluorocarbon (HCFC) gases or carbon dioxide. The HCFCs have been declared acceptable through the year 2035. Industrial ethylene oxide sterilizers are equipped with recovery systems whether they use 100% ethylene oxide or the HCFC-diluted material.

It is necessary to determine the bioburden and make cycle verification studies when ethylene oxide sterilization is used, as it is for other sterilization methods. The manufacturer of hospital sterilization equipment provides cycle recommendations based on the expected bioburden and the consideration of an appropriate safety factor. In ethylene oxide sterilization, it is necessary to determine if residues of the sterilant are absorbed by the sterilized article, and to examine the possible formation of other potentially toxic materials as a result of reaction with ethylene oxide.

The critical parameters of ethylene oxide sterilization are temperature, time, gas concentration, and relative humidity. The critical role of humidity has been demonstrated by a number of studies (11,18,19). Temperature, time, and gas concentration requirements are dependent not only on the bioburden, but also on the type of hardware and gas mixture used. If cycle development is not possible, as in the case of hospital sterilization, the manufacturer's recommendations should be followed.

Provisions must be made for allowing residues of the sterilant absorbed by the product to dissipate by using aeration cabinets that have forced-air circulation at elevated temperatures. The amount of remaining absorbed sterilant should be determined before releasing the sterilized articles. If, as in the case of hospital sterilization, such studies are not feasible, the recommendations of the manufacturers of the articles sterilized or of the aeration equipment should be obtained. The permissible residue concentrations are 10–250 ppm, depending on the type of article and on its intended use.

Biological monitoring of ethylene oxide sterilization is essential and is conducted using spores of *Bacillus subtilis*. See Table 2 for the required performance characteristics. Chemical indicators for ethylene oxide sterilization are usually of the color-change type and are capable of indicating the presence of ethylene oxide under some minimal conditions of temperature, time, and gas concentra-

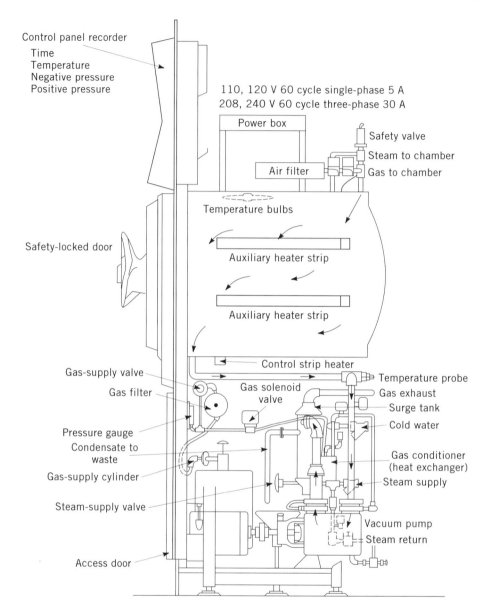

Fig. 2. Schematic of ethylene oxide sterilizer (1).

tion. A few types are also capable of indicating the absence of moisture. External processing indicators similar to autoclave tape, but containing ethylene oxide-responsive chemicals are also available.

General recommendations for instrumentation include monitoring gas concentration, temperature, time, and the moisture content of the chamber. Hospital sterilizers are not usually equipped with instrumentation providing direct display of gas concentration and moisture content. These rely instead on a specific sequence of steps performed automatically and the recording of pressure which when 100% ethylene oxide is used is a perfect measure for the concentration of this gas.

Other Sterilization Techniques

Ionizing Radiation. Radiation sterilization, as practiced in the 1990s, employs electron accelerators or radioisotopes (qv). Electrons have relatively low penetration ability, and the use of accelerators requires careful control. Gamma-radiation sterilization usually employs ^{60}Co and occasionally ^{137}Cs as the radioisotope source. A wide range of packaging materials can be used because gamma-rays possess a considerably greater penetrating ability. However, the materials must not be degraded to the point where the quality of the aseptic barrier is compromised.

Materials tend to undergo chemical changes when exposed to gamma-radiation. Therefore it is generally recommended that the effect of the radiation on any material be determined before gamma-ray sterilization is attempted. Exposure requirements are measured in terms of the delivered dose of radiations and the procedure becomes time independent. Bioburden determinations should be carried out to establish the effective dose. A dose of 2.5×10^4 Gy (2.5 Mrad) is selected for many articles, although lower doses are probably adequate. Radiation exposure can be monitored using counters and electronic measuring devices.

Chemical dosimeters based on ferrous sulfate, ferrous cupric sulfate, or ceric sulfate are generally used. Color-change process indicators are also used, but these cannot measure the radiation dose, only the extent of sterilization.

Filtration. The filtration process depends on the physical retardation of microorganisms from a fluid by a filter membrane or similarly effective medium. The effectiveness of this process is also influenced by the bioburden (6). Hollow-fiber membranes (qv) are most often employed. The resultant filtrate should be sterile, relatively free of particles, and not lose its effectiveness or safety. Filtration may be used for removal of both bacteria and viruses.

Liquid Sterilants. Formalin, a solution of methanal (formaldehyde (qv)) in water, has sterilizing properties (20), as does glutaraldehyde and hydrogen peroxide (qv) (21). The sterilizing action of these liquids is dependent on complete contact with the surface of the article to be sterilized, which can be prevented by grease, smudges, fingerprints, or other impurities. Liquid sterilants cannot penetrate a greasy layer, and careful precleaning is required before processing. The time of exposure is also critical. Although some sterilants can be used at temperatures higher than ambient, stability can be affected adversely. Changes in the pH of solutions may have similar adverse effects. The instructions of the sterilant manufacturer must be followed faithfully if liquid sterilants are to be effective.

Liquid sterilants are known to corrode the metal parts of articles and instruments that are to be sterilized, although articles composed exclusively of glass or certain type of corrosion-resistant metal alloys can be safely processed. Because the degree of corrosion is related to length of exposure, many articles are merely disinfected in a shorter exposure time. Disinfection may be suitable for certain applications. The safety of using liquid sterilants must be judged by a qualified microbiologist.

Other chemicals used historically as liquid sterilants include a solution based on hydrogen peroxide (qv). This was once offered commercially.

Processing in liquid sterilants results in wet products which require highly specialized packaging. Therefore, liquid sterilization should only be considered if the sterilized article is to be used almost immediately. Liquid sterilants or their residues can be harmful to living tissues. Therefore it is always necessary to rinse articles with sterile water or saline solution following treatment. Whereas liquid sterilization is an extremely useful method for articles that cannot withstand the conditions of steam sterilization, the problems associated with its use limit its application.

Other Sterilants.   Sterilization methods, developed in response to the requirements of a low temperature, noncorrosive sterilant and rapid turnaround time required by most hospitals, include use of hydrogen peroxide vapor, hydrogen peroxide plasma, and peroxy acetic acid. Acceptance of these methods was not universal as of this writing (ca 1996).

Other chemical agents or physical conditions that destroy microbial cells have limitations to general use. Ultraviolet light has sterilizing properties, but cannot penetrate many materials, whereas some substances are destroyed by exposure to a sterilizing dose. Some chemicals, such as 3-hydroxy-2-oxetanone (α-hydroxy-O-propiolactone), used in the past, have been found to be carcinogenic. Effectiveness of untested sterilants should be determined according to some acceptable test standard such as those listed in the Official Methods of the Association of Official Analytical Chemists (22). The manufacture and sale of chemical sterilants and disinfectants is regulated in the United States.

Sterilization Packaging

In rare instances it is possible to sterilize an article at the time and place of its use. However, sterilization generally takes place at one location prior to use of an article at another location. The main purpose of packaging (qv) is to protect the sterility of the contents. When an article is placed in its protective container and subsequently sterilized, the process is called terminal sterilization. When it is sterilized first and then placed in a presterilized container, the process is called sterile filling.

A sterile-filling operation requires an environment that excludes or diminishes the possibility of recontamination before the sterile product is sealed. Completely sealed units, such as glove boxes, are suitable. Specially constructed clean hoods or rooms which utilize laminar air flow having highly filtered air can be used for sterile filling. Personnel working in such environments have to wear special protective garments as well as masks, hair covers, etc. The packaging materials (qv) used in sterile filling can be of types that provide a hermetically sealed environment, such as glass, metal cans, or metal foil.

Packaging material used for terminal sterilization must permit full sterilant penetration as well as provide a microbial barrier. Consideration must also be given to the conditions to which the sterile package is to be exposed until used, such as storage, transportation, or frequency of handling.

Storage time by itself is not expected to affect the maintenance of sterility. However, longer storage times may increase the incidence of potentially harmful conditions. Frequent handling, wetness, and possible deterioration of the packaging material are typical examples of conditions that may compromise sterility and limit the shelf life of a package. The package contents may have a specific shelf life as well.

Most industrially prepared, presterilized packages contain inserts having statements that sterility is guaranteed only if the package is not opened or damaged. There is a wide choice of packaging materials and methods available for industrial processes. Using the appropriate packaging, sterilization methods, materials, and handling procedures, sterility can be protected for an indefinite length of time.

Hospital sterilization is more limited in the availability of sterilization methods and of packaging materials. Microbial invasion can occur particularly when articles are wrapped in traditional fabrics such as muslin (140-thread-count cotton). The expected shelf life of hospital-wrapped and sterilized articles is considered to be ca 21–30 days when a double-wrapping technique is used. Double-wrapping requires two successive wraps, each having a layer or layers of an approved packaging material.

Another type of hospital packaging system employs peel-open packages. These are constructed by heat sealing two webs of packaging material around the edges. One layer is usually a plastic film of composite construction, the other is a surgical-grade kraft paper designed to give an effective microbial barrier. Shelf life is extended to a time determined by need rather than sterility protection.

For any packaging method, provision must be made for opening of the package and retrieval of the sterilized article in a manner that does not compromise its sterility.

Related Techniques

Procedures less thorough than sterilization may be used for the preparation of foods and medical supplies. Some of these processes are capable of rendering an object microbiologically safe for a given purpose when employed using proper safeguards. For example, the cooking of food usually results in the reduction of spoilage because some of the potentially harmful organisms are destroyed, even though the process temperatures seldom exceed 100°C. Whereas the boiling of baby formula is not a sterilization procedure, it suffices in most cases.

Pasteurization, the heating of certain fluids, frequently milk or dairy products (see MILK AND MILK PRODUCTS), destroys potentially harmful organisms such as mycobacteria, *M. tuberculosis*, *M. bovis*, or *M. avium*. Pasteurization, carried out at 62°C for 30 min or at 72°C for 15 s, is not a sterilization procedure.

Disinfection destroys pathogenic organisms. This procedure can render an object safe for use. Disinfectants include solutions of hypochlorites, tinctures of iodine or iodophores, phenolic derivatives, quaternary ammonium salts, ethyl alcohol, formaldehyde, glutaraldehyde, and hydrogen peroxide (see DISINFECTANTS AND ANTISEPTICS). Effective use of disinfected materials must be judged by properly trained personnel.

Tyndalization, or fractional sterilization, is no longer considered acceptable for sterilization. Spores of vegetative organisms are the most difficult entities to destroy. In this procedure, rather than destroying the spores, spores are prompted to germinate and then destroyed by boiling water.

Bacteriostasis is the process of preventing the growth and reproduction of microorganisms. When the bacteriostat is removed or its power is exhausted, however, the organisms can resume growth. Bis(2-hydroxy-3,5,6-trichlorophenyl)methane (hexachlorophene) is a bacteriostatic agent.

Sanitization is a cleaning procedure that reduces microbial contaminants on certain surfaces to safe or relatively safe levels, as defined by the EPA or public health authorities. The article is usually cleaned with hot water and various germicidal detergents. Sanitization can be safe for a product in contact with intact skin or for food utensils, but it is not considered safe for articles to be inserted in the human body. Effective sanitization is a requirement in the processing of reusable medical supplies before packaging and sterilization. It is also a requirement in the maintenance of utensils and containers used for food preparation.

Decontamination is a procedure to render safe for handling, disposal, or the subsequent processing of an article that may contain a large amount of potentially infectious organisms. Decontamination and sterilization are similar procedures, except that in the former case the bioburden is higher. In both cases, all organisms present are destroyed. However, decontamination is not expected to result reliably in the 10^{-6} probability of microbial survival, as in sterilization, because of the higher bioburden. Decontamination may include sanitization and disinfection steps, but it most frequently involves sterilization procedures at exposure times two to four times longer than usual. Incineration is a frequent choice for decontamination of single-use articles (see INCINERATORS). Germicides are agents capable of killing some specific forms of microorganisms.

Sterilization in the Food-Processing Industry

The concept of heat processing of foods in hermetically sealed containers was introduced in 1810 (23). The role of microorganisms was unknown at the time, and only the so-called agents of putrefaction were eliminated.

The problem of microbial contamination of foods is twofold: foods may act as nutrients for, and carriers of, pathogenic organisms; additionally, foods may be spoiled by the action of certain organisms (see FOOD PROCESSING). Generally, four specific types of organisms are considered to be food poisons, although it is possible for a number of different organisms present in foods to cause disease. Salmonella and *Clostridium perfringens* require the ingestion of the organisms in large numbers. *Staphylococcus aureus* and *Clostridium botulinum* produce toxins and the organisms need not be ingested for the symptoms to occur. Salmonella is relatively easy to destroy by boiling, and so is *Staphylococcus aureus*, although the toxin of the latter is more heat resistant. The *Clostridia* are spore-forming organisms. The germinated cells can easily be destroyed, but the spores are heat resistant. *Escherichia coli*, a universal contaminant of the colon, has been found to be a cause of food-related disease outbreaks, but usually as the result of direct contamination by employees in restaurants in combination with the serving of undercooked food.

Food spoilage can be caused by enzymatic action or microorganisms. Certain preservation techniques, such as freezing, may retard spoilage by preventing the multiplication of microorganisms or the catalytic action of enzymes. Freezing, however, does not kill the organisms and should not be regarded as a sterilization method.

The most widely used sterilization method in the food industry is moist heat. The heat is usually supplied by high pressure steam, but because most foods already contain moisture the role of steam is to heat the food to the required temperature. The cooking and sterilization processes can frequently be combined into one. The food may be sealed into impervious containers of glass, metal, or plastic film and undergo terminal sterilization, or it may be presterilized in batches or in a continuous operation and then filled into a presterilized container. The latter process is called sterile filling.

The sterilizers or retorts used to process canned or prepackaged foods must be designed in such a way as to assure uniform temperature distribution throughout. Adequate venting permits complete air removal. The air vent is located at the opposite end from the steam inlet (24). The retorts may be horizontal or vertical in design.

The destruction of microorganisms in foods follows the same kinetic relationship as for other materials. The process is strongly influenced by the nature of the food, size of the container, and temperature. Industry-wide standards have been established by the National Canners Association (24).

The effect of various pHs has been well known for some time. Acidic foods such as fruits tend to retard microbial growth and resist certain types of contamination. For this reason, the standards adopted industry-wide have been based on the processing of foods of high acidity (low pH). In the United States, the FDA has regulatory responsibility over the preparation, sterilization, and distribution of foods.

Because of the large volume of food production, continuous processing offers economic advantages. The continuous sterilization of prefilled cans may be accomplished with relative ease because some cans are capable of withstanding the high pressures generated in order to attain the required sterilization temperatures. Such a process is, in reality, not truly continuous, because the prefilled containers represent a discontinuous phase. Time delays in heating full large-volume cans to sterilization temperatures place a limitation on the can sizes; presterilization followed by aseptic filling solves this problem. The containers are presterilized by heat, steam, radiation, etc. The food may be processed in large discontinuous batches, or in modern continuous retorts. Sterilization conditions are checked by appropriate instrumentation. A truly continuous cooking-sterilization process employs back-pressure valves or water towers to allow the attainment of pressures that result in required temperatures.

There are four types of food sterilization processes: terminal sterilization in prefilled containers in a batchwise process; terminal sterilization in prefilled containers of appropriate design heated to the required temperatures in a continuous process; aseptic filling following batchwise cooking in an appropriate retort; and aseptic filling in a continuous cooking system equipped with appropriate valves to allow the necessary pressures for attainment of the required sterilization temperatures.

BIBLIOGRAPHY

"Sterilization" in *ECT* 1st ed., Vol. 12, pp. 896–916, by E. L. Gaden, Jr., and E. J. Henley, Columbia University; in *ECT* 2nd ed., Vol. 18, pp. 805–829, by J. E. Doyle, Castle Co.; "Sterilization Techniques" in *ECT* 3rd ed., Vol. 21, pp. 626–644, by T. A. Augurt, Propper Manufacturing Co., Inc.

1. J. J. Perkins, *Principles and Methods of Sterilization in the Health Sciences*, 2nd ed., Charles C. Thomas, Springfield, Ill., 1969.
2. H. Chick, *J. Hyg.* **8**, 92 (1908).
3. L. L. Katzin, L. A. Sandholser, and M. E. Strong, *J. Bacteriol.* **45**, 265 (1943).
4. G. B. Phillips and W. Miller, *Industrial Sterilization*, Duke University Press, Durham, N.C., 1973, pp. 239–282.
5. G. B. Phillips and W. Miller, *Validation of Steam Sterilization*, Technical Monograph #1, Parenteral Drug Association, 1978.
6. *The United States Pharmacopeia XXIII* (USP XX–NF XV), The United States Pharmacopeial Convention, Inc., Rockville, Md., 1980, pp. 1037–1040.
7. Ref. 1, p. 163.
8. *Accreditation Manual for Hospitals*. Joint Commission on Accreditation of Hospitals, Chicago, Ill., 1996.
9. *Steam Sterilization and Sterility Assurance*, Association for Advancement of Medical Instrumentation, Arlington, Va., ANSI/AAMI ST46, 1993.
10. *Assoc. Oper. Room Nurses J.* **32**, 222 (Aug. 1980).
11. R. R. Ernst, *Industrial Sterilization: Ethylene Oxide Gaseous Sterilization for Industrial Applications*, Duke University Press, Durham, N.C., 1973.
12. Ref. 1, p. 165.
13. *Sterilization of Health Care Products—Chemical Indicators, Part 1: General Requirements*, AAMI 1114001-D, 1995-11-30, Association for the Advancement of Medical Instrumentation, Arlington, Va., 1995, proposed new American National Standard.
14. *Sterilization of Health Care Products—Chemical Indicators, Part 1: General Requirements*, ISO 11140-1:1995(E), International Standards Organization, Geneva, Switzerland, 1995.
15. U.S. Pat. 2,075,845 (1937), P. M. Gross and J. F. Dickson.
16. C. R. Phillips and S. Kaye, *Am. J. Hyg.* **50**, 270 (1949).
17. C. R. Phillips, *Am. J. Hyg.* **50**, 280 (1949).
18. S. Kaye and C. R. Phillips, *Am. J. Hyg.* **50**, 296 (1949).
19. R. R. Ernst and J. E. Doyle, *Biotechnol. Bioeng.* **10**, 1 (1963).
20. A. Heyl, *Surg. Bus.* **42** (Jan. 1972).
21. U.S. Pat. 3,016,328 (Jan. 9, 1962), R. E. Pepper and E. R. Lieberman (to Ethicon).
22. W. Horwitz, *Official Methods of Analysis of Association of Official Analytical Chemists*, AOAC, Washington, D.C.
23. *Chem. Week*, 40 (Mar. 10, 1982).
24. *Processes for Low Acid Foods*, National Canners Association, Washington, D.C., latest edition.

General References

S. S. Block, *Disinfection, Sterilization, and Preservation*, 2nd ed., Lea and Febinger, Philadelphia, Pa., 1977.

G. Sykes, *Disinfection and Sterilization*, 2nd ed., Spon, London, 1965.

S. Turco and R. E. King, *Sterile Dosage Forms, Their Preparation and Clinical Application*, Lea and Febinger, Philadelphia, Pa., 1974.

N. A. M. Eskin, H. M. Henderson, and R. S. Townsend, *Biochemistry of Foods*, Academic Press, Inc., New York, 1971.

National Conference of Spacecraft Sterilization, California Institute of Technology, U.S. National Aeronautics and Space Administration Technical Information Division, Calif., 1966.

Canned Food, Principles of Thermal Processing, Food Processing Institute, Washington, D.C.

Validation and Routine Monitoring of Moist Heat Sterilization Processes, ISO 11134, International Standards Organization, Geneva, Switzerland.

THOMAS A. AUGURT
Propper Manufacturing Co., Inc.

J.A.A.M. VAN ASTEN
National Institute for Public Health
and the Environment

STEROIDS

Steroids (**1**) are members of a large class of lipid compounds called terpenes that are biogenically derived from the same parent compound, isoprene, C_5H_8. Steroids contain or are derived from the perhydro-1,2-cyclopentenophenanthrene ring system (**1**) and are found in a variety of different marine, terrestrial, and synthetic sources. The vast diversity of the natural and synthetic members of this class depends on variations in side-chain substitution (primarily at C17), degree of unsaturation, degree and nature of oxidation, and the stereochemical relationships at the ring junctions.

There are many classes of natural and synthetic steroids best known for their wide array of biological activity. The naturally occurring steroids can be subdivided into several categories that include (*1*) nonhormonal, mammalian steroids such as sterols and bile acids; (*2*) vitamin D; (*3*) hormonal steroids such as the human sex hormones (androgen, estrogens, and progestins) and corticosteroids (glucocorticoids and mineralocorticoids); and (*4*) other naturally occurring steroids, such as plant steroids (eg, sapogenins, saponins, withanolides), marine steroids, ecdysteroids, cardiac steroids, steroid alkaloids, and steroid antibiotics (see HORMONES; VITAMINS, VITAMIN D; ANTIBIOTICS).

The biological activity of the naturally occurring steroids has been exploited in the development of therapeutic agents. Many of these synthetic steroids are the product of pharmaceutical research that has resulted in the development of several steroid-based medicines with a market value over 10 billion U.S. dollars (*2*). Among the medicinally important synthetic steroids are the antihormones, anesthetics, antiinflammatories, antiasthmatics, contraceptive drugs (qv), antibiotics, anticancer agents, cardiovascular agents (qv), and osteoporosis drugs (see ANTIASTHMATIC AGENTS; CHEMOTHERAPEUTICS, ANTICANCER).

History

Initial steroid research involved isolation of sterols and bile acids from natural sources. DeFourcroy is generally credited with the discovery of cholesterol [*57-88-5*] (**2**) in 1789 (*3*). In 1848, cholic acid [*81-25-4*] (**3**) was isolated from the saponification of ox bile and its elementary composition determined as $C_{24}H_{40}O_5$; 40 years later, Reintzer established the molecular formula of cholesterol as $C_{27}H_{46}O$. Degradative studies revealed the relationship of cholesterol and the bile acids (*4*).

(**2**)   (**3**)

Although many sterols and bile acids were isolated in the nineteenth century, it was not until the twentieth century that the structure of the steroid nucleus was first elucidated (*5*). X-ray crystallographic data first suggested that the steroid nucleus was a thin, lath-shaped structure (*6*). This perhydro-1,2-cyclopentenophenanthrene ring system was eventually confirmed by the identification of the Diels hydrocarbon [*549-88-2*] (**4**) and by the total synthesis of equilenin [*517-09-9*] (**5**) (*7*).

(4)   (5)

The period from the 1930s through the 1960s is often referred to as the golden age of steroid research. By the end of the 1930s, the isolation, structural determination, and synthesis of the sex hormones estrogen, progesterone, estradiol, and testosterone were completed (8). Also, investigations of steroid production in the adrenal cortex began in the 1930s. From 1936 to 1942, more than 30 crystalline steroids called corticosteroids were isolated from adrenal gland extracts. Included among these corticosteroids were the potent glucocorticoid, cortisol, and the mineralocorticoid, aldosterone.

Efforts toward producing synthetic steroids, particularly cortisol, expanded during World War II to enable researchers to explore the possibility of medicinal applications of corticosteroids. In 1948, the discovery that cortisone dramatically alleviates the symptoms of arthritis led to intensive research on the antiinflammatory properties of corticosteroids. The development of partial and total syntheses for the commercial preparation of cortisone, alternative methods for producing cortisone, and the search for more potent antiinflammatory analogues greatly stimulated both academic and industrial steroid research.

During this period of intense synthetic research, a search for inexpensive raw materials for the partial synthesis of steroids was initiated. Abundant quantities of the sapogenin diosgenin [512-04-9] were isolated from plant sources and used for the industrial preparation of steroids (9).

This explosion in steroid chemistry both stimulated and was aided by the development of conformational analysis (10). Many basic, physical organic chemistry principles were established as a result of the study of the logically predictable chemistry of the rigid perhydro-1,2-cyclopentenophenanthrene, steroid skeleton.

The 1950s and 1960s saw the development of orally active progestins based on the synthesis of steroids that lack the C19-angular methyl substituent (19-norsteroids). The commercial production of these compounds for the regulation of menstrual disorders began in 1957, and for oral contraception in 1960.

The 1970s were a vibrant period for steroid research. Beginning in the early 1970s, problems with Mexican supplies of the steroid raw material, diosgenin, prompted investigations into alternative sources of starting materials. Chemical methods to degrade the plant sterol stigmasterol [83-48-7] into useful starting materials were developed. In addition, new fermentation methods for the preparation of commercially important steroids from cholesterol and sitosterol were discovered. Also in the 1970s, studies on the control of cholesterol biosynthesis through selective enzymatic inhibition were begun. The active form of vitamin D was determined, and its biochemical effects delineated.

In the 1970s and 1980s, potent antiprogestins were discovered and used as contragestational agents, with possible applications for the treatment of various cancers. The synthesis of the antiprogestin RU-486 demonstrated a versatile way to functionalize the 11-position of a steroid nucleus.

In the 1980s, advances in biotechnology had a considerable impact on steroid research. During this period, the mechanism of steroid hormone-activated gene regulation became more clearly defined. These mechanistic studies still receive considerable attention in the primary literature.

Structure and Nomenclature

The long history of steroid nomenclature has been reviewed in "Definitive Rules for Nomenclature of Steroids" in 1972 (11). Most of these rules of steroid nomenclature have been adopted. A general definition of steroids is as follows. "Steroids are compounds possessing the skeleton of cyclopenta[a]phenanthrene or a skeleton derived therefrom by one or more bond scissions or ring expansions or contractions" (12). The position-numbering and ring-lettering conventions for steroids are shown in (1). Positions 18 and 19 are often angular methyl groups; in addition, position 19 can be a hydrogen and is not substituted when the A-ring is aromatic. Position 17 can be substituted, unsubstituted, and/or oxygenated. Compounds are systematically named as derivatives of the parent hydrocarbons shown in Figure 1. Substituents that extend below the plane of the steroid are referred to as α and are designated by a broken line; those attached to the plane of the steroid from above are called β and are shown by a bold or solid line. Substituents of unknown configuration are indicated by a wavy line. Generally, the ring junctions have an all-trans relationship with the hydrogen attached to C9 on the α-face, unless otherwise indicated. Changes in steroid nomenclature that have been introduced since 1972 include a wider use of the (R),(S)-system

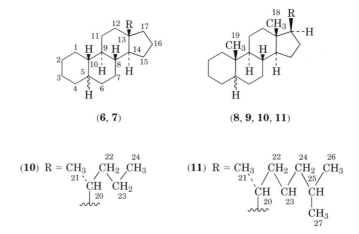

Fig. 1.   Nomenclature of the parent hydrocarbon ring skeletons. Gonane [4732-76-7] (**6**) R = H; estrane [24749-37-9] (**7**) R = CH$_3$; androstane [24887-75-0] (**8**) R = H; pregnane [24909-91-9] (**9**) R = C$_2$H$_5$; cholane [548-98-1] (**10**) R as shown; and cholestane [145-53-7] (**11**) R as shown.

for designating the stereochemistry in the side chain, and four new parent hydrocarbons, ie, campastane, portiferastane, stigmastane, and gorgostane, have been proposed but have yet to receive formal adaptation (13). Although the systematic nomenclature for steroids has been firmly established, the most common and most important steroids are often designated by trivial names.

Classification of Biologically Active, Natural Steroids

NONHORMONAL MAMMALIAN STEROIDS

Sterols and Cholesterol. Natural sterols are crystalline $C_{26}-C_{30}$ steroid alcohols containing an aliphatic side chain at C17. Sterols were first isolated as nonsaponifiable fractions of lipids from various plant and animal sources and have been identified in almost all types of living organisms. By far, the most common sterol in vertebrates is cholesterol (**2**). The total cholesterol content measured in all mammalian species that have been examined is between 1 and 2 g/kg. Cholesterol is found in nearly all tissue types in both an esterified and unesterified form. The bulk of this sterol is in the unesterified form; however, certain tissues, eg, plasma, skin, and hair, contain a considerable portion of cholesterol as an ester (14–18).

Cholesterol serves two principal functions in mammals. First, cholesterol plays a role in the structure and function of biological membranes. The Δ^5 bond, the flat alternating trans–antitrans stereochemistry at the ring junctures, and the C17β-side chain of cholesterol form a rigid nonpolar nucleus. Combined with the hydrophilic C3–OH group, this nucleus has appropriate dimensions to interdigitate between the amphipathic phospholipids in the bilayer of biomembranes. The incorporation of cholesterol into the membrane bilayer changes its physical properties in a way that optimizes the efficiency of the biomembrane functions. Functions such as structural integrity, permeability (eg, optimization of the efficiency of active and passive transport of metabolites), and overall shape are enhanced. Secondly, cholesterol serves as a central intermediate in the biosynthesis of many biologically active steroids, including bile acids, corticosteroids, and sex hormones (14–19).

Bile Acids and Alcohols. Bile acids have been detected in all vertebrates that have been examined and are a result of cholesterol metabolism. The C_{24} acid, 5β-cholanic acid [*546-18-9*] (**13**) is the structural derivative of the majority of bile acids in vertebrates. Most mammalian bile acids have a cis-fused A–B ring junction (**12**) resulting in a nonplanar steroid nucleus. Bile acids, like sterols, typically contain a C3α-hydroxyl group (lithocholic acid: 3α-hydroxycholanic acid

(**12**)

[*434-13-9*]). Along with the C3α-hydroxyl group, bile acids may contain a hydroxyl at C7α (chenodeoxycholic acid [*474-25-9*]), at C12α (7-deoxycholic acid [*83-44-3*], (**14**)), and at C7α and at C12α (cholic acid, (**3**)) to name but a few. Bile

(**13**)   (**14**)

acids with hydroxyl groups in the β-position at C3, C6, or C7 may occur, but in significantly reduced amounts compared to their α-counterparts. In an organism the hydroxyl groups may be progressively oxidized, leading to mixtures of keto acids and keto–hydroxy acids (17,20–23).

Bile salts, cholesterol, phospholipids, and other minor components are secreted by the liver. Bile salts serve three significant physiological functions. The hydrophilic carboxylate group, which is attached via an alkyl chain to the hydrophobic steroid skeleton, allows the bile salts to form water-soluble micelles with cholesterol and phospholipids in the bile. These micelles assist in the solvation of cholesterol. By solvating cholesterol, bile salts contribute to the homeostatic regulation of the amount of cholesterol in the whole body. Bile salts are also necessary for the intestinal absorption of dietary fats and fat-soluble vitamins (24–26).

VITAMIN D

Vitamin D refers to a group of seco-steroids that possess a common conjugated triene system of double bonds. Vitamin D_3 [*67-97-0*] (**16a**) and vitamin D_2 [*50-14-6*] (**16b**) are the best-known examples (Fig. 2). Vitamin D_3 (**16a**) is found primarily in vertebrates, whereas vitamin D_2 (**16b**) is found primarily in plants. The relationship of vitamin D_2 and vitamin D_3 to the more classical steroid nucleus is demonstrated by their immediate sterol precursors, ergosterol [*57-87-4*] (provitamin D_2, **15b**) and 7-dehydrocholesterol [*434-16-2*] (provitamin D_3, **15a**), respectively. Sunlight converts 7-dehydrocholesterol (**15a**) into vitamin D_3 (**16a**) in the skin. The term vitamin is a misnomer. Vitamin D_3 is a prohormone that is converted into physiologically active form, primarily 1,25-dihydroxyvitamin D_3 (**17**), by successive hydroxylations in the liver and kidney. This active form is part of a hormonal system that regulates calcium and phosphate metabolism in the target tissues. In addition, 1,25-dihydroxyvitamin D_3 [*32511-63-0*] (**17**) has a diverse range of biological actions, including effects on cell differentiation and proliferation and control of other hormonal systems (27,28). Rickets, a disease of early childhood characterized by faulty ossification of bone, and osteomalacia, a disease characterized by a failure of calcification of bone matrix, are due to a deficiency of vitamin D. Excessive doses of vitamin D can cause hypercal-

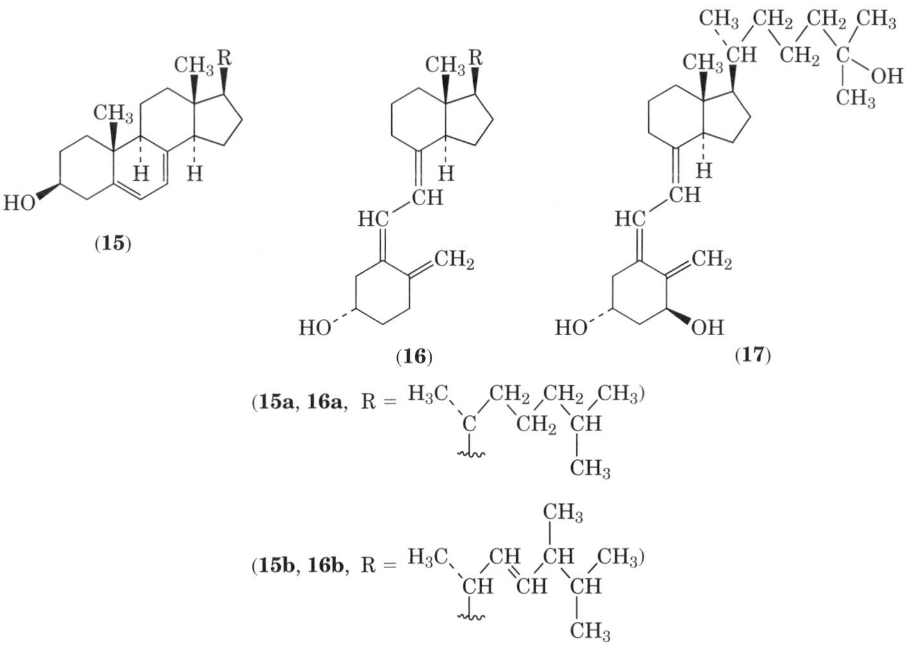

Fig. 2. Vitamin D: precursors (**15**), prohormones (**16**), and active hormone (**17**).

cemia and deposition of bone in the soft tissue (29). Only when skin irradiation is insufficient is there a true dietary requirement for vitamin D (30).

The irradiation pathway of precalciferol [50524-96-4] (**18**) is shown in Figure 3. When irradiated, ergosterol (**15b**) first undergoes a photochemically allowed conrotatory ring opening of the B-ring to form (**18**) as a central intermediate. Triene (precalciferol) (**18**) can undergo a thermally allowed 1,7-sigmatropic shift to form vitamin D$_2$. Precalciferol also forms tachysterol [115-61-7] (**19**) by photochemical isomerization of the central triene double bond and lumisterol [474-69-1] (**20**) by conrotatory ring closure (30–33). Although a true steady state of the photochemical reaction is never achieved owing to the formation of over-irradiated products, the experimentally determined quasistationary-state yields are in good agreement with the calculated steady-state values (34,35).

STEROID HORMONES

Generally, steroid hormones are metabolically short-lived steroids produced in small amounts by various endocrine glands. They serve as chemical messengers that regulate a variety of physiological and metabolic activities in vertebrates. Steroid hormones bind to soluble, intracellular receptor molecules. These steroid hormone receptors, members of the steroid–thyroid–retinoid superfamily, are large proteins that act as ligand-regulated transcription factors controlling specific gene expression. The mechanisms of gene activation of steroid hormones are much too complex for this review and differ slightly for various classes of steroids (36). On a cellular level, however, the mechanism of steroid hormone activation has several common features. The monomeric or untransformed steroid

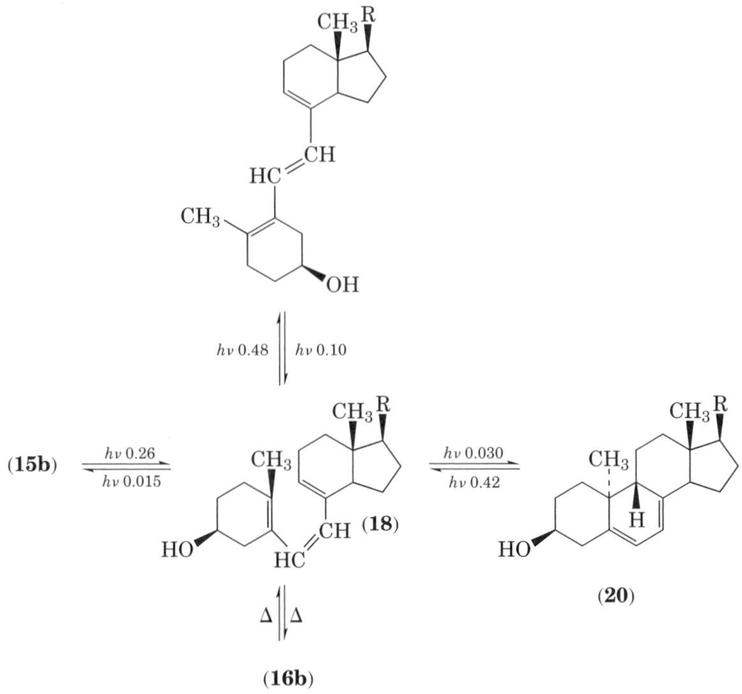

Fig. 3. Photochemical and thermal reactions of previtamin D_2 where the quantum yields for photochemical reactions are given by the arrow. R is as shown in Figure 2 for (**15b**) and (**16b**).

receptor is located in the cytoplasm or the nucleus of almost every living cell. This untransformed receptor is complexed to a variety of heat shock proteins (HSPs), including HSP-90. Upon binding the hormone, the receptors dissociate from the HSPs and dimerize. In the nucleus of the cell, these dimeric steroid receptors bind to DNA and, together with a heteromeric complex of proteins, regulate gene transcription (37). Molecules that interfere with steroid hormone gene regulation are called antagonists or antihormones (38).

Steroid hormones can be subdivided into sex hormones (androgens, estrogens, and progestins) and corticosteroids (glucocorticoids and mineralocorticoids).

Sex Hormones. Androgens, estrogens, and progestins are steroids that are secreted primarily by the genital glands. From a chemical point of view, the division of the sex hormones into these three groups is convenient; however, they may possess common physiological properties. Therefore, the sex hormones are organ-specific rather than sex-specific (39,40).

Androgens. These C_{19} steroids contain the basic perhydro-1,2-cyclopentenophenanthrene ring system with the C18 and C19 angular methyl group. A primary function of androgens is to maintain the male sex organs and secondary sex characteristics. Androgens were first isolated from the urine of males, females, and eunuchs. When androgens from testicular extracts were injected into castrated or immature males, restoration or development, respectively, of the male genital organs and secondary sexual characteristics were observed; hence the term male sex hormone. Examples of androgens are testosterone [58-22-0] (**21**), dihydrotestosterone (DHT) [521-18-6] (**22**), androsterone [53-41-8] (**24**), and dehydroepiandrosterone [53-43-0] (DHEA) (**23**). DHEA is one of the most

(**21**)   (**22**)   (**23**)

abundant steroids in human males; however, it is not a potent androgen. Androsterone (**24**) is not very important in humans, whereas 4-androstene-3,17-dione is important in females. DHT is one of the most potent androgens in humans. Inhibitors of the enzyme 5α-reductase that convert testosterone to DHT are used to treat benign prostate hyperplasia (41–45).

In addition to their masculinizing effects, androgens are anabolic (tissue-building) agents. Anabolic–androgenic steroids have been abused by athletes with the hope of improving their training, endurance, performance, or physique. Owing to the long-term toxicity of these natural and synthetic anabolic steroids in high doses and the potential for abuse, the U.S. Congress enacted the Anabolic Steroids Control Act in 1990. This Act requires the regulation of anabolic steroids in the same way that various opioid drugs, amphetamines, and barbiturates are regulated (46,47).

(**24**)   (**25**)   (**26**)

Estrogens. Estrogens were originally isolated between 1929 and 1935 and are characterized by having an aromatic A-ring and thus having a phenolic character. Estrogens stimulate the growth and development of the female reproductive organs and the secondary sex characteristics. Another primary function of estrogens along with progesterone, is to regulate the ovulatory cycle. Estrogens, as with all the steroid hormones, are important for healthy growth and development in women and men. The main production site of estradiol [50-28-2] (**25**) is the female ovary; however, small amounts of estrogens are produced in testes and the adrenal cortex, and significant amounts of estrogens are produced by peripheral aromatization of local and circulating androgens in skin and fat. Estrone [53-16-7] (**26**) and estriol [50-27-1] (**27**) are estrogenically active metabolites of estradiol. Synthetic and natural estrogens play an important role in the treatment of osteoporosis in post-menopausal women. Antiestrogens are important for the treatment of breast cancer (48–51).

(27)   (28)

Progestins.   Progesterone [*57-83-0*] (**28**), the principal progestin in mammals, is secreted primarily by the corpus luteum of the ovary. A main responsibility of progesterone, together with estrogen, is to prepare the endometrium for pregnancy. Synthetic progestins have wide applications in gynecology and contraception (52–57).

Corticosteroids.   Although the adrenal cortex secretes small amounts of androgens and estrogens, the major secretory steroids from this gland are called corticosteroids. Corticosteroids have several biological activities, including the regulation of electrolyte balance by mineralocorticoids and carbohydrate and protein metabolism by glucocorticoids. Over 30 steroids have been isolated from the adrenal cortex; however, the bulk of the biological activity of the corticosteroids has been attributed to only a few of these compounds. The division between glucocorticoids and mineralocorticoids can be confusing because many corticosteroids have contributing glucocorticoid and mineralocorticoid activities (58–61).

Glucocorticoids.   Natural, potent glucocorticoids possess a Δ^4-3-one group, an oxygen substituent at C11β (necessary for agonism), and a C17β-2-hydroxyethan-1-one sidechain. Cortisol [*50-23-7*] (**29**), corticosterone [*50-22-6*] (**30**), and cortisone [*53-06-5*] (**31**) are typical examples. The principal effects of

(29)   (30)   (31)

glucocorticoids are to mobilize fat and protein from tissues, utilize these nutrients to supply energy for the body, and decrease the rate of carbohydrate utilization for energy. Thus, they are diabetogenic and act as functional insulin antagonists. Glucocorticoids are only weakly active in respect to electrolyte metabolism. Glucocorticoids are also potent inhibitors of inflammation and have been the subject of intense synthetic and biological studies. They are used as therapeutics for a variety of different inflammatory diseases (62).

Mineralocorticoids. Aldosterone [*6251-69-0*] (**32**), the most potent natural mineralocorticoid, also possesses a Δ^4-3-one group, an oxygen substituent at C11β, and a C17β-2-hydroxyethan-1-one side chain. In addition, the C18 of

(**32**)

aldosterone is oxidized to an aldehyde. Mineralocorticoids, particularly aldosterone, act to retain sodium and to prevent the retention of excess potassium. Antimineralocorticoids have been used therapeutically as diuretics and as agents that regulate blood pressure (63–65).

OTHER NATURAL STEROIDS

Steroids are nearly ubiquitous to all living organisms and have a variety of structural variations. Herein a brief overview of a few natural steroids from both plant and animal sources that have interesting biological activities or industrial importance is given.

Sapogenins and Saponins. Steroids isolated from a variety of plant sources that contain a spiroketal between hydroxyl moieties at C16 and C26 and a carbonyl at C22 are called sapogenins (**33**).

(**33**)   (**34**)

In addition to this spiroketal moiety, sapogenins generally contain a 3β-hydroxyl group. Since the late nineteenth century, more than 200 sapogenin aglycones

have been isolated and characterized. Sapogenin aglycones, particularly diosgenin [512-04-9] (34), hecogenin [467-55-0] (35), and tigogenin [77-60-1] (36),
have been an important source of starting materials for the commercial steroid
industry owing to their relative abundance in easily cultivated plants and their
ease of isolation (66).

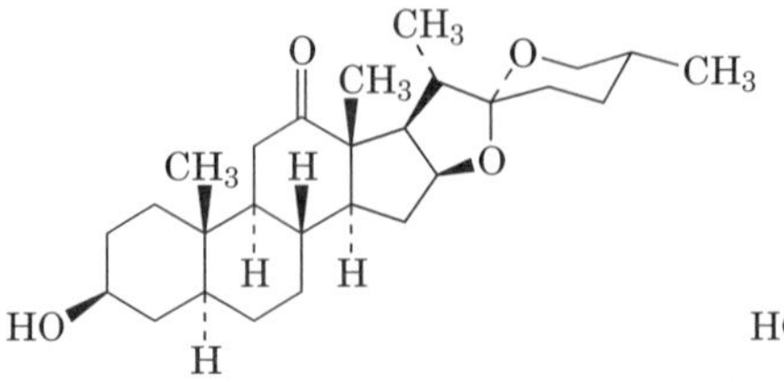

(35) 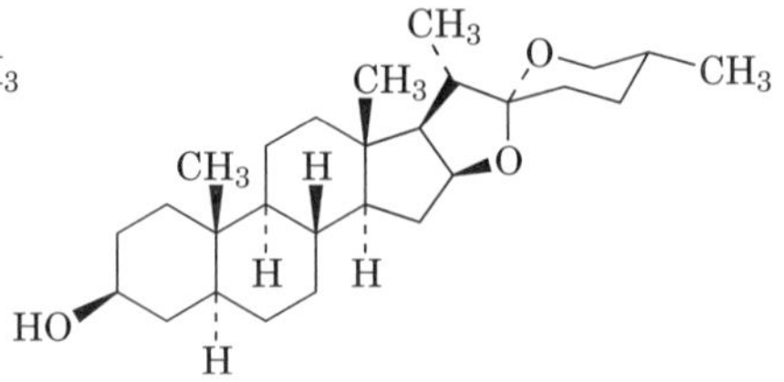   (36)

Saponins are widely distributed in plants and marine organisms and consist of a steroid or terpene skeleton attached to a saccharide (Fig. 4). In plants, for
example, many sapogenins contain sugar residues attached to the 3β-hydroxyl
group. Classical methods of isolation of saponins were inadequate for separating individual components; therefore, the characterization of most of the pure
saponins arose only in the late 1970s with the integration of silica gel column chromatography, semipreparative hplc, preparative tlc, and special isolation techniques adapted to particular situations.

Because of diversity in structure, pharmacology, and biological activities,
saponins have been studied for a number of different commercial applications.
Many of these plant glycosides form a soapy lather when shaken with water
and produce hemolysis when water solutions are injected into the blood stream.
Thus, saponins have been used as detergents, foaming agents, and fish toxins.
Although toxic to fish, saponins are nontoxic when ingested by humans, probably
because of nonabsorption by the intestine (67). Another commercial application
of saponins is in food flavoring. Depending on the structure, saponins can have

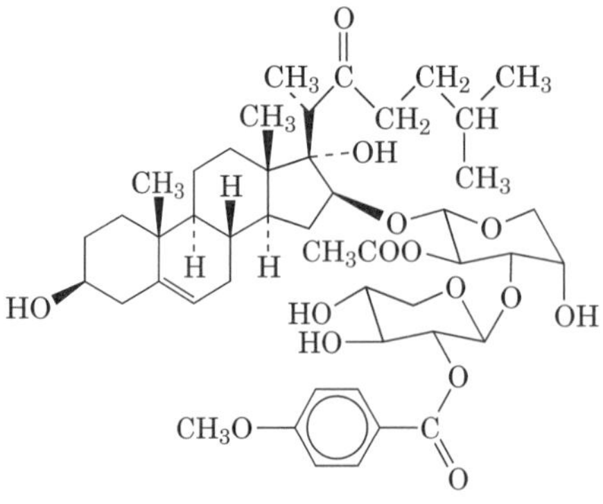

(37) 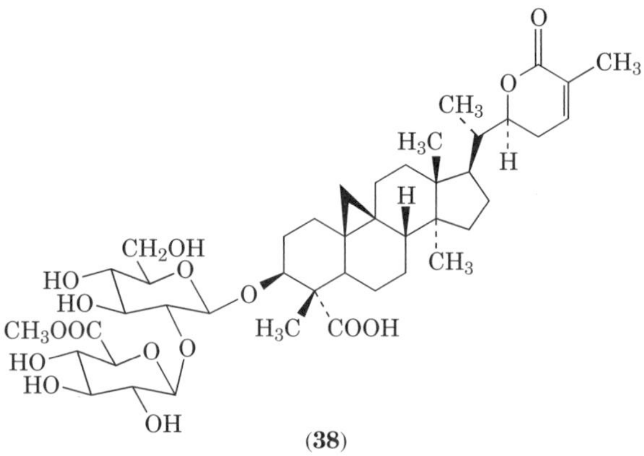   (38)

Fig. 4. Examples of saponins: OSW-1 [145075-81-6] (37) and abrusoside E6″-methyl
ester (38).

either a bitter or sweet taste. For example, (**38**) is a saponin derivative that is 150 times sweeter than sugar. Also, saponins have been studied for their antifungal and anticancer activities. By way of illustration, the saponin OSW-1 (**37**) is extremely toxic to cancer cells but has little toxicity to normal cells when assayed *in vitro* (68) (see Fig. 4).

Plant Sterols. Sterols have been identified in almost all types of living organisms and can be isolated, in varying quantities, from many different plants. Similar to cholesterol, plant sterols have a structural and functional role in biological systems and serve as intermediates in the biosynthesis of an assortment of biologically active steroids. The plant sterol brassinolide [*72962-43-7*] (**39**) is a member of ubiquitous plant steroids termed brassinosteroids. Brassinosteroids may play a role in the regulation of gene expression by light and therefore act as phytohormones (69). Stigmasterol [*83-48-7*] (**40**) and β-sitosterol [*83-46-5*] (**41**), isolated primarily from soybeans, have become an important source of starting materials for the commercial steroid industry (70).

(**39**)

(**40**, Δ22,23-)
(**41**, saturated 22,23-)

Steroid Alkaloids. Steroid alkaloids are compounds isolated from plants and some higher animals that possess the basic steroidal skeleton with nitrogen(s) incorporated as an integral part of the molecule. The nitrogen can be located within the perhydro-1,2-cyclopentenophenanthrene ring system or in a side chain. Because steroid alkaloids possess several different types of amine nitrogens and may also be conjugated to sugar residues, purification is often difficult. Preparative thin-layer chromatography and column chromatography with a variety of chromatographic materials and eluting solvents are the most common methods employed to isolate the steroidal bases.

Steroid alkaloids have been isolated from four families of terrestrial plant sources (*Solanaceae, Liliaceae, Apocynaceae,* and *Buxaceae*), two animal sources (*Salamandra* and *Phyllobates*), and several marine sources. Steroid alkaloids can be classified based on structure and fall into a variety of categories. The spirosolanes contain a C_{27} cholestane skeleton with a C20 spiroaminoketal moiety, as exemplified by the most abundant members of this class, veramine [*21059-48-3*] (**42**), tomatidine [*77-59-8*] (**43**), and solasodine [*126-17-0*] (**44**). Owing to

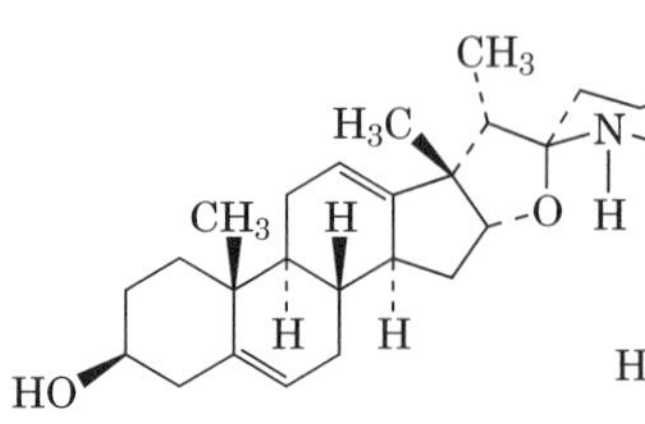

(42) 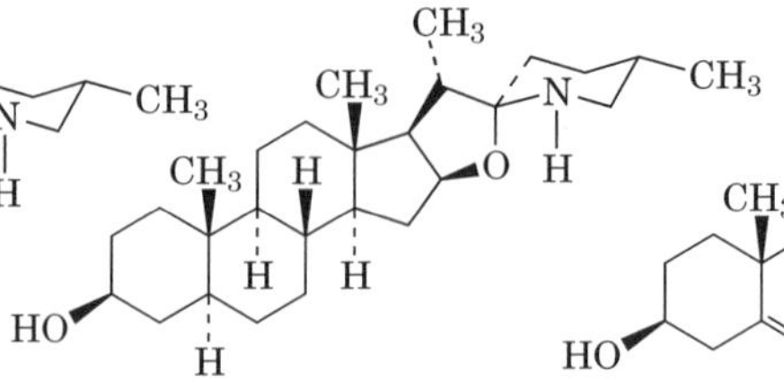   (43) 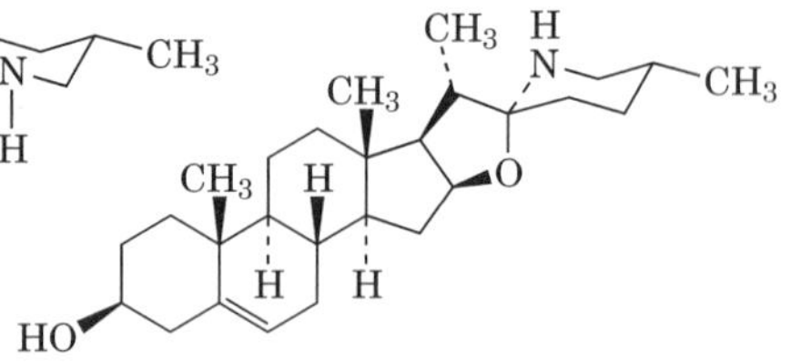   (44)

shortages in diosgenin, solasodine and tomatidine have gained importance as alternative raw materials for industrial steroid synthesis. Solanidine-type steroidal alkaloids are a small subclass consisting of only 34 compounds as of 1993. They have three skeletal variations that have a nitrogen-containing 5,6-bicylic ring fused to C16–C17 or C15–C16 of the steroidal skeleton as a common structural feature (Fig. 5). The largest subclass of steroidal alkaloids is the secosoline bases; (45) is a general secosolanidine. This class is characterized by a 2-ethyl-piperidine-like base attached to the steroid nucleus at either C16 or C17, exemplified by etioline [29271-49-6] (46). The pregnane-type alkaloids have one or more

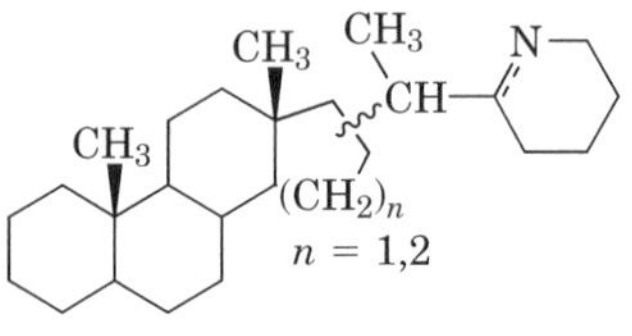

(45) 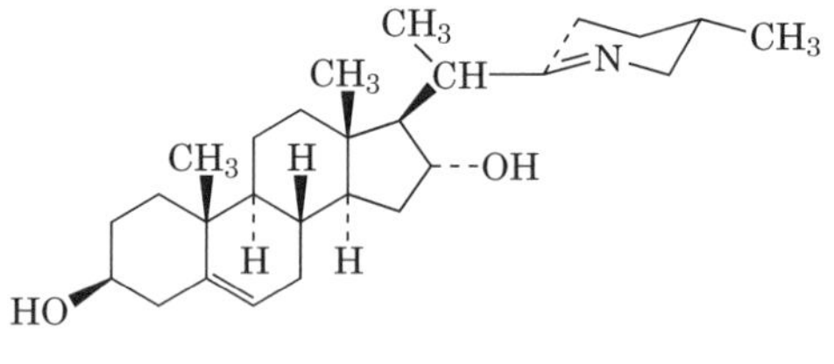   (46)

nitrogens attached to a pregnane skeleton, as illustrated by buxaprogenstine [113762-72-4] (47) and irehidiamine-A [3614-57-1] (48). The buxus alkaloids,

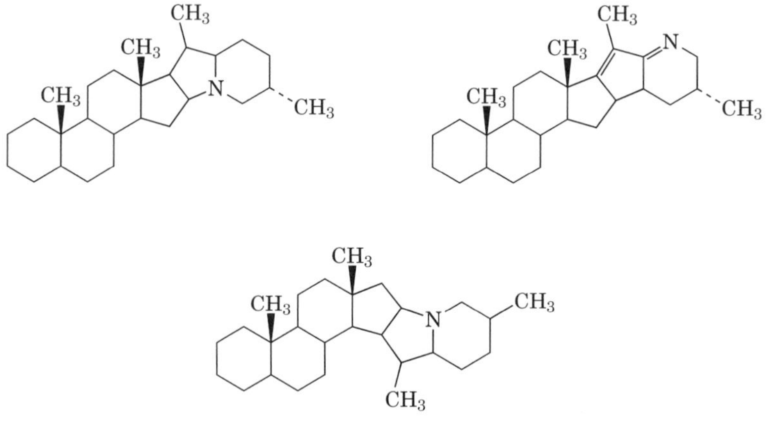

Fig. 5. Solanidine nuclei.

isolated from evergreen shrubs, contain carbon substitution at C4 and C14 and either a cyclopropane moiety between C9, C10, and C19 or the B-ring ex-

(47)   (48)

panded diene as demonstrated by cyclobuxine-D [2241-90-9] (49) and buxamine-E [14317-17-0] (50), respectively. The carbon substitution at C4 and C14 is

(49)   (50)

indicative of an intermediate in the biogenic scheme between lanosterol and cholesterol-type steroids. Buxus alkaloids have been used as folk remedies for a variety of disorders, including venereal disease, tuberculosis, cancer, and malaria. The samanine, jerveratrum, and ceveratrum-type compounds all have a structurally altered C_{27} steroid skeleton. The samanine alkaloids have an expanded A-ring with the formation of an isoxazoline ring system and a cis-A–B ring junction, as shown by the primary alkaloid of this group, samandarine [467-51-6] (51). The most abundant jerveratrum base, jervine [469-59-0] (52), illustrates the structural variation of this class, including a five-membered C-ring, a six-membered D-ring, and a piperidine ring system fused to C17 of the tetracyclic steroidal system via a spirotetrahydrofuran-linking group. The second-largest subclass of steroid alkaloids, ceveratrum, contains over 100 members. As exemplified by imperialine [61825-98-7] (53), ceveratrum alkaloids consist of a hexacyclic ring system and various degrees of oxidation at C3, C4, C6, C15,

(51)   (52)   (53)

C16, and C20 (71–76). Ritterazines and cephalostatins are among steroid al-kaloids recently isolated from marine sources that have received considerable attention. These compounds have two steroid nuclei linked at C2 and C3 by a pyrazine ring system, as shown by cephalostatin 1 [*11288-65-9*] (**54**). When as-sayed *in vitro*, cephalostatins are among the most potent cytotoxins ever screened by the National Cancer Institute (77).

(54)

Cardiac Steroids. Cardiac steroids (steroid lactones) and corresponding glycosides are characterized by their ability to exert a powerful inotropic (in-creasing the force of cardiac contraction) effect, and are used both for their inotropic and antiarrhythmic properties. The two most prevalent cardiac agly-cones are the cardenolides and bufadienolides. The cardenolides are C_{23} steroids that have a C17β-substituted five-membered lactone that is generally α,β-unsaturated, an unusual β-faced oxygen on C14, and a bile acid-like cis-A–B ring junction. Cardenolides are exemplified by digitoxigenin [*143-62-4*] (**55**) which is an active ingredient in digitalis. The bufadienolides differ in that they are C_{24}-steroids that possess a C17β-substituted six-membered lactone ring that generally has two degrees of unsaturation. Bufadienolides can be represented by bufalin [*465-21-4*] (**56**) which occurs in toad skin secretions.

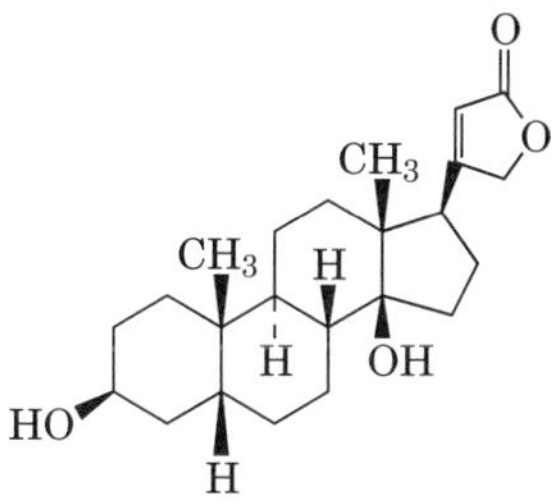

(**55**) 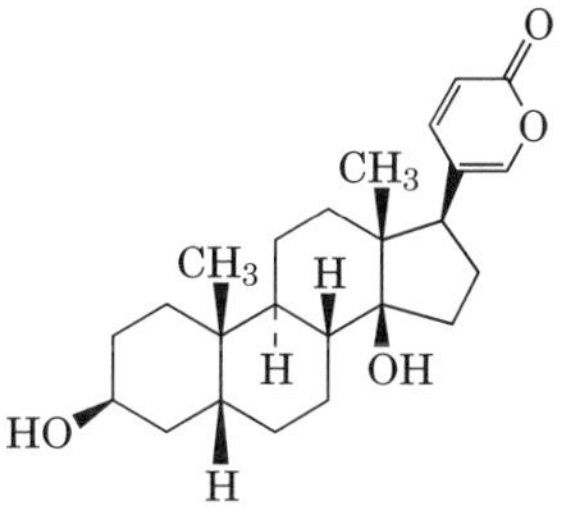   (**56**)

Other structural variations in both series are the stereochemistry at C3 and the degree of oxidation on the nucleus and side chains. Cardiac steroids probably exert their inotropic effects by acting as specific, noncompetitive inhibitors of Na^+–K^+-ATPases, known as sodium pumps, and thus increasing intracellular Na^+. Elevated levels of Na^+ lead to increases in Ca^{2+} via a sarcolemmal Na^+–Ca^{2+} exchange; hence, more Ca^{2+} becomes available for the contractile apparatus (78–83).

   Withanolides.   Withanolides are C_{28}-steroidal lactones that are isolated from the Solanaceae plant family. Withanolides are characterized by an ergostane-type skeleton, the C17-side chain of which is transformed into a six-member lactone ring (Fig. 6). The withanolides and the related ergostanes are the only known natural steroids obtained from the same family that have representatives with both α- and β-orientations of the C17 side chain, such as

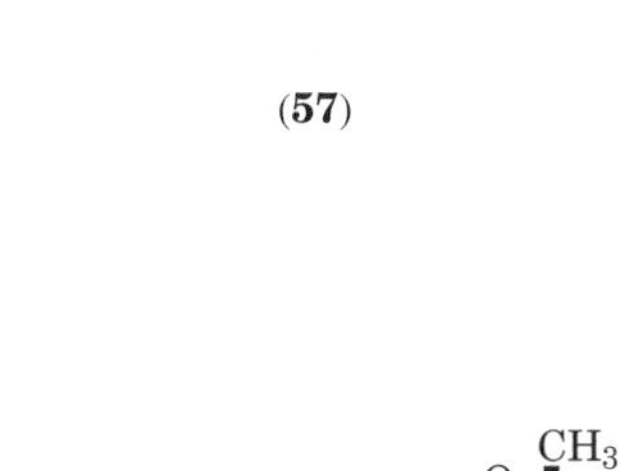

(**57**) 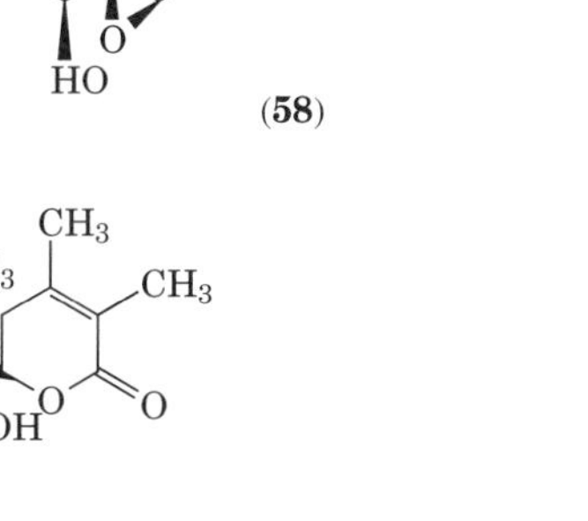   (**58**)

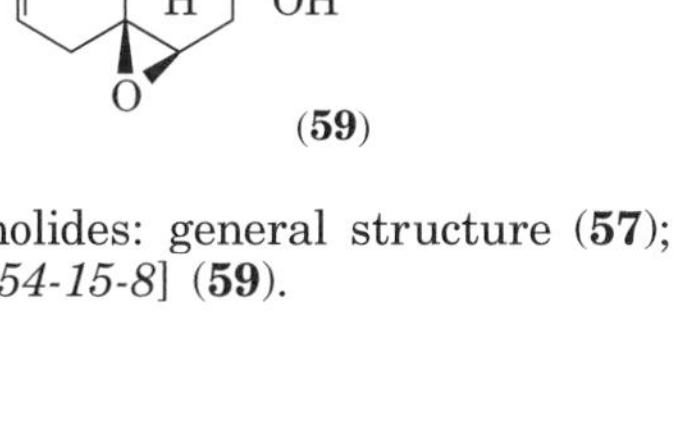

(**59**)

Fig. 6.   Examples of withanolides: general structure (**57**); withanolide D [*30655-48-2*] (**58**); and withanolide E [*38254-15-8*] (**59**).

withanolide D (**58**) and withanolide E (**59**) (84–88). Many biological properties of withanolides have been studied, including antitumor, antibiotic, immunomodulating, and insect antifeeding activities (89).

Ecdysteroids.   Ecdysteroids have been studied since the 1940s. They can be isolated from many species of the animal kingdom that belong to the phyla Protomia, eg, insects, worms, and arthropods, as well as a variety of different plant species. Ecdysteroids include the molting hormones; however, not all the over 60 ecdysteroids that have been isolated are active hormones. Ecdysteroids from animals are referred to as zooecdysteroids and from plants are referred to as phytoecdysteroids. Ecdysteroids all contain a cholestane skeleton with an A–B cis-ring junction and polyhydroxylation, including the 2β-, 3β-diol, 14α-hydroxy, and C17 side-chain hydroxylations, as exemplified by ecdysone [3604-87-3] (**60**) and 20-hydroxyecdysone [5289-74-7] (**61**). In addition, ecdysteroids are generally oxidized at C6 with a double bond at C7–C8 to form a cyclohexenone in the B-ring. Ecdysone (**60**) was the first insect hormone that was isolated (90) and characterized (91); however, its metabolite, 20-hydroxyecdysone (**61**),

(**60**, R=H)

(**61**, R=OH)

is recognized to be the universal molting hormone of insects and crustaceans. In addition to their molting properties, ecdysteroids have been studied for their antitumor and antimicrobial activities (92–96).

Marine Sterols.   Throughout the 1980s and 1990s, several hundred unique sterol structures have been elucidated from a variety of marine invertebrates. A single nucleus can be used to describe most terrestrial sterols, but no single template suffices for marine sterols. Basic substructures for marine sterol nuclei have been proposed, eg, (**62**), conventional; (**63**), 19-*nor*; (**64**), A-*nor* (Fig. 7). In addition, seco-sterols, such as 9,11-*seco*-sterols (**65**) and 8,9-*seco*-sterols (**66**), and highly degraded sterol nuclei have been observed. Much of the diversity of this group of sterols is incorporated into the C17 side chain. These variations include cyclopropane, cyclopropene, acetylene, allene, polyalkylated, 26-*nor*, and polyoxygenated side chains. Similar to cholesterol, marine sterols play a critical role in both the physiology and biochemistry of biological systems. In addition, marine sterols may have an ecological role in the marine environment (98–102).

Steroid Antibiotics.   The steroid antibiotics are a structurally diverse class of steroids that have a common biological function, ie, antibacterial, antifungal, antiviral, or antitumor activities. This group of compounds can overlap

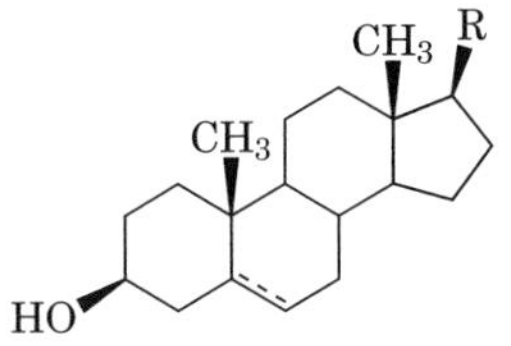

(**62**)

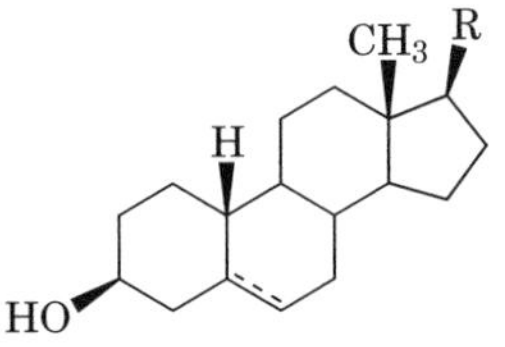

(**63**)

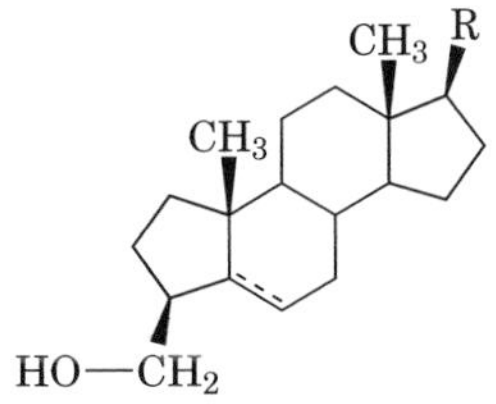

(**64**)

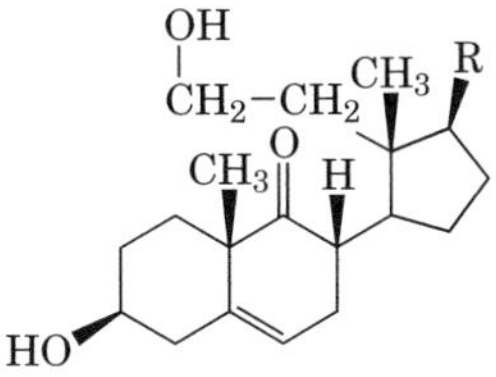

(**65**)

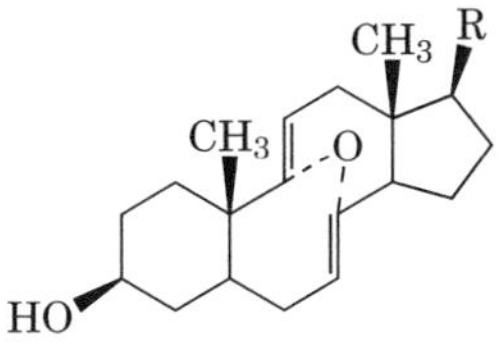

(**66**)

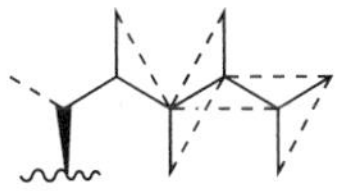

(**a**)

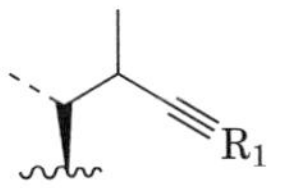

(**b**)

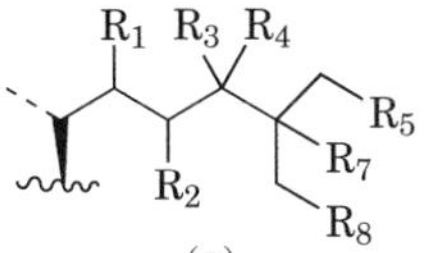

(**c**)

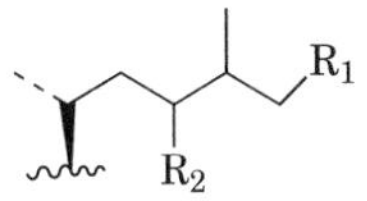

(**d**)

Fig. 7. Structural diversity of marine sterols where R = a variety of unique side chains, including (**a**) cyclopropa(e)ne, (**b**) acetylene, (**c**) polyalkylated, and (**d**) the 26-*nor* side chain.

with other steroid classes listed above. Fusidic acid [*6990-06-3*] (**67**), helvolic acid [*29400-42-8*] (**68**), and cephalosporin P_1 [*13258-72-5*] (**69**) exemplify a set of antibacterial steroids that contain a prolanostane skeleton with an unique trans–syn–trans–antitrans stereochemistry. This stereochemical relationship forces the B-ring into an unusual "boat" configuration (103). These compounds inhibit the growth of gram-positive bacteria by inhibiting protein synthesis, but have little activity against gram-negative bacteria (104,105).

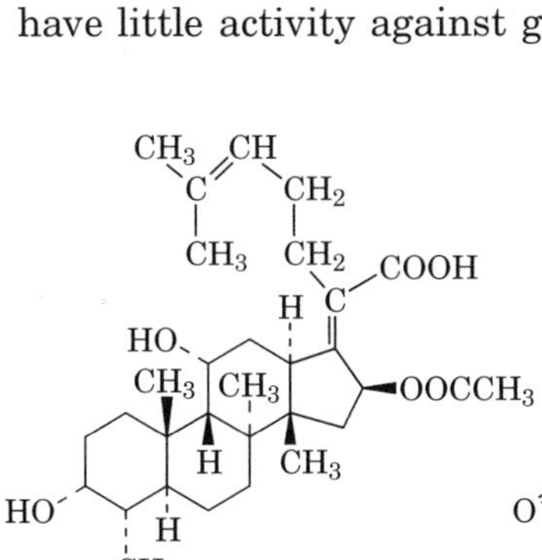

(**67**)

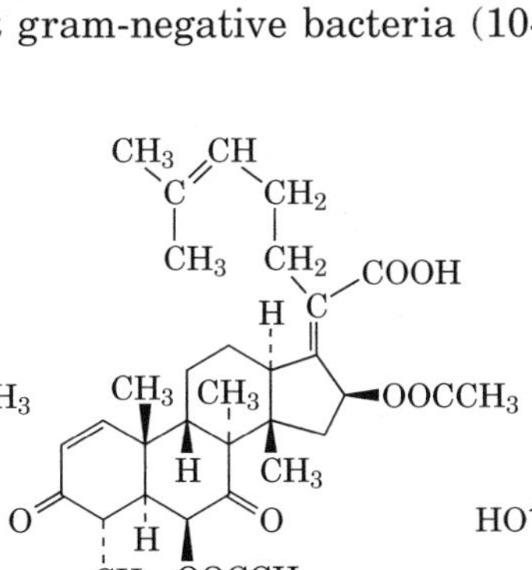

(**68**)

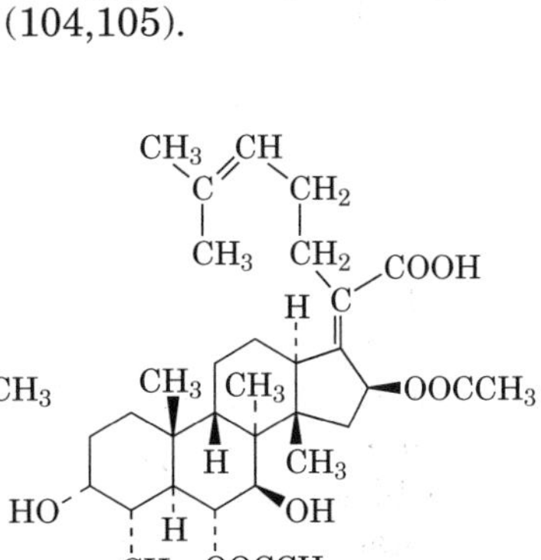

(**69**)

An antibiotic isolated from the tissues of the dogfish shark is the steroid alkaloid squalamine [*148717-90-2*] (**70**). Squalamine is a rare adduct of spermidine with an anionic bile acid intermediate. Squalamine is a broad-spectrum antibiotic that exhibits potent antimicrobial activity against fungi, protozoa, viruses, and both gram-negative and gram-positive bacteria (106). The precise mechanism of the antimicrobial activity of squalamine has not been identified. However, squalamine (**70**) could inhibit microbe growth by generating ion channels or pores in microbial membranes or by binding to microbial DNA. The total synthesis of squalamine from commercially available bile acids has been realized and the biological activity of squalamine has been mimicked by analogues that are more easily synthesized (**71**) (107,108).

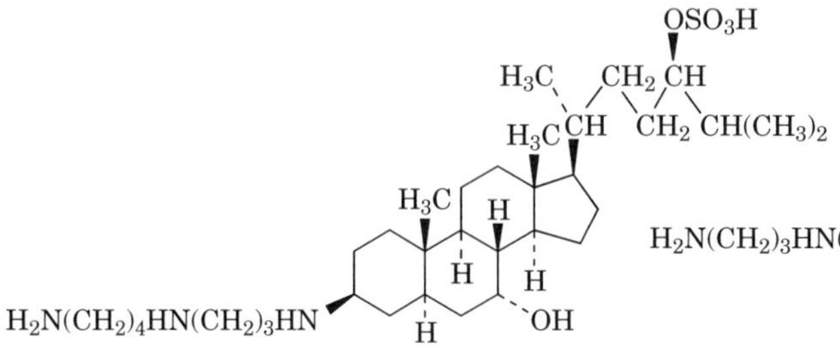

(**70**)

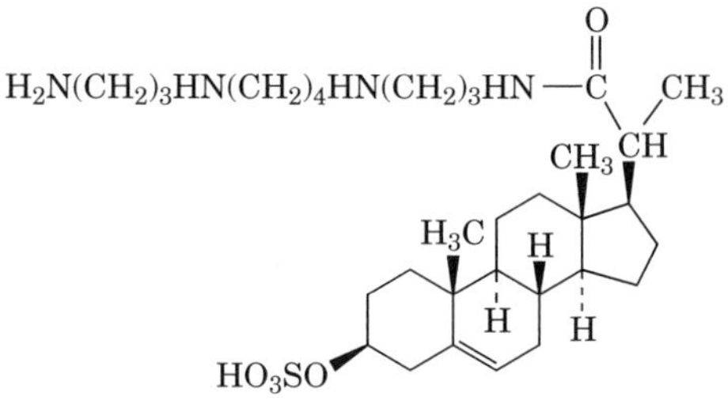

(**71**)

Biosynthesis

Steroids are members of a large class of lipid compounds called terpenes. Using acetate as a starting material, a variety of organisms produce terpenes by essentially the same biosynthetic scheme (Fig. 8). The self-condensation of two molecules of acetyl coenzyme A (CoA) forms acetoacetyl CoA. Condensation of acetoacetyl CoA with a third molecule of acetyl CoA, then followed by an NADPH-mediated reduction of the thioester moiety produces mevalonic acid [*150-97-0*] (**72**). Phosphorylation of (**72**) followed by concomitant decarboxylation and dehydration processes produce isopentenyl pyrophosphate. Isopentenyl pyrophosphate isomerase establishes an equilibrium between isopentenyl pyrophosphate and 3,3-dimethylallyl pyrophosphate (**73**). The head-to-tail addition of these isoprene units forms geranyl pyrophosphate. The addition of another isopentenyl pyrophosphate unit results in the sesquiterpene (C_{15}) farnesyl pyrophosphate (**74**). Both of these head-to-tail additions are catalyzed by prenyl transferase. Squalene synthetase catalyzes the head-to-head addition of two achiral molecules of farnesyl pyrophosphate, through a chiral cyclopropane intermediate, to form the achiral triterpene, squalene (**75**).

Stereospecific 2,3-epoxidation of squalene, followed by a nonconcerted carbocationic cyclization and a series of carbocationic rearrangements, forms lanosterol [*79-63-0*] (**77**) in the first steps dedicated solely toward steroid synthesis (109,110). Several biomimetic, cationic cyclizations to form steroids or steroid-

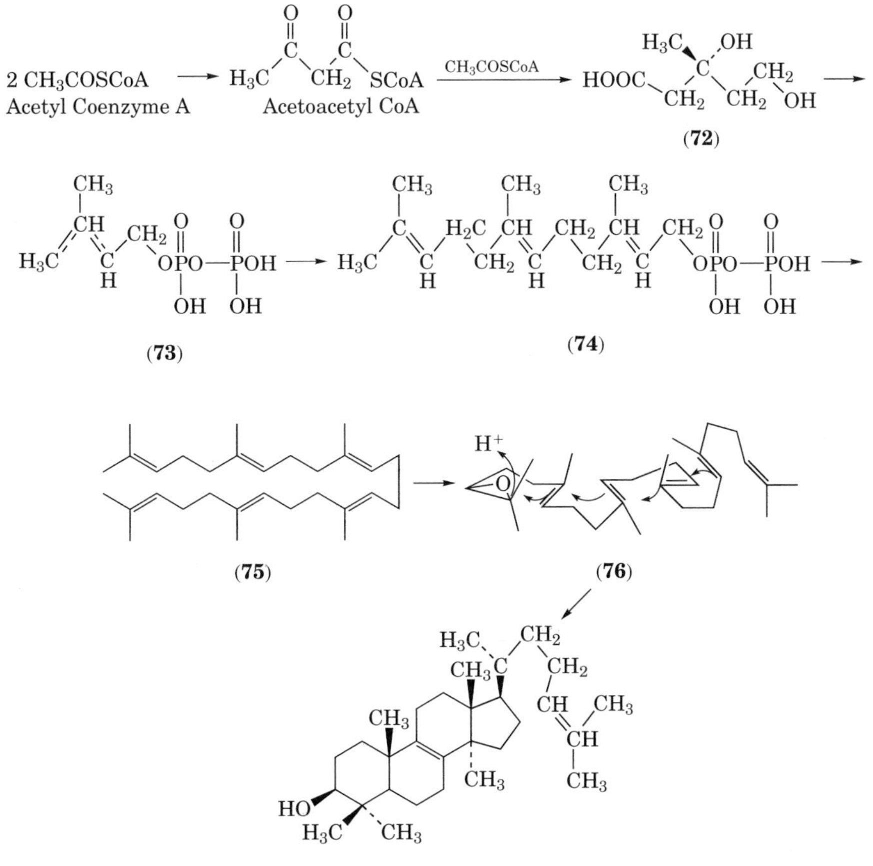

Fig. 8. Abbreviated terpene biosynthesis.

like nuclei have been observed in the laboratory (111), and the total synthesis of lanosterol has been accomplished by a carbocation–olefin cyclization route (112). Through a complex series of enzyme-catalyzed reactions, lanosterol is converted to cholesterol (**2**). Cholesterol is the principal starting material for steroid hormone biosynthesis in animals. The cholesterol biosynthetic pathway is composed of at least 30 enzymatic reactions. Lanosterol and squalene appear to be normal constituents, in trace amounts, in tissues that are actively synthesizing cholesterol.

The conversion of cholesterol (**2**) to pregnenolone [145-13-1] (**78**) is accomplished primarily through enzymatic systems in the adrenocortical and gonadal mitochondria. This conversion appears to be rate-limiting and therefore is regarded as the control point for the entire steroid hormone biosynthetic process. An abbreviated metabolic pathway for the corticosteroids and sex hormones is shown in Figure 9; however, many other metabolites play various roles in biosynthesis and excretion of steroids. The oxidation–reduction steps on the metabolic pathways are generally reversible, whereas the steps that include an oxidative C–C cleavage of the side chain are irreversible (40,113–119).

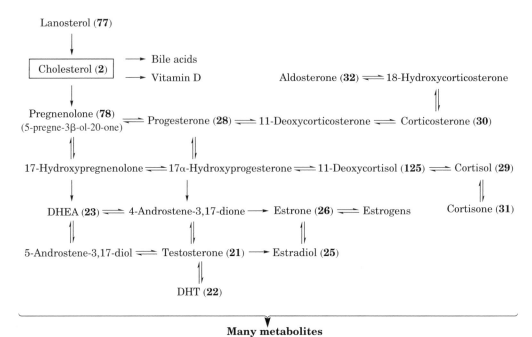

Fig. 9. Abbreviated steroid biosynthesis where DHEA = dehydroepiandrosterone and DHT = dihydrotestosterone.

Manufacture and Synthesis

There are three general processes that are used, as of ca 1996, worldwide for steroid production: (*1*) direct isolation from natural sources, (*2*) partial synthesis from steroid raw materials that have been isolated from plants and animals, and (*3*) total synthesis from nonsteroidal starting materials (120).

DIRECT ISOLATION

The two most important classes of steroid pharmaceuticals that are isolated directly from natural products are some estrogens and most cardiac steroids. Compounds with estrogenic activity have been isolated from different sources. Urine from pregnant women and from pregnant mares has been used for the commercial production of estrogens. The overall amount of estrogens from pregnant mare urine is generally 10 times greater than that of pregnant human urine; however, the hormone from pregnant mares is more highly conjugated than in humans and only 10–25% of the total is directly extractable. The isolation of estrogens from natural sources has been essentially the same since ca 1960 and has been reviewed (121).

Cardiac steroids occur in small amounts in various plants with a wide geographical distribution. The purple foxglove *Digitalis purpuras* has been used for centuries as both a drug and a poison. Isolation and characterization of the various cardiac steroids have been reviewed (122,123).

PARTIAL SYNTHESIS

Raw Materials and Extraction. The variety of natural sources of steroid raw materials is vast, and the exact details of manufacturing processes are ambiguous closely held industrial secrets. However, the most widely utilized raw materials for the partial synthesis of steroids appear to be the following: (*1*) the sapogenins, including diosgenin (**34**), hecogenin (**35**), and tigogenin (**36**); (*2*) the structurally related steroid alkaloids, including tomatidine (**43**) and solasodine (**44**); (*3*) sterols, such as cholesterol (**2**), stigmasterol (**40**), and β-sitosterol (**41**); and (*4*) bile acids, such as deoxycholic acid (**14**) (124).

Plants of the genus *Dioscorea*, which include *D. deltoidea*, *D. prazeri*, and *D. tubers*, are the most common source of diosgenin. This genus occurs abundantly in tropical and subtropical regions throughout the world. A variety of methods are used for the isolation of diosgenin. In a generalized process plant tubers are dried and powdered. This powdered material is first hydrolyzed with aqueous acid or enzymes, then extracted with an organic solvent such as petroleum ether. Diosgenin is isolated after recrystallization (125).

A pilot plant in India has been established to extract fiber, pulp, and juice from the leaves of sisal plants. The fiber is sold directly or used to manufacture rope, the crushed pulp is used in paper processing, and the juice is an excellent source of hecogenin. During a three- to five-day fermentation of the juice, partial enzymatic hydrolysis causes hecogenin to precipitate as the hemisaponin in the form of a fine sludge. This sediment is hydrolyzed with aqueous hydrochloric acid, neutralized, and filtered. This filter cake is washed with water and extracted with alcohol. The yield of hecogenin varies between 0.05 and 0.1% by the weight of the leaf (126).

Owing to periodic fluctuations in the price of diosgenin, alternative raw materials such as solasodine have been used for the synthesis of steroid drugs. Solasodine can be isolated from a medley of genera and species of plants found worldwide. Generally, solasodine appears in plants as a glycoside at the 3-position of the steroid. Solasodine is isolated by extraction of the fresh or dried glycoalkaloid-containing tissue with either alcohol or aqueous acids. Alcohol extraction yields more total solasodine but with significantly higher levels of contaminants. Therefore, overall yields of the two extraction processes are comparable owing to losses during the purification of the alcohol extracts (127).

In the United States, the plant sterols stigmasterol and β-sitosterol are a significant raw material for the synthesis of antiinflammatory glucocorticoids and other steroid hormones. Extracts from soybean oil by-products contain 12–25% stigmasterol along with large amounts of sitosterols. The industrial separation of these nearly identical sterols was accomplished in high yields and purity using a multistage countercurrent crystallization from selected solvents (70).

In addition to the isolation of steroid raw materials from whole plants, plant tissue cultures have been investigated as an alternative source of these steroids. Despite many advances (128), there are no industrial applications of plant cell cultures for the production of steroids (129,130).

Methods of Partial Synthesis. Partial syntheses are done typically by chemical degradation or fermentation/biotransformation.

Chemical Degradation. Initial efforts toward the synthesis of cortico-steroids in Europe and in the United States used 7-deoxycholic acid as a starting material. These syntheses included multistep oxygen transposition and C17 degradation to form various 11-oxy-steroids (131). The first significant break-through in the commercial synthesis of steroids was the chemical degradation of diosgenin. The Marker degradation became the principal method for commer-cial steroid synthesis in the 1940s and 1950s, and modifications of this process are still in use in the 1990s (Fig. 10). When diosgenin is heated to approxi-mately 200°C in acetic anhydride, elimination and acetylation of the oxygen in the F-ring produce the bis-acetylated enol ether (**79**). Oxidative cleavage of the enol ether of (**79**) with chromium trioxide followed by elimination of the C16-acyl-oxygen results in steroid (**80**) [*1162-53-4*]. Selective hydrogenation of the α,β-unsaturated ketone in the D-ring from the sterically less hindered α-face forms pregnenolone (**78**). Pregnenolone is readily converted into progesterone (**28**) under oxidative conditions (132).

This process was improved and expanded (133) to provide starting materi-als for the C19-sex hormones that include estrogens and androgens (see Fig. 10).

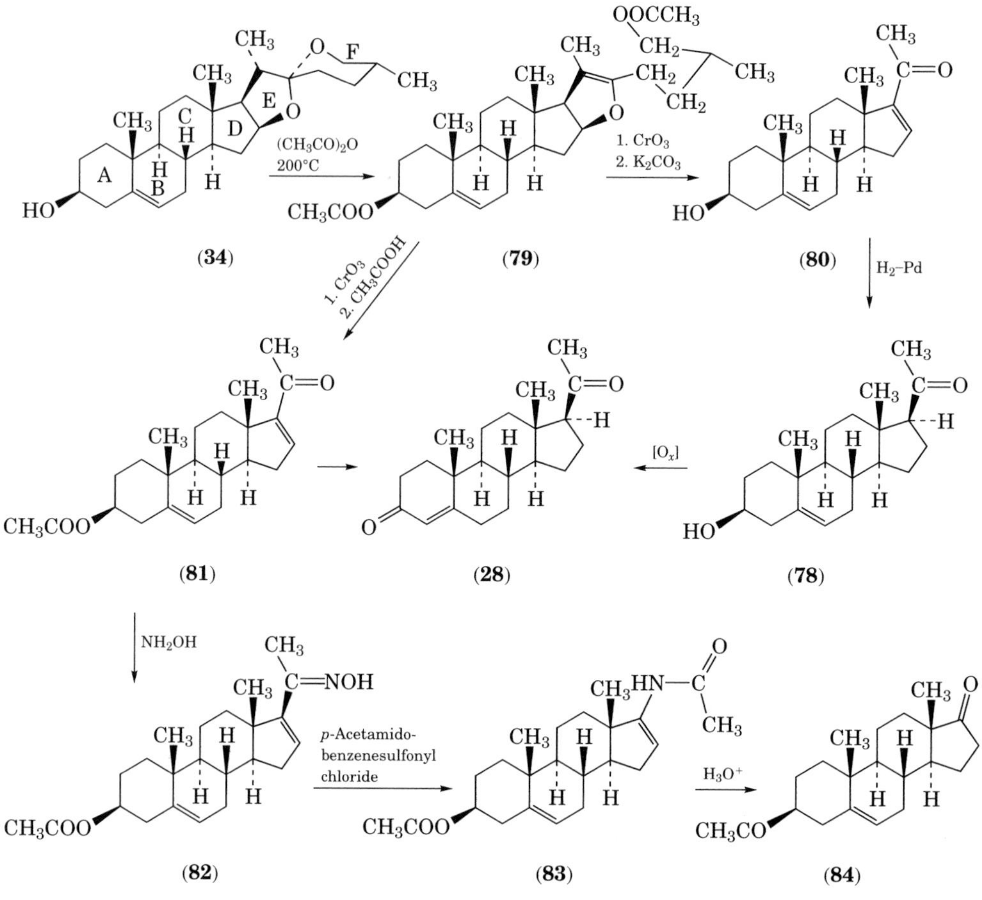

Fig. 10. Chemical degradation of diosgenin.

Oxidative cleavage of enol ether (**79**) with chromium trioxide followed by elimination of the C16-acyl-oxygen in hot acetic acid affords pregnenolone acetate (**81**) in over 80% yield. Pregnenolone acetate (**81**) can be converted to progesterone by methods similar to the Marker process. The cleavage of the C17 side chain begins with the treatment of (**81**) with hydroxylamine to afford the C20 oxime (**82**). Beckman rearrangement of (**82**) affords the ene-amide (**83**). Mild acid hydrolysis of (**83**) results in dehydroepiandrosterone acetate [*853-23-6*] (**84**) (133). The same processes have been applied to the structurally similar steroid alkaloids solasodine (**44**) and tomatidine (**43**).

Unlike diosgenin, hecogenin possesses a C12-keto group and is saturated at C5–C6. The keto group at C12 in the C-ring allows for a series of chemical steps to transpose this oxygen to the biologically important C11-position (Fig. 11). Bromination of hecogenin acetate (**85**) in either benzene, carbon tetrachloride, or dioxane produces the 11,23-dibromide (**86**) in over 50% recrystallized yield with 32% recovered, usable starting materials. Heating a biphasic mixture of (**86**) and sodium hydroxide in *t*-butyl alcohol and water, followed by acetylation and zinc-mediated debromination forms the bromine-free acetate (**87**) in approximately 80% yield. Finally, a dissolving metal reduction of (**87**) results in 11-ketotigogenin [*4802-74-8*] (**88**). The spiroketal ring can be degraded by methods similar to the degradations discussed above. The efficient synthesis of 11-ketotigogenin established a commercial production of cortisol from hecogenin (134).

Another commercial method that has been used for the production of progesterone is the chemical degradation of the side chain of stigmasterol (Fig. 12). Oxidation of the C3-hydroxyl of stigmasterol (**40**) with concomitant double-bond migration results in (**89**). The resultant C4–C5 double bond in the A-ring is electron-deficient and less reactive under the subsequent reaction conditions. Selective ozonation and oxidation cleavage of the side chain double bond of (**89**) result in aldehyde (**90**). Enol acetate formation of the C22-aldehyde (**91**), followed by a second ozonation and cleavage of the resultant side-chain double bond, yields progesterone (**28**) (135).

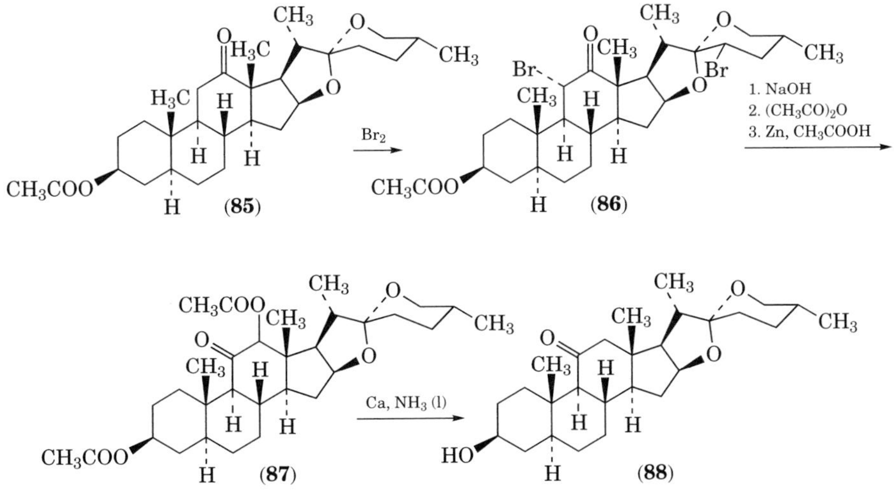

Fig. 11. Carbonyl transposition of hecogenin.

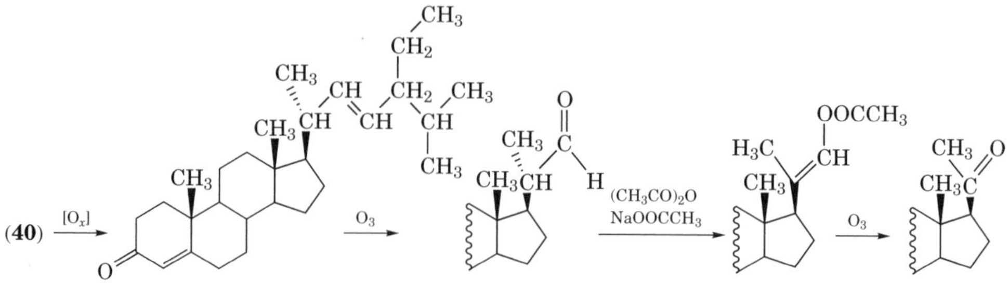

Fig. 12. Chemical degradation of stigmasterol.

Interest in the synthesis of 19-norsteroids as orally active progestins prompted efforts to remove the C19 angular methyl substituent of readily available steroid precursors. Industrial applications include the direct conversion of androsta-1,4-diene-3,17-dione [897-06-3] (**92**) to estrone [53-16-7] (**26**) by thermolysis in mineral oil at about 500°C (136), and reductive elimination of the angular methyl group of the 17-ketal of the dione [2398-63-2] (**93**) with lithium biphenyl radical anion to form the 17-ketal of estrone [900-83-4] (**94**) (137).

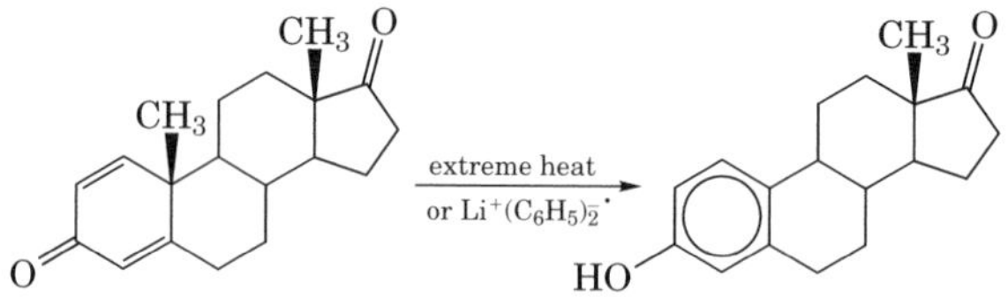

Another method to prepare 19-norsteroids is first to oxidize the C19 angular methyl substituent, followed by reductive decarboxylation or decarbonylation of the resultant C19 lactone, carboxylic acid, or aldehyde. All methods of oxidation of angular methyl groups proceed through high energy intermediates capable of oxidizing unactivated CH bonds. These high energy intermediates are generated from an intramolecular heteroatom in close proximity to the angular methyl group. Practical routes to 19-norsteroids are shown in Figure 13. The addition of hypohalous acid to Δ^5-steroids (**95**) gives 5α-halo-6β-carbinols (**96**). Cyclization of this 6β-alcohol onto the C19 angular methyl substituent can occur under a variety of different conditions. For example, ether (**97**) is produced by the treatment of (**96**) with lead tetraacetate under thermal or photolytic conditions. In addition, treatment of (**96**) with lead tetraacetate and iodine proceeds though an intermediate hypoiodite (138) that, after homolytic decomposition, results in ether (**97**) (139). Ether (**97**) can be either oxidized to lactone (**98**) or the C3-acetate hydrolyzed and oxidized to the C3-ketone (**99**). Elimination of the halogen of (**99**) followed by C6-oxygen reduction yields the C19 alcohol (**100**). Oxidation of (**100**) to the aldehyde or carboxylic acid (**101**), followed by decarbonylation or decarboxylation, respectively, results in the 19-norsteroid (**102**). Alternatively, acetate hydrolysis, C3-oxidation, and elimination of (**98**) forms lactone (**103**).

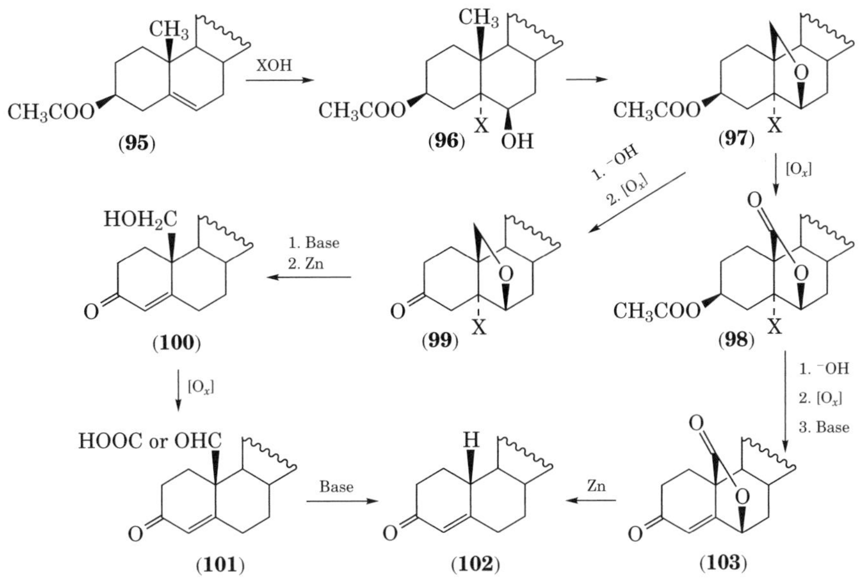

Fig. 13. Chemical degradation to 19-norsteroids where X = Cl, Br.

Concomitant C6-reduction and decarboxylation of (**103**) yield the 19-norsteroid (**102**) (140). In a similar process, the C19 and C18 angular methyl groups can be oxidized by photolytic activation of a nearly nitrite ester (141). A free-radical activation of the C18-angular methyl moiety has been exploited in a number of synthetic approaches to aldosterone (142,143).

Fermentation/Biotransformation.   In a search to decrease the cost and increase the efficiency of steroid synthesis, commercial biotechnology operations have focused on microbial agents for specific transformations of individual steroid substrates. The regio- and stereoselective hydroxylation of every site on virtually every steroid nucleus is possible (144). Many of these hydroxylation steps are of commercial importance. For example, the 9α-, 11α-, 11β-, and 16α-hydroxylations are key steps in the industrial synthesis of synthetic corticosteroid antiinflammatory drugs. These steps are accomplished almost exclusively by microbial transformations. In addition to hydroxylations, other useful microbial oxidations of steroids include alcohol oxidations, epoxidations, oxidative cleavage of carbon–carbon bonds (eg, C17 side-chain cleavage), introduction of double bonds, peroxidations, and heteroatom oxidations. Other invaluable microbial steroid transformations include reductions, degradations, A-ring aromatization, resolutions, isomerizations, conjugations, hydrolyses, heteroatom introduction, and sequential reactions (120,145,146).

There are two principal biotechnological applications dealing with steroids. Microbial agents are used for processing raw materials into useful intermediates for general steroid production and for specific transformations of steroids to advanced intermediates or finished products (120,145).

Processing Raw Materials.   Along with the aforementioned chemical methods of processing steroid raw materials, microbial transformations have been and are used in a number of commercial degradation processes. The microbial degradation of the C17 side chain of the two most common sterols, cholesterol (**2**) and β-sitosterol (**41**), is a principal commercial method for the preparation of starting materials in Japan and the United States.

Many microorganisms have been found that partially or completely degrade cholesterol (**2**) (147). Enzyme inhibitors or modified microbial agents have been used to control these degradations in order to form commercially useful steroidal intermediates (Fig. 14). When mixed with metal ions (Ni, Co, Pb, or Se) (148), chelating agents (149), or 8-hydroxyquinoline (150), various mycobacteria have been demonstrated to produce significant quantities of androsta-1,4-diene-3,17-dione [*63-05-8*] (**104**) from cholesterol (**2**). In a commercialized process for the microbial conversion of cholesterol to androsta-1,4-diene-3,17-dione, the 9α-hydroxylation step was inhibited by α,α-dipyridyl (151). In another commercial process, uv-generated mutations of *Mycobacterium* sp. have been used to produce androsta-4-ene-3,17-dione (**104**) and androsta-1,4-diene-3,17-dione [*897-06-3*] (**92**) from β-sitosterol (see Fig. 14) (152). In a similar industrial process, a mutant of *Mycobacterium fortuitum* degraded β-sitosterol to 9α-hydroxyandrosta-4-ene-3,17-dione [*560-62-3*] (**105**) (153). Dehydration of (**105**) to $\Delta^{9(11)}$-derivative (androsta-4,9-diene-3,17-dione [*1035-69-4*]) (**106**) provided starting material for corticosteroid synthesis (154). The rate of side-chain cleavage of sterols is limited by the low solubility of substrates and products and their low transport rates to and from cells. Cyclodextrins have been used to increase the solubilities of these compounds and to assist in their cellular transport. Cyclodextrins increase the rate and selectivity of side-chain cleavage of both cholesterol and β-sitosterol with no effect on cell growth. Optimal conditions have resulted in enhancement of molar yields of androsta-1,4-diene-3,17-dione (**92**) from 35–40% to >80% in the presence of cyclodextrins (120,145,146,155).

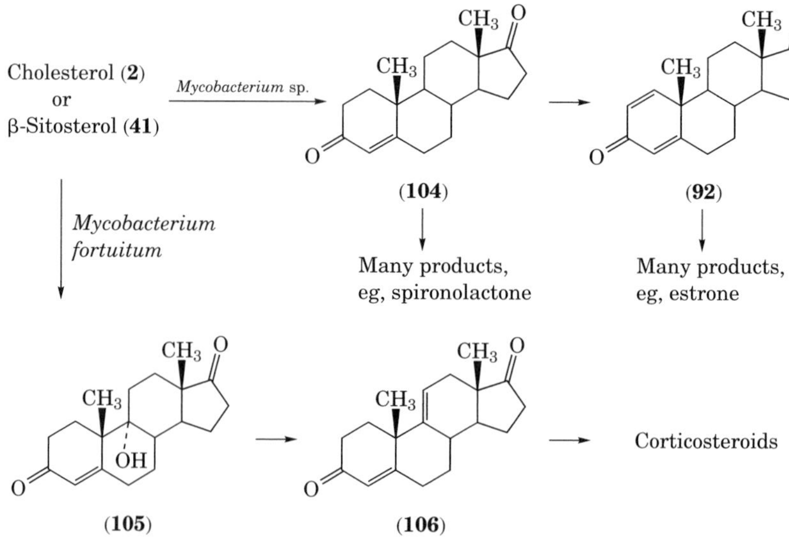

Fig. 14. Commercialized processes for the microbial transformation of readily available sterols to useful synthetic intermediates.

Besides the aforementioned chemical methods, microbial degradations have been used to remove the C19 angular methyl substituent of readily available steroid precursors. For example, 19-hydroxysterols, such as 3β-acetoxy-19-hydroxy-5-cholestene [750-59-4] (**107**), can be converted to estrone by *Nocardia* sp. in yields up to 70% (120,145,146).

(**107**)   (**26**)

Transformations of Steroids to Advanced Intermediates or Finished Products. The most difficult chemical step of the early chemical syntheses of cortisol or corticosteroid analogues was the introduction of an 11-hydroxy moiety. In 1949 the first successful biotransformation of 11-deoxycorticosterone to corticosterone using a perfusion of bovine adrenal glands was reported (156). Similar adrenal perfusions were used by G. D. Searle & Company for the preparation of adequate supplies of hydrocortisone for clinical evaluations (157). Using microorganisms, workers at Upjohn (1952) found that the specific *Mucorales* fungi grown in aerated cultures were capable of direct 11α-hydroxylation of progesterone in yields as high as 90% (158). Concurrently, workers at the Squibb Institute used *Aspergillus niger* to perform the same reaction on progesterone, 11-deoxycortisol, 11-deoxy-17α-hydroxycorticosterone, and 17α-hydroxyprogesterone (159). The 11α-hydroxy steroids were readily converted to the 11β-hydroxy steroids by first oxidation to the ketone with chromic acid, then sodium borohydride reduction from the sterically less hindered face. However, direct microbial transformation of 11-deoxy steroids to the 11β-hydroxy steroids was still desired. Shortly thereafter, Pfizer investigators found that *Curvularia lunata* converted progesterone, deoxycortisone, 11-deoxy-17α-hydroxycorticosterone, and 17α-hydroxyprogesterone directly to the 11β-hydroxy steroids (160,161).

Schering investigators uncovered a second significant breakthrough in microbial biotechnology of steroid production. They discovered that *Corynebacterium simplex* converted hydrocortisone (cortisol) (**29**) to prednisolone via a 1,2-dehydrogenation reaction. This $\Delta^{1,4}$-3-ketosteroid is a highly active antiinflammatory commercial product (162).

A third advancement in microbial biotechnology of steroid production was the ability to introduce a 16α-hydroxyl group microbiologically (163). Modifications of the 11β-hydroxylation, 16α-hydroxylation 1,2-dehydrogenation microbial processes are used for the synthesis of hydrocortisone, prednisolone, triamcinolone, and other steroid pharmaceuticals. A few microbial transformations that have been used to manufacture steroids are listed in Table 1 (164).

Representative Partial Syntheses. *Estranes.* The synthesis of 19-norsteroids was stimulated by the development of orally active progestins as birth control agents. The industrial synthesis of pure estrone (120,165) made conversion to 19-*nor*-A-ring-eneones by the Birch reduction economically feasible (166). An early industrial synthesis of the contraceptive agents norethynodrel (**111**) and norethindrone (**112**) is shown in Figure 15. The aromatic ring of estadiol-3-

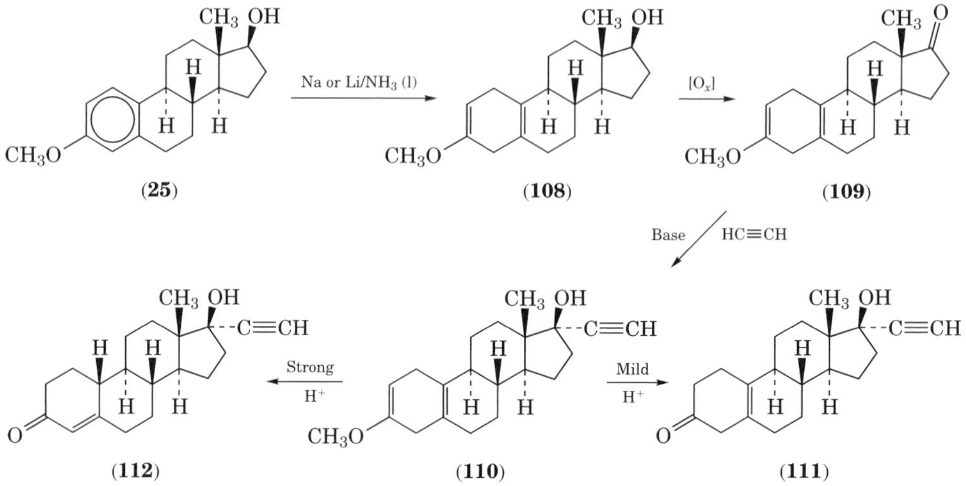

Fig. 15. Partial synthesis of norethynodrel (**111**) and norethindrone (**112**).

methyl ether is reduced to the 1,4-dihydroestrogen (**108**) with a dissolving metal reduction (167). Careful C17-oxidation of (**108**), followed by addition of acetylene to the resultant C17-ketone (**109**) under basic conditions, yields (**110**). The enol ether of (**110**) is cleaved with mild aqueous acid or strong aqueous acid to give norethynodrel [*68-23-5*] (**111**) and norethindrone [*68-22-4*] (**112**), respectively (168–170).

Most 19-norsteroids are produced through total synthesis from nonsteroidal starting materials. One notable exception is the production of the androgen agonist oxendolone (**113**) (Fig. 16). An aldol condensation of acetaldehyde with dehydroepiandrosterone acetate [*853-23-6*] (**114**) results in ene-one (**115**). Selective hydrogenation of the ene-one from α-face produces (**116**). Reduction of the C17-ketone from the less hindered α-face, followed by acetylation of the C3β- and resultant C17β-alcohols yields a diacetate. Addition of hypochlorous acid to this diacetate first forms an intermediate 5,6-α-chloronium ion, followed by diaxial ring opening produces (**117**). Degradation of the C19-methyl group proceeds as described previously (see Fig. 13), that is, (*1*) a lead tetraacetate-induced free-radical addition of the C6β-hydroxyl moiety to the C19-methyl substituent to form a cyclic ether, (*2*) selective saponification of the C3-acetate, (*3*) oxidation of the C3-alcohol to a ketone, (*4*) base-catalyzed elimination of the C5-chloride to form an A-ring enone, and (*5*) zinc-mediated reduction of the cyclic ether formed the C19 alcohol (**118**). Oxidation of the primary alcohol to an aldehyde, followed by base-catalyzed decarbonylation of this aldehyde and saponification of the C17-acetate, affords oxendolone (**113**) (171).

Table 1. Commercial Microbial Transformations Used To Produce Advanced Intermediates or Finished Products[a]

Substrate[b]	Transformation	Product	Organism
progesterone	11α-hydroxylation oxidation/ lactonization		*Rhizopus nigricans* *Cylindrocarpon radicicola*
11-deoxycortisol (17α-derivatives)	11β-hydroxylation	cortisol/derivatives	*Curvularia lunata*
6α-fluoro-16α-methyl-21-hydroxy-pregn-4-ene-3,20-dione	11β-hydroxylation	Paramethasone	*Curvularia lunata*
11-deoxy-16-methylene-cortisol	11β-hydroxylation	Prednylidene	*Curvularia lunata*
9α-fluorohydro-cortisone	1-dehydrogenation 16α-hydroxylation	Triamcinolone	*Arthrobacter simplex*
hydrocortisone	1-dehydrogenation	Prednisolone	*Arthrobacter simplex* or *Bacillus lentus*
6α-fluoro-16α-methyl cortico-sterone	1-dehydrogenation	Fluocortolone	
11β,21-dihydroxy-pregna-4,17(20)-dien-3-one	1-dehydrogenation		*Septomyxa affinis*
rac-3-methoxy-8,14-secoestra-1,3,5-(10),9(11)-tetra-ene-14,17-dione (Secosteroid)[c]	17-ketone reduction		*Saccharomyces uvarum*
androst-4-ene-3,17-dione[d]	17-ketone reduction		*Saccharomyces* sp.
21-acetoxy-17α-hydroxy-pregnenolone	Δ⁵-3β-alcohol dehydrogenase		*Flavobacterium dehydrogenans*
6α-fluoro-21-hydroxy-16α-methyl-pregn-4-ene-3,20-one	9α-hydroxylation		*Curvularia lunata*

[a]Refs. 120 and 165.
[b]Class is corticosteroid unless otherwise noted.
[c]Class is estrogen–progestin.
[d]Class is androgen.

Pregnanes. In 1944, Sarrett completed the first partial synthesis of cortisone (172). Like many of the early syntheses of corticosteroids, Sarrett began with the a bile acid, deoxycholic acid (**14**). Because bile acids are isolated from animal sources, their supply is by necessity limited (173). Following these early syntheses, several improvements and innovations have resulted in a number of industrial syntheses of cortisol and other corticosteroids.

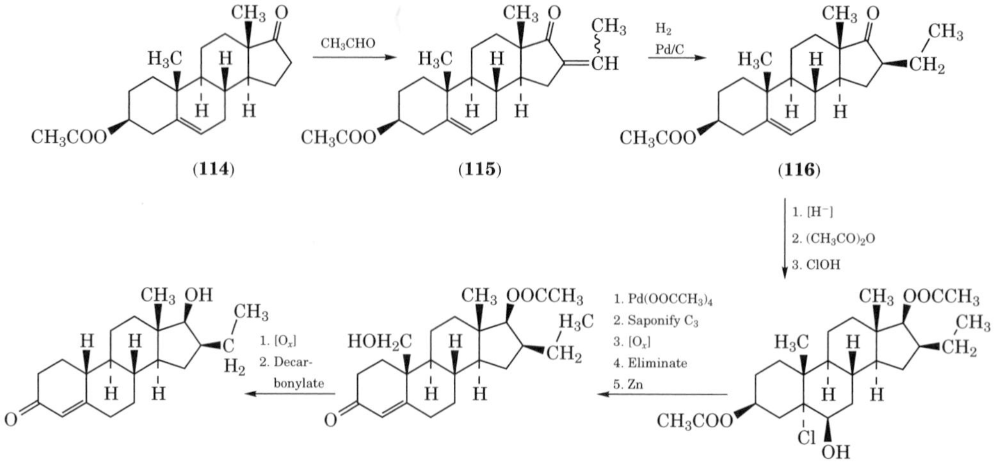

Fig. 16.   Partial synthesis of oxendolone [33765-68-3] (**113**).

Although there are many variations on industrial partial synthesis of corticosteroids, two basic processes are shown in Figure 17. Stigmasterol (**40**) is converted to 11-oxoprogesterone [516-15-4] (**119**) by methods that have been discussed. Under controlled conditions, base-catalyzed acylation of (**119**) forms a 21-monoglyoxylated compound as the primary product and a 2,21-bisglyoxylated compound (**120**) as a by-product. The preferred production process treats (**119**) with two or more moles of diethyl glyoxylate to form (**120**), exclusively. Treatment of (**120**) with bromine followed by alkaline cleavage of the glyoxylates produces intermediate (**121**). Without isolation of (**121**), Favorskii rearrangement forms (**122**). Direct debromination of (**122**) with Zn yields (**123**). In addition, (**122**) can be taken directly to the potent corticosteroid prednisolone in three chemical steps. Protection of the C3-ketone of (**123**) by formation of the ene-amine with pyrrolidine, followed by C11 and C21 reduction with lithium aluminum hydride and hydrolysis of the ene-amine results in (**124**). Osmium tetroxide-catalyzed oxidation of (**124**) yields cortisol (hydrocortisone, **29**) (174).

In another process, diosgenin is degraded to 16-dehydropregnenolone by chemical methods. Conversion of 16-dehydropregnenolone to 11-deoxycortisol (**125**) can be accomplished in 11 chemical steps. These steps result in hydroxylations at C21 and C17, oxidation at C3, and Δ^5 to Δ^4 double-bond isomerization (175). Microbial oxidation of (**125**) also produces cortisol (**29**).

Several cortisol analogues have become important therapeutically. Among the useful cortisol derivatives are the following: (1) dehydrogenation at C1−C2, (2) addition of fluorine to C6α and C9α, (3) addition of a methyl substituent to C6α, C16α, and C16β, and/or (4) hydroxylation at C16α (174). Commercial syntheses of a few of the C9α-fluoro analogues are shown in Figure 18. Tosylation of the C11α-hydroxyl substituent of (**126**), followed by elimination of this moiety in acidic acid/sodium acetate produces the $\Delta^{(9,11)}$-steroid (**127**). Bromohydrin

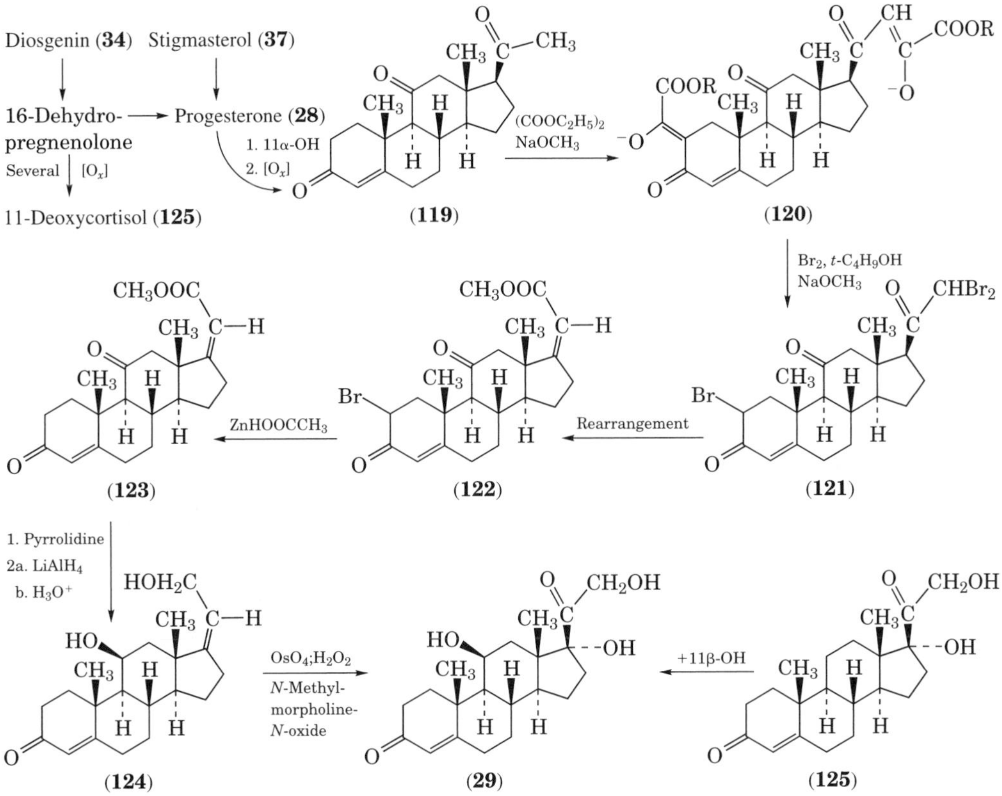

Fig. 17. Early industrial syntheses of cortisol (**29**).

formation on the $\Delta^{(9,11)}$-steroid (**127**) forms 9α-bromocortisol acetate. Treatment of this 9α-bromo-11β-carbinol with potassium acetate in boiling alcohol yields a $9\beta,11\beta$-epoxy steroid. Treatment of the $9\beta,11\beta$-epoxy steroid with hydrogen fluoride results in the corresponding 9α-fluorohydrin (9α-fluorohydro cortisone [*127-31-1*]) (**128**) (176,177). The other therapeutically important steroid syntheses shown in Figure 18 have been discussed in the biotransformations section.

Following the development of the efficient synthesis of the microbial degradation of cholesterol and sitosterol (177), the transformation of the C17-ketone of the resultant androstanes into the corticoid side chain were studied extensively (178). What appear to be among the most efficient of these pathways are shown in Figure 19. Treatment of androst-4-ene-3,17-dione (**106**) or androsta-1,4-diene-3,17-dione (**131**) with potassium cyanide in acetic acid forms the desired β-cyano-α-hydrin (**133**) through an equilibration of the two cyanohydrin epimers and selective crystallization of (**133**). The C17-hydroxyl moiety is protected and activated as the (chloromethyl)dimethylsilyl ether (**135**). Treatment of (**135**) with a strong base, eg, lithium diisopropylamide (LDA), results in deprotonation of the chloromethyl group and cyclization of this anion onto the adjacent nitrile to form a spirocyclic intermediate with a C20-imine. When the reaction mixture is quenched with aqueous acid, this C20-imine is hydrolyzed, the silyl ether is cleaved, and protodesilylation occurs resulting in (**140**) without

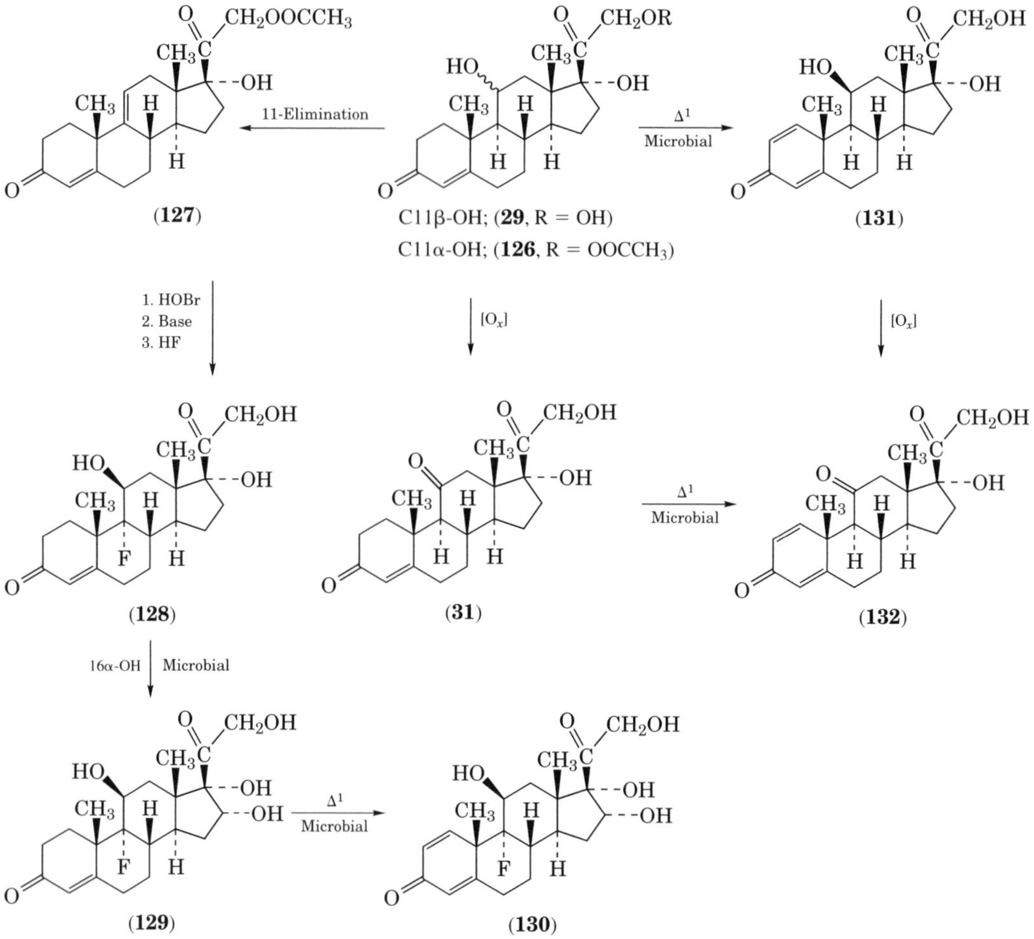

Fig. 18. Partial synthesis of corticosteroid drugs: 9α-fluoro-16α-hydroxyhydrocortisone [*337-02-0*] (**129**), triamcinolone [*124-94-7*] (**130**), prednisolone [*50-24-8*] (**131**), and prednisone [*53-03-2*] (**132**).

the need for a protecting group on the C3-ketone. The resultant C21-chloride of (**140**) can be displaced with acetate anion or reduced with zinc in acetic acid to produce (**141**) and (**139**), respectively. Alternatively if the C3-ketone is first protected, (**139**) is formed directly from (**135**) by reductive cyclization followed by hydrolysis (174,179).

A second industrial process for the conversion of androst-4-ene-3,17-dione (**106**) or androsta-1,4-diene-3,17-dione (**131**) into corticosteroid drugs is also shown in Figure 19. After selective protection of the C3-ketone, addition of the lithium compound (**132**) followed by an aqueous acid quench results in the chloro aldehyde (**134**). Treatment of (**134**) with potassium acetate and acetic anhydride forms the Δ¹⁶-corticoid (**137**). Steroid (**137**) is a versatile intermediate for the synthesis of a variety of the aforementioned, potent glucocorticoids. Conjugate reduction of the C16–C17 double bond of (**137**) with trialkylsilane yields enol ether (**136**). *meta*-Chloroperbenzoic acid (MCPBA) mediated epoxidation of (**136**)

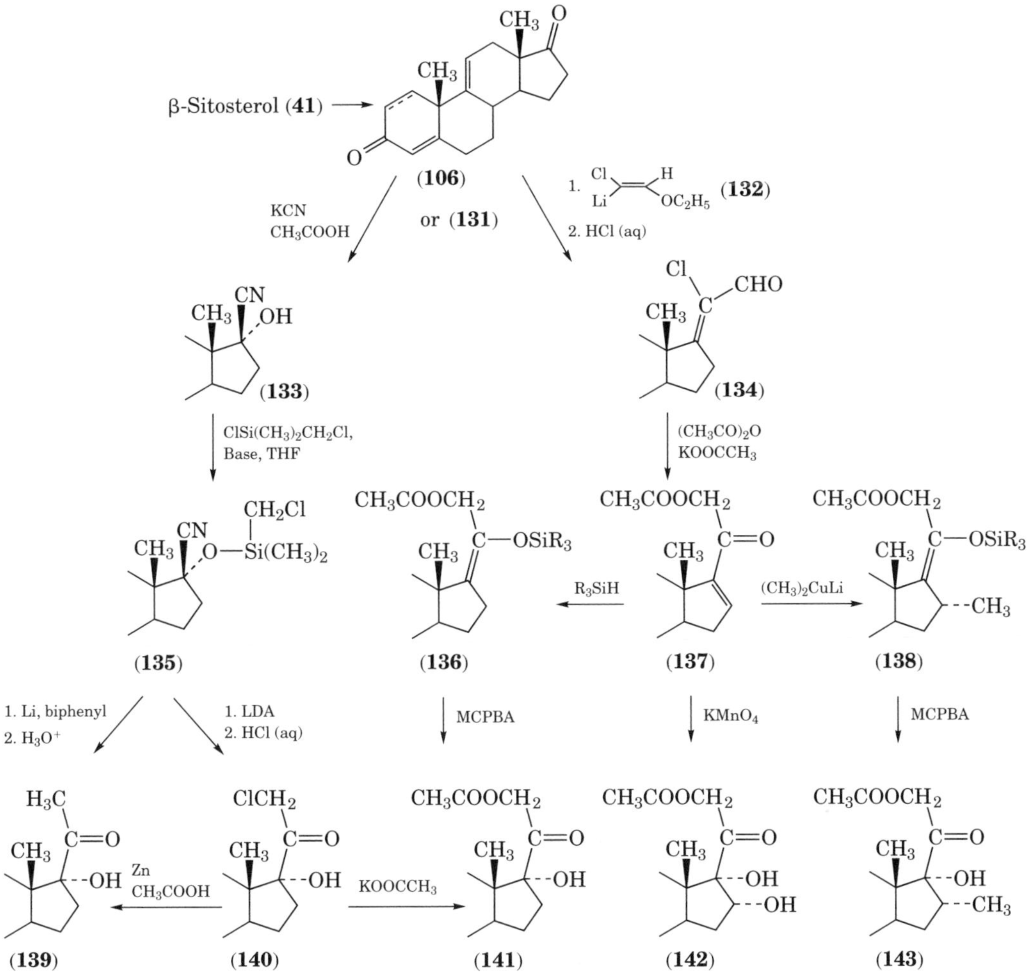

Fig. 19. C$_{17}$ chemistry.

produces a C17α,C20α-epoxide. Desilylation of the C20 silyl-enol-ether yields the C20-acylated glucocorticoid side chain (**141**). Treatment of (**137**) with potassium permanganate forms the α16,α17-diol (**142**). Conjugate addition of a methyl cuprate to the α-face of C16 of (**137**) followed by O-silylation of the resultant enolate results in the substituted enol ether (**138**). Epoxidation and desilylation, as before, yield (**143**) (174,180).

Another synthesis of the cortisol side chain from a C17-keto-steroid is shown in Figure 20. Treatment of a C3-protected steroid 3,3-ethanedyldimercapto-androst-4-ene-11,17-dione [*112743-82-5*] (**144**) with a trihaloacetate, zinc, and a Lewis acid produces (**145**). Addition of a phenol and potassium carbonate to (**145**) in refluxing butanone yields the aryl vinyl ether (**146**). Concomitant reduction of the C20-ester and the C11-ketone of (**146**) with lithium aluminum hydride forms (**147**). Deprotection of the C3-thioketal, followed by treatment of (**148**) with *meta*-chloroperbenzoic acid, produces epoxide (**149**). Hydrolysis of (**149**) under acidic conditions yields cortisol (**29**) (181).

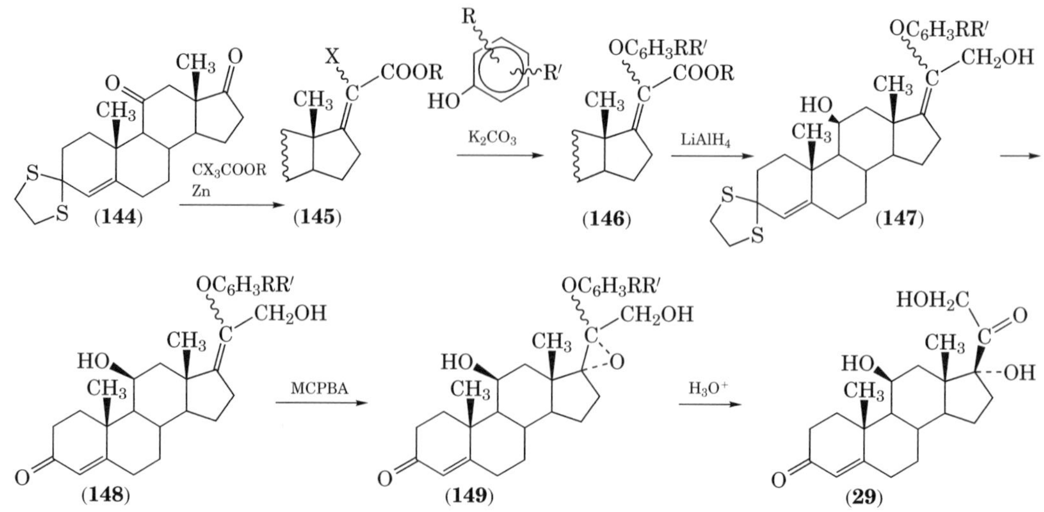

Fig. 20. Further C_{17} chemistry. The disubstituted phenol that reacts with (**145**) may be represented as $C_6H_3RR'OH$.

Mineralocorticoid antagonists are used as diuretics and antihypertensive agents. The introduction of a spirolactone function at C17 provides for the potent antimineralocorticoid activity. The commercial synthesis of the potassium-sparing diuretic spironolactone is outlined in Figure 21. The C17-ethynylated compound $17\beta H$-pregn-5-en-20-yne-3β-dio [*3604-60-2*] (**151**) is carboxylated with carbon dioxide and base to provide (**152**). Catalytic reduction of (**152**) followed by acid-catalyzed cyclization forms (**153**). A second catalytic hydro-

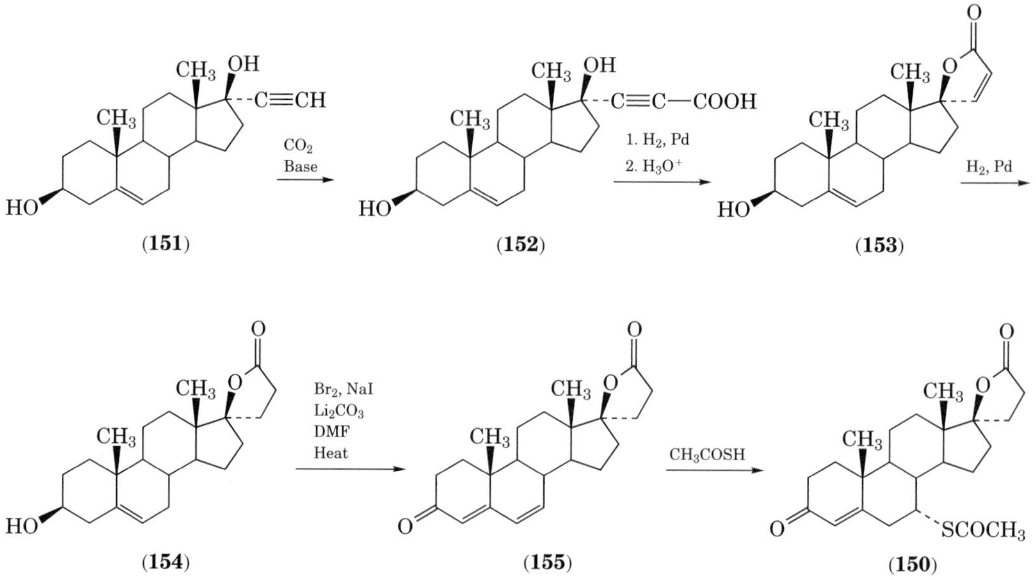

Fig. 21. Industrial synthesis of spironolactone [*52-01-7*] (**150**).

genation produces the spirolactone (**154**). Direct oxidation of (**154**) yields diene (**155**). Conjugate addition of thiolacetic acid to (**155**) affords the orally active spironolactone (**150**) (182).

In addition to its antimineralocorticoid activity, spironolactone (**150**) also possesses to some degree such undesirable effects as progestinal and antiandrogenic activity. Newer mineralocorticoid antagonists have been prepared in efforts to design agents free of these usually unwanted side effects (183). The synthesis of one such agent, spirorenone (**156**), is shown in Figure 22. This synthesis demonstrates an efficient method for C17-spirolactone construction. Microbiological oxidation of the dehydroepiandrosterone derivative (**157**) [67572-65-0] provides the 7β-hydroxylated compound (**158**). Selective acylation of the C3-hydroxyl group (**159**), followed by vanadium-catalyzed epoxidation affords (**160**) as the β-epoxide. Chlorination of the 7-hydroxyl moiety with triphenyl phosphine and carbon tetrachloride forms (**161**). Sequential reductive elimination (**162**) followed by saponification yields (**163**). Cyclopropanation of (**163**) with the Simmons-Smith reagent gave exclusively the β-cyclopropane (**164**). The C17-spirolactone construction begins with the addition of the dianion of propargyl alcohol to the C17-ketone of (**164**) to form (**165**). Palladium-catalyzed reduction

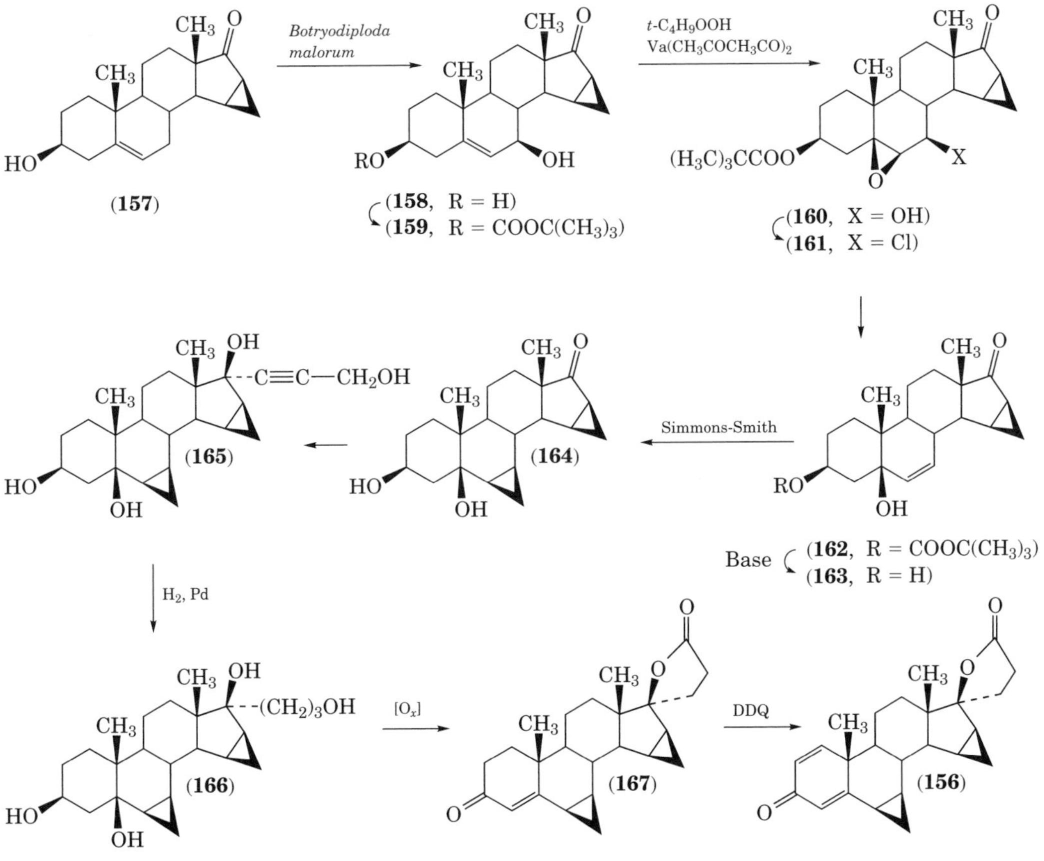

Fig. 22. Synthesis of spirorenone [74220-07-8] (**156**) where DDQ = 2,3-dichloro-5,6-dicyano-1,4-benzoquinone.

of (**165**) forms (**166**). Treatment of (**166**) with pyridinium chlorochromate simultaneously oxidizes the C3-alcohol to a ketone, oxidizes the primary alcohol to an acid, eliminates the 5-hydroxy-3-ketone to an ene-one, and cyclizes the D-ring hydroxy acid to a lactone to form (**167**). Dehydrogenation of the C1–C2 bond with 2,3-dichloro-5,6-dicyano-1,4-benzoquinone (DDQ) yields spirorenone (**156**) (184).

The stereocontrolled syntheses of steroid side chains for ecdysone, crustecdysone, brassinolide, withanolide, and vitamin D_3 have been reviewed (185). Also, other manuscripts, including reviews on the partial synthesis of steroids (186), steroid drugs (187–189), biologically active steroids (190), heterocyclic steroids (191), vitamin D (192), novel oxidations of steroids (193), and template-directed functionalization of steroids (194), have been published.

TOTAL SYNTHESIS

Estranes. Investigations into the total synthesis of steroids began in the 1930s shortly after the precise formula for cholesterol was established. The earliest studies focused on equilenin (**5**) because of its relative stereochemical simplicity when compared to other steroid nuclei. Initially, equilenin was synthesized in 20 chemical steps with an overall yield of 2.7%. This synthesis helped to confirm the perhydro-1,2-cyclopentenophenanthrene ring system of the steroid nucleus (195). Estrone was the second natural steroid to be synthesized from nonsteroidal starting materials in 0.1% overall yield in 18 steps (196). Many of the latter steps in this synthesis of estrone were essentially the same as those for the earlier synthesis of equilenin.

Since these original processes, a vast number of total syntheses of aromatic A-ring steroids have appeared (197,198). A highly efficient synthesis of (±)-9,11-dehydroestrone methyl ether (**177**), involving a tandem Cope-Claisen rearrangement, has been reported (199). Alkylation of the anion of methyl 2-methyl-2-cyclopentene-1-carboxylate [25662-31-1] (**169**) with bromide 4-bromomethyl-7-methoxy-1,2-dihydronaphthalene [83747-47-1] (**168**) produces ester (**170**, R = $COOCH_3$) in 94% yield. Reduction of (**170**) with lithium aluminum hydride yields (**171**, R = CH_2OH). Treatment of alcohol (**171**) with vinyl ethyl ether and mercury(II) acetate provides (**172**) in an overall yield of 87% from (**170**). Thermolysis of (**172**) at approximately 370°C undergoes a Cope

rearrangement to form (**173**), followed by a Claisen rearrangement producing a 2/1 mixture of diastereomeric aldehydes at C13. Separation of the major product from this rearrangement affords the desired diastereomer (**174**, X = CH_2; Y =

αH) in 35% yield. Ozonolysis of (**174**) results in tricarbonyl (**175**, X = O; Y = αH) in 70% yield. Epimerization of the hydrogen on the α-face of C8 of (**175**) with a methanol–sodium methoxide solution yields (**176**, X = O; Y = βH) in 69% yield. This compound contains the desired C8, C13, and C14 relative stereochemistry.

A McMurry coupling of (**176**, X = O; Y = βH) provides ($\pm$)-9,11-dehydroesterone methyl ether [*1670-49-1*] (**177**) in 56% yield. 9,11-Dehydroestrone methyl ether

(**177**) can be converted to estrone methyl ether by stereoselective reduction of the C_9–C_{11} double bond with triethyl silane in trifluoroacetic acid. In turn, estrone methyl ether can be converted to estradiol methyl ether by sodium borohydride reduction of the C17 ketone (199,200).

An asymmetric synthesis of estrone begins with an asymmetric Michael addition of lithium enolate (**178**) to the scalemic sulfoxide (**179**). Direct treatment of the crude Michael adduct with *meta*-chloroperbenzoic acid to oxidize the sulfoxide to a sulfone, followed by reductive removal of the bromine affords (**180**, X = α and βH; R =H) in over 90% yield. Similarly to the conversion of (**175**)

to (**176**), base-catalyzed epimerization of (**180**) produces an 85% isolated yield of (**181**, X = βH; R = H). C8 and C14 of (**181**) have the same relative and absolute stereochemistry as that of the naturally occurring steroids. Methylation of (**181**) provides (**182**). A (CH$_3$)$_2$CuLi-induced reductive cleavage of sulfone (**182**) followed by stereoselective alkylation of the resultant enolate with an allyl bromide yields (**183**). Ozonolysis of (**183**) produces (**184**) (wherein the aldehydric oxygen is

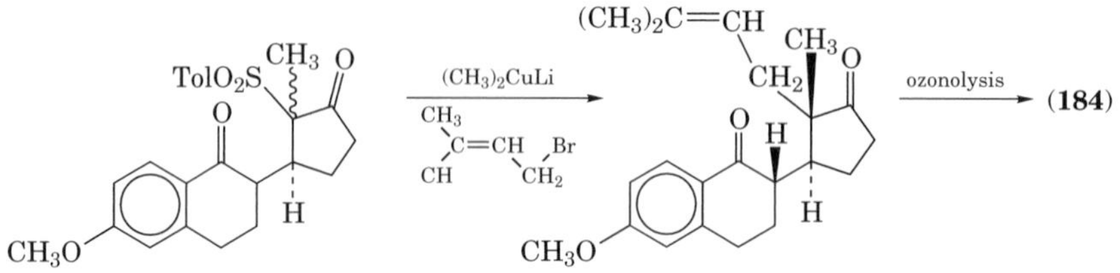

by isopropylidene) in 68% yield. Compound (**184**) is the optically active form of Ziegler's intermediate (**176**), and is converted to (+)-estrone in 6.3% overall yield and >95% enantiomeric excess (200).

The most recent, and probably most elegant, process for the asymmetric synthesis of (+)-estrone applies a tandem Claisen rearrangement and intramolecular ene-reaction (Fig. 23). Stereochemically pure (**185**) is synthesized from (2R)-1,2-O-isopropylidene-3-butanone in an overall yield of 86% in four

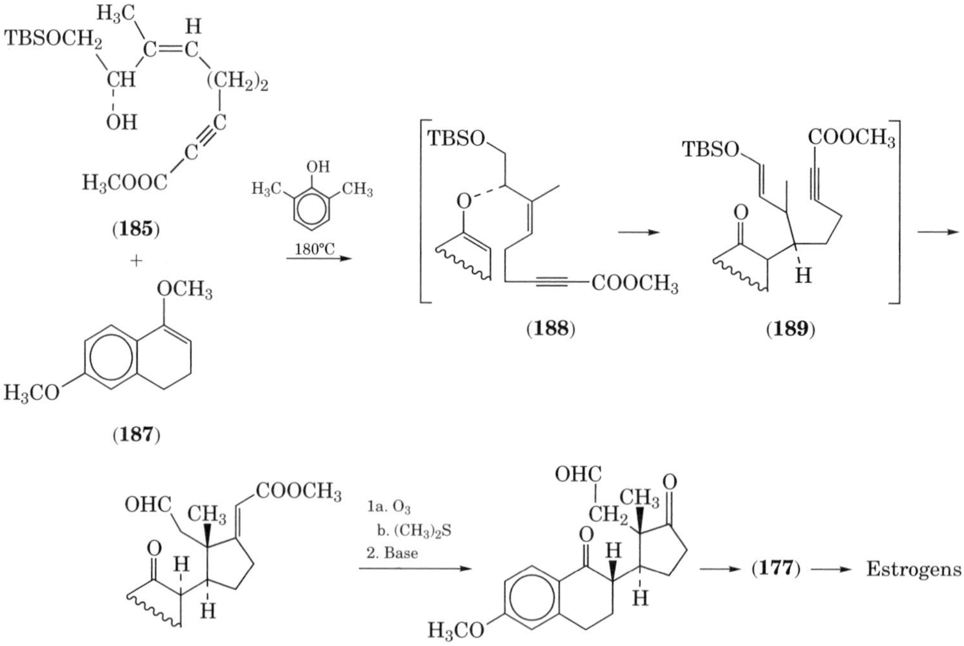

Fig. 23. Asymmetric total synthesis of estrogens, where TBS = *tert*-butyldimethylsilane.

chemical steps. Heating a toluene solution of (**185**), enol ether (**187**), and 2,6-dimethylphenol to 180°C in a sealed tube for 60 h produces (**190**) in 76% yield after purification. Ozonolysis of (**190**) followed by base-catalyzed epimerization of the C8α-hydrogen to a C8β-hydrogen (again similar to conversion of (**175**) to (**176**)) produces (**184**) in 46% yield from (**190**). Aldehyde (**184**) was converted to 9,11-dehydroestrone methyl ether (**177**) as discussed above. The overall yield of 9,11-dehydroestrone methyl ether (**177**) was 17% in five steps from 6-methoxy-1-tetralone (**186**) and (**185**) (201).

Most 19-norsteroid contraceptive agents are produced by total synthesis from nonsteroidal starting materials. A large number of syntheses of 19-norsteroids have been reported (202). An industrial synthesis of 19-norsteroids that is based on the classical Torgov (203) synthesis of aromatic steroids is shown in Figure 24. The addition of vinyl magnesium chloride (**192**) to 6-methoxy-1-tetralone (**186**) produces (**191**). The acidity of the 1,3-dione in 2-alkyl cyclopentadione (**200**, R = CH_3 or C_2H_5) catalyzes the addition of (**200**) to the vinyl carbinol (**191**), forming the tricyclic compound (**193**). The alkyl substituent or R-group of the cyclopentadione is either methyl or ethyl for the synthesis of estradiol and d-norgestrel (**209**), respectively. Microbial reduction of the prochiral secosteroid (**193**) produces the 13(R),17(S)-hydroxyketone (**194**). Acid-mediated cyclization

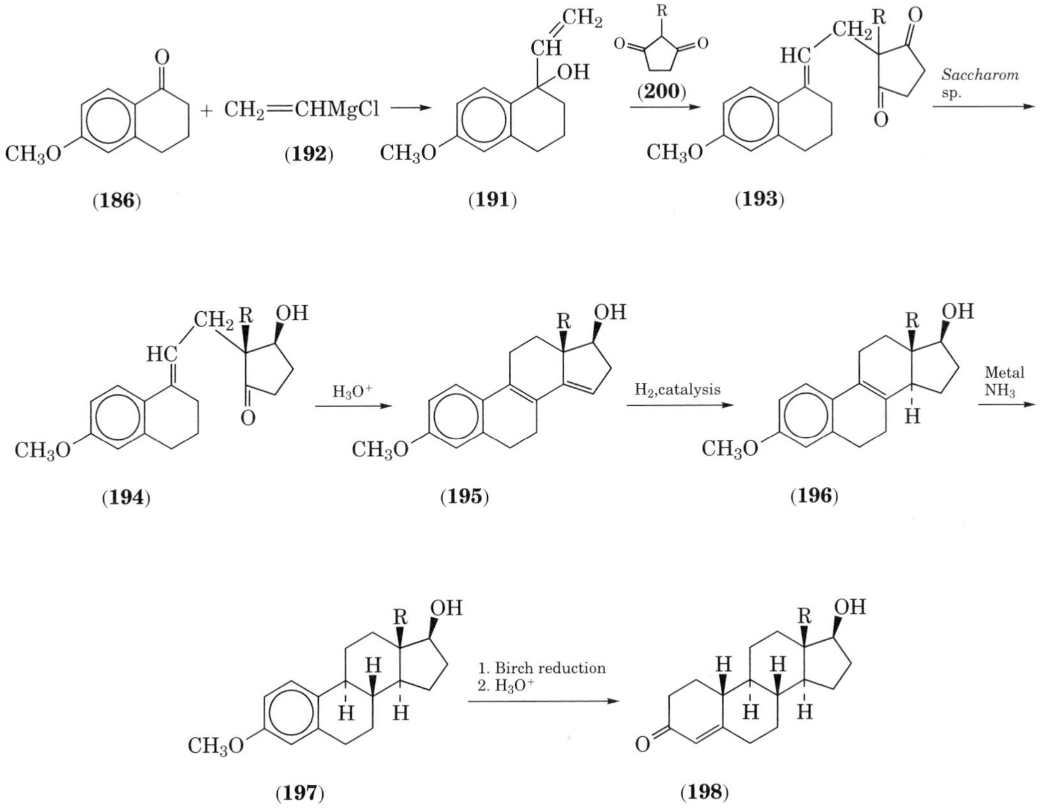

Fig. 24. Industrial total synthesis of gestrogens. For (**195**, R = CH_3) [4858-90-6]; (**195**, R = C_2H_5) [14507-45-0]. For (**196**, R = CH_3) [6733-79-5]; (**196**, R = C_2H_5) [7443-72-3].

of (**194**) forms steroids (**195**). Catalytic hydrogenation of the 14–15 double bond from the face opposite to the C18 substituent yields (**196**). Compound (**196**) contains the natural steroid stereochemistry around the D-ring. A metal-ammonia reduction of (**196**) forms the most stable product (**197**) thermodynamically. When R is equal to methyl, this process comprises an efficient total synthesis of estradiol methyl ester. Birch reduction of the A-ring of (**197**) followed by acid hydrolysis of the resultant enol ether allows access into the 19-norsteroids (**198**) (204).

Another efficient synthesis of d-norgestrel (**209**) begins with the condensation of 2-ethyl-1,3-cyclopentanedione (**200**, R = C_2H_5) with methyl vinyl ketone (**199**), producing (**201**). An asymmetric, intramolecular aldol condensation of (**201**) that is catalyzed by (S)-$(-)$-proline followed by an acid-catalyzed dehydration yields hydrindandione (**202**) in 97% optical purity (205). Condensation of (**202**) with formaldehyde and benzenesulfinic acid generates (**203**) in 85% yield.

(**200**) ⟶ (**199**) ⟶ (**201**)   1. (S)-$(-)$-Proline   2. H_3O^+ ⟶ (**202**) ⟶ (**203**)

In acid solution, the double bond of (**203**) is hydrogenated to the trans-fused sulfone (**204**). Presumably, this hydrogenation goes through a cis-fused intermediate that is rapidly epimerized to (**204**) under the acidic conditions of the reaction. Condensation of the sodium salt of 7,7-ethylenedioxy-3-oxooctanoate (**205**) with (**204**) produces (**206**). Crude (**206**) is cyclized, hydrolyzed, and decarboxylated,

(**204**) + (**205**)   NaH ⟶ (**206**)

producing the tricyclic compound (**207**). Hydrogenation of (**207**) followed by ketal hydrolysis and cyclization affords (**208**) in an overall yield of 35% from hydrindandione (**203**). d-Norgestrel [797-63-7] (**209**) is obtained by ethynylation of (**208**) (206).

(207)   (208)   (209)

Androstanes and Pregnanes.   The first total syntheses of nonaromatic steroids that contain the C19-angular methyl substituent were accomplished in the early 1950s. These syntheses all began with starting materials containing a two-ring system. The remaining two rings were appended through annulation sequences. For example, the Robinson (207) total synthesis began with the B/C-ring system followed by annulation of the A-ring and finally the D-ring. Although this synthesis consisted of approximately 40 chemical steps in an overall yield below $10^{-4}\%$, by demonstrating the possibility of the total synthesis of nonaromatic steroids it nevertheless has great historical value. Likewise, the basic plan of Sarrett's (208) first total synthesis of cortisol was based on the B–C to A–D-ring strategy in an overall yield of about 0.14% in approximately 30 chemical steps. The Woodward (209) total synthesis began with the C–D-ring system followed sequentially by the formation of the B- and A-ring systems (210).

A more recent ring annulation strategy for the total synthesis of steroids from the Stork group is shown in the following equation (**210 → 211 → 212**) and Figure 25. This synthesis begins with the formation of the C–D-ring system as a suitably functionalized indane. Condensation of the pyrrolidine enamine of cyclopentanone with ene-one (**210**) results in the bicyclic keto-ester (**211**) in 60–70% yield. Treatment of (**211**) with isopropenyl acetate and sulfuric acid produces a dienol acetate. Treatment of this dienol acetate with acetic anhydride and boron trifluoride etherate forms (**212**) as the major product. Reduction of

(210)   (211)   (212)

ene-one (**212**) to an allylic alcohol with sodium borohydride, followed by enol acetate cleavage and conjugation of the double bond produces (**213**). Conjugate addition of diethyl aluminum cyanide to (**213**), followed by protection of the ketone as the 1,3-dioxolane, affords (**214**) as a single diastereomer. Concomitant reduction of the cyano and ester moieties of (**214**) with lithium

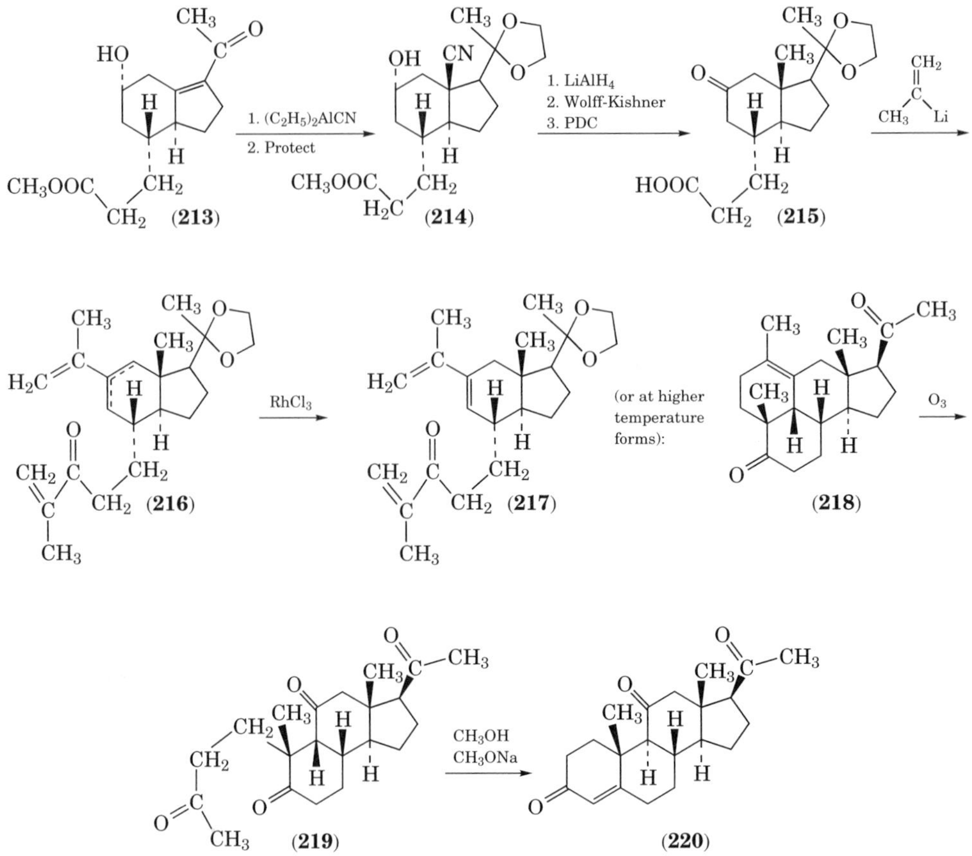

Fig. 25. Total synthesis of 11-keto-progesterone [516-15-4] (**220**).

aluminum hydride forms a diol with an angular imine moiety. Wolff-Kishner reduction of this resultant imine converts the angular imine to a methyl group. Oxidation with pyridinium dichromate (PDC) yields the keto acid (**215**). Treatment of (**215**) with isopropenyl lithium affords the isomeric diene (**216**) after dehydration. This mixture of dienes (**216**) is isomerized to diene (**217**) with catalytic ruthenium trichloride. Diene (**217**) is arranged for an intramolecular Diels-Alder reaction. More conveniently, treatment of dienes (**216**) with a catalytic amount of ruthenium trichloride in refluxing ethanol forms the endo Diels-Alder adduct (**218**) as the major isomer in 93% yield. Ozonolysis of (**218**) affords the trione (**219**) that contains the unnatural stereocenter at C9. Treatment of (**219**) with sodium methoxide in methanol epimerizes C9 to the thermodynamically most stable stereoisomer and cyclizes the A-ring to form 11-keto-progesterone (**220**) in 60% overall yield from indanonepropionic acid (**215**) (211).

A similar intramolecular Diels-Alder strategy was employed in an efficient synthesis to an appropriately functionalized hydrindanone nucleus (212). After functionalization, Diels-Alder cyclization, and appropriate functional group manipulation, this hydrindanone was converted into (±)-cortisone. The overall process afforded (±)-cortisone in a total of 18 chemical steps in approximately 3% yield.

A third variation of this strategy has been applied to an enantioselective total synthesis of cortisone. From an appropriately functionalized, scalemic hydrindan that possessed an 11-oxo-group and a masked corticoid side-chain, (+)-cortisol was produced in an 11-step total synthesis (213).

Several additional Diels-Alder cycloaddition strategies have been applied to the total synthesis of the steroid skeleton (214). For example, the first enantioselective synthesis of (+)-cortisone (**31**) was accomplished by the intramolecular [4+2] cycloaddition of an olefinic o-quinodimethane that contained an optically active stereodirecting group as the key chemical step. Condensation of aldehyde (**221**) with a lithiated oxathiane forms a 96% yield of alcohol (**222**, R' = OH; R'' = H); Swern oxidation of (**222**) produces ketone (**223**, R' = R'' = O). Addition of

(**221**)   (**222**)
                        (**223**)

2-propene magnesium bromide to ketone (**223**) yields (**224**) as a single stereoisomer at the isopropenyl alcohol and as a mixture of two diastereomers at the benzocyclobutenylic position; the absolute configuration at the benzocyclobutenylic position is inconsequential because this stereocenter is lost during thermolysis of (**224**). The thermal reaction of (**224**) proceeds through an intermediate o-quinodimethane (**225**), followed by a Diels-Alder [2+4] cycloaddition, producing the tricyclic compound (**226**) as the only detectable isomeric product in

(**224**)   (**225**)   (**226**)

quantitative yield. Protection of the C17 tertiary alcohol with methoxymethyl chloride (MOMCl) followed by oxidative hydrolysis of the C18 chiral auxiliary produces aldehyde (**227**, R = CHO). Reduction of (**227**) with sodium borohydride affords alcohol (**228**, R = CH_2OH). Birch reduction of (**228**) followed by acid-catalyzed hydrolysis of the resultant enol ether affords, in about 75% yield, the thermodynamically favored stereoisomer (**229**, R = H) as a single product. Diol (**229**) is converted into acetonide (**230**, R = isopropylidene), quantitatively. Allylic oxidation of (**230**) followed by pyridinium chlorochromate (PCC) mediated oxidation of the resultant mixture of allylic alcohols produces ene-dione (**231**). A 1,3-dipolar cycloaddition of diazomethane to (**231**)

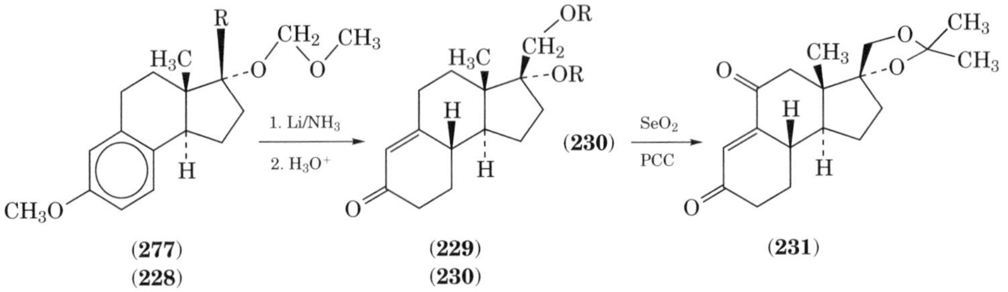
1. Li/NH3
2. H3O+
SeO2
PCC
(230)
(277)
(228)
(229)
(230)
(231)

followed by thermolysis forms (**232**). Compound (**232**) contains the C19 angular methyl substituent and an oxygen at C11. Stereoselective A-ring formation, through a known procedure (215), produces (**234**). The steroid (**234** [*128802-55-1*])

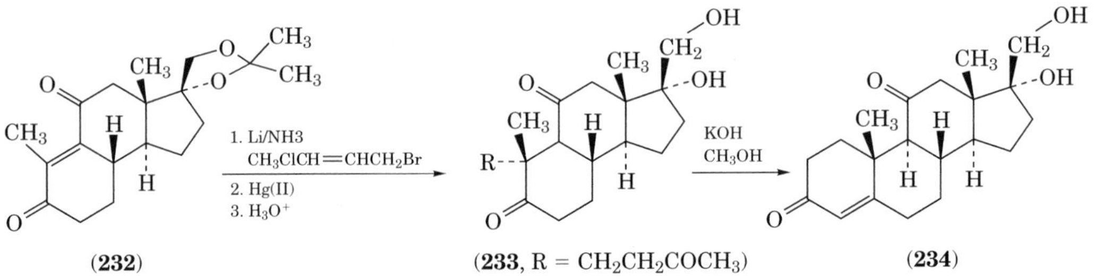
1. Li/NH3
CH3ClCH=CHCH2Br
2. Hg(II)
3. H3O+
KOH
CH3OH
(232)
(233, R = CH2CH2COCH3)
(234)

contains the basic skeleton of (+)-cortisone (**31**), therefore the total synthesis of (**31**) is completed by C3-ketone protection, side-chain manipulations, and final deprotection (**234**) → (**235**) → (**31**) (216,217).

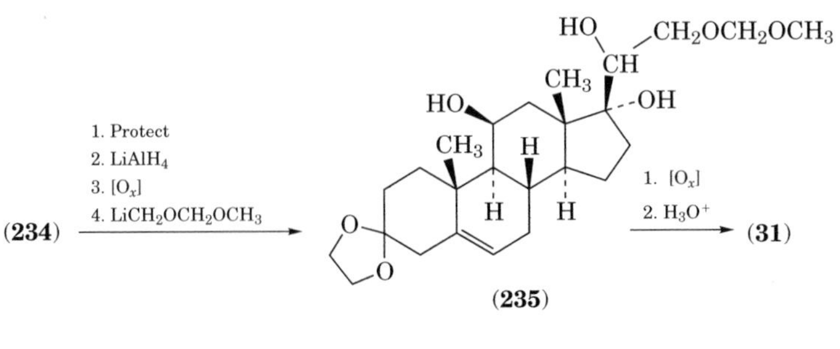
(234)
1. Protect
2. LiAlH4
3. [Ox]
4. LiCH2OCH2OCH3
1. [Ox]
2. H3O+
(31)
(235)

Other approaches to the stereoselective total synthesis of nonaromatic steroids include the carbocationic, biomimetic cyclization reactions. Generally, these cyclizations begin with the synthesis of an appropriately functionalized cyclopentenol. Acid-catalyzed cyclization forms the B–C–D rings of the steroid nucleus with the natural relative stereochemistry in a single step. Methods for the stereoselective synthesis of acyclic and monocyclic substrates and reaction conditions for their cyclization have been established (218). Optimization of these carbocationic cyclizations is accomplished by manipulating the initiating and the terminating groups. Figure 26 illustrates a carbocationic cyclization approach. The addition of diketal (**236**) to aldehyde (**237**) produces alcohol (**238**). Compound (**238**) is converted into aldehyde (**239**) after several functional group manipulations. Treatment of (**239**) with the anion of phosphonate (**240**) forms (**241**) as a 2:3 mixture of (*E*)- and (*Z*)-isomers. Compound (**241**) is converted to enone (**242**) and then to the desired carbinols (**243**) and (**244**) through established methodology (219). Compounds (**243**) and (**244**) contain an allylic cyclopentenol as a carbo-

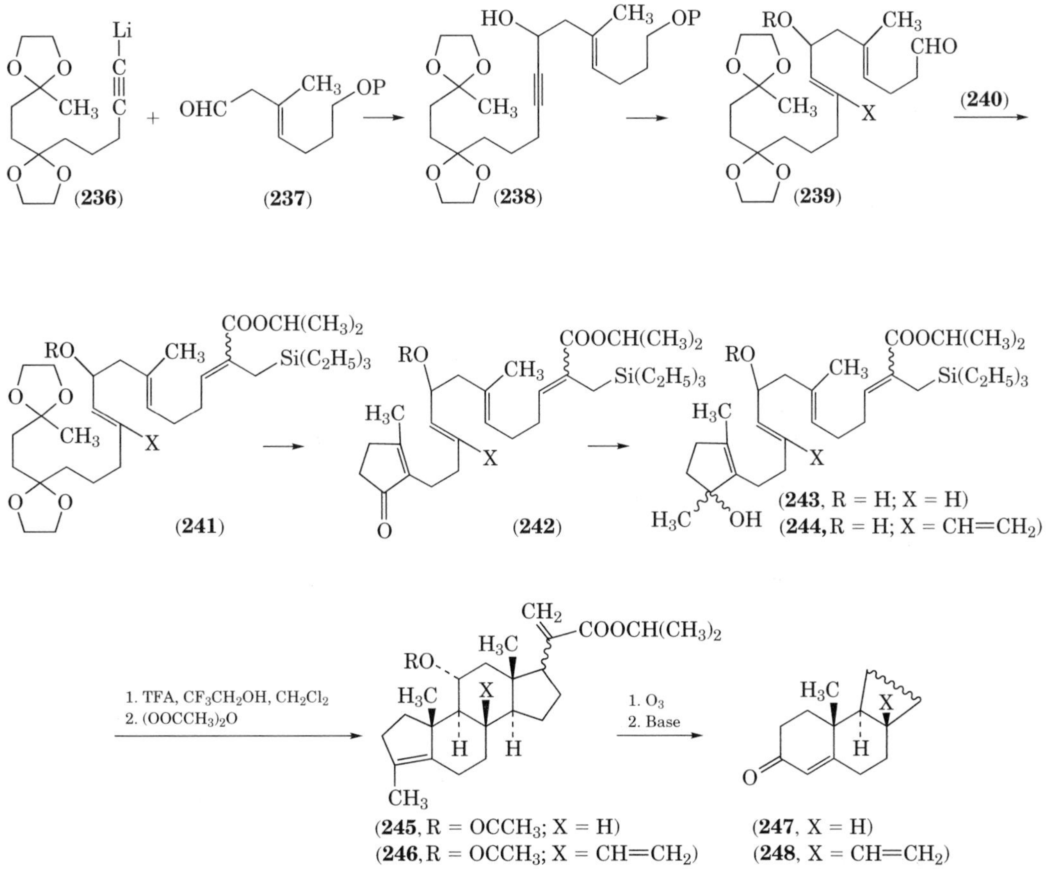

Fig. 26. Representative biomimetic cyclization where P = a protecting group, TFA = trifluoroacetic acid, and (**240**) = $(C_2H_5)_3SiCH_2\overset{|}{\underset{}{C}}H-\overset{O}{\overset{\|}{P}}(OC_2H_5)_2$

cation initiating group and allylic silanes as a terminating group. Treatment of (**243**) and (**244**) with trifluoroacetic acid (TFA) in 1:1 2,2,2-trifluororethanol and dichloromethane, followed by acetylation of the C17-alcohol, produces (**245**) and (**246**) in 20 and 80% yields, respectively (220). Previously, similar steroids that contain this methyl cyclopentene A-ring have been converted to the natural steroid A-ring by oxidative cleavage of the double bond, followed by base-catalyzed cyclization (221).

In the above carbocationic cyclizations (see Fig. 26), a C8β-vinyl substituent enhances the rate and overall yield of the reaction by providing a cation-stabilizing auxiliary. Chemical routes were explored to remove this C8β-vinyl substituent. These pathways required many chemical steps and were not very efficient. Therefore, fluoro-olefins were studied as removable cation-stabilizing auxiliaries (Fig. 27) (222). These studies culminated in the synthesis of *dl-β-*amyrin (**256**). Starting with mesityl oxide, fluoro-dienol (**249**) was prepared in nine steps in a 20% overall yield. Compound (**249**) was converted to cy-clopentenol (**250**) in approximately 18% overall yield in 15 chemical steps. Compound (**250**) contains an allylic cyclopentenol initiating group, a fluoro-olefin

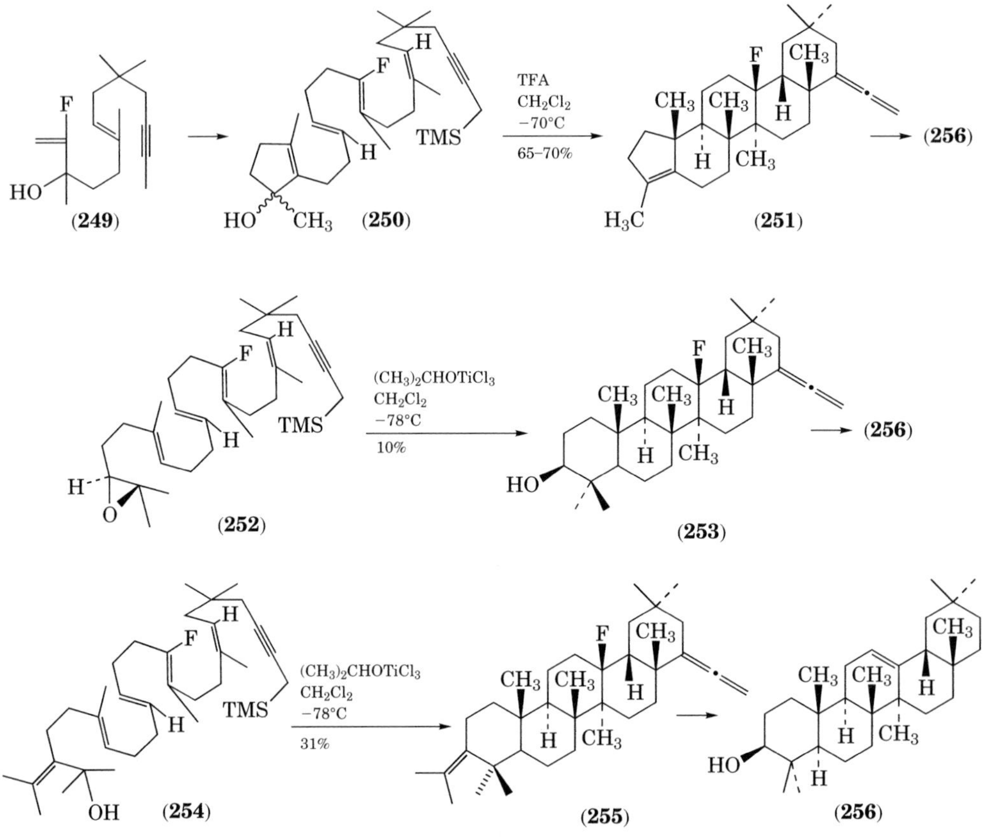

Fig. 27. Total synthesis of β-amyrin [*559-70-6*] (**256**) via a carbocationic cyclization, where TMS = trimethylsilyl.

cation-stabilizing auxiliary, and a propargyl silane-terminating group. Treatment of (**250**) with trifluoroacetic acid in dichloromethane produced the pentacyclic compound (**251**) in 65–70% yield. Four rings of (**251**) bearing seven chiral centers were formed during this cyclization step. Compound (**251**) was converted to *dl-β*-amyrin (**256**) by a series of chemical steps that included oxidative cleavage of the double bonds, base-catalyzed cyclization to form the six-membered A-ring, elimination of the C13-fluorine group, and dimethylation at C4 (223). The completion of the A-ring transformations, especially dimethylation at C4, was problematic. Therefore, other carbocation-initiating groups were studied. First, an epoxide was used as an initiating group. Treatment of polyene (**252**) with a Lewis acid in dichloromethane at −78°C provides pentacycle (**253**) in 10% yield. Steroid (**253**) contains the A-ring functionality of β-amyrin (**256**). In an improved process, an allylic alcohol-initiated polycyclization was investigated. Treatment of polyene (**254**) with a Lewis acid in dichloromethane at −78°C forms pentacycle (**255**) in 31% yield. This cyclization product (**255**) is suitably functionalized at C3, C13, and C22 to allow conversion to β-amyrin (**256**) with a minimum of synthetic manipulations (224).

Other, removable cation-stabilizing auxiliaries have been investigated for polyene cyclizations. For example, a silyl-assisted carbocation cyclization has been used in an efficient total synthesis of lanosterol. The key step, treatment of (**257**) with methyl aluminum chloride in methylene chloride at −78°C, followed by acylation and chromatographic separation, affords (**258**) in 55% yield (two steps). When this cyclization was attempted on similar compounds that did not contain the C7β-silicon substituent, no tetracyclic products were observed. Steroid (**258**) is converted to lanosterol (**77**) in three additional chemical steps (225).

In addition to cationic cyclizations, other conditions for the cyclization of polyenes and of ene-ynes to steroids have been investigated. Oxidative free-radical cyclizations of polyenes produce steroid nuclei with exquisite stereocontrol. For example, treatment of (**259**) and (**260**) with Mn(III) and Cu(II) afford the D-homo-5α-androstane-3-ones (**261**) and (**262**), respectively, in approximately 30% yield. In this cyclization, seven asymmetric centers are established in one chemical step (226,227). Another intramolecular cyclization reaction of iodo-ene

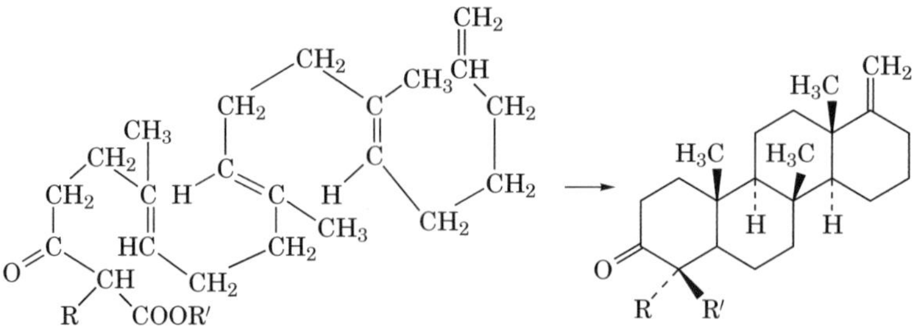

(**259**, R = CH$_3$; R′ = CH$_3$)

(**260**, R = H; R′ = C$_2$H$_5$)

(**261**, R = CH$_3$; R′ = COOCH$_3$)

(**262**, R = COOC$_2$H$_5$; R′ = H)

poly-ynes was reported using a carbopalladation cascade terminated by car-bonylation. This carbometalation–carbonylation cascade using CO at 111 kPa (1.1 atm) at 70°C converted an acyclic iodo–tetra-yne (**263**) to a D-homo-steroid nucleus (**264**) [162878-44-6] in approximately 80% yield in one chemical step (228). Intramolecular annulations between two alkynes and a chromium or

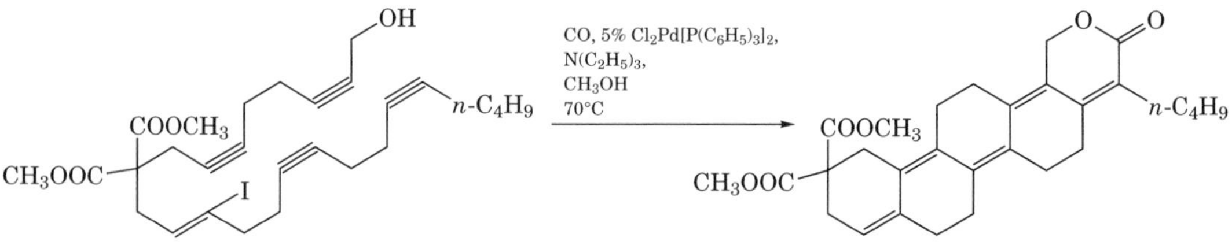

(**263**)

(**264**)

tungsten carbene complex have been examined for the formation of a variety of different fused-ring systems. A tandem Diels-Alder–two-alkyne annulation of a triynylcarbene complex demonstrated the feasibility of this strategy for the synthesis of steroid nuclei. Complex (**265**) was prepared in two steps from commercially available materials. Treatment of (**265**) with Danishefsky's diene in CH$_3$CN at room temperature under an atmosphere of carbon monoxide (101.3 kPa = 1 atm), followed by heating the reaction mixture to 110°C, provided (**266**) in

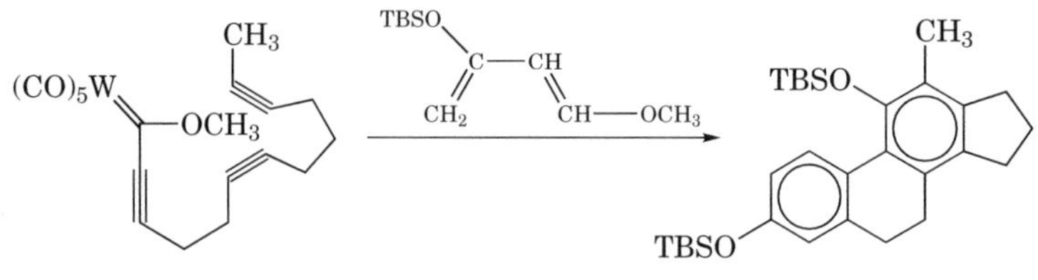

(**265**)

(**266**)

62% yield (TBS = *tert*-butyldimethylsilyl). In a second experiment, a sequential Diels-Alder–two-alkyne annulation of triynylcarbene complex (**267**) afforded a nonaromatic steroid nucleus (**269**) in approximately 50% overall yield from the acyclic precursors (**229**).

(**267**)   (**268**)

(**269**)

Besides the aforementioned A-ring aromatic steroids and contraceptive agents, partial synthesis from steroid raw materials has also accounted for the vast majority of industrial-scale steroid synthesis. One notable exception, however, was the first industrial-scale synthesis of optically active steroids performed by workers at Roussel-UCLAF. The linear synthesis began with a suitable B–C-ring synthon, 6-methoxy-1-tetralone (**186**). In a series of steps, tetralone (**186**) was converted to 2-methyl-2-cyanotetralone (**270**). Condensation of (**270**) with dimethyl succinate followed by carbonyl reduction, saponification, and resolution produced the optically active tricyclic acid (**271**). A series of

(**270**)   (**271**)

reductions, a decarboxylation, and a hydrolysis produced (**272**). Appendage of the A-ring functionality by alkylation produced intermediate (**273**). Compound (**273**) was used as a common intermediate for the synthesis of 19-norsteroids, estrogens, and corticosteroids (**230**).

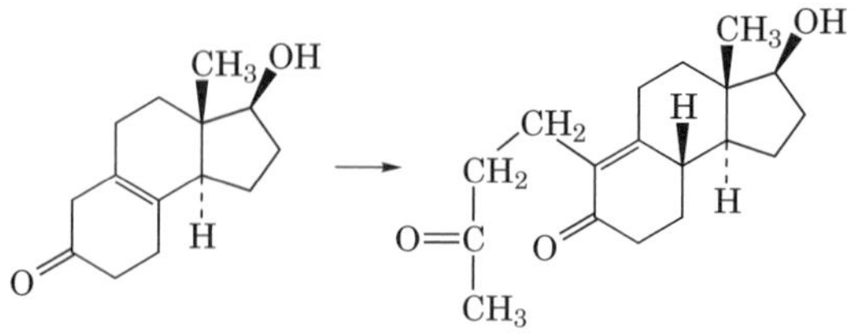

(272)   (273)

An interesting breakthrough in steroid endocrinology occurred with the discovery of a novel class of steroid antihormones. Several 11β-substituted 19-norsteroids display potent antiprogestinal activity. For example, RU-486 [84371-65-3] (**278**) is marketed in Europe as a contragestive agent. The synthesis of RU-486 demonstrates a unique method for functionalization of the 11β-position of a steroid nucleus. Condensation of the mono-ketal (**274**), available by either partial synthesis or total synthesis methods discussed above, with lithium propyne forms (**275**). Regiospecific epoxidation of (**275**) with hydrogen peroxide and hexafluoroacetone in methylene chloride produces epoxide (**276**) as a 3:1 mixture of 5α,10α-epoxide (**276**) and 5β,10β-epoxide, respectively. Pure epoxide (**276**) is isolated in approximately 60% yield after chromatographic purification. More recently, similar dienes have been epoxidized using a readily accessible pigment, Fe(II)-phthalocyane, as a catalyst to form a 10:1 mixture of a 5α,10α-epoxide and a 5β,10β-epoxide, respectively (231). In the key step, addition of a Grignard reagent to the unsaturated epoxide (**276**), via copper-catalyzed conjugate addition, produces the 11β-substituted steroid (**277**) in high yield. This reaction is extremely versatile for a variety of different copper-catalyzed organometallic reagents regardless of steric hindrance. Concomitant acid-catalyzed ketal hydrolysis and dehydration produces RU-486 (**278**) (232) (Fig. 28).

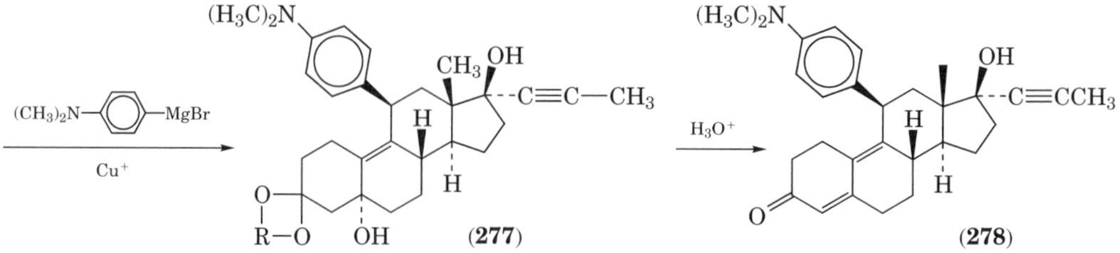

Fig. 28.   Industrial synthesis of RU-486.

Uses: Therapeutics and Toxicology

STEROID HORMONES

Sex Hormones. The largest economic impact of synthetic estrogen and progestin production has been for use as contraceptive agents and for treatment and prevention of osteoporosis. Mixtures of estrogens and progestins have been used as contraceptive agents since the early 1960s. The principal mode of steroid contraceptive action is exerted at the hypothalamic–pituitary–ovarian and uterine sites. Thus, contraceptive steroid mixtures have been used to treat a variety of related abnormal states including endometriosis, dysmenorrhea, hirsutism, polycystic ovarian disease, dysfunctional uterine bleeding, benign breast disease, and ovarian cyst suppression (233). One of two estrogens, ethinylestradiol [57-63-6] ((17α)-19-norpregna-1,3,5(10)-trien-20-yne-3,17-diol) or mestranol [72-33-3] ((17α)-3-methoxy-19-norpregna-1,3,5(10)-trien-20-yne-17-ol) is contained in most combination oral contraceptives. The progestin component in oral contraceptive pills is more variable. The progestin component can be progesterone derivatives that contain a C6-methyl group such as medroxyprogesterone [520-85-4] ((6α)-17-hydroxy-6-methyl-4-ene-3,20-dione) and megestrol acetate [595-33-5] ((17-hydroxy-6-methylpregna-4,6-diene-3,20-dione acetate). Also, 19-norsteroids, such as norethindrone [68-22-4] ((17α)-17-hydroxy-19-norpregn-4-en-20-yn-3-one) (**112**), norgestrel [6533-00-2] (13-ethyl-17-hydroxy-18,19-dinorpregn-4-en-20-yn-3-one) (**208**), norgestimate [35189-28-7] ((17α)-17-(acetyloxy)-13-ethyl-18,19-dinorpregn-4-20-yn-3-one) and gestodene [60282-87-3] ((17α)-13-ethyl-17-hydroxy-18,19-dinorpregnane-4,15-diene-20-yne-3-one) are used as orally active progestins. The C17-ethinyl moiety protects these steroids from metabolism and assists in their excellent pharmacokinetic profile (234–236).

Estrogens are routinely prescribed to post-menopausal women to prevent the development and exacerbation of osteoporosis because it can increase bone density and reduce fractures. Any increased risk of uterine cancer with the use of estrogen alone is practically eliminated by cyclic therapy with a progestinal agent. Estradiol (**25**) or conjugated estrogens, isolated from the urine of pregnant mares, are typical agents used for the prevention and treatment of osteoporosis (237).

Antiprogestins, such as RU-486 [84371-65-3] (17β-hydroxy-11β-(4-dimethylaminophenyl-1)-17α-(prop-1-ynyl)-estra-4,9-diene-3-one) (**278**) and ZK98299 [096346-61-1], (11β-(4-dimethylaminophenyl)-17α-hydroxy-17β-(3-hydroxypropyl-13α-methyl-4,9-gonadien-3-one) (**237**) represent a new class of drugs for fertility regulation. Also, these drugs have potential applications in the treatment of uterine cancer. An RU-486 and prostaglandin combination is accepted in Western Europe as a low resource method for early pregnancy termination (239).

During the 1960s and 1970s a wide range of estrogens and antiestrogens were synthesized primarily to study reproductive endocrinology. The focus of clinical applications of many of these antiestrogens has shifted to breast cancer therapy. These antiestrogens possess both steroidal, such as RU-58668 [151555-47-4] (11β-4-[5-(4,4,5,5,5-pentafluoropentylsulfonyl)pentyloxy]penyl]estra-1,3,5-(10)-triene-3,17β-diol (**279**) (240) and nonsteroidal, such as tamoxifen [10540-

29-1] (**280**, R = H), droloxifene [*82413-20-5*] (**281**, R = OH), and raloxifene [*84449-90-1*] (**282**), structures. Although structurally different, all of these

(**279**)   (**280**)   (**282**)

(**281**)

antiestrogens bind to the estrogen receptor in the breast cancer cell and exert a profound influence on cell replication. However, many of these compounds, eg, tamoxifene, may have other indirect mechanisms to control cell replication. Along with acting as estrogen antagonists in the breast, droloxifene and raloxifene are estrogen agonists in bone and are undergoing clinical trials for the treatment of osteoporosis (241,242).

Testosterone, alkylated testosterone, or testosterone esters are the primary anabolic–androgenic steroid drugs. Most of these synthetic testosterone derivatives were developed in the 1950s in failed attempts to separate the hormones' masculinizing (androgenic) and skeletal muscle-building (anabolic) effects. The difficulty in separating the androgenic and anabolic effects was probably experienced because both actions of these steroid derivatives are mediated by the same intracellular receptor. The medicinal uses for these drugs include treatment of certain types of anemias, hereditary angioedema, certain gynecological conditions, protein anabolism, certain allergic reactions, and use in replacement therapy in gonadal failure states. However, anabolic–androgenic steroids are best known for their nonmedical, and illegal, use to aid in body-building or to increase skeletal muscle size, strength, and endurance. The doses used by strength athletes often exceed testosterone-equivalent doses for replacement therapy by 10- to 100-fold. Consequently, a vast number of toxic and deleterious effects from these drugs have been reported. Among these adverse effects are impaired liver function including malignant hepatic tumors, impaired growth of prepubescent boys, altered sexual characteristics in both men and women, and prostatic enlargement. Other possible adverse effects include an increased risk for cardiovascular disease, colonic cancer, adenocarcinoma of the prostate, fatal rupture of a hepatic tumor, severe cystic acne, and psychosis (243).

Corticosteroids. The greatest portion of steroid drug production is aimed at the synthesis of glucocorticoids (244), which are highly effective agents for the treatment of chronic inflammation. Glucocorticoids exert their effects by binding to the cytoplasmic glucocorticoid receptor within the target cell and thus either increase or decrease transcription of a number of genes involved in the inflammatory process. Specifically, glucocorticoids down-regulate potential mediators of inflammation such as cytokines, certain cytokine receptors, inducible nitric oxide synthase (NOS), cyclooxygenase-2, endothelin-1, and phospholipase A_2 (PLA_2) and up-regulate agents that are potential inhibitors of inflammation such as lipocortin-1, β_2-adrenoreceptor, endonucleases, neutral endopeptidase (NEP), and angiotensin converting enzyme (ACE) (245).

The discovery that cortisone dramatically alleviates the symptons of arthritis led to intensive research on the antiinflammatory properties of corticosteroids. Glucocorticoids are used (ca 1996) to treat a variety of different diseases that are exacerbated by inflammation, such as arthritis, asthma, rhinitis, and skin irritations. Typical oral glucocorticoids used to treat rheumatoid arthritis are prednisone (**132**), 6α-methylprednisolone [83-43-2], and dexamethasone [50-02-2] (**283**). Systemic side effects of exogenous glucocorticoids include suppression of the pituitary–adrenal axis, resulting in a decreased response to stress which eventually normalizes after discontinuation, although the recovery can be delayed. An excess of exogenous glucocorticoids can result in symptoms of Cushing's disease, including weight gain, weakness, hypertension, and diabetes. In addition, excess exogenous glucocorticoid ingestion can cause increased bone reabsorption because of a negative calcium balance, delayed union fractures, secondary hyperparathyroidism, reduced bone formation, and sex hormone disturbances, all contributing to a significant increase in the risk of osteoporosis (246).

Corticosteroids are the most efficacious treatment available for the long-term treatment of asthma, and inhaled corticosteroids are considered to be a first-line therapy for asthma (247). In the early 1950s, cortisone (**31**) and cortisol (**29**) were used to treat asthma. However, drugs with fewer side effects and with

(**283**) (**284**)

a higher therapeutic index were sought. Prednisolone (**131**) and dexamethasone (**283**) were developed as orally active corticosteroids with increased glucocorticoid activity but decreased mineralocorticoid activity. Still, these drugs possessed unwanted systemic side effects. In 1972, betamethasone valerate and beclomethasone dipropionate [55340-19-8] (**284**) were introduced. These topically active corticosteroids had few side effects when administered by inhalation. Since 1972

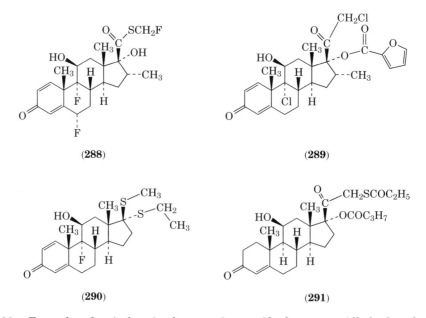

several corticosteroids have been designed to break down rapidly after reaching the systemic circulation. Therefore, these novel corticosteroids are less likely to suppress the hypothalamic–pituitary–adrenal axis. Some of these drugs that are in clinical development include fluticasone (**288**), mometasone (**289**), SQ-27239 (**290**), and JO-1222 (**291**) (Fig. 29) (248,249). Other corticosteroids include [*76-25-5*] (**285**, R = H triamcinolone acetonide), [*5611-51-8*] (**286**, R = OCCH$_2$C(CH$_3$)$_3$ triamcinolone hexacetonide), and dexflazacort [*14484-47-0*] (**287**).

Rhinitis is characterized by nasal stuffiness with partial or full obstruction, and itching of the nose, eyes, palate, or pharynx, sneezing, and rhinorrhoea. If left untreated it can lead to more serious respiratory diseases such as sinusitis or asthma. Although several types of drugs are available for treatment, nasal

Fig. 29.   Examples of topical antiasthma corticosteroids that are rapidly broken down in the circulation: fluticasone [*90566-53-3*] (**288**), monetasone [*83919-23-7*] (**289**), SQ-27239 [*85197-77-9*] (**290**), and JO-1222 [*98449-05-9*] (**291**).

spray topical corticosteroids are widely regarded as the reference standard in rhinitis therapy (250).

There are hundreds of topical steroid preparations that are available for the treatment of skin diseases. In addition to their aforementioned antiinflammatory effects, topical steroids also exert their effects by vasoconstriction of the capillaries in the superficial dermis and by reduction of cellular mitosis and cell proliferation especially in the basal cell layer of the skin. In addition to the aforementioned systemic side effects, topical steroids can have adverse local effects. Chronic treatment with topical corticosteroids may increase the risk of bacterial and fungal infections. A combination steroid and antibacterial agent can be used to combat this problem. Additional local side effects that can be caused by extended use of topical steroids are epidermal atrophy, acne, glaucoma and cataracts (thus the weakest concentrations should be used in and around the eyes), pigmentation problems, hypertrichosis, allergic contact dermatitis, perioral dermatitis, and granuloma gluteale infantum (251).

OTHER THERAPEUTIC STEROIDS

Saponins. Although the hypocholesterolemic activity of saponins has been known since the 1950s, their low potency and difficult purification sparked little interest in natural saponins as hypolipidemic agents. Synthetic steroids (**292**, **293**) that are structurally related to saponins have been shown to lower plasma cholesterol in a variety of different species (252). Steroid (**292**) is designated CP-88,818 [*99759-19-0*]. The hypocholesterolemic agent CP-148,623 [*150332-35-7*] (**293**) is not absorbed into the systemic circulation and does not inhibit enzymes involved in cholesterol synthesis, release, or uptake. Rather, (**293**) specifically inhibits cholesterol absorption into the intestinal mucosa (253). As of late 1996, CP-148,623 is in clinical trials as an agent that lowers blood concentrations of cholesterol (254).

(**292**, no carbonyl at C-11)
(**293**)

Heterocyclic Steroids. Steroid 5α-reductase (types 1 and 2) converts testosterone (**21**) to the physiologically more potent androgen dihydrotestosterone (DHT) (**22**). The type 1 isoform occurs in nongenital skin, whereas the type 2 isoform is the predominant form in the prostate (the type 1 isoform is present in a lesser extent) and genital skin fibroblasts. There has been much interest in developing inhibitors of steroid 5α-reductase as a therapy for a variety of disorders associated with elevated levels of DHT including benign prostatic hyperplasia (BPH), some prostatic cancers, certain skin disorders, and male pattern baldness. Among these inhibitors are 4-aza and 6-aza steroids. The 4-aza

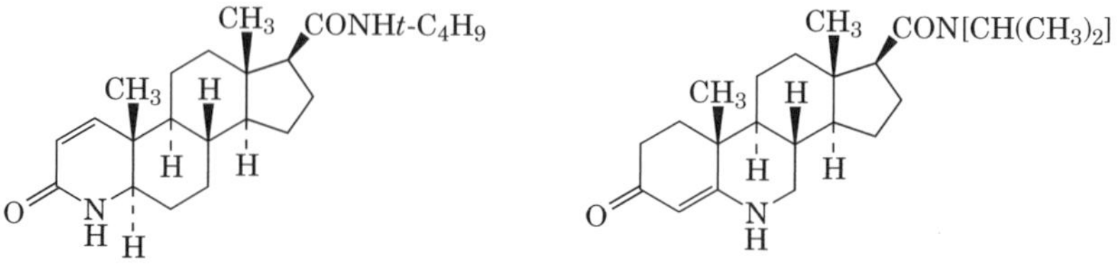

(294)

(295)

steroid, finasteride [*98319-26-7*] (**294**), shows selectivity for the human type 2 isoform and has been approved for the treatment of benign prostate hyperplasia (255).

Several steroids that contain C21 heterocycles (lazaroids) have been reported to be potent inhibitors of iron-dependent lipid peroxidation (Fig. 30). These compounds were evaluated as antioxidants, CNS neuroprotective agents, and antiasthmatic agents. This class of compounds does not appear to have any glucocorticoid activity. Thus, the precise mode of action for the antiasthmatic activity of lazaroids is under investigation. Tirilazad mesylate [*110101-67-2*] (U-74006F) (**296**) is undergoing clinical trials for head injury, subarachniod hemorrhage, and spinal cord trauma (256,257).

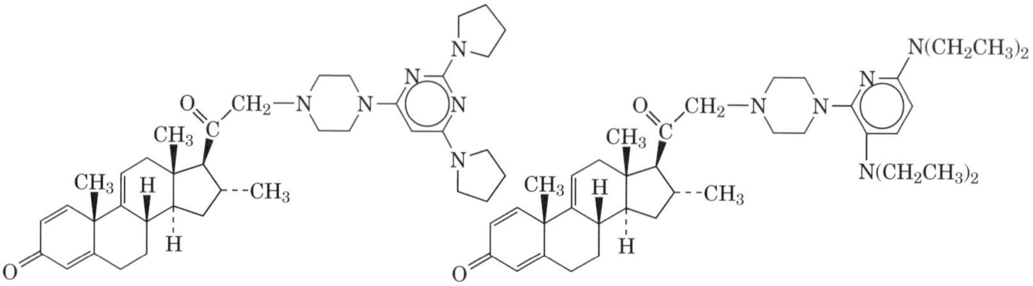

(296)

(297)

Fig. 30. C$_{21}$-Heterocyclic steroids that are inhibitors of iron-dependent lipid peroxidation: U-74006F [*110101-67-2*] (**296**) and U-74500A [*110101-65-0*] (**297**).

Economic Aspects

Raw Materials. In 1984, diosgenin accounted for at least 50% of the total steroid drug output worldwide (258); 1994 estimates suggest that approximately 60% of all steroids used as drugs are synthesized using diosgenin as the starting material (259). Until 1970, Mexico was the main source of diosgenin production. In the early 1970s the Mexican government's nationalization of the collection of *Dioscorea* plants together with a decrease in diosgenin content (6–4%) by overharvesting and increases in transportation costs, caused the price of a kilogram of diosgenin to rise from $10 to >$100 by the mid-1970s. Because of competition with other steroid raw materials, such as β-sitosterol, stigmasterol, and the solasidine alkaloids, and competition from other diosgenin-producing countries, the price of a kilogram of diosgenin fell to $25 by the early 1980s (260). As of 1991, the leading producers of diosgenin were China, Mexico, India, Guatemala, and Costa Rica. China, Brazil, and India are the leading exporters of hecogenin (261).

Therapeutics. The sale of steroid drugs is a multibillion dollar industry. The two largest selling groups of steroid drugs are the systemic sex hormones, including agents such as hormonal contraceptives, estrogens, progestins, and androgens, and the corticosteroids (Table 2). The 12-month period ending December 1994 saw systemic sex hormones account for approximately $5.4 billion whereas total corticosteroids accounted for about $5.7 billion in sales. Disregarding inflation, the sales of steroid drugs have climbed since the 1970s. For example, total worldwide sales of the systemic sex hormones and corticosteroids was around $7.4 billion in 1990 and approximately $11.2 billion in 1994.

In addition to the sex hormones and corticosteroids, other steroid drugs have substantial worldwide markets. For example, cytostatic hormones had worldwide sales of approximately $1.8 billion in 1994. Included in these $1.8 billion are several steroids or steroid-mimetics such as megestrol acetate and tamoxifen (**282**), respectively (262).

Table 2. Total Worldwide Sales of Systemic Sex Hormones and Corticosteroids

	Sales, $ \times 10^6$	
Steroid class	1990	1994
sex hormones, systemic	3,582	5,436
corticosteroids		
topical	1,558	1,891
systemic	903	1,181
respiratory	988	2,170
nasal	382	665
inhalants, systemic	606	1,505
steroids for sensory organs	396	507
Total	*7,427*	*11,185*

Analytical Methods

The field of steroid analysis includes identification of steroids in biological samples, analysis of pharmaceutical formulations, and elucidation of steroid structures. Many different analytical methods, such as ultraviolet (uv) spectroscopy, infrared (ir) spectroscopy, nuclear magnetic resonance (nmr) spectroscopy, x-ray crystallography, and mass spectroscopy, are used for steroid analysis. The constant development of these analytical techniques has stimulated the advancement of steroid analysis.

Data collected on the uv spectra of steroids are available in several books, spectrum atlases, and review articles (263). The most characteristic absorptions in steroid hormones include α,β-unsaturated ketones, conjugated dienes, and phenolic A-rings (264).

The most powerful method for structure elucidation of steroid compounds during the classical period of steroid chemistry (~1940–1950s) was ir-spectroscopy. As with the ultraviolet spectra, data collected on the infrared spectra of steroids are available in several books, spectrum atlases, and review articles (265,266). Unlike ultraviolet spectroscopy, even the least substituted steroid derivatives are relatively rich in characteristic absorption bands in infrared spectroscopy (264).

Generally, the most powerful method for structural elucidation of steroids is nuclear magnetic resonance (nmr) spectroscopy. There are several classical reviews on the one-dimensional (1-D) proton [1]H-nmr spectroscopy of steroids (267). ^{13}C-nmr, a technique used to observe individual carbons, is used for structure elucidation of steroids. In addition, ^{13}C-nmr is used for biosynthesis experiments with ^{13}C-enriched precursors (268). The availability of higher magnetic field instruments coupled with the arrival of 1-D and two-dimensional (2-D) techniques such as DEPT, COSY, NOESY, 2-D J-resolved, HOHAHA, etc, have provided powerful new tools for the structural elucidation of complex natural products including steroids (269).

A definitive method for structural determination is x-ray crystallography. Extensive x-ray crystal structure determinations have been done on a wide variety of steroids and these have been collected and listed (270). In addition, other analytical methods for steroid quantification or structure determination include, mass spectrometry (271), polarography, fluorimetry, radioimmunoassay (264), and various chromatographic techniques (272).

BIBLIOGRAPHY

"Sterols and Steroids" in *ECT* 1st ed., Vol. 12, pp. 917–947, by R. B. Turner, The Rice Institute, and L. I. Conrad, American Cholesterol Products, Inc.; "Steroids" in Suppl. Vol. 1, pp. 848–888, by G. Anner and A. Wettstein, Ciba, Ltd.; in *ECT* 2nd ed., Vol. 18, pp. 830–896, by D. Taub and T. B. Windholz, Merck, Sharp & Dohme; in *ECT* 3rd ed., Vol. 21, pp. 645–729, by G. R. Lenz, Searle Laboratories.

1. E. Heftman, *Steroid Biochemistry*, Academic Press, Inc., New York, London, 1970; L. F. Fieser and M. Fieser, *Steroids*, Reinhold Publishing Corp., New York, 1959; J. Fried and J. A. Edwards, eds., *Organic Reactions in Steroid Chemistry*, Vols. 1 and 2, Van Nostrand Rheinhold Co., New York, 1972; *Rodds's Chemistry of Carbon Compounds*, Vol. 2, Parts D and E, Elsevier Publishing Co., London, 1970; W. R. Butt, B. T. Rudd, and R. Morris, *Hormone Chemistry*, Vol. 2, Ellis Horwood Ltd., Chichester, U.K., 1976; J. A. Hogg, *Steroids*, **57**, 593 (1992); R. Hirschman, *Steroids*, **57**, 579 (1992); H. Herzog and E. P. Oliveto, *Steroids*, **57**, 617 (1992); F. B. Colton, *Steroids*, **57**, 624 (1992); C. Djerassi, *Steroids*, **57**, 631 (1992); K. Nakanishi, *Steroids*, **57**, 649 (1992); A. J. Birch, *Steroids*, **57**, 363 (1992); K. Bloch, *Steroids*, **57**, 378 (1992); J. Fried, *Steroids*, **57**, 384 (1992); P. A. Lehmann, *Steroids*, **57**, 403 (1992); E. R. H. Jones, *Steroids*, **57**, 357 (1992); A. Zaffaroni, *Steroids*, **57**, 403 (1992); E. R. H. Jones, *Steroids*, **57**, 357 (1992); A. Zaffaroni, *Steroids*, **57**, 642 (1992); M. Levitz, *Steroids*, **57**, 456 (1992); J. Gorski, *Steroids*, **57**, 240 (1992); G. Rosenkranz, *Steroids*, **57**, 409 (1992); S. Bernstein, *Steroids*, **57**, 392 (1992); C. Djerassi, *Steroid Reactions*,

Holden-Day, Inc., San Francisco, Calif., 1963; C. Djerassi, *Steroids Made it Possible*, American Chemical Society, Washington, D.C., 1990. For a list of steroid reviews prior to 1972 see G. P. Ellis, *Medicinal Chemistry Reviews*, Butterworth and Co. Ltd., Kent, U.K., 1972, pp. 71–80; J. H. Clark, W. T. Schrader, and R. T. O'Malley, in J. D. Wilson and D. W. Foster, eds., *Williams Textbook of Endocrinology*, W. B. Sanders Co., 1992; M. Sainsbury, ed., *Second Supplements to the 2nd Edition of Rodd's Chemistry of Carbon Compounds*, Vols. II B (Partial), C, D, and E, Elsevier Science BV, Amsterdam, the Netherlands, 1994.

2. Technical data, *The Pharmaceutical Market: World Review*, Vol. I, International, IMS Global Services, 1994, used with permission; C. Vezina, in *Basic Biotechnology*, J. Bu'Lock and B. Kristiansen, eds., Academic Press, Inc., New York, 1987, pp. 463–482.

3. M. deFourcroy, *Ann. Chim.*, 242 (1789).

4. H. Wieland, E. Dane, and E. Scholz, *Z. Physiol. Chem.* **211**, 261 (1932); A. Windaus, *Ber.* **36**, 3177 (1909).

5. H. Wieland, *Z. Physiol. Chem.* **80**, 287 (1912); J. D. Bernal, *Nature*, **129**, 277 (1932); J. D. Bernal, *Chem. Ind.* **51**, 466 (1932); H. Wieland and E. Dane, *Z. Physiol. Chem.* **210**, 268 (1932).

6. O. Rosenheim and H. King, *Nature*, **130**, 315 (1932); H. King, *Chem. Ind.* **51**, 954 (1932); H. King, **52**, 299 (1933).

7. O. Diels, W. Gädke, and P. Körding, *Ann.* **459**, 1 (1927); W. E. Bachmann, W. Cole, and A. L. Wilds, *J. Am. Chem. Soc.* **61**, 974 (1939); *Ibid.*, **62**, 824 (1940).

8. L. F. Fieser and M. Fieser, *Steroids*, Reinhold Publishing Corp., New York, 1959, p. 445.

9. P. A. Lehmann, A. Bolvar, and R. Quintero, *J. Chem. Ed.* **50**, 195 (1973); see also Ref. 1.

10. O. Hassel, *Tidsskr. Kjemi, Bergves.* **3**, 32 (1943); D. H. R. Barton, *Experimentia*, **6**, 316 (1950); W. Klyne and V. Prelog, *Experimentia*, **16**, 521 (1960).

11. IUPAC-IUB Committee on the Nomenclature of Organic Compounds, *Pure Appl. Chem.* **31**, 283 (1972).

12. IUPAC-IUB Joint Commission on Biochemical Nomenclature (JCBN), *Eur. J. Biochem.* **186**, 429 (1989); *Pure Appl. Chem.* **61**, 1783 (1989).

13. Nomenclature Committee of IUBMB (NC-IUBMB) and IUPAC-IUBMB JCBN: *Eur. J. Biochem.* **204**, 1 (1992); *Eur. J. Biochem.* **183**, 1 (1989); *Biochem. J.*, **165**, I-IV (1990); *Arch. Biochem. Biophys.* **272**, 262 (1989); *J. Am. Chem. Soc.* **82**, 5577 (1960); *Steroids*, **13**, 227 (1969); *J. Org. Chem.* **34**, 1547 (1969); *Arch. Biochem. Biophys.* **147**, 4 (1971).

14. Ref. 8, pp. 26–53 and 341–364.

15. G. F. Gibbons, K. A. Mitropoulos, and N. B. Myant, *Biochemistry of Cholesterol*, Elsevier Biomedical Press B.V., Amsterdam, the Netherlands, 1982.

16. R. P. Cook, ed., *Cholesterol*, Academic Press, Inc., New York, 1958.

17. H. Van Belle, *Cholesterol, Bile Acids, and Atherosclerosis*, North-Holland Publishing Co., Amsterdam, the Netherlands, 1965.

18. E. Heftmann, *Steroid Biochemistry*, Academic Press, Inc., New York, 1970, pp. 1–24.

19. L. Finegold, ed., *Cholesterol in Membrane Models*, CRC Press Inc., Boca Raton, Fla., 1993.

20. Ref. 8, pp. 421–443.

21. Ref. 18, pp. 51–62.

22. Ref. 15, pp. 189–202.

23. M. H. Briggs and J. Brotherton, *Steroid Biochemistry and Pharmacology*, Academic Press, London, 1970, pp. 183–188.

24. R. A. Davis, S. Dueland, and J. D. Trawick, in A. J. Lusis, J. I. Rotter, and R. S. Sparks, eds., *Molecular Genetics of Coronary Artery Disease: Candidate Genes and Processes in Atherosclerosis*, Basal, Karger, 1992, pp. 208–227.

25. G. Paumgartner, A. Stiehl, and W. Gerok, eds., *Bile Acids and Cholesterol in Health and Disease*, MTP Press Ltd, Boston, Mass., 1983.

26. E. Lund and I. Björkhem, *Acc. Chem. Res.* **28**, 241 (1995).

27. M. F. Holick and H. F. DeLuca, in D. E. M. Lawson, ed., *Vitamin D*, Academic Press, London, 1978, pp. 51–91.

28. M. R. Walters, *Endocrinology Rev.* **13**, 719 (1992); A. J. Brown, A. Dusso, and E. Slatopolsky, *Sem. Nephrol.* **14**(2), 156 (1994) and references therein; Y. Nishii, *Progress in Endocrinology*, Vol. 3, ICGSEM, 1993, pp. 391–395 and references therein.

29. Ref. 8, pp. 90–168.

30. D. E. M. Lawson, ed., *Vitamin D*, Academic Press, London, 1978.

31. M. F. Holick and H. F. DeLuca, in M. H. Briggs and G. A. Christie, eds., *Steroid Biochemistry and Pharmacology*, Academic Press, London, 1974, pp. 112–151.

32. Ref. 18, pp. 31–36.

33. Ref. 23, pp. 81–83.

34. E. Havinga, *Experimentia*, **29**, 1181 (1973).

35. F. Boomsma, H. J. C. Jacobs, E. Havinga, and A. Van der Gen, *Tetrahedron Lett.*, 427 (1975).

36. J. H. Clark, W. T. Schrader, and R. T. O'Malley, in J. D. Wilson and D. W. Foster, eds., *Williams Textbook of Endocrinology*, W.B. Sanders Co., 1992, pp. 35–90.

37. R. M. Evans, *Science*, **240**, 889 (1988); S. K. Nordeen, B. J. Bonna, C. A. Beck, D. P. Edwards, K. C. Borror, and D. B. DeFranco, *Steroids*, **60**, 97–104 (1995); A. P. H. Wright and co-workers, *J. Steroid Biochem. Molec. Biol.* **47**, 1–6 (1993); R. L. Miesfeld, *Crit. Rev. Biochem. Mol. Biol.* **24**(2), 101–117 (1989); C. Scheidereit and co-workers, *J. Steroid Biochem.* **24**(1), 19–24 (1986).

38. E. V. Jensen, *Ann. N.Y. Acad. Sci.* **761**, 1–4 (1995); M. K. Agerwal, *Drugs of Fut.* **20**, 903 (1995).

39. P. G. Marshall, *Rodds's Chemistry of Carbon Compounds*, Vol. 2, Parts D and E, Elsevier Publishing Co., London, 1970, pp. 197–267.

40. W. R. Butt, B. T. Rudd, and R. Morris, *Hormone Chemistry*, Vol. 2, Ellis Horwood Ltd, Chichester, U.K., 1976.

41. Ref. 8, pp. 503–576.

42. Ref. 40, pp. 118–146.

43. Ref. 18, pp. 115–130.

44. Ref. 23, pp. 121–138.

45. Z-X. Zhou, C-I. Wong, M. Sar, and E. M. Wilson, *Rec. Prog. Hormone Res.* **49**, 249 (1994).

46. J. M. Hoberman and C. E. Yesalis, *Sci. Am.*, 76 (1995).

47. S. E. Lukas *TIPS*, **14**, 61 (1993).

48. Ref. 40, pp. 81–117.

49. Ref. 8, pp. 444–502.

50. Ref. 18, pp. 131–146.

51. Ref. 23, pp. 86–120.

52. D. C. Collins, *Am. J. Obstet. Gynecol.* **170**(5:2), 1508 (1994).

53. N. Perone, in J. W. Goldzieher, ed., *Pharmacology of Contraceptive Steroids*, Raven Press Ltd, New York, 1994, pp. 5–19.

54. Ref. 40, pp. 60–80.

55. Ref. 8, pp. 539–576.

56. Ref. 18, pp. 63–74.

57. Ref. 23, pp. 86–120.
58. Ref. 40, pp. 147–190.
59. Ref. 8, pp. 600–726.
60. Ref. 18, pp. 97–114.
61. Ref. 23, pp. 139–170.
62. A. S. Brem and D. J. Morris, *Mol. Cell. Endo.* **97**, C1–C5 (1993).
63. E. P. Gómez-Sánchez, *Steroids*, **60**, 69–72 (1995).
64. J. Müller, *Steroids*, **60**, 2–9 (1995).
65. M. K. Agerwal, *Pharmacolog. Rev.*, 67 (1994).
66. S. B. Mahato, in M. Sainsbury, ed., *Second Supplements to the 2nd Edition of Rodd's Chemistry of Carbon Compounds*, Vol. II B (Partial), C, D and E, Elsevier Science BV, Amsterdam, the Netherlands, 1994, pp. 509–554.
67. Ref. 8, pp. 810–846.
68. A. M. Rouhi, *Chem. Eng. News*, 28–35 (Sept. 1995) and references therein.
69. J. Li, P. Nagpal, V. Vitart, T. C. McMorris, and J. Chory, *Science*, **272**, 398 (1996).
70. A. Poulus, J. W. Greiner, and G. A. Fevig, *Ind. Eng. Chem.* **53**, 949 (1961); U.S. Pat. 2,839,544 (1958), J. A. Greiner and G. A. Fevig.
71. A-U. Rahman and M. I. Choudhary, *Methods Plant Biochem.* **8**, 473 (1993).
72. A-U. Rahman and M. I. Choudhary, *Nat. Prod. Rep.*, 361 (1995).
73. R. Schakirov and M. S. Yunusov, *Nat. Prod. Rep.* **7**, 139 (1990).
74. D. M. Harrison, *Nat. Prod. Rep.* **1**, 219 (1984).
75. D. Gröger, *Cell, Culture, and Somatic Cell Genetics*, Vol. 5, Academic Press, Inc., New York, 1988, pp. 435–448.
76. Ref. 8, pp. 847–896.
77. S. Fukuzawa, S. Matsunaga, and N. Fusetani, *Tetrahedron*, **51**, 6707 (1995); C. H. Heathcock and S. C. Smith, *J. Org. Chem.* **59**, 6828 (1994) and references therein; J. U. Jeong, S. C. Sutton, S. Kim, and P. L. Fuchs, *J. Am. Chem. Soc.* **117**, 10,157 (1995).
78. T. Akizawa, T. Mukia, M. Matsukawa, M. Yoshioka, J. F. Morris, and V. P. Butler, *Chem. Pharm. Bull.* **42**(3), 754 (1994).
79. K. Shimada, K. Ohishi, H. Fukunaga, J. S. Ro, and T. Nambara, *J. Pharmacobio-Dyn.* **8**, 1054 (1988).
80. Ref. 8, pp. 727–809.
81. D. S. Fullerton and co-workers, *J. Med. Chem.* **27**, 256 (1984).
82. D. S. Fullerton and co-workers, in A. S. V. Burgan, G. C. K. Roberts, and M. S. Tute, eds., *Molecular Graphics and Drug Design*, Elsevier Science Publishers, New York, 1986, pp. 257–282.
83. K. R. H. Repke, R. Megges, J. Weiland, and R. Schön, *Angew. Chem. Int. Ed. Eng.* **34**, 282 (1995).
84. P. Christen, *Pharm. Acta. Helv.* **61**, 242 (1986).
85. G. Adam, N. Q. Chien, B. Goehring, and K. Dornberger, *Proceedings of the 3rd FECS International Conference on Chemistry and Biotechnology Act. Natural Products*, VCH, Weinheim, Germany, 1987, pp. 66–80.
86. E. Glotter, A. Goldman, and I. Kirkson, *Tennen Yuki Kagobusu Toronkai Koen Yoshishu*, **27**, 537 (1985).
87. A. B. Kundu, A. Mukherjee, and A. K. Day, *J. Sci. Ind. Res.* **35**, 616 (1976).
88. E. Glotter, *Nat. Prod. Rep.* **8**(4), 415 (1991).
89. I. Kirson and E. Glotter, *J. Nat. Prod.* **44**, 633 (1981).
90. A. Butenandt and K. Karlson, *Z. Naturforsch*, **9b**, 389 (1954).
91. P. Karlson, H. Hoffmeister, H. Hummel, H. Hocks, and G. Spiteller, *Chem. Ber.* **98**, 2394 (1965); R. Huber and W. Hoppe, *Chem. Ber.* **98**, 2403 (1965).
92. K. Nakanishi, *Steroids*, **57**, 649 (1992).

93. M. F. Feldlaufer, in J. Koolman, ed., *Ecdysone*, Thieme, Stuttgart, Germany, 1989, pp. 308–312.

94. K. H. Hoffmann, *Labor Praxis*, 86 (1993).

95. J. Koolman, *Zool. Sci.* **7**(4), 563 (1990).

96. H. Ishizaka, in Ref. 93, pp. 204–210.

97. R. M. K. Carlson, C. Tarchini, and C. Djerassi in S. N. Anachenko, ed., *Frontiers in Bioorganic Chemistry and Molecular Biology*, The Robert A. Welch Foundation, Houston, Tex., 1980, p. 211.

98. R. G. Kerr and B. J. Baker, *Nat. Prod. Rep.* **8**, 465 (1991).

99. M. V. D'Auria, L. G. Paloma, L. Minale, and R. Riccio, *Tetrahedron Lett.*, 2149 (1991).

100. B. J. Baker and R. G. Kerr, *Topics in Current Chemistry*, Vol. 167, Springer-Verlag, Berlin, 1993, pp. 1–31.

101. C. J. Djerassi and G. A. Doss, in A-U. Rahman, ed., *Studies in Natural Products Chemistry*, Vol. 9, Elsevier Science Publishers BV, Amsterdam, the Netherlands, 1991, pp. 35–50.

102. N. Ikekawa, in H. Danielsson and J. Sjovall, eds., *Steroids and Bile Acids*, Elsevier Science Publishers BV, Amsterdam, the Netherlands, 1985, pp. 199–230; C. Djerassi, in P. Krogsgaard-Larsen, S. Brogger Christensen, and H. Kofod, eds., *Alfred Benzon Symposium 20: Natural Products and Drug Development*, Munksgaard, Copenhagen, Denmark, 1984, pp. 164–176; N. W. Withers, in P. J. Scheuer, ed., *Marine Natural Products*, Vol. V, Academic Press, Inc., New York, 1983, pp. 87–130; C. Djerassi, *Pure Appl. Chem.* **53**, 873 (1981).

103. C. L. Harvey, C. J. Sih, and S. G. Knight, *Antibiotics*, **1**, 404 (1967).

104. N. Tanaka, T. Kinoshita, and H. Masukawa, *J. Biochem.* **65**, 459 (1969); N. Tanaka, H. Yamaki, and Y-C. Lin, *J. Antibiotics, Ser A*, 156 (1967); H. Hikino, Y. Asada, S. Arihara, and T. Takemoto, *Chem. Pharm. Bull.* **20**, 1067 (1972).

105. R. E. Ireland, P. Beslin, R. Giber, U. Hengartner, H. A. Kirst, H. Maag, *J. Org. Chem.* **42**, 1267 (1977), for synthetic studies.

106. K. S. Moore and co-workers, *Proc. Natl. Acad. Sci. USA*, **90**, 1354 (1993); S. L. Wehrli, K. S. Moore, H. Roder, S. Durrell, and M. Zasloff, *Steroids*, **58**, 370 (1993).

107. A. Sadownik and co-workers, *J. Am. Chem. Soc.* **117**, 6138 (1995); R. Mestel, *New Sci.* **2006**, 6 (1995).

108. R. M. Moriarty, S. M. Tuladhar, L. Guo, and S. Wehrli, *Tetrahedron Lett.* **35**, 8103 (1994); R. M. Moriarity and co-workers, *Tetrahedron Lett.* **36**, 5139 (1995).

109. I. Abe, M. Rohmer, and G. D. Prestwich, *Chem. Rev.* **93**, 2189 (1993).

110. E. J. Corey and co-workers, *J. Am. Chem. Soc.* **117**, 11,819 (1995); E. J. Corey and H. Cheng, *Tetrahedron Lett.* **37**(16), 2709 (1996).

111. W. S. Johnson, G. W. Daub, T. A. Lyle, and M. Niwa, *J. Am. Chem. Soc.* **102**, 7800 (1980); W. S. Johnson, B. F. Frel, and A. S. Goplan, *J. Org. Chem.* **46**, 1513 (1981); W. S. Johnson, T. A. Lyle, and G. W. Daub, *J. Org. Chem.* **47**, 163 (1982); W. S. Johnson, S. Escher, and B. W. Metcalf, *J. Am. Chem. Soc.* **98**, 1039 (1976); E. E. van Tamelen and J. R. Hwu, *J. Am. Chem. Soc.* **104**, 2490 (1983).

112. E. J. Corey, J. Lee, and D. R. Liu, *Tetrahedron Lett.* **35**, 9149 (1994).

113. T. A. Spencer, *Acc. Chem. Res.* **27**, 83–90 (1994).

114. K. M. Curnow, P. C. White, and L. Pascoe, *Curr. Opin. Endocrinol. Diabetes*, 10–15 (1994).

115. L. F. Fieser and M. Fieser, *Steroids*, Reinhold Publishing Corp., New York, 1959.

116. Ref. 18, pp. 1–30.

117. Ref. 25, pp. 52–87.

118. D. M. Harrison, *Nat. Prod. Rep.* **5**, 387 (1988), for biosynthesis of triterpenoids, steroids, and carotenoids.

119. P. Robel and E-E. Baulieu, *TEM*, **5**(1), 1–8 (1994).

120. L. L. Smith, in H.-J. Rehm and G. Reed, eds., *Biotechnology: A Comprehensive Treatise in 8 Volumes*, Vol. 6a, K. Kieslich, vol. ed., Deerfield Beach, Fla., 1984, pp. 31–78.

121. Ref. 115, pp. 444–502.

122. Ref. 115, pp. 727–809.

123. M. Wuchtl and M. Mangkudidjojo, *Pharmazeutische Zeitung*, **129**, 686/32 (1984).

124. D. Onken, *CLB Chemie in Labor und Biotechnik*, **43**(3), 134–136 (1992).

125. R. K. Jaggi and V. K. Kapoor, *Ind. Drugs*, **29**(5), 191 (1992); M. Ikram and G. A. Miana, *Sci. Ind.* **7**(1,2), 1 (1970).

126. M. S. Murphy, in *Recent Trends in Biotechnology*, C. Ayyanna, ed., Tata McGraw-Hill, New Delhi, India, 1993, pp. 475–80.

127. R. K. Jaggi and V. K. Kapoor, *J. Sci. Indrust. Res.* **53**, 34 (1994); J. D. Mann, *Adv. Agron.* **30**, 207 (1978).

128. A. W. Afermann, H. Spieler, and E. Reinhard, in B. Deus-Neumann, W. Barz, and E. Reinhard, eds., *Primary and Secondary Metabolism of Plant Cell Cultures*, Springer-Verlag, Berlin, 1985, pp. 316–322; R. G. Butenko, S. I. Demchenko, S. L. Karanova, I. A. Rappoport, and Z. B. Shamina, "Method of Producing the Diosgenin," *Otkritija Izobretenija*, N 31, Inventor's Sertificat N 440404, 1975.

129. A. Kh. Lipskii, *J. Biotechnol.* **26**(1), 83 (1991).

130. K. Kawaguchi, M. Hirotani, and T. Furuya, in Y. P. S. Bajaj, ed., *Biotechnology in Agriculture and Forestry*, Vol. 21, *Medicinal and Aromatic Plants IV*, Springer-Verlag, Berlin, 1993, for review of *in vitro* culture and production of cardenolides.

131. Ref. 8, pp. 634–659; T. F. Gallagher and W. P. Long, *J. Biol. Chem.* **162**, 496, 511, 521 (1946).

132. R. E. Marker and E. Rohrmann, *J. Am. Chem. Soc.* **61**, 3592 (1939); R. E. Marker and E. Rohrmann, *J. Am. Chem. Soc.* **62**, 518 (1940); R. E. Marker, T. Tsukamoto, and D. L. Turner, *J. Am. Chem. Soc.* **62**, 2525 (1940); R. E. Marker, R. B. Wagner, P. R. Ulshafer, E. L. Wittbecker, D. P. J. Goldsmith, and C. H. Ruof, *J. Am. Chem. Soc.* **69**, 2167 (1947).

133. G. Rosenkranz, O. Mancera, F. Sondheimer, and C. Djerassi, *J. Org. Chem.* **21**, 520 (1956); G. Rosenkranz, *Steroids*, **57**, 409 (1992).

134. J. Elks, G. H. Phillipps, T. Walker, and L. J. Wyman, *J. Chem. Soc.* **4330**, 4344 (1956).

135. F. W. Heyl, A. P. Centolla, and M. E. Herr, *J. Am. Chem. Soc.* **69**, 1957 (1947); A. P. Centolla, F. W. Heyl, and M. E. Herr, *J. Am. Chem. Soc.* **70**, 2953 (1948); F. W. Heyl and M. E. Herr, *J. Am. Chem. Soc.* **72**, 2617 (1950).

136. H. H. Imhoffen, *Angew. Chem.* **53**, 471 (1940); E. B. Hershberg, M. Rubin, and E. Schwenk, *J. Org. Chem.* **15**, 292 (1950); U.S. Pat. 3,994,938 (1976), K. Wakabayashi, Y. Chigira, and K. Fukuda.

137. H. L. Dryden, Jr., G. M. Webber, and J. J. Wieczorrek, *J. Am. Chem. Soc.* **86**, 742 (1964).

138. J. Kalvoda and K. Heusler, *Synthesis*, 501 (1971).

139. U.S. Pat. 3,176,014 (1965), R. Pappo and L. N. Nysted; A. Bowers, R. Villotti, J. A. Edwards, E. Denot, and O. Halpern, *J. Am. Chem. Soc.* **84**, 3204 (1962); H. Ueberwasser and co-workers, *Helv. Chim. Acta*, **34**, 344 (1963).

140. K. Heuslar and J. Kalvoda, in J. Fried and J. A. Edwards, eds., *Organic Reactions in Steroid Chemistry*, Vol. 2, Van Nostrand Rheinhold Co., New York, 1972, pp. 237–287 and references therein.

141. R. H. Hesse, *Adv. Free Radical Chem.* **3**, 83 (1969).

142. M. Miyano, *J. Org. Chem.* **46**, 1846 (1981) and references therein; D. H. R. Barton and J. M. Beaton, *J. Am. Chem. Soc.* **83**, 750 (1961); D. H. R. Barton, N. K. Basu, M. J. Day, R. H. Hesse, M. M. Pechet, and A. N. Starratt, *J. Chem. Soc. Perkin I*, 2243 (1975).

143. G. Majetich and K. Wheless, *Tetrahedron*, **51**, 7095 (1995) for a review on intramolecular free-radical functionalization.

144. W. Charney and H. L. Herzog, *Microbial Transformations of Steroids*, Academic Press, Inc., New York, 1967.

145. K. A. Martin, in Ref. 120, pp. 80–95.

146. S. B. Mahato and I. Majumdar, *Phytochemistry*, **34**(4), 883 (1993); K. E. Smith, F. Ahmed, and T. Antoniou, *Biochem. Soc. Trans.* **21**, 1077 (1993).

147. K. Arima, M. Nasgasawa, M. Bae, and G. Tamura, *Agr. Biol. Chem.* **33**(11), 1636 (1969).

148. W. F. van der Waard, J. Doodewaard, J. de Flines, and S. van der Weele, *Abh. Deut. Akad. Berlin Kl. Med.* **2**, 101 (1969).

149. U.S. Pat. 3,338,042 (1968), K. Arima and co-workers.

150. G. Wix, K. G. Buki, E. Tomorkeny, and G. Ambrus, *Steroids*, **11**, 401 (1968).

151. M. Nagasawa, H. Hashiba, N. Watanabe, M. Bae, G. Tamura, K. Arima, *Agr. Biol. Chem.* **34**(5), 801 (1970).

152. W. J. Marsheck, S. Kraychy, and R. D. Muir, *Appl. Microbiol.* **23**, 72 (1972).

153. M. G. Wovcha, F. J. Antosz, J. C. Knight, L. A. Kominek, and T. R. Pyke, *Biochim. Biopsy. Acta.* **539**, 308 (1978).

154. V. VanRheenen and K. P. Shepard, *J. Org. Chem.* **44**, 1582 (1979); U.S. Pat. 4,102,907 (1978), K. P. Shephard.

155. P. G. M. Hessenlink, S. V. Vliet, H. D. Vries, and B. Withholt, *Enzyme Microb. Technol.* **11**, 398 (1989).

156. O. Hechter and co-workers, *J. Am. Chem. Soc.* **71**, 3261 (1949).

157. O. K. Sebek and D. Perlman, in H. J. Peppler and D. Perman, eds., *Microbial Technology*, 2nd ed., Vol. 1, Academic Press, Inc., New York, 1979, pp. 483–496.

158. D. H. Peterson and H. C. Murray, *J. Am. Chem. Soc.* **74**, 4055 (1952); D. H. Peterson and co-workers, *J. Am. Chem. Soc.* **74**, 5933 (1952); S. H. Epstein and co-workers, *J. Am. Chem. Soc.* **75**, 408 (1953).

159. J. Fried, R. W. Thoma, J. R. Gerke, J. E. Herz, M. N. Donin, and D. Perlman, *J. Am. Chem. Soc.* **74**, 3962 (1952).

160. U.S. Pat. 2,658,023 (1953), G. M. Shull, D. A. Kita, and J. W. Davisson; G. M. Shull and D. A. Kita, *J. Am. Chem. Soc.* **77**, 763 (1955).

161. B. M. Bloom and G. M. Shull, *J. Am. Chem. Soc.* **77**, 5767 (1955); B. M. Bloom, E. J. Agnello, and G. D. Laubach, *Experientia*, **12**, 27 (1956).

162. J. Fried, R. W. Thoma, and A. Klingsberg, *J. Am. Chem. Soc.* **75**, 5764 (1953); H. L. Herzog and co-workers, *Tetrahedron*, **18**, 581 (1962); U.S. Pat. 2,837,464 (1958), A. Nobile; A. Nobile and co-workers, *J. Am. Chem. Soc.* **77**, 4184 (1955).

163. D. Perlman, E. Titus, and J. Fried, *J. Am. Chem. Soc.* **74**, 2126 (1952); R. W. Thoma, J. Fried, S. Bonanno, and P. Grabowich, *J. Am. Chem. Soc.* **79**, 4818 (1957).

164. S. B. Mathato, S. Banerjee, and S. Podder, *Phytochemistry*, **28**, 7 (1989); K. Kieslich, *Arzneim.-Forsch/Drug Res.* **36**(1) 888 (1986); L. Sedlaczek, J. Dlugonski, and A. Jaworski, *Appl. Microbiol. Biotechnol.* **20**, 166 (1984).

165. B. K. Lee, D. Y. Ryu, R. W. Thoma, and W. E. Brown, *Biotechnol. Bioeng.* **11**, 1255 (1969); K. Kieslish, H. Wieglepp, K. Petzoldt, and F. Hill, *Tetrahedron*, **27**, 445 (1971).

166. A. J. Birch, *Quart. Rev.* **4**, 69 (1950).

167. A. L. Wilds and N. A. Nelson, *J. Am. Chem. Soc.* **75**, 5360 and 5366 (1953).

168. U.S. Pat. 2,655,518 (1953), F. B. Colton; U.S. Pat. 2,691,028 (1954), F. B. Colton; U.S. Pat. 2,725,389 (1955), F. B. Colton.

169. C. Djerassi, L. Miramontes, G. Rosenkrantz, and F. Sondheimer, *J. Am. Chem. Soc.* **76**, 4092 (1954); J. Iriarte, C. Djerassi, and H. J. Ringold, *J. Am. Chem. Soc.* **81**, 436 (1959), for a similar procedure to norethindrone.

170. C. Djerassi, *Steroids*, **57**, 631 (1992); F. B. Colton, *Steroids*, **57**, 624 (1992).

171. G. Goto, K. Hiraga, T. Miki, and M. Sumi, *Yakugaku Zasshi*, **103**, 1042 (1983); D. Lednicer, L. A. Mitscher, and G. I. Georg, *The Organic Chemistry of Drug Synthesis*, Vol. 4, John Wiley and Sons, Inc., New York, 1990, pp. 65–77.

172. L. H. Sarrett, *J. Biol. Chem.* **162**, 601 (1946).

173. G. Rosenkranz and F. Sondheimer, *Chem. Org. Naturst.*, 275 (1953), for a review of cortisol syntheses before 1953.

174. J. A. Hogg, *Steroids*, **57**, 593 (1992) and references therein; J. A. Hogg and co-workers, *J. Am. Chem. Soc.* **77**, 4438 (1955).

175. H. J. Ringold, G. Rosenkrantz, and F. Sondheimer, *J. Am. Chem. Soc.* **78**, 820 (1956).

176. J. Fried and E. F. Sabo, *J. Am. Chem. Soc.* **75**, 2273 (1953); J. Fried and E. F. Sabo, *J. Am. Chem. Soc.* **76**, 1455 (1954); E. M. Hicks and E. S. Wallis, *J. Biol. Chem.* 162 (1946).

177. J. Y. Godard, in R. E. Banks and K. C. Lowe, eds., *Fluorine Medicines of the 21st Century*, Paper 10, University of Manchester, Institute of Science and Technology, Manchester, U.K., 1994, for a review introducing fluorine into corticosteroids.

178. J. Redpath and F. J. Zeelen, *Chem. Soc. Rev.* **12**, 75 (1983); I. Netta and H. Ueno, *Org. Synth. Chem.* **45**, 445 (1987); V. Van Rheenen and K. P. Shepard, *J. Org. Chem.* **44**, 1582 (1979); G. Neef, U. Ader, A. Seeger, and R. Wiechert, *Chem. Ber.* **113**, 1184 (1980); I. Nitta, S. Fujimori, and H. Ueno, *Bull. Chem. Soc. Jpn.* **58**, 978 and 981 (1985); A. R. Daniewski and J. Wojciechowska, *Synthesis*, 132 (1984); D. van Leusen and A. M. van Leusen, *Tetrahedron Lett.* **25**, 2581 (1984); D. H. R. Barton and S. Z. Zard, *J. Chem. Soc. Perkin Trans. I*, 2191 (1985); S. Solyom, K. Szilagy, and L. Toldy, *Liebigs Ann. Chem.* 153 (1987); P. G. Cattini, E. Morera, and G. Ottar, *Tetrahedron Lett.* **31**, 1889 (1990); J. G. Reid and T. Debiak-Krook, *Tetrahedron Lett.* **31**, 3669 (1990).

179. D. A. Livingston, J. E. Petre, and C. L. Bergh, *J. Am. Chem. Soc.* **112**, 6449 (1990); D. A. Livingston, *Adv. Med. Chem.* **1**, 137–174 (1992).

180. U.S. Pat. 4,216,159 (1980), E. J. Hessler and V. H. Van Rheenen.

181. Eur. Pat. 0 531 212 A2 (1993), F. Brion, J. Buendia, C. Diolez, and M. Vivat.

182. J. A. Cella, E. A. Brown, and R. R. Burtner, *J. Org. Chem.* **24**, 743 (1957); E. A. Brown, R. D. Muir, J. A. Cella, *J. Org. Chem.* **25**, 96 (1960).

183. G. E. Arth, H. Schwam, L. H. Sarett, and M. Glitzer, *J. Med. Chem.* **6**, 617 (1963); R. M. Weier and L. M. Hofmann, *J. Med. Chem.* **18**, 817 (1975); *Ibid.* **20**, 1304 (1977).

184. D. Bittler and co-workers, *Angew. Chem. Int. Ed. Engl.* **21** 696 (1982); D. Lednicer and L. A. Mitscher, *The Organic Chemistry of Drug Synthesis*, Vol. 3, John Wiley and Sons, Inc., New York, 1984, pp. 81–108.

185. T. Kametani, *Actual. Chim. Ther.* **15**, 131 (1988).

186. V. Cerny, *Nat. Prod. Rep.* **11**, 419 (1994); J. R. Hanson, *Nat. Prod. Rep.* **9**, 37 (1993); A. B. Turner, *Nat. Prod. Rep.* **8**, 17 (1992); A. B. Turner, *Nat. Prod. Rep.* **6**, 539 (1990); A. B. Turner, *Nat. Prod. Rep.* **5**, 311 (1989); J. Elks *Nat. Prod. Rep.* **3**, 515 (1987); J. R. Hanson, *Nat. Prod. Rep.* 567 (1995), for recent reviews on partial synthesis of steroids.

187. D. Lednicer and L. A. Mitscher, *The Organic Chemistry of Drug Synthesis*, Vols. 1–4, John Wiley and Sons, Inc., New York, 1977, 1980, 1984, 1990, for general references of partial synthesis of steroid drugs.

188. B. Camerino, *Prod. Chim. Aerosol Sel.* **25**(4), 26 (1984).

189. D. Onken, *CLB Chem. Labor Biotech.* **43**(5), 264 (1992); *Ibid.*, **43**(4), 200; *Ibid.*, **43**(3), 134.

190. P. Welzel and co-workers, *Izv. Khim.* **20**(1), 48 (1987).

191. P. Catsoulacos and D. Catsoulacos, *J. Heterocyclic Chem.* **30**(1), 1 (1993); P. Catsoulacos, G. Pairas, *Pharmakeutke*, **4**(1), 26 (1991); R. P. Reddy, A. Ravindranath, T. Ramaiah, T. Sundara, and M. Roa, *Ind. J. Chem.* **1**, 45 (1987); P. Catsoulacos, *Epitheor. Klin. Farmakol. Farmakokinet., Int. Ed.* **2**(2), 91 (1988).

192. G-D. Zhu and W. H. Okamura, *Chem. Rev.* **95**, 1877 (1995).

193. M. L. Sá e Melo, M.J.S.M. Moreno, S. C. Pereira da Costa, J. A. R. Salvador, and A. S. Campos Neves, *Ultrason. Sonochem.* **1**(1), S37 (1994).

194. R. Breslow, *Chemtracts: Org. Chem.* **1**(5), 333 (1988).

195. W. E. Bachmann, W. Cole, and A. L. Wilds, *J. Am. Chem. Soc.* **61**, 974 (1939); *Ibid.*, **62**, 824 (1940).

196. G. Anner and H. Miescher, *Helv. Chim. Acta.* **31**, 2173 (1948); *Ibid.*, **32**, 1957 (1949); *Ibid.*, **33**, 1379 (1950).

197. D. Taub, in J. A. Simon, ed., *The Total Synthesis of Natural Products*, Vol. 6, John Wiley & Sons, Inc., New York, 1984, pp. 1–50; K. P. C. Vollhardt, *Strategies and Tactics in Organic Synthesis*, Academic Press, Inc., New York, 1984, pp. 299–324, for reviews on total synthesis of aromatic steroids.

198. Torgov and co-workers, *Tetrahedron Lett.* 1553 (1963); Smith and co-workers, *J. Chem. Soc.* 5072 (1963); Johnson and co-workers, *J. Am. Chem. Soc.* **95**, 7501 (1973); Cohen and co-workers, *J. Org. Chem.* **40**, 681 (1975); Danishefsky and co-workers, *J. Am. Chem. Soc.* **98**, 4975 (1976); Kametani and co-workers, *J. Am. Chem. Soc.* **99**, 3461 (1977); Oppolzer and co-workers, *Helv. Chim. Acta.* **63**, 1703 (1980); Vollhardt and co-workers, *J. Am. Chem. Soc.* **102**, 5253 (1980); Quinkert and co-workers, *ACIE*, **19**, 1027 (1980); Grieco and co-workers, *J. Org. Chem.* **45**, 2247 (1980); Bryson and co-workers, *Tetrahedron Lett.* **21**, 2381 (1980); S. Djuric, T. Sarkar, and P. Magnus, *J. Am. Chem. Soc.* **102**, 6885 (1980); Y. Ito, M. Naktsuka, and T. Saegusa, *J. Am. Chem. Soc.* **103**, 476 (1981); J. H. Hutchinson and T. Money, *Tetrahedron Lett.* **26**, 1819 (1985); B. Barlaam, J. Boivin, L. Elkaim, S. Elton-Farr, and S. Z. Zard, *Tetrahedron*, **51**, 1675 (1995), for examples in total synthesis of aromatic A-ring steroids.

199. F. E. Ziegler and H. Lim, *J. Org. Chem.* **47**, 5229 (1982).

200. G. H. Posner and C. Switzer, *J. Am. Chem. Soc.* **108**, 1239 (1986).

201. K. Mikami, K. Takahashi, T. Nakai, and T. Uchimaru, *J. Am. Chem. Soc.* **116**, 10,948 (1994).

202. M. B. Groen and F. J. Zeelen, *Recl. Trav. Chim. Pays-Bas.* **105**, 465 (1986); N. Cohen, *Acc. Chem. Res.* 412 (1976).

203. S. N. Ananchenko and I. V. Torgov, *Doklady Akad. Nauk SSSR*, **127**, 553 (1959); S. N. Ananchenko and I. V. Torgov, *Tetrahedron Lett.*, 1553 (1963).

204. Th. B. Windholz and M. Windholz, *Angew. Chem.* **76**, 249 (1964); L. Velluz, J. Valis, and G. Nomine, *Angew. Chem.* **77**, 185 (1965); S. Schwarz, *Wiss. Zeitschr. THLM*, **29**(4), 444 (1987).

205. U. Ender, G. Sauer, and R. Wiechert, *Angew. Chem.* **83**, 492 (1971); Z. G. Hajos and D. R. Parrish, *J. Org. Chem.* **39**, 1612, 1615, 3239 (1974); *Ibid.* **38**, 3244 (1973).

206. G. Sauer, U. Eder, G. Haffer, G. Neef, and R. Wiechert, *Angew. Chem. Internat. Edit.* **14**, 417 (1975).

207. J. W. Cornforth and R. Robinson, *J. Chem. Soc.* 1855 (1949); J. W. Cornforth and R. Robinson, *Nature*, **160**, 737 (1947); H. M. E. Cardwell, J. W. Cornforth, S. R. Duff, H. Holtermann, and R. Robinson, *Chem. Ind.*, 389 (1951).

208. J. Sarret and co-workers, *J. Am. Chem. Soc.* **74**, 1393, 1397, 1401, 1406, 4974 (1952); *Ibid.*, **75**, 422, 1707, 2112 (1953); *Ibid.*, **76**, 1715, 5026, 5031 (1954); *Ibid.*, **77**, 1026, 3834 (1955).

209. R. B. Woodward and co-workers, *J. Am. Chem. Soc.*, **73**, 2403, 3547, 3548, 4057 (1951); *Ibid.*, **74**, 4223 (1952).

210. A. A. Akhrem and Y. A. Titov, *Total Steroid Synthesis*, Plenum Press, New York, 1970, for total synthesis of steroids before 1970.

211. G. Stork, G. Clark, and C. S. Shiner, *J. Am. Chem. Soc.* **103**, 4948 (1981); G. Stork and D. H. Sherman, *J. Am. Chem. Soc.*, **104**, 3758 (1982).

212. Y. Horiguchi, E. Nakamura, I. Kuwajima, *J. Org. Chem.*, **22**, 4323 (1986); Y. Horiguchi, E. Nakamura, and I. Kuwajima, *J. Am. Chem. Soc.*, **111**, 6257 (1989).

213. H. Shibasaki, T. Furuta, and Y. Kasuya, *Steroids*, **57**, 325 (1992); K. Minagawa, T. Furuta, Y. Kasuya, A. Fujino, M. A. Avery, and M. Tanabe, *J. Chem. Soc. Perkin Trans. 1*, 587 (1988).

214. G. Quinkert and M. Del Grosso, in Ottow, Schollkopf, and Schultz, eds., *Stereoselective Synthesis*, Springer-Verlag, Berlin, 1994, pp. 109–134; T. Kamettani and H. Nemoto, *Tetrahedron*, 3 (1981); C. Spino and J. Crawford, *Tetrahedron Lett.* **35**, 5559 (1984); H. Nemoto, M. Nagai, K. Fukumoto, and T. Kametani, *Tetrahedron Lett.* **26**, 4613 (1985).

215. G. Stork and E. W. Logusch, *J. Am. Chem. Soc.* **102**, 1219 (1980).

216. H. Nemoto, N. Matsuhashi, M. Imaizumi, M. Nagai, and K. Fukumoto, *J. Org. Chem.*, **55**, 5625 (1990).

217. H. Nemoto, A. Satoh, K. Fukumoto, and C. Kabuto, *J. Org. Chem.*, **60** (1995), for similar asymmetric synthesis of fluorosteroids.

218. W. S. Johnson, G. W. Daub, T. A. Lyle, and M. Niwa, *J. Am. Chem. Soc.*, **102**, 7800 (1980); W. S. Johnson, B. F. Frel, and A. S. Goplan, *J. Org. Chem.* **46**, 1513 (1981); W. S. Johnson, T. A. Lyle, and G. W. Daub, *J. Org. Chem.* **47**, 163 (1982); W. S. Johnson, S. Escher, and B. W. Metcalf, *J. Am. Chem. Soc.*, **98**, 1039 (1976); E. E. van Tamelen and J. R. Hwu, *J. Am. Chem. Soc.*, **104**, 2490 (1983).

219. W. S. Johnson, *Bioorg. Chem.*, **5**, 51–98 (1976).

220. W. S. Johnson, S. D. Lindell, and J. Steele, *J. Am. Chem. Soc.*, **109**, 5852 (1987).

221. W. S. Johnson, M. B. Gravestock, and B. E. McCarry, *J. Am. Chem. Soc.*, **93**, 4332 (1971); W. S. Johnson, S. Echer, and B. W. Metcalf, *J. Am. Chem. Soc.*, **98**, 1039 (1976).

222. W. S. Johnson, B. Chenera, F. S. Tham, and R. K. Kullnig, *J. Am. Chem. Soc.*, **115**, 493 (1993); W. S. Johnson, V. R. Fletcher, B. Chenera, W. R. Bartlett, F. S. Tham, and R. K. Kullnig, *J. Am. Chem. Soc.* **115**, 497 (1993); R. A. Buchanan, *Ibid.*, p. 504.

223. W. S. Johnson, M. S. Plummer, S. P. Reddy, and W. R. Bartlett, *J. Am. Chem. Soc.*, **115**, 515 (1993).

224. P. V. Fish and W. S. Johnson, *Tetrahedron Lett.*, **35**, 1469 (1994).

225. E. J. Corey, J. Lee, and D. R. Liu, *Tetrahedron Lett.*, **35**, 9149 (1994).

226. P. A. Zoretic, X. Weng, M. L. Caspar, and D. G. Davis, *Tetrahedron Lett.*, **32**, 4819 (1991).

227. Y. W. Andemichael, Y. Huang, and K. K. Wang, *J. Org. Chem.* **58**, 1651 (1993); P. A. Zoretic, Z. Shen, M. Wang, and A. A. Ribeiro, *Tetrahedron Lett.*, **36**, 2925 (1995), for other free-radical approaches to steroids.

228. T. Sugihara, C. Coperet, Z. Owezarczyk, L. S. Harring, and E. Negishi, *J. Am. Chem. Soc.*, **116**, 7923 (1994).

229. J. Bao, W. D. Wulff, V. Dragisich, S. Wenglowsky, and R. G. Ball, *J. Am. Chem. Soc.*, **116**, 7616 (1994).

230. L. Velluz, J. Valls, and G. Nomine, *Angew. Chem. Int. Ed. Eng.*, **4**, 181 (1965), and references therein.

231. R. Rohde, G. Neff, G. Sauer, and R. Wiechert, *Tetrahedron Lett.*, **26**, 2069 (1985).

232. A. Belanger, D. Philibert, and G. Teutsch, *Steroids*, **37**, 361 (1981); G. Teutsch, in M. K. Agarwal, ed., *Adrenal Steroid Antagonism*, Walter-deGruyer, Berlin, 1984, pp. 27–47; Eur. Pat. 0 057 115 A2 (1982b), G. Teutsch, G. Costerousse, D. Philibert, and R. Deraedt.

233. G. C. Starks, *J. Family Prac.*, **19**, 315 (1984).

234. H. A. Zcur and D. Stewart, *Curr. Opinion Obstet. Gynecol.* **4**, 365 (1992).

235. P. G. Stubblefield, *J. Reprod. Med.* **31**, 922 (1986); J. W. Goldzieher, *Am. J. Obstet. Gynecol.* **160**, 1260 (1989); B. E. McKenzie, *Toxicol. Pathol.*, **17**, 377 (1989).

236. L. J. Dorflinger, *Contraception*, **31**, 557 (1985), for relative potencies of synthetic progestins.

237. J. C. Robins, J. L. Ambrus, and C. M. Ambrus, *Steroids*, **42**, 3205 (1983); F. P. Cantatore, M. C. Loperfido, L. Mancini, and M. Carrozzo, *J. Rheumatology*, **19**(11), 1753 (1992); J. C. Robin, O. W. Suh, and J. L. Ambrus, *Steroids*, **40**, 125 (1982); "Script Reports," *Script Yearbook 1994*, Vol. I, PJB Publications Ltd., pp. 131–133.

238. J. T. Pento and J. Castañer, *Drugs of Fut.* **20**, 784 (1995).

239. R. Holt, *Law Med. Health Care*, **20**(3), 109 (1992); D. A. Grimes, D. R. Mishell, D. Shoupe, and M. Lacarra, *Am. J. Obstet. Gynecol.* **158**, 1307 (1988); E. E. Baulieu, A. Ulman, and D. Philibert, *Arch. Gynecol. Obstet.* **241**, 73 (1987); R. E. Gerfield and E. E. Baulieu, *Bailliere's Clin. Endocrinol. Metabol.* **1**(1), 207–221 (1987); A. Ulmann, *J. Steroid Biochem.* **27**, 1009 (1987); D. L. Healy, *Reprod. Fertil. Dev.* **2**, 477 (1990); K. B. Horowitz, L. Tung, and G. S. Taimoto, *J. Steroid Biochem.* **53**, 9 (1995).

240. F. Nique and P. Van de Velde, *Drugs of Fut.* **20**, 362 (1995).

241. V. C. Jordan and C. S. Murphy, *Endocrine Rev.* **11**, 578 (1990).

242. S. M. Sedlacek and K. B. Horwitz, *Steroids*, **44**, 467 (1984), for progestin use.

243. Council on Scientific Affairs, *JAMA*, **264**(22), 2923 (1990); A. D. Rogol and C. E. Yesalis, III, *J. Clin. Endo. Met.* **74**, 465 (1992); M. LaBree, *J. Sports Med. Phys. Fitness*, **31**, 618 (1991); K. L. Soe, M. Soe, and C. Gluud, *Liver*, **12**, 73 (1992); W. A. Narducci, J. C. Wagner, T. P. Hendrickson, and T. P. Jeffrey, *Clin. Toxicol.* **28**, 287 (1990); J. T. Laseter and J. A. Russel, *Med. Sci. Sports Exer.* **23**, 1 (1991); L. Uzych, *Can. J. Psych.* **37**, 23 (1992); F. Celotti and P. N. Cesi, *J. Steroid Biochem. Molec. Biol.* **43**, 469 (1992); G. Glazer, *Arch. Intern. Med.* **151**, 1925 (1991); D. A. Smith and P. J. Perry, *Ann. Pharmacother.* **26**, 520 (1992).

244. J. D. Mann, *Adv. Agron.* **30**, 207 (1978).

245. P. J. Barnes and I. Adcock, *Trends Pharmacol. Sci.* **14**(12), 436 (1993).

246. C. Montecucco, R. Caporali, P. Caprotti, M. Caprotti, and A. Notario, *J. Rheumatol.* **19**(12), 1895 (1992); O. D. Messina and co-workers, *J. Rheumatol.* **19**(10), 1520 (1992).

247. P. J. Barnes, *Brit. Med. Bull.* **48**, 231 (1991); L. Fabbri and co-workers, *Thorax*, **48** 817 (1993).

248. I. Pavord and A. Knox, *Clin. Pharmacokinet.* **25**(2), 126–135 (1993); I. M. Richards, S. K. Shields, R. L. Griffin, S. F. Fidler, and C. M. Dunn, *Clin. Exper. Aller.* **22**, 432 (1992).

249. J. L. Mobley, J. E. Chin, and I. M. Richards, *Exp. Opin. Invest. Drugs* **5**(7), 87 (1996).

250. C. F. LaForce and co-workers, *J. Family Prac.* **38**, 145 (1994); W. Busse, *J. Allerg. Clin. Immunol.* **82**, 890 (1988); R. P. Schleimer, *Ann. Rev. Pharmacol. Toxicol.* **25**, 381 (1985); S. C. Siegel, *J. Allerg. Clin. Immunol.* **81**, 984 (1988).

251. B. B. Furner, *Pharmacol.* **9**, 285 (1992); B. Giannotti and N. Pimpinelli, *Drugs*, **44**, 62 (1992).

252. H. J. Harwood and co-workers, *J. Lipid Res.* **34**, 377 (1993).

253. F.-W. Bangerter and co-workers, *Atherosclerosis*, **109**, 309 (1994); F.-W. Bangerter and co-workers, *Atherosclerosis*, **109**, 310 (1994).

254. F. J. Urban, B. S. Moore, and R. Breiter, *Tetrahedron Lett.* **31**, 4421 (1990); P. A. McCarthy, *J. Labeled Cpds. Radiopharm.* **28**, 4799 (1990); M. P. Zawistoski, J. P. Kiplinger, and P. A. McCarthy, *Tetrahedron*, **49**, 4799 (1993); L. M. Zaccaro and co-workers, *Atherosclerosis*, **109**, 315 (1994).

255. A. D. Abell and B. R. Henderson, *Curr. Med. Chem.* **2**, 583 (1995); J. M. Williams, G. Marchesini, R. A. Reamer, U-H. Dolling, and E. J. J. Grabowski, *J. Org. Chem.* **60**, 5337 (1995) and references therein; C. Haffner, *Tetrahedron Lett.* **36**, 4039 (1995) and references therein; G. H. Rasmusson and co-workers, *J. Med. Chem.* **29**, 2298 (1986).

256. E. J. Jacobsen and co-workers, *J. Med. Chem.* **33**, 1145 (1990); E. D. Hall, J. M. McCall, and E. D. Means, *Adv. Pharmacol.* **28**, 221 (1994).

257. G. L. Bundy and co-workers, *J. Med. Chem.* **38**, 4161 (1995), for recent work on nonsteroidal compounds with similar activities.
258. E. K. Janaki Ammal and P. Nagendra Prasad, *Curr. Sci.* **53**(11), 601 (1984).
259. Y. Chen and Y. Wu, *J. Herbs, Spices, Med. Plants*, **2**(3), 59 (1994).
260. D. Onken and D. Onken, *Pharmazie*, **35**, 193 (1980); C. Djerassi, *Proc. R. Soc. London Ser. B*, **195**, 175 (1976).
261. R. K. Jaggi and V. K. Kapoor, *Ind. Drugs*, **29**, 191 (1991).
262. *The Pharmaceutical Market: World Review*, Vol. I, International, IMS Global Services, 1990 and 1994.
263. L. Dorfman, *Chem. Rev.* **53**, 47 (1953); J. P. Duza, M. Heller, and S. Bernstein, in L. L. Engel, ed., *Physical Properties of Steroid Hormones*, Pergamon Press, Oxford, U.K., 1963, pp. 69–287; W. Neudert and H. Ropke, *Atlas of Steroid Spectra*, Springer-Verlag, Berlin, 1965.
264. S. Gorog, *Quantitative Analysis of Steroids*, Elsevier Scientific Publishing Co., Amsterdam, the Netherlands, and Akademiai Kiado, Budapest, Hungary, 1983.
265. K. Dobriner, E. R. Katzenellenbogen, and R. N. Jones, *Infrared Absorption Spectra of Steroids: An Atlas*, Vol. 1, Interscience Publishers, Inc., New York, 1953; G. Roberts, B. S. Gallagher, and R. N. Jones, Vol. 2, 1958.
266. E. Caspi and G. F. Scrimshaw, in H. Cartensen, ed., *Steroid Hormone Analysis*, Marcel Dekker, New York, 1967, pp. 55–92; M. F. Jayle, *Analyse des Steroids Hormonaux*, Vol. III, Masson, Paris, 1965.
267. N. S. Bhacca and D. H. Williams, *Applications of NMR Spectroscopy in Organic Chemistry*, Holden-Day, San Francisco, Calif., 1964; E. Caspi and Th. Wittstruck, in Ref. 266, pp. 93–133; C. C. Hinckley, in E. Heftman, ed., *Modern Methods in Steroid Analysis*, Academic Press, Inc., New York, 1973, pp. 265–291; D. P. Hollis, *Ibid.*, pp. 245–264.
268. G. J. Bird, D. J. Collins, F. W. Eastwood, and R. H. Exner, *Aust. J. Chem.* **32**, 797 (1979).
269. A. Rahman and M. M. Qureshi, *Pure Appl. Chem.* **62**, 1385 (1990); S. L. Wehrli, K. S. Moore, H. Roder, S. Durell, and M. Zasloff, *Steroids*, **58**, 370 (1993).
270. W. L. Duax and D. A. Norton, *Atlas of Steroid Structure*, IFI/Plenum Data, New York, 1975; *Terpenoids and Steroids*, Vol. 4, The Chemical Society, London, 1974, p. 531.
271. C. H. L. Shackleton and K. M. Straub, *Steroids*, **40**, 35 (1982).
272. J. C. Touchtone, in G. Zweig and J. Sherma, eds., *CRC Handbook of Chromatography: Steroids*, CRC Press Inc., Boca Raton, Fla., 1986; G. Szepesi, *J. Planar Chromatog.* **5**, 396 (1992).

BRADLEY P. MORGAN
MELINDA S. MOYNIHAN
Pfizer, Inc.

STILBENE DERIVATIVES. See FLUORESCENT WHITENING AGENTS; STILBENE DYES.

STILBENE DYES

Stilbene dyes of importance are mostly direct yellow dyes for cellulosic fibers, especially paper. There have been several red and blue stilbene-containing dyes reported, but they have not (ca 1996) been developed to commercial importance. There are brown leather dyes which are stilbene-based. The most important stilbene dyes are those known since the 1880s. The commercial importance of Direct Yellow 11 (1883), Direct Orange 15 (1888), Direct Yellow 4 (1886), and Direct Yellow 106 (1936) attest to the value, properties, and durability of stilbene yellow dyes.

Stilbene [103-30-0], $C_6H_5CH{=}CHC_6H_5$, is a crystalline hydrocarbon used in the manufacture of dyes; its name is derived from the Greek word *stilbein*, meaning to glitter (1). However, in all the references to commercial dyes, there is not a single one derived directly from stilbene itself nor is there a reference to anyone ever making a dye from stilbene. In most examples, the starting material for stilbene dyes is 4-nitrotoluene-2-sulfonic acid [121-03-9], which is oxidized under alkaline conditions to 4,4'-dinitro-2,2'-dinitrostilbenedisulfonic acid [128-42-7] (1) as the first descriptive substance. There are more than 100 stilbene dyes listed in the *Colour Index* which have been offered as commercial products (2); less than 12 are available, as listed in the latest *AATCC Buyer's Guide* (3).

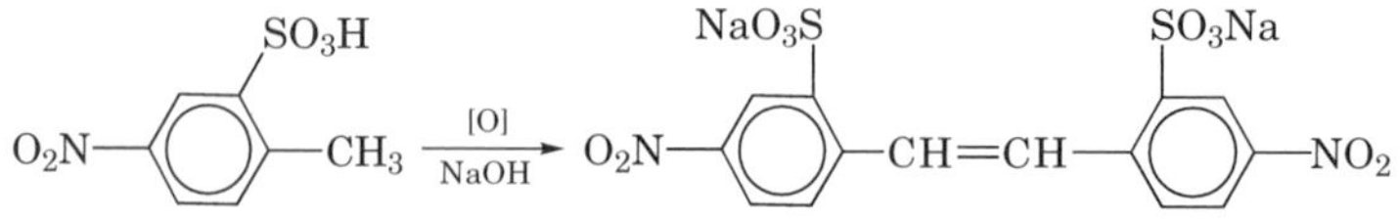

(1)

(1)

Stilbene dyes are classed as a subgroup of azo dyes having excellent color-fastness and typical direct dye wash fastness on cotton and are arranged into six categories by the Society of Dyers and Colourists (2), as described in the following.

(1) Self-condensation products of 4-nitrotoluene-2-sulfonic acid or its derivative 4,4'-dinitro-2,2'-stilbenedisulfonic acid or 4,4'-dinitro-2,2'-dibenzyldisulfonic acid [728-42-7] and products of their treatment with reducing or oxidizing agents. An example is Direct Yellow 11 (CI 40000) [1325-37-7] (2).

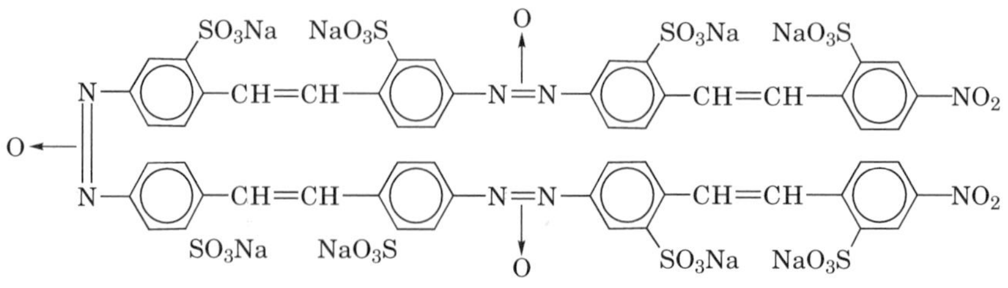

(2)

(*2*) Condensation products of 4-nitrotoluene-2-sulfonic acid or its derivatives together with phenols, naphthols, or aminophenols. An example here is Direct Yellow 19 (CI 40030) (**3**):

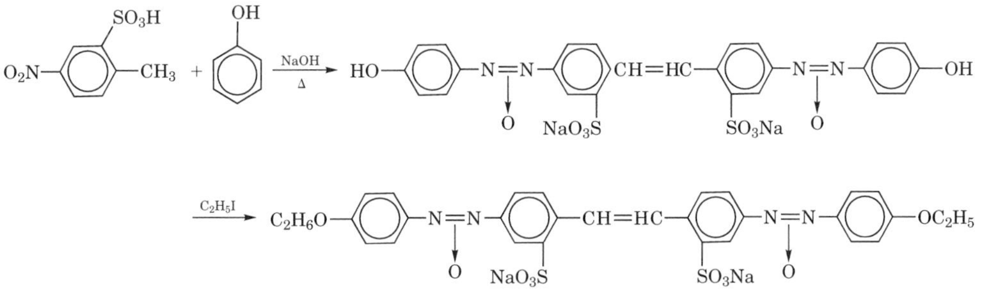

(**3**)

(*3*) Condensation products of 4-nitrotoluene-2-sulfonic acid or its derivatives together with aromatic amines. Direct Orange 28 (CI 40065) (**4**) is an example. The amine in this case is *para*-diaminobenzene.

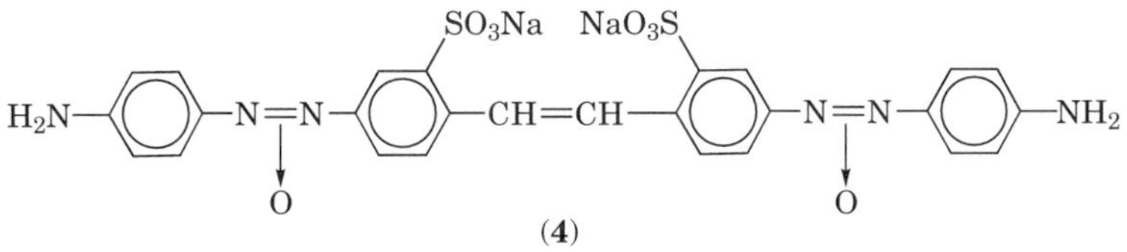

(**4**)

(*4*) Azo-stilbene dyes formed by condensation of 4,4′-dinitro-2,2′-stilbene-disulfonic acid or 4,4′-dinitro-2,2′-dibenzyldisulfonic acid (**1**) with aminoazo compounds. Direct Orange 34 (CI 40215) [*32651-66-4*] (**5**) is a representative:

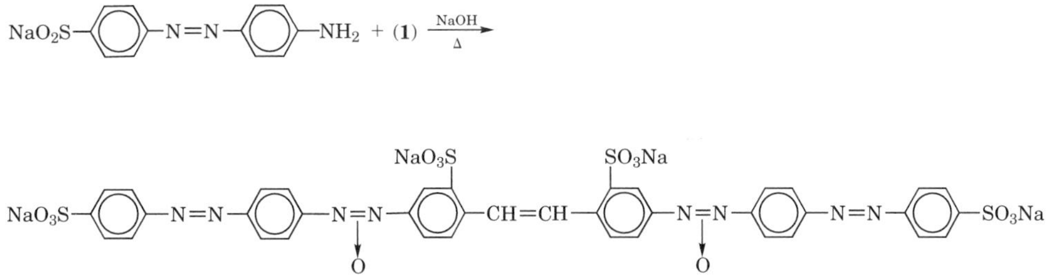

(**5**)

(*5*) Azo-stilbene dyes formed by diazotization of a condensation product containing primary amino groups and coupling with azo dye coupling components, eg, Direct Brown 29 (CI 40505) (**6**):

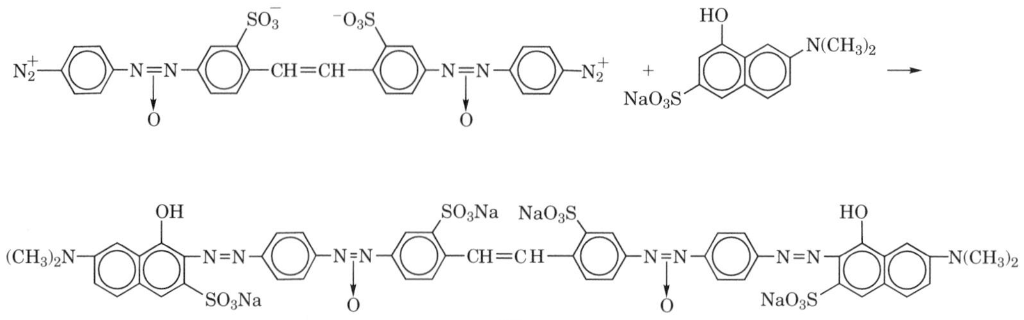

(6)

(6) Stilbene-azo dyes of more precise constitution prepared in the usual way by tetrazotization and coupling of 4,4'-diamino-2,2'-stilbenedisulfonic acid [81-11-8] (7). The product in this example is ZDirect Yellow 4 (CI 24890) [3051-11-4] (8):

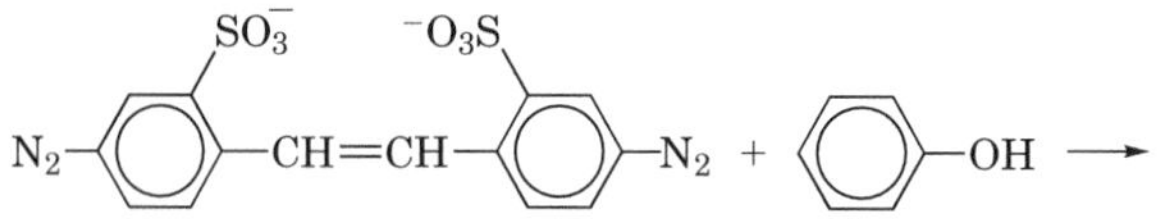

(7)

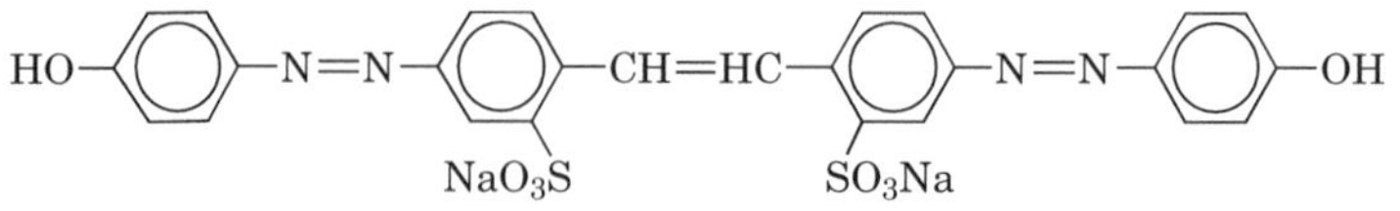

(8)

(6a) Stilbene-azo dyes from 4-nitro-4'-amino-2,2'-stilbenedisulfonic acid [119-72-2] include Direct Yellow 106 (CI 40300) [12222-60-5] (9). The synthetic route is shown in Figure 1.

Most stilbene dyes derived from 4-nitrotoluene-2-sulfonic acid are of non-definitive structure even though structures are proposed which are descriptive of the major components (4). 4-Nitrotoluene-2-sulfonic acid can be converted to 4,4'-dinitro-2,2'-dinitrostilbenedisulfonic acid (1) in yields of 60–80% by heating in dilute sodium hydroxide (3%) at about 50°C and slowly adding sodium hypochlorite (4), as shown in equation 1. The product obtained can be reduced to 4,4'-diamino-2,2'-stilbenedisulfonic acid (7), then tetrazotized and coupled to various phenols and amines to give dyes having reasonably definitive structures. These dyes are classed as normal azo dyes (2). As may be expected with most tetrazotizations and couplings, many decomposition and self-coupled by-products are obtained. 4,4'-Dinitro-2,2'-stilbenedisulfonic acid (1) can also be reduced with polysulfide to 4-nitro-4'-amino-2,2'-stilbenedisulfonic acid (10), which is a useful intermediate for dye synthesis.

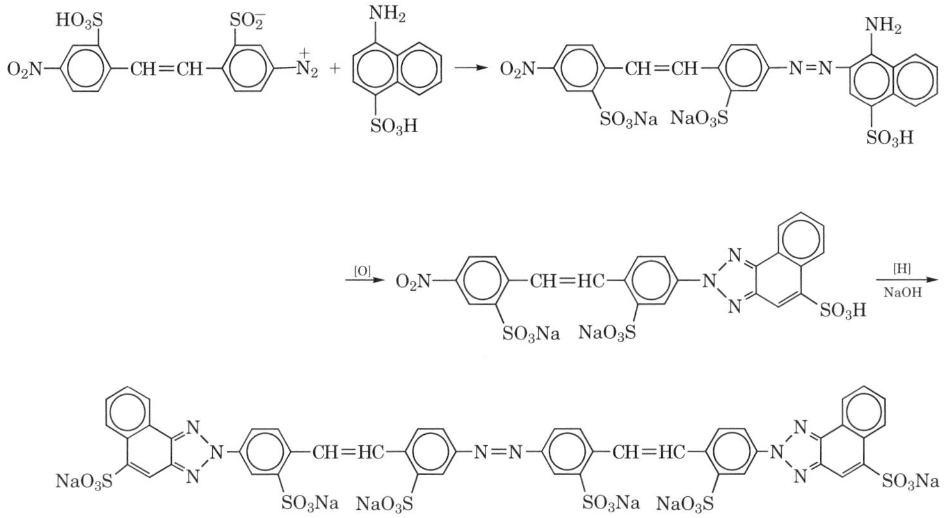

Fig. 1.   After diazo coupling, oxidation leads triazene ring formation. Alkaline reduction and coupling lead to (**9**).

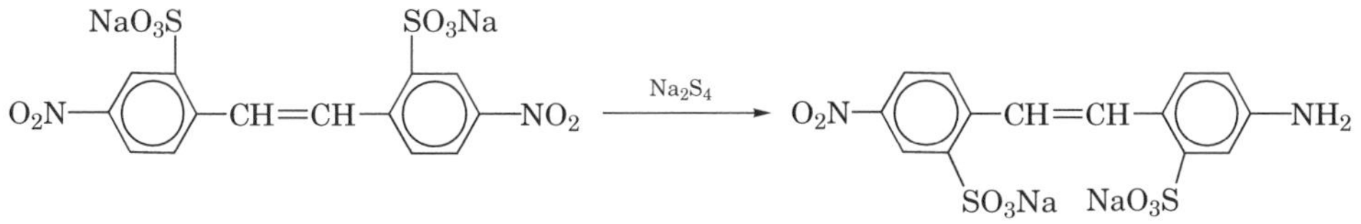

Stilbene dyes have generally been important as direct dyes and fluorescent brighteners for cellulosic fibers (4). Most stilbene dyes are yellow and orange, with some examples of reds and browns and even a few blues. Brown stilbene dyes have commercial value as leather dyes (4).

Direct Yellow 4 (CI 24890) (**11**) is the most familiar dye made from tetrazotized 4,4'-diamino-2,2'-stilbenedisulfonic acid [*57153-16-9*] (**7**) and phenol.

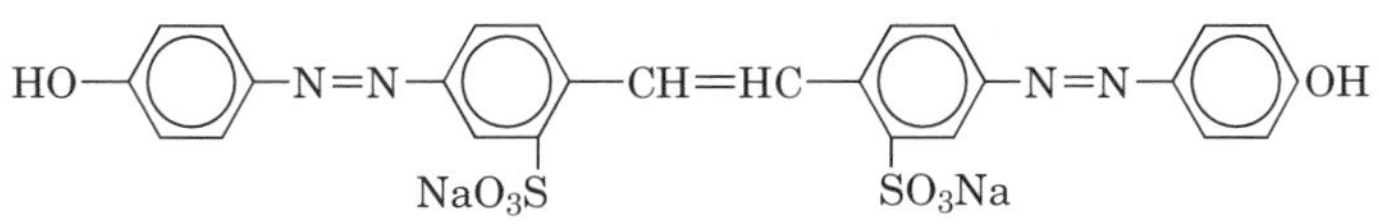

This dye has importance for dyeing paper and is also used as a pH indicator, changing to a red color under alkaline conditions.

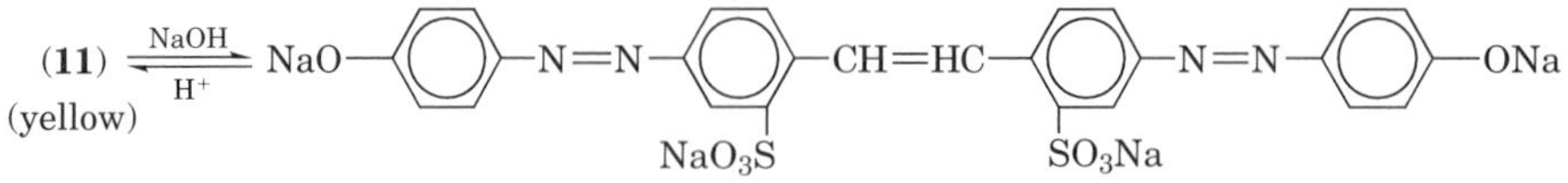

$$(11) \underset{H^+}{\overset{NaOH}{\rightleftharpoons}} NaO-\langle \bigcirc \rangle-N=N-\langle \bigcirc \rangle-CH=HC-\langle \bigcirc \rangle-N=N-\langle \bigcirc \rangle-ONa$$

(yellow) NaO_3S SO_3Na

(red)

An alkali-stable dye, Direct Yellow 12 (CI 24895) [*2870-32-8*] (**12**), is made by ethylating Direct Yellow 4 (**11**) with ethyl chloride or diethyl sulfate.

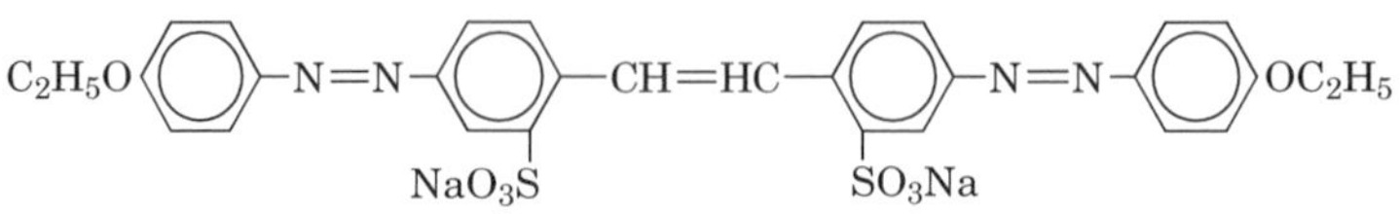

$$C_2H_5O\langle \bigcirc \rangle-N=N-\langle \bigcirc \rangle-CH=HC-\langle \bigcirc \rangle-N=N-\langle \bigcirc \rangle OC_2H_5$$

NaO_3S SO_3Na

(**12**)

Condensation dyes from 4-nitrotoluene-2-sulfonic acid are the most important of the stilbene dyes. Direct Yellow 11 (CI 40000), discovered in 1883 and commonly known as Sun Yellow [*1325-37-7*], is widely used in the paper industry (2,4).

Direct Orange 15 (CI 40003) [*1325-35-5*] (**13**), which is made by reducing the alkaline condensation product equivalent to Direct Yellow 11 with sulfide or formaldehyde, is an important paper dye used in dyeing brown paper for bags.

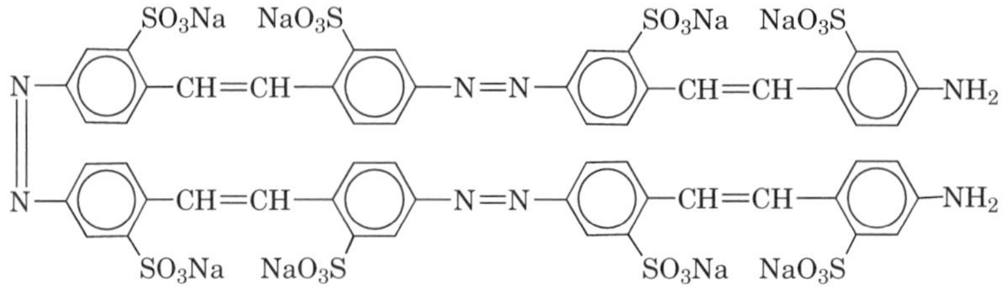

$$\begin{array}{c} SO_3Na \quad NaO_3S \qquad\qquad SO_3Na \quad NaO_3S \\ N-\langle \bigcirc \rangle-CH=CH-\langle \bigcirc \rangle-N=N-\langle \bigcirc \rangle-CH=CH-\langle \bigcirc \rangle-NH_2 \\ \| \\ N-\langle \bigcirc \rangle-CH=CH-\langle \bigcirc \rangle-N=N-\langle \bigcirc \rangle-CH=CH-\langle \bigcirc \rangle-NH_2 \\ SO_3Na \quad NaO_3S \qquad\qquad SO_3Na \quad NaO_3S \end{array}$$

(**13**)

The azoxy and nitro groups in Direct Yellow 11 are reduced to azo and amino groups in Direct Orange 15. Direct Yellow 6 (CI 40006) (**14**) is a greener and brighter shade of yellow than Direct Yellow 11 and is made by reductive azo formation from 4,4′-dinitro-2,2′-stilbenedisulfonic acid to an azo and azoxy dye.

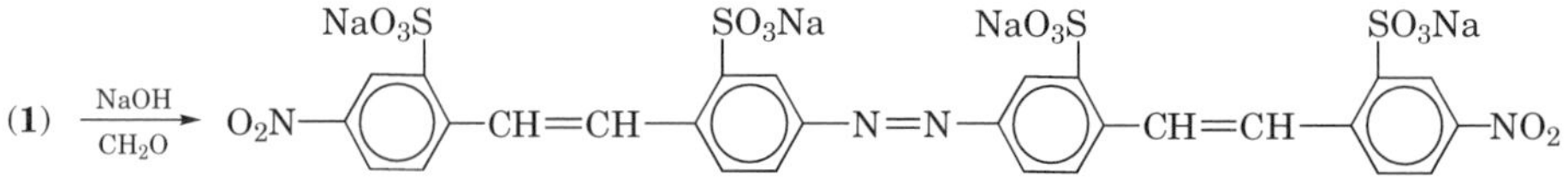

$$\begin{array}{c} NaO_3S \qquad\qquad SO_3Na \qquad NaO_3S \qquad\qquad SO_3Na \\ (1) \underset{CH_2O}{\overset{NaOH}{\longrightarrow}} O_2N-\langle \bigcirc \rangle-CH=CH-\langle \bigcirc \rangle-N=N-\langle \bigcirc \rangle-CH=CH-\langle \bigcirc \rangle-NO_2 \end{array}$$

(**14**)

This controlled synthesis gives a purer product than alkaline condensation of 4-nitrotoluene-2-sulfonic acid alone and is the reason for its brightness (4).

Activity in this class of compounds since the early 1980s has been mostly with stilbene fluorescent brighteners (see FLUORESCENT WHITENING AGENTS). There has, however, been some other activity in most of the six listed categories of dyes as well. 4-Nitrotoluene-2-sulfonic acid condensations to give stable solutions of Direct Yellow 11 type dyes for paper account for many efforts in this category (5–9). Reaction products of 4,4′-diamino-2,2′-stilbenedisulfonic acid tetrazo with various coupling components are claimed as dyes for paper (10–18) (Table 1).

The most widely reported developments have been in category 4, ie, 4,4′-dinitro-2,2′-stilbenedisulfonic acid (**1**) condensations with amino-containing azo components, some of which are copper complexes, to give dyes having excellent properties on leather (19–31) (Table 2).

Tetrakisazo dyes with good fastness properties prepared by tetrazotization of (**7**) coupling to 1-naphthylamine, retetrazotizing and coupling to, eg,

Table 1. Dyes Derived from Tetrazotized (7) Dye or Coupling Compound

Coupling component	CAS Registry Number	Ref.	Color
	[91779-63-4]	12	yellow
 R = CH$_3$CH$_2$— CH$_2$OHCH$_2$— CH$_3$CH$_2$CHOHCH$_2$—		15	blue
		16	yellow
o-cresol		18	yellow
	[118914-79-7]	11	brown
		11	brown

Table 2. Brown Dyes Derived From (1)

Coupling component	CAS Registry Number	Ref.
	[106199-75-1]	19,29
	[106564-02-7]	21
	[106222-76-8]	23
	[85895-92-7]	24
	[84373-09-1]	25
	[106769-40-8]	20,26
		27
	[102949-51-9]	30

2-naphthol, have been reported (14). Limited use is also reported for both basic and reactive stilbene-containing dyes (10,13,32,33) (Fig. 2). Although it is reported that basic dyes are suitable for dyeing paper, none are found in commerce.

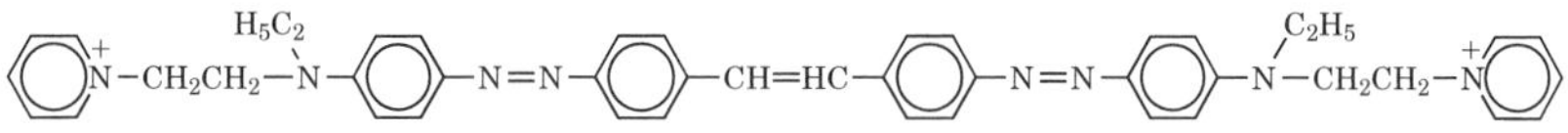

Fig. 2. Basic and reactive stilbene dyes: (**15**) [*122749-45-5*] (32); (**16**) [*122749-47-7*] (33); (**17**) [*89049-99-0*] (10); and (**18**) [*79146-01-3*] (13).

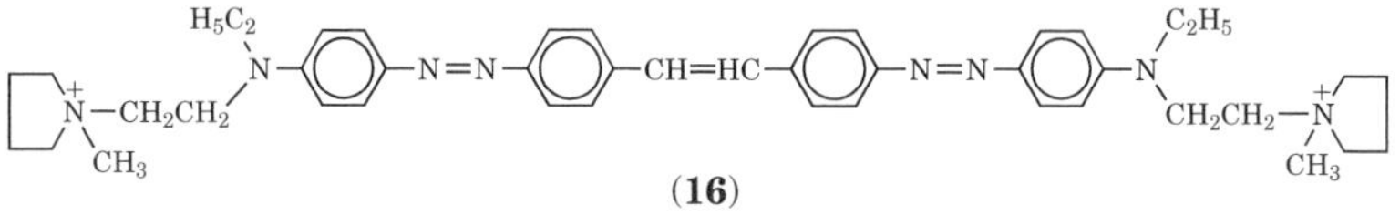

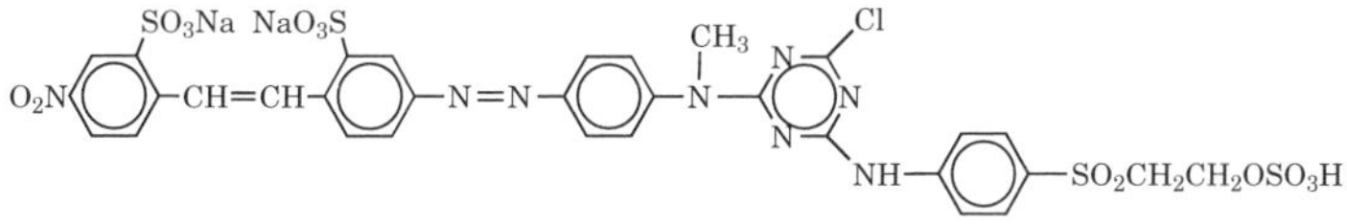

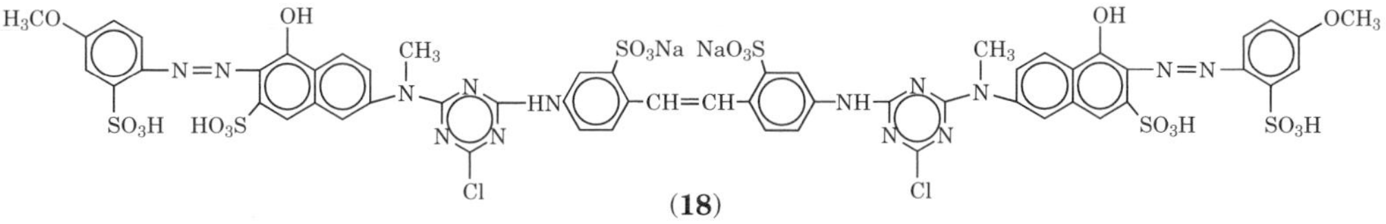

Coupling tetrazo (**7**) with *o*-cresol has been reported to give a dye (**19**) which is less alkali-sensitive than Direct Yellow 4 (CI 24890) and to have better cold-water solubility than Direct Yellow 12 (CI 24895) (18). One reference is made to a stilbene laser dye (10) and two each to dyes for light-polarizing films (34,35) and reprographic inks (36,37).

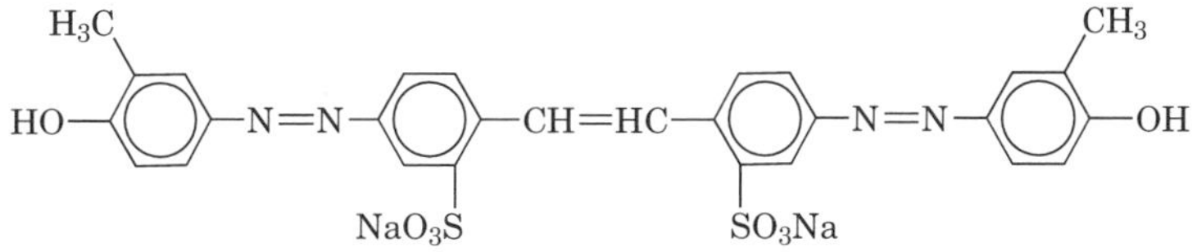

Economic Aspects

Direct Yellow 4, Direct Yellow 11, Direct Yellow 106, and Direct Orange 15 are the most important stilbene dyes, accounting for most sales of this type. Estimated volumes and values for liquid and powder forms appear in Table 3 (38).

Table 3. Estimated Production Volume and Prices of Stilbene Dyes

Dye	Volume, 10^3 kg		Average sales price, \$/kg		Manufacturer
	Liquid	Powder	Liquid	Powder	
Direct Yellow 4	455	23	2.20	6.60	BASF, DyStar, C & K
Direct Yellow 6	205	36	1.87	5.83	BASF, Ciba, C & K
Direct Yellow 11	2730	114	1.98	4.40	BASF, DyStar, Ciba, C & K
Direct Orange 15	136	23	2.75	7.15	DyStar, C & K
Direct Yellow 106		91		19.80	Ciba, C & K

Health and Safety Factors

Stilbene dyes are similar to azo dyes in their resistance to biological degradation. Typically the BOD is only a small percentage of the COD. Available toxicological data indicate relatively little personal or environmental hazard (39–42) (Table 4).

Table 4. Toxicological Data for Stilbene Dyes

Dye	LD_{50}, mg/kg[a]	LC_{50}, mg/L	BOD, mg O_2/g	COD, mg O_2/g
Direct Yellow 6	>5000	>1000 (96 h) (fat head minnow)	13	318
Direct Yellow 11	>5000	>280 mg/m (4 h) inhalation (rat)	100	6600
Direct Yellow 106	>7500	>1000 (48 h) (trout)	107	460
Direct Orange 15	>5000	>1000 (96 h) (blue gill)	200	6900

[a]Rat.

BIBLIOGRAPHY

"Stilbene Dyes" in *ECT* 1st ed., Vol. 12, pp. 949–955, by A. F. Plue, General Aniline and Film Corp.; in *ECT* 2nd ed., Vol. 19, pp. 1–13, by H. R. Schwander, J. R. Geigy AG and G. S. Dominguez, Geigy Chemical Corp.; in *ECT* 3rd ed., Vol. 21, pp. 729–746, by R. E. Farris, Sandoz Colors & Chemicals, Inc.

1. *Webster's New World Dictionary of the American Language*, College Edition, The World Publishing Company, Cleveland and New York, 1955.
2. *The Colour Index*, 3rd ed., Vol. 3, Society of Dyers and Colorists, Bradford, Yorkshire, U.K., 1971, pp. 4212–4214, 4365–4371, and Rev. 3rd ed., Vol. 6, 1st Suppl., p. 6400.
3. *Textile Chemist and Colorist*, Vol. 27, No. 7, American Association of Textile Chemists and Colorists, Research Triangle Park, N.C., July 1955.
4. K. Venkataraman, *The Chemistry of Synthetic Dyes and Pigments*, Vol. 1, Academic Press, Inc., New York, 1952, pp. 628–636.
5. Jpn. Pat. 60,118,753 (June 26, 1985), Mitsui Toatsu Chemicals.
6. Eur. Pat. 182,743 (May 28, 1986), A. Tzikas, P. Herzig, and J. Markert (to Ciba-Geigy).

7. Eur. Pat. 122,224 (Oct. 17, 1984), J. Markert (to Ciba-Geigy).
8. Ger. Pat. 3,046,450 (Aug. 27, 1981), R. Pedrazzi (to Sandoz).
9. Jpn. Pat. 56,065,061 (June 2, 1981), Pentel KK.
10. Jpn. Pat. 58,160,361 (Sept. 22, 1983), Mitsubishi Chemical Industries.
11. Ger. Pat. 3,710,007 (Oct. 6, 1988), K. Kunde (to Bayer).
12. U.S. Pat. 4,455,258 (June 19, 1984), D. A. Brode (to Crompton and Knowles).
13. USSR Pat. 834,045 (May 30, 1981), L. A. Surkina, K. V. Solodova.
14. Jpn. Pat. 61,053,362 (Mar. 17, 1986), Orient Kagaku Kogyo.
15. U.S. Pat. 4,314,816 (Feb. 9, 1982), V. Tullio (to Mobay Chem. Corp).
16. G. B. Pat. 1,601,232 (Oct. 28, 1981), F. Walker, J. Prince (to L. B. Holliday & Co.).
17. Ger. Pat. 3,007077 (Sept. 3, 1981), K. Kunde, P. Wild (to Bayer).
18. Belg. Pat. 885270 (Mar. 17, 1981), A. Brulard, A. Gerbaux (to Althouse Tertre SA).
19. East Ger. Pat. 236334 (June 4, 1986), W. Hepp and co-workers (to Bitterfeld).
20. East Ger. Pat. 236,338 (June 4, 1986), W. Hepp and co-workers (to Bitterfeld).
21. East Ger. Pat. 236,330 (June 4, 1986), G. Knoechel and co-workers (to Bitterfeld).
22. East Ger. Pat. 236,329 (June 4, 1986), G. Knoechel and co-workers (to Bitterfeld).
23. East Ger. Pat. 236,331 (June 4, 1986), G. Knoechel and co-workers (to Bitterfeld).
24. Rom. Pat. 79,485 (Apr. 29, 1983), I. Prejmereanu, D. Birlea, and C. Balour (to Interprenderea de Medicemente si Coloranti "Sintofarm").
25. Ger. Pat. 3,218,354 (Dec. 2, 1982), U. Schlesinger (to Ciba-Geigy).
26. East Ger. Pat. 236,335 (June 4, 1986), W. Hepp and co-workers (to Bitterfeld).
27. East Ger. Pat. 236,336 (June 4, 1986), W. Hepp and co-workers (to Bitterfeld).
28. East Ger. Pat. 236,335 (June 4, 1986), W. Hepp and co-workers (to Bitterfeld).
29. East Ger. Pat. 236,333 (June 4, 1986), W. Hepp and co-workers (to Bitterfeld).
30. East Ger. Pat. 236,328 (June 4, 1986), W. Hepp and co-workers (to Bitterfeld).
31. U.S. Pat. 4,742,162 (May 3, 1988), U. Schlesinger (to Ciba-Geigy).
32. Eur. Pat. 306,452 (Mar. 8, 1989), W. Stingelin (to Ciba-Geigy).
33. U.S. Pat. 4,940,738 (July 10, 1990), W. Stingelin (to Ciba-Geigy).
34. U.S. Pat. 5,272,259 (Dec. 21, 1993), U. Claussen and F. W. Kroock (to Bayer).
35. *J. Polym. Sci. Part A: Polym. Chem.* **28**, 1–13 (1990).
36. U.S. Pat. 5,254,159 (Oct. 19, 1993), K. Gundlach, G. A. R. Nobes, M. Breton, and R. Colt (to Xerox).
37. U.S. Pat. 5,360,472 (Nov. 1, 1994), E. Radigan, L. Isganitis, W. Solodar, and K. Gundlach (to Xerox).
38. Internal Market Survey 1995, Ciba-Geigy Corp., Greensboro, N.C., rev. Mar. 1996.
39. Material Safety Data Sheet, Ciba-Geigy Corp., Greensboro, N.C., rev. July 7, 1995.
40. Material Safety Data Sheet, Ciba-Geigy Corp., Greensboro, N.C., rev. June 23, 1989.
41. Material Safety Data Sheet, Ciba-Geigy Corp., Greensboro, N.C., rev. Mar. 15, 1995.
42. Material Safety Data Sheet, Ciba-Geigy Corp., Greensboro, N.C., rev. Aug. 26, 1986.

ROY E. SMITH
Ciba-Geigy Corporation

STIMULANTS

A variety of chemical agents have the capacity to stimulate the central nervous system (CNS) of mammals. Some have therapeutic uses; others are primarily of toxicological importance. The capacity of some of the agents to produce excessive CNS stimulation has greatly limited the usefulness of many in this class of compounds. Excessive CNS stimulation can lead to the production of convulsions which may have fatal consequences. Herein stimulants are separated into three more or less distinct pharmacological categories: analeptics, psychomotor stimulants, and antidepressants. The therapeutic uses of CNS stimulants have increased in the 1990s largely as a result of the widespread use of newer antidepressants (see PSYCHOPHARMACOLOGICAL AGENTS).

Analeptics

Analeptics are respiratory stimulants capable of stimulating respiratory and vasomotor centers in the medulla. These have been used to revive individuals poisoned by central nervous depressants such as barbiturates, alcohol, and general anesthetics (see ANESTHETICS; HYPNOTICS, SEDATIVES, ANTICONVULSANTS, AND ANXIOLYTICS). The action is not confined only to the medulla; at doses only slightly higher than those that stimulate the medulla, analeptics can stimulate the motor cortex and produce seizures. The initial CNS stimulation is followed, if larger doses are given, by CNS depression and ultimately by respiratory depression and cardiovascular collapse. As of this writing (1996), analeptics are seldom used to treat overdose by CNS depressants. Controlled studies (1) have shown that a higher incidence of mortality occurred when analeptics were administered to poisoned patients than when the drugs were not administered and the patients received only good nursing care.

Although the clinical usefulness is limited, analeptics continue to be valuable tools in the study of CNS neurotransmitters. A discussion of central neurotransmission is available in a number of textbooks in pharmacology and neuroscience, eg, Reference 2, and elsewhere in the *Encyclopedia* (see NEUROREGULATORS). A large number of chemical substances function in the mammalian CNS to regulate the transmission of information from one neuron to another. Neurotransmitters may be either excitatory or inhibitory, ie, may cause a depolarization or a hyperpolarization, respectively, of the neuronal membrane with which they interact. As a simplification, compounds that antagonize the actions of inhibitory transmitters tend to cause excitation and thus cause convulsions; compounds that facilitate inhibitory neurotransmission tend to be depressants. Conversely, agents that antagonize excitatory neurotransmission tend to be depressants and compounds that facilitate excitatory transmission tend to be convulsants.

Some naturally occurring analeptics have been known for centuries. Two of the best known and most thoroughly studied are strychnine [57-24-9] (**1**) and picrotoxin [124-87-8], a 1:1 combination of picrotoxinin [17617-45-7] (**2**) and picrotin [21416-53-5] (**3**). These continue to be of interest in the study of mammalian neurotransmission.

(1)   (2)   (3)

Strychnine (**1**), an alkaloid, was introduced into European medical practice in the early sixteenth century after being used as a rat poison (see ALKALOIDS). As of 1996, it was still used as a rodenticide. The total synthesis of this complex molecule is known (3,4).

There is good evidence that strychnine is a specific, competitive, post-synaptic antagonist of glycine in the CNS. Glycine, a known inhibitory transmitter in the mammalian CNS, is primarily located and functions at interneurons in the spinal cord. Glycine mediates inhibition of spinal cord neurons and is intimately involved in the regulation of spinal cord and brainstem reflexes. By directly antagonizing the inhibitory action of glycine, strychnine allows excitatory impulses to be greatly exaggerated, resulting in a characteristic seizure pattern known as opisthotonos. In humans, in whom the extensor muscles are normally dominant, a tonic hyperextension is observed, so that at its extreme, opisthotonos presents as a characteristic posture in which the back is arched and only the back of the head and the heels are touching the surface on which the patient is lying. In the presence of strychnine, all sensory stimuli produce exaggerated responses, and even slight sensory stimulation may precipitate convulsions. An important aspect of therapy is therefore to prevent the patient from receiving sensory stimulation.

Picrotoxin has been instrumental in establishing an inhibitory neurotransmitter role for the amino acid, gamma-aminobutyric acid (GABA), quantitatively the most important inhibitory neurotransmitter in the mammalian CNS. Whereas glycine is predominately localized in the spinal cord, GABA is more highly concentrated in the brain.

Picrotoxin, unlike strychnine and most other analeptics, is nonnitrogenous. The bitter principle is extracted from the Asian shrub, *Anamirta cocculus* L. Only picrotoxinin (**2**) has analeptic properties. Both picrotoxin and picrotin (**3**) have been synthesized by a multistep process starting with (−)-carvone (**5**); structures and absolute configurations have also been established (6). On GABAergic neurons in the CNS, there is a GABA binding site as well as a picrotoxin binding site, among other sites, surrounding the chloride channel (7). GABA acts to promote the influx of chloride into the cell by opening the chloride channel. When picrotoxinin is present, it binds to the picrotoxin binding site noncompetitively and acts to close the chloride channel, thereby antagonizing the ability of GABA to allow chloride to enter the cell and produce hyperpolarization. Other compounds that appear to function in a manner similar to picrotoxinin include some bicyclic cage compounds such as (^{35}S)t-butylbicyclophosphorothionate (TBPS) (**4**), the convulsant barbiturate isomer S(+)N-methyl-5-phenyl-5-propylbarbituric acid

($S(+)$MPPB) (**5**), and lindane [58-89-9] (**6**) (8). Pentylenetetrazol [54-95-5] (**7**), one of the first totally synthetic analeptics, is prepared by the reaction of cyclohexanone and hydrazoic acid (9). The compound, first introduced in the United States in 1927 as a treatment for barbiturate poisoning, was used to a limited extent for chemical shock therapy. As of the mid-1990s, it is used occasionally to enhance mental and physical activity in elderly patients and as a diagnostic aid, ie, as an electroencephalogram (EEG) activator, in epilepsy. Phentylenetetrazol is also an important laboratory tool for screening anticonvulsant drugs.

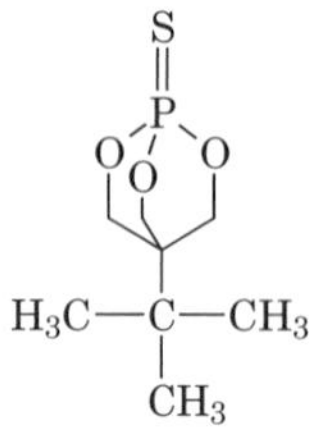

(**4**) 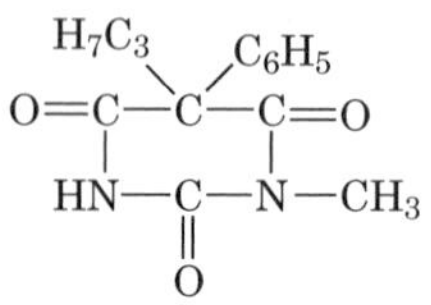   (**5**) 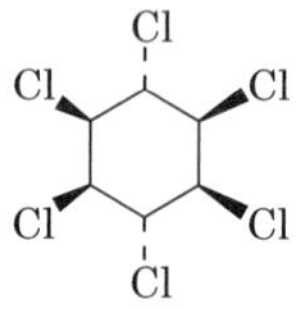   (**6**)

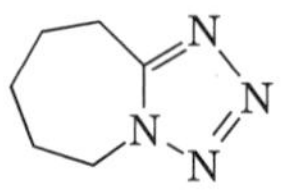

(**7**)

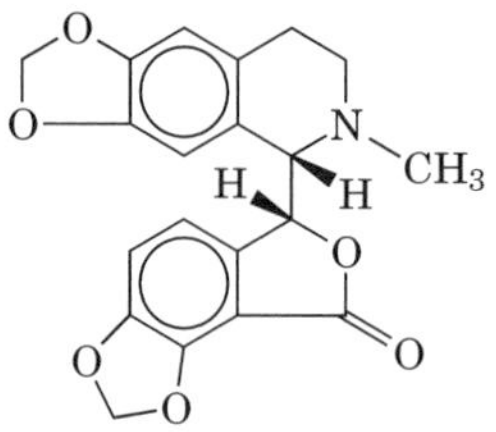

(**8**)

Bicuculline [485-49-4] (**8**) is another analeptic compound known to act by competitively antagonizing GABA at its receptor (9). There is no evidence that bicuculline has been evaluated in humans. Several other compounds, such as the steroid R5135 (**10**) or the arylaminopyridazines SR 95103 (**11**) and SR 95531 (**12**), appear to antagonize GABA competitively in a manner similar to that of bicuculline.

Flurothyl [333-36-8] (bis-(2,2,2-trifluoroethyl)ether) (**9**), an analeptic having strong convulsant properties, has been used for chemical shock therapy (13). The compound is unique in that it is a volatile fluorinated ether and its structure resembles those of many halogenated general anesthetics. Chemical shock therapy is rarely used.

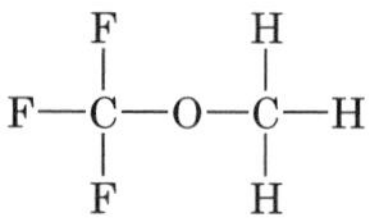

(**9**) 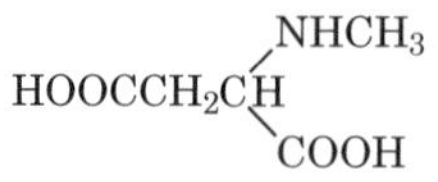   (**10**) 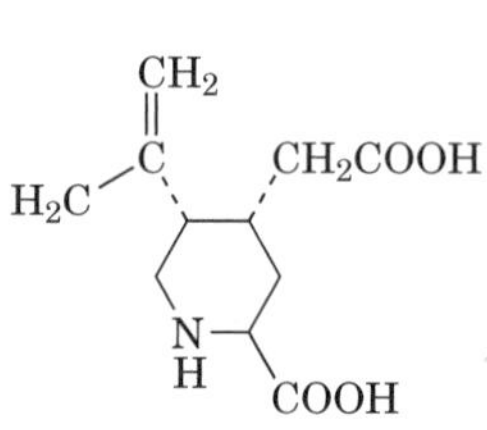   (**11**) 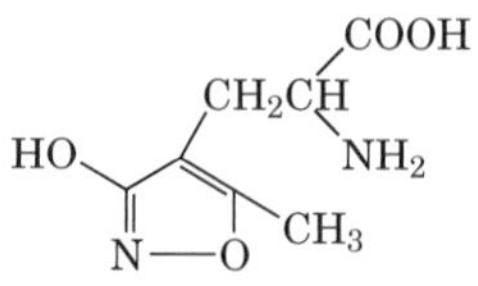   (**12**)

Compounds that have agonistic properties at glutamate or aspartate receptors are also CNS stimulants, readily cause convulsions, and presumably could also be employed as analeptics. Three separate excitatory amino acid receptor subtypes have been characterized pharmacologically, based on the relative potency of synthetic agonists. These three receptors are named for their respective prototypical agonists: N-methyl-D-aspartate [6384-92-5] (NMDA) (**10**), kainate from kainic acid [487-79-6] (**11**), and α-amino-3-hydroxy-5-methylisoxazole-4-propionate (AMPA) (**12**). All of the agonists, ie, NMDA, kainic acid, and AMPA, are stimulants and convulsants (14). These agents are used only experimentally.

Benzodiazepines have largely replaced barbiturates and barbiturate-like agents for use as anxiolytics and sedative–hypnotics. Because benzodiazepines rarely produce levels of CNS depression that require therapeutic intervention, the need for analeptics has decreased considerably. However, there are occasions in which the use of a respiratory stimulant may be warranted. By far the leading respiratory stimulant marketed in the United States is doxapram [309-29-5] (**13**), prepared by a unique rearrangement of the pyrrolidine [3471-97-4] (**14**) to the pyrrolidinone [3192-64-1] (**15**), followed by alkylation using morpholine (15).

(**14**)

(**15**)   (**13**)

Health and Safety Factors. Clinical side effects and LD_{50} values of most commercially available analeptics have been summarized (2). Overdoses produce symptoms of extreme CNS excitation, including restlessness, hyperexcitability, skeletal muscle hyperactivity, and in some cases convulsions.

Psychostimulants

Compounds having relatively specific cerebral stimulant properties are classified as psychostimulants or psychoanaleptics. Caffeine [58-08-2] (**16**), a mild psychostimulant, has been called the most widely used psychoactive substance on earth (16,17). Caffeine, theophylline [58-55-9] (**17**), and theobromine [83-67-0] (**18**) are three closely related alkaloids known as methylxanthines that occur in plants widely distributed throughout the world. The first two have CNS stimulant properties; the last is virtually inactive as a stimulant. The basis for the popularity of caffeine-containing beverages is their ability to elevate mood, decrease fatigue, and increase capacity for work. It is estimated that at least half the population of the world consumes tea on a regular basis; in the United States, coffee is the most important source of caffeine, and cola-flavored drinks seem a close second. The word caffeine is used exclusively herein even though some of the effects of caffeinated beverages may result from the theophylline content. Average caffeine contents per cup of brewed coffee (qv), instant coffee, tea (qv) (bagged), and cola beverages (12 oz (340 g)) are 110, 66, 46, and 47 mg, respectively (18) (see CARBONATED BEVERAGES).

The effects of low to moderate amounts of caffeine ingestion are generally salutary. At higher levels, however, more serious signs of CNS stimulation may be elicited. These may be expressed as nervousness, restlessness, insomnia, tremors, and anxiety. At higher doses, generalized convulsions may occur.

(**16**, R = R' = CH₃)

(**17**, R = CH₃; R' = H)

(**18**, R = H; R' = CH₃)

(**19**)

The mechanism by which the methylxanthines produce CNS stimulation is not clearly established. These agents may function, in part, to limit chloride channel activation in a manner similar to that of pentylenetetrazol (**7**) or biculculine (**8**). Another possibility is a specific antagonism of the inhibitory neurotransmitter adenosine [58-61-7] (**19**) (19).

Methylxanthines have a few valid therapeutic uses, including treatment of asthma and relief of dyspnea (see ANTIASTHMATIC AGENTS). The CNS stimulatory effects are also utilized for the treatment of the prolonged apnea that may be observed in premature infants. Theophylline may be combined with doxapram (**13**) for this use (20). The methylxanthine most widely used therapeutically is theophylline, although caffeine may also be used. For parenteral administration, a salt of theophylline is employed. There are several salts available, including theophylline ethylenediamine (aminophylline [317-34-0]) and oxtriphylline (choline theophyllinate). Other synthetic xanthines that are used include dyphylline [479-18-5] and enprofylline [41078-02-8] (21). Caffeine is obtained in pure form from tea waste, from the manufacture of caffeine coffee, and by total synthesis (22,23).

Sympathomimetics. Sympathomimetics are a group of mostly synthetic compounds that resemble the neurotransmitters epinephrine [51-43-4] and norepinephrine [*51-41-2*] pharmacologically and to some extent chemically (see EPINEPHRINE AND NOREPINEPHRINE). These agents have wide-ranging pharmacological effects, including, in some cases, profound CNS excitatory actions. Sympathomimetics that have selective central effects have been used in the treatment of narcolepsy, as an aid in weight reduction (see ANTIOBESITY DRUGS), and in the treatment of attention-deficit hyperactivity disorder (ADHD). The mechanism of action of the sympathomimetics in exerting their central effects is thought to be either directly, by interacting with an adrenoceptor (usually α-1-adrenoceptor) as an agonist, or indirectly, by causing the release of endogenous norepinephrine, which activates all adrenoceptors. Only a limited number of agents having sympathomimetic activity demonstrate CNS properties, ie, those that are able to penetrate the blood brain barrier by virtue of their lipid solubility.

The oldest of the centrally acting sympathomimetics is ephedrine [*299-42-3*] (**20**), used for over 5000 years in China before being introduced into Western medicine in 1924 (24). Ephedrine occurs in many varieties of plants of the genus *Ephedra* and may also be synthesized. Although formerly used extensively for its CNS effects as well as for its bronchodilator properties, ephedrine has been largely replaced by more effective and more selective agents. There are three other indirectly acting adrenomimetic compounds having CNS stimulant properties that have been employed clinically: amphetamine [*300-62-9*] (**21**), methamphetamine [*537-46-2*] (**22**), and methylphenidate [*113-45-1*] (**23**). D-Amphetamine is three to four times more potent in producing CNS stimulation than is L-amphetamine. Dextroamphetamine has been used to overcome fatigue, as an analeptic, as an aid in weight reduction, and in the treatment of ADHD. As of this writing, amphetamine is used to some extent in the therapy of narcolepsy, but is seldom used for weight reduction, even though it has marked effect on decreasing appetite (anorexic effect). Because tolerance develops rapidly to the anorexic effects, the weight loss is of limited duration (about two to three weeks). The development of tolerance, the abuse potential, and the insomnia and nervousness it causes have led to a marked reduction in amphetamine usage. Indeed, legal restrictions have markedly curtailed the production, as well as the availability, of both (**21**) and (**22**). Methylphenidate (**23**) is reported to have less abuse potential and as of 1996 is the drug of choice for ADHD (25).

R′
|
〈〉—CHCHNHR
|
CH₃

(**20**, R = CH₃; R′ = OH)
(**21**, R = R′ = H)
(**22**, R = CH₃; R′ = H)

〈〉—CH—〔N-H piperidine〕
O=C
|
OCH₃

(**23**)

C₆H₅   O   NH
       | |
       C   NH
       O

(**24**)

Pemoline [*2152-34-3*] (**24**), structurally dissimilar to amphetamine or methylphenidate, appears to share the CNS-stimulating properties. As a consequence, pemoline is employed in the treatment of ADHD and of narcolepsy. There are several other compounds that are structurally related to amphetamines, although not as potent and, presumably, without as much abuse potential. These compounds also have anorexic effects and are used to treat obesity. Some of the compounds available are phentermine [122-09-8], fenfluramine [458-24-2], and an agent that is available over-the-counter, phenylpropanolamine [14838-15-4] (26).

Side Effects and Abuse Potential. Sympathomimetics are one of the most abused classes of drugs marketed in the United States. Continued use for weight loss and insomnia relief leads to the development of tolerance and habituation. There is widespread use of amphetamine as well as other sympathomimetic compounds among recreational drug users. The abuse characteristics of amphetamines and related drugs are similar to that of cocaine [50-36-2], also a sympathomimetic. Acute intoxication from amphetamine-like drugs results in the patient exhibiting dizziness, confusion, tremor, irritability, hypertension, and cardiac palpitations. Acute paranoia and a state resembling schizophrenia may also be exhibited by individuals taking large amounts of amphetamines or cocaine (27). Because of the potential for abuse, the manufacture, distribution, and use of sympathomimetics are strictly controlled in the United States by the Drug Enforcement Agency (DEA).

Antidepressants

Depression. Disorders of mood or affect may be either a pathological state or a normal human emotion. The American Psychiatric Association has established diagnostic criteria that allow clinicians to distinguish between patients who require treatment and those who do not. It is estimated that about 5% of the adult population of the United States may be suffering from a mood disorder at any one time (28). The most common mood disorder, known as reactive depression, is commonly observed following adverse life events such as the death of a loved one. It may also be an accompaniment to other serious physical illnesses such as heart attacks, cancer, or debilitating diseases such as parkinsonism and alcoholism. In addition, depression may be an adverse effect of drugs, such as certain antihypertensives (see CARDIOVASCULAR AGENTS). Reactive depression is frequently expressed by depression, anxiety, or feelings of stress or guilt. Patients usually recover spontaneously, but medication and counseling frequently speed up the process. In contrast to reactive depression, the two principal mood disorders, unipolar disorder and bipolar disorder, appear to be genetically determined biochemical disorders which are lifelong diseases. Frequent swings between serious depression and reasonably normal behavior often result. These behavioral swings can make evaluation of therapeutic effectiveness extremely difficult (29).

The most common type of depression, known as unipolar disorder (major depression), accounts for about 25% of all depression. Signs include weight loss, loss of libido, alterations in sleep pattern, symptoms of negative self-image, suicidal thoughts, and overwhelming grief. The other type is known as bipolar disorder (manic–depressive disorder), and includes about 10–15% of depressions. Typically, the patient having bipolar disorder alternates between depression as seen in unipolar depression, and periods in which the symptoms are exactly the opposite: seemingly boundless energy, excessive talkativeness, increased libido, inflated self-esteem, and surges of creativity. The patient having bipolar illness cycles between depression and mania, in which the duration of each cycle is commonly measured in months. The patient having unipolar disorder, on the other hand, is usually in a state of constant depression (29).

Mechanism. The mechanisms involved in the etiology of mood disorders have been derived from an understanding of the mechanism of action of the drugs that are effective in the therapy of the disorders. The monoamine oxidase inhibitors (22–26) were the first effective drugs for the treatment of depression. Their mechanism of action is to elevate levels of those endogenous agents, eg, monoamines such as norepinephrine, dopamine [51-61-6], and serotonin [50-67-9], that are substrates for the enzyme monoamine oxidase (MAO). Compounds that decrease the concentrations of these same monoamines, such as reserpine [50-55-5], cause many of the signs and symptoms of depression. Many studies monitoring regulation of various monoamine receptors in the central nervous system have led to two important experimental findings. One is a downregulation of β-adrenoceptors following chronic administration of many drugs effective in depression. The other is an enhancement of transmission through a particular receptor, 5-HT_{1A}, after chronic administration of all clinically effective antidepressants and after electroconvulsive treatment. A detailed review of the neuropharmacology of antidepressants is available (30).

Treatment. Most, although not all, of the drugs effective in the treatment of depression are CNS stimulants. Until the middle of the twentieth century, pharmacological treatment was symptomatic, supportive, and frequently ineffective. Patients having suicidal tendencies, a common finding in depression, were isolated for their own protection. The development around the 1900s of psychotherapy and the discovery in the 1930s of the use of chemoshock, utilizing pentylenetetrazol or insulin treatment, and electroconvulsive therapy (ECT) gave the first indication that depression could be successfully treated. The discovery in the 1950s of agents known as monoamine oxidase inhibitors (MAOI) and later of the tricyclic antidepressants (TCA) has led to more effective and safer preparations.

Monoamine Oxidase Inhibitors. MAOIs inactivate the enzyme MAO, which is responsible for the oxidative deamination of a variety of endogenous and exogenous substances. Among the endogenous substances are the neurotransmitters, norepinephrine, dopamine, and serotonin. The prototype MAOI is iproniazid [54-92-2] (**25**), originally tested as an antitubercular drug and a close chemical relative of the effective antitubercular, isoniazid [54-85-3] (**26**). Tubercular patients exhibited mood elevation, although no relief of their tuberculosis, following chronic administration of iproniazid. In 1952 it was shown that the compound had potent MAO inhibitory activity, thus providing a biochemical basis for the effects (31). Iproniazid was introduced into psychiatry in 1957 (32,33). Since that time, many MAOI have been tested and marketed in the United States and in other countries. Because of toxicity, primarily of the liver, only three agents are available: isocarboxazid [59-63-2] (**27**), phenelzine [51-71-8] (**28**), and tranylcypromine [155-09-9] (**29**). The relative MAO inhibitory potencies of these agents as compared to iproniazid are listed in Table 1.

Table 1. Potency Data for MAO Inhibitors[a]

Generic name	Trade name	Manufacturer	Year of introduction	Relative MAO inhibitory potency[b]
iproniazid	Marsilid	Roche	1957	1
isocarboxazid	Marplan	Roche	1959	3.1
phenelzine	Nardil	Parke-Davis	1959	18
tranylcypromine	Parnate	Smith, Kline and French	1961	45

[a]Ref. 34.
[b]Rat brain MAO, tyramine as substrate.

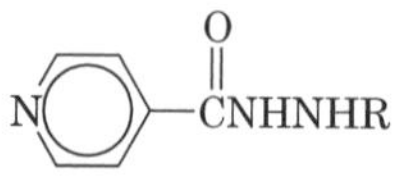

(**25**, R = CH(CH$_3$)$_2$)
(**26**, R = H)

(**30**) R = CH$_2$CH$_2$CNHCH$_2$C$_6$H$_5$

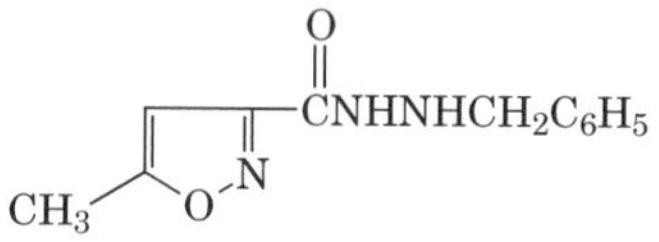

(**27**)

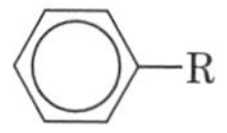

(**28**, R = CH$_2$CH$_2$NHNH$_2$)
(**31**, R = CH(CH$_3$)NHNH$_2$)

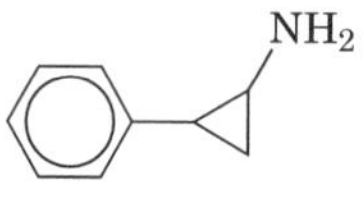

(**29**)

Nialamide [51-12-7] (**30**) and mebanazine [65-64-5] (**31**) are two MAO inhibitors marketed in Europe that have structural similarities to iproniazid and phenelzine, respectively. Both compounds are prepared by standard methods (35,36).

As of the mid-1990s, use of MAOIs for the treatment of depression is severely restricted because of potential side effects, the most serious of which is hypertensive crisis, which results primarily from the presence of dietary tyramine. Tyramine, a naturally occurring amine present in cheese, beer, wine, and other foods, is an indirectly acting sympathomimetic, that is, it potently causes the release of norepinephrine from sympathetic neurons. The norepinephrine that is released interacts with adrenoceptors and, by interacting with α-adrenoceptors, causes a marked increase in blood pressure; the resultant hypertension may be so severe as to cause death.

Normally, dietary tyramine is broken down in the gastrointestinal tract by MAO and is not absorbed. In the presence of MAOI, however, all of its potent sympathomimetic actions are seen. Other side effects of MAOI include excessive CNS stimulation, orthostatic hypotension, weight gain, and in rare cases hepatotoxicity. Because the monoamine oxidase inhibitors exhibit greater toxicity, yet no greater therapeutic response than other, newer agents, clinical use has been markedly curtailed. The primary use for MAOIs is in the treatment of atypical depressions, eg, those associated with increased appetite, phobic anxiety, hypersomnolence, and fatigues, but not melancholia (2).

Tricyclic Antidepressants. Imipramine [50-49-7] (32), which was the first tricyclic antidepressant to be developed, is one of many useful psychoactive compounds derived from systematic molecular modifications of the antihistamine promethazine [60-87-7] (see HISTAMINE AND HISTAMINE ANTAGONISTS). The sulfur atom of promethazine was replaced with an ethylene bridge and the dimethylamino group attached to an *n*-propyl group, rather than to an isopropyl one, of the side chain. The actual synthesis of (32) is typical of the compounds in this class (37).

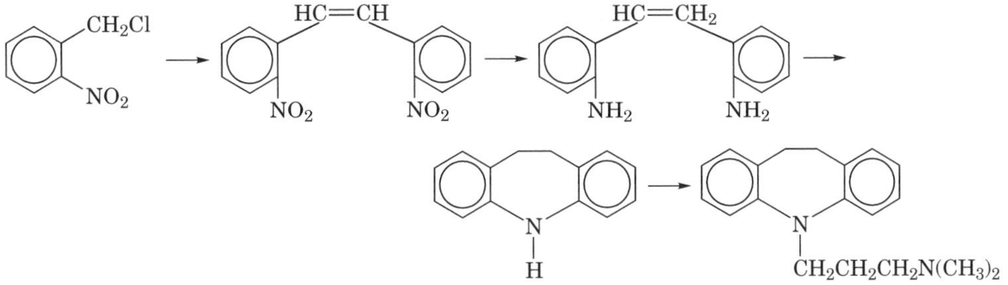

(32)

Selected for clinical trials as a compound to calm agitated patients, imipramine was relatively ineffective. However, it was observed to be effective in the treatment of certain depressed patients (38). Early studies on the mechanism of action showed that imipramine potentiates the effects of the catecholamines, primarily norepinephrine. This finding, along with other evidence, led to the hypothesis that the compound exerts its antidepressant effects by elevating norepinephrine levels at central adrenergic synapses. Subsequent studies have shown that the compound is a potent inhibitor of norepinephrine reuptake and, to a lesser extent, the uptake of serotonin, thus fitting the hypothesis that had been developed to explain the antidepressant actions of MAOIs.

Following the successful introduction of imipramine (32), many related compounds were prepared and clinically evaluated for antidepressant effects. Amitriptyline [50-48-6] (33), structurally related to imipramine but having a $C=CH$ group replacing the heterocyclic $N-CH_2$ fragment, has comparable pharmacological and clinical effects to imipramine. Both are particularly useful in the treatment of depressed patients exhibiting psychomotor agitation. Trimipramine [739-71-9] (34), which is also structurally similar to imipramine, differing only in the branched side chain, has similar activity as the latter, but is slightly less potent.

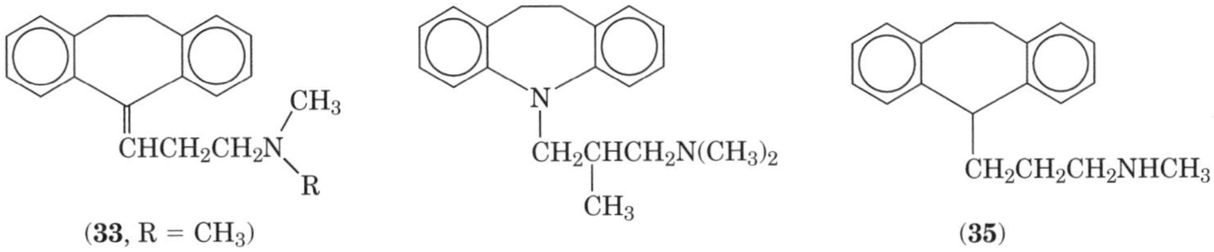

(33, R = CH$_3$)
(36, R = H)

(34)

(35)

Desipramine [50-47-5] (**35**) and nortriptyline [72-69-5] (**36**) are demethylated derivatives and principal metabolites of (**32**) and (**33**), respectively. Both compounds possess less sedative and stronger psychomotor effects than the tertiary amine counterparts, probably because tricyclics containing secondary amine groups generally show greater selectivity for inhibiting the reuptake of norepinephrine compared with the reuptake of serotonin. Protriptyline [438-60-8] (**37**), a structural isomer of nortriptyline, is another important secondary amine that displays a similar clinical profile.

(**37**)   (**38**)   (**39**)

(**40**)   (**41**)

Doxepin [1668-19-5] (**38**), unlike other commercially available tricyclics, has an oxygen atom in the bridge between the two aromatic rings. It is marketed as a cis–trans mixture (1:5) of isomers, both of which are active. This close relative of amitriptyline (**33**) has both sedative and anxiolytic properties associated with its antidepressant profile. Maprotiline [10262-69-8] (**39**) and amoxapine [14028-44-5] (**40**) are pharmacologically, although not chemically, similar to the tricyclic secondary amines. Clomipramine [303-49-1] (**41**) has similar pharmacological and antidepressant efficacy. However, clomipramine is approved by the U.S. FDA only for the treatment of obsessive–compulsive disorder. Representative brands of tricyclic antidepressants marketed in the United States are listed in Table 2.

Side Effects and Toxicity.   Adverse effects to the tricyclic antidepressants, primarily the result of the actions of these compounds on either the autonomic, cardiovascular, or central nervous systems, are summarized in Table 3. The most serious side effects of the tricyclics concern the cardiovascular system. Arrhythmias, which are dose-dependent and rarely occur at therapeutic plasma levels, can be life-threatening. In order to prevent adverse effects, as well as to be certain that the patient has taken enough drug to be effective, the steady-state serum levels of tricyclic antidepressant drugs are monitored as a matter of good

Table 2. Biological Data for U.S. Marketed Tricyclic Antidepressants

Generic name	Trade name	Manufacturer	Year of introduction	Uptake inhibition, IC$_{50}$, 10^{-7} M	
				Norepinephrine[a]	Serotonin[b]
imipramine	Tofranil	Geigy	1959	0.75	3.2
amitriptyline	Elavil	Merck, Sharp, and Dohme	1961	1.3	3.0
desipramine	Norpramine	Merrell-National	1964	0.014	18
nortriptyline	Aventyl	Lilly	1969	0.29	15
trimipramine	Surmontil	Ives	1979	77	260
protriptyline	Vivactil	Merck, Sharp and Dohme	1967	0.008	13
doxepin	Sinequan	Roerig	1969	5.5	39
amoxapine	Asendin	Lederle	1980		

[a]In mouse atria (39).
[b]In thrombocytes (40,41).

943

Table 3. Clinical Features of Antidepressants[a]

Agent or class	CNS effects[b]	Cardiovascular effects[c]		Anticholinergic effects[c]	Weight change
		Orthostatic hypotension	Arrhythmias		
MAOI	+	+++		++	+
TCA	−	+++	++	++++	+
SSRI[d]	+	0	0	+	−
buprion	+	0	0	+	0
trazadone	−	++	0	+	+

[a]Ref. 30.
[b]Stimulation, +; sedation, −.
[c]No effect, 0; increasing effect, +, ++, +++, and ++++.
[d]SSRI = selective serotonin reuptake inhibitor.

practice. A comprehensive review of structure–activity relationships among the tricyclic antidepressants is available (42).

Selective Serotonin Reuptake Inhibitors. In 1987, the FDA approved fluoxetine [54910-89-3] (**42**) for use in the treatment of major depression. Fluoxetine and related compounds, sertraline [79617-96-2] (**43**), and paroxetine [61869-08-7] (**44**) appear to inhibit selectively the reuptake of serotonin while having virtually no effect on the uptake of norepinephrine or dopamine. It is hypothesized that the elevated levels of serotonin that occur at the synapse as the result of decreased uptake in time produces a desensitization of certain serotonin receptors, eg, 5-HT$_{1A}$ and 5-HT$_{1B}$ (43). Ultimately, these events are believed to lead to a potentiation of serotonin neurotransmission at central synaptic sites. In addition to its effects on serotonin, sertraline has been shown to produce a down-regulation of β-adrenoceptors following chronic administration. These selective serotonin reuptake inhibitors (SSRIs) do not appear to be more effective than the tricyclics for the treatment of depression. However, the SSRIs do appear to lack many of the side effects associated with the tricyclics and other antidepressants, and are therefore both safer for and more readily accepted by the patient. Sexual dysfunction is a common complaint of most antidepressants. The principal side effects of the SSRIs appear to consist of headache, nausea, and restlessness. A summary of adverse effects of the SSRI drugs available in the United States as of 1996 is provided in Table 4.

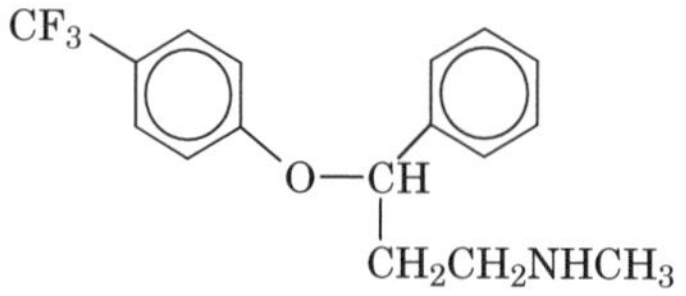

CF$_3$ ... O—CH ... CH$_2$CH$_2$NHCH$_3$

(**42**)

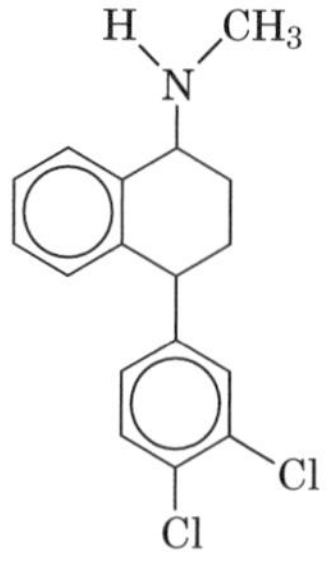

(**43**)

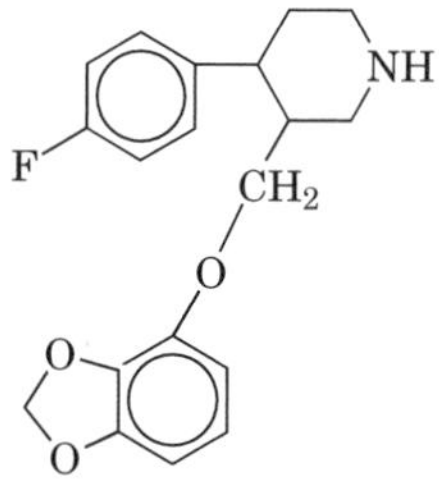

(**44**)

Table 4. Selective Serotonin Reuptake Inhibitors[a]

Generic name	Trade name	Manufacturer	Comment
fluoxetine	Prozac	Eli Lilly	possible side effects include rash, urticaria, nausea, diarrhea, CNS stimulation, insomnia, tremors, seizures, anorgasmia, ejaculatory delay, weight loss
sertraline	Zoloft	Pfizer	same possible side effects, but less CNS stimulation
paroxetine	Paxil	SmithKline Beecham	same possible side effects, but lower incidence of dry mouth

[a]Ref. 30.

Miscellaneous Antidepressants. There are a few agents that either chemically or pharmacologically do not fit neatly into any of the categorized antidepressant agents. Trazodone [*19794-93-5*] (**45**) was introduced as a safer, less toxic, and faster-acting antidepressant. It is effective in some patients, virtually ineffective in others. Trazodone, which appears to have effects at both serotonin and norepinephrine synapses, causes a high level of sedation, as well as dizziness, hypotension, and nausea. It is also reported to cause priapism, an uncommon but serious side effect. Nefazodone (**45a**) is similar in structure to trazodone and appears to share most of its clinical and pharmacological effects. Although priapism did not occur during early clinical studies, this is still a possibility considering the structural similarities with trazodone. Buprion [*34911-55-2*] (**46**) is devoid of inhibitory actions on both serotonin and norepinephrine uptake systems. However, it is a potent inhibitor of the uptake system for dopamine. Buprion is also not a monoamine oxidase inhibitor. It is well tolerated by patients of all ages, including the elderly, and is virtually devoid of cardiovascular and antimuscarinic side effects, particularly when compared with the tricyclics. However, buprion is a CNS stimulant and can cause convulsions at higher doses. More commonly, nervousness and insomnia are observed. These drugs are included in Table 3.

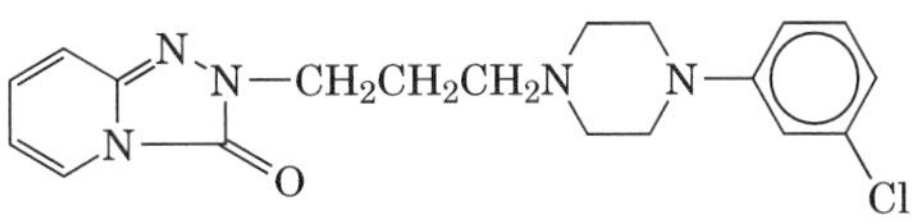

(**45**)

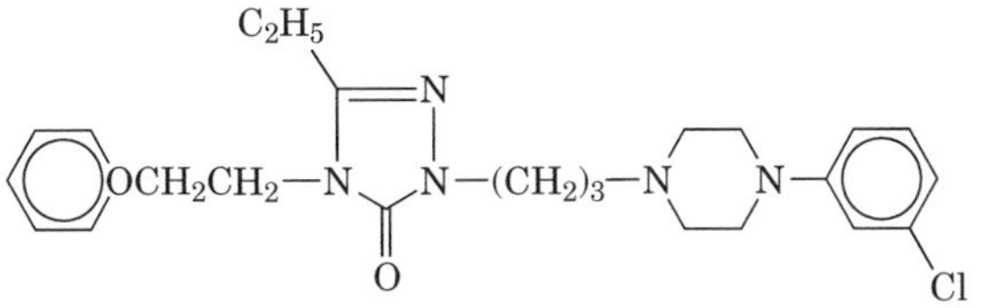

(**45a**)

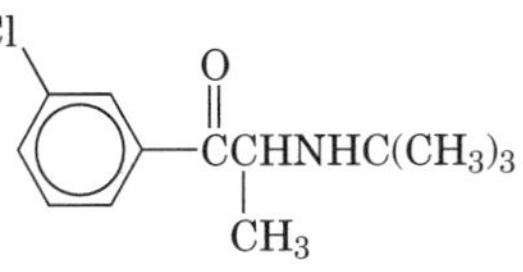

(**46**)

Other Drugs.   Agents not considered to be CNS stimulants yet employed for the treatment of certain types of depression includes lithium carbonate for the treatment of bipolar disorder. In most patients, lithium is the sole agent used to control manic behavior and is very effective (see PSYCHOPHARMACOLOGICAL AGENTS).

The market for antidepressant medication is very large. The market in the United States is ~\$6 × 10^9 per year.

BIBLIOGRAPHY

"Stimulants" in *ECT* 3rd ed., Vol. 21, pp. 747–761, by W. J. Welstead, Jr., A. H. Robins Co.

1. J. Kjaer-Larsen, *Lancet*, **271**, 967 (1956).
2. C. R. Craig, in C. R. Craig and R. E. Stitzel, eds., *Modern Pharmacology*, 4th ed., Little, Brown & Co., Boston, Mass., 1994, p. 328.
3. R. B. Woodward and co-workers, *J. Am. Chem. Soc.* **76**, 4749 (1954).
4. R. B. Woodward and co-workers, *Tetrahedron*, **19**, 247 (1963).
5. E. J. Corey and H. L. Pearce, *J. Am. Chem. Soc.* **101**, 5841 (1979).
6. L. A. Porter, *Chem. Rev.* **67**, 441 (1967).
7. D. R. Curtis, A. W. Duggan, D. Felix, and G.A.R. Johnston, *Nature*, **228**, 676 (1970).
8. R. F. Squires and co-workers, *Mol. Pharmacol.* **23**, 326 (1983).
9. K. F. Schmidt, *Ber.* **57**, 704 (1924).
10. P. Hunt and S. Clements-Jewery, *Neuropharmacol.* **20**, 357 (1981).
11. J. P. Chambon and co-workers, *Proc. Natl. Acad. Sci. USA*, **82**, 1832 (1985).
12. M. Heaulme and co-workers, *J. Neurochem.* **48**, 1677 (1987).
13. U.S. Pat. 3,363,006 (Jan. 9, 1968), J. F. Olin (to Pennsalt Chemicals Corp.).
14. D. D. Schoepp and co-workers, *Eur. J. Pharmacol.* **182**, 421 (1990).
15. C. D. Lunsford and co-workers, *J. Med. Chem.* **7**, 302 (1964).
16. S. H. Snyder and co-workers, *Science*, **211**, 1408 (1981).
17. S. H. Snyder and co-workers, *Proc. Nat. Acad. Sci.* **78**, 3260 (1981).
18. M. L. Bunker and M. McWilliams, *J. Am. Diet. Assoc.* **74**, 28 (1979).
19. T. W. Rall, *Pharmacologist*, **24**, 277, 1982.
20. K. J. Barrington and co-workers, *Pediatrics*, **80**, 22 (1987).
21. K.-E. Andersson and C. G. A. Persson, eds., *Anti-Asthma Xanthines and Adenosine*, Exerpta Medica, Amsterdam, the Netherlands, 1985.
22. U.S. Pat. 2,817,588 (Dec. 24, 1957), W. E. Barch (to Standard Brands).
23. W. Traube, *Chem. Ber.* **33**, 3052 (1900).
24. K. K. Chen and C. Schmidt, *Medicine, Baltimore*, **9**, 1 (1930).
25. J. Elia, *Drugs*, **46**, 863 (1993).
26. T. Silverstone, *Drug Alcohol Depend.* **17**, 151 (1986).
27. S. L. Satel, S. M. Southwick, and F. H. Gawin, *Am. J. Psychiatry*, **148**, 495 (1991).
28. E. Richelson, *Psych. Clin. N.A.* **16-3**, 461 (1993).
29. G. L. Gessa, W. Fratta, L. Pani, and G. Serra, *Depression and Mania From Neurobiology to Treatment*, Lippincott-Raven, Hagerstown, Md., 1995, p. 1.
30. A. J. Azzaro and H. E. Ward, in Ref. 2, p. 397.
31. E. A. Zeller and co-workers, *Experientia*, **8**, 349 (1952).
32. G. E. Crane, *Psychiatr. Res. Rep.* **8**, 142 (1957).
33. N. S. Kline, *J. Clin. Exp. Psychopathol.* **19**, 72 (1958).
34. D. R. Maxwell, W. R. Gray, and E. M. Taylor, *Br. J. Pharmacol.* **17**, 310 (1961).
35. U.S. Pat. 3,040,061 (June 10, 1962), B. M. Bloom and R. E. Carnham (to Pfizer).
36. C. G. Overberger and A. V. DiGiulio, *J. Am. Chem. Soc.* **80**, 6562 (1958).

37. W. Schindler and F. Haeflinger, *Helv. Chim. Acta*, **37**, 472 (1954).
38. R. Kuhn, *Am. J. Psychiat.* **115**, 459 (1958).
39. J. Hyttel, *Psychopharmacology*, **51**, 225 (1977).
40. G.V.R. Born and R. E. Gillson, *J. Physiol.* **146**, 472 (1977).
41. R. S. Stacey, *Br. J. Pharmacol.* **16**, 284 (1961).
42. F. J. Zeelen, in F. Hoffmeister and G. Stille, eds., *Psychotropic Agents*, Part I, Springer-Verlag, New York, 1980, p. 352.
43. R. J. Baldessarini, *J. Clin. Psychiat.* **50**, 117 (1989).

General Reference

R. H. Levy, R. H. Mattson, and B. S. Meldrum, eds., *Antiepileptic Drugs*, 4th ed., Raven Press, New York, 1995.

CHARLES R. CRAIG
West Virginia University

STONEWARE. See CERAMICS.

STORAX. See RESINS, NATURAL.

STOUT. See BEER.

STREPTOMYCIN AND RELATED ANTIBIOTICS. See ANTIBIOTICS.

STRESS CRACKING. See FRACTURE MECHANICS; MATERIALS RELIABILITY.

STRONTIUM AND STRONTIUM COMPOUNDS

Strontium

Strontium [*7440-24-6*], Sr, is in Group 2 (IIA) of the Periodic Table, between calcium and barium. These three elements are called alkaline-earth metals because the chemical properties of the oxides fall between the hydroxides of alkali metals, ie, sodium and potassium, and the oxides of earth metals, ie, magnesium, aluminum, and iron. Strontium was identified in the 1790s (1). The metal was first produced in 1808 in the form of a mercury amalgam. A few grams of the metal was produced in 1860–1861 by electrolysis of strontium chloride [10476-85-4].

Strontium forms 0.02–0.03% of the earth's crust and is present in igneous rocks in amounts averaging 375 ppm. It is the fifth most abundant metallic ion in seawater, occurring in quantities of ca 14 grams per metric ton. Strontium rarely forms independent minerals in igneous rocks but usually occurs as a minor constituent of the rock-forming minerals. Independent strontium minerals do develop in or near sedimentary rocks mainly associated with beds or lenses of gypsum, anhydrite, or rock salt; in veins associated with limestone and dolomite; disseminated in shales, marls, and sandstones; or as a gangue mineral in lead–zinc deposits.

Properties. Strontium is a hard white metal having physical properties shown in Table 1. It has four stable isotopes, atomic weights 84, 86, 87, and 88; and one radioactive isotope, strontium-90 [10098-97-2], which is a product of nuclear fission. The most abundant isotope is strontium-88.

The chemical properties of strontium are intermediate between those of calcium and barium. Strontium is more reactive than calcium, less reactive than barium. Strontium is bivalent and reacts with H_2 to form SrH_2 [13598-33-9] at reasonable speed at 300–400°C. It reacts with H_2O, O_2, N_2, F, S, and halogens to produce the expected compounds corresponding to its valence (+2). Strontium and the alkaline-earth metals are less active than the alkali metals, but all are strong reducing agents. At elevated temperatures, strontium reacts with CO_2 to form the carbide [12071-29-3] and strontium oxide [1314-11-0] (3):

$$5\,Sr + 2\,CO_2 \longrightarrow SrC_2 + 4\,SrO$$

Metallic strontium dissolves in liquid ammonia. The metal and its salts impart a brilliant red color to flames.

Table 1. Physical Properties of Strontium[a]

Property	Value
atomic weight	87.62
melting range, °C	768–791
boiling range, °C	1350–1387
density, g/cm^3	2.6
crystal system	face-centered cubic
lattice constant, pm	605
latent heat of fusion, kJ/kg[b]	104.7
electrical resistivity, $\mu\Omega$/cm	22.76
stable isotopes, % abundance	
^{84}Sr [15758-49-3]	0.56
^{86}Sr [13982-14-4]	9.86
^{87}Sr [13982-64-4]	7.02
^{88}Sr [14119-10-9]	82.56

[a] Refs. 1–3.
[b] To convert J to cal, divide by 4.184.

Occurrence. The principal strontium mineral is celestite, naturally occurring strontium sulfate. Celestite and celestine [7759-02-6] both describe this mineral. However, celestite is the form most widely used in English-speaking countries. Celestite has a theoretical strontium oxide content of 56.4 wt %, a hardness of 3–3.5 on Mohs' scale, and a specific gravity of 3.96. It is usually white or bluish white and has an orthorhombic crystal form.

Deposits of celestite in Gloucestershire, the United Kingdom, represented the main source of the world supply from 1884 to 1941 and provided up to 90% of the world strontium supply (4). During World War II, shipments to the United States and Western Europe from the United Kingdom were disrupted, and celestite deposits in Mexico and Spain were developed.

Strontianite is the naturally occurring form of strontium carbonate. It has a theoretical strontium oxide content of 70.2%, but no economically workable deposits are known. There are some naturally occurring strontium–barium and strontium–calcium isomorphs, but none has economic importance.

The scale of celestite mining has never been big enough to justify highly efficient mining methods or extensive exploration. Hand digging of the mineral is not uncommon and the ore is usually sorted by hand. Celestite used in the United States has an average strontium sulfate content of 90 wt % with a range of 88–93 wt %. As production has increased, more automated methods, including gravity separation and flotation, have been used more frequently.

Production. Metallic strontium was first successfully produced by the electrolysis of fused strontium chloride. Although many attempts were made to develop this process, the deposited metal has a tendency to migrate into the fused electrolyte and the method was not satisfactory. A more effective early method was that described in Reference 5. Strontium oxide is reduced thermally with aluminum according to the following reaction:

$$3\,SrO + 2\,Al \longrightarrow 3\,Sr + Al_2O_3$$

This reaction occurs in a vacuum and the gaseous metal is condensed in a cooler part of the apparatus. All strontium metal is produced commercially by the thermal reduction process in alloy steel retorts.

Uses. The main application for strontium is in the form of strontium compounds. The carbonate, used in cathode ray tubes (CRTs) for color televisions and color computer monitors, is used both in the manufacturing of the glass envelope of the CRT and in the phosphors which give the color.

Historically, strontium metal was produced only in very small quantities. Rapid growth of metal production occurred during the late 1980s, however, owing to use as a eutectic modifier in aluminum–silicon casting alloys. The addition of strontium changes the microstructure of the alloy so that the silicon is present as a fibrous structure, rather than as hard acicular particles. This results in improved ductility and strength in cast aluminum automotive parts such as wheels, intake manifolds, and cylinder heads.

Annual consumption of strontium metal has increased from a few kilograms to several hundred metric tons. The largest single producer worldwide of strontium metal is Timminco Metals, a division of Timminco Ltd. (Toronto, Canada), which commissioned a facility in 1987 in Westmeath, Ontario, for production of high purity strontium metal. Other producers of strontium metal include Calstron Corporation (Memphis, Tennessee) and Pechiney Electrometallurgie (France).

Prior to its addition to the aluminum casting alloys, the strontium metal is usually alloyed into the form of a master alloy. These master alloys are typically 10% Sr–90% Al or 90% Sr–10% Al, and improve the dissolution and handling characteristics of strontium in the foundry.

A second, more recently developed use for strontium metal is as an inoculant in ductile iron castings. Inoculants provide nuclei upon which graphite forms during the solidification of cast iron, thus preventing the formation of white cast iron. Elkem Metals Company has commercialized a range of fine-sized foundry inoculants for iron castings. These inoculants, called superseed, are ferrosilicon alloys containing 50 or 75% Si, 0.8% Sr. Most of the balance is iron.

Other newer applications developed for strontium metal include modifying the structure of 6000 series Al extrusion alloys, reducing the shrinkage microporosity in magnesium gravity-casting alloys, and as an addition to lead alloys used for starting, lighting, and ignition (SLI) batteries. None of these latest applications had reached any significant commercial usage as of 1996.

Strontium Compounds

Strontium has a valence of +2 and forms compounds that resemble the compounds of the other alkaline-earth metals (see BARIUM COMPOUNDS; CALCIUM COMPOUNDS). Although many strontium compounds are known, there are only a few that have commercial importance and, of these, strontium carbonate [1633-05-2], $SrCO_3$, and strontium nitrate [10042-76-9], $Sr(NO_3)_2$, are made in the largest quantities. The mineral celestite [7759-02-6], $SrSO_4$, is the raw material from which the carbonate or the nitrate is made.

Production. World production of strontium compounds averaged <20,000 t/yr throughout the 1950s and 1960s. Increased demand for strontium compounds led to a sharp rise in output between 1969 and 1971 as the use of strontium carbonate in the screens and glass envelopes of color televisions began to grow. Demand for strontium carbonate in the 1980s for computer monitors led to a steady rise in output of strontium compounds. Peak production of >291,000 t was reached in 1990. Demand then fell before stabilizing at 198,000 t in 1993.

As of 1996, Solvay Barium Strontium GmbH was the world's largest producer of strontium carbonate at its combined barium–strontium plant at Bad Hunningen (Germany). Cia Minera La Valencia (Mexico) was the next largest producer, followed by Chemical Products Corporation of the United States, which has two plants, one in the United States and one in Mexico. Dachan Specialty Chemicals (South Korea) was another principal producer. Together these four companies accounted for almost 80% of world capacity.

Mexico is the principal producer of strontium minerals, accounting for 36% of the world total in 1993. Iran and Turkey were primary suppliers of strontium to the former USSR, but the breakup of the USSR led to a significant reduction in demand. The dramatic decline in output of strontium minerals in Turkey was also partly a result of a drop in demand from the Solvay strontium carbonate plant in Germany. In order to secure supplies of strontium minerals, Solvay has opened a celestite mine in Spain. Mexico, Spain, Turkey, and Iran are the four largest-producing areas of strontium minerals as of this writing (ca 1996). Output from the deposits in the United Kingdom has dropped virtually to none. Deposits formerly mined in Cape Breton (Nova Scotia, Canada) are no longer in operation. Whereas strontium mineral deposits occur widely in the United States, these are uneconomical in comparison to the four principal producing countries. China, however, is becoming an increasingly important producer of strontium carbonate and had an estimated production capacity of 35,000 t/yr as of 1995.

Strontium carbonate is the principal strontium compound consumed. Other commercial compounds include strontium nitrate and strontium hydrate. The latter is available in an anhydrous form or as the dihydrate or octahydrate form and is called strontium hydrate.

There are two main processes for conversion of celestite, ie, strontium sulfate, to strontium carbonate. The principal process is the black ash process. Strontium nitrate is produced by dissolving celestite in nitric acid and purifying it. Most other strontium compounds are produced from strontium nitrate. To service this market, NOAH Technologies Corporation (San Antonio, Texas) has established a plant in Mexico to manufacture most commercial- and reagent-grade strontium compounds except strontium carbonate.

Economic Aspects. Celestite is mined and then shipped to countries where chemical processing takes place. Spain is the largest celestite-exporting country, as almost all of its output is exported. Mexico is the next largest exporter. About half of its annual output is exported. Most of celestite production from Turkey and Iran is also exported. Iranian exports are believed to be shipped to the CIS and other Eastern European countries.

Total exports of strontium carbonate have increased significantly in recent years, particularly those from Mexico. Between 1990 and 1993, Mexico overtook Germany as the principal exporting country.

Prices for strontium minerals and compounds are usually fixed directly between the producer and consumer. The use of other strontium compounds is growing, eg, strontium nitrate in pyrotechnics; strontium chloride, dental desensitizing; strontium hydroxide sugar refining and separating; strontium chromate, pigments and corrosion inhibitors; and strontium acetate, catalysts manufacturing. Typical prices for celestite ranged between U.S. $70–$140/t for minimum 92% $SrSO_4$ in the early to mid-1990s. Strontium carbonate prices were in the $900/t range in 1994. Strontium carbonate is sold in two grades: granular and powder. Granular grade, also referred to as glass grade, is made by calcining the powder grade. The powder grade, also known as chemical or ferrite grade, has an average particle size of 1 μm.

Uses. The United States and Japan account for about three quarters of the world demand for strontium carbonate, ca 150,000 t/yr. In 1993, 67% of U.S. and 73% of Japanese demand for strontium carbonate were accounted for by the manufacture of cathode ray tubes (CRTs). The other principal application for strontium carbonate is in permanent ceramic magnets, which represented 13% of U.S. strontium carbonate consumption and 17% of Japanese. Other applications for strontium compounds include pyrotechnics (qv), ca 11%; pigments (qv), 5%; and electrolytic zinc refining, 2%.

Health and Safety Factors.   The strontium ion has a low order of toxicity, and strontium compounds are remarkably free of toxic hazards. Chemically, strontium is similar to calcium, and strontium salts, like calcium salts, are not easily absorbed by the intestinal tract. Strontium carbonate has no commonly recognized hazardous properties. Strontium nitrate is regulated as an oxidizer that promotes rapid burning of combustible materials, and it should not be stored in areas of potential fire hazards.

Strontium Acetate.   Strontium acetate [543-94-2], $Sr(CH_3CO_2)_2$, is a white crystalline salt with a specific gravity of 2.1, and it is soluble to the extent of 36.4 g in 100 mL of water at 97°C. It crystallizes as the tetrahydrate or as the pentahydrate below 9.5°C. When heated, strontium acetate decomposes.

Strontium Carbonate.   Strontium carbonate, $SrCO_3$, occurs naturally as strontianite in orthorhombic crystals and as isomorphs with aragonite, $CaCO_3$, and witherite, $BaCO_3$. There are deposits in the United States in Schoharie County, New York; in Westphalia, Germany; and smaller deposits in many other areas. None is economically workable. Strontianite has a specific gravity of 3.7, a Mohs' hardness of 3.5, and it is colorless, gray, or reddish in color.

Strontium carbonate is a colorless or white crystalline solid having a rhombic structure below 926°C and a hexagonal structure above this temperature. It has a specific gravity of 3.70, a melting point of 1497°C at 6 MPa (60 atm), and it decomposes to the oxide on heating at 1340°C. It is insoluble in water but reacts with acids, and is soluble in solutions of ammonium salts.

Production.   In the commercial production of strontium carbonate, celestite ore is crushed, ground, and stored in bins before it is fed to rotary kilns. As the ground ore is being conveyed to the kilns, it is mixed with ground coke. In the kilns, the celestite is reduced to strontium sulfide [1314-96-1], known as black ash, according to the reaction:

$$SrSO_4 \text{ (s)} + 2\text{ C (s)} \longrightarrow SrS \text{ (s)} + 2\text{ CO}_2 \text{ (g)}$$

The product stream from the kilns is collected in storage bins. Black ash from the bins is fine-ground in a ball mill and fed to a leacher circuit, which is a system of stirred tanks, where it is dissolved in water and the muds are separated by countercurrent decantation. The solution from the decantation is passed through filter presses; the muds are washed, centrifuged, and discarded. The filtered product, a saturated solution containing 12–13 wt % strontium sulfide, is sent to an agitation tank where soda ash is added to cause precipitation of strontium carbonate crystals:

$$SrS \text{ (l)} + Na_2CO_3 \text{ (s)} \longrightarrow SrCO_3 \text{ (s)} + Na_2S \text{ (l)}$$

After precipitation is complete, the slurry is pumped to vacuum drum filters where a nearly complete liquid–solids separation is accomplished. The liquid is dilute sodium sulfide solution, which is concentrated by evaporation to a flaked 60 wt % sodium sulfide product. The filter cake is a 60 wt % strontium carbonate solid which is fed to a carbonate dryer. After drying, the strontium carbonate product is cooled, ground, and screened for packaging.

Strontium carbonate also precipitates from strontium sulfide solution with carbon dioxide. Hydrogen sulfide is generated as a by-product of this reaction and reacts with sodium hydroxide to produce sodium hydrosulfide, which is sold as by-product. The ability of the black ash process to produce a product exceeding 95% strontium carbonate, from ores containing ≤85% strontium sulfate, has led to its predominance.

In another process, strontium sulfate can be converted to strontium carbonate directly by a metathesis reaction wherein strontium sulfate is added to a solution of sodium carbonate to produce strontium carbonate and leave sodium sulfate in solution (6). Prior to this reaction, the finely ground ore is mixed with hydrochloric acid to convert the calcium carbonates and iron oxides to water-soluble chlorides.

Uses. The largest use of strontium carbonate is in the manufacture of glass faceplates for color-television tubes. It is present in glass at ca 12–14 wt % on a strontium oxide basis and functions as an x-ray absorber. Strontium carbonate is an effective x-ray barrier because strontium has a large atomic radius, and its presence is required in the relatively high voltage television sets used in the United States and Japan. The lower voltage Western European television faceplates contain the less expensive barium carbonate as an x-ray absorber. Strontium carbonate, when added to special glasses, glass frits, and ceramic glazes, increases the firing range, lowers acid solubility, and reduces pin-holing; strontium carbonate also has low toxicity. It is used in the production of high purity, low lead electrolytic zinc by a process patented by ASARCO, Inc. (7) and it is used in Australia, South Africa, the United States, and Japan. At a use level of 4.4–7.7 kg of strontium carbonate per ton of metal produced, it removes lead from the cathode zinc.

Strontium Chromate. Strontium chromate [7789-06-2], $SrCrO_4$, is made by precipitation of a water-soluble chromate solution using a strontium salt or of chromic acid using a strontium hydroxide solution. It has a specific gravity of 3.84 and is used as a low toxicity, yellow pigment and as an anticorrosive primer for zinc, magnesium, aluminum, and alloys used in aircraft manufacture (8) (see CORROSION AND CORROSION CONTROL).

Strontium Hexaferrite. Strontium hexaferrite [12023-91-5], $SrO \cdot 6Fe_2O_3$, is made by combining powdered ferric oxide, Fe_2O_3, and strontium carbonate, $SrCO_3$, and calcining the mixture at ca 1000°C in a rotary kiln (9). The material is crushed, mixed with a binder, and pressed or extruded into finished shapes and sintered at ca 1200°C. The sintered ferrite shapes are used as magnets in small electric motors, particularly in fractional horsepower motors for automobile windshield wipers and window risers. Flexible magnets for use as refrigerator door gaskets are made by blending ferrite powder with a polymer, melting the polymer, and extruding the mixture into the required shape.

Strontium Halides. Strontium halides are made by the reactions of strontium carbonate with the appropriate mineral acids. They are used primarily in medicines as replacements for other bromides and iodides.

Strontium bromide [10476-81-0], $SrBr_2$, forms white, needle-like crystals, which are very soluble in water (222.5 g in 100 mL water at 100°C) and soluble in alcohol. The anhydrous salt has a specific gravity of 4.216 and a melting point of 643°C.

Strontium chloride [10476-85-4], $SrCl_2$, is similar to calcium chloride but is less soluble in water (100.8 g in 100 mL water at 100°C). The anhydrous salt forms colorless cubic crystals with a specific gravity of 3.052 and a melting point of 873°C. Strontium chloride is used in toothpaste formulations (see DENTIFRICES).

Strontium fluoride [7783-48-4], SrF_2, forms colorless cubic crystals or a white powder with a specific gravity of 4.24 and a melting point of 1190°C. It is insoluble in water but is soluble in hot hydrochloric acid.

Strontium iodide [10476-86-5], SrI_2, forms colorless crystals which decompose in moist air. It is very soluble in water (383 g in 100 mL water at 100°C) and the anhydrous salt has a specific gravity of 4.549 and a melting point of 402°C.

Strontium Nitrate.   Strontium nitrate, $Sr(NO_3)_2$, in the anhydrous form is a colorless crystalline powder with a melting point of 570−645°C and a specific gravity of 2.986. It also exists as a tetrahydrate [13470-05-8] which has a density of 2.2 and a melting point of 100°C. The anhydrous salt is the commercially produced product. Strontium nitrate is made by the reaction of milled strontium carbonate with nitric acid. The nitrate slurry is filtered, crystallized, and centrifuged before it undergoes drying in a rotary dryer. The final product is screened and bagged for shipment. Strontium nitrate is classified as an oxidizer by the U.S. Department of Transportation and may react rapidly enough with easily oxidizable substances to cause ignition. The anhydrous material is normally stable but starts to decompose at ca 500°C with the evolution of nitrogen oxides.

The principal use of strontium nitrate is in the manufacture of pyrotechnics (qv) as it imparts a characteristic, brilliant crimson color to a flame. Railroad fusees and distress or rescue signaling devices are the main uses for strontium nitrate. It is also used to make red tracer bullets for the military.

Strontium Oxide, Hydroxide, and Peroxide.   Strontium oxide, SrO, is a white powder that has a specific gravity of 4.7 and a melting point of 2430°C. It is made by heating strontium carbonate with carbon in an electric furnace, or by heating celestite with carbon and treating the sulfide formed with caustic soda and then calcining the product (10). It reacts with water to form strontium hydroxide [18480-07-4] and is used as the source of strontium peroxide [1314-18-7].

Strontium hydroxide, $Sr(OH)_2$, resembles slaked lime but is more soluble in water (21.83 g per 100 g of water at 100°C). It is a white deliquescent solid with a specific gravity of 3.62 and a melting point of 375°C. Strontium soaps are made by combining strontium hydroxide with soap stocks, eg, lard, tallow, or peanut oil. The strontium soaps are used to make strontium greases, which are lubricants that adhere to metallic surfaces at high loads and are water-resistant, chemically and physically stable, and resistant to thermal breakdown over a wide temperature range (11).

Strontium peroxide, SrO_2, is a white powder with a specific gravity of 4.56 that decomposes in water. It is made by the reaction of hydrogen peroxide with strontium oxide and is used primarily in pyrotechnics and medicines.

Strontium Sulfate. Strontium sulfate, $SrSO_4$, occurs as celestite deposits in beds or veins in sediments or sedimentary rocks. Celestite has a specific gravity of ca 3.97, a Mohs' hardness of 3.0–3.5, and is colorless-to-yellow and often pale blue. Strontium sulfate forms colorless or white rhombic crystals with a specific gravity of 3.96 and an index of refraction of 1.622–1.631. It decomposes at 1580°C and has a solubility of 0.0113 g per 100 mL of water at 0°C.

Strontium Titanate. Strontium titanate [*12060-59-2*], $SrTiO_3$, is a ceramic dielectric material that is insoluble in water and has a specific gravity of 4.81. It is made from strontium carbonate and is used in the form of 0.5-mm thick disks as electrical capacitors in television sets, radios, and computers.

BIBLIOGRAPHY

"Strontium" under "Alkaline Earth Metals," in *ECT* 1st ed., Vol. 1, p. 463, by C. L. Mantell, Consulting Chemical Engineer; "Strontium Compounds" in *ECT* 1st ed., Vol. 13, pp. 113–118, by L. Preisman, Barium Reduction Corp., and D. M. C. Reilly, Food Machinery and Chemical Corp.; "Strontium" in *ECT* 2nd ed., Vol. 19, pp. 48–49, by L. M. Pidgeon, University of Toronto; "Strontium Compounds" in *ECT* 2nd ed., Vol. 19, pp. 50–54, by L. Preisman, Pittsburgh Plate Glass Co.; in *ECT* 3rd ed., Vol. 21, pp. 762–769, by A. F. Zeller, FMC Corp.

1. J. W. Mellor, *A Comprehensive Treatise on Inorganic and Theoretical Chemistry*, Vol. III, John Wiley & Sons, Inc., New York, 1922.
2. *Metals Handbook*, 9th ed., Vol. 2, American Society for Metals, Metals Park, Ohio, 1979.
3. *Gmelins Handbuch der Anorganischen Chemie*, System Number 29, Verlag Chemie, Weinheim/Bergstr., Germany, 1964.
4. *Celestite*, Mineral Dossier No. 6, Mineral Resources Consultative Committee, HMSO, London, 1973.
5. M. Guntz and M. Galliot, *Compt. Rend.* **151**, 813 (1910).
6. D. L. Stein, in T. J. Gray, ed., *Conference on Strontium Containing Compounds*, Nova Scotia Technical College, Halifax, Nova Scotia, Canada, 1973, pp. 1–9.
7. U.S. Pat. 2,539,681 (Jan. 30, 1951), R. P. Yeck and Y. E. Lebedeff (to American Smelting and Refining Co.).
8. T. J. Gray and R. J. Rontil, "Strontium-Based Pigments and Glazes," in Ref. 6.
9. T. J. Gray and R. J. Routil, "Strontium Hexaferrite," in Ref. 6.
10. G. D. Parkes, ed., *Mellor's Modern Inorganic Chemistry*, John Wiley & Sons, Inc., New York, 1967.
11. T. J. Gray and T. Betancourt, "Strontium Soaps and Greases" in Ref. 6, pp. 137–148.

STEPHEN G. HIBBINS
Timminco Metals

STRUCTURAL FOAMS. See FOAMED PLASTICS.

STRUCTURE-ACTIVITY RELATIONSHIPS. See PHARMACODYNAMICS.

STYRENE

Styrene [*100-42-5*] (phenylethene, vinylbenzene, phenylethylene, styrol, cinnamene), $C_6H_5CH=CH_2$, is the simplest and by far the most important member of a series of aromatic monomers. Also known commercially as styrene monomer (SM), styrene is produced in large quantities for polymerization. It is a versatile monomer extensively used for the manufacture of plastics, including crystalline polystyrene, rubber-modified impact polystyrene, expandable polystyrene, acrylonitrile–butadiene–styrene copolymer (ABS), styrene–acrylonitrile resins (SAN), styrene–butadiene latex, styrene–butadiene rubber (qv) (SBR), and unsaturated polyester resins (see ACRYLONITRILE POLYMERS; STYRENE PLASTICS).

Styrene was first isolated in the nineteenth century from the distillation of storax, a natural balsam. Although it was known to polymerize, no commercial applications were attempted for many years because the polymers were brittle and readily cracked. The development of dehydrogenation processes by I. G. Farben in Germany and Dow Chemical in the United States during the 1930s was the first step toward the modern styrene technology. Several plants were built in Germany before World War II to produce styrene, primarily for making synthetic rubber. It also became a material of strategic importance in the United States when the supply of nature rubber from South Asia was cut off from the Allied countries's access, and large-scale plants were built. After the war the demand for styrene monomer continued to grow, but its main use has shifted from synthetic rubber to polystyrene. Polystyrene (PS) as of ca 1996 accounts for 65% of the total styrene demand. The production of styrene in the United States was 2.0 million metric tons in 1970, which increased to 3.2×10^6 t in 1980 and 5.0×10^6 t in 1995. Rapid growth has also been seen in Western Europe and Japan and, since the 1980s in the Pacific Rim. The total worldwide production was approximately 16.5×10^6 t in 1995 and is projected to grow at a rate of more than 4% annually until the year 2005. Many factors contribute to its growth: it is a liquid that can be handled easily and safely, it can be polymerized and copolymerized under a variety of conditions by common methods of plastics technology to a large number of polymers of different properties and applications, polystyrene is easy to extrude and mold and is one of the least expensive thermoplastics volumetrically, the raw materials benzene and ethylene are produced in very large quantities in refineries and can be supplied to styrene plants through pipelines, and the manufacturing technologies are efficient and plants can be built on a large scale to produce styrene at low cost.

Two process routes are used commercially for the manufacture of styrene: dehydrogenation and coproduction with propylene oxide. Both routes use ethylbenzene as the intermediate; ethylbenzene is made from benzene and ethylene. The manufacture of styrene via ethylbenzene consumes more than 50% of the commercial benzene in the world. The great majority of the ethylbenzene and styrene plants are based on licensed technologies which are available at rather modest licensing fees. Styrene is a commodity chemical traded in large volumes domestically and internationally. The product specifications are largely dictated by the market. The minimum purity is usually 99.8%, which can be easily met in a well-operated plant of modern design. Some producers choose to have their plants designed to produce high purity styrene at a small incremental investment to gain marketing advantages and in anticipation of its future demand. Production of 99.95% styrene could become routine as a result of recent advances in manufacturing technology.

The commodity nature of the product and the easy access to the licensed processes enable new producers, particularly in developing countries, to enter the global styrene merchant market with little experience in styrene technology. Access to ethylene, which cannot be easily transported by means other than pipelines, is a key factor in considering new styrene facilities. Timing, or luck, is even more important because the supply and demand of styrene are seldom in balance and the price fluctuates broadly and rapidly as a result. Most of the time, the producers either suffer losses (1981–1985, 1991–1993) or enjoy handsome profits (1987–1990, 1994–mid-1995). Investments in styrene plants are known to have been recovered in less than a year, but prosperity encourages over-investment and lean years may follow.

Properties

Styrene is a colorless liquid with an aromatic odor. Important physical properties of styrene are shown in Table 1 (1). Styrene is infinitely soluble in acetone, carbon tetrachloride, benzene, ether, n-heptane, and ethanol. Nearly all of the commercial styrene is consumed in polymerization and copolymerization processes. Common methods in plastics technology such as mass, suspension, solution, and emulsion polymerization can be used to manufacture polystyrene and styrene copolymers with different physical characteristics, but processes relating to the first two methods account for most of the styrene polymers currently (ca 1996) being manufactured (2–8). Polymerization generally takes place by free-radical reactions initiated thermally or catalytically. Polymerization occurs slowly even at ambient temperatures. It can be retarded by inhibitors.

Styrene undergoes many reactions of an unsaturated compound, such as addition, and of an aromatic compound, such as substitution (2,8). It reacts with various oxidizing agents to form styrene oxide, benzaldehyde, benzoic acid, and other oxygenated compounds. It reacts with benzene on an acidic catalyst to form diphenylethane. Further dehydrogenation of styrene to phenylacetylene is unfavorable even at the high temperature of 600°C, but a concentration of about 50 ppm of phenylacetylene is usually seen in the commercial styrene product.

Ethylbenzene Manufacture

Styrene is manufactured from ethylbenzene. Ethylbenzene [100-41-4] is produced by alkylation of benzene with ethylene, except for a very small fraction that is recovered from mixed C_8 aromatics by superfractionation. Ethylbenzene and styrene units are almost always installed together with matching capacities because nearly all of the ethylbenzene produced commercially is converted to styrene. Alkylation is exothermic and dehydrogenation is endothermic. In a typical ethylbenzene–styrene complex, energy economy is realized by advantageously integrating the energy flows of the two units. A plant intended to produce ethylbenzene exclusively or mostly for the merchant market is also not considered viable because the merchant market is small and sporadic.

Table 1. Physical Properties of Styrene Monomer[a]

Property	Value
boiling point at 101.3 kPa = 1 atm, °C	145.0
freezing point, °C	−30.6
flash point (fire point), °C	
Tag open-cup	34.4 (34.4)
Cleveland open-cup	31.4 (34.4)
autoignition temperature, °C	490.0
explosive limits in air, %	1.1–6.1
vapor pressure, kPa, Antoine equation[b]	$\log_{10} P = 6.08201 - \dfrac{1445.58}{209.43 + t°C}$
critical pressure, P_c, MPa[c]	3.81
critical temperature, t_c, °C	369.0
critical volume, V_c, cm^3/g	3.55
refractive index, $n_D{}^{20}$	1.5467

	at 0°C	20°C	60°C	100°C	140°C
viscosity, mPa·s(=cP)	1.040	0.763	0.470	0.326	0.243
surface tension, mN/m (=dyn/cm)	31.80	30.86	29.01	27.15	25.30
density, g/cm^{3} [d]	0.9237	0.9059	0.8702	0.8346	
specific heat (liquid), J/(g·K)[e]	1.636	1.690	1.811	1.983	2.238
specific heat (vapor) at 25°C, C_p, J/(g·K)[e]			1.179		

Property	Value
latent heat of vaporization, ΔH_v, J/g[e]	
at 25°C	428.44
145°C	354.34
heat of combustion (gas at constant pressure) at 25°C, ΔH_c, kJ/mol[e]	4262.78
heat of formation (liquid) at 25°C, ΔH_f, kJ/mol[e]	147.36
heat of polymerization, kJ/mol[e]	74.48
Q value	1.0
e value	−0.8
volumetric shrinkage upon polymerization, vol %	17.0
cubical coefficient of expansion, °C^{-1}	
at 20°C	9.710×10^{-4}
40°C	9.902×10^{-4}
solubility (at 25°C), wt %	
monomer in H_2O	0.032
H_2O in monomer	0.070

[a]Ref. 1.
[b]To convert from $\log_{10} P_{kPa}$ to $\log_{10} P_{mm\ Hg}$, add 0.8751 to the constant.
[c]To convert MPa to psi, multiply by 145.
[d]Density at 150°C is 0.7900 g/cm^3.
[e]To convert J to cal, divide by 4.184.

The reaction of benzene and ethylene takes place on acidic catalysts and

$$C_6H_6 + CH_2{=}CH_2 \rightleftharpoons C_6H_5CH_2CH_3$$

can be carried out either in the liquid or vapor phase. Benzene in excess of the stoichiometric ratio to ethylene is used. The forward reaction is highly favored thermodynamically; the equilibrium conversion of ethylene is nearly complete. Alkylation does not stop at monoethylbenzene. Part of the ethylbenzene is further alkylated to diethylbenzenes, triethylbenzenes, tetraethylbenzenes, etc, collectively referred to as polyethylbenzenes (PEB). A high benzene-to-ethylene ratio in the feed mixture gives a low polyethylbenzenes-to-ethylbenzene ratio in the reaction product, but requires larger equipment and more energy consumption to recover the additional unreacted benzene by distillation for recycle to the reactor. The quantity of polyalkylbenzenes formed relative to ethylbenzene is also affected by the catalyst and reaction conditions. The polyethylbenzenes are recovered by distillation and transalkylated with benzene to produce additional ethylbenzene. For example, the transalkylation of diethylbenzene proceeds as follows:

$$CH_3CH_2C_6H_4CH_2CH_3 + C_6H_6 \rightleftharpoons 2\ C_6H_5CH_2CH_3$$

The transalkylation reaction is essentially isothermal and is reversible. A high ratio of benzene to polyethylbenzene favors the transalkylation reaction to the right and retards the disproportionation reaction to the left. Although alkylation and transalkylation can be carried out in the same reactor, as has been practiced in some processes, higher ethylbenzene yield and purity are achieved with a separate alkylator and transalkylator, operating under different conditions optimized for the respective reactions.

A number of side reactions occur in both the alkylator and the transalkylator, including oligomerization, cracking, dehydrogenation, isomerization, alkylation of alkanes by ethylene, alkylation of toluene by ethylene, alkylation of benzene by higher olefins, etc. The reactor effluent can be shown by high sensitivity chromatography to contain a large number of aromatic and nonaromatic by-products, most of them in parts per million or less. Isomerization causes formation of xylenes. Oligomerization is the first step in the formation of most of the by-products and is the side reaction that is most harmful to the yield and quality of the product and the stability of the catalyst. Some of the oligomers boil close to ethylbenzene and remain as nonaromatic impurities in the product. Other oligomers react further to form cumene, n-propylbenzene, butylbenzenes, and other higher alkylbenzenes. These alkylbenzenes either become impurities in the product or are recycled with the polyethylbenzene stream to the transalkylator. Depending on the process used, the alkylbenzenes that cannot be transalkylated to ethylbenzene are either decomposed or converted to heavies on repeated recycle. The heavies, consisting of diphenyl compounds and heavy aromatics and nonaromatics, are rejected from the process as residue, which is usually used as fuel. Control of side reactions and by-products is the most important consideration in the development of catalysts and processes for ethylbenzene production.

The benzene feedstock contains C_6 nonaromatics, ranging from 50 to 2000 ppm, depending on the source of supply. The C_6 nonaromatics do not directly contaminate the product because their boiling points are typically 40–60°C lower than that of ethylbenzene. However, part of these nonaromatics are alkylated by ethylene in certain processes to form higher nonaromatics, which may contaminate the product. The nonaromatics may also be cracked in the reactors, and the resulting components react with benzene to form various alkylbenzenes, such as cumene, which may also contaminate the product. The only aromatic impurity in a significant concentration in the typical benzene feedstock is toluene, ranging from 50 to 1000 ppm. It reacts with ethylene to form ethyltoluene in the alkylator; ethyltoluene is converted back to toluene in the transalkylator. Toluene is distilled together with ethylbenzene and becomes a product impurity that is innocuous downstream in the dehydrogenation process. However, any ethyltoluene that is not separated from ethylbenzene contaminates the styrene product as ethyltoluene and vinyltoluene.

The ethylene feedstock used in most plants is of high purity and contains 200–2000 ppm of ethane as the only significant impurity. Ethane is inert in the reactor and is rejected from the plant in the vent gas for use as fuel. Dilute gas streams, such as treated fluid-catalytic cracking (FCC) off-gas from refineries with ethylene concentrations as low as 10%, have also been used as the ethylene feedstock. The refinery FCC off-gas, which is otherwise used as fuel, can be an attractive source of ethylene even with the added costs of the treatments needed to remove undesirable impurities such as acetylene and higher olefins. Its use for ethylbenzene production, however, is limited by the quantity available. Only large refineries are capable of delivering sufficient FCC off-gas to support an ethylbenzene–styrene plant of an economical scale.

Commercial ethylbenzene is manufactured almost exclusively for captive use to produce styrene; only a small fraction is traded. The specifications of merchant ethylbenzene are usually negotiated between the seller and the buyer. An assay as low as 99% can be acceptable, provided that most of the impurities are benzene, which is strictly a distillation consideration, and toluene, which is largely dependent on the benzene feedstock. They are usually inconsequential in the dehydrogenation of ethylbenzene which makes far more benzene and toluene by side reactions. However, they must not be excessive, because the distillation train in the styrene unit is designed to accept specified quantities of benzene and toluene. Other possible impurities in ethylbenzene are nonaromatics in the C_7–C_{10} range and aromatics in the C_8–C_{10} range. Their nature and quantities depend on the feedstocks, the process used, and the design and operation of the plant. The C_7–C_{10} nonaromatics originate, in some processes, from oligomerization of ethylene and alkylation of the C_6 nonaromatics contained in benzene. The C_8–C_{10} aromatics impurities include xylenes, cumene, n-propylbenzene, ethyltoluenes, and butylbenzenes.

Diethylbenzenes are not considered by-products of the ethylbenzene process because they are recycled and transalkylated to ethylbenzene. Nonetheless, their concentrations in the ethylbenzene product must be minimized. Diethylbenzenes are converted in the dehydrogenation reactor to divinylbenzenes, which are exceedingly reactive, forming cross-linked polymers and causing polymer buildup in pipes, columns, and tanks. In severe cases, a styrene plant can be forced to shut down to clean out the polymers. The diethylbenzene content in the ethylbenzene product is generally specified to be 10 ppm maximum, largely because more accurate analytical instruments have become readily available only in recent years. The new plants are designed to limit this content to ≤1 ppm, which is not difficult to achieve by distillation because the relative volatility of ethylbenzene and diethylbenzenes is very high. Chlorides are poisons to the dehydrogenation catalyst. This fact rarely poses a problem in the manufacturing process, but possible contamination in storage and shipping must be prevented.

Most of the ethylbenzene plants built before 1980 are based on use of aluminum chloride catalysts. Aluminum chloride is an effective alkylation catalyst but is corrosive; its use as the catalyst requires glass-lined, brick-lined, or Hastelloy reactors, and thus results in high capital costs for the reactor section. The necessity of washing and neutralizing the reactor effluent to remove the spent catalyst adds to the complexity of both the plant itself and its operation. It also creates an aqueous waste that is environmentally objectionable and becoming increasingly difficult to dispose of. The corrosivity also increases plant downtime and maintenance cost. For these reasons, the aluminum chloride-based processes have not been used for new plants in the 1990s. However, many older plants based on aluminum chloride are still in operation and together accounted for about 40% of the worldwide ethylbenzene capacity as of 1995. The newer plants are based on zeolite catalysts.

Zeolite-Based Alkylation. Zeolites have the obvious advantages of being noncorrosive and environmentally benign. They have been extensively researched as catalysts for ethylbenzene synthesis. Earlier efforts were unsuccessful because the catalysts did not have sufficient selectivity and activity and were susceptible to rapid coke formation and deactivation.

The Mobil-Badger vapor-phase ethylbenzene process was the first zeolite-based process to achieve commercial success. It is based on a synthetic zeolite catalyst, ZSM-5, developed and optimized by Mobil Oil Corporation for the synthesis of ethylbenzene (9–18) (see MOLECULAR SIEVES). It has the desirable characteristics of high activity, low oligomerization, and low coke formation. The catalyst is regenerable by controlled burning of the coke deposited on the catalyst. A process based on this catalyst was developed at The Badger Company (now Raytheon Engineers & Constructors). In 1976, the Alkar unit at Cosden's Big Spring, Texas, refinery was converted to the Mobil-Badger process operating on dilute ethylene feedstock. In early 1980, a single-train plant of one billion pounds (450,000 t) annual capacity was brought on-stream for American Hoechst at Bayport, Texas. By the end of 1995, the Mobil-Badger process had been licensed to 35 plants with a combined capacity of nearly 10,000,000 t/yr, including all of the new units built in the United States since 1980. Because alkylation is carried out in the vapor phase, the Mobil-Badger process can accept dilute ethylene, such as treated FCC off-gas, as well as polymer-grade ethylene as a feedstock. Two commercial plants based on this process are in operation with dilute ethylene feeds (18–20), by Shell (Stanlow, U.K.) and Dow (Terneuzen, the Netherlands).

Since its first commercialization, the Mobil-Badger process has undergone several important improvements. The Third Generation Technology, which is offered currently (ca 1996), extends the catalyst cycle length between regenerations to two years and eliminates the spare reactor in the earlier generations. It achieves a yield of greater than 99.5% and reduces the xylene, which is the only significant impurity in the Mobil-Badger product, by more than 50%. Oligomerization is negligible in this process and the product does not contain nonaromatics. A unique feature of this process is its ability to decompose nonaromatics to light gases and remove them in the vent gas, thus avoiding the need to purge part of the recycled benzene to remove the nonaromatics. This process is also unique in its ability to remove higher alkylbenzenes such as cumene by dealkylation and to recover the benzene value. A high nonaromatics content in the benzene feedstock and a high propylene content in the ethylene feedstock have little effect on the yield and purity of the product.

A simplified flow diagram of the Mobil-Badger vapor-phase process is shown in Figure 1. Fresh and recycled benzene are vaporized and preheated to the desired temperature and fed to a multistage fixed-bed reactor. Ethylene is distributed to the individual stages. Alkylation takes place in the vapor phase at 350–450°C and the pressure that is deemed to be most economical for the individual plant, usually 0.69–2.76 MPa (100–400 psig). Separately, the polyethylbenzene stream from the distillation section is mixed with benzene, vaporized and heated, and fed to the transalkylator, where polyethylbenzenes react with benzene to form additional ethylbenzene. The combined reactor effluent, consisting of excess benzene, ethylbenzene, polyethylbenzenes, and trace impurities, is distilled in the benzene column. Benzene is condensed in the overhead for recycle to the reactors. Light hydrocarbons consisting of the ethane contained in the ethylene feed and components decomposed from the nonaromatic contained in the fresh benzene feed are removed as vent gas for use as fuel. The bottoms from the benzene column are distilled in the ethylbenzene column to recover the ethylbenzene product in the overhead. The bottoms stream from the ethylbenzene column is further distilled in the polyethylbenzene column to remove a small quantity of residue, the latter being a free-flowing liquid usually used as fuel. The overhead polyethylbenzene stream is recycled to the reactor section for transalkylation to ethylbenzene. The process heat, including the heat of reaction, is recovered by generating steam for use in the styrene unit. The heat-exchange system, not shown in Figure 1, is designed to optimize the overall efficiency of the ethylbenzene–styrene complex, consistent with the utility costs and constraints of the individual location.

A liquid-phase process based on an ultraselective Y (USY)-type zeolite catalyst developed by Unocal in the 1980s (21,22) and licensed by ABB Lummus Crest was commercialized at Nippon Styrene Monomer Corporation in 1990. Three commercial plants based on this process were in operation, all in Japan, as of mid-1995. This process is now called the Lummus-UOP process. The flow scheme of this liquid-phase process, shown schematically in Figure 2, is similar to the Mobil-Badger vapor-phase process. The differences are primarily in the catalysts, reaction conditions, reactor sizes, yields, and product specifications. To maximize the productivity of the catalyst in the Unocal process, the reactors operate at close to the critical temperatures of the reaction mixtures, ie, up to 270°C, and at pressures sufficient to maintain the reaction mixtures in the liquid state, ie, approximately 3.79 MPa (550 psig). A compressor is required unless the supply pressure of ethylene is sufficient to deliver it to the reactor. Large quantities of catalyst are required in two multistage alkylators and a

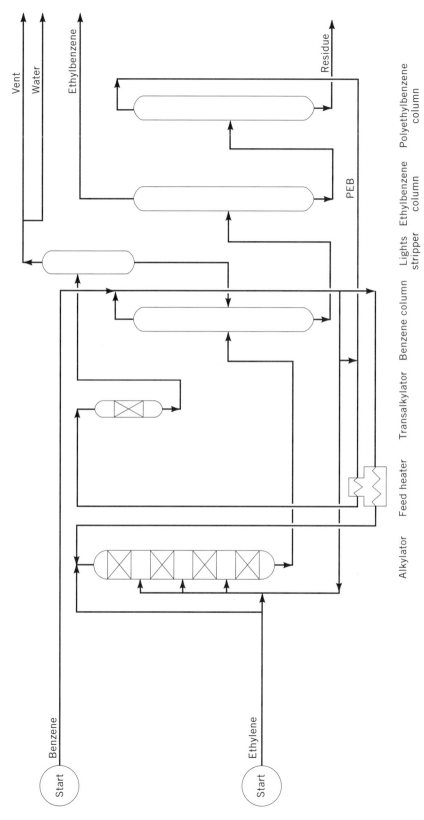

Fig. 1. Mobil-Badger vapor-phase ethylbenzene process where PEB = polyethylbenzene.

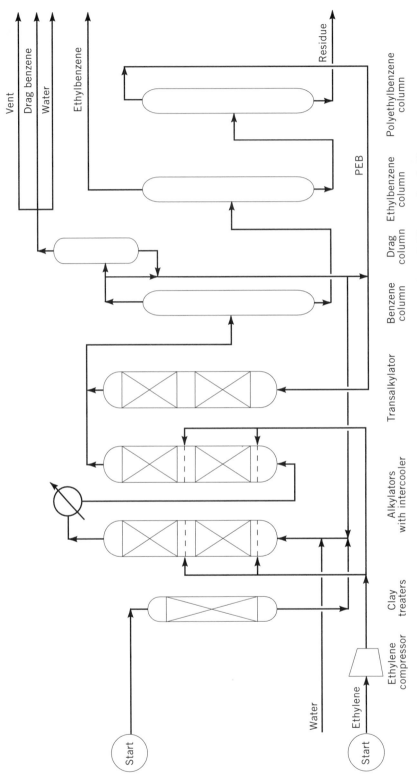

Fig. 2. Lummus–Unocal–UOP liquid-phase ethylbenzene process where PEB = polyethylbenzene.

964

transalkylator. To suppress ethylene oligomerization on the USY catalysts, water is injected into the feed stream to moderate the catalyst activity. Because the catalyst is susceptible to poisoning by basic materials, especially nitrogen compounds, the benzene feed is pretreated in clay guard chambers. Some of the C_6 nonaromatics contained in the fresh benzene are decomposed to components which may react with benzene to form various alkylbenzenes, or are alkylated to higher nonaromatics which may contaminate the product. The remaining C_6 nonaromatics are removed by purging a fraction of the recycled benzene called benzene drag. The overall yield to ethylbenzene is estimated to be 98–98.5%, with the residue accounting for most of losses. The ethylbenzene produced from this process contains low concentrations of xylene, but substantial quantities of other aromatic impurities (mainly cumene, n-propylbenzene, and ethyltoluene) and nonaromatic impurities. Because complete dissolution of ethylene is essential to liquid-phase alkylation, this process is not suitable for low concentration ethylene feeds, such as FCC off-gas; in order to provide sufficient liquid to dissolve the diluents (mainly ethane, methane, and hydrogen), the benzene recycle rate would have to be raised to an uneconomical level.

A two-phase process developed by CDTech in the 1980s (22,23) and licensed also by ABB Lummus Crest is claimed to be applicable to dilute ethylene feed as well as polymer-grade ethylene feed. The alkylation reaction is carried out in a catalytic distillation column with the liquid and vapor in countercurrent flow. The catalyst is packaged into bales, which are usually fiber glass containers having a cylindrical shape. The bales are arranged in the column to provide separate paths for the liquid to move downward by gravity and for the vapor to move upward by pressure. The complexity of the reactor and the use of baled catalyst, which does not flow freely, complicate the loading, unloading, and regeneration of the catalyst. The mechanical design of this column is an adaptation of CDTech's process for the etherification of methanol and isobutene to methyl *tert*-butyl ether (MTBE). In the MTBE application, catalytic distillation permits partial separation of the reactants and product, thus shifting the etherification reaction toward formation of MTBE, which is otherwise limited by thermodynamic equilibrium. The heat of reaction is removed by evaporation of some reactants in the column. This feature is beneficial to etherification, which is thermodynamically more favorable at a lower temperature. However, these benefits of catalytic distillation do not apply to the alkylation reaction. The alkylation of benzene is virtually irreversible, the conversion of ethylene being over 99.9999% in thermodynamic equilibrium. The sole purpose of using a catalytic distillation column for alkylation is to suppress oligomerization of ethylene on the catalyst. Oligomerization is a higher order reaction than alkylation with respect to the ethylene concentration. A high benzene-to-ethylene ratio would give a low ethylene concentration in the reaction mixture and restrain oligomerization, but it carries an economic penalty in the recovery and recycle of benzene. Catalytic distillation offers an alternative way to reduce the local ethylene concentration. The segregation of ethylene in the vapor phase and aromatics in the liquid phase creates mass-transfer barriers to hinder the transport of ethylene to the catalyst, which is in contact with the liquid, thus retarding oligomerization of ethylene. The mass-transfer barriers, however, also retard the conversion of ethylene to ethylbenzene, thus requiring a large catalyst volume and a complex reactor system to avoid a high ethylene loss. In addition to the alkylator, a finishing reactor of the conventional packed bed design is required to complete the conversion of ethylene. Although removal of the heat of reaction by evaporation in principle permits injection of the entire ethylene feed to a single catalytic bed, in practice the catalyst is divided into several beds to allow multiple injections of ethylene to further reduce the local ethylene concentration. The countercurrent flow of liquid and vapor results in a condition wherein the highest concentrations of ethylene and alkylated benzenes meet at the bottom layer of the catalyst. It favors the formation of polyethylbenzenes and adds duty to the transalkylation system. A distillation column is still required to recover the unreacted benzene in the alkylator and transalkylator effluents. One design option is to place the alkylator on top of the benzene column. This option, however, makes the loading and unloading of the baled catalyst even more difficult. A commercial plant based on this technology was reported to have been commissioned at Mitsubishi Chemical in Japan in late 1994.

A vapor-phase process primarily for FCC off-gas feeds was developed by Sinopec Technology Company based on a zeolite catalyst of the Pentasil type (24,25). It relies on frequent regeneration of the catalyst to minimize pretreatment of the FCC off-gas and allows the impurities in the feed gas to react with benzene to form by-products. Consequently, the product yield and purity are low. Joint licensing by ABB Lummus Crest and Sinopec was announced in 1994.

All of the processes described above require more benzene recycle than the aluminum chloride-based processes. Retrofitting an existing aluminum chloride-based plant to a zeolite-based plant requires not only replacement of the reaction section but also additional investment in the distillation train. Several producers have chosen to replace their old plants during the 1980s and early 1990s with new ones based on the Mobil-Badger process, at expanded capacities.

More recently, a new alkylation technology named EBMax was developed by Mobil Oil Corporation and Raytheon Engineers & Constructors and commercialized in Japan in late 1995 at Chiba Styrene Monomer Company, in a unit originally based on the Lummus-UOP liquid-phase process. The EBMax technology is based on a new Mobil zeolite catalyst called MCM-22. MCM-22 overcomes the oligomerization problem that plagues other liquid-phase alkylation processes. Because the catalyst is highly active for alkylation but inactive for oligomerization and cracking, it permits operation at low benzene-to-ethylene ratios, while achieving the highest yield and product purity among the ethylbenzene processes (26).

Aluminum Chloride-Based Alkylation.   The earlier alkylation processes were variations of the Friedel-Craft reaction on an aluminum chloride catalyst complex in a liquid-phase reactor (27), including those developed by Dow Chemical, BASF, Monsanto, and Union Carbide in cooperation with Badger. The Union Carbide-Badger process was the one most widely used during the 1960s and 1970s, with 20 plants built worldwide.

An improved aluminum chloride-based process was developed by Monsanto (28) in the 1970s. Using a presynthesized aluminum chloride complex and operating the reactor at higher temperature and pressure, the catalyst inventory is reduced to below its solubility in the reaction mixture. The reactants and the catalyst complex are mixed in the reactor to form a homogeneous liquid. The transalkylation of polyethylbenzenes is carried out separately. These improvements result in a higher yield. The higher reactor temperature permits recovery of the heat of reaction as useful steam. The reduced inventory of the catalyst, however, also reduces its capacity to resist deactivation by water and other poisons. A simplified flow diagram of the homogeneous alkylation process is shown in Figure 3. Alkylation and transalkylation are carried out either in two separate reactors or in a single vessel with two compartments. The catalyst complex is synthesized separately. It enters into an upflow alkylator together with benzene that has been dried in a dehydration column. Ethylene is sparged into the liquid in the reactor. The reaction progresses as the liquid moves upward, and the reaction mixture circulates through a heat exchanger where the exothermic heat of reaction is recovered as low pressure steam. The effluent from the alkylator is mixed with the polyethylbenzene (PEB) stream recovered in the distillation section. This mixture passes through the transalkylator where part of the polyethylbenzenes react with benzene to form additional ethylbenzene. In a plant operating in the steady state, the polyethylbenzenes made in the alkylator are offset by the polyethylbenzenes converted in the transalkylator, so that the recycle rate remains constant. The effluent from the transalkylator is flashed to remove the light components which then go to the vent gas. The liquid is cooled, water-washed, and neutralized with caustic solution to remove the aluminum catalyst. The washed organic phase is separated into four fractions in a series of three distillation columns. The first column recovers unreacted benzene for recycle to the reactor; the second recovers and purifies the ethylbenzene product; and the third separates the polyethylbenzene stream from the residue for recycle to the reactor section.

Other Technologies.   Ethylbenzene can be recovered from mixed C_8 aromatics by superfractionation. This technology was first practiced by Cosden Oil & Chemical Company in 1957 (Big Spring, Texas), based on a design developed jointly with The Badger Company. Several superfractionation plants were built in the United States, Europe, and Japan around 1960. This route is no longer competitive, for several reasons. The quantity that can be recovered is limited by the supply of mixed C_8 aromatics and by the low ethylbenzene content (ca 20%). High capital and energy costs resulting from the small boiling-point difference of 1.8°C between ethylbenzene and *para*-xylene and low product purity are some other factors. A process for recovery of ethylbenzene from C_8 aromatic streams by selective adsorption called EBEX was developed by UOP in the 1970s.

The Alkar process (29–31), developed by UOP and commercialized in 1960 at Cosden, uses boron trifluoride on alumina support as the catalyst. It has been used for polymer-grade as well as dilute ethylene feeds. This process produces a high purity product, but the high corrosivity of the fluoride and the susceptibility of the catalyst to poisoning have prevented it from gaining widespread acceptance. The last plant based on this process was built by Shell at Moerdijk, the Netherlands, in 1979. It was decommissioned and replaced in 1992 by a new plant based on the Mobil-Badger process. Another process for ethylbenzene production developed by UOP based on a solid phosphoric acid catalyst (32) was commercialized at El Paso Natural Gas Company's (Odessa, Texas) plant in the 1960s using a dilute ethylene feed. The conversion of ethylene and the selectivity to ethylbenzene are low. This technology is similar to UOP's SPA process, which

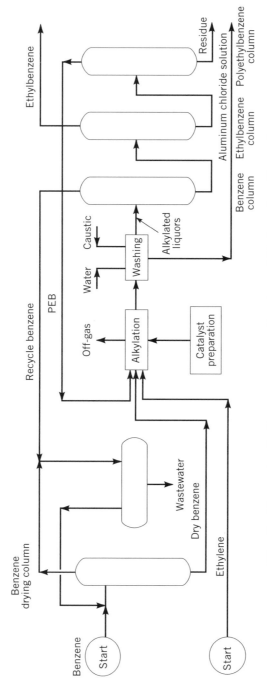

Fig. 3. Monsanto ethylbenzene process. Courtesy of *Hydrocarbon Processing*.

968

was widely used for the production of cumene, but it did not meet with the same degree of success.

Production of ethylbenzene from butadiene has been investigated by many researchers. It consists of two steps: cyclodimerization of 1,3-butadiene to 4-vinylcyclohexene and dehydrogenation of the vinylcyclohexene to ethylbenzene.

Styrene Manufacture

Styrene manufacture by dehydrogenation of ethylbenzene is simple in concept and has the virtue of being a single-product technology, an important consideration for a product of such enormous volume. This route is used for nearly 90% of the worldwide styrene production. The rest is obtained from the coproduction of propylene oxide (PO) and styrene (SM). The PO–SM route is complex and capital-intensive in comparison to dehydrogenation of ethylbenzene, but it still can be very attractive. However, its use is limited by the mismatch between the demands for styrene and propylene oxides (qv). The global demand for propylene oxide is only 2/10 of that of styrene by weight, while the ratio of propylene oxide to styrene from the PO–SM route is more than 4:10. The propylene oxide from PO–SM must also compete with that obtained from other routes. Consequently, the propylene oxide market dictates the PO–SM capacity that can be profitably installed, and the larger-volume styrene is thus reduced to the status of a by-product. As such, this styrene can be priced elastically and profitably at the same time. The PO–SM operators are thus the envy of the styrene industry when propylene oxide is in demand. However, the styrene production in the PO–SM plant cannot be increased alone to take advantage of a rising styrene market, and viewed in this light their position becomes less enviable. The PO–SM processes, unlike some of the dehydrogenation processes, are not available for licensing.

Dehydrogenation. The dehydrogenation of ethylbenzene to styrene takes

$$C_6H_5CH_2CH_3 \rightleftharpoons C_6H_5CH{=\!=}CH_2 + H_2$$

place on a promoted iron oxide–potassium oxide catalyst in a fixed-bed reactor at the 550–680°C temperature range in the presence of steam. The reaction is limited by thermodynamic equilibrium. Low pressure favors the forward reaction, as is to be expected from the stoichiometric relation that one mole of reactant dissociates into two moles of products. Earlier dehydrogenation reactors were designed for operation above atmospheric pressure, at about 138 kPa (20 psia), so that a compressor would not be required to remove hydrogen from the condensed reactor effluent. Since the 1970s, vacuum design has become standard because the benefits of high conversion and selectivity achievable and low dilution steam required at low pressure outweigh the cost of the compressor. Pressures of 41 kPa (6 psia) or lower at the reactor outlet have been designed. Dehydrogenation is an endothermic reaction. The reaction rate is highly temperature-dependent. High temperature favors dehydrogenation both kinetically and thermodynamically but also increases by-products from side reactions and decreases the styrene selectivity.

The main by-products in the dehydrogenation reactor are toluene and benzene. The formation of toluene accounts for the biggest yield loss, ie, approximately 2% of the styrene produced when a high selectivity catalyst is used. Toluene is formed mostly from styrene by catalytic reactions such as the following:

$$C_6H_5CH{=}CH_2 + 2\ H_2 \longrightarrow C_6H_5CH_3 + CH_4$$

$$C_6H_5CH{=}CH_2 \longrightarrow C_6H_5CH_3 + C$$

The formation of benzene accounts for a yield loss of up to 1% of the styrene produced, mostly the result of thermal decomposition of ethylbenzene and styrene. Other by-products include carbon dioxide, methane, ethylene, phenylacetylene, α-methylstyrene, allylbenzene, vinyltoluenes, xylenes, cumene, n-propylbenzene, ethyltoluenes, butylbenzenes, and heavy aromatics. Although they are rather insignificant in terms of the related yield losses, these compounds can affect the cost of purification and the quality of the styrene product. The alkylbenzene and alkenylbenzene by-products are formed by free-radical and catalytic reactions from ethylbenzene and styrene. With the exception of phenylacetylene, these compounds may also originate from impurities in the ethylbenzene feedstock. Phenylacetylene is formed from dehydrogenation of styrene thermally and catalytically. It is largely temperature-dependent but is also affected by catalyst formulation. Its concentration relative to styrene increases typically from 50 ppm at the start-of-run of a high selectivity catalyst to over 100 ppm at the end-of-run as the temperature is raised to compensate for the declining activity of the catalyst. Phenylacetylene is undesirable in certain styrene uses, such as SBR, because of its propensity to cross-link with styrene and to terminate free-radical reactions during polymerization. If necessary, it can be reduced to parts per million in the styrene product by hydrogenation on a noble metal catalyst under mild conditions. Technologies such as Fina-Badger's PAR technology are available for this treatment. Among the aromatic by-products, α-methylstyrene has the highest concentration in crude styrene. It is partially separated from styrene by distillation and removed in the residue stream.

A small fraction of the hydrocarbons decompose and deposit on the catalyst as carbon. Although the effect is minute in terms of yield losses, this carbon can still significantly reduce the activity of the catalyst. The carbon is formed from cracking of alkyl groups on the aromatic ring and of nonaromatics present in certain ethylbenzene feedstocks. It can be removed by the water gas reaction, which is catalyzed by potassium compounds in the catalyst. Steam, which is

$$C + 2\ H_2O \longrightarrow CO_2 + 2\ H_2$$

co-fed with ethylbenzene to the reactor, serves three purposes: as an energy carrier, it supplies the endothermic heat of reaction; as a diluent, it shifts the equilibrium toward the formation of styrene; and as a reactant in the water gas reaction, it retards and limits carbon deposition on the catalyst. However, as important as steam is in the dehydrogenation reactor, its use must be minimized for economic reasons. The steam-to-hydrocarbon ratio, or SHR, is an important consideration in the selection of catalysts and processes.

Dehydrogenation catalysts usually contain 40–90% Fe_2O_3, 5–30% K_2O, and promoters such as chromium, cerium, molybdenum, calcium, and magnesium oxides. Criterion Catalysts (partnership company of Shell and American Cyanamid) and the Sud-Chemie Group (including Sud-Chemie in Germany, United Catalysts in the United States, and Nissen Girdler in Japan) are the principal catalyst developers and manufacturers for the styrene industry. Dow and BASF manufacture proprietary catalysts, primarily for captive use. During the 1970s, Shell-105 catalyst, called C-105 as of ca 1996 and marketed by Criterion, was widely used because of its high activity and high chemical and mechanical stability (34), despite a modest selectivity in the 89–93% range. As raw material costs for styrene production increased rapidly during the 1970s, the need for development of high selectivity catalysts became urgent. The earlier catalysts of this type suffer from poor mechanical integrity and low activity. The high selectivity catalysts generally require higher reaction temperature to effect the same ethylbenzene conversion. The higher temperature imposes additional demands on the equipment metallurgy, and increases the formation of phenylacetylene. During the 1980s, the mechanical integrity of the high selectivity catalyst was greatly improved. The catalyst manufacturers have also had considerable success in increasing the activity while preserving the selectivity. Criterion (34) offers a series of high selectivity catalysts, including C-025A, C-045, VersiCat, and IronCat. The Sud-Chemie Group (35) offers the G-64, G-84, and Styromax series of catalysts.

An important property of the dehydrogenation catalyst is its steam stability, which refers to the ability of the catalyst to sustain the desired ethylbenzene conversion under a given set of operating conditions, ie, pressure, temperature, and especially steam-to-hydrocarbon ratio (SHR). Steam instability is characterized by accelerated decrease in the ethylbenzene conversion as a result of rapid carbon deposition on the catalyst when the SHR is too low. Most catalysts require a minimum SHR in the 8–10 range. Since the steam-to-hydrocarbon ratio has a large impact on the energy efficiency of a styrene plant, the search for a high selectivity catalyst that is stable at a low ratio has been a topic of intensive research efforts. A recently developed catalyst has been shown to be stable at a SHR of 7 or lower while retaining a high selectivity in pilot-plant operations.

The quantity of catalyst used for a given plant capacity is related to the liquid hourly space velocity (LHSV), ie, the volume of liquid hydrocarbon feed per hour per volume of catalyst. To determine the optimal LHSV for a given design, several factors are considered: ethylene conversion, styrene selectivity, temperature, pressure, pressure drop, SHR, and catalyst life and cost. In most cases, the LHSV is in the range of 0.4–0.5 h/L. It corresponds to a large quantity of catalyst, approximately 120 m^3 or 120–160 t depending on the density of the catalyst, for a plant of 300,000 t/yr capacity.

All of the dehydrogenation catalysts deteriorate, or age, in use, causing the ethylbenzene conversion to decrease with time. This phenomenon is intrinsic in the catalytic reaction, though the aging rate may also be affected by poisons and impurities in the feed. The iron oxide–potassium oxide catalyst is susceptible to poisoning by chlorides, but this is a rare occurrence in commercial plants, particularly when the ethylbenzene feedstock comes from a zeolite-based process. In most cases catalyst aging is caused by carbon deposition on the catalyst, chemical decomposition of the active ingredient in the catalyst, and mechanical disintegration of catalyst particles. In commercial practice, the reaction temperature is increased to compensate for the decrease in the activity of the catalyst and to maintain the desired conversion. This procedure continues until the temperature reaches the limit allowed by the mechanical design of the reactor system or when the styrene selectively is deemed too low to be economical. The useful lives of dehydrogenation catalysts vary from one to four years, and in most cases 18–24 months, depending on the nature of the catalyst, the design and operation of the reactor system, and the quality of the feedstock. Catalyst is a significant cost in the production of styrene owing to the large quantity required. There is considerable incentive to extend the useful life of the catalyst. Even more important than savings in catalyst cost is the reduction in downtime and increase in production that result from an extended catalyst life. Carbon deposition can be reversed to a large degree by steaming the catalyst, but the effect is only short-term; frequent steaming is not deemed to be economical. Steaming is ineffective for extending catalyst life because it does not prevent chemical decomposition of the catalyst. A method of preventing decomposition and disintegration of the dehydrogenation catalysts called Catalyst Stabilization Technology (CST) was invented in 1995 at Weymouth Laboratory of Raytheon Engineers and Constructors (36).

Dehydrogenation is carried out either isothermally or adiabatically. In principle, isothermal dehydrogenation has the duel advantage of avoiding a very high temperature at the reactor inlet and maintaining a sufficiently high temperature at the reactor outlet. In practice, these advantages are negated by formidable heat-transfer problems. First, to facilitate heat transfer from an external source to provide the reaction heat, expensive tubular reactors are used. Second, the reaction temperature exceeds the stable temperatures of the molten salts commonly used as heat-transfer media in tubular reactors. Flue gas is used, but its heat capacity and heat-transfer rate are low, requiring a large number of tubes and multiple reactor trains for large plants. Isothermal processes are practiced by BASF in Europe and Asahi in Japan. An isothermal reactor system that uses a molten salt mixture of sodium, lithium, and potassium carbonates as the heating medium has been offered for licensing by Lurgi, Montedison, and Denggendarfer. A demonstration unit was built in 1985 by Montedison at Montova, Italy. High conversion and selectivity are claimed but no commercial units have been built.

In an adiabatic reactor, the endothermic heat of reaction is supplied by the preheated steam that is mixed with ethylbenzene upstream of the reactor. As the reaction progresses, the temperature decreases. To obtain a high conversion of ethylbenzene to styrene, usually two, and occasionally three, reactors are used in series with a reheater between the reactors to raise the temperature of the reaction mixture. The reactors are fixed bed in design with the catalyst confined between two concentric screens in each reactor, which is cylindrical and vertical. The reaction mixture flows radially from the inner screen through the catalyst and outside screen. The radial design provides a large flow area and minimizes the pressure drop through the catalyst in an economical way. Large plants with a single train capable of producing more than 500,000 t/yr of styrene are in operation.

Other than the reactor system, the distillation column that separates the unconverted ethylbenzene from the crude styrene is the most important and expensive equipment in a styrene plant. It is expensive because the relative volatility between ethylbenzene and styrene is small, requiring a large number of distillation stages and a high reflux ratio. Its design must also take into consideration that styrene polymerizes in the liquid phase even at the ambient temperature and the rate of polymerization increases rapidly with the temperature and the concentration. In addition to reducing the styrene yield, polymerization could cause operating difficulties. Therefore, the distillation is carried out at a high vacuum to keep the temperature low, and inhibitors are used to retard polymerization. Many efforts have been made to minimize the pressure drop in the distillation column, and thus to reduce the temperatures in the stripping section and in the reboiler. The Linde sieve tray (37), developed in the 1960s, succeeded in reducing the pressure drop, decreasing the liquid holdup, and increasing the tray efficiency. Another significant advancement in the styrene distillation technology was introduced in the 1980s by the development of structural packings (38). This results in further reduction of pressure drop and increases of separation efficiency and throughput. Structural packings are offered by several equipment manufacturers, including Glitch, Koch, Norton, and Sultzer.

To minimize yield losses and to prevent equipment fouling by polymer formation, which may lead to a reduced production rate and plant shutdown, polymerization inhibitors are used in the distillation train, product storage, and more recently in vent gas compressors. Inhibitor usage is a significant cost in the production of styrene. Sulfur was used extensively as polymerization inhibitor until the mid-1970s. It is effective but the residue leaving the distillation train is contaminated with sulfur and is not environmentally acceptable as fuel. New inhibitors, mostly nitrogen-containing organic compounds such as dinitrophenol and dinitrocresol, have been developed to replace sulfur. They are more expensive and some are highly toxic, but it is acceptable to use the residue containing inhibitors of this type as fuel. The styrene product also needs inhibitor to retard polymerization in storage and shipping. The most commonly used inhibitor for this service is 4-tert-butylcatechol [98-29-3] (TBC), which is colorless, usually in 10–15 ppm concentration. The term inhibitor is used in the styrene industry often indiscriminately to denote substances ranging from true inhibitor to retarder. True inhibitors prevent the initiation of polymerization by reacting rapidly with free radicals; retarders reduce the rate of polymerization. New inhibitor systems that are claimed to be capable of dramatically reducing polymer formation have been developed by Uniroyal Chemical under the trade name SFR, by Betz Process Chemicals under the trade name STYREX, and by Ciba-Geigy.

The qualities of the styrene product and toluene by-product depend primarily on three factors: the impurities in the ethylbenzene feedstock, the catalyst used, and the design and operation of the dehydrogenation and distillation units. Other than benzene and toluene, the presence of which is usually inconsequential, possible impurities in ethylbenzene are C_7–C_{10} nonaromatics and C_8–C_{10} aromatics. Parts of the nonaromatics are cracked and dehydrogenated. The C_7 and some of the C_8 nonaromatics remain in the toluene stream as alkane and alkene impurities and decrease its by-product value. These impurities are also reported to be the likely cause of color problems in the toluene by-product (39). The remaining nonaromatics remain in the styrene product. Among the C_8 aromatics, m- and p-xylene can be recycled to extinction; o-xylene remains in the styrene product. The C_9-C_{10} aromatics include cumene, n-propylbenzene, ethyltoluenes, butylbenzenes, etc. They pass through the dehydrogenation reactor partly unchanged and partly converted to their corresponding dehydrogenated compounds. For example, cumene is partially converted to α-methylstyrene. Part of these aromatic and dehydrogenated compounds remain in the styrene product.

The development of the high selectivity catalyst is the most important advancement in the styrene manufacturing technology during the quarter century prior to the mid-1990s. Other important improvements include better design of the reactor system to assure uniform composition, temperature, and distribution of the reaction mixture, to minimize thermal reactions, and to prevent localized catalyst aging; low reactor pressure to enhance styrene selectivity and catalyst stability; efficient heat exchange systems to reduce the energy cost; structural packings to reduce the pressure drops in distillation columns; polymerization inhibitors that are more effective and less environmentally objectionable; use of high purity ethylbenzene feedstock to increase the purity of the styrene product; and improved design of the recovery and purification system to minimize polymer formation and styrene loss to the residue.

A modern adiabatic dehydrogenation unit with two reactors and an interreactor reheater, operating with a high selectivity catalyst and at a pressure of 41 kPa (6 psia) at the second reactor outlet, can give a styrene selectivity as high as 97% at an ethylbenzene conversion of 60–70%. The choice of a desired conversion is a matter of economic optimization. A low conversion gives a high selectivity but increases energy consumption; a high conversion decreases the energy cost but increases the raw material cost.

Most of the styrene plants in operation or under construction in the world as of ca 1996 are based on one of two adiabatic processes: the Fina-Badger process licensed by Badger Technology Division of Raytheon Engineers & Constructors and the Monsanto process licensed by ABB Lummus Crest. Dow Chemical, a primary styrene producer, uses its own adiabatic process. These processes are similar to one another in principle, but differ in details. In selecting a process, the licensee usually considers the capital cost, yield, product quality, energy efficiency, inhibitor consumption, plant reliability, operating flexibility, and other factors that are of significance to the project. The details are important. For example, operating reliability and flexibility have proved to be crucial to the profitability of a plant because the supply and demand of styrene are seldom in balance. From 1980 to the mid-1990s its price has fluctuated by as much as a factor of three or more within one or two years, whereas the raw material and energy costs have remained relatively stable.

Figure 4 illustrates a typical dehydrogenation unit. Fresh and recycled ethylbenzene are combined and mixed with steam, then heat-exchanged with the reactor effluent. The vapor mixture of ethylbenzene and steam is raised to the desired reaction temperature by mixing with more steam that has been superheated to 800°C or higher in a fired heater. This mixture is fed to the reactors, two or three stages in series with interstage reheaters. The reactor effluent goes through heat exchangers to preheat the ethylbenzene–steam feed mixture and

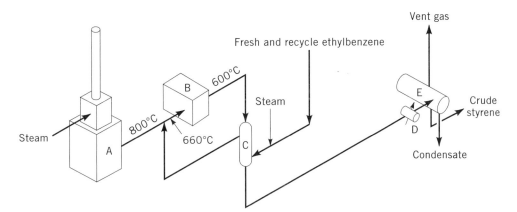

Fig. 4. Manufacture of styrene by adiabatic dehydrogenation of ethylbenzene: A, steam superheater; B, reactor section; C, feed–effluent exchanger; D, condenser; and E, settling drum. Courtesy of The Badger Company, Inc.

to generate steam for use in the distillation train. The condensed reactor effluent is separated in the settling drum into vent gas (mostly hydrogen), process water, and organic phase. The vent gas is removed by a compressor for use as fuel or for hydrogen recovery. The process water is stripped of organics and reused. The organic phase with polymerization inhibitor added is pumped to the distillation train. In an optional energy recovery scheme offered in conjunction with the Fina-Badger process, after the high level energy in the reactor effluent is recovered, the latent heat is recovered by using the condensing reactor effluent to vaporize a ethylbenzene–water azeotrope, which is then compressed and used as the reactor feed. This option is highly energy efficient but requires an incremental investment. It has been used in the Far East.

Figure 5 illustrates a typical distillation train in a styrene plant. Benzene and toluene by-products are recovered in the overhead of the benzene–toluene column. The bottoms from the benzene–toluene column are distilled in the ethylbenzene recycle column, where the separation of ethylbenzene and styrene is effected. The ethylbenzene, containing up to 3% styrene, is taken overhead and recycled to the dehydrogenation section. The bottoms, which contain styrene, by-products heavier than styrene, polymers, inhibitor, and up to 1000 ppm ethylbenzene, are pumped to the styrene finishing column. The overhead product from this column is purified styrene. The bottoms are further processed in a residue-finishing system to recover additional styrene from the residue, which consists of heavy by-products, polymers, and inhibitor. The residue is used as fuel. The residue-finishing system can be a flash evaporator or a small distillation column. This distillation sequence is used in the Fina-Badger process and the Dow process.

In the Monsanto process (now called Lummus-UOP Classic process, as distinct from the SMART process to be described later), ethylbenzene, together with benzene and toluene, is separated from styrene in the first column. The overhead condensate from this column then goes to the benzene–toluene column and is redistilled to recover the benzene–toluene fraction in the overhead and ethylbenzene in the bottoms. It has been argued that energy is wasted in this scheme

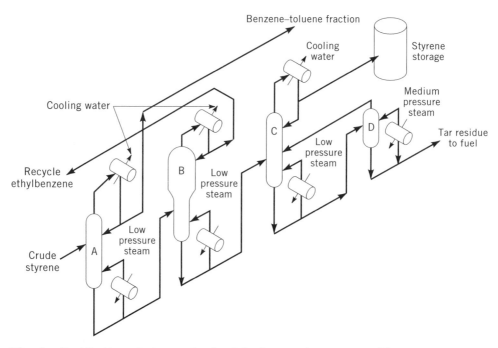

Fig. 5. Purification of styrene in the dehydrogenation reactor effluent in the Fina-Badger styrene process: A, benzene–toluene column; B, ethylbenzene recycle column; C, styrene finishing column; and D, residue finishing. Courtesy of The Badger Company, Inc.

by distilling the benzene–toluene fraction twice, once together with ethylbenzene and once away from ethylbenzene. Actually, this extra energy consumption, which can be computed, is small. It has also been argued that less polymer is formed in having the styrene bypass the benzene–toluene column. Actually, any possible reduction of polymerization resulting from this arrangement, which has never been substantiated, is insignificant because polymerization of the styrene in solution with ethylbenzene through a small benzene–toluene column is negligible in comparison with polymerization of the high concentration styrene in the much larger ethylbenzene–styrene distillation column. Moreover, the higher vapor and liquid loadings required by having to distill benzene and toluene together with ethylbenzene may slightly increase styrene polymerization. In any case, these effects are insignificantly small and every process owner continues to use its own design very likely for historical reasons. An optional scheme to utilize the latent heat of the ethylbenzene vapor in the column overhead to vaporize an ethylbenzene–water azeotrope for use as reactor feed is offered in conjunction with the Monsanto process. This scheme recovers less energy than the Fina-Badger scheme but does not require a compressor. However, the condensing temperature of the ethylbenzene vapor must be raised by having the ethylbenzene–styrene column operate at higher temperatures. The inhibitor usage and polymer formation increase substantially because the rate of polymerization increases rapidly with the temperature. Consequently, the quantity and viscosity of the residue are high. A wipe-film evaporator is then used downstream of the styrene finishing column to minimize yield losses. This scheme has also been used in the Far East. It has been reported that the increased polymerization has caused difficult operating problems and reduced operating flexibility. The economics of the optional energy recovery schemes are marginal at best in most situations insofar as the energy savings are negated by the incremental investment, extra inhibitor usage and yield losses, and reduced operating flexibility. Neither of them is used by the producers in the United States.

The benzene–toluene fraction is further fractionated in a small column, not shown in Figure 5, to recover benzene for recycle to the alkylation unit and toluene for sale. This toluene can be converted to benzene by hydrodealkylation but the high selectivity catalyst has reduced the formation of toluene in the dehydrogenation reactor to the point where the cost of installing a hydrodealkylation unit is difficult to justify even in a large styrene plant.

PO–SM Coproduction. The coproduction of propylene oxide and styrene (40–49) includes three reaction steps: (1) oxidation of ethylbenzene to ethylbenzene hydroperoxide, (2) epoxidation of ethylbenzene hydroperoxide with propylene to form α-phenylethanol and propylene oxide, and (3) dehydration of α-phenylethanol to styrene.

$$C_6H_5CH_2CH_3 + O_2 \longrightarrow C_6H_5CH(CH_3)OOH$$

$$C_6H_5CH(CH_3)OOH + CH_2{=}CHCH_3 \longrightarrow C_6H_5CH(CH_3)OH + \underset{\underset{O}{\diagdown\diagup}}{CH_2CHCH_3}$$

$$C_6H_5CH(CH_3)OH \longrightarrow C_6H_5CH{=}CH_2 + H_2O$$

The oxidation step is similar to the oxidation of cumene to cumene hydroperoxide that was developed earlier and is widely used in the production of phenol and acetone. It is carried out with air bubbling through the liquid reaction mixture in a series of reactors with decreasing temperatures from 150 to 130°C, approximately. The epoxidation of ethylbenzene hydroperoxide to α-phenylethanol and propylene oxide is the key development in the process. It is carried out in the liquid phase at 100–130°C and catalyzed by a soluble molybdenum naphthenate catalyst, also in a series of reactors with interreactor coolers. The dehydration of α-phenylethanol to styrene takes place over an acidic catalyst at about 225°C. A commercial plant (50,51) was commissioned in Spain in 1973 by Halcon International in a joint venture with Enpetrol based on these reactions, in a process that became known as the Oxirane process, owned by Oxirane Corporation, a joint venture of ARCO and Halcon International. Oxirane Corporation merged into ARCO in 1980 and this process is now generally known as the ARCO process. It is used by ARCO at its Channelview, Texas, plant and in Japan and Korea in joint ventures with local companies. A similar process was developed by Shell (52–55) and commercialized in 1979 at its Moerdijk plant in the Netherlands. The Shell process uses a heterogeneous catalyst of titanium oxide on silica support in the epoxidation step. Another plant by Shell is under construction in Singapore (ca 1996).

The recovery and purification facilities in these processes are also complex. One reason is that oxygenated by-products are made in the reactors.

Oxygenates, particularly aldehydes, are known to hinder polymerization of styrene and to cause color instability. Elaborate purification is required to remove the oxygenates. The styrene assay becomes very high as a result, but its quality as a monomer relative to the styrene produced by the dehydrogenation technology has been a matter of differing opinion among polymer producers.

Other Technologies.   As important as dehydrogenation of ethylbenzene is in the production of styrene, it suffers from two theoretical disadvantages: it is endothermic and is limited by thermodynamic equilibrium. The endothermicity requires heat input at high temperature, which is difficult. The thermodynamic limitation necessitates the separation of the unreacted ethylbenzene from styrene, which are close-boiling compounds. The obvious solution is to effect the reaction oxidatively:

$$C_6H_5CH_2CH_3 + {}^1\!/_2\,O_2 \longrightarrow C_6H_5CH{=}CH_2 + H_2O$$

Oxidative dehydrogenation is exothermic and is irreversible for all practical purposes. It could greatly reduce the cost of styrene production. The theoretical elegance and the potential benefit of the concept are so seductive that it has been a topic of extensive research and development since the 1940s not only in the petrochemical industry but also in universities. Molecular oxygen as well as various oxygen carriers, such as sulfur trioxide and ferric oxide, have been used as oxidants in fixed-bed and fluid-bed reactors. These research efforts have not resulted in commercial processes mainly because the selectivity to styrene is too low, in comparison with the dehydrogenation technology, to be economical.

A similar but somewhat less ambitious approach is to carry out dehydrogenation of ethylbenzene and oxidation of the hydrogen product alternately in separate reactors containing different catalysts:

Dehydrogenation	$C_6H_5CH_2CH_3 \rightleftharpoons C_6H_5CH{=}CH_2 + H_2$
Oxidation	$H_2 + {}^1\!/_2\,O_2 \longrightarrow H_2O$

The oxidation reactor replaces the reheater in the adiabatic dehydrogenation technology. The oxidation of hydrogen not only reheats the reaction mixture from a dehydrogenation reactor but also shifts the reaction equilibrium, thus allowing a higher conversion of ethylbenzene in the following dehydrogenation reactor. An ethylbenzene conversion of greater than 80% is claimed to be obtainable with three dehydrogenation reactors and two oxidation reactors. Such an oxidative reheat technology using a palladium catalyst in the oxidation reactor was developed by UOP (56) and implemented in a demonstration-scale unit at Mitsubishi Chemical in Japan. It was offered for licensing as a new styrene process in the mid-1980s under the trade name of StyroPlus. It has not attained commercial acceptance, probably because of the high cost of the palladium catalyst and the necessity of diluting the oxygen with steam prior to its injection into the reaction mixture. There are also concerns about the safety of the oxygen mixing step, potential damage to the dehydrogenation catalyst by oxygen, and the possibility of contaminating the styrene product with oxygenates. It is now called oxidative reheat technology and offered as a way of increasing existing plant capacity. It is also called SMART (styrene monomer advanced reactor technology) in combination with the Monsanto dehydrogenation process licensed by ABB Lummus Crest (55–58).

A three-step process involving the oxidation of acetophenone, hydrogenation of the ketone to α-phenylethanol, and dehydration of the alcohol to styrene was practiced commercially by Union Carbide (59) until the early 1960s. Other technologies considered during the infancy of the styrene industry include side-chain chlorination of ethylbenzene followed by dehydrochlorination or followed by hydrolysis and dehydration.

Technologies for producing styrene from alternative raw materials have also been seriously investigated. Toluene is priced lower than benzene historically and its use as a starting raw material in lieu of benzene received much attention during the 1970s. Side-chain alkylation of toluene with methanol or formaldehyde was studied by Monsanto and others (60,61). Monsanto and others (62–64) also undertook programs to develop oxidative coupling of toluene to stilbene, followed by disproportionation of stilbene with ethylene to form styrene:

$$2\ C_6H_5CH_3 + O_2 \longrightarrow C_6H_5CH\!=\!CHC_6H_5 + 2\ H_2O$$
$$C_6H_5CH\!=\!CHC_6H_5 + CH_2\!=\!CH_2 \longrightarrow 2\ C_6H_5CH\!=\!CH_2$$

This route uses toluene in lieu of benzene and cuts the ethylene consumption in half in comparison with the commercial alkylation–dehydrogenation technology via ethylbenzene. It affords a potential cost savings of 20% in raw materials. It would also avoid the costly separation of ethylbenzene and styrene. In actuality, however, these toluene-based technologies suffer from low yields which negate the potential advantages. They were not brought to the commercial stage, and the research programs were discontinued in the early 1980s. A technology for cracking diphenylethane to styrene and benzene originally developed by Nippon Petrochemicals is reported to have been used to reclaim the diphenylethane contained in the residue from a ethylbenzene plant based on the Unocal alkylation process. Recovery of styrene from various petroleum streams, such as pyrolysis gasoline, has been considered but is deemed uneconomical.

Production of styrene from butadiene has also been extensively investigated. Recently, Dow announced licensing a process involving cyclodimerization of 1,3-butadiene to 4-vinylcyclohexene, followed by oxidative dehydrogenation of the vinylcyclohexene to styrene (65,66). The cyclodimerization step takes place in

$$2\ C_4H_6 \longrightarrow C_6H_9CH\!=\!CH_2$$

the liquid phase at about 100°C and 2.76 MPa (400 psig) over a copper–zeolite catalyst. The vinylcyclohexene is converted to styrene by oxidative dehydrogenation with molecular oxygen diluted with steam in the vapor phase at about 400°C

$$C_6H_9CH\!=\!CH_2 + O_2 \longrightarrow C_6H_5CH\!=\!CH_2 + 2\ H_2O$$

and 138 kPa (20 psig) on a mixed metal oxide catalyst. A three-step process from butadiene to styrene has also been extensively investigated (67–69). The first step also involves cyclodimerization of butadiene to vinylcyclohexene, using an iron dinitrosyl chloride–zinc catalyst complex which is very active and selective. The reaction is carried out in tetrahydrofuran solvent at about 75°C and 0.52 MPa (75 psig). The second step involves dehydrogenation of vinylcyclohexene to ethylbenzene in the vapor phase at 400°C and 0.69 MPa (100 psig) or

$$C_6H_9CH{=}CH_2 \longrightarrow C_6H_5CH_2CH_3 + H_2$$

higher pressure on a noble metal catalyst. The third step is the conventional dehydrogenation of ethylbenzene to styrene. Recent interest in the butadiene route is prompted by the projection of a worldwide surplus of butadiene as a by-product of ethylene plants. The potential of this route is limited by the supply of butadiene as a low priced by-product in a quantity sufficient to support a styrene plant of an economical size. This limitation can be readily seen from the fact that a world-scale naphtha cracker that produces 450,000 t/yr of ethylene generates only enough by-product butadiene to support a styrene plant of 50,000 t/yr capacity. In contrast, it requires only 84,000 t/yr of ethylene to support a world-scale styrene plant of 300,000 t/yr capacity. Nevertheless, it is conceivable that the butadiene from a big naphtha cracker can be economically converted to ethylbenzene, which would then be used to supplement the ethylbenzene from alkylation for the production of styrene on an economical scale.

Economic Aspects

Capacity. The producers of styrene in the United States and their capacities (69–71), together with the technologies used, are listed in Table 2. It is a mature industry and the capacity in the United States is not expected to increase rapidly. The annual capacity of 5.3×10^6 t already substantially exceeds the domestic demand. Approximately 20% of U.S. production is exported. The U.S.

Table 2. U.S. Styrene Producers and Capacities, 1995

Company (location)	Capacity, 10^3 t/yr	Technologies Ethylbenzene	Technologies Styrene
Amoco (Texas City, Tex.)	364	AlCl$_3$	dehydro
ARCO (Channelview, Tex.)	511	Monsanto AlCl$_3$	ARCO PO/SM
	636	Mobil-Badger ZSM-5	ARCO PO/SM
Chevron (St. James, La.)	682	Mobil-Badger ZSM-5	Fina-Badger dehydro
Cos-Mar (Carville, La.)	864	Mobil-Badger ZSM-5	Fina-Badger dehydro
Dow (Freeport, Tex.)	636	Dow AlCl$_3$	Dow dehydro
Huntsman (Bayport, Tex.)	568	Mobil-Badger ZSM-5	Fina-Badger dehydro
Rexene (Odessa, Tex.)	145	UOP Alkar	UOP dehydro
Sterling (Texas City, Tex.)	727	Monsanto AlCl$_3$	Monsanto dehydro
Westlake (Lake Charles, La.)	159	Mobil-Badger ZSM-5	Fina-Badger dehydro
Total	*5292*		

styrene producers benefit from the huge ethylene capacity and the relatively low energy cost present in the Gulf Coast.

Worldwide styrene capacity, approximately 18×10^6 t/yr, is shown by geographic region in Table 3. Styrene is also a mature industry in Western Europe and Japan. In the other areas of Pacific Rim, particularly in Korea, the capacity has been increasing rapidly since the early 1980s. A large number of new construction projects are in progress or in planning in Korea, Taiwan, South Asia, China, and India. The styrene capacity in the Asia/Pacific region, including Japan, is expected to surpass that in North America by 1998.

Supply and Demand. Styrene is traded in large volumes internationally. Its supply and demand must therefore be viewed in a global perspective. The supply, demand, and net trade by geographic region in 1995 are estimated in Table 4. The worldwide production in 1995 was estimated to be approximately 17 $\times 10^6$ t. The demand is expected to grow at 2.5% per year in the United States and at a somewhat lesser rate in Western Europe and Japan. The demand in the rest of the Asia/Pacific region is expected to grow at a high rate, up to 10% per year. There was a huge shortage of styrene production in the Asia/Pacific region in the decade from 1985–1995. It is expected to continue into the early part of the twenty-first century in spite of the rapid capacity expansion. This regional supply/demand imbalance is presently covered mostly by imports from North America. North America and, increasingly, the Middle East will thus continue to be the exporters. Worldwide, the demand of styrene is expected to grow at an average rate of 4.5% annually until the year 2005.

Price. Styrene is a global commodity chemical. Its price is fairly uniform worldwide, except for short-term fluctuations on the regional spot market. However, the price of styrene has been highly cyclic. During the worldwide recession

Table 3. Worldwide Styrene Capacity, 1995

Regions	Capacity, 10^3 t/yr
North America	6,200
South America	400
Western Europe	4,200
Eastern Europe	1,200
Asia/Pacific	5,500
Middle East/Africa	600
Total	*18,100*

Table 4. Worldwide Styrene Supply and Demand, 1995

Regions	Production, 10^3 t/yr	Demand, 10^3 t/yr	Net trade, 10^3 t/yr
North America	5,900	4,900	1,000
South America	400	600	(200)
Western Europe	4,000	4,000	
Eastern Europe	1,100	1,300	(200)
Asia/Pacific	5,300	6,300	(1,000)
Middle East/Africa	600	200	400
Total	*17,300*		

in the early 1980s, it went below \$450/metric ton, which was less than the cost of production. With the economic recovery in the mid-1980s, it increased rapidly, reaching \$1500/t or higher on the spot market in the Far East in 1988. The 1987–1989 period was extraordinarily profitable years for styrene producers. Many new plants were constructed and excessive production capacity followed. The price then dropped to about \$500/t during 1991–1993 period. It again surged in early 1994, reaching \$1350/t on the spot market during the second quarter of 1995, but then plunged to \$500/t within a few months. As of September 1996, contract prices were ca \$660/t but spot prices were still under \$550/t.

Specifications and Analysis

Styrene specifications are largely dictated by the market. There is no formal agreement in the industry. Specifications for a typical polymerization-grade styrene product are shown in Table 5. The typical assay increased from 99.6% to 99.8% during the 1980s. As of 1996, an assay of 99.8% is still generally accepted. Newer plants in the United States are designed for 99.90% purity and in some cases as high as 99.95%. Some producers choose to have their plants designed to produce high purity styrene at a small incremental investment to gain marketing advantages and in anticipation of its future demand.

The advance (ca 1996) in alkylation technology enables the production of ultrahigh purity ethylbenzene at a low cost. With this ethylbenzene as the intermediate, a dehydrogenation unit of the present design will be able to produce styrene of 99.95% purity routinely. It may prompt a new standard in the styrene industry.

Methods for the analysis of styrene monomer are also shown in Table 5. The freezing point measurement was the standard method for the determination of styrene assay until the 1970s, but it has been largely replaced by gas chromatography. Modern gas chromatography gives a great amount of detail on the impurities and is accurate to parts per million or lower; styrene assay is determined by difference. However, certain impurities and properties are not amenable to chromatographic analysis. Color is measured spectrophotometrically and registered

Table 5. Specifications for Typical Polymerization-Grade Styrene Monomer Product

Parameter	Specification[a]	Test method
purity	99.80[b]	chromatography
hydrocarbon impurities	2000	chromatography
color	10[c]	ASTM D1209
polymer	10	ASTM D2121
aldehydes (as benzaldehyde)	50	ASTM D2119
peroxides (as hydrogen peroxide)	30	ASTM D2340
TBC[d] (inhibitor)	10–50	ASTM D2120
chlorides (as chlorine)	1	ASTM D4929
sulfur	1	ASTM D2747

[a]Value shown is in ppm by wt max, unless otherwise specified.
[b]Wt % min.
[c]Pt–Co scale, max.
[d]4-*tert*-Butylcatechol.

on the APHA or the platinum–cobalt scale. The TBC (4-*tert*-butylcatechol, the inhibitor) is extracted with aqueous sodium hydroxide and the resulting reddish quinone is measured photometrically. The polymer content is precipitated by addition of methanol to the monomer and the resulting mixture is measured with a turbidity meter. Other possible impurities, such as chlorides, sulfur, aldehydes (expressed as benzaldehyde), and peroxides (expressed as hydrogen peroxide) are measured by the methods commonly used in the chemical industry, also listed in Table 5.

Health and Safety Factors

Styrene is mildly toxic, flammable, and can be made to polymerize violently under certain conditions. However, handled according to proper procedures, it is a relatively safe organic chemical. Styrene vapor has an odor threshold of 50–150 ppm (72,73).

Styrene is listed in the U.S. Toxic Substance Control Act (TSCA) Inventory of Chemicals. It is not confirmed as a carcinogen but is considered a suspect carcinogen. The recommended exposure limits are OSHA PEL 50 ppm, ACGIH TLV 50 ppm. For higher concentrations, NIOSH/MSHA-approved respiratory protection devices should be used. For skin protection, use of protective garments and gloves of Viton, Nitrile, or PVA construction should be made. The acute effects of overexposure to styrene are shown in Table 6 (74).

Styrene liquid is inflammable and has sufficient vapor pressure at slightly elevated temperatures to form explosive mixtures with air, as shown in Figure 6. Uninhibited styrene polymerizes slowly at room temperature, and the polymerization rate increases with the temperature. Styrene polymerization is exothermic and this can cause the reaction to run awry. Polymerization can be inhibited by the presence of 10 ppm or more of TBC. Oxygen is necessary for effective action of the inhibitor (75). It is recommended that air be added periodically to the storage vessels, which are blanketed with an inert gas. Properly inhibited and attended, styrene can be stored for an extended period of time. In climates where the average daily temperature in excess of 27°C is common, refrigeration of bulk storage is recommended (76). Copper and copper alloys should not be used in the handling or storage of styrene, because copper can dissolve in styrene to cause discoloration and interferes with polymerization.

Table 6. Acute Effects of Overexposure to Styrene[a]

Area affected	Effect
eye	slight to moderate irritation
skin	slight irritation; repeated exposure may produce severe irritation including blistering
inhalation	may cause headache, nausea, dizziness, muscle weakness; produces central nervous system depression; irritates nose, throat, and lungs
ingestion	may cause nausea, headache, dizziness, muscle weakness; produces central nervous system depression, and diarrhea; may be aspirated into lungs if swallowed, resulting in pulmonary edema and chemical pneumonitis

[a]Ref. 74.

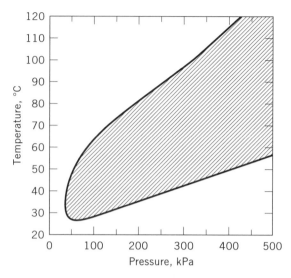

Fig. 6. Approximate explosive limits of styrene monomer vapor in equilibrium with liquid styrene in air, where ▨ represents the explosive region. To convert kPa to mm Hg, multiply by 7.5. Courtesy of the Dow Chemical Company.

Uses

Commercial styrene is used almost entirely for the manufacture of polymers. Polystyrene accounts for 64% of the worldwide demand for styrene. The rest is for manufacture of copolymers: ABS, 9%; SB latex, 7%; UPR, 5%; SBR, 4%; others, 11%.

Polystyrene (PS). Common applications include packaging, food containers, and disposable tableware; toys; furniture, appliances, television cabinets, and sports goods; and audio and video cassettes. For some of these applications, PS is modified by blending or graft polymerization with SBR to form impact polystyrene, which is less sensitive to breakage. Expandable polystyrene is widely used in construction for thermal insulation.

Acrylonitrile–Butadiene–Styrene Copolymer (ABS). Uses for ABS are in sewer pipes, vehicle parts, appliance parts, business machine casings, sports goods, luggage, and toys.

Styrene–Butadiene Latex. SB latex is used in coatings, carpet backing, paper adhesives, cement additives, and latex paint.

Styrene–Butadiene Rubber (SBR). This elastomer is used primarily in tires, vehicle parts, and electrical components.

Unsaturated Polyester Resins (UPR). The principal uses are in putty, coatings, and adhesives. Glass-reinforced UPR is used for marine, construction, and vehicle materials, as well as for electrical parts.

Derivatives

A large number of compounds related to styrene have been reported in the literature. Those having the vinyl group $CH_2=CH-$ attached to the aromatic ring are referred to as styrenic monomers. Several of them have been used for manufacturing small-volume specialty polymers. The specialty styrenic monomers that are manufactured in commercial quantities are vinyltoluene, *para*-methylstyrene, α-methylstyrene, and divinylbenzene. In addition, 4-*tert*-butylstyrene [*1746-23-2*] (TBS) is a specialty monomer that is superior to vinyltoluene and *para*-methylstyrene in many applications. It is manufactured by Amoco in a complex process and its use has been limited by its high price. Deltech is market-testing a TBS monomer that is produced in a much simpler process that can substantially reduce its price and widen its use. Other styrenic monomers produced in small quantities include chlorostyrene and vinylbenzene chloride.

With the exception of α-methylstyrene, which is a by-product of the phenol–acetone process, these specialty monomers are more difficult and expensive to manufacture than styrene. Their more complex molecular structures give rise to the formation of more types and greater quantities of by-products in chemical reactions, and the yields are low in comparison with styrene synthesis. As a result, the recovery and purification facilities are more complex. The high boiling points of these monomers require that distillation be carried out at high temperatures, which would cause high yield losses to polymerization, or at low pressures, which would increase the investment and operating costs. These difficulties, together with the lack of economy in large-scale production, keep the cost of these monomers high and limit their uses.

Vinyltoluene. This is a specialty monomer with properties similar to those of styrene (77). It was originally developed to compete with styrene when benzene was in short supply after World War II. It is more difficult to manufacture than styrene. With the development of new technologies for producing benzene from petroleum, notably catalytic reforming and naphtha cracking, styrene retained its lead position and became a large-volume monomer. Vinyltoluene [*25013-15-4*] has continued to be produced for special applications. Its copolymers are more heat-resistant than the corresponding styrene copolymers. The increased solubility in aliphatic solvents and the faster reaction rates resulting from the added methyl group make vinyltoluene better suited than styrene as a specialty monomer for paint, varnish, and polyester applications.

Vinyltoluene (VT) is a mixture of *meta*- and *para*-vinyltoluenes, typically in the ratio of 60:40. This isomer ratio results from the ratio of the corresponding ethyltoluenes in thermodynamic equilibrium. Physical properties and chemical analysis of a typical vinyltoluene product are shown in Tables 7 and 8, respectively. Vinyltoluene monomer is produced by Dow Chemical Company and Fina Oil & Chemical Company. The worldwide consumption is estimated to be approximately 100,000 t/yr.

Ethyltoluene is manufactured by aluminum chloride-catalyzed alkylation similar to that used for ethylbenzene production. All three isomers are formed. A typical analysis of the reactor effluent is shown in Table 9. After the unconverted toluene and light by-products are removed, the mixture of ethyltoluene isomers and polyethyltoluenes is fractionated to recover the meta and para isomers (bp 161.3 and 162.0°C, respectively) as the overhead product, which typically contains 0.2% or less ortho isomer (bp 165.1°C). This isomer separation is difficult but essential because *o*-ethyltoluene undergoes ring closure to form

Table 7. Physical Properties of a 60:40 Mixture of *m*- and *p*-Vinyltoluene[a]

Property	Value
molecular weight	118.17
refractive index, n_D^{20}	1.5422
viscosity at 20°C, mPa·s (=cP)	0.837
surface tension at 20°C, mN/m (=dyn/cm)	31.66
density at 20°C, g/cm^3	0.8973
boiling point at 101.3 kPa (=1 atm), °C	172
freezing point, °C	−77
flash point (Cleveland open-cup), °C	60
fire point (Cleveland open-cup), °C	68.3
autoignition temperature, °C	575
explosive limits, % in air	1.9–6.1
vapor pressure, kPa[b]	
at 20°C	0.15
60°C	1.76
160°C	74.66
critical pressure, P_c, MPa[c]	4.19
critical temperature, t_c, °C	382
critical volume, V_c, mL/g	3.33
critical density, d_c, g/mL	0.30
specific heat of vapor, C_p, at 25°C, J/(g·K)[d]	1.2284
latent heat of vaporization, ΔH_v, J/g[d]	
at 25°C	426.10
boiling point	349.24
heat of combustion (ΔH_c), kJ/mol[d], gas at constant pressure at 25°C	4816.54
heat of formation (ΔH_f), kJ/mol[d], liquid at 25°C	115.48
heat of polymerization, kJ/mol[d]	+66.9±0.2
Q value	0.95
e value	−0.89
shrinkage upon polymerization, vol %	12.6
cubical coefficient of expansion at 20°C	9.361×10^{-4}
solubility, wt % in H$_2$O at 25°C	0.0089
H$_2$O in monomer at 25°C	0.47
solvent compatibility[e]	

[a] Ref. 1.
[b] To convert kPa to mm Hg, multiply by 7.5.
[c] To convert MPa to psi, multiply by 145.
[d] To convert J to cal, divide by 4.184.
[e] Infinitely soluble in acetone, carbon tetrachloride, benzene, diethyl ether, *n*-heptane, and ethanol.

indan and indene in the subsequent dehydrogenation process. These compounds are even more difficult to remove from vinyltoluene, and their presence in the monomer results in inferior polymers. The *o*-ethyltoluene and polyethyltoluenes are recovered and recycled to the reactor for isomerization and transalkylation to produce more ethyltoluenes. Fina uses a zeolite-catalyzed vapor-phase alkylation process to produce ethyltoluenes.

Table 8. Chemical Analysis of Commercial Vinyltoluene[a]

Assay	ppm[b]
purity	99.6[c]
polymer	none
phenylacetylene	58
aldehydes as CHO	10
peroxides as H_2O_2	5
chlorides as Cl	5
TBC	12
m-vinyltoluene	60[c]
p-vinyltoluene	40[c]

[a]Ref. 1.
[b]Unless otherwise indicated.
[c]Percent.

Table 9. Composition of Unrefined Product Obtained in the Manufacture of Ethyltoluene

Compound	Wt %
benzene and lights	0.2
toluene	48.3
ethylbenzene and xylenes	1.2
p-ethyltoluene	11.9
m-ethyltoluene	19.3
o-ethyltoluene	3.8
polyethyltoluenes and other by-products	14.4
tar	0.9

The dehydrogenation of the mixture of m- and p-ethyltoluenes is similar to that of ethylbenzene, but more dilution steam is required to prevent rapid coking on the catalyst. The recovery and purification of vinyltoluene monomer is considerably more difficult than for styrene owing to the high boiling point and high rate of thermal polymerization of the former and the complexity of the reactor effluent, which contains a large number of by-products. Pressures as low as 2.7 kPa (20 mm Hg) are used to keep distillation temperatures low even in the presence of polymerization inhibitor. The finished vinyltoluene monomer typically has an assay of 99.6%.

***para*-Methylstyrene.** PMS is the para isomer of vinyltoluene in high purity. It is probably the only compound that has the potential to compete against styrene in the large commodity market as a styrenic monomer. p-Methylstyrene [627-97-9] has an apparent raw material cost advantage over styrene, and the characteristics of its polymers and copolymers are superior to those of the corresponding styrene polymers and copolymers in some important aspects. PMS is made by alkylation of toluene with ethylene to p-ethyltoluene, followed by dehydrogenation of p-ethyltoluene. Stoichiometrically, it takes 78 t of toluene and 24 t of ethylene to make 100 t of PMS, whereas it takes 75 t of benzene and

25 t of ethylene to make 100 t of styrene. Toluene is priced lower than benzene by weight, and ethylene is more expensive than both. The use of toluene in lieu of benzene and the lower ethylene requirement together give PMS a potential savings of 10–20%. It has an additional cost advantage of 4% as a monomer resulting from poly-PMS having a lower density than polystyrene, as polymers are usually sold (and used) by volume rather than by weight.

The most difficult problem in manufacturing PMS is in the synthesis of the *p*-ethyltoluene precursor in high purity. A conventional alkylation process makes all three isomers, and isolation of the para isomer from this mixture by distillation or any other technique would be prohibitively expensive. The manufacture of PMS as a large-volume monomer became feasible with the development by Mobil of a ZSM-5 catalyst which is selective to the para isomer in alkylation (78). It attracted much attention in the styrene industry in the early 1980s. PMS was billed as "high-tech styrene" and forecasts were made of its gradually replacing styrene as the dominant styrenic monomer. This did not happen for several reasons. Aside from the isomer selectivity, the production of PMS is more difficult than that of styrene with higher investment, lower yield, and higher energy consumption in both the alkylation and dehydrogenation steps. In reality, the production cost advantage is far less than the apparent raw material cost savings. Moreover, the markets for both the monomer and its polymers must be developed on a large scale for it to compete on the basis of both quality and price. This requires a large investment, which would involve considerable risk in view of the well-established position of styrene monomer in the plastics industry. It was deemed to be too risky, and the expectation of PMS becoming a large-volume monomer faded. It is now manufactured by Deltech as a specialty monomer at Baton Rouge, Louisiana. The typical product contains 97% *p*-vinyltoluene, 3% *m*-vinyltoluene, and 0.2% *o*-vinyltoluene.

Divinylbenzene. This is a specialty monomer used primarily to make cross-linked polystyrene resins. Pure divinylbenzene (DVB) monomer is highly reactive polymerically and is impractical to produce and store. Commercial DVB monomer (76–79) is generally manufactured and supplied as mixtures of *m*- and *p*-divinylbenzenes and ethylvinylbenzenes. DVB products are designated by commercial grades in accordance with the divinylbenzene content. Physical properties of DVB-22 and DVB-55 are shown in Table 10. Typical analyses of DVB-22 and DVB-55 are shown in Table 11. Divinylbenzene [*1321-74-0*] is readily polymerized to give brittle insoluble polymers even at ambient temperatures. The product is heavily inhibited with TBC and sulfur to minimize polymerization and oxidation.

Small quantities of DVB copolymerized with styrene yield polymers which have the appearance of polystyrene but possess higher heat distortion temperatures, greater hardness, and slightly better impact and tensile strengths. The increased resistance to thermal distortion allows the plastics to be machined more easily and to have broader use in electrical insulation applications. Beads of styrene–divinylbenzene resins made by suspension polymerization are used as the basis of ion-exchange resins. (The ionic sites are formed afterward by substitution on the aromatic rings.) The cross-links resulting from divinylbenzene help stabilize the bead structure and minimize swelling. The largest use

Table 10. Physical Properties of Two Divinylbenzene Mixtures[a]

Property	Value DVB-22[b]	DVB-55[b]
molecular weight	130.08	130.18
refractive index, n_D^{25}	1.5326	1.5585
viscosity at 25°C, mPa·s (=cP)	0.883	1.007
surface tension at 25°C, mN/m (=dyn/cm)	30.55	32.10
density at 20°C, g/cm^3	0.8979	0.9126
boiling point, °C	180[c]	195[c]
freezing point, °C		−45
flash point (Cleveland open-cup), °C	57	74
fire point (Cleveland open-cup), °C	57	74
explosive limits, % in air	1.1–6.2	≥1.1[d]
critical pressure, P_c, MPa[e]	2.45[c]	2.45[c]
critical temperature, t_c, °C	348[c]	369[c]
latent heat of vaporization, ΔH_v, at boiling point, J/g[f]	320.49	350.62
solubility at 25°C, % in H_2O	0.0065	0.0052
H_2O in monomer at 25°C	0.051	0.054
solvent compatibility[g]		

[a]Ref. 1.
[b]Dow Chemical Co.'s designations for 22 and 55% divinylbenzene, respectively.
[c]Calculated.
[d]Could not be measured at 130°C.
[e]To convert MPa to psi, multiply by 145.
[f]To convert J to cal, divide by 4.184.
[g]Both types are infinitely soluble in acetone, carbon tetrachloride, benzene, and ethanol.

Table 11. Chemical Analysis of Two Divinylbenzene Mixtures[a]

Assay	Value DVB-22	DVB-55
polymer, ppm	100	100
aldehydes as CHO, ppm	40	40
peroxides as H_2O_2, ppm	5	5
sulfur as S, ppm	20	230
TBC, ppm	1000	1000
total unsaturation[b]	83.3	149.4
divinylbenzene, %		
meta	17.1	36.4
para	8.2	18.6
total	*25.3*	*55.0*
ethylvinyltoluene, %		
meta	23.1	25.0
para	10.0	13.0

[a]Ref. 1.
[b]As ethylvinylbenzene.

of DVB is in ion-exchange resins for domestic and industrial water softening. Ion-exchange resins are also used as solid acid catalysts for certain reactions, such as esterification.

Divinylbenzene is manufactured by dehydrogenation of diethylbenzene, which is an internal product in the alkylation plant for ethylbenzene production. This internal product is normally transalkylated to produce more ethylbenzene. A stream of the diethylbenzene can be diverted for divinylbenzene production. The technology for dehydrogenation of diethylbenzene is similar to that for dehydrogenation of ethylbenzene. The presence of two ethyl groups on the aromatic ring, however, greatly increases the number of possible products and by-products. Light by-products include benzene, toluene, xylene, ethylbenzene, styrene, ethyltoluene, and vinyltoluene. *Meta-* and *para-*diethylbenzenes are converted to their corresponding ethylvinylbenzenes and divinylbenzenes. The unconverted diethylbenzenes and part of the ethylvinylbenzenes are removed together with the light by-products and, after further fractionation, recycled to the reactor. *ortho*-Ethylbenzene [*135-01-3*] is dehydrogenated and cyclized to naphthalene, which is removed together with other heavy by-products as the residue from the product, which is a mixture of divinylbenzenes and ethylvinylbenzenes. This last distillation is carried out at a very low pressure of 1.33–2 kPa (10–15 mm Hg) absolute and in the presence of inhibitors to keep the temperature low and to prevent polymerization. Further isolation of the product components is neither necessary nor desirable. For most commercial applications, divinylbenzene is used in low concentrations as a cross-linking agent. Ethylvinylbenzene itself is a monomer much like vinyltoluene and is easily incorporated into polymers of styrene.

α-Methylstyrene.   This compound is not a styrenic monomer in the strict sense. The methyl substitution on the side chain, rather than the aromatic ring, moderates its reactivity in polymerization. It is used as a specialty monomer in

ABS resins, coatings, polyester resins, and hot-melt adhesives. As a copolymer in ABS and polystyrene, it increases the heat-distortion resistance of the product. In coatings and resins, it moderates reaction rates and improves clarity. Physical properties of α-methylstyrene [*98-83-9*] are shown in Table 12.

Production of α-methylstyrene (AMS) from cumene by dehydrogenation was practiced commercially by Dow until 1977. It is now produced as a by-product in the production of phenol and acetone from cumene. Cumene is manufactured by alkylation of benzene with propylene. In the phenol–acetone process, cumene is oxidized in the liquid phase thermally to cumene hydroperoxide. The hydroperoxide is split into phenol and acetone by a cleavage reaction catalyzed by sulfur dioxide. Up to 2% of the cumene is converted to α-methylstyrene. Phenol and acetone are large-volume chemicals and the supply of the by-product

Table 12. Physical Properties of α-Methylstyrene[a]

Property	Value
molecular weight	118.18
refractive index, n_D^{20}	1.53864
viscosity at 20°C, mPa·s (=cP)	0.940
surface tension at 20°C, mN/m (=dyn/cm)	32.40
density at 20°C, g/cm^3	0.9106
boiling point, °C	165
freezing point, °C	−23.2
flash point (Cleveland open-cup), °C	57.8
fire point (Cleveland open-cup), °C	57.8
explosive limits, % in air	0.7–3.4
vapor pressure, kPa[b]	
at 20°C	0.253
60°C	2.400
100°C	13.066
160°C	88.660
critical pressure, P_c, MPa[c]	4.36
critical temperature, t_c, °C	384
critical volume, V_c, mL/g	3.26
critical density, d_c, g/mL	0.29
specific heat of liquid, J/(g·K)[d]	
at 40°C	2.0460
100°C	2.1757
specific heat of vapor at 25°C, J/(g·K)[d]	1.2357
latent heat of vaporization, ΔH_v, J/g[d]	
at 25°C	404.55
boiling point	326.35
heat of combustion, ΔH_c, gas at constant pressure at 25°C, kJ/mol[d]	4863.73
heat of formation, ΔH_f, liquid at 25°C, kJ/mol[d]	112.97
heat of polymerization, kJ/mol[d]	39.75
Q value	0.76
e value	−1.17
cubical coefficient of expansion at 20°C	9.774×10^{-4}
solubility in H$_2$O at 25°C	0.056
H$_2$O in monomer at 25°C	0.010
solvent compatibility[e]	

[a]Ref. 1.
[b]To convert kPa to mm Hg, multiply by 7.5.
[c]To convert MPa to psi, multiply by 145.
[d]To convert J to cal, divide by 4.184.
[e]Infinitely soluble in acetone, carbon tetrachloride, benzene, diethyl ether, n-heptane, and ethanol.

α-methylstyrene is well in excess of its demand. Producers are forced to hydrogenate it back to cumene for recycle to the phenol–acetone plant. Estimated plant capacities of the U.S. producers of α-methylstyrene are listed in Table 13 (80).

Table 13. U.S. α-Methylstyrene Producers and Capacities, 1995

Company	Location	Capacity, t/yr
AlliedSignal	Frankford, Pa.	25,000
Aristech	Haverhill, Ohio	17,300
JLM Chemicals	Blue Island, Ill.	2,300
Georgia-Gulf	Pasadena, Tex.	3,200
	Plaquemine, La.	6,800
Texaco	El Dorado, Kans.	1,400
Total		*56,000*

BIBLIOGRAPHY

"Styrene" in *ECT* 1st ed., Vol. 13, pp. 119–146, by A. L. Ward and W. J. Roberts, Pennsylvania Industrial Chemical Corp.; in *ECT* 2nd ed., Vol. 19, pp. 55–85, by K. E. Coulter, H. Kehde, and B. F. Hiscock, The Dow Chemical Co.; in *ECT* 3rd ed., Vol. 21, pp. 770–801, by P. J. Lewis, C. Hagopian, and P. Koch, The Badger Co., Inc.

1. *Styrene-Type Monomers*, Technical Bulletin No. 170-151B-3M-366, The Dow Chemical Company, Midland, Mich. (no date).
2. R. H. Boundy and R. F. Boyer, eds., *Styrene, Its Polymers, Copolymers, and Derivatives*, Reinhold Publishing Corp., New York, 1952.
3. H. Ohlinger, *Polystyrol*, Springer-Verlag, Berlin, Germany, 1955.
4. J. Elly, R. N. Haward, and W. Simpson, *J. Appl. Chem.* **2**, 347 (1951).
5. C. H. Basdekis, in W. M. Smith, ed., *Manufacturing of Plastics*, Reinhold Publishing Corp., New York, 1964.
6. H. Fikentscher, H. Gerrens, and H. Schuller, *Angew. Chem.* **72**, 856 (1960).
7. C. E. Schildknecht, *Polymer Processes*, Interscience Publishers, Inc., New York, 1956, Chapt. 4.
8. K. F. Coulter, H. Kehde, and B. F. Hiscock, in E. C. Leonard, ed., *Vinyl Monomers*, Wiley-Interscience, New York, 1969.
9. U.S. Pat. 3,751,506 (Aug. 7, 1973), G. T. Burress (to Mobil Oil Corp.).
10. U.S. Pat. 4,107,224 (Aug. 15, 1978), F. G. Dwyer (to Mobil Oil Corp.).
11. H. W. Grote and C. F. Gerald, *Chem. Eng. Prog.* **56**(1), 60 (1960).
12. H. W. Grote, *Oil Gas J.* **56**(13), 73 (1958).
13. L. L. Hegedus and co-workers, *Catalyst Design*, John Wiley & Sons, Inc., New York, 1987.
14. N. Y. Chen, T. F. Degnan, and C. M. Smith, *Molecular Transport and Reaction in Zeolite*, VCH Publishers, New York, 1994.
15. F. G. Dwyer, P. J. Lewis, and F. M. Schneider, *Chem. Eng.* **83**, 90 (Jan. 5, 1976).
16. P. J. Lewis and F. G. Dwyer, *Oil Gas J.* **75**(40), 55 (1977).
17. *Hydrocarbon Process.* **74**(3), 116 (1995).
18. K. J. Fallon and S. Ram, *J. Japan Aro. Ind. Assoc.*, 45 (1993).
19. *Chem. Mark. Rep.*, **240**(20), 7 (Nov. 18, 1991).
20. K. J. Fallon, H. K. H. Wang, and C. R. Venkat, *Oil Gas J.* **93**(16), 50 (1995).
21. *Hydrocarbon Process.* **74**(3), 114 (1995).
22. J. A. Valentine and co-workers, *Increase Competitiveness in the Styrene Market*, 1992, Sud-Chemie International Styrene Symposium, Ohita, Japan, Nov. 9, 1992.
23. J. Chen, *CDTECH Aromatic Alkylation Technologies*, Worldwide Solid Acid Process Conference, Houston, Tex., Nov. 14, 1993.
24. *Chem. Mark. Rep.* **246**(16), 27 (Oct. 17, 1994).
25. D. Gandi and T. Mortimer, *Direct Ethylbenzene Process*, 1994 Sud-Chemie International Styrene Symposium, Louisville, Ky., Nov. 2–4, 1994.

26. B. Maerz, S. S. Chen, C. R. Venkat, and D. Mazzone, EBMax: *Leading Edge Ethylben-zene Technology from Mobil/Badger*, 1996 DeWitt Petrochemical Review, Houston, Tex., Mar. 19–21, 1996.

27. G. A. Olah, ed., *Friedel-Crafts and Related Reactions*, Wiley-Interscience, New York, 1964.

28. A. C. MacFarlane, *Oil Gas J.* **74**(6), 55 (1977).

29. U.S. Pat. 3,183,233 (May 11, 1965), H. S. Bloch (to Universal Oil Products Co.).

30. U.S. Pat. 3,200,163 (Aug. 10, 1965), E. R. Fenske (to Universal Oil Products Co.).

31. *Chem. Week* **98**(7), 80 (1966).

32. E. K. Jones, *Oil Gas J.* **58**(9), 80 (1960).

33. Technical Bulletin on Shell-005, Shell Chemical Co., Houston, Tex., 1978.

34. Product Bulletin, Criterion Catalysts, Houston, Tex., 1994.

35. Product Bulletin, United Catalysts, Louisville, Ky., 1994.

36. U.S. Pat. 5,461,179 (Oct. 24, 1995), S. S. Chen, S. Y. Hwang, S. A. Oleksy, and S. Ram (to Raytheon Engineers & Constructors).

37. J. C. Frank, G. R. Geyer, H. Kehde, *Chem. Eng. Prog.* **65**(2), 79 (1969).

38. D. B. McMullen and co-workers, *Chem. Eng. Prog.* **87**(7), 187 (1991).

39. S. E. Knipling, *Toluene Color-More Questions than Answers*, 1994 Sud-Chemie International Styrene Symposium, Louisville, Ky., Nov. 2–4, 1994.

40. Br. Pats. 1,122,732 and 1,122,731 (Aug. 7, 1978), C. Y. Choo (to Halcon International, Inc.).

41. Br. Pat. 1,128,150 (Sept. 25, 1968), C. Y. Choo and R. L. Golden (to Halcon International, Inc.).

42. U.S. Pat. 3,459,810 (Aug. 5, 1969), C. Y. Choo and R. L. Golden (to Halcon International, Inc.).

43. Ger. Pat. 2,631,016 (July 10, 1975), M. Becker (to Halcon International, Inc.).

44. U.S. Pat. 3,987,115 (Oct. 19, 1976), J. G. Zajacek and F. J. Hilbert (to Atlantic Richfield Co.).

45. U.S. Pat. 4,066,706 (Jan. 3, 1978), J. P. Schmidt (to Halcon International, Inc.).

46. S. Af. Pat. 66 05,917 (Apr. 1, 1968), H. B. Pell and E. I. Korchak (to Halcon International Inc.).

47. Fr. Pat. 1,548,198 (Nov. 29, 1968), T. W. Stein, H. Gilman, and R. L. Bobeck (to Halcon International, Inc.).

48. U.S. Pat. 3,849,451 (Nov. 19, 1974), T. W. Stein, H. Gilman, and R. L. Bobeck (to Halcon International, Inc.).

49. Ger. Pat. 1,939,791 (Feb. 26, 1970), M. Becker and S. Khoobiar (to Halcon International, Inc.).

50. *Chem. Week* **99**(6), 19 (1966).

51. *Chem. Week* **99**(5), 49 (1966).

52. Ger. Pat. 2,165,027 (July 6, 1973), J. J. Coyle (to Shell Oil Corp.).

53. U.S. Pat. 3,829,392 (Aug. 13, 1974), H. P. Wulff (to Shell Oil Corp.).

54. U.S. Pat. 3,873,578 (Mar. 25, 1975), C. S. Bell and H. P. Wulff (to Shell Oil Corp.).

55. U.S. Pat. 4,059,598 (Nov. 22, 1977), J. J. Coyle (to Shell Oil Corp.).

56. U.S. Pat. 4,435,607 (May 6, 1984), T. Imai (to UOP).

57. K. Egawa and co-workers, *Aromatics* **43**, 5–6 (1991).

58. *Hydrocarbon Process.*, **74**(3), 147 (1995).

59. H. J. Sanders, H. F. Keag, and H. S. McCullough, *Ind. Eng. Chem.* **45**(1), 2 (1953).

60. U.S. Pat. 3,965,206 (June 22, 1976), P. D. Montgomery, P. N. Moore, and W. R. Knox (to Monsanto Co.).

61. U.S. Pat. 4,247,729 (Jan. 27, 1981), H. Susumu, M. Yoshiyuhi, T. Hideyuki (to Mitsubishi Petrochemical Co.).

62. U.S. Pat. 4,115,424 (Sept. 19, 1978), M. L. Unland and G. E. Barker (to Monsanto Co.).

63. U.S. Pat. 4,140,726 (Feb. 20, 1978), M. L. Unland and G. E. Barker (to Monsanto Co.).
64. *Styrene/Ethylbenzene*, PERP report 94/95-8, Chem Systems, Tarrytown, N.Y., Mar. 1996.
65. U.S. Pat. 4,144,278 (Mar. 13, 1979), D. J. Strope (to Phillips).
66. U.S. Pat. 3,903,185 (Sept. 2, 1975), H. H. Vogel, H. M. Weitz, E. Lorenz, and R. Platz (to BASF).
67. U.S. Pat. 5,276,257 (Jan. 4, 1994), R. W. Diesen (to Dow Chemical Co.).
68. U.S. Pat. 5,329,057 (July 12, 1994), R. W. Diesen, R. S. Dixit, and S. T. King (to Dow Chemical Co.).
69. *Styrene from Butadiene*, PERP report 93S3, Chem Systems, Tarrytown, N.Y., (Mar. 1995).
70. *Styrene/Ethylbenzene*, PERP Report 91-9, Chem Systems, Tarrytown, N.Y., (Oct. 1992).
71. *Chem. Mark. Rep.* **248**(5), 41 (July 24, 1995).
72. F. A. Fazzalari, ed., *Compilation of Odor and Taste Threshold Values Data*, American Society for Testing and Materials, Philadelphia, Pa., 1977.
73. *Threshold Limits Values for Chemical Substances in Workroom Air*, American Conference of Governmental Industrial Hygienists, Cincinnati, Ohio, 1981.
74. *Styrene Material Data Sheet*, Phillips 66 Co. (Jan. 28, 1991).
75. *Storage and Handling of Styrene-Type Monomers*, Form No. 115-575-79, Organic Chemicals Dept., Dow Chemical USA, Midland, Mich., 1979.
76. *Specialty Monomers Product Stewardship Manual*, Form No. 505-0007-1290JB, Dow Chemical USA, Midland, Mich., 1990.
77. U.S. Pat. 2,763,702 (Sept. 18, 1956), J. L. Amos and K. E. Coulter (to Dow Chemical Co.).
78. U.S. Pat. 4,100,217 (July 11, 1978), L. B. Young (to Mobil Oil Corp.).
79. Product Bulletin, Deltech Corp., Baton Rouge, La., 1994.
80. *Chem. Mark. Rep.* **248**(2), 41 (July 10, 1995).

SHIOU-SHAN CHEN
Raytheon Engineers & Constructors

STYRENE–BUTADIENE RUBBER

Attempts to produce synthetic rubber have been carried out since the 1800s. The introduction of automobiles in the early 1900s gave added impetus to find a substitute for natural rubber, the price of which tripled from $2.16/kg in 1900 to $6.73/kg in 1910 (1). The advent of World War I gave Germany incentive to start a crash program on an alternative to natural rubber. From this work, products based on dimethylbutadiene were used, but these were not found to be good substitutes.

In the late 1920s Bayer & Company began reevaluating the emulsion polymerization process of polybutadiene as an improvement over their Buna technology, which was based on sodium as a catalyst. Incorporation of styrene (qv) as a comonomer produced a superior polymer compared to polybutadiene. The product Buna S was the precursor of the single largest-volume polymer produced in the 1990s, emulsion styrene–butadiene rubber (ESBR).

When the United States entered World War II, access to the source of natural rubber was eliminated and the federal government set up a consortium of American rubber manufacturers under the auspices of the Office of Rubber Reserve. A total of 15 styrene–butadiene rubber (SBR) plants were constructed between 1941 and 1942. These plants, devoted to the production of a natural rubber substitute, were run by the four principal rubber companies in the United States: Goodyear, U.S. Rubber, B.F. Goodrich, and Firestone. To increase the number of rubber companies participating in the program, two additional companies were formed from smaller rubber producers: Copolymer Corporation and National Synthetic Rubber Corporation (2). The product agreed upon was called GR-S, an abbreviation for Government Rubber–Styrene, and was based on Buna S. It was the standard general-purpose SBR manufactured until right after World War II, when Germany's work on a low temperature system to generate free radicals for cold emulsion polymerization of SBR became known. The properties of this product were so much better than the hot GR-S that the Office of Rubber Reserve ordered all of the consortium's plants to install refrigeration equipment and begin phasing in the cold-polymerized GR-S. This product has been the basis of the emulsion polymer industry in the United States ever since.

In the mid-1950s, the Nobel Prize-winning work of K. Ziegler and G. Natta introduced anionic initiators which allowed the stereospecific polymerization of isoprene to yield high cis-1,4 structure (3,4). At almost the same time, another route to stereospecific polymer architecture by organometallic compounds was announced (5).

In the 1960s, anionic polymerized solution SBR (SSBR) began to challenge emulsion SBR in the automotive tire market. Organolithium compounds allow control of the butadiene microstructure, not possible with ESBR. Because this type of chain polymerization takes place without a termination step, an easy synthesis of block polymers is available, whereby glassy (polystyrene) and rubbery (polybutadiene) segments can be combined in the same molecule. These thermoplastic elastomers (TPE) have found use in nontire applications. Styrene–butadiene–styrene blocks are formed by the anionic solution polymerization technique. The long styrene block portion of the polymer chain imparts a rigidity similar to that of vulcanized rubber. Being thermoplastic, the polymers are easily worked on conventional equipment. Also, because the styrene segments soften on heating, the polymer scrap can be reworked, unlike vulcanized rubber.

Physical Properties

Desirable properties of elastomers include elasticity, abrasion resistance, tensile strength, elongation, modulus, and processibility. These properties are related to and dependent on the average molecular weight and mol wt distribution, polymer macro- and microstructure, branching, gel (cross-linking), and glass-transition temperature (T_g) (see ELASTOMERS, SYNTHETIC).

Emulsion polymerization gives SBR polymer of high molecular weight. Because it is a free-radical-initiated process, the composition of the resultant chains is heterogeneous, with units of styrene and butadiene randomly spaced throughout. Unlike natural rubber, which is polyisoprene of essentially all cis-1,4 configuration, giving an ordered structure and hence crystallinity, ESBR is amorphous. For cold ESBR polymerized at 4–10°C, the typical microstructure of the butadiene portion is 72% trans-1,4, 12% cis-1,4, and 16% 1,2. In hot polymerized ESBR (51.5°C), the typical ratio is changed somewhat to 65% trans-1,4, 18.5% cis-1,4, and 16.5% 1,2 (6).

Unlike SSBR, the microstructure of which can be modified to change the polymer's T_g, the T_g of ESBR can only be changed by a change in ratio of the monomers. Glass-transition temperature is that temperature where a polymer experiences the onset of segmental motion (7).

For ESBR polymerized at 50°C, knowing the percentage of bound styrene in the copolymer allows estimation of the T_g by the following, where S is the weight fraction of the styrene (% bound styrene) (8).

$$T_g \ (°C) = (-85 + 135 \ S)/(1 - 0.5 \ S) \tag{1}$$

Similarly, for ESBR made at 5°C, the T_g is given by equation 2:

$$T_g \ (°C) = (-78 + 128 \ S)/(1 - 0.5 \ S) \tag{2}$$

The glass-transition temperatures for solution-polymerized SBR as well as ESBR are routinely determined by nuclear magnetic resonance (nmr), differential thermal analysis (dta), or differential scanning calorimetry (dsc).

Among the techniques employed to estimate the average molecular weight distribution of polymers are end-group analysis, dilute solution viscosity, reduction in vapor pressure, ebulliometry, cryoscopy, vapor pressure osmometry, fractionation, hplc, phase distribution chromatography, field flow fractionation, and gel-permeation chromatography (gpc). For routine analysis of SBR polymers, gpc is widely accepted. Table 1 lists a number of physical properties of SBR (random) compared to natural rubber, solution polybutadiene, and SB block copolymer.

Advantages of NR/IR are high resilience and strength, and abrasion-resistance. BR shows low heat buildup in flexing, good resilience, and abrasion-resistance. Random SBR is low in price, wears well, and bonds easily. Block SBR is easily injection-molded, and is not cross-linked. Applications of NR/IR include tires, tubes, belts, bumpers, tubing, gaskets, seals, foamed mattresses, and padding. BR is used in tire treads and mechanical goods, as is random SBR. Block SBR is used in toys, rubber bands, and mechanical goods.

Raw Materials

The monomers, butadiene (qv) and styrene (qv), are the most important ingredients in the manufacture of SBR polymers. For ESBR, the largest single material is water; for solution SBR, the solvent.

Table 1. Properties and Applications of Cross-Linked Rubber Compounds[a]

Property	NR/IR[b]	BR[c]	SBR random[d]	SBR block[e]
	Gum stock[f]			
density, g/cm^3	0.93	0.93	0.94	0.94–1.03
tensile strength, MPa[g]	17–21	1.4–7	1.4–2.8	11.7–25.5
resistivity, Ω·cm, log	15–17		15	13
dielectric constant at 1 kHz	2.3–3.0	2.3–3.0	3.0	3.4
dissipation factor at 1 kHz	0.0002–0.0003	0.0002–0.0003		
dielectric strength, MV/m[h]				18.9
	Reinforced stock			
tensile strength, MPa[g]	21–28	14–24	14–24	7–21
elongation at break, %	300–700	300–700	300–700	500–1000
hardness, Shore A	20–100	30–100	40–100	40–85
resilience	excellent	excellent	good	excellent
stiffening temperature, °C	−30 to −45	−35 to −50	−20 to −45	−50 to −60
brittle temperature, °C	−60	−70	−60	−70
continuous high temperature limit, °C	100	100	110	65
	Resistance to solvents and conditions			
acid	good	good	good	good
alkali	good	good	good	good
gasoline and oil	poor	poor	poor	poor
aromatic hydrocarbons	poor	poor	poor	poor
ketones	good	good	good	poor
chlorinated solvents	poor	poor	poor	poor
oxidation	good	good	good	good
ozone	poor	poor	poor	poor
γ-radiation	good	good	good	good

[a]Ref. 9.
[b]*cis*-1,4-Polyisoprene, natural rubber (NR), also made synthetically (IR).
[c]*cis*-1,4-Polybutadiene.
[d]Styrene–butadiene random copolymer, 25 wt % styrene.
[e]Styrene–butadiene block copolymer, ~25% SBR styrene (YSBR).
[f]Cross-linked and unfilled.
[g]To convert MPa to psi, multiply by 145.
[h]To convert MV/m to kV/in., multiply by 25.65.

The quality of the water used in emulsion polymerization has long been known to affect the manufacture of ESBR. Water hardness and other ionic content can directly affect the chemical and mechanical stability of the polymer emulsion (latex). Poor latex stability results in the formation of coagulum in the polymerization stage as well as other parts of the latex handling system.

In converting ESBR latex to the dry rubber form, coagulating chemicals, such as sodium chloride and sulfuric acid, are used to break the latex emulsion. This solution eventually ends up as plant effluent. The polymer crumb must also be washed with water to remove excess acid and salts, which can affect the cure properties and ash content of the polymer. The requirements for large amounts of good-quality fresh water and the handling of the resultant effluent are of utmost importance in the manufacture of ESBR and directly impact on the plant operating costs.

Solution polymerization can use various solvents, primarily aliphatic and aromatic hydrocarbons. The choice of solvent is usually dictated by cost, availability, solvency, toxicity, flammability, and polymer structure. SSBR polymerization depends on recovery and reuse of the solvent for economical operation as well as operation under the air-quality permitting of the local, state, and federal mandates involved.

Styrene.   Commercial manufacture of this commodity monomer depends on ethylbenzene, which is converted by several means to a low purity styrene, subsequently distilled to the pure form. A small percentage of styrene is made from the oxidative process, whereby ethylbenzene is oxidized to a hydroperoxide or alcohol and then dehydrated to styrene. A popular commercial route has been the alkylation of benzene to ethylbenzene, with ethylene, after which the crude ethylbenzene is distilled to give high purity ethylbenzene. The ethylbenzene is directly dehydrogenated to styrene monomer in the vapor phase with steam and appropriate catalysts. Most styrene is manufactured by variations of this process. A variety of catalyst systems are used, based on ferric oxide with other components, including potassium salts, which improve the catalytic activity (10).

In 1990, the annual U.S. capacity to manufacture styrene monomer was 4,273,000 t/yr, and production was 3,636,000 t/yr (11). Polystyrene resin is the dominant user of styrene monomer. SBR use is about 7% of U.S. domestic styrene monomer production. Worldwide production in 1995 was projected to be 77% of capacity as demand increased just under 5% per year, from 1990 consumption of 13,771,000 to 17,000,000 metric tons in 1995.

Butadiene.   Although butadiene was produced in the United States in the early 1920s, it was not until the start of World War II that significant quantities were produced to meet the war effort. A number of processes were investigated as part of the American Synthetic Rubber Program. Catalytic dehydrogenation of n-butenes and n-butanes (Houdry process) and thermal cracking of petroleum hydrocarbons were chosen (12).

Economic considerations in the 1990s favor recovering butadiene from by-products in the manufacture of ethylene. Butadiene is a by-product in the C4 streams from the cracking process. Depending on the feedstocks used in the production of ethylene, the yield of butadiene varies. For use in polymerization, the butadiene must be purified to 99+%. Crude butadiene is separated from C_3 and C_5 components by distillation. Separation of butadiene from other C_4 constituents is accomplished by salt complexing/solvent extraction. Among the solvents used commercially are acetonitrile, dimethylacetamide, dimethylformamide, and *N*-methylpyrrolidinone (13). Based on the available crude C_4 streams, the worldwide forecasted production is as follows: 1995, 6,712,000; 1996, 6,939,000; 1997, 7,166,000; and 1998, 7,483,000 metric tons (14). As of January 1996, the 1995 actual total was 6,637,000 t.

Soap. A critical ingredient for emulsion polymerization is the soap (qv), which performs a number of key roles, including production of oil (monomer) in water emulsion, provision of the loci for polymerization (micelle), stabilization of the latex particle, and impartation of characteristics to the finished polymer.

Fatty acid soap was first used for ESBR. Its scarcity prompted the investigation of rosin acids from gum and wood as substitutes (1). The discovery of the disproportionation of rosin allowed rosin acid soaps to overcome the polymerization inhibition of untreated rosin acids. Rosin acid soaps gave the added benefit of tack to the finished polymer. In the 1990s, both fatty acid and rosin acid soaps, mainly derived from tall oil, are used in ESBR.

Polymerization

ESBR and SSBR are made from two different addition polymerization techniques: one radical and one ionic. ESBR polymerization is based on free radicals that attack the unsaturation of the monomers, causing addition of monomer units to the end of the polymer chain, whereas the basis for SSBR is by use of ionic initiators (qv).

Free-radical initiation of emulsion copolymers produces a random polymerization in which the trans/cis ratio cannot be controlled. The nature of ESBR free-radical polymerization results in the polymer being heterogeneous, with a broad molecular weight distribution and random copolymer composition. The microstructure is not amenable to manipulation, although the temperature of the polymerization affects the ratio of trans to cis somewhat.

In solution-based polymerization, use of the initiating anionic species allows control over the trans/cis microstructure of the diene portion of the copolymer. In solution SBR, the alkyllithium catalyst allows the 1,2 content to be changed with certain modifying agents such as ethers or amines. The use of anionic initiators to control the molecular weight, molecular weight distribution, and the microstructure of the copolymer has been reviewed (15).

Emulsion Polymerization. Emulsion SBR was commercialized and produced in quantity while the theory of the mechanism was being debated. Harkins was among the earliest researchers to describe the mechanism (16); others were Mark (17) and Flory (18). The theory of emulsion polymerization kinetics by Smith and Ewart is still valid, for the most part, within the framework of monomers of limited solubility (19). There is general agreement in the modern theory of emulsion polymerization that the process proceeds in three distinct phases, as elucidated by Harkins (20): nucleation (initiation), growth (propagation), and completion (termination).

The nucleation stage for sparingly water-soluble monomers, such as styrene and butadiene, lasts up to 15 to 20% conversion of monomer to polymer. Free radicals from some source initiate polymerization in one of four loci: (*1*) in the water phase where a free radical attacks a soluble monomer molecule (homolytic nucleation); (2) at some part of the monomer-swollen micelle (micellar nucleation); (3) at the combined surfactant molecule and solvated monomer oligomer in the dispersed phase (nucleation by hydrophobic association); and (4) at the surface of the stabilized monomer droplet. In the first case, the growing oligomer reaches a finite chain length and collapses to form a polymer particle. Available soap then stabilizes this particle. For SBR, the most important source of polymer generation is the micelle. Nucleation by hydrophobic association, similar to homogenous nucleation, forms particles as the oligomer chain reaches a finite length and precipitates out of solution as a particle stabilized with a layer of soap. In the fourth case, because the number of monomer droplets is small in relation to the number of available monomer-swollen micelles, polymer formation in the droplet is not statistically significant on account of the great difference in surface area of micelles compared to the monomer droplets and is therefore generally ignored.

The onset of the growth stage is characterized by the disappearance of the micelles. At this point, the growing micelle undergoes a metamorphosis, becoming a polymer particle, now surrounded by a charged layer of surfactant. During this phase, monomer from the shrinking droplets continues to diffuse through the water phase. Polymerization proceeds at a constant rate under steady-state conditions. It is during this stage that free surfactant is made available. The surfactant arises from the following sources: the shrinking monomer droplets, released surfactant from agglomeration of latex particles, and released surfactant from the growing polymer particles which need less surfactant coverage at a larger size because of the surface-to-volume ratio. If enough soap becomes available then more micelles can form. This results in another particle population that gives a bimodal particle distribution. This stage for SBR ends at 55–65% conversion.

The completion stage is identified by the fact that all the monomer has diffused into the growing polymer particles (disappearance of the monomer droplet) and reaction rate drops off precipitously. Because the free radicals that now initiate polymerization in the monomer-swollen latex particle can more readily attack unsaturation of polymer chains, the onset of gel is also characteristic of this third stage. To maintain desirable physical properties of the polymer formed, emulsion SBR is usually terminated just before or at the onset of this stage.

Advantages of emulsion polymerization of SBR include the following: (1) high molecular weight polymer at high rates; (2) relatively low viscosity of the latex formed to give good heat transfer of the generated exotherm; (3) good control of the polymerization temperature; (4) ease of removing unreacted monomers and their recovery; and (5) ease of processing the finished latex into dry polymer product.

The original emulsion SBR was run at high temperature (50°C) using a fatty acid soap system (Table 2). This was because the available free-radical source was potassium peroxydisulfate, $K_2S_2O_4$, which depends on elevated temperature for the production of free radicals (thermal decomposition). It was found early on that dodecyl mercaptan played a key role in the polymerization initiation, as well as in the regulation of the polymer molecular weight. It was proposed

Table 2. Typical SBR Recipes[a]

Elastomer	Hot SBR 1000	Cold SBR 1500	Cold SBR 1502
Composition, phm			
butadiene	71	71	71
styrene	29	29	29
potassium peroxydisulfate	0.3		
p-menthane hydroperoxide (PMHP)		0.12	0.12
n-dodecyl mercaptan (DDM)	0.5		
t-dodecyl mercaptan (TDM)		0.2	0.18
Emulsifier makeup, phm			
water[b] monomer/water	200	200	200
disproportionated tall oil rosin acid soap	4.5–5	4.5	1.35
hydrogenated tallow fatty acid soap	4.5		3.15
potassium chloride	0.3	0.3	0.3
DARVAN WAQ (secondary emulsifier)		0.1	0.1
Versene Fe-3 (iron complexing agent)		0.01	0.01
sodium dithionate (oxygen scavenger)[c]		0.025	0.025
Sulfoxylate activator makeup, phm			
ferrous sulfate heptahydrate		0.04	0.04
Versene Fe-3		0.06	0.06
sodium formaldehyde sulfoxylate (SFS)		0.06	0.06
Shortstop makeup, phm			
methyl namate	0.05	0.05	0.05
diethylhydroxylamine	0.015	0.015	0.015
Polymerization conditions			
temperature, °C	50	5	5
final conversion, %	72	60–65	60–65
coagulation	acid/amine	acid/amine	acid/amine
antioxidant, ~0.5–0.75 on rubber	stainer	stainer	nonstainer
organic acid content, by weight	5–7	5–7	5–7
styrene content, by weight	24	24	24
mooney viscosity ML-4, min at 100°C	48	46–58	46–58

[a]Ref. 21.
[b]Monomer/water ratio adjusted to 1:2.
[c]pH of solution to 10–10.5.

that the mercaptan radical formed from the action of the persulfate radical as the initiating species for polymerization (22). More recently it has been shown that the enhancement of the mercaptan promoting effect is specific to fatty acid soap containing persulfate-initiated diene systems (23).

Originally, the production of emulsion SBR was conducted by the batch method, which involves the charging of the entire recipe ingredients at the same time. As the polymerization advances, the amount of solids (polymer) increases. The percentage of solids is related to the percentage of conversion of monomer to polymer. At the proper conversion rate, in this case 75%, a dilute solution of a chemical called a shortstopping agent is added to the reactor. Hydroquinone, at low levels, is effective. The termination of polymerization is accomplished in two ways. First, shortstopping agents, ie, reducing agents, combine with the initiating species, ie, an oxidizing agent, and effectively destroy the source of free radicals. The second step is to trap and neutralize the free radicals already generated. In the case of hydroquinone, it reduces the persulfate and effectively destroys the generation of any free radicals (24). As it is oxidized to the quinone structure, it becomes an effective free-radical acceptor, scavenging any remaining free radicals.

The shortstopped latex is transferred to one of a series of large vessels, referred to as blowdown tanks, which hold two batches in addition to adequate vapor space. The hot latex is then transferred to large horizontal tanks by pressure differential. These tanks are referred to as flash tanks, for as the latex enters the tank it degasses, that is, releases the residual butadiene, gaseous monomer. The degassed latex is then pumped to the top of a plate column where it is contacted by steam entering the bottom, with the column operating under vacuum. This countercurrent steam distillation of the hot latex efficiently removes the residual styrene monomer. Because SBR contains a high percentage of unsaturation (double bonds) from the butadiene, it is susceptible to degradation by heat, light (uv energy), and O_2 (oxidation). Certain chemicals called antioxidants effectively protect polymers from this degradation. The finished latex containing antioxidant is then subjected to a two-stage coagulation that breaks the emulsion, resulting in a slurry of rubber crumb. The crumb is then passed through a series of washes and dewatering, finally reground and then dried to a low moisture content in large, continuous-belt dryers heated by direct gas-fired burners or steam pipes. The crumb exiting the dryer is then baled in a hydraulic baler in standard 80-lb bales (~36 kg), bagged or film-wrapped, and boxed for shipment in standardized containers.

Through the years, improvements in the finishing process eliminated the need of two-stage coagulation, ie, creaming the latex with sodium chloride and coagulating this agglomerated latex with sulfuric acid. The same results are achieved by mixing the salt and acid and coagulating in one step. To aid the complete coagulation of the SBR latex, small amounts of coagulant aids are added. These coagulant aids are usually based on polyamines. In the mid-1970s, many commercial ESBR manufacturers replaced the brine (NaCl) by increasing the level of polyamine coagulants. A significant benefit from the change was in the reduction of ash residue in the polymer. Single-screw dewatering presses replaced vacuum filters for effective dewatering of the crumb. The screw dewaterers express large amounts of water at its boiling point as a result of the work put into the polymer as it passes through the machine. Soluble salts are released with the water, resulting in lower ash content of the polymer. Figures 1 and 2 illustrate a modern production flow sheet for manufacturing solid ESBR.

Initially, all of the SBR polymer known as GR-S produced during World War II was by the batch process. Later, it was thought that a higher volume of polymer would be needed for the war effort. The answer was found in switching from batchwise to continuous production. This was demonstrated in 1944 at the Houston, Texas synthetic rubber plant operated by The Goodyear Tire & Rubber Company. One line, consisting of 12 reactors, was lined up in a continuous

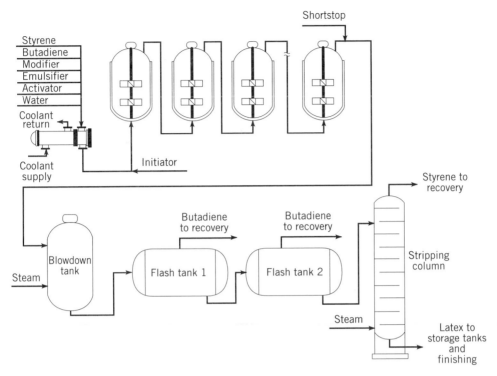

Fig. 1. Typical continuous emulsion SBR polymerization.

mode, producing GR-S that was more consistent than the batch-produced polymer (25). In addition to increased productivity, improved operation of the recovery of monomers resulted because of increased (20%) reactor capacity as well as consistent operation instead of up and down, as by batchwise polymerization.

Continuous polymerization is accomplished by making up all recipe ingredients in solutions, which are mixed with the monomer streams, brought to desired temperature, and initiated by addition of the free-radical source. Of critical importance is the accurate makeup of the solutions and precise metering of all the streams. Modern ESBR plants depend on computer control (distributive control systems). The reactors are connected in series so that the liquid flow is into the bottom and out the top of each agitated vessel. The residence time in the reactor train is controlled by either a variable volume last reactor or a series of large-diameter pipes (displacement columns), whereby at the correct solids percentage (conversion), the shortstop is added to the exiting latex (see Fig. 1).

After World War II, it was reported that the Germans had been working on a cold emulsion polymerization based on a redox initiation system that gave far superior SBR to the conventional hot polymerization (see Table 2). The principal differences were a polymerization temperature of 5°C instead of 50°C, monomer-to-polymer conversion of 60–65% instead of 75%, and the use of a reduction/oxidation system (redox) to supply free radicals by chemical rather than thermal means. The superiority of cold over hot SBR was attributed, in part, to the greater linearity and less gel of the cold polymer. It was shown that in the cross-linking reaction of butadiene, the ratio of the rate of cross-

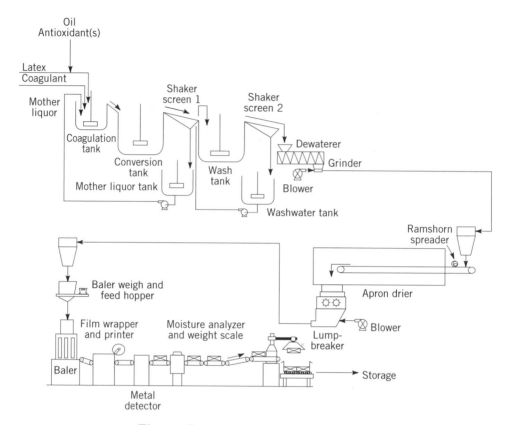

Fig. 2. Typical emulsion SBR finishing.

linking to the rate of propagation is constant up to the point where the monomer phase disappears (26). This ratio decreases as the polymerization temperature is lowered. Thus, there are fewer cross-linking sites at lower temperatures. At the point of monomer depletion (Smith-Ewart third stage), cross-linking increases rapidly. By lowering the conversion rate at which the polymerization is terminated, that point is not reached.

The impact of cold GR-S was quite pronounced. The U.S. government edicted that all of the emulsion SBR plants switch to the cold process. This required addition of refrigeration capacity in these plants as well as other significant changes, such as insulation of reactors, improved vacuum to reduce oxygen that retards polymerization, and the heating of latex in blowdown tanks to aid in the disengagement of butadiene when transferred to the flash tanks.

It was found that cold emulsion SBR could be made at very high molecular weight, combined with petroleum-based oils at a very high level. Although the polymer could take up oil to the extent of 60 parts of oil per 100 parts of polymer, a standard level of 37.5 parts of oil per 100 parts of polymer was established because the plasticizing power of the oil would require extremely high polymer molecular weight at the 60 part level. The oil was added to the latex as an emulsion and cocoagulated in the standard system. The resultant oil-extended polymer had the benefits of very high molecular weight (improved stress–strain properties) with the processability of medium Mooney viscosity SBR. Extending pure polymer with oil increased the amount of product over 25%. In addition, the economics of oil-extended SBR was quite attractive considering that a 36-kg bale contained about 10 kg of a low cost process oil.

Several different petroleum process oils are used as extenders for SBR. These are classified as highly aromatic, aromatic, naphthenic, and paraffinic. The highly aromatic oils give the most desirable physical properties to the polymer and are widely used in tire treadstock. However, these oils impart a black color to the polymer and produce staining, ie, discoloration of light-colored articles that come in contact with products containing the oil. For light-colored articles and products in contact with light-colored materials, naphthenic or paraffinic oils are preferred.

The properties of emulsion SBR are comparable to those of natural rubber only when SBR is reinforced with carbon black. Carbon black can be slurried in water and then combined with the SBR latex and cocoagulated to give a product called black masterbatch. The black masterbatch would require considerably less time and energy to get the carbon black dispersed in the nonproductive compound than if it were dry-mixed with the polymer. Other advantages of handling black masterbatch are the elimination of inventorying and handling carbon black, less energy demand in mixing stocks, and better dispersability of the carbon black in the compound.

Black masterbatch can be made with or without oil-extended SBR. Commercially there are available, worldwide, 11 numbered ESBR cold black masterbatches and 15 ESBR cold oil-black masterbatches (27). These types range in black type and content as well as oil type and amount. Of course, not every listed product is available from every supplier.

Solution SBR. Addition polymerization is accomplished in three steps: initiation, propagation, and termination. The alkyllithium-initiated polymerization in hydrocarbon solvent has no termination step, resulting in living polymer. The mechanism for initiation is thought to be dependant on the addition of the lithium alkyl across the vinyl double bond giving an organolithium compound. Propagation is effected by the further addition of monomer as a stepwise reaction, reforming the same type of carbon–lithium bond (28).

Without the termination step in the alkyllithium polymerization, the batch polymerization results in a copolymer composition of styrene and butadiene that markedly changes with polymer conversion. This produces blockiness, ie, long segments of one constituent in the copolymer chains. An explanation for this behavior is that the butadiene reactivity is considerably greater than that of styrene in the copolymerization and has been shown to be caused by the high cross-propagation rate of the styryllithium anion (29). This mechanism actually allows great flexibility in producing very different types of polymers. The so-called tapered block polymer is made by polymerizing styrene and butadiene monomers with butyllithium as the initiator. By addition of certain chemicals such as ethers, tertiary amines, and phosphites, random distribution of the comonomers is achieved. Certain of these additives also affect the microstructure of the butadiene portion. It is thus possible to control polymer composition, monomer distribution sequence, microstructure (trans-1,4/cis-1,4 ratio constant at variable vinyl level), molecular weight, molecular weight distribution, and polymer chain structure. Figure 3 shows a flow diagram for the continuous production of solution SBR. The kinetics of alkyllithium solution copolymerization

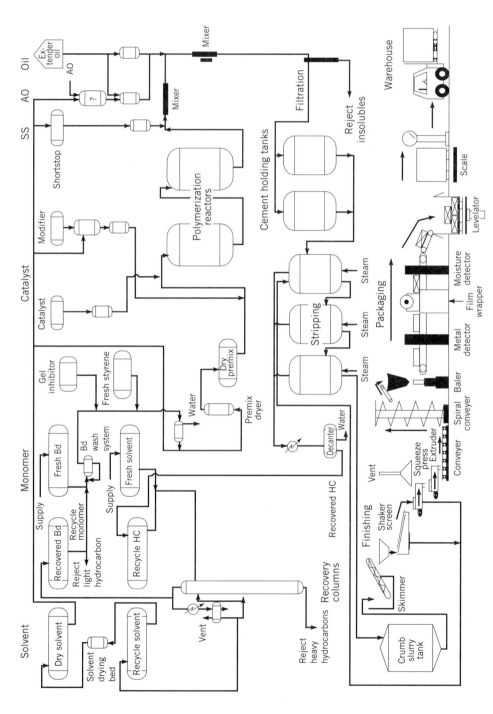

Fig. 3. Solution SBR manufacture by continuous process where Bd = butadiene, HC = hydrocarbon, AO = antioxidant, and SS = shortstop.

of styrene and butadiene in hydrocarbon solvent, along with techniques to randomize the resultant polymer, have been reviewed (30).

Once the ability to control the randomization of solution SBR was understood and established, the commercialized SSBR began to displace some of the ESBR in tire applications. It was found that the SSBR alone could achieve the same desired properties that blends of ESBR and polybutadiene gave, thus simplifying the compound recipes. The combination of low rolling resistance and high wet traction without sacrificing abrasion resistance (tread wear) is attainable with SSBR but not with ESBR alone. Also, by controlling the vinyl microstructure to a low level, a lower T_g is achieved than for ESBR at the same styrene content. This allows for a higher level of oil and carbon black to be added (31).

Because the styrene segments of block copolymers soften on heating, properties that depend on stiffness, hardness, abrasion, etc, may suffer. Conversely, the polymers flow much more readily on molding, extruding, etc. This property allows them to be used in shoe soling and wire insulation applications. Until more recently, the use of block SSBR in tire tread applications was not considered because of the significantly higher running temperature associated with compounds containing them. A Japanese patent claims that the use of a low molecular weight random block copolymer of styrene and butadiene (A–B–A block SSBR) as part of a tire tread composition imparts improved grip (adhesion) and abrasion resistance to the tire (32).

SBR Compounding and Processing

The initial production of GR-S rubber required a learning period in processing before rubber workers were comfortable with the new synthetic rubber. Although softer and more plastic initially, the GR-S did not break down as much as natural rubber. Once familiar with the differences, processing GR-S was handled quite comfortably. The same was true when the cold GR-S was introduced, followed by the oil-extended version, and SSBR.

Development of the Mooney viscometer gave compounders an indication of the processibility of different lots of the uncompounded polymer. This machine measures the torque resistance encountered by a rotor revolving in a chamber surrounded by polymer at a constant temperature. The resulting Mooney number describes the toughness of the polymer and is an indirect measure of molecular weight.

Gel, the insoluble fraction of polymer formed by high conversion of monomer to polymer, excessive mixing, poor antioxidant protection, etc, increases the difficulty of mixing, reduces tensile strength, increases modulus, and decreases the resistance to cut growth. Polymers with significant gel content have reduced elastic memory (but lower shrinkage and lower surface roughness) and improved dimensional stability, two important qualities in extrusion and calendering operation.

Vulcanization was discovered by Goodyear in 1839 (33). By incorporating elemental sulfur with natural rubber and heating the mass, he found that the resultant compound yielded a tough, elastic material, markedly different from the unvulcanized state. Modern rubber compounding (qv) has come a long way since the discovery of vulcanization, but its explanation still remains quite theoretical (34). As a class, SBR is slower curing than natural rubber. This is thought to be caused by the lower unsaturation in SBR.

The art of compounding requires extensive experience and knowledge of the many compound ingredients. A typical rubber compound in addition to polymer contains one or more ingredients from the following general classes: vulcanizing agents, accelerators, accelerator activators, antioxidants, pigments, and softeners (see RUBBER CHEMICALS) (35).

The vulcanizing agent, which supplies the bridge between the polymer chains, is still furnished predominantly by the sulfur molecule in commercial formulations. Peroxide vulcanizers that produce carbon-to-carbon cross-links are also important. Thiuram disulfides are used in special applications, such as improved heat resistance. Other agents are of more academic interest.

Accelerators are chemical compounds that increase the rate of cure and improve the physical properties of the compound. As a class, they are as important as the vulcanizing agent itself. Without the accelerator, curing requires hours or even days to achieve acceptable levels. Aldehyde amines, thiocarbamates, thiuram sulfides, guanidines, and thiazoles are all classified as accelerators. By far, the most widely used are the thiazoles, represented by mercaptobenzothiazole (MBT) and benzothiazyl disulfide (MBTS).

Accelerator activators are chemicals required to initiate the acceleration of the curing process. They also improve the polymer compound quality. This class includes zinc oxide (pure) and stearic acid. Other compounds that have been in use are litharge, magnesium oxide, amines, and amine soaps.

Antioxidants (qv) are routinely added to the compounds over and above those contained in the polymer at manufacture. Finished products are subjected to a wide variety of activities causing degradation. Molecules with the propensity to neutralize free radicals by destruction, absorption, or disproportionation can accomplish this phenomenon in different ways. Incorporation of two or more different chemical categories often enhances protection much better (synergism) than the use of higher levels of individual additives. The most popular chemical classes shown to have good antioxidant properties are phenols, phosphites, thioesters, and amines. Because many antioxidant compounds are highly colored, they may tint the finished article. Amines and quinolines have the propensity to stain, that is, to transfer some of the color they impart from the compound in which they are used to a surface in contact with that compound, and are classified as stainers. These characteristics must be kept in mind when choosing the antioxidant system.

Antiozonants (qv) prevent or reduce polymer degradation by the active ozone molecule. Some antioxidant compounds, such as the *para*-phenylenediamines, are excellent as antiozonants (36). The protection by these compounds is thought to be either a reaction with the ozone before it can react with the surface of the rubber or an aid in reuniting chains severed by ozone (37).

Pigments (qv) improve or change polymer properties as well as lower product costs. Reinforcement of SBR by carbon blacks allows this family of polymers to compete with natural rubber (see CARBON, CARBON BLACK). It is the most important attribute of the pigment in SBR processing. Two other important groups of pigments are the silicates and silicas. Other materials such as zinc oxide, clays, and calcium carbonate act more as inert fillers to lower the overall compound cost rather than to impart reinforcement. Softeners, ie, plasticizers, reinforcing agents, extenders, lubricants, tackifiers, and dispersing aids, are used as processing aids to enhance mixing of uncured stocks and soften cured compounds.

SBR rubber compounds are prepared in two stages: the nonproductive and the productive phases. In the nonproductive phase the compound ingredients are mixed, usually in internal mixers (Banbury). The mixing time is usually short and the compound temperature is in the 160–195°C range. This stock is discharged from the mixer to equipment that allows cooling and a convenient storage form, such as a mill or an extruder/die plate that yields a sheet or pelletized form. Usually the material is coated with a slurry of clay, calcium carbonate, or zinc stearate to prevent self-adhesion.

The productive stock, ie, the curable compound, is made up by mixing the nonproductive stock in the Banbury once more with the curative package (sulfur, accelerators, etc). This time the drop temperature is lower, in the range of 95–112°C. The productive stock is then sheeted or pelletized and coated with the dip coat, cooled, and finally stored, ready for further processing for final fabrication.

Economic Aspects and Uses

Styrene–butadiene elastomers, emulsion and solution types combined, are reported to be the largest-volume synthetic rubber, with 28.7% of the world consumption of all synthetic rubber in 1994 (38). This percentage has decreased steadily since 1973 when SBR's market share was 57% (39). The decline has been attributed to the switch to radial tires (longer milage) and the growth of other synthetic polymers, such as polyethylene, polypropylene, polyester, and polystyrene. Since 1985, production of SBR has been flat (Table 3).

Worldwide long-term consumption of SBR is projected to increase 2.5% per year through 1999 (Table 4). SBR is forecast to remain the dominant elastomer of

Table 3. Synthetic Rubber Production[a,b], 10^3t

Type[c]	1985	1990	1992	1994	% Change, 1993/1994
SBR					
solid	2284	2398	2358	2433	5.0
latex[d]	289	264	300	299	−2.3
polybutadiene elastomer (BR)	978	1164	1264	1452	5.9
polyisoprene rubber (IR)	110	142	115	111	16.8
chloroprene rubber (CR)	321	306	260	276	15.0
ethylene propylene diene monomer (EPDM)	453	622	617	672	11.3
nitrile rubber (NBR)	213	246	255	277	9.1
others	563	574	606	543[e]	−11.3
Total	*5211*	*5716*	*5775*	*6063[e]*	*4.5*

[a]Ref. 40. [b]Excludes Central Europe, Socialist Countries of Asia, CIS. [c]For all polymers latices are included, except for SBR where they are shown separately. [d]Excludes carboxylated latices. [e]The change reflects a change in the International Institute of Synthetic Rubber Producers (IISRP) reporting base; growth rates not available in 1994.

Table 4. Worldwide Long-Term New Rubber Consumption By Elastomer Type[a], 10³ t

Elastomer	1994	1995[b]	1999[b]
SBR			
solid	2,589	2,774	3,085
latex	400	413	467
carboxylated SBR latex	1,473	1,509	1,670
polybutadiene	1,427	1,541	1,726
ethylene propylene diene monomer (EPDM)	653	679	750
chloroprene rubber (CR)	252	258	293
nitrile rubber (NBR)	319	337	391
other synthetics[c]	1,176	1,369	1,549
Total Central Europe SR	*250*	*268*	*351*
Total Asia CPEC SR	*485*	*519*	*680*
Total synthetics	*9,022*	*9,666*	*10,962*
natural rubber	5,425	5,595	6,264
Total new rubber	*14,447*	*15,289*	*17,226*
synthetic, %	62.4	63.2	63.6

[a]Ref. 40.
[b]1995 and 1999 values are forecast.
[c]Includes isoprene–isobutyl rubber (IIR), polyisoprene rubber (IR), and other synthetic rubbers (SR).

Table 5. Worldwide Long-Term New Rubber Consumption By Area and Type[a], 10³ t

Elastomer	Western Europe			North America		
	1994[b]	1995[c]	1999[c]	1994[b]	1995[c]	1999[c]
SBR						
solid	568	586	620	852	860	883
latex[d]	125	125	125	79	80	83
carboxylated SBR	570	584	645	634	646	699
polybutadiene elastomer (BR)	280	288	312	500	507	510
ethylene propylene diene monomer (EPDM)	225	236	266	264	270	274
chloroprene rubber (CR)	66	66	67	76	77	79
nitrile rubber (NBR) (solid and latex)	89	92	100	122	124	125
other synthetics[e]	241	251	278	408	420	426
Total synthetic rubber	*2164*	*2228*	*2413*	*2935*	*2984*	*3080*
natural rubber	900	930	1010	1106	1111	1103
Total new rubber	*3064*	*3159*	*3423*	*4041*	*4094*	*4183*
synthetic, %	70.6	70.5	70.5	72.6	72.9	73.6
thermoplastic elastomer (TPE)	299	314	371	358	387	496

[a]Ref. 40.
[b]Estimated December 1994.
[c]Forecast prepared during first quarter of 1995.
[d]Noncarboxylated.
[e]Includes isoprene–isobutyl rubber (IIR), polyisoprene rubber (IR), and others.

all synthetic polymers in the same time frame. In 1993, use of SBR encompassed the following: tires and tire-related products, including tread rubber, 80%; mechanical goods, 11%; other automotive uses, 6%; and adhesives, chewing gum base, shoe products, flooring, etc, for the remaining 3% (41).

North America has led the world in consumption of SBR and will continue for the rest of the twentieth century. As shown in Table 5, total demand for SBR, including emulsion (solid) and solution (solid), was 852,000 metric tons in 1994, up from 811,000 metric tons in 1993. Since the early 1990s growth has averaged about 3.2% per year. Projected growth in the near future, based on the tight supply of natural rubber, is 4–5%, through 1999.

Health and Safety Factors

Air quality and plant effluent have been monitored and more or less regulated from the inception of SBR manufacture. Between the 1970s and 1990s, regulatory restrictions on plant operations have increased significantly. The ever-decreasing limits on exposure levels include styrene and butadiene monomers as well as their dimers and by-products. Most local and state governments have strict discharge permits that limit what kind of chemicals and how much of it can be emitted into the environment.

The American Conference of Governmental Industrial Hygienists (ACGIH) has set a time-weighted average (TWA) of 50 ppm for styrene monomer (42). For butadiene monomer, ACGIH has set the exposure limit of TWA at 10 ppm (43). As of this writing (1996), the Occupational Safety and Health Administration

Table 5. (Continued)

Asia and Oceania			Africa and Middle East		
1994[b]	1995[c]	1999[c]	1994[b]	1995[c]	1999[c]
684	714	832	60.7	63.2	71.5
141	146	164	4.0	4.1	6.0
231	241	280			
445	464	540	28.0	28.7	31.5
141	146	165	5.0	5.3	7.0
84	87	101	6.0	6.1	6.5
77	80	94	3.2	3.2	3.7
217	227	267	21.0	21.8	25.0
2020	*2105*	*2443*	*127.9*	*132.5*	*151.3*
2070	2120	2340	188.0	194.9	215.2
4090	*4224*	*4783*	*315.8*	*327.4*	*366.5*
49.4	49.8	51.0	40.5	40.5	41.3
148	154	179	5.5	5.9	7.7

(OSHA) is reviewing their TWA, which is set at 1000 ppm for butadiene. It is likely that the new TWA will be set at 1 or 2 ppm. Both styrene and butadiene are considered suspect carcinogens.

Title V of the Clean Air Act Amendments of 1990 covers federally approved state operating permits for manufacturing facilities. One requirement of this regulation is that manufacturers must report emissions information of identified hazardous air pollutants specific to their operation from a list of 189 named in the Clean Air Act Amendments. Rubber and tire manufacturers had to meet this requirement by the end of 1995. The Rubber Manufacturers Association has begun an industrywide project to develop accurate and reliable emissions data to aid manufacturers to comply with these requirements (44).

In 1988, the German government issued technical regulation for hazardous materials, TRGS 552. Among other things, 12 volatile nitrosamines were specified and an orientation value of 2.5 μg/m^3 was set for the rubber industry. In 1992 the regulation was redefined. The following levels/action are being enforced in Germany: (*1*) up to 0.1 μg/m^3 constitutes no exposure; (*2*) if levels exceed 0.25 μg/m^3, action needs to be taken; (*3*) if level exceeds 0.65 μg/m^3, respirators and medical monitoring must be offered; and (*4*) if levels exceed 1.00 μg/m^3, respirators are mandatory.

This regulation is only concerned with volatile nitrosamines in the workplace air. A principal problem in enforcement is in the detection method. Only certain analytical laboratories are certified and reproducibility is difficult to achieve. Epidemiological studies have shown volatile nitrosamines to be carcinogenic in animals (45). Volatile nitrosamines are formed when secondary amine compounds break down and are nitrosated. In rubber this occurs primarily during the vulcanization stage, where accelerators, which are predominantly secondary amine compounds, decompose, forming lower molecular weight compounds, and are nitrosated either from oxides of nitrogen in the air or from nitrate–nitrite salts in the vulcanization process. Other sources of these secondary amines are as contaminants in compounding ingredients and as trace amounts in emulsion SBR, from the residue of certain shortstopping chemicals used in its manufacture.

There is an industry trend to supply SBR certifiably free of volatile nitrosamines or nitrosatable compounds. This has generally been accomplished by replacing shortstop systems based on carbamates and hydroxyl amines with products that are not based on secondary amines or are secondary amines of high molecular weight, such as dibenzyldithiocarbamate. A more recently issued patent for ESBR shortstop is based on isopropylhydroxylamine, a primary amine that does not form nitrosamine (46).

Of primary concern to local, state, and federal governments is the growing stockpile of scrap tires. It was estimated by the Rubber Manufacturer's Association that for the year 1994, 242 million tires were added to the staggering number stockpiled haphazardly throughout the United States. The threat of huge piles of scrap tires catching fire is cited as a principal concern. Past experience has shown how such fires pollute the air and threaten groundwater as the large quantities of oil released in the incomplete burning become a serious runoff problem. In 1991 the federal government instituted the Intermodal Surface Transportation Act, which, among other things, mandates the use of ground rubber from scrap tires in asphalt paving. Being an unfunded mandate, it was ignored by most states and the Federal Highway Administration has not been authorized money to enforce this provision (47). There is heated debate on whether or not this technology is in fact worthwhile (48). Although use of scrap tires is projected to increase rapidly in the next three to five years, the only economically feasible use has been as a fuel or fuel supplement in utility and industrial applications (see RECYCLING, RUBBER) (49).

BIBLIOGRAPHY

"Styrene Resins and Plastics" in *ECT* 1st ed., Vol. 13, pp. 146–179, by J. A. Struthers, R. F. Boyer, and W. C. Goggin, The Dow Chemical Co.; "Styrene Plastics" in *ECT* 2nd ed., Vol. 19, pp. 85–134, by H. Keskkula, A. E. Platt, and R. F. Boyer, The Dow Chemical Co.; in *ECT* 3rd ed., Vol. 21, pp. 801–847, by A. E. Platt and T. C. Wallace, The Dow Chemical Co.

1. R. F. Dunbrook, in G. S. Whitby, ed., *Synthetic Rubber*, John Wiley & Sons, Inc. New York, 1954, p. 34.
2. J. W. Livingston and J. T. Cox, Jr., in Ref. 1, p. 180.
3. K. Ziegler and co-workers, *Angew. Chem.* **67**, 541 (1955).
4. G. Natta, *J. Polym. Sci.* **16**, 143 (1955).
5. F. W. Stavely and co-workers, *Ind. Eng. Chem.* **48**, 778 (1956).
6. S. G. Turly and H. Keshkula, *Polymer*, **21**, 466 (1980).
7. F. Rodriguez, *Principles of Polymer Systems*, 2nd Ed., Hemisphere Publishing Corp., Washington, D.C., 1982, p. 37.
8. L. A. Wood, *J. Polym. Sci.* **28**, 319 (1958).
9. Ref. 7, pp. 540–545.
10. D. H. James, J. B. Gardner, and E. C. Mueller, in J. I. Kroschwitz, ed., *Encyclopedia of Polymer Science and Engineering*, 2nd ed., Vol. 16, John Wiley & Sons, Inc., New York, 1989, pp. 6–10.
11. K. L. Ring, *CEH Marketing Research Report Styrene, Chemical Economics Handbook*, SRI International, Menlo Park, Calif., 1992, p. 694.3000D.
12. C. E. Morrell, in Ref. 1, pp. 59–61.
13. D. P. Tate and T. W. Bethea, in Ref. 10, Vol. 2, p. 548.
14. Chemical Market Associates, Inc., *CMAI Monomers Market Report 1994*, Houston, Tex.
15. R. N. Cooper, *AIChE Symp. Ser.* **69**(135), 83–85 (1973).
16. W. D. Harkins, *J. Am. Chem. Soc.* **69**, 1428–1444 (1947).
17. H. Mark and R. Raff, *High Polymeric Reactions*, Interscience, New York, 1941.
18. P. J. Flory, *J. Am. Chem. Soc.* **59**, 241–253 (1937).
19. W. V. Smith and R. H. Ewart, *J. Chem. Phys.* **16**, 592–599 (1948).
20. D. C. Blackley, *Emulsion Polymerization Theory and Practice*, Applied Science Publishers Ltd., London, 1975, pp. 58–72.
21. *The Vanderbilt Rubber Handbook*, 13th ed., R. T. Vanderbilt Co., Inc., Norwalk, Conn., 1990, pp. 58–59.
22. I. M. Kolthoff and A. I. Medalic, *J. Polymer Sci.* **6**, 189–207 (1951).
23. E. M. Verdurmen, J. M. Geurts, and A. L. German, *Makromol. Chem.* **195**(2), 641–645 (Feb. 1994).
24. S. A. Sundet, *Government Synthetic Rubber Report*, Vol. 28, CR-650, May 10, 1945.
25. J. W. Livingston and J. T. Cox, Jr., in Ref. 1, p. 209.
26. M. Morton and P. P. Salatiello, *J. Polym. Sci.* **8**, 111–121, 215–224 (1951).

27. *The Synthetic Rubber Manual*, 12th ed., International Institute of Synthetic Rubber Producers, Inc., Houston, Tex., 1992, pp. 19–20, 26–28.

28. S. Bywater and D. J. Worsfold, *ACS Advances in Chemistry*, No. 52, ACS, Washington, D.C., 1966, p. 37.

29. M. Morton and F. Ellis, *J. Polym. Sci.* **61**, 25 (1962).

30. T. C. Bouton and S. Futamura, *Rubber Age*, **106**(3), 33–35 (Mar. 1974).

31. *Cariflex Technical Manual*, GPR 1.2, Shell Chemicals, Houston, Tex., 1983.

32. U.S. Pat. 4,396,743 (July 2, 1983), Fujimaki and co-workers (to Bridgestone Tire Co, Ltd.).

33. L. E. Oneacre, in M. Morton, ed., *Introduction to Rubber Technology*, Reinhold Publishing Corp., New York, 1956, p. 93.

34. W. A. Wilson and D. C. Grimm, *Molecular Structures from Polymerization and Vulcanization*, Southern Rubber Group, Knoxville, Tenn., 1994.

35. B. S. Garvey, Jr., in Ref. 33, pp. 34–40.

36. *Antioxidants and Antiozonants for Rubber and Rubber-like Products*, The Goodyear Tire & Rubber Co., Akron, Ohio, 1988.

37. F. W. Billmeyer, Jr., *Textbook of Polymer Science*, 3rd ed., John Wiley & Sons, Inc., New York, 1984, p. 517.

38. *Worldwide Forecast-1995*, International Institute of Synthetic Rubber Producers, Inc., Houston, Tex., 1995.

39. *Chem. Ind. Newsletter*, 11 (Nov./Dec. 1993).

40. Technical data, International Institute of Synthetic Rubber Producers, Inc., Houston, Tex., Feb. 24, 1995.

41. *Chemical Marketing Reporter*, **22**, 45 (May 30, 1994).

42. *Suspect Chemicals Sourcebook*, Master Index 3, Roytech Publications, Bethesda, Md., 1993, p. 144.

43. Ref. 42, p. 159.

44. *Rubber World*, **210**(6), 11 (Sept. 1994).

45. P. N. Magee and J. M. Barnes, *Brit. J. Cancer*, **10**, 114–122, (1956).

46. U.S. Pat. 5,384,372 (Jan. 24, 1995), R R. Lattime (to The Goodyear Tire & Rubber Co.).

47. *Akron Beacon J.* (224-94) B-15 (Nov. 24, 1994).

48. M. Moore, *Tire Bus.* **11**(24), 3/20, (Mar. 21, 1994).

49. J. Paul, in Ref. 10, Vol. 14, p. 788.

RICHARD R. LATTIME
The Goodyear Tire & Rubber Company

STYRENE–BUTADIENE SOLUTION COPOLYMERS. See
ELASTOMERS, SYNTHETIC.

STYRENE PLASTICS

Polystyrene [9003-53-6] (PS), the parent of the styrene plastics family, is a high molecular weight linear polymer which, for commercial uses, consists of ~1000 styrene units. Its chemical formula (**1**), where $n = $ ~1000, tells little of its properties.

$$-\left(CH-CH_2\right)_n-$$

(**1**)

The main commercial form of PS is amorphous and hence PS is highly transparent. The polymer chain stiffening effect of the pendent phenyl groups raises the glass-transition temperature (T_g) to slightly over 100°C. Therefore, under ambient conditions, the polymer is a clear glass, whereas above the T_g it becomes a viscous liquid which can be easily fabricated, with only slight degradation, by extrusion or injection-molding techniques. This ease with which PS can be converted into useful articles accounts for the very high volume (>9 million tons/yr) used in world commerce. Even though crude oil is the source of the polymer, the energy savings and environmental impact accrued during fabrication and use, compared to alternative materials, more than offset the short life of many PS articles (1).

Most PS is manufactured using continuous bulk polymerization plants. Generally, the key problems associated with manufacture of the polymer are removal of the heat of polymerization and pumping highly viscous solutions. Conversion of the monomer to the polymer is energetically very favorable and occurs spontaneously on heating without the addition of initiators. Because it is a continuous polymerization process, material-handling problems are minimized during manufacture. By almost any standard, the polymer produced is highly pure and is usually greater than 99 wt % PS; for particular applications, however, processing aids are often deliberately added to the polymer. Methods for improving the toughness, solvent resistance, and upper use temperature have been developed. Addition of butadiene-based rubbers increases impact resistance, and copolymerization of styrene with acrylonitrile produces heat- and solvent-resistant plastics (see ACRYLONITRILE POLYMERS). Uses for these plastics are extensive. Packaging applications, eg, disposable tumblers, television cabinets, meat and food trays, and egg cartons, are among the largest area of use for styrene plastics. Rigid foam insulation in various forms is being used increasingly in the construction industry, and modified styrene plastics are replacing steel or aluminum parts in automobiles; both applications result in energy savings beyond the initial investment in crude oil. The cost of achieving a given property, eg, impact strength, is among the lowest for styrene plastics as compared to other competitive materials.

Properties

The general mechanical properties of styrene polymers are given in Table 1. Considerable differences in performance can be achieved by using the various styrene plastics. Within each group, additional variation is expected. In choosing an appropriate resin for a given application, other properties and polymer behavior during fabrication must be considered. These factors depend on the combination of inherent polymer properties, the fabrication technique, and the devices, eg, a mold, used for obtaining the final object. Accordingly, consideration must be given to such factors as the surface appearance of the part and the development of anisotropy, and the effect of anisotropy on mechanical strength, ie, long-term resistance of the molding to external strain.

Physical. An extensive compilation of physical properties of PS has been given (2,3). In general, a polymer must have a weight-average molecular weight (M_w) about 10 times higher than its chain entanglement molecular weight (M_e) to have optimal strength. Below M_e, the strength of a polymer changes rapidly with M_w. However, at about 10 times M_e, the strength reaches a plateau region (Fig. 1). For PS, the M_e is 18,100 (4). Thus PS having an $M_w < 180,000$ is generally too brittle to be useful. This indicates why no general-purpose molding and extrusion grades of PS are sold commercially that have $M_w < 190,000$.

Stress–Strain Properties. The strain energy, derived from the area under the stress–strain curve, is considered to indicate the level of toughness of a polymer. High impact PS (HIPS) has a higher strain energy than an acrylonitrile–butadiene–styrene (ABS) plastic, as shown in Figure 2. Based on different impact-testing techniques, ABS materials are generally tougher than HIPS materials (6,7). The failure of the stress–strain curve to reflect this ductility can be related to the fact that ABS polymers tend to show only localized flow or necking tendency at low rates of extension, and therefore fail at low elongation. HIPS extends uniformly during such tests, and the test specimen whitens over all of its length and extends well beyond the yield elongation. At higher testing speeds, ABS polymers also deform more uniformly and give high elongations (7).

Tensile strengths of styrene polymers vary with temperature. Increased temperature lowers the strength. However, tensile modulus in the temperature region of most tests (−40 to 50°C) is affected only slightly. The elongations of PS and styrene copolymers do not vary much with temperature (−40 to 50°C), but the elongation of rubber-modified polymers first increases with increasing temperature and ultimately decreases at high temperatures.

The molecular orientation of the polymer in a fabricated specimen can significantly alter the stress–strain data as compared with the data obtained for an isotropic specimen, eg, one obtained by compression molding. For example, tensile strengths as high as 120 MPa (18,000 psi) have been reported for PS films and fibers (8). PS tensile strengths below 14 MPa (2000 psi) have been obtained in the direction perpendicular to the flow.

Creep, Stress Relaxation, and Fatigue. The long-term engineering tests on plastics in intended-use environments and temperatures are required for predicting the overall performance of a polymer in a given application. Creep tests involve the measurement of deformation as a function of time at a constant stress or load. For styrene-based plastics, many such studies have been carried

Table 1. Mechanical Properties of Injection-Molded Specimens of Main Classes of Styrene-Based Plastics[a]

Property	Polystyrene (PS)	Poly(styrene-co-acrylonitrile) (SAN)[b]	Glass-filled PS[c]	High impact PS[d]	HIPS	Acrylonitrile–butadiene–styrene terpolymer (ABS)[d]		Standard ABS	Super ABS
						Type 1	Type 2		
CAS Registry Number	[9003-53-6]	[9003-54-7]			[9003-53-6]			[9003-56-9]	
specific gravity	1.05	1.08	1.20	1.05	1.05	1.05	1.05	1.04	1.04
Vicat softening point, °C	96	107	103	103	95	99	108	103	108
tensile yield, MPa[e]	42.0	68.9	131	39.6	29.6	31.0	53.8	41.4	34.5
elongation, rupture, %	1.8	3.5	1.5	15	58	55	10	20	60
modulus, MPa[e]	3170	3790	7580	2690	2140	2620	2620	2070	1790
impact strength (notched Izod), J/m[f]	21	21	80	96	134	193	187	267	428
dart-drop impact strength	very low	very low	medium high	low	medium high	medium high	high	very high	very high
relative ease of fabrication	excellent	excellent	poor	excellent	excellent	excellent	good	good	medium good

[a]Ref. 2.
[b]24 wt % acrylonitrile.
[c]20% glass fibers.
[d]Medium mol wt.
[e]To convert MPa to psi, multiply by 145.
[f]To convert J/m to ft·lbf/in., divide by 53.38.

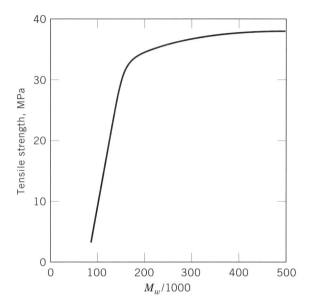

Fig. 1.   PS tensile strength vs M_w (5). To convert MPa to psi, multiply by 145.

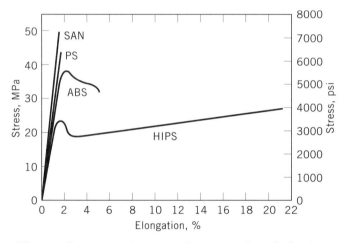

Fig. 2.   Stress–strain curves for styrene-based plastics.

out (9,10). Creep curves for styrene and its copolymers at room temperature show low elongation having only small variation with stress, whereas the rubber-modified polymers exhibit a low elongation region, followed by crazing and increasing elongation, usually to ca 20%, before failure (Fig. 3).

Creep tests are ideally suited for the measurement of long-term polymer properties in aggressive environments. Both the time to failure and the ultimate elongation in such creep tests tend to be reduced. Another test to determine plastic behavior in a corrosive atmosphere is a prestressed creep test in which the specimens are prestressed at different loads, which are lower than the creep load, before the final creep test (11).

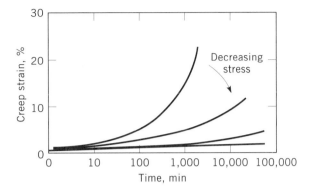

Fig. 3. Typical creep behavior for rubber-modified styrene polymers.

Stress-relaxation measurements, where stress decay is measured as a function of time at a constant strain, have also been used extensively to predict the long-term behavior of styrene-based plastics (9,12). These tests have also been adapted to measurements in aggressive environments (13). Stress-relaxation measurements are further used to obtain modulus data over a wide temperature range (14).

Fatigue is another property of considerable interest to the design engineer. Cyclic deflections of a predetermined amplitude, short of giving immediate failure, are applied to the specimen, and the number of cycles to failure is recorded. In addition to mechanically induced periodic stresses, fatigue failure can be studied when developing cyclic stresses by fluctuating the temperature.

Fatigue in polymers has been reviewed (15); detailed theory and practice of fatigue testing are covered. Fatigue tests, conducted both in air and in aggressive environments (16), are carried out for two main reasons: to learn the inherent fatigue resistance of the material and to study the relationship between specimen design and fatigue failure.

Melt Properties. The melt properties of PS at temperatures between 120 and 260°C are important because it is in this temperature range that PS is extruded to make sheets, foams, and films, or molded into parts. Generally it is desirable to make parts having high strength from materials having low melt viscosity for easy melt processing. However, polymer molecular weight increases both strength and melt viscosity. The melt viscosity of PS can be decreased to improve its melt processibility by the addition of a plasticizer such as mineral oil. However, the addition of a plasticizer has a penalty, ie, the heat distortion temperature is lowered. In applications where heat resistance is important, melt processibility can be influenced, without a significant effect on heat resistance, by control of the polydispersity (17), by branching (18), or by the introduction of pendent ionic groups, eg, sodium sulfonate (19,20).

Impact Strength. PS and styrene copolymers are brittle polymers under normal use conditions. A high speed blow at temperatures below T_g causes catastrophic failure without significant deformation, crazing, or yielding in the polymer. Rubber-modified styrene polymers, however, are significantly more impact-resistant. These polymers are characterized by significant whitening of the specimen during the test as a result of craze formation, separation of the rubber phase from the matrix polymer, and cracks (21,22). The mechanism by which the dispersed rubber particles cause such increased toughness continues to be debated (23–25). Craze initiation and termination are controlled by the particles. Under tensile stress, crazes are initiated near the equators of the particles, ie, at the point of maximum stress, and propagate outward. The highly oriented PS fibrils with the surrounding voids constitute the craze (26). Because of the molecular orientation, such material is load-bearing and ductile. The large number of particles leads to a large number of small crazes, and on a microscopic scale the plastic is ductile instead of brittle. Creation of the craze matter absorbs energy and, so long as the stress does not cause the crazes to become true cracks, the plastic can recover after the stress is removed. Particle size sensitivity probably results from the ineffectiveness of very small particles in stopping craze growth. They may also be inefficient in producing the necessary stress level around the particle to initiate crazing. Although similar considerations apply to some ABS plastics, the observation of stress whitening or crazing and necking indicates shear yielding (27).

A brittle fracture of a styrene polymer can be brought about by producing uniaxially oriented moldings. Measuring the strength in these moldings across the flow direction or by biaxial loading, such as in a dart drop test, shows the embrittlement of an otherwise tough polymer. Injection moldings of HIPS produced at low temperatures have been shown to be particularly prone to brittle fracture, whereas the ABS polymers in general are less subject to fabrication-induced anisotropy and the consequent embrittlement (28). However, it has been shown that tough moldings of PS can be obtained through the introduction of balanced, multiaxial orientation (29). One way this orientation can be achieved is by molding objects with rotational symmetry at low molding temperatures and by rotating one half of the mold during and after filling, and until the polymer in the mold is cool enough to resist molecular relaxation. In addition to enhanced toughness, craze resistance is improved in such moldings.

Embrittlement of otherwise tough rubber-modified styrene polymers occurs through aging of these polymers (30). The effects of outdoor aging of rubber-modified thermoplastics have been studied (31). Outdoor aging was simulated by laminating a brittle film onto a virgin rubber-modified polymer molding. These experiments showed that aging reduces the energy of crack initiation, so that the final impact strength is determined by the inherent crack-propagation energy of the rubber-modified polymer.

Surface Appearance.   HIPS materials can have a glossy or a dull surface appearance. This surface appearance is a function of surface roughness caused by rubber particles disrupting the surface regularity. The rubber particles near the surface can disrupt the surface by causing either depressions or elevations. These irregularities are caused not only by the nature of the rubber particles, eg, size and shape, but also by processing conditions. For example, during injection molding the surface of the polymer is pressed against a very smooth polished mold surface and quenched, locking in the smooth surface. However, when HIPS is extruded into sheets and thermoformed into parts, it is allowed to cool slowly, giving the rubber particles near the surface time to relax. Polybutadiene rubber particles shrink more upon cooling than the PS matrix, thus causing depressions (32).

Material Types

General-Purpose Polystyrene. Polystyrene is a high molecular weight ($M_w = 2 - 3 \times 10^5$), crystal-clear thermoplastic that is hard, rigid, and free of odor and taste. Its ease of heat fabrication, thermal stability, low specific gravity, and low cost result in moldings, extrusions, and films of very low unit cost. In addition, PS materials have excellent thermal and electrical properties that make them useful as low cost insulating materials (see INSULATION, ELECTRIC; INSULATION, THERMAL).

Commercial polystyrenes are normally rather pure polymers. The amount of styrene, ethylbenzene, styrene dimers and trimers, and other hydrocarbons is minimized by effective devolatilization or by the use of chemical initiators (33). Polystyrenes with low overall volatiles content have relatively high heat-deformation temperatures. The very low content of monomer and other solvents, eg, ethylbenzene, in PS is desirable in the packaging of food. The negligible level of extraction of organic materials from PS is of crucial importance in this application.

When additional lubricants, eg, mineral oil and butyl stearate, are added to PS, easy-flow materials are produced. Improved flow is usually achieved at the cost of lowering the heat-deformation temperature. Stiff-flow PS has a high molecular weight and a low volatile level and is useful for extrusion applications. Typical levels of residuals in PS grades are listed in Table 2. Differences in molecular weight distribution are illustrated in Figure 4.

Table 2. Residuals in Typical Polystyrene, wt %

	Grade	
Polystyrene	Extrusion	Injection-molding
styrene	0.04	0.1
ethylbenzene	0.02	0.1
styrene dimer	0.04	0.1
styrene trimer	0.25	0.8

Specialty Polystyrenes. These include ionomers and PS of specified tacticity, as well as stabilized PS.

Ionomers. PS ionomers are typically prepared by copolymerizing styrene with an acid functional monomer, eg, acrylic acid, or by sulfonation of PS followed by neutralization of the pendent acid groups with monovalent or divalent alkali metals. The introduction of ionic groups into PS leads to a significant modification of both solid-state and melt properties. The introduction of ionic interactions in PS leads to increasing T_g, rubbery modulus, and melt viscosity with increasing ion content (20). For the sodium salt of sulfonated PS, it has been shown that the mode of deformation changes from crazing to shear deformation as the ion content increases (34,35).

Tactic Polystyrenes. Isotactic (IPS) and syndiotactic (SPS) polystyrenes can be obtained by the polymerization of styrene with stereospecific catalysts of the Ziegler-Natta type. Aluminum-activated $TiCl_3$ yields IPS, whereas soluble Ti complexes, eg, $(\eta^5\text{-}C_5H_5)TiCl_3$, in combination with a partially hydrolyzed alkylaluminum, eg, methylalumoxane, yield SPS. The discovery of the SPS

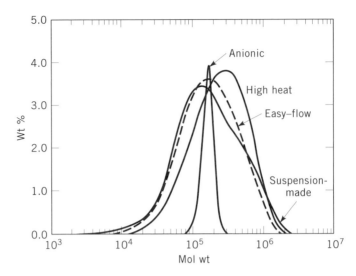

Fig. 4. Molecular weight distribution curves for representative polystyrenes.

catalyst system was first reported in 1986 (36). As a result of the regular tactic structure, both IPS (phenyl groups cis) and SPS (phenyl groups alternating trans) can be crystallized. Samples of IPS quenched from the melt are amorphous but become crystalline if annealed for some time at a temperature slightly below the crystalline melting point. The IPS rate of crystallization is relatively slow compared to that of SPS and those of other crystallizable polymers, eg, polyethylene or polypropylene. This slow rate of crystallization is what has kept IPS from becoming a commercially important polymer even though the material has been known for over 40 years. SPS, on the other hand, crystallizes rapidly from the melt and, as of this writing (1996), is in the process of being developed for commercial use in Japan by Idemitsu Petrochemical Company and in the United States by The Dow Chemical Company. In the amorphous state, the properties of IPS and SPS are very similar to those of conventional atactic PS. Crystalline IPS has a melting temperature of around 240°C, whereas SPS melts at about 270°C (37). In the crystalline state, both IPS and SPS are opaque and insoluble in most common PS solvents.

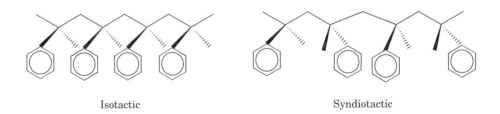

Isotactic   Syndiotactic

Stabilized Polystyrenes. Stabilized polystyrenes are materials with added stabilizers, eg, uv-screening agents, antioxidants (qv), and synergistic agents.

Early stabilization systems for PS included alkanolamines and methyl salicylate (38). Improved stabilizing systems have been developed; these involve a uv-radiation absorber, eg, Tinuvin P (Ciba-Geigy Corp.) with a phenolic antioxidant. Iron as a contaminant, even at a very low concentration, can cause color formation during fabrication. However, this color formation can be appreciably retarded by using tridecyl phosphite as a costabilizer with the uv-radiation absorber and the antioxidant (39). Rubber-modified styrene polymers are heat-stabilized with nonstaining rubber antioxidants, eg, Irganox 1076 (CIBA-GEIGY Corp.). Typically stabilizer formulations for PS are designed by trial and error. However, more recently a predictive model was developed for PS photodegradation, which allows the prediction of weatherability of PS containing a certain concentration of a uv absorber (40) (see Uv STABILIZERS).

Polymers containing flame retardants (qv) have been developed. The addition of flame retardants does not make a polymer noncombustible, but rather increases the polymer's resistance to ignition and reduces the rate of burning with minor fire sources. The primary commercial developments are in the areas of PS foams (see FOAMED PLASTICS) and television and computer housings. Both inorganic (hydrated aluminum oxide, antimony oxide) and organic (alkyl and aryl phosphates) additives have been used (41). Synergistic effects between halogen compounds and free-radical initiators have been reported (42). Several new halogenated compounds and corrosion inhibitors are effective additives (see CORROSION AND CORROSION INHIBITORS) (43). The polymer manufacturer's recommendations with regard to maximum fabrication temperature should be carefully observed to avoid discoloration of the molded part or corrosion of the mold or the machine.

Antistatic polystyrenes have been developed in terms of additives or coatings to minimize primarily dust collecting problems in storage (see ANTISTATIC AGENTS). Large lists of commercial antistatic additives have been published (41). For styrene-based polymers, alkyl and/or aryl amines, amides, quaternary ammonium compounds, anionics, etc, are all used.

Styrene Copolymers. Acrylonitrile, butadiene, α-methylstyrene, acrylic acid, and maleic anhydride have been copolymerized with styrene to yield commercially significant copolymers. Acrylonitrile copolymer with styrene (SAN), the largest-volume styrenic copolymer, is used in applications requiring increased strength and chemical resistance over PS. Most of these polymers have been prepared at the cross-over or azeotropic composition, which is ca 24 wt % acrylonitrile (see ACRYLONITRILE POLYMERS; COPOLYMERS).

Copolymers are typically manufactured using well-mixed continuous-stirred tank reactor (CSTR) processes, where the lack of composition drift does not cause loss of transparency. SAN copolymers prepared in batch or continuous plug-flow processes, on the other hand, are typically hazy on account of composition drift. SAN copolymers with as little as 4% by wt difference in acrylonitrile composition are immiscible (44). SAN is extremely incompatible with PS; as little as 50 ppm of PS contamination in SAN causes haze. Copolymers with over 30 wt % acrylonitrile are available and have good barrier properties. If the acrylonitrile content of the copolymer is increased to >40 wt %, the copolymer becomes ductile. These copolymers also constitute the rigid matrix phase of the ABS engineering plastics.

Butadiene copolymers are mainly prepared to yield rubbers (see STYRENE–BUTADIENE RUBBER). Many commercially significant latex paints are based on styrene–butadiene copolymers (see COATINGS; PAINT). In latex paint the weight ratio S:B is usually 60:40 with high conversion. Most of the block copolymers prepared by anionic catalysts, eg, butyllithium, are also elastomers. However, some of these block copolymers are thermoplastic rubbers, which behave like cross-linked rubbers at room temperature but show regular thermoplastic flow at elevated temperatures (45,46). Diblock (styrene–butadiene (SB)) and triblock (styrene–butadiene–styrene (SBS)) copolymers are commercially available. Typically, they are blended with PS to achieve a desirable property, eg, improved clarity/flexibility (see POLYMER BLENDS) (46). These block copolymers represent a class of new and interesting polymeric materials (47,48). Of particular interest are their morphologies (49–52), solution properties (53,54), and mechanical behavior (55,56).

Maleic anhydride readily copolymerizes with styrene to form an alternating structure. Accordingly, equimolar copolymers are normally produced, corresponding to 48 wt % maleic anhydride. However, by means of CSTR processes, copolymers with random low maleic anhydride contents can be produced (57). Depending on their molecular weights, these can be used as chemically reactive resins, eg, epoxy systems and coating resins, for PS–foam nucleation, or as high heat-deformation molding materials (58).

It has been discovered that styrene forms a linear alternating copolymer with carbon monoxide using palladium II–phenanthroline complexes. The polymers are syndiotactic and have a crystalline melting point ~280°C (59). Shell Oil Company is commercializing carbon monoxide α-olefin plastics based on this technology (60).

Polymers of Styrene Derivatives. Many styrene derivatives have been synthesized and the corresponding polymers and copolymers prepared (61). Glass-transition temperatures for a series of substituted styrene polymers are shown in Table 3. The highest T_g is that of poly(α-methylstyrene), which can be prepared by anionic polymerization. Because it has a low ceiling temperature (61°C), depolymerization can occur during fabrication with the formed monomer acting as a plasticizer and lowering the heat distortion to 110–125°C (62). The

Table 3. Glass-Transition Temperatures of Substituted Polystyrene

Polymer	CAS Registry Number	T_g, °C
polystyrene	[9003-53-6]	100
poly(o-methylstyrene)	[25087-21-2]	136
poly(m-methylstyrene)	[25037-62-1]	97
poly(p-methylstyrene)	[24936-41-2]	106
poly(2,4-dimethylstyrene)	[25990-16-3]	112
poly(2,5-dimethylstyrene)	[34031-72-6]	143
poly(p-tert-butylstyrene)	[26009-55-2]	130
poly(p-chlorostyrene)	[24991-47-7]	110
poly(α-methylstyrene)	[25014-31-7]	170

polymer, which is difficult to fabricate because of its high melt viscosity, is more brittle than PS but can be toughened with rubber.

Some polymers from styrene derivatives seem to meet specific market demands and to have the potential to become commercially significant materials. For example, monomeric chlorostyrene is useful in glass-reinforced polyester recipes because it polymerizes several times as fast as styrene (61). Poly(sodium styrenesulfonate) [9003-59-2], a versatile water-soluble polymer, is used in water-pollution control and as a general flocculant (see WATER, INDUSTRIAL WATER TREATMENT; FLOCCULATING AGENTS) (63,64). Poly(vinylbenzyl ammonium chloride) [70504-37-9] has been useful as an electroconductive resin (see ELECTRICALLY CONDUCTIVE POLYMERS) (65).

Rubber-Modified Polystyrene. Rubber is incorporated into PS primarily to impart toughness. The resulting materials, commonly called high impact polystyrene (HIPS), are available in many different varieties. In rubber-modified PS, the rubber is dispersed in the PS matrix in the form of discrete particles. The mechanism of rubber particle formation and rubber reinforcement, as well as several reviews of HIPS and other heterogeneous polymers, have been published (21,22,66–70). The photomicrographs in Figure 5 show the different morphologies possible in HIPS materials prepared using various types of rubbers (71,72). If the particles are much larger than 5–10 micrometers, poor surface appearance of moldings, extrusions, and vacuum-formed parts are usually noted. Although most commercial HIPSs contain ca 3–10 wt % polybutadiene or styrene–butadiene copolymer rubber, the presence of PS occlusions within the rubber particles gives rise to a 10–40% volume fraction of the reinforcing rubber phase (22,73). Accordingly, a significant portion of the PS matrix is filled with rubber particles. Techniques have been published for evaluating the morphology of HIPS (72,74,75).

For effective toughening of otherwise brittle PS with rubbers, the following generalizations can be made. In order to have good impact strength over a wide temperature range, the T_g of the rubber must be below $-50°C$, as measured, eg, by torsion pendulum at 1 Hz. The use of butadiene rubbers is particularly effective when the rubber is present during the polymerization of styrene. Grafting of some styrene to rubber takes place, and occlusion of PS extends the volume fraction of the dispersed, reinforcing rubber phase. The rubber phase in the final product is cross-linked to some degree for the most effective reinforcement. Because the rubber phase exists in the form of discrete rubber particles, the degree of cross-linking does not significantly influence the melt flow, which is that of a linear, ie, uncross-linked, thermoplastic polymer. A variation in the degree of cross-linking may be needed to optimize product properties for different applications. Depending on the process and rubber concentration used, there is some latitude in the size of the rubber particles that results in a good balance of properties. This range may extend from <1 to >5 μm.

Rubber-Modified Copolymers. Acrylonitrile–butadiene–styrene polymers have become important commercial products since the mid-1950s. The development and properties of ABS polymers have been discussed in detail (76) (see ACRYLONITRILE POLYMERS). ABS polymers, like HIPS, are two-phase systems in which the elastomer component is dispersed in the rigid SAN copolymer matrix. The electron photomicrographs in Figure 6 show the difference in morphology

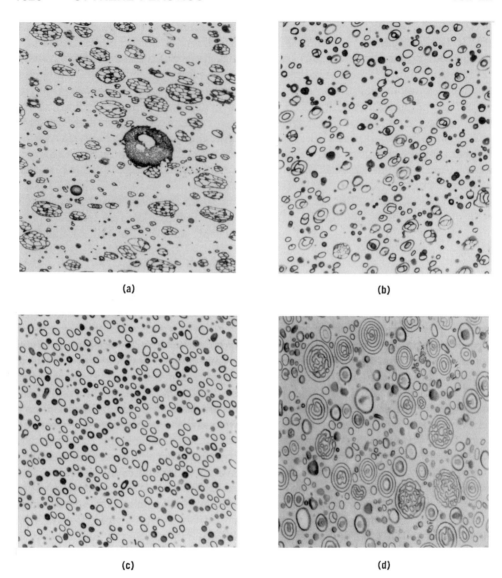

Fig. 5.   Electron photomicrographs of several HIPS resins prepared using different types of rubbers.

of mass vs emulsion ABS polymers. The differences in structure of the dispersed phases are primarily a result of differences in production processes, types of rubber used, and variation in rubber concentrations.

Because of the possible changes in the nature and concentration of the rubber phase, a wide range of ABS polymers is available. Generally, they are rigid, showing a modulus at room temperature of 1.8–2.6 GPa ((2.6–3.8) × 10^5 psi), and have excellent notched impact strength at room temperature, ca 135–4.00 J/m (2.5–7.5 ft lb/in.), and at lower (eg, −40°C) temperatures, 50–140 J/m (0.94–2.6 ft lb/in.). This combination of stiffness, impact strength, and solvent resistance makes ABS polymers particularly suitable for demanding applications. Another important attribute of several ABS polymers is their

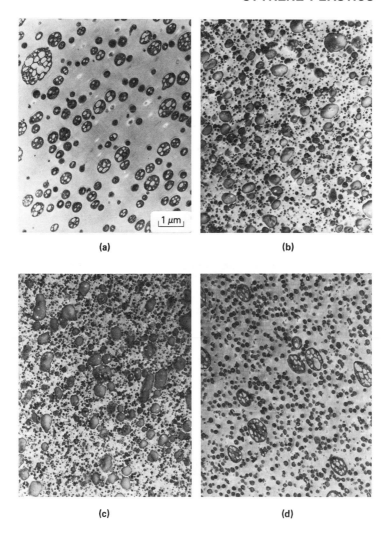

(a)   **(b)**

(c)   **(d)**

Fig. 6. Electron photomicrographs of some commercial ABS resins produced by bulk, emulsion, or a mixture of the two: (**a**) Dow ABS 340; (**b**) Borg-Warner Cycolac T-1000; (**c**) United States Steel Kralastic 606ED, ACFXS53972; and (**d**) Monsanto Lustrex I-448. Scale in (**a**) applies to all photomicrographs.

minimum tendency to orient or develop mechanical anisotropy during molding (28,77). Accordingly, uniform tough moldings are obtained. In addition, ABS polymers exhibit good ease of fabrication and produce moldings and extrusions with excellent gloss, which can be decorated by many techniques, eg, lacquer painting, vacuum metallizing, and electroplating (78–81). In the case of electroplating, the strength of the molded piece is significantly improved (77). When an appropriate decorative coating or a laminated film is applied, ABS polymers can be used outdoors (82).

ABS can be blended with bisphenol A polycarbonate resins to make a material having excellent low temperature toughness. The most important application of this blend is for automotive body panels.

When inherent outdoor weatherability is important, acrylonitrile–ethylene–styrene (AES) or acrylonitrile–styrene–acrylate (ASA) materials are typically used. These materials utilize ethylene–propylene (EP) copolymers and poly(butyl acrylate), respectively, as the rubber phase. EP and poly(butyl acrylate) rubbers are inherently more weatherable than polybutadiene because they are more saturated, leaving fewer labile sites for oxidation (83). Even though AES and ASA polymers are more weatherable than ABS, additives are needed to stabilize the materials against outdoor photochemical degradation for prolonged periods. Additive packages for AES and ASA weatherable materials generally contain both uv absorbers and hindered amine light stabilizers. Typical outdoor applications for weatherable polymers include recreational vehicles, camper tops, and swimming pool accessories. The extremely hostile environments where these stabilized AES materials are utilized still take their toll on the materials. Over prolonged periods of use, the surface gains a chalky appearance as the polymer degrades. Serious discoloration can also result and has been found to be caused by degradation of the additives (84).

High heat ABS resins are produced by adding a third monomer to the styrene and acrylonitrile to stiffen the polymer backbone, thus raising the T_g. Two monomers used commercially for this purpose are α-methylstyrene (85) and N-phenylmaleimide (86).

Not only are ABS polymers useful engineering plastics, but some of the high rubber compositions are excellent impact modifiers for poly(vinyl chloride) (PVC). Styrene–acrylonitrile-grafted butadiene rubbers have been used as modifiers for PVC since 1957 (87).

New Rubber-Modified Styrene Copolymers.   Rubber modification of styrene copolymers other than HIPS and ABS has been useful for specialty purposes. Transparency has been achieved with the use of methyl methacrylate as a comonomer; styrene–methyl methacrylate copolymers have been successfully modified with rubber. Improved weatherability is achieved by modifying SAN copolymers with saturated, aging-resistant elastomers (88).

Glass-Reinforced Styrene Polymers.   Glass reinforcement of PS and SAN markedly improves mechanical properties. The strength, stiffness, and fracture toughness are generally doubled at least. Creep and relaxation rates are significantly reduced and creep rupture times are increased. The coefficient of thermal expansion is reduced by more than one half, and generally response to temperature changes is minimized (89). Normally, ca 20 wt % glass fibers, eg, 6-mm long, 0.009-mm dia E glass, can be used to achieve these improvements. PS, SAN, HIPS, and ABS have been used with glass reinforcement.

Four approaches are available in the 1990s for producing glass-reinforced parts: use of preblended, reinforced molding compound; blending of reinforced concentrates with virgin resin; a direct process, in which the glass is cut and weighed automatically and blended with the polymer at the molding machine; and general inplant compounding (77). The choice of any of these processes depends primarily on the size of the operation and the corresponding economics. The use of concentrates, eg, 80 wt % glass–20 wt % PS, for the subsequent blending to ca 20 wt % glass in the final product has many attractive features and seems to be appropriate for a medium-sized operation, whereas the direct process is emphasized in very large-scale operations.

Degradation

Like almost all synthetic polymers, styrene plastics are susceptible to degradation by heat, oxidation, uv radiation, high energy radiation, and shear, although
in normal use only uv radiation imposes any real limit on the general usefulness of these plastics (90,91). Thus, it is generally recommended that the use of
styrene plastics in outdoor applications be avoided.

Typically, PS loses about 10% of its M_w when it is fabricated. A significant
amount of research has been carried out to determine the nature of the weak
links in PS (92–95). Various modes of initiation of degradation mechanisms
have been proposed: (*1*) chain-end initiation, (*2*) random scission initiation, and
(*3*) scission of weak links in the polymer backbone. It has been suggested that
chain-end initiation is the predominant mechanism at 310°C, whereas random
scission produces stable molecules. Evidence for weak-link scissions comes mainly
from studies showing loss of molecular weight vs degradation time. These plots
usually show a rapid initial drop in molecular weight indicating initial rapid
weak-link scission. However, the picture is also complicated by the fact that the
mechanism of degradation is temperature-dependent. It appears that weak-link
scissions taking place at high temperatures initiate depolymerization whereas
at lower temperatures scissions simply cause a decrease in molecular weight.
In any case, a clear difference in thermal stability has been shown between PS
produced using free-radical (FRPS) and anionic (APS) polymerization (Fig. 7).
This difference is due mainly to the initiator-derived fragments that remain in
the polymer after isolation.

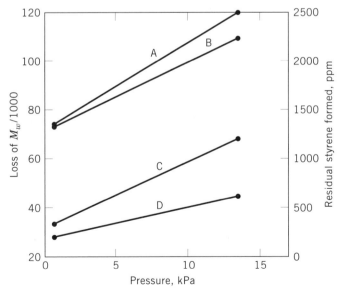

Fig. 7. Comparison of the thermal stability, ie, formation of monomer and loss of M_w, of
FRPS and APS upon heating for 2.5 h at 285°C in glass tubes sealed at 0.67 or 13.3 kPa. A
is FRPS residual styrene formed; B, FRPS M_w loss; C, APS M_w loss; and D, APS residual
styrene formed. To convert kPa to mm Hg, multiply by 7.5.

Poly(α-methylstyrene) unzips to monomer exclusively. Figure 8 is a comparison of the thermal stability of several copolymers. Thermal oxidative degradation of PS occurs much faster, leading to additional volatile components consisting of aldehydes and ketones, yellowing of the polymer with a very dramatic drop in molecular weight, and some cross-linking. Rates and yields are highly oxygen- and temperature-sensitive. Figure 9 shows the magnitude of oxidative attack on PS and the extent to which an antioxidant can protect the polymer.

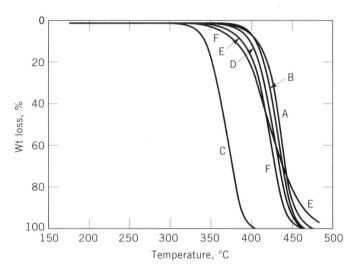

Fig. 8. Thermogravimetric analysis of polymers and copolymers of styrene in nitrogen at 10°C/min: A represents PS; B, poly(vinyltoluene); C, poly(α-methylstyrene); D, poly(styrene-*co*-acrylonitrile), with 71.5% styrene; E, poly(styrene-*co*-butadiene), with 80% styrene; and F, poly(styrene-*co*-methylstyrene), with 75% styrene.

Environmental Considerations

Polystyrene has received much public and media attention and has been described as being nondegradable, nonrecyclable, toxic when burned, landfill-choking, ozone-depleting, wildlife-killing, and even carcinogenic. These misconceptions regarding PS have resulted in boycotts and bans across North America. In reality, PS comprises less than 0.5% of the solid waste going to landfills (Fig. 10) (96).

The approach that the plastics industry has taken in managing plastics waste is an integrated one of source reduction, recycling, incineration for energy recovery, reduction of litter, development of photodegradable plastics for specific litter-prone applications, the development of biodegradable plastics, and increasing public awareness of the recyclability and value of PS. The balance of these approaches varies from year to year depending on public concerns, political pressures, legislation, technological advancement, and the development of an understanding of the actual contribution of plastics to total solid-waste generation. In the mid-1990s, recycling is the most heavily researched and developed. The National Plastics Recycling Company (NPRC) was established in 1989

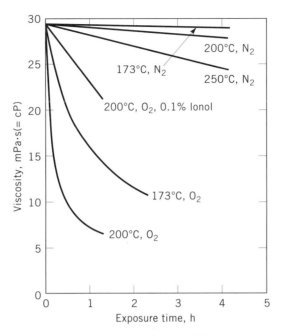

Fig. 9. Thermal and thermooxidative degradation of PS. Viscosity is for a 10% solution in toluene at 25°C.

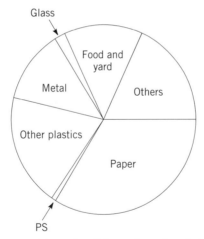

Fig. 10. Approximate composition of solid waste going into landfills in the United States (96).

through the combined efforts of the top eight U.S. PS producers: Huntsman, Dow, Polysar, Fina, Arco, Mobil, Chevron, and Amoco. Its charter is to facilitate plastic recycling, with the ultimate objective of recycling 25% of the PS produced in the United States each year (97).

Environmental Degradation.   Natural sunlight emits energy only in wavelengths above 290 nm. Therefore any polymer that does not absorb light energy at wavelengths above 290 nm should not be photodegradable. The activation spectrum of PS vs the intensity of the solar spectrum is shown in Figure 11. Polystyrene, the degradation of which can be activated by radiation at wavelengths above 290 nm, is quite photodegradable and must be stabilized by the addition of uv-absorbing additives if it is to be used in outdoor applications where durability is important.

Even though PS is naturally quite photodegradable, there have been considerable efforts to accelerate the process to produce so-called photodegradable PS (98–103). The approach is to add photosensitizers, typically ketone containing molecules, that absorb sunlight, eg, benzophenone. The absorbed light energy is then transferred to the polymer to cause backbone scission via an oxidation mechanism (Fig. 12). Photodegradable PS is useful in litter-prone applications, eg, fast food packaging.

A more effective approach to enhancing the rate of photodegradation of PS is to copolymerize styrene with a small amount of a ketone-containing monomer. Thus the ketone groups are attached to the polymer during its manufacture (Fig. 13) (98,100).

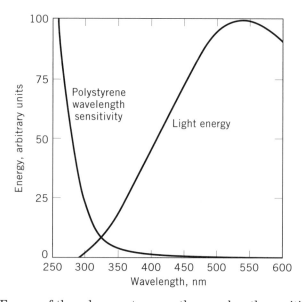

Fig. 11. Energy of the solar spectrum vs the wavelength sensitivity of PS.

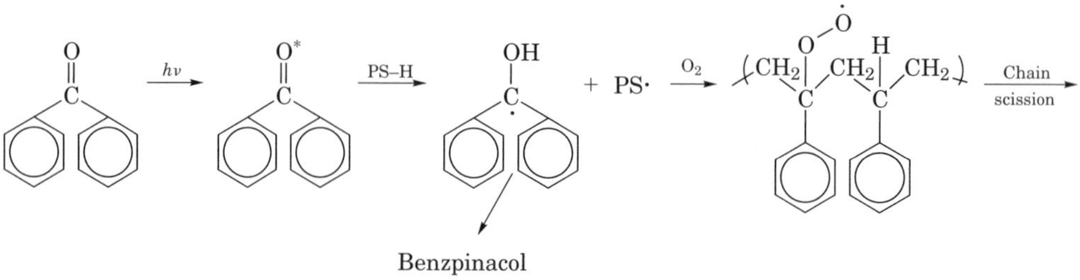

Fig. 12. Chemistry of molecular weight breakdown of PS during outdoor exposure.

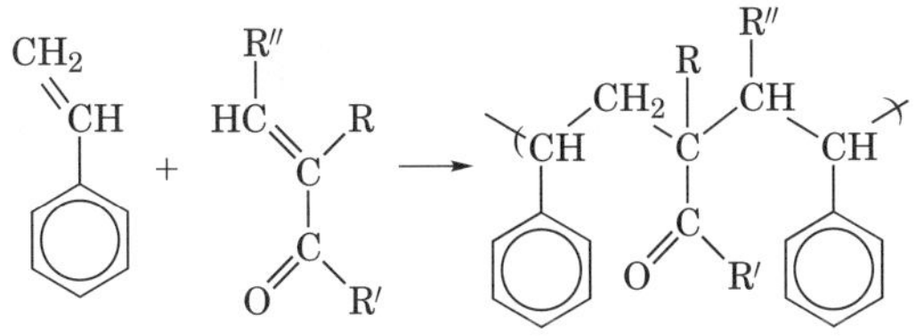

Fig. 13. Incorporation of photosensitive ketone groups into PS during manufacture by copolymerization of styrene and vinyl ketone.

Attaching the ketone groups to the polymer backbone is more efficient on a chain scission/ketone basis because some of the light energy that the pendent ketone absorbs leads directly to chain scission via the Norrish type II mechanism, as well as photooxidation via the Norrish type I mechanism (see POLYMERS, ENVIRONMENTALLY DEGRADABLE).

A key problem with the manufacture of photodegradable PS containing low levels of methyl vinyl ketone and methyl isopropenyl ketone is their human toxicity. This problem has been solved by adding the β-ketoalcohol intermediate, formed during vinyl ketone manufacture, to the styrene polymerization and generating the α,β-unsaturated ketone *in situ* (102,103). A concern in the use of photodegradable PS is the environmental impact of the products of photooxidation. However, photodegraded PS is expected to be more susceptible to biodegradation because the molecular weight has been reduced, the PS chains have oxidized end groups, the incorporation of oxygen as alcohol and ketone groups has increased the hydrophilicity of the PS fragments, and the surface area has increased (104–106).

Biodegradation. PS is inherently resistant to biodegradation mainly on account of its hydrophobicity. Efforts have been made to enhance the biodegradability of PS by inserting hydrolyzable linkages, eg, ester and amide, into its backbone. This was accomplished by adding monomers to the polymerization which are capable of undergoing ring-opening copolymerization (Fig. 14) (107).

Blowing Agents. Until the mid-1980s, the most common blowing agents for extruded PS foams were chlorofluorocarbons (CFCs) (see FLUORINE COMPOUNDS, ORGANIC–FLUORINATED ALIPHATIC COMPOUNDS). However, when these materials were shown to contribute to the ozone depletion problem, a considerable effort to find alternative blowing agents for the manufacture of extruded PS foam emerged. Most of the research has focused on the development of carbon

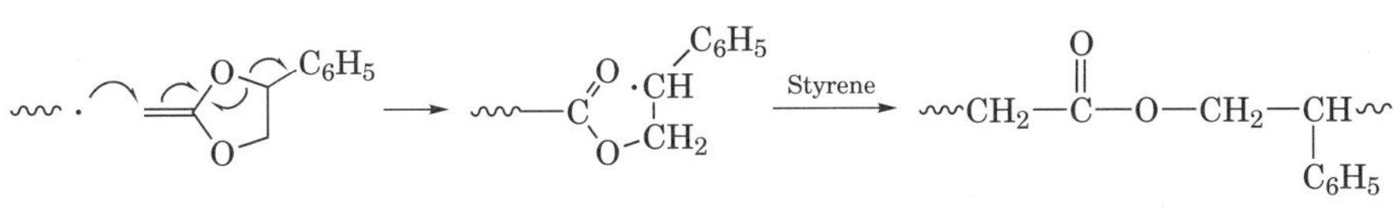

Fig. 14. Preparation of biodegradable PS by incorporating ester linkages into the backbone via ring-opening copolymerization of styrene with a cyclic ketene acetal.

dioxide foaming technology for PS (108). By contrast, PS bead foam uses hydrocarbon blowing agents, eg, pentane. Hydrocarbon blowing agents are extremely flammable and add volatile organic compounds to the atmosphere.

Polymerization

Styrene [*100-42-5*] and most of its derivatives are among the few monomers that can be polymerized by all four distinct mechanisms, ie, anionic, cationic, free-radical, and Ziegler-Natta. These include processes dependent on electromagnetic radiation, which is usually a free-radical mechanism, or high energy radiation, which is either a cationic or free-radical mechanism, depending on the water content of the system. All mechanisms, other than Ziegler-Natta, generally yield polymers with a high degree of random placement of the phenyl group relative to the backbone, ie, the polymers are classified as atactic and amorphous. Anionically made PS is usually atactic and amorphous, but in some cases, eg, at low temperatures, isotactic PS has been prepared.

Each of the mechanisms used to polymerize styrene has its own unique advantages and disadvantages. In the anionic mechanism, initiation, propagation, and termination steps are sequential, resulting in the formation of narrow polydispersity ($M_w/M_n < 1.1$). The termination step is controlled, allowing control of end-group structure. However, the polymerization feed must be purified. High molecular weight polymers are difficult to make by the cationic mechanism on account of instability of the polystyryl carbocation giving fast termination. In addition, polymerization feed must be purified. Free-radical initiation, propagation, and termination steps are simultaneous, resulting in the formation of broad polydispersity ($M_w/M_n > 2$). Multiple termination paths lead to a variety of end groups. However, the polymerization feed need not be purified. In the Ziegler-Natta method, metal complexes allow stereospecific polymerization, resulting in the formation of high melting crystalline tactic PS. The polymerization feed must be purified.

Free-radical polymerization is the preferred industrial route both because monomer purification is not required (109) and because initiator residues need not be removed from polymer for they have minimal effect on polymer properties.

Styrene–butadiene block copolymers are made with anionic chain carriers, and low molecular weight PS is made by a cationic mechanism (110). Analytical standards are available for PS prepared by all four mechanisms (see INITIATORS).

FREE-RADICAL POLYMERIZATION

The styrene family of monomers are almost unique in their ability to undergo spontaneous or thermal polymerization merely by heating to >100°C. Styrene in essence acts as its own initiator. The mechanism by which this spontaneous polymerization occurs has challenged researchers since the 1940s. Two mechanisms explaining spontaneous styrene polymerization have been proposed and supported by considerable circumstantial evidence. The oldest mechanism, first postulated by Flory (111), involves a bond-forming reaction between two molecules of styrene (S) to form a 1,4-diradical. However, experiments to test the mechanism showed that there was no significant difference in the molecular weight of PS initiated by monoradicals compared with the spontaneously initiated polymerizations taken to the same monomer conversion (112). It thus became clear that the initiating species was not a diradical. Evidences favoring this mechanism include the identification of *cis-* and *trans-*1,2-diphenylcyclobutanes as principal dimers (113,114), and the large differences between spontaneous and chemically initiated (azobisisobutyronitrile) styrene polymerizations in the presence of the free-radical scavenger 1, 1′-diphenyl-2-picrylhydrazyl (DPPH). The rate of consumption of DPPH is 25 times faster than that expected from polymerization rate measurements. This difference was explained by the spontaneous formation of diradicals, many of which become self-terminated before initiating polymer radicals (115).

The second mechanism, proposed by Mayo (116), involves the Diels-Alder reaction of two styrene molecules to form a reactive dimer (DH) followed by a molecular assisted homolysis between DH and another styrene molecule.

The Mayo mechanism has been generally preferred even though critical reviews (113,117) have pointed out that the mechanism is only partially consistent with the available data. Also, the postulated intermediate DH has never been isolated. Evidences supporting the mechanism include kinetic investigations (118,119), isotope effects (117), and isolation/structure determination of oligomers (117,120). Even though the reactive dimer intermediate DH has never been isolated, the aromatized derivative DA has been detected in PS (120). Also, D· has been indicated as an end-group in PS using ^{1}H-nmr and uv spectroscopy (121).

The Flory and Mayo proposals can be combined by the common diradical ·D·, which collapses to either DH or 1,2-diphenylcyclobutane. Nonconcerted Diels-Alder reactions are permissible for two nonpolar reactants (122).

The spontaneous polymerization of styrene was studied in the presence of various acid catalysts (123) to see if the postulated reactive intermediate DH could be intentionally aromatized to form inactive DA. The results showed that the rate of polymerization of styrene is significantly retarded by acids, eg, camphorsulfonic acid, accompanied by increases in the formation of DA. This finding gave further confirmation of the intermediacy of DH because acids would have little effect on the cyclobutane dimer intermediate in the Flory mechanism.

DH $\xrightarrow{\text{aromatization}}$

C_6H_5

DA

A potentially important commercial benefit of adding an acid catalyst to the spontaneous free-radical polymerization of styrene is that a significant shift results in the rate–molecular weight curve for PS. This shift (Fig. 15) is most pronounced at high molecular weights, thus allowing formation of high molecular weight PS at much faster polymerization rates. An explanation for this phenomenon is that the rate of formation of initiating radicals is reduced in the presence of acid. To return the rate of initiating free-radical formation back to a higher level, the temperature must be increased. The increased temperature increases the rate of propagation. The main mechanism of termination is radical coupling (124), the rate of which is most affected by radical concentration. Because the polymerization temperature can be raised in the presence of acid without increasing free-radical concentration, the propagation rate is increased relative to termination rate, thereby raising the molecular weight. Other mechanisms of termination include chain-transfer with solvent and with the Diels-Alder dimer (DH), and disproportionation between two polystyryl radicals.

PS produced by the spontaneous initiation mechanism is typically contaminated by dimers and trimers (1–2 wt/wt). These oligomers are somewhat volatile and cause problems during extrusion (vapors) and molding (mold sweat) operations. The use of chemicals to generate initiating free radicals significantly reduces the formation of the oligomers. Oligomer production is reduced because the

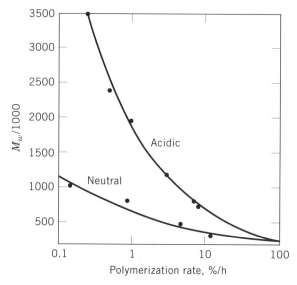

Fig. 15. Polymerization rate vs molecular weight relationship for spontaneous bulk styrene polymerization under neutral and acidic conditions.

polymerization temperature can be lowered to slow down the Diels-Alder dimerization reaction. A wide range of free-radical initiators are available. They differ mainly in the temperature at which each generate free radicals at a useful rate. Typically, it is best to polymerize styrene at about the 1-h half-life temperature of the initiator it contains.

Initiators (qv) that have been utilized to initiate styrene polymerization can be generally categorized into three types: peroxides, azo, and carbon–carbon (125). Peroxides are thermally unstable, decomposing by homolysis of the O–O bond and resulting in the formation of two oxy radicals. Azo compounds decompose by concerted homolysis of the N–C bonds on either side of the azo linkage, resulting in extrusion of nitrogen gas and the formation of two carbon-centered radicals. Carbon–carbon initiators decompose by homolysis of a sterically strained C–C bond, resulting in the formation of two carbon-centered radicals. The radicals that initiate styrene polymerization end up attached to the chain ends and may have an effect on polymer stability (126–129).

Because most of the bulk PS reactors were originally designed to produce spontaneously initiated PS in the 100–170°C temperature range, the peroxide initiators used generally have 1-h half-lives in the range of 90–140°C. If the peroxide decomposes too rapidly, a runaway polymerization could result; if it decomposes too slowly, peroxide exits the reactor. Because organic peroxides are significantly more expensive than styrene monomer (5–20 times), it is economically prudent to choose an initiator that is highly efficient and entirely consumed during the polymerization.

Another economically driven objective is to utilize initiators that increase the rate of styrene polymerization to form PS having the desired molecular weight. The commercial weight average molecular weight (M_w) range for general-purpose PS is 200,000–400,000. For spontaneous polymerization, the M_w is inversely proportional to polymerization rate (Fig. 16).

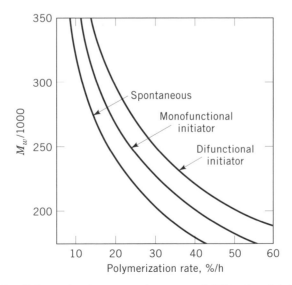

Fig. 16. Polymerization rate advantage of difunctional initiator.

The main reason that the M_w decreases as the polymerization temperature increases is the increase in the initiation and termination reactions, which leads to a decrease in the kinetic chain length (Fig. 17). At low temperature, the main termination mechanism is polystyryl radical coupling, but as the temperature increases, radical disproportionation becomes increasingly important. Termination by coupling results in higher M_w PS than any of the other termination modes.

There are typically several different product grades produced in a single polymerization reactor; transition between these products in the minimum time

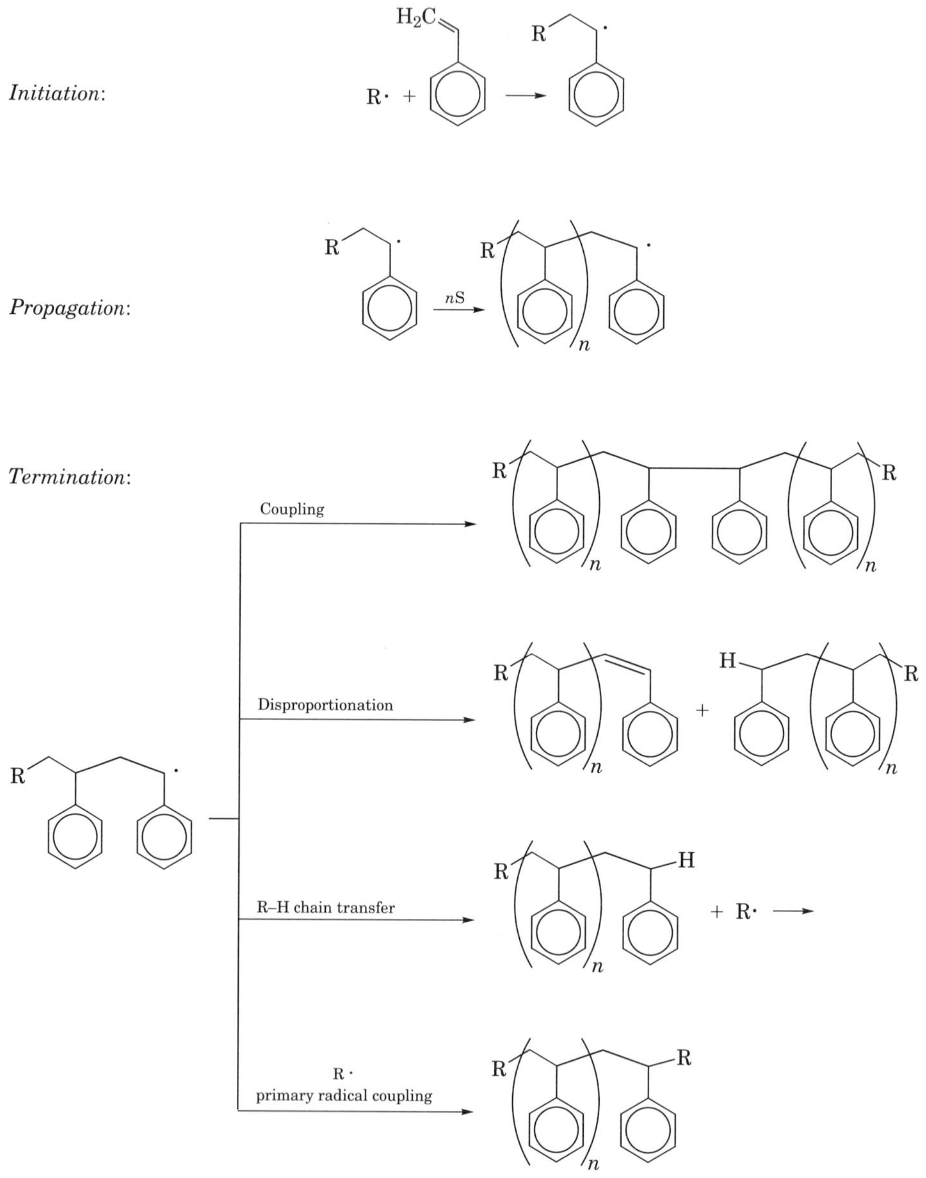

Fig. 17.   General chemistry of free-radical styrene polymerization.

maximizes production yield. Most PS producers rely on the use of kinetic modeling and computer simulation to aid in the manufacture of PS to minimize transition time between product grades. Kinetic models have been developed for styrene polymerization without added initiators (130–132), using one (133–137) or two (138) monofunctional initiators with different half-lives; and using symmetrical difunctional (Fig. 18**a**, **b**, and **c**) or unsymmetrical difunctional initiators (Fig. 18**d** and **e**). These models clearly show the polymerization rate advantage of using initiators for the manufacture of PS using continuous bulk polymerization processes (see Fig. 16).

One of the key reasons for the polymerization rate advantage using difunctional initiators is their theoretical ability to form initiator fragments, which can initiate polymer growth from two different sites within the same fragment and ultimately lead to double-ended PS. If double-ended PS chains are produced, provided that the main mechanism of termination is chain coupling, higher M_w PS can be produced at faster rates using difunctional initiators (145).

Below 80°C, radical combination is the primary termination mechanism (146); above it, both disproportionation and chain transfer with the Diels-Alder dimer are increasingly important. The gel or Trommsdorff effects, as manifested by a period of accelerating rate concomitant with increasing molecular weight, is apparent at below 80°C in styrene polymerization, although subtle changes

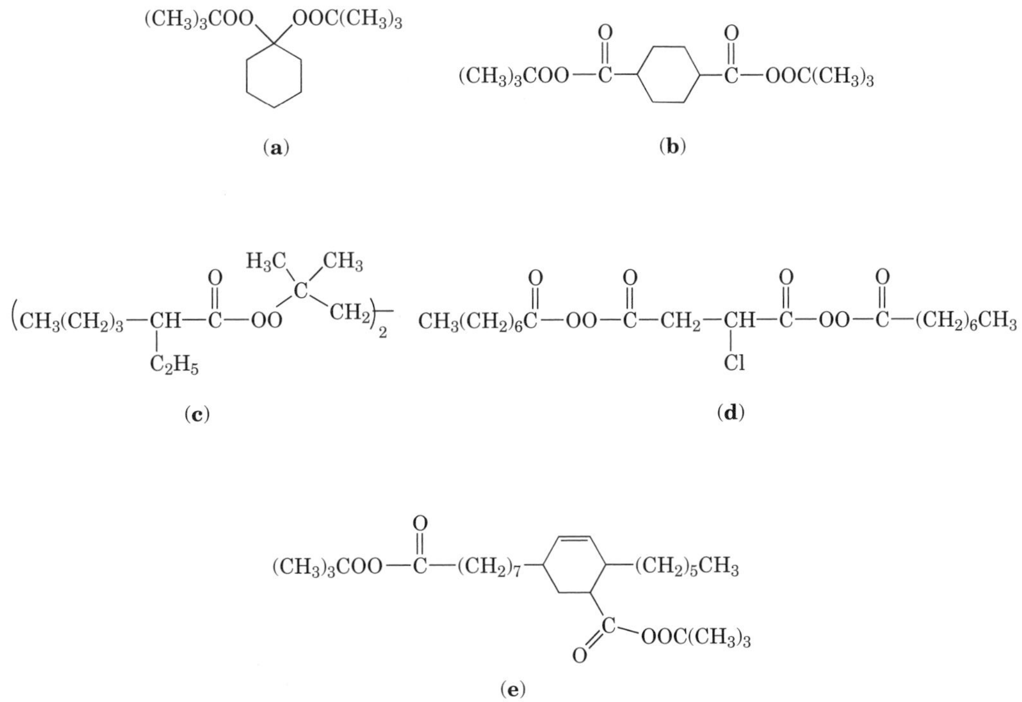

Fig. 18. Difunctional initiators for styrene polymerization: (**a**) (139), (**b**) (139), and (**c**) (140) are symmetrical difunctional; (**d**) (141,142) and (**e**) (143,144) are unsymmetrical difunctional initiators.

during the polymerization at higher temperature may be attributed to variation of the specific rate constants with viscosity (146,147).

In some cases, inhibition of polymerization can be regarded as a special type of chain transfer. This is of importance in commercial-scale operations involving styrene storage for extended periods. The majority of inhibitors are of the phenolic/quinone family. All of these species function as inhibitors only in the presence of oxygen. 4-*tert*-Butylcatechol (TBC) at 12–50 ppm is the most universally used inhibitor for protecting styrene. At ambient conditions and with a continuous supply of air, TBC has a half-life of 6–10 weeks (148). The requirement of oxygen causes complex side reactions, resulting in significant oxidation of the monomer, which causes yellow coloration, especially in the vapor phase. An inert gas blanket reduces this problem as well as flammability hazards, but precautions must be taken to ensure an adequate level of dissolved oxygen in the liquid phase. Another family of inhibitors is that characterized by the N–O bond, which does not seem to require oxygen for effectiveness. This family includes a variety of nitrophenol compounds, hydroxylamine derivatives, and nitrogen oxides (149,150). Nitric oxide is particularly useful. Presumably, the function of this family depends on the unpaired electron associated with the N–O bond.

Other miscellaneous compounds that have been used as inhibitors are sulfur and certain sulfur compounds (qv), picrylhydrazyl derivatives, carbon black, and a number of soluble transition-metal salts (151). Both inhibition and acceleration have been reported for styrene polymerized in the presence of oxygen. The complexity of this system has been clearly demonstrated (152). The key reaction is the alternating copolymerization of styrene with oxygen to produce a polyperoxide, which at above 100°C decomposes to initiating alkoxy radicals. Therefore, depending on the temperature, oxygen can inhibit or accelerate the rate of polymerization.

Chain Transfer. Chain-transfer (CT) agents are occasionally added to styrene to reduce the molecular weight of the polymer, although for many applications this is unnecessary. Polymerization temperature alone is generally sufficient to achieve molecular weight control. Diluents, ie, materials with little or no chain-transfer activity, are sometimes used to reduce engineering problems. Some typical chain-transfer agents for styrene polymerization are listed in Table 4. The CT agents of commercial significance are α-methylstyrene dimer, terpinolene, dodecane-1-thiol, and 1,1-dimethyldecane-1-thiol. Chain transfer to styrene monomer has been reported, but some work strongly suggests that this

Table 4. Chain-Transfer Constants (K_{ct}) in Free-Radical Styrene Polymerization

Compound	K_{ct}	Temperature, °C
benzene	0.00002	100
toluene	0.00005	100
ethylbenzene	0.0002	100
isopropylbenene	0.0002	100
α-methylstyrene dimer	0.3	120
1-dodecanethiol	13	130
1,1-dimethyl-1-decanethiol	1.1	120
1-hexanethiol	15	100

reaction is negligible and transfer with the Diels-Alder dimer is the actual inherent transfer reaction (153,154). Chain transfer to PS has received some attention, but experiments indicate that it is minimal (155,156). If it did occur at a significant extent, then the polymer would be branched. The CT constants of several common solvents and CT agents are also shown in Table 4.

High levels of CT agents can be used to control the termination process. If the CT agent has a functional group attached to it, the functional group ends up becoming attached to the end of the PS chain. If the functionalized CT agent operates by donating an H atom to the polystyryl radical, the functional group ends up becoming attached to only one end of the PS chain. However, this technique does not quantitatively functionalize the polymer chains because not all chains get initiated by the functionalized CT agent fragment formed by loss of the H atom. However, if both the initiator and the CT agent contain the functional group, high purity one-end functional PS can be produced (157).

Another approach to control end-group structure is the use of CT agents that operate by an addition-fragmentation mechanism. This approach can lead to the formation of PS having functional groups at both ends if both the initiating and terminating fragments contain functional groups (Fig. 19). It has been found that the common CT agent α-methylstyrene dimer operates by an addition-fragmentation mechanism (158). The use of addition-fragmentation CT agents also places a reactive double bond on one end of the polymer chain and thus yields a macromonomer. Copolymerization of the chain-end double bond with more styrene leads to branched PS.

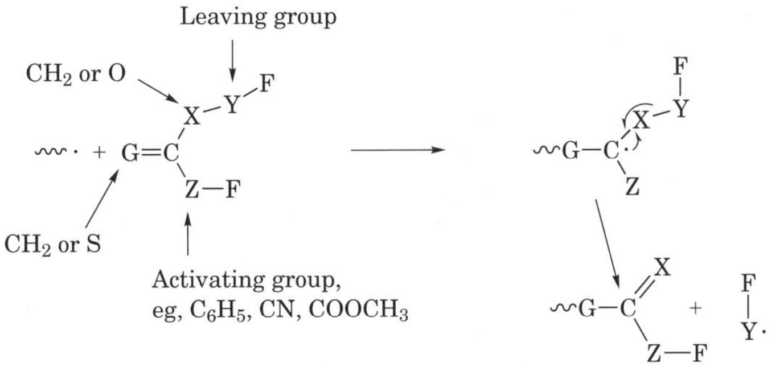

Fig. 19. Generic addition fragmentation CT structure and the mechanism of action, where F = functional group.

IONIC POLYMERIZATION

Instead of a neutral unpaired electron, styrene polymerization can proceed with great facility through a positively charged species (cationic polymerization) or a negatively charged species (anionic polymerization). The polymerization reaction is more sensitive to impurities than the free-radical system, and pretreatment of the monomer is generally required (159). n-Butyllithium (NBL) is the most widely used initiator for anionic polymerization of styrene. In solution, it exists as six-membered aggregates, and a key step in the initiation sequence is dissociation yielding at least one isolated molecule.

If the initiation reaction is much faster than the propagation reaction, then all chains start to grow at the same time. Because there is no inherent termination step, the statistical distribution of chain lengths is very narrow. The average molecular weight is calculated from the mole ratio of monomer-to-initiator sites. Chain termination is usually accomplished by adding proton donors, eg, water or alcohols, or electrophiles such as carbon dioxide.

Anionic polymerization, if carried out properly, can be truly a living polymerization (160). Addition of a second monomer to polystyryl anion results in the formation of a block polymer with no detectable free PS. This technique is of considerable importance in the commercial preparation of styrene–butadiene block copolymers, which are used either alone or blended with PS as thermoplastics.

Anionic polymerization offers fast polymerization rates on account of the long life-time of polystyryl carbanions. Early studies have focused on this attribute, most of which were conducted at short reactor residence times (<1 h), at relatively low temperatures (10–50°C), and in low chain-transfer solvents (typically benzene) to ensure that premature termination did not take place. Also, relatively low degrees of polymerization (DP) were typically studied. Continuous commercial free-radical solution polymerization processes to make PS, on the other hand, operate at relatively high temperatures (>100°C), at long residence times (>1.5 h), utilize a chain-transfer solvent (ethylbenzene), and produce polymer in the range of 1000–1500 DP.

Two companies, Dow and BASF, have devoted significant research efforts to develop continuous anionic polymerization processes for PS. However, no PS as of this writing (1996) is produced commercially using these processes. The Dow Chemical Company researchers utilized conventional free-radical polymerization reactors of the CSTR type (161–165) to study the anionic polymerization of styrene, whereas BASF focused on continuous reactors of the linear flow type (166). In an anionic CSTR process, initiation, propagation, and termination are occurring simultaneously and thus the polydispersity of the resulting PS is ~2. A CT solvent (ethylbenzene) and relatively high temperatures (80–140°C) were also used. Under these conditions, chain transfer to solvent (CTS) is extremely high because the high monomer conversions (>99%) achieved under steady-state operation result in a large ratio (typically 500:1) of solvent to monomer. Figure 20 depicts the ebullient CSTR anionic process investigated by Dow researchers.

The result of high CTS is that very high yields of PS based on NBL (the most costly raw material) and PS having high clarity (low color) are produced (Fig. 21). Under stringent feed purification conditions, as high as 8000% yield based on NBL initiator (109) can be achieved. According to Figure 21, without high levels of CTS, it would be impossible to make a PS having sufficient clarity to meet the 1990s color requirements to be sold as prime resin. However, with no CTS, 640 ppm of NBL is required to produce a PS of 1000 DP.

One of the key benefits of anionic PS is that it contains much lower levels of residual styrene monomer than free-radical PS (167). This is because free-radical polymerization processes only operate at 60–80% styrene conversion, whereas anionic processes operate at >99% styrene conversion. Removal of unreacted styrene monomer from free-radical PS is accomplished using continuous devolatilization at high temperature (220–260°C) and vacuum. This process leaves about 200–800 ppm of styrene monomer in the product. Taking the styrene to

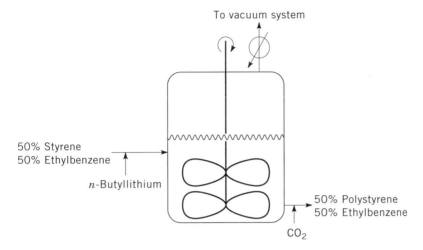

Fig. 20. Dow ebullient CSTR anionic polymerization process.

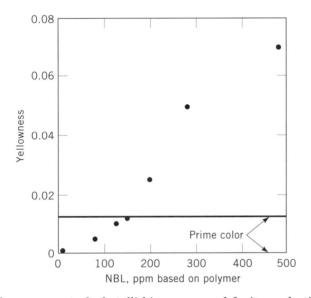

Fig. 21. PS color vs amount of n-butyllithium consumed for its production in a CSTR.

a lower level requires special devolatilization procedures such as steam stripping (168).

More recent process research aimed at anionic PS is that of BASF AG. Unlike the Dow Process, the BASF process utilizes continuous linear-flow reactors (LFR) with no back-mixing to make narrow polydispersity resins. This process consists of a series alternating reactors and heat exchangers (Fig. 22). Inside the reactors, the polymerization exotherm carries the temperature from 30°C at the inlet to 90°C at the outlet. The heat exchangers then take the temperature back down to 30°C. This process, which requires no solvent, results in the formation of narrow polydispersity PS.

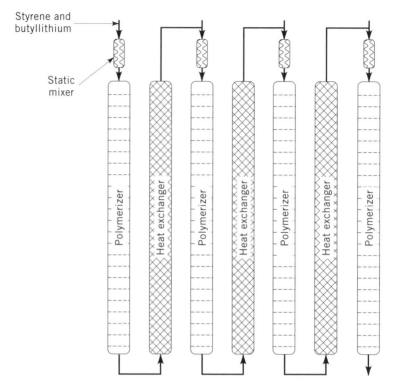

Fig. 22. BASF continuous anionic process.

Continuous anionic polymerization conducted above the ceiling temperature (61°C) is also useful for making thermally stable styrene-co-α-methylstyrene (SAMS) having high α-methylstyrene (AMS) content. Preparation of SAMS having >20 wt % of AMS units using bulk free-radical polymerization leads to very slow polymerization rates, low molecular weight polymer, and the formation of high levels of oligomers. The preparation of SAMS using an anionic CSTR polymerizer operating at 90–100°C allows the production of SAMS having up to about 70 wt % AMS units (Fig. 23) (169). By conducting the polymerization above the ceiling temperature, no AMS polyads of more than two units are possible. SAMS copolymers having polyads of greater than two units in length are thermally unstable (170).

Cationic polymerization of styrene can be initiated either by strong acids, eg, perchloric acid, or by Friedel-Crafts reagents with a proton-donating activator, eg, boron trifluoride or aluminum trichloride with a trace of a protonic acid or water. The solvent again plays an important role, and chain-transfer reactions are very common where the reactants are polymer, monomer, solvent, and counterion. As a result, high molecular weights are more difficult to achieve and molecular weight distributions are often comparable to those obtained from free-radical polymerizations. Commercial use of cationic styrene polymerization is reported only where low molecular weight polymers are desired (110).

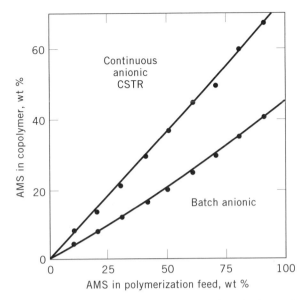

Fig. 23.   Comparison of continuous (CSTR) and batch anionic production of SAMS.

Considerable advances have taken place in the 1990s with regard to cationic polymerization of styrene. Its uses to make block copolymers and even living cationic polymerization have been reported (171).

Ziegler-Natta.   Ziegler-Natta-initiated styrene polymerization yields stereoregular tactic PS. The tacticity can be isotactic (phenyl rings cis to each other) or syndiotactic (phenyl rings trans to each other), depending on the initiator structure. All commercially important styrenic plastics in the 1990s are amorphous because the vast excess of these polymers is made by a free-radical mechanism. However, Dow Chemical and Idemitsu Petrochemical Companies are working together to commercialize crystalline syndiotactic PS. SPS is produced using a bulk polymerization process. Because the SPS crystallizes as it forms, the polymerization process is challenging, requiring mixing and removal of the heat of polymerization from a sticky solid. Proper reactor design is required to keep the sticky mass from solidifying.

LIVING STYRENE POLYMERIZATION

The requirements for a polymerization to be truly living are that the propagating chain ends must not terminate during polymerization. If the initiation, propagation, and termination steps are sequential, ie, all of the chains are initiated and then propagate at the same time without any termination, then monodisperse (ie, $M_w/M_n = 1.0$) polymer is produced. In general, anionic polymerization is the only mechanism that yields truly living styrene polymerization and thus monodisperse PS. However, significant research has been conducted to achieve living styrene polymerization using cationic, free-radical, and Ziegler-Natta mechanisms.

Since living free-radical polymerization (LFRP) was first proposed in 1982 (172,173), there has emerged a multitude of examples (174–182). However, significant controversy exist over the nature of the living chain-end and whether the process can really be called living. Instead, terms such as pseudo-living, quasi-living, resuscitatable, and dormant have been used to better describe the process. The process generally involves the addition of relatively stable free radicals to the polymerization. The growing polystyryl radicals are intercepted and terminated by coupling with the stable radicals, thereby minimizing termination by chain transfer and the coupling of polystyryl radicals with themselves. Prior to the discovery of LFRP, stable free radicals were only added to styrene as polymerization inhibitors. They acted as radical scavengers to stop the propagation process. However, the bond formed between the polystyryl radical and the stable free radical is somewhat labile and styrene can insert into the bond once it is activated by heat or photolysis. The main evidences used to support the living nature of the polymerization are the increase of molecular weight with monomer conversion, the formation of narrow-polydispersity PS, and the ability to prepare block copolymers. Normally free-radical polymerizations show relatively flat molecular weight vs conversion plots. Although the preparation of truly monodisperse ($M_w/M_n = 1.0$) PS using LFRP has yet been demonstrated, polydispersities approaching 1.1 have been reported for very low molecular weight PS prepared in the presence of stable nitroxyl radicals, ie, $M_w = 20,000$.

Most of the LFRP research in the 1990s is focused on the use of nitroxides as the stable free radical. The main problems associated with nitroxide-mediated styrene polymerizations are slow polymerization rate and the inability to make high molecular weight narrow-polydispersity PS. This inability is likely to be the result of side reactions of the living end leading to termination rather than propagation (183). The polymerization rate can be accelerated by the addition of acids to the process (184). The mechanism of the accelerative effect of the acid is not certain.

COPOLYMERIZATION

Styrene readily copolymerizes with many other monomers spontaneously. The styrene double bond is electronegative on account of the donating effect of the phenyl ring. Monomers that have electron-withdrawing substituents, eg, acrylonitrile and maleic anhydride, tend to copolymerize most readily with styrene because their electropositive double bonds are attached to the electronegative styrene double bond. Spontaneous copolymerization experiments of many different monomer pair combinations indicate that the mechanism of initiation changes with the relative electronegativity difference between the monomer pairs (185).

Copolymerization makes possible dramatic improvements in one or more properties. There is a vast amount of literature on styrene-containing copolymers (186). The reactivity ratios for the monomer pairs are listed in Table 5. Styrene–acrylonitrile (SAN) copolymers are large-volume thermoplastics having improved mechanical properties and heat and solvent resistance with some loss of color stability (see ACRYLONITRILE POLYMERS). The disparity of reactivity ratios shown in Table 5 implies an unequal insertion rate of the monomers into the copolymer. Figure 24 illustrates manifestations of this effect. Most SAN copolymers are manufactured at or near the azeotrope or cross-over point (A in Fig. 24) to avoid significant drift in composition (Fig. 25) (187,188). SAN copoly-

Table 5. **Free-Radical Copolymerization Reactivity Ratios, r**

Monomer 2	r_1	r_2	Temperature, °C
acrylonitrile	0.4	0.04	60
butadiene	0.5	1.40	50
methyl methacrylate	0.54	0.49	60
m-divinylbenzene	0.65	0.60	60

[a]Styrene = monomer 1.

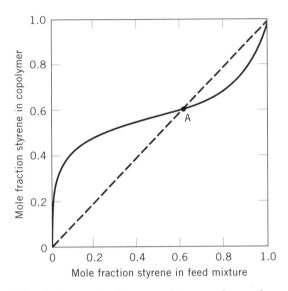

Fig. 24. Relationship between feed composition and copolymer composition of styrene–acrylonitrile copolymerization. See text.

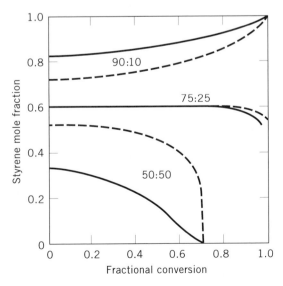

Fig. 25. Drift in monomer composition (——) and copolymer composition (– – –) with conversion for three initial monomer mixtures. Ratios are based on wt %.

mers having as little as a difference of 4 wt % in AN content are not miscible (44). Therefore the preparation of SAN under conditions where composition drift can occur leads to haze. Reactors have been designed to overcome composition drift problems on a commercial scale (189). All SAN copolymers produced in the United States are made in a mass or solution continuous process. The greater tendency of these copolymers to develop yellow coloration requires more attention to techniques of polymerization and devolatilization (190–193). The discoloration mechanism has been linked to both AN sequence distribution in the polymer backbone (194) and the thermal decomposition of oligomers (195–198).

Copolymers with butadiene, ie, those containing at least 60 wt % butadiene, are an important family of rubbers. In addition to synthetic rubber, these compositions have extensive uses as paper coatings, water-based paints, and carpet backing. Because of unfavorable reaction kinetics in a mass system, these copolymers are made in an emulsion polymerization system, which favors chain propagation but not termination (199). The result is economically acceptable rates with desirable chain lengths. Usually such processes are run batchwise in order to achieve satisfactory particle size distribution.

Styrene–maleic anhydride (SMA) copolymers are used where improved resistance to heat is required. Processes similar to those used for SAN copolymers are used. Because of the tendency of maleic anhydride to form alternating copolymers with styrene, composition drift is extremely severe unless the polymerization is carried out in CSTR reactors having high degrees of back-mixing.

Divinylbenzene copolymers with styrene are produced extensively as supports for the active sites of ion-exchange resins and in biochemical synthesis. About 1–10 wt % divinylbenzene is used, depending on the required rigidity of the cross-linked gel, and the polymerization is carried out as a suspension of the monomer-phase droplets in water, usually as a batch process. Several studies have been reported on the reaction kinetics (200,201).

Polymerization of styrene in the presence of polybutadiene rubber yields a much tougher material than PS. Although the material is usually opaque and has a somewhat lower modulus, the gain in impact resistance has placed this plastic in wide commercial usage. The polymerization is complex because the rubber has sufficient reactivity, ie, at the allylic hydrogen and 1:2 addition units, to graft and cross-link, and the polymerization system is comprised of two phases. The graft copolymer of PS on polybutadiene acts as an emulsifier, stabilizing the dispersion against coalescence. Events taking place during this polymerization can be followed with the ternary phase diagram shown in Figure 26. For example, for a feed mixture made up of 8 wt % rubber in styrene (point A) on an increment of conversion to PS (to D), phase separation occurs. Small droplets of styrene–PS, which are stabilized by the graft copolymer, are dispersed in the styrene–polybutadiene solution. The composition of the two phases is given by points B and C for the PS-rich phase and polybutadiene-rich phase, respectively. The phase–volume ratio, ie, rubber phase/PS phase, is given by the ratio DB/DC. As the reaction proceeds along the line AE, the phase composition and phase–volume ratio can be read from the tie lines. Larger droplets of PS solution form and the small, original ones remain. When the phase–volume ratio is about unity (point F), provided that there is sufficient shearing agitation, the larger PS-containing droplets, which have mostly coalesced into large pools, be-

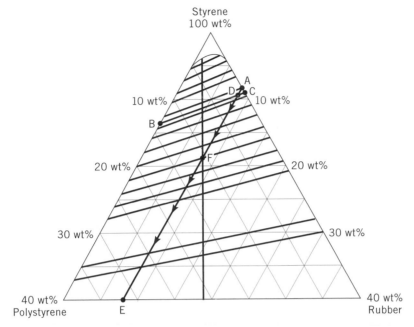

Fig. 26.   Ternary-phase diagram for the system styrene–PS–polybutadiene rubber.

come the continuous phase and polybutadiene-containing droplets are dispersed therein (202,203). However, the latter contain the small, original PS particles as occlusions.

The particle size and size distribution are largely controlled by the applied shear rate during and after phase inversion, the viscosity of the continuous phase, the viscosity ratio of the two phases, and the interfacial tension between the phases (51,202). The viscosity parameters depend on phase compositions, polymer molecular weights, and temperatures; the interfacial tension is largely controlled by the amount and structure of the graft copolymer available at the particle surface. Phase-contrast micrographs of this phase-inversion sequence, wherein the PS phase is dark, are shown in Figure 27. If shearing agitation is insufficient, then phase inversion does not occur and the product obtained is a network of cross-linked rubber with PS and its properties are much inferior. The morphology of the dispersed rubber phase remains much as formed after phase inversion, unless there is a step change in phase concentration, eg, by blending streams. Rubber-phase volume, including the occlusions, largely controls performance. Izod impact strength depends on rubber particle size and varies from 48 J/m (0.9 ft·lbf/in.) at an average particle diameter of 0.6 μm to ~100 J/m (1.9 ft·lbf/in.) at 3.5 μm diameter (204).

Once the desired particle size and morphology with sufficient graft to provide adequate adhesion between the rubber particles and the PS phase have been achieved, the integrity of the particle must be protected against deformation or destruction during fabrication by cross-linking of the polybutadiene chains. These chemical changes taking place on the rubber begin with polymerization. Either anionic or Ziegler-Natta-polymerized polybutadiene is used at 5–10 wt %

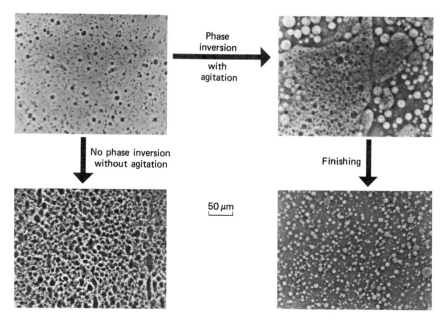

Fig. 27. Phase contrast photomicrographs showing particle formation via phase inversion.

styrene. If peroxide initiators are used, about one half of the rubber is grafted by the time phase inversion occurs; less is grafted in the absence of peroxides (205).

Hydrogen abstraction by alkoxy radicals at any of the allylic sites yields a site for styrene polymerization, resulting in a graft (205,206). Model studies based on styrene–butadiene block copolymers show a strong influence of graft structure on particle morphology (51,52). The double bonds in polybutadiene are much less reactive than those in styrene but as conversion exceeds ca 80% and temperature approaches or exceeds 200°C, copolymerization through the polybutadiene (1,2 units) chain seems to occur and leads to cross-linking and hence gel formation (207).

Commercial Processes

There are two problems in the manufacture of PS: removal of the heat of polymerization (ca 700 kJ/kg (300 Btu/lb)) of styrene polymerized and the simultaneous handling of a partially converted polymer syrup with a viscosity of ca 10^5 mPa(=cP). The latter problem strongly aggravates the former. A wide variety of solutions to these problems have been reported; for the four mechanisms described earlier, ie, free radical, anionic, cationic, and Ziegler, several processes can be used. Table 6 summarizes the processes which have been used to implement each mechanism for liquid-phase systems. Free-radical polymerization of styrenic systems, primarily in solution, is of principal commercial interest. Details of suspension processes, which are declining in importance, are available (208,209), as are descriptions of emulsion processes (210) and summaries of the historical development of styrene polymerization processes (208,211,212).

Table 6. Process vs Mechanism for Styrene Polymerization

Process	Mechanism			
	Free radical	Anionic	Cationic	Zeigler
mass/bulk	used for all styrene plastics	yes	yes	
solution	used for all styrene plastics	styrene–butadiene block copolymers	yes	yes
suspension	used for all styrene plastics			
precipitation	yes			
emulsion	used for styrene–butadiene latices and ABS plastics			

Two types of reactors are used for continuous solution polymerization: the linear-flow reactor (LFR), approximating in the ideal case a plug-flow reactor, and the continuous-stirred tank reactor (CSTR), which ideally is isotropic in composition and temperature (Fig. 28) (see REACTOR TECHNOLOGY). LFR usually involve conductive heat transfer to many tubes through which a heat-transfer fluid flows. Multiple temperature zones can easily be achieved in a single reactor; agitation is provided by a rotating shaft with arms down the central axis of the tubular vessel. Reactors of this type operate for long periods and handle very high viscosity partial polymers with reliable heat-transfer coefficients. CSTRs are either of the recirculated coil configuration and rely on conduction to the cooled jacket for heat removal, or ebullient and makes use primarily of evaporative cooling to achieve temperature control. The limitation of this reactor occurs at high viscosity, when the rate of vapor disengagement is exceeded by the rate of vapor generation and the ability to mix in the condensed vapor. Typically, this limit occurs at 60–70 wt % solids. Figures 29 and 30 illustrate typical processes based on these two types of reactors. Most general-purpose, ie,

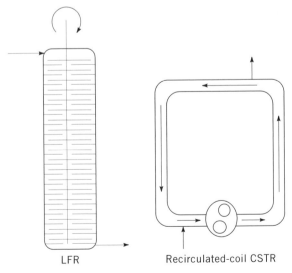

Fig. 28. Linear and CSTR reactor configuration used commercially for PS manufacture (see also Fig. 20).

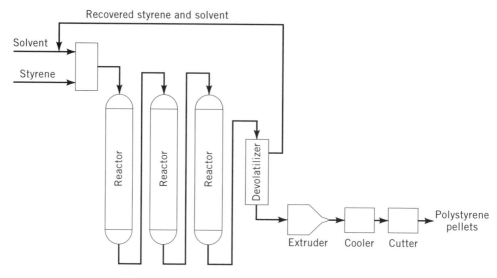

Fig. 29. Typical LFR polymerization process.

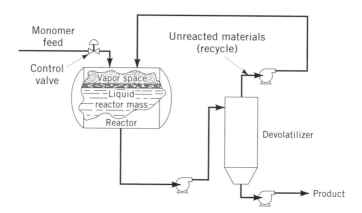

Fig. 30. Typical CSTR polymerization process.

unmodified PS is manufactured in the United States in such facilities. The LFR process (see Fig. 29) typically uses up to 20 wt % ethylbenzene mainly to reduce viscosity. The three reactors may each be subdivided into three temperature-control zones, which allow for flexible control of the molecular weight distribution of the polymer as well as optimum output per reactor volume. Temperatures are 100–170°C. At ca 80 wt % solids, the partial polymer is heated from 180 to 240°C before being devolatilized upon entering a vacuum tank. The recovered solvent and monomer are continuously recycled to the feed. The molten polymer leaves the vacuum tank and is stranded, cooled, and chopped into pellets for shipping. The CSTR process (see Fig. 30) differs in that it is typically a single reactor and, therefore, involves a single temperature zone, although more can be added, and operates at ca 60 wt % solids, requiring a large polymer heater to achieve adequate devolatilization (213,214).

There has been some research aimed at continuously converting monomer completely to polymer using reactive extrusion (215). However, this process has not yet been commercialized as of this writing.

Rubber-modified PS manufacture places several additional demands: dissolving ca 5–10 wt % polybutadiene in styrene, often a 10–20-h process; shear conditions to achieve phase inversion and the desired particle size for a given product; control of phase compositions to produce the desired particle morphology; and sufficient time and temperature at the end of the process to achieve the necessary cross-linking, gel formation, and swelling index. Graft copolymer formation occurs throughout the polymerization, and sufficient amount of graft copolymer for phase stability is necessary at phase inversion. In the case of an LFR system, only addition of a rubber dissolver is needed to meet the above requirements. For a CSTR-based system, multiple reactors in series are required to avoid rapid shifts in phase compositions on mixing of one reactor effluent with the next reaction contents with loss of desirable morphology (216,217). Figure 31 shows a multizone CSTR reactor system. Phase inversion and particle sizing are accomplished in the first-stage reactor with the effluent entering a single-tank, baffled reactor operating as five stages with a common vapor space. An alternative is three CSTR reactors in series, followed by an adiabatic reactor to achieve high solids content (216). Multiple stages are necessary to achieve favorable rubber particle morphology with CSTRs; LFR systems inherently avoid sudden phase composition changes, unless severe internal back-mixing is allowed to occur.

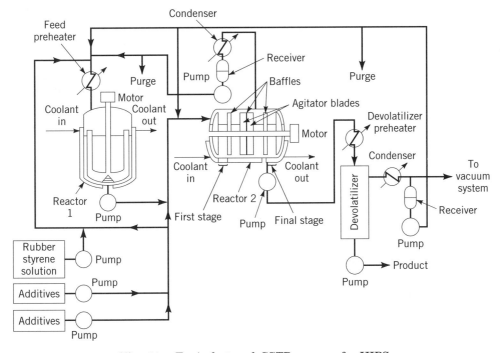

Fig. 31. Typical staged CSTR process for HIPS.

Generally, phase inversion and particle sizing occur in the first reactor. Grafting rate is highest initially, and the amount formed is adequate to stabilize the particles. The shear field in the reactor is critical to control of particle size as well as to meet the minimum shear rate for phase inversion, without which the product consists of a continuous cross-linked rubber network with markedly inferior properties (202,203). After the particles are formed and sized, only gentle agitation is applied to aid heat transfer without disturbing the particle structure. Conditions that produce the desired PS-phase molecular weight are used. Prior to devolatilization, the partial polymer is heated to ca 200°C to cross-link the rubber chain in order to protect the rubber particles from the very high shear fields during fabrication. Stranding, cooling, and chopping into pellets follow.

Polybutadiene rubbers are used almost exclusively as reinforcing agents for styrenic plastics. The very low glass-transition temperature (-80°C or lower) gives good low temperature properties, and the allylic hydrogen atom and weakly active double bonds can provide the desired degree of grafting and cross-linking. Other rubbers have been used; acrylates, ethylene–propylene–diene rubbers, polyisoprene, and polyethylene have been reported, but with limited commercial success because of their low chemical reactivity, not withstanding their acceptable glass-transition temperature (218).

Fabrication

Injection Molding. There are two basic types of injection-molding machines in use: the reciprocating screw and the screw preplasticator (219). Their simple design, uniform melting temperature, and excellent mixing characteristics make them the preferred choice for injection molding. Machines with shot capacities up to 25 kg for solid injection-molded parts and 65 kg for structural foam parts are available (220,221). Large solid moldings include automotive dash panels, television cabinets, and furniture components. One-piece structural foam parts weighing 35 kg or more are molded for increased rigidity, strength, and part weight reduction (222) (see PLASTICS PROCESSING).

The injection-molding process is basically the forcing of melted polymer into a relatively cool mold where it freezes and is removed in a minimum time. The shape of a molding is defined by the cavity of the mold. Quick entry of the material into the mold followed by quick setup results in a significant amount of orientation in the molded part. The polymer molecules and, in the case of heterogeneous rubber-modified polymers, the rubber particles tend to be highly oriented at the surface of the molding. Orientation at the center of the molding tends to be significantly less because of the relaxation of the molten polymer.

The anisotropy that develops during the molding operation is detrimental to the performance of the fabricated part in several ways. First, highly oriented moldings, which form particularly if low melt temperatures are used, exhibit good gloss, have an abnormally narrow use-temperature range that causes early warping, and, perhaps most importantly, tend to be brittle even though the material is inherently capable of producing tough parts (223). However, this development of polymer orientation during molding can also be used to advantage, as in the case of rotational orientation during the molding operation (224).

The achievement of isotropic moldings is also important when the molded part is to be decorated, eg, painted or metallized. Highly oriented parts that have a high frozen-in internal stress memory tend to give rise to rough or distorted surfaces as a result of the relaxing effect of the solvents and/or heat. For electroplating ABS moldings, it is particularly important to obtain isotropic moldings. Isotropy can be achieved by the use of a high melt temperature, a slow fill speed, a low injection pressure, and a high mold temperature (80,81,225,226).

Injection molding of styrene-based plastics is usually carried out at 200–300°C. For ABS polymers, the upper limit may be somewhat less, because these polymers tend to yellow somewhat if too high a temperature and/or too long a residence time are imposed.

To obtain satisfactory moldings with good surface appearance, contamination, including that by moisture, must be avoided. For good molding practice, particularly with the more polar styrene copolymers, drying must be part of the molding operation. A maximum of 0.1 wt % moisture can be tolerated before surface imperfections appear.

For achieving appropriate economics, injection-molding operations are highly automated and require few operating personnel (219). Loading of the hopper is usually done by an air-conveying system; the pieces are automatically ejected, and the rejects and sprues are ground and reused with the virgin polymer. Also, hot probes or manifold dies are used to eliminate sprues and runners.

Extrusion. Extrusion of styrene polymers is one of the most convenient and least expensive fabrication methods, particularly for obtaining sheet, pipe, irregular profiles, and films. Relatively small extruders, eg, 11.5-cm dia, can produce well over 675 kg/h of polymer sheet (227). Extrusion is also the method for plasticizing the polymer in screw injection-molding machines and is used to develop the parison for blow molding. Extrusion of plastics is one of the most economical methods of fabrication because it is a continuous method involving relatively inexpensive equipment. The extrusion process has been studied in great detail (228,229). Single-screw extruders work extremely well with styrene-based plastics. Machines are available with length-to-diameter (L/D) ratios of 36:1 or more. Some of the longer L/D extruders are used with as many as three vent zones for removal of volatiles, often eliminating the necessity for predrying as is practiced with hygroscopic materials, eg, SAN and ABS. Where venting is inadequate, these polymers must be predried to a maximum moisture content of 0.03–0.05 wt % to obtain high quality sheet (see FILM AND SHEETING MATERIALS).

Many rubber-modified styrene plastics are fabricated into sheet by extrusion primarily for subsequent thermoforming operations. Much consideration has been given to the problem of achieving good surface quality in extruded sheet (230,231). Excellent surface gloss and sheet uniformity can be obtained with styrene-based polymers.

Considerable work has been done on mathematic models of the extrusion process, with particular emphasis on screw design. Good results are claimed for extrusion of styrene-based resins using these mathematical methods (229,232). With the advent of low cost computers, closed-loop control of the extrusion system has become commonplace. More uniform gauge control at higher output rates is achievable with many commercial systems (233,234).

Lamination of polymer films, both styrene-based and other polymer types, to styrene-based materials can be carried out during the extrusion process for protection or decorative purposes. For example, an acrylic film can be laminated to ABS sheet during extrusion for protection in outdoor applications. Multiple extrusion of styrene-based plastics with one or more other plastics has grown rapidly from the mid-1980s to the mid-1990s.

Thermoforming and Orientation. Thermoforming of HIPS and ABS extruded sheet is of considerable importance in several industries. In the refrigeration industry, large parts are obtained by vacuum-forming extruded sheet. Vacuum forming of HIPS sheet for refrigerator-door liners was one of the most significant early developments promoting the rapid growth of the whole family of HIPS. When a thermoplastic polymer film or sheet is heated above its glass-transition temperature, it can be formed or stretched. Under controlled conditions, new shapes can be controlled; also, various amounts of orientation can be imparted to the polymer film or sheet for altering its mechanical behavior.

Thermoforming is usually accomplished by heating a plastic sheet above its softening point and forcing it against a mold by applying vacuum, air, or mechanical pressure. On cooling, the contour of the mold is reproduced in detail. In order to obtain the best reproduction of the mold surface, carefully determined conditions for the plastic-sheet temperature, ie, heating time and mold temperature, must be maintained.

Several modifications of thermoforming plastic sheet have been developed. In addition to straight vacuum forming, there are vacuum snapback forming, drape forming, and plug-assisted-pressure-and-vacuum forming. Some combinations of these techniques are also practiced. Such modifications are usually necessary to achieve more uniform wall thickness in the finished deep-draw sections. Vacuum forming can also be continuous by using the sheet as it is extruded. An example of this technology is practiced with several high speed European lines in operation in the United States. Precise temperature conditioning allows carefully controlled levels or orientation in the finished part (235).

Thermoforming is perhaps the process with the lowest unit cost. Examples of thermoformed articles are refrigerator-door and food-container liners, containers for dairy products, and luggage. Some of the largest formed parts are camper/trailer covers and liners for refrigerated-railroad-car doors (236).

Orientation of styrene-based copolymers is usually carried out at temperatures just above T_g. Biaxially oriented films and sheet are of particular interest. Such orientation increases tensile properties, flexibility, toughness, and shrinkability. PS produces particularly clear and sparkling film after being oriented biaxially for envelope windows, decoration tapes, etc. Oriented films and sheet of styrene-based polymers are made by the bubble process and by the flat-sheet or tentering process. Fibers and films can be produced by uniaxial orientation (237) (see FILM AND SHEETING MATERIALS).

Blow Molding. Blow molding is a multistep fabrication process for manufacturing hollow symmetrical objects. The granules are melted and a parison is obtained by extrusion or by injection molding. The parison is then enclosed by the mold, and pressure or vacuum is applied to force the material to assume the contour of the mold. After sufficient cooling, the object is ejected.

Styrene-based plastics are used somewhat in blow molding but not as much as linear polyethylene and PVC. HIPS and ABS are used in specialty bottles, containers, and furniture parts. ABS is also used as one of the impact modifiers for PVC. Clear, tough bottles with good barrier properties are blow-molded from these formulations.

PS or copolymers are used extensively in injection blow molding. Tough and craze-resistant PS containers have been made by multiaxially oriented injection-molded parisons (238). This process permits the design of blow-molded objects with a high degree of controlled orientation, independent of blow ratio or shape.

Additives. Processing aids, eg, plasticizers and mold-release agents, are often added to PS (see RELEASE AGENTS). Even though PS is an inherently stable polymer, other compounds are sometimes added to give extra protection for a particular application. Rubber-modified polymers containing unreacted allylic groups are very susceptible to oxidation and require carefully considered antioxidant packages for optimum long-term performance. Ziegler-Natta-initiated polybutadiene rubbers are especially sensitive in this respect, for they often contain organocobalt residues from the catalyst complex. For food contact applications, the additives must be FDA-approved. Important additives used in styrene plastics are listed in Table 7.

Economic Aspects

Most of the styrene monomer manufactured globally goes into the manufacture of PS and its copolymers, thus the price of the two tend to parallel each other (Fig. 32).

PS is a global product, of which North America, Western Europe, and Southeast Asia are the principal consumers (Fig. 33). Global PS production capacity generally parallels the demand for the material (Fig. 34). However, the trend since early 1980s has been toward narrowing the gap between capacity and demand in an effort to maximize the profitability of the business.

Characterization

Four modes of characterization are of interest: chemical analyses, ie, qualitative and quantitative analyses of all components; mechanical characterization, ie, tensile and impact testing; morphology of the rubber phase; and rheology at a range of shear rates. Other properties measured are stress crack resistance, heat distortion temperatures, flammability, creep, etc, depending on the particular application (239).

Because plastics are almost invariably modified with one or more additives, there are three components of chemical analysis: the high molecular weight portion, ie, the polymer; the additives, ie, plasticizer and mold-release agent; and the residuals remaining from the polymerization process. The high molecular weight portion is best characterized using gpc, which gives molecular weight and polydispersity information. If the gpc is equipped with a diode array detector, chromophores responsible for polymer discoloration can be located (125). The additives in the polymer are best characterized by extracting them from the polymer and analyzing the extract using high performance liquid chromatography (240). The relevant characterization methods are listed in Table 8. Figure 4 illustrates typical molecular weight distribution curves from gpc. Capillary gas chromatogry is used to determine residuals. Mechanical testing is carried out according to standard ASTM methods.

Table 7. Additives Used in Styrene Plastics

Type	Compounds	Examples	Amount, wt %	Comments
plasticizers (qv)	mineral oil phthalate esters adipate esters		<4	all cause loss of heat distortion temperature
mold-release agents	steric acid metal sterates sterate esters silicones amide waxes		0.1 3	many mold-release agents cause yellowness of haze
antioxidants (qv)	alkylated phenols phosphite esters	ionol tris(nonylphenyl) phosphite	1 1	synergism some-times achieved with multiple use
	thioesters	dilaurylthio-dipropionate	1	
antistatic agents (qv)	quaternary ammonium compounds		2	
uv stabilizers (qv)	benzotriazoles benzophenones	Tinuvin P	0.25	
ignition suppression agents	halogenated compounds, antimony oxide, aluminum oxide, phosphate esters			synergism some-times achieved with multiple use

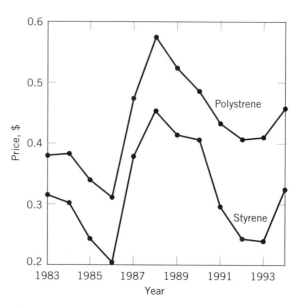

Fig. 32. Historical comparison of styrene monomer and general-purpose PS pricing.

1058

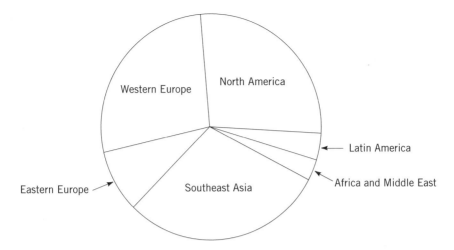

Fig. 33. Global consumption of PS by geographical area (96).

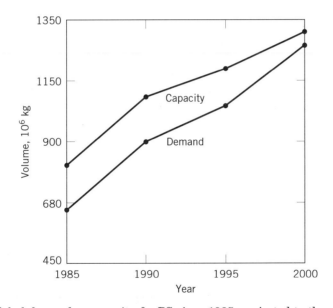

Fig. 34. Global demand vs capacity for PS since 1985, projected to the year 2000 (96).

Rubber particle size distribution is usually measured with a Coulter counter or directly from electron photomicrographs (72,241); the latter also gives details of particle morphology. Rheological studies are made using a modified tensile tester with capillary rheometer (ASTM D1238-79) or with the powerful Rheometrics mechanical spectrometer (242). For product specifications, a simple melt-flow test is used (ASTM D1238 condition G) with a measurement of the heat distortion temperature (ASTM D1525). These two tests, with solution viscosity as a measure of molecular weight, have been used historically to characterize PS. However, molecular weight distribution and residuals analyses, in general, have replaced them.

Table 8. Chemical Characterization Techniques

High mol wt component		Additives		Residuals	
Variable	Method[a]	Variable	Method[a]	Variable	Method[a]
mol wt	solution viscometry	mineral oil	gc/hplc	styrene	gc
mol wt distribution	gpc	phthalate esters	gc	solvents	gc
copolymer composition	ir	stearates	ir	oligomers	gc
rubber content	ir	other	hplc		
graft yield	solubility				

[a]gpc = gel-permeation chromatography; gc = gas chromatography; hplc = high performance liquid chromatography; ir = infrared analysis.

Health and Safety Factors

In pellet form, styrene-based plastics have a very low degree of toxicity (243). Under normal conditions of handling and use, they should pose no unusual problems from ingestion, inhalation, or eye and skin contact. Heating of these polymers usually results in the release of some vapors. The vapors contain some styrene (TLV = 100 ppm (1974); human detection, 5–10 ppm) and, in the case of styrene–acrylonitrile copolymers, some acrylonitrile (TLV = 2 ppm (1978)) as well as very low levels of degraded and oxidized hydrocarbons and additives. The warning properties of styrene and any oxidized organics are such that the vapors normally can be detected by humans at very low levels, and hazardous concentrations are usually irritating and obnoxious. Nevertheless, adequate ventilation should be provided.

Styrene polymers burn under the right conditions of heat and oxygen supply. This can occur even when they are modified by addition of ignition suppression chemicals. Once ignited, these polymers may burn rapidly and produce a dense black smoke. Combustion products from any burning organic material should be considered toxic. Ignition temperatures of several styrenic polymers are listed in Table 9. Fires can be extinguished by conventional means, ie, water and water fog.

Table 9. Ignition Temperatures and Burning Rates of Styrene-Based Polymers[a]

Polymer	Minimum flash-ignition temperature, °C	Minimum self-ignition temperature, °C	Burning rate[b], cm/s
polystyrene			
general-purpose	345–360	488–496	<0.06
rubber-modified	385–399	435–474	<0.06
styrene–acrylonitrile copolymer	366	454	<0.06
acrylonitrile–butadiene–styrene	404	>404	<0.06

[a]Test specimens for burning rate data were 1.27 × 15.24 × 0.318 cm³. Descriptions of burning rate and other flammability characteristics developed from small-scale laboratory testing do not reflect hazards presented by these or any other materials under actual fire conditions.
[b]In a horizontal position.

and raw material usage reduction. Syntactic foams are combinations of foamable beads, eg, expandable PS or styrene–acrylonitrile copolymers. These are mixed with a resin, usually thermosetting, which has a large exotherm during curing, eg, epoxy or phenolic resins (253). The mixture is then placed in a mold and the exotherm from the resin cure causes the expandable particles to foam and forces the resin to the surface of the mold. A typical example is expandable PS in a flexible polyurethane foam matrix used in cushioning applications (254). Sandwich panels are made either by foaming beads between the skin materials or by adhering skins to planks cut to precise dimensions. Foamed PS beads are admixed with concrete for lightweight masonry structures (255).

Extruded Rigid Foam.   In addition to low temperature thermal insulation, foamed PSs are used for insulation against ambient temperatures in the form of perimeter insulation and insulation under floors and in walls and roofs. The upside-down roof system has been patented (256), in which foamed plastic such as Styrofoam (Dow) plastic foam is applied above the tar-paper vapor seal, thereby protecting the tar paper from extreme thermal stresses that cause cracking. The foam is covered with gravel or some other wear-resistant topping (see ROOFING MATERIALS).

In addition to such thermal insulation uses in buildings, there is a potentially tremendous new area in the form of highway underlayment to prevent frost damage. Damage to roadways, roadbeds, and airfields because of frost action is a costly and aggravating problem. Conventional treatments to prevent such damage are expensive and unreliable. During the 1960s, an improved and more economical solution to the problem was developed (257). It is based on the use of thermal insulation to reduce the heat loss in the frost-susceptible subgrade soil so that no freezing occurs. Many miles of insulated pavements have been built in the United States, Canada, Europe, and Japan. The concept has proved valid, and the performance of extruded PS foam has been completely satisfactory. It is expected that the use of this concept will continue to grow as natural road-building materials continue to become more scarce and expensive and as the weight and speed of land and airborne vehicles continue to increase.

A 2.54-cm Styrofoam plastic foam with thermal conductivity of ca 0.03 W/(m·K) (0.21 (Btu·in.)/(ft·h·°F)) is equivalent to 61 cm of gravel. Any synthetic foam having compressive strength sufficiently high and thermal conductivity sufficiently low is effective. However, the resistance of PS-type foams to water, frost damage, and microorganisms in the soil makes them especially desirable. An interesting and important application of this concept was the use of Styrofoam in the construction of the Alaska pipeline. In this case, the foam was used to protect the permafrost.

Rigid Foam from Foaming-in-Place Beads.   Expandable PS (EPS) has been in wide use in packaging since its introduction. Applications range from pallet shipping containers for computer terminals, material-handling stacking trays, to packages for fresh fruits and vegetables. Expandable PS serves as a cushioning material, a package insert for blocking and bracing, a flotation item, or as insulation.

The basic resin for EPS is in the form of beads that are expanded to a desired density before molding. Densities for packaging parts are typically 20–40 kg/m^3. Once expanded, the beads are fused in a steam-heated mold to

Uses

PS Foams. The early history of foamed PS is available (244), as cussions of the theory of plastic foams (245). Foamable PS beads were d in the 1950s by BASF under the trademark of STYROPOR (246–24ξ beads, made by suspension polymerization in the presence of blowin such as pentane or hexane, or by post-pressurization with the same agents, have had an almost explosive growth, with 200,000 metric ton: 1980. Some typical physical properties of PS foams are listed in Tabl FOAMED PLASTICS).

The following are commercially significant foamed PSs. Extrude and boards that are largely flame-retardant and in the density range of ζ are used for low temperature thermal insulation, buoyancy, floral disp elty, packaging (qv), and construction purposes. Foamed boards an from foaming-in-place (FIP) beads (density 16–32 kg/m^3) are used for ing, buoyancy, insulation, and numerous other applications. Batch m boards and shapes as well as automated molding of continuous planks : Extruded foamed PS sheet 17-mm thick with densities of 64–160 kg/m^3 largely in packaging applications. Extremely fine cell size is required fo and strength. Special nucleating agents are employed for this purpose, mers of styrene and maleic anhydride, citric acid–sodium bicarbonate I and talc (249). High density extruded planks (density 35–64 kg/m^3), for heavy-duty structural applications. Styrene copolymer foams contai acrylonitrile for gasoline resistance, are made either by extrusion or fro High density (480–800 kg/m^3) injection-molded objects are made from of FIP beads and PS granules or more commonly HIPS or ABS with a blowing agent, eg, azodicarbonamide. These moldings have a high den of essentially PS and a foamed core. Foamable ABS systems, eg, lamin ABS skins and a heat-foamable core, are used for structural applicatic car body parts (250). Of growing commercial significance is coextrud core ABS pipe (251,252). Reducing the density of the foam core by ι half and using a chemical blowing agent result in a ∼20% overall pi

Table 10. Characteristic Properties of Some Polystyrene Foams

Property	Styrofoam^a, extruded	Bead, board-molded	Foam sheet
density, kg/m$^{3\,b}$	35	32	96
compressive strength, kPa^c	310	207–276	290
tensile strength, kPa^c	517	310–379	2070–345(
flexural strength, MPa^d	1138	379–517	
thermal conductivity, W/(m·K)^e	0.030	0.035	0.035
heat distortion, °C	80	85	85

^{a}Registered trademark of The Dow Chemical Company.
^{b}To convert kg/m^3 to lb/ft^3, multiply by 0.0624.
^{c}To convert kPa to psi, multiply by 0.145.
^{d}To convert MPa to psi, multiply by 145.
^{e}To convert W/(m·K) to (Btu·in.)/(h·ft^2·°F), multiply by 6.933.

form a specific shape. Most parts are molded of standard-white resins, although several pastel colors are available.

Converters who manufacture EPS parts and components for packaging are called shape molders. Others who mold large billets, eg, measuring $0.6 \times 1.2 \times 2.4$ m^3, are called block molders. The package user can obtain EPS parts without the benefits of a mold; parts can be fabricated from billets by hot-wire cutting. Whether molded or fabricated, EPS packages and their components are typically designed by careful consideration of the compression and cushioning properties of expanded PS (258).

A large number of factors that influence the foaming of foamable PS beads, eg, the molecular weight of the polymer, polymer type, blowing agent type and content, and bead size, have been analyzed (259). It is suggested that at least part of the pentane blowing agent is present in microvoids in the glassy PS. Thus, the density of beads containing 5.7 wt % n-pentane is 1080 kg/m^3, compared to a value of 1050 kg/m^3 for pure PS beads and a calculated density of 1020 kg/m^3 for a simple mixture in which n-pentane is dissolved in the PS. If all of the pentane were in voids, the calculated density would be 1120 kg/m^3. About 60% of the pentane is held in voids, with the balance presumably dissolved. n-Pentane present in voids has a reduced vapor pressure, which depends on the effective microcapillary diameter. Dissolved n-pentane in PS also has a reduced vapor pressure, by an amount determined by the thermodynamic interaction between solvent and polymer.

An unspecified experimental styrene copolymer, possibly with acrylonitrile, shows a greatly reduced tendency to lose blowing agent on aging at 23°C (259). FIP beads from chlorostyrene likewise have a greatly enhanced ability to retain blowing agent, as indicated in Table 11. The higher heat distortion of this polymer requires steam pressures of 210–340 kPa (30–50 psi) for blowing.

Table 11. Loss of Isopentane from Foaming-in-Place Chlorostyrene Beads[a] in Open Storage, wt %[b]

	After 20 h		After 300 h	
Temperature, °C	Polystyrene	Poly(chlorostyrene)	Polystyrene	Poly(chlorostyrene)
25	18	1.3	40	10
50	38	12	72	33
80	48	24	80	54

[a]420-mm beads.
[b]Initial concentration of isopentane = 6 wt %.

Various other factors that influence the foaming of PS, especially cross-linking, have been described (255). Reference 255 discusses foaming as a vis-coelastic process in which there is competition between the blowing pressure, which increases with temperature, and the thinning and rupture of cell walls with consequent collapse of the foam. The presence of cross-links reduces the tendency for cell wall rupture. Figure 35 shows foam volumes attained as a function of temperature for various amounts of divinylbenzene. The effects of temperature and cross-linking on the kinetics of foaming have also been discussed (255).

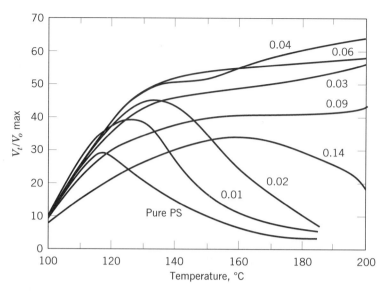

Fig. 35.   Maximum foaming volume of styrene divinylbenzene copolymers containing 8.8 wt % CO_2 as a function of divinylbenzene content and temperature. Numbers beside curves indicate wt % divinylbenzene, V_t/V_o is the ratio of final volume to initial volume at temperature t (258).

Another important factor affecting the foaming of PS beads is the differential rate of diffusion of n-pentane compared to steam and air. It is estimated that the quantity of n-pentane normally used is sufficient to produce a foaming volume of 30-fold and hence a foam density of no more than 32 kg/m³. However, the steam and/or air that diffuse into the beads contribute to further expansion, so that foaming volumes of 60-fold or more are commonly achieved. This aspect of diffusion has been discussed in some detail (260). In cycle foaming in air, beads are allowed to equilibrate with air after each heat-foaming step through a series of stages. Foaming volumes of 200:1 have been achieved in this manner, with densities down to 3.2–4.8 kg/m³.

Lightweight concrete is made from prefoamed EPS beads, Portland cement, and organic binders. Precast shapes are being used to provide structural strength, thermal insulation, and sound deadening.

PS as a Raw Material for Rigid Thermoplastic Foams.   The low cost of PS and its high thermal stability, which provides good recycle efficiency, as well as the possibility of using low cost blowing agents, are important to its use in rigid thermoplastic foams. The heat distortion point of ca 80°C, in contrast to 65°C for PVC foam, permits its use in most construction applications. The low sensitivity of the polymer to moisture is another important factor. Because of the above and possibly for other reasons, PS has been the dominant material used in rigid thermoplastic foams. Poly(vinyl chloride) is economical, has good physical properties, and is flame-retarding, but it has marginal heat distortion properties for some uses and does not readily lend itself to the continuous extrusion process used for PS foam. No FIP beads system based on PVC has been perfected as of this writing. This could well be a result of its high vapor barrier compared to PS.

Polyethylene is too permeable to retain a blowing agent and hence is ruled out as an FIP bead material, but it is commercially significant as extruded foam.

Flame Retardants. The growing use of PS foams in the construction industry has provided impetus for continuing research on flame-retardant additives (see FLAME RETARDANTS). Organic bromine compounds, eg, pentabromomonochlorocyclohexane, have long been used for this purpose in extruded PS foam. Use of injection-molded structural foams, based on HIPS and ABS and containing flame-retardant additives, has grown rapidly since the late 1970s. A typical additive system to impart the required flame-retardant properties is bis(pentabromophenyl) oxide and a synergist, eg, antimony trioxide. Because of environmental concerns over the use of halogenated flame-retardant chemicals, research is focusing on the development of halogen-free systems. Phosphorus-containing compounds are the most heavily researched alternative in the 1990s.

Foamed Sheet. PS foamed sheet is used for foamed trays, egg cartons, disposable dinnerware, and packaging. Foam sheet manufacturing techniques, manufacturers' logistics, and markets have been described (261). Choice of the correct blowing agent for foam sheet is critical in determining ultimate density and physical properties. The choice of blowing agent may also be dictated by safety considerations, eg, flammability of hydrocarbons, or environmental concerns, eg, chlorofluorocarbons. These two categories of blowing agents have been discussed (262).

Oriented PS Film and Sheet. Oriented PS film, in addition to being heat-shrinkable and having the lowest cost of any of the rigid plastic materials, offers a high degree of optical clarity, high surface gloss, and excellent dimensional stability, particularly with regard to relative humidity. It is not a barrier film; in fact, one of its largest uses, that of packaging field-fresh produce, depends on its being highly permeable to oxygen and water vapor (Table 12). A review of PS film as part of a larger study on films of all types is available (263,264). Other relevant reviews have also been published (265,266).

Although PS is normally considered a rather brittle material, biaxial orientation imparts some extremely desirable properties, particularly in regard to an increase in elongation. Thus, the 1.5–2% of elongation normally associated with unoriented PS can become as high as 10%, depending on the exact conditions of preparation (267,268).

There have been several methods utilized for preparing biaxially oriented PS film. The bubble process is used commercially. The high softening point, coupled with the tendency of oriented film to shrink above 85°C, makes heat sealing difficult. Solvent, adhesive, and ultrasonic sealing are used in most applications. Scratch resistance, impact strength, and crease resistance are low. Although the bubble process superficially resembles the common bubble process

Table 12. Gas and Water-Vapor Permeabilities[a] of PS Films

Composition	Oxygen	Carbon dioxide	Water
unmodified polystyrene	1,560	6,240	714,000
72:28 styrene-*stat*-acrylonitrile	40	2,060	1,290,000

[a]Gas permeability units = nmol/m·s·GPa.

for making polyethylene, polypropylene, and Saran film, there are a number of substantive differences; the PS process is considerably more difficult to achieve. In the case of polyethylene, for example, crystallinity stabilizes the blown bubble, whereas with a true thermoplastic, eg, PS, the blown film must be stabilized by cooling below the T_g. This involves extremely careful control of temperature in the blowing process.

As a general rule, there is an economic break-even point at ca 0.08 mm, which coincides with the defined difference between film and sheet. Film is made more economically by the bubble method and sheet by the tenter-frame method. The exact thickness for break-even depends on technological improvements, which can be made in both processes, in the degree of control used in regulating them and in quality requirements.

PS film contains no plasticizers, absorbs negligible moisture, and exhibits exceptionally good dimensional stability. It does not become brittle with age nor distort when exposed to low or high humidity. Excellent clarity, stability, and machinability, ie, ability to pass through packaging machinery at high speeds and be cut and sealed, etc, are central factors in the use of oriented PS in window envelopes. These same characteristics have also led to applications of film for window cartons. An antifog film 0.025–0.032-mm thick has been used extensively for the windows of cartons for bacon. Substantial amounts of film are also employed as sheet protectors and as inserts in wallets. A thickness of 0.060–0.130 mm is customary for these applications. PS film can be printed by means of flexographic, rotogravure, and silkscreen methods. A lamination of reverse-printed film and paper has been used for an attractive, high sparkle package for hand soap. The film is an excellent base for metallizing because of the almost complete absence of volatiles. Oriented PS film is also widely used for lamination to PS foamed sheet for preprinted decoration and/or property enhancement (269).

For PS sheet, the use responsible for its rapid growth is in meat trays, where the merchandising value is important, ie, the transparency of the package, as well as its dimensional stability and resistance to moisture. A smaller but growing use is as a photographic film base where dimensional stability at all humidities and low gel count are both important. The thermoplastic nature of PS and the memory effects built into it through biaxial orientation make it especially suitable for packaging applications. It must be processed under pressure rather than by vacuum-forming equipment to avoid the shrinkage that would otherwise result from the high level of orientation.

Important uses of PS are as windows in envelopes and as produce overwrap. High optical clarity is required for mechanized reading of characters through envelope windows, a significant technological trend that is certain to enhance the use of PS film. In regard to produce overwrap, the transmission characteristics of PS for water vapor and oxygen coincide more or less with the metabolic requirements of the produce being packaged, eg, lettuce, so that in addition to giving mechanical protection, the overwrap allows the produce to metabolize in a normal manner (270). Low cost and high optical clarity are important. Furthermore, the feel of PS film gives an impression of crispness to the product. PS film also has potential use as a type of synthetic paper (see PULP, SYNTHETIC). Whereas HIPS is reasonably opaque, general-purpose PS must be treated so that it is opaque; this can take the form of pigmentation, mechanical abrasion, or surface treatment, eg, chemical etch. PS film can be given a suitable coating, eg, a latex-clay coating, of the kind that is used on ordinary paper for the purpose of achieving a good printable surface. General discussions of gas permeabilities in plastic films are available (6,8,9,271,272) (see BARRIER POLYMERS).

BIBLIOGRAPHY

"Styrene Resins and Plastics" in *ECT* 1st ed., Vol. 13, pp. 146–179, by J. A. Struthers, R. F. Boyer, and W. C. Goggin, The Dow Chemical Co.; "Styrene Plastics" in *ECT* 2nd ed., Vol. 19, pp. 85–134, by H. Keskkula, A. E. Platt, and R. F. Boyer, The Dow Chemical Co.; in *ECT* 3rd ed., Vol. 21, pp. 801–847, by A. E. Platt and T. C. Wallace, The Dow Chemical Co.

1. M. Hocking, *Science*, **251**, 504 (1991).
2. R. F. Boyer, H. Keskkula, and A. E. Platt, in N. M. Bikales, ed., *Encyclopedia of Polymer Science and Technology*, Vol. 13, John Wiley & Sons, New York, 1970, pp. 128–447; E. R. Moore and co-workers, *ibid.*, Vol. 13, 1989, pp. 1–246.
3. J. F. Rudd, in J. Brandrup and E. H. Immergut, eds., *Polymer Handbook*, 3rd ed., John Wiley & Sons, Inc., 1989, p. V-81–86.
4. W. W. Graessley, *Adv. Polym. Sci.* **16**, 3 (1974).
5. M. Szwarc, M. Levy, and G. Milkovich, *J. Am. Chem. Soc.* **78**, 2656 (1956).
6. H. Keskkula, G. M. Simpson, and F. L. Dicken, *Soc. Plast. Eng. Preprints Ann. Tech. Conf.* **12**, XV-2 (1966).
7. F. J. Furno, R. S. Webb, and N. P. Cook, *Prod. Eng.* **35**, 87 (1964); V. E. Malpass, *Soc. Plast. Eng. Preprints Ann. Tech. Conf.* **13**, 618 (1969).
8. W. E. Brown, *Proceedings from International Plastics Congress*, 1966, N.V. t'Raedthuys, Utrecht, Netherlands, 1967.
9. R. L. Bergen and W. E. Wolstenholme, *SPE J.* **16**, 1235 (1960).
10. G. B. Jackson and J. L. McMillan, *SPE J.* **19**, 203 (1963).
11. S. G. Turley and H. Keskkula, *Polym. Eng. Sci.* **7**, 1 (1967).
12. W. E. Brown, *Performance of Plastics in Building*, No. 1004, Building Research Institute, Inc., Washington, D.C., 1963.
13. R. McFedries, Jr., *Plast. World*, **24**, 34 (Oct. 1963).
14. E. Scalco, T. W. Huseby, and L. L. Blyler, Jr., *J. Appl. Polym. Sci.* **12**, 1343 (1968).
15. E. H. Andrews, in J. V. Schmitz and W. E. Brown, eds., *Testing of Polymers*, Vol. 4, Wiley-Interscience, New York, 1969, p. 237.
16. W. H. Haslett, Jr. and L. A. Cohen, *SPE J.* **20**, 246 (1964).
17. U.S. Pat. 4,585,825 (Apr. 29, 1986), M. A. Wesselmann (to Dow Chemical Co.).
18. J. Roovers, *Macromolecules*, **17**, 1196 (1984).
19. C. G. Bazuin, A. Eisenberg, and M. Kamal, *J. Polym. Sci. Part B: Polym. Phys.* **24**, 1155 (1986).
20. M. Hara, P. Jar, and J. A. Sauer, *Polymer*, **32**, 1622 (1991).
21. C. B. Bucknall and R. R. Smith, *Polymer*, **6**, 437 (1965).
22. J. A. Schmitt, *J. Appl. Polym. Sci.* **12**, 533 (1968).
23. W. Retting, *Angew. Makromol. Chem.* **58/59**, 133 (1977).
24. I. Katime, J. R. Quintana, and C. Price, *Mater. Lett.* **22**, 297 (1995).
25. C. B. Bucknall, *Toughened Plastics*, Applied Science Publishers, London, 1977, Chapts. 7 and 10.
26. D. G. Gilbert and A. M. Donald, *J. Mater. Sci.* **21**, 1819 (1986).
27. Y. Okamoto and co-workers, *Macromolecules*, **24**, 5639 (1991).

28. H. Keskkula, G. M. Simpson, and F. L. Dicken, *Soc. Plast. Eng. Preprints Annu. Tech. Conf.* **12**, XV-2 (1966).

29. K. J. Cleereman, *SPE J.* **23**, 43 (1967); **25**, 55 (1969).

30. L. C. Stirik, *Ann. N.Y. Acad. Sci.* **279**, 78 (1976).

31. C. B. Bucknall and D. G. Street, *J. Appl. Polym. Sci.* **2**, 289 (1959).

32. A. Hall and I. Burnstein, *J. Mater. Sci.* **29**, 6523 (1994).

33. D. B. Priddy, *Adv. Polym. Sci.* **114**, 69 (1994).

34. M. A. Bellinger, J. A. Sauer, and M. Hara, *Polymer*, **35**, 5478 (1994).

35. M. A. Bellinger, J. A. Sauer, and M. Hara, *Macromolecules*, **27**, 1407 (1994).

36. N. Ishihara and co-workers, *Macromolecules*, **19**, 2464 (1986).

37. A. J. Pasztor, B. G. Landes, and P. J. Karjala, *Thermochemica Acta*, **177**, 187 (1991).

38. U.S. Pat. 2,287,188 (June 23, 1942), L. A. Matheson and R. F. Boyer (to Dow Chemical Co.).

39. R. L. Miller and L. E. Nielsen, *J. Polym. Sci.* **55**, 643 (1961).

40. D. S. Allan and co-workers, *Macromolecules*, **27**, 7621 (1994).

41. *Modern Plastics Encyclopedia*, Vol. 45, 1968, p. 43.

42. J. Eichhorn, *J. Appl. Polym. Sci.* **8**, 2497 (1964).

43. U.S. Pat. 3,324,076 (June 6, 1967), M. E. Elder, R. T. Dickerson, and W. F. Tousignant (to Dow Chemical Co.).

44. G. E. Molau, *J. Polym. Sci. Polym. Phys. Ed.* **3**, 1007 (1965).

45. U.S. Pat. 3,265,765 (Aug. 9, 1966), G. Holden and R. Milkovich (to Shell Oil Co.).

46. U.S. Pat. 4,267,284 (May 12, 1981), A. G. Kitchen (to Phillips Petroleum Co.).

47. G. E. Molau, *J. Polym. Sci. Polym. Chem. Ed.* **3**, 1267, 4235 (1965).

48. J. Moacanin, G. Holden, and N. W. Tschoegl, *J. Polym. Sci. Polym. Symp.* **26** (1969).

49. G. Reiss and co-workers, *J. Macrotool. Sci. Phys.* **B17**, 355 (1980).

50. R. Xie and co-workers, *Macromolecules*, **27**, 3444 (1994).

51. A. Echte, *Angew. Makrotool. Chem.* **58/59**, 175 (1977).

52. H. Feng, Z. Feng, and L. Shen, *Macromolecules*, **27**, 7840 (1994).

53. J. Stock and R. Stadler, *Makromol. Chem. Macromol. Symp.* **76**, 127 (1993).

54. M. R. Ambler, *Chromatogr. Sci. Ser.* **19**, 29 (1981).

55. T. Guo and W. Yang, *J. Damage Mech.* **2**, 364 (1993).

56. H. L. Hsieh and co-workers, *Chemtech*, **11**, 626 (1981).

57. U.S. Pat. 3,336,267 (Apr. 15, 1967), R. I. Zimmerman and W. E. O'Connor (to Dow Chemical Co.).

58. U.S. Pat.. 3,231,524 (Jan. 25, 1966), D. W. Simpson (to Dow Chemical Co.).

59. M. Barsacchi and co-workers, *Angew. Chem. Int. Ed. Engl.* **30**, 989 (1991).

60. E. Drent, J. A. M. van Broekhoven, and M. J. Doyle, *J. Organomet. Chem.* **417**, 235 (1991).

61. R. H. Boundy, R. F. Boyer, and S. Stoesser, *Styrene, Its Polymers, Copolymers and Derivatives*, ACS, Monograph No. 115, Reinhold Publishing Corp., New York, 1952.

62. H. W. McCormick, *J. Polym. Sci.* **25**, 488 (1957).

63. U.S. Pat. 3,206,445 (Sept. 14, 1965), H. Volk (to Dow Chemical Co.).

64. U.S. Pat. 3,340,238 (Sept. 5, 1967), W. E. Smith and H. Volk (to Dow Chemical Co.).

65. U.S. Pat. 3,011,918 (Dec. 5, 1961), L. H. Silvernail and M. W. Zembal (to Dow Chemical Co.).

66. G. E. Molau and H. Keskkula, *J. Polym. Sci. Polymer Chem. Ed.* **4**, 1595 (1966).

67. S. Strella, in E. Baer, ed., *Engineering Design for Plastics*, Reinhold Publishing Corp., New York, 1964, pp. 795–814.

68. R. F. Boyer, *Polym. Eng. Sci.* **8**, 161 (1968).

69. C. B. Bucknall, *Toughened Plastics*, Applied Science Publishers, London, 1977.

70. H. Willersinn, *Makromol. Chem.* **101**, 297 (1966).

71. H. Keskkula, S. G. Turley, and R. F. Boyer, *J. Appl. Polym. Sci.* **15**, 351 (1971).

72. K. Kato, *J. Electron Microsc.* **14**, 220 (1965); *Polym. Eng. Sci.* **7**, 38 (1967).
73. D. A. Walker, in Ref. 41, pp. 334.
74. H. Keskkula and P. A. Traylor, *J. Appl. Polym. Sci.* **11**, 2361 (1967).
75. R. J. Williams and R. W. A. Hudson, *Polymer*, **8**, 643 (1967).
76. Ref. 2, Vol. 1, p. 388.
77. *Br. Plast.*, **38**, 708 (Dec. 1965).
78. G. M. Kraynak, *Soc. Plast. Eng. Annu. Tech. Conf.* **13**, 896 (1967).
79. *Vacuum Metallizing Cycolac*, Technical Report P135, Borg-Warner Corp., Chicago, Ill., 1980.
80. E. N. Hildreth, *Modern Plastics Encyclopedia*, Vol. 43, 1966, p. 991.
81. K. Stoeckhert, *Kunststoffe*, **55**, 857 (1965).
82. *Mod. Plast.* **45**, 84 (Aug. 1968).
83. N. L. Maecker and D. B. Priddy, *J. Appl. Polym. Sci.* **42**, 21 (1991).
84. B. Bell and co-workers, *J. Appl. Polym. Sci.* **54**, 1605 (1994).
85. U.S. Pat. 4,874,829 (Oct. 17, 1989), C. E. Schwier and W. C. Wu (to Dow Chemical Co.).
86. U.S. Pat. 5,091,470 (Feb. 25, 1992), H. W. Wolsink, J. J. Plommer, and T. D. Traugott (to Dow Chemical Co.).
87. U.S. Pat. 2,802,809 (Aug. 13, 1957), R. A. Hayes (to Firestone Tire and Rubber Co.).
88. E. Zahn, *Appl. Polym. Syrup.* **11**, 209 (1969).
89. W. E. Brown, J. D. Striebel, and D. C. Fuccella, *Proceedings from Automotive Engineering Congress*, paper 680059, 1968.
90. C. H. Bamford and C. F. M. Tipper, eds., *Chemical Kinetics*, Vol. 14, Elsevier Science Inc., New York, 1975.
91. N. Grassie, ed., *Developments in Polymer Degradation*, Vols. 1 and 2, Applied Science Publishers, London, 1977.
92. M. Guaita, O. Chiantore, and L. Costa, *Polym. Deg. Stab.* **12**, 315 (1985).
93. O. Chiantore, M. Guaita, and N. Grassie, *Polym. Deg. Stab.* **12**, 141 (1985).
94. G. G. Cameron, W. A. Bryce, and L. T. McWalter, *Eur. Polym. J.* **20**, 563 (1984).
95. R. S. Lehrle, R. E. Peakman, and J. C. Robb, *Eur. Polym. J.* **18**, 517 (1982).
96. J. R. Stoppert, *Polystyrene Deserves More Respect*, Form No. 301-01772-1291, The Dow Chemical Co., Midland, Mich., 1991.
97. G. A. Mackey, R. C. Westphal, and R. Coughanour, in R. J. Ehrig, ed., *Plastics Recycling*, Hanser-Gardner Publishers, Cincinnati, Ohio, 1992.
98. K. Sikkema and co-workers, *Polym. Degr. Stabil.* **38**, 119 (1992).
99. K. Sikkema and co-workers, *Polym. Degr. Stabil.* **38**, 113 (1992).
100. M. J. Hanner and co-workers, *Polym. Degr. Stabil.* **39**, 235 (1993).
101. U.S. Pat. 5,187,249 (Feb. 16, 1993), V. A. Dais and D. B. Priddy (to Dow Chemical Co.).
102. J. J. O'Brien, P. B. Smith, and D. B. Priddy, *Polym. Degr. Stabil.* **39**, 69 (1993).
103. U.S. Pat. 5,194,527 (Mar. 16, 1993), J. J. O'Brien, K. Sikkema, and D. B. Priddy (to Dow Chemical Co.).
104. J. E. Potts and co-workers, *Polym. Sci. Technol.* **3**, 61 (1973).
105. J. E. Guillet, M. Heskins, and L. R. Spencer, *Polym. Mater. Sci. Eng.* **58**, 80 (1988).
106. S. Karlsson, O. Ljungquist, and A. Albertsson, *Polym. Degr. Stabil.* **21**, 237 (1988).
107. W. J. Bailey, *Polym. J.* **17**, 85 (1985).
108. PCT Int. Appl. WO 9322371 (Nov. 11, 1993), G. C. Welsh, B. D. Dalke, and E. K. C. Lee (to Dow Chemical Co.).
109. D. B. Priddy and M. Pirc, *J. Appl. Polym. Sci.* **40**, 41 (1990).
110. U.S. Pat. 4,087,599 (May 2, 1978), J. M. Roe and D. B. Priddy (to Dow Chemical Co.).
111. P. J. Flory, *J. Am. Chem. Soc.* **59**, 241 (1937).
112. K. E. Russel and A. Tobolsky, *J. Am. Chem. Soc.* **76**, 395 (1954).

113. W. A. Pryor, *ACS Symp. Ser.* **69**, 33 (1978).

114. W. Brown, *Makromol. Chem.* **128**, 130 (1969).

115. N. J. Barr and co-workers, *Eur. Polym. J.* **14**, 245 (1978).

116. F. R. Mayo, *J. Am. Chem. Soc.* **90**, 1289 (1968).

117. W. A. Pryor and L. Lasswell, *Adv. Free Radical Chem.* **5**, 27 (1975).

118. F. R. Mayo, *J. Am. Chem. Soc.* **75**, 6133 (1953).

119. R. R. Hiatt and P. D. Bartlett, *J. Am. Chem. Soc.* **81**, 1149 (1959).

120. R. Kaiser and co-workers, *Angew. Makromol. Chem.* **12**, 25 (1970).

121. Y. K. Chong, E. Rizzardo and D. H. Solomon, *J. Am. Chem. Soc.* **105**, 7761 (1983).

122. J. Mulzer and co-workers, *Chem. Ber.* **121**, 2231 (1988).

123. W. C. Buzanowski and co-workers, *Polym. Prepts*, **32**, 220 (1991).

124. D. R. Hensley and co-workers, *Macromolecules*, **27**, 2351 (1994).

125. D. B. Priddy, *Adv. Polym. Sci.* **111**, 69 (1994).

126. G. Moad, D. H. Solomon, and R. I. Willing, *Macromolecules*, **21**, 855 (1988).

127. J. F. McKellar and N. F. Allen, *Photochemistry of Man Made Polymers*, Applied Science, London, 1978.

128. H. Susuki, Y. Tanaka, and Y. Ishii, *Sen'i Gakkaishi*, **35**, 296 (1979).

129. D. Bellus and co-workers, *J. Polym. Sci. Part C*, **22**, 629 (1969).

130. A. D. Schmidt and W. H. Ray, *Chem. Eng. Sci.* **36**, 1401 (1981).

131. J. W. Hammer, T. A. Akramov, and W. A. Ray, *Chem. Eng. Sci.* **36**, 1897 (1981).

132. K. Ito and T. Aoyama, *Eur. Polym. J.* **23**, 955 (1987).

133. L. Blavier and J. Villermaux, *Chem. Eng. Sci.* **39**, 101 (1984).

134. C. H. Bamford, *Polymer*, **31**, 1720 (1990).

135. J. Bogunjoko and B. W. Brooks, *Makromol. Chem.* **184**, 1603 (1983).

136. K. Y. Hsu and S. A. Chen, *Polym. Eng. Sci.* **24**, 1253 (1984).

137. F. L. Marten and A. E. Hamielec, *J. Appl. Polym. Sci.* **27**, 489 (1982).

138. K. J. Kim and K. Y. Choi, *Polym. Eng. Sci.* **31**, 333 (1991).

139. M. A. Villalobos, A. E. Hamielec, and P. E. Wood, *J. Appl. Polym. Sci.* **42**, 629 (1991).

140. K. Y. Choi, W. R. Liang, and G. D. Lei, *J. Appl. Polym. Sci.* **35**, 1562 (1988).

141. K. J. Kim, W. Liang, and K. Y. Choi, *Ind. Eng. Chem. Res.* **28**, 131 (1989).

142. K. J. Kim and K. Y. Choi, *Chem. Eng. Sci.* **44**, 297 (1989).

143. K. J. Kim and K. Y. Choi, *Chem. Eng. Sci.* **43**, 965 (1988).

144. K. Y. Choi and G. D. Lei, *AIChE J.* **33**, 2067 (1987).

145. V. R. Kamath, *Mod. Plast.* **58**, 106 (1981).

146. J. C. Bevington, H. W. Melville, and R. P. Taylor, *J. Polym. Sci.* **12**, 449 (1954).

147. A. V. Tobolsky, C. E. Rogers, and R. D. Brickman, *J. Am. Chem. Soc.* **28**, 1277 (1960).

148. Form No. 115-575-79, Dow Chemical Company, Midland, Mich., 1979.

149. U.S. Pat. 4,177,110 (Dec. 4, 1979), J. M. Watson (to Cosden).

150. U.S. Pat. 4,086,147 (Apr. 25, 1978), J. M. Watson (to Cosden).

151. M. H. George, in G. E. Ham, ed., *Vinyl Polymerization*, Marcel Dekker, Inc., New York, 1967, pp. 186–188.

152. A. A. Miller and F. R. Mayo, *J. Am. Chem. Soc.* **78**, 1017 (1956).

153. O. F. Olaj, H. F. Kauffmann, and J. W. Breitenback, *Makromol. Chem.* **178**, 2707 (1977).

154. Ref. 3, pp. 11–61.

155. E. Niki and Y. Kamiya, *J. Org. Chem.* **38**, 1403 (1975).

156. E. Niki and Y. Kamiya, *J. Chem. Soc. Perkin Trans.* **1975**, 1221 (1975).

157. U.S. Pat. 4,910,274 (Mar. 20, 1990), D. B. Priddy (to Dow Chemical Co.).

158. Y. Watanabe and co-workers, *Chem. Lett. (Japan)*, **1975**, 1089 (1993).

159. D. B. Priddy and M. Pirc, *J. Appl. Polym. Sci.* **40**, 41 (1990).

160. M. Szwarc, *Living Polymers and Electron Transfer Processes*, Wiley-Interscience, New York, 1968.

161. D. B. Priddy, M. Pirc, and B. J. Meister, *Polym. React. Eng.* **1**, 343 (1993).

162. D. B. Priddy and M. Pirc, *J. Appl. Polym. Sci.* **37**, 1079 (1989).

163. D. B. Priddy and M. Pirc, *J. Appl. Polym. Sci.* **37**, 393 (1989).

164. D. B. Priddy and M. Pirc, *Polym. Prepr.* **29**, 340 (1988).

165. U.S. Pat. 4,859,748 (Aug. 22, 1989), D. B. Priddy and M. Pirc (to Dow Chemical Co.).

166. R. Thiele, *Chem. Eng. Technol.* **17**, 127 (1994).

167. Ger. Pat. DE 4235980 (Apr. 28, 1994), F. Brandstetter, H. Gausepohl, and R. Thiele (to BASF AG).

168. U.S. Pat. 3,773,740 (Nov. 20, 1973), T. T. Szabo (to Union Carbide).

169. U.S. Pat. 4,647,632 (Mar. 3, 1987), D. B. Priddy (to Dow Chemical Co.).

170. D. B. Priddy and co-workers, *J. Appl. Polym. Sci.* **41**, 383 (1990).

171. K. Miyashita and co-workers, *Macromolecules*, **27**, 1093 (1994).

172. T. Otsu, M. Yoshida, and T. Tazaki, *Makromol. Chem. Rapid Commun.* **3**, 133 (1982).

173. T. Otsu and M. Yoshida, *Makromol. Chem. Rapid Commun.* **3**, 127 (1982).

174. T. Otsu and A. Kuriyama, *Polym. Bull.* **11**, 135 (1984).

175. T. Otsu and A. J. Kuriyama, *Macromol. Sci. Chem.* **A21**, 961 (1984).

176. T. Otsu, *Kobunshi*, **37**, 248 (1988).

177. T. Otsu, T. Tazaki, and M. Yoshioka, *Chem. Express*, **5**, 801 (1990).

178. E. Rizzardo, *Chemistry in Australia*, 32 (Jan./Feb. 1987).

179. B. R. Smirnov, *Vysokomol. Soedin. Ser. A*, **32**, 583 (1990).

180. D. Mardare and K. Matyjaszewski, *Polym. Prepr.* **35**, 778 (1994).

181. M. K. Georges and co-workers, *Macromolecules*, **26**, 2987 (1993).

182. K. Endo, K. Murata, and T. Otsuk, *Macromolecules*, **25**, 5554 (1992).

183. I. Li and co-workers, *Macromolecules*, **28**, 6692 (1995).

184. M. K. Georges and co-workers, *ACS Polym. Prepr.* **35**, 870 (1994).

185. H. K. Hall, *Angew. Chem. Int. Ed. Engl.* **22**, 440 (1983).

186. Ref. 3, pp. II-303–333.

187. G. E. Molau, *J. Polym. Sci.* **B3**, 1007 (1965).

188. H. J. Karam, in K. Solc, ed., *Polymer Compatibility and Incompatibility*, MMI Press, Midland, Mich., 1981.

189. A. W. Hansen and R. L. Zimmermann, *Ind. Eng. Chem.* **49**, 1803 (1957).

190. Ger. Pat. 2,138,176 (July 30, 1971) (Badishe Anilin u. Soda-Fabrik).

191. U.S. Pat. 4,243,781 (July 6, 1981), R. W. Kent, Jr. (to Dow Chemical Co.).

192. U.S. Pat. 4,268,652 (May 19, 1981), R. W. Kent, Jr. (to Dow Chemical Co.).

193. U.S. Pat. 4,206,293 (June 3, 1980), R. L. Kruse (to Monsanto Co.).

194. D. B. Priddy, *Adv. Polym. Sci.* **121**, 123 (1995).

195. D. S. Allan and co-workers, *Macromolecules*, **26**, 6068 (1993).

196. B. Bell and co-workers, *Macromol. Reports*, **A29**, 1 (1992).

197. D. L. Hasha and co-workers, *Macromolecules*, **25**, 3046 (1992).

198. E. Stark and co-workers, *J. Macromol. Sci., Macromol. Reports*, **A29**, 1 (1992).

199. F. A. Miller, in R. R. Meyers and J. S. Long, eds., *Treatise on Coatings*, Vol. 1, Part 2, Marcel Dekker, Inc., New York, 1968, pp. 1–57.

200. O. Okay and co-workers, *Macromolecules*, **28**, 2728 (1985).

201. O. Okay, *Polymer* **35**, 796 (1994).

202. G. F. Freeguard, *Br. Polym. J.* **6**, 205 (1974).

203. U.S. Pat. 2,694,692 (Nov. 16, 1954), J. L. Amos, J. L. McCurdy, and O. R. McIntire (to Dow Chemical Co.).

204. U.S. Pat. 4,214,056 (July 22, 1980), R. E. Lavengood (to Monsanto Co.).

205. A. Brydon, G. M. Burnett, and G. G. Cameron, *J. Polym. Sci. Chem. Ed.* **11**, 3255 (1973); **12**, 1011 (1974); **18**, 2143 (1980).

206. T. Kotaka, *Makromol Chem.* **177**, 159 (1976).

207. D. J. Stein, G. Fahrbach, and H. Adler, *Adv. Chem. Ser.* **142**, 148 (1975).

208. R. H. M. Simon and D. C. Chappelear, in J. N. Henderson and T. C. Bouton, eds., *Polymerization Reactors and Processes*, ACS Symposium Series 104, American Chemical Society, Washington, D.C., 1979, pp. 71–112.

209. R. B. Bishop, *Practical Polymerization for Polystyrene*, Cahners Books, Boston, Mass., 1971.

210. D. C. Blackley, *Emulsion Polymerization: Theory and Practice*, Applied Science Publishers Ltd., London, 1975.

211. J. L. Amos, *Polym. Eng. Sci.* **14**, 1 (1974).

212. R. F. Boyer, *J. Macromol. Sci. Chem.* **A15**, 1411 (1981).

213. U.S. Pat. 3,884,766 (May 20, 1975), W. G. Bir and J. Novack (to Monsanto Co.).

214. U.S. Pat. 3,966,538 (June 29, 1976), C. G. Hagberg (to Monsanto Co.).

215. W. Michaeli and co-workers, *J. Appl. Polym. Sci.* **48**, 871 (1993).

216. U.S. Pat. 4,011,284 (Mar. 8, 1977), G. Gawne and C. Ouwerkerk (to Shell Oil Co.).

217. U.S. Pat. 3,903,202 (Sept. 2, 1975), D. E. Carter and R. H. M. Simon (to Monsanto Co.).

218. E. Martuscelli, R. Palumbo, and M. Kryszewski, *Polymer Blends*, Plenum Press, New York, 1980.

219. J. Lignon, *Mod. Plast.* **57**, 317 (1980).

220. C. Milacron, *Mod. Plast.* **57**, 332 (1980).

221. Hoover Universal Sales Literature, HV 450, Manchester, Mich., undated.

222. *Plastics Design Forum*, 22 (May/June 1980).

223. H. Keskkula and J. W. Norton, Jr., *J. Appl. Polym. Sci.* **2**, 289 (1959).

224. K. J. Cleereman, *SPE J.* **23**, 43 (Oct. 1967); **25**, 55 (Jan. 1969).

225. P. A. M. Ellis, *Plast. Inst. Trans. J.* **35**, 537 (1967).

226. H. R. Jacobi, *Screw Extrusion of Plastics*, Life Books Ltd., London, 1963.

227. G. A. Kruder, *SPE J.* **28**, 56 (Oct. 1972).

228. J. F. T. Pittman and R. Sander, *Int. Polym. Process.* **9**, 326 (1994).

229. H. Shidara and M. M. Denn, *J. Non-Newtonian Fluid Mech.* **48**, 101 (1993).

230. J. Fredos, ed., *Plastic Engineering Handbook*, Van Nostrand Reinhold Co., Inc., New York, 1976.

231. E. P. Weaver, *Polym. Eng. Sci.* **6**, 172 (1966).

232. C. Guerrero and P. J. Carreau, *J. Polym. Eng.* **12**, 155 (1993).

233. R. W. Brand and R. L. Keiks, *Plast. Technol.* **37** (Feb. 1972).

234. N. Fountas, *Plast. World*, **37**, 40 (Nov. 1979).

235. U.S. Pat. 4,039,609 (Aug. 2, 1977), A. W. Thief and H. Hell (to Bellaplast GmbH).

236. Cycolac Brand, *ABS Polymers Sales Bulletin*, Marbon Chemical Division of Borg-Warner Corp., Chicago, Ill., 1980.

237. J. T. Seitz, *Encyclopedia of Polymer Science and Engineering*, 2nd ed., Vol. 16, Wiley-Interscience, New York, 1989, p. 156.

238. K. J. Cleereman, W. J. Schrenk, and L. S. Thomas, *SPE J.* **24**, 27 (1968).

239. R. A. Bubeck and co-workers, *Polym. Eng. Sci.* **21**, 624 (1981).

240. N. E. Skelly and co-workers, *Polym. Mater. Sci. Eng.* **59**, 23 (1988).

241. D. E. James, *Polym. Eng. Sci.* **8**, 241 (1968).

242. T. Shimada, P. L. Horng, and R. S. Porter, *J. Rheology*, **24**, 78 (1980).

243. *Product Stewardship–Styrene Plastics*, Form. No. 304-109-1280, The Dow Chemical Co., Midland, Mich., 1980.

244. R. N. Kennedy, in R. J. Bender, ed., *Handbook of Foamed Plastics*, Lake Publishing Co., Libertyville, Ill., 1965.

245. K. C. Frisch and J. H. Saunders, *Plastic Foams*, Vol. 1, Part 1, Marcel Dekker, Inc., New York, 1972.

246. U.S. Pat. 2,681,321 (June 15, 1954), F. Stasny and R. Gaeth (to Badishe Anilin u. Soda-Fabrik).

247. U.S. Pat. 2,744,291 (May 8, 1956), F. Stasny and K. Buchholtz (to Badishe Anilin u. Soda-Fabrik).
248. U.S. Pat. 2,787,809 (Apr. 9, 1957), F. Stasny (to Badishe Anilin u. Soda-Fabrik).
249. U.S. Pat. 3,231,524 (Jan. 25, 1966), D. W. Simpson and C. H. Pottenger (to Dow Chemical Co.).
250. D. C. Wollard, *Proceedings from the 12th Annual SPI Conference*, Washington, D.C., 1967.
251. U.S. Pat. 4,249,875 (Feb. 8, 1981), E. Hart and R. Rutledge (to Cosden).
252. F. R. Bush and G. C. Rollerson, *Proceedings from the 38th SPE-ANTEC*, New York, 1980.
253. *Low Temperature Systems*, Form No. 179-2086-77, The Dow Chemical Co., Midland, Mich., 1977.
254. B. Brooks and L. G. Rey, *J. Cell. Plast.* **9**, 232 (1973).
255. L. C. Rubens, *J. Cell. Plast.* **1**, 3 (1965).
256. U.S. Pat. 3,411,256 (Nov. 19, 1968), J. B. Best (to Dow Chemical Co.).
257. U.S. Pat. 3,250,188 (May 10, 1966), G. A. Leonards (to Dow Chemical Co.).
258. R. L. Chatman, *Package Eng.* **26**, 56 (July 1981).
259. A. R. Ingram and H. A. Wright, *Mod. Plast.* **41**, 152 (Nov. 1963).
260. S. J. Skinner, S. Baxter, and P. J. Grey, *Plast. Inst. Trans. J.* **32**, 180 (1964).
261. R. Martino, *Mod. Plast.* **55**, 34 (Aug. 1978).
262. J. G. Burt, *J. Cell. Plast.* **15**, 158 (1979).
263. O. Sweeting, ed., *Science and Technology of Polymer Films*, Vols. 1 and 2, Wiley-Interscience, New York, 1968–1969.
264. J. Pinsky, in Ref. 263, Vol. 2.
265. W. R. R. Park, ed., *Plastics Film Technology*, Reinhold Publishing Corp., New York, 1969.
266. C. Saikaishi, *Jpn. Chem. Quart.* **1**, 49 (1968).
267. Ref. 237, Vol. 11, 1989, p. 159.
268. H. J. Karam, in Ref. 263, Vol. 1, pp. 227–253.
269. *Lamination of Trycite Plastic Films to Polystyrene Foam Sheet*, Bulletin No. 500-898-79, The Dow Chemical Co., Midland, Mich., 1979.
270. C. R. Scott, F. J. Butt, and J. Eichhorn, *Mod. Packag.* **38**, 135 (1965).
271. H. Yasuda, H. G. Clark, and V. Stannett, in N. M. Bikales, ed., *Encyclopedia of Polymer Science and Technology*, Vol. 9, Wiley-Interscience, New York, 1968, pp. 794–807.
272. C. E. Rogers, in E. Baer, ed., *Engineering Design for Plastics*, Reinhold Publishing Corp., New York, 1964, pp. 609–688.

DUANE B. PRIDDY
The Dow Chemical Company

SUBERIC ACID. See DICARBOXYLIC ACIDS.

SUCCINIC ACID AND SUCCINIC ANHYDRIDE

Succinic acid [*110-15-6*] (butanedioic acid; 1,2-ethanedicarboxylic acid; amber acid), $C_4H_6O_4$, occurs frequently in nature as such or in the form of its esters. It can be found in animal tissues (1), in vegetables and fruit (2,3), or in spring water (4), and has also been identified in meteorites (5). It is formed in alcoholic fermentation (6) and in the chemical and biochemical oxidation of fats. Succinic acid is present in amber (7) (*Succinum*) and can be obtained by distillation, by which method it was first isolated by Georgius Agricola in 1550.

Succinic anhydride [*108-30-5*] (3,4-dihydro-2,5-furandione; butanedioic anhydride; tetrahydro-2,5-dioxofuran; 2,5-diketotetrahydrofuran; succinyl oxide), $C_4H_4O_3$, was first obtained by dehydration of succinic acid. In the 1990s anhydride is produced by hydrogenation of maleic anhydride and the acid by hydration of the anhydride, by hydrogenation of aqueous solutions of maleic acid, or as a by-product in the manufacture of adipic acid (qv) (see MALEIC ANHYDRIDE, MALEIC ACID, AND FUMARIC ACID).

From the chemical point of view, succinic acid and its anhydride are characterized by the reactivity of the two carboxylic functions and of the two methylene groups. Uses range from pharmaceuticals to food, detergents, cosmetics, plastics and resins, plant growth regulators, textiles, photography, and gas and water treatment.

Physical Properties

The acid occurs both as colorless triclinic prisms (α-form) and as monoclinic prisms (β-form) (8). The β-form is triboluminescent and is stable up to 137°C; the α-form is stable above this temperature. Both forms dissolve in water, alcohol, diethyl ether, glacial acetic acid, anhydrous glycerol, acetone, and various aqueous mixtures of the last two solvents. Succinic acid sublimes with partial dehydration to the anhydride when heated near its melting point.

Succinic acid is absorbed from aqueous solutions by anion-exchange resins or active carbon (9–11). Succinic anhydride forms rhombic pyramidal or bipyramidal crystals. It is relatively insoluble in ether, but soluble in boiling chloroform and ethyl acetate. Succinic anhydride reacts with water and alcohols, giving the acid and monoesters, respectively.

Physical properties of the acid and its anhydride are summarized in Table 1. Other references for more data on specific physical properties of succinic acid are as follows: solubility in water at 278.15–338.15 K (12); water-enhanced solubility in organic solvents (13); dissociation constants in water–acetone (10 vol %) at 30–60°C (14), water–methanol mixtures (10–50 vol %) at 25°C (15,16), water–dioxane mixtures (10–50 vol %) at 25°C (15), and water–dioxane–methanol mixtures at 25°C (17); nucleation and crystal growth (18–20); calculation of the enthalpy of formation using semiempirical methods (21); enthalpy of solution (22,23); and enthalpy of dilution (23). For succinic anhydride, the enthalpies of combustion and sublimation have been reported (24).

Table 1. Physical Properties of Succinic Acid and Succinic Anhydride

Property	Succinic acid	Succinic anhydride
molecular weight	118.09	100.08
melting point, °C	188	119.6
boiling point at 101.3 kPa (=1 atm), °C	dehydration, 235	261
boiling/sublimation point, Pa[a], in °C		
133.3		92
267	156–157	
667		115
1,333		128
6,667		169
13,332		189
density, g/cm³		
solid, 20°C		1.2
solid, 25°C	1.572	
solubility, g/100 g solvent		
in water at 0°C	2.8	
in water at 100°C	121	
96% ethanol at 15°C	10	
ethyl ether at 15°C	1.2	
methylene chloride at bp	insoluble	6.6
chloroform at bp	insoluble	3.7
in water at 25°C		
pK_1	4.16	
pK_2	5.61	
enthalpy of combustion, kJ/mol[b]	−1491	−1537.9
enthalpy of formation at 298.15 K, kJ/mol[b]	−940.5	−607.8
heat capacity at 298.15 K, J/(mol·K)[b]	153	
enthalpy of solution, kJ/mol[b]	−27.3	
enthalpy of sublimation, kJ/mol[b]		80.7
enthalpy of fusion, kJ/mol[b]		20.41
dielectric constant at 3–97°C, 5 kHz	2.29–2.90	
flammability point, °C		147

[a]To convert Pa to mm Hg, divide by 133.3.
[b]To convert kJ/mol to kcal/mol, divide by 4.184.

Chemical Properties

Succinic acid and anhydride undergo most of the reactions characteristic of dicarboxylic acids and cyclic acid anhydrides, respectively. Other interesting reactions take place at the active methylene groups.

Heat.   When heated, succinic acid loses water and forms an internal anhydride with a stable ring structure. Dehydration starts at 170°C and becomes rapid at 190–210°C (25). Further heating of succinic anhydride causes decarboxylation and the formation of the dilactone of gamma ketopimelic acid (26) (eq. 1). The same reaction takes place at lower temperatures in the presence of alkali.

$$2 \quad \text{[anhydride]} \xrightarrow{\text{Na}^+} \text{[spiro dilactone]} + CO_2 \qquad (1)$$

At higher temperatures the presence of alkali causes an explosive reaction that does not stop at the bimolecular stage. Precautions must therefore be taken to exclude traces of alkali when handling succinic anhydride.

Succinic anhydride is stabilized against the deteriorative effects of heat by the addition of small amounts (0.5 wt %) of boric acid (27), the presence of which also decreases the formation of the dilactone of gamma ketopimelic acid (28). Compared with argon, CO_2 has an inhibiting effect on the thermal decomposition of succinic acid, whereas air has an accelerating effect (29,30).

Hydration and Dehydration.   Succinic anhydride reacts slowly with cold water and rapidly with hot water to give the acid. For this reason it must be carefully stored in anhydrous conditions. Succinic acid can be dehydrated to the anhydride by heating at 200°C, optionally in the presence of a solvent (31). Dehydration can also be performed with clay catalysis in the presence of isopropenyl acetate under microwave irradiation (32) or with bis(trichloromethyl) carbonate at room temperature (33).

Esterification.   Succinic anhydride reacts readily with alcohols to give monoesters of succinic acid, which are readily esterified to diesters by the usual methods (34–37).

Dimethyl succinate [*106-65-0*] (mp 19°C, bp 196°C at atmospheric pressure) can be produced from methanol and the anhydride or the acid, or by hydrogenation of dimethyl maleate (38,39). The same methods can be used to prepare diethyl succinate [*123-25-1*] (mp −18°C, bp 216.5°C at atmospheric pressure) and diisopropyl succinate [*924-88-9*].

Succinic acid diesters are also obtained by one-step hydrogenation (over Pd on charcoal) and esterification of maleic anhydride dissolved in alcohols (40); carbonylation of acrylates in the presence of alcohols and Co complex catalysts (41–43); carbonylation of ethylene in alcohol in the presence of Pd or Pd–Cu catalysts (44–50); hydroformylation of acetylene with Mo and W complexes in the presence of butanol (51); and a biochemical process from dextrose/corn steep liquor, using *Anaerobiumspirillum succiniciproducens* as a bacterium (52).

An important use of dialkyl succinates is in the preparation of dialkyl succinyl succinates (35,53–56), which are intermediates in the manufacture of quinacridone pigments. The reaction is carried out in the presence of alkali metal alkoxides (eq. 2).

$$2 \quad \text{[diester]} \xrightarrow{\text{MOR}} \text{[cyclohexane derivative]} + 2\,ROH \qquad (2)$$

In many applications succinic acid and anhydride are esterified with poly-hydric compounds, ie, polyols (57–59), cellulose (60), or starch (61–64).

One of the methods used to isolate succinic acid from the waste stream of the adipic acid process is esterification of the mixture of succinic, glutaric, and adipic acid followed by fractionation (65–69).

Oxidation. Succinic acid reacts with hydrogen peroxide, giving differ-ent products that depend on the experimental conditions: peroxysuccinic acid [2279-96-1] $(CH_2COOOH)_2$, oxosuccinic acid [328-42-7] (oxaloacetic acid); malonic acid [141-82-2], or a mixture of acetaldehyde, malonic acid, and malic acid [6915-15-7]. Succinic anhydride in dimethylformamide (DMF) with H_2O_2 gives monoperoxysuccinic acid [3504-13-0], $HOOCCH_2CH_2COOOH$, mp 107°C (70).

Potassium permanganate oxidizes succinic acid to a mixture of malic and tartaric acid [133-37-9]. 3-Hydroxypropionic acid [503-66-2] is obtained with sodium perchlorate. Cerium(IV) sulfate in sulfuric acid medium oxidizes succinic acid to oxaloacetic acid (71).

Hydrogenation. Gas-phase catalytic hydrogenation of succinic anhydride yields γ-butyrolactone [96-48-0] (GBL), tetrahydrofuran [109-99-9] (THF), 1,4-butanediol (BDO), or a mixture of these products, depending on the experimen-tal conditions. Catalysts mentioned in the literature include copper chromites with various additives (72), copper–zinc oxides with promoters (73–75), and ruthenium (76). The same products are obtained by liquid-phase hydrogenation; catalysts used include Pd with various modifiers on various carriers (77–80), Ru on C (81) or Ru complexes (82,83), Rh on C (79), Cu–Co–Mn oxides (84), Co–Ni–Re oxides (85), Cu–Ti oxides (86), Ca–Mo–Ni on diatomaceous earth (87), and Mo–Ba–Re oxides (88). Chemical reduction of succinic anhydride to GBL or THF can be performed with 2-propanol in the presence of ZrO_2 cata-lyst (89,90).

Halogenation. Succinic acid and succinic anhydride react with halogens through the active methylene groups. Succinic acid heated in a closed vessel at 100°C with bromine yields 2,3-dibromosuccinic acid almost quantitatively. The yield is reduced in the presence of excess water as a result of the formation of brominated hydrocarbons. The anhydride gives the mono- or dibromo derivative, depending on the equivalents of bromine used.

Succinyl chloride [543-20-4] is obtained from phosphorous pentachloride and succinic acid, from thionyl chloride and succinic acid or anhydride (91,92), or from phosgene and succinic anhydride (93).

Condensation with Aldehydes and Ketones. Succinic anhydride and suc-cinic esters in the presence of different catalysts react in the gas phase with formaldehyde to give citraconic acid or anhydride and itaconic acid (94–96). Di-alkyl acyl succinates are obtained by reaction of dialkyl succinates with C_{1-4} aldehydes over peroxide catalysts (97).

Succinic esters condense with aldehydes and ketones in the presence of bases, eg, sodium alkoxide or piperidine, to form monoesters of alkylidene-succinic acids, eg, condensation of diethyl succinate with acetone yields ethyl 2-isopropylidenesuccinate (eq. 3). This reaction, known as Stobbe condensation, is specific for succinic esters and substituted succinic esters (98,99).

$$\text{(3)}$$

Aromatic aldehydes (100), eg, cinnamaldehyde, and ketones (101) react in a similar manner (eq. 4). Ketones containing reactive methyl or methylene groups give with succinates, in the presence of sodium hydride, both the Stobbe condensation and the formation of diketones by a Claisen mechanism (102) (eq. 5).

$$\text{(4)}$$

$$\text{(5)}$$

Friedel-Crafts Reactions. In the presence of Friedel-Crafts catalysts, succinic anhydride reacts with alkyl benzenes to form alkylbenzoylpropionic acids (103), eg, the reaction with indane gives a 97% yield of 4-oxo-(4,5-indanyl)butyric acid (eq. 6).

$$\text{(6)}$$

Friedel-Crafts acylation of unsaturated fatty acids can be carried out with succinic anhydride in the presence of alkylaluminium halide (104).

Reactions with Nitrogen Compounds. Succinimide [123-56-8], mp 126°C, can be prepared by reaction of aqueous solutions of the acid with ammonia (105) or urea (106) (eq. 7). The solution is heated until water and ammonia are no longer evolved and the molten crude succinimide is purified by fractionation. Alternatively, the crude product can be recrystallized from water (105).

$$\begin{array}{c} \text{CH}_2\text{—C(=O)—ONH}_4 \\ \text{CH}_2\text{—C(=O)—ONH}_4 \end{array} \xrightarrow{\Delta} \text{(succinimide)NH} + 2\,\text{H}_2\text{O} + \text{NH}_3 \qquad (7)$$

N-Alkyl or *N*-aryl succinimides can be prepared from the corresponding amines (107) or from succinic anhydride, ammonia, and the corresponding alcohol (108). Succinimides are also obtained by vapor-phase hydrogenation of the corresponding maleimides in the presence of a catalyst (109).

Halogen-substituted succinimides are a class of products with important applications. *N*-Bromosuccinimide [*128-08-5*], mp 176–177°C, is the most important product in this group, and is prepared by addition of bromine to a cold aqueous solution of succinimide (110,111) or by reaction of succinimide with NaBrO$_2$ in the presence of HBr (112). It is used as a bromination and oxidation agent in the synthesis of cortisone and other hormones. By its use it is possible to obtain selective bromine substitution at methylene groups adjacent to double bonds without addition reactions to the double bond (113).

N-Chlorosuccinimide [*128-09-6*], mp 150–151°C, forms orthorhombic crystals and has a chlorine-like odor; it is prepared from succinimide and hypochlorous acid (114,115). Because of its powerful germicide properties, it is used in disinfectants for drinking water. Like its bromine derivative, it is also a halogenating agent.

Diamines react with two moles of succinic anhydride to give *N*,*N'*-disuccinimides (eq. 8).

$$\text{H}_2\text{NCH}_2\text{CH}_2\text{NH}_2 + 2\;(\text{succinic anhydride}) \xrightarrow{\Delta} (\text{succinimide})\text{NCH}_2\text{CH}_2\text{N}(\text{succinimide}) + 2\,\text{H}_2\text{O} \qquad (8)$$

Succinic acid reacts with urea in aqeous solution to give a 2:1 compound having mp 141°C (116,117), which has low solubility in water. A method for the recovery of succinic acid from the wastes from adipic acid manufacture is based on this reaction (118,119). The monoamide succinamic acid [*638-32-4*], NH$_2$COCH$_2$CH$_2$COOH, is obtained from ammonia and the anhydride or by partial hydrolysis of succinimide. The diamide succinamide [*110-14-5*], (CH$_2$CONH$_2$)$_2$, mp 268–270°C, is obtained from succinyl chloride and ammonia or by partial hydrolysis of succinonitrile. Heating succinimide with a primary amine gives *N*-alkylsuccinimides (eq. 9).

$$\text{(9)}$$

One-step manufacture of 2-pyrrolidinone can be effected by reaction of succinic acid or anhydride with NH_3 and H_2 in the presence of $Pd-Al_2O_3$ catalyst in water or ether (120).

Reactions with Sulfur Compounds. Thiosuccinic anhydride [3194-60-3] is obtained by reaction of diethyl or diphenyl succinate [621-14-7] with potassium hydrogen sulfide followed by acidification (eq. 10). Thiosuccinic anhydride is also obtained from succinic anhydride and hydrogen sulfide under pressure (121).

$$\text{(10)}$$

Sulfur trioxide reacts with both methylene groups to yield 2,3-disul-fosuccinic acid [54060-35-4] (eq. 11) (see SULFUR COMPOUNDS).

$$\text{(11)}$$

Miscellaneous Reactions. Radiolysis at room temperature of diluted aqueous solutions of succinic acid produces 1,2,3,4-butane tetracarboxylic acid [1703-58-8] (122), which has numerous industrial and agricultural applications (eq. 12).

$$\text{(12)}$$

Degradation. Heating of succinic acid or anhydride yields γ-ketopimelic dilactone, cyclohexane-1,4-dione, and a mixture of decomposition products that include acetic acid, propionic acid, acrylic acid, acetaldeide, acrolein, oxalic acid, cyclopentanone, and furane. In argon atmosphere, thermal degradation of succinic anhydride takes place at 340°C (123). Electrolysis of succinic acid produces ethylene and acetylene.

Manufacture and Processing

Succinic anhydride is manufactured by catalytic hydrogenation of maleic anhydride [108-31-6]. In the most widely used commercial process this reaction is performed in the liquid phase, at temperatures of 120–180°C and at moderate pressures, in the range of 500–4000 kPa (72–580 psi). Catalysts mentioned in the patent literature include nickel (124), Raney nickel (125,126), palladium on different carriers (127,128), and palladium complexes (129). The hydrogenation of the double bond is exothermic: $\Delta H = -133.89$ kJ/mol (-32 kcal/mol) (130).

The reactor is designed to provide efficient heat removal and a good contact of the gas (hydrogen) with the solid catalyst and the liquid reaction mixture. Depending on plant capacity, hydrogenation can be batch (124) or continuous. Continuous hydrogenation is performed in three-phase fixed-bed reactors (128) or in a cascade of two or three continuous-stirred-tank reactors with the catalyst suspended as a powder in the liquid reaction mixture. Standard precautions for hydrogenation reactions, eg, inertization with nitrogen at startup and shutdown and on-line continuous oxygen analyzers connected to interlocks, must be taken to prevent, at any time, the simultaneous presence in the hydrogenation reactor of oxygen and hydrogen. The yield of the hydrogenation reaction is virtually theoretical. Impurities detected in raw succinic anhydride include small amounts (300–2000 ppm each) of unconverted maleic anhydride, γ-butyrolactone, butyric acid, and propionic acid.

After separation of the catalyst by filtration, raw succinic anhydride is purified by distillation under reduced pressure, ie, 4–13 kPa (30–98 mm Hg), and flaked. The material of construction of the plant is stainless steel. Typical specific consumptions for the production of one metric ton of succinic anhydride are maleic anhydride at 1050 kg; hydrogen, 300 m^3; steam, 4500 kg; cooling water, 100 m^3; electricity, 350 kW; nitrogen, 100 m^3; and methane, 100 m^3. Effluents to be disposed of are hydrogen vent, low and high boiling by-products from the distillation unit, and exhausted catalyst.

In the early 1990s, processes were developed for the production of 1,4-butanediol and γ-butyrolactone by gas-phase catalytic hydrogenation of maleic anhydride (131–134). Succinic anhydride is obtained as a partial hydrogenation by-product in these processes. It can be recycled to complete the hydrogenation to the desired products, or be separated and purified. This process could in the future become a significant commercial route for succinic anhydride.

The simplest route to succinic acid is by hydration of its anhydride. Pure succinic anhydride is dissolved in hot water, succinic acid is formed, separated as crystals upon cooling, filtered, and dried.

Succinic acid can also be produced by catalytic hydrogenation of aqueous solutions of maleic or fumaric acid in the presence of noble metal catalysts, ie, palladium, rhodium, ruthenium, or their mixtures, on different carriers (135–139) or on Raney nickel (140).

A mixture of succinic (15–25 wt %), glutaric (45–55 wt %), and adipic acid (25–35 wt %) is obtained as a by-product in the oxidation of cyclohexane to adipic acid. In 1993, the production of adipic acid by this process was in the range of two million metric tons, which corresponds to a production of about 100,000 metric tons of the mixture of the three acids.

Various techniques have been proposed for the recovery of pure succinic acid, including extraction (141–145), selective crystallization (146–151), heating to dehydrate the acid and subsequent recovery of succinic anhydride by distillation (152), esterification followed by fractionation of the mixture of the esters (65–69), and separation as urea adduct (118,119).

Other preparations of succinic acid mentioned in the literature are electrochemical reduction of maleic or fumaric acid (153,154), ultrasound-promoted Zn–acetic acid reduction of maleic or fumaric acid (155), reduction of maleic acid with H_3PO_2 at room temperature (156), electrochemical reduction of CO_2 in the presence of ethylene (157), oxidation of furfural (158,159), oxidation of 1,4-butanediol catalyzed by $Pd-C/Pb(OOCCH_3)_2$ system (160), oxidation of coal by RuO_4 (161,162), carbonylation of acrylic acid in the presence of cobalt (163) or rhodium (164), CO, and H_2, and fermentation of glucose from wet milling of corn (165). None of these methods has found commercial application.

Succinic anhydride can be prepared from succinic acid by dehydration; it operates in high boiling solvent (31), in the presence of clays as a catalyst (32), or at room temperature with triphosgene (33).

Production and Shipment

The total consumption of succinic acid and succinic anhydride in 1990 was 1,500 t in the United States, 2,500 t in Europe, 7,500 t in Japan, and 1,500 t in other countries. Production was 500 t in the United States, 2,500 t in Europe, and 11,000 t in Japan. The total installed capacity is in the 18,000–20,000-t/yr range. The total consumption decreased slightly between 1990 and 1994 mainly because of the replacement of succinic acid by fumaric acid in bath preparations, which is one of the main uses of succinic acid in Japan. The principal producers are Buffalo Color in the United States, Lonza SpA and Chemie Linz in Europe, Kawasaki Kasei, Nippon Shokubai, Takeda Chemical, Kyowa Hakko, and New Japan Chemical in Japan.

Succinic acid and succinic anhydride are sold in 25-kg net polyethylene (PE) bags having cardboard box protection for the anhydride, in 70-liter (50-kg net) fiber drums, and in 55-gallon (275-lb; 125-kg net) drums. The two products must be stored in a fresh, dry, ventilated area. Succinic anhydride must be carefully protected from moisture during transportation and storage to avoid hydrolysis to succinic acid.

In 1994 the price of ton lots of succinic acid was in the $2–3/kg range, depending on the end use. The price of succinic anhydride was $2.5/kg, showing little variation over the previous five years and down from the 1981 price of $3.5/kg (166). The two products are not controlled by such transport regulations as International Maritime Dangerous Goods code (IMDG), International Civil Aviation Organization/International Air Transport Association (ICAO/IATA), Règlement

International Dangereuses (RID), or European Agreement of International transport of Dangerous goods by Road (ADR).

Specifications and Analysis

Commercial specifications of succinic acid and succinic anhydride are given in Table 2.

Various methods can be used to analyze succinic acid and succinic anhydride, depending on the characteristics of the material. Methods generally used to control specifications of pure products include acidimetric titration for total acidity or purity; comparison with Pt–Co standard calibrated solutions for color; oxidation with potassium permanganate for unsaturated compounds; subtracting from the total acidity the anhydride content measured by titration with morpholine for content of free acid in the anhydride; atomic absorption or plasma spectroscopy for metals; titration with $AgNO_3$ or $BaCl_2$ for chlorides and sulfates, respectively; and comparison of the color of the sulfide solution of the metals with that of a solution with a known Pb content for heavy metals.

Techniques used for the determination of small concentrations of succinic acid or anhydride in various substances include gc or capillary gc in fruit and vegetables (167), eggs (168), rain (169), and liquors from wood (170); ion chromatography in rainwater (171), streams from the adipic acid process (172),

Table 2. Specifications of Succinic Acid and Succinic Anhydride

Property	Succinic acid	Succinic anhydride
melting point, °C, min	185.0	
solidification point, °C, min		118.3
physical appearance	small white crystals	white flakes
color (APHA), molten, max		100
total acidity (purity), %, min	99.5[a]	99.5[b]
free acidity (as succinic acid), %, max		1
water content (Karl Fisher), %, max	1.0	
unsaturated compounds, %, max	0.2[c]	0.2[d]
ash, %, max	0.1	
chlorides, ppm, max		100
iron, ppm, max	5	
lead, ppm, max	5	
arsenic, ppm, max		3
sulfates (as SO_4), ppm, max	500	400
heavy metals (as Pb), ppm, max	10	10
fineness, <2 mm (<10 mesh), %	100	
appearance of solution (1 g in 10-mL H_2O at 50°C)	clear, colorless	
organic impurity (1 g in 10-mL H_2SO_4 concentration)	clear, colorless	

[a]As succinic acid.
[b]As succinic anhydride.
[c]As maleic acid.
[d]As maleic anhydride.

and food (173); gc/ms in amber (174); hplc in wine and champagne (175,176), vinegar (177), beverages (178), apples (179), as well as milk and cheese (180,181); and polarography in γ-butyrolactone (182). Succinic acid can be separated from other organic acids by liquid chromatography by using cation-exchange resins as a stationary phase (183).

Health and Safety Factors

Succinic acid is Generally Recognized As Safe (GRAS) by the U.S. FDA (184) and is approved as a flavor enhancer, as a pH control agent in condiments, and for use in meat products. It causes irritation to the eyes (185), skin, mucous membranes, and upper respiratory tract. LD_{50} in rat is 2260 mg/kg. Succinic acid, like most materials in powder form, can cause dust explosion.

Succinic anhydride is extremely irritating to the eyes. It causes skin, mucous membranes, and respiratory tract irritation. It may be a sensitizer. There is no evidence of carcinogenic activity in male or female rats given 50 or 100 mg/kg succinic anhydride (186); the Ames test is negative (187). LD_{50} in rat 1510 mg/kg. There are no established exposure limits for ACGIH TLV or TWA.

Succinic acid and anhydride should be handled with rubber or plastic gloves; safety goggles and a dust filter are recommended when handling the products in powder form. A full-face gas mask with a type A (brown) filter cartridge should be worn when handling molten products.

Incineration in an approved combustion plant is the preferred method of disposal. Wastewater from succinic acid processes is suitable for biological degradation by activated sludge (188). Polymeric sorbents (189) and ferric chloride treatment processes (190) can also be used for wastes containing succinic acid. Chemical oxygen demand has been determined by the permanganate method (191).

Uses

Table 3 summarizes many of the uses mentioned in the literature. The main use of succinic acid in Japan is for bath preparations (314–322). This application in 1994 accounted for nearly 80% of total consumption. After recording a more than 10% yearly increase in the late 1980s, the growth of this application has slowed down, and consumption is decreasing on account of the replacement of succinic acid by fumaric acid for economic reasons. This trend is expected to continue in the coming years.

An important emerging use of succinic acid and anhydride is the production of inherently degradable polymers (388–395), which are enjoying a very high rate of growth: over 10%/yr in the early 1990s (see POLYMERS, ENVIRONMENTALLY DEGRADABLE). Other important applications are in the fields of food additives (qv), detergents, cosmetics (qv), pigments (qv), toners, cement additives, soldering fluxes, as well as in the synthesis of pharmaceutical products.

Table 3. Uses of Succinic Acid and Succinic Anhydride

Use	Reference
Adhesives and sealants	
hot-melt adhesives	192
adhesive primers for aluminized surfaces	193
Agriculture	
plant growth regulators	194–197
herbicide compositions	194, 198, 199
fungicides	107, 200, 201
pesticidal effervescent granules	202
insecticides	203
mosquito attractants	204
Building and construction	
cement additives and cement compositions	205–215
asphalt paving materials	216
Ceramics	
manufacture of porous titanium oxide	217
manufacture of sinterable boehmite powder	218
manufacture of porous ceramics	219
Coating, pigments, dyes, inks	
coating compositions	220–225
automobile topcoats	226, 227
powder coating	228, 229
radiations, uv-curable coating compositions	224, 230, 231
photocurable ink compositions	232
toners	233, 234
quinacridone pigments	35, 53–56, 235, 236
dye intermediates	237, 238
Corrosion inhibitors	
for steel, copper, and alloys	239–241
for coatings	242, 243
for antifreeze	244
for metalworking oils	245
for lead in perchloric acid	246
boiler-water treatment	247
Detergents, cleaning agents, emulsifiers, antiscaling agents	
dishwashing detergent compositions	248, 249
fabric incrustation prevention and softeners	250, 251
liquid laundry detergents	252, 253
hard-surface detergent compositions	254–259
acid cleaning compositions	260–263
bleaching compositions	264–266
biodegradable aqueous filter cleaning compositions	267
solid soaps	268
detergent powders and tablets	269–271
granular detergent compositions	272
Electric and electronic	
solution for electrolytic capacitors	273–275

Table 3. (*Continued*)

Use	Reference
cleaner for printed circuit boards	276
gettering compounds	277

Electrochemistry

Use	Reference
electrochemical stripping of silver	278
electrodialysis	279
electrochemical graining of aluminum	280

Food

Use	Reference
disodium succinate used as food seasoning	281
condiments	282
food preservative	283
carbon dioxide-generating compositions for cold drinks	284
beverages prepared from apricot or plum purees	285

Household products

Use	Reference
deodorants for air	286–290
deodorization, sanitation of toilets	291–295

Lithography

Use	Reference
photosensitive lithographic plates	296–299
diazo photosensitive compositions	300
aqueous developer for lithographic printing plates	301

Lubricants

Use	Reference
synthetic lubricants	37
metalworking lubricating oils	302
lubricants for thermoplastics	303
lubricant compositions for electrical contacts	304

Metallurgy

Use	Reference
ore processing	305
etching agents	306, 307
manufacture of nickel powder	308
bath for polishing steel	309, 310
binder compositions in sintering of steels	311

Paper

Use	Reference
deinking agents for recycling waste paper	312
paper sizing composition	313

Personal care

Use	Reference
effervescent bath tablets	314–322
denture cleaners and mouthwashes	323–327
cosmetic compositions and moisturizers	328–332
adhesive sheets for application to the skin	333
hair rinses	334, 335
cleansing solution for contact lenses	336
deodorant napkins for hands	337

Pharmaceuticals

Use	Reference
effervescent tablets, programmed drug release	339–345
antiseptic tablets	346, 347
disinfectants and antibacterials	348, 349

Table 3. (*Continued*)

Use	Reference
ischemia protection	350
treatment of wounds	351
bioabsorbable sutures	352
hardening composition for dental materials	353
drug detection reagent	354
buffer for serum	355
derivatives in pharmaceuticals	356–360

Photography

in bleaching or bleach fixing solutions	361–364

Plating

electroplating baths for Ni, Zn, Cr, Au,	365–370
electroless plating for Ag, Ni, Ni–P, Ni–Mo–P, Co, Cu, Pd	371–383
electroless plating of magnetic materials	384–386
electroless Ni coating of piezoceramics	387

Polymers and resins

biodegradable polymers	388–395
biodegradable packaging foams	396
curing agent for epoxy resins	397–401
epoxy elastomers	402
preparation of liquid crystals	403
modified polyamides	404
water-soluble polymers	405

Reagent in chemical synthesis

dehydrating agent	406
acylating agent	407–409
oligonucleotide synthesis	410, 411

Soldering

pastes and fluxes	412–428

Textiles and fibers

finishes for silk or cellulosic fabrics	429–432
dyeing aids for cotton or polyamide fibers	433–435
succinylation of silk	436
wet strengthening agents	437
adhesive composition for textile laminates	438
textile bleaching agents	439

Water and gas treatment

bactericide tablets for swimming pools	440
algicide compositions	441
treatment of water for use in beauty parlor	442
flue gas desulfurization	443, 444
air purification and disinfection	445–447

Miscellaneous

catalyst manufacture	448
fireproofing aids	449
wood treatment	450

Table 3. (*Continued*)

Use	Reference
Miscellaneous	
superconductor manufacture	451
electrolyte for lithium batteries	452–455
leather manufacture	456
animal feeds	457

BIBLIOGRAPHY

"Succinic Acid and Succinic Anhydride" in *ECT* 1st ed., Vol. 13, pp. 180–191, by W. H. Gardner and L. H. Flett, Allied Chemical & Dye Corp.; in *ECT* 2nd ed., Vol. 19, pp. 134–150, by P. Turi, Sandoz, Inc.; in *ECT* 3rd ed., Vol. 21, pp. 848–864, by L. O. Winstrom, Consultant.

1. W. Zurburg, *Adattamenti Biochim. Anim. Acquatici Carenza Ossigeno, Atti Incontro Biochim Mar.* **1**, 39–66 (1982).
2. I. Calvarano, M. Calvarano, and G. Di Giacomo, *Essenze Deriv. Agrum.* **59**(1), 5–17 (1989).
3. D. H. Picha, *J. Agric. Food Chem.* **33**(4), 743–745 (1985).
4. K. Kawamura and A. Nissenbaum, *Org. Geochem.* **18**(4), 469–476 (1992).
5. E. T. Peltzer and co-workers, *Adv. Space Res.* **4**(12), 69–74 (1984).
6. F. Radler, *Microbiol. Ind. Aliment. Ann. Congr. Int.* **2**, 5–16 (1980).
7. J. S. Mills, R. White, and L. J. Gough, *Chem. Geol.* **47**(1–2), 15–39 (1984).
8. N. N. Petropavlov and S. B. Yarantsev, *Kristallografiya*, **28**(6), 1132–1139 (1983).
9. C. Y. C. Lee, O. E. Pedram, and A. L. Hines, *J. Chem. Eng. Data*, **31**(2), 133–136 (1986).
10. H. Li and co-workers, *Taiyuan Gongxueyuan Xuebao*, (2), 59–67 (1984).
11. Y. Onal and Z. Tez, *Doga: Turk Kim. Derg.* **13**(1), 49–56 (1989).
12. A. Apelblat and E. Manzurola, *J. Chem. Thermodyn.* **19**(3), 317–320 (1987).
13. J. N. Starr and C. J. King, *Energy Res. Abstr.* **17**(8), Abstr. No. 21954 (1992).
14. M. F. Amira, S. A. El-Shazly, and M. M. Khalil, *Thermochim. Acta*, **115**, 1–10 (1987).
15. G. Papanastasiou, I. Ziogas, and D. Jannakoudakis, *Chem. Chron.* **15**(3), 147–160 (1986).
16. G. Papanastasiou, G. Stalidis, and D. Jannakoudakis, *Bull. Soc. Chim. Fr.* 1984, (9–10) 255–259 (1984).
17. G. Papanastasiou and I. Ziogas, *Bull. Soc. Chim. Fr.* (5), 725–730 (1985).
18. Y. Qiu and A. C. Rasmuson, *AIChE J.* **37**(9), 1293–1304 (1991).
19. Y. Qiu and A. C. Rasmuson, *AIChE J.* **36**(5), 665–676 (1990).
20. J. W. Mullin and M. J. L. Whiting, *Ind. Eng. Chem. Fundam.* **19**(1), 117–121 (1980).
21. C. B. Aakeroy, *Theochem*, **100**(2–3), 259–267 (1992).
22. A. Apelblat, *J. Chem. Thermodyn.* **18**(4), 351–357 (1986).
23. S. Taniewska-Osinska, M. Tkaczyk, and A. Apelblat, *J. Chem. Thermodyn.* **22**(7), 715–720 (1990).
24. M. Y. Yang and G. Pilcher, *J. Chem. Thermodyn.* **22**(9), 893–898 (1990).
25. Technical data, Lonza S.p.A., Italy, 1996.
26. U.S. Pat. 2,302,321 (Nov. 17, 1942), H. Hopff and H. Griesshaber (to Alien Property Custodian).
27. U.S. Pat. 4,257,958 (Mar. 24, 1981), J. C. Powell (to Texaco Inc.).
28. Technical data, Lonza S.p.A., Italy, 1996.
29. K. Muraishi and Y. Suzuki, *Thermochim. Acta*, **232**(2), 195–203 (1994).

30. K. Muraishi, Y. Suzuki, and A. Kikuchi, *Thermochim. Acta,* **239**(1–2), 51–59 (1994).

31. Ger. Offen. 2,319,574 (Nov. 14, 1974), W. Mesch and A. Wittwer (to BASF AG).

32. D. Villemin, L. Bouchta, and A. Loupy, *Synth. Commun.* **23**(4), 419–424 (1993).

33. R. Kocz, J. Roestamadji, and S. Mobashery, *J. Org. Chem.* **59**(10), 2913–2914 (1994).

34. H. J. Bart and co-workers, *Int. J. Chem. Kinet.* **26**(10), 1013–1021 (1994).

35. Jpn. Kokai Tokkyo Koho, JP 04091055 (Mar. 24, 1992), K. Osada and S. Nishama (to Shinnippon Rika K.K.).

36. M. M. Amirkhanyan and M. F. Elanyan, *Tr. IREA,* **46**, 7–10 (1984).

37. I. A. El-Magly, E. S. Nasr, and M. S. El-Samanoudy, *J. Synth. Lubr.* **7**(2), 89–103 (1990).

38. Austrian Pat. AT 383,115 (May 25, 1987), H. Hornich and J. Schaller (to Chemie Linz AG).

39. Ger. Offen. DE 3,503,485 (Aug. 7, 1986), J. Schaller and H. Hornich (to Lentia GmbH).

40. Hung. Pat. 41727 (May 28, 1987), L. Torkos and co-workers (to Nitroil Vegyipari Termelo-Fejleszto Kozos Vallalat).

41. Ger. Offen. DE 3,332,018 (March 21, 1985), J. G. Reuvers, W. Richter, and R. Kummer (to BASF AG).

42. Jpn. Kokai Tokkyo Koho JP 06199736 (July 19, 1994), A. Matsuda and co-workers (to Kogyo Gijutsuin).

43. Jpn. Kokai Tokkyo Koho JP 01254641 (Oct. 11, 1989), N. Okada and O. Takahashi (to Idemitsu Kosan Co., Ltd.).

44. Jpn. Kokai Tokkyo Koho JP 81,110,645 (Sept. 1, 1981) (to Ube Industries Ltd).

45. Eur. Pat. Appl. EP 163,442 (Dec. 4, 1985), R. A. Lucy and G. E. Morris (to British Petroleum Co. PLC).

46. Eur. Pat. Appl. EP 231,044 (Aug. 5, 1987), E. Drent and A. J. M. Breed (to Shell Internationale Research Maatschappij B.V.).

47. U.S. Pat. 4,827,023 (May 2, 1989), C. Y. Hsu (to Sun Refining and Marketing Co.).

48. U.S. Pat. 4,667,053 (May 19, 1987), J. J. Lin (to Texaco Inc.).

49. G. E. Morris and co-workers, *J. Chem. Soc., Chem. Commun.* (6), 410–411 (1987).

50. Jpn. Kokai Tokkyo Koho JP 57042653 (Mar. 10, 1982) (to Asahi Chem. Ind. Co.).

51. L. G. Bruk and co-workers, *Izv. Akad. Nauk SSSR, Ser. Khim,* (1), 232–234 (1983).

52. Eur. Pat. Appl. EP 389,103 (Sept. 26, 1990), D. A. Glassner and R. Datta (to Michigan Biotechnology Institute).

53. Ger. Offen. DE 3320415 (Dec. 6, 1984), H. Heise, M. Hintzmann, and K. Brueckmann (to Hoechst AG).

54. Eur. Pat. Appl. EP 536083 (Apr. 7, 1993), C. D. Campbell and E. E. Jaffe (to CIBA-GEIGY AG).

55. Ger. Offen. DE 3104644 (Aug. 19, 1982), M. Rolf and co-workers (to Bayer AG).

56. Jpn. Kokai Tokkyo Koho JP 58 85,843 (May 23, 1983) (to Mitsui Toatsu Chemicals, Inc.).

57. U.S. Pat. 3,329,635 (July 4, 1967), T. J. Miranda (to O'Brien Corp.).

58. K. Geckeler and E. Bayer, *Polym. Bull.* **3**(6–7), 347 (1980).

59. Ger. Offen. 2,012,526 (Oct. 7, 1971), D. Stoye and co-workers (to Chemische Werke Huels AG).

60. J. A. Cuculo, *Text. Res. J.* **41**(5), 375 (1971).

61. Ger. Offen. 2,048,350 (May 27, 1971), F. J. Germino and co-workers (to CPC International, Inc.).

62. P. C. Trubiano, in O. B. Wurzburg, ed., *Modified Starches: Properties and Uses*, CRC Press, Inc., Boca Raton, Fla., pp. 131–147, 1987.

63. U.S. Pat. 4,061,611 (Dec. 6, 1977), R. C. Glowsky and co-workers (to Sherwin-Williams Co.).

64. U.S. Pat. 4,231,803 (Nov. 4, 1980), E. M. Bovier and co-workers (to Anheuser-Busch, Inc.).

65. U.S. Pat. 4,105,856 (Aug. 8, 1978), C. A. Newton (to El Paso Products Co.).

66. U.S. Pat. 4,316,775 (Feb. 23, 1982), W. D. Nash (to El Paso Products Co.).

67. U.S. Pat. 4,365,080 (Dec. 21, 1982), N. F. Cywinski (to El Paso Products Co.).

68. U.S. Pat. 4,442,303 (Apr. 10, 1984), S. S. Mims (to El Paso Products Co.).

69. U.S. Pat. 3,991,100 (Nov. 9, 1976), S. Hochberg (to E. I. du Pont de Nemours & Co., Inc.).

70. Brit. Pat. 1,139,507 (Jan. 8, 1969), E. G. E. Hawkins (to Distiller Co. Ltd.).

71. B. M. Rao, T. P. Sastry, and T. S. Parekh, *Z. Phys. Chem.* (*Leipzig*), **264**(5), 906–919 (1983).

72. Eur. Pat. Appl. EP 373,947 (June 20, 1990), S. Suzuki, H. Inagaki, and H. Ueno (to Tonen Co., Ltd.).

73. Jpn. Kokai Tokkyo Koho JP 02233627 (Sept. 17, 1990), S. Sadakatsu, H. Inagaki, and H. Ueno (to Tonen Co. Ltd.).

74. Jpn. Kokai Tokkyo Koho JP 47023294 (June 29, 1972), M. Yamaguchi and T. Okano (to Mitsubishi Rayon Co., Ltd.).

75. U.S. Pat. 4,810,807 (Mar. 7, 1989), J. R. Budge and S. E. Pedersen (to Standard Oil Co.).

76. U.S. Pat. 4,301,077 (Nov. 17, 1981), F. A. Pesa and A. M. Graham (to Standard Oil Co.).

77. Jpn. Kokai Tokkyo Koho JP 05,148,254 (June 15, 1993), T. Fuchigami and co-workers (to Tosoh Corp.; Sagami Chem. Res.).

78. Jpn. Kokai Tokkyo Koho JP 05,222,022 (Aug. 31, 1993), T. Fuchigami and co-workers (to Tosoh Corp.; Sagami Chem. Res.).

79. PCT Int. Appl. WO 92 02,298 (Feb. 20, 1992), R. E. Ernst and J. B. Michel (to E. I. du Pont de Nemours and Co., Inc.).

80. Jpn. Kokai Tokkyo Koho JP 62,111,974 (May 22, 1987), H. Wada and co-workers (to Mitsubishi Chemical Industries Co., Ltd.).

81. Jpn. Kokai Tokkyo Koho JP 06,157,491 (June 3, 1994), T. Fuchigami and co-workers (to Tosoh Corp.).

82. Eur. Pat. Appl. EP 420062 (Apr. 3, 1991), C. Miyazawa and co-workers (to Mitsubishi Kasei Corp.).

83. Eur. Pat. Appl. EP 453,948 (Oct. 30, 1991), Y. Hara and H. Inagaki (to Mitsubishi Kasei Corp.).

84. Ger. Offen. DE 3,726,509 (Feb. 16, 1989), R. Fisher and co-workers (to BASF AG).

85. U.S. Pat. 5,086,030 (Feb. 4, 1992), G. Bjornson and J. J. Stark (to Phillips Petroleum Co.).

86. U.S. Pat. 4,929,777 (May 29, 1990), G. Irick Jr., P. N. Mercer, and K. E. Simmons (to Eastman Kodak Co.).

87. Jpn. Kokai Tokkyo Koho JP 01,143,865 (June 6, 1989), S. Takigawa and co-workers (to Nissan Chemical Industries, Ltd.; Nissan Girdler Catalyst Co., Ltd.).

88. Jpn. Kokai Tokkyo Koho 63,088,045 (Apr. 19, 1988), K. Hasegawa, T. Aoki, and S. Seo (to Mitsubishi Petrochemical Co., Ltd.).

89. Jpn. Kokai Tokkyo Koho JP 04,154,776 (May 27, 1992), K. Takahashi, M. Shibagaki, and H. Matsushita (to Japan Tobacco, Inc.).

90. K. Takahashi, M. Shibagaki, and H. Matsushita, *Bull. Chem. Soc. Jpn.* **65**(1), 262–266 (1992).

91. Faming Zhuanli Shenquing Gongkai Shoumingshu CN 1,062,346 (July 1, 1992), J. Zhu, X. Sima, and Z. Wang (to Shaanxi Provincial Institute of Medical Industry).

92. C. G. Calin and co-workers, *Ind. Usoara* **36**(9), 407–411 (1989).

93. U.S. Pat. 3,810,940 (May 14, 1974), C. F. Hauser (to Union Carbide Corp.).

94. Ger. Offen. 2,352,468 (May 30, 1974), R. Berg and co-workers (to Pfizer, Inc.).

95. Jpn. Kokai JP 49101327 (Sept. 25, 1974), T. Shimizui and C. Fujii (to Denki Kagaku Kogyo KK).

96. Jpn. Kokai JP 49101326 (Sept. 25, 1974), T. Shimizui and C. Fujii (to Denki Kagaku Kogyo KK).

97. Jpn. Kokai Tokkyo Koho JP 61,243,047 (Oct. 29, 1986), H. Yoshida and co-workers (to Nippon Shokubai Kagaku Kogyo Co., Ltd.).

98. H. Stobbe, *Ber.* **26**, 2312 (1893).

99. W. S. Johnson and G. H. Daub, in R. Adams and co-workers, eds., *Organic Reactions*, Vol. 6, John Wiley & Sons, Inc., New York, 1951, pp. 2–73.

100. J. M. Lawlor and M. B. McNamee, *Tetrahedron Lett.* **24**(21), 2211–2212 (1983).

101. Czech. Pat. CS 225,399 (June 30, 1985), J. Krepelka and M. Simonova.

102. G. H. Daub and W. S. Johnson, *J. Am. Chem. Soc.* 1950, **72**, 501–504 (1950).

103. G. Peto, in G. A. Olah, ed., *Friedel-Krafts and Related Reactions*, Vol. 3, Pt. 1, Wiley-Interscience, New York, 1964, pp. 550–663.

104. U. Biermann and J. O. Metzger, *Fett. Wiss. Technol.* **94**(9), 329–332 (1992).

105. Jpn. Kokai Tokkyo Koho JP 04,282,361 (Oct. 7, 1992), M. Fukuhara and S. Oda (to Daihaci Chemical Industry Co., Ltd.).

106. Israel Pat. 38,852 (Nov. 29, 1974), A Stern and co-workers (to Makhteshim Chemical Works Ltd.).

107. Jpn. Kokai Tokkyo Koho JP 60 06,659 (Jan. 14, 1985) (to Showa Denko KK).

108. U.S. Pat. 4,841,069 (June 20, 1989), R. J. Olsen (to Amoco Corp.).

109. U.S. Pat. 4,851,546 (July 25, 1989), A. M. Graham and T. G. Attig (to Standard Oil Co.).

110. L. Horner and E. H. Winkelmann, *Angew. Chem.* **71**, 349 (1959).

111. Jpn. Kokai Tokkyo Koho JP 06056772 (Mar. 1, 1994), A. Eto and T. Sakai (to Sogo Yatsuko KK).

112. S. Kajigaeshi and co-workers, *Bull. Chem. Soc. Jpn.* **58**(2), 769–770 (1985).

113. J. H. Incremona, *Diss. Abstr.* **B27**(7), 2295 (1967).

114. Jpn. Kokai Tokkyo Koho JP 04,282,362 (Oct. 7, 1992), M. Fukuara and S. Oda (to Daihachi Chemical Industry Co., Ltd.).

115. D. Matte and co-workers, *Can. J. Chem.* **70**(1), 89–99 (1992).

116. C. H. Walker, *Sep. Sci.* **2**(3), 399 (1967).

117. H. Wiedenfeld and F. Knock, *Acta Crystallogr. Sect. C: Cryst. Struct. Commun.* **C46**(6), 1038–1040 (1990).

118. Jpn. Kokai Tokkyo Koho JP 81, 10,897 (Mar. 11, 1981), (to Asahi Chemical Industry Co., Ltd.).

119. Jpn. Kokai Tokkyo Koho JP 81, 10,898 (Mar. 11, 1981), (to Asahi Chemical Industry Co., Ltd.).

120. U.S. Pat. 4,904,804 (Feb. 27, 1990), M. S. Matson (to Phillips Petroleum Co.).

121. T. Takido and co-workers, *Yuki Gosei Kagaku Kyokai Shi*, **31**(10), 826 (1973).

122. A. Negron-Mendoza and G. Albarran, *Radiat. Phys. Chem.* **42**(4–6), 973–976 (1993).

123. T. Ando, Y. Fujimoto, and S. Morisaki, *J. Hazard. Mater.* **28**(3), 251–280 (1991).

124. Faming Zhuanli Shenqing Gongkai Shuomingshu CN 1,078,716 (Nov. 24, 1993), X. Ma (to Weinan Science and Technology Information Institute).

125. Faming Zhuanli Shenqing Gongkai Shuomingshu CN 1,063,484 (Aug. 12, 1992), X. Tang.

126. Czech. Pat. CS 218,083 (June 15, 1984), J. Sraga and P. Hrnciar.

127. Rus. Pat. 721,406 (Mar. 15, 1980), A. Avots,V. Kuplenieks, and D. Kreille (to Institute of Organic Synthesis Academy of Sciences Latvian SSR).

128. P. Ruiz and co-workers, *Chemical Catalyst Reaction Modeling*, ACS, Washington, D.C., 1984, pp. 15–36.

129. Rus. Pat. SU 1,541,210 (Feb. 7, 1990), G. A. Tolstikov and co-workers (to Bashkir Institute of Chemistry).

130. S. Minoda and M. Miyajima, *Hydrocarbon Process.* **49**, 176–178 (Nov. 1970).

131. G. L. Castiglioni and co-workers, *Chem. Ind.* 510–511 (July 1993).

132. G. L. Castiglioni and co-workers, *Erdoel Kohle*, **47**(4), 146–149 (1994).

133. G. L. Castiglioni and co-workers, *Erdoel Khole*, **47**(9), 337–341 (1994).

134. U.S. Pat. 5,196,602 (Mar. 23, 1993), J. R. Budge, T. G. Attig, and A. M. Graham (to Standard Oil Co.).

135. Ger. Auslegeschrift 1,259,869 (Feb. 1, 1968), W. Schmitz (to C. Still).

136. J. Hanika, J. Horych, and V. Ruzicka, *Sb. Vys. Sk. Chem.-Technol. Praze, Org. Chem. Technol.* **C28**, 35–47 (1983).

137. Jpn. Kokai Tokkyo Koho JP 61,204,148 (Sept. 10, 1986), K. Matsuzaki, S. Ikemoto, and M. Narita (to Kawasaki Kasei Chemicals, Ltd.).

138. Jpn. Kokai Tokkyo Koho JP 61,204,149 (Sept. 10, 1986), K. Matsuzaki and co-workers (to Kawasaki Kasei Chemicals, Ltd.).

139. Jpn. Kokai Tokkyo Koho JP 02,121,946 (May 9, 1990), K. Fujita, M. Kokubu, and M. Takeda (to Kyowa Yuka Co., Ltd.).

140. Rom. Pat. RO 79,020 (June 30, 1982), S. D. Paucescu and co-workers (to Intreprinderea Chimica Dudesti).

141. U.S. Pat. 2,840,607 (June 6, 1958), E. C. Attane Jr. and co-workers (to Union Oil Co.).

142. U.S. Pat. 2,870,203 (Jan. 20, 1959), A. D. Cyphers Jr. and A. A. Gruber (to E. I. du Pont de Nemours & Co., Inc.).

143. U.S. Pat. 3,329,712 (July 4, 1967), D. E. Danley and G. L. Whitesell (to Monsanto Co.).

144. Brit. Pat. 1,216,844 (Dec. 23, 1970), W. Bowyer and co-workers (to ICI Ltd.).

145. Czech. Pat. 139,128 (Nov. 15, 1970), M. Polievka and co-workers.

146. U.S. Pat. 3,338,959 (Aug. 29, 1967), C. T. Sciance and L. S. Scott (to E. I. du Pont de Nemours & Co., Inc.).

147. Brit. Pat. 1,366,933 (Sept. 18, 1974), J. E. Lambert (to ICI Ltd.).

148. Pol. Pat. 51,539 (July 20, 1966), L. Dworakowski and co-workers (to Spoldzielnia Pracy Chemikow Argon).

149. Pol. Pat. 63,909 (Nov. 20, 1971), S. Ciborowski (to Instytut Chemii Przemyslowej).

150. Rus. Pat. 405,861 (Nov. 5, 1973), I. Ya. Lubyanitskii and co-workers.

151. Jpn. Kokai 77 19,618 (Feb. 15, 1977), (to BP Chemicals Ltd).

152. PCT Int. Appl. WO 88 04,650 (June 30, 1988), J. B. Copper (to BP Chemicals Ltd.).

153. Pol. PL 148,885 (Mar. 31, 1990), J. Gora, K. Smigielski, and J. Kula (to Politechnika Lodzka).

154. R. Fujita and T. Sekine, *Denki Kagaku oyobi Kogyo Butsuri Kagaku* **50**(6), 491–493 (1982).

155. A. P. Marchand and G. M. Reddy, *Synthesis*, (3), 198–200 (1991).

156. X. Wang and Y. Guan, *Chin. Chim. Lett.* **4**(5), 407–408 (1993).

157. G. I. Shul'zhenko and Y. B. Vasil'ev, *Izv. Akad. Nauk SSSR, Ser. Khim.* (6), 1377–1380 (1991).

158. L. A. Badovskaya and co-workers, *Khim. Sel'sk Khoz.* (1), 59–60 (1990).

159. M. J. N. Garcia and M. B. Iglesias, *Acta Cient. Compostelana*, **22**(2–4), 809–817 (1985).

160. M. Akada and co-workers, *Bull. Chem. Soc. Jpn.* **66**(5), 1511–1515 (1993).

161. L. M. Stock and S. H. Wang, *Fuel*, **65**(11), 1552–1562 (1986).

162. E. S. Olson, J. W. Diehl, and M. L. Froehlich, *Fuel Process. Technol.* **15**, 319–326 (1987).

163. Jpn. Kokai Tokkyo Koho JP 02,06,427 (Jan. 10, 1990), N. Okada and O. Takahashi (to Idemitsu Kosan Co., Ltd.).

164. Eur. Pat. Appl. EP 188,209 (July 23, 1986), P. M. Burke (to E. I. du Pont de Nemours and Co., Inc.).
165. *Chem. Mark. Rep.* **246**, 5 (Oct. 10, 1994).
166. *Chem. Mark. Rep.* **232**, 14 (Mar. 30, 1981).
167. M. Morvai and I. Molnar-Perl, *Chromatographia*, **34**(9–10), 502–504 (1992).
168. S. Littmann, E. Schulte, and L. Acker, *Z. Lebensm.-Unters. Forsch.* **175**(2), 101–105 (1982).
169. K. Kawamura, S. Steinberg, and I. R. Kaplan, *Int. J. Environ. Anal. Chem.* **19**(3), 175–188 (1985).
170. R. Alen, K. Niemela, and E. Sjostrom, *J. Chromatogr.* **301**(1), 273–276 (1984).
171. M. Matsumoto, *Taiki Osen Gakkaishi*, **23**(1), 64–71 (1988).
172. J. C. Tompsen and R. H. Smith, *Process Control Qual.* **2**(1), 55–58 (1992).
173. I. Yoshida, K. Hayakawa, and M. Miyazaki, *Eisei Kagaku*, **31**(5), 317–323 (1985).
174. G. C. Galletti and R. Mazzeo, *Rapid. Commun. Mass Spectrom.* **7**(7), 646–650 (1993).
175. J. Wang and co-workers, *Sepu*, **11**(3), 183–185 (1993).
176. D. Tusseau and C. Benoit, *J. Chromatogr.* **395**, 323–333 (1987).
177. Y. Maeda and co-workers, *Shizuoka-ken Eisei Kankyo Senta Hokoku*, **28**, 47–50 (1985).
178. R. M. Marce and co-workers, *Chromatographia*, **29**(1–2), 54–58 (1990).
179. G. D. Blanco and co-workers, *Chromatographia*, **25**(12), 1054–1058 (1988).
180. K. Shimazaki and co-workers, *Obihiro Chikusan Daigaku Gakujutsu Kenkyu Hokoku, Dai-1-Bu*, **15**(2), 101–105 (1987).
181. G. Panari, *Milchwissenschaft*, **41**(4), 214–216 (1986).
182. D. Sun and X. Li, *Fenxi Huaxue*, **11**(2), 138–139 (1983).
183. Jpn. Kokai Tokkyo Koho JP 05,38,455 (Feb. 19, 1993), S. Ishiguro and S. Shioda (to Showa Denko KK).
184. *Fed. Reg.* **44**(68), 20656 (Apr. 6, 1979).
185. C. P. Carpenter and H. F. Smyth, *Am. J. Ophtalmo.* **29**, 1363–1372 (1946).
186. R. Melnick, Report NTP-TR-373, NIH/PUB-90-2828, National Toxicology Program, Research Triangle Park, N.C., 1990.
187. M. Ishidate Jr., T. Sonufi, and K. Yoshikawa, *GANN Monograph on Cancer Research*, **27**, 95–108 (1981).
188. G. W. Malaney and co-workers, *Water Pollut. Control Fed. Pt. 2*, **41**(2), R-18 (1969).
189. M. Wojaczynska and co-workers, *Metody Fizikochem Oczszczania Wody Sciekow, Mater. Conf.*, 2nd Nauk Miedzynar Konf., Vol. 1, 1979.
190. Jpn. Kokai Tokkyo Koho JP 80 20,652 (Feb. 14, 1980), T. Ogasa and co-workers (to Sumitomo Metal Mining Co., Ltd.).
191. H. Matsumoto, S. Miyajima, and R. Matsumoto, *Eisei Kagaku*, **35**(1), 86–92 (1989).
192. U.S. Pat. 4,325,853 (Apr. 20, 1982), V. Acharya and P. R. Lakshmanan (to Gulf Oil Corp.).
193. Jpn. Kokai Tokkyo Koho JP 58,154,763 (Sept. 14, 1983) (to Tokyo Cellophane Paper Co., Ltd.).
194. Eur. Pat. Appl. EP 176,294 (Apr. 2, 1986), G. J. Farquharson, K. G. Watson, and G. J. Bird (to ICI Australia, Ltd.).
195. PCT Int. Appl. WO 86 07,354 (Dec. 18, 1986), G. Matolcsy and co-workers (to Reanal Finomvergyszergyar).
196. K. Takeno, R. L. Legge, and R. P. Pharis, *Plant Phisiol.* **67**(suppl.), 581 (1981).
197. A. M. Reinbol'd, G. S. Pasechnik, and D. P. Popa, *Izv. Akad Nauk Mold. SSR, Ser. Biol. Khim. Nauk*, (4), 55–59 (1989).
198. Jpn. Kokai Tokkyo Koho JP 04,297,403 (Oct. 21, 1992), Y. Ogawa and co-workers (to Ishihara Sangyo Kaisha, Ltd.).
199. Jpn. Kokai Tokkyo Koho JP 04,297,404 (Oct. 21, 1992), Y. Ogawa and co-workers (to Ishihara Sangyo Kaisha, Ltd.).

200. Jpn. Kokai Tokkyo Koho JP 60,146,806 (Aug. 2, 1985), H. Tomioka and co-workers (to Sumitomo Chemical Co., Ltd.).

201. Eur. Pat. Appl. EP 253,501 (Jan. 20, 1988), V. M. Anthony, C. J. Urch, and P. A. Worthington (to Imperial Chemical Industries PLC).

202. Eur. Pat. Appl. EP 391,851 (Oct. 10, 1990), J. M. Zellweger (to GIBA-GEIGY AG).

203. Jpn. Kokai Tokkyo Koho JP 04,279,503 (Oct. 5, 1992), Y. Ogawa and co-workers (to Ishihara Sangyo Kaisha, Ltd.).

204. U.S. Pat. 4,818,526 (Apr. 4, 1989), R. A. Wilson and co-workers (to International Flavors and Fragrances Inc.; University of Florida).

205. Jpn. Kokai Tokkyo Koho JP 01,111,761 (Apr. 28, 1989), O. Imamura, T. Ito, and N. Katsuki (to Sanko Colloid Kagaku KK).

206. Jpn. Kokai Tokkyo Koho JP 04,119,953 (Apr. 21, 1992), T. Yamakawa, S. Nakamura, and Y. Muto (to Shin-Etsu Chemical Industry Co., Ltd.).

207. Jpn. Kokai Tokkyo Koho JP 06,032,639 (Feb. 8, 1994), Y. Sasagawa, K. Kaizaki, and S. Ogawa (to Denki Kagaku Kogyo KK).

208. Jpn. Kokai Tokkyo Koho JP 06,032,640 (Feb. 8, 1994), Y. Sasagawa, K. Kaizaki, and S. Ogawa (to Denki Kagaku Kogyo KK).

209. Jpn. Kokai Tokkyo Koho JP 06,032,641 (Feb. 8, 1994), Y. Sasagawa, K. Kaizaki, and S. Ogawa (to Denki Kagaku Kogyo KK).

210. Jpn. Kokai Tokkyo Koho JP 06,032,642 (Feb. 8, 1994), Y. Sasagawa, K. Kaizaki, and S. Ogawa (to Denki Kagaku Kogyo KK).

211. Jpn. Kokai Tokkyo Koho JP 06,144,904 (May 24, 1994), Y. Sasagawa, K. Shimada, and M. Nishama (to Denki Kagaku Kogyo KK).

212. Jpn. Kokai Tokkyo Koho JP 06,144,905 (May 24, 1994), Y. Sasagawa, K. Shimada, and M. Nishama (to Denki Kagaku Kogyo KK).

213. Jpn. Kokai Tokkyo Koho JP 06,157,096 (June 3, 1994), Y. Sasagawa, K. Shimada, and M. Nishama (to Denki Kagaku Kogyo KK).

214. Jpn. Kokai Tokkyo Koho JP 06,157,095 (June 3, 1994), Y. Sasagawa, K. Shimada, and M. Nishama (to Denki Kagaku Kogyo KK).

215. Jpn. Kokai Tokkyo Koho JP 06,157,099 (June 3, 1994), Y. Sasagawa, K. Shimada, and M. Nishama (to Denki Kagaku Kogyo KK).

216. J. L. Boucher, I. H. Wang, and R. A. Romine, *Prepr. Am. Chem. Soc., Div. Pet. Chem.* **35**(3), 556–561 (1990).

217. Jpn. Kokai Tokkyo Koho JP 02,196,029 (Aug. 2, 1990), M. Saiki and co-workers (to Sanyo Color Works, Ltd.).

218. Jpn. Kokai Tokkyo Koho JP 01,275,421 (Nov. 6, 1989), Y. Oguri, J. Saito, and M. Wakabayashi (to Mistubishi Kasei Corp.).

219. Jpn. Kokai Tokkyo Koho JP 01,308,888 (Dec. 13, 1989), S. Kojima and Y. Taniguchi (to Asahi Optical Co., Ltd.).

220. Jpn. Kokai Tokkyo Koho JP 06 32,925 (Feb. 8, 1994), H. Kuramochi (to Polytec Design KK).

221. Jpn. Kokai Tokkyo Koho JP 05,112,583 (May 7, 1993), T. Yoshida (to Toshiba Silicone).

222. Jpn. Kokai Tokkyo Koho JP 05,098,168 (Apr. 20, 1993), M. Oooka, G. Iwamura, and S. Takezawa (to Dainippon Ink & Chemicals).

223. Jpn. Kokai Tokkyo Koho JP 05,112,662 (May 7, 1993), T. Mizobuchi and co-workers (to Mitsubishi Rayon Co.).

224. Jpn. Kokai Tokkyo Koho JP 05,117,429 (May 14, 1993), Y. Mizobuchi and co-workers (to Mitsubishi Rayon Co.).

225. Ger. Offen. DE 3,818,050 (Dec. 15, 1988), S. Miyabayshi and co-workers (to Takeda Chemical Industries, Ltd.).

226. Jpn. Kokai Tokkyo Koho JP 05 39,452 (Feb. 19, 1993), M. Saiba and co-workers (to Aishin Kako K.k.; Toyota Motor Co., Ltd.).

227. Jpn. Kokai Tokkyo Koho JP 05 39,454 (Feb. 19, 1993), K. Nakamura and co-workers (to Aishin Kako K.k.; Toyota Motor Co., Ltd.).

228. Jpn. Kokai Tokkyo Koho JP 03 52,968 (Mar. 7, 1991), T. Agawa and H. Takeda (to Dainippon Ink and Chemicals, Inc.).

229. Eur. Pat. Appl. EP 355676 (Feb. 28, 1990), D. Oberkobusch, W. Gress, and B. Wegemund (to Henkel K. GaA).

230. Eur. Pat. Appl. EP 329,298 (Aug. 23, 1989), T. P. Klun, A. F. Robbins, and M. Ali (to Minnesota Mining and Manufacturing Co.).

231. Jpn. Kokai Tokkyo Koho JP 05,247,373 (Sept. 24, 1993), N. Hosokawa and K. Hayama (to Mitsubishi Petrochemical Co.).

232. Jpn. Kokai Tokkyo Koho JP 02,153,909 (June 13, 1990), K. Isobe, K. Matsumoto, and K. Nawata (to Nippon Kayaku Co., Ltd.).

233. Jpn. Kokai Tokkyo Koho JP 03 87,756 (Apr. 12, 1991), J. Aoshima and H. Fujita (to Tomoegawa Paper Co., Ltd.).

234. Jpn. Kokai Tokkyo Koho JP 02 03,076 (Jan. 8, 1990), H. Yushina, K. Koizumi, and T. Ito (to Mitsubishi Kasei Corp.).

235. Jpn. Kokai Tokkyo Koho JP 62,106,062 (May 16, 1987), T. Sonoda and F. Shimoyama (to Nippon Shokubai Kagaku Kogyo Co., Ltd.).

236. Jpn. Kokai Tokkyo Koho JP 61,286,343 (Dec. 16, 1986), T. Sonoda and F. Shimoyama (to Nippon Shokubai Kagaku Kogyo Co., Ltd.).

237. Jpn. Kokai Tokkyo Koho JP 61,161,243 (July 21, 1986), H. Yoshida and co-workers (to Nippon Shokubai Kagaku Kogyo Co., Ltd.).

238. Jpn. Kokai Tokkyo Koho JP 81,166,155 (Dec. 21, 1981) (to Sumitomo Chemical Co., Ltd.).

239. Jpn. Kokai Tokkyo Koho JP 58 84,981 (May 21, 1983) (to Nippon Steel Corp.).

240. Pol. Pat. PL 154,449 (Nov. 29, 1991), A. Kotlinski and co-workers (to Instytut Chemii Przemyslowej).

241. Eur. Pat. Appl. EP 295,108 (Dec. 14, 1988), A. Lenack (to Exxon Chemical Patents, Inc.).

242. A. Braig, *FATIPEC-Kongr.* **3**(19), 401–421 (1988).

243. Jpn. Kokai Tokkyo Koho JP 03,232,982 (Oct. 16, 1991), Y. Takahashi and co-workers (to Mitsubishi Oil Co., Ltd.).

244. Eur. Pat. Appl. EP 348,303 (Dec. 27, 1989), B. Plassin.

245. Eur. Pat. Appl. EP 464,473 (Jan. 8, 1992), W. Ritschel, H. Lorke, and G. Kremer (to Hoechst AG).

246. S. Sankarapapavinasam, F. Pushpanaden, and M. F. Ahmed, *Bull. Electrochem.* **5**(5), 319–323 (1989).

247. Eur. Pat. Appl. EP 603,811 (June 29, 1994), S. Taya, T. Ueda, and M. Itoh (to Kurita Water Industries Ltd.).

248. Jpn. Kokai Tokkyo Koho JP 04,334,347 (Nov. 20, 1992), K. Oi and co-workers (to Taiyo Kagaku Co., Ltd.).

249. Eur. Pat. Appl. EP 429,124 (May 29, 1991), L. X. Huynh (to Procter and Gamble Co.).

250. Eur. Pat. Appl. EP 463,802 (Jan. 2, 1992), S. B. Kong (to Clorox Co.).

251. Eur. Pat. Appl. EP 234,082 (Sept. 2, 1987), J. L. Copeland (to Ecolab, Inc.).

252. Ger. Offen. DE 3,625,268 (Feb. 5, 1987), T. Ouhadi and L. Dehan (to Colgate-Palmolive Co.).

253. Eur. Pat. Appl. EP 264,977 (Apr. 27, 1988), I. Herbots, J. P. Johnston, and F. De Buzzaccarini (to Procter and Gamble Co.; Procter and Gamble European Technical Center).

254. Eur. Pat. Appl. EP 151,517 (Aug. 14, 1985), Y. J. Nedonchelle (to Uniliver NV).

255. Eur. Pat. Appl. EP 379,256 (July 25, 1990), W. J. Cook and co-workers (to Colgate-Palmolive Co.).

256. Eur. Pat. Appl. EP 589,761 (Mar. 30, 1994), L. Regis, M. Marchal, and C. Blanvalet (to Colgate-Palmolive Co.).

257. Eur. Pat. Appl. EP 411,708 (Feb. 6, 1991), M. Thomas, G. Blandiaux, and B. Valange (to Colgate-Palmolive Co.).

258. Eur. Pat. Appl. EP 336,878 (Oct. 11, 1989), M. Thomas, B. Valance, and G. Blandiaux (to Colgate-Palmolive Co.).

259. Eur. Pat. Appl. EP 368,146 (May 16, 1990), M. Loth, C. Blanvalet, and B. Valange (to Colgate-Palmolive Co.).

260. U.S. Pat. 5,019,288 (May 28, 1991), S. M. Garcia (to Chem. Shield, Inc.).

261. Can. Pat. Appl. CA 2077398 (Mar. 7, 1993), W. J. Cook and co-workers (to Colgate-Palmolive Co., USA).

262. Jpn. Kokai Tokkyo Koho JP 60,150,899 (Aug. 8, 1985), I. Ito and K. Mizuno (to Kurita Water Industries, Ltd.).

263. U.S. Pat. 4,970,015 (Nov. 13, 1990), S. M. Garcia (to Chem. Shield, Inc.).

264. Ger. Offen. DE 4,029,297 (Mar. 19, 1992), H. Lueders (to Huels AG).

265. Jpn. Kokai Tokkyo Koho JP 02,238,098 (Sept. 20, 1990), A. Soi, N. Nomura, and K. Ohira (to Kao Corp.).

266. Brit. Pat. Appl. GB 2,193,510 (Feb. 10, 1988), I. C. Callaghan and B. N. Love (to British Petroleum Co. PLC).

267. U.S. Pat. 5,324,443 (June 28, 1994), S. Arif and B. B. Sandel (to Olin Corp.).

268. Jpn. Kokai Tokkyo Koho JP 04,130,198 (May 1, 1992), T. Nozaki (to Kao Corp.).

269. Eur. Pat. Appl. EP 242,141 (Oct. 21, 1987), S. D. Liem and F. A. Veer (to Unilever PLC; Unilever NV).

270. Eur. Pat. Appl. EP 242,138 (Oct. 21, 1987), G. J. Huijben and co-workers (to Unilever PLC; Unilever NV).

271. Jpn. Kokai Tokkyo Koho JP 62 30,198 (Feb. 9, 1987), T. Okubo, K. Mukoyama, and K. Umehara (to Lion Corp.).

272. Jpn. Kokai Tokkyo Koho JP 62 62,899 (Mar. 19, 1987), K. Saito and co-workers (to Kao Corp.).

273. Jpn. Kokai Tokkyo Koho JP 03,136,223 (June 11, 1991), K. Morita (to Sanyo Electric Co., Ltd.).

274. Czech. Pat. CS 260,892 (May 15, 1989), M. Bubenicek.

275. Jpn. Kokai Tokkyo Koho JP 61,050,321 (Mar. 12, 1986), Y. Yokoyama and T. Ito (to Nippon Chemi-Con Corp.).

276. PCT Int. Appl. WO 90 08,206 (July 26, 1990), M. E. Tadros, C. A. Matzdorf, and T. K. Shah (to Martin-Marietta Corp.).

277. PCT Int. Appl. WO 94 19,275 (Sept. 1, 1994), K. E. Howard and J. R. Moyer (to Dow Chemical Co.).

278. Jpn. Kokai Tokkyo Koho JP 02,104,699 (Apr. 17, 1990), T. Nakamura and T. Shimomura (to C. Uyemura and Co., Ltd.).

279. Ger. Offen. DE 4,137,377 (May 19, 1993), N. Rish and co-workers (to Fa. Rudolph Jatzke).

280. Eur. Pat. Appl. EP 401,601 (Dec. 12, 1990), O. Gobbetti (to Diaprint S.r.I.).

281. Jpn. Kokai Tokkyo Koho JP 62 06,535 (Feb. 12, 1987), K. Ogawa, T. Toyoda, and K. Katagiri (to Takeda Chemical Industries, Ltd.).

282. Jpn. Kokai Tokkyo Koho JP 62 87,070 (Apr. 21, 1987), E. Tsuburaya, S. Omori, and H. Masai (to Nakano Vinegar Co., Ltd.).

283. Jpn. Kokai Tokkyo Koho JP 58,183,467 (Oct. 26, 1983), (to Toppan Printing Co., Ltd.).

284. Jpn. Kokai Tokkyo Koho JP 02,172,810 (July 4, 1990), K. Saegusa and T. Mori (to Kao Corp.).

285. L. A. Badovskaya, N. A. Severina, and N. A. Yatsun, *Izv. Vyssh. Uchebn. Zaved. Pishch. Tekhnol.* (1), 19–21 (1992).

286. Jpn. Kokai Tokkyo Koho JP 04156851 (May 29, 1992), M. Shigemitsu and M. Ariyoshi (to Toyo Ink Mfg. Co., Ltd.).

287. Jpn. Kokai Tokkyo Koho JP 01,158,960 (June 22, 1989), T. Ueda and co-workers (to Nippon Zeon Co., Ltd.).

288. Jpn. Kokai Tokkyo Koho JP 01,166,760 (June 30, 1989), T. Saijo (to Shoko Kagaku Kenkyusho K.K.).

289. Jpn. Kokai Tokkyo Koho JP 57,200,159 (Dec. 8, 1982) (to Kyodo Milk Industry Co., Ltd. Ayugawa Taizo).

290. Jpn. Kokai Tokkyo Koho JP 58,173,551 (Oct. 12, 1983) (to Ayukawa, Yazuso Sankyo K.K.).

291. Jpn. Kokai Tokkyo Koho JP 05 004095 (Jan. 14, 1993), E. Takemura, H. Irie, and T. Kaneko (to Nippon Soda Co., Ltd.).

292. Jpn. Kokai Tokkyo Koho JP 03,224,922 (Oct. 3, 1991), N. Miyakoshi, E. Takemura, and T. Kaneko (to Nippon Soda Co., Ltd.).

293. Jpn. Kokai Tokkyo Koho JP 03,223,386 (Oct. 2, 1991), N. Miyakoshi, E. Takemura, and H. Suzuki (to Nippon Soda Co., Ltd.).

294. Jpn. Kokai Tokkyo Koho JP 02,140,300 (May 29, 1990), K. Okada, H. Itaya, and M. Kayama (to Pola Chemical Industries, Inc.).

295. Brit. Pat. Appl. GB 2,078,522 (Jan. 13, 1982) (to Antec AH International Ltd.).

296. Jpn. Kokai Tokkyo Koho JP 63,276,047 (Nov. 14, 1988), H. Nakai and co-workers (to Konika Co.; Mitsubishi Kasei Corp.).

297. Jpn. Kokai Tokkyo Koho JP 01,307,741 (Dec. 12, 1989), H. Nakai and co-workers (to Konika Co.; Mitsubishi Kasei Corp.).

298. Jpn. Kokai Tokkyo Koho JP 02 44,359 (Feb. 14, 1990), Y. Maeda (to Mitsubishi Kasei Corp.).

299. Eur. Pat. Appl. EP 508457 (Oct. 14, 1992), J. Yamada and S. Shinohara (to Mitsubishi Paper Mills, Ltd.).

300. Jpn. Kokai Tokkyo Koho JP 63,214,746 (Sept. 7, 1988), T. Asano and S. Natori (to Fuji Pharmaceutical Industries Co., Ltd.).

301. U.S. Pat. 5,279,927 (Jan. 18, 1994), J. E. Walls and co-workers (to Eastman Kodak Co.).

302. U. J. Moeller, *Mineraloeltechnik*, **38**(2), 31 (1993).

303. Ger. Offen. DE 3,630,783 (Mar. 24, 1988), K. Worschech and co-workers (to Neynaber Chemie GmbH).

304. Jpn. Kokai Tokkyo Koho JP 59,142,293 (Aug. 15, 1984) (to Alps Electric Co., Ltd.).

305. S. I. Pol'kin and co-workers, in A. M. Gol'man and I. L. Dmitrieva, eds., *Flotatsionnye Reagenty* Nauka, Moscow, 1986, pp. 104–109.

306. S. A. Merritt and M. Dagenais, *J. Electrochem. Soc.* **140**(9), L138–L139 (1993).

307. A. J. Tang, K. Sadra, and B. G. Streetman, *J. Electrochem. Soc.* **140**(5), L82–L83 (1993).

308. Jpn. Kokai Tokkyo Koho JP 06,049,557 (Feb. 22, 1994), Y. Kenmochi and J. Pponda (to Japan Metals & Chem. Co., Ltd.).

309. Jpn. Kokai Tokkyo Koho JP 63,229,262 (Sept. 26, 1988), H. Kobayashi and R. Kato (to Tipton Mfg. Corp.).

310. Rus. Pat. SU 1,525,236 (Nov. 30, 1989), V. I. Kolomenskov, V. D. Selinavov, and L. I. Nikolaev.

311. PCI Int. Appl. WO 9322469 (Nov. 11, 1993), S. Luk (to Hoeganaes Corp.).

312. Jpn. Kokai Tokkyo Koho JP 04,202,881 (July 23, 1992), K. Hamaguchi and H. Urushibata (to Kao Corp.).

313. Finn. Pat. FI 67,736 (Jan. 31, 1985), A. Juopperi, B. Lax, and L. Frilund (to Osakeyhtio Kasvioljy-Vaxtolje AB).

314. Eur. Pat. Appl. EP 339,276 (Nov. 2, 1989), Y. Ichii and co-workers (to Kao Corp.).

315. Eur. Pat. Appl. EP 229,616 (July 22, 1987), H. Yorozu and co-workers (to Kao Corp.).
316. Jpn. Kokai Tokkyo Koho JP 61,229,815 (Oct. 14, 1986), Y. Eguchi and H. Yorozu (to Kao Corp.).
317. Jpn. Kokai Tokkyo Koho JP 04,029,926 (Jan. 31, 1992), J. Ichii, W. Ookawa, and H. Yorozu (to Kao Corp.).
318. Jpn. Kokai Tokkyo Koho JP 63,264,518 (Nov. 1, 1988), S. Unoki and co-workers (to Wakamoto Pharmaceutical Co., Ltd.).
319. Jpn. Kokai Tokkyo Koho JP 01,290,622 (Nov. 22, 1989), Y. Ora (to Sunpalco Co., Ltd.).
320. Jpn. Kokai Tokkyo Koho JP 02,121,917 (May 9, 1990), S. Tachibana and H. Sakai (to Lion Corp.).
321. Jpn. Kokai Tokkyo Koho JP 03,058,920 (Mar. 14, 1991), S. Tachibana and H. Sakai (to Lion Corp.).
322. Jpn. Kokai Tokkyo Koho JP 02,218,608 (Aug. 31, 1990), T. Uramoto and co-workers (to Pola Chemical Industries Inc.).
323. Ger. Off. DE 4,222,056 (Jan. 5, 1994), A. Hamann, W. Neumann, and R. Staeck (to Chemie A.-G. Bitterfeld-Wolfen).
324. U.S. Pat. 5,229,103 (July 20, 1993), S. Eagle and G. Barker (to Hydrodent Laboratories, Inc.).
325. Jpn. Kokai Tokkyo Koho JP 01,125,316 (May 17, 1989), N. Komyama and K. Sugawara (to Lion Corp.).
326. Jpn. Kokai Tokkyo Koho JP 61,176,518 (Aug. 8, 1986), T. Sema, H. Watanabe, and M. Yogo (to Lion Corp.).
327. Eur. Pat. Appl. EP 133,354 (Feb. 20, 1985), N. E. Banks, G. J. Hignett, and E. Smith (to Interox Chemicals Ltd.).
328. Jpn. Kokai Tokkyo Koho JP 04,149,113 (May 22, 1992), H. Kikuchi and co-workers (to Kanebo, Ltd.).
329. Eur. Pat. Appl. EP 414,608 (Feb. 27, 1991), H. Noel (to Roussel-UCLAF).
330. Jpn. Kokai Tokkyo Koho JP 05194153 (Aug. 3, 1993), Y. Natori and J. Nishida (to Lion Corp.).
331. Eur. Pat. Appl. EP 327,379 (Aug. 9, 1989), K. Coupland and P. J. Smith (to Croda International PLC).
332. Jpn. Kokai Tokkyo Koho JP 05,279,239 (Oct. 26, 1993), E. Kobayashi and co-workers (to Hotsukaido Sooda KK; Kawaken Fine Chemicals Co.).
333. Jpn. Kokai Tokkyo Koho JP 63,277,619 (Nov. 15, 1988), M. Inoue, T. Tanaka, and H. Doi (to Nippon Oils and Fats Co., Ltd.).
334. Jpn. Kokai Tokkyo Koho JP 05,221,839 (Aug. 31, 1993), S. Iki and Y. Iwamoto (to Kanebo Ltd.).
335. Jpn. Kokai Tokkyo Koho JP 05,201,835 (Aug. 10, 1993), S. Iki and Y. Iwamoto (to Kanebo Ltd.).
336. Eur. Pat. Appl. EP 471,352 (Feb. 19, 1992), K. Ogata, K. Ushio, and H. Nakayama (to Senju Pharmaceutical Co., Ltd.).
337. Jpn. Kokai Tokkyo Koho JP 57,200,161 (Dec. 8, 1982) (to Kyodo Milk Industry Co., Ltd. Konde K.K.).
338. Y. Jin and W. Yue, *Yaoxue Tongbao* **18**(2) 100–102 (1983).
339. PCT Int. Appl. WO 91 04,757 (Apr. 18, 1991), F. Wehling, S. Schuehle, and N. Madamala (to Cima Labs, Inc.).
340. PCT Int. Appl. WO 93 09,781 (May 27, 1993), M. Okada and co-workers (to SS Pharmaceutical Co., Ltd.).
341. PCT Int. Appl. WO 9404135 (Mar. 3, 1994), A. Yamada and co-workers (to Teikoku Seiyaku Kabushiki Kaisha).
342. C. Van der Veen, H. Buitendijk, and C. F. Lerk, *Eur. J. Pharm. Bipharm.* **37**(4), 238–242 (1991).

343. K. Thoma and T. Zimmer, *Pharm. Ind.* **51**(1), 98–101 (1989).

344. S. R. Desai and co-workers, *Drug Dev. Ind. Pharm.* **15**(5), 671–689 (1989).

345. PCT Int. Appl. WO 93 00,898 (Jan. 21, 1993), D. Clapham (to SmithKline Beecham PLC).

346. Jpn. Kokai Tokkyo Koho JP 01,26,502 (Jan. 27, 1989), S. Nomura, Y. Oka, and T. Shimamura (to Shikoku Chemicals Corp.).

347. Jpn. Kokai Tokkyo Koho JP 63,239,203 (Oct. 5, 1988), S. Nomura and Y. Oka (to Shikoku Chemicals Corp.).

348. PCT Int. Appl. WO 93 13,793 (July 22, 1992), P. Blackburn, S. J. Projan, and E. B. Edward (to Applied Microbiology, Inc.).

349. H. H. Moussa and L. M. Chabaka, *Egypt. J. Chem.* **26**(3), 267–273 (1983).

350. R. Takayanagi, F. Umeda, and M. Sakamoto, *Kassei Sanso Furi Rajikaru*, **2**(6), 779–783 (1991).

351. Jpn. Kokai Tokkyo Koho JP 06,065,041 (Mar. 8, 1994), I. Sasaki and co-workers (to Kosei Kk).

352. U.S. Pat. 4,481,353 (Nov. 6, 1984), E. Nyilas and T. H. Chiu (to Children's Hospital Medical Center, Boston).

353. Ger. Offen. DE 4,141,174 (June 17, 1992), H. Ohno, T. Satou, and H. Okamoto (to Tokuyama Soda KK).

354. Eur. Pat. Appl. EP 386,644 (Sept. 12, 1990), S. J. Salamone and S. Vitone (to F. Hoffmann-La Roche and Co. AG).

355. Ger. Offen. DE 3,324,626 (Jan. 17, 1985), H. Rehner, K. Bartl, and R. Portenhauser (to Boehringer Mannheim GmbH).

356. B. M. Khadilkar and S. R. Bhayade, *Indian J. Chem. Sect. B.* **32B**(3), 338–342 (1993).

357. D. C. Phan, N. H. Do, and L. T. Phan, *Tap Chi Duoc Hoc*, (5), 10–12 (1992).

358. Jpn. Kokai Tokkyo Koho JP 60,06,686 (Jan. 14, 1985) (to Kissei Pharmaceutical Co., Ltd.).

359. V. F. Konev and co-workers, *Farm. Zh. (Kiev)*, (5), 39–42 (1988).

360. G. J. Quallich, M. T. Williams, and R. C. Friedmann, *J. Org. Chem.* **55**(16), 4971–4973 (1990).

361. Eur. Pat. Appl. EP 563,571 (Oct. 6, 1993), Y. Ueda and H. Yamashita (to Konica Co.).

362. Jpn. Kokai Tokkyo Koho JP 06,03,788 (Jan. 14, 1994), W. Satake (to Konishiroku Photo Ind.).

363. Eur. Pat. Appl. EP 556782 (Aug. 25, 1993), H. Yamashita and Y. Ueda (to Konica Corp.).

364. Eur. Pat. Appl. EP 574829 (Dec. 22, 1993), S. Kuse and co-workers (to Konica Co.).

365. A. V. Zvyagintseva and co-workers, *Izv. Vyssh. Uchebn. Zaved. Khim. Khim. Tekhnol.* **31**(12), 91–95 (1988).

366. Z. Zuo and G. Wang, *Cailiao Baohu*, **26**(1), 4–9 (1993).

367. Z. A. Miroshnik and A. I. Falicheva, *Izv. Vyssh. Uchebn. Zaved. Khim. Khim. Tekhnol.* **32**(8), 63–65 (1989).

368. Eur. Pat. Appl. EP 494,431 (July 15, 1992), N. Totsuka and co-workers (to Kawasaki Steel Corp.).

369. B. Radziuniene and V. Skominas, *Liet. TSR Mokslu Akad. Darb. Ser. B.* (4), 25–32 (Russ.) (1986).

370. U.S. Pat. 4,755,264 (July 5, 1988), J. A. Lochet and R. B. Patel (to Vanguard Research Associates, Inc.).

371. Eur. Pat. Appl. EP 400349 (Dec. 5, 1990), B. V. Sodervall and T. Lundeberg (to Adtech Holding).

372. Ger. Pat. DD 272,194 (Oct. 4, 1989), B. Wilcke and K. Richter (to VEB Elektronische Bauelemente).

373. Eur. Pat. Appl. EP 376,494 (July 4, 1990), I. S. Mahmoud (to International Business Machines Corp.).
374. R. Tarozaite and A. Luneckas, *Chemija*, (1), 83–90 (1990).
375. Jpn. Kokai Tokkyo Koho JP 02,66,175 (Mar. 6, 1990), K. Kojima (to Shimadzu Corp.).
376. Jpn. Kokai Tokkyo JP 05,295,557 (Nov. 9, 1993), H. Yamamoto and T. Shimazaki (to Hitachi Chemical Co., Ltd.).
377. A. M. T. van der Putten and J. W. G. Bakker, *J. Electrochem. Soc.* **140**(8), 2229–2235 (1993).
378. Z. Guo, *Cailiao Baohu*, **25**(5), 46–48 (1993).
379. J. Bielinski and E. Skudlarska, *Oberflaeche-Surf.* **26**(3), 76–82 (1985).
380. PCT Int. Appl. WO 90,09,467 (Aug. 23, 1990), I. G. McDonald and P. J. Archer (to Polymetals Technology Ltd.).
381. PCT Int. Appl. WO 90,09,468 (Aug. 23, 1990), I. G. McDonald and P. J. Archer (to Polymetals Technology Ltd.).
382. Ger. Pat. DE 4,111,558 (Jan. 9, 1992), L. Stein (to Schering AG).
383. Ger. Pat. DE 4,316,679 (July 28, 1994), L. Stein, H. Mahlkow, and W. Strache (to Atotech Deutschland GmbH).
384. Jpn. Kokai Tokkyo Koho JP 03,17,278 (Jan. 25, 1991), F. Goto (to NEC Corp.).
385. Jpn. Kokai Tokkyo Koho JP 03,53,078 (Mar. 7, 1991), T. Watanabe and co-workers (to Nippon Kokan KK).
386. T. Osaka, *Hyomen Gijutsu*, **44**(10), 767–773 (1993).
387. Ger. Offen. DE 3,402,494 (July 25, 1985), H. Januschowetz and H. Laub (to Siemens AG).
388. Eur. Pat. Appl. EP 565,235 (Oct. 13, 1993), E. Takiyama and co-workers (to Showa Highpolymer Co., Ltd.).
389. Jpn. Kokai Tokkyo Koho JP 05,295,098 (Nov. 9, 1993), E. Takyama and co-workers (to Showa Highpolymer).
390. Jpn. Kokai Tokkyo Koho JP 05,295,099 (Nov. 9, 1993), E. Takyama, T. Fujimaki, and N. Harigai (to Showa Highpolymer).
391. Jpn. Kokai Tokkyo Koho JP 06,172,624 (June 21, 1994), M. Imaizumi, M. Kotani, and E. Takyama (to Showa Highpolymer, Showa Denko KK).
392. Jpn. Kokai Tokkyo Koho JP 06 80,872 (Mar. 22, 1994), T. Fujimaki and co-workers (to Showa Highpolymer).
393. Jpn. Kokai Tokkyo Koho JP 06,145,313 (May 24, 1994), Y. Iwaya and T. Ikeda (to Unitika Ltd.).
394. Jpn. Kokai Tokkyo Koho JP 06,145,283 (May 24, 1994), Y. Iwaya and T. Ikeda (to Unitika Ltd.).
395. Jpn. Kokai Tokkyo Koho JP 06,157,703 (June 7, 1994), R. Hasegawa, K. Fukunaga, and N. Harada (to Nippon Kayaku KK).
396. PCT Int. Appl. WO 93,08014 (Apr. 29, 1993), H. J. Jeffs (to Bio-Products International).
397. Belg. Pat. BE 895,533 (May 2, 1983) (to Ashland Oil, Inc.).
398. Jpn. Kokai Tokkyo Koho JP 05,51,518 (Mar. 2, 1993), M. Oooka and K. Yamamura (to Dainippon Ink & Chemicals).
399. Jpn. Kokai Tokkyo JP 59,174,615 (Oct. 3, 1984) (to Toshiba Corp.).
400. Czech. Pat. CS 224,137 (Nov. 1, 1984), K. Jelinek and co-workers.
401. Ger. Pat. DD 248,598 (Aug. 12, 1987), J. Klee and H. H. Hoerhold (to Friedrich Schiller Universität).
402. Ger. Offen. DE 3,525,695 (Jan. 30, 1986), S. Yagishita and co-workers (to Nippon Zeon Co., Ltd.; Toyoda Gosei Co., Ltd.).
403. Ger. Offen. DE 4,011,812 (Oct. 17, 1991), R. Eidenschink (to Consortium Fuer Elektrochemische Industrie GmbH).

404. Jpn. Kokai Tokkyo Koho JP 05,93,063 (Apr. 16, 1993), H. Kawakami and H. Minematsu (to Teijin Ltd.).
405. R. Devarajan and co-workers, *J. Appl. Polym. Sci.* **48**(5), 921–930 (1992).
406. Jpn. Kokai Tokkyo Koho JP 63,258,865 (Oct. 25, 1988), H. Higuchi and co-workers (to Nippon Oils and Fats Co., Ltd.).
407. Eur. Pat. Appl. EP 488,682 (June 3, 1992), S. Koyama and co-workers (to Chisso Corp.).
408. P. Auger and co-workers, *J. Labelled Comp. Radiopharm.* **33**(4), 263–276 (1993).
409. A. Patel, R. C. Poller, and E. B. Rathbone, *Appl. Organomet. Chem.* **1**(4), 325–329 (1987).
410. PCT Int. Appl. WO 93,20,092 (Oct. 14, 1993), M. J. McLean and co-workers (to Zeneca Ltd.).
411. PCT Int. Appl. WO 93,20,091 (Oct. 14, 1993), M. J. McLean and A. J. Garman (to Zeneca Ltd.).
412. Jpn. Kokai Tokkyo Koho JP 04,200,992 (July 21, 1992), M. Takemoto, T. Onishi, and M. Aihara (to Nippodenso Co., Ltd.; Harima Chemicals, Inc.).
413. U.S. Pat. 5,004,508 (Apr. 2, 1991), E. W. Mace, J. Sickler, and V. Srinivasan (to International Business Machines Corp.).
414. Eur. Pat. Appl. EP 413,312 (Feb. 20, 1991), T. Gomi, H. Ota, and K. Kaneko (to Yuho Chemicals, Inc.).
415. Eur. Pat. Appl. EP 510,539 (Oct. 28, 1992), T. Gomi and Y. Douzaki (to Yuho Chemicals, Inc.).
416. Jpn. Kokai Tokkyo Koho JP 05,146,891 (June 15, 1993), H. Aoki and co-workers (to Aoki Metal).
417. Jpn. Kokai Tokkyo Koho JP 04,147,792 (May 21, 1992), M. Sakamoto and co-workers (to Asahi Chemical Research Laboratory Co., Ltd).
418. U.S. Pat. 5,176,749 (Jan. 5, 1993), B. J. Costello, J. Langan, and R. Harkins (to Argus International).
419. U.S. Pat. 5,004,509 (Apr. 2, 1991), S. V. Bristol (to Delco Electronics Corp.).
420. U.S. Pat. 5,297,721 (Mar. 29, 1994), A. F. Schneider, D. B. Blumel, and J. V. Tomczak (to Frys Metals, Inc.).
421. U.S. Pat. 5,141,568 (Aug. 25, 1992), R. L. Turner, K. E. Johnson, and L. L. Kimmel (to Hughes Aircraft Co.).
422. Eur. Pat. Appl. EP 458,161 (Nov. 27, 1991), R. L. Turner (to Hughes Aircraft Co.).
423. Eur. Pat. Appl. EP 483,762 (May 6, 1992), R. L. Turner, K. E. Johnson, and L. L. Kimmel (to Hughes Aircraft Co.).
424. Eur. Pat. Appl. EP 483,763 (May 6, 1992), R. L. Turner, K. E. Johnson, and L. L. Kimmel (to Hughes Aircraft Co.).
425. U.S. Pat. 5,281,281 (Jan. 25, 1994), K. Stefanowski (to Litton Systems Inc.).
426. Jpn. Kokai Tokkyo Koho JP 04,37,497 (Feb. 7, 1992), H. Minahara and co-workers (to Mec K.K.).
427. USSR SU 1,604,536 (Nov. 7, 1990), L. V. Moroz and co-workers (to Odessa State University).
428. Eur. Pat. Appl. EP 379,290 (July 25, 1990), M. R. Oddy and A. M. Wagland (to Cookson Group PLC).
429. Jpn. Kokai Tokkyo Koho JP 01,132,877 (May 25, 1989), M. Tsukada (to Japan Ministry of Agricultural and Forestry, Sericultual Experiment Station).
430. Jpn. Kokai Tokkyo Koho JP 01,052,875 (Feb. 28, 1989), K. Iwabori, H. Shinagawa, and Y. Nomura (to Toyobo Co., Ltd.).
431. Eur. Pat. Appl. EP 390,109 (Oct. 3, 1990), T. Sasakura and Y. Anasako (to Nitto Boseki Co., Ltd.).
432. Eur. Pat. Appl. EP 308,914 (Mar. 29, 1989), L. Jaeckel, R. Kleber, and B. Mees (to Hoechst AG).

433. B. A. K. Andrews, E. J. Blanchard, and R. M. Reinhardt, *Am. Dyest. Rep.* **79**(9), 48, 50, 52, 54, 94 (1990).

434. U.S. Pat. 4,836,828 (June 6, 1989), S. Hussamy (to Burlington Industries, Inc.).

435. PCT Int. Appl. WO 91,00,318 (Jan. 10, 1991), T. C. Hemling and H. Stitzel (to Olin Corp.).

436. H. Shiozaki, M. Tsukada, and M. Matsumura, *Nippon Sanshigaku Zasshi*, **57**(2), 165–168 (1988).

437. U.S. Pat. 5,104,923 (Apr. 14, 1992), P. J. Steinwand and D. P. Stack (to Union Oil Co. of California).

438. Jpn. Kokai Tokkyo Koho JP 61,243,875 (Oct. 30, 1986), Y. Toyoshima and M. Tanaka (to Sumimoto Chemical Co., Ltd.).

439. Ger. Offen. DE 3,438,529 (Apr. 24, 1986), M. Dankowski (to Degussa AG).

440. Jpn. Kokai Tokkyo Koho JP 62,89,616 (Apr. 24, 1987), T. Matsumoto, Y. Ishikawa, and K. Abe (to Kao Corp.).

441. Jpn. Kokai Tokkyo Koho JP 02,174,704 (July 6, 1990), K. Ono, H. Odanaka, and K. Kodama (to Nippon Shokubai Kagaku Kogyo Co., Ltd.).

442. Jpn. Kokai Tokkyo Koho JP 01,215,391 (Aug. 29, 1989), K. Kamasaka.

443. J. D. Mobley, O. W. Hargrove Jr., and J. D. Colley, *Proc. Annu. Meet. Air Pollut. Control Assoc.* **75**(4) (1982).

444. J. D. Colley and co-workers, *Proceedings of the 8th Symposium on Flue Gas Desulfurization*, Palo Alto, Calif., 1983.

445. Jpn. Kokai Tokkyo Koho JP 06,007,418 (Jan. 18, 1994), A. Nakayama and E. Tanaka (to Kuraray Chemical Kk.).

446. Jpn. Kokai Tokkyo Koho JP 04 99,571 (Mar. 31, 1992), H. Moriguchi, T. Matsushima, and T. Watanabe (to Mitsubishi Materials K.k.).

447. Jpn. Kokai Tokkyo Koho JP 05,322,218 (Dec. 7, 1993), Y. Kodaira and H. Inagaki (to Ee Pii Esu Kk).

448. Jpn. Kokai Tokkyo Koho JP 04,260,442 (Sept. 16, 1992), J. Kanai (to Sumitomo Metal Mining Co., Ltd.).

449. Jpn. Kokai Tokkyo Koho JP 06,128,190 (May 10, 1994), T. Nakamura, S. Komatsu, and Y. Goto (to Nippon Oils & Fats Co., Ltd.).

450. E. Dunnigham and G. Parker, *FRI Bull.* **176**, 58–66 (1992).

451. Jpn. Kokai Tokkyo Koho JP 04,182,315 (June 29, 1992), N. Sadakata and co-workers (to Fujikura Ltd.).

452. Jpn. Kokai Tokkyo Koho JP 05,82,168 (Apr. 2, 1993), R. Ooshita and co-workers (to Sanyo Electric Co.).

453. Jpn. Kokai Tokkyo Koho JP 01,134,872 (May 26, 1989), Y. Toyoguchi and co-workers (to Matsushita Electric Industrial Co., Ltd.).

454. Jpn. Kokai Tokkyo Koho JP 04,355,065 (Dec. 9, 1992), Y. Mifuji and co-workers (to Matsushita Electric Industrial Co., Ltd.).

455. Jpn. Kokai Tokkyo Koho JP 01,30,178 (Feb. 1, 1989), T. Hirai and co-workers (to Nippon Telegraph and Telephone Public Corp.).

456. H. Zhang, Y. Sun, and B. Han, *Pige Keji*, (4), 27–29 (1987).

457. PCT Int. Appl. WO 83,01,559 (May 11, 1983), J. Beres, I. Klenczner, and L. Puskas (to Vetomagtermelteto es Ertekesito Vallalat).

CARLO FUMAGALLI
LONZA SpA

NORTHERN MICHIGAN UNIVERSITY

3 1854 006 155 330